Continuum Electromechanics

To Janet Damman Melcher

Continuum Electromechanics

James R. Melcher

The MIT Press
Cambridge, Massachusetts, and London, England

65323051

Printed and bound in the United States of America

Library of Congress Cataloging in Publication Data

Melcher, James R.
Continuum electromechanics.

Includes index.
1. Electric engineering. 2. Electrodynamics.
3. Continuum mechanics. 4. Electromagnetic fields.
I. Title.
TK145.M616 621.3 81-1578
ISBN 0-262-13165-X AACR2

Preface

The three stages in which this text came into being give some insight as to how the material has matured. As "notes" written in the early 1960's, it was intended to serve as an introduction to the subject of electrohydrodynamics. Thus, it reflected the author's early research interests. During this period, the author had the privilege of collaborating with Herbert H. Woodson (now University of Texas) on the development of an undergraduate subject, "Fields, Forces and Motion". That effort resulted in the text Electromechanical Dynamics (Wiley, 1968). There has also been a strong influence from Hermann A. Haus, with whom the author has collaborated for a number of years in the development and teaching of an undergraduate electromagnetic field theory subject. Both Woodson, with his interests in rotating machinery and magnetohydrodynamics, and Haus, who then worked in areas ranging from electron beam engineering and plasmas to the electrodynamics of continuous media, stimulated the notion that there was a set of fundamental ideas that permeated many different "specialty areas". To be taught were widely applicable basic laws, approaches to modeling and mathematical techniques for disclosing what the models had to say.

The text took its second form in 1972-1973, when the objective was to achieve this broader and more enduring aspect of the material. Much of the writing was done while the author was on a Guggenheim Fellowship and a Fellow of Churchill College, Cambridge University, England. During that year, as a guest of George Batchelor's Department of Applied Mathematics and Theoretical Physics, and with the privilege of working with Sir Geoffrey Taylor, there was the opportunity to further broaden the perspective. Here, the influences were toward the disciplines of continuum mechanics.

Unfortunately, the manuscript resulting from this second writing was more in the nature of two books than one. More integration and culling of material was required if the self-imposed objective was to be achieved of helping to define a discipline rather than simply covering a number of interrelated topics.

The third version, this text, would probably not have come into being had it not been for the active encouragement of Aina Sils. Her editorial help and typewriter artistry provided teaching material that was immediately sufficiently attractive to serve as an incentive to commit nights and weekends to yet another rewrite.

As a close colleague who has been instrumental in establishing as an area the continuum electromechanics of biological systems, Alan J. Grodzinsky has been both a source of technical insight and an inspiration to complete the publication of material that for so many years had been referenced in theses as "notes."

Research carried out by still other colleagues at MIT will be seen to have influenced the scope and content. The Electric Power Systems Engineering Laboratory, directed by Gerald L. Wilson, is an example with its activities in superconducting machinery (James L. Kirtley, Jr.) and its model power system (Steven D. Umans). Others are the High Voltage Laboratory (John G. Trump and Chathan M. Cooke), the National Magnet Laboratory (Ronald R. Parker and Richard D. Thornton), the Research Laboratory of Electronics (Paul Penfield, Jr. and David H. Staelin), the Materials Processing Center (Merton C. Flemings), the Energy Laboratory (Janos M. Beer and Jean F. Louis), the Polymer Processing Program (Nam P. Suh), and the Laboratory for Insulation Research, (Arthur R. Von Hippel and William B. Westphal).

A great satisfaction and motivation has come from seeing the ideas promolgated here serve the needs of industry. The author's consulting activities, for more than 30 different companies, provided many useful examples. In the face of an increasing awareness of the importance of energy to our societal institutions and our way of life, it has been satisfying to see the concepts presented here applied not only to the development of new energy systems, but to the conflicting problem of environmental control as well.

Where possible, examples have intentionally been chosen that can be illustrated with generally available films. Referenced in Appendix C, these are in two series. The series from the National Committee on Fluid Mechanics Films was being developed at the Education Development Center while the author was active in making three films in the series from the National Committee on Electrical Engineering Films. Interaction with such individuals as Ascher H. Shapiro and J. A. Shercliff fostered an interest in using films to enliven and undergird classroom education.

While graduate students involved with the subject or carrying out their PhD theses, a number of people have made substantial contributions. Some of these are James F. Hoburg (Secs. 8.17 and 8.18), Jose Ignacio Perez Arriaga (Secs. 4.5 and 4.8), Peter W. Dietz (Sec. 5.17), Richard S. Withers (Secs. 5.8 and 5.9), Kent R. Davey (Sec. 8.5), and Richard M. Ehrlich (Sec. 5.9).

Problems at the ends of chapters were typed by Eleanor J. Nicholson. Figures were drawn by the author.

Solutions to the problems have been prepared in the form of a manual. Intended as an aid to those either presenting this material in the classroom or using it for self-study, this manual is available for the cost of reproduction from the author. Requests should be over the signature of either a member of a university faculty or the industrial equivalent.

James R. Melcher

Cambridge, Massachusetts
January, 1981

Contents

Continuum Electromechanics

1

Introduction to Continuum Electromechanics

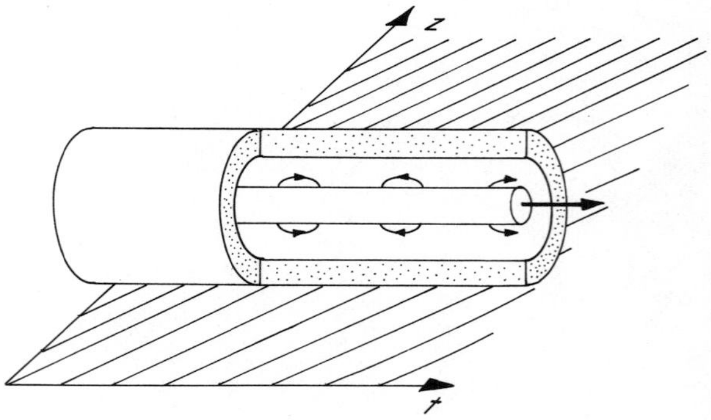

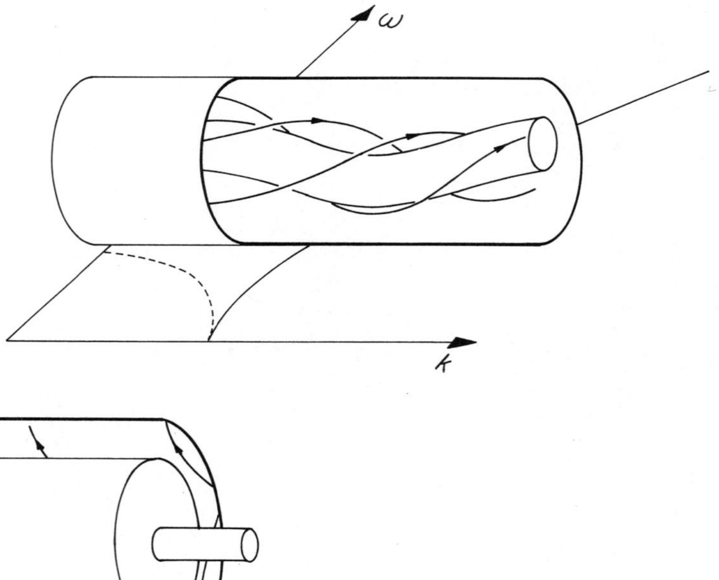

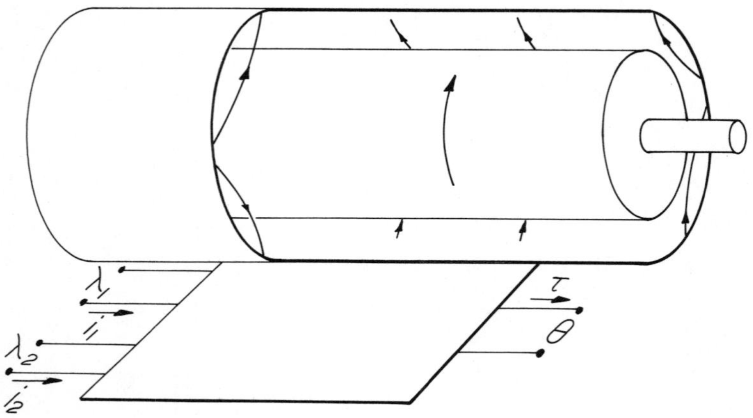

1.1 Background

There are two branches to the area of electromagnetics. One is primarily concerned with electromagnetic waves. Typically of interest are guided and propagating waves ranging from radio to optical frequencies. These may propagate through free space, in plasmas or through optical fibers. Although the interaction of electromagnetic waves with media of great variety is of essential interest, and indeed the media modify these waves, it is the electromagnetic wave that is at center stage in this branch. Dynamical phenomena of interest to this branch are typified by times, τ, shorter than the transit time of an electromagnetic wave propagating over a characteristic length of the "system" being considered. For a characteristic length ℓ and wave velocity c (in free space, the velocity of light), this transit time is ℓ/c.

In the chapters that follow, it is the second branch of electromagnetics that plays the major role. In the sense that electromagnetic wave transit times are short compared to times of interest, the electric and magnetic fields are quasistatic: $\tau \gg \ell/c$. The important dynamical processes relate to conduction phenomena, to the mechanics of ponderable media, and to the two-way interaction created by electromagnetic forces as they elicit a mechanical response that in turn alters the fields.

Because the mechanics can easily upstage the electromagnetics in this second branch, it is likely to be perceived in terms of a few of its many parts. For example, from the electromagnetic point of view there is much in common between issues that arise in the design of a synchronous alternator and of a fusion experiment. But, on the mechanical side, the rotating machine, with its problems of vibration and fatigue, seems to have little in common with the fluid-like plasma continuum. So, the two areas are not generally regarded as being related.

In this text, the same fundamentals bear on a spectrum of applications. Some of these are reviewed in Sec. 1.2. The unity of these widely ranging topics hinges on concepts, principles and techniques that can be traced through the chapters that follow. By way of a preview, Secs. 1.3-1.7 are outlines of these chapters, based on themes designated by the section headings.

Chapters 2 and 3 are concerned with fundamentals. First the laws and approximations are introduced that account for the effect of moving media on electromagnetic fields. Then, the force densities and associated stress tensors needed to account for the return influence of the fields on the motion are formulated.

Chapter 4 takes up the class of devices and phenomena that can be described by models in which the distributions (or the relative distributions) of both the material motion and of the field sources are constrained. This subject of electromechanical kinematics embraces lumped parameter electromechanics. The emphasis here is on using the field point of view to determine the relationship between the lumped parameters and the physical attributes of devices, and to determine the distribution of stress and force density.

Chapters 5 and 6 retain the mechanical kinematics, but delve into the self-consistent evolution of fields and sources. Motions of charged microscopic and macroscopic particles entrained in moving media are of interest in their own right, but also underlie the limitations of commonly used conduction constitutive laws. These chapters both introduce basic concepts, such as the Method of Characteristics and temporal and spatial modes, and model practical devices ranging from the electrostatic precipitator to the linear induction machine.

Chapters 7-11 treat interactions of fields and media where not only the field sources are free to evolve in a way that is consistent with the effect of deforming media, but the mechanical systems respond on a continuum basis to the electric and magnetic forces.

Chapter 7 introduces the basic laws and approximations of fluid mechanics. The formulation of laws, deduction of boundary conditions and use of transfer relations is a natural extension of the viewpoint introduced in the context of electromagnetics in Chap. 2.

Chapter 8 is concerned with electromechanical static equilibria and the dynamics resulting from perturbing these equilibria. Illustrated are a range of electromechanical models motivated by Chaps. 5 and 6. It is here that temporal instability first comes to the fore.

Chapter 9 is largely devoted to electromechanical flows. Included is a discussion of flow development, understood in terms of the same physical processes represented by characteristic times in the previous four chapters. Flows that display super- and sub-critical behavior presage causal effects of wave propagation taken up in Chap. 11. The last half of this chapter is an introduction to "direct" thermal-to-electric energy conversion.

Chapter 10 is divided into parts that are each concerned with diffusion processes. Thermal diffusion, together with convective heat transfer, is considered first. Electrical dissipation accompanies almost all electromechanical processes, so that heat transfer often poses an essential limitation on invention and design. Because fields are often used for dielectric or induction heating, this is a subject in its own right. This part begins with examples where the coupling is "one-way" and ends by considering some of the mechanisms for two-way coupling between the thermal and electromechanical subsystems. The second part of this chapter serves as an introduction to electromechanical processes that occur on a spatial scale small enough that molecular diffusion processes come into play. Here introduced is the interplay between electric and mechanical stresses that makes it possible for particles to undergo electrophoresis rather than migrate in an electric field. The concepts introduced in this second part are applicable to physicochemical systems and point to the electromechanics of biological systems.

Chapter 11 brings together models and concepts from Chaps. 5-10, emphasizing streaming interactions, in which ordered kinetic energy is available for participation in the energy conversion process. Included are fluid-like continua such as electron beams and plasmas.

1.2 Applications

Transducers and rotating machines that are described by the lumped parameter models of Chap. 4 are so pervasive a part of modern day technology that their development might be regarded as complete. But, with new technologies outside the domain of electromechanics, there come new needs for electromechanical devices. The transducers used to drive high-speed computer print-outs are an example. New devices in other areas also result in electromechanical innovations. For example, high power solid-state electronics is revolutionizing the design and utilization of rotating machines.

As energy needs press the capabilities of electric power systems, rotating machines continue to be the mainstay of energy conversion to electrical form. Synchronous generators are subject to increasingly stringent demands. To improve capabilities, superconducting windings are being incorporated into a new class of generators. In these synchronous alternators, magnetic materials no longer play the essential role that they do in conventional machines, and new design solutions are required.

The Van de Graaff machine also considered in Chap. 4 should not be regarded as a serious approach to bulk power generation, but nevertheless represents an important approach to the generation of extremely high potentials. It is also the grandfather of proposed energy conversion approaches. An example is the electrogasdynamic "thermal-to-electric" energy converter of Chap. 9, Sec. 9.

Chapters 5 and 6 begin to hint at the diversity of applications outside the domain of lumped parameter electromechanics. The behavior of charged particles in moving fluids is important for understanding liquid insulation in transformers and cables. Again, in the area of power generation and distribution, ions and charged macroscopic particles contribute to the contamination of high-voltage insulators. Also related to the overhead line transmission of electric power is the generation of audible noise. In this case, the charged particles considered in Chap. 5 contribute to the transduction of electrical energy into acoustic form, the result being a sufficient nuisance that it figures in the determination of rights of way.

Some examples in Chap. 5 are intended to give basic background relevant to the control of particulate air pollution. The electrostatic precipitator is widely used for air pollution control. Gases cleaned range from the recirculating air within a single room to the exhaust of a utility. With industries of all sorts committed to the use of increasingly dirtier fuels, new devices that also exploit electrical forces are under development. These include not only air pollution control equipment, but devices for painting, agricultural spraying, powder deposition and the like.

Image processing is an application of charged particle dynamics, as are other matters taken up in later chapters. Charged droplet printing is under development as a means of marrying the computer to the printed page. Xerographic and aerosol printing of considerable variety exploit electrical forces on particles.

A visit to a printing plant, to a paper mill or to a textile factory makes the importance of charges and associated electrical forces on moving materials obvious. The charge relaxation processes considered in Chap. 5 are fundamental to understanding such phenomena.

The induction machines considered in Chap. 6 are the most common type of rotating motor. But related interactions between moving conductors and magnetic fields also figure in a host of other applications. The development of high-speed ground transportation has brought into play the linear induction machine as a means of propulsion, and induced magnetic forces as a means of producing magnetic lift. Even if these developments do not reach maturity, the induction type of interaction would remain important because of its application to material transport in manufacturing processes, and to melting, levitation and pumping in metallurgical operations. The application of induced magnetic forces to the sorting of refuse is an example of how such processes can figure in seemingly unrelated areas.

Chapter 7 plays a role relative to fluid mechanics that Chap. 2 does with respect to electromagnetics. Without a discourse on the applications of this material in its own right, consider the relevance of topics that are taken up in the subsequent chapters.

Fields can be used to position, levitate and shape fluids. In many cases, a static equilibrium is desired. Examples treated in Chap. 8 include the levitation of liquid metals for metallurgical purposes, shaping of interfaces in the processing of plastics and glass, and orientation of ferrofluid seals and of cryogenic liquids in zero gravity environments.

The electromechanics of systems having a static equilibrium is often dominated by instabilities. The insights gained in Chap. 8 are a starting point in understanding atomization processes induced by means of electric fields. Here, droplets formed by means of electric fields figure in electrostatic paint spraying and corona generation from conductors under foul weather conditions. Internal instabilities also taken up in Chap. 8 are basic to mixing of liquids by electrical means and for electrical control of liquid crystal displays. Both two-phase (boiling and condensation) and convective heat transfer can be augmented by electromechanical coupling, usually through the mechanism of instability. Perhaps not strictly in the engineering domain is thunderstorm electrification. The stability of charged drops and the electrohydrodynamics of air entrained collections of charged drops are topics touched upon in Chap. 8 that have this meteorological application.

The statics and dynamics of hydromagnetic equilibria is now a subject in its own right. Largely because of its relevance to fusion machines, the discussion of hydromagnetic waves and surface instabilities serves as an introduction to an area of active research that, like other applications, has important implications for the energy posture. Internal modes taken up in Chap. 8 also have counterparts in hydromagnetics.

Magnetic pumping of liquid metals, taken up in Chap. 9, has found application in nuclear reactors and in metallurgical operations. Electrically induced pumping of semi-insulating and insulating liquids, also discussed in Chap. 9, has seen application, but in a range of modes. A far wider range of fluids have properties consistent with electric approaches to pumping and hence there is the promise of innovation in manufacturing and processing.

Magnetohydrodynamic power generation is being actively developed as an approach to converting thermal energy (from burning coal) to electrical form. The discussion of this approach in Chap. 9 is not only intended as an introduction to MHD energy conversion, but to the general issues confronted in any approach to thermal-to-electrical energy conversion, including turbine-generator systems. The electrohydrodynamic converter also discussed there is an alternative to the MHD approach that sees periodic interest. For that reason, its applicability is a matter that needs to be understood.

Inductive and dielectric heating, even of materials at rest and with no electromechanical considerations, are the basis for important technologies. These topics, as well as the generation and transport of heat in electromechanical systems where thermal effects often pose primary design limitations, are part of the point of the first half of Chap. 10. But, thermal effects can also be central to the electromechanical coupling itself. Examples where thermally induced property inhomogeneities result in such coupling include electrothermally induced convection of liquid insulation.

Electromechanical coupling seated in double layers, also taken up in Chap. 10, relates to processes (such as electrophoretic particle motions) that see applications ranging from the painting of automobiles to the chemical analysis of large molecules. One of the reasons for including electrokinetic and electrocapillary interactions is the suggestion it gives of mechanisms that can come into play in biological systems, a subject that draws heavily on physicochemical considerations. The purely electromechanical models considered here serve to identify this developing area.

The electromechanics of streaming fluids and fluid-like systems, taken up in Chap. 11, has perhaps its best known applications in the domain of electron beam engineering. Klystrons, traveling-wave tubes, resistive-wall amplifiers and the like are examples of interactions between streams of charged particles (electrons) and various types of structures. The space-time issues of Chap. 11 have general application to problems ranging from the stimulation of liquid jets used to form drops, to electromechanical processes for making synthetic fibers, to understanding liquid flow through "wall-less" pipes (in which electric or magnetic fields play the role of a duct wall), to beam-plasma interactions that result in instabilities that are used as a mechanism for heating plasmas.

1.3 Energy Conversion Processes

A theme of the chapters to follow is conversion of energy between electrical and mechanical forms. The relation between electromechanical power flow and the product of electric or magnetic stress and material velocity is first emphasized in Chap. 4. Rotating machines deserve to be highlighted in this basic sense, because for bulk power generation they are a standard for comparison. But, even where kinematic systems are superseded by those involving self-consistent interactions, there is value in considering the kinematic examples. They make clear the basic objectives governing the engineering of materials and fields even when the objectives are achieved by more devious methods. For example, the synchronous interactions with constrained charged particles are not directly applicable to practical devices, but highlight the basically electroquasistatic electric shear stress interaction that underlies electron beam interactions in Chap. 11.

The classification of energy conversion processes made in Chap. 4 provides a frame of reference for many of the self-consistent interactions described in later chapters. Thus, d-c rotating machines from Chap. 4 have counterparts with fluid conductors in Chap. 9, and the Van de Graaff generator is a prototype for the gasdynamic models developed in Chaps. 5 and 9. Electric and magnetic induction machines, respectively taken up in Chaps. 5 and 6, are a prototype for induction interactions with fluids in Chap. 9, And, the synchronous interactions of Chap. 4 motivate the self-consistent electron beam interactions of Chap. 11.

1.4 Dynamical Processes and Characteristic Times

Rate processes familiar from electrical circuits are the discharge of a capacitor (C) or an inductor (L) through a resistor (R), or the oscillation of energy between a capacitor and an inductor. One way to characterize the dynamics is in terms of the times RC, L/R and $\sqrt{LC}$, respectively.

Characteristic times describing rate processes on a continuum basis are a recurring theme. The electromagnetic times summarized in Table 1.4.1 are the field analogues of those familiar from circuit theory. Rather than defining the variables, reference is made to the section where the characteristic times are introduced. Some of the mechanical and thermal ones also have lumped parameter counterparts. For example, the viscous diffusion time, which represents the mechanical damping of ponderable material, is the continuum version of the damping rate for a dash-pot connected to a mass.

The electromechanical characteristic times represent the competition between electric or magnetic forces and viscous or inertial forces. In specialized areas, they may appear in a different guise. For example, with the electric field intensity $\vec{E}$ that due to the bunching of electrons in a plasma, the electro-inertial time is the reciprocal plasma frequency. In a highly conducting fluid stressed by a magnetic field intensity H, the magneto-inertial time is the transit time for an Alfvén wave.

Especially in fluid mechanics, these characteristic times are often brought into play as dimensionless ratios of times. Table 1.4.2 gives some of these ratios, again with references to the sections where they are introduced.

1.5 Models and Approximations

There are three classes of approximation, used repeatedly in the following chapters, that should be recognized as a recurring theme. Formally, these are based on time-rate, space-rate and amplitude-parameter expansions of the relevant laws.

The time-rate approximation gives rise to a quasistatic model, and exploits the fact that temporal rates of change of interest are slow compared to one or more times characterizing certain dynamical processes. Some possible times are given in Table 1.4.1. Both for electroquasistatics and magnetoquasistatics, the critical time is the electromagnetic wave transit time, τ_{em} (Sec. 2.3).

Space-rate approximations lead to quasi-one-dimensional (or two-dimensional) models. These are also known as long-wave models. Here, fields or deformations in a "transverse" direction can be approximated as being slowly varying with respect to a "longitidunal" direction. The magnetic field in a narrow but spatially varying air gap and the flow of a gas through a duct of slowly varying cross section are examples.

Amplitude parameter expansions carried to first order result in linearized models. Often they are used to describe dynamics departing from a static or steady equilibrium. Long-wave and linearized models are discussed and exemplified in Sec. 4.12, and are otherwise used repeatedly without formality.

Table 1.4.1. Characteristic times for systems having a typical length ℓ.

Time	Nomenclature	Section reference
	Electromagnetic	
$\tau_{em} = \ell/c$	Electromagnetic wave transit time	2.3
$\tau_e = \varepsilon/\sigma$	Charge relaxation time	2.3, 5.10
$\tau_m = \mu\sigma\ell^2$	Magnetic diffusion time	2.3, 6.2
$\tau_{mig} = \ell/bE$	Particle migration time	5.9
	Mechanical and thermal	
$\tau_a = \ell/a$	Acoustic wave transit time	7.11
$\tau_v = \rho\ell^2/\eta$	Viscous diffusion time	7.18, 7.24
$\tau_c = \eta/\rho a^2$	Viscous relaxation time	7.24
$\tau_D = \ell^2/\kappa$	Molecular diffusion time	10.2
$\tau_T = \ell^2\rho c_v/k_T$	Thermal diffusion time	10.2
	Electromechanical	
$\tau_{EV} = \eta/\varepsilon E^2$	Electro-viscous time	8.7
$\tau_{MV} = \eta/\mu H^2$	Magneto-viscous time	8.6
$\tau_{EI} = \ell\sqrt{\rho/\varepsilon E^2}$	Electro-inertial time	8.7
$\tau_{MI} = \ell\sqrt{\rho/\mu H^2}$	Magneto-inertial time	8.6

Table. 1.4.2. Dimensionless numbers as ratios of characteristic times. The material transit or residence time is $\tau = \ell/U$, where U is a typical material velocity.

Number	Symbol	Nomenclature	Sec. ref.
		Electromagnetic	
$\tau_e/\tau = \varepsilon U/\ell\sigma$	R_e	Electric Reynolds number	5.11
$\tau_m/\tau = \mu\sigma\ell U$	R_m	Magnetic Reynolds number	6.2
		Mechanical and thermal	
$\tau_a/\tau = U/a$	M	Mach number	9.19
$\tau_v/\tau = \rho\ell U/\eta$	R_y	Reynolds number	7.18
$\tau_D/\tau = \ell U/\kappa$	R_D	Molecular Peclet number	10.2
$\tau_T/\tau = \rho c_p\ell U/k_T$	R_T	Thermal Peclet number	10.2
$\tau_D/\tau_v = \eta/\rho k_D$	P_D	Molecular-viscous Prandtl number	10.2
$\tau_T/\tau_v = c_p\eta/k_T$	P_T	Thermal-viscous Prandtl number	10.2
		Electromechanical	
$\sqrt{\tau_m/\tau_{MV}} = \mu H\ell\sqrt{\sigma/\eta}$	H_m	Magnetic Hartmann number	8.6
$\sqrt{\tau_{mig}/\tau_{EV}}$; τ_e/τ_{EV}	H_e	Electric Hartmann number	9.12
$\tau_m/\tau_v = \eta\mu\sigma/\rho$	P_m	Magnetic-viscous Prandtl number	8.6

1.6 Transfer Relations and Continuum Dynamics of Linear Systems

Fields, flows and deformations in systems that are uniform in one or more "longitudinal" directions can have the dependence on the associated coordinate represented by complex amplitudes, Fourier series, Fourier transforms, or the appropriate extension of these in various coordinate systems. Typically, configurations are nonuniform in the remaining "transverse" coordinate. The dependence of variables on this direction is represented by "transfer relations." They are first introduced in Chap. 2 as flux-potential relations that encapsulate Laplacian fields in coordinate systems for which Laplace's equation is variable separable.

At the risk of having a forbidding appearance, most chapters include summaries of transfer relations in the three common coordinate systems. This is done so that they can be a resource, helping to obviate tedious manipulations that tend to obscure what is essential in the derivation of a model. The transfer relations help in organizing a development. Once the way in which they represent the space-time dynamics of a given medium is appreciated, they are also a way of quickly communicating the physical nature of a continuum.

Applications in Chap. 4 begin to exemplify how the transfer relations can help to organize the representation of configurations involving piece-wise uniform media. The systems considered there are spatially periodic in the "longitudinal" direction.

With each of the subsequent chapters, the application of the transfer relations is broadened. In Chap. 5, the temporal transient response is described in terms of the temporal modes. Then, spatial transients for systems in the temporal sinusoidal steady state are considered. In Chap. 6, magnetic diffusion processes are represented in terms of transfer relations, which take a form equally applicable to thermal and particle diffusion.

Much of the summary of fluid mechanics given in Chap. 7 is couched in terms of transfer relations. There, the variables are velocities and stresses. In a wealth of electromechanical examples, coupling between fields and media can be represented as occurring at boundaries and interfaces, where there are discontinuities in properties. Thus, in Chap. 8, the purely mechanical relations of Chap. 7 are combined with the electrical relations from Chap. 2 to represent electromechanical systems. More specialized are electromechanical transfer relations representing charged fluids, electron beams, hydromagnetic systems and the like, derived in Chaps. 8-11.

A feature of many of the examples in Chap. 8 is instability, so that again the temporal modes come to the fore. But with effects of streaming brought into play in Chap. 11, there is a question of whether the instability is absolute in the sense that the response becomes unbounded with time at a given point in space, or convective (amplifying) in that a sinusoidal steady state can be established but with a response that becomes unbounded in space. These issues are taken up in Chap. 11.

2

Electrodynamic Laws, Approximations and Relations

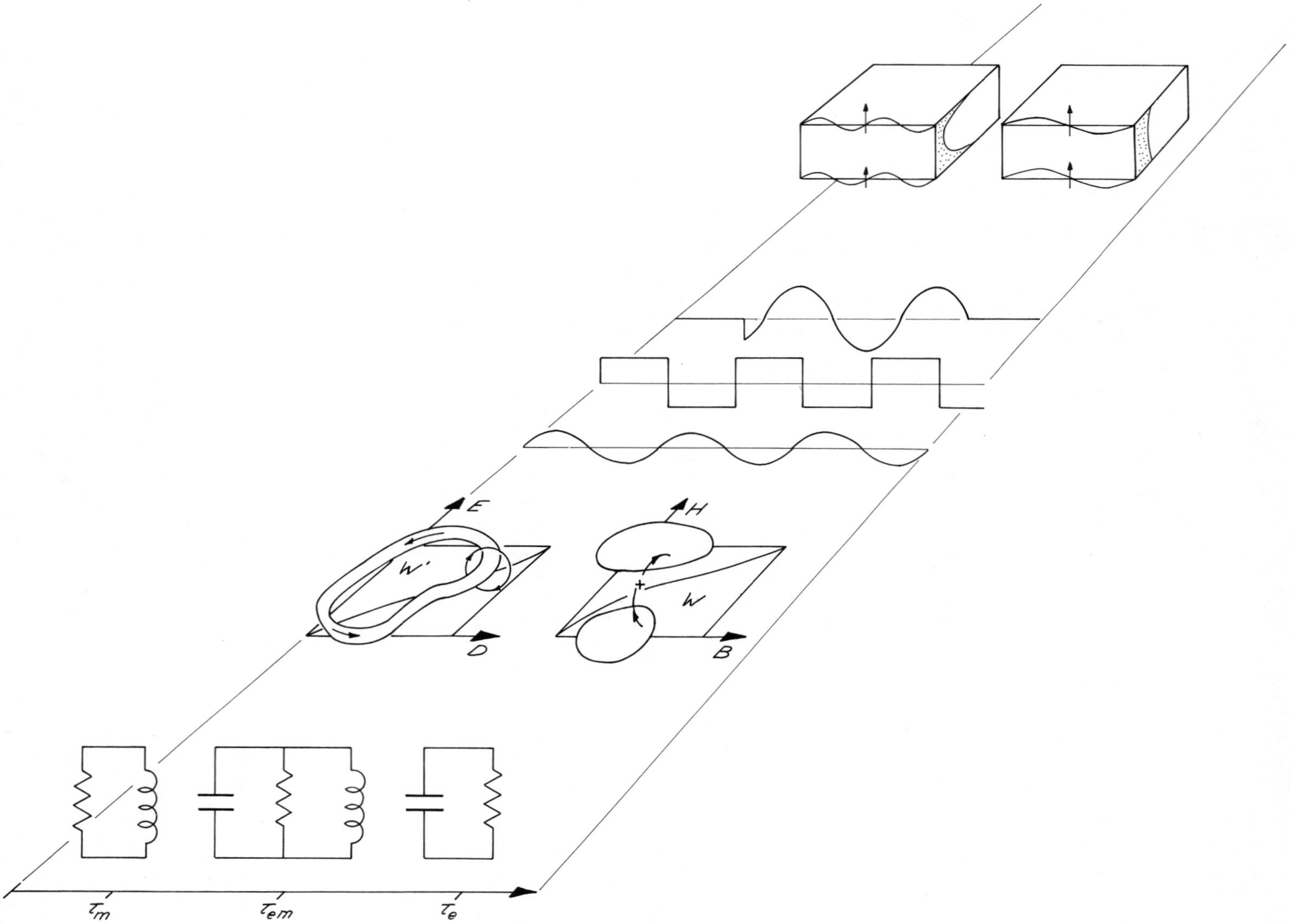

2.1 Definitions

Continuum electromechanics brings together several disciplines, and so it is useful to summarize the definitions of electrodynamic variables and their units. Rationalized MKS units are used not only in connection with electrodynamics, but also in dealing with subjects such as fluid mechanics and heat transfer, which are often treated in English units. Unless otherwise given, basic units of meters (m), kilograms (kg), seconds (sec), and Coulombs (C) can be assumed.

Table 2.1.1. Summary of electrodynamic nomenclature.

Name	Symbol	Units
Discrete Variables		
Voltage or potential difference	v	[V] = volts = m^2 kg/C sec^2
Charge	q	[C] = Coulombs = C
Current	i	[A] = Amperes = C/sec
Magnetic flux	λ	[Wb] = Weber = m^2 kg/C sec
Capacitance	C	[F] = Farad C^2 sec^2/m^2 kg
Inductance	L	[H] = Henry = m^2 kg/C^2
Force	f	[N] = Newtons = kg m/sec^2
Field Sources		
Free charge density	ρ_f	C/m^3
Free surface charge density	σ_f	C/m^2
Free current density	$\vec{J}_f$	A/m^2
Free surface current density	$\vec{K}_f$	A/m
Fields (name in quotes is often used for convenience)		
"Electric field" intensity	$\vec{E}$	V/m
"Magnetic field" intensity	$\vec{H}$	A/m
Electric displacement	$\vec{D}$	C/m^2
Magnetic flux density	$\vec{B}$	Wb/m^2 (tesla)
Polarization density	$\vec{P}$	C/m^2
Magnetization density	$\vec{M}$	A/m
Force density	$\vec{F}$	N/m^3
Physical Constants		
Permittivity of free space	$\varepsilon_o = 8.854 \times 10^{-12}$	F/m
Permeability of free space	$\mu_o = 4\pi \times 10^{-7}$	H/m

Although terms involving moving magnetized and polarized media may not be familiar, Maxwell's equations are summarized without prelude in the next section. The physical significance of the unfamiliar terms can best be discussed in Secs. 2.8 and 2.9 after the general laws are reduced to their quasistatic forms, and this is the objective of Sec. 2.3. Except for introducing concepts concerned with the description of continua, including integral theorems, in Secs. 2.4 and 2.6, and the discussion of Fourier amplitudes in Sec. 2.15, the remainder of the chapter is a parallel development of the consequences of these quasistatic laws. That the field transformations (Sec. 2.5), integral laws (Sec. 2.7), splicing conditions (Sec. 2.10), and energy storages are derived from the fundamental quasistatic laws, illustrates the important dictum that internal consistency be maintained within the framework of the quasistatic approximation.

The results of the sections on energy storage are used in Chap. 3 for deducing the electric and magnetic force densities on macroscopic media. The transfer relations of the last sections are an important resource throughout all of the following chapters, and give the opportunity to explore the physical significance of the quasistatic limits.

2.2 Differential Laws of Electrodynamics

In the Chu formulation,[1] with material effects on the fields accounted for by the magnetization density $\vec{M}$ and the polarization density $\vec{P}$ and with the material velocity denoted by $\vec{v}$, the laws of electrodynamics are:

Faraday's law

$$\nabla \times \vec{E} = -\mu_o \frac{\partial \vec{H}}{\partial t} - \mu_o \frac{\partial \vec{M}}{\partial t} - \mu_o \nabla \times (\vec{M} \times \vec{v}) \tag{1}$$

---

1. P. Penfield, Jr., and H. A. Haus, Electrodynamics of Moving Media, The M.I.T. Press, Cambridge, Massachusetts, 1967, pp. 35-40.

Ampère's law

$$\nabla \times \vec{H} = \varepsilon_o \frac{\partial \vec{E}}{\partial t} + \frac{\partial \vec{P}}{\partial t} + \nabla \times (\vec{P} \times \vec{v}) + \vec{J}_f \tag{2}$$

Gauss' law

$$\varepsilon_o \nabla \cdot \vec{E} = -\nabla \cdot \vec{P} + \rho_f \tag{3}$$

divergence law for magnetic fields

$$\mu_o \nabla \cdot \vec{H} = -\mu_o \nabla \cdot \vec{M} \tag{4}$$

and conservation of free charge

$$\nabla \cdot \vec{J}_f + \frac{\partial \rho_f}{\partial t} = 0 \tag{5}$$

This last expression is imbedded in Ampère's and Gauss' laws, as can be seen by taking the divergence of Eq. 2 and exploiting Eq. 3. In this formulation the electric displacement $\vec{D}$ and magnetic flux density $\vec{B}$ are defined fields:

$$\vec{D} = \varepsilon_o \vec{E} + \vec{P} \tag{6}$$

$$\vec{B} = \mu_o (\vec{H} + \vec{M}) \tag{7}$$

2.3 Quasistatic Laws and the Time-Rate Expansion

With a quasistatic model, it is recognized that relevant time rates of change are sufficiently low that contributions due to a particular dynamical process are ignorable. The objective in this section is to give some formal structure to the reasoning used to deduce the quasistatic field equations from the more general Maxwell's equations. Here, quasistatics specifically means that times of interest are long compared to the time, τ_{em}, for an electromagnetic wave to propagate through the system.

Generally, given a dynamical process characterized by some time determined by the parameters of the system, a quasistatic model can be used to exploit the comparatively long time scale for processes of interest. In this broad sense, quasistatic models abound and will be encountered in many other contexts in the chapters that follow. Specific examples are:

(a) processes slow compared to wave transit times in general; acoustic waves and the model is one of incompressible flow, Alfvén and other electromechanical waves and the model is less standard;

(b) processes slow compared to diffusion (instantaneous diffusion models). What diffuses can be magnetic field, viscous stresses, heat, molecules or hybrid electromechanical effects;

(c) processes slow compared to relaxation of continua (instantaneous relaxation or constant-potential models). Charge relaxation is an important example.

The point of making a quasistatic approximation is often to focus attention on significant dynamical processes. A quasistatic model is by no means static. Because more than one rate process is often imbedded in a given physical system, it is important to agree upon the one with respect to which the dynamics are quasistatic.

Rate processes other than those due to the transit time of electromagnetic waves enter through the dependence of the field sources on the fields and material motion. To have in view the additional characteristic times typically brought in by the field sources, in this section the free current density is postulated to have the dependence

$$\vec{J}_f = \sigma(\vec{r})\vec{E} + \vec{J}_v(\vec{v}, \rho_f, \vec{H}) \tag{1}$$

In the absence of motion, $\vec{J}_v$ is zero. Thus, for media at rest the conduction model is ohmic, with the electrical conductivity σ in general a function of position. Examples of $\vec{J}_v$ are a convection current $\rho_f \vec{v}$, or an ohmic motion-induced current $\sigma(\vec{v} \times \mu_o \vec{H})$. With an underbar used to denote a normalized quantity, the conductivity is normalized to a typical (constant) conductivity σ_o:

$$\sigma = \sigma_o \, \underline{\sigma}(\vec{r}, t) \tag{2}$$

To identify the hierarchy of critical time-rate parameters, the general laws are normalized. Coordinates are normalized to one typical length ℓ, while τ represents a characteristic dynamical time:

$$(x,y,z) = (\ell\underline{x}, \ell\underline{y}, \ell\underline{z}); \quad t = \tau\underline{t} \tag{3}$$

In a system sinusoidally excited at the angular frequency ω, $\tau = \omega^{-1}$.

The most convenient normalization of the fields depends on the specific system. Where electromechanical coupling is significant, these can usually be categorized as "electric-field dominated" and "magnetic-field dominated." Anticipating this fact, two normalizations are now developed "in parallel," the first taking $\mathcal{E}$ as a characteristic electric field and the second taking $\mathcal{H}$ as a characteristic magnetic field:

$$\vec{E} = \mathcal{E}\underline{\vec{E}},\ \vec{P} = \varepsilon_o\mathcal{E}\underline{\vec{P}},\ \vec{v} = (\ell/\tau)\underline{\vec{v}},\ \vec{J}_v = \frac{\varepsilon_o\mathcal{E}}{\tau}\underline{\vec{J}}_v,\quad \rho_f = \frac{\mathcal{E}\varepsilon_o}{\ell}\underline{\rho}_f,\ \vec{H} = \frac{\varepsilon_o\ell\mathcal{E}}{\tau}\underline{\vec{H}},\ \vec{M} = \frac{\varepsilon_o\ell\mathcal{E}}{\tau}\underline{\vec{M}}$$

$$\vec{H} = \mathcal{H}\underline{\vec{H}},\ \vec{M} = \mathcal{H}\underline{\vec{M}},\ \vec{v} = (\ell/\tau)\underline{\vec{v}},\ \vec{J}_v = \frac{\mathcal{H}}{\ell}\underline{\vec{J}}_v,\quad \vec{E} = \frac{\mu_o\ell\mathcal{H}}{\tau}\underline{\vec{E}},\ \rho_f = \frac{\varepsilon_o\mu_o\mathcal{H}}{\tau}\underline{\rho}_f,\ \vec{P} = \frac{\varepsilon_o\mu_o\ell\mathcal{H}}{\tau}\underline{\vec{P}} \qquad (4)$$

It might be appropriate with this step to recognize that the material motion introduces a characteristic (transport) time other than τ. For simplicity, Eq. 4 takes the material velocity as being of the order of ℓ/τ.

The normalization used is arbitrary. The same quasistatic laws will be deduced regardless of the starting point, but the normalization will determine whether these laws are "zero-order" or higher order in a sense to now be defined.

The normalizations of Eq. 4 introduced into Eqs. 2.2.1-5 result in

Electric normalization:

$$\nabla\cdot\vec{E} = -\nabla\cdot\vec{P} + \rho_f$$

$$\nabla\cdot\vec{H} = -\nabla\cdot\vec{M}$$

$$\nabla\times\vec{H} = \frac{\tau}{\tau_e}\sigma\vec{E} + \vec{J}_v + \frac{\partial\vec{E}}{\partial t} + \frac{\partial\vec{P}}{\partial t} + \nabla\times(\vec{P}\times\vec{v})$$

$$\nabla\times E = -\beta\left[\frac{\partial\vec{H}}{\partial t} + \frac{\partial\vec{M}}{\partial t} + \nabla\times(\vec{M}\times\vec{v})\right]$$

$$\nabla\cdot\sigma\vec{E} + \frac{\tau_e}{\tau}\left[\nabla\cdot\vec{J}_v + \frac{\partial\rho_f}{\partial t}\right] = 0$$

Magnetic normalization:

$$\nabla\cdot\vec{E} = -\nabla\cdot\vec{P} + \rho_f \qquad (\underline{5})$$

$$\nabla\cdot\vec{H} = -\nabla\cdot\vec{M} \qquad (\underline{6})$$

$$\nabla\times\vec{H} = \frac{\tau_m}{\tau}\sigma\vec{E} + \vec{J}_v + \beta\left[\frac{\partial E}{\partial t} + \frac{\partial P}{\partial t} + \nabla\times(\vec{P}\times\vec{v})\right] \qquad (\underline{7})$$

$$\nabla\times\vec{E} = -\frac{\partial\vec{H}}{\partial t} - \frac{\partial\vec{M}}{\partial t} - \nabla\times(\vec{M}\times\vec{v}) \qquad (\underline{8})$$

$$\nabla\cdot\sigma\vec{E} + \frac{\tau}{\tau_m}\nabla\cdot\vec{J}_v + \beta\frac{\tau}{\tau_m}\frac{\partial\rho_f}{\partial t} = 0 \qquad (\underline{9})$$

where underbars on equation numbers are used to indicate that the equations are normalized and

$$\tau_m \equiv \mu_o\sigma_o\ell^2,\quad \tau_e \equiv \varepsilon_o/\sigma_o$$

and

$$\beta = \left(\frac{\tau_{em}}{\tau}\right)^2;\quad \tau_{em} \equiv \sqrt{\mu_o\varepsilon_o}\,\ell = \ell/c \qquad (10)$$

In Chap. 6, τ_m will be identified as the magnetic diffusion time, while in Chap. 5 the role of the charge-relaxation time τ_e is developed. The time required for an electromagnetic plane wave to propagate the distance ℓ at the velocity c is τ_{em}. Given that there is just one characteristic length, there are actually only two characteristic times, because as can be seen from Eq. 10

$$\sqrt{\tau_m\tau_e} = \tau_{em} \qquad (11)$$

Unless τ_e and τ_m, and hence τ_{em}, are all of the same order, there are only two possibilities for the relative magnitudes of these times, as summarized in Fig. 2.3.1.

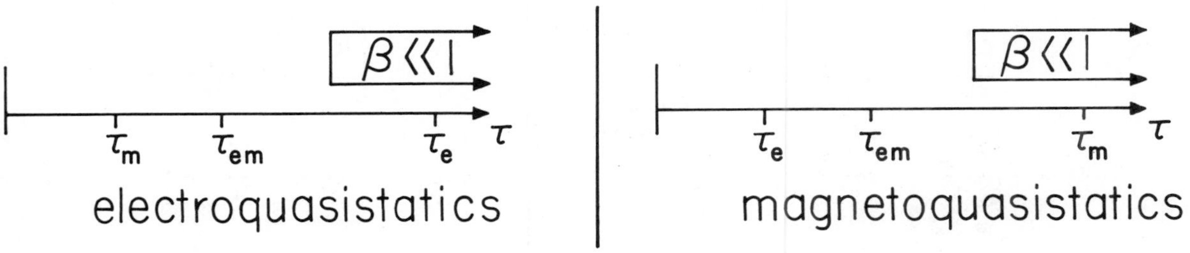

Fig. 2.3.1. Possible relations between physical time constants on a time scale τ which typifies the dynamics of interest.

By electroquasistatic (EQS) approximation it is meant that the ordering of times is as to the left and that the parameter $\beta = (\tau_{em}/\tau)^2$ is much less than unity. Note that τ is still arbitrary relative to τ_e. In the magnetoquasistatic (MQS) approximation, β is still small, but the ordering of characteristic times is as to the right. In this case, τ is arbitrary relative to τ_m.

To make a formal statement of the procedure used to find the quasistatic approximation, the normalized fields and charge density are expanded in powers of the <u>time-rate parameter</u> β.

$$\vec{E} = \vec{E}_o + \beta\vec{E}_1 + \beta^2\vec{E}_2 + \cdots$$

$$\vec{H} = \vec{H}_o + \beta\vec{H}_1 + \beta^2\vec{H}_2 + \cdots$$

$$\vec{J}_v = (\vec{J}_v)_o + \beta(\vec{J}_v)_1 + \beta^2(\vec{J}_v)_2 + \cdots$$

$$\rho_f = (\rho_f)_o + \beta(\rho_f)_1 + \beta^2(\rho_f)_2 + \cdots$$

(12)

In the following, it is assumed that constitutive laws relate $\vec{P}$ and $\vec{M}$ to $\vec{E}$ and $\vec{H}$, so that these densities are similarly expanded. The velocity $\vec{v}$ is taken as given. Then, the series are substituted into Eqs. 5-9 and the resulting expressions arranged by factors multiplying ascending powers of β. The "zero order" equations are obtained by requiring that the coefficients of β^0 vanish. These are simply Eqs. 5-9 with $\beta = 0$:

$$\nabla\cdot\vec{E}_o = -\nabla\cdot\vec{P}_o + (\rho_f)_o \qquad\qquad \nabla\cdot\vec{E} = -\nabla\cdot\vec{P}_o + (\rho_f)_o \quad (13)$$

$$\nabla\cdot\vec{H}_o = -\nabla\cdot\vec{M}_o \qquad\qquad\qquad \nabla\cdot\vec{H}_o = -\nabla\cdot\vec{M}_o \quad (14)$$

$$\nabla\times\vec{H}_o = \frac{\tau}{\tau_e}\sigma\vec{E}_o + (\vec{J}_v)_o + \frac{\partial\vec{E}_o}{\partial t} \qquad \nabla\times\vec{H}_o = \frac{\tau_m}{\tau}\sigma\vec{E}_o + (\vec{J}_v)_o \quad (15)$$

$$+ \frac{\partial\vec{P}_o}{\partial t} + \nabla\times(\vec{P}_o\times\vec{v})$$

$$\nabla\times\vec{E}_o = 0 \qquad\qquad \nabla\times E_o = -\frac{\partial\vec{H}_o}{\partial t} - \frac{\partial\vec{M}_o}{\partial t} - \nabla\times(\vec{M}_o\times\vec{v}) \quad (16)$$

$$\nabla\cdot\sigma\vec{E}_o + \frac{\tau_e}{\tau}\left[\nabla\cdot(\vec{J}_v)_o + \frac{\partial(\rho_f)_o}{\partial t}\right] = 0 \qquad \nabla\cdot\sigma\vec{E}_o + \frac{\tau}{\tau_m}\nabla\cdot(\vec{J}_v)_o = 0 \quad (17)$$

The zero-order solutions are found by solving these equations, augmented by appropriate boundary conditions. If the boundary conditions are themselves time dependent, normalization will turn up additional characteristic times that must be fitted into the hierarchy of Fig. 2.3.1.

Higher order contributions to the series of Eq. 12 follow from a sequential solution of the equations found by making coefficients of like powers of β vanish. The expressions resulting from setting the coefficients of β^n to zero are:

$$\nabla\cdot\vec{E}_n + \nabla\cdot\vec{P}_n - (\rho_f)_n = 0 \qquad\qquad \nabla\cdot\vec{E}_n + \nabla\cdot\vec{P}_n - (\rho_f)_n = 0 \quad (18)$$

$$\nabla\cdot\vec{H}_n + \nabla\cdot\vec{M}_n = 0 \qquad\qquad\qquad \nabla\cdot\vec{H}_n + \nabla\cdot\vec{M}_n = 0 \quad (19)$$

$$\nabla\times\vec{H}_n - \frac{\tau}{\tau_e}\sigma\vec{E}_n - (\vec{J}_v)_n - \frac{\partial\vec{E}_n}{\partial t} \qquad \nabla\times\vec{H}_n - \frac{\tau_m}{\tau}\sigma\vec{E}_n - (\vec{J}_v)_n =$$

$$- \frac{\partial\vec{P}_n}{\partial t} - \nabla\times(\vec{P}_n\times\vec{v}) = 0 \qquad \left[\frac{\partial\vec{E}_{n-1}}{\partial t} + \frac{\partial\vec{P}_{n-1}}{\partial t} + \nabla\times(\vec{P}_{n-1}\times\vec{v})\right] \quad (20)$$

$$\nabla\times E_n = -\left[\frac{\partial\vec{H}_{n-1}}{\partial t} + \frac{\partial M_{n-1}}{\partial t} + \nabla\times(\vec{M}_{n-1}\times\vec{v})\right] \qquad \nabla\times\vec{E}_n + \frac{\partial\vec{H}_n}{\partial t} + \frac{\partial\vec{M}_n}{\partial t} + \nabla\times(\vec{M}_o\times\vec{v}) = 0 \quad (21)$$

$$\nabla\cdot\sigma\vec{E}_n + \frac{\tau_e}{\tau}\left[\nabla\cdot(\vec{J}_v)_n + \frac{\partial(\rho_f)_n}{\partial t}\right] = 0 \qquad \nabla\cdot\sigma\vec{E}_n + \frac{\tau}{\tau_m}\nabla\cdot(\vec{J}_v)_n = -\frac{\tau}{\tau_m}\frac{\partial(\rho_f)_{n-1}}{\partial t} \quad (22)$$

To find the first order contributions, these equations with n=1 are solved with the zero order solutions making up the right-hand sides of the equations playing the role of known driving functions. Boundary conditions are satisfied by the lowest order fields. Thus higher order fields satisfy homogeneous boundary conditions.

Once the first order solutions are known, the process can be repeated with these forming the "drives" for the n=2 equations.

In the absence of loss effects, there are no characteristic times to distinguish MQS and EQS systems. In that limit, which set of normalizations is used is a matter of convenience. If a situation represented by the left-hand set actually has an EQS limit, the zero order laws become the quasistatic laws. But, if these expressions are applied to a situation that is actually MQS, then first-order terms must be calculated to find the quasistatic fields. If more than the one characteristic time τ_{em} is involved, as is the case with finite τ_e and τ_m, then the ordering of rate parameters can contribute to the convergence of the expansion.

In practice, a formal derivation of the quasistatic laws is seldom used. Rather, intuition and experience along with comparison of critical time constants to relevant dynamical times is used to identify one of the two sets of zero order expressions as appropriate. But, the use of normalizations to identify critical parameters, and the notion that characteristic times can be used to unscramble dynamical processes, will be used extensively in the chapters to follow.

Within the framework of quasistatic electrodynamics, the unnormalized forms of Eqs. 13-17 comprise the "exact" field laws. These equations are reordered to reflect their relative importance:

Electroquasistatic (EQS)	Magnetoquasistatic (MQS)	
$\nabla\cdot\varepsilon_o\vec{E} = -\nabla\cdot\vec{P} + \rho_f$	$\nabla\times\vec{H} = \vec{J}_f$	(23)
$\nabla\times\vec{E} = 0$	$\nabla\cdot\mu_o\vec{H} = -\nabla\cdot\mu_o\vec{M}$	(24)
$\nabla\cdot\vec{J}_f + \frac{\partial\rho_f}{\partial t} = 0$	$\nabla\times\vec{E} = -\frac{\partial\mu_o\vec{H}}{\partial t} - \frac{\partial\mu_o\vec{M}}{\partial t} - \mu_o\nabla\times(\vec{M}\times\vec{v})$	(25)
$\nabla\times\vec{H} = \vec{J}_f + \frac{\partial\varepsilon_o\vec{E}}{\partial t} + \frac{\partial\vec{P}}{\partial t} + \nabla\times(\vec{P}\times\vec{v})$	$\nabla\cdot\vec{J}_f = 0$	(26)
$\nabla\cdot\mu_o\vec{H} = -\nabla\,\mu_o\vec{M}$	$\nabla\cdot\varepsilon_o\vec{E} = -\nabla\cdot\vec{P} + \rho_f$	(27)

The conduction current $\vec{J}_f$ has been reintroduced to reflect the wider range of validity of these equations than might be inferred from Eq. 1. With different conduction models will come different characteristic times, exemplified in the discussions of this section by τ_e and τ_m. Matters are more complicated if fields and media interact electromechanically. Then, $\vec{v}$ is determined to some extent at least by the fields themselves and must be treated on a par with the field variables. The result can be still more characteristic times.

The ordering of the quasistatic equations emphasizes the instantaneous relation between the respective dominant sources and fields. Given the charge and polarization densities in the EQS system, or given the current and magnetization densities in the MQS system, the dominant fields are known and are functions only of the sources at the given instant in time.

The dynamics enter in the EQS system with conservation of charge, and in the MQS system with Faraday's law of induction. Equations 26a and 27a are only needed if an after-the-fact determination of $\vec{H}$ is to be made. An example where such a rare interest in $\vec{H}$ exists is in the small magnetic field induced by electric fields and currents within the human body. The distribution of internal fields and hence currents is determined by the first three EQS equations. Given $\vec{E}$, $\vec{P}$, and $\vec{J}_f$, the remaining two expressions determine $\vec{H}$. In the MQS system, Eq. 27b can be regarded as an expression for the after-the-fact evaluation of ρ_f, which is not usually of interest in such systems.

What makes the subject of quasistatics difficult to treat in a general way, even for a system of fixed ohmic conductivity, is the dependence of the appropriate model on considerations not conveniently represented in the differential laws. For example, a pair of perfectly conducting plates, shorted on one pair of edges and driven by a sinusoidal source at the opposite pair, will be MQS at low frequencies. The same pair of plates, open-circuited rather than shorted, will be electroquasistatic at low frequencies. The difference is in the boundary conditions.

Geometry and the inhomogeneity of the medium (insulators, perfect conductors and semiconductors) are also essential to determining the appropriate approximation. Most systems require more than one characteristic dimension and perhaps conductivity for their description, with the result that more than two time constants are often involved. Thus, the two possibilities identified in Fig. 2.3.1 can in principle become many possibilities. Even so, for a wide range of practical problems, the appropriate field laws are either clearly electroquasistatic or magnetoquasistatic.

Problems accompanying this section help to make the significance of the quasistatic limits more substantive by considering cases that can also be solved exactly.

2.4 Continuum Coordinates and the Convective Derivative

There are two commonly used representations of continuum variables. One of these is familiar from classical mechanics, while the other is universally used in electrodynamics. Because electromechanics involves both of these subjects, attention is now drawn to the salient features of the two representations.

Consider first the "Lagrangian representation." The position of a material particle is a natural example and is depicted by Fig. 2.4.1a. When the time t is zero, a particle is found at the position $\vec{r}_o$. The position of the particle at some subsequent time is $\vec{\xi}$. To let $\vec{\xi}$ represent the displacement of a continuum of particles, the position variable $\vec{r}_o$ is used to distinguish particles. In this sense, the displacement $\vec{\xi}$ then also becomes a continuum variable capable of representing the relative displacements of an infinitude of particles.

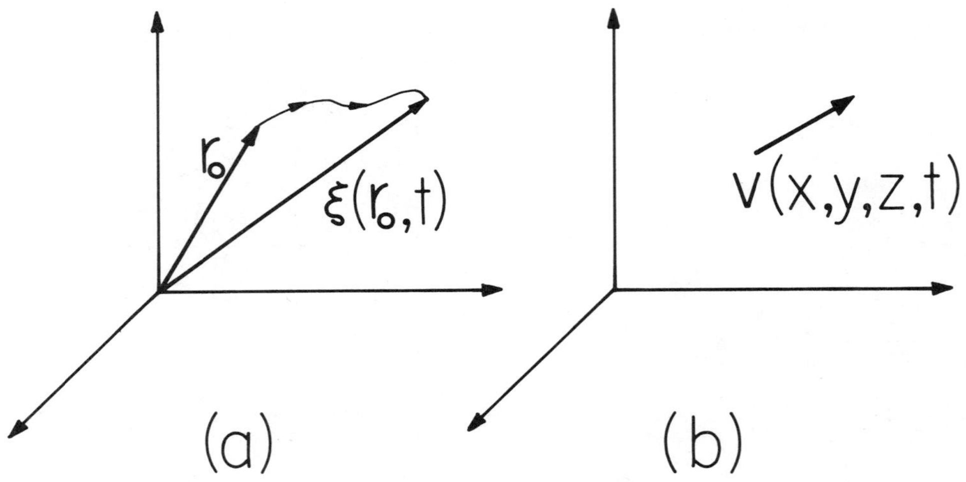

Fig. 2.4.1. Particle motions represented in terms of (a) Lagrangian coordinates, where the initial particle coordinate $\vec{r}_o$ designates the particle of interest, and (b) Eulerian coordinates, where (x,y,z) designates the spatial position of interest.

In a Lagrangian representation, the velocity of the particle is simply

$$\vec{v} = \frac{\partial \vec{\xi}}{\partial t} \tag{1}$$

If concern is with only one particle, there is no point in writing the derivative as a partial derivative. However, it is understood that, when the derivative is taken, it is a particular particle which is being considered. So, it is understood that $\vec{r}_o$ is fixed. Using the same line of reasoning, the acceleration of a particle is given by

$$\vec{a} = \frac{\partial \vec{v}}{\partial t} \tag{2}$$

The idea of representing continuum variables in terms of the coordinates (x,y,z) connected with the space itself is familiar from electromagnetic theory. But what does it mean if the variable is mechanical rather than electrical? We could represent the velocity of the continuum of particles filling the space of interest by a vector function $\vec{v}(x,y,z,t) = \vec{v}(\vec{r},t)$. The velocity of particles having the position (x,y,z,) at a given time t is determined by evaluating the function $\vec{v}(\vec{r},t)$. The velocity appearing in Sec. 2.2 is an example. As suggested by Fig. 2.4.1b, if the function is the velocity evaluated at a given position in space, it describes whichever particle is at that point at the time of interest. Generally, there is a continuous stream of particles through the point (x,y,z).

Computation of the particle acceleration makes evident the contrast between Eulerian and Lagrangian representations. By definition, the acceleration is the rate of change of the velocity computed for a given particle of matter. A particle having the position (x,y,z) at time t will be found an instant Δt later at the position $(x + v_x\Delta t, y + v_y\Delta t, z + v_z\Delta t)$. Hence the acceleration is

$$\vec{a} = \lim_{\Delta t \to 0} \frac{\vec{v}(x + v_x\Delta t, y + v_y\Delta t, z + v_z\Delta t, t + \Delta t) - \vec{v}(x,y,z,t)}{\Delta t} \tag{3}$$

Expansion of the first term in Eq. 3 about the initial coordinates of the particle gives the <u>convective derivative</u> of $\vec{v}$:

$$\vec{a} = \frac{\partial \vec{v}}{\partial t} + v_x \frac{\partial \vec{v}}{\partial x} + v_y \frac{\partial \vec{v}}{\partial y} + v_z \frac{\partial \vec{v}}{\partial z} \equiv \frac{\partial \vec{v}}{\partial t} + \vec{v}\cdot\nabla\vec{v} \tag{4}$$

The difference between Eq. 2 and Eq. 4 is resolved by recognizing the difference in the significance of the partial derivatives. In Eq. 2, it is understood that the coordinates being held fixed are the initial coordinates of the particle of interest. In Eq. 4, the partial derivative is taken, holding fixed the particular point of interest in space.

The same steps show that the <u>rate of change of any vector variable $\vec{A}$, as viewed from a particle having the velocity $\vec{v}$,</u> is

$$\frac{D\vec{A}}{Dt} \equiv \frac{\partial \vec{A}}{\partial t} + (\vec{v}\cdot\nabla)\vec{A}; \quad \vec{A} = \vec{A}(x,y,z,t) \tag{5}$$

The time rate of change of any scalar variable for an observer moving with the velocity $\vec{v}$ is obtained from Eq. 5 by considering the particular case in which $\vec{A}$ has only one component, say $\vec{A} = f(x,y,z,t)\vec{i}_x$. Then Eq. 5 becomes

$$\frac{Df}{Dt} \equiv \frac{\partial f}{\partial t} + \vec{v}\cdot\nabla f \tag{6}$$

Reference 3 of Appendix C is a film useful in understanding this section.

2.5 <u>Transformations between Inertial Frames</u>

In extending empirically determined conduction, polarization and magnetization laws to include material motion, it is often necessary to relate field variables evaluated in different reference frames. A given point in space can be designated either in terms of the coordinate $\vec{r}$ or of the coordinate $\vec{r}'$ of Fig. 2.5.1. By "<u>inertial reference frames</u>," it is meant that the relative velocity between these two frames is constant, designated by $\vec{u}$. The positions in the two coordinate systems are related by the <u>Galilean transformation</u>:

$$\vec{r}' = \vec{r} - \vec{u}t; \quad t' = t \tag{1}$$

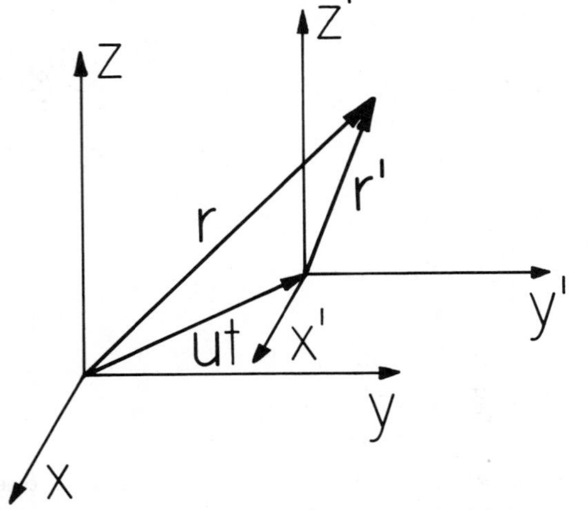

Fig. 2.5.1

Reference frames have constant relative velocity $\vec{u}$. The coordinates $\vec{r} = (x,y,z)$ and $\vec{r}' = (x',y',z')$ designate the same position.

It is a familiar fact that variables describing a given physical situation in one reference frame will not be the same as those in the other. An example is material velocity, which, if measured in one frame, will differ from that in the other frame by the relative velocity $\vec{u}$.

There are two objectives in this section: one is to show that the quasistatic laws are invariant when subject to a Galilean transformation between inertial reference frames. But, of more use is the relationship between electromagnetic variables in the two frames of reference that follows from this proof. The approach is as follows. First, the postulate is made that the quasistatic equations take the same form in the primed and unprimed inertial reference frames. But, in writing the laws in the primed frame, the spatial and temporal derivatives must be taken with respect to the coordinates of that reference frame, and the dependent field variables are then fields defined in that reference frame. In general, these must be designated by primes, since their relation to the variables in the unprimed frame is not known.

For the purpose of writing the primed equations of electrodynamics in terms of the unprimed coordinates, recognize that

$$\nabla' \rightarrow \nabla$$

$$\frac{\partial \vec{A}}{\partial t'} \rightarrow \left(\frac{\partial}{\partial t} + \vec{u}\cdot\nabla\right)\vec{A} \equiv \frac{\partial \vec{A}}{\partial t} + \vec{u}\nabla\cdot\vec{A} - \nabla \times (\vec{u} \times \vec{A}) \qquad (2)$$

$$\frac{\partial \psi}{\partial t'} \rightarrow \left(\frac{\partial}{\partial t} + \vec{u}\cdot\nabla\right)\psi \equiv \frac{\partial \psi}{\partial t} + \nabla\cdot\vec{u}\psi$$

The left relations follow by using the chain rule of differentiation and the transformation of Eq. 1. That the spatial derivatives taken with respect to one frame must be the same as those with respect to the other frame physically means that a single "snapshot" of the physical process would be all required to evaluate the spatial derivatives in either frame. There would be no way of telling which frame was the one from which the snapshot was taken. By contrast, the time rate of change for an observer in the primed frame is, by definition, taken with the primed spatial coordinates held fixed. In terms of the fixed frame coordinates, this is the convective derivative defined with Eqs. 2.4.5 and 2.4.6. However, $\vec{v}$ in these equations is in general a function of space and time. In the context of this section it is specialized to the constant $\vec{u}$. Thus, in rewriting the convective derivatives of Eq. 2 the constancy of $\vec{u}$ and a vector identity (Eq. 16, Appendix B) have been used.

So far, what has been said in this section is a matter of coordinates. Now, a physically motivated postulate is made concerning the electromagnetic laws. Imagine one electromagnetic experiment that is to be described from the two different reference frames. The postulate is that provided each of these frames is inertial, the governing laws must take the same form. Thus, Eqs. 23-27 apply with $[\nabla \rightarrow \nabla', \partial(\ )/\partial t \rightarrow \partial(\ )/\partial t']$ and all dependent variables primed. By way of comparing these laws to those expressed in the fixed-frame, Eqs. 2 are used to rewrite these expressions in terms of the unprimed independent variables. Also, the moving-frame material velocity is rewritten in terms of the unprimed frame velocity using the relation

$$\vec{v}' = \vec{v} - \vec{u} \qquad (3)$$

Thus, the laws originally expressed in the primed frame of reference become

$\nabla\cdot\varepsilon_o\vec{E}' = -\nabla\cdot\vec{P}' + \rho_f'$	$\nabla \times \vec{H}' = \vec{J}_f'$	(4)
$\nabla \times \vec{E}' = 0$	$\nabla\cdot\mu_o\vec{H}' = -\nabla\cdot\mu_o\vec{M}'$	(5)
$\nabla\cdot(\vec{J}_f' + \vec{u}\rho_f') + \dfrac{\partial \rho_f'}{\partial t} = 0$	$\nabla\times(\vec{E}' - \vec{u} \times \mu_o\vec{H}') = -\dfrac{\partial \mu_o\vec{H}'}{\partial t} - \dfrac{\partial \mu_o\vec{M}'}{\partial t} - \mu_o\nabla \times (\vec{M}' \times \vec{v})$	(6)
$\nabla \times (\vec{H}' + \vec{u} \times \varepsilon_o\vec{E}') = (\vec{J}_f' + \vec{u}\rho_f') + \dfrac{\partial \varepsilon_o\vec{E}'}{\partial t} + \dfrac{\partial \vec{P}'}{\partial t} + \nabla \times (\vec{P}' \times \vec{v})$	$\nabla\cdot\vec{J}_f' = 0$	(7)
$\nabla\cdot\mu_o\vec{H}' = -\nabla\cdot\mu_o\vec{M}'$	$\nabla\cdot\varepsilon_o\vec{E}' = -\nabla\cdot\vec{P}' + \rho_f'$	(8)

In writing Eq. 7a, Eq. 4a is used. Similarly, Eq. 5b is used to write Eq. 6b. For the one experiment under consideration, these equations will predict the same behavior as the fixed frame laws, Eqs. 2.3.23-27, if the identification is made:

EQS	MQS	
$\vec{E}' = \vec{E}$	$\vec{H}' = \vec{H}$	(9)
$\vec{P}' = \vec{P}$	$\vec{M}' = \vec{M}$	(10)
$\rho_f' = \rho_f$	$\vec{J}_f' = \vec{J}_f$	(11)
$\vec{J}_f' = \vec{J}_f - \vec{u}\rho_f$	$\vec{E}' = \vec{E} + \vec{u} \times \mu_o\vec{H}$	(12)
$\vec{H}' = \vec{H} - \vec{u} \times \varepsilon_o\vec{E}$		(13)
and hence, from Eq. 2.2.6	and hence, from Eq. 2.2.7	
$\vec{D}' = \vec{D}$	$\vec{B}' = \vec{B}$	(14)

The primary fields are the same whether viewed from one frame or the other. Thus, the EQS electric field polarization density and charge density are the same in both frames, as are the MQS magnetic field, magnetization density and current density. The respective dynamic laws can be associated with those field transformations that involve the relative velocity. That the free current density is altered by the relative motion of the net free charge in the EQS system is not surprising. But, it is the contribution of this same convection current to Ampère's law that generates the velocity dependent contribution to the EQS magnetic field measured in the moving frame of reference. Similarly, the velocity dependent contribution to the MQS electric field transformation is a direct consequence of Faraday's law.

The transformations, like the quasistatic laws from which they originate, are approximate. It would require Lorentz transformations to carry out a similar procedure for the exact electrodynamic laws of Sec. 2.2. The general laws are not invariant in form to a Galilean transformation, and therein is the origin of special relativity. Built in from the start in the quasistatic field laws is a self-consistency with other Galilean invariant laws describing mechanical continua that will be brought in in later chapters.

2.6 Integral Theorems

Several integral theorems prove useful, not only in the description of electromagnetic fields but also in dealing with continuum mechanics and electromechanics. These theorems will be stated here without proof.

If it is recognized that the gradient operator is defined such that its line integral between two endpoints (a) and (b) is simply the scalar function evaluated at the endpoints, then[1]

$$\int_{\vec{a}}^{\vec{b}} \nabla\psi\cdot\vec{d\ell} = \psi(\vec{b}) - \psi(\vec{a}) \qquad (1)$$

Two more familiar theorems[1] are useful in dealing with vector functions. For a closed surface S, enclosing the volume V, Gauss' theorem states that

$$\int_V \nabla\cdot\vec{A}dV = \oint_S \vec{A}\cdot\vec{n}da \qquad (2)$$

while Stokes's theorem pertains to an open surface S with the contour C as its periphery:

$$\int_S \nabla \times \vec{A}\cdot\vec{n}da = \oint_C \vec{A}\cdot\vec{d\ell} \qquad (3)$$

In stating these theorems, the normal vector is defined as being outward from the enclosed volume for Gauss' theorem, and the contour is taken as positive in a direction such that $\vec{d\ell}$ is related to $\vec{n}$ by the right-hand rule. Contours, surfaces, and volumes are sketched in Fig. 2.6.1.

A possibly less familiar theorem is the generalized Leibnitz rule.[2] In those cases where the surface is itself a function of time, it tells how to take the derivative with respect to time of the integral over an open surface of a vector function:

---

1. Markus Zahn, Electromagnetic Field Theory, a problem solving approach, John Wiley & Sons, New York, 1979, pp. 18-36.
2. H. H. Woodson and J. R. Melcher, Electromechanical Dynamics, Vol. 1, John Wiley & Sons, New York, 1968, pp. B32-B36.(See Prob. 2.6.2 for the derivation of this theorem.)

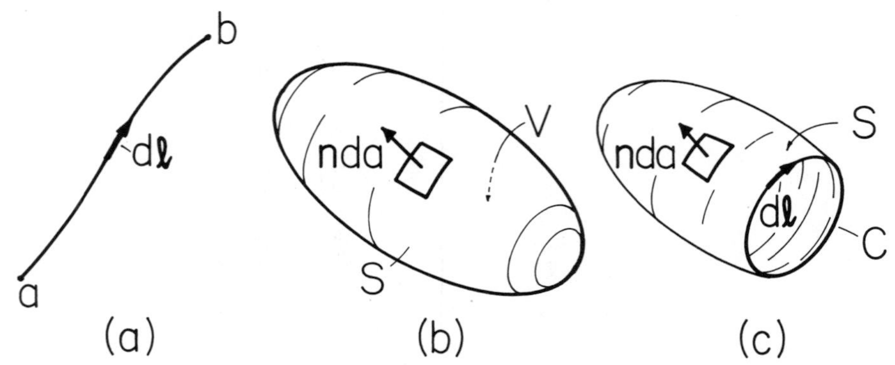

Fig. 2.6.1. Arbitrary contours, volumes and surfaces: (a) open contour C; (b) closed surface S, enclosing volume V; (c) open surface S with boundary contour C.

$$\frac{d}{dt}\int_S \vec{A}\cdot\vec{n}da = \int_S [\frac{\partial\vec{A}}{\partial t} + (\nabla\cdot\vec{A})\vec{v}_s]\cdot\vec{n}da + \oint_C (\vec{A}\times\vec{v}_s)\cdot\vec{d\ell} \qquad (4)$$

Again, C is the contour which is the periphery of the open surface S. The velocity $\vec{v}_s$ is the velocity of the surface and the contour. Unless given a physical significance, its meaning is purely geometrical.

A limiting form of the generalized Leibnitz rule will be handy in dealing with closed surfaces. Let the contour C of Eq. 4 shrink to zero, so that the surface S becomes a closed one. This process can be readily visualized in terms of the surface and contour sketch in Fig. 2.6.1c if the contour C is pictured as the draw-string on a bag. Then, if $\zeta \equiv \nabla\cdot\vec{A}$, and use is made of Gauss' theorem (Eq. 2), Eq. 4 becomes a statement of how to take the time derivative of a volume integral when the volume is a function of time:

$$\frac{d}{dt}\int_V \zeta dV = \int_V \frac{\partial\zeta}{\partial t}dV + \oint_S \zeta\vec{v}_s\cdot\vec{n}da \qquad (5)$$

Again, $\vec{v}_s$ is the velocity of the surface enclosing the volume V.

2.7 Quasistatic Integral Laws

There are at least three reasons for desiring Maxwell's equations in integral form. First, the integral equations are convenient for establishing jump conditions implied by the differential equations. Second, they are the basis for defining lumped parameter variables such as the voltage, charge, current, and flux. Third, they are useful in understanding (as opposed to predicting) physical processes. Since Maxwell's equations have already been divided into the two quasistatic systems, it is now possible to proceed in a straightforward way to write the integral laws for contours, surfaces, and volumes which are distorting, i.e., that are functions of time. The velocity of a surface S is $\vec{v}_s$.

To obtain the integral laws implied by the laws of Eqs. 2.3.23-27, each equation is either (i) integrated over an open surface S with Stokes's theorem used where the integrand is a curl operator to convert to a line integration on C and Eq. 2.6.4 used to bring the time derivative outside the integral, or (ii) integrated over a closed volume V with Gauss' theorem used to convert integrations of a divergence operator to integrals over closed surfaces S and Eq. 2.6.5 used to bring the time derivative outside the integration:

$$\oint_S (\varepsilon_o\vec{E} + \vec{P})\cdot\vec{n}da = \int_V \rho_f dV \qquad\qquad \oint_C \vec{H}\cdot\vec{d\ell} = \int_S \vec{J}_f\cdot\vec{n}da \qquad (1)$$

$$\oint \vec{E}\cdot\vec{d\ell} = 0 \qquad\qquad \oint_S \mu_o(\vec{H} + \vec{M})\cdot\vec{n}da = 0 \qquad (2)$$

$$\oint_S \vec{J}_f'\cdot\vec{n}da + \frac{d}{dt}\int_V \rho_f dV = 0 \qquad\qquad \oint_C \vec{E}'\cdot\vec{d\ell} = -\frac{d}{dt}\int_S \mu_o(\vec{H} + \vec{M})\cdot\vec{n}da - \oint_C \mu_o\vec{M}\times(\vec{v}-\vec{v}_s)\cdot\vec{d\ell} \qquad (3)$$

$$\oint_C \vec{H}'\cdot d\vec{\ell} = \int_S \vec{J}_f'\cdot\vec{n}da + \frac{d}{dt}\int_S (\varepsilon_o\vec{E} + \vec{P})\cdot\vec{n}da + \oint_C \vec{P} \times (\vec{v} - \vec{v}_s)\cdot d\vec{\ell} \qquad \oint_S \vec{J}_f\cdot\vec{n}da = 0 \tag{4}$$

$$\oint_S \mu_o(\vec{H} + \vec{M})\cdot\vec{n}da = 0 \qquad \oint_S (\varepsilon_o\vec{E} + \vec{P})\cdot\vec{n}da = \oint_V \rho_f dV \tag{5}$$

where

$$\vec{J}_f' = \vec{J}_f - \vec{v}_s\rho_f$$

$$\vec{H}' = \vec{H} - \vec{v}_s \times \varepsilon_o\vec{E}$$

where

$$\vec{E}' \doteq \vec{E} + \vec{v}_s \times \mu_o\vec{H}$$

The primed variables are simply summaries of the variables found in deducing these equations. However, these definitions are consistent with the transform relationships found in Sec. 2.5, and the velocity of these surfaces and contours, $\vec{v}_s$, can be identified with the velocity of an inertial frame instantaneously attached to the surface or contour at the point in question. Approximations implicit to the original differential quasistatic laws are now implicit to these integral laws.

2.8 Polarization of Moving Media

Effects of polarization and magnetization are included in the formulation of electrodynamics postulated in Sec. 2.2. In this and the next section a review is made of the underlying models.

Consider the electroquasistatic systems, where the dominant field source is the charge density. Not all of this charge is externally accessible, in the sense that it cannot all be brought to some position through a conduction process. If an initially neutral dielectric medium is stressed by an electric field, the constituent molecules and domains become polarized. Even though the material retains its charge neutrality, there can be a local accrual or loss of charge because of the polarization. The first order of business is to deduce the relation of such polarization charge to the polarization density.

For conceptual purposes, the polarization of a material is pictured as shown in Fig. 2.8.1.

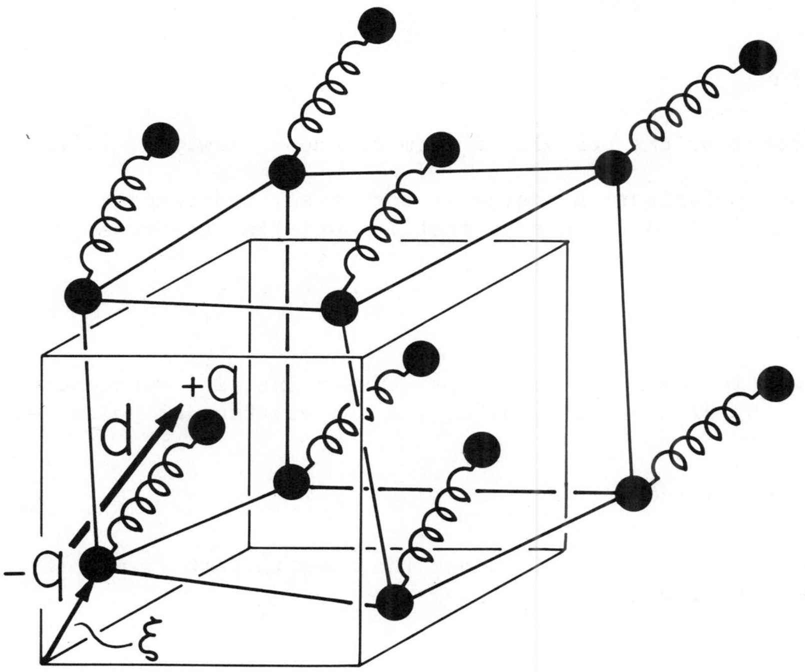

Fig. 2.8.1. Model for dipoles fixed to deformable material. The model pictures the negative charges as fixed to the material, and then the positive halves of the dipoles fixed to the negative charges through internal constraints.

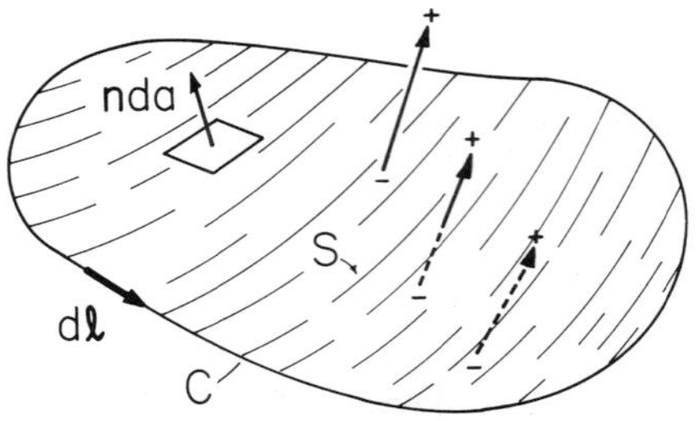

Fig. 2.8.2

Polarization results in net charges passing through a surface.

The molecules or domains are represented by dipoles composed of positive and negative charges $\pm q$, separated by the vector distance $\vec{d}$. The dipole moment is then $\vec{p} = q\vec{d}$, and if the particles have a number density n, the polarization density is defined as

$$\vec{P} = nq\vec{d} \tag{1}$$

In the most common dielectrics, the polarization results because of the application of an external electric field. In that case, the internal constraints (represented by the springs in Fig. 2.8.1) make the charges essentially coincident in the absence of an electric field, so that, on the average, the material is (macroscopically) neutral. Then, with the application of the electric field, there is a separation of the charges in some direction which might be coincident with the applied electric field intensity. The effect of the dipoles on the average electric field distribution is equivalent to that of the medium they model.

To see how the polarization charge density is related to the polarization density, consider the motion of charges through the arbitrary surface S shown in Fig. 2.8.2. For the moment, consider the surface as being closed, so that the contour enclosing the surface shown is shrunk to zero. Because polarization results in motion of the positive charge, leaving behind the negative image charge, the net polarization charge within the volume V enclosed by the surface S is equal to the negative of the net charge having left the volume across the surface S. Thus,

$$\int \rho_p dV = - \oint_S nq\vec{d}\cdot\vec{n}da = - \oint_S \vec{P}\cdot\vec{n}da \tag{2}$$

Gauss' theorem, Eq. 2.6.2, converts the surface integral to one over the arbitrary volume V. It follows that the integrand must vanish so that

$$\rho_p = - \nabla\cdot\vec{P} \tag{3}$$

This polarization charge density is now added to the free charge density as a source of the electric field intensity in Gauss' law:

$$\nabla\cdot\varepsilon_o\vec{E} = \rho_f + \rho_p \tag{4}$$

and Eqs. 3 and 4 comprise the postulated form of Gauss' law, Eq. 2.3.23a.

By definition, polarization charge is conserved, independent of the free charge. Hence, the polarization current $\vec{J}_p$ is defined such that it satisfies the conservation equation

$$\nabla\cdot\vec{J}_p + \frac{\partial\rho_p}{\partial t} = 0 \tag{5}$$

To establish the way in which $\vec{J}_p$ transforms between inertial reference frames, observe that in a primed frame of reference, by dint of Eq. 2.5.2c, the conservation of polarization charge equation becomes

$$\nabla\cdot[\vec{J}_p' + \vec{u}\rho_p'] + \frac{\partial\rho_p'}{\partial t} = 0 \tag{6}$$

It has been shown that $\vec{P}$, and hence ρ_p, are the same in both frames (Eq. 2.5.10a). It follows that the required transformation law is

$$\vec{J}_p' = \vec{J}_p - \vec{u}\rho_p \tag{7}$$

If the dipoles are attached to a moving medium, so that the negative charges move with the same velocity $\vec{v}$ as the moving material, the motion gives rise to a current which should be included in Ampère's law as a source of magnetic field. Even if the material is fixed, but the applied field is time-varying so as to induce a time-varying polarization density, a given surface is crossed by a net charge and there is a current caused by a time-varying polarization density. The following steps determine the current density $\vec{J}_p$ in terms of the polarization density and the material velocity.

The starting point is the statement

$$\int_S \vec{J}_p' \cdot \vec{n}da = \frac{d}{dt}\int_S \vec{P}\cdot\vec{n}da \qquad (8)$$

The surface S, depicted by Fig. 2.8.2, is attached to the material itself. It moves with the negative charges of the dipoles. Integrated over this deforming surface of fixed identity, the polarization current density evaluated in the frame of reference of the material is equal to the rate of change with respect to time of the net charge penetrating that surface.

With the surface velocity identified with the material velocity, Eq. 2.6.4 and Eq. 3 convert Eq. 8 to

$$\int_S \vec{J}_p' \cdot \vec{n}da = \int_S \left(\frac{\partial\vec{P}}{\partial t} - \rho_p\vec{v}\right)\cdot\vec{n}da + \oint_C \vec{P}\times\vec{v}\cdot d\vec{\ell} \qquad (9)$$

On the left, $\vec{J}_p'$ is replaced by Eq. 7 evaluated with $\vec{u} = \vec{v}$, while on the right Stokes's theorem, Eq. 2.6.3, is used to convert the line integral to a surface integral. The result is an equation in surface integrals alone. Although fixed to the deforming material, the surface S is otherwise arbitrary and so it follows that the required relation between $\vec{J}_p$ and $\vec{P}$ for the moving material is

$$\vec{J}_p = \frac{\partial\vec{P}}{\partial t} + \nabla\times(\vec{P}\times\vec{v}) \qquad (10)$$

It is this current density that has been added to the right-hand side of Ampère's law, Eq. 2.3.26a, to complete the formulation of polarization effects in the electroquasistatic system.

2.9 Magnetization of Moving Media

It is natural to use polarization charge to represent the effect of macroscopic media on the macroscopic electric field. Actually, this is one of two alternatives for representing polarization. That such a choice has been made becomes clear when the analogous question is asked for magnetization. In the absence of magnetization, the free current density is the source of the magnetic field, and it is therefore natural to represent the macroscopic effects of magnetizable media on $\vec{H}$ through an equivalent magnetization current density. Indeed, this viewpoint is often used and supported by the contention that what is modeled at the atomic level is really a system of currents (the electrons in their orbits). It is important to understand that the use of equivalent currents, or of equivalent magnetic charge as used here, if carried out self-consistently, results in the same predictions of physical processes. The choice of models in no way hinges on the microscopic processes accounting for the magnetization. Moreover, the magnetization is often dominated by dynamical processes that have more to do with the behavior of domains than with individual atoms, and these are most realistically pictured as small magnets (dipoles). With the Chu formulation postulated in Sec. 2.2, the dipole model for representing magnetization has been adopted.

An advantage of the Chu formulation is that magnetization is developed in analogy to polarization. But rather than starting with a magnetic charge density, and deducing its relation to the polarization density, think of the magnetic material as influencing the macroscopic fields through an intrinsic flux density $\mu_o\vec{M}$ that might be given, or might be itself induced by the macroscopic $\vec{H}$. For lack of evidence to support the existence of "free" magnetic monopoles, the total flux density due to all macroscopic fields must be solenoidal. Hence, the intrinsic flux density $\mu_o\vec{M}$, added to the flux density in free space $\mu_o\vec{H}$, must have no divergence:

$$\nabla\cdot\mu_o(\vec{H}+\vec{M}) = 0 \qquad (1)$$

This is Eq. 2.3.24b. It is profitable to think of $-\nabla\cdot\mu_o\vec{M}$ as a source of $\vec{H}$. That is, Eq. 1 can be written to make it look like Gauss' law for the electric field:

$$\nabla\cdot\mu_o\vec{H} = \rho_m; \quad \rho_m = -\nabla\cdot\mu_o\vec{M} \qquad (2)$$

The magnetic charge density ρ_m is in this sense the source of the magnetic field intensity.

Faraday's law of induction must be revised if magnetization is present. If $\mu_o\vec{M}$ is a magnetic flux density, then, through magnetic induction, its rate of change is capable of producing an induced electric field intensity. Also, if Faraday's law of induction were to remain valid without alteration, then its divergence must be consistent with Eq. 1; obviously, it is not.

To generalize the law of induction to include magnetization, it is stated in integral form for a contour C enclosing a surface S fixed to the material in which the magnetized entities are imbedded. Then, because $\mu_o(\vec{H} + \vec{M})$ is the total flux density,

$$\oint_C \vec{E}'\cdot d\vec{\ell} = -\frac{d}{dt}\int_S \mu_o(\vec{H} + \vec{M})\cdot\vec{n}da \tag{3}$$

The electric field $\vec{E}'$ is evaluated in the frame of reference of the moving contour. With the time derivative taken inside the temporally varying surface integrals (Eq. 2.6.4) and because of Eq. 1,

$$\oint_C \vec{E}'\cdot d\vec{\ell} = -\int_S \frac{\partial}{\partial t}[\mu_o(\vec{H} + \vec{M})]\cdot\vec{n}da + \int_S \nabla \times [\vec{v} \times \mu_o(\vec{H} + \vec{M})]\cdot\vec{n}da \tag{4}$$

The transformation law for $\vec{E}$ (Eq. 2.5.12b with $\vec{u} = \vec{v}$) is now used to evaluate $\vec{E}'$, and Stokes's theorem, Eq. 2.6.3, used to convert the line integral to a surface integral. Because S is arbitrary, it then follows that the integrand must vanish:

$$\nabla \times \vec{E} = -\frac{\partial}{\partial t}[\mu_o(\vec{H} + \vec{M})] + \nabla \times (\vec{v} \times \mu_o\vec{M}) \tag{5}$$

This generalization of Faraday's law is the postulated equation, Eq. 2.3.25b.

2.10 Jump Conditions

Systems having nonuniform properties are often modeled by regions of uniform properties, separated by boundaries across which these properties change abruptly. Fields are similarly often given a piece-wise representation with jump conditions used to "splice" them together at the discontinuities. These conditions, derived here for reference, are implied by the integral laws. They guarantee that the associated differential laws are satisfied through the singular region of the discontinuity.

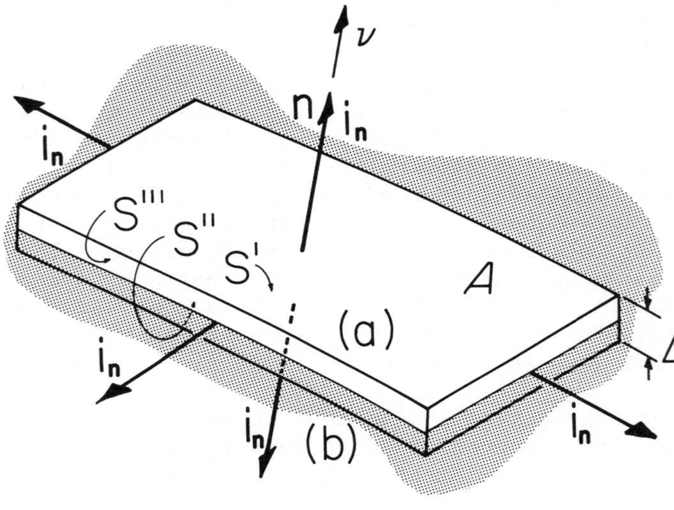

Fig. 2.10.1. Volume element enclosing a boundary. Dimensions of area A are much greater than Δ.

Electroquasistatic Jump Conditions: A section of the boundary can be enclosed by a volume element having the thickness Δ and cross-sectional area A, as depicted by Fig. 2.10.1. The linear dimensions of the cross-sectional area A are, by definition, much greater than the thickness Δ. Implicit to this statement is the assumption that, although the surface can be curvilinear, its radius of curvature must be much greater than a characteristic thickness over which variations in the properties and fields take place.

The normal vector $\vec{n}$ used in this section is a unit vector perpendicular to the boundary and directed from region b to region a, as shown in Fig. 2.10.1. Since this same symbol is used in connection with integral theorems and laws to denote a normal vector to surfaces of integration, these latter vectors are denoted by $\vec{i}_n$.

First, consider the boundary conditions implied by Gauss' law, Eq. 2.3.23a, with Eq. 2.8.3 used to introduce ρ_p. This law is first multiplied by ν^m and then integrated over the volume V:

$$\int_V \nu^m \nabla\cdot\varepsilon_o\vec{E}dV = \int_V \nu^m\rho_f dV + \int_V \nu^m\rho_p dV \qquad (1)$$

Here, ν is a coordinate (like x,y, or z) perpendicular to the boundary and hence in the direction of $\vec{n}$, as shown in Fig. 2.10.1.

First, consider the particular case of Eq. 1 with m = 0. Then, the integration gives

$$\vec{n}\cdot[\![\varepsilon_o E]\!] = \sigma_f + \sigma_p \qquad (2)$$

where $[\![\vec{A}]\!] \equiv \vec{A}^a - \vec{A}^b$ and $[\![\psi]\!] \equiv \psi^a - \psi^b$ and the free surface charge density σ_f and polarization surface charge density σ_p have been defined as

$$\sigma_f = \lim_{A\to 0}\frac{1}{A}\int\rho_f dV, \quad \sigma_p = \lim_{A\to 0}\frac{1}{A}\int\rho_p dV \qquad (3)$$

The relationship between the surface charge and the electric field intensity normal to the boundary can be pictured as shown in Fig. 2.10.2b.

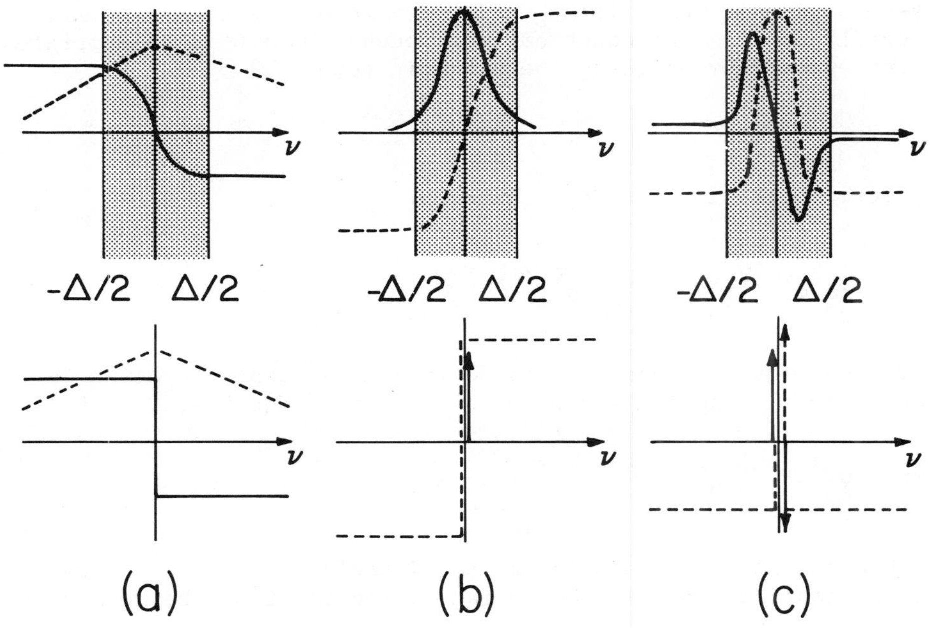

Fig. 2.10.2. Sketches of the charge distribution represented by the solid lines, and the electric field intensity normal to the boundary represented by broken lines. Sketches at the top represent actual distributions, while those below represent idealizations appropriate if the thickness Δ of the region over which the electric field intensity makes its transition is small compared to other dimensions of interest: (a) volume charge density to either side of interface but no surface charge; (b) surface charge; (c) double layer.

In view of Eq. 2, the normal electric field intensity is continuous at the interface unless there is a singularity in charge. Thus, with volume charges to either side of the interface, there is an abrupt change in the rate of change of the electric field intensity normal to the boundary, but the field is itself continuous. On the other hand, as illustrated by the sketches of Fig. 2.10.2b, if there is an appreciable charge per unit area within the boundary, the electric field intensity is discontinuous, and undergoes a step discontinuity.

A somewhat less familiar situation is that of Fig. 2.10.2c. Within the boundary there are regions of large positive and negative charge concentrations with an associated intense electric field between. In the limit where the boundary becomes very thin, a component of the surface charge density becomes a doublet, and the electric field becomes an impulse.

The double layer can be pictured as being positive surface charges disposed on one side of the boundary, and negative surface charges distributed on the other, with an internal component of the electric field originating on the positive charges and terminating on the negative ones. The magnitude of the double layer is equal to the product of the positive surface charge density and the distance between these layers, Δ. In the limit where the layer thickness becomes infinitely thin while the double-layer magnitude remains constant, the electric field within the double layer must approach infinity. Thus, associated with the doublet of charge density, there is an impulse in the electric field intensity, as sketched in Fig. 2.10.2c.

The boundary condition to be used in connection with a double layer is found from Eq. 1 by letting m = 1. The left-hand side of Eq. 1 can be integrated by parts, so that it becomes

$$\int_V \nabla\cdot(\varepsilon_o \nu\vec{E})dV - \int_V \varepsilon_o\vec{E}\cdot\nabla\nu dV = \int_V \nu(\rho_f + \rho_p)dV \tag{4}$$

For the incremental volume, the surface double layer density is defined as

$$p_\Sigma = \lim_{A\to 0}\frac{1}{A}\int \nu(\rho_f + \rho_p)dV = \int_{\nu_-}^{\nu_+}\nu(\rho_f + \rho_p)d\nu \tag{5}$$

and so the right-hand side of Eq. 4 is Ap_Σ. The origin of the Δ axis remains to be defined but $\Delta \equiv \nu_+ - \nu_-$. To glean a jump condition from the equation, the second EQS law is incorporated. That $\vec{E}$ is irrotational, Eq. 2.3.24a, is represented by defining the electric potential

$$\vec{E} = -\nabla\Phi \tag{6}$$

Thus, the second term on the left in Eq. 4 becomes

$$\int_V \varepsilon_o E\cdot\nabla\nu dV = -\int_V \varepsilon_o\nabla\Phi\cdot\nabla\nu dV = -\int_V \varepsilon_o\nabla\cdot(\Phi\nabla\nu)dV + \int_V \varepsilon_o\Phi\nabla^2\nu dV \tag{7}$$

Evaluation of $\nabla^2\nu$ gives nothing because ν is defined as a local Cartesian coordinate. The last integral vanishes, and with the application of Gauss' theorem, Eq. 2.6.2, it follows that Eq. 4 becomes

$$\oint_S \varepsilon_o\nu\vec{E}\cdot\vec{i}_n da + \oint_S \varepsilon_o\Phi\nabla\nu\cdot\vec{i}_n da = A\,p_\Sigma \tag{8}$$

Provided that within the layer, $\vec{E}$ parallel to the interface and Φ are finite (not impulses in the limit $\Delta\to 0$), Eq. 8 only has contributions to the surface integrals from the regions to either side of the interface. Thus,

$$A\varepsilon_o(\nu_+\vec{E}^a - \nu_- E^b)\cdot\vec{n} + A\varepsilon_o[\![\Phi]\!] = A\,p_\Sigma \tag{9}$$

The origin of the ν axis is adjusted to make the first term vanish. The required boundary condition to be associated with Eqs. 2.3.23a and 2.3.23b is

$$\varepsilon_o[\![\Phi]\!] = p_\Sigma \tag{10}$$

The gradient of Eq. 10 within the plane of the interface converts the jump condition to one in terms of the electric field:

$$\varepsilon_o[\![\vec{E}_t]\!] = -\nabla_\Sigma p_\Sigma \tag{11}$$

Here ∇_Σ is the surface gradient and t denotes components tangential to the interfacial plane.

In the absence of a double-layer surface density, these last two boundary conditions are the familiar statement that the tangential electric field intensity at a boundary must be continous. The statement given in Eq. 10 that the potential must be continuous at a boundary is another way of stating this requirement on the tangential electric field intensity. With a double layer, the tangential electric field intensity is discontinuous, as is also the potential.

Equations 10 and 11 could also be derived using the condition that the line integral of the electric field intensity around a closed loop intersecting the boundary vanish. Usually, the tangential electric field is continuous because there is no contribution to this line integral from those segments of the contour passing through the boundary. However, with the double layer, the electric field intensity within the boundary is infinite; so, even though the segments of the line integral across the boundary vanish as $\Delta \to 0$, there is a net contribution from these segments of the integration.

It is clear that higher order singularities could also be handled by considering values of m in Eq. 1 greater than unity. However, the doublet is as singular a charge distribution as of interest physically.

There are two reasons for wishing to include the doublet charge distribution, one mathematical and one physical. Just as the surface charge density is a singularity in the volume charge density which can be used to terminate a normal electric field intensity at a boundary, the double layer is a termination of a tangential electric field. On the physical side, there are many situations in which a double layer actually exists within a very thin region of material. Double layers abound at interfaces between liquids and metals and between metals. The double-layer concept is useful for modeling electromechanical coupling involving these interfacial regions.

So far, those EQS laws have been considered that do not explicitly involve time rates of change. Conservation of charge does involve a dynamic term. Its associated boundary conditions can therefore be derived only by making further stipulations as to the nature of the boundary. It is now admitted that the boundary can, in general, be one which is deforming. Because time did not appear explicitly in the previous derivations of this section, the conditions derived are automatically appropriate, even if the boundary is moving.

The integral form of charge conservation, Eq. 2.7.3a, is written for a volume V and surface S tied to the material itself. Thus, with $\vec{v}_s \to \vec{v}$,

$$\oint_S (\vec{J}_f - \rho_f \vec{v})\cdot\vec{i}_n da = -\frac{d}{dt}\int_V \rho_f dV \tag{12}$$

As seen in Fig. 2.10.1, the volume of integration always encloses material of fixed identity and intersects the boundary. Implicit to this statement is the assumption that the boundary is one of demarcation between material regions. The material velocity is presumed to at most have a step singularity across the boundary. (It is important to recognize that there are other types of boundaries. For example, the boundary could be a shock front, with a gas moving through from one side of the interface to the other. In that case, the boundary conditions thus far derived would remain correct, because no mention has yet been made of the physical nature of the boundary.)

The left-hand side of Eq. 12 can be handled in a manner similar to that already illustrated, since it does not involve time rates of change. The integration is divided into two parts: one over the upper and lower surfaces of the volume, the other over the parts of the surface which intersect the boundary. The contributions to a current flow through these side surfaces comes from a surface current. It follows by using a two-dimensional form of Gauss' theorem, Eq. 2.6.2, that the left-hand side of Eq. 12 is

$$\int_{S'+S''} (\vec{J}_f - \rho_f\vec{v})\cdot\vec{i}_n da + \int_{S'''} (\vec{J}_f - \rho_f\vec{v})\cdot\vec{i}_n da = A\{\vec{n}\cdot[\![\vec{J}_f - \vec{v}\rho_f]\!] + \nabla_\Sigma\cdot(\vec{K}_f - \sigma_f\vec{v}_t)\} \tag{13}$$

Here, A is the area of intersection between the volume element and the boundary. The right-hand side of Eq. 12 is, by the definition of Eq. 3,

$$\frac{d}{dt}\int_V \rho_f dV = \frac{d}{dt}\int_A \sigma_f da \tag{14}$$

Note that, if the volume of integration V, and hence the area of integration A, is one always fixed to the material, then the area A is time-varying. The surface charge density is a function only of the two dimensions within the plane of the interface. Thus, the term on the right in Eq. 14 is a time derivative of a two-dimensional integral. This is a two-dimensional special case of the situation described by the generalized Leibnitz rule, Eq. 2.6.5, which stated how the time derivative of a volume integral could be represented, even if the volume of integration were time-varying. Thus, Eq. 14 becomes

$$\frac{d}{dt}\int_V \rho_f dV = A\left[\frac{\partial\sigma_f}{\partial t} + \nabla_\Sigma\cdot(\vec{v}_t\sigma_f)\right] \tag{15}$$

Finally, with the use of Eqs. 13 and 15, Eq. 12 becomes the required jump condition representing charge conservation:

$$\vec{n}\cdot[\![\vec{J}_f - \rho_f\vec{v}]\!] + \nabla_\Sigma\cdot\vec{K}_f = -\frac{\partial\sigma_f}{\partial t} \tag{16}$$

By contrast with Eqs. 10 and 11, the expression is specialized to interfaces that do not support charge distributions so singular as a double layer. In using Eq. 16, note that a partial derivative with respect to time is usually defined as one taken holding the spatial coordinates constant. A review of the derivation of Eq. 16 will make it clear that such is not the significance of the partial derivative on the right in Eq. 16. The surface charge density is not defined throughout the three-dimensional space. Thus, this derivative means the partial derivative with respect to time, holding the coordinates within the plane of the interface constant.

The component of current normal to the boundary represented by the first term in Eq. 16 will be recognized as the free current density in a frame of reference moving with the boundary. A good question would be, "why is it that the normal current density appears in Eq. 16 evaluated in the primed frame of reference, while the surface free current density is not?" The answer points to the physical situation for which Eq. 16 is appropriate. As the material boundary moves in the normal direction, the material ahead and behind carries a charge distribution along, but one that never reaches the boundary. By contrast, materials can flow in and out within the surface of the volume of interest, and carry with them a surface charge density of a convective nature. Thus, the surface divergence appearing in the second term of Eq. 16 can include both a conduction surface current and a convection surface current.

Magnetoquasistatic Jump Conditions: The integral forms of Ampère's law and Gauss' law for magnetic fields incorporate no time rates of change. Hence, the jump conditions implied by these laws are familiar from elementary electrodynamics. Ampère's law, Eq. 2.7.1b, is integrated over the surface S and around the contour C enclosing the boundary, as sketched in Fig, 2.10.3, to obtain

$$\vec{n}\times[\![\vec{H}]\!] = \vec{K}_f \tag{17}$$

where $\vec{K}_f$ is the surface current density. Although it is entirely possible to consider a doublet of current density as a model, this impulsive singularity in the distribution of free current density is of as high an order as necessary to model MQS electromechanical situations of general interest.

From Gauss' law for magnetic fields, Eq. 2.7.2b, applied to the incremental volume enclosing the interface, Fig. 2.10.1, the jump condition is

$$\vec{n}\cdot[\![\mu_o(\vec{H}+\vec{M})]\!] = 0 \tag{18}$$

Faraday's law of induction brings into play the time rate of change, and it is expected that motion of the boundary leads to an addition to the jump condition not found for stationary media. According to Eq. 2.7.3b, the integral form of Faraday's law, for a contour fixed to the material (of fixed identity) so that $\vec{v}_s \to \vec{v}$, is

$$\oint_C (\vec{E}+\vec{v}\times\mu_o\vec{H})\cdot d\vec{\ell} = -\frac{d}{dt}\int_S \mu_o(\vec{H}+\vec{M})\cdot\vec{n}da \tag{19}$$

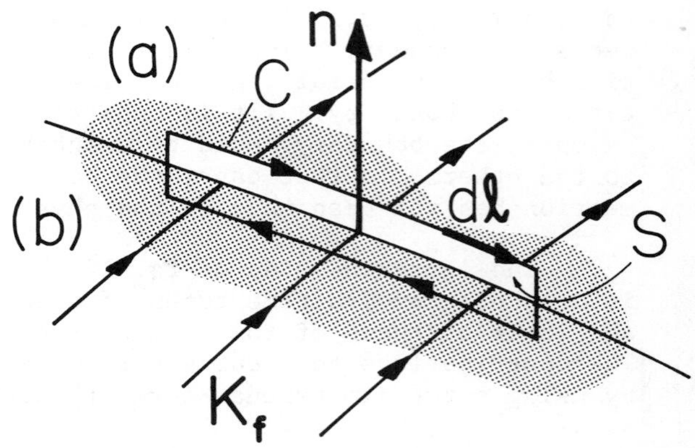

Fig. 2.10.3. Contour of integration C enclosing a surface S that intersects the boundary between regions (a) and (b).

With Eq. 19, it has already been assumed that the boundary is a material one. Consistent with Eq. 17 is the assumption that it can be carrying a surface current with it as it deforms. If the surface S were not one of fixed identity, this would mean that the surface integral on the right could be a step function of time as the boundary passed through the surface of integration. The result would be a temporal impulse on the right which would make a contribution to the boundary condition even in the limit where the surface S becomes vanishingly small. By contrast, because the surface S is one of fixed identity, in the limit where the surface area vanishes, the right-hand side of Eq. 19 makes no contribution.

With the assumption that fields and velocity are at most step functions across the boundary, the integral on the left in Eq. 19 gives

$$\vec{n}\times[\![\vec{E}+\vec{v}\times\mu_o\vec{H}]\!] = 0 \tag{20}$$

This expression is what would be expected, in view of the transformation law for the electric field in the MQS system. It states that $\vec{E}'_t$ is continuous across the interface.

Summary of Electroquasistatic and Magnetoquasistatic Conditions: Table 2.10.1 summarizes the jump conditions.

Table 2.10.1. Quasistatic jump conditions; $[\![\vec{A}]\!] \equiv \vec{A}^a - \vec{A}^b$.

EQS	MQS	
$\vec{n}\cdot[\![\varepsilon_o\vec{E}+\vec{P}]\!]=\sigma_f$ $\vec{n}\cdot[\![\vec{P}]\!]=-\sigma_p$	$\vec{n}\times[\![\vec{H}]\!]=\vec{K}_f$	(21)
$\varepsilon_o[\![\Phi]\!]=\sigma_d$ $\varepsilon_o[\![\vec{E}_t]\!]=-\nabla_\Sigma\sigma_d$	$\vec{n}\cdot\mu_o[\![\vec{H}+\vec{M}]\!]=0$ $\vec{n}\cdot\mu_o[\![\vec{M}]\!]=-\sigma_m$	(22)
$\vec{n}\cdot[\![\vec{J}_f-\rho_f\vec{v}]\!]+\nabla_\Sigma\cdot\vec{K}_f=-\dfrac{\partial\sigma_f}{\partial t}$	$\vec{n}\times[\![\vec{E}+v\times\mu_o\vec{H}]\!]=0$	(23)
$\vec{n}\times[\![\vec{H}-\vec{v}\times\varepsilon_o\vec{E}]\!]=\vec{K}_f-\sigma_f\vec{v}_t$	$\vec{n}\cdot[\![\vec{J}_f]\!]=0$	(24)

Included in the summary are several that are either rarely used, are matters of definition, or are obvious. That the surface polarization charge and surface magnetic charge are related to $\vec{P}$ and $\vec{M}$ respectively follows from Eqs. 2.8.3 and 2.9.2 used in conjunction with Gauss' theorem and the elemental volume of Fig. 2.10.1. Similarly, Eq. 24b follows from the solenoidal nature of the MQS current density. Finally, Eq. 24a follows from the EQS form of Ampère's law, integrated over the surface S of Fig. 2.10.3, following the line of reasoning used in connection with Eq. 20.

2.11 Lumped Parameter Electroquasistatic Elements

Lumped parameter electromechanical models are sufficiently practical that they warrant detailed examination.[1] Even though the electromechanical coupling may be of a definitely continuum and distributed nature, it is most often the case that interest is in inputs and outputs at discrete terminal pairs. This section reviews the definition of energy storage elements in EQS systems.

An abstract representation of a system of perfectly conducting electrodes, each having a potential v_i relative to a reference electrode, is shown in Fig. 2.11.1. Not only are the electrodes and their connecting leads perfectly conducting, but the environment surrounding them is perfectly insulating.

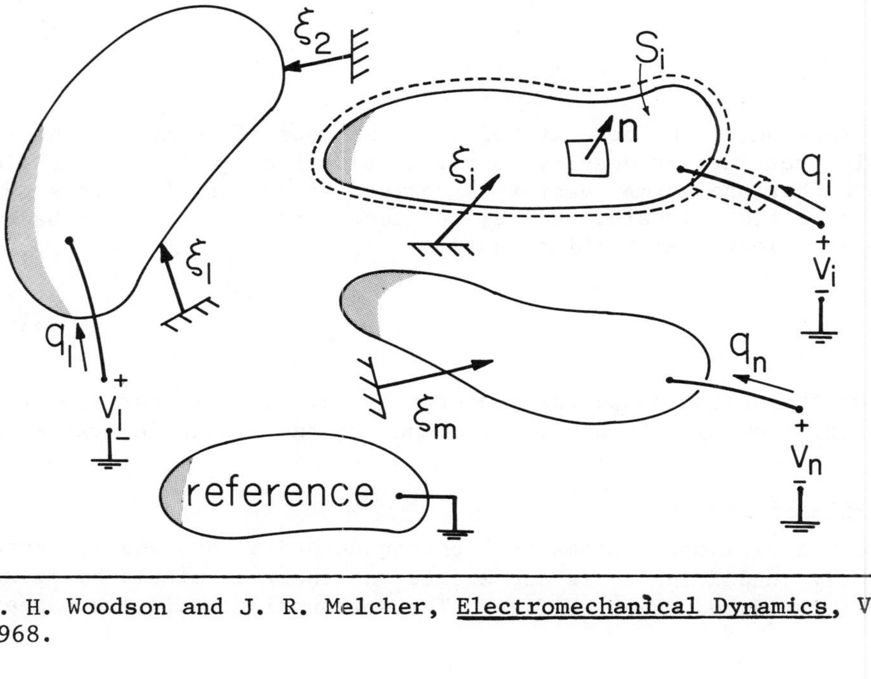

Fig. 2.11.1

Schematic view of an electrode system consisting of n electrodes composed of perfect conductors and immersed in a perfectly insulating medium.

1. H. H. Woodson and J. R. Melcher, Electromechanical Dynamics, Vol. I, John Wiley & Sons, New York, 1968.

The charge on each of the n electrodes is the free charge density integrated over a volume enclosing the electrode:

$$q_i \equiv \int_{V_i} \rho_f dV = \oint_{S_i} \vec{D}\cdot\vec{n}da \tag{1}$$

The total charge on an electrode is indicated by an arrow pointing toward the electrode from the terminal pair attached to that electrode. The associated voltage is defined in terms of the electric field and potential by

$$v_i = -\int_{ref}^{(i)} \vec{E}\cdot d\vec{\ell} = \Phi_i - \Phi_{ref} = \Phi_i \tag{2}$$

This relation is justified because the electric field is irrotational and hence the negative gradient of of Φ.

Given the geometry of the electrodes at a certain instant in time, displacements $\xi_1\cdots\xi_j\cdots\xi_m$ are known, and the condition that the field be irrotational and satisfy Gauss' law leads to equations that can in principle be used to determine the charges on the individual electrodes at a given instant:

$$q_i = q_i(v_1\cdots v_n, \xi_1\cdots\xi_m) \tag{3}$$

If the dielectrics are electrically linear in the sense that $\vec{D} = \varepsilon\vec{E}$, where ε is a function of position but not of time or the field, then it is useful to define a capacitance

$$C_{ij} = \left.\frac{q_i}{v_j}\right|_{v_{i\neq j}=0} = \frac{\oint_{S_i} \varepsilon\vec{E}\cdot\vec{n}da}{-\int_{ref}^{j}\vec{E}\cdot d\vec{\ell}} \tag{4}$$

The capacitance of the ith electrode relative to the jth electrode is the charge on the ith electrode per unit voltage on the jth electrode, with all other electrodes held at zero voltage. The capacitance is useful as a parameter because the charge on an electrode in a linear dielectric is proportional to the voltage itself; hence, the capacitance is purely a function of the electrical properties of the system and the geometry:

$$q_i = \sum_{j=1}^{n} C_{ij}v_j, \quad C_{ij} = C_{ij}(\xi_1\cdots\xi_m) \tag{5}$$

To define the capacitance as with Eqs. 4 and 5, no reference is required to the time rate of change. In these relations q_i, v_i, and ξ_i can all be functions of time. The dynamics enter by virtue of conservation of charge, which can be written for a volume including the ith electrode as (Eq. 2.7.3a):

$$\oint_{S_i} \vec{J}_f'\cdot\vec{n}da = -\frac{d}{dt}\int_{V_i} \rho_f dV \tag{6}$$

The quantity on the right in this expression is the negative of the time rate of change of the total free charge on the ith electrode. The only free current density normal to a surface enclosing the electrode is that through the wire itself. Note that the normal vector is defined as outward from this surface, while a positive current through the wire flows inward. Hence, the left-hand side of Eq. 6 becomes the negative of the total current at the ith electrical terminal pair:

$$i_i = \frac{dq_i}{dt} \tag{7}$$

With the charge given as a function of the voltages and the geometry by Eq. 3, or in particular by Eq. 5, Eq. 7 can be used to compute the current flowing into a given terminal of the electrode system.

2.12 Lumped Parameter Magnetoquasistatic Elements

An extremely practical idealization of lumped parameter magnetoquasistatic systems is sketched schematically in Fig. 2.12.1. Perfectly conducting coils are excited at their terminals by currents i_i and, in general, coupled together by the induced magnetic flux. The surrounding medium is magnetizable

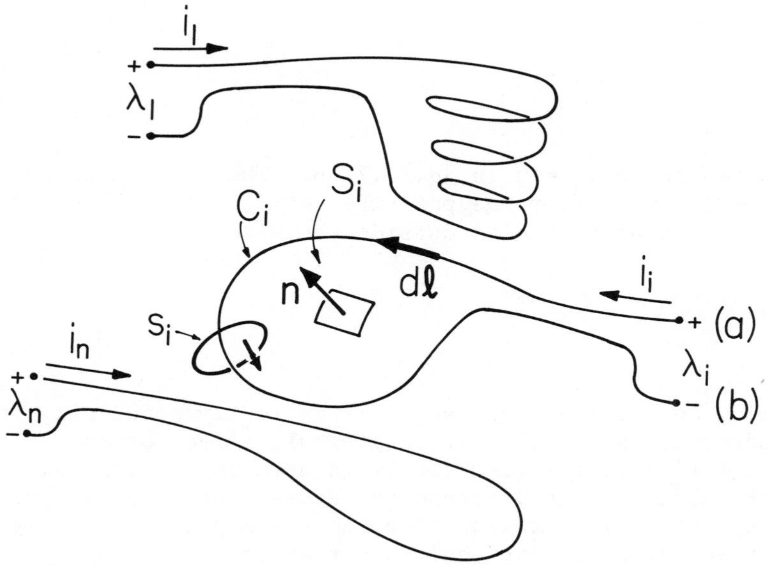

Fig. 2.12.1

Schematic representation of a system of perfectly conducting coils. The ith coil is shown with the wire assuming the contour C_i enclosing a surface S_i. There is a total of n coils in the system.

but free of electrical losses. The total flux λ_i linked by the ith coil is a terminal variable, defined such that

$$\lambda_i = \int_{S_i} \vec{B}\cdot\vec{n}da \tag{1}$$

A positive λ is determined by first assigning the direction of a positive current i_i. Then, the direction of the normal vector (and hence the positive flux) to the surface S_i, enclosed by the contour C_i followed by the current i_i, has a direction consistent with the right-hand rule, as Fig. 2.12.1 illustrates.

Because the MQS current density is solenoidal, the same current flows through the cross section of the wire at any point. Thus, the terminal current is defined by

$$i_i = \int_{s_i} \vec{J}_f\cdot\vec{i}_n\, da \tag{2}$$

where the surface s_i intersects all of the cross section of the wire at any point, as illustrated in the figure.

The first two MQS equations are sufficient to determine the flux linkages as a function of the current excitations and the geometry of the coil. Thus, Ampère's law and the condition that the magnetic flux density be solenoidal are solved to obtain relations having the form

$$\lambda_i = \lambda_i(i_1\cdots i_n,\xi_1\cdots\xi_m) \tag{3}$$

If the materials involved are magnetically linear, so that $\vec{B} = \mu\vec{H}$, where μ is a function of position but not of time or the fields, then it is convenient to define inductance parameters which depend only on the geometry:

$$L_{ij} = \frac{\lambda_i}{i_j}\bigg|_{i_{i\neq j}=0} = \frac{\int_{S_i} \mu\vec{H}\cdot\vec{n}da}{\int_{s_j} \vec{J}_f\cdot\vec{i}_n da} \tag{4}$$

The inductance L_{ij} is the flux linked by the ith coil per unit current in the jth coil, with all other currents zero. For the particular cases in which an inductance can be defined, Eq. 3 becomes

$$\lambda_i = \sum_{j=1}^{n} L_{ij}i_j,\ L_{ij} = L_{ij}(\xi_1\cdots\xi_m) \tag{5}$$

The dynamics of a lumped parameter system arise through Faraday's integral law of induction, Eq. 2.7.3b, which can be written for the ith coil as

$$\oint_{C_i} \vec{E}'\cdot d\vec{\ell} = -\frac{d}{dt}\int_{S_i} \vec{B}\cdot\vec{n}da \tag{6}$$

Here the contour is one attached to the wire and so $\vec{v}_s = \vec{v}$ in Eq. 2.7.3b. The line integration can be broken into two parts, one of which follows the wire from the positive terminal at (a) to (b), while the other follows a path from (b) to (a) in the insulating region outside the wire

$$\oint_{C_i} \vec{E}'\cdot d\vec{\ell} = \int_a^b \vec{E}'\cdot d\vec{\ell} + \int_b^a \vec{E}'\cdot d\vec{\ell} \tag{7}$$

Even though the wire is in general deforming and moving, because it is perfectly conducting, the electric field intensity $\vec{E}'$ must vanish in the conductor, and so the first integral called for on the right in Eq. 7 must vanish. By contrast with the EQS fields, the electric field here is not irrotational. This means that the remaining integration of the electric field intensity between the terminals must be carefully defined. Usually, the terminals are located in a region in which the magnetic field is sufficiently small to take the electric field intensity as being irrotational, and therefore definable in terms of the gradient of the potential. With the assumption that such is the case, the remaining integral of Eq. 7 is written as

$$\int_b^a \vec{E}'\cdot d\vec{\ell} = -\int_b^a \nabla\Phi\cdot d\vec{\ell} = -(\Phi_a - \Phi_b) \equiv -v_i \tag{8}$$

Thus it follows from Eq. 6, combined with Eqs. 1 and 8, that the voltage at the coil terminals is the time rate of change of the associated flux linked:

$$v_i = \frac{d\lambda_i}{dt} \tag{9}$$

With λ_i given by Eq. 3 or Eq. 5, the terminal voltage follows from Eq. 9.

2.13 Conservation of Electroquasistatic Energy

This and the next section develop a field picture of electromagnetic energy storage from fundamental definitions and principles. Results are a first step in the derivation of macroscopic force densities in Chap. 3. Energy storage in a conservative EQS system is considered first, followed by a statement of power flow. In this and the next section the macroscopic medium is at rest.

Thermodynamics: Whether in electric or magnetic form, energy storage follows from the definition of the electric field as a force per unit charge. The work required to transport an element of charge, δq, from a reference position to a position p in the presence of the electric field intensity is

$$\delta w = -\int_{ref}^{p} \delta q\vec{E}\cdot d\vec{\ell} \tag{1}$$

The integral is the work done by the external force on the electric subsystem in placing the charge at p. If this process can be reversed, it can be said that the work done results in a stored energy equal to Eq. 1. In an electroquasistatic system, the electric field is irrotational. Hence, $\vec{E} = -\nabla\Phi$. Then, if Φ_{ref} is defined as zero, it follows that Eq. 1 becomes

$$\delta w = \int_{ref}^{p} \delta q\nabla\Phi\cdot d\vec{\ell} = \delta q\Phi \tag{2}$$

where use has been made of the gradient integral theorem, Eq. 2.6.1. Consider now energy storage in the system abstractly represented by Fig. 2.13.1. The system is perfectly insulating, except for the perfectly conducting electrodes introduced into the volume of interest, as in Sec. 2.11. It will be termed an "electroquasistatic thermodynamic subsystem."

The electrodes have terminal variables as defined in Sec. 2.11; voltages v_i and total charges q_i. But, in addition, the volume between the electrodes supports a free charge density ρ_f. By definition, the energy stored in assembling these charges is equal to the work required to carry the charges from a reference position to the positions of interest. Thus, the incremental energy storage associated with incremental changes in the electrode charges, δq_i, or in the charge density, $\delta\rho_f$, in a given neighborhood on the insulator, is

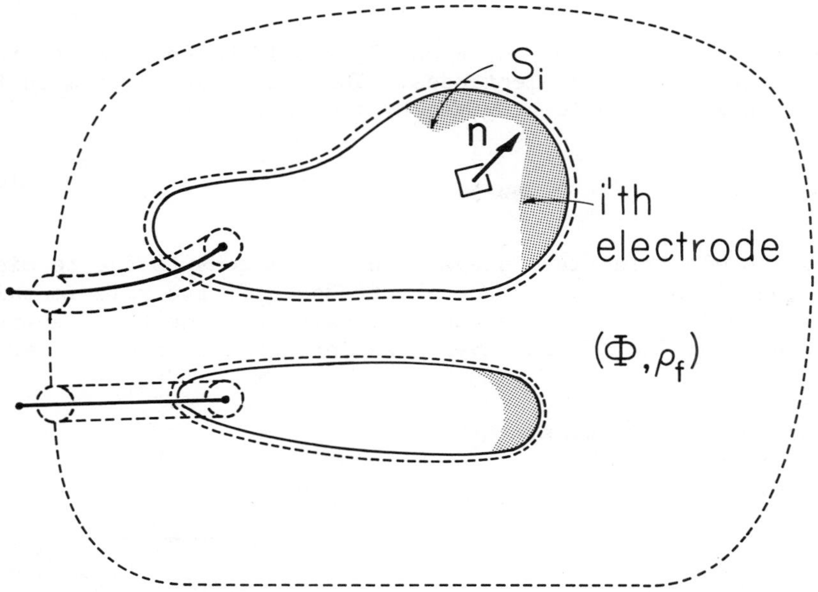

Fig. 2.13.1

Schematic representation of electroquasistatic system composed of perfectly conducting electrodes imbedded in a perfectly insulating dielectric medium.

$$\delta w = \sum_{i=1}^{n} v_i \delta q_i + \int_{V'} \Phi \delta \rho_f dV \tag{3}$$

The volume V' is the volume excluded by the electrodes. Note that the reference electrode is not included in the summation, because the electric potential on that electrode is, by definition, zero. The work required to place a free charge at its final position correctly accounts for the polarization, because the polarization charges induced in carrying the free charges to their final position are reflected in the potential.

Consider now the field representation of the electroquasistatic stored energy. From Gauss' law (Eq. 2.3.23a), the contribution of the summation in Eq. 3 can be represented in terms of an integral over the surfaces S_i of the electrodes:

$$\delta w = \sum_{i=1}^{n} \oint_{S_i} \Phi_i \delta \vec{D} \cdot \vec{n} da + \int_{V'} \Phi \delta \rho_f dV \tag{4}$$

Here, Φ_i is the potential on the surface S_i. The surfaces enclosing the electrodes can be joined together at infinity, as shown in Fig. 2.13.1. The resulting simply connected surface encloses all of the electrodes, the wires as they extend to infinity, with the surface completed by a closure at infinity. Thus, the surface integration called for with the first term on the right in Eq. 4 can be represented by an integration over a closed surface. Gauss' theorem is then used to convert this surface integral to a volume integration. However, note that the normal vector used in Eq. 4 points into the volume V' excluded by the electrodes and included by the surface at infinity. Thus, in using Gauss' theorem, a minus sign is introduced and Eq. 4 becomes

$$\delta w = -\int_{V'} \nabla \cdot (\Phi \delta \vec{D}) dV + \int_{V'} \Phi \delta \rho_f dV = \int_{V'} [-\Phi \nabla \cdot \delta \vec{D} - \delta \vec{D} \cdot \nabla \Phi + \Phi \delta \rho_f] dV \tag{5}$$

In rewriting the integral, the identity $\nabla \cdot \psi \vec{C} = \vec{C} \cdot \nabla \psi + \psi \nabla \cdot \vec{C}$ has been used.

From Gauss' law, $\delta \rho_f = \delta \nabla \cdot \vec{D} = \nabla \cdot \delta \vec{D}$. It follows that the first and last terms in Eq. 5 cancel. Also, the electric field is irrotational ($\vec{E} = -\nabla \Phi$). So Eq. 5 becomes

$$\delta w = \int_V \vec{E} \cdot \delta \vec{D} dV \tag{6}$$

There is no $\vec{E}$ inside the electrodes, so the integration is now over all of the volume V.

The integrand in Eq. 6 is an energy density, and it is therefore appropriate to define the incremental change in electric energy density as

$$\delta W = \vec{E} \cdot \delta D \tag{7}$$

The field representation of the energy, as given by Eqs. 6 and 7, should be compared to that for lumped parameters. Suppose all of the charge resided on electrodes. Then, the second term in Eq. 3 would be zero, and the incremental change in energy would be given by the first term:

$$\delta w = \sum_{i=1}^{n} v_i \delta q_i \tag{8}$$

Comparison of Eqs. 6 and 8 suggests that the electric field plays a role analogous to the terminal voltage, while the displacement vector is the analog of the charge on the electrodes. If the relationship between the variables $\vec{E}$ and $\vec{D}$, or v and q, is single-valued, then the energy density and the total energy in the continuum and lumped parameter systems can be viewed, respectively, as integrals or areas under curves as sketched in Fig. 2.13.2.

If it is more convenient to have all of the voltages, rather than the charges, as independent variables, then Legendre's dual transformation can be used. That is, with the observation that

$$v_i \delta q_i = \delta v_i q_i - q_i \delta v_i \tag{9}$$

Eq. 8 becomes

$$\delta w' = \sum_{i=1}^{n} q_i \delta v_i; \quad w' \equiv \sum_{i=1}^{n} (v_i q_i - w) \tag{10}$$

with w' defined as the coenergy function.

In an analogous manner, a coenergy density, W', is defined by writing $\vec{E}\cdot\delta\vec{D} = \delta(\vec{E}\cdot\vec{D}) - \vec{D}\cdot\delta\vec{E}$ and thus defining

$$\delta W' = \vec{D}\cdot\delta\vec{E}; \quad W' \equiv \vec{E}\cdot\vec{D} - W \tag{11}$$

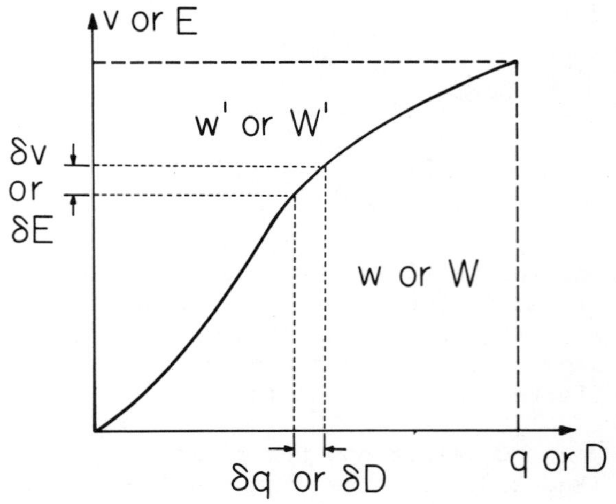

Fig. 2.13.2. Geometric representation of energy w, coenergy w', energy density W, and coenergy density W' for electric field systems.

The coenergy and coenergy density functions have the geometric relationship to the energy and energy density functions, respectively, sketched in Fig. 2.13.2. In those systems in which there is no distribution of charge other than on perfectly conducting electrodes, Eqs. 6 and 8 can be regarded as equivalent ways of computing the same incremental change in electroquasistatic energy. If the charge is distributed throughout the volume, Eq. 6 remains valid.

With the notion of electrical energy storage goes the concept of a conservative subsystem. In the process of building up free charges on perfectly conducting electrodes or slowly conducting charge to the bulk positions (one mechanism for carrying out the process pictured abstractly by Eq. 3), the work is stored much as it would be in cocking a spring. The electrical energy, like that of the spring, can later be released (discharged). Included in the subsystem is storage in the polarization. For work done on polarizable entities to be stored, this polarization process must also be reversible. Here, it is profitable to think of the dipoles as internally constrained by spring-like nondissipative elements, capable of releasing energy when the polarizing field is turned off. Mathematically, this restriction on the nature of the polarization is brought in by requiring that $\vec{P}$ and hence $\vec{D}$ be a single-valued function of the instantaneous $\vec{E}$, or that $\vec{E} = \vec{E}(\vec{D})$. In lumped parameter systems, this is tantamount to q = q(v) or v = v(q).

<u>Power Flow</u>: The electric and polarization energy storage subsystem is the field theory generalization of a capacitor. Just as practical circuits involve a capacitor interconnected with resistors and other types of elements, in any actual physical system the ideal energy storage subsystem is imbedded with and coupled to other subsystems. The field equations, like Kirchhoff's laws in circuit theory, encompass all of these subsystems. The following discussion is based on forming quadratic expressions from the field laws, and hence relate to the energy balance between subsystems.

For a geometrical part of the ith subsystem, having the volume V enclosed by the surface S, a statement of power flow takes the integral form

$$\oint_S \vec{S}_i\cdot\vec{n}da + \int_V \frac{\partial W_i}{\partial t} dV = \int_V \phi_i dV \tag{12}$$

Here, S_i is the power flux density, W_i is the energy density, and ϕ_i is the dissipation density.

Different subsystems can occupy the same volume V. In Eq. 12, V is arbitrary, while i distinguishes the particular physical processes considered. The differential form of Eq. 12 follows by applying Gauss' theorem to the first term and (because V is arbitrary) setting the integrand to zero:

$$\nabla\cdot\vec{S}_i + \frac{\partial W_i}{\partial t} = \phi_i \tag{13}$$

This is a canonical form which will be used to describe various subsystems. In a given region, W_i can increase with time either because of the volumetric source ϕ_i or because of a power flux $-\vec{n}\cdot\vec{S}_i$ into the region across its bordering surfaces.

For an electrical lumped parameter terminal pair, power is the product of voltage and current. This serves as a clue for finding a statement of power flow from the basic laws. The generalization of the voltage is the potential, while conservation of charge as expressed by Eq. 2.3.25a brings in the free current density. So, the sum of Eqs. 2.3.25a and the conservation of polarization charge equation, Eq. 2.8.5, is multiplied by Φ to obtain

$$\Phi[\nabla\cdot(\vec{J}_f+\vec{J}_p) + \frac{\partial}{\partial t}(\rho_f + \rho_p)] = 0 \tag{14}$$

With the objective an expression having the form of Eq. 13, a vector identity (Eq. 15, Appendix B) and Gauss' law, Eq. 2.3.23a, convert Eq. 14 to

$$\nabla\cdot[\Phi(\vec{J}_f + \vec{J}_p)] + \vec{E}\cdot(\vec{J}_f + \vec{J}_p) + \Phi\frac{\partial}{\partial t}\nabla\cdot\varepsilon_o\vec{E} = 0 \tag{15}$$

In the last term the time derivative and divergence are interchanged and the vector identity used again to obtain the expression

$$\nabla\cdot\vec{S}_e + \frac{\partial W_e}{\partial t} = \phi_e \tag{16}$$

where, with Eq. 2.8.10 used for $\vec{J}_p$,

$$\vec{S}_e \equiv \Phi\left(\vec{J}_f + \vec{J}_p + \frac{\partial\varepsilon_o\vec{E}}{\partial t}\right) = \Phi[\vec{J}_f + \frac{\partial\vec{D}}{\partial t} + \nabla \times (\vec{P}\times\vec{v})]$$

$$W_e \equiv \frac{1}{2}\varepsilon_o\vec{E}\cdot\vec{E}$$

$$\phi_e \equiv -\vec{E}\cdot(\vec{J}_f + \vec{J}_p) = -\vec{E}\cdot[\vec{J}_f + \frac{\partial\vec{P}}{\partial t} + \nabla\times(\vec{P}\times\vec{v})]$$

Which terms appear where in this expression is a matter of what part of a physical system (which subsystem) is being described. Note that W_e does not include energy stored by polarizing the medium. Also, it can be shown that $\nabla\cdot\vec{S}_e = \nabla\cdot(\vec{E}\times\vec{H})$, so that $\vec{S}_e$ is the poynting vector familiar from conventional classical electrodynamics. In the dissipation density, $\vec{E}\cdot\vec{J}_f$ can represent work done on an external mechanical system due to polarization forces or, if the polarization process involves dissipation, heat energy given up to a thermal subsystem.

The polarization terms in ϕ_e can also represent energy storage in the polarization. This is illustrated by specializing Eq. 16 to describe a subsystem in which $\vec{P}$ is a single-valued function of the instantaneous $\vec{E}$, the free current density is purely ohmic, $\vec{J}_f = \sigma\vec{E}$, and the medium is at rest. Then, the polarization term from ϕ_e can be lumped with the energy density term to describe power flow in a subsystem that includes energy storage in the polarization:

$$\nabla\cdot\vec{S}_E + \frac{\partial W_E}{\partial t} = \phi_E \tag{17}$$

where

$$\vec{S}_E \equiv \Phi[\sigma\vec{E} + \frac{\partial\vec{D}}{\partial t}];\quad W_E \equiv \int_o^{\vec{D}} \vec{E}\cdot\delta\vec{D};\quad \phi_E \equiv -\sigma\vec{E}\cdot\vec{E}$$

Note that the integral defining the energy density W_E, which is consistent with Eq. 7, involves an integrand $\vec{E}$ which is time dependent only through the time dependence of $\vec{D}$: $\vec{E} = \vec{E}[\vec{D}(t)]$. Thus, $\partial W_E/\partial t = \vec{E}\cdot(\partial\vec{D}/\partial t)$.

With the power flux density placed on the right, Eq. 17 states that the energy density decreases because of electrical losses (note that $\phi_E < 0$) and because of the divergence of the power density.

2.14 Conservation of Magnetoquasistatic Energy

Fundamentally, the energy stored in a magnetic field involves the same work done by moving a test charge from a reference position to the position of interest as was the starting point in Sec. 2.13. But, the same starting point leads to an entirely different form of energy storage. In a magnetoquasistatic system, the net free charge is a quantity evaluated after the fact. A self-consistent representation of the fields is built upon a statement of current continuity, Eq. 2.3.26b, in which the free charge density is ignored altogether. Yet, the energy stored in a magnetic field is energy stored in charges transported against an electric field intensity. The apparent discrepancy in these statements is resolved by recognizing that the charges of interest in a magnetoquasistatic system are at least of two species, with the charge density of one species alone far outweighing the net charge density.

Thermodynamics: Because the free current density is solenoidal, a current "tube" can be defined as shown in Fig. 2.14.1. This tube is defined with a cross section having a normal $\vec{i}_n$ in the direction of the local current density, and a surrounding surface having a normal perpendicular to the local current density. An example of a current tube is a wire surrounded by insulation and hence carrying a total current i which is the same at one cross section as at another.

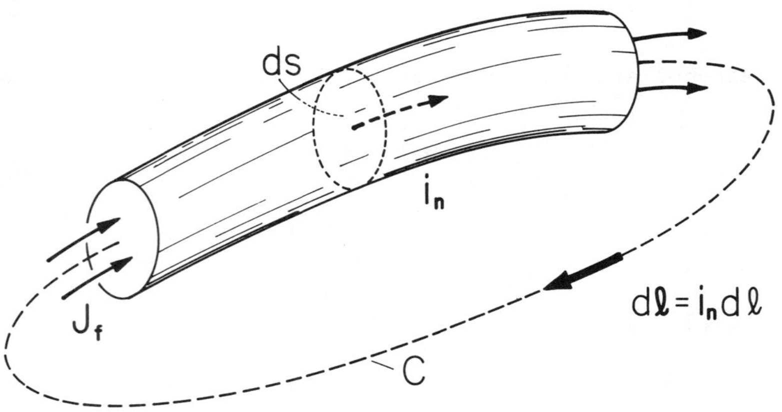

Fig. 2.14.1

Current tube defined as having cross-sectional area ds perpendicular to the local current density, and an outside surface with a normal vector perpendicular to the current density.

For bipolar conduction, and a stationary medium, the current density within the tube is related to the charge density by the expression

$$\vec{J}_f = \rho_+\vec{v}_+ - \rho_-\vec{v}_- \tag{1}$$

Here the conduction process is visualized as involving two types of carriers, one positive, with a charge density ρ_+, and the other negative, with a magnitude ρ_-. The carriers then have velocities which are, respectively, $\vec{v}_+$ and $\vec{v}_-$. Even though there is a current density, in the magnetoquasistatic system there is essentially no net charge: $\rho_f = \rho_+ - \rho_- \simeq 0$. In an increment of time δt, the product of the respective charge densities and net displacements is $\rho_+\vec{v}_+\delta t$ and $-\rho_-\vec{v}_-\delta t$. The work done on the charges as they undergo these displacements is the energy stored in magnetic form. This work is computed by recognizing that the force on each of the charged species is the product of the charge density and the electric field intensity. Hence, the energy stored in the field by a length of the current tube $d\ell$ is to first order in differentials $d\ell$ and ds,

$$-(\rho_+\vec{v}_+ - \rho_-\vec{v}_-)\cdot\vec{E}\delta t ds d\ell = -\vec{J}_f\cdot\vec{E}\delta t ds d\ell \tag{2}$$

The expression for the free current density, Eq. 1, is used on the right to restate the energy stored in the increment of time δt. The unit vector $\vec{i}_n$ is defined to be the direction of $\vec{J}_f$. Thus, $\vec{J}_f = (\vec{J}_f\cdot\vec{i}_n)\vec{i}_n$. Because the current density is solenoidal, it follows closed paths. The product $\vec{J}_f\cdot\vec{i}_n ds$ is, by definition, constant along one of these paths, and if $\vec{i}_n d\ell$ is defined as an increment of the line integral, it then follows from Eq. 2 that the energy stored in a single current tube is

$$-\vec{J}_f\cdot\vec{i}_n ds\left(\oint_C \vec{E}\cdot\vec{i}_n d\ell\right)\delta t \tag{3}$$

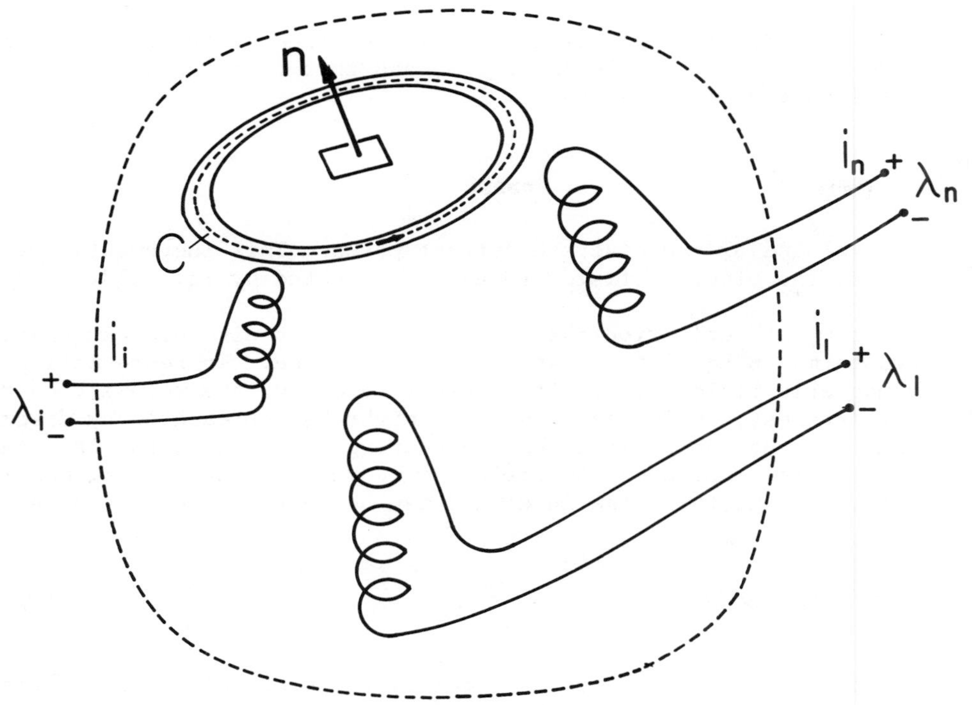

Fig. 2.14.2

Schematic representation of a magnetoquasistatic energy storage system. Currents are either distributed in current loops throughout the volume of interest, or confined to one of n possible contours connected to the discrete terminal pairs.

By contrast with the electroquasistatic system, in which the electric field intensity is induced by the charge density (Gauss' law), the electric field intensity in Eq. 3 is clearly rotational. This emphasizes the essential role played by Faraday's law of magnetic induction.

It is helpful to have in mind at least the abstraction of a physical system. Figure 2.14.2 shows a volume of interest in which the currents are either distributed throughout the volume or confined to particular contours (coils), the latter case having been discussed in Sec. 2.12.

First, consider the energy stored in the current paths defined by coils having cross-sectional area ds. From Eq. 3, this contribution to the total energy is conveniently written as

$$-\vec{J}_f\cdot\vec{i}_n ds\Big(\oint_{C_i} \vec{E}\cdot\vec{i}_n d\ell\Big)\delta t = i_i\delta\lambda_i \tag{4}$$

Faraday's law and the definition of flux linkage, Eqs. 2.12.1 and 2.12.6, are the basis for representing the line integral as a change in the flux linkage.

Because the free current density is solenoidal, the distribution of free currents within the volume V excluded by the discrete coils can be represented as the superposition of current tubes. From Eq. 4 and the integral form of Faraday's law, Eq. 2.7.3b with $\vec{v}_s = v = 0$ (the medium is fixed), it follows that the energy stored in a current tube is

$$\delta w_{\text{current tube}} = \vec{J}_f\cdot\vec{i}_n ds\Big(\int_{S_{tube}} \delta\vec{B}\cdot\vec{n}da\Big) \tag{5}$$

The magnetic flux density is also solenoidal, and for this reason it is convenient to introduce the magnetic vector potential $\vec{A}$, defined such that $\vec{B} = \nabla \times \vec{A}$, so that the magnetic flux density is automatically solenoidal. With this representation of the flux density in terms of the vector potential, Stokes's theorem, Eq. 2.6.3, converts Eq. 5 to

$$\vec{J}_f\cdot\vec{i}_n ds \oint_{C_{tube}} \delta\vec{A}\cdot\vec{i}_n d\ell = \oint_{C_{tube}} (\vec{J}_f\cdot\delta\vec{A})dsd\ell = \int_{V_{tube}} \vec{J}_f\cdot\delta\vec{A}dV \tag{6}$$

Here, $\vec{J}_f$ is by definition in the direction of $\vec{i}_n$, so that $\vec{J}_f\cdot\delta\vec{A}$ takes the component of $\delta\vec{A}$ in the $\vec{i}_n$ direction. The second equality is based upon recognition that the product $\vec{ds}\cdot\vec{d\ell}$ is a volume element of the current tube, and the line integration constitutes an integration over the volume, V_{tube}, of the tube.

To include all of the energy stored in the distributed current loops, it is necessary only that the integral on the right in Eq. 6 be extended over all of the volume occupied by the tubes. The combination of the incremental energy stored in the discrete loops, Eq. 4, and that from the distributed current loops, Eq. 6, is the incremental total energy of the system

$$\delta w = \sum_{i=1}^{n} i_i \delta\lambda_i + \int_V \vec{J}_f \cdot \delta\vec{A}dV \tag{7}$$

In this expression, V is the volume excluded by the discrete current paths. This incremental magnetic energy storage is analogous to that for the electric field storage represented by Eq. 2.13.3.

In retrospect, it is apparent from the derivation that the division into discrete and distributed current paths, represented by the two terms in Eq. 7, is a matter of convenience. In representing the incremental energy in terms of the magnetic fields alone, it is handy to extend the volume V over all of the currents within the volume of interest, including those that might be represented by discrete terminal pairs. With this understanding, the incremental change in energy, Eq. 7, is the last term only, with V extended over the total volume. Moreover, Ampere's law represents the current density in terms of the magnetic field intensity, and, in turn, the integrand can be rewritten by use of a vector identity (Eq. 8, Appendix B):

$$\delta w = \int_V \nabla \times \vec{H} \cdot \delta\vec{A}dV = \int_V [\vec{H} \cdot \nabla \times \delta\vec{A} + \nabla \cdot (\vec{H} \times \delta\vec{A})]dV \tag{8}$$

The last term in Eq. 8 can be converted to a surface integral by using Gauss' theorem. With the understanding that the system is closed in the sense that the fields fall off rapidly enough at infinity so that the surface integration can be ignored, the remaining volume integration on the right in Eq. 8 can be used to obtain a field representation of the incremental energy change. With the curl of the vector potential converted back to a flux density, Eq. 8 becomes

$$\delta w = \int_V \vec{H} \cdot \delta\vec{B}dV \tag{9}$$

The integrand of Eq. 9 is defined as an incremental magnetic energy density

$$\delta W = \vec{H} \cdot \delta\vec{B} \tag{10}$$

It is helpful to note the clear analogy between this energy density and the incremental total energy represented by lumped parameters. In the absence of volume free current densities that cannot be represented by discrete terminal pairs, Eq. 7 reduces to the lumped parameter form

$$\delta w = \sum_{i=1}^{n} i_i \delta\lambda_i \tag{11}$$

The magnetic field intensity plays the continuum role of the discrete terminal currents, and the magnetic flux density is the continuum analog of the lumped parameter flux linkages. The situation in this magnetic case is, of course, analogous to the electrical incremental energy storages in continuum and in lumped parameter cases, as discussed with Eqs. 7 and 8 of Sec. 2.13.

Just as it is often convenient in dealing with electrical lumped parameters to use the voltage as an independent variable, so also in magnetic field systems it is helpful to use the terminal currents as independent variables. In that case, the coenergy function w' is conveniently introduced as an energy function

$$\delta w' = \sum_{i=1}^{n} \lambda_i \delta i_i \tag{12}$$

In an analogous way, the co-energy density, w', is defined such that

$$\delta W' = \vec{B} \cdot \delta\vec{H}; \quad W' = \vec{H} \cdot \vec{B} - W \tag{13}$$

<u>Power Flow</u>: Thus far, the storage of energy in magnetic form has been examined. The postulate has been that all work done in moving the charges against an electric field is stored. In any system as a whole this is not likely to be the case. The general magnetoquasistatic laws enable a deduction of an equation representing the flow of power, and the rate of change of the stored energy. This places

the energy storage in the context of a more general system.

A clue as to how an energy conservation statement might be constructed from the differential magnetoquasistatic laws is obtained from Eq. 2, which makes it clear that the product of the free current density and the electric field intensity are closely connected with the statement of conservation of energy. The dot product of the electric field and Ampère's law, Eq. 2.3.23b, is

$$\vec{E}\cdot[\nabla \times \vec{H} - \vec{J}_f] = 0 \tag{14}$$

Use of a vector identity (Eq. 8, Appendix B) makes it possible to rewrite this expression as

$$\vec{H}\cdot\nabla \times \vec{E} - \nabla\cdot(\vec{E}\times\vec{H}) = \vec{E}\cdot\vec{J}_f \tag{15}$$

With the additional use of Faraday's law to represent $\nabla \times \vec{E}$, Eq. 15 takes the form of Eq. 2.13.16, with

$$S_e \equiv \vec{E}\times\vec{H}$$

$$W_e \equiv \frac{1}{2}\mu_o\vec{H}\cdot\vec{H} \tag{16}$$

$$\phi_e \equiv -\vec{E}\cdot\vec{J}_f - \vec{H}\cdot\frac{\partial\mu_o\vec{M}}{\partial t} - \vec{H}\cdot\nabla\times(\mu_o\vec{M}\times\vec{v})$$

These quantities have much the same physical significances discussed in connection with Eq. 2.13.16.

To place the magnetic energy storage identified with the thermodynamic arguments in the context of an actual system, consider a material which is ohmic and fixed so that $\vec{v} = 0$ and $\vec{J}_f = \sigma\vec{E}$. Then the second term on the right in Eq. 16c is in the form of a time rate of change of magnetization energy density. Hence, the power flow equation assumes the form of Eq. 2.13.17, with

$$\vec{S}_E = \vec{E}\times\vec{H}$$

$$W_E = \int_o^{\vec{B}} \vec{H}\cdot\delta\vec{B} \tag{17}$$

$$\phi_E = -\sigma\vec{E}\cdot\vec{E}$$

Implicit is the assumption that $\vec{H}$ is a single-valued function of the instantaneous $\vec{B}$. The resulting energy density includes magnetization energy and is consistent with Eq. 2.14.10.

2.15 Complex Amplitudes; Fourier Amplitudes and Fourier Transforms

The notion of a continuum network is introduced for the first time in the next section. The associated transfer relations illustrated there are a theme throughout the chapters which follow. Among several reasons for their use is the organization they lend to the representation of complicated, largely linear, systems. In this chapter, the continuum networks represent electromagnetic fields. Later, they represent fluid and (to some degree) solid mechanics, heat and mass transfer, and electromechanical continua in general. These networks make it possible to set aside one part of a given problem, derive the associated relations once and for all and accumulate these for later use. Such relations will be picked up over and over in solving different problems and, properly understood, are a useful reference.

Complex Amplitudes: In many practical situations, excitations are periodic in one or two spatial directions, in time or in space and time. The complex amplitude representation of fields, useful in dealing with these situations, is illustrated by considering the function $\Phi(z,t)$ which has dependence on z given explicitly by

$$\Phi(z,t) = \mathrm{Re}\,\tilde{\Phi}(t)e^{-jkz} \tag{1}$$

With the wavenumber k real, the spatial distribution is periodic with wavelength $\lambda = 2\pi/k$ and spatial phase determined by the complex amplitude $\tilde{\Phi}$. For example, if $\tilde{\Phi} = \tilde{\Phi}_o(t)$ is real and k is real, then $\Phi(z,t) = \Phi_o(t)\cos kz$.

The spatial derivative of Φ follows from Eq. 1 as

$$\frac{\partial\Phi}{\partial z} = \mathrm{Re}\,[-jk\tilde{\Phi}(t)e^{-jkz}] \tag{2}$$

The following identifications can therefore be made:

$$[\Phi(z,t), \frac{\partial\Phi}{\partial z}(z,t)] \Leftrightarrow [\tilde{\Phi}(t), -jk\tilde{\Phi}(t)] \tag{3}$$

with it being understood that even though complex amplitudes are being used, the temporal dependence is arbitrary. There will be occasions where the time dependence is specified, but the space dependence is not. For example, complex amplitudes will take the form

$$\Phi(z,t) = \text{Re}\,\hat{\Phi}(z)e^{j\omega t} \tag{4}$$

where $\hat{\Phi}(z)$ is itself perhaps expressed as a Fourier series or transform (see Sec. 5.16).

Most often, complex amplitudes will be used to represent both temporal and spatial dependences:

$$\Phi(z,t) = \text{Re}\hat{\Phi}e^{j(\omega t-kz)} \tag{5}$$

The (angular) frequency ω can in general be complex. If Φ is periodic in time with period T, then $T = 2\pi/\omega$. For complex amplitudes $\hat{\Phi}$, the identifications are:

$$[\Phi(z,t), \frac{\partial\Phi}{\partial z}(z,t), \frac{\partial\Phi}{\partial t}(z,t)] \Leftrightarrow [\hat{\Phi}, -jk\hat{\Phi}, j\omega\hat{\Phi}] \tag{6}$$

If ω and k are real, Eq. 5 represents a traveling wave. At any instant, its wavelength is $2\pi/k$, at any position its frequency is ω and points of constant phase propagate in the +z direction with the phase velocity ω/k.

Fourier Amplitudes and Transforms: The relations between complex amplitudes are identical to those between Fourier amplitudes or between Fourier transforms provided that these are suitably defined. For a wide range of physical situations it is the spatially periodic response or the temporal sinusoidal steady state that is of interest. Simple combinations of solutions represented by the complex amplitudes then suffice, and there is no need to introduce Fourier concepts. Even so, it is important to recognize at the outset that the spatial information required for analysis of excitations with arbitrary spatial distributions is inherent to the transfer relations based on single-complex-amplitude solutions.

The Fourier series represents an arbitrary function periodic in z with fundamental periodicity length ℓ by a superposition of complex exponentials. In terms of complex Fourier coefficients $\tilde{\Phi}_n(t)$, such a series is

$$\Phi(z,t) = \sum_{n=-\infty}^{\infty} \tilde{\Phi}_n(t)e^{-jk_n z}; \quad k_n \equiv 2n\pi/\ell; \quad \tilde{\Phi}_n^* = \tilde{\Phi}_{-n} \tag{7}$$

where the condition on $\tilde{\Phi}_n$ insures that Φ is real. Thus, with the identification $\tilde{\Phi} \rightarrow \tilde{\Phi}_n$ and $k \rightarrow k_n$, each complex exponential solution of the form of Eq. 1 can be taken as one term in the Fourier series. The mth Fourier amplitude $\tilde{\Phi}_m$ follows by multiplying Eq. 7 by the complex conjugate function $\exp(jk_m z)$ and integrating over the length ℓ to obtain only one term on the right. This expression can then be solved for $\tilde{\Phi}_m$ to obtain the inverse relation

$$\tilde{\Phi}_m = \frac{1}{\ell}\int_z^{z+\ell} \Phi(z,t)e^{jk_m z}\,dz \tag{8}$$

If the temporal dependence is also periodic, with fundamental period T, the Fourier series can also be used to represent the time dependence in Eq. 7:

$$\Phi(z,t) = \sum_{m=-\infty}^{+\infty}\sum_{n=-\infty}^{+\infty} \hat{\Phi}_{mn}\, e^{j(\omega_m t-k_n z)}; \quad \hat{\Phi}_{mn}^* = \hat{\Phi}_{-m-n} \tag{9}$$

where the condition on the amplitudes insures that $\Phi(z,t)$ is real. One component out of this double summation is the traveling-wave solution represented by the complex amplitude form, Eq. 5. The rules given by Eqs. 3 and 6 pertain either to the complex amplitudes or the Fourier coefficients.

The Fourier transform is convenient if the dependence is not periodic. With the Fourier transform $\tilde{\Phi}(k,t)$ given by

$$\tilde{\Phi}(k,t) = \int_{-\infty}^{+\infty} \Phi(z,t)e^{jkz}\,dz$$

the functional dependence on z is a superposition of the complex exponentials

$$\Phi(z,t) = \int_{-\infty}^{+\infty} \tilde{\Phi}(k,t)e^{-jkz}\,\frac{dk}{2\pi} \tag{11}$$

The relation between the transform and the transform of the derivative can be found by taking the transform of $\partial\Phi/\partial z$ using Eq. 11 and integrating by parts. Recall that $\int v\,du = uv - \int u\,dv$ and identify $du \to \partial\Phi/\partial z\,dz$ and $v \to \exp jkz$, and it follows that

$$\int_{-\infty}^{\infty} \frac{\partial\Phi}{\partial z} e^{jkz}\,dz = \Phi e^{jkz}\Big|_{-\infty}^{+\infty} - jk\int_{-\infty}^{+\infty} \Phi e^{jkz}\,dz \tag{12}$$

For properly bounded functions the first term on the right vanishes and the second is $-jk\tilde{\Phi}(k,t)$. The transform of $\partial\Phi/\partial z$ is simply $-jk\tilde{\Phi}$ and thus the Fourier transform also follows the rules given with Eq. 3.

Extension of the Fourier transform to a second dimension results in the transform pair

$$\begin{aligned}\Phi(z,t) &= \int_{-\infty}^{\infty}\int_{-\infty}^{\infty} \hat{\Phi}(k,\omega)e^{j(\omega t-kz)}\,\frac{dk}{2\pi}\frac{d\omega}{2\pi}\\ \hat{\Phi}(k,\omega) &= \int_{-\infty}^{+\infty}\int_{-\infty}^{+\infty} \Phi(z,t)e^{-j(\omega t-kz)}\,dt\,dz\end{aligned} \tag{13}$$

which illustrates how the traveling-wave solution of Eq. 5 can be viewed as a component of a complicated function. Again, relations between complex amplitudes are governed by the same rules, Eq. 6, as are the Fourier amplitudes $\hat{\Phi}(k,\omega)$.

If relationships are found among quantities $\tilde{\Phi}(t)$, then the same relations hold with $\tilde{\Phi} \to \hat{\Phi}$ and $\partial(\;)/\partial t \to j\omega$, because the time dependence $\exp(j\omega t)$ is a particular case of the more general form $\tilde{\Phi}(t)$.

<u>Averages of Periodic Functions</u>: An identity often used to evaluate temporal or spatial averages of complex-amplitude expressions is

$$\left\langle \mathrm{Re}\,\tilde{A}\,e^{-jkz}\;\mathrm{Re}\,\tilde{B}\,e^{-jkz}\right\rangle_z = \frac{1}{2}\,\mathrm{Re}\,\tilde{A}\,\tilde{B}^* \tag{14}$$

where $\langle\;\rangle_z$ signifies an average over the length $2\pi/k$ and it is assumed that k is real. This relation follows by letting

$$\mathrm{Re}\,\tilde{A}\,e^{-jkz}\;\mathrm{Re}\,\tilde{B}\,e^{-jkz} = \frac{1}{2}\left[\tilde{A}\,e^{-jkz} + \tilde{A}^*e^{jkz}\right]\frac{1}{2}\left[\tilde{B}\,e^{-jkz} + \tilde{B}^*e^{jkz}\right] \tag{15}$$

and multiplying out the right-hand side to obtain

$$\frac{1}{4}\left[\tilde{A}\,\tilde{B}\,e^{-2jkz} + \tilde{A}^*\tilde{B}^*e^{2jkz}\right] + \frac{1}{4}\left[\tilde{A}\,\tilde{B}^* + \tilde{A}^*\tilde{B}\right] \tag{16}$$

The first term is a linear combination of cos 2kz and sin 2kz and hence averages to zero. The second term is constant and identical to the right-hand side of Eq. 14.

A similar theorem simplifies evaluation of the average of two periodic functions expressed in the form of Eq. 7:

$$\begin{aligned}\langle AB\rangle_z &= \left\langle \sum_{n=-\infty}^{+\infty} \tilde{A}_n(t)\,e^{-jk_nz} \sum_{m=-\infty}^{+\infty} \tilde{B}_m(t)\,e^{-jk_mz}\right\rangle_z\\ &= \sum_{n=-\infty}^{+\infty} \tilde{A}_n\tilde{B}_{-n} = \sum_{n=-\infty}^{+\infty} \tilde{A}_n\tilde{B}_n^*\end{aligned} \tag{17}$$

Of course, either the complex amplitude theorem of Eq. 14 or the Fourier amplitude theorem of Eq. 17 applies to time averages with $kz \rightarrow -\omega t$.

2.16 Flux-Potential Transfer Relations for Laplacian Fields

It is often convenient in the modeling of a physical system to divide the volume of interest into regions having uniform properties. Surfaces enclosing these regions are often planar, cylindrical or spherical, with the volume then taking the form of a planar layer, a cylindrical annulus or a spherical shell. Such volumes and bounding surfaces are illustrated in Tables 2.16.1-3. The question answered in this section is: given the potential on the bounding surfaces, what are the associated normal flux densities? Of immediate interest is the relation of the electric potentials to the normal displacement vectors. But also treated in this section is the relation of the magnetic potential to the normal magnetic flux densities. First the electroquasistatic fields are considered, and then the magnetoquasistatic relations follow by analogy.

Electric Fields: If any one of the regions shown in Tables 2.16.1-3 is filled with insulating charge-free ($\rho_f = 0$) material of uniform permittivity ε,

$$\vec{P} = (\varepsilon - \varepsilon_o)\vec{E}, \; \vec{D} = \varepsilon\vec{E} \tag{1}$$

the governing field equations are Gauss' law, Eq. 2.3.23a,

$$\nabla\cdot\vec{D} = 0 \tag{2}$$

and the condition that $\vec{E}$ be irrotational, Eq. 2.3.24a. The latter is equivalent to

$$\vec{E} = -\nabla\Phi \tag{3}$$

Thus, the potential distribution within a volume is described by Laplace's equation

$$\nabla^2\Phi = 0 \tag{4}$$

In terms of Φ,

$$\vec{D} = -\varepsilon\nabla\Phi \tag{5}$$

Magnetic Fields: For magnetoquasistatic fields in an insulating region ($J_f = 0$) of uniform permeability

$$\vec{M} = (\frac{\mu}{\mu_o} - 1)\vec{H}; \; \vec{B} = \mu\vec{H} \tag{6}$$

Thus, from Ampère's law, Eq. 2.3.23b, $\vec{H}$ is irrotational and it is appropriate to define a magnetic potential Ψ:

$$\vec{H} = -\nabla\Psi \tag{7}$$

In addition, there is Eq. 2.3.24b:

$$\nabla\cdot\vec{B} = 0 \tag{8}$$

Thus, the potential again satisfies Laplace's equation

$$\nabla^2\Psi = 0 \tag{9}$$

and in terms of Ψ, the magnetic flux density is

$$\vec{B} = -\mu\nabla\Psi \tag{10}$$

Comparison of the last two relations to Eqs. 4 and 5 shows that relations now derived for the electric fields can be carried over to describe the magnetic fields by making the identification $(\Phi,\vec{D},\varepsilon) \rightarrow (\Psi,\vec{B},\mu)$.

Planar Layer: Bounding surfaces at $x = \Delta$ and $x = 0$, respectively denoted by α and β, are shown in Table 2.16.1. So far as developments in this section are concerned, these are not physical boundaries. They are simply surfaces at which the potentials are respectively

$$\Phi(\Delta,y,z,t) = \text{Re}\,\tilde{\Phi}^\alpha(t)\exp[-j(k_y y+k_z z)]; \; \Phi(0,y,z,t) = \text{Re}\,\tilde{\Phi}^\beta(t)\exp[-j(k_y y+k_z z)] \tag{11}$$

Table 2.16.1. Flux-potential transfer relations for planar layer in terms of electric potential and normal displacement (Φ, D_x). To obtain magnetic relations, substitute $(\Phi, D_x, \varepsilon) \rightarrow (\Psi, B_x, \mu)$.

Planar layer

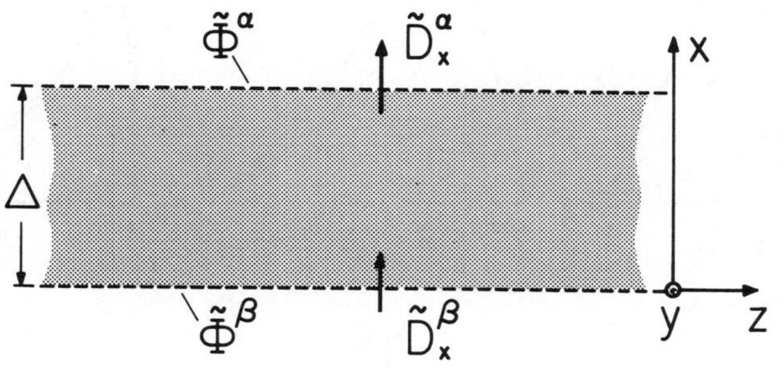

$$\Phi = \text{Re}\, \tilde{\Phi}(x,t)\, e^{-j(k_y y + k_z z)}$$

$$\gamma \equiv \sqrt{k_y^2 + k_z^2}$$

$$\begin{bmatrix} \tilde{D}_x^\alpha \\ \tilde{D}_x^\beta \end{bmatrix} = \varepsilon\gamma \begin{bmatrix} -\coth(\gamma\Delta) & \dfrac{1}{\sinh(\gamma\Delta)} \\ \dfrac{-1}{\sinh(\gamma\Delta)} & \coth(\gamma\Delta) \end{bmatrix} \begin{bmatrix} \tilde{\Phi}^\alpha \\ \tilde{\Phi}^\beta \end{bmatrix} \qquad \text{(a)}$$

$$\begin{bmatrix} \tilde{\Phi}^\alpha \\ \tilde{\Phi}^\beta \end{bmatrix} = \frac{1}{\varepsilon\gamma} \begin{bmatrix} -\coth(\gamma\Delta) & \dfrac{1}{\sinh(\gamma\Delta)} \\ \dfrac{-1}{\sinh(\gamma\Delta)} & \coth(\gamma\Delta) \end{bmatrix} \begin{bmatrix} \tilde{D}_x^\alpha \\ \tilde{D}_x^\beta \end{bmatrix} \qquad \text{(b)}$$

These will be recognized as generalizations of the complex amplitudes introduced with Eq. 2.15.1. That the potentials at the α and β surfaces can be quite general follows from the discussion of Sec. 2.15, which shows that the following arguments apply when $\tilde{\Phi}$ is a spatial Fourier amplitude or a Fourier transform.

In view of the surface potential distributions, solutions to Eq. 4 are assumed to take the form

$$\Phi = \text{Re}\, \tilde{\Phi}(x,t) \exp[-j(k_y y + k_z z)] \qquad (12)$$

Substitution shows that

$$\frac{d^2\tilde{\Phi}}{dx^2} - \gamma^2\tilde{\Phi} = 0; \quad \gamma = \sqrt{k_y^2 + k_z^2} \qquad (13)$$

Solutions of this equation are linear combinations of $e^{\pm\gamma x}$ or alternatively of $\sinh \gamma x$ and $\cosh \gamma x$. With $\tilde{\Phi}_1$ and $\tilde{\Phi}_2$ arbitrary functions of time, the solution takes the form

$$\tilde{\Phi} = \tilde{\Phi}_1 \sinh \gamma x + \tilde{\Phi}_2 \cosh \gamma x \qquad (14)$$

The two coefficients are determined by requiring that the conditions of Eq. 11 be satisfied. For the simple situation at hand, an instructive alternative to performing the algebra necessary to evaluate $(\tilde{\Phi}_1, \tilde{\Phi}_2)$ consists in recognizing that a linear combination of the two solutions in Eq. 14 is $\sinh \gamma(x - \Delta)$. Thus, the solution can be written as the sum of solutions that are individually zero on one or the other of the bounding surfaces. By inspection, it follows that

$$\tilde{\Phi} = \tilde{\Phi}^\alpha \frac{\sinh \gamma x}{\sinh \gamma\Delta} - \tilde{\Phi}^\beta \frac{\sinh \gamma(x - \Delta)}{\sinh \gamma\Delta} \qquad (15)$$

From Eqs. 5 and 15, $\vec{D}$ can be determined:

$$\vec{D}_x = -\varepsilon \frac{\partial\Phi}{\partial x} = -\varepsilon\, \text{Re}\, \gamma \left[\tilde{\Phi}^\alpha \frac{\cosh \gamma x}{\sinh \gamma\Delta} - \tilde{\Phi}^\beta \frac{\cosh \gamma(x-\Delta)}{\sinh \gamma\Delta}\right] e^{-j(k_y y + k_z z)} \qquad (16)$$

Evaluation of this equation at $x = \Delta$ gives the displacement vector normal to the α surface, with complex amplitude $\tilde{D}_x^\alpha$. Similarly, evaluated at $x = 0$, Eq. 16 gives $\tilde{D}_x^\beta$. The components of the "flux" $(\tilde{D}_x^\alpha, \tilde{D}_x^\beta)$ are now determined, given the respective potentials $(\tilde{\Phi}^\alpha, \tilde{\Phi}^\beta)$. The transfer relations, Eq. (a) of Table 2.16.1, summarize what is found. These relations can be solved for any pair of variables as a function of the remaining pair. The inverse transfer relations are also summarized for reference in Table 2.16.1, Eq. (b)

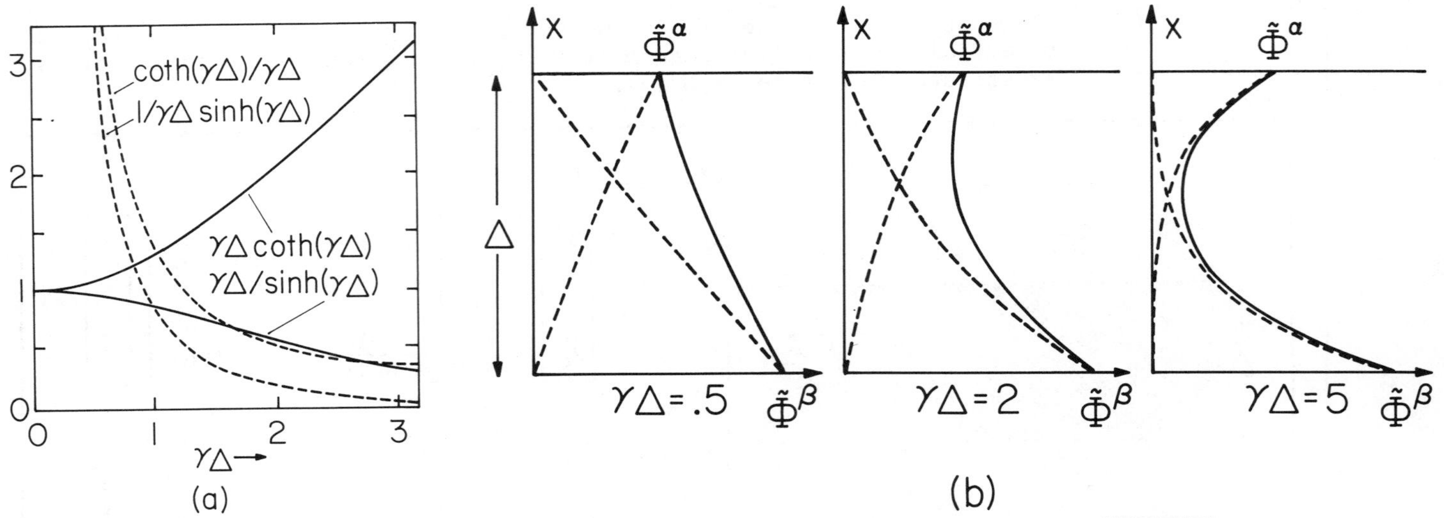

Fig. 2.16.1. (a) Transfer coefficients as a function of $\Delta\gamma \equiv \Delta\sqrt{k_y^2 + k_z^2}$.
(b) Distribution of Φ across layer.

That the layer is essentially a distributed capacitance (inductance) is emphasized by drawing attention to the analogy between the transfer relations and constitutive laws for a system of linear capacitors (inductors). For a two-terminal-pair system, Eq. 2.11.5 comprises two terminal charges (q_1,q_2) expressed as linear functions of the terminal voltages (v_1,v_2). Analogously, the (D_x^α,D_x^β) (which have units of charge per unit area and an arbitrary time dependence) are given as linear functions of the potentials by Eq. (a) of Table 2.16.1. A similar analogy exists between Eq. 2.12.5, expressing (λ_1,λ_2) as functions of (i_1,i_2), and the transfer relations between (B_x^α,B_x^β) (units of flux per unit area) and the magnetic potentials (ψ^α,ψ^β).

According to Eq. (a) of Table 2.16.1, D_x is induced by a "self term" (proportional to the potential at the same surface) and a "mutual term." The coefficients which express this self- and mutual-coupling have a dependence on $\Delta\gamma$ ($2\pi/\gamma$ the wavelength in the y-z plane) shown in Fig. 2.16.1a. Written in the form of Eq. 15, the potential has components, excited at each surface, that decay to zero, as shown in Fig. 2.16.1b, at a rate that is proportional to how rapidly the fields vary in the y-z plane. For long waves the decay is relatively slow, as depicted by the case $\Delta\gamma = 0.5$, and the mutual-field is almost as great as the self field. But as the wavelength is shortened relative to Δ ($\Delta\gamma$ increased), the surfaces couple less and less.

In this discussion it is assumed that γ is real, which it is if k_y and k_z are real. In fact, the transfer relations are valid and useful for complex values of (k_y,k_z). If these numbers are purely imaginary, the field distributions over the layer cross section are periodic. Such solutions are needed to satisfy boundary conditions imposed in an x-y plane.

Cylindrical Annulus: With the bounding surfaces coaxial cylinders having radii α and β, it is natural to use cylindrical coordinates (r, θ, z). A cross section of this prototype region and the coordinates are shown in Table 2.16.2. On the outer and inner surfaces, the potential has the respective forms

$$\Phi(\alpha,\theta,z,t) = \mathrm{Re}\,\tilde{\Phi}^\alpha(t)\,e^{-j(m\theta+kz)};\quad \Phi(\beta,\theta,z,t) = \mathrm{Re}\,\tilde{\Phi}^\beta(t)e^{-j(m\theta+kz)} \tag{17}$$

Hence, it is appropriate to assume a bulk potential

$$\Phi = \mathrm{Re}\,\tilde{\Phi}(r,t)e^{-j(m\theta+kz)} \tag{18}$$

Substitution in Laplace's equation (see Appendix A for operations in cylindrical coordinates), Eq. 4, then shows that

$$\frac{d^2\tilde{\Phi}}{dr^2} + \frac{1}{r}\frac{d\tilde{\Phi}}{dr} - (k^2 + \frac{m^2}{r^2})\tilde{\Phi} = 0 \tag{19}$$

By contrast with Eq. 13, this one has space-varying coefficients. It is convenient to categorize the solutions according to the values of (m,k). With m = 0 and k = 0, the remaining terms are a perfect differential which can be integrated twice to give the solutions familiar from the problem of the field

Table 2.16.2. Flux-potential relations for cylindrical annulus in terms of electric potential and normal displacement (Φ, D_r). To obtain magnetic relations, substitute $(\Phi, D_r, \varepsilon) \rightarrow (\Psi, B_r, \mu)$.

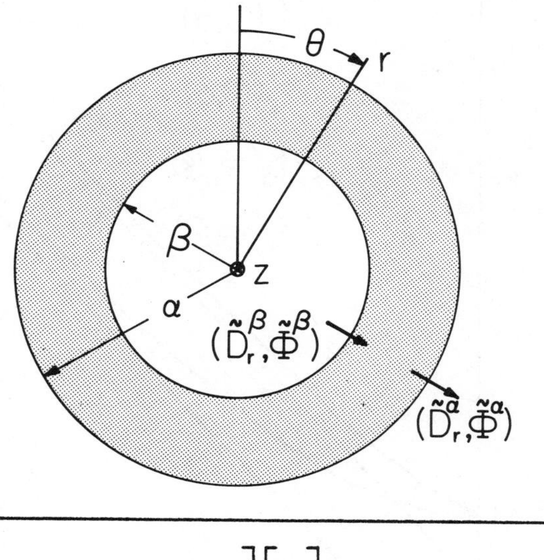

$$\Phi = \text{Re}\, \tilde{\Phi}(r,t) e^{-j(m\theta + kz)}$$

$$\begin{bmatrix} \tilde{D}_r^\alpha \\ \tilde{D}_r^\beta \end{bmatrix} = \varepsilon \begin{bmatrix} f_m(\beta,\alpha) & g_m(\alpha,\beta) \\ g_m(\beta,\alpha) & f_m(\alpha,\beta) \end{bmatrix} \begin{bmatrix} \tilde{\Phi}^\alpha \\ \tilde{\Phi}^\beta \end{bmatrix} \quad \text{(a)}$$

$k = 0,\ m = 0$

$$f_o(x,y) = \frac{1}{y}\Big/\ln\left(\frac{x}{y}\right);\quad g_o(x,y) = \frac{1}{x}\Big/\ln\left(\frac{x}{y}\right)$$

$k = 0,\ m = 1,2,\cdots$

$$f_m(x,y) = \frac{m}{y}\frac{\left[\left(\frac{x}{y}\right)^m + \left(\frac{y}{x}\right)^m\right]}{\left[\left(\frac{x}{y}\right)^m - \left(\frac{y}{x}\right)^m\right]}$$

$$g_m(x,y) = \frac{2m}{x}\frac{1}{\left[\left(\frac{x}{y}\right)^m - \left(\frac{y}{x}\right)^m\right]}$$

$k \neq 0,\ m = 0,1,2\cdots$*

$$f_m(x,y) = \frac{jk[H_m(jkx)J_m'(jky) - J_m(jkx)H_m'(jky)]}{[J_m(jkx)H_m(jky) - J_m(jky)H_m(jkx)]}$$

$$g_m(x,y) = \frac{-2j}{\pi x[J_m(jkx)H_m(jky) - J_m(jky)H_m(jkx)]}$$

$$f_m(x,y) = \frac{k[K_m(kx)I_m'(ky) - I_m(kx)K_m'(ky)]}{[I_m(kx)K_m(ky) - I_m(ky)K_m(kx)]}$$

$$g_m(x,y) = \frac{1}{x[I_m(kx)K_m(ky) - I_m(ky)K_m(kx)]}$$

$$\begin{bmatrix} \tilde{\Phi}^\alpha \\ \tilde{\Phi}^\beta \end{bmatrix} = \frac{1}{\varepsilon} \begin{bmatrix} F_m(\beta,\alpha) & G_m(\alpha,\beta) \\ G_m(\beta,\alpha) & F_m(\alpha,\beta) \end{bmatrix} \begin{bmatrix} \tilde{D}_r^\alpha \\ \tilde{D}_r^\beta \end{bmatrix} \quad \text{(b)}$$

$k = 0,\ m = 0$

No inverse

$k = 0,\ m = 1,2,\cdots$

$$F_m(x,y) = \frac{y}{m}\frac{\left[\left(\frac{x}{y}\right)^m + \left(\frac{y}{x}\right)^m\right]}{\left[\left(\frac{x}{y}\right)^m - \left(\frac{y}{x}\right)^m\right]}$$

$$G_m(x,y) = \frac{2y}{m}\frac{1}{\left[\left(\frac{x}{y}\right)^m - \left(\frac{y}{x}\right)^m\right]}$$

$k \neq 0,\ m = 0,1,2,\cdots$*

$$F_m(x,y) = \frac{1}{jk}\frac{[J_m'(jkx)H_m(jky) - H_m'(jkx)J_m(jky)]}{[J_m'(jky)H_m'(jkx) - J_m'(jkx)H_m'(jky)]}$$

$$G_m(x,y) = \frac{-2}{j\pi k(kx)[J_m'(jky)H_m'(jkx) - J_m'(jkx)H_m'(jky)]}$$

$$F_m(x,y) = \frac{1}{k}\frac{[I_m'(kx)K_m(ky) - K_m'(kx)I_m(ky)]}{[I_m'(ky)K_m'(kx) - I_m'(kx)K_m'(ky)]}$$

$$G_m(x,y) = \frac{1}{k(kx)[I_m'(ky)K_m'(kx) - I_m'(kx)K_m'(ky)]}$$

$\beta \rightarrow 0$

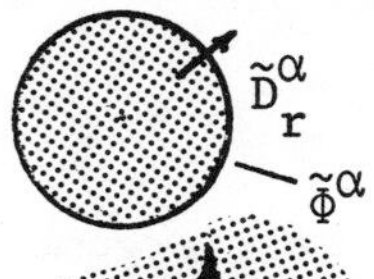

$$\tilde{D}_r^\alpha = \varepsilon f_m(0,\alpha)\tilde{\Phi}^\alpha;\quad f_m(0,\alpha) = -\frac{kI_m'(k\alpha)}{I_m(k\alpha)} \quad \text{(c)}$$

$\alpha \rightarrow \infty$

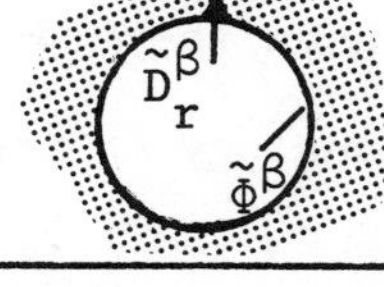

$$\tilde{D}_r^\beta = \varepsilon f_m(\infty,\beta)\tilde{\Phi}^\beta;\quad f_m(\infty,\beta) = -\frac{kK_m'(k\beta)}{K_m(k\beta)} \quad \text{(d)}$$

*See Prob. 2.17.2 for proof that $H_m(jkx)J_m'(jkx) - J_m(jkx)H'_m(jkx) = -2/(\pi kx)$ and $K_m(kx)I_m'(kx) - I_m(kx)K_m'(kx) = 1/kx$ incorporated into g_m and G_m.

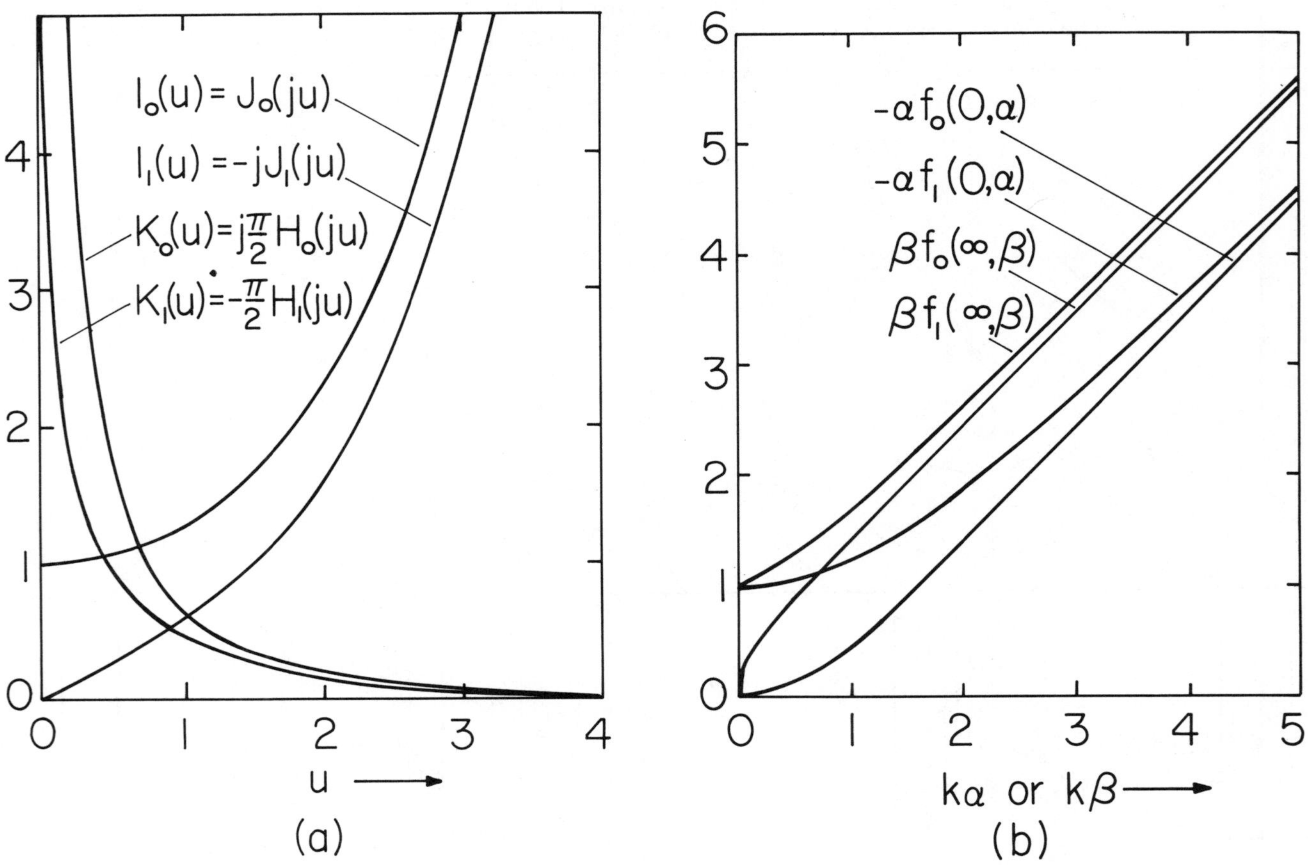

Fig. 2.16.2. (a) Modified Bessel functions. (b) Self-field coefficients of cylindrical transfer relations in limits where surfaces do not interact.

between coaxial circular conductors. In view of the boundary conditions at $r = \alpha$ and $r = \beta$,

$$\tilde{\Phi} = \tilde{\Phi}^\alpha \frac{\ln(\frac{r}{\beta})}{\ln(\frac{\alpha}{\beta})} - \tilde{\Phi}^\beta \frac{\ln(\frac{r}{\alpha})}{\ln(\frac{\alpha}{\beta})} = \tilde{\Phi}_\alpha + (\tilde{\Phi}_\beta - \tilde{\Phi}_\alpha)\frac{\ln(\frac{r}{\alpha})}{\ln(\frac{\beta}{\alpha})}; \quad (m,k) = (0,0) \tag{20}$$

For situations that depend on θ, but not on z (polar coordinates) so that $k = 0$, substitution shows the solutions to Eq.19 are $r^{\pm m}$. By inspection or algebraic manipulation, the linear combination of these that satisfies the conditions of Eq. 17 is

$$\tilde{\Phi} = \tilde{\Phi}^\alpha \frac{[(\frac{\beta}{r})^m - (\frac{r}{\beta})^m]}{[(\frac{\beta}{\alpha})^m - (\frac{\alpha}{\beta})^m]} + \tilde{\Phi}^\beta \frac{[(\frac{r}{\alpha})^m - (\frac{\alpha}{r})^m]}{[(\frac{\beta}{\alpha})^m - (\frac{\alpha}{\beta})^m]}; \quad (m,k) = (m,0) \tag{21}$$

For k finite, the solutions to Eq. 19 are the modified Bessel functions $I_m(kr)$ and $K_m(kr)$. These play a role in the circular geometry analogous to $\exp(\pm\gamma x)$ in Cartesian geometry. The radial dependences of the functions of order $m = 0$ and $m = 1$ are shown in Fig. 2.16.2a. Note that I_m and K_m are respectively singular at infinity and the origin.

Just as the exponential solutions could be determined from Eq. 13 by assuming a power series in x, the Bessel functions are determined from an infinite series solution to Eq. 19. Like γ, k can in general be complex. If it is, it is customary to define two new functions which, in the special case where k is real, have imaginary arguments:

$$J_m(jkr) \equiv j^m I_m(kr), \; H_m(jkr) = \frac{2}{\pi} j^{-(m+1)} K_m(kr) \tag{22}$$

These are respectively the Bessel and Hankel functions of first kind. For real arguments, I_m and K_m are real, and hence J_m and H_m can be either purely real or imaginary, depending on the order.

Large real-argument limits of the functions I_m and K_m reinforce the analogy to the Cartesian exponential solutions:

$$\lim_{u\to\infty} I_m(u) = \frac{1}{\sqrt{2\pi u}}\exp(u);\ \lim_{u\to\infty} K_m(u) = \sqrt{\frac{\pi}{2u}}\ \exp(-u) \tag{23}$$

Useful relations in the opposite extreme of small arguments are

$$\lim_{u\to 0} jH_o(ju) = \frac{2}{\pi}\ln\left(\frac{2}{1.781072u}\right);\ \lim_{u\to 0} J_m(ju) = \frac{(ju)^m}{m!2^m}$$

$$\lim_{u\to 0} H_m(ju) = \frac{(m-1)!2^m}{j\pi(ju)^m};\ m \neq 0 \tag{24}$$

By inspection or algebraic manipulation, the linear combination of J_m and H_m satisfying the boundary conditions of Eq. 17 is

$$\tilde{\Phi} = \tilde{\Phi}^\alpha \frac{[H_m(jk\beta)J_m(jkr) - J_m(jk\beta)H_m(jkr)]}{[H_m(jk\beta)J_m(jk\alpha) - J_m(jk\beta)H_m(jk\alpha)]} + \tilde{\Phi}^\beta \frac{[J_m(jk\alpha)H_m(jkr) - H_m(jk\alpha)J_m(jkr)]}{[J_m(jk\alpha)H_m(jk\beta) - H_m(jk\alpha)J_m(jk\beta)]} \tag{25}$$

The evaluation of the surface displacements (D_r^α, D_r^β) using Eqs. 20, 21, or 25 is now accomplished using the same steps as for the planar layer. The resulting transfer relations are summarized by Eq. (a) in Table 2.16.2. Inversion of these relations, to give the surface potentials as functions of the surface displacements, results in the relations summarized by Eq. (b) of that table. Primes denote derivatives with respect to the entire specified argument of the function. Useful identities are:

$$uI_m'(u) = mI_m(u) + uI_{m+1}(u);\ uI_m'(u) = -mI_m(u) + uI_{m-1}(u)$$

$$uK_m'(u) = mK_m(u) - uK_{m+1}(u)$$

$$R_o'(u) = -R_1(u) \tag{26}$$

$$uR_m'(u) = -mR_m(u) + uR_{m-1}(u);\ uR_m'(u) = mR_m(u) - uR_{m+1}(u)$$

where R_m can be J_m, H_m, or the function N_m to be defined with Eq. 29.

Two useful limits of the transfer relations are given by Eqs. (c) and (d) of Table 2.16.2. In the first, the inner surface is absent, while in the second the outer surface is removed many wavelengths $2\pi/k$. The self-field coefficients $f_m(0,\alpha)$ and $f_m(\infty,\beta)$ are sketched for m=0 and m=1 in Fig. 2.16.2b. Again, it is useful to note the analogy to the planar layer case where the appropriate limit is $k\Delta \to \infty$. In fact, for $k\alpha$ or $k\beta$ reasonably large, the k dependence and the signs are the same as for the planar geometry:

$$\lim_{k\alpha\to\infty} \alpha f_m(0,\alpha) \to -k\alpha;\ \lim_{k\beta\to\infty} \beta f_m(\infty,\beta) \to k\beta \tag{27}$$

For small arguments, these functions become

$$\lim_{k\alpha\to 0} \alpha f_o(0,\alpha) \to -\frac{(k\alpha)^2}{2};\ \lim_{k\beta\to 0} \beta f_o(\infty,\beta) \to \frac{1}{\ln\left[\frac{2}{1.781072k\beta}\right]}$$

$$\lim_{k\alpha\to 0} \alpha f_m(0,\alpha) \to -m \text{ for } m \neq 0;\ \lim_{k\beta\to 0} \beta f_m(\infty,\beta) \to m \text{ for } m \neq 0 \tag{28}$$

In general, k can be complex. In fact the most familiar form for Bessel functions is with k purely imaginary. In that case, J_m is real but H_m is complex. By convention

$$H_m(u) \equiv J_m(u) + jN_m(u) \tag{29}$$

where, if u is real, J_m and N_m are real and Bessel functions of first and second kind. As might be expected from the planar analogue, the radial dependence becomes periodic if k is imaginary. Plots of the functions in this case are given in Fig. 2.16.3.

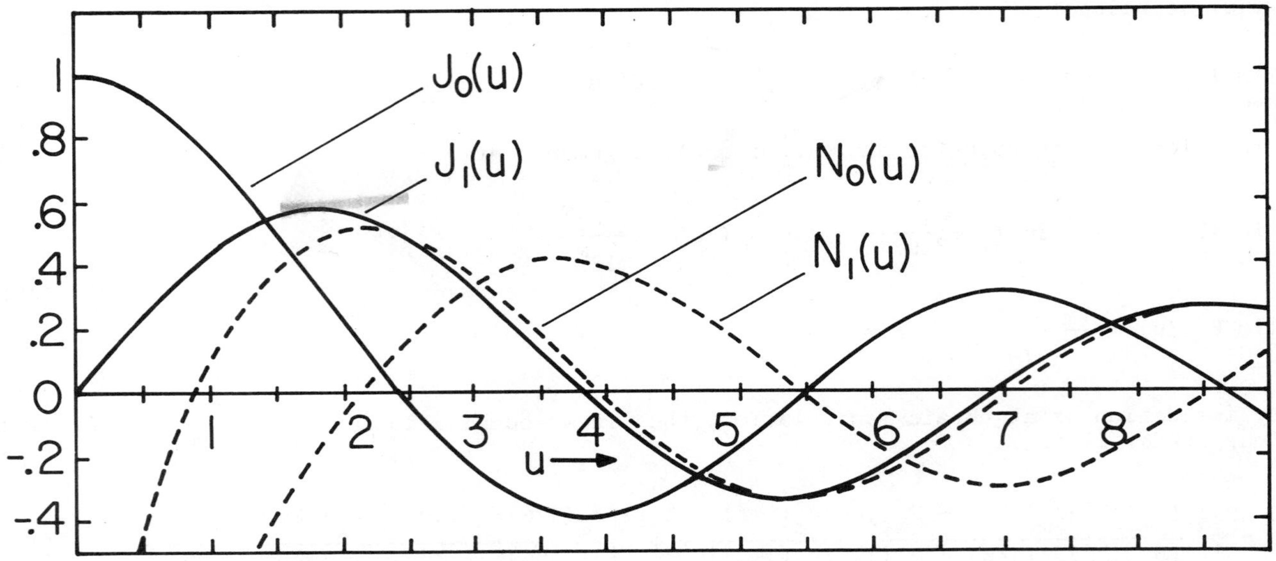

Fig. 2.16.3. Bessel functions of first and second kind and real arguments. References for the Bessel and related functions should be consulted for more details concerning their properties and numerical values.[1-4]

Spherical Shell: A region between spherical surfaces having outer and inner radii α and β, respectively, is shown in the figure of Table 2.16.3. In the volume, the potential conveniently takes the variable separable form

$$\Phi = \text{Re}\, \tilde{\Phi}(r,t)\, \Theta(\theta) e^{-jm\phi} \tag{30}$$

where (r,θ,ϕ) are spherical coordinates as defined in the figure. Substitution of Eq. 30 into Laplace's equation, Eq. 4, shows that the ϕ dependence is correctly assumed and that the (r,θ) dependence is determined from the equations

$$\frac{1}{\sin\theta\Theta}\frac{d}{d\theta}[\sin\theta \frac{d\Theta}{d\theta}] - \frac{m^2}{\sin^2\theta} = -K^2$$

$$\frac{1}{\tilde{\Phi}}\frac{d}{dr}(r^2\frac{d\tilde{\Phi}}{dr}) = K^2 \tag{31}$$

where the separation coefficient K^2 is independent of (r,θ). With the substitutions

$$u = \cos\theta, \sqrt{1-u^2} = \sin\theta \tag{32}$$

Eq. 31a is converted to

$$(1-u^2)\frac{d^2\Theta}{du^2} - 2u\frac{d\Theta}{du} + (K^2 - \frac{m^2}{1-u^2})\,\Theta = 0 \tag{33}$$

For $K^2 = n(n+1)$ and n an integer, solutions to Eq. 33 are

$$\Theta = P_n^m(u) \tag{34}$$

---

1. F. B. Hildebrand, Advanced Calculus for Applications, Prentice-Hall, Englewood Cliffs, N.J., 1962, pp. 142-165.

2. S. Ramo, J. R. Whinnery and T. Van Duzer, Fields and Waves in Communication Electronics, John Wiley and Sons, New York, 1965, pp. 207-218.

3. M. Abramowitz and I. A. Stegun, Handbook of Mathematical Functions with Formulas, Graphs, and Mathematical Tables, National Bureau of Standards, Applied Mathematics Series 55, U.S. Government Printing Office, Washington D. C. 20402, 1964, pp. 355-494.

4. E. Jahnke and F. Emde, Table of Functions with Formulae and Curves, Dover Publications, New York 1945, pp. 128-210.

Table 2.16.3. Flux-potential transfer relations for spherical shell in terms of electric potential and normal displacement (Φ, D_r). To obtain magnetic relations, substitute $(\Phi, D_r, \varepsilon) \rightarrow (\Psi, B_r, \mu)$.

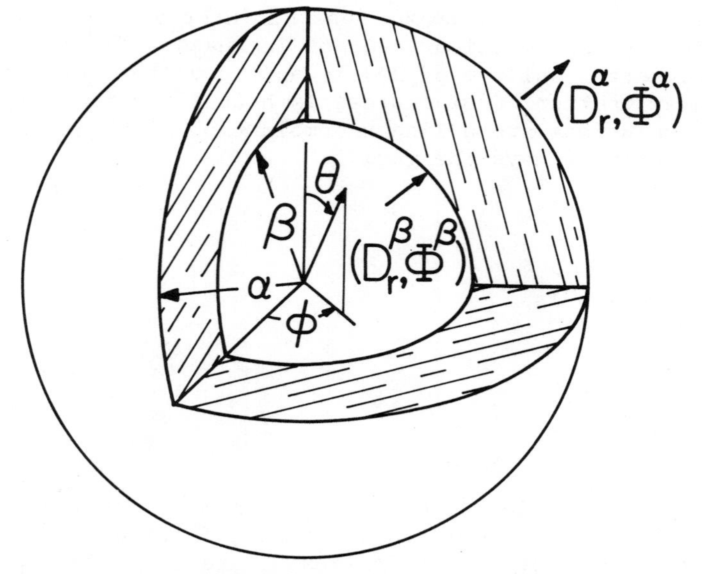

$$\Phi = \text{Re}\, \tilde{\Phi}(r,t) P_n^m(\cos\theta) e^{-jm\phi}$$

$$P_n^m = (1 - x^2)^{m/2} \frac{d^m P_n}{dx^m}$$

$$P_o = 1,\ P_1 = x,\ P_2 = \frac{1}{2}(3x^2 - 1)$$

$$P_3 = \frac{1}{2}(5x^3 - 3x)$$

$$P_4 = \frac{1}{8}(35x^4 - 30x^2 + 3)$$

m	P_o^m	P_1^m	$P_1^m \cos m\phi$	P_2^m	$P_2^m \cos m\phi$	P_3^m	$P_3^m \cos m\phi$
0	1	$\cos\theta$	+ −	$\frac{1}{2}(3\cos^2\theta - 1)$	+ − +	$\frac{1}{2}(5\cos^3\theta - 3\cos\theta)$	+ − + −
1	0	$\sin\theta$	+ − +	$3\sin\theta\cos\theta$	+ − + − + −	$\frac{3}{2}\sin\theta(5\cos^2\theta - 1)$	+ − + − + − + − +
2	0	0		$3\sin^2\theta$	+ − + − +	$15\sin^2\theta\cos\theta$	+ − + − + − + − + −
3	0	0		0		$15\sin^3\theta$	+ − + − + − +

$$\begin{bmatrix} \tilde{D}_r^\alpha \\ \tilde{D}_r^\beta \end{bmatrix} = \varepsilon \begin{bmatrix} f_n(\beta,\alpha) & g_n(\alpha,\beta) \\ g_n(\beta,\alpha) & f_n(\alpha,\beta) \end{bmatrix} \begin{bmatrix} \tilde{\Phi}^\alpha \\ \tilde{\Phi}^\beta \end{bmatrix} \qquad \text{(a)}$$

$$f_n(x,y) = \frac{[n(\frac{y}{x})^n + (n+1)(\frac{x}{y})^{n+1}]}{[x(\frac{x}{y})^n - y(\frac{y}{x})^n]}$$

$$g_n(x,y) = \frac{(2n+1)}{x^2[\frac{1}{y}(\frac{x}{y})^n - \frac{1}{x}(\frac{y}{x})^n]}$$

$$\begin{bmatrix} \tilde{\Phi}^\alpha \\ \tilde{\Phi}^\beta \end{bmatrix} = \frac{1}{\varepsilon} \begin{bmatrix} F_n(\beta,\alpha) & G_n(\alpha,\beta) \\ G_n(\beta,\alpha) & F_n(\alpha,\beta) \end{bmatrix} \begin{bmatrix} \tilde{D}_r^\alpha \\ \tilde{D}_r^\beta \end{bmatrix} \qquad \text{(b)}$$

$$F_n(x,y) = \frac{y}{x} \frac{[\frac{1}{n}(\frac{y}{x})^n + \frac{1}{n+1}(\frac{x}{y})^{n+1}]}{[\frac{1}{y}(\frac{x}{y})^n - \frac{1}{x}(\frac{y}{x})^n]}$$

$$G_n(x,y) = \frac{y}{x}\frac{(2n+1)}{n(n+1)} \frac{1}{[\frac{1}{y}(\frac{x}{y})^n - \frac{1}{x}(\frac{y}{x})^n]}$$

$\beta \rightarrow 0$

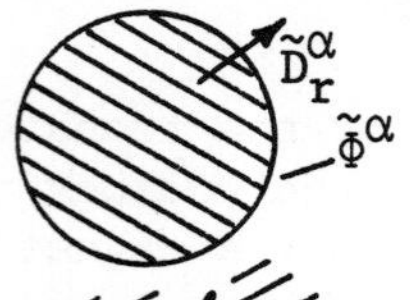

$$\tilde{D}_r^\alpha = -\frac{\varepsilon n}{\alpha}\tilde{\Phi}^\alpha \qquad \text{(c)}$$

$\alpha \rightarrow \infty$

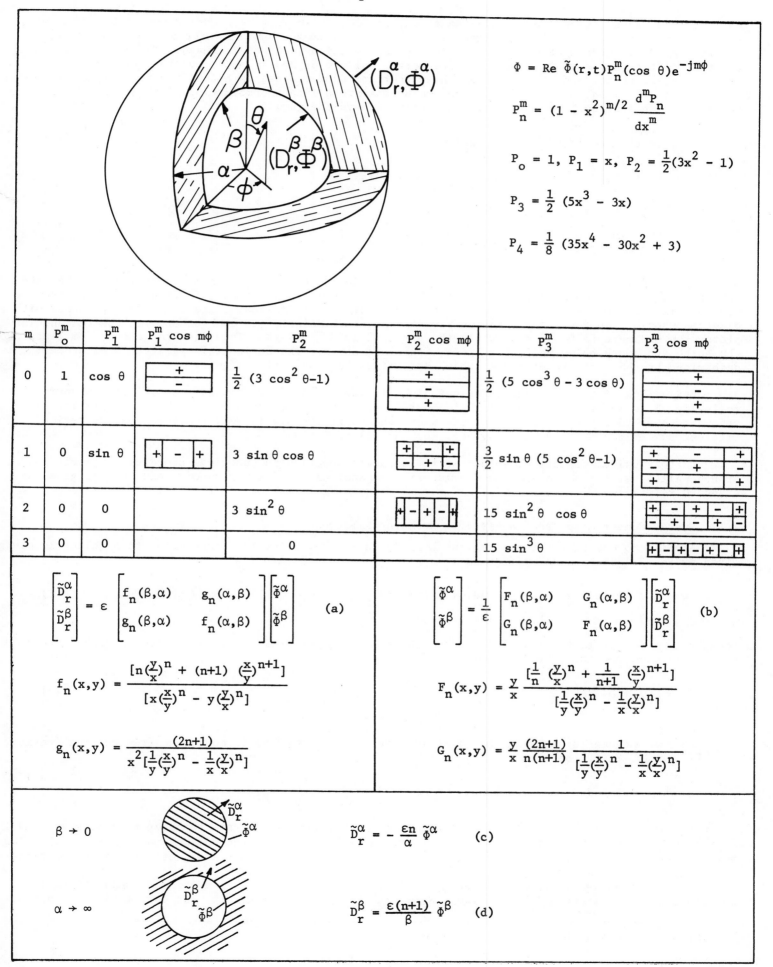

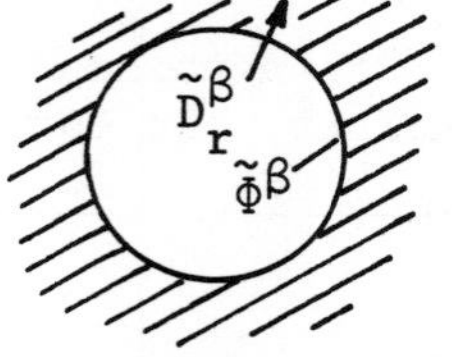

$$\tilde{D}_r^\beta = \frac{\varepsilon(n+1)}{\beta}\tilde{\Phi}^\beta \qquad \text{(d)}$$

where P_n^m are the associated Legendre functions of the first kind, order n and degree m. In terms of the Legendre polynomials P_n, these functions are summarized in Table 2.16.3. Note that these solutions are closed. They do not require infinite series for their representation.

To the second order differential equation, Eq. 33, there must be a second set of solutions Q_n^m. Because these are singular in the interval $0 \leq \theta \leq \pi$, and situations of interest here include the entire spherical surface at any given radius, these solutions are not included. The functions P_n^m play the role of exp(jkz)(say) in cylindrical geometry, while exp(jmϕ) is analogous to exp(jmθ). The radial dependence, which is much of the bother in cylindrical coordinates, is actually quite simple in spherical coordinates. From Eq. 31b it is seen that solutions are a linear combination of r^n and $r^{-(n+1)}$. With the assumption that surface potentials respectively have the form

$$\Phi(^{\alpha}_{\beta},\theta,\phi,t) = \text{Re}\ \tilde{\Phi}^{\alpha}_{\beta}(t)P_n^m(\cos\theta)\exp(jm\phi) \tag{35}$$

it follows that the appropriate linear combination is

$$\tilde{\Phi} = \tilde{\Phi}^{\alpha}\frac{[(\frac{r}{\beta})^n - (\frac{\beta}{r})^{n+1}]}{[(\frac{\alpha}{\beta})^n - (\frac{\beta}{\alpha})^{n+1}]} + \tilde{\Phi}^{\beta}\frac{[(\frac{r}{\alpha})^n - (\frac{\alpha}{r})^{n+1}]}{[(\frac{\beta}{\alpha})^n - (\frac{\alpha}{\beta})^{n+1}]} \tag{36}$$

The complex amplitudes $(\tilde{\Phi}^{\alpha},\tilde{\Phi}^{\beta})$ determine the combination of cos mϕ and sin mϕ, constituting the distribution of Φ with longitudinal distance. For a real amplitude, the distribution is proportional to cos mϕ. In the summary of Table 2.16.3, the lowest orders of P_n^m (cos θ) are tabulated, together with diagrams showing the zones that are positive and negative relative to each other. In the rectangular plots, the ordinate is $0 \leq \theta \leq \pi$, while the abscissa is $0 \leq \phi \leq 2\pi$. Thus, the top and bottom lines are the north and south poles while the lines within are nodes. The horizontal register of each diagram is determined by the complex amplitude, which determines the phase of exp(jmϕ).

Evaluation of the transfer relations given in Table 2.16.3 by Eqs. (a) and (b) is now carried out following the same procedure as for the planar layer. From these relations follow the limiting situations of a solid spherical region or one where the outer surface is well removed from the region of interest summarized for reference by Eqs. (c) and (d) of Table 2.16.3.

Further useful aspects of solutions to Laplace's equation in spherical coordinates, including orthogonality relations that permit Fourier-like expansions and evaluation of averages, are given in standard references.[5]

2.17 Energy Conservation and Quasistatic Transfer Relations

Applied to one of the three regions considered in Sec. 2.16, the incremental total electric energy given by Eq. 2.13.6, can be written as

$$\delta w = -\int_V \nabla\Phi\cdot\delta\vec{D}dV = -\int_V \nabla\cdot(\Phi\delta\vec{D})dV + \int_V \Phi\nabla\cdot\delta\vec{D}dV \tag{1}$$

Because $\rho_f = 0$, the last integral is zero. The remaining integral is converted to a surface integral by Gauss' theorem, and the equation reduces to

$$\delta w = -\oint_S \Phi\delta\vec{D}\cdot\vec{n}da \tag{2}$$

Similar arguments apply in the magnetic cases. Because there is no volume free current density, $\vec{H} = -\nabla\Psi$ and Eq. 2.14.9 becomes

$$\delta w = -\oint_S \Psi\delta\vec{B}\cdot\vec{n}da \tag{3}$$

Consider now the implications of these last two expressions for the transfer relations derived in Sec. 2.16. Discussion is in terms of the electrical relations, but the analogy made in Sec. 2.16 clearly pertains as well to Eqs. 2 and 3, so that the arguments also apply to the magnetic transfer relations.

Suppose that the increment of energy δw is introduced through S to a volume bounded by sections of the α and β surfaces extending one "wavelength" in the surface dimensions. In Cartesian coordinates,

---

5. F. B. Hildebrand, loc. cit., pp. 159-165.

this volume is bounded by (y,z) surfaces extending one wavelength in the y and z directions. In cylindrical coordinates, the volume is a pie-shaped cylinder subtended by outside and inside surfaces having length $2\pi/k$ in the z direction and $2\pi\alpha/m$ and $2\pi\beta/m$ respectively in the azimuthal direction. In spherical coordinates, the volume is a sector from a sphere with $\theta = 2\pi/m$ radians along the equator, θ extending from $0 \rightarrow \pi$ and the surfaces at $r = \alpha$ and $r = \beta$. In any of these cases, conservation of energy, as expressed by Eq. 2, requires that

$$\delta w = -a^{\alpha}\left\langle\!\left\langle \Phi^{\alpha}\delta D_n^{\alpha}\right\rangle\!\right\rangle + a^{\beta}\left\langle\!\left\langle \Phi^{\beta}\delta D_n^{\beta}\right\rangle\!\right\rangle \tag{4}$$

The $\langle\!\langle\ \rangle\!\rangle$ indicate averages over the respective surfaces of excitation. The areas (a^{α},a^{β}) are in particular

$$a^{\beta}_{\alpha} = \begin{cases} (2\pi)^2/k_y k_z & \text{Cartesian} \\ [(2\pi)^2/mk]\binom{\alpha}{\beta} & \text{cylindrical} \\ (4\pi/m)\binom{\alpha^2}{\beta^2} & \text{spherical} \end{cases} \tag{5}$$

In writing Eq. 2 as Eq. 4, contributions of surfaces other than the α and β surfaces cancel because of the spatial periodicity. It is assumed that (k_y,k_z), (m,k) and m are real numbers.

The transfer relations developed in Sec. 2.16 take the general form

$$\begin{bmatrix} \tilde{\Phi}^{\alpha} \\ \tilde{\Phi}^{\beta} \end{bmatrix} = \begin{bmatrix} -A_{11} & A_{12} \\ -A_{21} & A_{22} \end{bmatrix} \begin{bmatrix} \tilde{D}_n^{\alpha} \\ \tilde{D}_n^{\beta} \end{bmatrix} \tag{6}$$

The coefficients A_{ij} are real. Hence, for the purpose of deducing properties of A_{ij}, there is no loss in generality in taking $(\tilde{D}_n^{\alpha},\tilde{D}_n^{\beta})$ and hence $(\tilde{\Phi}^{\alpha},\tilde{\Phi}^{\beta})$ as being real. Then, Eq. 4 takes the form

$$\delta w = C[-a^{\alpha}\tilde{\Phi}^{\alpha}\delta\tilde{D}_n^{\alpha} + a^{\beta}\tilde{\Phi}^{\beta}\delta\tilde{D}_n^{\beta}) \tag{7}$$

where C is 1/2 in the Cartesian and cylindrical cases and is a positive constant in the spherical case.

With the assumption that $w = w(\tilde{D}^{\alpha},\tilde{D}^{\beta})$, the incremental energy can also be written as

$$\delta w = \frac{\partial w}{\partial \tilde{D}_n^{\alpha}}\delta\tilde{D}_n^{\alpha} + \frac{\partial w}{\partial \tilde{D}_n^{\beta}}\delta\tilde{D}_n^{\beta} \tag{8}$$

where $(\tilde{D}_n^{\alpha},\tilde{D}_n^{\beta})$ constitute independent electrical "terminal" variables. Thus, from Eqs. 7 and 8,

$$-a^{\alpha}\tilde{\Phi}^{\alpha} = \frac{\partial w}{\partial \tilde{D}_n^{\alpha}}\ ;\ a^{\beta}\tilde{\Phi}^{\beta} = \frac{\partial w}{\partial \tilde{D}_n^{\beta}} \tag{9}$$

A reciprocity condition is obtained by taking derivatives of these expressions with respect to $\tilde{D}_n^{\beta}$ and $\tilde{D}_n^{\alpha}$, respectively, and eliminating the energy function. In view of the transfer relations, Eq. 6,

$$a^{\alpha}A_{12} = a^{\beta}A_{21} \tag{10}$$

Thus, in the planar layer where the areas a^{α} and a^{β} are equal, the mutual coupling terms $A_{12} = A_{21}$. That the relations are related by Eq. 10 in the spherical case is easily checked, but the complicated expressions for the cylindrical case simplify the mutual terms (footnote to Table 2.16.2).

The energy can be evaluated by integrating Eq. 7 using the "constitutive" laws of Eq. 6. The integration is first carried out with $\tilde{D}^{\beta} = 0$, raising $\tilde{D}^{\alpha}$ to its final value. Then, with $\tilde{D}^{\alpha} = \tilde{D}^{\alpha}$, $\tilde{D}^{\beta}$ is raised to its final value

$$w = C\left[\frac{1}{2}a^{\alpha}A_{11}(\tilde{D}_n^{\alpha})^2 - a^{\beta}A_{21}\tilde{D}_n^{\alpha}\tilde{D}_n^{\beta} + \frac{1}{2}a^{\beta}A_{22}(\tilde{D}_n^{\beta})^2\right] \tag{11}$$

With either excitation alone, w must be positive and so from this relation it follows that

$$A_{11} > 0, \quad A_{22} > 0 \tag{12}$$

These conditions are also met by the relations found in Sec. 2.16.

2.18 Solenoidal Fields, Vector Potential and Stream Function

Irrotational fields, such as the quasistatic electric field, are naturally represented by a scalar potential. Not only does this reduce the vector field to a scalar field, but the potential function evaluated on such surfaces as those of "perfectly" conducting electrodes becomes a lumped parameter terminal variable, e.g., the voltage.

Solenoidal fields, such as the magnetic flux density $\vec{B}$, are for similar reasons sometimes represented in terms of a vector potential $\vec{A}$:

$$\vec{B} = \nabla \times \vec{A} \tag{1}$$

Thus, $\vec{B}$ automatically has no divergence. Unfortunately, the vector field $\vec{B}$ is represented in terms of another vector field $\vec{A}$. However, for important two-dimensional or symmetric configurations, a single component of $\vec{A}$ is all required to again reduce the description to one involving a scalar function. Four commonly encountered cases are summarized in Table 2.18.1.

The first two are two-dimensional in the usual sense. The field $\vec{B}$ lies in the x-y (or r-θ) plane and depends only on these coordinates. The associated vector potential has only a z component. The third configuration, like the second, is in cylindrical geometry, but with $\vec{B}$ independent of θ and hence with $\vec{A}$ having only an $\vec{i}_\theta$ component. The fourth configuration is in spherical geometry with symmetry about the z axis and the vector potential directed along ϕ_θ.

Like the scalar potential used to represent irrotational fields, the vector potential is closely related to lumped parameter variables. If $\vec{B}$ is the magnetic flux density, it is convenient for evaluation of the flux linkage λ (Eq. 2.12.1). For an incompressible flow, where $\vec{B}$ is replaced by the fluid velocity $\vec{v}$, the vector potential is conveniently used to evaluate the volume rate of flow. In that application, A and Λ become "stream functions."

The connection between the flux linked and the vector potential follows from Stokes's theorem, Eq. 2.6.3. The flux Φ_λ through a surface S enclosed by a contour C is

$$\Phi_\lambda = \int_S \vec{B}\cdot\vec{n}da = \int_S \nabla \times \vec{A}\cdot\vec{n}da = \oint_C \vec{A}\cdot d\vec{\ell} \tag{2}$$

In each of the configurations of Table 2.18.1, Eq. 2 amounts to an evaluation of the surface integral. For example, in the Cartesian two-dimensional configuration, contributions to the integration around a contour C enclosing a surface having length ℓ in the z direction, only come from the legs running in the z direction. Along these portions of the contour, denoted by (a) and (b), the coordinates (x,y) are constant. Hence, the flux through the surface is simply ℓ times the difference A(a) - A(b), as summarized in Table 2.18.1.

In the axisymmetric cylindrical and spherical configurations, r and r sin θ dependences are respectively introduced, so that evaluation of Λ essentially gives the flux linked. For example, in the spherical configuration, the flux linked by a surface having inner and outer radii r cos θ evaluated at (a) and (b) is simply

$$\Phi_\lambda = \oint_C \frac{\Lambda(r,\theta)}{r \sin\theta}\vec{i}_\phi\cdot d\vec{\ell} = \frac{\Lambda}{r\sin\theta}2\pi(r\sin\theta)\Big|_b^a = 2\pi[\Lambda(a) - \Lambda(b)] \tag{3}$$

Used in fluid mechanics to represent incompressible fluid flow, Λ is the Stokes's stream function. Note that the flux is positive if directed through the surface in the direction of $\vec{n}$, which is specified in terms of the contour C by the right-hand rule.

2.19 Vector Potential Transfer Relations for Certain Laplacian Fields

Even in dealing with magnetic fields in regions where $\vec{J}_f = 0$, if the flux linkages are of interest, it is often more convenient to develop a model in terms of transfer relations specified in terms of a vector rather than scalar potential. The objective in this section is to summarize these relations for the first three configurations identified in Table 2.18.1.

Table 2.18.1. Important configurations having solenoidal field $\vec{B}$ represented by single components of vector potential $\vec{A}$.

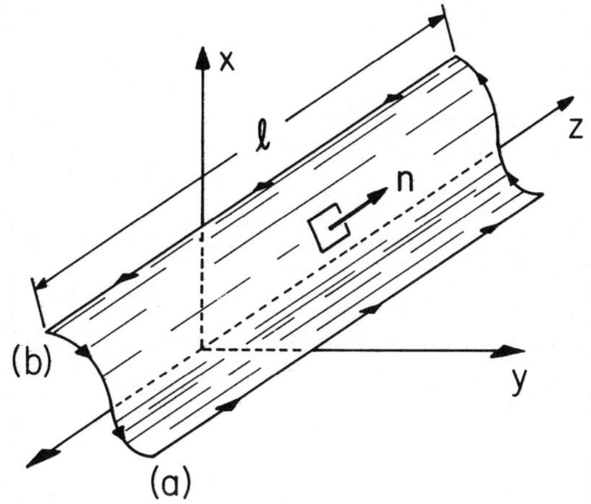

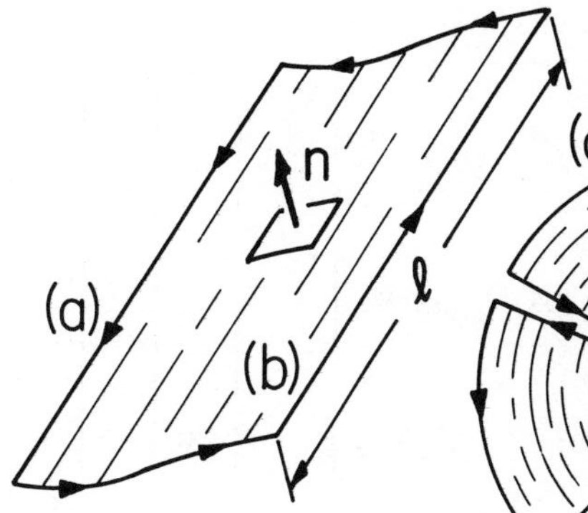

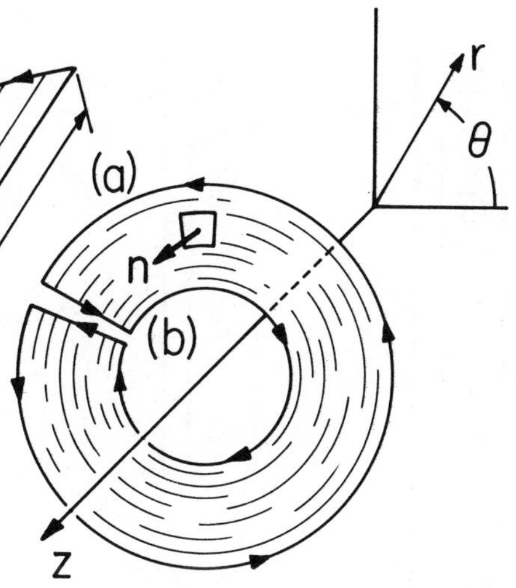

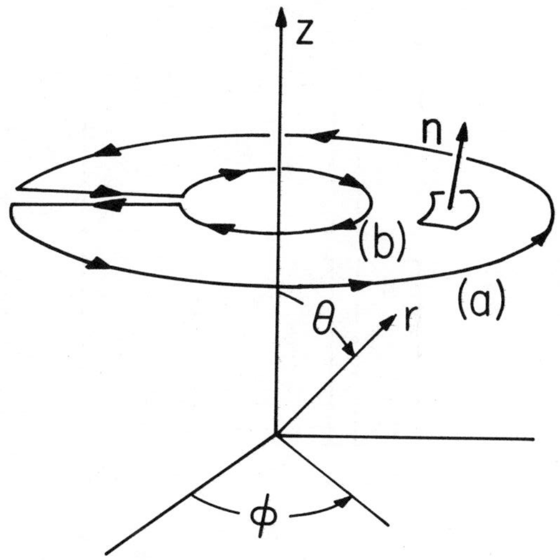

Two-dimensional Cartesian	Polar	Axisymmetric cylindrical	Axisymmetric spherical
$\vec{A} = A(x,y)\vec{i}_z$ (a)	$\vec{A} = A(r,\theta)\vec{i}_z$ (d)	$\vec{A} = \frac{\Lambda(r,z)}{r}\vec{i}_\theta$ (g)	$\vec{A} = \frac{\Lambda(r,\theta)}{r \sin\theta}\vec{i}_\phi$ (j)
$\vec{B} = \frac{\partial A}{\partial y}\vec{i}_x - \frac{\partial A}{\partial x}\vec{i}_y$ (b)	$\vec{B} = \frac{1}{r}\frac{\partial A}{\partial \theta}\vec{i}_r - \frac{\partial A}{\partial r}\vec{i}_\theta$ (e)	$\vec{B} = -\frac{1}{r}\frac{\partial \Lambda}{\partial z}\vec{i}_r + \frac{1}{r}\frac{\partial \Lambda}{\partial r}\vec{i}_z$ (h)	$\vec{B} = \frac{1}{r\sin\theta}[\frac{1}{r}\frac{\partial\Lambda}{\partial\theta}\vec{i}_r - \frac{\partial \Lambda}{\partial r}\vec{i}_\theta]$ (k)
$\Phi_\lambda = \ell[A(a) - A(b)]$ (c)	$\Phi_\lambda = \ell[A(a) - A(b)]$ (f)	$\Phi_\lambda = 2\pi[\Lambda(a) - \Lambda(b)]$ (i)	$\Phi_\lambda = 2\pi[\Lambda(a) - \Lambda(b)]$ (l)

Table 2.19.1. Vector potential transfer relations for two-dimensional or symmetric Laplacian fields.

Two-dimensional Cartesian	Polar	Axisymmetric cylindrical
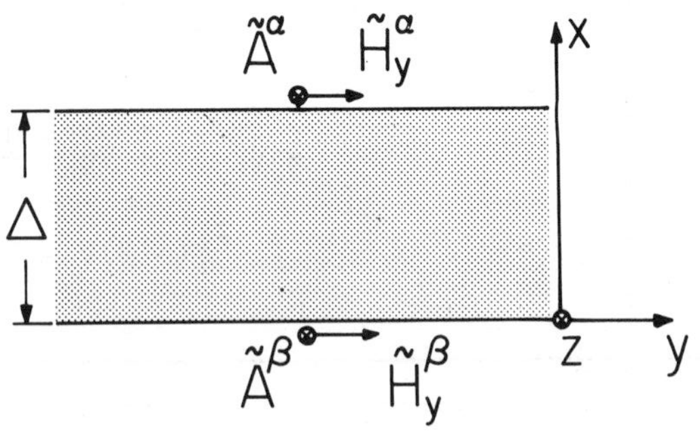	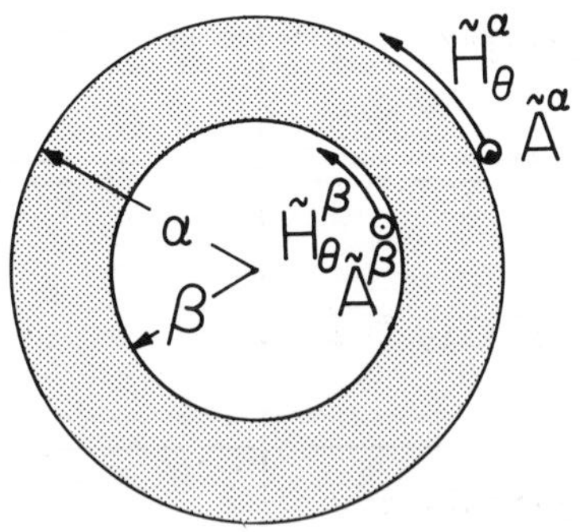	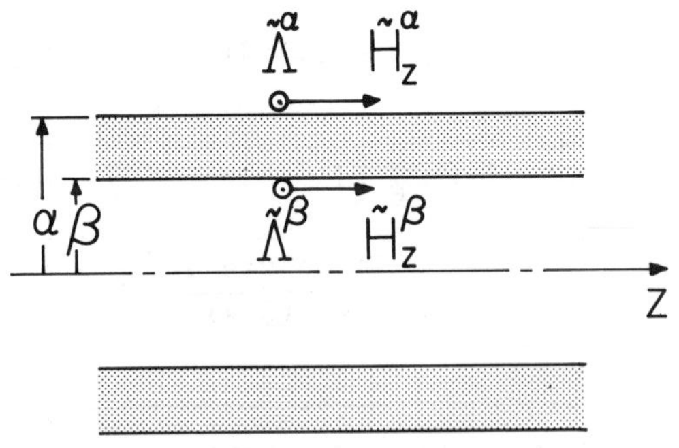
$\vec{A} = \vec{i}_z \operatorname{Re} \tilde{A}(x) \exp(-jky)$ $\begin{bmatrix} \tilde{H}_y^\alpha \\ \tilde{H}_y^\beta \end{bmatrix} = \frac{k}{\mu} \begin{bmatrix} -\coth(k\Delta) & \frac{1}{\sinh(k\Delta)} \\ \frac{-1}{\sinh(k\Delta)} & \coth(k\Delta) \end{bmatrix} \begin{bmatrix} \tilde{A}^\alpha \\ \tilde{A}^\beta \end{bmatrix}$ (a) $\begin{bmatrix} \tilde{A}^\alpha \\ \tilde{A}^\beta \end{bmatrix} = \frac{\mu}{k} \begin{bmatrix} -\coth(k\Delta) & \frac{1}{\sinh(k\Delta)} \\ \frac{-1}{\sinh(k\Delta)} & \coth(k\Delta) \end{bmatrix} \begin{bmatrix} \tilde{H}_y^\alpha \\ \tilde{H}_y^\beta \end{bmatrix}$ (b)	$\vec{A} = \vec{i}_z \operatorname{Re} \tilde{A}(r) \exp(-jm\theta)$ $\begin{bmatrix} \tilde{H}_\theta^\alpha \\ \tilde{H}_\theta^\beta \end{bmatrix} = \frac{1}{\mu} \begin{bmatrix} f_m(\beta,\alpha) & g_m(\alpha,\beta) \\ g_m(\beta,\alpha) & f_m(\alpha,\beta) \end{bmatrix} \begin{bmatrix} \tilde{A}^\alpha \\ \tilde{A}^\beta \end{bmatrix}$ (c) $\begin{bmatrix} \tilde{A}^\alpha \\ \tilde{A}^\beta \end{bmatrix} = \mu \begin{bmatrix} F_m(\beta,\alpha) & G_m(\alpha,\beta) \\ G_m(\beta,\alpha) & F_m(\alpha,\beta) \end{bmatrix} \begin{bmatrix} \tilde{H}_\theta^\alpha \\ \tilde{H}_\theta^\beta \end{bmatrix}$ (d) For f_m, g_m, F_m, G_m see Table 2.16.2, $k = 0$, $m \neq 0$	$\vec{A} = \vec{i}_\theta \operatorname{Re} \tilde{A}(r)\exp(-jkz)$; $\tilde{\Lambda} = \tilde{A}r$ $\begin{bmatrix} \tilde{H}_z^\alpha \\ \tilde{H}_z^\beta \end{bmatrix} = \frac{-k^2}{\mu} \begin{bmatrix} F_o(\beta,\alpha) & G_o(\alpha,\beta) \\ G_o(\beta,\alpha) & F_o(\alpha,\beta) \end{bmatrix} \begin{bmatrix} \frac{\tilde{\Lambda}^\alpha}{\alpha} \\ \frac{\tilde{\Lambda}^\beta}{\beta} \end{bmatrix}$ (e) $\begin{bmatrix} \frac{\tilde{\Lambda}^\alpha}{\alpha} \\ \frac{\tilde{\Lambda}^\beta}{\beta} \end{bmatrix} = -(\frac{\mu}{k^2}) \begin{bmatrix} f_o(\beta,\alpha) & g_o(\alpha,\beta) \\ g_o(\beta,\alpha) & f_o(\alpha,\beta) \end{bmatrix} \begin{bmatrix} \tilde{H}_z^\alpha \\ \tilde{H}_z^\beta \end{bmatrix}$ (f) For F_o, G_o, f_o, g_o see Table 2.16.2, $m = 0$, $k \neq 0$
$\begin{bmatrix} \hat{A}^\alpha \\ \hat{A}^\beta \end{bmatrix} = \frac{j}{k} \begin{bmatrix} \hat{B}_x^\alpha \\ \hat{B}_x^\beta \end{bmatrix}$	$\begin{bmatrix} \hat{A}^\alpha \\ \hat{A}^\beta \end{bmatrix} = \frac{j}{m} \begin{bmatrix} \alpha\hat{B}_r^\alpha \\ \beta\hat{B}_r^\beta \end{bmatrix}$	$\begin{bmatrix} \frac{\hat{\Lambda}^\alpha}{\alpha} \\ \frac{\hat{\Lambda}^\beta}{\beta} \end{bmatrix} = \frac{-j}{k} \begin{bmatrix} \hat{B}_r^\alpha \\ \hat{B}_r^\beta \end{bmatrix}$

With $\vec{B}$ represented in terms of $\vec{A}$ by Eq. 2.18.1, Ampère's law (Eq. 2.3.23) requires that in a region of uniform permeability μ,

$$\nabla \times \nabla \times \vec{A} = \mu\vec{J}_f \tag{1}$$

For a given magnetic flux density $\vec{B}$, curl $\vec{A}$ is specified. But to make $\vec{A}$ unique, its divergence must also be specified. Here, the divergence of $\vec{A}$ is defined as zero. Thus, the vector identity $\nabla \times \nabla \times \vec{A} = \nabla(\nabla\cdot\vec{A}) - \nabla^2\vec{A}$ reduces Eq. 1 to the vector Poisson's equation:

$$\nabla^2\vec{A} = -\mu\vec{J}_f; \quad \nabla\cdot\vec{A} = 0 \tag{2}$$

The vector Laplacian is summarized in Appendix A for the three coordinate systems of Table 2.18.1. Even though the region described in the following developments is one where $\vec{J}_f = 0$, the source term on the right has been carried along for later reference.

Cartesian Coordinates: In the Cartesian coordinate system of Table 2.18.1 it is the z component of Eq. 2 that is of interest. The z component of the vector Laplacian is the same operator as for the scalar Laplacian. Thus, the situation is analogous to that outlined by Eqs. 2.16.11 to 2.16.16 with $\Phi \rightarrow A$. With solutions of the form $A = \text{Re}\, A(x,t)\exp(-jky)$ so that $\gamma \rightarrow k_y \equiv k$, the appropriate linear combination of solutions is

$$\tilde{A} = \tilde{A}^{\alpha}\frac{\sinh kx}{\sinh k\Delta} - \tilde{A}^{\beta}\frac{\sinh k(x-\Delta)}{\sinh k\Delta} \tag{3}$$

Because $\vec{H} = \vec{B}/\mu$, the associated tangential field intensity is given by Eq. (b), Table 2.18.1,

$$H_y = -\frac{1}{\mu}\frac{\partial A}{\partial x} \tag{4}$$

Expressed in terms of Eq. 3 and evaluated at the surfaces $x = \alpha$ and $x = \beta$, respectively, Eq. 4 gives the first transfer relations, Eq. (a), of Table 2.19.1. Inversion of these relations gives Eqs. (b).

Polar Coordinates: In cylindrical coordinates with no z dependence, it is again the z component of Eq. 2 that is pertinent. The configuration is summarized in Table 2.18.1. Solutions take the form $A = \text{Re}\, A(r,t)\exp(-jm\theta)$ and are analogous to Eq. 2.16.21 with Φ replaced by A:

$$\tilde{A} = \tilde{A}^{\alpha}\frac{\left[\left(\frac{\beta}{r}\right)^m - \left(\frac{r}{\beta}\right)^m\right]}{\left[\left(\frac{\beta}{\alpha}\right)^m - \left(\frac{\alpha}{\beta}\right)^m\right]} + \tilde{A}^{\beta}\frac{\left[\left(\frac{r}{\alpha}\right)^m - \left(\frac{\alpha}{r}\right)^m\right]}{\left[\left(\frac{\beta}{\alpha}\right)^m - \left(\frac{\alpha}{\beta}\right)^m\right]} \tag{5}$$

The tangential field is then evaluated from Eq. (e), Table 2.18.1:

$$H_\theta = -\frac{1}{\mu}\frac{\partial A}{\partial r} \tag{6}$$

Evaluation at the respective surfaces $r = \alpha$ and $r = \beta$ gives the transfer relations, Eqs. (c) of Table 2.19.1. Inversion of these relations gives Eqs. (d).

Axisymmetric Cylindrical Coordinates: By contrast with the two-dimensional configurations so far considered, where the vector Laplacian of A_z is the same as the scalar Laplacian, the vector nature of Eq. 2 becomes apparent in the axisymmetric cylindrical configuration. The θ component of Eq. 2 is the scalar Laplacian of A_θ plus $(-A_\theta/r^2)$ (see Appendix A). With $A_\theta \equiv A$,

$$\frac{\partial^2 A}{\partial r^2} + \frac{1}{r}\frac{\partial A}{\partial r} - \frac{A}{r^2} + \frac{\partial^2 A}{\partial z^2} = -\mu J_\theta \tag{7}$$

Even though solutions do not have a θ dependence, so that

$$A = \text{Re}\, \tilde{A}(r,t)e^{-jkz} \tag{8}$$

equation 7 reduces to a form of Bessel's equation to which solutions are Bessel's and Hankel's functions of order unity:

$$\frac{\partial^2\tilde{A}}{\partial r^2} + \frac{1}{r}\frac{\partial\tilde{A}}{\partial r} - \left(k^2 + \frac{1}{r^2}\right)\tilde{A} = -\mu\tilde{J} \tag{9}$$

(Compare Eq. 9 to Eq. 2.16.19.) It follows that solutions are of the form of Eq. 2.16.25 with $\tilde{\Phi} \rightarrow \tilde{A}$ and m = 1:

$$\tilde{\Lambda} \equiv r\tilde{A} = \tilde{A}^{\alpha} \frac{H_1(jk\beta)[rJ_1(jkr)] - J_1(jk\beta)[rH_1(jkr)]}{H_1(jk\beta)J_1(jk\alpha) - J_1(jk\beta)H_1(jk\alpha)}$$

$$+ \tilde{A}^{\beta} \frac{J_1(jk\alpha)[rH_1(jkr)] - H_1(jk\alpha)[rJ_1(jkr)]}{J_1(jk\alpha)H_1(jk\beta) - H_1(jk\alpha)J_1(jk\beta)} \tag{10}$$

The tangential field intensity follows from Eq. 10 and Eq. (h) of Table 2.18.1:

$$H_z = \frac{1}{\mu r}\frac{\partial \Lambda}{\partial r} \tag{11}$$

In performing the differentiation, observe from Eq. 2.16.26d that whether R_m is J_m or H_m,

$$\frac{d}{dr}[rR_1(jkr)] = jkrR_o(jkr) \tag{12}$$

Evaluation of H_z at the respective surfaces $r = \alpha$ and $r = \beta$ gives the transfer relations, Eqs. (e) of Table 2.19.1. Inversion of these relations gives Eqs. (f).

2.20 Methodology

As descriptions of subregions composing a heterogeneous system, transfer relations (illustrated for quasistatic fields in Sec. 2.16) are building blocks for describing complicated interactions. By appropriate identification of variables, the same relations can be used to describe different regions.

As an example, three planar regions are shown in Fig. 2.20.1. The symbols in parentheses denote positions adjacent to the surfaces demarking subregions. At the surfaces, variables can be discontinuous. Hence it is necessary to distinguish variables evaluated on adjacent sides of a boundary. The transfer relations describe the fields within the subregions and not across the boundaries.

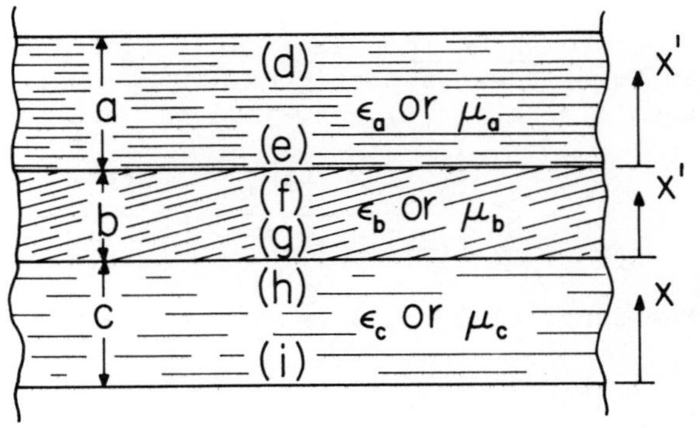

Fig. 2.20.1. Convention used to denote surface variables.

The transfer relations of Table 2.16.1 can be applied to the upper region by identifying $(\alpha) \rightarrow (d)$, $(\beta) \rightarrow (e)$, $\Delta \rightarrow a$ and ε or $\mu \rightarrow \varepsilon_a$ or μ_a. Similarly, for the middle region, $(\alpha) \rightarrow (f)$, $(\beta) \rightarrow (g)$, $\Delta \rightarrow b$, and ε or $\mu \rightarrow \varepsilon_b$ or μ_b. Boundary conditions and jump relations across the surfaces then provide coupling conditions on the surface variables. Once the surface variables have been self-consistently determined, the field distributions within the region can be evaluated using the bulk distributions evaluated in terms of the surface coefficients. With appropriate surface amplitudes and $x \rightarrow x'$, where the latter is defined for each region in Fig. 2.20.1, Eq. 2.16.15 describes the potential distribution.

This approach will be used not only in other geometries but in representing mechanical and electromechanical processes.

Problems for Chapter 2

For Section 2.3:

Prob. 2.3.1 Perfectly conducting plane parallel plates are shorted at $z = 0$ and driven by a distributed current source at $z = -\ell$, as shown in Fig. P2.3.1.

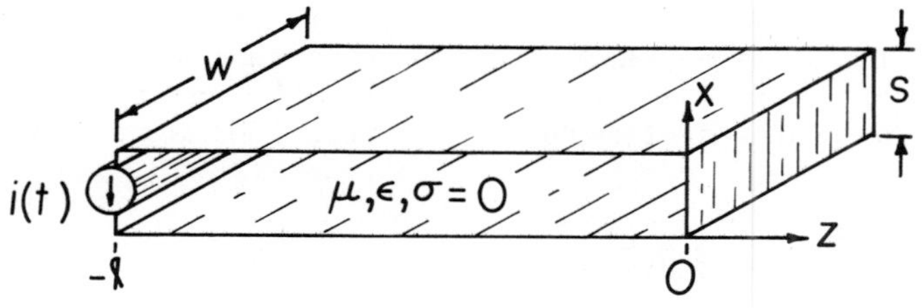

Fig. P2.3.1

(a) Apply the normalization of Eq. 4b to Maxwell's equations used to represent the fields between the plates. There is no material between the plates, so magnetization, polarization and conduction between the plates are ignorable.

(b) Simplify these equations by assuming that $\underline{\vec{E}} = \underline{E}_x(\underline{z},\underline{t})\vec{i}_x$ and $\underline{\vec{H}} = \underline{H}_y(\underline{z},\underline{t})\vec{i}_y$.

(c) The driving current is $i(t) = \text{Re}\, I_o \exp j\omega t$. Find $\underline{E}_x$, $\underline{H}_y$, the surface current and surface charge on the lower plate to second order.

(d) Convert the results of (c) to dimensional expressions.

(e) Solve for the exact fields and expand in β to check the results of (d).

Prob. 2.3.2 The parallel plates of Prob. 2.3.1 are now driven along their left edges by a voltage source v(t). They are open along their right edges. Carry out the steps analogous to those of Prob. 2.3.1. A normalization that makes the EQS limit the zero order approximation is appropriate.

Prob. 2.3.3 Perfectly conducting plane parallel electrodes in the planes $x = a$ and $x = 0$ "sandwich" and make electrical contact with a layer of material having conductivity σ and thickness a. These plates are driven along their edges so that the surface current is $\text{Re}\, \hat{K} \exp(j\omega t)\vec{i}_z$ in the lower plate at $z = -\ell$ and the negative of this in the upper plate. The edges of the plates at $z = 0$ are "open-circuit." In the conductor, fields take the form $E_x(z,t)$, $H_y(z,t)$.

(a) Show that all of Maxwell's equations are satisfied if

$$\frac{d^2\hat{H}_y}{d^2z} + k^2\hat{H}_y = 0; \quad k \equiv \sqrt{\omega^2\mu_o\varepsilon_o - j\omega\mu_o\sigma}; \quad \hat{E}_x = \frac{-1}{(\sigma + j\omega\varepsilon_o)}\frac{d\hat{H}_y}{dz}$$

(b) Show that

$$H_y = \text{Re}\, \hat{K}\frac{e^{-jkz} - e^{jkz}}{e^{jk\ell} - e^{-jk\ell}}e^{j\omega t}; \quad E_x = \frac{\text{Re}\, \hat{K}jk(e^{-jkz} + e^{jkz})e^{j\omega t}}{(\sigma + j\omega\varepsilon_o)(e^{jk\ell} - e^{-jk\ell})}$$

(c) In Fig. 2.3.1, $\tau \rightarrow 1/\omega$ and provided $\tau_e \neq \tau_m$, there are two possibilities:

(i) $\omega\tau_{em} \ll 1$ and $\omega\tau_m \ll 1$. Show that in this case $k\ell \ll 1$ and

$$E_x \rightarrow \text{Re}\, \frac{\hat{K}\, e^{j\omega t}}{(\sigma + j\omega\varepsilon_o)\ell}$$

so that the system is equivalent to a capacitor shorted by a resistor (what values?).

(ii) $\omega\tau_{em} \ll 1$, $\omega\tau_e \ll 1$. Show that in this case $k \rightarrow (-1 + j)/\delta_m$, where the skin depth

$\delta_m \equiv \sqrt{2/\omega\mu\sigma}$, and that H_y is the superposition of "skin-effect" waves decaying in the direction of phase propagation.

(d) Now, consider the EQS model from the outset. Under what conditions are the laws (Eqs. 23a - 27a) valid? Show that the solution for E_x is consistent with part (c).

(e) Consider the magnetoquasistatic laws (Eqs. 23b - 27b) from the outset and show that the result is consistent with part (c). For what conditions are these laws valid?

Prob. 2.3.4 Given the EQS laws, Eqs. 23a - 25a, together with conduction and polarization constitutive laws and the material motions, $\vec{E}$, $\vec{P}$ and ρ_f can be determined. This is generally possible because the constitutive laws do not typically involve $\vec{H}$. Then, if $\vec{H}$ is required, Eqs. 26a and 26b, together with a magnetization constitutive law, can be used. It is clear that these relations uniquely define $\vec{H}$, because they stipulate both $\nabla \times \vec{H}$ and $\nabla \cdot \vec{H}$. Consider now the analogous question of uniquely determining $\vec{E}$ in an MQS system. In such a system the conduction and magnetization constitutive laws respectively take the form

$$\vec{J}_f = \sigma(\vec{r},t)(\vec{E} + \vec{v}\times\mu_o\vec{H}) \quad ; \quad \vec{M} = \vec{M}(\vec{H},\vec{v})$$

and Eqs. 23b - 25b together with a knowledge of the material motion can be used to find $\vec{H}$ and $\vec{M}$. Show that $\vec{E}$ is then uniquely specified and that recourse to Gauss' Law is made only to make an "after the fact" evaluation of the charge density.

For Section 2.4:

Prob. 2.4.1 A material suffers a rigid-body rotation about the z axis with constant angular velocity Ω. The particle at the position (r_o,θ_o) when $t = 0$ is found at

$$\vec{\xi}(r_o,\theta_o,t) = r_o\cos(\Omega t + \theta_o)\vec{i}_x + r_o\sin(\Omega t + \theta_o)\vec{i}_y$$

at a subsequent time t. This Lagrangian description is pictured in Fig. P2.4.1. Use Eqs. 2.4.1 and 2.4.2 to show that the velocity and acceleration are respectively

$$\vec{v} = r_o\,\Omega[-\sin(\Omega t + \theta_o)\vec{i}_x + \cos(\Omega t + \theta_o)\vec{i}_y]$$

$$\vec{a} = -\Omega^2\,\vec{\xi}$$

Fig. P2.4.1. Specific example in which rigid-body steady rotation is represented in (a) Lagrangian coordinates and (b) Eulerian coordinates.

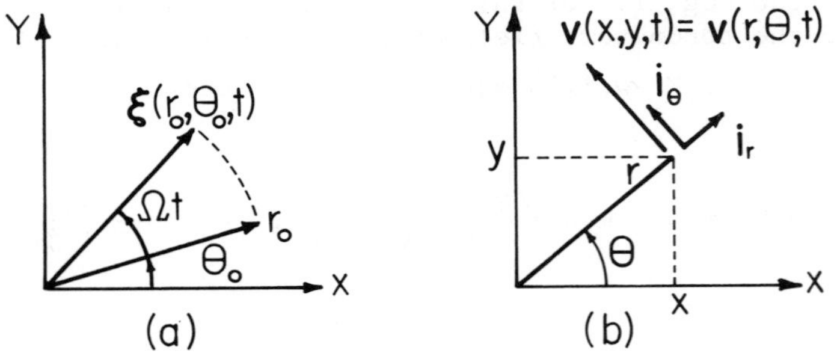

Prob. 2.4.2 One incentive for using an Eulerian representation is that motions which are time dependent in Lagrangian coordinates can become independent of time. To illustrate, consider the alternative representation of the rigid body rotation of Prob. 2.4.1.

The material velocity at a given point (r,θ) or (x,y) is

$$\vec{v} = \vec{i}_\theta\Omega r = \Omega(-r\sin\theta\vec{i}_x + r\cos\theta\vec{i}_y) = \Omega(-y\vec{i}_x + x\vec{i}_y)$$

i.e., the velocity is independent of time. Clearly the acceleration is not obtained by taking the partial derivative with respect to time, as might be suggested by the misuse of Eq. 2.4.2. Use Eq. 2.4.4 to find $\vec{a}$ and compare to the result of Prob. 2.4.1.

For Section 2.5:

Prob. 2.5.1 A scalar function takes the traveling-wave form $\Phi = \mathrm{Re}\hat{\Phi}(x,y)\exp^{j(\omega t-kz)}$ in the frame of reference $(\vec{r},t)$. The primed frame moves in the z direction relative to the unprimed frame with the velocity U. Use the convective derivative to find the rate of change of Φ for an observer moving with the velocity $U\vec{i}_z$. Compute this same time rate of change by expressing $\Phi = \Phi(x',y',z',t')$ and finding $\partial/\partial t'$. Use these results to deduce the transformation $\omega' = \omega - kU$. If $\omega' = 0$, $\omega = kU$. Explain in physical terms.

Prob. 2.5.2 A vector function $\vec{A}(x,y,z,t)$ can also be evaluated as $\vec{A}(x',y',z',t')$ where the prime coordinates are related to the unprimed ones by Eq. 2.5.1. Show that Eq. 2.5.2b holds.

For Section 2.6:

Prob. 2.6.1 The one-dimensional form of Leibnitz' rule pertains to taking an integral between end-points (b) and (a) which are themselves a function of time, as sketched in Fig. P2.6.1.

Fig. P2.6.1. One-dimensional form of Leibnitz' rule specifies how derivative can be taken of the integral between time-varying endpoints.

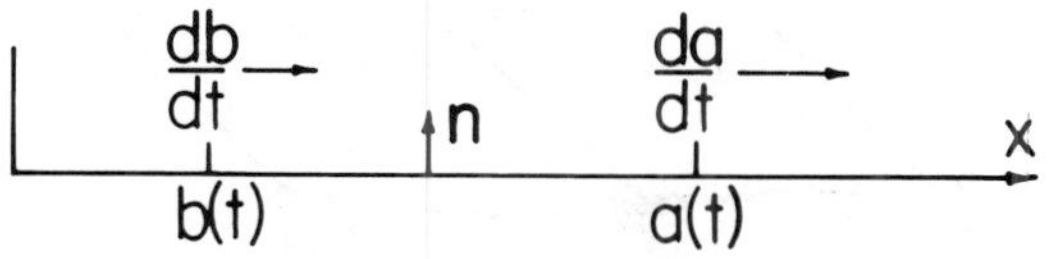

Define $\vec{A} = f(x,t)\vec{i}_z$ and use Eq. 2.6.4 with a suitable surface to show that, for the one-dimensional case, Leibnitz' rule becomes

$$\frac{d}{dt}\int_{b(t)}^{a(t)} f(x,t)\,dx = \int_b^a \frac{\partial f}{\partial t}\,dx + f(a,t)\frac{da}{dt} - f(b,t)\frac{db}{dt}$$

Prob. 2.6.2 The following steps lead to a derivation of the generalized Leibnitz rule, Eq. 2.6.4 where S is pictured as S_2, and S_1 at the times $t + \Delta t$ and t, respectively. The vector function $\vec{A}$ depends on both space and time. However, for convenience, the spatial dependence is not explicitly indicated in the following. By definition:

$$\frac{d}{dt}\int_S \vec{A}\cdot\vec{n}\,da = \lim_{\Delta t\to 0}\frac{1}{\Delta t}\left(\int_{S_2}\vec{A}(t+\Delta t)\cdot\vec{n}da - \int_{S_1}\vec{A}(t)\cdot\vec{n}da\right) \tag{1}$$

so the first integral in brackets on the right must be evaluated to first order in Δt. To that end,

(a) Apply Gauss theorem to the volume V swept out by S during the time Δt. Note that $\vec{n}$ is the normal to the open surface S and show that to first order in Δt,

$$\int_V \nabla\cdot\vec{A}dV = \int_{S_2}\vec{A}(t)\cdot\vec{n}da - \int_{S_2}\vec{A}(t)\cdot\vec{n}da - \Delta t\oint_{C_1}\vec{A}\cdot\vec{v}\times d\vec{\ell} \tag{2}$$

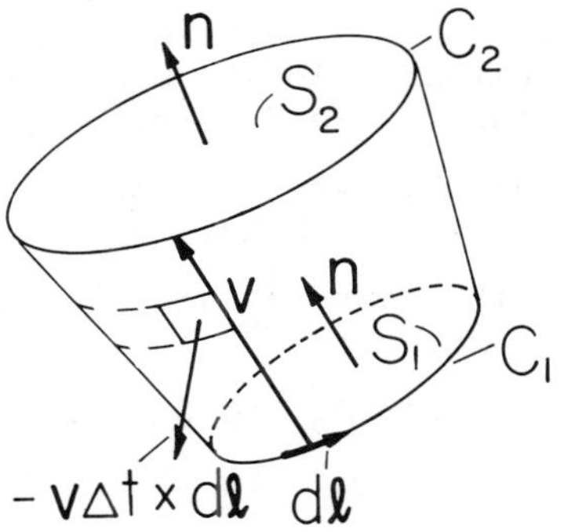

Fig. P2.6.2

(b) Argue that also to first order in Δt,

$$\int_{S_2}\vec{A}(t+\Delta t)\cdot\vec{n}da \simeq \int_{S_2}\vec{A}(t)\cdot\vec{n}da + \int_{S_1}\frac{\partial\vec{A}}{\partial t}(t)\Delta t\cdot\vec{n}da + \cdots \tag{3}$$

(c) Finally, show that the volume element dV, called for in evaluating the left side of Eq. 2, is $dV = \Delta t\vec{v}\cdot\vec{n}da$.

(d) Combine these results to evaluate the right-hand side of Eq. 1 and deduce Eq. 2.6.4.

Prob. 2.6.3 It is sometimes necessary to evaluate the time rate of change of a line integral of a vector variable having time-varying end points. The problem is to evaluate the derivative

$$\frac{d}{dt}\int_{\vec{a}(t)}^{\vec{b}(t)} \vec{A}\cdot d\vec{\ell} = \lim_{\Delta t \to 0} \frac{\left[\int_{\vec{a}(t+\Delta t)}^{\vec{b}(t+\Delta t)} \vec{A}(t+\Delta t)\cdot d\vec{\ell} - \int_{\vec{a}(t)}^{\vec{b}(t)} \vec{A}(t)\cdot d\vec{\ell}\right]}{\Delta t}$$

Here a and b denote time-dependent vector positions in space. What is meant by the line integration is indicated by Fig. P2.6.3.

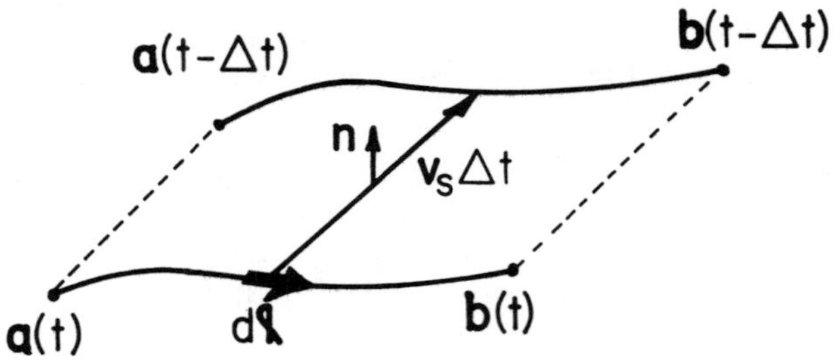

Fig. P2.6.3. Time-varying contour of line integration.

The contour of integration at the time t is instantaneously sketched. At that instant each point on the contour has a velocity $\vec{v}_s$ so that in a time Δt the contour has moved by an amount $\vec{v}_s\Delta t$. By definition, the velocity of the end point is $\vec{v}_s$ evaluated at the end point.

The theorem to be derived shows how the integration can be carried out after the time derivative has been taken. Thus it is analogous to the generalized Leibnitz rule for differentiation of a surface integral having time-varying geometry. The desired theorem states that

$$\frac{d}{dt}\int_{\vec{a}(t)}^{\vec{b}(t)} \vec{A}\cdot d\vec{\ell} = \int_{\vec{a}(t)}^{\vec{b}(t)} \frac{\partial \vec{A}}{\partial t}\cdot d\vec{\ell} + \vec{A}(\vec{b},t)\cdot v_s(\vec{b},t) - \vec{A}(\vec{a},t)\cdot v_s(\vec{a},t) + \int_{\vec{a}}^{\vec{b}} (\nabla\times\vec{A})\times\vec{v}_s\cdot d\vec{\ell}$$

Show that this rule can be derived following steps motivated by those used in the derivation of the generalized Leibnitz rule for a time-varying surface integration.

For Section 2.8:

Prob. 2.8.1 To illustrate how the steady-state motion of dipoles results in a $\vec{J}_p$ and hence an induced magnetic field, consider a slab of material extending to infinity in the y and z directions between infinitely permeable surfaces at $x = \pm a$. The slab has a thickness 2a, moves in the y direction with uniform velocity U and supports the polarization $\vec{P} = -(\rho_o a/\pi)\sin(\pi x/a)\vec{i}_x$, where ρ_o is a given constant. Fields are in the steady state and there is no free current density.

(a) Observe that Ampere's law, Eq. 2.2.2, and the boundary conditions are satisfied by making $\vec{H} = \vec{P}\times\vec{v}$. What is $\vec{H}$?

(b) Compute $\vec{J}_p$ and then use Ampere's law to find $\vec{H}$ in much the same way as if $\vec{J}_p$ were a free current density.

(c) Find ρ_p and show that in this case $\vec{J}_p$ is simply the result of polarization charge in motion

For Section 2.9:

Prob. 2.9.1 To someone not appreciating the importance of keeping field transformations consistent with the fundamental laws, it might appear that Faraday's law written in the Chu formulation (Eq. 2.2.1) would imply that a magnetized and conducting material set into motion would automatically support an electric field that would drive a free current density. In fact, there is an $\vec{E}$, but no $\vec{J}_f$. Consider as a specific case a magnetized slab, having $\vec{M} = -(\rho_o a/\pi\mu_o)\sin(\pi x/a)\vec{i}_x$, extending to infinity in the y and z directions, having boundaries at $x = \pm a$ in the x direction and suffering a uniform y-

Prob. 2.9.1 (continued)

directed translation with velocity U. Perfectly conducting walls bound the slab at $x = \pm a$. Steady state conditions prevail.

(a) Find the $\vec{H}$ induced by the given magnetization.

(b) Use Faraday's law to deduce $\vec{E}$.

(c) Now, if the material also has a conductivity σ, so that an observer at rest in the conductor can apply Ohm's law in the form $\vec{J}_f' = \sigma \vec{E}'$, because $\vec{J}_f = \vec{J}_f'$ but $\vec{E}' = \vec{E} + \vec{v}\times\mu_o\vec{H}$ (Eqs. 2.5.11 and 2.5.12), $\vec{J}_f = \sigma(\vec{E} + \vec{v}\times\mu_o\vec{H})$. Show that in fact $\vec{J}_f = 0$.

For Section 2.11:

Prob. 2.11.1 A plane parallel capacitor with electrodes at potentials v_1 and v_2 is used to impose a field on a third electrode that is grounded and free to move either longitudinally or transversely with displacements (ξ_1, ξ_2). The electrodes, shown in Fig. P2.11.1, have depth d into paper. Ignore fringing fields and find the capacitance matrix relating the charges (q_1,q_2) to the voltages (v_1,v_2).

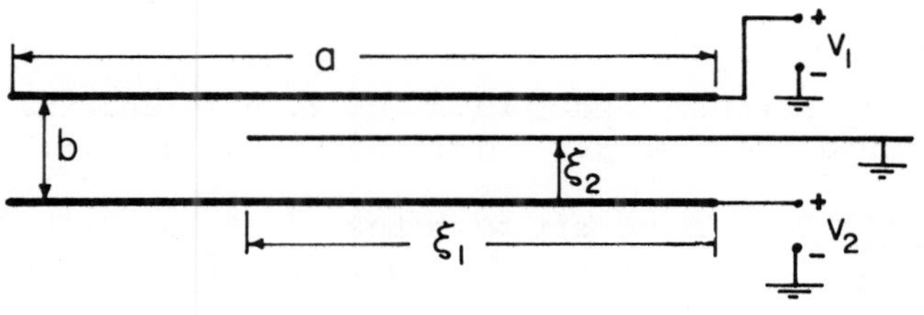

Fig. P2.11.1

For Section 2.12:

Prob. 2.12.1 A pair of perfectly conducting coaxial one-turn coils have the shape of circular cylinders of radius a and ξ, each with a length $d >> a$. Currents i_1 and i_2 are fed to the coils through parallel electrodes having a spacing that is negligible compared to other dimensions of interest. Determine the inductance matrix, Eq. 2.12.5, relating (λ_1, λ_2) to (i_1,i_2).

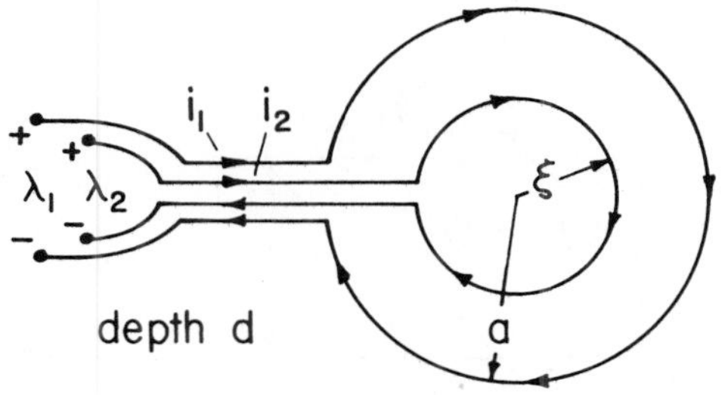

Fig. P2.12.1

For Section 2.13:

Prob. 2.13.1 For the system of Prob. 2.11.1, find the total coenergy storage $w'(v_1,v_2,\xi_1,\xi_2)$ by integrating Eq. 2.13.10.

Prob. 2.13.2 The dielectric slab shown in Fig. P2.13.2 is composed of material having the constitutive law $\vec{D} = \varepsilon_o\vec{E} + \vec{E}/\alpha_1 \sqrt{\alpha_2^2 + E^2}$. The slab has depth d into the paper. Under the assumption that $\rho_f=0$ in the dielectric and that its edges remain well removed from the fringing fields, find the dependence of the coenergy on (v,ξ).

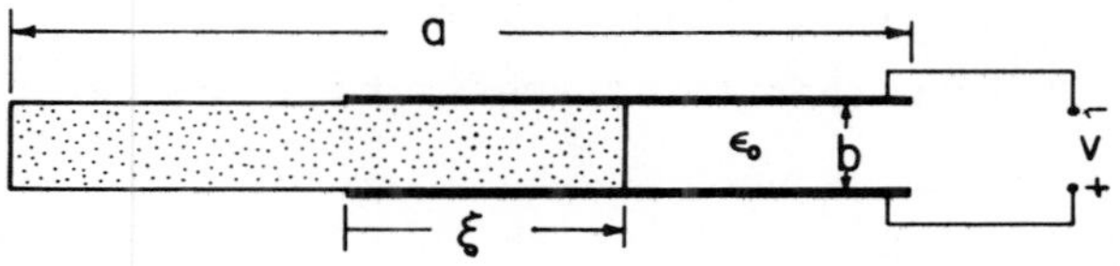

Fig. P2.13.2

For Section 2.14:

Prob. 2.14.1 For the system described in Prob. 2.12.1,

(a) Find the energy, $w = w(\lambda_1,\lambda_2, \xi)$, (b) the coenergy $w' = w'(i_1,i_2,\xi)$.

For Section 2.15:

Prob. 2.15.1 Show that the Fourier coefficients given by Eq. 2.15.8 follow from the procedure outlined in the paragraph following Eq. 2.15.7.

Prob. 2.15.2 A function $\Phi(z,t)$ is a square-wave function of z with magnitude $V_o(t)$. That is, $\Phi = V_o(t)$, $-\ell/4 < z < \ell/4$ and $\Phi = -V_o(t)$, $\ell/4 < z < 3\ell/4$. Show that the Fourier coefficients are

$$\tilde{\Phi}_m = 0,\ m \text{ even and } \tilde{\Phi}_m = 4V_o(t)\sin\left(\frac{k_m\ell}{4}\right)/(k_m\ell),\ m \text{ odd}$$

Prob. 2.15.3 A function $\Phi(z,t)$ is zero except in the interval $-\ell/2 < z < \ell/2$, where it is $V_o(t)$. Show that its Fourier transform is $\tilde{\Phi}(k,t) = \ell V_o(t)\ \sin(\frac{k\ell}{2})/(k\ell/2)$.

Prob. 2.15.4 Carry out the spatial average of the product of two Fourier series, as called for in completing Eq. 2.15.17.

For Section 2.16:

Prob. 2.16.1 Start with Eq. 2.16.14 and the relation between potential and flux, Eq. 2.16.5 and deduce the transfer relations of Table 2.16.1 for a planar layer.

Prob. 2.16.2 Start with Eqs. 2.16.20, 2.16.21 and 2.16.25 and deduce the transfer relations of Table 2.16.2. Use the properties of the Bessel functions as $r \to 0$ and $r \to \infty$ to deduce the limiting cases of Eqs. c and d.

Prob. 2.16.3 Start with Eq. 2.16.36 and deduce the transfer relations of Table 2.16.3. Evaluate the appropriate limits to arrive at Eqs. c and d.

Prob. 2.16.4 A region of free space is bounded by fictitious parallel planes at $x = \Delta$ and $x = 0$, as shown in Fig. P2.16.4.

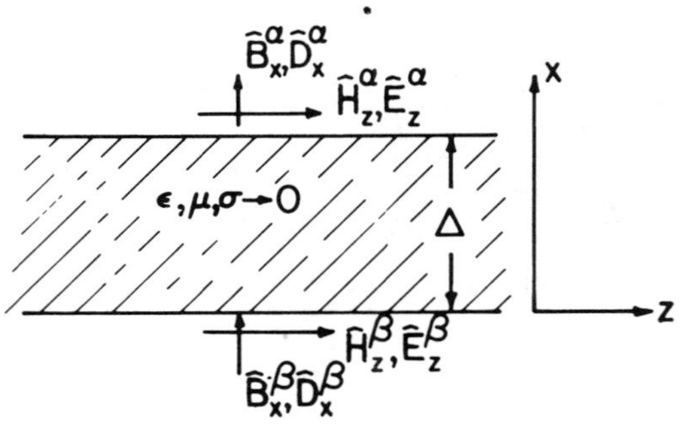

Fig. P2.16.4

Fields take the form
$\vec{E} = \text{Re}\,\hat{\vec{E}}(x)\ e^{j(\omega t-kz)}$;
$\vec{H} = \text{Re}\,\hat{\vec{H}}(x)\ e^{j(\omega t-kz)}$

so that there is no dependence on y and the time dependence is explicitly taken as exp $(j\omega t)$. The objective is to obtain transfer relations between tangential and perpendicular field components at the α and β surfaces without the quasistatic approximation.

(a) With fields taking the given form, show that all components of $\vec{E}$ and $\vec{H}$ can be written in terms of the axial components of $\hat{E}_z$ and $\hat{H}_z$. (This follows from Ampere's and Faraday's laws). Also show that E_z and H_z satisfy the wave equation.

(b) Write $\hat{E}_z$ and $\hat{H}_z$ in terms of the amplitudes $\hat{E}_z^\alpha$, $\hat{E}_z^\beta$ and $\hat{H}_z^\alpha$, $\hat{H}_z^\beta$ defined as these quantities evaluated on the respective surfaces.

(c) Show that the transfer relation for the layer is

$$\begin{bmatrix} \varepsilon\hat{E}_x^\alpha \\ \varepsilon\hat{E}_x^\beta \\ \mu\hat{H}_x^\alpha \\ \mu\hat{H}_x^\beta \end{bmatrix} = \begin{bmatrix} j\frac{\varepsilon k}{\gamma}\coth(\gamma\Delta) & -j\frac{\varepsilon k}{\gamma}\frac{1}{\sinh(\gamma\Delta)} & 0 & 0 \\ j\frac{\varepsilon k}{\gamma}\frac{1}{\sinh(\gamma\Delta)} & -j\frac{\varepsilon k}{\gamma}\coth(\gamma\Delta) & 0 & 0 \\ 0 & 0 & j\frac{\mu k}{\gamma}\coth(\gamma\Delta) & -j\frac{\mu k}{\gamma}\frac{1}{\sinh(\gamma\Delta)} \\ 0 & 0 & j\frac{\mu k}{\gamma}\frac{1}{\sinh(\gamma\Delta)} & -j\frac{\mu k}{\gamma}\coth(\gamma\Delta) \end{bmatrix} \begin{bmatrix} \hat{E}_z^\alpha \\ \hat{E}_z^\beta \\ \hat{H}_z^\alpha \\ \hat{H}_z^\beta \end{bmatrix}$$

where the other components of $\vec{E}$ and $\vec{H}$ are found from

$$\hat{H}_y = \frac{\omega\varepsilon_o}{k}\hat{E}_x\ ,\quad \hat{E}_y = \frac{-\omega\mu_o}{k}\hat{H}_x\ ,\quad \text{and } \gamma \equiv \sqrt{k^2 - \omega^2\mu\varepsilon}$$

Prob. 2.16.4 (continued)

(d) Show that in the quasistatic limit the relation reduces to the electroquasistatic and magnetoquasistatic transfer relations of Table 2.16.1 with appropriate identification of variables for the electric and magnetic relations.

(e) To make a connection with TE and TM modes in a plane parallel plate waveguide, let the α and β surfaces be perfectly conducting electrodes. Thus, the boundary conditions are

$$\hat{E}_z^{\alpha} = \hat{E}_z^{\beta} = 0 \qquad \text{TM modes}$$

$$\hat{B}_x^{\alpha} = \hat{B}_x^{\beta} = 0 \qquad \text{TE modes}$$

where the transverse magnetic and transverse electric modes can be separated because of the form taken by the transfer relations. Use these relations to argue that fields within that satisfy these homogeneous boundary conditions must also satisfy the dispersion equations

$$\omega^2\mu\varepsilon = k^2 + \left(\frac{n\pi}{\Delta}\right)^2 \quad ; \qquad n = 1, 2, 3 \ldots$$

Prob. 2.16.5 A planar region, shown in Table 2.16.1, is filled by an inhomogeneous dielectric, with a permittivity that depends on x:

$$\varepsilon(x) = \varepsilon_\beta \exp 2\eta x, \quad \eta \equiv \ln(\varepsilon_\alpha/\varepsilon_\beta)/2\Delta$$

The free charge density is zero.

(a) Show that the potential distribution is

$$\tilde{\Phi} = \tilde{\Phi}^{\alpha} e^{-\eta(x-\Delta)} \frac{\sinh\lambda x}{\sinh\lambda\Delta} - \tilde{\Phi}^{\beta} e^{-\eta x} \frac{\sinh\lambda(x-\Delta)}{\sinh\lambda\Delta}$$

where

$$\lambda \equiv \sqrt{k^2 + \eta^2}$$

(b) Show that the transfer relations are

$$\begin{bmatrix} \tilde{D}_x^{\alpha} \\ \\ \tilde{D}_x^{\beta} \end{bmatrix} = \varepsilon_\beta \lambda \begin{bmatrix} \left(\frac{\eta}{\lambda} - \coth\lambda\Delta\right) e^{\eta 2\Delta} & \frac{e^{\eta\Delta}}{\sinh\lambda\Delta} \\ \\ \frac{-e^{\eta\Delta}}{\sinh\lambda\Delta} & \frac{\eta}{\lambda} + \coth\lambda\Delta \end{bmatrix} \begin{bmatrix} \tilde{\Phi}^{\alpha} \\ \\ \tilde{\Phi}^{\beta} \end{bmatrix}$$

Prob. 2.16.6 A planar region, shown in Table 2.16.1, is filled by an anisotropic material having the constitutive law $D_i = \varepsilon_{ij}E_j$. The permittivity coefficients are uniform throughout. Determine the transfer relations in the form of Eqs. (a) of Table 2.16.1.

For Section 2.17:

Prob. 2.17.1 In developing conditions on coefficients in the transfer relations with the potentials expressed as functions of the "flux" variables, it is natural to use the energy function as exemplified in this section. The coenergy function is more convenient in dealing with the potentials as the independent variables. For the transfer relations of Sec. 2.16 written in the form

$$\begin{bmatrix} \tilde{D}_n^{\alpha} \\ \\ \tilde{D}_n^{\beta} \end{bmatrix} = \begin{bmatrix} -B_{11} & B_{12} \\ \\ -B_{21} & B_{22} \end{bmatrix} \begin{bmatrix} \tilde{\Phi}^{\alpha} \\ \\ \tilde{\Phi}^{\beta} \end{bmatrix}$$

derive conditions analogous to those of Eqs. 2.17.10 and 2.17.12.

Prob. 2.17.2 Use the reciprocity condition, Eq. 2.17.10 to show

$$kx[H_m(jkx)\,J_m'(jkx) - J_m(jkx)\,H_m'(jkx)] = \text{constant}$$

Use Eqs. 2.16.22 and 2.16.23 to establish that the constant is $2/\pi$. Thus, the numerators of the functions g_m and G_m in the cases $k \neq 0$ of Table 2.16.2 are considerably simplified from what is obtained by direct evaluation.

Prob. 2.17.3 With Eq. 2.17.7, it is assumed that the excitations on the α and β surfaces are in spatial phase, and that the A_{ij} are real. By allowing the excitations to have arbitrary phase, it is possible to learn more about these coefficients. In general, the expression replacing Eq. 2.17.7 in Cartesian or cylindrical geometry is

$$\delta w = \frac{1}{2}\, C\, \mathrm{Re}[-a^{\alpha}\tilde{\Phi}^{\alpha}\delta(\tilde{D}_n^{\alpha})^* + a^{\beta}\tilde{\Phi}^{\beta}\delta(\tilde{D}_n^{\beta})^*]$$

Because $\mathrm{Re}\,\tilde{u}\,\delta\tilde{V} = \tilde{u}_r\delta\tilde{V}_r + \tilde{u}_i\delta\tilde{V}_i$, this expression becomes

$$\delta w = \frac{1}{2}\, C[-a^{\alpha}\tilde{\Phi}_r^{\alpha}\delta\tilde{D}_{nr}^{\alpha} - a^{\alpha}\tilde{\Phi}_i^{\alpha}\delta\tilde{D}_{ni}^{\alpha} + a^{\beta}\tilde{\Phi}_r^{\beta}\delta\tilde{D}_{nr}^{\beta} + a^{\beta}\tilde{\Phi}_i^{\beta}\delta\tilde{D}_{ni}^{\beta}]$$

That is, the real and imaginary parts of the excitations on each surface are independent variables. Use the fact that the energy is a state variable: $w = w(\tilde{D}_{nr}^{\alpha}, \tilde{D}_{ni}^{\alpha}, \tilde{D}_{nr}^{\beta}, \tilde{D}_{ni}^{\beta})$ and show that

$$-a^{\alpha}\tilde{\Phi}_r^{\alpha} = \frac{\partial w}{\partial \tilde{D}_r^{\alpha}} \quad ; \quad -a^{\alpha}\tilde{\Phi}_i^{\alpha} = \frac{\partial w}{\partial \tilde{D}_i^{\beta}} \quad ; \quad a^{\beta}\tilde{\Phi}_r^{\beta} = \frac{\partial w}{\partial \tilde{D}_r^{\beta}} \quad ; \quad a^{\beta}\tilde{\Phi}_i^{\beta} = \frac{\partial w}{\partial \tilde{D}_i^{\beta}}$$

From these relations, derive reciprocity relations between the derivatives of $(\tilde{\Phi}_r^{\alpha}, \tilde{\Phi}_i^{\alpha}, \tilde{\Phi}_r^{\beta}, \tilde{\Phi}_i^{\beta},)$ with respect to $(\tilde{D}_{nr}^{\alpha}, \tilde{D}_{ni}^{\alpha}, \tilde{D}_{nr}^{\beta}, \tilde{D}_{ni}^{\beta})$. Assume that the A_{ij} can have real and imaginary parts, and show from these reciprocity relations that A_{11} and A_{22} must be real and that $a^{\alpha}A_{12} = a^{\beta}A^*_{21}$.

Prob. 2.17.4 Use the results of Prob. 2.17.1 to show that the transfer relations of Prob. 2.16.5 satisfy the reciprocity relations.

For Section 2.18:

Prob. 2.18.1 For the axisymmetric cylindrical case of Table 2.18.1, show that Eq. (h) follows from Eq. (g) and that Eq. 2.18.2 can be used to deduce the expression for the total flux, Eq. (i).

Prob. 2.18.2 Show that Eq. (k) of Table 2.18.1 follows from Eq. (j).

For Section 2.19;

Prob. 2.19.1 Derive Eqs. (e) and (f) of Table 2.19.1.

3

Electromagnetic Forces, Force Densities and Stress Tensors

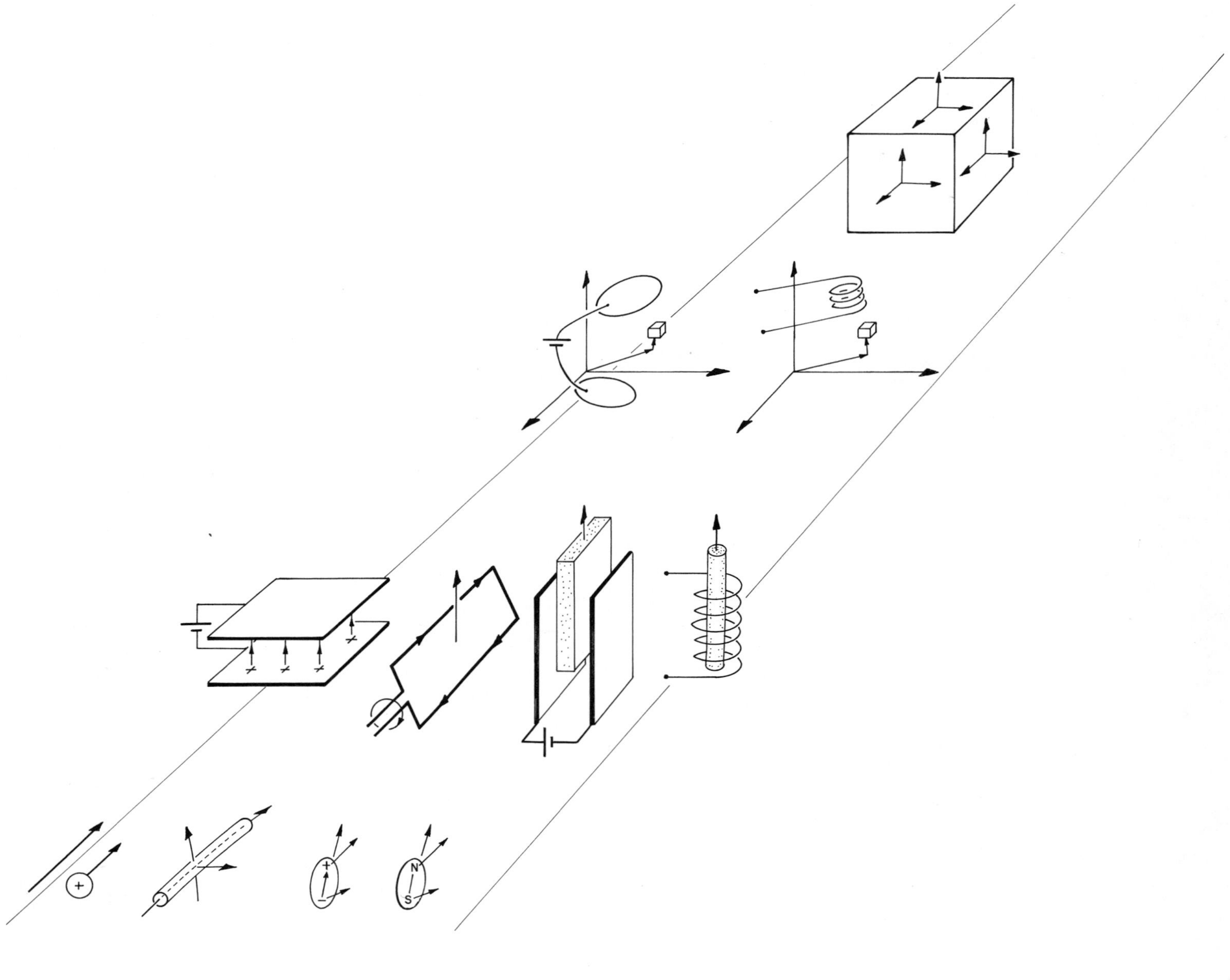

3.1 Macroscopic versus Microscopic Forces

Most important in this chapter is the distinction between forces on fundamental particles and forces on macroscopic media. It is common to speak of the "force on a charge" or the "force on a current" even though what is meant is the force on ponderable material. Interest might actually be in electric and magnetic forces acting on collections of fundamental charge carriers. (Motions of electron beams in vacuum are an example. The charged particles in that case constitute the continuum, in the sense that it is the electron inertia that enters into the equation of motion.) But, more commonly, the charged particles are imbedded in media, and it is the resulting force on the material that is of interest. Examples are as obvious as the electrical force of attraction between the capacitor plates of an electrostatic voltmeter or the magnetic torque exerted on current-carrying conductors in a meter movement.

Section 3.2 develops a specific model to illustrate how momentum imparted to charged particles by the fields is transferred to the neutral media that support those particles. That macroscopic forces are more than simply an average over the forces on fundamental charges is further emphasized by considering the practical cases of polarization and magnetization forces. Force densities of engineering significance exist even in regions where the free charge and free current (and for that matter polarization charge or magnetization charge) are absent. Such forces can be associated with a microscopic picture, discussed in Sec. 3.6, in which electrical forces on dipoles are transferred to the media.

Although the dipole model is useful for forming a microscopic picture of electric polarization forces, it is restricted to cases where the dipoles do not significantly interact. In the pursuit of a less restricted force density, developments in Secs. 3.7-3.8 are based on such measured macroscopic parameters as the permittivity and permeability. It is the business of thermodynamics to convert that information into the desired force densities. In its own way, the line of reasoning presented in Secs. 3.5, 3.7 and 3.8 exemplifies a more basic point of view than one geared to a particular microscopic model. Thermodynamic concepts provide a means for replacing detailed and specialized derivations by carefully defined physical measurements.

The stress-tensor representation of electromagnetic forces which concludes this chapter will see continual application in the following chapters. The tensor concept itself, introduced in Sec. 3.9, will also be applied to the formulation of continuum mechanical and electromechanical equations.

3.2 The Lorentz Force Density

Although macroscopic forces were the first measured in the development of electricity and magnetism, it is now normally accepted that the fundamental force is that on a "test" charge. This charge might be a single electron in free space. If the charged particle has a total charge q and moves with a velocity $\vec{v}_p$, then the Lorentz force acting on the particle supporting the charge is

$$\vec{f} = q\vec{E} + q\vec{v}_p \times \mu_o\vec{H} \tag{1}$$

This statement, like the electrodynamic laws summarized in Chap. 2, is an empirical one. In most of the areas of continuum electromechanics, it is forces due to many charges that are of interest, and it is therefore appropriate to sum the individual forces of Eq. 1 over the charges within a given unit of volume to arrive at the Lorentz force density

$$\vec{F} = \rho_f\vec{E} + \vec{J}_f \times \mu_o\vec{H} \tag{2}$$

Incremental volumes of interest have dimensions much greater than the characteristic distances between particles. But also, for the average electrical field to have meaning, it must be primarily due to sources external to the differential volume of interest. This ensures that, over an incremental volume, each particle experiences essentially the same electric field. The contribution to the field of the charges within the differential volume is negligible. Similar arguments apply to the magnetic field intensity, which must be produced over a given differential volume largely by currents outside the volume.

Equation 2 represents the force density acting on a ponderable medium if means are available for the force on the particles to be transmitted to the medium. The mechanisms by which this happens are diverse, and implicit to the conduction process. Whether the fundamental carriers are electrons in a metal, holes and electrons in a semiconductor or ions in a liquid or gas, the average motions of fundamental charge carriers are superimposed on random motions. The flights of fundamental carriers are interrupted by collisions with lattice molecules (in a solid) or molecules that are themselves in a Brownian equilibrium (in a liquid or gas) with a frequency that is usually extremely high compared to reciprocal times of interest. These collisions transfer momentum from the fundamental charge carriers to the ponderable medium.

To more fully appreciate the transition from the force acting on fundamental carriers, Eq. 1, to that on a material, Eq. 2, it is helpful to make a formal derivation. Although the discussion leads to rather general conclusions, only two families of carriers are now considered, one positive with charge per particle q_+ and number density n_+ and the other negative with a magnitude of charge q_- and number density n_-. The average Lorentz force, Eq. 1, is in equilibrium with an average force representing the effect of collisions on the net migration of the particles:

$$\begin{aligned} q_+\vec{E} + q_+(\vec{v}_+ + \vec{v}) \times \mu_o\vec{H} &= m_+\nu_+\vec{v}_+ \\ -q_-\vec{E} - q_-(\vec{v}_- + \vec{v}) \times \mu_o\vec{H} &= m_-\nu_-\vec{v}_- \end{aligned} \tag{3}$$

The retarding forces on the right are much as would be conceived for a swarm of macroscopic particles moving through a viscous liquid. The average carrier velocities $\vec{v}_\pm$ are measured relative to the medium, which itself has the velocity $\vec{v}$. Hence, on the right it is relative velocities of particles and medium that appear, while in the Lorentz force it is total particle velocities that are appropriate. The coefficients for the collisional forces are written as the product of the particle masses $m_\pm$ and collision frequencies $\nu_\pm$ as a matter of convention. Note that the inertial force on the carriers is ignored compared to that due to collisions. This approximation would be invalidated in a plasma if the frequency of an applied electric field intensity were extremely high. But, in many conductors and certainly in the most usual electromechanical situations, the inertial effects of the charge carriers can be ignored (see (Problem 3.3.1.).

The charge density and current density are written in terms of the microscopic variables as

$$\rho_f = n_+q_+ - n_-q_- \tag{4}$$

$$\begin{aligned} \vec{J}_f &= n_+q_+(\vec{v}_+ + \vec{v}) - n_-q_-(\vec{v}_- + \vec{v}) \\ &= n_+q_+\vec{v}_+ - n_-q_-\vec{v}_- + \rho_f\vec{v} \end{aligned} \tag{5}$$

The average force density acting on the ponderable medium is the sum of the right-hand sides of Eq. 3, respectively, multiplied by the particle densities $n_\pm$:

$$\vec{F} = n_+m_+\nu_+\vec{v}_+ + n_-m_-\nu_-\vec{v}_- \tag{6}$$

The point in writing this equation is to formalize the statement that, through some collisional process, the force on the fundamental carriers becomes the force on the medium. It is evident from the next step that, at least in so far as the Lorentz force density is concerned, the details of the collisional equilibrium are not important. The left-hand sides of Eq. 3 (regardless, for example, of whether $m_\pm\nu_\pm$ are functions of $v_\pm$ or are constant) are substituted for the respective terms in Eq. 6 to obtain

$$\vec{F} = (n_+q_+ - n_-q_-)\vec{E} + [(n_+q_+\vec{v}_+ - n_-q_-\vec{v}_-) + (n_+q_+ - n_-q_-)\vec{v}] \times \mu_o\vec{H} \tag{7}$$

In view of the definitions given by Eqs. 4 and 5, this expression is the Lorentz force density of Eq. 2. Its validity hinges on there being an instantaneous equilibrium between the forces on the fundamental carriers and the "collisions" with the ponderable medium, but not on the details of that interaction.

3.3 Conduction

There are three objectives in this section. The first is to have a microscopic picture of the carrier motions to associate with ohmic or unipolar conduction models. The second is to illustrate how constitutive laws for media in motion can be derived from models based on particular microscopic models, or (on the basis of the field transformations) found by generalizing empirically determined laws established in the laboratory for materials at rest. Finally, a byproduct of the discussion is an introduction to Hall effect.

Consider the carrier motions represented by Eqs. 3.2.3, with the magnetic field $\vec{H} = H_o\vec{i}_x$ externally imposed. The components of these equations then respectively become

$$\begin{bmatrix} 1 & 0 & 0 \\ 0 & 1 & \mp b_\pm\mu_oH_o \\ 0 & \pm b_\pm\mu_oH_o & 1 \end{bmatrix} \begin{bmatrix} v_{x\pm} \\ v_{y\pm} \\ v_{z\pm} \end{bmatrix} = \begin{bmatrix} \pm b_\pm E_x \\ \pm b_\pm E_y \pm b_\pm v_z\mu_oH_o \\ \pm b_\pm E_z \mp b_\pm v_y\mu_oH_o \end{bmatrix} \tag{1}$$

where particle mobilities are defined as $b_{\pm} = q_{\pm}/m_{\pm}\nu_{\pm}$.

These three equations can be inverted to find the relative carrier velocities in terms of $(\vec{E},\vec{H},\vec{v})$:

$$\begin{bmatrix} v_{x\pm} \\ v_{y\pm} \\ v_{z\pm} \end{bmatrix} = \frac{1}{\Delta_{\pm}} \begin{bmatrix} \frac{\pm b_{\pm}}{\Delta_{\pm}} & 0 & 0 \\ 0 & \pm b_{\pm} & b_{\pm}^2 \mu_o H_o \\ 0 & -b_{\pm}^2 \mu_o H_o & \pm b_{\pm} \end{bmatrix} \begin{bmatrix} E_x \\ E_y + v_z \mu_o H_o \\ E_z - v_y \mu_o H_o \end{bmatrix} \tag{2}$$

where $\Delta_{\pm} = 1 + (\mu_o H_o b_{\pm})^2$.

These velocity components can now be introduced into Eq. 3.2.5 to express the free current density as

$$\vec{J}_f = (n_+q_+b_+ + n_-q_-b_-)E'_x\vec{i}_x + \left(\frac{n_+q_+b_+}{\Delta_+} + \frac{n_-q_-b_-}{\Delta_-}\right)(E'_y\vec{i}_y + E'_z\vec{i}_z)$$
$$+ \left(\frac{n_+q_+b_+^2}{\Delta_+} - \frac{n_-q_-b_-^2}{\Delta_-}\right)\mu_o\vec{E}' \times \vec{H}_o + \rho_f\vec{v} \tag{3}$$

where $\vec{E}' \equiv \vec{E} + \vec{v} \times \mu_o\vec{H}$ is the electric field in a frame of reference moving with the material (for a magnetoquasistatic system).

From Eq. 3, it is clear that there are two components to the current density, one in the direction of the imposed electric field and the second perpendicular to it. The latter term is called the Hall current and is due to the tendency of the particles to move perpendicular to their own velocity and to the imposed magnetic field intensity. This last term is ignorable if

$$\mu_o H_o b_{\pm} << 1 \tag{4}$$

A typical magnetic flux density is $\mu_o H_o = 1$ (10,000 gauss, which is in the range where magnetic materials saturate). Electrons in copper have a mobility on the order of 3×10^{-3} m^2/volt sec, so that the parameter on the left is then much less than 1. Ions in liquids have mobilities that are typically 5×10^{-8} m^2/volt sec and the approximation is even better. But in silicon or germanium, where the electron mobility is in the range of 10^{-1} m^2/volt sec, the Hall effect is coming into play by the time $\mu_o H_o$ is of the order of unity. With the inequality of Eq. 4 satisfied, Eq. 3 reduces to the familiar form

$$\vec{J}_f = (n_+q_+b_+ + n_-q_-b_-)\vec{E}' + \rho_f\vec{v} \tag{5}$$

If the number density of charge carriers n_+ and/or n_- remains essentially the same in spite of the application of $\vec{E}$, then the factor multiplying $\vec{E}$ in Eq. 5 is usefully regarded as a parameter characterizing the material, the electrical conductivity σ. This case of ohmic conduction is displayed by materials ranging from metallic conductors, where the carriers are electrons and essentially immobile ions, to electrolytes, where ions of at least two species participate in the conduction. In any of these cases, for the ohmic model to be valid, the conduction must involve at least two species with both n_+q_+ and n_-q_- greatly exceeding the net charge ρ_f. By introducing the conductivity as a parameter, the detailed analysis necessary to determine the self-consistent distributions of the individual carriers is avoided. But to examine the conditions under which the conductivity model is valid, it is necessary to formulate the laws that govern the self-consistent carrier motions. This is best done in the context of molecular diffusion (Chap. 10) so that other important limitations on the model can also be identified.

Even though in accounting for conduction it is useful to have in mind microscopic mechanisms, it is also important to recognize the far-reaching implications of empirical relations. Given any conduction law based on laboratory measurements made with a fixed sample, effects of material motion can be brought in by using the transformation laws. For example, if it is known that the conductor obeys Ohm's law when stationary, then in a primed inertial frame moving with the velocity $\vec{v}$ of the conductor, the experiment shows that

$$\vec{J}'_f = \sigma\vec{E}' \tag{6}$$

In an electroquasistatic system, including polarization, $\vec{J}'_f = \vec{J}_f - \rho_f\vec{v}$ (Eq. 2.5.12a) and $\vec{E}' = \vec{E}$ (Eq. 2.5.9a). Hence, Eq. 6 becomes Eq. 5. In a magnetoquasistatic system, including magnetization,

$\vec{J}_f' = \vec{J}_f$ (Eq. 2.5.11b) and $\vec{E}' = \vec{E} + \vec{v} \times \mu_o\vec{H}$ (Eq. 2.5.12b). Substitution in Eq. 6 now gives Eq. 5, except for the charge convection term $\rho_f\vec{v}$. In a magnetoquasistatic system, this term is second-order, as will be argued in the next section.

Fundamental to the use of an empirical law determined for the stationary material is the assumption that material acceleration and deformation do not influence the conduction. In any case, if acceleration did effect the conduction, the close tie between conduction and the Lorentz force density, illustrated in this and the previous section, calls into question the notion that the electromechanics can be modeled by a single continuum subject to the Lorentz force density.

3.4 Quasistatic Force Density

The Lorentz force density, Eq. 3.2.2, is composed of what will be termed, respectively, an electric force density and a magnetic force density

$$\vec{F} = \rho_f\vec{E} + \vec{J}_f \times \mu_o\vec{H} \tag{1}$$

It is found in a wide range of applications that the force density is predominantly one or the other of these contributions. Polarization and magnetization force densities, not included in Eq. 1, are similarly identified with the respective quasistatic systems. In this section, dimensional arguments are given that demonstrate that the electric force density generally dominates in electroquasistatic systems, while the magnetic force density dominates in magnetoquasistatic systems.

The line of reasoning is an extension of that introduced in Sec. 2.2. The force density is normalized in accordance with Eq. 2.3.4 and the free current density is represented as having the form of Eq. 2.3.1. Thus,

$$\vec{F} = \frac{\varepsilon_o\mathscr{E}^2}{\ell}[\rho_f\,\vec{E} + \frac{\tau_m}{\tau}(\sigma\,\vec{E} + \frac{\tau_e}{\tau}\vec{J}_v) \times \vec{H}] \qquad \text{EQS} \tag{2}$$

$$\vec{F} = \frac{\mu_o\mathscr{H}^2}{\ell}[(\frac{\tau_{em}}{\tau})^2\rho_f\,\vec{E} + (\frac{\tau_m}{\tau}\vec{E} + \vec{J}_v) \times \vec{H}] \qquad \text{MQS} \tag{3}$$

The relative values of the time constants are summarized by Fig. 2.3.1. In the electroquasistatic system, $\tau_m/\tau \ll 1$ and $\tau_m\tau_e/\tau^2 = (\tau_{em}/\tau)^2 \ll 1$. Hence, the free charge density term is zero-order in Eq. 1, and the magnetic term is consistently ignored.[1] In the magnetoquasistatic force density of Eq. 3, $(\tau_{em}/\tau)^2 \ll 1$, and the free charge force density is negligible compared to the magnetic term. Hence, the second term of Eq. 1 is used to the exclusion of the first in magnetoquasistatic systems.

3.5 Thermodynamics of Discrete Electromechanical Coupling

In this section, the thermodynamic electric and magnetic energy storage subsystems are expanded to include the possibility of a finite number of discrete mechanical displacements of macroscopic material. Based on the notion of an energy function and a thermodynamic equilibrium, the force of electrical origin associated with each of these displacements is determined. Typically, the method exploits a knowledge of the electrical terminal relations to determine the forces. The approach is generalized in Secs. 3.7 and 3.8, where constitutive laws are the basis for finding the force density of electric origin. Except for mathematical manipulations, the derivations now reviewed draw upon all of the demanding issues confronted later in deriving force densities.

Electroquasistatic Coupling: An example of a lumped-parameter electroquasistatic system is given with Fig. 2.11.1, including a schematic representation of a finite number of mechanical displacements. Associated with each of the displacements is an electromechanical force tending to displace a lumped element by an amount $\delta\xi_i$.

Conservation of energy for the system with the geometry fixed is expressed by Eq. 2.13.8. Now, an incremental increase in the total energy caused by placing an increment of charge δq_i on an electrode having the voltage v_i can be diminished by an amount equal to the work done on the external environment by the forces of electrical origin acting through the displacements of the associated mechanical entities. Thus, energy conservation requires that

$$\delta w = \sum_{i=1}^{n} v_i\delta q_i - \sum_{j=1}^{m} f_j\delta\xi_j; \quad w = w(q_1\cdots q_n,\xi_1\cdots\xi_m) \tag{1}$$

---

1. Electrons in vacuum can have a velocity approaching that of light. In that case an imposed magnetic field can have a crucial effect on the EQS dynamics (See Sec. 11.2).

Given the charges $q_1 \cdots q_n$ and the displacement $\xi_1 \cdots \xi_m$ as independent variables, the energy function is uniquely determined. The "displacements" should be recognized as generalized variables in that they could just as well be angular deflections, in which case the associated "forces" would be torques.

To determine w, constitutive relations $v_i(q_1 \cdots q_n, \xi_1 \cdots \xi_n)$ must be known so that Eq. 1 can be integrated. The integration is a line integral in a state-space composed of the independent variables. Because the f_j's are not known, and are defined as equal to zero in the absence of electrical excitations, integration on the mechanical variables ξ_j is carried out first. This gives no contribution because as the displacements are brought to their final values, $f_j = 0$ (no work is required to assemble the system with the q_j's = 0). Then, the integration on successive electrical variables is carried out, first on q_1 with all other q_j's = 0, then on q_2 with q_1 at its final value and all others zero, etc. Formally, the integration of Eq. 1 gives

$$w = \sum_{j=1}^{n} \int_{o}^{q_j} v_j(q_1 \cdots q_j', 0 \cdots 0, \xi_1, \xi_2 \cdots \xi_m)\delta q_j' \tag{2}$$

Because the energy function is a state function specified by the independent variables, an incremental change in the total energy can also be written as

$$\delta w = \sum_{i=1}^{n} \frac{\partial w}{\partial q_i} \delta q_i + \sum_{j=1}^{m} \frac{\partial w}{\partial \xi_j} \delta \xi_j \tag{3}$$

If the q's and the ξ's are independent variables in the sense that Eqs. 1 and 3 hold for arbitrary combinations of incremental changes in these electrical and mechanical variables, then

$$v_i = \frac{\partial w}{\partial q_i} \; ; \quad f_j = - \frac{\partial w}{\partial \xi_j} \tag{4}$$

Note that the q's and ξ's are not necessarily independent of each other unless the system is isolated from the total system in which it is imbedded. Given w from Eq. 2, the electrical forces are determined.

A consequence of the conservation of energy expressed by Eq. 1 is the reciprocity condition between pairs of terminal variables. For example, derivatives of Eq. 4a, first with respect to q_j and then of the same equation but with i replaced by j, and with respect to q_i, are related by

$$\frac{\partial v_i}{\partial q_j} = \frac{\partial^2 w}{\partial q_i \partial q_j} = \frac{\partial v_j}{\partial q_i} \tag{5}$$

Other reciprocity conditions follow from Eq. 4 by taking cross-derivatives to relate forces and voltages to each other.

In dealing with practical lumped-parameter systems, it is often convenient to use the voltages rather than the charges as independent variables. If all of the voltages are to be independent variables, it is appropriate to recognize that

$$\sum_{i=1}^{n} v_i \delta q_i = \sum_{i=1}^{n} [\delta(v_i q_i) - q_i \delta v_i] \tag{6}$$

so that substitution into Eq. 1 gives

$$\delta w' = \sum_{i=1}^{n} q_i \delta v_i + \sum_{j=1}^{m} f_j \delta \xi_j \tag{7}$$

where a coenergy function has been defined in terms of the energy function as

$$w'(v_1 \cdots v_n, \xi_1 \cdots \xi_m) \equiv \sum_{i=1}^{n} v_i q_i - w \tag{8}$$

The coenergy function is a particular case of an arbitrarily large number of functions that can be defined. Any combination of charges and voltages can be independent variables, and a hybrid energy function, appropriately defined as a state function of this combination. With the voltages as independent variables, an equation similar to Eq. 2 is found with the charges replaced by the voltages, and the voltages and displacements the independent variables:

$$q_i = \frac{\partial w'}{\partial v_i} \; ; \quad f_j = \frac{\partial w'}{\partial \xi_j} \tag{9}$$

The coenergy function, like the energy function, is found from purely electrical considerations, as described in Sec. 2.13.

Magnetoquasistatic Coupling: Lumped-parameter electromechanical coupling in a magnetic field system, described schematically by Fig. 2.12.1, can be given the same thermodynamic representation as that outlined for electroquasistatic systems. The statement of conservation of energy for the system of discrete coils and mechanical displacements is the generalization of Eq. 2.14.11, with the addition of the mechanical work done as an electrical force f_j causes an incremental displacement $\delta\xi_j$:

$$\delta w = \sum_{i=1}^{n} i_i \delta\lambda_i - \sum_{j=1}^{m} f_j \delta\xi_j \tag{10}$$

All of the arguments given for the electric systems follow for the magnetic field systems if variables are identified:

$$q_i \rightarrow \lambda_i,\ v_i \rightarrow i_i \tag{11}$$

$$w = w(\lambda_1 \cdots \lambda_n,\ \xi_1 \cdots \xi_m);\ w' = w'(i_1 \cdots i_n,\ \xi_1 \cdots \xi_m)$$

The magnetic force is the negative partial derivative of the magnetic energy with respect to the appropriate associated displacement, with the other displacements and all of the flux linkages held constant. Similarly, the force can be found from the coenergy function by taking the derivative with respect to the associated displacement with the other displacements and the currents held constant.

3.6 Polarization and Magnetization Force Densities on Tenuous Dipoles

Forces due to polarization and magnetization lend further emphasis to the importance of making a distinction between forces on microscopic charged particles and macroscopic forces on materials supporting those charges. The experiment depicted by Fig. 3.6.1 makes it clear that (1) there is more to the force density than accounted for by the Lorentz force density, and (2) the additional force density is not $\rho_p\vec{E}$ (or in the magnetic analogue, $\rho_m\vec{H}$).

A pair of capacitor plates are dipped into a dielectric liquid. With the application of a potential difference v, it is found experimentally that the liquid rises between the plates.* To make it clear that the issues involved can be understood in terms of lumped-parameter concepts, the liquid between the plates is replaced by a solid dielectric material having the same polarizability as the liquid, so that the problem is reduced to one of a solid dielectric slab rising between the plates as it is pulled from the liquid below.

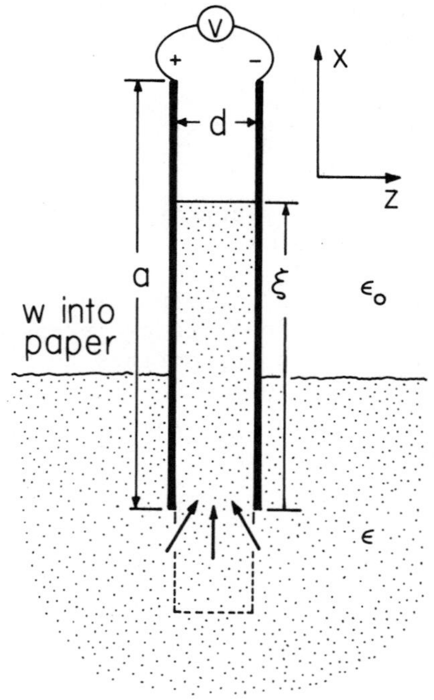

Fig. 3.6.1. Experiment demonstrating the existence of polarization forces that are not explicable in terms of forces on single charges.

Recall that if the interface is well removed from the edges of the plates, an exact solution satisfying the quasistatic differential equations and boundary conditions in the neighborhood of the interface is $\vec{E} = (v/d)\vec{i}_z$. Of course, there is a fringing field in the neighborhood of the edges of the capacitor plates. However, because the slab and the liquid have the same dielectric constant and $\rho_f = 0$, the fringing field has the same distribution as if the dielectric were not present.

It might be tempting to take the force as being the product of the net charge at any given point and the local electric field, or $\rho_p\vec{E}$. However, everywhere in the dielectric bulk the polarization density is proportional by the same constant to the electric field (Eq. 2.16.1). Because $\rho_f = 0$, it follows from Gauss' law that $\vec{E}$ and hence $\vec{P}$ have no divergence, and so there is also no polarization charge in the dielectric. Furthermore, because the electric field is uniform and tangential to the interface, there is not even a polarization surface charge density at the interface (Eq. 2.10.21). Throughout the dielectric, on the interface and in the bulk, there is no polarization charge. Clearly, the force which makes the dielectric rise between the plates cannot be accounted for by a polarization charge density.

---

* In an experiment, a-c voltage is used with a sufficiently high frequency that the material responds only to the rms field and free charge cannot accumulate in the bulk.

If the polarized material is composed of individual dipoles, each subject to an electrical force, and each transmitting this electrical force to the neutral medium, it is clear that there is really no reason to expect that the force density should take the same form as that for free charges. With free charges, it is the individual charges that transmit their forces to the neutral medium through mechanisms discussed in Sec. 3.2. Now concern is with the force on individual dipoles which transmit that force to the neutral medium, either because they are tied to a lattice structure (Fig. 2.8.1) or through collisional mechanisms similar to those discussed for charge carriers in Sec. 3.2.

In the following, it is assumed that the dipoles are subject to an electric field that is the average, or macroscopic, electric field. The development ignores the distortion of the electric field intensity at one dipole because of the neighboring dipoles. For this reason, the result is designated a force density acting on tenuous dipoles.

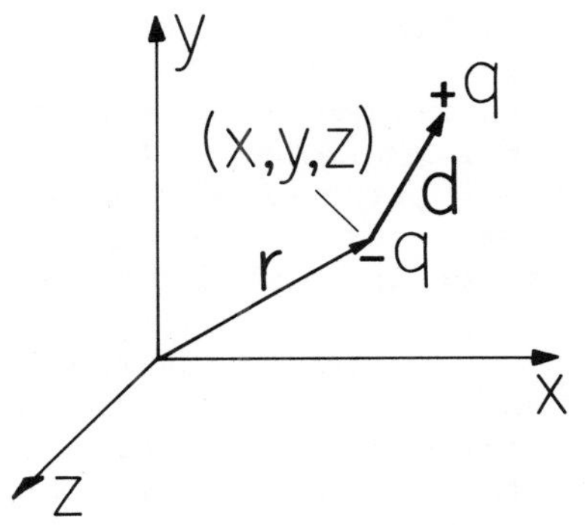

Fig. 3.6.2. Definition of displacement and charge locations for dipole.

A single dipole is shown in Fig. 3.6.2. The dipole can be pictured as a pair of oppositely signed charges having the vector separation $\vec{d}$. The negative charge is located at $\vec{r}$. With the assumption that the force on the dipole is transmitted to the medium, the procedure is to compute the force on a single dipole, and then to average this force over all the dipoles. The net force in the ith direction on the pair of charges taken as a unit is

$$f_i = \lim_{\substack{\vec{d}\to 0 \\ q\to\infty}} q[E_i(\vec{r} + \vec{d}) - E_i(\vec{r})] \tag{2}$$

The limit is one in which the spacing of the charges becomes extremely small compared to other distances of interest and, at the same time, the magnitude of the charges becomes very large, so that the product $q\vec{d} \equiv \vec{p}$ remains finite. The dipole moment is defined as $\vec{p}$. The required limit of Eq. 2 becomes

$$f_i = \lim_{\substack{\vec{d}\to 0 \\ q\to\infty}} q[E_i(\vec{r}) + \frac{\partial E_i}{\partial x_j} d_j - E_i(\vec{r})] = p_j \frac{\partial E_i}{\partial x_j} \tag{3}$$

Thus, there is a net force on each dipole given in vector notation by

$$\vec{f} = \vec{p}\cdot\nabla\vec{E} \tag{4}$$

Note that implicit to this vector representation is the definition of what is meant by the operator $\vec{A}\cdot\nabla\vec{B}$

By assumption, the net force on each dipole is transmitted to the macroscopic medium and it is appropriate then to think of averaging these polarization forces over all dipoles within the medium. In general, this average would have to be taken with recognition that the microscopic dipoles could assume a spectrum of polarizations in a given electric field intensity. For present purposes, the average can simply be represented as the multiplication of Eq. 4 by the number of dipoles, n, per unit volume. With the definition of the polarization density as $\vec{P} = n\vec{p}$, the <u>Kelvin polarization force density</u> is found:

$$\vec{F} = \vec{P}\cdot\nabla\vec{E} \tag{5}$$

Can the force density given by Eq. 5 be used to explain the rise of the dielectric between the plates in Fig. 3.6.1? Certainly, there is no force density in material regions of uniform electric field, because then the spatial derivatives called for with Eq. 5 vanish. However, in the fringing field at the lower edges of the plates, the electric field intensity does vary rapidly. In that region, the permittivity is a constant, and for a linear dielectric, where $\vec{D} = \varepsilon\vec{E}$, Eq. 5 becomes [in dealing with vectors and tensors, a term in which a subscript appears twice is to be summed 1 to 3 (unless otherwise indicated)]

$$F_i = (\varepsilon - \varepsilon_o)E_j \frac{\partial E_i}{\partial x_j} = (\varepsilon - \varepsilon_o)E_j \frac{\partial E_j}{\partial x_i} = (\varepsilon - \varepsilon_o) \frac{\partial}{\partial x_i} (\frac{1}{2} E_j E_j) \tag{6}$$

where the irrotational nature of $\vec{E}$ is exploited, $\partial E_i/\partial x_j = \partial E_j/\partial x_i$. In vector notation, Eq. 6 becomes

$$\vec{F} = \nabla[\frac{1}{2} (\varepsilon - \varepsilon_o)\vec{E}\cdot\vec{E}] \tag{7}$$

Remember, this relation pertains only to regions of a linear dielectric in which the permittivity is constant, and is simply a means of visualizing the distribution of the Kelvin force density. In such regions, the force density has the direction of maximum rate of increase of the electric energy storage. Typical force vectors, sketched in Fig. 3.6.1, tend to push the dielectric upward between the plates. It is important not to overgeneralize from Eq. 7. In any configuration in which there is a component of $\vec{E}$ perpendicular to an interface, there is a singular component of the Kelvin force density acting at the interface -- a surface force density. Such a component would be incorrectly inferred from Eq. 7, which is not valid through the interfacial region.

Consider now the force density acting on a continuum of dilute magnetic dipoles that, like the analogous electric dipoles just considered, pass along a force of electric origin to a macroscopic medium via collisions or lattice constraints. It is not possible to use the Lorentz force law as a starting point unless magnetic monopoles and an analogous force law on these magnetic "charges" is postulated. Without introducing such notions, the Kelvin magnetization force density can be deduced as follows.

Electroquasistatic and magnetoquasistatic systems are pictured abstractly in Fig. 3.6.3. A volume enclosing the region occupied by a dipole having the position $\vec{\xi}$ has a surface S and includes neither free charge in the EQS system nor free current in the MQS system. Hence the fields are governed by

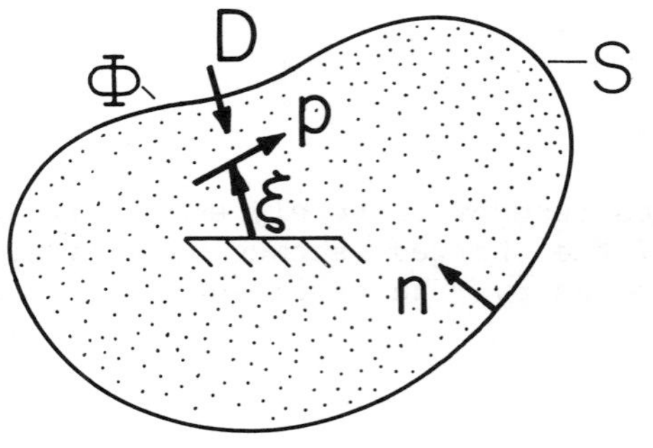

Fig. 3.6.3a. EQS system

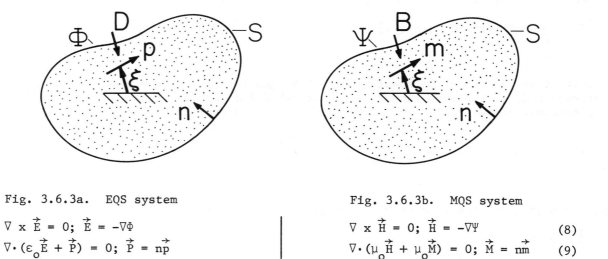

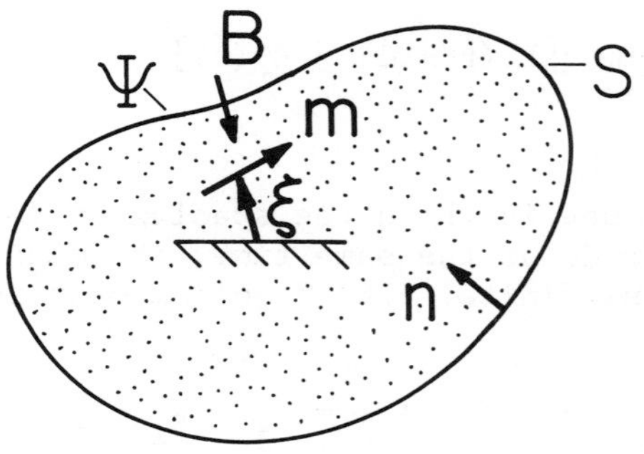

Fig. 3.6.3b. MQS system

$$\nabla \times \vec{E} = 0; \quad \vec{E} = -\nabla\Phi \qquad\Big|\qquad \nabla \times \vec{H} = 0; \quad \vec{H} = -\nabla\Psi \tag{8}$$

$$\nabla\cdot(\varepsilon_o\vec{E} + \vec{P}) = 0; \quad \vec{P} = n\vec{p} \qquad\Big|\qquad \nabla\cdot(\mu_o\vec{H} + \mu_o\vec{M}) = 0; \quad \vec{M} = n\vec{m} \tag{9}$$

Statements that the input of electric energy either goes into increasing the total energy stored or into doing work on the dipoles are (see Eqs. 3.5.1 and 2.13.4 or Eq. 3.5.10 and Eq. 2.14.9 integrated by parts):

$$\oint_S \Phi\delta\vec{D}\cdot\vec{n}da = \delta w + \vec{f}\cdot\delta\vec{\xi} \qquad\Big|\qquad \oint_S \Psi\delta\vec{B}\cdot\vec{n}da = \delta w + \vec{f}\cdot\delta\vec{\xi} \tag{10}$$

To find the force on the dipole, the energy would be determined as a function of the electrical excitations and $\vec{\xi}$. Then, with the understanding that the derivative is taken with the quantities $\vec{D}\cdot\vec{n}$ and $\vec{B}\cdot\vec{n}$, respectively, held fixed on the surface S, the respective forces follow as

$$f_i = -\frac{\partial w}{\partial \xi_i} \qquad\Big|\qquad f_i = -\frac{\partial w}{\partial \xi_i} \tag{11}$$

Now, what would be obtained if this procedure were carried through for the electric case is already known to be given by Eq. 4. Moreover, there is a complete analogy between every aspect of the electric and magnetic systems. The calculation in the magnetic case need not be repeated once the electric one is carried out. Rather, an identification of variables suffices to give the answer, $\vec{E} \to \vec{H}$, $\vec{P} \to \mu_o\vec{M}$. Hence, it follows that Eq. 5 is replaced by the Kelvin magnetization force density

$$\vec{F} = \mu_o\vec{M}\cdot\nabla\vec{H} \tag{12}$$

The Kelvin force densities, Eqs. 5 and 12, suffer the weakness that they do not take into account the interaction between dipoles. Moreover, is the average over the spectrum of dipole moments $\vec{p}$ or $\vec{m}$ leading to the polarization and magnetization densities consistent with the usage of these densities in Chap. 2? These difficulties are overcome by a derivation based on thermodynamic principles. Because force densities are then based on electrically measured constitutive laws, consistency with definitions already introduced is insured.

3.7 Electric Korteweg-Helmholz Force Density

The thermodynamic technique used in this section for deducing the electric force density with combined effects of free charge and polarizarion is a generalization of that used in determining discrete forces in Sec. 3.5. This principle of virtual work is exploited because it is not practical to predict the relationship between microscopic and macroscopic fields.

In any derivation of a force density, it is important to be clear about (a) what empirically determined information is required, and (b) what postulates or assumptions are incorporated into the derivation or are implicit to an application of the force density. Generally, empirically determined information can be used to replace assumptions. As derived here, the only empirical information required is an electrical constitutive law relating the macroscopic electric field to the polarization density $\vec{P}$ (or displacement $\vec{D}$). This relationship is typically determined by making electrical measurements on homogeneous samples of the material. These amount to measurements of the terminal characteristics of capacitor-like configurations incorporating samples of the material. (In the lumped-parameter systems of Sec. 3.5, the analogous empirical information was the electrical terminal relation.) With so little empirical information, the force density can only be identified if the system considered is a conservative thermodynamic subsystem. Thus, the force density is derived picturing the system as having no dissipation mechanisms. (The same conservative system is considered in Sec. 3.5 to find discrete forces.) The <u>assumption</u> is then made that the force density remains valid even in modeling systems with dissipation. If dissipation mechanisms were to be incorporated into the system considered, then a virtual power principle could be exploited to find the force density, but additional empirical information would be required.

Experiments show that, for a wide range of materials, electrical constitutive laws take the form of state functions

$$\vec{E} = \vec{E}(\alpha_1 \cdots \alpha_m, \vec{D}) \quad \text{or} \quad \vec{D} = \vec{D}(\alpha_1 \cdots \alpha_m, \vec{E}) \tag{1}$$

The α's are properties of the material. Thus, if measurements are made on a homogeneous sample of the material, the α's are varied by changing the composition of the sample. For example, α_1 might be the concentration of dipoles of a given species, or the concentration of one liquid in another. The number of α's used depends on the specific application. Most important for now is the distinction between changing $\vec{E}$ in Eq. 1 by changing the material and hence changing α's, and doing so by changing $\vec{D}$. Some special cases of Eq. 1 are given in Table 3.7.1.

Table 3.7.1. Constitutive laws having the general form of Eq. 1a.

Law	Description
$\vec{E} = \varepsilon^{-1}(\alpha_1 \cdots \alpha_m)\vec{D}$	Electrically linear and (fields) collinear
$E_i = s_{ij}(\alpha_1 \cdots \alpha_m)\vec{D}_j$	Electrically linear and anisotropic
$\vec{E} = \varepsilon^{-1}(\alpha_1 \cdots \alpha_m, D^2)\vec{D}$	Electrically nonlinear and (fields) collinear
$E_i = s_{ij}(\alpha_1 \cdots \alpha_m, D_1, D_2, D_3)D_j$	Electrically nonlinear and anisotropic

The third case of the table might represent a material in which dipoles are in Brownian equilibrium with a nonpolar liquid. An applied field tends to line up the dipoles and hence give rise to a polarization density and hence to a contribution to $\vec{D}$. In terms of two properties (α_1,α_2), a model including the saturation effect, resulting as all dipoles become aligned with the field, might be

$$\varepsilon = \frac{\alpha_1}{\sqrt{1 + \alpha_2^2\, \vec{E}\cdot\vec{E}}} + \varepsilon_o \tag{2}$$

Built into this example, and the general relation, Eq. 1, is the assumption that the constitutive law is a state function. It does not depend on rates of change, and it is a single-valued function of the variables and hence not dependent on the path followed to arrive at the given state.

The continuum now considered is not homogeneous, in that at any given instant the α's can vary from one position to another. Moreover, for the electromechanical subsystem considered, the properties are tied to the material. As the material moves, properties change. For material within a volume of fixed identity,

$$\int_V \alpha_i dV = \text{constant} \tag{3}$$

By definition, the volume V is always composed of the same material. By definition, the α's must satisfy Eq. 3 when the subsystem is considered to be isolated from other subsystems.

The finite number of mechanical degrees of freedom for the discrete coupling of Sec. 3.5 is now replaced by an infinite number of degrees of freedom. The mechanical continuum, perhaps a fluid, perhaps a solid, is capable of undergoing the vector deformations $\delta\vec{\xi}$. These incremental displacements are viewed as small departures from an equilibrium mechanical configuration which is precisely that for which the force density is required.

Since the time derivative of Eq. 3 vanishes, the generalized Leibnitz rule, Eq. 2.6.5, gives

$$\frac{d}{dt}\int_V \alpha_i dV = \int_V \frac{\partial \alpha_i}{\partial t} dV + \oint_S \alpha_i \frac{\partial \vec{\xi}}{\partial t}\cdot\vec{n}da = 0 \tag{4}$$

where by definition the velocity of the surface S is equal to that of the material $(\vec{v}_s \rightarrow \frac{\partial \vec{\xi}}{\partial t})$. Gauss' theorem converts the second integral to a volume integral. Although of fixed identity, the volume is arbitrary, and so it follows from Eq. 4 that changes in the property α_i are linked to the material deformations by an expression that is equivalent to Eq. 3:

$$\delta\alpha_i = -\nabla\cdot(\alpha_i \delta\vec{\xi}) \tag{5}$$

The framework has now been established for stating and exploiting conservation of energy for the electromechanical subsystem. The procedure is familiar from Sec. 3.5. With electrical excitations absent, a system, such as shown in Fig. 2.13.1, is assembled mechanically. Because the force density of electrical origin is by definition zero during the process, no work is required. The system now consists of rigid electrodes for producing part or all of the electrical excitations and a mechanical continuum in the intervening space. This material is described by Eq. 1. With the mechanical deformations fixed ($\delta\vec{\xi} = 0$), the electrical excitations are next raised by placing bulk charges at the positions of interest in the material and by raising the potentials on the electrodes. The result is a stored electrical energy given by Eq. 2.13.6:

$$w = \int_V W dV; \quad W = \int_0^{\vec{D}} \vec{E}(\alpha_1\cdots\alpha_m,\vec{D}')\cdot\delta\vec{D}' \tag{6}$$

Here, V is the volume occupied by the material and the fields, and hence excluding the electrodes.

Now, with the net charge on each electrode constrained to be constant, consider variations in the energy caused by incremental displacements of the material. A statement of energy conservation accounting for work done on the external mechanical world by the force density of electrical origin is

$$\int_V [\delta W + \vec{F}\cdot\delta\vec{\xi}]dV = 0 \tag{7}$$

There are two consequences of the incremental displacement. First, the mechanical deformation carries the properties with it, as already stated by Eq. 5. Second, there is a redistribution of the free charge. Because the system is conservative, the free charge is constrained to move with the material. The charge within a volume always composed of the same material particles is constant. Thus, Eq. 3 also holds with $\alpha_i \rightarrow \rho_f$, and it follows that an expression similar to Eq. 5 can be written for the change in charge density at a given location caused by the material displacement $\delta\vec{\xi}$:

$$\delta\rho_f = -\nabla\cdot(\rho_f\delta\vec{\xi}) \tag{8}$$

It is extremely important to recognize the difference between δW in Eq. 7, and δW in Sec. 2.13. In Eq. 7, the change in energy is caused by material displacements $\delta\vec{\xi}$, whereas in Sec. 2.13 it is due to changes in the electrical excitations. The energy W is assumed to be a state function of the same variables as used to express the constitutive law, Eq. 1. Hence,

$$\delta W = \sum_{i=1}^{m} \frac{\partial W}{\partial \alpha_i}\delta\alpha_i + \frac{\partial W}{\partial \vec{D}}\cdot\delta\vec{D} \tag{9}$$

where

$$\frac{\partial W}{\partial \vec{D}}\cdot\delta\vec{D} \equiv \sum_{i=1}^{3}\frac{\partial W}{\partial D_i}\delta D_i$$

With the understanding that the partial derivative is taken with the α's held fixed, it follows from Eq. 6 that

$$\frac{\partial W}{\partial D_j} = E_j \tag{10}$$

Hence, the last term in Eq. 9 is written using Eq. 10 with $\vec{E}$ in turn replaced by $-\nabla\Phi$. Then, integration by parts* gives

$$\int_V \frac{\partial W}{\partial \vec{D}} \cdot \delta\vec{D}dV = -\oint_S \Phi\delta\vec{D}\cdot\vec{n}da + \int_V \Phi(\nabla\cdot\delta\vec{D})dV \tag{11}$$

The part of the surface coincident with the electrode surfaces gives a contribution from each electrode equal to the electrode potential multiplied by the change in electrode charge. Because the electrode charges are held fixed while the material is deformed, this integration gives no contribution. The remaining part of the surface integration is sufficiently well removed from the region of interest that the fields have fallen off sufficiently to make a negligible contribution. Thus, the first term on the right vanishes and, because of Gauss' law, Eq. 11 becomes

$$\int \frac{\partial W}{\partial \vec{D}} \cdot \delta\vec{D}dV = \int \Phi\delta\rho_f dV \tag{12}$$

It is now possible to write Eq. 7 with effects of $\delta\vec{\xi}$ represented explicitly. Substitution of Eq. 8 into 12 and then Eqs. 12 and 5 into 9, and finally of Eq. 9 into 7, gives

$$\int_V [-\sum_{i=1}^{m} \frac{\partial W}{\partial \alpha_i} \nabla\cdot(\alpha_i\delta\vec{\xi}) - \Phi\nabla\cdot(\rho_f\delta\vec{\xi}) + \vec{F}\cdot\delta\vec{\xi}]dV = 0 \tag{13}$$

With the objective of writing the integrand in the form $(\quad)\cdot\delta\vec{\xi}$, the first two terms are integrated by parts. Because the surface integrations are either on the rigid electrode surfaces where $\delta\vec{\xi}\cdot\vec{n} = 0$, or at infinity where the fields have decayed to zero, and $\vec{E} = -\nabla\Phi$, Eq. 13 becomes

$$\int_V [\sum_{i=1}^{m} \alpha_i\nabla(\frac{\partial W}{\partial \alpha_i}) - \rho_f\vec{E} + \vec{F}]\cdot\delta\vec{\xi}dV = 0 \tag{14}$$

It is tempting, and in fact correct, to set the integrand of this expression to zero. But the justification is not that the volume V is arbitrary. To the contrary, the volume V is a special one enclosing all of the region occupied by the deformable medium and fields. (The volume integration plays the role of a summation over the mechanical variables for the lumped-parameter systems of Sec. 3.5.) The integrand is zero because $\delta\vec{\xi}$ (like the lumped-parameter displacements) is an independent variable. The equation must hold for any deformation, including one confined to any region where $\vec{F}$ is to be evaluated:

$$\vec{F} = \rho_f\vec{E} - \sum_{i=1}^{m} \alpha_i\nabla(\frac{\partial W}{\partial \alpha_i}) \tag{15}$$

It is most often convenient to write the second term so that it is clear that it consists of a force density concentrated where there are property gradients and the "gradient of a pressure":

$$\vec{F} = \rho_f\vec{E} + \sum_{i=1}^{m} \frac{\partial W}{\partial \alpha_i}\nabla\alpha_i - \nabla[\sum_{i=1}^{m} \alpha_i \frac{\partial W}{\partial \alpha_i}] \tag{16}$$

The implications of Eq. 16 and the method of its derivation are appreciated by considering three commonly encountered limiting cases and then writing Eq. 16 in such a way that its relation to the Kelvin force density is clear.

Incompressible Media: Deformations are then such that

$$\nabla\cdot\delta\vec{\xi} = 0 \tag{17}$$

Because $\delta\vec{\xi}\cdot\vec{n} = 0$ on the rigid electrode surfaces that comprise part of the surface S enclosing V in Eq. 7, any pressure function π that approaches zero with sufficient rapidity at infinity to make the surface integration there negligible will satisfy the relation

*Integration by parts in three dimensions amounts to

$$\int_V \Psi\nabla\cdot\vec{A}dV = \int_V \nabla\cdot(\Psi\vec{A})dV - \int_V \vec{A}\cdot\nabla\Psi dV = \oint_S \Psi\vec{A}\cdot\vec{n}da - \int_V \vec{A}\cdot\nabla\Psi dV$$

$$\oint_S \pi\delta\vec{\xi}\cdot\vec{n}da = \int_V \nabla\cdot(\pi\delta\vec{\xi})dV = 0 \tag{18}$$

Thus, Eq. 14 remains valid even if the volume integral of Eq. 18 is added to it. But, for incompressible deformations as defined with Eq. 17, $\nabla\cdot(\pi\delta\vec{\xi}) = \nabla\pi\cdot\delta\vec{\xi}$. Thus, the term added to Eq. 14, like those already appearing in its integrand, can be written with $\delta\vec{\xi}$ as a factor. It follows that for incompressible deformations, the gradient of any scalar pressure, π, can be added to the force density of Eq. 16. For example, π might be $\vec{P}\cdot\vec{E}$, since this function decays with distance from the system sufficiently rapidly to make the contribution of the surface integration at infinity vanish. On the basis of this apparent arbitrariness in the force density, the following observation is now made for the first time, and will be emphasized again in Chap. 8. Two force densities differing by the gradient of a scalar pressure will give rise to the same incompressible deformations. Physically this is so because in modeling a continuum as incompressible, the pressure becomes a "left-over" variable. It becomes whatever it must be to make Eq. 17 valid. Whatever the $\nabla\pi$ added to the force density of electrical origin, π can be absorbed into the "mechanical" pressure of the continuum-force equation.

For incompressible deformations, where the force density is arbitrary to within the gradient of a pressure, the gradient term can be omitted from Eq. 16, which then takes the convenient form

$$\vec{F} = \rho_f\vec{E} + \sum_{i=1}^{m}\frac{\partial W}{\partial \alpha_i}\nabla\alpha_i \tag{19}$$

This expression concentrates the force density where there are property gradients. In a charge-free system composed of regions having uniform properties, the force density is thus confined to interfaces between regions.

<u>Incompressible and Electrically Linear</u>: For an incompressible material having the constitutive law

$$\vec{D} = \varepsilon_o(1 + \chi_e)\vec{E} \equiv \varepsilon\vec{E} \tag{20}$$

the susceptibility χ_e is conserved by a volume of fixed identity. That is, α_1 can be taken as χ_e in Eq. 3 and m = 1. Then, from Eq. 6,

$$W = \frac{1}{2}\frac{D^2}{\varepsilon_o(1+\chi_e)};\ \frac{\partial W}{\partial \chi_e} = -\frac{\varepsilon_o}{2}E^2 \tag{21}$$

and because $\nabla\chi_e = \nabla[(1+\chi_e)]$, it follows that the force density of Eq. 19 specializes to

$$\vec{F} = \rho_f\vec{E} - \frac{1}{2}E^2\nabla\varepsilon \tag{22}$$

<u>Electrically Linear with Polarization Dependent on Mass Density Alone</u>: Certainly a possible parameter α_1 is the mass density ρ, since then Eq. 3 is satisfied. For a compressible medium it is possible that the susceptibility χ_e in Eq. 20 is only a function of ρ. Then,

$$\alpha_1 = \rho,\ \chi_e = \chi_e(\rho),\ W = \frac{1}{2}\frac{D^2}{\varepsilon_o[1+\chi_e(\rho)]};\ \frac{\partial W}{\partial\rho} = -\frac{\varepsilon_o}{2}E^2\frac{\partial\chi_e}{\partial\rho} \tag{23}$$

and, because $(\partial\varepsilon/\partial\rho)\nabla\rho = \nabla\varepsilon$, the force density given by Eq. 16 becomes

$$\vec{F} = \rho_f\vec{E} - \frac{1}{2}E^2\nabla\varepsilon + \nabla[\frac{1}{2}\rho\frac{\partial\varepsilon}{\partial\rho}E^2] \tag{24}$$

Because the last term is associated with volumetric changes in the material, it is called the electrostriction force density.

<u>Relation to the Kelvin Force Density</u>: Because $W = W(\alpha_1,\alpha_2\cdots\alpha_m,\vec{D})$, the kth component of the gradient of W is

$$(\nabla W)_k = \sum_{i=1}^{m}\frac{\partial W}{\partial\alpha_i}\frac{\partial\alpha_i}{\partial x_k} + \frac{\partial W}{\partial D_j}\frac{\partial D_j}{\partial x_k} \tag{25}$$

In view of Eq. 10, it follows that

$$\sum_{i=1}^{m}\frac{\partial W}{\partial\alpha_i}\frac{\partial\alpha_i}{\partial x_k} = \frac{\partial W}{\partial x_k} - \frac{\partial}{\partial x_k}(\vec{E}\cdot\vec{D}) + D_j\frac{\partial E_j}{\partial x_k} \tag{26}$$

This expression can be substituted for the second term in Eq. 16, which with some manipulation then becomes

$$\vec{F} = \rho_f\vec{E} + \vec{P}\cdot\nabla\vec{E} + \nabla[\frac{1}{2}\varepsilon_o\vec{E}\cdot\vec{E} + W - \vec{E}\cdot\vec{D} - \sum_{i=1}^{m}\alpha_i\frac{\partial W}{\partial\alpha_i}] \tag{27}$$

In this form, the force density is the sum of a free charge force density, the Kelvin force density (Eq. 3.6.5) and the gradient of a pressure. This last term can consistently be ignored in predicting the deformations of an incompressible continuum. For such situations, the Kelvin force density or the Korteweg-Helmholtz force density in the form of Eq. 19 will give rise to the same deformations. Note that they have very different distributions.

Apparently the last term in Eq. 27 represents the interaction between dipoles omitted from the derivation of the Kelvin force density. In fact, this term vanishes when the constitutive law takes a form consistent with the polarization being due to noninteracting dipoles. In that case, the susceptibility should be linear in the mass density so that $\chi_e = c\rho$, where c is a constant. In Eq. 23, $\partial\chi_e/\partial\rho = c$, and evaluation shows that, indeed, the last term in Eq. 27 does vanish.

3.8 Magnetic Korteweg-Helmholtz Force Density

Thermodynamic techniques for determining the magnetization force density are analogous to those outlined for the polarization force density in Sec. 3.7. In fact, if there were no free current density, the magnetic field intensity, like the electric field intensity, would be irrotational. It would then be possible to make a derivation that would be the complete analog of that for the polarization force density. However, in the following the force density due to free currents is included and hence $\vec{H}$ is not irrotational.

The constitutive law takes the form

$$\vec{H} = \vec{H}(\alpha_1,\alpha_2\cdots\alpha_m,\vec{B}) \text{ or } \vec{B} = \vec{B}(\alpha_1,\alpha_2\cdots\alpha_m,\vec{H}) \tag{1}$$

with specific possibilities given in Table 3.7.1 with $\varepsilon \to \mu$, $\vec{E} \to \vec{H}$ and $\vec{D} \to \vec{B}$. A conservative electromechanical subsystem is assembled mechanically, with no electrical excitations, so that it assumes a configuration identical to the one for which the force density is required. By the definition of the subsystem, this process requires no energy. Then, with the mechanical system fixed (the α's fixed), electrical excitations are applied so as to establish the free currents in excitation coils and in the medium itself, with the distribution that for which the force density is required. This procedure is formalized in Sec. 2.12 and a system schematic is shown in Fig. 2.14.2. As was shown in Sec. 2.14, currents in excitation coils are conveniently regarded as part of the total distribution of free current density. Hence, the volume of interest now includes all of the region permeated by the magnetic field.

Now, with the electrical excitations established, a statement of conservation of energy, with the electrical excitations held fixed but the material undergoing an incremental displacement, is Eq. 3.7.7, where now W is the magnetic energy density given from Eq. 2.14.10 by

$$W = \int_o^{\vec{B}} \vec{H}(\alpha_1,\alpha_2\cdots\alpha_m,\vec{B}')\cdot\delta\vec{B}' \tag{2}$$

The following steps, leading to a deduction of the force density, are analogous to those taken in Sec. 3.7. The link between the α's and $\delta\xi$ is given by Eq. 3.7.5. What is the connection between $\vec{J}_f$ and $\delta\vec{\xi}$?

Actually, it is a link between the flux linkage and $\vec{\xi}$ that is appropriate. If the medium is to both support a free current density and be conservative, the material must be idealized as having an infinite conductivity. This means that any open material surface S (surface of fixed identity) must link a constant flux:

$$\delta\int_S \vec{B}\cdot\vec{n}da = 0 \tag{3}$$

One way to make this deduction is to use the integral form of Faraday's law for a contour C enclosing a surface S of fixed identity, Eq. 2.7.3b, with $\vec{v} = \vec{v}_s$. Because the medium is perfectly conducting, $\vec{E}' = 0$ and what remains of Faraday's law is Eq. 3. From the generalized Leibnitz rule, Eq. 2.6.4, Eq. 3 and the solenoidal nature of $\vec{B}$ require that

$$\int_S \delta\vec{B}\cdot\vec{n}da + \oint_C(\vec{B}\times\delta\vec{\xi})\cdot\vec{d\ell} = 0 \tag{4}$$

Stokes's theorem, Eq. 2.6.3, converts the contour integral to a surface integral. Because this surface is arbitrary, the sum of the integrands must vanish. If it is further recognized that $\delta\vec{B} = \nabla \times \delta\vec{A}$, then it follows that

$$\delta\vec{A} = \delta\vec{\xi} \times \vec{B} \tag{5}$$

Thus, there is established the link between material deformations and the alterations of the field that are required if the deformations are to be flux-conserving.

The change in W associated with the material deformation, called for in the conservation of energy equation, Eq. 3.7.7, is in general

$$\delta W = \sum_{i=1}^{n} \frac{\partial W}{\partial \alpha_i} \delta\alpha_i + \frac{\partial W}{\partial \vec{B}} \cdot \delta\vec{B} \tag{6}$$

where, in view of Eq. 2,

$$\frac{\partial W}{\partial B_j} = H_j \tag{7}$$

It is the integral over the total volume V of δW that is of interest. The integral of the last term in Eq. 6 is

$$\int_V \frac{\partial W}{\partial \vec{B}} \cdot \delta\vec{B}dV = \int_V \vec{H}\cdot\delta\vec{B}dV = \int_V \vec{H}\cdot\nabla \times \delta\vec{A}dV \tag{8}$$

Because the fields decay to zero sufficiently rapidly at infinity that the surface integral vanishes and because Ampere's law, Eq. 2.3.23b, gives $\nabla \times \vec{H} = \vec{J}_f$, integration of the last term in Eq. 8 by parts gives

$$\int_V \frac{\partial W}{\partial \vec{B}} \cdot \delta\vec{B}dV = \int_V \nabla\cdot(\delta\vec{A} \times \vec{H})dV + \int_V \delta\vec{A}\cdot\nabla \times \vec{H}dV = \oint_S \delta\vec{A} \times \vec{H}\cdot\vec{n}da + \int_V \vec{J}\cdot\delta\vec{A}dV = \int_V \delta\vec{A}\cdot\vec{J}_f dV \tag{9}$$

Substitution for $\delta\vec{A}$ from Eq. 5 finally gives an expression explicitly showing the $\vec{\xi}$ dependence:

$$\int_V \frac{\partial W}{\partial \vec{B}} \cdot \delta\vec{B}dV = \int_V \delta\vec{\xi} \times \vec{B}\cdot\vec{J}_f dV = -\int_V \vec{J}_f \times \vec{B}\cdot\delta\vec{\xi}dV \tag{10}$$

Finally, the energy conservation statement, Eq. 3.7.7, is written with δW given by Eq. 6 and in turn, $\delta\alpha_i$ given by Eq. 3.7.5 and the last term given by Eq. 10:

$$\int_V [-\sum_{i=1}^{n} \frac{\partial W}{\partial \alpha_i} \nabla\cdot(\delta\vec{\xi}\alpha_i) - \vec{J}_f \times \vec{B}\cdot\delta\vec{\xi} + \vec{F}\cdot\delta\vec{\xi}]dV = 0 \tag{11}$$

With the objective of writing the first term as a dot product with $\delta\vec{\xi}$, the first term is integrated by parts (exactly as in going from Eq. 3.7.13 to Eq. 3.7.14) to obtain

$$\int_V [\sum_{i=1}^{n} \alpha_i \nabla \frac{\partial W}{\partial \alpha_i} - \vec{J}_f \times \vec{B} + \vec{F}]\cdot\delta\vec{\xi}dV = 0 \tag{12}$$

The integrand must be zero, not because the volume is arbitrary (it includes all of the system involved in the electromechanics) but rather because the virtual displacements $\delta\vec{\xi}$ are arbitrary in their distribution. Hence, the force density is

$$\vec{F} = \vec{J}_f \times \vec{B} - \sum_{i=1}^{n} \alpha_i \nabla \frac{\partial W}{\partial \alpha_i} \tag{13}$$

The special cases considered in Sec. 3.7 have analogs that similarly follow from Eq. 13. Because what is involved in deriving these forms involves the magnetization term in Eq. 13, and not the free current force density, these expressions can be written down by direct analogy.

Incompressible Media: The convenient form emphasizing the importance of regions where there are property gradients is

$$\vec{F} = \vec{J}_f \times \vec{B} + \sum_{i=1}^{n} \frac{\partial W}{\partial \alpha_i} \nabla\alpha_i \tag{14}$$

Incompressible and Electrically Linear: With a constitutive law

$$\vec{B} = \mu_o(1 + \chi_m)\vec{H} = \mu\vec{H} \tag{15}$$

the force density of Eq. 13 reduces to

$$\vec{F} = \vec{J}_f \times \vec{B} - \frac{1}{2} H^2\nabla\mu \tag{16}$$

Electrically Linear with Magnetization Dependent on Mass Density Alone: With the constitutive law in the form of Eq. 15, but $\chi_m = \chi_m(\rho)$, where ρ is the mass density, the force density is the sum of Eq. 14 and a magnetostrictive force density taking the form of the gradient of a pressure:

$$\vec{F} = \vec{J}_f \times \vec{B} - \frac{1}{2} H^2\nabla\mu + \nabla\left(\frac{1}{2} \rho \frac{\partial\mu}{\partial\rho} H^2\right) \tag{17}$$

Relation to Kelvin Force Density: With the stipulation that $W = W(\alpha_1,\alpha_2\cdots\alpha_m,\vec{B})$ is a state function, Eq. 13 becomes the sum of a Lorentz force density due to the free current density, the Kelvin force density and the gradient of a pressure:

$$\vec{F} = \vec{J}_f \times \mu_o\vec{H} + \mu_o\vec{M}\cdot\nabla\vec{H} + \nabla\left[\frac{1}{2}\mu_o\vec{H}\cdot\vec{H} + W - \vec{H}\cdot\vec{B} - \sum_{i=1}^{m} \alpha_i \frac{\partial W}{\partial \alpha_i}\right] \tag{18}$$

The discussion of Sec. 3.7 is as appropriate for understanding these various forms of the magnetic force density as it is for the electric force density.

3.9 Stress Tensors

Most of the force densities of concern in this text can be written as the divergence of a stress tensor. The representation of forces in terms of stresses will be used over and over again in the chapters which follow. This section is intended to give a brief summary of the differential and integral properties of the stress tensor.

Suppose that the ith component of a force density can be written in the form

$$F_i = \frac{\partial T_{ij}}{\partial x_j}; \quad (\vec{F} = \nabla\cdot\vec{T}) \tag{1}$$

Here, the Einstein summation convection is applicable, so that because the j's appear twice in the same term, they are to be summed from one to three. An alternative notation, in parentheses, represents the same operation in vector notation. Much of the convenience of recognizing the stress tensor representation of a force density comes from then being able to convert an integration of the force density over a volume to an integration of the stress tensor over a surface enclosing the volume. This generalization of Gauss' theorem is easily shown by fixing attention on the ith component (think of i as given) and defining a vector such that

$$\vec{G}_i = T_{i1}\vec{i}_1 + T_{i2}\vec{i}_2 + T_{i3}\vec{i}_3 \tag{2}$$

Then the right-hand side of Eq. 1 is simply the divergence of $\vec{G}_i$. Gauss' theorem then shows that

$$\int_V F_i dV = \int_V \nabla\cdot\vec{G}_i dV = \oint_S \vec{G}_i\cdot\vec{n}da \tag{3}$$

or, in index notation and using the definition of $\vec{G}_i$ from Eq. 2,

$$\int_V F_i dV = \oint_S T_{ij}n_j da \tag{4}$$

This tensor form of Gauss' theorem is the integral counterpart of Eq. 1. Physically, Eq. 4 states that an alternative to integrating the force density in some Cartesian direction over the volume V is an integration of the integrand on the right over a surface completely enclosing that volume V. The integrand of the surface integral can therefore be interpreted as a force/unit area acting on the enclosing surface in the ith direction. To distinguish it from a surface force density, it will be referred to as the "traction." It does not act on a physical surface and has physical significance only when integrated over a closed surface. It is simply the force/unit area that must be integrated over the entire surface to find the net force due to the volume force density

$$T_i = T_{ij}n_j; \quad \vec{T} = \vec{\vec{T}}\cdot\vec{n} \tag{5}$$

In vector notation and in terms of the traction $\vec{T}$, Eq. 4 is written as

$$\int_V \vec{F}dV = \oint_S \vec{T}\cdot\vec{n}da \tag{6}$$

Figure 3.9.1 shows the general relationship of the traction and normal vector. The traction can act in an arbitrary direction relative to the surface.

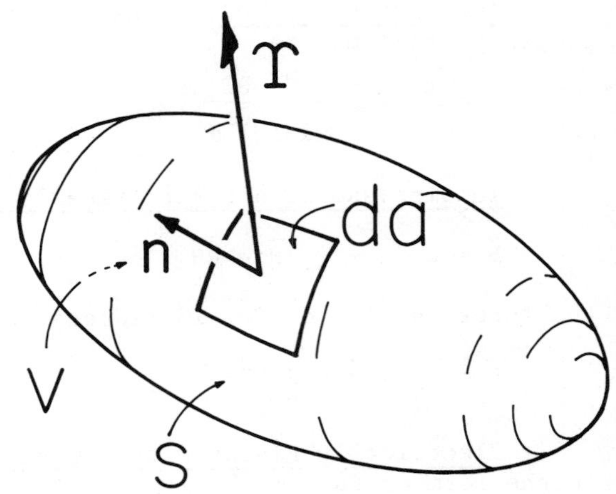

Fig. 3.9.1. Schematic view of volume V enclosed by surface S, showing traction acting on elements of surface.

To develop a physical interpretation of the stress tensor components, it is helpful to consider a particular volume V and surface S with surfaces having normals in the Cartesian coordinate directions. The cube shown in Fig. 3.9.2 is such a volume. Suppose that interest is in determining the net force on the cube in the x direction, from Eq. 4. The required surface integration can then be broken into separate integrations over each of the cube's surfaces. For the integration on the right face, the normal vector has only an x component, so the only contribution to that surface integration is from T_{xx}. Similarly, on the left surface, the normal vector is in the -x direction, and the integral over that surface is of $-T_{xx}$. The minus sign is represented by directing the stress arrow in the minus x direction in Fig. 3.9.2. On the top and bottom surfaces, the normal vector is in the y direction, and the integration is of plus and minus T_{xy}. Similarly, on the front and back surfaces, the only terms contributing to the traction are T_{xz}. The stress tensor components represent normal stresses if the indices are equal, and shear stresses if they are unequal. In either case, the stress component acting in the ith direction on a surface having its normal in the jth direction is T_{ij}.

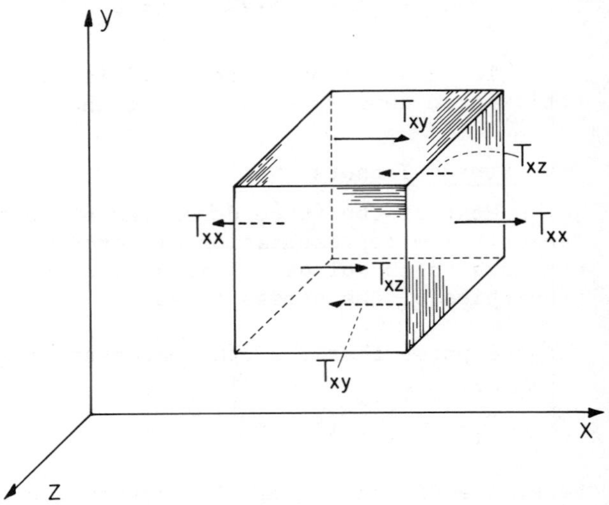

Fig. 3.9.2. Stress components acting on cube in the x direction.

Orthogonal components are a familiar way of representing a vector $\vec{F}$. In the coordinate system (x_1,x_2,x_3) the components are denoted by F_j. What is meant by a vector is implicit to how these components decompose into the components of the vector expressed in a second orthogonal coordinate system $(x_1',x_2'.x_3')$ pictured in Fig. 3.9.3. The two coordinate systems are related by the transformation

$$x_k' = a_{k\ell}x_\ell; \quad \frac{\partial x_k'}{\partial x_\ell} = a_{k\ell} \tag{7}$$

where $a_{k\ell}$ is the cosine of the angle between the x_k' axis and the x_ℓ axis.

A component of the vector in the primed frame in the ith direction is then given by

$$F_i' = a_{ij}F_j \tag{8}$$

For example, suppose that i = 1. Then, Eq. 8 gives the x_1' component of $\vec{F}'$ as the projections of the components in the x_1, x_2, x_3 directions onto the x_1' direction. Equation 8 summarizes how a vector transforms from one coordinate system onto another, and could be used to define what is meant by a "vector."

Similarly, the components of a tensor transform from the unprimed to the primed coordinate system in a way that can be used to define what is meant by a "tensor." To deduce the transformation, begin with Eq. 8 using the divergence of a stress tensor to represent each of the force densities (Eq. 1):

$$\frac{\partial T'_{ik}}{\partial x'_k} = a_{ij}\frac{\partial T_{i\ell}}{\partial x_\ell} \tag{9}$$

Now, if use is made of the chain rule for differentiation, and Eq. 7, it follows that

$$\frac{\partial T'_{ik}}{\partial x'_k} = a_{ij}\frac{\partial T_{j\ell}}{\partial x'_k}\frac{\partial x'_k}{\partial x_\ell} = a_{ij}a_{k\ell}\frac{\partial T_{j\ell}}{\partial x'_k} \tag{10}$$

Thus, the tensor transformation follows as

$$T'_{ik} = a_{ij}a_{k\ell}T_{j\ell} \tag{11}$$

Useful conditions on the direction cosines a_{ij} are obtained by recognizing that the transformation from the primed frame to the unprimed frame, given generally by

$$F_j = b_{ji}F'_i \tag{12}$$

involves the same direction cosines, because b_{ji}, defined as the cosine of the angle between the x_j axis and the x'_i axis, is equal to a_{ij}. Thus, Eqs. 12 and 8 together show that

$$F'_i = a_{ik}F_k = a_{ik}a_{\ell k}F'_\ell \tag{13}$$

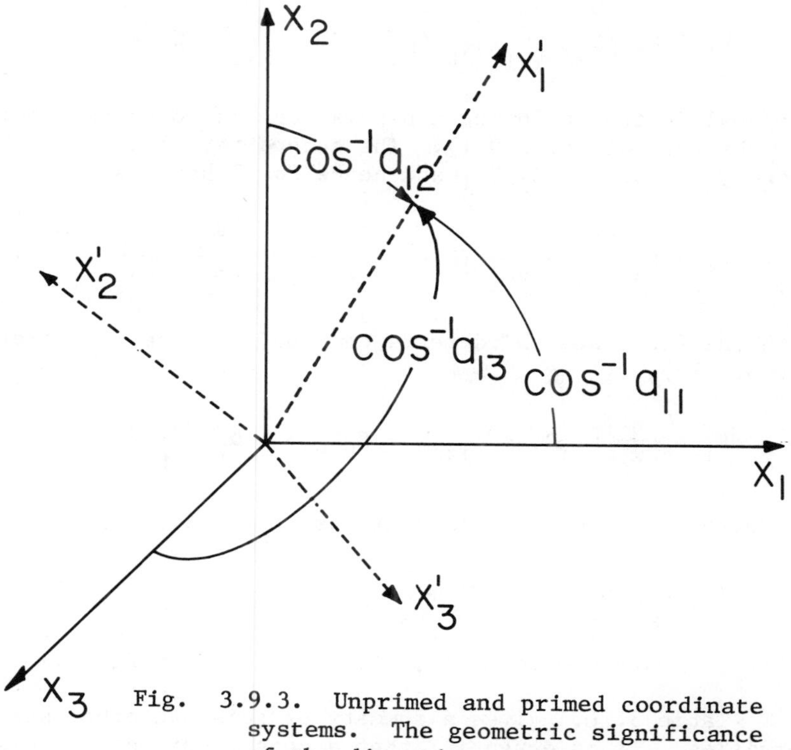

Fig. 3.9.3. Unprimed and primed coordinate systems. The geometric significance of the direction cosine a_{ij} is shown.

and it follows that the direction cosines satisfy the condition that

$$a_{ik}a_{\ell k} = \delta_{i\ell} \tag{14}$$

where the Kronecker delta function δ_{ik} by definition takes the values

$$\delta_{ik} = \begin{cases} 1 & i = k \\ 0 & i \neq k \end{cases} \tag{15}$$

Finally, suppose that a total torque rather than a total force is to be computed. By way of analogy to Eq. 6, is there a way in which the integration of the torque density can be converted to an integration over the enclosing surface? With respect to the origin, the total torque on material within the volume V is

$$\vec{\tau} = \int_V \vec{r} \times \vec{F}dV \tag{16}$$

where $\vec{r}$ is the vector distance from the origin. With $\vec{F}$ given as the divergence of a stress tensor, Eq. 1, and provided that $\bar{\bar{T}}$ is symmetric ($T_{ij} = T_{ji}$), the tensor form of Gauss' theorem can be used to show that

$$\vec{\tau} = \oint_S \vec{r} \times (\vec{T}\cdot\vec{n})da \tag{17}$$

The net torque is the integral over the enclosing surface of a surface torque density $\vec{r} \times \vec{T}$ (see Problem 3.9.1).

3.10 Electromechanical Stress Tensors

The objectives in this section are to illustrate how the stress tensor associated with any one of the force densities in Secs. 3.7 and 3.8 is determined, and to summarize the stress tensors for future reference.

The ith component of the Korteweg-Helmholtz force density, Eq. 3.7.16, written using Gauss' law to eliminate ρ_f, is

$$F_i = E_i \frac{\partial D_j}{\partial x_j} + \sum_{k=1}^{m} \frac{\partial W}{\partial \alpha_k}\frac{\partial \alpha_k}{\partial x_i} - \frac{\partial}{\partial x_i}\left[\sum_{k=1}^{m} \alpha_k \frac{\partial W}{\partial \alpha_k}\right] \tag{1}$$

The goal in the following manipulations is to express this equation in the form of a tensor divergence (in the form of Eq. 3.9.1). The second term can be replaced by Eq. 3.7.26. Also, because $\vec{E}$ is irrotational, $\partial E_i/\partial x_j = \partial E_j/\partial x_i$ and hence Eq. 1 becomes

$$F_i = E_i \frac{\partial D_j}{\partial x_j} + \frac{\partial}{\partial x_i}(W - E_k D_k) + D_j \frac{\partial E_i}{\partial x_j} - \frac{\partial}{\partial x_i}\left[\sum_{k=1}^{m} \alpha_k \frac{\partial W}{\partial \alpha_k}\right] \tag{2}$$

With the first and third terms combined and the Kronecker delta function δ_{ij} introduced (see Eq. 3.9.15),

$$F_i = \frac{\partial}{\partial x_j}\left[E_i D_j + \delta_{ij}\left(W - E_k D_k - \sum_{k=1}^{m} \alpha_k \frac{\partial W}{\partial \alpha_k}\right)\right] \tag{3}$$

It follows from a comparison of Eqs. 2 and 3.9.1 that the required stress tensor is

$$T_{ij} = E_i D_j - \delta_{ij}\left(W' + \sum_{k=1}^{m} \alpha_k \frac{\partial W}{\partial \alpha_k}\right) \tag{4}$$

where the coenergy density, W', is defined by Eq. 2.13.11.

Table 3.10.1 gives a summary of this and other stress tensors together with the associated force densities. It is essential that a consistent pair be used.

Table 3.10.1. Summary of force densities and associated stress tensors.

Equation	Force density	Stress tensors
3.7.16	$\vec{F} = \rho_f \vec{E} + \sum_{k=1}^{m} \frac{\partial W}{\partial \alpha_k} \nabla \alpha_k - \nabla\left[\sum_{k=1}^{m} \alpha_k \frac{\partial W}{\partial \alpha_k}\right]$	$T_{ij} = E_i D_j - \delta_{ij}\left(W' + \sum_{k=1}^{m} \alpha_k \frac{\partial W}{\partial \alpha_k}\right)$
3.8.13	$\vec{F} = \vec{J}_f \times \vec{B} + \sum_{k=1}^{m} \frac{\partial W}{\partial \alpha_k} \nabla \alpha_k - \nabla\left[\sum_{k=1}^{m} \alpha_k \frac{\partial W}{\partial \alpha_k}\right]$	$T_{ij} = H_i B_j - \delta_{ij}\left(W' + \sum_{k=1}^{m} \alpha_k \frac{\partial W}{\partial \alpha_k}\right)$
	Incompressible media	
3.7.19	$\vec{F} = \rho_f \vec{E} + \sum_{k=1}^{m} \frac{\partial W}{\partial \alpha_k} \nabla \alpha_k$	$T_{ij} = E_i D_j - \delta_{ij} W'$
3.8.14	$\vec{F} = \vec{J}_f \times \vec{B} + \sum_{k=1}^{m} \frac{\partial W}{\partial \alpha_k} \nabla \alpha_k$	$T_{ij} = H_i B_j - \delta_{ij} W'$
	Incompressible and electrically linear: $\vec{D} = \varepsilon \vec{E}, \vec{B} = \mu \vec{H}$	
3.7.22	$\vec{F} = \rho_f \vec{E} - \frac{1}{2} E^2 \nabla \varepsilon$	$T_{ij} = \varepsilon E_i E_j - \frac{\varepsilon}{2} \delta_{ij} E_k E_k$
3.8.14	$\vec{F} = \vec{J}_f \times \vec{B} - \frac{1}{2} H^2 \nabla \mu$	$T_{ij} = \mu H_i H_j - \frac{\mu}{2} \delta_{ij} H_k H_k$
	Electrically linear, ε and μ dependent on mass density ρ only	
3.7.24	$\vec{F} = \rho_f \vec{E} - \frac{1}{2} E^2 \nabla \varepsilon + \nabla\left(\frac{1}{2} \rho \frac{\partial \varepsilon}{\partial \rho} E^2\right)$	$T_{ij} = \varepsilon E_i E_j - \frac{\varepsilon}{2} \delta_{ij} E_k E_k \left(1 - \frac{\rho}{\varepsilon}\frac{\partial \varepsilon}{\partial \rho}\right)$
3.8.17	$\vec{F} = \vec{J}_f \times \vec{B} - \frac{1}{2} H^2 \nabla \mu + \nabla\left(\frac{1}{2} \rho \frac{\partial \mu}{\partial \rho} H^2\right)$	$T_{ij} = \mu H_i H_j - \frac{\mu}{2} \delta_{ij} H_k H_k \left(1 - \frac{\rho}{\mu}\frac{\partial \mu}{\partial \rho}\right)$
	Kelvin force density and stress tensor	
3.6.5	$\vec{F} = \rho_f \vec{E} + \vec{P} \cdot \nabla \vec{E}$	$T_{ij} = E_i D_j - \frac{1}{2} \delta_{ij} \varepsilon_o E_k E_k$
3.5.12	$\vec{F} = \vec{J}_f \times \mu_o \vec{H} + \mu_o \vec{M} \cdot \nabla \vec{H}$	$T_{ij} = H_i B_j - \frac{1}{2} \delta_{ij} \mu_o H_k H_k$

The stress tensor makes it possible to compute the total force on an object by integrating over an enclosing surface S in accordance with Eq. 3.9.6. For an isolated object in free space, this force is the same regardless of the particular force density used. If the force is considered as the integral of the force density over the volume of the object, this fact is by no means obvious. But, note that in free space the stress tensors of Table 3.10.1 all agree. Because the enclosing surface S is in this free space region, the same total force will result from integrating Eq. 3.9.6 regardless of the force density associated with the stress tensor.

3.11 Surface Force Density

In many systems, the electric or magnetic force density is concentrated in a thin layer, usually comprising the interface between two regions. If the thickness of this layer is small compared to the dimensions of the adjacent regions and other lengths of interest, then the force per unit area on the interface may be used to describe the layer. An interfacial section is enclosed by the incremental volume of thickness Δ and area $A = \delta x \delta y$, shown in Fig. 3.11.1. The surface force density is defined as a force per unit area of the interface in a limit in which first Δ and then A approach zero. The integration of the electric force density throughout the control volume is conveniently carried out using the appropriate stress tensor T_{ij} integrated over the enclosing surface. With $\vec{n}$ defined as the unit normal to the interface and $\vec{i}_n$ the unit normal to the control surface, the surface force density is

$$\vec{T} = \lim_{\substack{\Delta\to 0\\ A\to 0}} \frac{1}{A} \oint_S \vec{\vec{T}}\cdot\vec{i}_n da = [\![\vec{\vec{T}}]\!] \cdot \vec{n} + \lim_{A\to 0} \frac{1}{A} \oint_C \int_{0^-}^{0^+} \vec{\vec{T}}\cdot\vec{i}_n d\nu d\ell \qquad (1)$$

Integration is divided into two parts. The first is the contribution from the surfaces external to the layer, having normals $\vec{n}$ and $-\vec{n}$, respectively. The second accounts for the "edges" of the volume where the surface cuts through the double layer. If fields within the layer are of the same order as those outside, contributions of the second integral vanish as $\Delta \to 0$. In electroquasistatic systems, the double layer presents a case where the internal fields are sufficiently intense that the second term not only makes a contribution but one that can dominate the first term. The remainder of this section is devoted to converting this contribution to a more useful form.

The distance normal to the interface is y, with (μ,ξ) orthogonal coordinates in the local interfacial plane, as shown in Fig. 3.11.1. In the absence of a double layer, the electric field is of the same order of magnitude throughout, and hence in the limit $\Delta \to 0$, the second term in Eq. 1 becomes negligible compared to the first. With the double layer, the stress contributions from the edges of the control volume are of the same order as those from the exterior surfaces.

As discussed in Sec. 2.10, the tangential electric field suffers a discontinuity through the double layer. However, the tangential field within the layer is of the same order as the external field. Because the thickness Δ over which the interior stresses act is much smaller than the linear dimensions $\delta\xi$ and $\delta\mu$, the internal stress contributions to the integrations around the periphery of the control volume are ignorable unless the double-layer charges are themselves responsible for a substantially larger internal field than external field. This double-layer-generated field is directed normal to the interface and dominates in determining the interior stresses. The stress taken now as represented by Eq. 3.7.19b of Table 3.10.1 is

$$T_{ij} = E_i D_j - \delta_{ij} W' \qquad (2)$$

where, in the case of a linearly polarized dielectric, the coenergy density W' is simply $\varepsilon E^2/2$. Stress components associated with the dominant field in the double layer interior are essentially

$$\begin{aligned} &T_{\xi\xi} \to T_{\mu\mu} \to -W' \\ &T_{ij} \to 0; \quad i \neq j \end{aligned} \qquad (3)$$

The traction acting on the periphery of the control volume is therefore approximately

$$\int_{0^-}^{0^+} T\cdot\vec{i}_n d\nu = -\int_{0^-}^{0^+} W' d\nu \vec{i}_n \equiv \gamma_E \vec{i}_n \qquad (4)$$

The normal vector $\vec{i}_n$ can be written as $-\vec{n}x d\vec{\ell}$, so that Eq. 1 becomes

$$\vec{T} = [\![\vec{\vec{T}}]\!] \cdot \vec{n} - \lim_{A\to 0} \frac{1}{A} \oint_C \gamma_E \vec{n} x d\vec{\ell}$$

In the limit $A\to 0$, the contour integral in Eq. 5 need only be evaluated to first order in $\delta\xi\cdot\delta\mu$. Expansion about the origin, denoted by the subscript o, gives an approximate expression for the integral

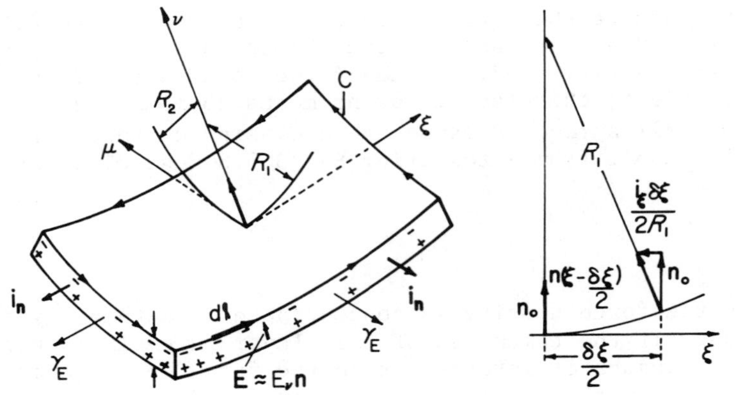

Fig. 3.11.1

(a) Volume enclosing section of interface. Thickness Δ is sufficient to include double layer but small compared to linear dimensions of A. (b) Cross-sectional view of interface showing relation of radius of curvature R to n and dℓ.

that becomes exact in the limit. The contour C is taken as rectangular with edges parallel to the (ξ,μ) axes. The segment of length $\delta\mu$ at $\xi = \delta\xi/2$ has $-\vec{n}\times d\vec{\ell} \approx \delta\mu(\vec{i}_\xi + \vec{n}_o\delta\xi/R_1)$ and gives a contribution to the contour integral

$$\left\{[\gamma_E]_o + [\frac{\partial\gamma_E}{\partial\xi}]_o \frac{\delta\xi}{2}\right\}\left\{\vec{i}_\xi + \frac{\vec{n}_o\delta\xi}{R_1}\right\}\delta\mu \tag{6}$$

The three additional sides of the rectangular contour give similar contributions, so that altogether,

$$-\lim_{A\to 0}\frac{1}{A}\oint_C \gamma_E\vec{n}\times d\vec{\ell} = \lim_{\delta\xi\delta\mu\to 0}\frac{1}{\delta\xi\delta\mu}\Bigg\{\left\{[\gamma_E]_o + [\frac{\partial\gamma_E}{\partial\xi}]_o\frac{\delta\xi}{2}\right\}\left\{\vec{i}_\xi + \frac{\vec{n}_o}{R_1}\frac{\delta\xi}{2}\right\}\delta\mu$$

$$+\left\{[\gamma_E]_o - [\frac{\partial\gamma_E}{\partial\xi}]_o\frac{\delta\xi}{2}\right\}\left\{-\vec{i}_\xi + \frac{\vec{n}_o}{R_1}\frac{\delta\xi}{2}\right\}\delta\mu + \left\{[\gamma_E]_o + [\frac{\partial\gamma_E}{\partial\mu}]_o\frac{\delta\mu}{2}\right\}\left\{\vec{i}_\mu + \frac{\vec{n}_o}{R_2}\frac{\delta\mu}{2}\right\}\delta\xi$$

$$+\left\{[\gamma_E]_o - [\frac{\partial\gamma_E}{\partial\mu}]_o\frac{\delta\mu}{2}\right\}\left\{-\vec{i}_\mu + \frac{\vec{n}_o}{R_2}\frac{\delta\mu}{2}\right\}\delta\xi\Bigg\}$$

$$= \vec{n}\gamma_E[\frac{1}{R_1} + \frac{1}{R_2}] + \nabla_\Sigma\gamma_E \tag{7}$$

Here, R_1 and R_2 are radii of curvature for the interface, reckoned in the orthogonal planes defined respectively by the normal and ξ and the normal and μ. Note that the sign of each curvature term is taken as positive if the center of curvature is on the side of the interface toward which $\vec{n}$ is directed. The surface force density associated with surface tension takes this same form. However, the convention used in Chap. 7 is with the radii of curvature the negatives of R_1 and R_2. With the understanding that R_1 and R_2 are radii of curvature taken as positive if the center of curvature is on the side of the interface out of which $\vec{n}$ is directed, Eqs. 1, 4, and 7 give the surface force density, with the double-layer contribution represented by the function γ_E,

$$\vec{T} = [\![\vec{\vec{T}}]\!]\cdot\vec{n} - \vec{n}\gamma_E[\frac{1}{R_1} + \frac{1}{R_2}] + \nabla_\Sigma\gamma_E \tag{8}$$

where

$$\gamma_E \equiv \int_{0^-}^{0^+} W'd\nu$$

It is shown in Sec. 7.6 that the second term in Eq. 8 can also be expressed as $-\gamma_E(\nabla\cdot\vec{n})\vec{n}$.

The double layer surface force density is exemplified in Chap. 10.

3.12 Observations

The force densities and associated stress tensors of Table 3.10.1 are of two origins. The Kelvin force densities, the last two in the table, come from a microscopic picture of particles and dipoles subject to electric or magnetic forces which, through the agent of a kinetic equilibrium, are passed along to the ponderable continuum. The Korteweg-Helmholz force densities, all of the others in the table, are based on an energy conservation principle. The connection between micro and macro fields, needed to apply this principle, is made using electrical measurements of constitutive laws to interrelate the macroscopic fields $\vec{D}$ and $\vec{E}$ or $\vec{B}$ and $\vec{H}$.

The arguments underlying each type of force density envoke certain assumptions which point to possible inadequacies. The Kelvin force densities picture the force acting on each dipole and each point charge in isolation and this force as being that transmitted to the ponderable media. This does not allow for the possibility that the micro fields of one dipole contribute to the force on a neighboring dipole.

This shortcoming is obviated by the energy method, which is based on a statement of energy conservation for an electromechanical subsystem. The resulting Korteweg-Helmholz force densities [1] are of course also restricted. On the one hand, they are more broadly applicable than might be concluded from the derivations. For example, the MQS continuum is viewed as "perfectly conducting," but the free current force density is certainly applicable in cases where the conductivity is finite. This is evident from its agreement with the Lorentz force density of Sec. 3.1, because the later model includes a finite mobility and hence electrical dissipation.

One way to derive a force density without ambiguity as to the validity of the result in nonconservative systems is to replace statements of energy conservation with those of power flow.[2] However, the principle of virtual power requires information beyond that required by the principle of virtual work used here. In addition to the constitutive laws relating the macroscopic field variables is the requirement for the power flux density, which must either be assumed or measured.

Underlying all of the discussions in this chapter has been the presumption that a clear distinction can be made between electric or magnetic force densities and those of other origins. This is tantamount to being able to isolate electromagnetic energy storage from other forms of energy storage. Piezoelectric coupling is an example where it is not fruitful to make this distinction. In that area, the stress and force density generally represent combined electric and mechanical electromechanical effects.

---

1. J. A. Stratton, Electromagnetic Theory, McGraw-Hill Book Co. Inc., New York, 1941, pp. 137-159.
2. P. Penfield, Jr., and H. H. Haus, Electrodynamics of Moving Media, The M.I.T. Press, Cambridge, Massachusetts, 1967, pp. 35-40.

Problems for Chapter 3

For Section 3.3:

Prob. 3.3.1 In writing Eq. 3.2.3, the inertia of the charge carriers is ignored. Add inertial terms to the equations, assume that the magnetic field is zero and consider an imposed electric field $\vec{E} = \text{Re}\,\hat{E}\exp(j\omega t)$. Show that the effects of inertia are negligible if $\omega \ll \nu_{\pm}$. For copper, the electron mobility is about 3×10^{-3} m^2/volt sec, while $q_-/m_- = 1.76 \times 10^{11}$ m^2/sec^2 volt. What must the frequency be to make the electron inertia significant?

For Section 3.5:

Prob. 3.5.1 For the system of Probs. 2.11.1 and 2.13.1,

(a) Show that the reciprocity condition requires that $C_{21} = C_{12}$.

(b) Find the electrical forces (f_1, f_2) in terms of (v_1, v_2, ξ_1, ξ_2) that tend to displace the movable plate in the directions $(\xi_1, \xi_2,)$ respectively.

Prob. 3.5.2 In Fig. 3.6.1, a dielectric slab is pictured as being pulled upward between plane parallel electrodes from a dielectric fluid having the same permittivity as the slab.

(a) What is the total coenergy, $w'(v,\xi)$? (Ignore fringing fields.)

(b) Use the force-energy relation, Eq. 3.5.9, to find the polarization force tending to make the slab rise.

Prob. 3.5.3 Determine the electrical force tending to increase the displacement ξ of the saturable dielectric material of Prob. 2.13.2.

Prob. 3.5.4 For the MQS configuration described in Probs. 2.12.1 and 2.14.1,

(a) Find the radial surface force density T_r by using the coenergy function to obtain $T_r(i_1, i_2, \xi)$.

(b) Compare the operations necessary to obtain $T_r(\lambda_1, \lambda_2, \xi)$ using the energy function w to those using w'. Even though the coenergy formulation is more convenient for this problem, the energy function is more convenient if one or more flux linkages are constrained.

(c) If the inner coil is shorted at a time when its flux linkage is $\lambda_2 = 0$, what is $T_r(\lambda_1, \xi)$?

For Section 3.6:

Prob. 3.6.1 In a fluid at rest, external force densities are held in equilibrium by the gradient of the fluid pressure p. Hence, force equilibrium for each incremental volume of the fluid subject to a force density $\vec{F}$ is represented by

$$\nabla p = \vec{F}$$

Suppose that the bottom of the dielectric slab pictured in Fig. 3.6.1 is well above the lower edges of the electrodes, so that the fringing field, and hence the ∇E^2, is confined to the liquid dielectric. Then there is no Kelvin force density acting on the slab, and the force density of Eq. 3.6.7 prevails in the liquid. Use Eq. 3.6.7 in Eq. 3.6.1 and integrate from the exterior free surface to the bottom of the slab to find the fluid pressure acting on the bottom of the slab. Show that this pressure, acting over the bottom of the slab, gives a net upward force that is consistent with the result of Prob. 3.5.2.

Prob. 3.6.2 Use arguments similar to those leading to Eq. 3.6.4 to show that the torque on an electric dipole is

$$\vec{\tau} = \vec{P} \times \vec{E}$$

Based on arguments similar to those used in deducing Eq. 3.6.12 from Eq. 3.6.5, argue that the torque on a magnetic dipole is

$$\vec{\tau} = \mu_o \vec{m} \times \vec{H}$$

For Section 3.7:

Prob. 3.7.1 Show that the last paragraph in Sec. 3.7 is correct.

For Section 3.9:

Prob. 3.9.1 One way to show that Eq. 3.9.17 can be used to compute $\vec{\tau}$ is to write Eq. 3.9.16 in Cartesian coordinates and use the symmetry of the stress tensor to bring the components of $\vec{r}$ inside the spatial derivatives. Carry out these steps and then use the tensor form of Gauss' theorem to obtain Eq. 3.9.17.

For Section 3.10:

Prob. 3.10.1 For certain purposes, the electric force density in an incompressible liquid with no free charge density might be represented as

$$\vec{F} = \frac{1}{2}\varepsilon\nabla(\vec{E}\cdot\vec{E})$$

where ε is a function of the spatial coordinates. Show that this differs from Eq. 3.7.22 by the gradient of a pressure and that the accompanying stress components are

$$T_{ij} = \varepsilon E_i E_j$$

Prob. 3.10.2 A fluid has the electrical constitutive law

$$\vec{D} = \alpha_1\vec{E} + \alpha_2(\vec{E}\cdot\vec{E})\vec{E}$$

It is inhomogeneous, so that α_1 and α_2 are functions of the spatial coordinates. There is no free charge density and the fluid can be assumed incompressible. Integrate the conservation of coenergy equations to show that the coenergy density is

$$W' = \frac{1}{2}\alpha_1\vec{E}\cdot\vec{E} + \frac{\alpha_2}{4}(\vec{E}\cdot\vec{E})^2$$

Find the force density $\vec{F}$ in terms of $\vec{E}$, α_1 and α_2. Find the stress tensor T_{ij} associated with this force density. Prove that $\vec{F}$ can be written in the form $\vec{F} = \vec{P}\cdot\nabla\vec{E} + \nabla\pi$, where $\vec{P}$ is the polarization density.

Prob. 3.10.1 For certain purposes, the electrical force density in an incompressible liquid with no free charge density might be represented as

$$\vec{F} = \frac{1}{2}\varepsilon\nabla(\vec{E}\cdot\vec{E})$$

where ε is a function of the spatial coordinates. Show that this differs from Eq. 3.7.22 by the gradient of a pressure, and that the accompanying stress components are

$$T_{ij} = \varepsilon E_i E_j$$

Prob. 3.10.2 A fluid has the electrical constitutive law

$$\vec{D} = (\varepsilon_o+\alpha_1)\vec{E} + \alpha_2(\vec{E}\cdot\vec{E})\vec{E}$$

It is inhomogeneous, so that α_1 and α_2 are functions of the spatial coordinates. There is no free charge density and the fluid can be assumed incompressible. Integrate the conservation of coenergy equations to show that the coenergy density is

$$W' = \frac{1}{2}(\varepsilon_o+\alpha_1)\vec{E}\cdot\vec{E} + \frac{\alpha_2}{4}(\vec{E}\cdot\vec{E})^2$$

Find the force density $\vec{F}$ in terms of $\vec{E}$, α_1 and α_2. Find the stress tensor T_{ij} associated with this force density. Prove that $\vec{F}$ can be written in the form

$$\vec{F} = \vec{P}\cdot\nabla\vec{E} + \nabla\pi$$

where $\vec{P}$ is the polarization density.

Prob. 3.10.3 Fig. P3.10.3 shows a circular cylindrical tube of inner radius a into which a second tube of outer radius b projects half way. On top of this inner tube is a "blob" of liquid metal (shown inside the broken-line box) having an arbitrary shape, but having a base radius equal to that of the inner tube. The outer and inner tubes, as well as the blob, are all essentially perfectly conducting on the time scale of interest. When $t=0^-$, there are no magnetic fields. When $t=0^+$, the outer tube is used to produce a magnetic flux which has density $B_o\vec{i}_z$ a distance $\ell >> a$ above the end of the inner tube. What is the magnetic flux density over the cross section of the annulus between tubes a distance ℓ ($\ell >> a$) below the end of the inner tube? Sketch the distribution of surface current on the perfect conductors (outer and inner tubes and blob), indicating the relative densities. Use qualitative arguments to state whether the vertical magnetic force on the blob acts upward or downward. Use the stress tensor to find the magnetic force acting on the blob in the z direction. This expression should be exact if $\ell >> a$, and be written in terms of a, b, B_o and the permeability of free space μ_o.

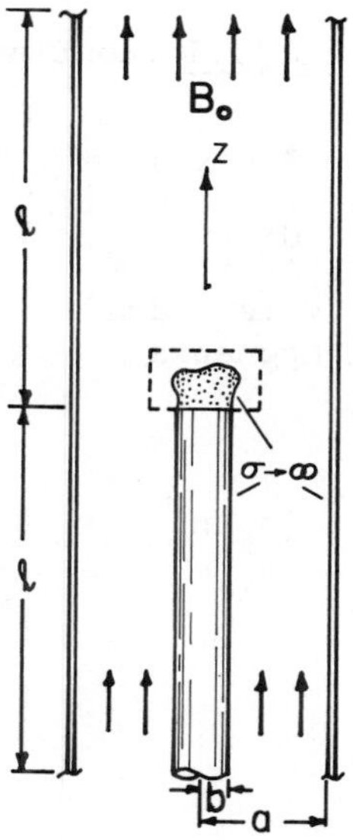

Fig. P3.10.3

Prob. 3.10.4 The mechanical configuration is as in Prob. 3.10.3. But, instead of the magnetic field, an electric field is produced by making the outer cylinder have the potential V_o relative to the inner one. Sketch the distribution of the electric field, and give qualitative arguments as to whether the electrical force on the blob is upward or downward. What is the electric field in the annulus at points well removed from the tip of the inner cylinder? Use the electric stress tensor to determine the z-directed electric force on the blob.

Prob. 3.10.5 In an EQS system with polarization, the force density is not $\vec{F} = \rho_p\vec{E} + \rho_f\vec{E}$, where ρ_p is the polarization charge. Nevertheless, this force density can be used to correctly determine the total force on an object isolated in free space. The proof follows from the argument given in the paragraph following Eq. 3.10.4. Show that the stress tensor associated with this force density is

$$T_{ij} = \varepsilon_o E_i E_j - \frac{1}{2}\delta_{ij}\varepsilon_o E_k E_k$$

Show that the predicted total force will agree with that found by any of the force densities in Table 3.10.1.

Prob. 3.10.6 Given the force density of Eq. 3.8.13, show that the stress tensor given for this force density in Table 3.10.1 is correct. It proves helpful to first show that

$$[(\nabla \times \vec{H}) \times \vec{B}]_i = \left(\frac{\partial H_i}{\partial x_j} - \frac{\partial H_j}{\partial x_i}\right) B_j$$

Prob. 3.10.7 Given the Kelvin force density, Eq. 3.5.12, derive the consistent stress tensor of Table 3.10.1. Note the vector identity given in Prob. 3.10.6.

Prob. 3.10.8 Total forces on objects can sometimes be found by the energy method "ignoring" fringing fields and yet obtaining results that are "exact." This is because the change in total energy caused by a virtual displacement leaves the fringing field unaltered. There is a "theorem" than any configuration that can be described in this way by an energy method can also be described by integrating the stress tensor over an appropriately defined surface. Use Eqs. 3.7.22 of Table 3.10.1 to find the force derived in Prob. 2.13.2.

For Section 3.11:

Prob. 3.11.1 An alternative to the derivation represented by Eq. 3.11.7 comes from exploiting an integral theorem that is analogous to Stokes's theorem.[1]

---

1. C. E. Weatherburn, Advanced Vector Analysis, G. Bell and Sons, Ltd., London, 1966, p. 126.

Prob. 3.11.1 (continued)

$$\oint_C \vec{V} \times d\vec{\ell} = \int_S [\vec{n}\nabla\cdot\vec{V} - \vec{n}\cdot(\vec{\nabla}\nabla)]da \qquad (1)$$

Here $\vec{\nabla}\nabla$ is a dyadic operator defined in Cartesian coordinates such that, "premultiplied" by $\vec{n}$, it has the components

$$[n_x \; n_y \; n_z] \begin{bmatrix} \frac{\partial V_x}{\partial x} & \frac{\partial V_x}{\partial y} & \frac{\partial V_x}{\partial z} \\ \frac{\partial V_y}{\partial x} & \frac{\partial V_y}{\partial y} & \frac{\partial V_y}{\partial z} \\ \frac{\partial V_z}{\partial x} & \frac{\partial V_z}{\partial y} & \frac{\partial V_z}{\partial z} \end{bmatrix} \qquad (2)$$

Hence,

$$\begin{aligned} \vec{n}\cdot\vec{\nabla}\nabla = \vec{i}_x &\left[n_x \frac{\partial V_x}{\partial x} + n_y \frac{\partial V_y}{\partial x} + n_x \frac{\partial V_z}{\partial x}\right] \\ \vec{i}_y &\left[n_x \frac{\partial V_x}{\partial y} + n_y \frac{\partial V_z}{\partial y} + n_z \frac{\partial V_z}{\partial y}\right] \\ \vec{i}_z &\left[n_x \frac{\partial V_x}{\partial z} + n_y \frac{\partial V_y}{\partial z} + n_z \frac{\partial V_z}{\partial z}\right] \end{aligned} \qquad (3)$$

Show that if $\vec{V} = \gamma_E\vec{n}$, it follows that

$$-\oint_C \gamma_E\vec{n} \times d\vec{\ell} = \int_S [-\vec{n}\gamma_E(\nabla\cdot\vec{n}) - \vec{n}(\vec{n}\cdot\nabla\gamma_E) + \nabla\gamma_E]da \qquad (4)$$

Thus if it is recognized that

$$\vec{n}\gamma_E \; \nabla\cdot\vec{n} = \vec{n}\gamma_E(\frac{1}{R_1} + \frac{1}{R_2})$$

(see Sec. 7.6) and that

$$\nabla_\Sigma\gamma_E \equiv \nabla\gamma_E - \vec{n}(\vec{n}\cdot\nabla\gamma_E)$$

then Eq. 3.11.7 follows.

Prob. 3.11.2 A force density is concentrated in interfacial regions where it can be represented by a surface force density $\vec{T}$. The total force on any material supporting this surface force density is then found by integrating the surface force density over the surface upon which it acts:

$$\vec{f} = \int_S \vec{T}\, da \qquad (1)$$

Suppose that the surface S is closed and that the external stress contributions to the surface force density are negligible, so that it is given by the second and third terms in Eq. 3.11.8. Use the integral theorem given in Prob. 3.11.1 to show that the resulting net force is zero.

4

Electromechanical Kinematics: Energy-Conversion Models and Processes

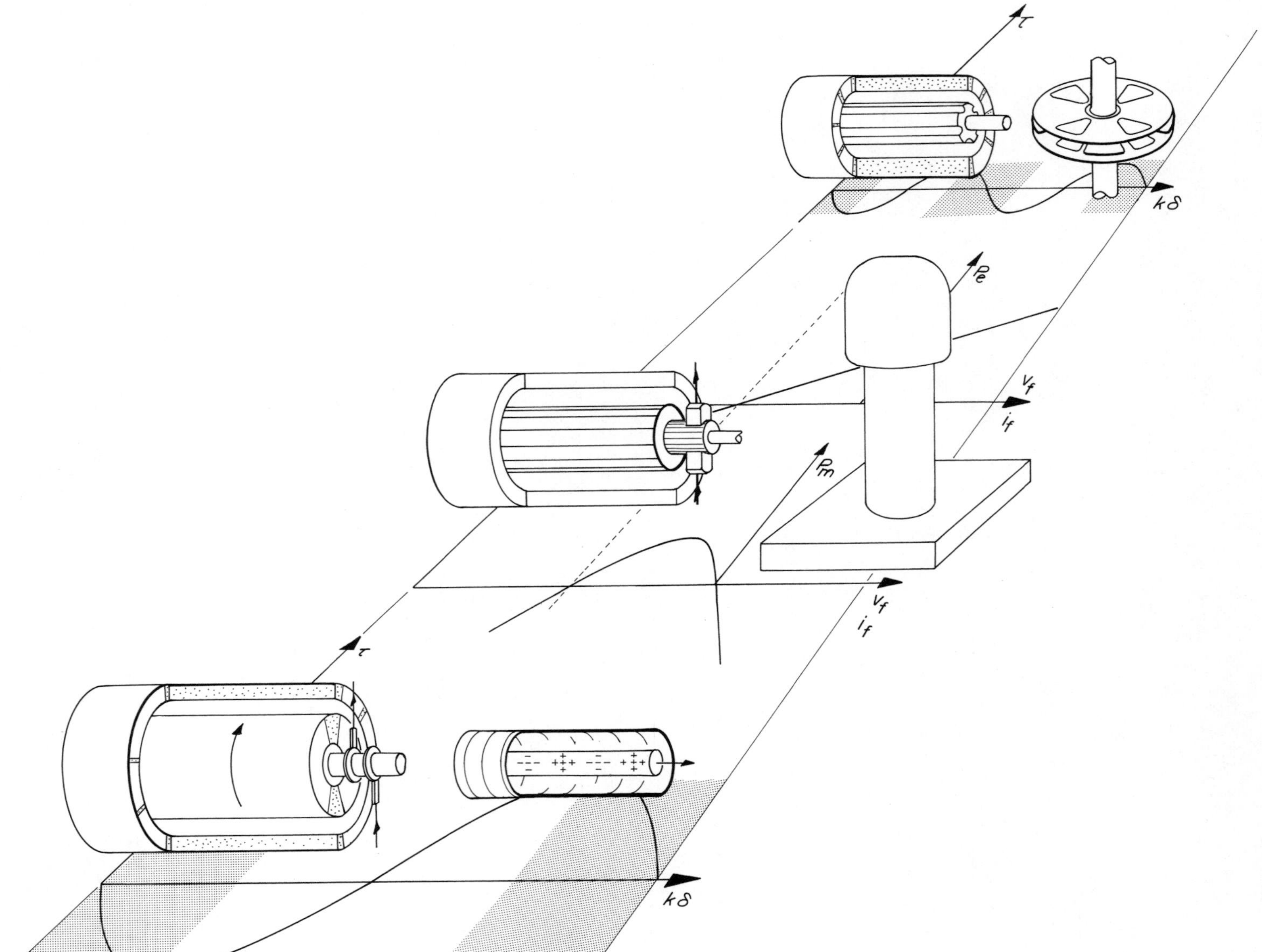

4.1 Objectives

Beginning with this chapter, progressively more electromechanical "degrees of freedom" are considered. The subject of electromechanical kinematics is first because then the relative mechanical motions as well as the paths and trajectories of charges and currents are known from the outset. The mechanics involves rigid-body translations or rotations, while charges and currents might be constrained by electrodes and wires. Processes in this category can be represented by lumped-parameter models. The field approach of this chapter provides the basis for conceptualizing and interrelating such interactions, for appreciating energy conversion limitations, and for deriving the parameters used in lumped-parameter models.

The representation of total forces and torques in terms of Maxwell stresses is developed in Sec. 4.2, followed in Sec. 4.3 by a classification of common types of energy converters, based on the fundamental field interactions. An extension of the transfer relations found in Secs. 2.16 and 2.19 to describe regions occupied by specified distributions of charge and current is made in Secs. 4.5 and 4.8. Although this chapter is concerned with modeling specific interactions, it is the technique for representing these systems that is the message. Section 4.4 exemplifies the notation and strategy underlying the methodical formulation of complex systems in not only this chapter, but those to follow. Of the remaining sections, only one does not pertain to a specific class of devices. Section 4.12 lends some formality to the philosophy underlying quasi-one-dimensional models. Such approximations retain nonlinear interactions and are illustrated in Secs. 4.13 and 4.14. By contrast, Secs. 4.4, 4.6 - 4.9 and 4.11 are concerned with field models that are naturally linear, or are linearized. Formally, the linearized model, in which products of amplitudes are ignored compared to terms that are linear in the amplitudes, is the zero-order approximation in an amplitude-parameter expansion for the exact solution. Similarly, the quasi-one-dimensional model is a zero-order approximation to an expansion in a space-rate parameter.

The analogies that exist between electric and magnetic field interactions is a theme throughout the chapter. This is clear in Sec. 4.3. But a thoughtful comparison of the characteristics of the d-c magnetic machine, considered in more detail in Sec. 4.10, with those of the Van de Graaff machine in Sec. 4.14 is worth while.

An overview of the chapter is given in Sec. 4.15.

4.2 Stress, Force and Torque in Periodic Systems

The configurations shown in Fig. 4.2.1 typify devices exploiting force or torque producing interactions between spatially periodic excitations on a "stator" structure and spatially periodic constrained or induced sources on a "rotor." In each of these, the interaction is across an air gap, a region having the electromagnetic characteristics of free space. The planar configuration of Fig. 4.2.1a might represent a linear motor or generator with the relevant force between "stator" (above) and "rotor" (below) z-directed, or it might be a developed model for the cylindrical geometry of Fig. 4.2.1c (appropriate in the limit where the air-gap spacing is small compared to the radius of the rotor). Figure 4.2.1b shows the cross section of either a planar "slab" with the interaction across two air gaps, or a cylindrical structure having an annular air gap. In either case the relevant net force is z-directed.

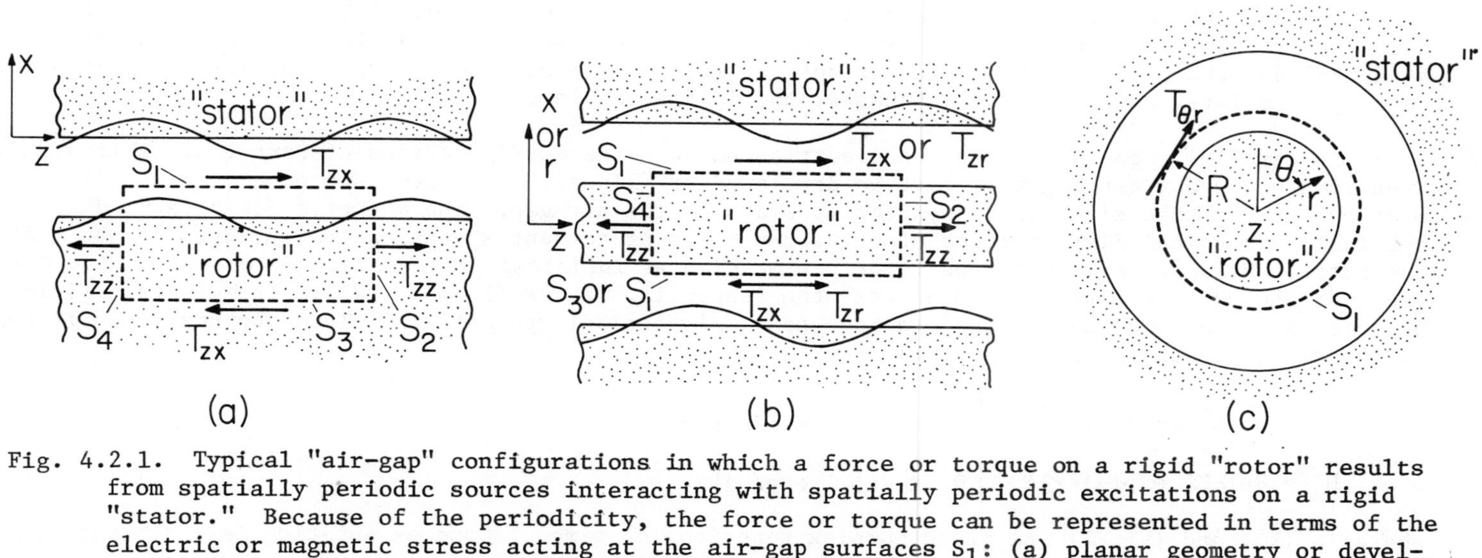

Fig. 4.2.1. Typical "air-gap" configurations in which a force or torque on a rigid "rotor" results from spatially periodic sources interacting with spatially periodic excitations on a rigid "stator." Because of the periodicity, the force or torque can be represented in terms of the electric or magnetic stress acting at the air-gap surfaces S_1: (a) planar geometry or developed model; (b) planar or cylindrical beam; (c) cylindrical rotor.

The total force acting in the z-direction on the "rotor" of Fig. 4.2.1a is conveniently determined by integrating the Maxwell stress, in accordance with Eq. 3.9.4, over the surface S enclosing a portion of the rotor having one fundamental length of periodicity. The portion S_1 of this surface is at an arbitrary plane x = constant in the air gap. Because the fields and hence the stress components T_{zz} are periodic in z, the contributions to the integration of the stress over surfaces S_2 and S_4 cancel regardless of where S_1 is located in the air gap. The contribution to the integration over S_3 can vanish for several reasons. The rotor may be perfectly permeable, of infinite permittivity or infinitely conducting, in which case $\vec{H}$ or $\vec{E}$ is zero on S_3. In Cartesian coordinates, the fields associated with excitations that are periodic in the z-direction decay in the x direction and if S_3 is well removed from the air gap, the contribution on S_3 asymptotically vanishes. Yet another possibility is that the planar model really is a developed model for the cylindrical configuration of Fig. 4.2.1c, in which case the surface S is "pie" shaped and the section S_3 does not exist. In any of these cases, the z-directed force acting on the rotor of Fig. 4.2.1a is simply

$$f_z = A\left\langle T_{zx}\right\rangle_z\Big|_{S_1} \tag{1}$$

where A is the y-z area of the air gap and T_{zx} is the magnetic or electric stress tensor, as the case may be. The brackets indicate a spatial average is taken, as discussed in Sec. 2.15.

There is no question as to which of the stress tensors in Table 3.10.1 should be used. As discussed in Sec. 3.10, in the free-space region of the air gap, all of the magnetic and all of the electric stress tensors agree.

If Fig. 4.2.1b represents a planar layer, then there are stress contributions from surfaces S_1 and S_3, and the net force acting on a section of the layer having area A in the y-z plane is

$$f_z = A\left[\left\langle T_{zx}\right\rangle_z\Big|_{S_1} - \left\langle T_{zx}\right\rangle_z\Big|_{S_3}\right] \tag{2}$$

On the other hand, if the "rotor" in that figure is a cylinder, then the net force takes the form of Eq. 1, with A the area of an enclosing cylindrical surface and appropriate shear stress $T_{zx} \rightarrow T_{zr}$ evaluated on that surface.

In computing the net torque on the rotor of Fig. 4.2.1c, it is tempting to multiply the space-average shear stress $\langle T_{\theta r}\rangle_\theta$ by the lever arm R and the area A of a cylindrical enclosing surface having radius R:

$$\tau_z = RA\left\langle T_{\theta r}\right\rangle_\theta\Big|_{S_1} \tag{3}$$

Because the stress is symmetric, this notion is rigorous, as can be seen by applying Eq. 3.9.16 to the surface S_1 of Fig. 4.2.1c.

4.3 Classification of Devices and Interactions

Based on the developed or linear air-gap configuration of Fig. 4.2.1a, this section begins with illustrative simplified examples of "synchronous" and "d-c" magnetic and electric interactions. Then, a general discussion is given of the various classes of machines, some having lumped-parameter models developed in later sections of this chapter and in the problems.

In parallel, consider first the electric and magnetic configurations of Part 1 of Table 4.3.1. Even though the devices might in fact be developed or "linear," the terms stator and rotor will be used to refer to the elements on respective sides of the air gap. The magnetic field is produced by spatially sinusoidal distributions of current modeled as current sheets on the surfaces of the stator and rotor. Because the stator and rotor are modeled as infinitely permeable, $\vec{H} = 0$ outside the air gap and the surface currents "terminate" the tangential fields (Eq. 2.10.21). The electric field is produced by electrodes constrained to have spatially periodic potentials. Thus, boundary conditions at the air-gap boundaries (s) and (r) are

$$\begin{array}{l|l} H_z^s = \text{Re}[\tilde{K}^s \exp(-jkz)] & \Phi^s = \text{Re}[\tilde{V}^s \exp(-jkz)] \\ H_z^r = \text{Re}[-\tilde{K}^r \exp(-jkz)] & \Phi^r = \text{Re}[\tilde{V}^r \exp(-jkz)] \end{array} \tag{1}$$

where $(\tilde{K}^s,\tilde{K}^r)$ and $(\tilde{V}^s,\tilde{V}^r)$ are given complex functions of time. (Complex notation is introduced in Sec. 2.15.)

With the surface S_1 taken as the rotor surface, (r), it follows from Eq. 4.2.1 and the average theorem, Eq. 2.15.14, that the force on a section of the rotor having area A is

Table 4.3.1. Basic configurations illustrating classes of electromechanical interactions and devices. MQS and EQS systems respectively in left and right columns.

sources imposed on moving member

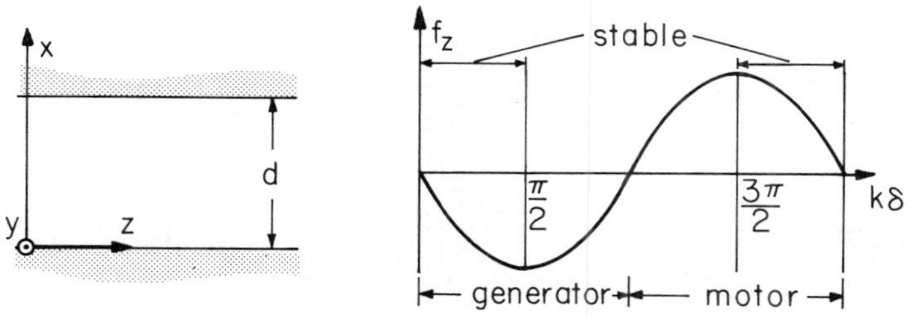

1. currents (potentials) constrained on both windings (electrodes)

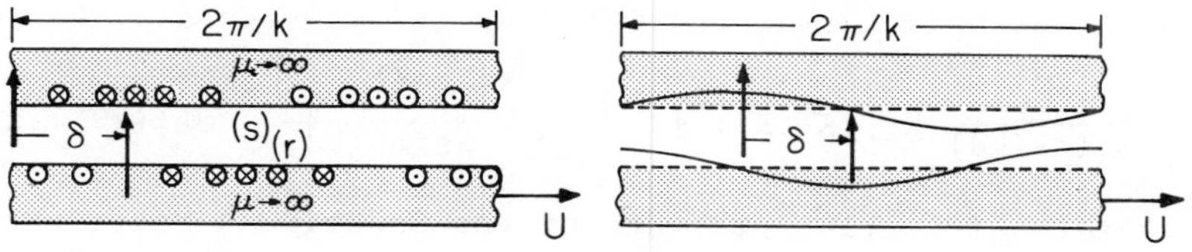

2. current (potential) constrained on "stator" and permanent magnetization (polarization) on "rotor"

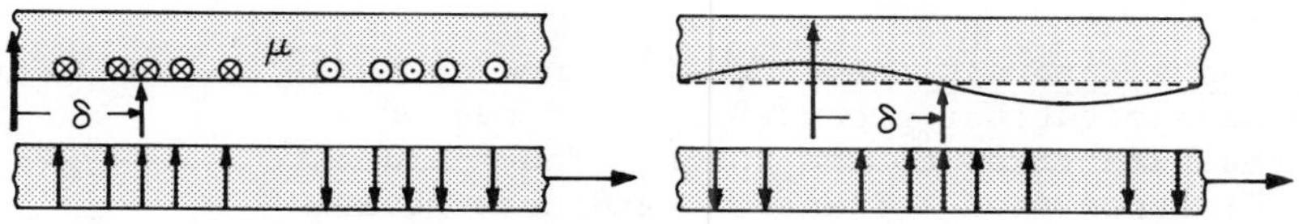

3. current (potential) constrained on "stator" and flux (charge) constrained on "rotor"

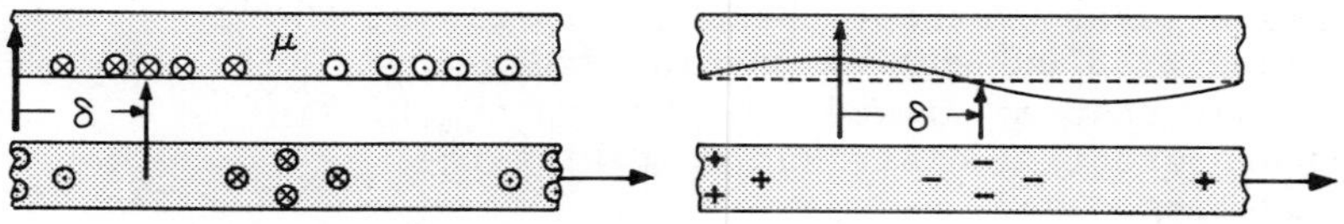

sources instantaneously induced on nonuniform moving member

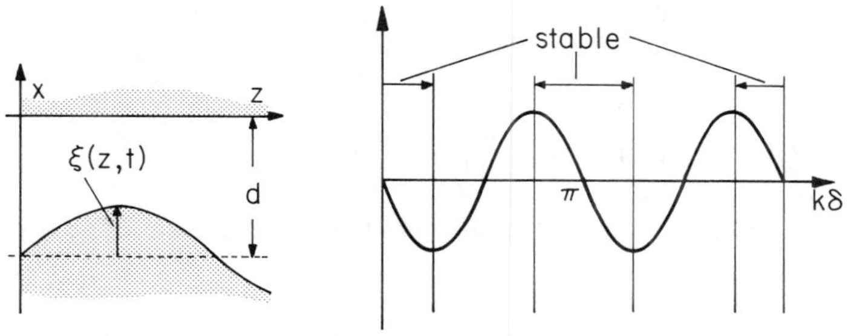

4. current (potential) constrained on "stator" and magnetization (polarization) induced on "rotor" having saliency

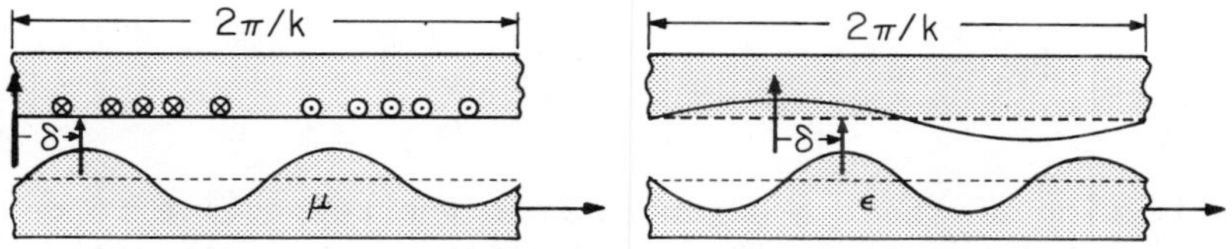

5. current (potential) constrained on "stator" and currents (charges) induced on "rotor" having saliency

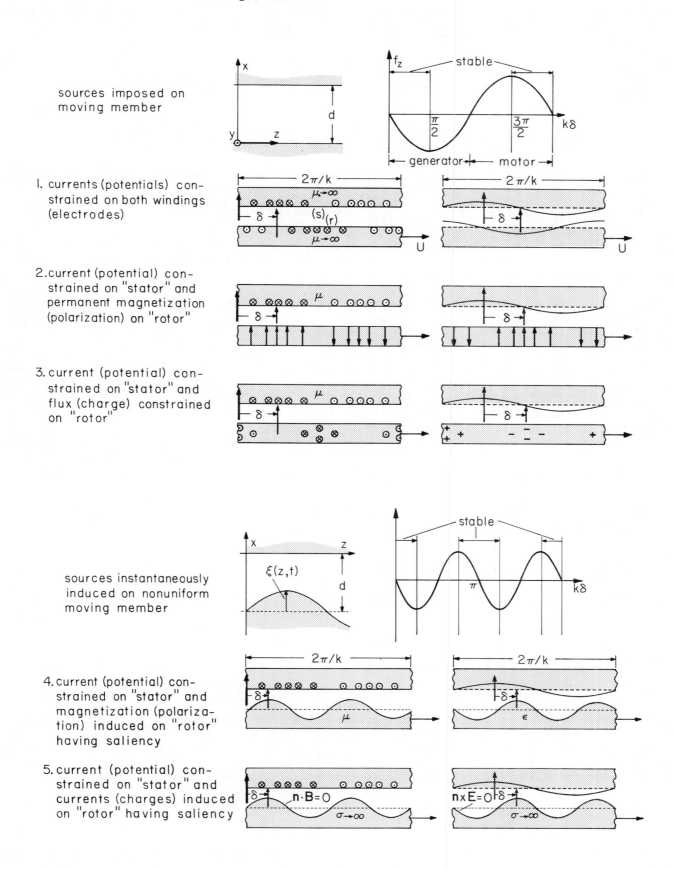

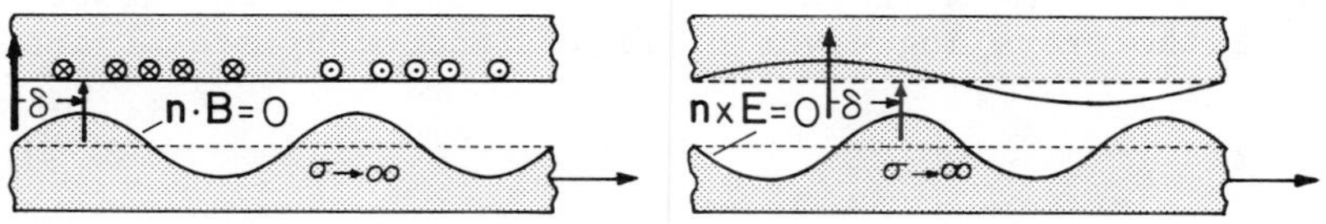

$$f_z = \frac{A}{2}\,\mathrm{Re}\,\mu_o \tilde{H}_x^r (\tilde{H}_z^r)^* = \frac{A}{2}\,\mathrm{Re}\,\mu_o \tilde{H}_x^r (-\tilde{K}^r)^* \qquad \Bigg| \qquad f_z = \frac{A}{2}\,\mathrm{Re}\,\varepsilon_o \tilde{E}_x^r (\tilde{E}_z^r)^* = \frac{A}{2}\,\mathrm{Re}\,\varepsilon_o \tilde{E}_x^r (jk\tilde{V}^r)^* \tag{2}$$

The gap transfer relations, Eq. (a) of Table 2.16.1, give the normal fluxes at (s) and (r) in terms of the potentials there. In the magnetic case, $H_z = jk\tilde{\Psi}$ and because of the boundary conditions, Eq. 1, these relations become

$$\begin{bmatrix} \mu_o \tilde{H}_x^s \\ \\ \mu_o \tilde{H}_x^r \end{bmatrix} = \mu_o k \begin{bmatrix} -\coth(kd) & \dfrac{1}{\sinh(kd)} \\ \\ \dfrac{-1}{\sinh(kd)} & \coth(kd) \end{bmatrix} \begin{bmatrix} \dfrac{\tilde{K}^s}{jk} \\ \\ \dfrac{-\tilde{K}^r}{jk} \end{bmatrix} \qquad \Bigg| \qquad \begin{bmatrix} \varepsilon_o \tilde{E}_x^s \\ \\ \varepsilon_o \tilde{E}_x^r \end{bmatrix} = \varepsilon_o k \begin{bmatrix} -\coth(kd) & \dfrac{1}{\sinh(kd)} \\ \\ \dfrac{-1}{\sinh(kd)} & \coth(kd) \end{bmatrix} \begin{bmatrix} \tilde{V}^s \\ \\ \tilde{V}^r \end{bmatrix} \tag{3}$$

Substitution of the normal flux densities at (r) expressed by Eqs. 3 into Eqs. 2 gives the desired forces

$$f_z = -\frac{A\mu_o}{2\sinh(kd)}\,\mathrm{Re}[j\tilde{K}^s(\tilde{K}^r)^*] \qquad \Bigg| \qquad f_z = \frac{A\varepsilon_o}{2\sinh(kd)}\,\mathrm{Re}[j(k\tilde{V}^s)(k\tilde{V}^r)^*] \tag{4}$$

Note that the terms involving products of the individual rotor excitations do not contribute. (They are imaginary and hence dropped in taking the real part.) Physically, this is expected because such terms represent the rotor self-field interactions.

Synchronous Interactions: Consider now systems with the rotor excitations produced by windings or electrodes that are fixed to the rotor. The coordinate z' measures distance from a frame of reference moving with the velocity U of the rotor, as sketched in Fig. 4.3.1. Fixed and moving frame coordinates are related in the figure. Perhaps through slip rings, the rotor is excited by a current of angular frequency ω_r, in such a way that as viewed from the rotor there is a current or potential distribution taking the form of a traveling wave:

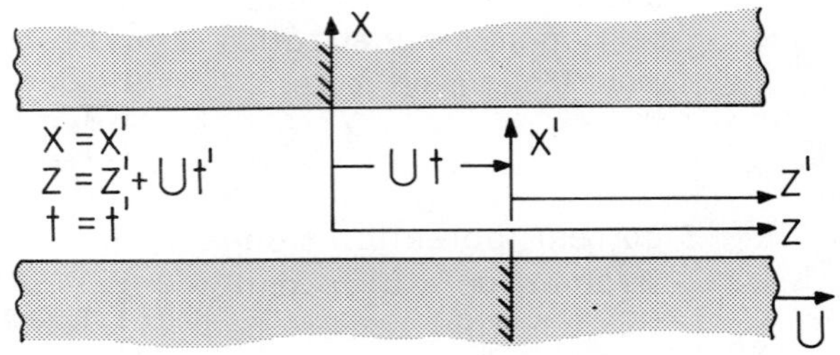

Fig. 4.3.1. Rotor and stator reference frames z' and z.

$$K^r = K_o^r \sin[\omega_r t - k(z' - \delta)] \qquad \Bigg| \qquad V^r = -V_o^r \cos[\omega_r t - k(z' - \delta)] \tag{5}$$

On the stator, a similar arrangement of windings or electrodes, with excitations at the angular frequency ω_s, give the traveling waves:

$$K^s = K_o^s \sin\,[\omega_s t - kz] \qquad \Bigg| \qquad V^s = V_o^s \cos\,[\omega_s t - kz] \tag{6}$$

Because $z' = z - Ut$, Eqs. 5 and 6 can be written in terms of complex amplitudes:

$$\begin{aligned} \tilde{K}^r &= -jK_o^r\, e^{j(\omega_r + kU)t}\, e^{jk\delta} \\ \tilde{K}^s &= -jK_o^s\, e^{j\omega_s t} \end{aligned} \qquad \Bigg| \qquad \begin{aligned} \tilde{V}^r &= -V_o^r\, e^{j(\omega_r + kU)t}\, e^{jk\delta} \\ \tilde{V}^s &= V_o^s\, e^{j\omega_s t} \end{aligned} \tag{7}$$

Substitution of these amplitudes into the respective force relations of Eq. 4 gives forces with sinusoidal time dependences. The frequencies are in each case $\omega_s - \omega_r - kU$. Only if this frequency is zero will these forces have time-average values. Division of the resulting frequency condition by k shows that these time-average forces exist because, as viewed from the stator frame of reference, the velocities of the traveling waves of field induced by stator and rotor sources are equal:

$$\frac{\omega_s}{k} = \frac{\omega_r}{k} + U \tag{8}$$

Usually, the rotor is d-c excited so that $\omega_r = 0$ and the phase velocity of the stator traveling wave, ω_s/k, is equal to the rotor velocity U. Under the synchronous condition, the substitution of Eqs. 7 into Eqs. 4 gives the forces as functions of the relative spatial phase $k\delta$ between traveling waves:

$$f_z = -\frac{A\mu_o K_o^s K_o^r}{2\sinh kd}\sin k\delta \qquad \Bigg| \qquad f_z = -\frac{A\varepsilon_o (kV_o^s)(kV_o^r)}{2\sinh kd}\sin k\delta \tag{9}$$

The sketches of the stator and rotor excitations in Part 1 of Table 4.3.1 (at the instant $t = 0$) show the relative distributions with $\delta = \lambda/4$, and hence $k\delta \equiv 2\pi(\delta/\lambda) = \pi/2$. According to Eqs. 9, it is at this spatial phase that the greatest retarding force acts on the rotor. The observation is consistent with what would be expected intuitively for the sketched distributions. Under the synchronous conditions the relative distribution of stator and rotor field sources is invariant. The stator current distribution gives rise to a normal flux density that peaks at the current null. This is the stator magnetic axis, indicated by the vertical arrow on the stator. This field interacts with the rotor current to produce the time-average force in the -z direction. Stator and rotor magnetic axes tend to line up. Similarly, in regions of positive and negative electrode potential there are positive and negative surface charges (although not exactly in phase with the potential). Thus, the retarding electric force results from the attraction of neighboring opposite charges. The rotor and stator axes, denoted by the vertical arrows, also tend to line up.

The classic force (or torque) phase-angle diagram, the graphical representation of Eqs. 9, is shown at the top of Fig. 4.3.1. Angles of positive and negative force can respectively give motor and generator operation. But, operation is generally restricted to the shaded regions because then a change in relative phase, $k\delta$, results in a force that tends to return the rotor to its original angle.

Parts 2 and 3 of Table 4.3.1 illustrate other types of excitations that result in synchronous interactions. In each of these, the rotor sources are "attached" to the rotor and hence the synchronous condition of Eq. 8 reduces to $\omega_s/k = U$. Each has a force with the same dependence on relative phase $k\delta$ illustrated by Eqs. 9.

Small machines having permanent magnet rotors are common, but electric analogues having permanent polarization (Sec. 4.4) are not. By contrast, electric synchronous interactions between traveling waves of charge and potential are common, whereas, devices making use of a trapped rotor flux are not. The former, a kinematic model for electron beam devices, will be considered further in Sec. 4.6.

D-C Interactions: The family of magnetic devices called d-c machines has as an electric field analogue devices of the Van de Graaff type. The configurations shown in Table 4.3.1, Part 1, can also be used to illustrate this class of devices, provided the sketched current and potential distributions are understood to be time-varying in amplitude but stationary in space. Currents are supplied to the rotor windings through brushes and commutator segments in such a way that even though the rotor moves, the rotor's relative current distribution is stationary. The stator current distribution is similarly stationary in space and shifted by the distance δ. The stationary distribution of rotor potential in the electric analogue is an approximation to the potential associated with charge placed on a moving belt at one fixed location and removed at another. Excitations therefore take the form

$$K^r = \text{Re}[-jK_o^r(t)e^{jk\delta}]e^{-jkz} = -K_o^r(t)\sin k(z-\delta) \qquad \Bigg| \qquad V^r = \text{Re}[-V_o^r(t)e^{jk\delta}]e^{-jkz} = -V_o^r(t)\cos k(z-\delta)$$

$$K^s = \text{Re}[-jK_o^s(t)]e^{-jkz} = -K_o^s(t)\sin kz \qquad \Bigg| \qquad V^s = \text{Re}\, V_o^s(t)e^{-jkz} = V_o^s(t)\cos kz \tag{10}$$

Note that the complex amplitudes multiplying $\exp(-jkz)$, now arbitrary functions of time, are as required to evaluate Eqs. 4. The resulting forces are in fact the same as given by Eqs. 9, provided it is understood that (K_o^s, K_o^r) and (V_o^s, V_o^r) are now arbitrary real functions of time.

The magnetic version of the d-c machine is modeled in Sec. 4.10, while the Van de Graaff machine is taken up in Sec. 4.14.

Synchronous Interactions with Instantaneously Induced Sources: Common examples of devices that exploit instantaneously induced magnetization forces on a moving member are variable-reluctance or salient-pole machines. Electric field members of this family of devices include variable-capacitance machines. (By contrast with magnetic and electric "induction" interactions, naturally taken up in the next two chapters, the rotor sources induced by the stator excitations move synchronously with the material. Geometry rather than a rate process, such as magnetic diffusion or charge relaxation, is involved.)

Linear or developed salient-pole models are shown in Part 4 of Table 4.3.1. The rotor, which in the magnetic case is perhaps highly magnetizable magnetically soft iron, has surface saliencies. In a two-pole rotating machine, the rotor represented by this model (with $2\pi/k$ the circumference of the stator) could be a squashed cylinder protruding toward the stator at two positions and away from it at two others. The conventional method for finding the magnetic force on the moving member is to use the energy method of Sec. 3.5 and knowledge of the inductance or capacitance of the stator windings or electrodes. Because of the rotor saliency, the stator terminal relations clearly depend on the rotor position, and hence so also does the magnetic or electric energy storage.

With the objective of fitting this type of interaction into the field point of view, the development is in terms of the magnetic interaction. Similitude then makes it possible to apply the results to the polarization case. In the limit where the material is highly magnetizable, $\vec{H}$ is excluded from the rotor so that on the rotor surface the tangential field vanishes. As a result, the magnetic traction acts normal to the surface of the rotor. That is, in a local Cartesian coordinate system on the rotor surface, having the axis n in the normal direction, any of the stress tensors (Table 3.10.1) evaluated in free space next to the rotor surface give a traction

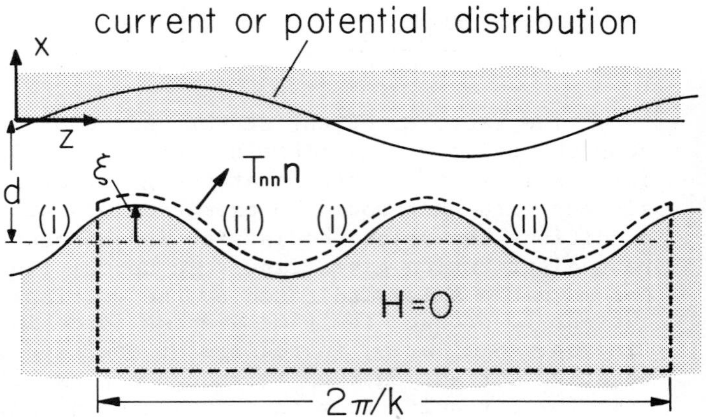

Fig. 4.3.2. Traction $\vec{\vec{T}}\cdot\vec{n} = T_{nn}\vec{n}$ acts normal to rotor surface.

$$\vec{T} = \vec{\vec{T}}\cdot\vec{n} = T_{nn}\vec{n} \tag{11}$$

Although not convenient for mathematical derivations, the surface enclosing one periodicity length $2\pi/k$ of the rotor, shown in Fig. 4.3.2, helps in understanding how the magnetic traction gives rise to a net force on the rotor. The traction acting normal to the surface has a value $T_{nn} = \mu_o H_n^2/2$ and hence is positive. No matter what the excitation from the stator winding, it is clear that at positions (i), where the slope of the stator surface is positive, the magnetic field tends to pull the rotor to the left while at point (ii) the pull is to the right. It is the spatial phase relationship between the stator current distribution and the rotor saliencies that makes one or the other of these forces dominant. It is clear, for example, that if the rotor surface wavelength matched that of the stator current there could be no net force. The z-directed traction acting at any given point would then be cancelled by that acting at a point on the rotor surface a half-wavelength away.

In deriving the relation of the excitation and rotor geometry to the net force, the rotor surface is taken as being at

$$x = -d + \xi(z,t) = -d + \text{Re}\,\hat{\xi}\, e^{-j(2k)(z-Ut)} \tag{12}$$

The rotor travels with the linear velocity $U = \omega/k$ and hence its surface, with wavelength π/k half that of the stator excitation, moves in synchronism with the traveling wave of stator surface current:

$$\vec{K} = \text{Re}\hat{K}^s e^{j(\omega t - kz)}\vec{i}_y \tag{13}$$

A surface, represented by $F(x,y,z,t) = x + d - \xi = 0$, has a normal vector

$$\vec{n} = \frac{\nabla F}{|\nabla F|} = \frac{\vec{i}_x - \frac{\partial\xi}{\partial z}\vec{i}_z}{|\nabla F|} \tag{14}$$

As a reminder that this is a familiar relation, the surface might be one of zero potential ($F \to \Phi$), with $\vec{n}$ the negative of the electric field intensity normalized so that it has unit magnitude. The condition that there be no tangential field on the rotor surface is then

$$[\vec{n} \times \vec{H}]_y = 0 \Rightarrow H_z = -H_x\frac{\partial\xi}{\partial z} \text{ at } x = -d + \xi \tag{15}$$

To match this boundary condition is in general difficult. In this section, it is assumed that ξ is small, so that Eq. 15 is evaluated approximately (to first order in ξ) at the "equilibrium" position of the rotor surface, $x = -d$. With H_x evaluated at $x = -d$ rather than at $x = -d + \xi$, the right-hand side of Eq. 15 is already written to first order in ξ:

$$H_z(x = -d + \xi) = H_z(x = -d) + \frac{\partial H_z}{\partial x}(x = -d)\xi \tag{16}$$

If it is further recognized that because $\vec{H}$ is irrotational, $\partial H_z/\partial x = \partial H_x/\partial z$, then to first order in ξ, Eq. 15 becomes a boundary condition to be evaluated at $x = -d$, defined as the position (r):

$$H_z^r = -\frac{\partial}{\partial z}(H_x^r\xi) \tag{17}$$

What must be used in evaluating H_x^r is the zero-order field. This is the field that would be found with $\xi = 0$, with the rotor presenting a planar surface to a gap excited on the stator side by the current sheet given by Eq. 13. Thus, Eq. 17 takes the form

$$\begin{aligned} H_z^r &= -\frac{\partial}{\partial z}\left[\mathrm{Re}\hat{H}_x^r e^{j(\omega t-kz)}\mathrm{Re}\hat{\xi}e^{-2jk(z-Ut)}\right] \\ &= -\frac{\partial}{\partial z}\left\{\frac{1}{2}\left[\hat{H}_x^r e^{j(\omega t-kz)} + (\hat{H}_x^r)^* e^{-j(\omega t-kz)}\right]\frac{1}{2}\left[\hat{\xi}\, e^{-2jk(z-Ut)} + \hat{\xi}^*\, e^{2jk(z-Ut)}\right]\right\} \end{aligned} \tag{18}$$

Because of the synchronism condition, $\omega = kU$, multiplying out this expression gives a term having the same spatial frequency as the stator current and a term at three times that frequency:

$$H_z^r = -\frac{\partial}{\partial z}\left[\mathrm{Re}\hat{\psi}_k e^{j(\omega t-kz)} + \mathrm{Re}\hat{\psi}_{3k}e^{3j(\omega t-kz)}\right];\ \hat{\psi}_k \equiv \frac{1}{2}(\hat{H}_x^r)^*\hat{\xi},\ \hat{\psi}_{3k} \equiv \frac{1}{2}(\hat{H}_x^r)\hat{\xi} \tag{19}$$

Note that this expression takes the form $\vec{H} = -\nabla\Psi$. With the surface S_1 of Fig. 4.2.1a taken as contiguous with the stator, the desired space-average rotor force is

$$f_z = A\langle T_z\rangle_z = A\left\langle \mu_o H_x^s \mathrm{Re}\hat{K}^s e^{j(\omega t-kz)}\right\rangle_z \tag{20}$$

Note that the terms in Eq. 19 are written in the standard complex form, with the quantity in brackets the magnetic potential Ψ. The amplitudes at the stator and rotor surfaces (at s and r) are therefore related by the transfer relation (Eqs. (a) of Table 2.16.1):

$$\begin{bmatrix} \mu_o\hat{H}_x^s \\ \mu_o\hat{H}_x^r \end{bmatrix} = \mu_o k \begin{bmatrix} -\coth(kd) & \dfrac{1}{\sinh(kd)} \\ \dfrac{-1}{\sinh(kd)} & \coth(kd) \end{bmatrix} \begin{bmatrix} \dfrac{\hat{K}^s}{jk} \\ \hat{\Psi}_k \end{bmatrix} \tag{21}$$

for components with dependence $\exp[j(\omega t - kz)]$ and

$$\begin{bmatrix} \mu_o H_x^s \\ \mu_o H_s^r \end{bmatrix} = \mu_o 3k \begin{bmatrix} -\coth(3kd) & \dfrac{1}{\sinh(3kd)} \\ \dfrac{-1}{\sinh(3kd)} & \coth(3kd) \end{bmatrix} \begin{bmatrix} 0 \\ \hat{\Psi}_{3k} \end{bmatrix} \tag{22}$$

for components with dependence $\exp 3j(\omega t - kz)$. The infinitely permeable material backing the stator current sheet requires that the third harmonic tangential field at the stator in Eq. 22a vanish.

The normal flux density $\mu_o H_x^s$ in Eq. 20 is a superposition of the components found using Eqs. 21a and 22a. Because it multiplies $\hat{\xi}$, $\hat{H}_x^r$ on the right in these expressions need only be evaluated to zero order in ξ. Thus, $\hat{H}_x^r$ is given by Eq. 21b with $\hat{\xi} = 0$, and hence $\hat{\Psi}_k = 0$. The second term in Eq. 19 also excites a field at the stator surface given by Eq. 22a. But, inserted into Eq. 20, this higher harmonic gives no space-average contribution and hence can be dropped. Thus, Eq. 20 becomes

$$f_z = A\left\langle \mathrm{Re}\left\{ j\mu_o\coth(kd)\hat{K}^s + \frac{\mu_o k}{2}\left[\frac{-j(\hat{K}^s)^*\hat{\xi}}{\sinh^2(kd)}\right]\right\} e^{j(\omega t-kz)}\ \mathrm{Re}\left[\hat{K}^s e^{j(\omega t-kz)}\right]\right\rangle_z \tag{23}$$

The averaging theorem, Eq. 2.15.14, can now be applied to Eq. 23 to obtain the first of these relations:

$$f_z = \frac{\mu_o kA}{4\sinh^2(kd)}\mathrm{Re}\left[(\hat{K}^s)^2 j\hat{\xi}^*\right] \quad\Bigg|\quad f_z = \frac{-\varepsilon_o kA}{4\sinh^2(kd)}\mathrm{Re}\left[(k\hat{V}^s)^2 j\hat{\xi}^*\right] \tag{24}$$

The second expression pertains to the electric configuration of Part 4, Table 4.3.1, and has been obtained by recognizing that, in terms of the magnetic and electric potentials, the air-gap fields are analogous. The only difference is that in the magnetic case the stator magnetic potential is $\hat{K}^s/jk$, while in the electric case, the stator electric potential is $\hat{V}^s$. Hence, the electric time average force is found (using the complete analogy discussed at the beginning of Sec. 2.16) by replacing $\mu_o \rightarrow \varepsilon_o$ and $\hat{K}^s \rightarrow jk\hat{V}^s$ in Eq. 24a to obtain Eq. 24b.

As specific examples having the stator excitations and rotor position when t = 0 shown in Part 4 of Table 4.3.1, let

$$\xi = \xi_o \cos 2k[Ut - (z - \delta)] = \mathrm{Re}\,\xi_o e^{2jk\delta} \exp[2jk(Ut - z)] \tag{25}$$

and

$$K^s = K_o^s \sin(\omega t - kz) = \mathrm{Re}(-jK_o^s) \exp[j(\omega t - kz)] \quad \Big| \quad V^s = V_o^s \cos(\omega t - kz) = \mathrm{Re}V_o^s \exp[j(\omega t - kz)] \tag{26}$$

where ξ_o, K_o^s and V_o^s are taken as real. Then, Eqs. 24 take the specific forms

$$f_z = -\frac{-\mu_o k(K_o^s)^2 \xi_o A}{4\sinh^2(kd)} \sin(2k\delta) \quad \Big| \quad f_z = \frac{-\varepsilon_o k(kV_o^s)^2 \xi_o A}{4\sinh^2(kd)} \sin(2k\delta) \tag{27}$$

The dependence of these forces on the spatial phase of stator excitations and rotor position, sketched in Table 4.3.1, is typical of salient-pole synchronous devices. That $\langle T_z \rangle_z$ has twice the periodicity in $k\delta$, obtained with the rotor excited directly by sources having the same periodicity as the stator excitations, is a direct consequence of the induced nature of the magnetization or polarization. Because the surface traction is proportional to the square of the local field, the same force is obtained if the rotor is shifted in relative position by $\delta = \pi/k$. The $[\sinh(kd)]^{-2}$ dependence of the force on the gap dimension d results because the only excitation is on the stator. By contrast with the synchronous interactions between excited stators and rotors [with (d) dependence $\sinh(kd)^{-1}$], here there is a round-trip attenuation of the excitation field, first in reaching the rotor surface and then in being reflected back to the stator.

Of the many configurations in the general family of "salient-pole" devices, two more are shown in Part 5 of Table 4.3.1. The magnetic case is considered in the problems, while the electric one is formally the same as if the rotor were perfectly polarizable. Hence it is also described by Eqs. 24b and 27b.

Practical devices make use of large amplitude saliency. One approach to obtaining an appropriate model is developed in Secs. 4.12 and 4.13, where the variable capacitance machine is considered in more detail.

4.4 Surface-Coupled Systems: A Permanent Polarization Synchronous Machine

With field sources modeled by surface charges or surface currents, it is natural to generalize the approach taken in Sec. 4.3 to the description of a wide class of complex electromechanically kinematic systems. The technique involves breaking the region of interest into source-free subregions that have uniform properties and hence can be described by the transfer relations of Sec. 2.16. Sources are then relegated to boundaries between subregions and are taken into account in the boundary conditions used to splice fields together. It is the objective in this section to illustrate the systematic approach that can be taken with such models by developing the lumped-parameter mechanical and electrical terminal relations for the rotating machine shown in Fig. 4.4.1.

The rotor consists of a material having polarization density that is uniform and permanent:

$$\vec{P} = P_o[\vec{i}_r \cos(\theta - \theta_r) - \vec{i}_\theta \sin(\theta - \theta_r)] = \mathrm{Re}P_o(\vec{i}_r - j\vec{i}_\theta)e^{-j(\theta-\theta_r)} \tag{1}$$

Field coordinates are (r,θ) while $\theta_r(t)$ is the rotor axis. Thus, the polarization density is uniform and directed collinear with the rotor axis at the angle $\theta_r(t)$. The region between the rotor (with radius R) and the stator (radius R_o) is an air gap. Stator electrodes shown in the figure have respective potentials $\pm v(t)$ and are imbedded in a dielectric having permittivity ε_s. The length of the device in the z direction, ℓ, is considered large compared to the radial dimensions.

Within the rotor, there is no free charge density. Moreover, because the permanent polarization is uniform and hence has no divergence, Gauss' law (Eq. 2.3.27) reduces to

$$\nabla \cdot \varepsilon_o \vec{E} = 0 \tag{2}$$

Within the rotor, as well as in the air gap and in the surrounding dielectric of the stator, the fields are Laplacian. The transfer relations of Sec. 2.16 are directly applicable to describing the bulk fields.

Boundary Conditions: The potential at $r = R_o$ is constrained to be $\pm v(t)$ on the respective portions of the stator surface covered by the electrodes. The potential between the electrodes on the dielectric surface at $r = R_o$ is approximated by the continuous linear distribution shown in Fig. 4.4.2.

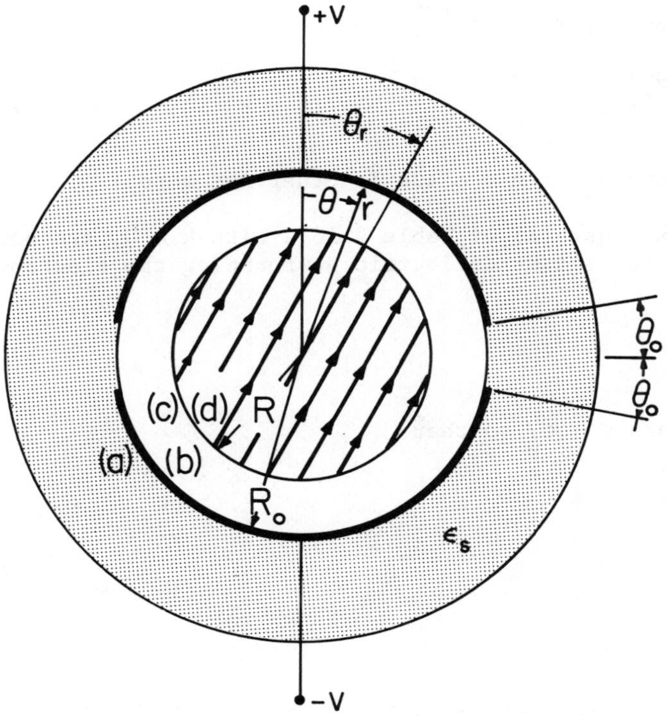

Fig. 4.4.1

Cross-sectional view of permanent polarization rotating machine.

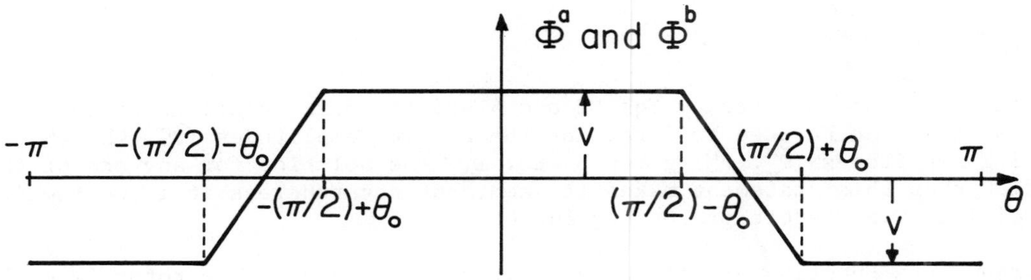

Fig. 4.4.2. Distribution of stator potential used to model the device shown in Fig. 4.4.1.

In Fig. 4.4.1, the notation (a)...(d) is used to denote positions adjacent to interfaces between regions. (This convention is introduced in Sec. 2.20.) Thus, the potential distribution of Fig. 4.4.2 is both Φ^a and Φ^b. In anticipation of the Laplacian solutions used to describe the bulk fields in cylindrical geometry, the potential of Fig. 4.4.2 is now expanded in a Fourier series (see Sec. 2.15 for a discussion of Fourier series):

$$\Phi^a = \Phi^b = \sum_{\substack{m=-\infty \\ (\text{odd})}}^{+\infty} \tilde{\Phi}^a_m(t)\, e^{-jm\theta}; \quad \tilde{\Phi}^a_m = \frac{2v(t)}{m\pi} \frac{\sin(m\theta_o)}{\theta_o m} \sin\left(\frac{m\pi}{2}\right) \tag{1}$$

In the following it is assumed that the dielectric surrounding the rotor is of sufficient radius compared to R_o, that fields decay to zero before reaching the outer surface of the dielectric.

At the rotor air-gap interface the tangential $\vec{E}$ and hence the potential must be continuous. Thus the Fourier amplitudes are related by

$$\tilde{\Phi}^c_m = \tilde{\Phi}^d_m \tag{2}$$

In addition, Gauss' law (Eq. 2.10.21a) and Eq. 1 require that

$$\vec{n}\cdot\varepsilon_o [\![E]\!] = -\vec{n}\cdot [\![P]\!] \Rightarrow \varepsilon_o E_r^c - \varepsilon_o E_r^d = \mathrm{Re}(P_o e^{j\theta_r}) e^{-j\theta} \quad (3)$$

This latter expression relates the Fourier amplitudes by

$$\varepsilon_o \tilde{E}_{rm}^c - \varepsilon_o \tilde{E}_{rm}^d = \frac{P_o}{2}\left[\delta_{1m} e^{j\theta_r} + \delta_{-1m} e^{-j\theta_r}\right] \quad (4)$$

where δ_{nm}, Kronecker's delta function, is unity for $n = m$ and is otherwise zero.

Bulk Relations: The transfer relations, Eqs. (a) of Table 2.16.2 with $k = 0$, are now used to represent the fields at the boundaries. In the stator dielectric surrounding the electrodes ($r > R_o$), $\alpha \to \infty$ and $\beta = R_o$, while $\varepsilon \to \varepsilon_s$:

$$\varepsilon_s \tilde{E}_{rm}^a = \varepsilon_s f_m(\infty, R_o)\tilde{\Phi}^a \quad (5)$$

In the air gap ($R_o > r > R$), $\alpha \to R_o$, $\beta \to R$ and $\varepsilon \to \varepsilon_o$ so that

$$\begin{bmatrix} \varepsilon_o \tilde{E}_{rm}^b \\ \varepsilon_o \tilde{E}_{rm}^c \end{bmatrix} = \varepsilon_o \begin{bmatrix} f_m(R, R_o) & g_m(R_o, R) \\ g_m(R, R_o) & f_m(R_o, R) \end{bmatrix} \begin{bmatrix} \tilde{\Phi}_m^b \\ \tilde{\Phi}_m^c \end{bmatrix} \quad (6)$$

Finally, within the rotor ($r < R$) the relations are used with $\alpha = R$, $\beta \to 0$ and $\varepsilon \to \varepsilon_o$:

$$\varepsilon_o \tilde{E}_{rm}^d = \varepsilon_o f_m(0, R)\tilde{\Phi}_m^d \quad (7)$$

The boundary conditions given by Eqs. 2 and 4 and the bulk relations of Eqs. 5, 6 and 7 comprise six expressions that can be used to determine the Fourier amplitudes ($\tilde{\Phi}_m^c$, $\tilde{\Phi}_m^d$, $\tilde{E}_{rm}^c$, $\tilde{E}_{rm}^d$, $\tilde{E}_{rm}^a$, $\tilde{E}_{rm}^b$) with the driving amplitudes ($\tilde{\Phi}_m^a$, $\tilde{\Phi}_m^b$) given by Eq. 1. The solution for any one of the amplitudes is usually much easier than this statement makes it seem, but nevertheless it is worthwhile to have the objective of the model in view before proceeding further.

Torque as a Function of Voltage and Rotor Angle (v, θ_r): The rotor is enclosed by a surface at the radial position (c) in the air gap. The method using the Maxwell stress to compute the torque is as outlined in connection with Eq. 4.2.3. With the fields represented by Fourier series, Eq. 2.15.17 reduces the average of the shear stress over the enclosing surface to a summation on the products of the Fourier amplitudes:

$$\tau_z = R(2\pi R\ell)\langle D_r^c E_\theta^c \rangle_\theta = 2\pi R^2 \ell \sum_{m=-\infty}^{+\infty} (\varepsilon_o \tilde{E}_{rm}^c)^* \left(\frac{jm}{R}\tilde{\Phi}_m^c\right) \quad (8)$$

Substitution for $\varepsilon_o \tilde{E}_{rm}^c$ from Eq. 6b introduces the stator field, which is given by Eq. 1, and the same field $\tilde{\Phi}_m^c$ as already appears in Eq. 8. On physical grounds it is expected that this latter "self-field" term should not make a contribution. This is indeed the case, because f_m is an even function of m so that terms in $|\tilde{\Phi}_m^c|^2$ cancel out of the sum. The mth term is cancelled by the -mth term. Thus, Eq. 8 reduces to

$$\tau_z = 2\pi R^2 \ell \sum_{m=-\infty}^{\infty} \varepsilon_o g_m(R, R_o)(\tilde{\Phi}_m^b)^* \left(\frac{jm}{R}\tilde{\Phi}_m^c\right) \quad (9)$$

and all that is required to determine the torque is an evaluation of $\tilde{\Phi}_m^c$.

With this objective, substitution of Eqs. 6b and 7 into Eq. 4 with Eq. 2 used to replace $\tilde{\Phi}_m^d$ with $\tilde{\Phi}_m^c$ gives an expression that can be solved for $\tilde{\Phi}_m^c$:

$$\tilde{\Phi}_m^c = \frac{\frac{P_o}{2}[\delta_{1m} e^{j\theta_r} + \delta_{-1m} e^{-j\theta_r}] - \varepsilon_o g_m(R, R_o)\tilde{\Phi}_m^b}{\varepsilon_o [f_m(R_o, R) - f_m(0, R)]} \quad (10)$$

This expression and Eq. 1 in turn can be used to evaluate the torque, Eq. 9. (Again, because g_m and f_m

are even in m, the self-field terms sum to zero):

$$\tau_z(v,\theta_r) = \frac{-4R\ell g_1(R,R_o)}{f_1(R_o,R) - f_1(0,R)} \frac{\sin\theta_o}{\theta_o} v(t)P_o \sin\theta_r \tag{11}$$

In a lumped parameter model for the device, with v(t) and $\theta_r(t)$ functions of time determined by the external electrical and mechanical constraints, this relation represents the electrical-to-mechanical coupling. The reciprocal mechanical-to-electrical coupling completes the model.

Electrical Terminal Relations: To describe the electrical terminals, the total charge q on the respective electrodes is required, again as a function of the terminal variables (v,θ_r). The charge on the upper electrode is

$$q = \ell \int_{-\frac{\pi}{2}+\theta_o}^{\frac{\pi}{2}-\theta_o} (\varepsilon_s E_r^a - \varepsilon_o E_r^b) R_o d\theta = \ell \int_{-\frac{\pi}{2}+\theta_o}^{\frac{\pi}{2}-\theta_o} \sum_{m=-\infty}^{+\infty} (\varepsilon_s \tilde{E}_{rm}^a - \varepsilon_o \tilde{E}_{rm}^b) e^{-jm\theta} R_o d\theta$$

$$= \ell R_o \sum_{m=-\infty}^{+\infty} \frac{2}{m}(\varepsilon_s \tilde{E}_{rm}^a - \varepsilon_o \tilde{E}_{rm}^b)\sin m(\tfrac{\pi}{2} - \theta_o) \tag{12}$$

The electric flux normal to the outer and inner surfaces of the electrode are computed from Eqs. 5 and 6a, respectively:

$$\varepsilon_s \tilde{E}_{rm}^a - \varepsilon_o \tilde{E}_{rm}^b = \varepsilon_s f_m(\infty,R_o)\tilde{\Phi}^a - \varepsilon_o f_m(R,R_o)\tilde{\Phi}_m^b - \varepsilon_o g_m(R_o,R)\tilde{\Phi}_m^c \tag{13}$$

The amplitudes $(\tilde{\Phi}_m^a, \tilde{\Phi}_m^b)$ are given in terms of v(t) by Eq. 2, while $\tilde{\Phi}_m^c$ is given by Eq. 10. Thus Eq. 13 is evaluated in terms of (v,θ_r):

$$q = C_s v(t) - A_r P_o \cos\theta_r(t) \tag{14}$$

where C_s, the stator self-capacitance, is independent of θ_r and is

$$C_s = \frac{4\ell R_o}{\pi} \sum_{\substack{m=-\infty \\ \text{odd}}}^{+\infty} \frac{\sin m(\frac{\pi}{2} - \theta_o)}{m^2} \frac{\sin m\theta_o}{m\theta_o} \sin(\tfrac{m\pi}{2}) \Big[\varepsilon_s f_m(\infty,R_o) - \varepsilon_o f_m(R,R_o) + \frac{\varepsilon_o g_m(R_o,R) g_m(R,R_o)}{f_m(R_o,R) - f_m(0,R)}\Big] \tag{15}$$

and A_r is a constant having the units of area

$$A_r = \frac{2\ell R_o g_1(R_o,R)}{f_1(R_o,R) - f_1(0,R)} \cos\theta_o \tag{16}$$

The required electrical terminal relation is Eq. 14.

For reasons that stem from the approximations made in the field description, the model represented by Eqs. 11 and 14 is not self-consistent. At the dielectric air-gap interface between electrodes, the potential is continuous, but $\vec{n}\cdot[\![\vec{D}]\!]$ is not. In physical terms, this means that the fields are as though segmented electrodes existed at $r = R_o$ in these transition regions having the linear potential distribution of Fig. 4.4.2 and supporting a surface charge that can be computed from Eq. 13. This charge is not included in Eq. 14 and might for some purposes be ignored. But, if the mechanical and electrical terminal relations are used as stated, the electromechanical system, which after all does not include energy dissipating elements, is given a model that does not conserve energy. In fact, once the torque is known, energy conservation formalisms introduced in Sec. 3.5 not only provide an alternative to computing the electrical terminal relations, but lead to a self-consistent model and a recognition that Eq. 15 can be considerably simplified.

In terms of lumped parameters, the system can be pictured as having the terminal pairs of Fig. 4.4.3. The electrical terminal pairs are interconnected so that $v_1 = -v_2 = v$ and by symmetry,

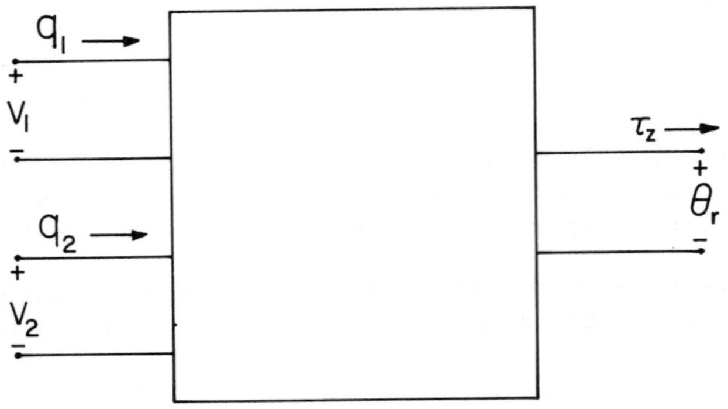

Fig. 4.4.3. Three-terminal pair lumped parameter system representing system of Fig. 4.4.1.

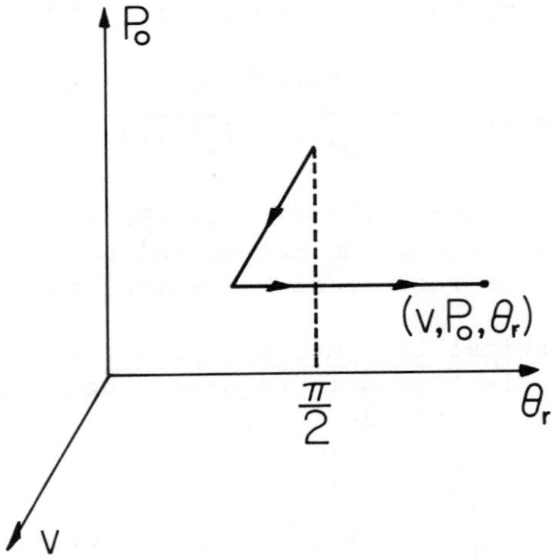

Fig. 4.4.4. State space integration contour.

$q_1 = -q_2 = q$. Thus, the incremental energy conservation equation is

$$\delta w = 2v\delta q - \tau_z d\theta_r \tag{17}$$

Not accessible through the external electrical terminals is the electric energy storage due to the permanent polarization. In Eq. 17 it is understood that P_o is held fixed. Transformation to a hybrid energy function $w''(v,P_o,\theta_r)$ is made by replacing $v\delta(2q) \to \delta(2qv) - 2q\delta v$ and defining $w'' = 2qv - w$, so that

$$\delta w'' = 2q\delta v + \tau_z d\theta_r \tag{18}$$

This expression is integrated on the state-space contour shown in Fig. 4.4.4. First, with the rotor at $\theta_r = \pi/2$, the polarization is brought up to its final state. Then the voltage is raised. Finally, with P_o and v held fixed, the rotor is turned to the angle θ_r of interest. With the rotor at $\theta_r = \pi/2$, the net charge induced on the upper electrode because of the polarization is zero. Hence, the net charge on the upper stator electrode is computed from Eq. 13, but with $\varepsilon_o E_r^b$ determined as if the rotor were not present. From Eq. 6,

$$\varepsilon_o \tilde{E}_{rm}^b = \varepsilon_o f_m(0,R_o)\tilde{\Phi}_m^b \tag{19}$$

Hence, Eq. 12 gives

$$q = C_s v; \quad C_s = \frac{4\ell R_o}{\pi} \sum_{\substack{m=-\infty \\ \text{odd}}}^{\infty} \frac{\sin m(\frac{\pi}{2} - \theta_o)}{m^2} \frac{\sin m\theta_o}{m\theta_o} \sin(\frac{m\pi}{2})[\varepsilon_s f_m(\infty,R_o) - \varepsilon_o f_m(0,R_o)] \tag{20}$$

In view of Eqs. 20 and 11, the integration of Eq. 18 on v and then on θ_r leads to

$$w'' = 2[\frac{1}{2} C_s v^2] + \left[\frac{4R\ell g_1(R,R_o)}{f_1(R_o,R) - f_1(0,R)} \frac{\sin \theta_o}{\theta_o}\right] vP_o \cos \theta_r \tag{21}$$

Finally, because $w'' = w''(v,P_o,\theta_r)$, the required terminal charge follows as

$$q = \frac{1}{2} \frac{\partial w''}{\partial v} = C_s v - A_r P_o \cos \theta_r \tag{22}$$

where

$$A_r = \frac{-2R\ell g_1(R,R_o)}{f_1(R_o,R) - f_1(0,R)} \frac{\sin \theta_o}{\theta_o} \tag{23}$$

and C_s is given by Eq. 20. Simplification of Eq. 15 leads to Eq. 20, but for the reasons discussed, Eqs. 16 and 23 differ by the factor $[\sin \theta_o/\theta_o]/\cos \theta_o$. The use of Eqs. 22 and 23 for the electrical terminal relation has the advantage that the model is then self-consistent in its representation of energy flow. The same advantage would exist if the energy relations were used to compute the electrical torque from the electrical terminal relations. This more conventional technique would make use of Eq. 14 and an integration of Eq. 18 in the sequence, P_o, θ_r and v. To carry out the second leg of this integration without making a contribution requires that symmetry be used to argue that there is no electrical torque even though the rotor is polarized.

4.5 Constrained-Charge Transfer Relations

For field sources constrained in their relative distribution, the transfer relation approach can not only be used for sources confined to boundaries, but can also be used to describe interactions with sources distributed through the bulk of a subregion. The objective in this section is to develop the principles underlying this generalization of the transfer relations for electroquasistatic fields and to summarize useful relations. The method is extended to certain magnetoquasistatic systems in Sec. 4.7.

In a region having a given net charge density ρ and uniform permittivity ε, Gauss' law and the requirement of irrotationality for $\vec{E}$ (Eqs. 2.3.23a and 2.3.23b) show that the electric potential Φ must satisfy Poisson's equation:

$$\nabla^2\Phi = -\frac{\rho}{\varepsilon} \tag{1}$$

In solving this linear equation, consider the solution to be a superposition of a homogeneous part Φ_H satisfying Laplace's equation and a particular solution Φ_P which, at each point in the volume of interest, has a Laplacian $-\rho/\varepsilon$:

$$\Phi = \Phi_H + \Phi_P \tag{2}$$

It is this latter component that balances the "drive" provided by the charge density when the total solution Φ is inserted into Eq. 1. By definition

$$\nabla^2\Phi_P = -\frac{\rho}{\varepsilon} \tag{3}$$

$$\nabla^2\Phi_H = 0 \tag{4}$$

In the three standard coordinate systems, the particular solution can be written as a superposition of the same variable-separable solutions used in Sec. 2.16 for the homogeneous solution. Thus,

$$\Phi_P = \begin{cases} \text{Re}\ \tilde{\Phi}_P(x,t)\exp[-j(k_y y + k_z z)] & \text{(Cartesian)} \\ \text{Re}\ \tilde{\Phi}_P(r,t)\exp[-j(m\theta + kz)] & \text{(cylindrical)} \\ \text{Re}\ \tilde{\Phi}_P(r,t)\,P_n^m(\cos\theta)\exp[-jm\phi] & \text{(spherical)} \end{cases} \tag{5}$$

With n used to denote the normal component at the respective bounding surfaces of the region described by the transfer relations, the homogeneous transfer relations of Tables 2.16.1, 2.16.2 and 2.16.3 relate the components of the homogeneous part of the solutions evaluated at the respective surfaces. Thus, in these relations, the substitution is made

$$\begin{aligned} &\tilde{\Phi}^\alpha \to \tilde{\Phi}_H^\alpha = \tilde{\Phi}^\alpha - \tilde{\Phi}_P^\alpha; \quad \tilde{\Phi}^\beta \to \tilde{\Phi}_H^\beta = \tilde{\Phi}^\beta - \tilde{\Phi}_P^\beta \\ &\tilde{D}_n^\alpha \to \tilde{D}_{nH}^\alpha = \tilde{D}_n^\alpha - \tilde{D}_{nP}^\alpha; \quad \tilde{D}_n^\beta \to \tilde{D}_{nH}^\beta = \tilde{D}_n^\beta - \tilde{D}_{nP}^\beta \end{aligned} \tag{6}$$

The transfer relations, which take the general form of Eq. 2.17.6, therefore relate the new surface variables and the particular solution evaluated at the surfaces:

$$\begin{bmatrix} \tilde{\Phi}^\alpha - \tilde{\Phi}_P^\alpha \\ \tilde{\Phi}^\beta - \tilde{\Phi}_P^\beta \end{bmatrix} = \begin{bmatrix} -A_{11} & A_{12} \\ -A_{21} & A_{22} \end{bmatrix} \begin{bmatrix} \tilde{D}_n^\alpha - \tilde{D}_{nP}^\alpha \\ \tilde{D}_n^\beta - \tilde{D}_{nP}^\beta \end{bmatrix} \tag{7}$$

Multiplied out, the transfer relations for regions with a bulk distribution of charge are

$$\begin{bmatrix} \tilde{\Phi}^\alpha \\ \tilde{\Phi}^\beta \end{bmatrix} = \begin{bmatrix} -A_{11} & A_{12} \\ -A_{21} & A_{22} \end{bmatrix} \begin{bmatrix} \tilde{D}_n^\alpha \\ \tilde{D}_n^\beta \end{bmatrix} + \begin{bmatrix} \tilde{h}^\alpha \\ \tilde{h}^\beta \end{bmatrix} \tag{8}$$

where

$$\begin{bmatrix} \tilde{h}^\alpha \\ \\ \tilde{h}^\beta \end{bmatrix} = \begin{bmatrix} \tilde{\Phi}_P^\alpha \\ \\ \tilde{\Phi}_P^\beta \end{bmatrix} + \begin{bmatrix} A_{11}\tilde{D}_{nP}^\alpha - A_{12}\tilde{D}_{nP}^\beta \\ \\ A_{21}\tilde{D}_{nP}^\alpha - A_{22}\tilde{D}_{nP}^\beta \end{bmatrix} \tag{9}$$

Associated with the surface variables related by these transfer relations are the bulk distributions of potential. These are obtained from the distributions of potential for no charge density by again using the substitutions summarized by Eq. 6. For example, in Cartesian coordinates, the potential distribution is the sum of Eq. 2.16.15 with $(\tilde{\Phi}^\alpha, \tilde{\Phi}^\beta)$ replaced by $(\tilde{\Phi}^\alpha - \tilde{\Phi}_P^\alpha, \tilde{\Phi}^\beta - \tilde{\Phi}_P^\beta)$ and the particular solution.

$$\tilde{\Phi} = (\tilde{\Phi}^\alpha - \tilde{\Phi}_P^\alpha)\frac{\sinh \gamma x}{\sinh \gamma\Delta} - (\tilde{\Phi}^\beta - \tilde{\Phi}_P^\beta)\frac{\sinh \gamma(x - \Delta)}{\sinh \gamma\Delta} + \tilde{\Phi}_P(x) \tag{10}$$

The same substitution generalizes the cylindrical coordinate potentials, Eqs. 2.16.20, 2.16.21 and 2.16.25 as well as those in spherical coordinates, Eq. 2.16.36.

Particular Solutions (Cartesian Coordinates): Any Φ_P having the form of Eq. 5 can be used in Eqs. 8 and 9. "Inspection" yields solutions in many cases. However, it is often true that the most useful solutions belong to a class that can be generated by the procedure now illustrated in Cartesian coordinates.

Within the planar region (shown in Table 2.16.1) there is a charge distribution that has an arbitrary dependence on the transverse coordinate x but the y-z dependence of Eq. 5a for complex amplitude, Fourier series or Fourier transform representations:

$$\rho = \text{Re} \sum_{i=0}^{\infty} \tilde{\rho}_i(t)\Pi_i(x)e^{-j(k_y y + k_z z)} \tag{11}$$

Here, the distribution has been represented as a superposition of modes $\Pi_i(x)$ having individual complex amplitudes $\tilde{\rho}_i(t)$. These as yet to be determined modes are defined such that the particular solution can be written as a superposition of the same modes:

$$\Phi_P = \text{Re} \sum_{i=0}^{\infty} \tilde{\Phi}_i(t)\Pi_i(x)e^{-j(k_y y + k_z z)} \tag{12}$$

The same functions are used for both ρ and Φ_P because then substitution into Poisson's equation, Eq. 3, shows that a particular solution has been found, provided that the modes satisfy the Helmholtz equation:

$$\frac{d^2\Pi_i}{dx^2} + \nu_i^2\Pi_i = 0; \quad \nu_i^2 = \frac{\tilde{\rho}_i}{\varepsilon\tilde{\Phi}_i} - k_y^2 - k_z^2 \tag{13}$$

It follows from Eq. 13 that Π_i is a linear combination of $\sin(\nu_i x)$ and $\cos(\nu_i x)$. Boundary conditions, selected as a matter of convenience and to give orthogonal modes that can be used to expand an arbitrary charge distribution in a quickly convergent series, complete the specification of the modes. For example, the transfer relations, Eqs. 8 and 9, are simplified if

$$\tilde{D}_{nP}^\alpha \equiv -\varepsilon \left.\frac{d\tilde{\Phi}_P}{dx}\right|_\alpha = 0; \quad \tilde{D}_{nP}^\beta \equiv -\varepsilon \left.\frac{d\tilde{\Phi}_P}{dx}\right|_\beta = 0 \tag{14}$$

so these will be used as boundary conditions in solving Eq. 13. It follows that for a layer with α and β surfaces at $x = \Delta$ and $x = 0$, respectively,

$$\Pi_i = \cos \nu_i x; \quad \nu_i = \frac{i\pi}{\Delta}; \quad i = 0,1,2,\cdots \tag{15}$$

From the definition of ν_i, Eq. 13, the potential and charge-density amplitudes called for in Eqs. 11 and 12 are related by

$$\tilde{\Phi}_i = \frac{\tilde{\rho}_i}{\varepsilon(\nu_i^2 + k_y^2 + k_z^2)} \tag{16}$$

The charge-density amplitudes are determined from a given distribution Re $\tilde{\rho}(x,t)\exp[-j(k_y y + k_z z)]$ by a Fourier analysis. That is, Eq. 11 is multiplied by Π_k, integrated from $0 \to \Delta$, solved for $\tilde{\rho}_k$ and $k \to i$:

$$\tilde{\rho}_i = \frac{2}{\Delta}\int_o^\Delta \tilde{\rho}(x,t)\Pi_i(\nu_i x)dx;\ i \neq 0:\ \tilde{\rho}_o = \frac{1}{\Delta}\int_o^\Delta \tilde{\rho}\ (x,t)dx \tag{17}$$

The associated transfer relations, Eqs. 8 and 9 evaluated using Eqs. 12, 15 and 16, with A_{ij}'s from Table 2.16.1, become

$$\begin{bmatrix} \tilde{\Phi}^\alpha \\ \\ \tilde{\Phi}^\beta \end{bmatrix} = \frac{1}{\varepsilon\gamma}\begin{bmatrix} -\coth\gamma\Delta & \dfrac{1}{\sinh\gamma\Delta} \\ \\ \dfrac{-1}{\sinh\gamma\Delta} & \coth\gamma\Delta \end{bmatrix}\begin{bmatrix} \tilde{D}_x^\alpha \\ \\ \tilde{D}_x^\beta \end{bmatrix} + \sum_{i=0}^{+\infty}\frac{\tilde{\rho}_i}{\varepsilon(\nu_i^2+\gamma^2)}\begin{bmatrix} (-1)^i \\ \\ 1 \end{bmatrix} \tag{18}$$

The potential distribution is given in terms of these amplitudes and the particular solution (Eqs. 12, 15 and 16) by Eq. 10. Note that to make use of Eq. 10 the origin of the x axis need not be coincident with the β surface. The equation applies to a region with the β surface at $x = a$ if the substitution is made $x \to x + a$.

Cylindrical Annulus: In cylindrical coordinates, the given charge distribution and particular solution take the form

$$\rho = \text{Re}\sum_{i=0}^{\infty}\tilde{\rho}_i(t)\Pi_i(r)e^{-j(m\theta+kz)};\quad \Phi_P = \text{Re}\sum_{i=0}^{\infty}\tilde{\Phi}_i(t)\Pi_i(r)e^{-j(m\theta+kz)} \tag{19}$$

Thus, Poisson's equation, Eq. 1, requires that

$$\frac{d^2\Pi_i}{dr^2} + \frac{1}{r}\frac{d\Pi_i}{dr} + (\nu_i^2 - \frac{m^2}{r^2})\,\Pi_i = 0;\ \nu_i^2 \equiv \frac{\tilde{\rho}_i}{\varepsilon\tilde{\Phi}_i} - k^2 \tag{20}$$

and the potential amplitudes are related to the charge density amplitudes by

$$\tilde{\Phi}_i = \frac{\tilde{\rho}_i}{\varepsilon(\nu_i^2 + k^2)} \tag{21}$$

Boundary conditions used in selecting solutions to Eq. 20 might be selected analogous to those of Eq. 14. This would simplify the transfer relations, but require solution of a relatively complicated transcendental equation for the ν_i's. Instead, the particular solution is required to vanish on the outer surface only and solutions that are singular at the origin are excluded. In cylindrical coordinates this is sufficient to result in a complete set of orthogonal modes:

$$\tilde{D}_{rP}^\alpha = -\varepsilon\left.\frac{d\tilde{\Phi}_P}{dr}\right|_\alpha = 0 \tag{22}$$

Comparison of Eq. 20 to Eq. 2.16.19 shows that the solutions that are not singular at the origin are Bessel's functions of first kind and order m:

$$\Pi_i = J_m(\nu_i r) \tag{23}$$

To satisfy the boundary condition, Eq. 22, the ν_i's must be roots of

$$\nu_i J_m'(\nu_i\alpha) = 0 \tag{24}$$

In now evaluating the transfer relations, Eqs. 8 and 9, the normal flux density is zero at the α surface, but otherwise all of the particular solution entries make a contribution:

$$\begin{bmatrix}\tilde{\Phi}^\alpha\\ \\ \tilde{\Phi}^\beta\end{bmatrix} = \frac{1}{\varepsilon}\begin{bmatrix}F_m(\beta,\alpha) & G_m(\alpha,\beta)\\ \\ G_m(\beta,\alpha) & F_m(\alpha,\beta)\end{bmatrix}\begin{bmatrix}\tilde{D}_r^\alpha\\ \\ \tilde{D}_r^\beta\end{bmatrix} + \sum_{i=0}^{\infty}\frac{\tilde{\rho}_i}{\varepsilon(\nu_i^2+k^2)}\begin{bmatrix}J_m(\nu_i\alpha)+\nu_i G_m(\alpha,\beta)J_m'(\nu_i\beta)\\ \\ J_m(\nu_i\beta)+\nu_i F_m(\alpha,\beta)J_m'(\nu_i\beta)\end{bmatrix} \tag{25}$$

An important limiting case is $\beta \to 0$ so that the region is a "solid" cylinder. This limit is most conveniently taken by first using the limiting form of the transfer relation, Eq. (b) of Table 2.16.2, which becomes

$$\tilde{\Phi}^\alpha - \tilde{\Phi}_P^\alpha = \frac{1}{\varepsilon}F_m(0,\alpha)[\tilde{D}_r^\alpha - \tilde{D}_{rP}^\alpha] \tag{26}$$

Put in the form of Eq. 25, the transfer relation for a solid cylinder is

$$\tilde{\Phi}^\alpha = \frac{1}{\varepsilon}F_m(0,\alpha)\tilde{D}_r^\alpha + \sum_{i=0}^{\infty}\frac{\tilde{\rho}_i}{\varepsilon(\nu_i^2+k^2)}J_m(\nu_i\alpha) \tag{27}$$

The charge-density amplitudes $\tilde{\rho}_i$ are evaluated in terms of the given charge distribution by exploiting the orthogonality of the Π_i's.

<u>Orthogonality of Π_i's and Evaluation of Source Distributions</u>: The given transverse distribution of ρ is used to evaluate the mode amplitudes, $\Pi_i(x)$ or $\Pi_i(r)$ and hence $\tilde{\rho}_i$. Because the particular solutions are in each case a superposition of solutions to the Helmholtz equation, with appropriate boundary conditions, the eigenmodes Π_i are orthogonal. In the Cartesian coordinate cases, this means that

$$\int_0^\Delta \Pi_i(\nu_i x)\Pi_j(\nu_j x)dx = \frac{\Delta}{2}\delta_{ij} \tag{28}$$

This relation is the basis for evaluating the Fourier coefficients, for example Eq. 17. Proof of orthogonality and determination of the coefficients is possible in this case by direct integration. But, in the circular geometry, a more powerful method is needed, one based on the properties of $\Pi_i(\nu_i r)$ that can be deduced from the differential equation and boundary conditions. The proof of orthogonality and determination of the normalizing factor is as follows.

Multiply Eq. 20 by $r\Pi_j$ and integrate from the origin to the outer radius. The first term can then be integrated by parts to obtain

$$r\Pi_j(\nu_j r)\left.\frac{d\Pi(\nu_i r)}{dr}\right|_o^\alpha - \int_o^\alpha r\frac{d\Pi_i(\nu_i r)}{dr}\frac{d\Pi_j(\nu_j r)}{dr}dr + \int_o^\alpha r(\nu_i^2 - \frac{m^2}{r^2})\Pi_i\Pi_j dr = 0 \tag{29}$$

This expression also holds with i and j reversed. The latter equation, subtracted from Eq. 29, gives

$$(\nu_i^2 - \nu_j^2)\int_o^\alpha r\Pi_i\Pi_j\,dr = r\Pi_i\left.\frac{d\Pi_j}{dr}\right|_o^\alpha - r\Pi_j\left.\frac{d\Pi_i}{dr}\right|_o^\alpha \tag{30}$$

Thus, it is clear that either for $\Pi_i = 0$ or $d\Pi_i/dr = 0$ at $r = \alpha$, the functions Π_i and Π_j are orthogonal in the sense that the integral appearing in Eq. 30 vanishes provided $i \neq j$.

The value of the integral for $i = j$ is required in evaluating the coefficients in the charge density expansion, and is deduced by taking the limit where $\nu_j \to \nu_i$, or $\Delta\nu \to 0$ in $(\nu_j = \nu_i + \Delta\nu)$

$$\Pi_j(\nu_j r) = \Pi_j[\nu_i r + (\Delta\nu)r] \simeq \Pi_j(\nu_i r) + [\Pi_j'(\nu_i r)]r\Delta\nu \tag{31}$$

Again, the prime indicates a derivative with respect to the argument $(\nu_j r)$. Expansion of Eq. 31 to first order in $\Delta\nu$ shows that in the limit $\Delta\nu \to 0$,

$$\int_o^\alpha r\Pi_i\Pi_j dr = \delta_{ij}\frac{\alpha^2}{2}\left\{[\Pi_i'(\nu_i\alpha)]^2 + [1 - \frac{m^2}{(\nu_i\alpha)^2}]\Pi_i^2(\nu_i\alpha)\right\} \tag{32}$$

In obtaining this result, the fact that Π_i satisfies Bessel's equation, Eq. 20, has again been used to substitute for Π_i'' in terms of Π_i and Π_i'.

An example exploiting the cylindrical constrained-charge transfer relations and orthogonality relations is developed in Sec. 4.6.

4.6 Kinematics of Traveling-Wave Charged-Particle Devices

Synchronous interactions between a "stator" potential wave and a traveling wave of charge are abstracted in Part 3 of Table 4.3.1. In the most common practical devices exploiting such electric interactions, the space-charge wave is itself created by the electromechanical interaction between a structure potential and a uniformly charged beam. These examples are not "kinematic" in the sense that the relative distribution of space charge cannot be prescribed. Nevertheless, by representing the interaction as though independent control can be obtained over the beam and structure traveling waves, the energy conversion principles are highlighted. In addition, this section illustrates how the constrained-charge transfer relations of Sec. 4.5 are put to work. Self-consistent interactions through electrical stresses will be developed in Chaps. 5 and 8.

In the model shown in Fig. 4.6.1, the space-charge wave has the shape of a circular cylinder of radius R and charge density

$$\rho = -\rho_B \cos(\omega t - kz + k\delta) = \text{Re}\,\tilde{\rho}\exp(-jkz);\ \tilde{\rho} \equiv [-\rho_B \exp(jk\delta)]\exp(j\omega t) \qquad (1)$$

where ρ_B is a constant.

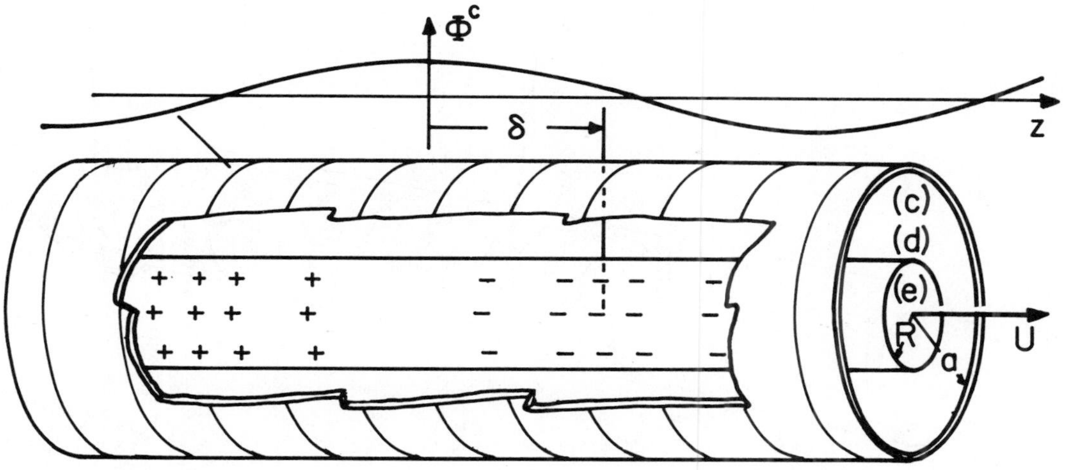

Fig. 4.6.1. Regions of positive and negative charge represent concentrations and rarefactions in the local charge density of an initially uniformly charged beam moving in the z direction with the velocity U.

In an electron beam device,[1] the stream is initially of uniform charge density. But, perhaps initiated by means of a modulating field introduced upstream, the particles become bunched. The resulting space charge can be viewed as the superposition of uniform and periodic space-charge components. The uniform component gives rise to an essentially radial field which tends to spread the beam. (Through the $q\vec{v} \times \vec{B}$ force attending any radial motion of the particle, a longitudinal magnetic field is often used to confine the beam and prevent its spreading. In any case, here the effect of this radial field is considered negligible.)

In traveling-wave beam devices, the interaction is with a traveling wave of potential on a slow-wave (perhaps helical) structure, such as that shown schematically in Fig. 4.6.2a. The structure is designed to propagate an electromagnetic wave with velocity less than that of light, so that it can be in synchronism with the space-charge wave. For the present purposes, this potential is imposed on a wall at r = c:

$$\Phi^c = V_o \cos(\omega t - kz) = \text{Re}\tilde{V}_o e^{-jkz};\ \tilde{V}_o = V_o e^{j\omega t} \qquad (2)$$

In the kinematic model of Fig. 4.6.1, the coupling can either retard or accelerate the beam, depending on whether operation is akin to a generator or motor (Table 4.3.1). Traveling-wave electron beam amplifiers and oscillators are generators, in that they convert the steady kinetic energy of the beam to an a-c electrical output. The result of the interaction is a time-average retarding force that tends

1. Basic electron beam electromechanics are discussed in the text Field and Wave Electrodynamics, by Curtis C. Johnson, McGraw-Hill Book Company, New York, 1965, p. 275.

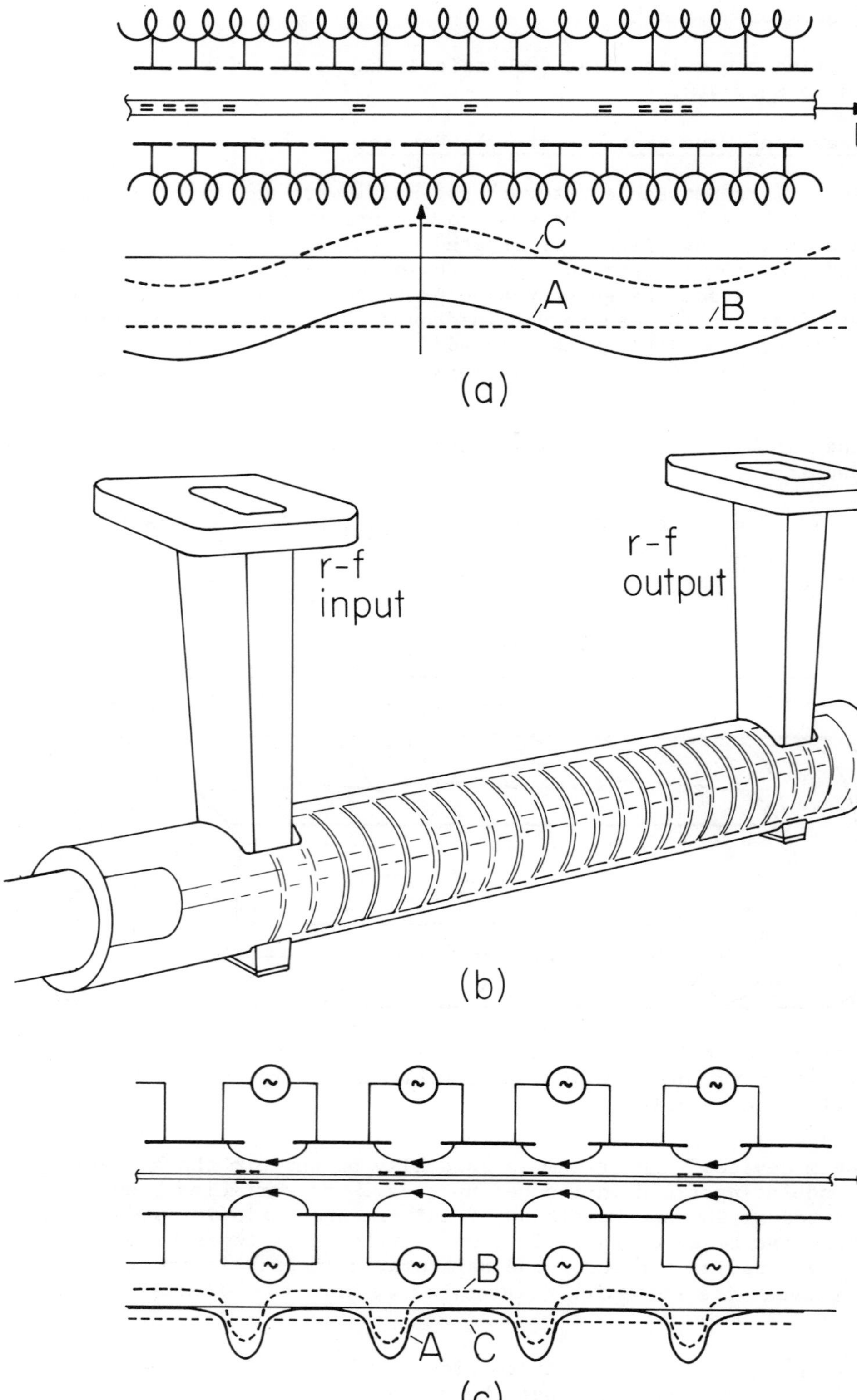

Fig. 4.6.2. (a) Schematic representation of traveling-wave electron beam device with slow-wave structure modeled by distributed circuit coupled to beam through the electric field. Below structure is distribution of space charge in the beam (A), and the equivalent distribution of a uniform charge density (B) and a periodic distribution (C). (b) Combination cutaway and phantom view of low-noise low-power traveling-wave tube that operates in part of the frequency range 2 to 40 GHz. (c) Schematic of linear accelerator designed so that oscillating gap voltages "kick" particles as they pass. Shown below are "bunches" of particles and hence space charge (A) and the equivalent superposition of periodic and uniform parts (B) and (C).

to slow the beam.

The "motor" of particle beam devices is the particle accelerator typified by Fig. 4.6.2c. Here, the object is to accelerate bunches of particles to extremely high velocities by subjecting them to alternating electric fields phased in such a way that when a bunch arrives at an accelerating gap, the fields tend to give it an additional "kick" in the axial direction.[2] The complex fields associated with the traveling particle bunches and accelerating fields are typically represented as traveling waves, as suggested by Fig. 4.6.2c. The principal periodic component of the space-charge wave is represented in the model of Fig. 4.6.1.

In this section it is presumed that the particle velocities are unaffected by the interaction; U is a constant. In fact, the object of the generator is to slow the beam, and of the accelerator is to increase the velocity; a more refined analysis is likely to be required for particular design purposes.

In yet another physical situation, the constraints on mechanical motion and wall potentials assumed in this section are imposed. At low frequencies and velocities, it is possible to deposit charge on a moving insulating material. Then, the relative charge velocity is known. Moreover, at low frequencies it is possible to use segmented electrodes and voltage sources to impose the postulated potential distribution.

As will be seen, at low velocities it is difficult to achieve competitive energy conversion densities using macroscopic electric forces. So, at low frequencies, the class of devices discussed in this section might be used as high-voltage generators rather than as generators of bulk power.

The net force on a section of the beam having length ℓ is found by integrating the stress over a surface adjacent to the outer wall (see Fig. 4.2.1b for detailed discussion of this step):

$$f_z = 2\pi a\ell\left\langle D_r^c E_z^c\right\rangle_z = \pi a\ell \mathrm{Re}[(\tilde{D}_r^c)^* jk\tilde{V}_o] \tag{3}$$

To compute $\tilde{D}^c$, and hence f_z, the potential is related to the normal electric flux and charge density by the transfer relation for a "solid" cylinder of charge, Eq. 4.5.27 with m = 0:

$$\tilde{\Phi}^\alpha = \frac{1}{\varepsilon_o} F_o(0,\alpha)\tilde{D}_r^\alpha + \sum_{i=0}^{\infty} \frac{\tilde{\rho}_i J_o(\nu_i\alpha)}{\varepsilon_o(\nu_i^2 + k^2)} \tag{4}$$

Table 2.16.2 summarizes $F_o(0,\alpha)$.

Single-Region Model: It is instructive to consider two alternative ways of representing the fields. First, consider that the beam and the surrounding annular region comprise a single region with a charge density distribution as sketched in Fig. 4.6.3. Then, in Eq. 4, the radius $\alpha = a$ and the position $(\alpha) \rightarrow (c)$. Multiplication of Eq. 4.5.19a by $r\Pi_j(\nu_j r)$ and integration $0 \rightarrow a$ then gives

$$\int_o^R \tilde{\rho} r J_o(\nu_j r)dr = \sum_{i=0}^{\infty} \tilde{\rho}_i \int_o^a rJ_o(\nu_i r)J_o(\nu_j r)dr \tag{5}$$

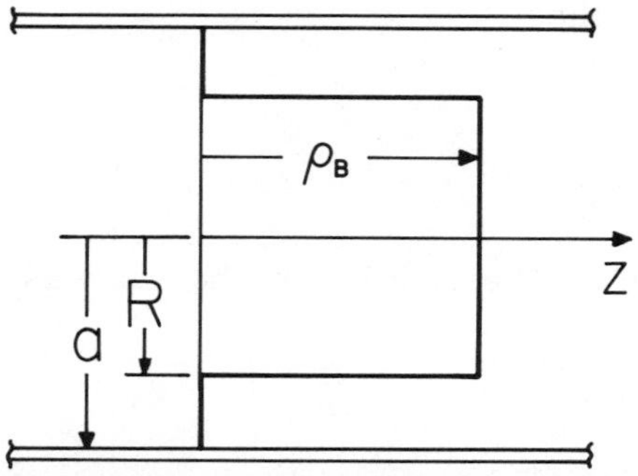

Fig. 4.6.3

Radial distribution of charge density.

The right-hand side is integrated using Eq. 4.5.32, while the left-hand side is an integral that can be evaluated from tables or by using the fact that $J_o(\nu_i r)$ satisfies Eq. 4.5.20 with m = 0 and Eq. 2.16.26c holds for J_o:

$$\tilde{\rho}\frac{R}{\nu_i} J_1(\nu_j R) = \tilde{\rho}_j \frac{a^2}{2} J_o^2(\nu_j a) \Rightarrow \tilde{\rho}_i = \frac{2\tilde{\rho}RJ_1(\nu_i R)}{\nu_i a^2 J_o^2(\nu_i a)}; \; i \neq 0 \tag{6}$$

2. A discussion of synchronous-type particle accelerators is given in Handbook of Physics, E. U. Condon and H. Odishaw, eds., McGraw-Hill Book Company, New York, 1958, pp. 9-156.

The root $\nu_i = 0$ to Eq. 4.5.24 is handled separately in integrating Eq. 5. In that case $J_o = 1$ and $\tilde{\rho}_o = R^2\tilde{\rho}/a^2$.

Because $\tilde{\Phi}^c = \tilde{V}_o$, Eq. 4 can now be solved for $\tilde{D}_r^c$:

$$\tilde{D}_r^c = \varepsilon F_o^{-1}(0,a)\left\{\tilde{V}_o - \tilde{\rho}\left[\frac{R^2}{\varepsilon(ak)^2} + \sum_{i=1}^{\infty}\frac{2RJ_1(\nu_i R)}{\varepsilon\nu_i a^2(\nu_i^2 + k^2)J_o(\nu_i a)}\right]\right\} \tag{7}$$

It follows from Eq. (3) that, for the distribution of charge and structure potential given by Eqs. 1 and 2, the required force on a length ℓ of the beam is

$$f_z = -(\pi R^2\ell)(kV_o\rho_B \sin k\delta)L_1 \tag{8}$$

where

$$L_1 = -\frac{a}{R}\left\{\left(\frac{R}{a}\right)\frac{1}{(ak)^2} + \sum_{i=1}^{\infty}\frac{2J_1[(\nu_i a)\frac{R}{a}]}{(\nu_i a)[(\nu_i a)^2 + (ak)^2]J_o(\nu_i a)}\right\}aF_o^{-1}(0,a)$$

Hence, the force has the characteristic dependence on the spatial phase shift between structure potential and beam space-charge waves identified for synchronous interactions in Sec. 4.3.

Two-Region Model: Consider next the alternative description. The region is divided into a part having radius R and described by Eq. 4 (with the position $\alpha \to e$ and radius $\alpha \to R$) and an annulus of free space. Because the charge density is uniform over the inner region, only the $i = 0$ term (having the eigenvalue $\nu_o = 0$) in the series of Eq. 4.5.1 is required to exactly describe the charge and potential distributions. With variables labeled in accordance with Fig. 4.6.1, Eq. 4 becomes

$$\tilde{\Phi}^e = \frac{D_r^e F_o(0,R)}{\varepsilon} + \frac{\tilde{\rho}}{\varepsilon k^2} \tag{9}$$

The annular region of free space is described by Eqs. (a) of Table 2.16.2:

$$\begin{bmatrix}\tilde{D}_r^c \\ \tilde{D}_r^d\end{bmatrix} = \varepsilon\begin{bmatrix} f_o(R,a) & g_o(a,R) \\ g_o(R,a) & f_o(a,R)\end{bmatrix}\begin{bmatrix}\tilde{\Phi}^c \\ \tilde{\Phi}^d\end{bmatrix} \tag{10}$$

Boundary conditions splice the regions together:

$$\tilde{\Phi}^c = \tilde{V}_o,\quad \tilde{\Phi}^d = \tilde{\Phi}^e,\quad \tilde{D}_r^d = \tilde{D}_r^e \tag{11}$$

In view of these conditions, Eqs. 9 and 10b combine to show that

$$\tilde{\Phi}^d = \frac{g_o(R,a)\tilde{V}_o + \tilde{\rho}\,F_o^{-1}(0,R)\varepsilon^{-1}k^{-2}}{F_o^{-1}(0,R) - f_o(a,R)} \tag{12}$$

From Eq. 10a $\tilde{D}_r^c$ can be found and the force, Eq. 3, evaluated. The result is the same as Eq. 8 except that L_1 is replaced by

$$L_2 = \frac{[ag_o(a,R)][aF_o^{-1}(0,R)]}{(ka)^2(\frac{R}{a})^2[aF_o^{-1}(0,R) - af_o(a,R)]} = \frac{1}{kR}\frac{I_o'(kR)}{I_o(ka)} \tag{13}$$

To obtain the second expression, note that the reciprocity condition, Eq. 2.17.10, requires that $ag_o(a,R) = -Rg_o(R,a)$.

Numerically, Eqs. 8 and 13 are the same. They are identical in form in the limit where the charge completely fills the region r<a, as can be seen by taking the limit $R \to a$ in each expression

$$L_1 \to L_2 = -\frac{aF_o^{-1}(0,a)}{(ak)^2} \tag{14}$$

In the example considered here the second representation gives the simpler result. But, if the splicing approach exemplified by Eq. 13 were used to represent a more complicated radial distribution of charge, the clear advantage would be with the single region representation illustrated by Eq. 8.

The dependence of L_2 on the wavenumber normalized to the wall radius is shown in Fig. 4.6.4. As would be expected, the coupling to the wall becomes weaker with increasing k (decreasing wavelength). The part of the coupling represented by L_2 also becomes smaller as the beam becomes more confined to the center. Note however that there is an R^2 factor in Eq. 8 that makes the effect of decreasing R much stronger than reflected in L_2 (or L_1) alone.

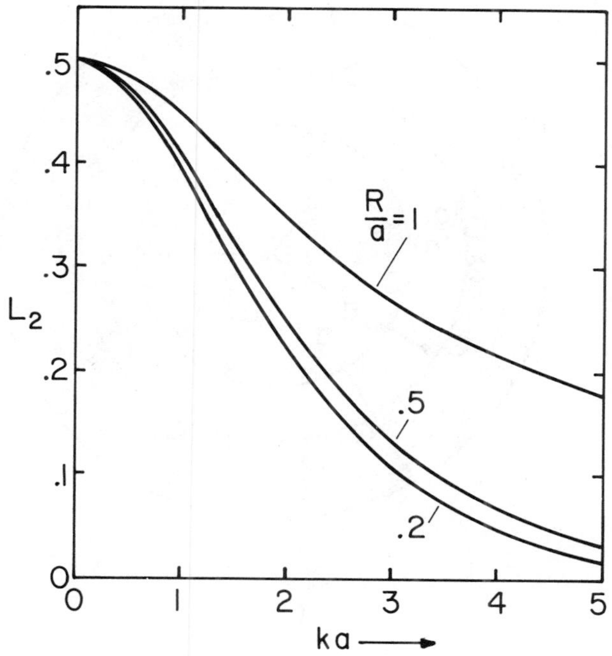

Fig. 4.6.4. Function L_2 defined by Eq. 4.6.8.

4.7 Smooth Air-Gap Synchronous Machine Model

A specific result in this section is the terminal relations that constitute the lumped-parameter model for a three-phase two-pole smooth air-gap synchronous machine. The derivations are aimed at exemplifying the pattern that can be followed in describing a wide class of magnetic field devices modeled by coupling at surfaces.

In the cross-sectional view of the smooth air-gap machine shown in Fig. 4.7.1a, the stator structure consists of a laminated circular cylindrical material having permeability μ_s with outside radius a and inner radius b. Imbedded in slots on this inner surface are three windings, having turns densities that vary sinusoidally with θ. These slots are typically as shown in Fig. 4.7.2b, where the laminations used for construction of rotor and stator for the generator of Fig. 4.7.2a are pictured. Only one of these stator windings is shown in Fig. 4.7.1, the "a" phase with its magnetic axis at $\theta = -90°$. The "b" and "c" phases are similarly distributed but rotated so that their magnetic axes are respectively at the angles 30° and 150°. Thus the peak surface current density for the respective windings comes at the angles $\theta = 0$, $\theta = 120°$, and $\theta = 240°$. These stator windings have peak turns densities N_a, N_b, N_c, respectively, and carry the terminal currents (i_a, i_b, i_c). Because the stator windings essentially form a current sheet at the radius b, their contribution to the field is modeled by the surface current density

$$K_z^s = i_a(t)N_a \cos\theta + i_b(t)N_b \cos(\theta - \tfrac{2\pi}{3}) + i_c(t)N_c \cos(\theta - \tfrac{4\pi}{3})$$
$$= \text{Re}\,\tilde{K}^s e^{-j\theta};\ \tilde{K}^s = i_aN_a + i_bN_b e^{j(\frac{2\pi}{3})} + i_cN_c e^{j(\frac{4\pi}{3})} \tag{1}$$

There is only one phase on the rotor, consisting of sinusoidally distributed windings of peak turns density N_r excited through slip rings by the terminal current i_r. With the rotor angular position denoted by θ_r, the rotor current is modeled by a surface current density at $r = c$ of

$$K_z^r = i_r(t)N_r \cos(\theta - \theta_r) = \text{Re}\,\tilde{K}^r e^{-j\theta};\ \tilde{K}^r = i_rN_r e^{j\theta_r} \tag{2}$$

These excitations have been written in the complex amplitude notation. Fields in each region are described by the polar coordinate transfer relations of Table 2.19.1 with m = 1.

The objective in the following calculations is to relate the electrical and mechanical terminal relations so that electromechanical coupling, represented schematically in Fig. 4.7.3, is specified in the form

$$\begin{bmatrix}\lambda_a\\ \lambda_b\\ \lambda_c\\ \lambda_r\end{bmatrix} = \begin{bmatrix}L_{aa} & L_{ab} & L_{ac} & L_{ar}\\ L_{ba} & L_{bb} & L_{bc} & L_{br}\\ L_{ca} & L_{cb} & L_{cc} & L_{cr}\\ L_{ra} & L_{rb} & L_{rc} & L_{rr}\end{bmatrix}\begin{bmatrix}i_a\\ i_b\\ i_c\\ i_r\end{bmatrix} \tag{3}$$

$$\tau_z = \tau_z(i_a, i_b, i_c, i_r, \theta_r) \tag{4}$$

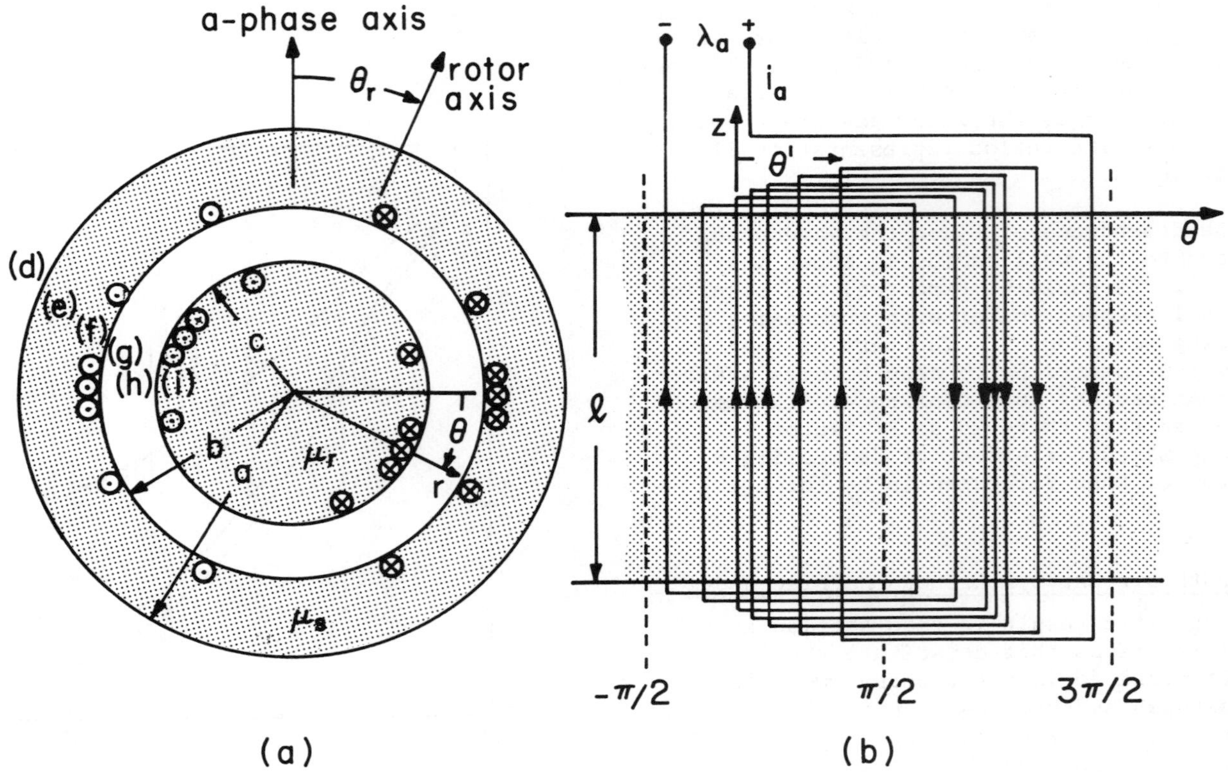

Fig. 4.7.1. (a) Cross-sectional view of smooth air-gap synchronous machine showing only one of three phases on stator. (b) Distribution of "a"-phase windings on stator as seen looking radially inward.

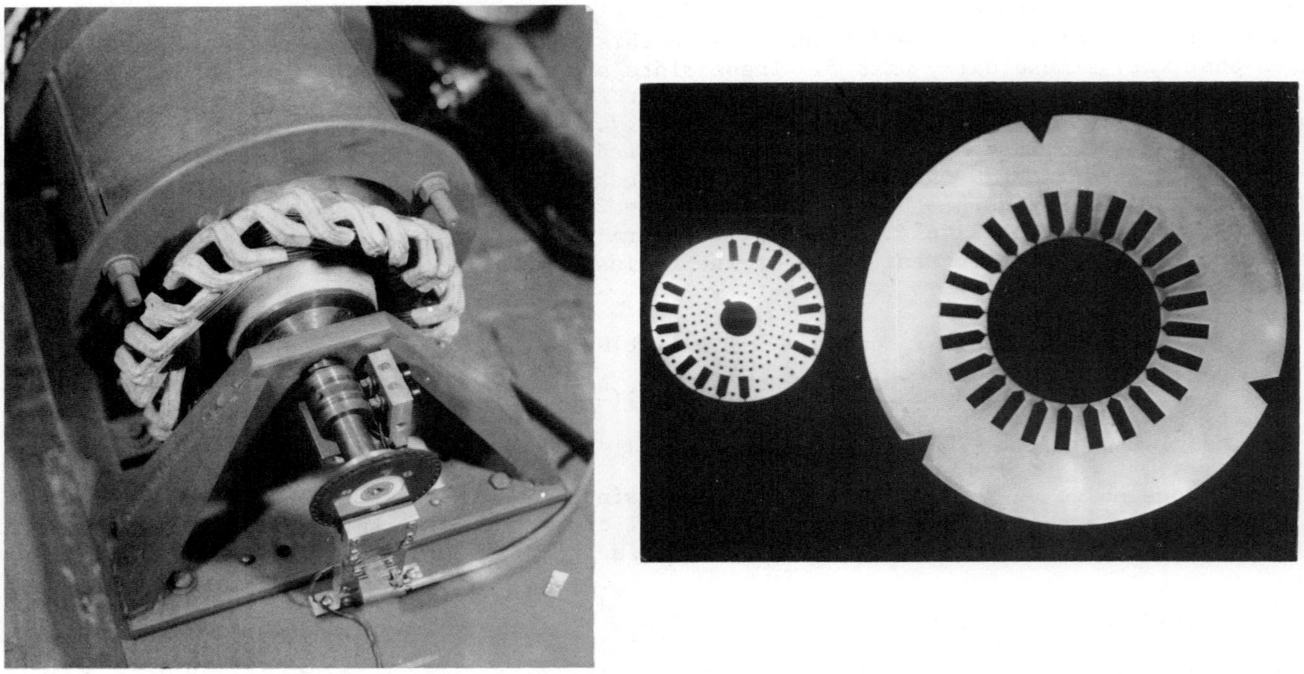

(a) (b)

Fig. 4.7.2. (a) Model synchronous alternating having rating of about one kVA and modeling 900 MVA machine. Unit is one of several used in MIT Electric Power Systems Engineering Laboratory as part of model power system. Slip rings for supplying field current are on shaft near bearing. Disk with holes is for measurement of angular position of rotor. (b) Rotor and stator laminations used for model machine of (a). Rectangular slots carry windings. Conducting rods inserted through the circular holes in the rotor are shorted at the ends of the rotor to simulate transient eddy-current (induction machine) effects in full-scale machine. The scaling requires that the model have extremely narrow air gap of about 0.23 mm, as compared to the gap of about 7 cm in the full-scale machine.

Boundary Conditions: The field excitations represented by Eqs. 1 and 2, written in complex-amplitude notation, can be matched by single components of the fields represented in each region by the polar coordinate transfer relations of Table 2.19.1. In view of the θ dependence of the current sheets, m = 1.

Positions adjacent to the boundaries between current-free regions of uniform permeability in Fig. 4.7.1a are denoted by (d) - (i). Fields are assumed to vanish far from the outer surface. At each surface, the normal flux density is continuous (Eq. 2.10.22). This means that the vector potential is continuous, and hence

$$\tilde{A}^d = \tilde{A}^e \tag{5}$$

$$\tilde{A}^f = \tilde{A}^g \tag{6}$$

$$\tilde{A}^h = \tilde{A}^i \tag{7}$$

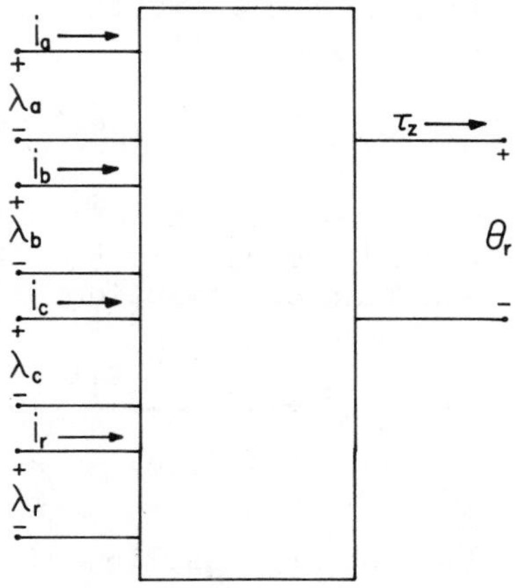

Fig. 4.7.3. Electromechanical coupling network for system of Fig. 4.7.1.

The jump in the tangential field intensity is equal to the surface current density (Eq. 2.10.21), and hence

$$\tilde{H}_\theta^d - \tilde{H}_\theta^e = 0 \tag{8}$$

$$\tilde{H}_\theta^f - \tilde{H}_\theta^g = \tilde{K}^s \tag{9}$$

$$\tilde{H}_\theta^h - \tilde{H}_\theta^i = \tilde{K}^r \tag{10}$$

Bulk Relations: Each of the uniform regions is described by Eq. (c) of Table 2.19.1. In the exterior region, $\alpha \to \infty$, $\beta = a$, and $\mu = \mu_o$

$$\tilde{H}_\theta^d = \frac{1}{\mu_o} f_1(\infty, a)\tilde{A}^d \tag{11}$$

In the stator, $\alpha = a$, $\beta = b$, and $\mu = \mu_s$:

$$\begin{bmatrix} \tilde{H}_\theta^e \\ \tilde{H}_\theta^f \end{bmatrix} = \frac{1}{\mu_s}\begin{bmatrix} f_1(b,a) & g_1(a,b) \\ g_1(b,a) & f_1(a,b) \end{bmatrix}\begin{bmatrix} \tilde{A}^e \\ \tilde{A}^f \end{bmatrix} \tag{12}$$

In the air gap, $\alpha = b$, $\beta = c$, and $\mu = \mu_o$:

$$\begin{bmatrix} \tilde{H}_\theta^g \\ \tilde{H}_\theta^h \end{bmatrix} = \frac{1}{\mu_o}\begin{bmatrix} f_1(c,b) & g_1(b,c) \\ g_1(c,b) & f_1(b,c) \end{bmatrix}\begin{bmatrix} \tilde{A}^g \\ \tilde{A}^h \end{bmatrix} \tag{13}$$

and finally, in the rotor, $\alpha = c$, $\beta \to 0$, and $\mu = \mu_r$:

$$\tilde{H}_\theta^i = \frac{1}{\mu_r} f_1(0,c)\tilde{A}^i \tag{14}$$

Torque as a Function of Terminal Currents and Rotor Angle: With the surface of integration for the stress tensor just inside the stator, it follows from Eq. 4.2.3 that the rotor torque is

$$\tau_z = (2\pi b^2 \ell)\,\frac{1}{2}\,\mathrm{Re}\left[(\tilde{H}_\theta^g)^*\tilde{B}_r^g\right] = \pi b^2 \ell\,\mathrm{Re}\left[(\tilde{H}_\theta^g)^*\left(\frac{-j\tilde{A}^g}{b}\right)\right] \tag{15}$$

It will be seen shortly that the electrical terminal relations can be computed from $\tilde{A}^g$. It is therefore convenient to also express Eq. 15 in terms of $\tilde{A}^g$ and the given surface currents. To this end, Eqs. 5 and 8 are used to replace (d) → (e) in Eq. 11, while Eqs. 6 and 9 are used to replace $\tilde{H}_\theta^f$ and $\tilde{A}^f$ in Eq. 12b. Thus, Eqs. 12 can be solved for $\tilde{H}_\theta^g$ as a function of $\tilde{K}^s$ and $\tilde{A}^g$:

$$\tilde{H}_\theta^g = -\tilde{K}^s + \frac{\tilde{A}^g}{\mu_s}\left\{ f_1(a,b) + \frac{g_1(b,a)g_1(a,b)}{\left[\frac{\mu_s}{\mu_o} f_1(\infty,a) - f_1(b,a)\right]}\right\} \tag{16}$$

Because the geometric quantity multiplying $\tilde{A}^g$ is real, it is clear that substitution of Eq. 16 into Eq. 15 gives only

$$\tau_z = \pi b \ell \mathrm{Re}[(\tilde{K}^s)^* j\tilde{A}^g] \tag{17}$$

To evaluate $\tilde{A}^g$ in terms of $\tilde{K}^s$ and $\tilde{K}^r$ (and hence in terms of the terminal currents and θ_r), Eqs. 7 and 10 are used in Eq. 14, which is solved for $\tilde{H}^h$. This latter quantity is substituted into Eq. 13b. Simultaneous solution of Eqs. 13 then gives a second expression for $\tilde{H}^g_\theta$:

$$\tilde{H}^g_\theta = \frac{\tilde{K}^r g_1(b,c)}{f_1(b,c) - \frac{\mu_o}{\mu_r} f_1(0,c)} + \frac{\tilde{A}^g}{\mu_o}\left\{ f_1(c,b) + \frac{g_1(b,c)g_1(c,b)}{\frac{\mu_o}{\mu_r} f_1(0,c) - f_1(b,c)} \right\} \tag{18}$$

By equating Eqs. 16 and 18, it is now possible to solve for $\tilde{A}^g$ in terms of the surface currents:

$$\tilde{A}^g = \frac{\mu_o}{D}\tilde{K}^s + \frac{\mu_o g_1(b,c)}{D[f_1(b,c) - \frac{\mu_o}{\mu_r} f_1(0,c)]}\tilde{K}^r \tag{19}$$

where

$$D \equiv \frac{\mu_o}{\mu_s}\left\{ f_1(a,b) + \frac{g_1(b,a)g_1(a,b)}{\left[\frac{\mu_s}{\mu_o} f_1(\infty,a) - f_1(b,a)\right]} \right\} - \left\{ f_1(c,b) + \frac{g_1(b,c)g_1(c,b)}{\left[\frac{\mu_o}{\mu_r} f_1(0,c) - f_1(b,c)\right]} \right\}$$

A methodical approach to solving the boundary and bulk relations is suited to those comfortable with the reduction of determinants or inclined to use matrix computations. Following this alternative, the boundary conditions, Eqs. 5 to 10, are used to eliminate the "d", "f", and "i" variables in the bulk relations, Eqs. 11 to 14. These latter equations are then written in the form

$$\begin{bmatrix} -1 & \frac{1}{\mu_o} f_1(\infty,a) & 0 & 0 & 0 & 0 \\ -1 & \frac{1}{\mu_s} f_1(b,a) & 0 & \frac{1}{\mu_s} g_1(a,b) & 0 & 0 \\ 0 & \frac{1}{\mu_s} g_1(b,a) & -1 & \frac{1}{\mu_s} f_1(a,b) & 0 & 0 \\ 0 & 0 & -1 & \frac{1}{\mu_o} f_1(c,b) & 0 & \frac{1}{\mu_o} g_1(b,c) \\ 0 & 0 & 0 & \frac{1}{\mu_o} g_1(c,b) & -1 & \frac{1}{\mu_o} f_1(b,c) \\ 0 & 0 & 0 & 0 & -1 & \frac{1}{\mu_r} f_1(0,c) \end{bmatrix} \begin{bmatrix} \tilde{H}^e_\theta \\ \tilde{A}^e \\ \tilde{H}^g_\theta \\ \tilde{A}^g \\ \tilde{H}^h_\theta \\ \tilde{A}^h \end{bmatrix} = \begin{bmatrix} 0 \\ 0 \\ \tilde{K}^s \\ 0 \\ 0 \\ -\tilde{K}^r \end{bmatrix} \tag{20}$$

Cramer's rule is then used to deduce $\tilde{A}^g$, Eq. 19.

Substitution of Eq. 19 into the torque expression, Eq. 17, shows that

$$\tau_z = \frac{\pi b \ell \mu_o}{D[f_1(b,c) - \frac{\mu_o}{\mu_r} f_1(0,c)]} \mathrm{Re}[j\tilde{K}^r(\tilde{K}^s)^*] \tag{21}$$

It follows from Eqs. 1 and 2 that the torque, expressed in terms of the terminal currents, is

$$\tau_z = \frac{-\pi b \ell \mu_o g_1(b,c)}{D[f_1(b,c) - \frac{\mu_o}{\mu_r} f_1(0,c)]} i_r N_r [i_a N_a \sin\theta_r + i_b N_b \sin(\theta_r - \tfrac{2\pi}{3}) + i_c N_c \sin(\theta_r - \tfrac{4\pi}{3})] \tag{22}$$

Electrical Terminal Relations: The flux linked by one turn of the "a"-phase coil running in the +z direction at $\theta = \theta'$ and returning in the -z direction at $\theta = \theta' + \pi$ is

$$\Phi_\lambda = \ell[A(b,\theta') - A(b,\theta' + \pi)] = \ell\mathrm{Re}\tilde{A}^g[e^{-j\theta'} - e^{-j(\theta'+\pi)}] \tag{23}$$

Here, use has been made of the relation between the vector potential and the flux, as described in Sec. 2.18 (Eq. (f) of Table 2.18.1).

The flux linked by the turns in the azimuthal interval $bd\theta'$ is then $\Phi_\lambda(bd\theta' N_a \cos\theta')$, and the total flux linked by the "a" phase is

$$\lambda_a = -b\ell N_a \int_{-\pi/2}^{\pi/2} \mathrm{Re}\,\frac{1}{2}\,[e^{j\theta'} + e^{-j\theta'}]\tilde{A}^g[e^{j\pi} - 1]e^{-j\theta'}d\theta' = \ell b N_a \pi \mathrm{Re}\tilde{A}^g \tag{24}$$

Substitution of $\tilde{A}^g$ from Eq. 19 and the surface currents from Eqs. 1 and 2 then gives the terminal relation for the "a" phase, in the form of Eq. 3a, where

$$L_{aa} = \frac{\pi\ell b\mu_o N_a^2}{D},\; L_{ab} = -\frac{\pi\ell b\mu_o N_a N_b}{2D},\; L_{ac} = -\frac{\pi\ell b\mu_o N_a N_c}{2D},\quad L_{ar} = L_o\ell b N_a N_r \cos\theta_r;$$

$$L_o \equiv \frac{\pi\mu_o g_1(b,c)}{D[f_1(b,c) - \frac{\mu_o}{\mu_r} f_1(0,c)]} \tag{25}$$

By symmetry, the inductances for the "b" and "c" phases are obtained without carrying out the evaluation by simply replacing indices in Eq. 25. For the "b" phase, replace indices $a \to b$, $b \to c$, $c \to a$, and $\theta_r \to \theta_r - 2\pi/3$ and for the "c" phase, $a \to c$, $b \to a$, $c \to b$, and $\theta_r \to \theta_r - 4\pi/3$.

The remaining flux linkage, λ_r, is computed by first recognizing that the flux linked by one turn on the rotor winding running in the z direction at θ' and returning at $\theta' + \pi$ is

$$\Phi_\lambda = -\ell\mathrm{Re}\tilde{A}^h[e^{j\pi} - 1]e^{-j\theta'} \tag{26}$$

Hence, the total flux linking the rotor winding is

$$\lambda_r = \int_{\theta_r - \frac{\pi}{2}}^{\theta_r + \frac{\pi}{2}} N_r \cos(\theta' - \theta_r)\Phi_\lambda c\,d\theta' = N_r c\ell\pi\mathrm{Re}\tilde{A}^h e^{j\theta_r} \tag{27}$$

The vector potential amplitude required to evaluate this expression follows from Eqs. 7, 10, 13b, and 14:

$$\tilde{A}^h = \frac{g_1(c,b)\tilde{A}^g - \mu_o\tilde{K}^r}{\frac{\mu_o}{\mu_r} f_1(0,c) - f_1(b,c)} \tag{28}$$

where $\tilde{A}^g$ is again Eq. 19, and the surface currents are evaluated in terms of the terminal currents using Eqs. 1 and 2. Thus, with the use of the transfer function reciprocity relation, $cg_1(c,b) = -bg_1(b,c)$, Eq. 2.17.10,

$$L_{ra} = L_o\ell b N_r N_a \cos\theta_r,\; L_{rb} = L_o\ell b N_r N_b \cos(\theta_r - \frac{2\pi}{3}),\; L_{rc} = L_o\ell b N_r N_c \cos(\theta_r - \frac{4\pi}{2}),$$

$$L_{rr} = L_o\ell b N_r^2 \left\{\frac{g_1(b,c)}{D[f_1(b,c) - \frac{\mu_o}{\mu_r} f_1(0,c)]} - \frac{1}{g_1(c,b)}\right\} \tag{29}$$

Energy Conservation: Because the electromechanical coupling network represented by Fig. 4.7.3 is conservative, there is considerable redundancy in the terminal relations that have been derived. Conservation of energy requires that (Eq. 3.5.7 applied to a magnetic system)

$$\delta w' = \lambda_a \delta i_a + \lambda_b \delta i_b + \lambda_c \delta i_c + \lambda_r \delta i_r + \tau_z \delta\theta_r \tag{30}$$

From the assumption that w' is a state function, it follows that (see Eq. 3.5.4)

$$\lambda_k = \frac{\partial w'}{\partial i_k};\; k = a,b,c,r;\; \tau_z = \frac{\partial w'}{\partial \theta_r} \tag{31}$$

Lumped-parameter reciprocity conditions are generated by taking cross-derivatives of these relations:

$$\frac{\partial \lambda_k}{\partial i_\ell} = \frac{\partial \lambda_\ell}{\partial i_k};\; \frac{\partial \tau_z}{\partial i_k} = \frac{\partial \lambda_\ell}{\partial \theta_r};\; \begin{cases} k = a,b,c,r \\ \ell = a,b,c,r \end{cases} \tag{32}$$

The four relations among the electrical terminal variables show that

$$L_{k\ell} = L_{\ell k} \tag{33}$$

and these conditions are met by the results summarized by Eqs. 25 and the subsequent substitution of indices and Eq. 29. The reciprocity conditions between the torque and the flux linkages, Eq. 32, is also satisfied by Eqs. 22 and Eqs. 25 and 29. Note that to make it clear that the lumped-parameter reciprocity relations are satisfied, the reciprocity condition for the air-gap transfer relations was used in writing Eq. 29.

4.8 Constrained-Current Magnetoquasistatic Transfer Relations

By way of exemplifying how transfer relations can be used to represent fields in bulk regions, including volume distributions of known current density, these relations are derived in this section for one important class of physical situations. The current density (which is typically the result of exciting distributions of wire) is z-directed, while the magnetic field is in the (r,θ) plane. Thus, the relations are directly applicable to rotating machines with negligible end effects. Such an application is taken up in the next section.

In a broad sense, the objective in this section is to magnetic field systems what the objective in Sec. 4.5 was to electric field systems. But, the solution of the vector Poisson's equation, Eq. 2.19.2, is more demanding than the scalar Poisson's equation, Eq. 4.5.1, and hence the technique now illustrated is limited to certain configurations in which only one component of the vector potential describes the fields. Such cases are discussed in Sec. 2.18 and the associated transfer relations for a region of free space are derived in Sec. 2.19. The following discussion relates to the polar-coordinate situations of Tables 2.18.1 and 2.19.1.

In the two-dimensional cylindrical coordinates, the vector Poisson's equation (Eq. 2.19.2) has only a z component and the Laplacian is the same as the scalar Laplacian:

$$\nabla^2 A = -\mu J_z \tag{1}$$

Following the line of attack used in Sec. 4.5, the solution is divided into homogeneous and particular parts,

$$A = A_H + A_P \tag{2}$$

defined such that

$$\nabla^2 A_P = -\mu J_z;\; \nabla^2 A_H = 0 \tag{3}$$

The imposed current is now represented in the complex amplitude form

$$J_z = \mathrm{Re}\tilde{J}(r,t)e^{-jm\theta} \tag{4}$$

Of course, by superposition, such solutions could be the basis for a Fourier representation of an arbitrary current distribution. Substitution of Eq. 4 into Eq. 3 shows that $\tilde{A}_p$ must satisfy the equation

$$\frac{d^2\tilde{A}_P}{dr^2} + \frac{1}{r}\frac{d\tilde{A}_P}{dr} - \frac{m^2}{r^2}\tilde{A}_P = -\mu\tilde{J}(r) \tag{5}$$

The particular solution can be any solution to Eq. 5. The magnetic field associated with this particular solution is, by the definition of the vector potential (Eq. 2.18.1),

$$H_{\theta P} = -\frac{1}{\mu}\frac{d\tilde{A}_P}{dr};\ \tilde{B}_{rP} = -\frac{jm}{r}\tilde{A}_P \tag{6}$$

From Eq. 2 it follows that the homogeneous solution is the total solution with the particular solution subtracted off. That is,

$$\tilde{A}_H = \tilde{A} - \tilde{A}_P;\ \tilde{H}_{\theta H} = \tilde{H}_\theta - \tilde{H}_{\theta P} \tag{7}$$

The homogeneous parts are related by the transfer relations, Eqs. (d) of Table 2.19.1, so that substitution from Eq. 7 shows that

$$\begin{bmatrix} \tilde{A}^\alpha - \tilde{A}_P^\alpha \\ \\ \tilde{A}^\beta - \tilde{A}_P^\beta \end{bmatrix} = \mu \begin{bmatrix} F_m(\beta,\alpha) & G_m(\alpha,\beta) \\ \\ G_m(\beta,\alpha) & F_m(\alpha,\beta) \end{bmatrix} \begin{bmatrix} \tilde{H}_\theta^\alpha - \tilde{H}_{\theta P}^\alpha \\ \\ \tilde{H}_\theta^\beta - \tilde{H}_{\theta P}^\beta \end{bmatrix} \tag{8}$$

These relations, multiplied out, are the transfer relations for the cylindrical annulus supporting a given distribution of z-directed current density:

$$\begin{bmatrix} \tilde{A}^\alpha \\ \\ \tilde{A}^\beta \end{bmatrix} = \mu \begin{bmatrix} F_m(\beta,\alpha) & G_m(\alpha,\beta) \\ \\ G_m(\beta,\alpha) & F_m(\alpha,\beta) \end{bmatrix} \begin{bmatrix} \tilde{H}_\theta^\alpha \\ \\ \tilde{H}_\theta^\beta \end{bmatrix} + \begin{bmatrix} \tilde{A}_P^\alpha \\ \\ \tilde{A}_P^\beta \end{bmatrix} - \mu \begin{bmatrix} F_m(\beta,\alpha) & G_m(\alpha,\beta) \\ \\ G_m(\beta,\alpha) & F_m(\alpha,\beta) \end{bmatrix} \begin{bmatrix} \tilde{H}_{\theta P}^\alpha \\ \\ \tilde{H}_{\theta P}^\beta \end{bmatrix} \tag{9}$$

Following the format used in Sec. 4.5, it would be natural to now proceed to generate particular solutions that form a complete set of orthogonal functions which are solutions to the Helmholtz equation. Such an approach to evaluating the particular solutions in Eq. 9 is required if an arbitrary radial distribution of current density is to be represented. The approach parallels that presented in Sec. 4.5.

In important physical configurations, to which the remainder of this section is confined, the radial distribution is uniform:

$$\tilde{J}(r) = \tilde{J} \tag{10}$$

Fortunately, inspection of Eq. 5 in this case yields simple particular solutions:

$$\tilde{A}_P = \mu\tilde{J}\begin{cases} \dfrac{r^2}{m^2 - 4}; & m \neq 2 \\ -\dfrac{1}{4}r^2 \ln r; & m = \pm 2 \end{cases} \tag{11}$$

Thus, for the case of a radially uniform current density distribution, substitution of Eq. 11 into Eq. 9 yields the transfer relations

$$\begin{bmatrix} \tilde{A}^\alpha \\ \\ \tilde{A}^\beta \end{bmatrix} = \mu_o \begin{bmatrix} F_m(\beta,\alpha) & G_m(\alpha,\beta) \\ \\ G_m(\beta,\alpha) & F_m(\alpha,\beta) \end{bmatrix} \begin{bmatrix} \tilde{H}_\theta^\alpha \\ \\ \tilde{H}_\theta^\beta \end{bmatrix} + \mu_o\tilde{J} \begin{bmatrix} h_m(\alpha,\beta) \\ \\ h_m(\beta,\alpha) \end{bmatrix} \tag{12}$$

where

$$h_m(x,y) = \begin{cases} \dfrac{1}{m^2 - 4}[x^2 + 2xF_m(y,x) + 2yG_m(x,y)]; & m \neq 2 \\ \dfrac{x}{8}[x + g_m(x,y)y^2 \ln(\dfrac{x}{y})]; & m = \pm 2 \end{cases}$$

and the functions F_m, G_m, and g_m are defined in Table 2.16.2 with k = 0.

The radial distribution of A within the volume of the annular region described by Eq. 12 is obtained by adding to the homogeneous solution, which is Eq. 2.19.5 with $\tilde{A}^\alpha \rightarrow \tilde{A}^\alpha - \tilde{A}_p^\alpha$, and $\tilde{A}^\beta \rightarrow \tilde{A}^\beta - \tilde{A}_p^\beta$, the particular solution $\tilde{A}_p$:

$$\tilde{A} = (\tilde{A}^\alpha - \tilde{A}_p^\alpha)\frac{\left[(\frac{\beta}{r})^m - (\frac{r}{\beta})^m\right]}{\left[(\frac{\beta}{\alpha})^m - (\frac{\alpha}{\beta})^m\right]} + (\tilde{A}^\beta - \tilde{A}_p^\beta)\frac{\left[(\frac{r}{\alpha})^m - (\frac{\alpha}{r})^m\right]}{\left[(\frac{\beta}{\alpha})^m - (\frac{\alpha}{\beta})^m\right]} + \tilde{A}_p \qquad (13)$$

For Eq. 12, the particular solution is given by Eq. 11, so the associated volume distribution is evaluated using Eq. 11.

The constrained-current transfer relations are applied to a specific problem in the next section.

4.9 Exposed Winding Synchronous Machine Model

The structure shown in cross section in Fig. 4.9.1 consists of a stator supporting three windings (a,b,c) and a rotor with a single winding (r). It models a three-phase two-pole synchronous alternator, and is similar to the configuration taken up in Sec. 4.7. The difference is that the windings on both rotor and stator are not embedded in slots of highly permeable material and take up a radial thickness that is appreciable compared to the air gap. As a result, the surface current model used in Sec. 4.7 is not appropriate.

The configuration considered here is an example to which the constrained-current transfer relations of Sec. 4.8 can be applied. It closely resembles models that have been developed for synchronous alternators making use of superconducting field (rotor) windings.[1] With superconductors, it is possible to generate magnetic fields that more than saturate magnetizable materials. As a result, the magnetic materials in which conductors are embedded in conventional machines can be dispensed with. This makes it possible to design for greater voltages than would be possible in a conventional machine, where the slot material in which a conductor is embedded must be grounded. But, because the conductors are exposed to the full magnetic force, methods of construction must be radically altered. A machine built

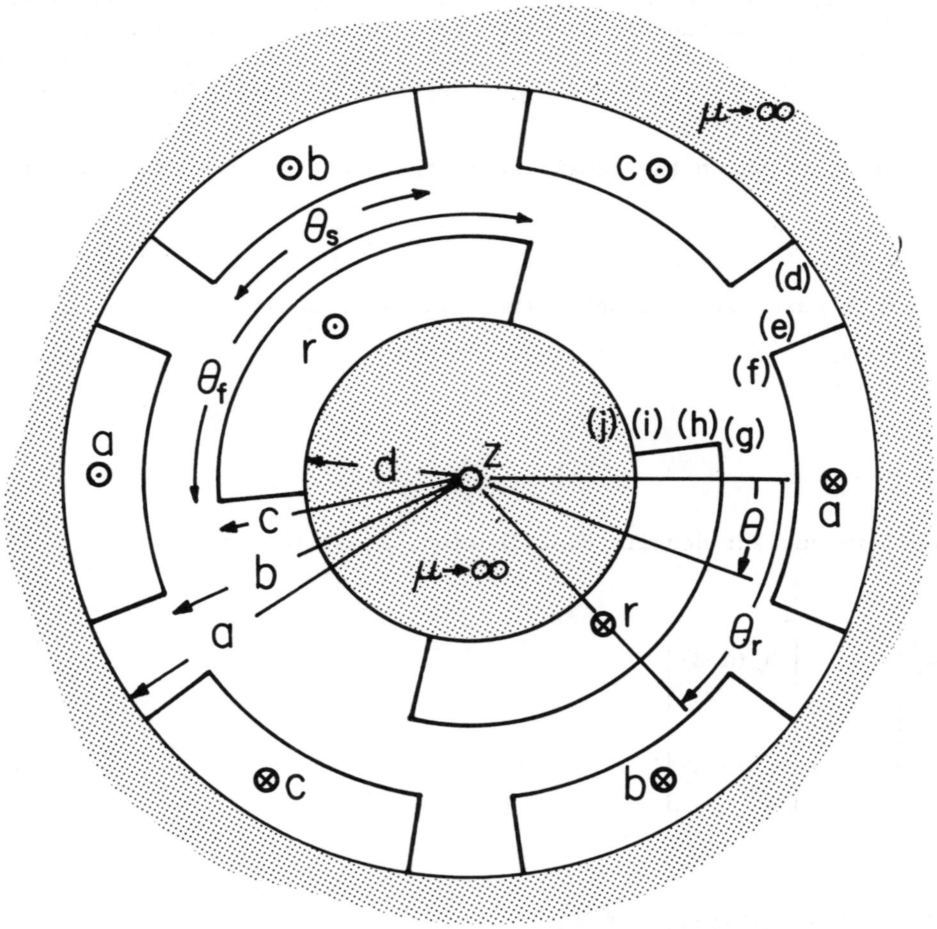

Fig. 4.9.1.

Cross section of synchronous machine model typifying structure used in superconducting field alternator.

1. J. L. Kirtley, Jr., "Design and Construction of an Armature for an Alternator with a Superconducting Field Winding," Ph.D. Thesis, Department of Electrical Engineering, Massachusetts Institute of Technology, Cambridge, Mass., 1971; J. L. Kirtley, Jr., and M. Furugama, "A Design Concept for Large Superconducting Alternators," IEEE Power Engineering Society, Winter Meeting, New York, Jan. 1975.

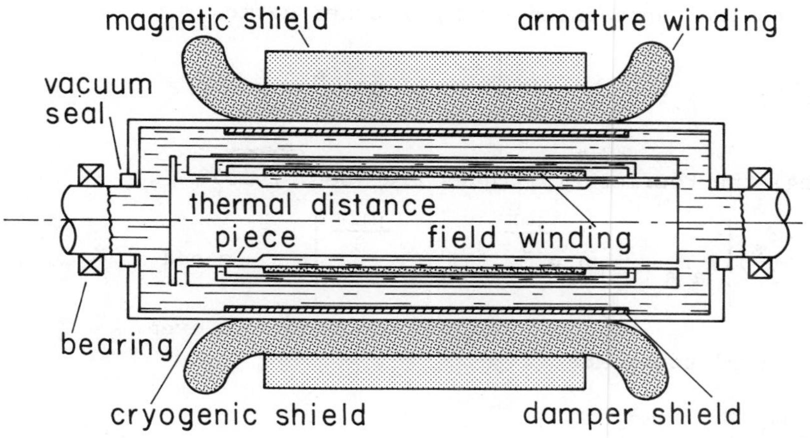

Fig. 4.9.2. Cross section of superconducting field alternator projected in design for 1000 and 10,000 MVA machines on basis of M.I.T. experiments on 2-3 MVA.[1] Not included in model of this section is conducting shell between rotor and stator to help prevent time-varying fields due to transients from reaching superconductors. Also, magnetic core of rotor used to simplify model in this section is not present in machine shown. Phenolic materials are used in projected design to construct stator and rotor.

to test approaches to constructing a rotating "refrigerator" required if the field is to be superconducting is shown in Fig. 4.9.2.

In the configuration considered here, it is assumed that surrounding the stator is a highly permeable shield material with inner radius (a) equal to the outer radius of the stator windings. Similarly, the rotor windings are bounded from inside by a "perfectly" permeable core. The magnetic materials are introduced into the model to make the example reasonably free of algebraic complications. In a machine having a superconducting field, a magnetic core would not be used. Development of a model without the magnetic rotor core follows the same pattern as now described.

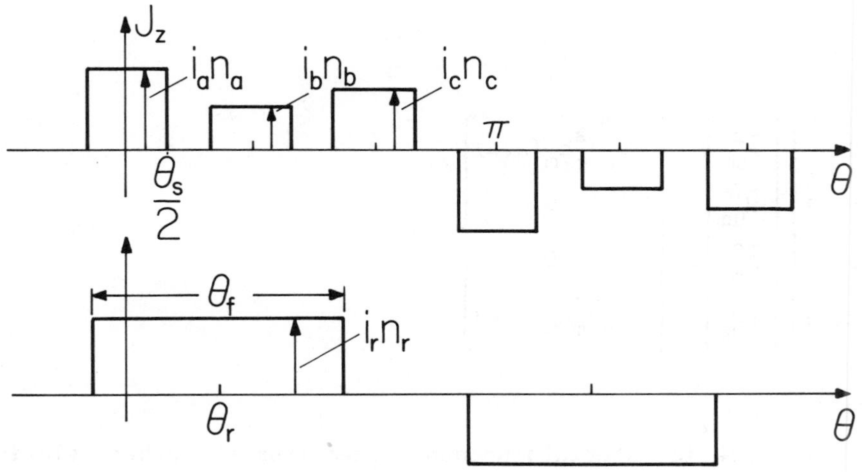

Fig. 4.9.3

Azimuthal current density distribution on stator and rotor.

The distribution of stator and rotor current densities with azimuthal position is shown in Fig. 4.9.3. The turns densities (n_a,n_b,n_c,n_r) (conductors per unit area) respectively carry the terminal currents (i_a,i_b,i_c,i_r). The conductors are unformly distributed. Hence, these current density distributions can be represented by the Fourier series

$$J_z^s = \sum_{m=-\infty}^{+\infty} \tilde{J}_m^s e^{-jm\theta}, \quad b < r < a; \quad J_z^r = \sum_{m=-\infty}^{+\infty} \tilde{J}_m^r e^{-jm\theta}, \quad d < r < c \tag{1}$$

For the stator winding, the Fourier amplitudes are (Sec. 2.15)

$$\tilde{J}_m^s = \begin{cases} \frac{2}{\pi m} \sin\left(\frac{m\theta_s}{2}\right)\left[i_a n_a + i_b n_b e^{\frac{jm\pi}{3}} + i_c n_c e^{\frac{jm2\pi}{3}}\right]; & m \text{ odd} \\ 0 \qquad ; & m \text{ even} \end{cases} \tag{2}$$

while on the rotor the amplitudes are

$$\tilde{J}_m^r = \begin{cases} \frac{2}{\pi m} \sin\left(\frac{m\theta_f}{2}\right) i_r n_r e^{jm\theta_r}; & m \text{ odd} \\ 0 \qquad ; & m \text{ even} \end{cases} \tag{3}$$

The constrained-current distribution is now as assumed in the previous section, Eqs. 4.8.4 and 4.8.10. The associated transfer relations relate the Fourier amplitudes of the tangential magnetic field intensities and vector potentials at the surfaces of the annular regions comprising the stator, the air gap and the rotor winding with designations (d) - (j) shown in Fig. 4.9.1.

Boundary Conditions: There are no surface currents in the model, so the tangential magnetic fields are continuous between regions and vanish on the stator and rotor magnetic materials. The normal flux density is continuous, and this requires that the vector potential be continuous:

$$\begin{aligned} &\tilde{H}_{\theta m}^d = 0; \ \tilde{H}_{\theta m}^e = \tilde{H}_{\theta m}^f; \ \tilde{H}_{\theta m}^g = \tilde{H}_{\theta m}^h; \ \tilde{H}_{\theta m}^i = 0 \\ &\tilde{A}_m^e = \tilde{A}_m^f; \ \tilde{A}_m^g = \tilde{A}_m^h \end{aligned} \tag{4}$$

Bulk Relations: The transfer relations, Eq. 4.8.12, are now applied in succession to the stator, the air gap and the rotor regions. In writing these expressions, the conditions of Eq. 4 are used to eliminate (e,h) variables in favor of the (f,g) variables:

$$\tilde{A}_m^d = \mu_o G_m(a,b)\tilde{H}^f{}_m + \mu_o \tilde{J}_m^s h_m(a,b)$$

$$\begin{bmatrix} -1 & \mu_o F_m(a,b) & 0 & 0 \\ -1 & \mu_o F_m(c,b) & 0 & \mu_o G_m(b,c) \\ 0 & \mu_o G_m(c,b) & -1 & \mu_o F_m(b,c) \\ 0 & 0 & -1 & \mu_o F_m(d,c) \end{bmatrix} \begin{bmatrix} \tilde{A}_m^f \\ \tilde{H}_{\theta m}^f \\ \tilde{A}_m^g \\ \tilde{H}_{\theta m}^g \end{bmatrix} = \begin{bmatrix} -\mu_o \tilde{J}_m^s h_m(b,a) \\ 0 \\ 0 \\ -\mu_o \tilde{J}_m^r h_m(c,d) \end{bmatrix} \tag{5}$$

$$\tilde{A}_m^i = \mu_o G_m(d,c)\tilde{H}_{\theta m}^g + \mu_o \tilde{J}_m^r h_m(d,c)$$

Because the boundary conditions on the magnetic materials uncouple them from the other relations, the first and last of these relations are written separately.

Torque as a Function of Terminal Variables: The torque is computed by integrating the Maxwell stress over the surface at (g) on the rotor side of the air gap (sec. 4.2). Because $B_r = (1/r)(\partial A/\partial\theta)$, the torque becomes (Eqs. 4.2.3 and 2.15.17):

$$\tau_z = 2\pi\ell c^2 \sum_{m=-\infty}^{+\infty} \left(\frac{-jm}{c}\tilde{A}_m^g\right)\left(\tilde{H}_{\theta m}^g\right)^* \tag{6}$$

To evaluate this expression, the amplitudes $\tilde{A}_m^g$ and $\tilde{H}_{\theta m}^g$ are found from the matrix equation of Eq. 5, using Cramer's rule:

$$\begin{aligned} \tilde{A}_m^g &= \tilde{J}_m^s C_1 + \tilde{J}_m^r C_2 \\ \tilde{H}_{\theta m}^g &= \tilde{J}_m^s C_3 + \tilde{J}_m^r C_4 \end{aligned} \tag{7}$$

where

$$C_1 = \frac{\mu_o^3}{D} h_m(b,a)G_m(c,b)F_m(d,c)$$

$$C_2 = \frac{\mu_o^3}{D} h_m(c,d)[F_m(a,b)F_m(b,c) - F_m(c,b)F_m(b,c) + G_m(c,b)G_m(b,c)]$$

$$C_3 = \frac{\mu_o^2}{D} h_m(b,a)G_m(c,b)$$

$$C_4 = \frac{\mu_o^2}{D} h_m(c,d)[F_m(a,b) - F_m(c,b)]$$

$$D = \mu_o^2\{G_m(c,b)G_m(b,c) - [F_m(c,b) - F_m(a,b)][F_m(b,c) - F_m(d,c)]\}$$

In using Eqs. 7 to evaluate Eq. 6, observe that $\tilde{J}^s_m(\tilde{J}^s_m)^*$ and $\tilde{J}^r(\tilde{J}^r_m)^*$ are even in m, as are also the functions h_m, F_m, and G_m. Because of the latter, the C_i's are also even in m. Thus, the summations of the self-field terms in $|\tilde{J}^s_m|^2$ and $|\tilde{J}^r_m|^2$ are odd functions of m and result in no contribution. The mth terms are canceled by the -mth terms. Only the cross terms appear, as Eq. 6 becomes

$$\tau_z = 2\pi \ell c \sum_{m=-\infty}^{+\infty} (-jm)[\tilde{J}^r_m(\tilde{J}^s_m)^* C_2C_3 + \tilde{J}^s_m(\tilde{J}^r_m)^* C_1C_4] \tag{8}$$

Substitution of Eqs. 2 and 3 therefore gives the torque as

$$\tau_z = \frac{16\ell c}{\pi} i_r n_r \sum_{\substack{m=1\\(\text{odd})}}^{\infty} \frac{(C_2C_3 - C_1C_4)}{m} \sin\left(\frac{m\theta_r}{2}\right) \sin\left(\frac{m\theta_s}{2}\right)[i_a n_a \sin m\theta_r + i_b n_b \sin m(\theta_r - \frac{\pi}{3}) + i_c n_c \sin m(\theta_r - \frac{2\pi}{3})] \tag{9}$$

where

$$(C_2C_3 - C_1C_4) = \frac{\mu_o^5}{D^2} h_m(c,d)h_m(b,a)G_m(c,b)[F_m(a,b)F_m(b,c) - F_m(c,b)F_m(b,c) + G_m(c,b)G_m(b,c) + F_m(d,c)F_m(c,b) - F_m(d,c)F_m(a,b)]$$

Electrical Terminal Relations: Each of the three phase windings of the stator, as well as the rotor winding, can be represented by the coil shown cross-sectionally in Fig. 4.9.4. For the "a" phase of the stator, variables are identified as $\theta_1 = \theta_s/2, \theta_2 = -\theta_s/2, \alpha = a, \beta = b$. For the rotor, $\theta_1 = \theta_r + \theta_f/2, \theta_2 = \theta_r - \theta_f/2$, $\alpha = c$, $\beta = d$.

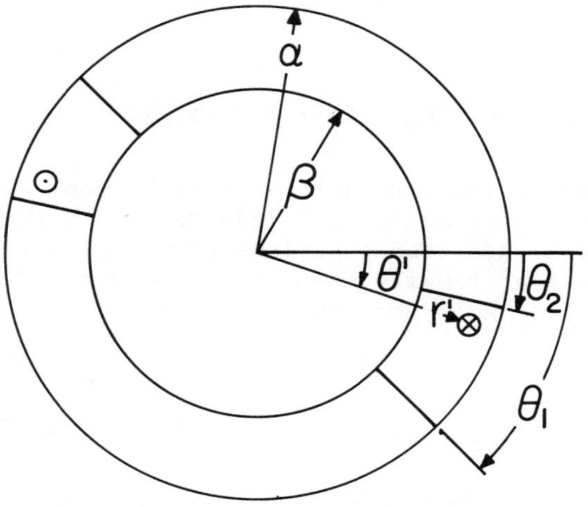

Fig. 4.9.4

Prototype coil representing each of the four in Fig. 4.9.1.

The flux linked by a single turn of the coil carrying current in the z direction at (r',θ') and returning it at $(r',\theta' + \pi)$ is conveniently evaluated in terms of the vector potential (Eq. (f) of Table 2.18.1):

$$\Phi_\lambda = \ell[A(r',\theta') - A(r',\theta' - \pi)] \tag{10}$$

With n defined as the turns per unit cross-sectional area, there are $nr'd\theta'dr'$ turns in a differential area and hence the total flux linked by the coil is

$$\lambda = \ell \int_\beta^\alpha \int_{\theta_2}^{\theta_1} \sum_{m=-\infty}^{+\infty} [\tilde{A}_m(r')e^{-jm\theta'} - \tilde{A}_m(r')e^{-jm(\theta'-\pi)}]nr'd\theta'dr' \tag{11}$$

The integration on θ' can be carried out directly to reduce Eq. 11 to

$$\lambda = 2j\ell n \sum_{\substack{m=-\infty \\ \text{odd}}}^{+\infty} \frac{\left(e^{-jm\theta_1} - e^{-jm\theta_2}\right)}{m} \int_\beta^\alpha \tilde{A}_m(r')r'dr' \tag{12}$$

To complete the radial integration, Eq. 4.8.13 is used to express $\tilde{A}_m$, while for the case being considered, $\tilde{A}_p$ is given by Eq. 4.8.11:

$$\lambda = 2j\ell n \sum_{\substack{m=-\infty \\ \text{odd}}}^{+\infty} \frac{\left(e^{-jm\theta_1} - e^{-jm\theta_2}\right)}{m} [\tilde{A}_m^\alpha M_m(\alpha,\beta) - \tilde{A}_m M_m^\beta(\beta,\alpha) - \mu_o \tilde{J}S_m(\alpha,\beta)] \tag{13}$$

where

$$M_m(x,y) = \frac{1}{2}[x^2 - m^2 h_m(x,y)]$$

$$S_m(x,y) = \begin{cases} \dfrac{x^2}{m^2-4} M_m(x,y) - \dfrac{y^2}{m^2-4} M_m(y,x) - \dfrac{1}{4}\dfrac{(x^4-y^4)}{m^2-4}, & m \neq \pm 2 \\[2ex] -\dfrac{1}{4}x^2 \ln x M_m(x,y) + \dfrac{1}{4}y^2 \ln y M_m(y,x) + \dfrac{1}{16}[x^4(\ln x - \tfrac{1}{4}) - y^4(\ln y - \tfrac{1}{4})], & m = \pm 2 \end{cases}$$

By appropriate identification of variables, Eq. 13 can now be used to compute the flux linked by each of the four electrical terminal pairs. The procedure is illustrated by considering the field winding. Then, variables are identified:

$$\lambda = \lambda_r,\ d \to c,\ \beta \to d,\ \theta_1 = \theta_r - \frac{\theta_f}{2},\ \theta_2 = \theta_r + \frac{\theta_f}{2},\ n = n_r,\ \tilde{A}^\alpha \to \tilde{A}^g,\ \tilde{A}^\beta \to \tilde{A}^i,\ \tilde{J} = \tilde{J}_m^r \tag{14}$$

The amplitudes $(\tilde{A}_m^g, \tilde{A}_m^i)$ are respectively evaluated from Eqs. 7a and the combination of Eqs. 5f and 7b. Thus, identified with the field winding, Eq. 13 becomes

$$\lambda_r = -\mu_o 4\ell n_r \sum_{\substack{m=-\infty \\ \text{odd}}}^{+\infty} \frac{1}{m}\sin\left(\frac{m\theta_f}{2}\right)e^{-jm\theta_r}\Big\{\tilde{J}_m^s[C_1M_m(\alpha,\beta) - \mu_oG_m(d,c)C_3M_m(\beta,\alpha)]$$
$$+ \tilde{J}_m^r[C_2M_m(\alpha,\beta) - \mu_oG_m(d,c)C_4M_m(\beta,\alpha) + \mu_oM_m(\beta,\alpha)h_m(d,c) - \mu_oS_m(\alpha,\beta)]\Big\} \tag{15}$$

The current density amplitudes are in turn related to the terminal currents by Eqs. 2 and 3. Thus, Eq. 15 is expressed in terms of three mutual inductances and a self-inductance, in the form of Eq. 4.7.3d. In writing these inductances, observe that F_m and G_m are even functions of m. It follows that h_m and hence M_m and S_m are also even functions of m, and that finally the coefficients of $(\tilde{J}_m^s, \tilde{J}_m^r)$ in Eq. 15 are even in m. Thus, the summation can be converted to one on positive values of m:

$$\begin{bmatrix} L_{ra} \\ L_{rb} \\ L_{rc} \end{bmatrix} = -\frac{16\ell n_r}{\pi} \sum_{\substack{m=1 \\ \text{odd}}}^{\infty} \frac{\sin\left(\frac{m\theta_f}{2}\right)}{m} \frac{\sin\left(\frac{m\theta_s}{2}\right)}{m} [C_1M_m(\alpha,\beta) - \mu_oG_m(d,c)C_3M_m(\beta,\alpha)] \begin{bmatrix} n_a \cos m\,\theta_r \\ n_b \cos m(\theta_r + \frac{\pi}{3}) \\ n_c \cos m(\theta_r + \frac{2\pi}{3}) \end{bmatrix} \tag{16}$$

$$L_{rr} = -\frac{8\ell n_r^2}{\pi}\sum_{\substack{m=1\\ odd}}^{\infty}\frac{\sin(\frac{m\theta_f}{2})}{m}\frac{\sin(\frac{m\theta_s}{2})}{m}[C_2M_m(\alpha,\beta) - \mu_o G_m(d,c)C_4M_m(\beta,\alpha)$$
$$+ \mu_o M_m(\beta,\alpha)h_m(d,c) - \mu_o S_m(\alpha,\beta)] \quad (17)$$

Because of the energy-conserving nature of the electromechanical coupling, there is redundancy of information in the electrical and mechanical terminal relations. Reciprocity, as expressed by Eq. 4.7.32b, can be made the basis for finding the θ_r dependent parts of the mutual inductances from the torque, Eq. 9. (Here, there are rotor positions at which each of the mutual inductances vanish, and hence Eq. 9 uniquely specifies the mutual inductances.) The reciprocity condition shows that an alternative to the coefficient used to express the mutual inductances in Eq. 16 is

$$[C_1M_m(\alpha,\beta) - \mu_o G_m(d,c)C_3M_m(\beta,\alpha)] = c[C_2C_3 - C_1C_4] \quad (18)$$

where the quantity on the right is given with Eq. 9.

With the reciprocity relations in view, one efficient approach to determining the complete lumped-parameter terminal relations is to first find the torque, Eq. 9, then use the reciprocity conditions to find the mutual inductances and finally compute the self-inductances from Eq. 13. This last step only requires evaluation of $(\tilde{A}_m^\alpha,\tilde{A}_m^\beta)$ with self-current excitations (with currents in other windings removed).

A more conventional approach is to compute the full inductance matrix from Eq. 13 and use the lumped-parameter energy method (Sec. 3.5) to find the torque.

4.10 D-C Magnetic Machines

The wide use of the d-c rotating machine justifies the model development undertaken in this section. But, these devices are also a prototype for a family of "conduction" machines which includes the homopolar generator[1] and magnetohydrodynamic energy convertors, to be taken up in Chap. 9. Analogous electric field devices are the Van de Graaff generator, considered in Sec. 4.14, and electrogas dynamic pumps and generators, described in Chaps. 5 and 9.

The developed model for the d-c machine given in Sec. 4.3 (Table 4.3.1, Part 3) is given a more complete characterization in Figs. 4.10.1 through 4.10.4. What is by convention termed the "field" winding is on the stator, which consists of a highly permeable structure wound with a total of $2n_f$ turns excited through the terminal pair (i_f,v_f). The "armature" is the rotor, with a winding connected through the commutator to the terminal pair (i_a,v_a), so that the distribution of current is essentially stationary in space. The θ dependence is shown in Fig. 4.10.2. The rotor core, like the stator magnetic circuit, is modeled here as being infinitely permeable.

With the assumption that the stator is infinitely permeable, it is clear that the magnetic potential on the stator surface, Ψ^f, is constant for those points at $r = R_o$ contiguous with the stator. Integration of Ampere's integral law, Eq. 2.7.1b, over any contour passing between the pole faces through the field winding and closing through the air gap shows that the pole faces differ in Ψ by $2n_fi_f$. The horizontal mid-plane is defined as the reference $\Psi = 0$. As an approximation that specifies the fringing field in the ranges of θ between pole faces, the magnetic potential is taken as the linear interpolation shown in Fig. 4.10.2a. Because the rotor is modeled as infinitely permeable, the tangential magnetic field at the rotor surface is equal to the surface current density K_z, as shown in Fig. 4.10.2b (an application of Eq. 2.10.21). The number of turns per unit azimuthal length on the rotor is N_a.

The commutator, which consists of conducting segments that are sequentially connected to the armature terminals through brushes, as shown in Fig. 4.10.3a,[2] is attached to one end of the rotor. Thus it rotates with the same angular velocity Ω (defined as positive in the positive θ direction) as the rotor. The model now developed does not include "end effects," in that the rotor is assumed to have a length ℓ that is much greater than the air gap R_o-R.

The boundary conditions, pictured graphically in Fig. 4.10.2, are first represented by Fourier series (Eqs. 2.15.7 and 2.15.8 with $k_n z \to n_\theta$ and $\ell \to 2\pi R$). Thus, with (f) denoting the radial position $r=R_o$,

$$\Psi^f = \sum_{\substack{m=-\infty\\ (odd)}}^{\infty}\tilde{\Psi}_m^f e^{-jm\theta}; \quad \tilde{\Psi}_m^f = \frac{2n_fi_f \sin m\theta_o}{m\pi(\theta_o m)} je^{\frac{jm\pi}{2}} \quad (1)$$

1. H. H. Woodson and J. R. Melcher, Electromechanical Dynamics, Part I, John Wiley & Sons, New York, 1968, p. 312.

2. A. E. Fitzgerald, Ch. Kingsley, Jr., and A. Kusko, Electric Machinery, McGraw-Hill Book Company, New York, 1971, p. 192.

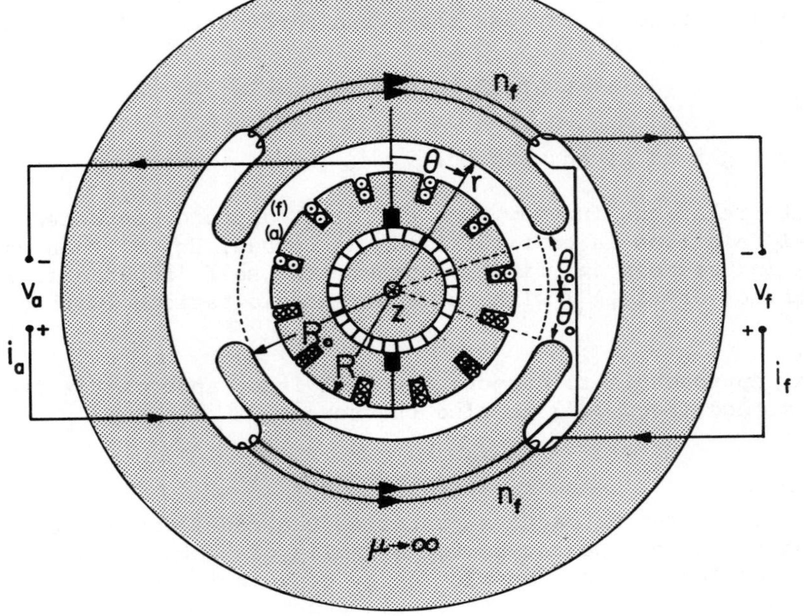

Fig. 4.10.1. Cross section of d-c machine.

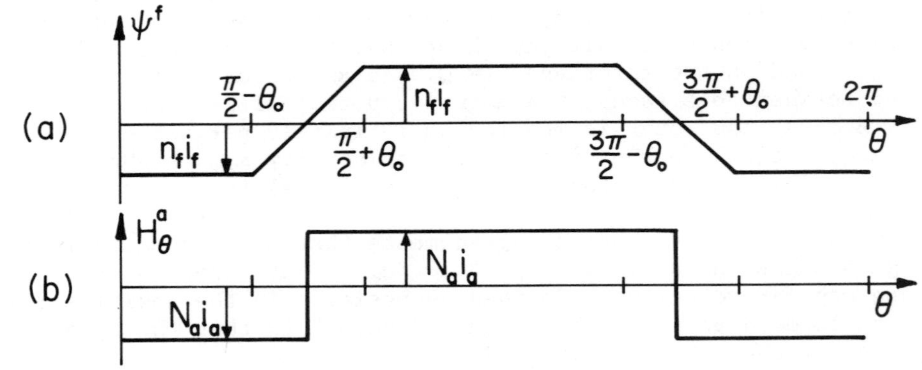

Fig. 4.10.2. Circumferential distribution of magnetic potential at $r = R_o$ and tangential magnetic field intensity at $r = R$.

and at the rotor surface where r = R,

$$\tilde{H}^a_\theta = \sum_{\substack{m=-\infty\\(\text{odd})}}^{\infty} \tilde{H}^a_{\theta m} e^{-jm\theta}; \quad \tilde{H}^a_{\theta m} = \frac{2N_a i_a}{m\pi} j e^{\frac{jm\pi}{2}} \tag{2}$$

Fields in the air gap are represented by the transfer relations, Eqs. (a) of Table 2.16.2 with k = 0. Hence, with positions (α) → (f) and (β) → (a) and with radii $\alpha \to R_o$ and $\beta \to R$,

$$\begin{bmatrix} \tilde{B}^f_{rm} \\ \tilde{B}^a_{rm} \end{bmatrix} = \mu_o \begin{bmatrix} f_m(R,R_o) & g_m(R_o,R) \\ g_m(R,R_o) & f_m(R_o,R) \end{bmatrix} \begin{bmatrix} \tilde{\Psi}^f_m \\ R\tilde{H}^a_{\theta m}/jm \end{bmatrix} \tag{3}$$

where $\tilde{H}^a_{\theta m}$ has been introduced by using $H_\theta = -(\nabla\Psi)_\theta$.

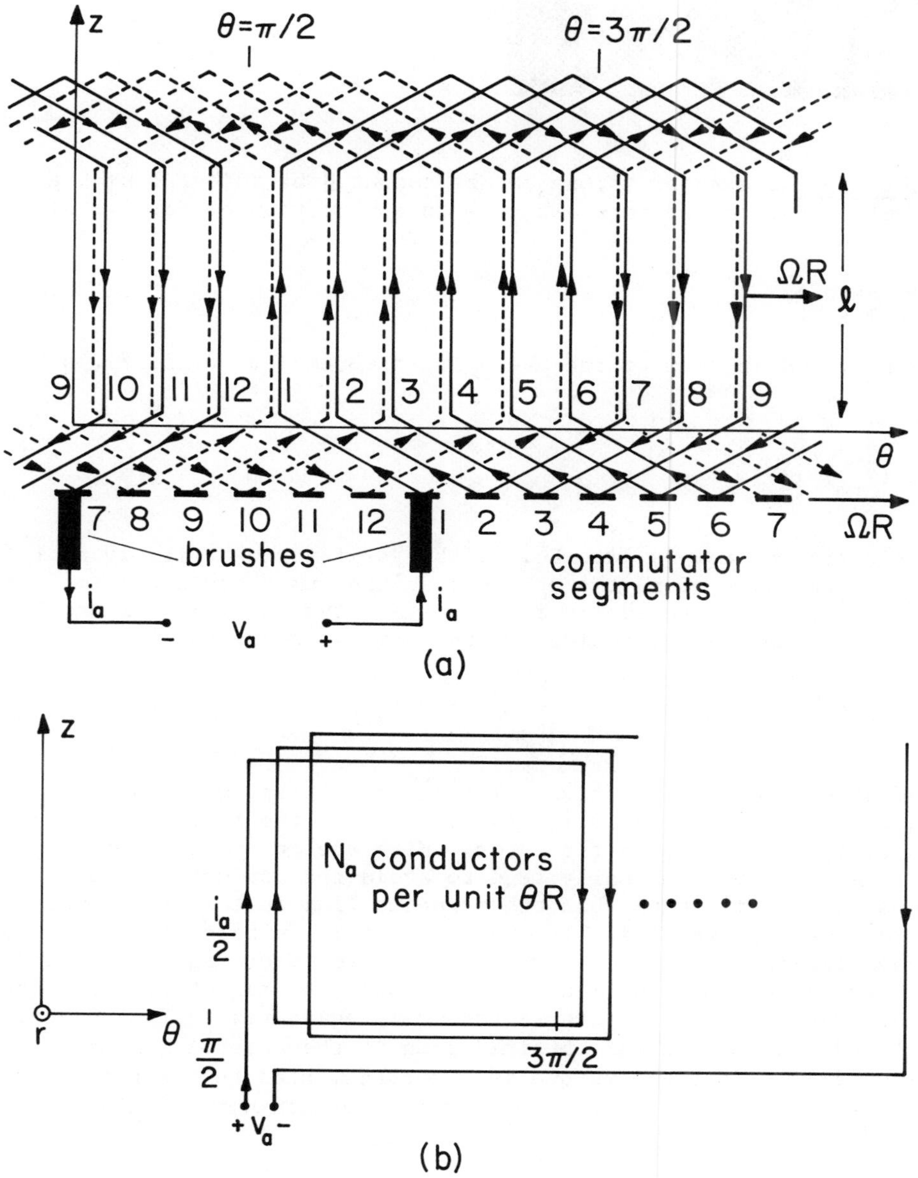

Fig. 4.10.3. (a) Typical winding scheme for armature of d-c machine shown in Fig. 4.10.1. The r axis is directed out of the paper. Brushes make contact with commutator segments which move to the right with armature conductors.[2] (b) Winding distribution of solid wires.

Fig. 4.10.4

This venerable d-c machine, of historical interest because it generated electric power for Boston at the turn of the century, has the advantage of putting the commutator segments and brushes in clear view. The pole faces surrounding the rotor at the upper right have a shape similar to that shown in Fig. 4.10.1, but the associated magnetic circuit is driven by armature coils wrapped on a horse-shoe magnetic circuit closing above the rotor. This is one of the first machines made after Thomas A. Edison moved from New York City to Schenectady in 1886.

Mechanical Equations: The rotor torque can be computed by integrating the Maxwell stress over a surface at $r = R_o$ just inside the stator. This is an application of Eq. 4.2.3:

$$\tau = (2\pi R_o \ell) R_o \left\langle B_r^f H_\theta^f \right\rangle_\theta \tag{4}$$

Because $\tilde{H}_{\theta m}^f = \tilde{\Psi}_m^f (jm/R_o)$, and in view of the averaging theorem (Eq. 2.15.17), substitution of Eqs. 1 and 2 converts Eq. 4 to

$$\tau = 2\pi R_o^2 \ell \sum_{m=-\infty}^{\infty} (\tilde{B}_{rm}^f)^* \left(\frac{jm}{R_o}\right) \tilde{\Psi}_m^f \tag{5}$$

With the substitution of Eq. 3a into Eq. 5, the "self-torque" (involving $\tilde{\Psi}_m^f (\tilde{\Psi}_m^f)^*$) sums to zero. (Because f_m/m is an odd function of m, the mth term in the sum cancels the -mth term.) The remaining expression is a sum on $\tilde{H}_{\theta m}^a \tilde{\Psi}_m^f$. These amplitudes are evaluated using Eqs. 1 and 2. The resulting magnetic torque is thus expressed as a function of the terminal currents:

$$\tau = -G_m i_f i_a ; G_m = \frac{16}{\pi} R R_o \ell \mu_o N_a n_f \sum_{\substack{m=1\\ (odd)}}^{+\infty} \frac{g_m(R_o, R)}{m^2} \frac{\sin(m\theta_o)}{m\theta_o} \tag{6}$$

The speed coefficient, G_m, is positive. This is consistent with the $(\vec{J} \times \vec{B})$ density expected with i_f and i_s positive, as shown in Fig. 4.10.1. But the use of the force density $\vec{J} \times \vec{B}$ misrepresents the actual distribution of force density on the rotor. With the conductors embedded in slots of highly permeable material, the flux lines actually tend to avoid the conductors and pass through the rotor surface between the slots. This means that the magnetic flux in the region where there is a current density tends to zero as the permeability becomes infinite. In fact, the magnetic torque is largely the result of the magnetization force density acting on the rotor magnetic material between the slots. Fortunately, the stress tensor used to find Eq. 6 includes the magnetization force density, so the deductions are sound. But, because the stress tensor is evaluated in free space, the same calculations would be carried out and the same answer obtained even if the essential role of the magnetization force density were not recognized. That the torque is not transmitted to the rotor through the conductor is important, because it alleviates problems encountered in maintaining insulation in the face of mechanical stress and vibration.

In terms of the electrical and mechanical terminal variables (i_f, i_a, τ_θ), Eq. 6 represents the electrical-to-mechanical coupling.

Electrical Equations: To complete the model, it is necessary to express the mechanical-to-electrical coupling in terms of the terminal variables. This is done by taking advantage of Faraday's law, written for a contour of integration that is fixed in the laboratory frame of reference and passes through the appropriate winding:

$$\oint_C (\vec{E} - \vec{v} \times \mu_o \vec{M}) \cdot d\vec{\ell} = -\int_S \frac{\partial \vec{B}}{\partial t} \cdot \vec{n}da \tag{7}$$

For the armature, the circuit C is composed of whatever is externally connected to the terminals (v_a, i_a) and the armature windings. The brushes are idealized as making continuous contact with the moving conductors. A particular possible winding that would give the uniform distribution of rotor current density is shown in Fig. 4.10.3.

The fixed frame electric field integrated on the left in Eq. 7 is related to the conductor current density $\vec{J}$ by Ohm's Law, Eqs. 3.3.6, 2.5.11b, and 2.5.12b. Hence, $\vec{E} = \vec{J}/\sigma - \vec{v} \times \mu_o \vec{H}$ and

$$\vec{E} - \vec{v} \times \mu_o \vec{M} = \frac{\vec{J}}{\sigma} - \vec{v} \times \vec{B} \tag{8}$$

where $\vec{v} = \Omega R \vec{i}_\theta$ is the velocity of the moving conductors. At a given instant, the armature winding amounts to a superimposed parallel pair of windings connected through the brushes to the armature terminals. One of the pair is shown in Fig. 4.10.3b. The other coil, represented by the dotted wires of Fig. 4.10.3a, links the same flux. Each of these windings carries half of the armature current and has the turns density N_a.

For the "solid" windings, Eq. 7 becomes

$$-v_a + \int_{\text{wire}} \frac{\vec{J}}{\sigma} \cdot d\vec{\ell} + \int_{\text{wire}} \Omega R B_r \vec{i}_z \cdot d\vec{\ell} = -\frac{d}{dt}\int_S B_r da \tag{9}$$

where S is an integration over the surface enclosed by the contour C composed of the wire. The integration of $\vec{E}$ between the terminals external to the machine gives the term $-v_a$.

The current density in the wire is the net current $i_a/2$ divided by the cross-sectional area of the wire, A_a. Hence, the second term in Eq. 9 becomes

$$\int_{\text{wire}} \frac{\vec{J}}{\sigma} \cdot d\vec{\ell} = \frac{i_a}{2A_a} \frac{1}{\sigma_a} \ell_a = R_a i_a; \quad R_a \equiv \frac{\ell_a}{2A_a \sigma_a} \tag{10}$$

where A_a is the cross-sectional area of the wire and ℓ_a is the total length of the wire joining the brushes at the given instant (the total length of the "solid" wire in Fig. 4.10.3a). Hence, R_a is the d-c resistance "seen" at the armature terminals.

The third term in Eq. 9 is evaluated by recognizing that those conductors between θ and $\theta + d\theta$ number $(N_a R)d\theta$, and therefore give a contribution $\Omega R B_r(\theta) N_a R d\theta$. This integrand makes a positive contribution in the interval $\pi/2<\theta<3\pi/2$, where the contour is in the positive z direction, and a negative contribution in the interval $-\pi/2<\theta<\pi/2$ where the wires are returning in the -z direction:

$$\begin{aligned}\int_{\text{wire}} \Omega R B_r^a \vec{i}_z \cdot d\vec{\ell} &= \ell \int_{\pi/2}^{3\pi/2} \Omega R^2 N_a B_r^a d\theta - \ell \int_{-\pi/2}^{\pi/2} \Omega R^2 N_a B_r^a d\theta \\ &= -4\Omega \ell R^2 N_a \sum_{\substack{m=-\infty \\ (\text{odd})}}^{+\infty} \frac{\tilde{B}_{rm}^a}{m} j e^{-j\frac{m\pi}{2}}\end{aligned} \tag{11}$$

The second equality results from substitution of the Fourier series and carrying out the integration. It follows from substitution for B_{rm}^a using Eq. 3b with Eqs. 1 and 2 used to relate $\tilde{\Psi}_m^f$ and $\tilde{H}_{\theta m}^a$ to the terminal currents that

$$\int_{\text{wire}} \Omega R B_r^a \vec{i}_z \cdot d\vec{\ell} = -\Omega G_m i_f \tag{12}$$

where G_m is the same as defined with Eq. 6. To complete these steps, observe that f_m/m^3 is an odd function of m, so that the contribution that is proportional to i_a sums to zero. Also, $Rg_m(R,R_o) = -R_o g_m(R_o,R)$, as can be seen from the definition in Table 2.16.2 or by application of the reciprocity condition, Eq. 2.17.10. There is no contribution to Eq. 12 of the part of B_r^a induced by the armature current because this "self-field" contribution to $\vec{v} \times \vec{B}$ at a winding location θ is cancelled by that at −θ.

To evaluate the right-hand side of Eq. 9, first observe that the flux linked by the coils having their left edges in the range dθ' in the neighborhood of θ' is the product of the flux linked by one turn and the number of turns in that range of θ':

$$-\left[\ell \int_{\theta'}^{\theta'+\pi} B_r^a R d\theta\right] N_a R d\theta' \tag{13}$$

As a result, the total flux linked by all of the turns is

$$\int_S B_r da = -\int_{\pi/2}^{3\pi/2} \left[\ell \int_{\theta'}^{\theta'+\pi} B_r^a R d\theta\right] N_a R d\theta' \tag{14}$$

Again, substitution of the Fourier series for B_r^a and evaluation of the integrals gives

$$\int_S B_r da = 4\ell N_a R^2 \sum_{\substack{m=-\infty \\ (\text{odd})}}^{+\infty} \frac{\tilde{B}_{rm}^a}{m^2} e^{-j\frac{m\pi}{2}} \tag{15}$$

Further evaluation, using Eqs. 3b, 1 and 2, with the observation that g_m/m^3 is an odd function of m so that the contribution proportional to i_f vanishes, gives

$$\int_S B_r da = L_a i_a; \quad L_a \equiv \frac{16\ell N_a^2 \mu_o R^3}{\pi} \sum_{\substack{m=1 \\ \text{odd}}}^{\infty} \frac{f_m(R_o,R)}{m^4} \tag{16}$$

That i_f makes no contribution to the <u>net</u> flux linked by the armature winding is evident from Fig. 4.10.1. The armature and field magnetic axes are perpendicular. Thus, with the substitution of Eqs. 10, 12 and 16, the armature circuit equation, Eq. 9, becomes

$$v_a = R_a i_a - \Omega G_m i_f + L_a \frac{di_a}{dt} \tag{17}$$

where R_a, G_m and L_a are given by Eqs. 10, 6 and 16.

The circuit equation for the field winding is similarly found by applying Faraday's integral law, Eq. 7, to a contour composed of the field winding. The right-hand side of Eq. 7 is <u>approximated</u>. by the flux contribution over the surfaces of the respective poles:

$$\int_S B_r da = n_f \ell \int_{-\frac{\pi}{2}+\theta_o}^{\frac{\pi}{2}-\theta_o} B_r^f R_o d\theta - n_f \ell \int_{\frac{\pi}{2}+\theta_o}^{\frac{3\pi}{2}-\theta_o} B_r^f R_o d\theta \tag{18}$$

Substitution of the Fourier series for B_r^f and integration gives

$$\int_S B_r^f da = 4 n_f \ell R_o \sum_{\substack{m=1 \\ (\text{odd})}}^{\infty} \frac{e^{-j\frac{m\pi}{2}}}{(-jm)} \tilde{B}_r^f \cos m\theta_o \tag{19}$$

This expression can now be evaluated using first Eq. 3a and then Eqs. 1 and 2. Because g_m/m^3 is an odd function of m, the term proportional to i_a sums to zero with the result

$$\int_S B_r^f da = L_f i_f; \quad L_f \equiv -\frac{16 n_f^2 \ell R_o \mu_o}{\pi} \sum_{\substack{m=1 \\ (\text{odd})}}^{\infty} \frac{\cos m\theta_o}{m^2} \frac{\sin m\theta_o}{m\theta_o} f_m(R,R_o) \tag{20}$$

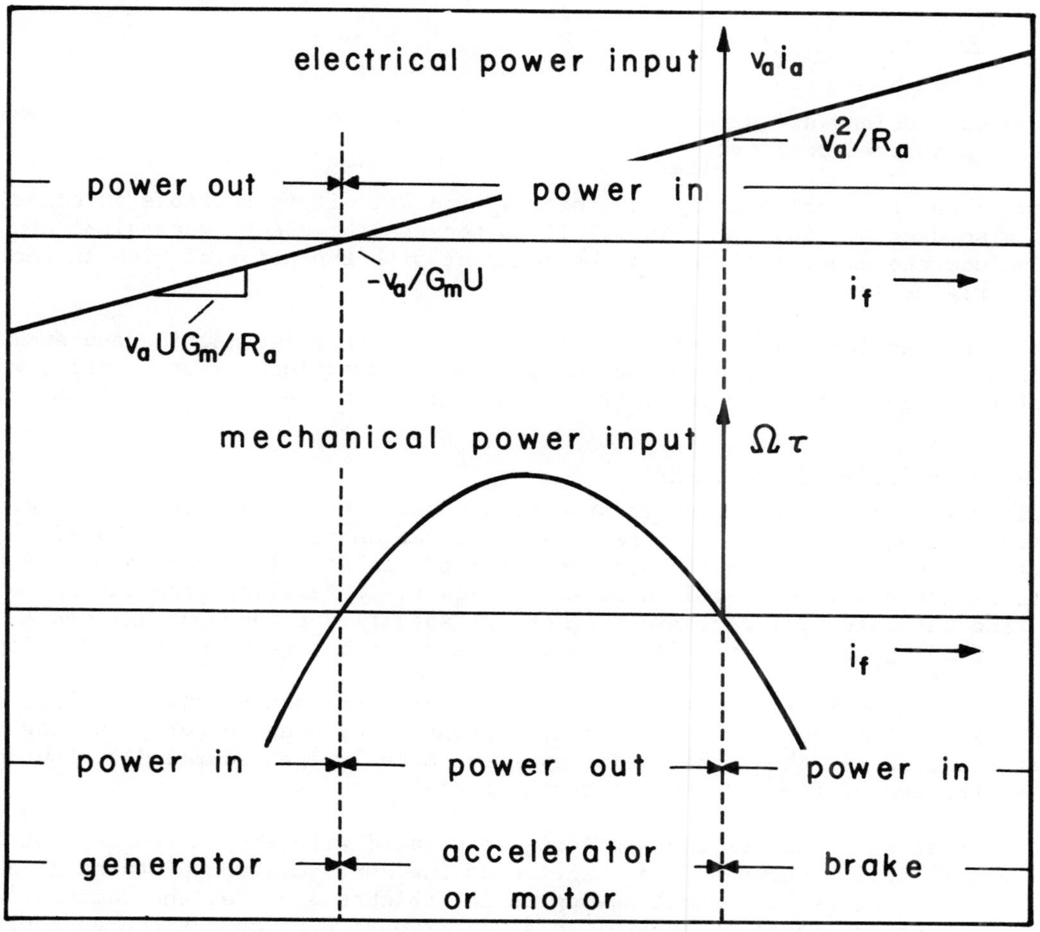

Fig. 4.10.5. Regimes of energy conversion for a d-c magnetic field type interaction. Armature voltage v_a is fixed and field current i_f is varied. With the identification of variables $i_f \rightarrow v_f$, $v_a \rightarrow i_a$, $R_a \rightarrow R_a^{-1}$, $G_m \rightarrow G_e$, the power characteristics also represent the Van de Graaff type of device developed in Sec. 4.14.

Note from the definition of f_m in Table 2.16.2 or the energy relation, Eq. 2.17.12, that $f_m(R,R_o) < 0$, so that L_f is positive. The left-hand side of Eq. 7 is evaluated as for the armature except that the conductor is fixed. Hence, Eq. 7 becomes the required circuit equation for the field:

$$v_f = R_f i_f + L_f \frac{di_f}{df} \tag{21}$$

The total resistance of the field winding is $R_f = A_f \ell_f/\sigma_f$, and L_f is given by Eq. 20.

<u>The Energy Conversion Process</u>: Simple consideration of Eqs. 6 and 17 relates the discrete electrical and mechanical terminal variables to the energy conversion process. Consider the field excitation current i_f and the armature voltage v_a as constrained by external sources. The steady-state dependence of the armature current and the magnetic torque on the constrained variables implied by Eqs. 6 and 17 is then

$$i_a = \frac{v_a}{R_a} + \frac{\Omega G_m}{R_a} i_f \tag{22}$$

$$\tau = -G_m i_f \left[\frac{v_a}{R_a} + \frac{\Omega G_m}{R_a} i_f\right] \tag{23}$$

The electrical power input to the device follows from Eq. 22 as

$$i_a v_a = \frac{v_a}{R_a}[v_a + \Omega G_m i_f] \tag{24}$$

while the mechanical power output is given by Eq. 23 multiplied by the angular velocity

$$\Omega\tau = -\frac{\Omega G_m}{R_a} i_f[v_a + G_m i_f] \tag{25}$$

These last two expressions are sketched in Fig. 4.10.5 to show the power-flow dependence on the field current i_f with Ω assumed positive.

In view of the physical significance of $i_a v_a$ and $\Omega\tau$, it is possible to classify the regimes of operation as also sketched in Fig. 4.10.5. It is because the electromechanical coupling has been defined to include the electrical losses (by contrast with the point of view in Sec. 4.9, for example) that the brake regime is possible.

The power conversion characteristics exemplified by this d-c machine and summarized in Fig. 4.10.5 are in common to the family of d-c or conduction type interactions. For example, with appropriate redefinition of variables, the same characteristics pertain to the Van de Graaff machine of Sec. 4.14.

4.11 Green's Function Representations

In dealing with fields that are related to sources (the charge density or current density) through linear differential equations, it is possible to use yet another approach that is based on the fact that superposition of sources implies superposition of fields. This approach, which is an alternative applicable to situations illustrated in Secs. 4.5 - 4.9, is familiar from the use of the superposition integral to find the potential response from charge specified throughout all space or from the Biot-Savart law for finding the magnetic field, given the distribution of current density throughout space.

Volume source distributions can often be considered the sum of distributions of surface charge or surface current. The transfer relations are a convenient vehicle for obtaining the response to such singular sources. By then integrating over the actual given source distribution, the field is represented as the sum of field responses to the surface sources.

The determination of the fields and force associated with the charge beam of Sec. 4.6 illustrates the method. Figure 4.11.1 shows a cross section of the configuration pictured in Fig. 4.6.1, but with the only volume charge in a shell having radial thickness dr' at the radius r', where the density is $\rho(r')$. The fields due to an arbitrary radial distribution of charge can be constructed once the response to this surface charge, having density $\tilde{\rho}(r')dr'$, is determined. At the outset, consider the field to be a superposition of fields due to the potential $\tilde{V}_o$ imposed at the surface r = a and to the distribution of charge in the volume. The latter is determined by using the boundary conditions

$$\tilde{\Phi}^c = 0,\ \tilde{\Phi}^d = \tilde{\Phi}^e,\ \tilde{D}_r^d - \tilde{D}_r^e = \tilde{\rho}(r')dr' \tag{1}$$

Implicit is the understanding that there is no θ dependence, and that the z dependence is exp(-jkz).

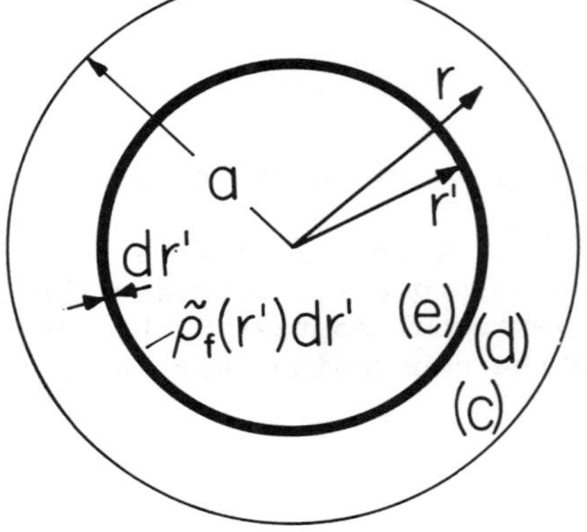

Fig. 4.11.1

Shell having surface-charge density $\tilde{\rho}_f(r')dr'$ gives rise to fields that can be summed to determined field due to arbitrary charge distribution.

In the region r > r', the flux-potential relations, Eq. (a) of Table 2.16.2, apply:

$$\begin{bmatrix} \tilde{D}_r^c \\ \tilde{D}_r^d \end{bmatrix} = \varepsilon \begin{bmatrix} f_o(r',a) & g_o(a,r') \\ g_o(r',a) & f_o(a,r') \end{bmatrix} \begin{bmatrix} \tilde{\Phi}^c \\ \tilde{\Phi}^d \end{bmatrix} \tag{2}$$

whereas in the inner region, r < r', the limiting form of Eq. (c) is appropriate:

$$\tilde{D}_r^e = \varepsilon f_o(0,r')\tilde{\Phi}^e \tag{3}$$

Subtraction of Eq. 3 from Eq. 2b and use of the boundary conditions of Eq. 1 gives

$$\tilde{\Phi}^d = \tilde{\Phi}^e = \frac{\tilde{\rho}(r')dr'}{\varepsilon[f_o(a,r') - f_o(0,r')]} \tag{4}$$

By the judicious use of these amplitudes and the potential distribution given for a canonical annular region by Eq. 2.16.25, it is now possible to write the radial distribution of $\tilde{\Phi}$ for an arbitrary distribution of charge density. There are three terms. The first is simply the potential due to the voltage $\tilde{V}_o$ applied at the outer wall. For this part, Eq. 2.16.25 is evaluated with $\beta \to 0$ and $\tilde{\Phi}^\alpha = \tilde{V}_o$. The second term comes from evaluating Eq. 2.16.25 for the potential at r due to the charge shell at $r' < r$ (so that $\alpha = a$, $\beta = r'$, $\tilde{\Phi}^\alpha = \tilde{\Phi}^c = 0$ and $\tilde{\Phi}^\beta = \tilde{\Phi}^d$) and adding up all contributions attributable to charge inside the radius of observation r. Finally, the third term is written by again using Eq. 2.16.25 to express the potential, but this time due to charge at a greater radius than the r, at $r < r'$ (so that $\alpha = r'$, $\beta \to 0$ and $\tilde{\Phi}^\alpha = \tilde{\Phi}^d$) and integrating over the distribution outside the observation position r:

$$\tilde{\Phi}(r) = \tilde{V}_o \frac{J_o(jkr)}{J_o(jka)} + \int_o^r \frac{[J_o(jka)H_o(jkr) - H_o(jka)J_o(jkr)]}{[J_o(jka)H_o(jkr')-H_o(jka)J_o(jkr')]} \frac{\tilde{\rho}(r')dr'}{\varepsilon[f_o(a,r') - f_o(0,r')]}$$
$$+ \int_r^a \frac{J_o(jkr)}{J_o(jkr')} \frac{\tilde{\rho}(r')dr'}{\varepsilon[f_o(a,r') - f_o(0,r')]} \tag{5}$$

To find the axial force acting on the entire beam, it is only the normal flux density at the outer wall that is required. This can be found from Eq. 5, but is more easily determined directly from Eqs. 2a, used first with $\tilde{\Phi}^c = \tilde{V}_o$ and (d) $\to$ 0 to find the flux density due to the wall potential alone and then with $\tilde{\Phi}^c = 0$ and $\tilde{\Phi}^d$ given by Eq. 4 to find the part due to the volume charge. The latter is summed over the total distribution of charge.

$$\tilde{D}_r^c = \varepsilon f_o(0,a)\tilde{V}_o + \int_o^a \frac{g_o(a,r')\tilde{\rho}(r')dr'}{[f_o(a,r') - f_o(0,r')]} \tag{6}$$

The force is thus determined by substituting this expression into Eq. 4.6.3. Equation 6 holds for an arbitrary charge distribution, but consider the uniform distribution of charge inside the radius R. Then the integration needs only be carried out from 0 to R. With $\tilde{V}_o$ and $\tilde{\rho}(r')$ selected consistent with Eqs. 4.6.1 and 4.6.2, it follows that the force is given by Eq. 4.6.8 with L_1 replaced by L_3, where

$$L_3 = \frac{a}{R^3} \int_o^R \frac{g_o(a,r')dr'}{[f_o(a,r') - f_o(0,r')]} = \frac{1}{(kR)^2} \int_o^{kR} (kr') \frac{I_o(kr')}{I_o(ka)} d(kr') \tag{7}$$

The integral is carried out by recognizing that $I_o(kr')$ is a solution to Eq. 2.16.19 with $r \to r'$ and $m = 0$:

$$\frac{d}{dr'}\left(r' \frac{dI_o(kr')}{dr'}\right) = k^2 r I_o(kr') \tag{8}$$

Hence, Eq. 7 gives the same result, Eq. 4.6.13, as found in Sec. 4.6 using the "splicing approach."

The same procedure applies if the charge has θ dependence $\exp(-jm\theta)$. Thus, by making use of a Fourier series representation in θ and z, the method can be used to describe fields associated with arbitrary dependence on θ and z.

The Green's function approach exemplified here is applicable to modeling the synchronous machines developed in Secs. 4.7 and 4.8.[1]

4.12 Quasi-One-Dimensional Models and the Space-Rate Expansion

The "narrow-air-gap" model for rotating machines and long-wave models for electromagnetic wave propagation are examples of quasi-one-dimensional models. The following sections illustrate the use of such models in the kinematic description of electromechanical interactions. Extensive use will be made in later chapters of models that similarly exploit a relatively slow variation of distributed quantities in a "longitudinal" direction relative to "transverse" directions.

---

1. This is the method used by Kirtley in "Design and Construction of an Armature for an Alternator with a Superconducting Field Winding," Ph.D. Thesis, Department of Electrical Engineering, MIT, Cambridge, Mass., 1971, for a configuration closely resembling that considered in Sec. 4.8.

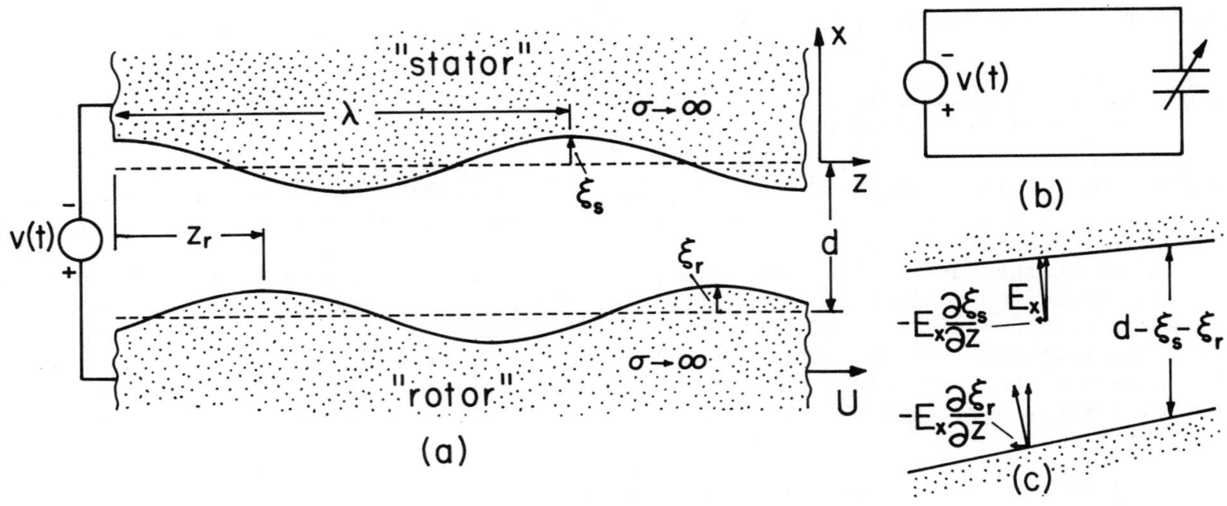

Fig. 4.12.1. (a) Cross-sectional view of synchronous electric field energy converter with stator and rotor composed of perfectly conducting materials constrained by a time-varying voltage source. The stator geometry is static, while the rotor moves to the right. (b) Interaction represented by time-varying capacitance. (c) Detail of air gap showing components of E_z to satisfy boundary conditions.

An example is shown in Fig. 4.12.1. Perfectly conducting surfaces having the potential difference v(t), vary from the planes x = 0 and x = -d by the amounts $\xi_s(z,t)$ and $\xi_r(z,t)$, respectively. What are the fields in the gap? This configuration is the basis for the study of the variable-capacitance machine in Sec. 4.13. Fields in the gap can be approximated by two techniques. If ξ_s and ξ_r are small compared to d, the boundary conditions can be linearized, and the fields found approximately. This is the approach used in Sec. 4.3 for describing the salient pole interactions (Eq. 4.3.16). It formally amounts to expanding the fields in an amplitude parameter expansion with the zero-order fields those with ξ_s and ξ_r equal to zero, the first-order terms those given by keeping only linear terms in (ξ_s,ξ_r) and so on. Thus, the validity of the model hinges on the amplitudes (ξ_s,ξ_r) being small.

In quasi-one-dimensional models, amplitudes are not necessarily small. Rather, certain spatial rates of change are small. In the configuration of Fig. 4.12.1, the distance λ typifying variations in the z direction is long compared to the distance d, $\gamma \equiv (d/\lambda)^2 \ll 1$.

The relationship between linearized and quasi-one-dimensional models is illustrated in Fig. 4.12.2. Linearized quasi-one-dimensional models must be consistent with the long-wave limit of the linearized model. In establishing complex models, this fact is often used to motivate the appropriate "zero-order" approximation which is the starting point in developing a quasi-one-dimensional model.

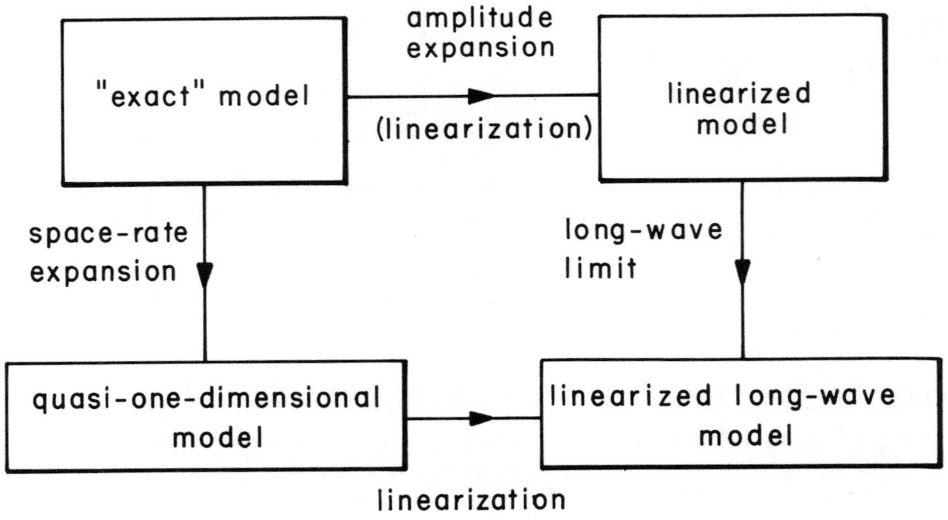

Fig. 4.12.2. Schematic characterization of relationships among three-dimensional, quasi-one-dimensional and linearized models.

Usually, quasi-one-dimensional models are motivated by physical reasoning, with little need for formality. This is partly because higher order terms are seldom used. But, at least once, it is worhwhile to see how higher order terms are found, and that the approximation used is the lowest order term in an expansion in powers of a space-rate parameter, in the example of Fig. 4.12.1, of $\gamma = (d/\lambda)^2$.

The procedure here is analogous to that of Sec. 2.3 on quasistatics. The spatial coordinate z, in which variables evolve slowly, plays the role of time. The physical idea that this slow variation ought to make one field component dominate the other is built into the normalization of variables. If modulations of the electrodes are slowly varying compared to the transverse distance d, each section of the electrodes tends to form a parallel-plate capacitor. With E_o a typical electric field in the x direction (the "dominant" field component), d taken as the typical length in the x direction, but λ as that length in the z direction, the appropriate normalization is

$$E_x = E_o \underline{E}_x \qquad x = d\underline{x}$$
$$E_z = E_o(d/\lambda)\underline{E}_z \qquad z = \lambda\underline{z} \tag{1}$$
$$\xi_r = d\underline{\xi}_r,\ \xi_s = d\underline{\xi}_s \qquad v = (E_o d)\underline{v}$$

In the gap, $\vec{E}$ is irrotational and solenoidal. In terms of the normalized variables, these conditions are

$$\frac{\partial E_x}{\partial z} - \frac{\partial E_z}{\partial x} = 0$$
$$\frac{\partial E_x}{\partial x} = -\gamma\frac{\partial E_z}{\partial z} \tag{2}$$

where the space-rate parameter $\gamma \equiv (d/\lambda)^2$. To complete the formulation in terms of normalized variables, boundary conditions at the scalloped perfect conductors are that the potential difference be v(t) and the tangential fields vanish:

$$E_z = -\frac{\partial \xi_s}{\partial z}E_x(x = \xi_s);\ E_z = -\frac{\partial \xi}{\partial z}E_x(x = \xi_r - 1);\ \int_{\xi_r - 1}^{\xi_s} E_x dx = v \tag{3}$$

Only two of these three expressions are independent.

The normalized field components are now expanded in series of the form

$$E_x = E_{xo} + \gamma E_{x1} + \gamma^2 E_{x2} + \cdots$$
$$E_z = E_{zo} + \gamma E_{z1} + \gamma^2 E_{z2} + \cdots \tag{4}$$

Note that only one dimensionless parameter is involved, so for the particularly simple case at hand, there is no ambiguity as to what lengths are most critical.

Substitution of the series of Eq. 4 into Eqs. 2 gives a pair of expressions which are polynomial in γ. Coefficients of each order in γ must vanish; thus, the zero-order terms involve only the zero-order fields

$$\frac{\partial E_{xo}}{\partial z} - \frac{\partial E_{zo}}{\partial x} = 0$$
$$\frac{\partial E_{xo}}{\partial x} = 0 \tag{5}$$

but the first order expressions are "driven" by the zero order fields

$$\frac{\partial E_{x1}}{\partial z} - \frac{\partial E_{z1}}{\partial x} = 0$$
$$\frac{\partial E_{x1}}{\partial x} = -\frac{\partial E_{zo}}{\partial z} \tag{6}$$

It follows from Eqs. 3c and 5b that E_{xo} is quasi-one-dimensional. It only depends on (z,t):

$$E_{xo} = E_{xo}(z,t) = \frac{v}{\xi_s + 1 - \xi_r} \tag{7}$$

What has been deduced as the zero-order E_x is just the voltage divided by the distance between conductors. If variations with z are sufficiently slow, each section of the system forms a plane-parallel capacitor. To find the other component of the zero-order field, note that E_{xo} is only a function of (z,t), so Eq. 5a can be integrated to obtain

$$E_{zo} = x \frac{\partial E_{xo}}{\partial z} + f(z,t) \tag{8}$$

where f(z,t) is an integration function. This function is determined by substitution of Eq. 8 into Eq. 3a:

$$E_{zo} = x \frac{\partial E_{xo}}{\partial z} - \frac{\partial}{\partial z}(E_{xo}\xi_s) \tag{9}$$

Substitution now shows that the tangential field on the lower surface is zero, Eq. 3b is satisfied. The zero-order fields are represented in dimensionless form by Eqs. 7 and 9.

The first-order fields are predicted by Eqs. 6, now that the zero-order fields are known. From Eqs. 6b and 9,

$$-\frac{\partial E_{x1}}{\partial x} = -\frac{\partial E_{zo}}{\partial z} = -x\frac{\partial^2 E_{xo}}{\partial z^2} + \frac{\partial^2}{\partial z^2}(E_{xo}\xi_s) \tag{10}$$

The functional dependence on x on the right in this expression is explicit, and therefore integration gives

$$E_{x1} = -\frac{x^2}{2}\frac{\partial^2 E_{xo}}{\partial z^2} - x\frac{\partial^2}{\partial z^2}(E_{xo}\xi_s) + g(z,t) \tag{11}$$

Because the zero-order E_x already satisfies the boundary condition, E_{xo} integrates to v across the gap (Eq.3c), the same integral of Eq. 11 must vanish and that serves to determine the integration function g(z,t). At this point, two terms in the series of Eq. 4a have been found, and they are sufficient to show what is meant by the expansion

$$E_x = \frac{v}{(1+\xi_s-\xi_r)} + \gamma\left\{\frac{\partial^2 E_{xo}}{\partial z}\left[-\frac{x^2}{2} + \frac{1}{6}\frac{\xi_s^3 + (1-\xi_r)^3}{\xi + (1-\xi_r)}\right] + \frac{\partial^2}{\partial z^2}(E_{xo}\xi_s)\left(x - \frac{1}{2}[\xi_s - (1-\xi_r)]\right)\right\} \tag{12}$$

By the definition of ℓ used in normalizing z, $\partial^2 E_{xo}/\partial z^2$ is on the order of E_{xo}. Hence, the first term in Eq. 12 gives an accurate picture of the field, provided $\gamma << 1$.

The procedure outlined is mainly of conceptual value. Certainly the quasi-one-dimensional modeling of a complex problem begins with a physically motivated approximation: here, Eq. 7. Because no more than the zero-order solutions are usually required, the formalism of normalizing the variables and identifying dimensionless space parameters is not usually required.

In retrospect, the zero-order fields have a dependence on the transverse direction (x) that is the lowest order polynomial in x consistent with the boundary conditions. Thus, E_{xo} varies as x^o (it is independent of x); while E_{zo} can satisfy the boundary conditions only if it includes a linear dependence on x.

4.13 Variable-Capacitance Machines

A model for one of the most commonly discussed "electrostatic" synchronous machines (which are themselves rather uncommon) is shown in Fig. 4.12.1a. Both the fixed and moving members have saliency and consist essentially of perfectly conducting material. The time-varying voltage between stator and rotor can either be the source of electrical power for producing a synchronous force in the z direction on the rotor, or it can serve as the voltage of a bus representing an energy sink for the device acting

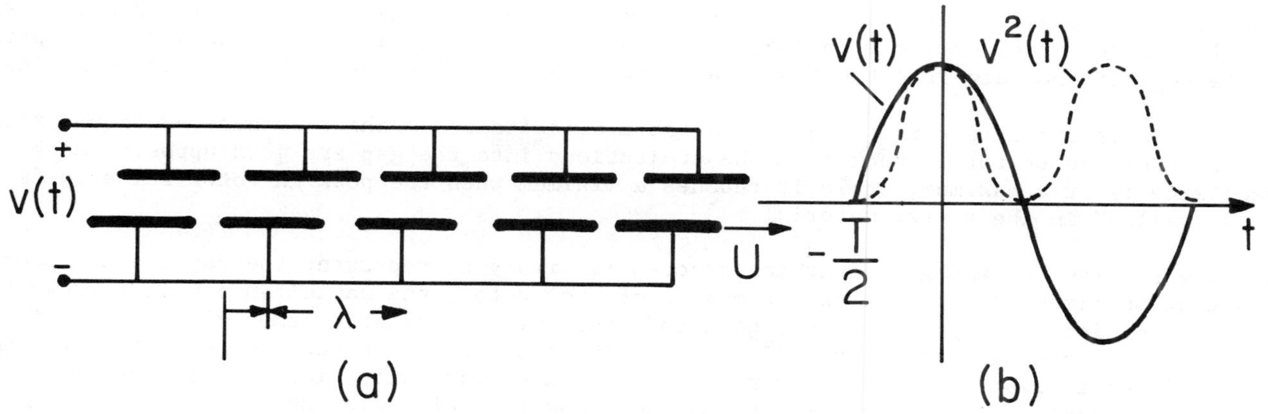

Fig. 4.13.1. Physical realization of variable-capacitance machine modeled in Fig. 4.12.1. (a) Stator and rotor structure consisting of vanes. (b) Sinusoidal voltage supplied through slip rings together with $v^2(t)$, showing temporal dependence of instantaneous force.

Fig. 4.13.1c. Variable-capacitance generator designed for use with vacuum insulation. Estimated output at 30,000 rpm is 6 kW at 20 kV (courtesy Goodrich High Voltage Corp.). Development of variable-capacitance machines was attempted for the generation of high-voltage power with application to ion propulsion in the space program. In space, vacuum insulation is easily obtained. See reports for Contract No. AF33(616)-7230 from Goodrich-High Voltage Astronautics, Inc., Burlington, Mass., to Aeronautical Systems Division, Air Force Systems Command, U.S. Air Force, Wright-Patterson Air Force Base, Ohio. For example, Phase II report by A. S. Denholm et al., 1961.

as a generator. In practice, the stator and rotor members might consist of metallic fins, as shown in Fig. 4.13.1. In the model, regions on the stator and rotor that project into the air gap represent the fins, while regions that dip into the stator and rotor material represent the gaps between fins.

The device is often referred to as a "variable-capacitance" machine because, when the relative position of rotor and stator is such that the projections into the gap are just opposite each other, the capacitance is at a maximum, while it reaches a minimum when the peak in rotor saliency falls just opposite a "valley" in the stator material.

One way to view the energy conversion process is simply to represent the capacitance seen by the voltage source as time-varying. Given the motion of the rotor, the capacitance C is a known function of time, and the electrical problem comes down to determining a suitable temporal variation for C, relative to a time-varying voltage, v. If power is supplied to the voltage source, it must come from the mechanical forces responsible for making the capacitance vary with time. Thus, the other side of the energy conversion process raises the question: How is a time-average force produced on the rotor by the combination of the salient configuration and the time-varying applied voltage? In this section, we will take up the second question first. What is the electrical force in the direction of motion on the moving member?

The field point of view taken here results in the relation between geometry and capacitance needed to model an actual system, even if the circuit point of view is taken. But also, it makes the example useful in conceptualizing electromechanical interactions that cannot be given a lumped-parameter model. For example, suppose that the undulations on the "rotor" were in fact material deformations produced by the field itself. This type of self-consistent electromechanical coupling is not kinematic and will be taken up in Chap. 9.

Synchronous Condition: With a sinusoidal voltage v(t) having period T, applied between the rotor and stator by means of a slip-ring, a time-average electrical force can act in the z direction on the rotor only if there is a synchronism between the applied voltage and the rotor motion. To this end, consider the physical origins of this force in terms of the model shown in Fig. 4.12.1. Regardless of the field polarity, at any position on the rotor surface there is an electric force per unit area that is directed perpendicular to the surface and into the air gap. This latter fact makes it clear that without the surface undulations, there can be no electrical force in the z direction.

To make a synchronous motor, on the time average, fields acting to the right over regions of the rotor surface with a negative slope must produce a greater force than those acting to the left on the regions where the slope is positive. What is the relationship between the excitation period T and the rotor velocity U that could result in there being a time-average electrical force? In terms of the displacement z_r of Fig. 4.12.1, a maximum in the force to the right is obtained with z_r in the neighborhood of $\lambda/4$. Thus, with the rotor in this position, the applied v^2 should be at its maximum. By the time the rotor is at $z_r = 3\lambda/4$, the force produced is in the wrong direction, and hence v^2 should be near a null. By the time $z_r = 5\lambda/4$, v^2 should be peaking again. It is concluded that in the time T/2, the rotor should move one wavelength: $UT/2 = \lambda$. Thus, the synchronism condition is met if

$$z_r = Ut + \delta; \; U = \frac{2\lambda}{T} \tag{1}$$

Here, δ is a spatial phase-angle determined by the mechanical load on a motor or the electrical load on a generator.

The quasi-one-dimensional electric field is given by Eqs. 4.12.7 and 4.12.9 un-normalized:

$$E_x = \frac{v}{d + \xi_s - \xi_r}; \; E_z = (x + d)\frac{\partial E_x}{\partial z} - \frac{\partial}{\partial z}(\xi_r E_x) \tag{2}$$

The force on a section of the rotor one wavelength long and a length ℓ in the y direction is found by integrating the Maxwell stress tensor over an enclosing surface as pictured in Fig. 4.2.1a. The only surface giving a contribution is the one of constant x in the air gap:

$$f_z = \ell \int_z^{z+\lambda} \varepsilon_o E_x E_z dz \tag{3}$$

This integral can be evaluated using the fields of Eq. 2. That it does not matter what x = constant plane is used in carrying out the integration (except for physical reasons, to have the assurance that the surface does not cut through one of the electrode inward peaks) is evident from the fact that

$$\int_z^{z+\lambda} \varepsilon_o E_x(x+d)\frac{\partial E_x}{\partial z}dz = \varepsilon_o(x+d)\int_z^{z+\lambda}\frac{\partial}{\partial z}(\tfrac{1}{2}E_x^2)dz = \varepsilon_o(x+d)[E_x^2(z+\lambda) - E_x^2(z)] = 0 \qquad (4)$$

The final deduction follows from the spatial periodicity of the structure. The remaining contributions to the integral are expressed using the normalization

$$z = \lambda\underline{z},\ \xi_s = d\underline{\xi}_s,\ \xi_r = d\underline{\xi}_r,\ \delta = \lambda\underline{\delta},\ z_r = \lambda\underline{z}_r \qquad (5)$$

With $f_z \equiv (\varepsilon_o \ell v^2/d)\underline{f}_z$, Eq. 3 becomes

$$\underline{f}_z = -\int_{\underline{z}}^{\underline{z}+1}\frac{1}{1+\underline{\xi}_s-\underline{\xi}_r}\frac{\partial}{\partial \underline{z}}\left[\frac{\underline{\xi}_r}{1+\underline{\xi}_s-\underline{\xi}_r}\right]d\underline{z} \qquad (\underline{6})$$

Carrying out the differentiation in the integrand gives

$$\underline{f}(\underline{z}_r) \equiv -\int_{\underline{z}}^{\underline{z}+1}\frac{(1+\underline{\xi}_s)\frac{\partial \underline{\xi}_r}{\partial \underline{z}} - \underline{\xi}_r\frac{\partial \underline{\xi}_s}{\partial \underline{z}}}{(1+\underline{\xi}_s-\underline{\xi}_r)^3}d\underline{z} \qquad (\underline{7})$$

Once the integral is completed, the function f depends on the amplitudes of ξ_s and ξ_r and on their relative displacement $\underline{z}_r$. The time-average force is then computed by specifying this relative displacement in terms of Eq. 1. In normalized variables, with $t = T\underline{t}$

$$\langle f_z\rangle_t = \frac{\varepsilon_o \ell}{d}\int_{\underline{t}}^{\underline{t}+1} v^2(\underline{t})\underline{f}(2\underline{t}+\underline{\delta})d\underline{t} \qquad (8)$$

As an example, consider stator and rotor electrodes having sinusoidal shapes of equal amplitude and a sinusoidal excitation voltage (note that Eqs. 7 and 8 are general in regard to these specifications):

$$\underline{\xi}_s = \underline{\xi}_o \cos 2\pi\underline{z},\ \underline{\xi}_r = \underline{\xi}_o\cos 2\pi(\underline{z}-\underline{z}_r),\ v(t) = V\cos 2\pi\underline{t} \qquad (9)$$

Numerical integration of Eq. 7 then gives the dependence on relative displacement and amplitude shown in Fig. 4.13.2a. To highlight the nonlinear effects of ξ_o, f is normalized to $\underline{\xi}_o^2$ so that much of the dependence on the electrode amplitudes is suppressed.

The electrodes make their closest approach to each other with $\underline{z}_r = 0.5$ and are furthest apart when $\underline{z}_r = 0$. Thus, for a given voltage, the fields tend to be more intense in the range $0.25<\underline{z}_r<0.5$ than they are in the range $0<\underline{z}_r<0.25$. This nonlinear efffect is reflected in the tendency of the force to be skewed toward relative deflections in the former range. As would be expected from the singularity in the denominator of Eq. 7, as the electrodes tend to touch ($\underline{\xi}_o \to 0.5$), the force tends to approach infinity just to the left of $\underline{z}_r = 0.5$. The function $f(z_r)$ is then used to numerically integrate Eq. 8, with the result the normalized time-average force shown as a function of relative displacement phase δ and amplitude $\underline{\xi}_o$ in Fig. 4.13.2b. Again, the dependence on ξ_o is partially suppressed in the normalization.

The electromechanical model exemplified by Eqs. 7 and 8 is nonlinear, in the sense that the electrode deflections can be of arbitrary amplitude in the range $0 < \underline{\xi}_o < 0.5$. The fact that the time-average force becomes infinite as $\underline{\xi}_o \to 0.5$ is to be expected. At some instant, the electrodes are then at the point of touching and the associated field is becoming extremely large where the electrodes are nearly in contact. (Physically, electrical breakdown would of course present a limit on the validity of the theory.) Within the validity of an air-gap dielectric that does not permit electrical breakdown, the procedure which has been followed is an example of the left vertical leg in Fig. 4.12.2.

Further linearization, based on $\xi_s << d$ and $\xi_r << d$, demonstrates what is meant by a "linearized quasi-one-dimensional" model and by the completion of the step represented by the lower horizontal leg in Fig. 4.12.2.

For small amplitudes, $(1+\xi_s-\xi_r)^{-3} \simeq 1 - 3(\xi_s - \xi_r)$, and hence Eq. 7 becomes

$$\underline{f}(\underline{z}_r) \to \int_{\underline{z}}^{\underline{z}+1}\left[(1+\underline{\xi}_s)\frac{\partial \underline{\xi}_r}{\partial \underline{z}} - \underline{\xi}_r\frac{\partial \underline{\xi}_s}{\partial \underline{z}} - 3(\underline{\xi}_s-\underline{\xi}_r)\frac{\partial \underline{\xi}_r}{\partial \underline{z}} + \cdots\right]d\underline{z} = \underline{\xi}_o\pi \sin 2\pi\underline{z}_r \qquad (\underline{10})$$

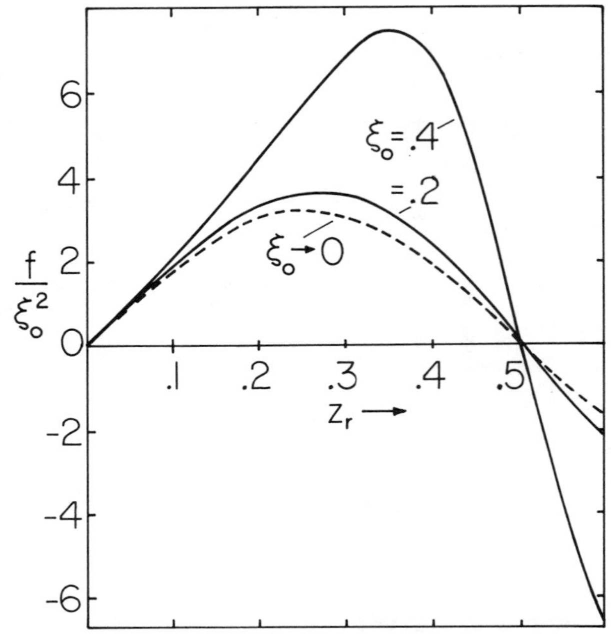

Fig. 4.13.2a. Electrical force on rotor of variable-capacitance machine (Fig. 4.12.1) as a function of normalized relative displacement $\underline{z}_r = z_r/\lambda$, with amplitude of electrodes as a parameter.

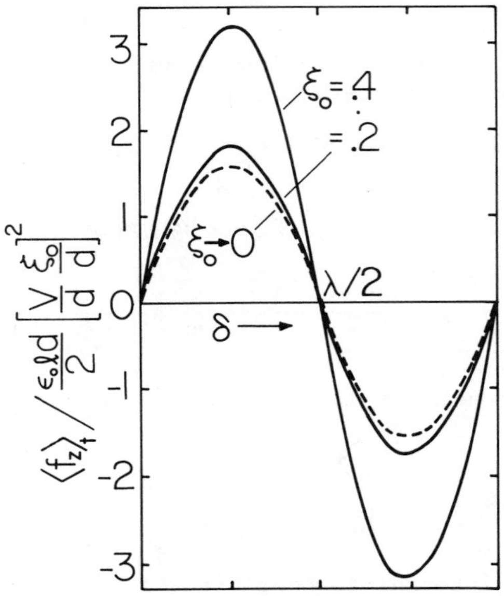

Fig. 4.13.2b. Normalized time-average force as a function of relative phase of sinusoidal excitation and rotor position.

(In carrying out this and the next integration it is helpful to represent the expressions of Eq. 9 in complex notation and make use of the averaging theorem, Eq. 2.15.14.) In turn, the time average called for by Eq. 8 can now be evaluated:

$$\langle f_z \rangle_t = \frac{\varepsilon_o \ell V^2 \xi_o^2 \pi}{d} \int_{\underline{t}}^{\underline{t}+1} \cos^2 2\pi \underline{t} \sin 2\pi(2\underline{t} + \underline{\delta}) d\underline{t} \tag{11}$$

Carrying out this integration gives

$$\langle f_z \rangle_t = (\ell\lambda)\left[\varepsilon_o\left(\frac{V}{d}\right)^2\right] \frac{d}{\lambda} \left(\frac{\xi_o}{d}\right)^2 \frac{\pi}{4} \sin\left(\frac{2\pi\delta}{\lambda}\right) \tag{12}$$

This approximation to the time-average force is shown by the broken curve of Fig. 4.13.2b.

Note that the small-amplitude force of Eq. 12 takes the form of the area $\ell\lambda$ multiplied by the electric pressure $\varepsilon_o(V/d)^2$ times factors representing the fraction of this product obtained by dint of the geometry and the relative phase of the rotor and the driving voltage.

The variable-capacitance machine is closely related to the salient-pole machine described in Sec. 4.3 (Case 4b of Table 4.3.1). In that example, the stator is "smooth" with electrodes constrained by a traveling wave of potential. The effect of having a stator with saliencies driven by a simple voltage source (which is likely to be more convenient) is to produce a similar time-average force.

Linearized from the outset, the variable-capacitance machine of this section could also be viewed in terms of an interaction between the rotor traveling wave and one of two stator waves, the sum of which is equivalent to the physical stator structure considered. The result of such an analysis would be a model without restrictions as to the gap width relative to the wavelength. For the related example of Sec. 4.3, Eq. 4.3.27b retains information (represented by the denominator, $\sinh^2(kd)$) about the effect of the air gap in the limit where d becomes large. This result, restricted to small amplitude but valid for arbitrary air-gap spacing, is typical of the amplitude parameter expansion or linearization modeling step of Fig. 4.12.2. Taking the long-wave limit for the example from Sec. 4.3 constitutes taking the limit of Eq. 4.3.27b, $kd \ll 1$. Following this route of first linearizing and then taking the long-wave limit for the variable-capacitance machine considered in this section is an alternative derivation of Eq. 12, and is considered in the problems.

4.14 Van de Graaff Machine

A cross-sectional view of a Van de Graaff generator is shown in Fig. 4.14.1. An insulating belt is charged to one polarity as it passes over the lower pulley. This charge is carried upward to the essentially field-free region under the high-voltage terminal dome where it is removed and replaced by charge of opposite polarity, which then makes the return trip on the downward moving portion of the belt. Surrounding the belt are equipotential rings which help in controlling the field distribution by supporting much of the charge imaging that on the belt. The electric field consists of a generated field that is essentially vertical and a self-field associated with the charge on the belt. The equipotential rings help to insure that the self-field is essentially perpendicular to the belt surface and hence does not reinforce the generated field. To achieve relatively high electric stress (exceeding 10^7 V/m), the machine is operated in electronegative gases at elevated pressure.

An objective in this section, achieved while developing a lumped-parameter model for the simplified Van de Graaff generator shown in Fig. 4.14.2, is to further illustrate the use of quasi-one-dimensional models. This makes it possible to point out the analogies between d-c magnetic machines, Sec. 4.10, and what might be termed "d-c electric machines."

In several regards, the model shown in Fig. 4.14.2 does not include features of the machine shown in Fig. 4.14.1. To avoid undue complexity, the equipotential rings are uniformly distributed between the high-voltage dome and the ground at the bottom. In the machine pictured in Fig. 4.14.1, charging is by means of a corona discharge (ion impact charging). An alternative scheme, which has the advantage of being more easily related to a physical model, makes use of induction charging of a belt consisting of conductors linked by insulators.[1] For the present purposes, the belt (having thickness d) is considered to carry metallic segments that are insulated from each other. "Field" voltage sources v_f are used to induce belt charges of opposite polarity at the top and bottom. As the belt passes over the lower pulley, successive segments contact a grounded brush and hence form essentially plane-parallel capacitors having a voltage v_f across the belt thickness d. With the assumption that the belt electrodes essentially cover all of the belt surface, the belt surface charge is related to the field voltage by

$$\sigma_f = \frac{\varepsilon v_f}{d} \tag{1}$$

The current i_a' both supplies the charge carried upward by the belt and neutralizes that coming downward. Hence, for a pulley angular velocity Ω and radius R,

$$i_a' = 2\sigma_f \ell(\Omega R) = \frac{2\ell R\varepsilon}{d}\Omega v_f \tag{2}$$

<u>Quasi-One-Dimensional Fields</u>: In the ideal, the generated field is uniformly distributed with respect to the z axis. To achieve this ideal, in spite of the metal pressure vessel, the equipotential rings are tapped onto a distributed bleeder resistance running from the dome to the ground plane. At least under steady-state conditions, this insures that the ring potential $\Phi_r(z)$ has the required linear distribution consistent with a uniform z-directed electric field. The following developments identify the implications of having time-varying terminal variables, (v_a, i_a) and (v_f, i_f).

The transverse field components are determined as though any local region along the z axis is one in which the x-directed fields are independent of z. Thus, in the region between rings and pressure vessel,

$$E_{x3} = \frac{\Phi_r}{c} \tag{3}$$

The fields E_{x2} and E_{x1} (Fig. 4.14.2) must satisfy Gauss' law at the belt surface and be consistent with the potential being the same on the ring where it faces the belt on the right and at the same z location on the left. Hence, with fields defined positive if they are as shown in Fig. 4.14.2,

$$\varepsilon_o(E_{x1} + E_{x2}) = \sigma_f \tag{4}$$

$$-2bE_{x2} + 2aE_{x1} = 0 \tag{5}$$

Here, E_{x2} is approximated as being uniform over the width of the belt, even though the rings are cylindrical and the belt is flat. The distance b is an average spacing. Simultaneous solution of these

---

1. W. D. Allen and N. G. Joyce, "Studies of Induction Charging Systems for Electrostatic Generators: The Laddertron," J. Electrostatics <u>1</u>, 71-89 (1975).

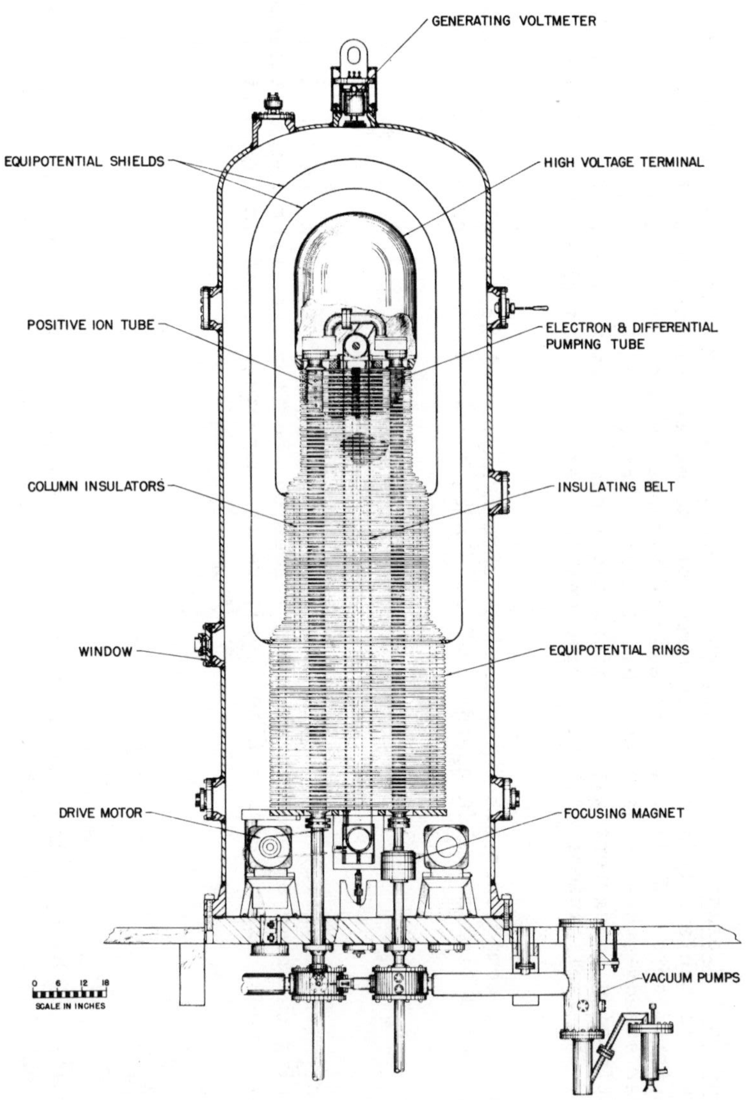

Fig. 4.14.1. Cross-sectional view of Van de Graaf high-voltage generator at M.I.T., High Voltage Research Laboratory. Device, used to provide accelerator potentials in medical research, operates at terminal voltages up to 5 MV.

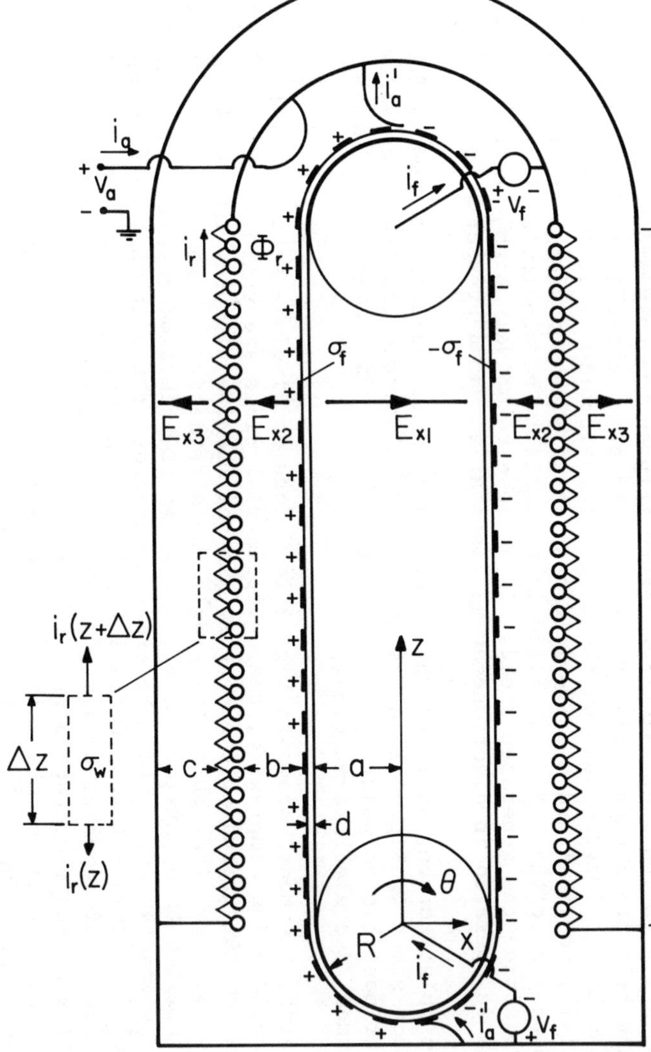

Fig. 4.14.2. Model for simple Van de Graaff machine exploiting inductive charging of belt carrying metal segments.

last two expressions shows that

$$E_{x2} = \frac{\sigma_f}{\varepsilon_o(1 + \frac{b}{a})} \tag{6}$$

These transverse fields make it possible to now write expressions that determine the field dependence on z. A section of the ring structure having incremental length Δz is shown in Fig. 4.14.2. Conservation of charge for this incremental section, which takes the form of a ring-shaped volume enclosing rings in the length Δz, is written be defining a ring charge per unit length (in the z direction), λ_r:

$$i_r(z + \Delta z) - i_r(z) = \frac{\partial(\lambda_r \Delta z)}{\partial t} \tag{7}$$

In the limit $\Delta z \to 0$, Eq. 7 becomes

$$\frac{\partial i_r}{\partial z} = \frac{\partial \lambda_r}{\partial t} \tag{8}$$

By symmetry, the contribution to the ring structure charge from the field inside (the images of the belt charges) cancel. What negative charge there is on the rings at the left imaging the positive belt charge on the upward-moving belt is canceled by the positive charge on the right imaging the downward moving negative belt charge. Hence, the only contribution to λ_r in Eq. 8 comes from the fields between the ring structure and the pressure vessel wall, approximated by Eq. 3; $\lambda_r \simeq 2\ell\varepsilon_o\Phi_r/c$. Thus, Eq. 8 becomes

$$\frac{\partial i_r}{\partial z} = \frac{2\ell\varepsilon_o}{c}\frac{\partial \Phi_r}{\partial t} \tag{9}$$

A second law is required to determine the distribution of (i_r, Φ_r). This is simply Ohm's law relating the z component of the electric field to the current carried by the bleeder resistance. With R_a the total resistance, and hence R_a/L the resistance per unit length, it follows that

$$\frac{\partial \Phi_r}{\partial z} = \frac{R_a}{L} i_r \tag{10}$$

Quasistatics: There is now enough of the model developed that a meaningful discussion can be made of two quasistatic approximations implicit to a lumped parameter model for the Van de Graaff machine.

First, Eq. 1 is misleading in that it implies that the belt charge is instantaneously established in proportion to the field voltage over the full length of the belt. Of course, an abrupt change in v_f would result in a "wave" of surface charge carried to the high-voltage dome by the moving belt. In the model developed here, temporal variations are presumed to be long compared to a transport time $L/\Omega R$. With this caveat as to the dynamic range of the resulting model, the belt charge is taken as proportional to the field voltage over the full length of the machine. The machine dynamics are quasistatic relative of the time required for the belt to traverse the distance between pulleys.

A second quasistatic approximation is necessary to approximate the field distribution governed by Eqs. 9 and 10 in a way that leads to a lumped-parameter model. Elimination of i_r between these equations results in the diffusion equation. The potential (and hence the ring charge) diffuses in the z direction, and the resulting dynamics are not in general representable in lumped-parameter terms. The subject of charge diffusion on heterogeneous structures is taken up in Sec. 5.15. Here, the quasistatic concepts of Sec. 2.3 are revisited to obtain a low-frequency lumped parameter model. But, now the critical rate process is represented by a charge diffusion time, not an electromagnetic wave transit time.

If the fields were truly static, Eq. 9 shows that the current would be independent of z. Thus, the zero-order current is $i_r = i_{ro}(t)$. The associated potential distribution can then be found by integrating Eq. 10:

$$\Phi_{ro} = v_a \frac{z}{L}; \quad v_a = R_a i_{ro} \tag{11}$$

This is the desired potential distribution. It assures a uniform generated field (z-directed) over the region of the moving belt.

Because the voltages (v_f, v_a) are in general time-varying, there is an additional capacitative current. The capacitance is distributed between the high-voltage terminal and ground, and is deduced by considering the first-order current i_{r1}, determined from Eq. 9 with the zero-order voltage (given by Eq. 11) introduced for Φ_r. (Note that the procedure followed here is an informal version of that outlined in Sec. 2.3.):

$$\frac{\partial i_{r1}}{\partial z} = \frac{2\ell\varepsilon_o}{c}\frac{dv_a}{dt}\frac{z}{L} \tag{12}$$

The z dependence is given explicitly, so this expression can be integrated to obtain

$$i_{r1} = \frac{\ell\varepsilon_o}{cL}\frac{dv_a}{dt}z^2 + f(t) \tag{13}$$

with f(t) an integration function to be determined shortly by boundary conditions. Introduction of Eq. 13 on the right in Eq. 10 gives an expression for Φ_{r1} that is similarly integrated to obtain

$$\Phi_{r1} = \frac{R_a}{L}\left[\frac{\ell\varepsilon_o}{cL}\frac{dv_a}{dt}\frac{z^3}{3} + f(t)z\right] \tag{14}$$

Because $\Phi_r = 0$ at $z = 0$, the second integration function has been set equal to zero.

The total voltage and current distributions consist of the sum of zero and first order parts. Because the zero-order distributions already satisfy the correct boundary conditions, the first order voltage must vanish at z = L. This serves to evaluate f(t) in Eq. 14. If f(t) is then introduced into Eq. 13, and that expression evaluated at z = L, the current $i_r(L,t)$ has been found:

$$i_r \simeq i_{ro} + i_{r1} = \frac{v_a}{R_a} + \frac{2\ell L\varepsilon_o}{3c}\frac{dv_a}{dt} \tag{15}$$

Note that because of the essentially linear distribution of voltage over the length of the structure, the equivalent capacitance is 1/3 what it would be if the structure formed a plane-parallel capacitor with the vessel wall. (This same equivalent capacitance can be computed with much less trouble and much less insight by simply finding the total electric energy storage and setting it equal to $\frac{1}{2} C_{eq} v_a^2$.)

Electrical Terminal Relations: The high-voltage terminal has a total current i_a which is the sum of $-i_a'$ given by Eq. 2, the ring-structure current i_r from Eq. 15, and a current required to charge the dome. With the last of these modeled as charging half of a spherical capacitor, the high-voltage terminal relation has the form

$$i_a = \frac{v_a}{R_a} - G_e \Omega v_f + C_a \frac{dv_a}{dt} \tag{16}$$

where

$$G_e \equiv \frac{2\ell R\varepsilon}{d}; \quad C_a \equiv \frac{2\ell L\varepsilon_o}{3c} + 2\pi\varepsilon_o(a + b)$$

The field terminal relations depend on details of the specific geometry in the region of the pulleys. They take the form

$$i_f = \frac{v_f}{R_f} + C_f \frac{dv_f}{dt} \tag{17}$$

where R_f is the resistance of the belt material and the pulley mounting and C_f is the capacitance of the pulley relative to ground or to the high-voltage terminal.

Mechanical Terminal Relations: The electrical torque acting in the θ direction on the lower pulley is computed by simply multiplying the z-directed force per unit area, $\sigma_f E_z$, by the total belt area $2\ell L$ and the lever arm R. In view of Eq. 1 for σ_f and the fact that $E_z = -v_a/L$,

$$\tau = -G_e v_f v_a \tag{18}$$

where the coefficient G_e is the same as defined with Eq. 16.

Analogy to the Magnetic Machine: The terminal relations summarized by these last three equations have a canonical form not only found to describe other electric machines of quite different configuration, but also to describe magnetic d-c machines. For example, compare these relations to Eqs. 4.10.17 4.10.21, and 4.10.6. The analogy is complete provided that the identification is $i_f \to v_f$, $v_a \to i_a$, $R_a \to R_a^{-1}$, $G_m \to G_e$.

The Energy Conversion Process: Modes of energy conversion are explored by considering the machine constrained in such a way that the high-voltage terminal current i_a is fixed, as is also the angular velocity Ω. Then, the machine is made to pass from one energy conversion regime to another by varying the field voltage v_f.

Under steady-state conditions, the electrical power input is expressed by solving Eq. 16 for v_a and multiplying by i_a:

$$v_a i_a = R_a i_a (i_a + \Omega G_e v_f) \tag{19}$$

The mechanical power output is also written in terms of (v_f, i_a) by substituting for v_a in Eq. 18 and multiplying by Ω:

$$\Omega\tau = -\Omega G_e R_a v_f (i_a + \Omega G_e v_f) \tag{20}$$

With the appropriate identification of variables, plots of these expressions, and the implied modes of energy conversion, are as shown in Fig. 4.10.5.

4.15 Overview of Electromechanical Energy Conversion Limitations

This chapter has two broad objectives. On the one hand, examples are chosen to illustrate techniques for using a field description in deducing lumped-parameter models. On the other hand, the examples convey an overview of systems that are electromechanically kinematic while providing a background for understanding the kinematic systems taken up in Chaps 5 and 6 and the coupling to deformable media developed in later chapters.

The Maxwell stress acting on a "control volume" enclosing the moving material, introduced in Sec. 4.2 as a convenient way to relate the fields to the total force or torque, is also useful in obtaining a qualitative perception of basic limitations on the energy conversion processes. These volumes are represented in an abstract way by Fig. 4.15.1. The longitudinal direction, denoted by (ℓ), generally represents the direction of material motion. Perpendicular to this is the transverse direction denoted by (t).

The net magnetic or electric force on the volume in general has contributions from both the transverse and longitudinal surfaces, A_ℓ and A_t. But, in all of the examples of this chapter, shear stresses rather than normal stresses contribute to the energy conversion process. To exploit this fact, the active volume of the devices has a longitudinal dimension that is large compared to transverse dimensions. For example, in rotating machines, maximum use of the magnetic or electric stress is made by having an "air gap" that is narrow compared to the circumference of the rotor. In the Van de Graaff machine, the same considerations lead to a "slender" configuration with the belt charges producing an electric field E_t across a narrow gap and the generated field being E_ℓ.

In all of these "shearing" types of electromechanical energy converters, the mechanical power output takes the form

$$P_m = UA_t K \langle \mu H_\ell H_t \rangle \quad \Big| \quad P_m = UA_t K \langle \varepsilon E_\ell E_t \rangle \tag{1}$$

Here, U and A_t are respectively the material velocity and an effective transverse area, e.g., the rotor surface velocity and area respectively in a rotating machine. The largest possible net contribution of the magnetic or electric shear stress contribution, $\langle \mu H_\ell H_t \rangle$ and $\langle \varepsilon E_\ell E_t \rangle$ respectively, is obtained if stress contributions to one of the surfaces of the control volume are minimized. Generally, this is accomplished by designing field sources into the volume. The factor K in Eq. 1 reflects geometry, material properties and phase angles. In a synchronous machine, it accounts for the air-gap spacing, the sinusoidal spatial dependence of the excitations and the relative phase of stator and rotor excitations. In the variable-capacitance machine of Sec. 4.13, this factor (which represents the "cut" of the ideal power output that is obtained) is also proportional to the product of the saliency amplitudes on rotor and stator.

Because of their higher energy conversion density, it is generally recognized that conventional magnetic electromechanical energy conversion systems are more practical than their electric counterparts. This predisposition has its basis in the extreme disparity between electric and magnetic shear stresses that can be produced under ordinary conditions.

In conventional magnetic equipment, the limit on the magnetic flux density, set by the saturation of magnetic materials, is in the range of 1-2 tesla (10 - 20 kgauss). The electric field intensity in air at atmospheric pressure (over macroscopic dimensions in the range of 1 mm to 10 cm usually of interest) is limited to less than the breakdown strength, 3×10^6 V/m. Thus under conventional conditions, the ratio of powers converted by electric and magnetic devices having the same

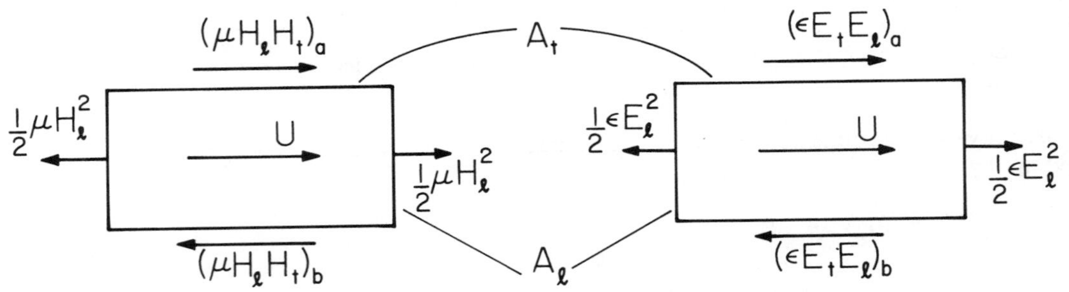

Fig. 4.15.1. Abstraction of regions of active electromechanical coupling in magnetic and electric field systems.

velocity U, effective area A_t and factor K is (from Eq. 1) the ratio of the respective shear stresses. Using as typical numbers, B = 1 and E = 10^6 V/m, this ratio is

$$\frac{(P_m)_{electric}}{(P_m)_{magnetic}} \simeq \frac{\varepsilon_o E_\ell E_t}{B_\ell B_t/\mu_o} \simeq 10^{-5} \tag{2}$$

The disadvantage inherent to electric energy conversion devices can be made up by increasing the velocity, the effective area, or the electrical breakdown strength. Now, illustrated by some examples is the way in which rough estimates of the energy converted can be made with Eqs. 1, provided the factors are evaluated with some appreciation for the underlying engineering limitations.

Synchronous Alternator: A large synchronous machine, driven by a turbine in a modern power plant, would have the typical parameters:

rotor radius $b \simeq 0.5$ m

rotor surface velocity $U = 2\pi 60 b = 188$ m/sec

rotor length $\ell = 7$ m

air gap transverse and longitudinal flux densities $\simeq 1$ tesla

These figures are typical of the full-scale generator modeled by the machine shown in Fig. 4.7.1c. An upper bound on the factor K in Eq. 1 to take into account the sinusoidal field distributions on rotor and stator, is reasonably taken as 1/2. Thus, from Eq. 1a, the mechanical power requirement (and with reasonable efficiency, therefore the maximum electrical power output) is expected to be approximately

$$P_m = (188)[(2\pi)(0.5)(7)](0.5)(1)/4\pi \times 10^{-7} = 1.6 \times 10^9 \text{ watts} \tag{3}$$

This is about 50% more than the power rating of existing equipment having roughly the parameters used.

Superconducting Rotating Machine: The limit on practical magnetic shear stress set by the saturation of magnetic materials more basically arises from the Ohmic heating limit on current density. A synchronous machine like that described in Sec. 4.9 but with no magnetic materials is in principle not limited by saturation. But it is limited by the current density consistent with available means for removing the heat from the windings. (A current density of 3×10^6 A/m^2 is projected for the normal conducting armature of the machine shown in Fig. 4.9.2.) The incremental increase in magnetic field associated with increasing the current density once the magnetic materials have been saturated makes conventional operation in this range generally unattractive.

One way to obtain higher field intensities than are practical using conventional conductors is to make use of superconductors. In time-varying fields, superconductors in fact have losses and are difficult to stabilize. But, for slowly varying and d-c fields they can be used to produce magnetic field intensities greater than the 1-2 tesla range of conventional equipment. Under balanced synchronous conditions, the field winding is only subject to d-c fields, while the armature winding carries a-c currents and is subject to a-c fields. Thus, in the machine of Fig. 4.9.2, the rotor winding is superconducting while the stator is composed of normal conductors. With that machine, the projected (rotor) field is in the range of 5-6 tesla and the area A_t required for a given power conversion accordingly reduced. For example, a two-pole 60 Hz machine having $B_r = 1$ tesla, $B_\theta = 5$ tesla and rotor length and radius $\ell = 5$ m and R = 0.3 m, respectively, has an estimated mechanical power input of $A_t T_{\theta r} R\Omega = (2\pi\ell R)(B_r B_\theta/2\mu_o)(R)(2\pi f) \simeq 2 \times 10^9$ watts. These are representative of the parameters for a projected

2000 MVA superconducting alternator.[1]

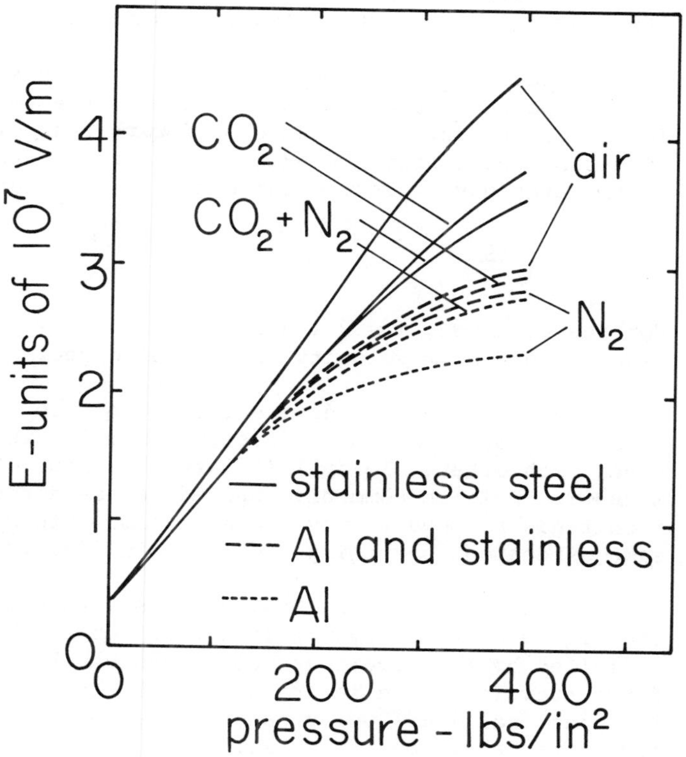

Fig. 4.15.2. Breakdown strength of common gases as a function of gas pressure for several different electrode combinations.[2]

Variable-Capacitance Machine: In machines exploiting electrical shear stresses, the limit on power converted posed by electrical breakdown can be pushed back by either making the insulation an electronegative gas under pressure, or vacuum. Typical improvements in breakdown strength with increasing pressure above atmospheric are shown in Fig. 4.15.2.[2] In principle the field intensity can be increased to more than 3×10^7 V/m, and hence the electric shear stress can be increased by a factor of more than 100 over that used in calculating Eq. 2.

The machine shown in Fig. 4.13.1c is designed for operation in vacuum. Here, the mean free path is very long compared to the distance between electrodes. As a result, breakdown results as particles are emitted from the electrode surfaces, accelerating until impacting the opposite electrode where they can produce further catastrophic results. Because the voltage difference between electrodes determines the velocity to which particles are accelerated, breakdown is voltage-dependent. Put another way, the breakdown field that can be supported by vacuum is a decreasing function of the gap distance. It also depends on the electrodes. Using steel electrodes having exposed areas of 20 cm², a typical breakdown strength under practical conditions appears to be 4×10^7 volts across a 1-mm gap.[3]

The electric machines illustrate how the power conversion density can be increased by dividing the device volume into active subregions. In an electric machine, current densities are small and as a result little conducting material is required to make an electrode function as an equipotential. By making stator and rotor blades (as well as intervening vacuum gaps) thin, it is possible to pack a larger amount of area A_t into a given volume. The limitation on the thickness and hence on the degree of reticulation that can be achieved in practice comes from the mechanical strength and stability of the rotor. Because of material creep and fracture, centrifugal forces pose a limit on the rotational velocity; but more important in this case, if a blade passes through a high-field region slightly off center, the result can be a transverse deflection that is reinforced by the next pulsation. The tendency for the blades to undergo transverse vibrations as they respond parametrically to the pulsating electric stress on each of their surfaces limits the effective area.

As numbers typical of the machine shown in Fig. 4.13.1c (where there are six gaps), consider:

R = mean radius of blades = 0.2 m
blade length = 0.12 m

U = mean blade velocity at 30,000 rpm = 630 m/sec (an extremely high velocity)

$E = 5 \times 10^6$ V/m

$A_t = (0.2)(2\pi)(0.12) = 0.9\ \text{m}^2$

Remember that the maximum electric field appears where the electrodes have their nearest approach, so the average field used is considerably less than the maximum possible. According to Eq. 1b with K=1, the power output is then at most 125 kW. Actually, the factor K significantly modifies this rough estimate. According to Fig. 4.13.2b, for $\xi_o/d = 0.4$ and a $\lambda/4$ phase,

$$K = (3.2)\left[\frac{d}{2\lambda}\left(\frac{\xi_o}{d}\right)^2\right] \tag{4}$$

1. J. L. Kirtley, Jr., and M. Furugama, "A Design Concept for Large Superconducting Alternators," IEEE Power Engineering Society, Winter Meeting, New York, January 1975.

2. J. G. Trump et al., "Influence of Electrodes on D-C Breakdown in Gases at High Pressure," Electrical Engineering, November (1950).

3. A. S. Denholm, "The Electrical Breakdown of Small Gaps in Vacuum," Can. J. Phys. 36, 476 (1959).

For $d/\lambda \simeq 0.1$, $K \simeq 2.5 \times 10^{-2}$, and the fraction of the ideal energy conversion is not very large. Instead of 125 kW, the postulated machine is predicted to produce 3 kW.

Electron-Beam Energy Converters: One class of electric field energy convertors that often have very respectable energy conversion densities make use of electrons themselves as the moving material. The model of Sec. 4.6 is developed with this class of devices in mind. A high-energy conversion density can result from the extremely large electron velocities that are easily obtained. For example, an electron having mass m and charge q accelerated to the potential Φ has the velocity

$$U = \sqrt{\frac{2q\Phi}{m}} \tag{5}$$

For the electron, $m = 9.1 \times 10^{-31}$ kg and $q = 1.6 \times 10^{-19}$ C. Thus, an accelerating potential of 10 kV results in a beam velocity of 6×10^7 m/sec!

In electron-beam devices, the electric shear stress is not usually limited by electrical breakdown, but rather by the necessity for maintaining columnated electrons in spite of their tendency to repel each other. To inhibit lateral motion of the charged particles due to their space charge, a magnetic field is commonly imposed in the direction of electron streaming. The Lorentz force, Eq. 3.1.1, then tends to convert any radial motion into an orbital motion, while letting electrons stream in the same direction as the imposed magnetic field.[4]

Electron beams are typically used to convert d-c electrical energy to high-frequency a-c. In fact, the high beam velocity requires that for a synchronous interaction, the frequency f is the beam velocity U divided by the wavelength of charge bunches; $f = U/\lambda$. Hence, for a wavelength $\lambda = 6$ cm, the frequency for a traveling-wave interaction with the 10 kV beam would be essentially $f = 6 \times 10^7/6 \times 10^{-2} = 10^9$ Hz. The practical limit on how short λ can be while obtaining useful coupling between beam and traveling-wave structure is evident from Sec. 4.6.

The kinematic picture for the beam is useful for making the electroquasistatic origins of the coupling clear and to identify the nature of the synchronous interaction upon which devices like the traveling-wave tube depend. But, because the electron bunching takes place self-consistently with the coupling fields, it is necessary, in engineering electron-beam devices, to treat the electrons as a continuum in their own right.[4] Such examples are taken up in Chap. 11.

Both electron-beam devices and synchronous alternators convert mechanical to electrical energy. As a reminder rather than a revelation, note that the synchronous alternator is of far more fundamental importance for human welfare, because when attached to the shaft of a turbine driven by a thermal heat cycle, it is capable of converting low-grade thermal energy to a high-grade electrical form. Its conversion of energy naturally fits into schemes for production of energy from natural basic sources. By contrast, the electron-beam devices only convert d-c electrical energy to a high-frequency electrical form.

---

4. M. Chodorow and C. Susskind, Fundamentals of Microwave Electronics, McGraw-Hill Book Company, New York, 1964.

For Section 4.3:

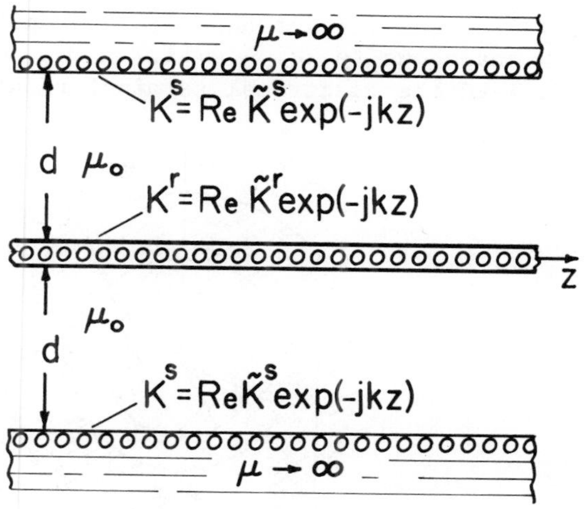

Fig. P4.3.1

Prob. 4.3.1 The cross section of a "double-sided machine" is shown in Fig. P4.3.1. The "rotor" is modeled as a current sheet.

(a) Find the force f_z acting in the z direction on an area A of the sheet.

(b) Now take the excitations as given by Eqs. 4.3.5a and 4.3.6a for synchronous interactions and evaluate f_z

(c) For a d-c interaction, the excitations are given by Eqs. 4.3.10a. Find f_z.

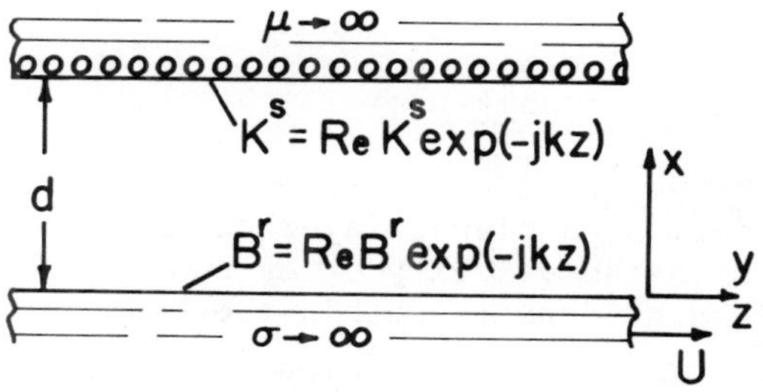

Fig. P4.3.2

Prob. 4.3.2 The developed model for a "trapped flux" synchronous machine is shown in Fig. P4.3.2. (See case 3a of Table 4.3.1). The stator surface current is specified as in Eq. 4.3.5a. The "rotor" consists of a perfectly conducting material. When t=0, the currents in this material have a pattern such that the flux normal to the rotor surface is $B_x^r = B_o^r \cos k[Ut-(z-\delta)]$, where U is the velocity of the rotor. Find f_z first in terms of $\tilde{K}^s$ and $\tilde{B}^r$ and then in terms of K_o^s and B_o^r. In practice, such a synchronous force would exist as a transient provided the initial current distribution diffused away, as described in Sec. 6.6, on a time scale long compared to that of interest.

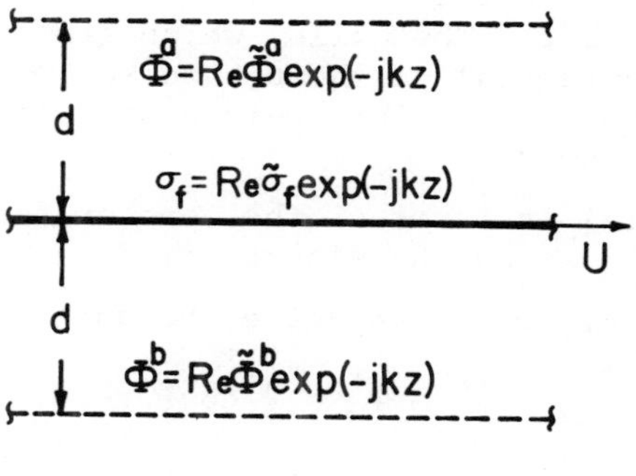

Fig. P4.3.3

Prob. 4.3.3 The moving member of an EQS device takes the form of a sheet, supporting the surface charge σ_f and moving in the z direction, as shown in Fig. P4.3.3. Electrodes on the adjacent walls constrain the potentials there.

(a) Find the force f_z on an area A of the sheet in terms of $(\hat{\Phi}^a, \hat{\sigma}_f, \hat{\Phi}^b)$.

(b) For a synchronous interaction, $\omega/k = U$. The surface charge is given by $-\sigma_o \cos[\omega t - k(z-\delta)]$ and $\Phi^a = V_o \cos(\omega t - kz)$. For even excitations $\Phi^b = \Phi^a$. Find f_z.

(c) An example of a d-c interaction is the Van de Graaf machine taken up in Sec. 4.14. With the excitations $\Phi^a = \Phi^b = -V_o \cos kz$ and $\sigma_f = \sigma_o \sin kz$, find f_z.

For Section 4.4:

Prob. 4.4.1 This problem is intended to give the opportunity to follow through the approach to developing a lumped parameter model illustrated in Sec. 4.4. However, for best efficiency in determining the electrical terminal relations, it will be helpful to use the transfer relations of Sec. 2.19, and study of Sec. 4.7 is recommended in this regard.

The cross section of a model for a permanent-magnetization rotating magnetic machine is shown in Fig. P4.4.1. The magnetization density in the rotor is uniform and of magnitude M_o. The stator is wound with a uniform turn density N, so that the surface current density over $2\theta_o$, the span of the turns, is Ni(t).

(a) Show that in the rotor volume, $\vec{B}$ is both solenoidal and irrotational so that the transfer relations of Table 2.19.1 apply provided that μH_θ is taken as B_θ.

(b) Show that boundary conditions at the rotor interface implied by the divergence condition on $\vec{B}$ and Ampere's law are

$$\vec{n} \cdot [\![\vec{B}]\!] = 0 \quad ; \quad \vec{n} \times [\![\vec{B}]\!] = \mu_o \vec{K}_f + \mu_o \vec{n} \times [\![\vec{M}]\!]$$

Prob. 4.4.1 (continued)

(c) Find the instantaneous torque on the rotor as a function of (θ_r, i). (Your result should be analogous to Eq. 4.4.11.)

(d) Find the electrical terminal relation $\lambda(\theta_r, i, M_o)$. (This result is analogous to Eq. 4.4.14.)

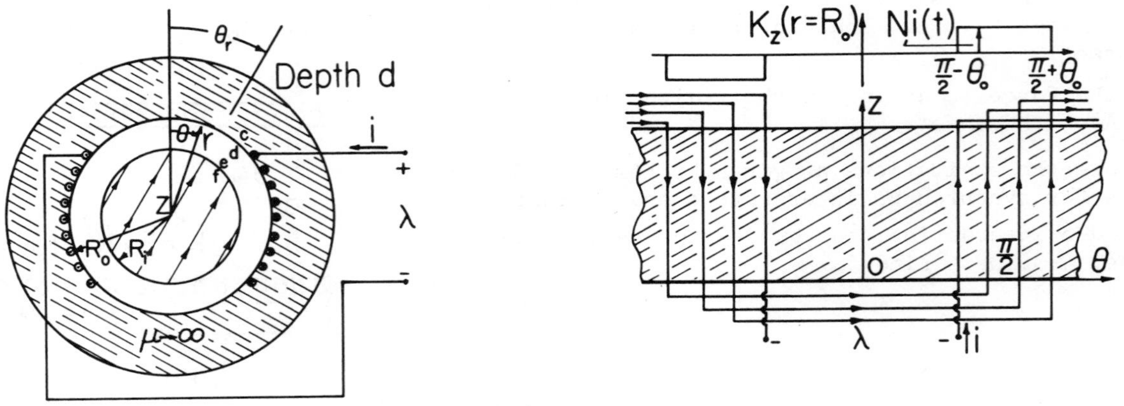

Fig. P4.4.1

For Section 4.6:

Prob. 4.6.1 A charged particle beam takes the form of a planar layer moving in the z direction with the velocity U, as shown in Fig. P4.6.1. The charge density within the beam is

$$\rho = \text{Re}\,\tilde{\rho}_o\, e^{-jkz}$$

Thus the density is uniform in the x direction within the beam, i.e., in the region $-b/2 < x < b/2$. The walls, which are constrained in potential as shown, are separated from the beam by planar regions of free space of thickness d.

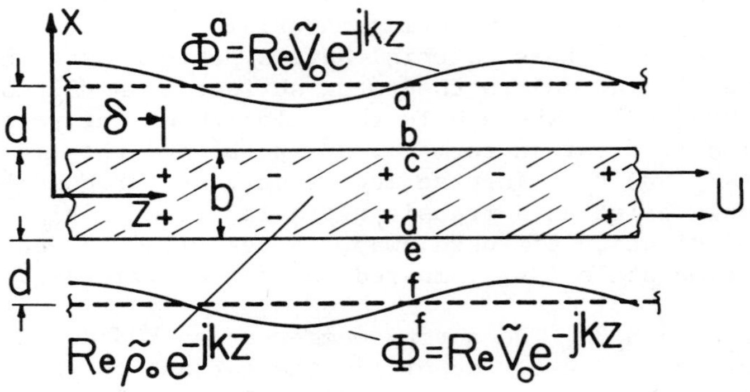

Fig. P4.6.1

(a) In terms of the complex functions of time $\tilde{V}_o$ and $\tilde{\rho}_o$, find the electrical force acting on an area A (in the y-z plane) of the beam in the z direction.

(b) Now, specialize the analysis by letting

$$\Phi^a = \Phi^f = V_o \cos(\omega t - kz)$$

$$\rho = -\rho_o \cos[\omega t - k(z-\delta)]$$

Given that the charged particles comprising the beam move with velocity U, and that k is specified what is ω? Evaluate the force found in (a) in terms of the phase displacement δ and the amplitudes V_o and ρ_o.

(c) Now consider the same problem from another viewpoint. Consider the entire region $-(d+\frac{b}{2}) < x < (d+\frac{b}{2})$ as one region and find alternative expressions for parts (a) and (b).

For Section 4.8:

Prob. 4.8.1 Transfer relations are developed here that are the Cartesian coordinate analogues of those in Sec. 4.8.

(a) With variables taking the form $A = \text{Re}\,\tilde{A}(x,t)e^{-jky}$ and $H_y = \text{Re}\,\tilde{H}_y(x,t)e^{-jky}$ and a volume current density (in the z direction) $J = \text{Re}\,\tilde{J}(x,t)e^{-jky}$, start with Eq. (b) of Table 2.19.1 and show

Prob. 4.8.1 (continued)

that the transfer relations take the form

$$\begin{bmatrix}\tilde{A}^\alpha \\ \tilde{A}^\beta\end{bmatrix} = \frac{\mu}{k}\begin{bmatrix}-\coth k\Delta & \frac{1}{\sinh k\Delta} \\ \frac{-1}{\sinh k\Delta} & \coth k\Delta\end{bmatrix}\begin{bmatrix}\tilde{H}^\alpha_y \\ \tilde{H}^\beta_y\end{bmatrix} - \frac{\mu}{k}\begin{bmatrix}-\coth k\Delta & \frac{1}{\sinh k\Delta} \\ \frac{-1}{\sinh k\Delta} & \coth k\Delta\end{bmatrix}\begin{bmatrix}\tilde{H}^\alpha_{yp} \\ \tilde{H}^\beta_{yp}\end{bmatrix} + \begin{bmatrix}\tilde{A}^\alpha_p \\ \tilde{A}^\beta_p\end{bmatrix}$$

(b) The bulk current density and particular solution for A are represented in terms of modes $\Pi_i(x)$:

$$J = \text{Re}\sum_{i=0}^{\infty}\tilde{J}_i(t)\Pi_i(x)e^{-jky} \quad ; \quad A_p = \text{Re}\sum_{i=0}^{\infty}\tilde{A}_i(t)\Pi_i(x)e^{-jky}$$

Show that if the modes are required to have zero derivatives at the surfaces, the transfer relations become

$$\begin{bmatrix}\tilde{A}^\alpha \\ \tilde{A}^\beta\end{bmatrix} = \frac{\mu}{k}\begin{bmatrix}-\coth k\Delta & \frac{1}{\sinh k\Delta} \\ \frac{-1}{\sinh k\Delta} & \coth k\Delta\end{bmatrix}\begin{bmatrix}\tilde{H}^\alpha_y \\ \tilde{H}^\beta_y\end{bmatrix} + \sum_{i=0}^{\infty}\frac{\mu\tilde{J}_i}{(\frac{i\pi}{\Delta})^2 + k^2}\begin{bmatrix}(-1)^i \\ 1\end{bmatrix}$$

For Section 4.9:

Prob. 4.9.1 A developed model for an exposed winding machine is shown in Fig. P4.9.1. The infinitely permeable stator structure has a winding that is modeled by the surface current $K^s = \text{Re}\,\tilde{K}^s e^{-jky}$. The rotor consists of a winding that completely fills the air gap and is backed by an infinitely permeable material. At a given instant, the current distribution in the rotor windings is uniform over the cross section of the gap; it is a square wave in the y direction, as shown. That is, the winding density (n wires per unit area) is uniform. Use the result of Prob. 4.8.1 to find the force per unit y-z area in the y direction acting on the rotor (note Eq. 2.15.17). Express this force for the synchronous interaction in which $K^s = K^s_o\cos(\omega t - \frac{2\pi}{\ell}y)$.

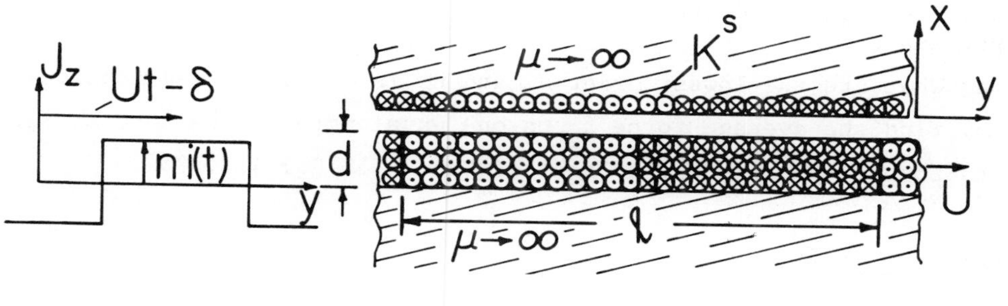

Fig. P4.9.1

For Section 4.10:

Prob. 4.10.1 A developed model for a d-c machine is shown in Fig. P4.10.1. The field winding is represented by a surface current distribution at x = b that is a positive impulse at z = 0 and a negative one at z = ℓ, each of magnitude $n_f i_f$ as shown. Following the outline given in Sec. 4.10, develop the mechanical and electrical terminal relations analogous to Eqs. 4.10.6, 4.10.17 and 4.10.21. (See

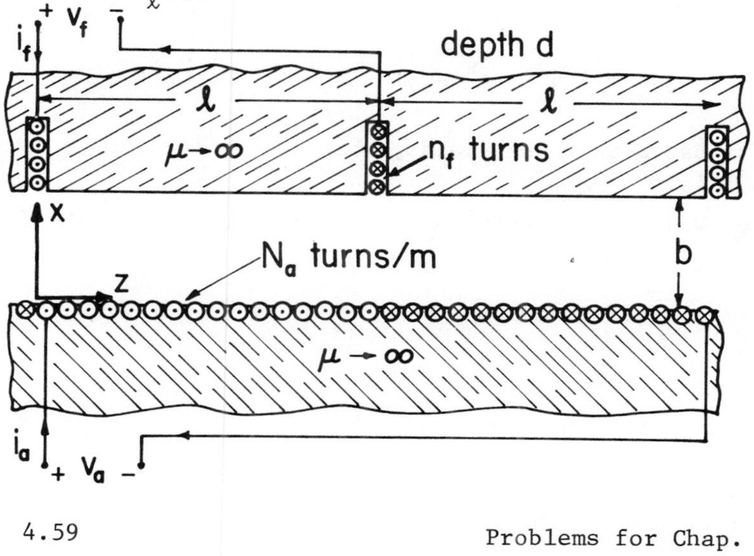

Fig. P4.10.1

Prob. 4.10.1 (continued)

Prob. 4.14.1 for a different approach with results that suggest simplification of those found here.)

For Section 4.12:

Prob. 4.12.1 The potential along the axis of a cylindrical coordinate system is $\Phi(z)$. The system is axisymmetric, so that $E_r = 0$ along the z axis. Show that fields in the vicinity of the z axis can be approximated in terms of $\Phi(z)$ by $E_z = -d\Phi/dz$ and

$$E_r = -\frac{r}{2}\frac{d^2\Phi}{dz^2}$$

For Section 4.13:

Prob. 4.13.1 An alternative to the quasi-one-dimensional model developed in this section is a "linearized" model, based on the stator and rotor amplitudes being small compared to the mean spacing d. In the context of a salient-pole machine, this approach is illustrated in Sec. 4.3. Assume at the outset that $\xi_r/d \ll 1$ and $\xi_s/d \ll 1$ but that the wavelength λ is arbitrary compared to d. Find the time-average force acting on one wavelength of the rotor. Take the limit $2\pi d/\lambda \ll 1$, and show that this force reduces to Eq. 4.13.12.

Prob. 4.13.2 A developed model for a salient pole magnetic machine is shown in Fig. P4.13.2. A set of distributed windings on the stator surface impose the surface current

$$K_y = K_o^s \sin(\omega t - kz)$$

and the geometry of the rotor surface is described by

$$\xi = \xi_o \cos 2k[Ut-(z-\delta)]$$

Both the rotor and stator are infinitely permeable.

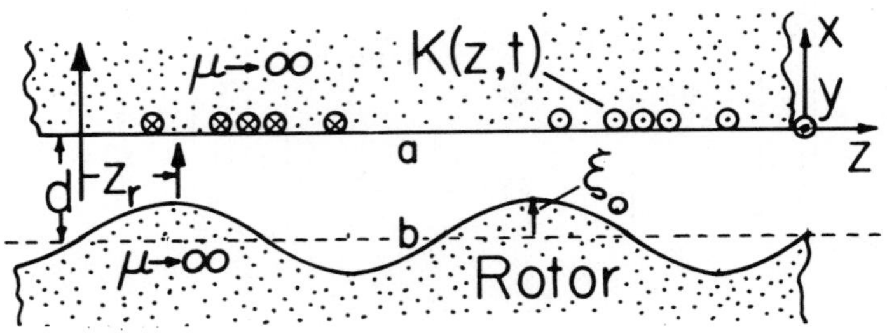

Fig. P4.13.2

(a) What are the lowest order H_x and H_z in a quasi-one-dimensional model?

(b) Find the average force f_z on one wavelength in the form of Eq. 4.13.8.

(c) Compare your result to that of Sec. 4.3, Eq. 4.3.27.

For Section 4.14:

Prob. 4.14.1

(a) For the magnetic d-c machine described in Prob. 4.10.1, show that the quasi-one-dimensional fields in the gap (based on $\ell \gg d$) are

$$H_x = \pm \frac{N_a i_a}{b}\left(z - \begin{matrix}\ell/2 \\ 3\ell/2\end{matrix}\right) + \frac{n_f i_f}{2b} \qquad \begin{matrix} 0 < z < \ell \\ \ell < z < 2 \end{matrix} \tag{1}$$

$$H_z = \pm N_a i_a \left(\frac{x}{b} - 1\right) \tag{2}$$

(b) Based on these fields, what is the force on a length, 2ℓ, of the armature written in the form $f_z = -G_m i_f i_a$?

(c) Write the electrical terminal relations in the form of Eqs. 4.10.17 and 4.10.21.

5

Charge Migration, Convection and Relaxation

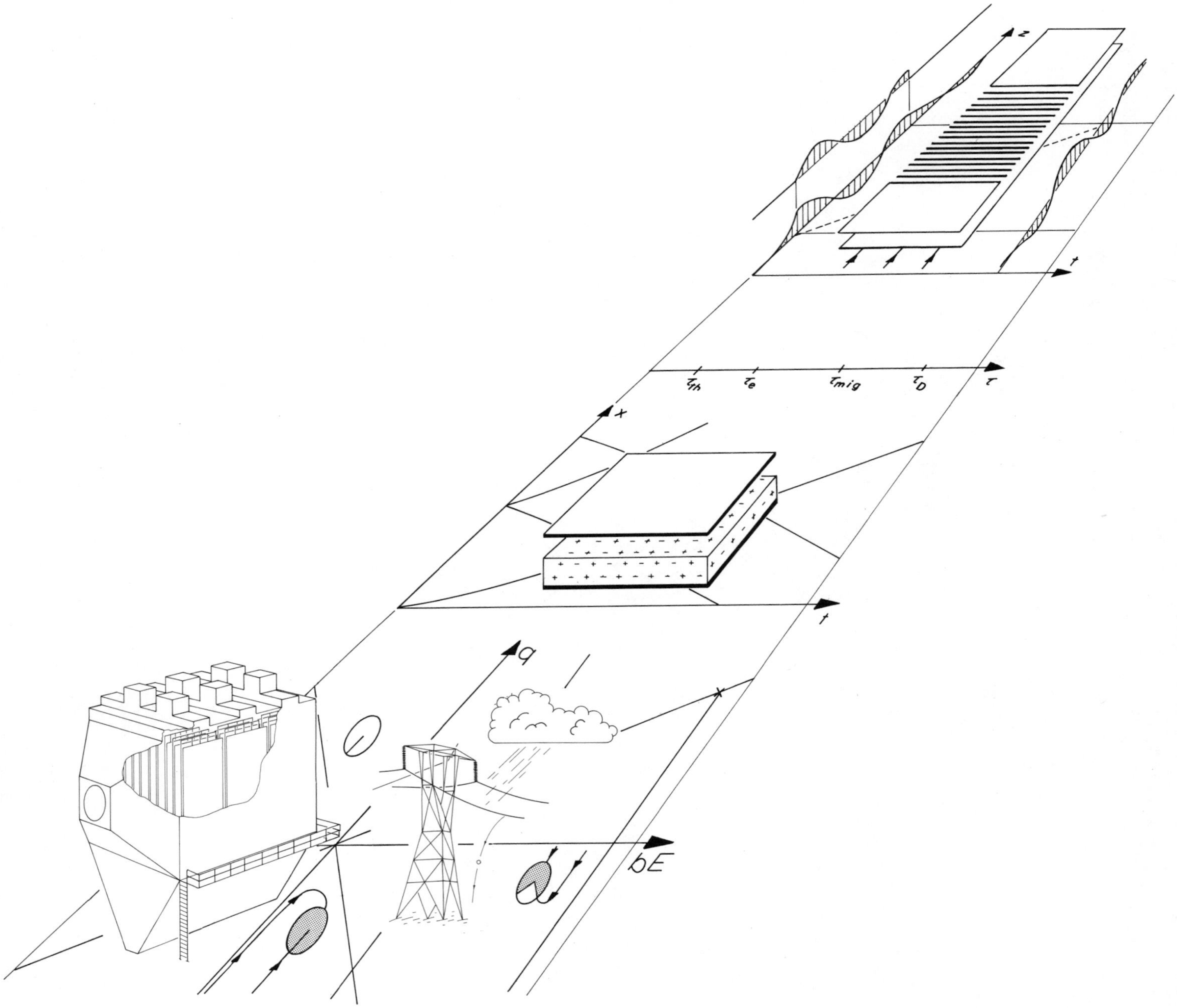

5.1 Introduction

In Chap. 4, the subject is electromechanical kinematics. Field sources are physically constrained to have predetermined spatial distributions and the relative motion is prescribed. As a result, in a typical example, the electromechanical dynamics can be incapsulated in a lumped-parameter model. In this and the next chapter, the mechanics remain kinematic, in that the material deformations are again prescribed. However, now material may be suffering relative deformations, represented by a given velocity field $\vec{v}(\vec{r},t)$. More important, in this and the next chapter, electrodes and wires are no longer used to constrain the "free" field sources. Rather, the distribution of free charge and current is now determined by the field laws themselves, augmented by conservation laws and constitutive relations.

The physical situations now considered are electroquasistatic and the sources are therefore charge densities. In Chap. 6, magnetoquasistatic systems are of interest, the relevant sources are the free current density and magnetization density, and the subject is magnetic diffusion in the face of material convection.

In the next section, equations are deduced that represent the fate of each species of charge. Throughout this chapter, the charge carriers are dominated in their motions by collisions with neutral particles and with each other. On the average, collisions are so frequent that the inertia of each carrier can be ignored. Such collision-dominated carrier motions are introduced in Secs. 3.2 and 3.3, where the observation is made that it is only if the particle inertia is ignorable that the electrical force on the carrier can be taken as instantaneously transmitted to the media through which it moves. If the carrier inertia is important, the carrier densities constitute mechanical continua in their own right. Such examples are the electron beam in vacuum and the ions and electrons that constitute a "cold" plasma. These models are therefore appropriately included in Chaps. 7 and 8, where fluids and fluid-like continua are studied.

The conservation of charge equations, together with the electroquasistatic field laws and the specified material deformation, constitute a description of the way in which the fields and their sources self-consistently evolve. Whether to gain insights concerning the implications of these equations, or to solve these equations in a specific situation, characteristic coordinates are valuable. Thus, the characteristic approach to partial differential equations is introduced in the context of charge-charrier migration, relaxation and convection. The method of characteristics will be used extensively to describe other phenomena involving propagation in later sections and chapters.

Examples treated in Secs. 5.4 and 5.5, which illustrate "imposed field and flow" dynamics of systems of carriers, involve a space charge due to the charge carriers that is ignorable in its contribution to the field. The impact charging of macroscopic particles treated in Sec. 5.5 results in a model widely used in atmospheric sciences, macroscopic particle physics and air-pollution control.

When space-charge effects are significant, it is necessary to be more specialized in the treatment. In Sec. 5.6 only one species of charge carrier is presumed to be significant. The unipolar carriers might be ions injected by a corona discharge into a neutral gas or into a highly insulating liquid. They might also be charged macroscopic particles carrying a constant charge per particle and migrating through a gas or liquid. Section 5.7 considers steady-flow one-dimensional unipolar conduction and its relation to the d-c family of energy converters.

Bipolar conduction, discussed in Secs. 5.8 and 5.9, has as a limiting model ohmic conduction. These sectiona have two major objectives, to illustrate charge migration and convection phenomena with more than one species of carrier, and to put the ohmic conduction model in perspective. In Sec. 5.10, charge relaxation is described in general terms by again resorting to the method of characteristics. The remaining sections are based on the ohmic conduction model.

The transfer relations for regions of uniform conductivity are discussed in Sec. 5.12 and applied to important illustrative physical situations in Secs. 5.13 and 5.14. These case studies are profitably contrasted with their magnetic counterparts developed in Secs. 6.4 and 6.5.

Temporal transients, initiated from spatially periodic initial conditions, are considered in Sec. 5.15. Just as the natural modes are closely related to the driven response of lumped-parameter linear systems, the natural modes of the continuum systems discussed in terms of their responses to spatially periodic drives in Secs. 5.13 and 5.14 are found to be closely related to the natural modes for distributed systems. This section, which is the first to illustrate the third category of response for linear systems that are uniform in at least one direction, as presaged in Sec. 1.2, also illustrates how heterogeneous systems of uniform ohmic conductors (which support a charge relaxation process in each bulk region) can display charge diffusion in the system taken as a whole. This type of diffusion should be discriminated from diffusion at the carrier (microscopic) level. Diffusion in the latter sense is included in Sec. 5.2 so that the domain of validity of migration and convection processes in which diffusion is neglected can be appreciated. Molecular diffusion and its effect on charge evolution, introduced in Sec. 5.2, is largely delayed until Chap. 10.

Finally, in Sec. 5.16, the response of an Ohmic moving sheet is used to introduce the fourth type of continuum linear response eluded to in Sec. 1.2, a spatially transient response to a drive that is temporarily in the sinusoidal steady state.

5.2 Charge Conservation with Material Convection

With the objective of deriving a law obeyed by each species of charge carrier in its self-consistent evolution, consider a volume V of the deforming material having a fixed identity. That is, in a macroscopic sense, the surface S enclosing this volume is always composed of the same material particles: S = S(t). The ith species of charge carrier is defined as having a number density n_i (particles per unit volume), charge magnitude q_i (per particle) and hence a magnitude of charge density $\rho_i = n_i q_i$. Positive and negative charge or charge density will be denoted explicitly by upper and lower signs respectively.

A statement that the total charge of the ith species is lost from V at a rate determined by the net outward current flux and accrued at a rate determined by the net effect of volumetric processes is

$$\frac{d}{dt}\int_V \pm\rho_i dV = -\oint_S \vec{J}_i'\cdot\vec{n}da \pm \int_V (G - R)dV \tag{1}$$

Generation and recombination of the carriers within the volume are represented by G and R, respectively, which have the units of charge/unit volume/sec. Because S is fixed relative to the media, $\vec{J}_i'$ is defined as the ith species current density measured in the materials frame of reference.

The generalized Leibnitz rule for differentiation of an integral over a time-varying volume, Eq. 2.6.5, makes it possible to take the time derivative inside the integral on the left in Eq. 1. In using Eq. 2.6.5 for this purpose, note that the velocity of the surface S is the material velocity $\vec{v}$. Thus Eq. 1 is converted to

$$\int_V \frac{\partial \pm\rho_i}{\partial t} dV + \oint_S \pm\rho_i \vec{v}\cdot\vec{n}da = -\oint_S \vec{J}_i'\cdot\vec{n}da \pm \int_V (G - R)dV \tag{2}$$

By Gauss' theorem, Eq. 2.6.2, the surface integrations are converted to volume integrations. Because the volume V is arbitrary, it follows that

$$\frac{\partial\rho_i}{\partial t} + \nabla\cdot[\rho_i\vec{v} \pm \vec{J}_i'] = G - R \tag{3}$$

To make use of this differential law, the current density must be related to the charge density, and the rates of generation and recombination must be specified.

Carriers, dominated by collisions in their motion through a neutral medium, are usually described by the current density

$$\vec{J}_i' = n_i b_i q_i \vec{E} \mp K_{Di}\nabla(q_i n_i) \equiv b_i\rho_i\vec{E} \mp K_{Di}\nabla\rho_i \tag{4}$$

The term proportional to $q_i\vec{E}$ represents migration and is familiar from Sec. 3.2. Because of the electric field, a charged particle sustains a net migration as it undergoes frequent thermally induced collisions with neutral particles. These collisions are so frequent that on the time scale of interest there is an instantaneous equilibrium between the electrical force and an effective drag force. In terms of a friction coefficient $(m_i\nu_i)$, this force equilibrium is expressed by

$$\pm q_i\vec{E} = (m_i\nu_i)\vec{v}_i \tag{5}$$

The particle velocity $\vec{v}_i$ relative to the neutral medium is expressed in terms of the mobility b_i as

$$\vec{v}_i = \pm b_i\vec{E} \tag{6}$$

where $b_i \equiv q_i/m_i\nu_i$. Thus, the first term in Eq. 4 is the product of the charge density $\pm\rho_i$ and the particle velocity $\vec{v}_i$. Large molecules and macroscopic particles in gases[1] and liquids[2] are often modeled as being spherical and obeying Stokes's law (Sec. 7.21), in which case the friction factor is $m_i\nu_i = 6\pi\eta a$, where η and a are the fluid viscosity and particle radius respectively. For such particles, the mobility is

$$b_i = \frac{q_i}{6\pi\eta a} \tag{7}$$

The second term in Eq. 4 recognizes that because of the thermally induced motions of the particles, on the average there will be a particle flux away from regions of high concentration. This flux is proportional to the spatial rate of change of concentration.

As might be expected from their common origins in the thermal particle motions, the diffusion coefficient K_{Di} and the mobility are related properties of the medium through which given particles migrate and diffuse. For ideal gases and liquids, K_{Di} and b_i are linked by the Einstein relation

$$K_{Di} = \left(\frac{kT}{q_i}\right)b_i;\ \frac{kT}{e} = 26.6 \times 10^{-3} \text{ volts at } T = 20^\circ\text{C} \tag{8}$$

where k is the Boltzmann constant, T is the absolute temperature in degrees Kelvin and q_i is the particle charge. The quantity kT/q is measured in volts and at room temperature for q equal to the electron charge, e, has the value given with Eq. 8.

Physical examples to which Eq. 4 applies are given in Table 5.2.1, together with typical values for the mobility and diffusion coefficient.

In inserting Eq. 4 into the charge conservation equation, Eq. 3, it is now assumed that the material deformations of interest are incompressible in the sense that $\nabla\cdot\vec{v} = 0$, so that

$$\frac{\partial\rho_i}{\partial t} + (\vec{v} \pm b_i\vec{E})\cdot\nabla\rho_i = \nabla\cdot(K_{Di}\nabla\rho_i) \mp \rho_i\nabla\cdot b_i\vec{E} + G - R \tag{9}$$

Each of n species contributing to the transfer of charge is described by an expression of the form of Eq. 9. The evolution of one species is linked to the others through Gauss' law, which recognizes that the net charge from all of the species is the source for the electric field:

$$\nabla\cdot\varepsilon\vec{E} = \sum_{i=1}^{n} \pm\rho_i \tag{10}$$

Of course, in the electroquasistatic approximation $\vec{E}$ is irrotational, a condition that is automatically met by requiring that

$$\vec{E} = -\nabla\Phi \tag{11}$$

Given appropriate source and recombination functions G and R, and the material velocity distribution $\vec{v}(r,t)$, Eqs. 9-11 constitute n+1 scalar expressions and one vector equation describing n charge densities, Φ and the vector $\vec{E}$.

In the remainder of this chapter, certain of the physical implications of these relations are explored, with emphasis on the interplay of the material convection and the charge transport processes. Approximations are necessary if practical use is to be made of these relations. In this regard, the relative importance of the migration and diffusion contributions to the current density, Eq. 4, is important. To approximate the ratio of diffusion and migration terms for a given species, the charge density gradient is characterized by ρ_i/ℓ, where ℓ is a typical length. For media described by the Einstein relation, Eq. 8,

$$\frac{\text{diffusion current density}}{\text{migration current density}} = \frac{kT/q_i}{\ell|\vec{E}|} \tag{12}$$

Suppose that each carrier supports one electronic charge. Then if $|\vec{E}| = 1$ V/m, the influence of diffusion equals or exceeds that of migration for length scales shorter than about 2.5 cm. But, for

1. C. Orr, Jr., Particle Technology, Macmillan Company, New York, 1966, p. 296.

2. F. Daniels and R. A. Alberty, Physical Chemistry, 3rd ed., John Wiley & Sons, New York, 1967, pp. 405-406.

Table 5.2.1. Typical mobilities of various charged particles.

Macroscopic Particles in Fluids

Charged to saturation by ion impact, the particle charge is given by Eq. 5.5.1. Introduced into Eq. 7, this charge implies the mobility

$$b = \frac{2\varepsilon_o aE}{\eta} \qquad \text{(a)}$$

where a is the particle radius, E is the electric field in which the charging occurs, and η is the viscosity of the gas or liquid. In air under standard conditions this expression is valid for radii down to about 0.5 μm, below which the finite mean free path of air molecules and diffusional charging become important.[3] For air, this expression becomes 8.8×10^{-7} aE, so that for a = 1 μm and E = 10^6 V/m the mobility is 10^{-7} (m/sec)/(V/m).

Ions in Gases

At atmospheric pressure, ions are typically generated by a corona discharge. Ions drawn from the discharge by an electric field are usually not distinguished. Reported ion mobilities distinguish among various gases, but do not specify the type of ion. Some published values, unless otherwise indicated for atmospheric pressure and 20°C, are:

Gas	Air (dry)	CCl_4	CO_2	H_2	H_2O (100°C)	H_2S	N_2	N_2 Very pure	O_2	SO_2
b_+ (units of 10^{-4} m^2/V sec)	1.36	0.30	0.84	5.9	1.1	0.62	1.27	1.28	1.31	0.41
b_- (units of 10^{-4} m^2/V sec)	2.1	0.31	0.98	8.15	0.95	0.56	1.84	145	1.8	0.41

Low mobilities in impure gases are thought to result from formation of "clusters," while extremely high negative mobilities are attributed to an "ion" spending part of its time as a free electron.[4]

Ions in Highly Insulating Liquids

Approximate formulas relate mobility to the viscosity,

$$b_+ \simeq 1.5 \times 10^{-11}/\eta; \quad b_- \simeq 3 \times 10^{-11}/\eta \qquad \text{(b)}$$

Thus, for a liquid having the viscosity of water, $\eta \simeq 10^{-3}$, mobilities are 1.5×10^{-8} and 3×10^{-8} respectively. For a careful evaluation with liquid and type of ion specified see Adamczewski.[5]

Ions in Water at 25°C Forming an Electrolyte at Infinite Dilution[6]

Ion	Na^+	K^+	H^+	Cl^-	I^-	OH^-	Ca^{2+}	SO_4^{2-}	NO_3^-
$b_\pm$ (units of 10^{-8} m^2/V sec)	5.20	7.62	36.3	7.90	7.96	20.5	6.16	8.27	7.40

3. H. J. White, Industrial Electrostatic Precipitation, Addison-Wesley Publishing Company, Reading, Mass., 1963, p. 137.
4. Handbook of Physics, E. U. Condon and H. Odishaw, Eds., McGraw-Hill Book Company, New York, 1958, pp. 4-161.
5. I. Adamczewski, Ionization, Conductivity and Breakdown in Dielectric Liquids, Taylor & Francis, London, 1969, pp. 224-225.
6. Ref. 2, p. 395.

fields of the order of 10^4 V/m, the length scale must be shorter than 2.5 μm for this to be true. In relatively conducting materials, such as electrolytes, fields of interest might be no more than 1 V/m. But, motions of ions in insulating liquids and gases, with fields typically exceeding 10^4 V/m, are not influenced by diffusion except in accounting for certain processes in the immediate vicinity of boundaries.

5.3 Migration in Imposed Fields and Flows

In this section, the spatial scale of interest is such that the diffusion current can be considered negligible compared to the migration current. In addition, the medium is one in which generation and recombination of the charged species is negligible. Hence, the first and last two terms in Eq. 5.2.9 can be dropped. For carriers having a constant mobility, what remains on the right in Eq. 5.2.9 is proportional to the divergence of the electric field. By Gauss' law, this term is therefore proportional to the net space charge. If the density of carriers is small, Gauss' law, Eq. 5.2.4, requires that $\vec{E}$ be solenoidal:

$$\nabla\cdot\vec{E} = 0 \tag{1}$$

and Eq. 5.2.9 therefore reduces to

$$\frac{\partial \rho_i}{\partial t} + (\vec{v} \pm b_i\vec{E})\cdot\nabla\rho_i = 0 \tag{2}$$

In this "imposed field" approximation, the electric field is essentially determined by charges outside of the region of interest. Typically, these charges reside on boundaries and, in terms of the potential, $\vec{E}$ is governed by Laplace's equation. Thus, as an example, if the potentials of all boundaries were constrained, $\vec{E}$ would be determined by solving Laplaces equation subject to these boundary conditions, and that value of $\vec{E}$ "imposed" in Eq. 2. For such a physical situation, each species migrates independently of the others, as is evident from the fact that the coupling between species afforded by Gauss' law is now absent.

The assumption that the electric field distribution is not appreciably affected by the migrating species says that the net charge density is small but not necessarily zero. In general there is an electrical force density acting throughout the moving medium. As in all of this chapter, it is assumed that the effect of this force density on the relative velocity distribution $\vec{v}(\vec{r},t)$ is negligible. In this sense, the flow is also "imposed."

The imposed field and flow approximation gives the opportunity to study the effect of convection on the migration of charged particles. As can be seen from Table 5.2.1, ions moving in a field of 10^5 V/m through air have a migration velocity $b_i E$ on the order of 20 m/sec. Thus, an air velocity on this order could have a large influence on an ion trajectory. Macroscopic charged particles, such as dust in an electrostatic precipitator, typically have a considerably lesser mobility, and are therefore strongly influenced by modest motions of the gas. Although typical velocities of a liquid are likely to be less than for a gas, because of the relatively lower mobilities of ions and macroscopic particles in highly insulating liquids, the effects of convection can again be appreciable.

With the replacement of the velocity by the ion velocity $\vec{v} \pm b_i\vec{E}$, Eq. 2 takes the form of a convective derivative. It states that the time rate of change of the species charge density as viewed by a charged particle of fixed identity is zero (see Sec. 2.4 for a discussion of the physical significance of the convective derivative):

$$\frac{d\rho_i}{dt} = 0 \tag{3}$$

on

$$\frac{d\vec{r}}{dt} = \vec{v} \pm b_i\vec{E} \tag{4}$$

In what amounts to a rederivation of the convective derivative, consider the transition from Eq. 2 to the representation of Eqs. 3 and 4 in a somewhat more formal way. The three spatial coordinates and time constitute a four-dimensional space. Each set of coordinates $(\vec{r},t)$ in this space has an associated solution $\rho_i(\vec{r},t)$. An incremental change in the coordinates therefore leads to a change in ρ_i given by

$$d\rho_i = dt\,\frac{\partial\rho_i}{\partial t} + dx\,\frac{\partial\rho_i}{\partial x} + dy\,\frac{\partial\rho_i}{\partial y} + dz\,\frac{\partial\rho_i}{\partial z} \tag{5}$$

As it stands, this expression is nothing more than a prescription for computing $d\rho_i$ for a given change

$(d\vec{r},dt)$ in the coordinates of the $(\vec{r},t)$ space. But, can these incremental changes be specified so that Eq. 2 reduces to an ordinary differential equation? Division of Eq. 5 by dt and comparison to Eq. 2 shows that the desired specification is Eq. 4. Along a given characteristic line, represented by Eq. 4, Eq. 2 becomes Eq. 3. These lines have the physical significance of being the trajectories of the carriers.

If the evolution of the charge species is to be determined within a given volume V, then the charge density of each species must be specified where the associated characteristic line "enters" the volume of interest. The "direction" of a characteristic line is one of increasing time. Formally, with $\vec{n}$ taken as positive if directed outward from the volume of interest, the boundary condition is imposed on the ith species wherever

$$\vec{n}\cdot(\vec{v} \pm b_i\vec{E}) < 0 \tag{6}$$

Boundary conditions consistent with causality seem obvious in the transient case, but Eqs. 4 and 5 pertain also to steady flows in which rates of change with respect to time for an observer at a fixed location are zero.

Steady Migration with Convection: In the laboratory frame of reference, $\vec{v},\vec{E}$ and the boundary conditions represented by Eq. 6 are all invariant. Even so, the time rate of change for the particle, as expressed by Eq. 4, is finite. Explicit expressions for the particle trajectories can be found in a wide class of physically interesting situations, following the approach now illustrated.

Both $\vec{v}$ and $\vec{E}$ are solenoidal, and hence can be represented in terms of vector potentials. The discussion of Sec. 2.18 centers around four common configurations in which only a single component of these vector potentials is required to describe the vector functions. By way of illustration, the polar and axisymmetric spherical configurations are now considered, with the results applied to specific problems in the next two sections.

In polar coordinates, define vector potentials such that

$$\begin{bmatrix}\vec{E}\\ \vec{v}\end{bmatrix} = \begin{bmatrix}\vec{i}_r\frac{1}{r}\frac{\partial}{\partial\theta} - \vec{i}_\theta\frac{\partial}{\partial r}\end{bmatrix}\begin{bmatrix}A_E\\ A_v\end{bmatrix} \tag{7}$$

as suggested by Table 2.18.1. Similarly, in spherical coordinates

$$\begin{bmatrix}\vec{E}\\ \vec{v}\end{bmatrix} = \frac{1}{r\sin\theta}\begin{bmatrix}\vec{i}_r\frac{1}{r}\frac{\partial}{\partial\theta} - \vec{i}_\theta\frac{\partial}{\partial r}\end{bmatrix}\begin{bmatrix}\Lambda_E\\ \Lambda_v\end{bmatrix} \tag{8}$$

In terms of these functions, in the respective configurations, Eq. 4 becomes

Polar	Axisymmetric spherical	
$\frac{dr}{dt} = \frac{1}{r}\frac{\partial}{\partial\theta}(A_v \pm b_iA_E)$	$\frac{dr}{dt} = \frac{1}{r\sin\theta}[\frac{1}{r}\frac{\partial}{\partial\theta}](\Lambda_v \pm b_i\Lambda_E)$	(9)
$r\frac{d\theta}{dt} = -\frac{\partial}{\partial r}(A_v \pm b_iA_E)$	$r\frac{d\theta}{dt} = -\frac{1}{r\sin\theta}\frac{\partial}{\partial r}(\Lambda_v \pm b_i\Lambda_E)$	(10)

Remember that steady-state conditions prevail, so that the quantities on the right are independent of time. Time is therefore eliminated as a parameter by solving each of these expressions for dt and setting the respective equations equal to each other

$\frac{\partial}{\partial r}(A_v \pm b_iA_E)dr + \frac{\partial}{\partial\theta}(A_v \pm b_iA_E)d\theta = 0$	$\frac{\partial}{\partial r}(\Lambda_E \pm b_i\Lambda_v)dr + \frac{\partial}{\partial\theta}(\Lambda_v \pm b_i\Lambda_E)d\theta = 0$	(11)

Because there is no time dependence to the potential functions, these expressions constitute total derivatives, and can be just as well written as

$d(A_v \pm b_iA_E) = 0$	$d(\Lambda_v \pm b_i\Lambda_E) = 0$	(12)

The lines along which a species charge density is constant are implicitly given by

$A_v \pm b_iA_E = \text{constant}$	$\Lambda_v \pm b_i\Lambda_E = \text{constant}$	(13)

Quasistationary Migration with Convection: To integrate the particle equations of motion, and thus arrive at Eqs. 13, it is necessary to require that the particles be in essentially the same field and flow distribution throughout their motions through the volume of interest. In that sense, the motions are steady. But the particle transit times may be brief compared to a dynamical time of interest, perhaps that required for a surface upon which the particles impinge to charge, and hence change the electric field intensity. Thus, over a longer time scale, the flow and field distribution, hence the functions (A_E,A_v) and (Λ_E,Λ_v), may be functions of time. This is often the situation during impact charging of macroscopic particles, discussed in Sec. 5.5.

For unipolar migration, the assumption that the electric field is solenoidal (that space charge has a negligible effect on the electric field distribution) is equivalent to the postulate of quasistationary migration (that the transit time for a particle through the volume of interest is short compared to the time required to charge a boundary). This point is best made in Sec. 5.6 after a quasistationary process is considered in Sec. 5.5.

5.4 Ion Drag Anemometer

The example of this section is intended to illustrate how charged particle trajectories can be computed using the approximations introduced in Sec. 5.3. A pair of electrodes is embedded in the wall bounding a fluid moving uniformly to the right, as shown in Fig. 5.4.1. A potential, V, applied to the right electrode gives an electric field intensity which terminates on the left electrode. In the neighborhood of the coordinate origin this field can be approximated as azimuthal. Thus, the imposed velocity and electric field intensity distributions are

$$\vec{v} = U[\sin\theta\vec{i}_r + \cos\theta\vec{i}_\theta] \qquad (1)$$

$$\vec{E} = -\frac{V}{\pi r}\vec{i}_\theta \qquad (2)$$

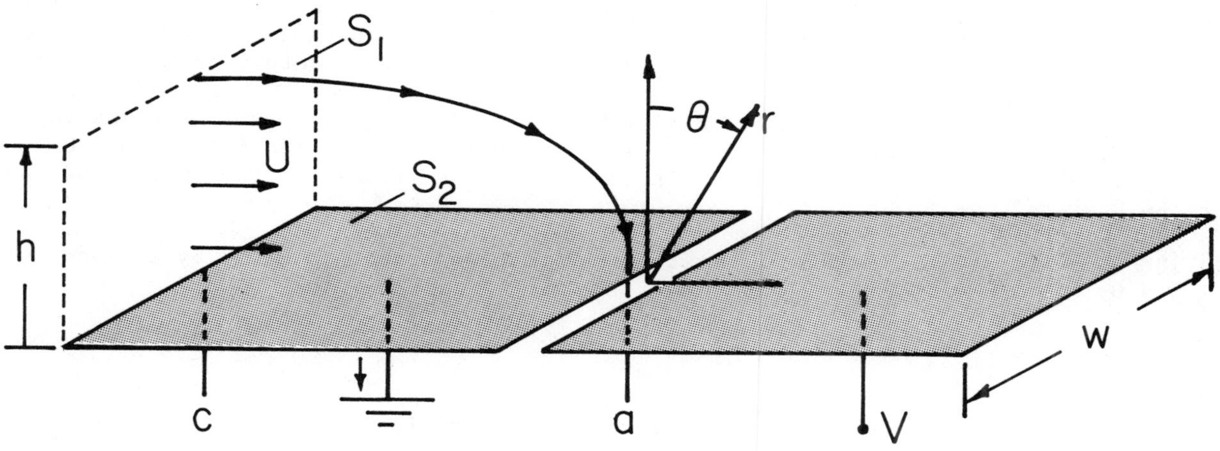

Fig. 5.4.1. Electrodes embedded in a smooth wall have the potential difference V. Ions enter from the left, entrained in the uniform velocity U. With a positive V, the left electrode intercepts some of the ions from the flow.

Fluid flow is represented as inviscid, and hence uniform right up to the electrode surfaces. Positive ions, present in the stream entering from the left, are sampled by the electrodes. The flux of ions to the left electrode caused by applying a positive voltage V to the right electrode is to be computed with a view toward obtaining the associated current i as a way of measuring the gas velocity.[1]

It follows from Eqs. 5.3.6 that

$$A_E = \frac{V}{\pi}\ln\left(\frac{r}{a}\right) \qquad (3)$$

$$A_v = -Ur\cos\theta \qquad (4)$$

The characteristic lines, along which the charge density is constant, are given by Eq. 5.3.13, which in view of Eqs. 3 and 4 becomes

$$-Ur\cos\theta + \frac{bV}{\pi}\ln\left(\frac{r}{a}\right) = \text{constant} \qquad (5)$$

1. K. J. Nygaard, "Anemometric Characteristics of a Wire-to-"Plane" Electrical Discharge," Rev. Sci. Instr. 36, 1771 (1965).

The constant is evaluated by fixing attention on the characteristic line entering at an altitude h over the left edge of the left electrode. Thus, at $r \sin\theta = -c$, $r\cos\theta = h$ and $r = \sqrt{h^2 + c^2}$. Then, Eq. 5 becomes

$$\underline{V} \ln\left[\frac{\sqrt{\underline{h}^2 + 1}}{\underline{r}}\right] = \underline{h} - \underline{r}\cos\theta \qquad (6)$$

where normalization of (V,h,r) is introduced:

$$\underline{V} = \frac{bV}{\pi Uc}, \quad \underline{h} = \frac{h}{c}, \quad \underline{r} = \frac{r}{c}$$

The quantity on the right is the distance downward (toward the electrode) measured from the initial altitude, $\underline{h}$, of a particle. Hence, the particle trajectories can be simply plotted by specifying the normalized voltage $\underline{V}$ and $\underline{h}$ for the trajectory of interest. With compass in hand, a graphical construction of a trajectory is obtained by picking a normalized radial coordinate $\underline{r}$, computing the left-hand side of Eq. 6, and finding the azimuthal angle θ at which the distance downward from the initial height $\underline{h}$, is as computed.

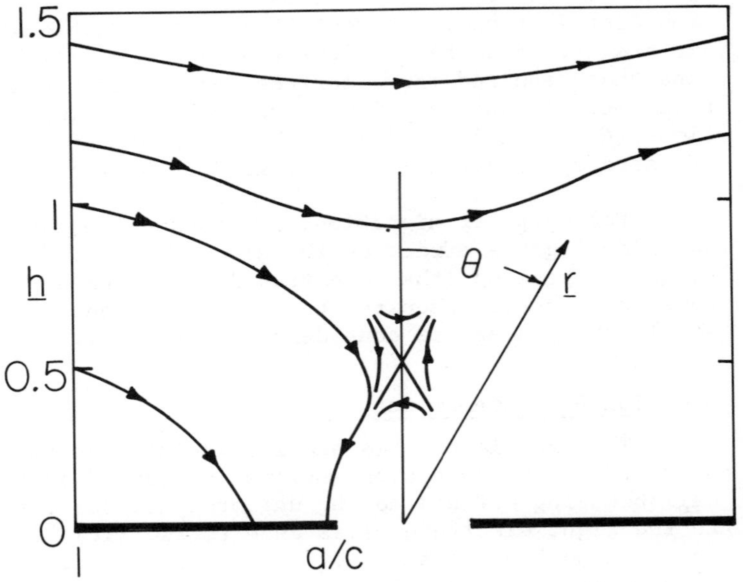

Fig. 5.4.2. Characteristic (force) lines for the physical configuration of Fig. 5.4.1. Vertical and horizontal distances have been normalized to c, with the left electrode then extending from 1→a/c. In this sketch, $\underline{V}$=0.5.

Typical plots are shown in Fig. 5.4.2. Concern is with positive ions only so that characteristic lines emanating from the wall to the right of the origin enter the volume of interest where there is no source of charge. Hence, the constant charge density to be associated with those lines is zero. On lines entering from the left, the charge density is a constant determined by conditions to the left.

The point (r,θ) = (0.5,θ) shown in Fig. 5.4.2 is one of zero force. Setting the r and θ components of $\overline{v} + b\overline{E}$ to zero shows that this critical point is at $\underline{r} = \underline{V}$ and θ = 0. At this point, characteristic lines entering from the left split into those that remain in the stream and those that reach the plane θ = −π/2.

The characteristic line passing through the critical point is found by evaluating Eq. 5 at $\underline{r} = \underline{V}, \theta = 0$:

$$-\underline{r}\cos\theta + \underline{V}\ln\left(\underline{r}\frac{c}{a}\right) = -\underline{V} + \underline{V}\ln\left(\underline{V}\frac{c}{a}\right) \qquad (7)$$

The position $\underline{r} = \underline{r}^*$ on the surface θ = −π/2 where this critical characteristic line impinges then follows by evaluating Eq. 7 with θ = −π/2:

$$\underline{r}^* = \frac{\underline{V}}{e} \qquad (8)$$

Thus, the critical characteristic line impinges on the electrode if $\underline{r}^* > (a/c)$, i.e., if

$$\underline{V} > \left(\frac{a}{c}\right)e \qquad (9)$$

For lesser values of $\underline{V}$, all of the electrode surface collects particles entering from the left, and the total current i is the integral of $-\rho bE_\theta$ over the entire electrode surface:

$$i = w\int_a^c \frac{\rho bV}{\pi}\frac{dr}{r} = (\rho Ucw)\underline{V}\ln\left(\frac{c}{a}\right) \qquad (10)$$

This dependence of i on V is presented graphically in Fig. 5.4.3, valid so long as $\underline{V} < \frac{a}{c}e$.

If V is increased beyond this value, only that portion of the electrode to the left of $\underline{r} = \underline{r}^*$ collects particles. The rest intercepts characteristic lines carrying no charge because they originate on the boundary θ = π/2 to the right. Thus, the current is

$$i = w\int_{(c\underline{V}/e)}^{c} \frac{\rho bV}{\pi}\frac{dr}{r} = \rho Ucw\,\underline{V}(1 - \ln\underline{V}) \qquad (11)$$

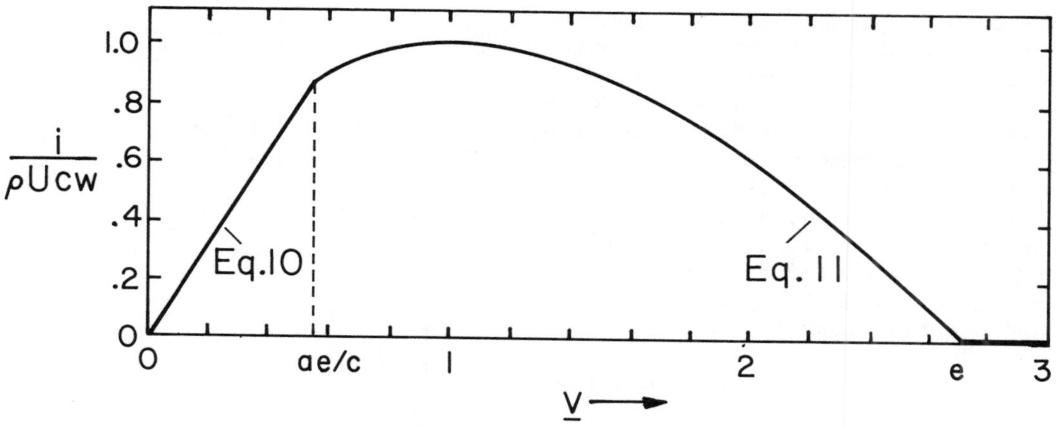

Fig. 5.4.3. Normalized current to electrode in Fig. 5.4.1 as function of normalized voltage $\underline{V} = b\overline{V}/\pi Uc$.

With V beyond the value e, all of the characteristic lines reaching the electrode surface originate to the right where there is no source of particles. For voltages greater than this, the electric field diverts the particles completely before they can reach the electrode, and i = 0. The current dependence given by Eq. 11 is also summarized in Fig. 5.4.3.

It should be clear from the i-V characteristic summarized by Fig. 5.4.3 that there are many ways in which practical use could be made of the charged particle collection process. The peak current is a measure of U, while the voltage at which the curve peaks, or cuts off, gives a measure of either the velocity or the mobility.

5.5 Impact Charging of Macroscopic Particles: The Whipple and Chalmers Model

Electrostatic precipitators, used for the collection of particulate from gases in air-pollution control systems, make use of ion impact charging. A typical configuration is shown in Fig. 5.5.1. Dust laden gas enters the metallic tube from the bottom, and the object is to separate the dust from the gas before the latter leaves at the top. The high-voltage wire supported at the center of the tube sustains a corona discharge, a type of electrical breakdown that remains localized around the wire. Within this corona discharge, both positive and negative ions are created. Positive ions are drawn outside the immediate vicinity of the corona where they migrate along the lines of force toward the grounded coaxial electrode.

A particle of dust that interrupts the electric field also interrupts the ion migration. As a result the particle becomes charged.

Once charged, it too is subject to an electrical force and hence also tends to migrate to the cylindrical wall. The final stage of particle collection consists in rapping the electrode so that compacted dust falls from the walls into a hopper below.[1]

Provided that the contribution of the migrating ions to the electric field in the immediate vicinity of the particle is negligible, the model developed in this section describes the charging process. As the particle acquires charge, its own contribution to the field is altered, so this example gives the opportunity to exemplify the quasistationary migration presaged in Sec. 5.3.

Typical electric fields in an electrostatic precipitator are 5×10^5 V/m. Thus, ions having a mobility of about 2×10^{-4} (m/sec)/(V/m) have velocities $bE \simeq 100$ m/sec. Typical gas velocities are only 1-2 m/sec, so the effects of convection on the charging process are usually not significant.

But convection is an important factor in other situations to which the impact charging model pertains. It is well known that as drops of water fall through the atmosphere, they become charged because of interactions with ions. In a thunderstorm, a system of ions and drops can be subject to a significant electric field.

The particle shown in Fig. 5.5.2 is taken as spherical with an "imposed" electric field E that is locally uniform and if positive directed as shown. As envisioned by meteorologists, the particle is a

1. H. J. White, Industrial Electrostatic Precipitation, Addison-Wesley Publishing Company, Reading, Mass., 1963, pp. 33-48.

water drop falling through the atmosphere, so from the frame of reference of the particle, there is an ambient gas velocity U directed upward in the -z direction. For the meteorologist the question is, given ions of a certain density carried by the combined field and flow, what is the charging law for the particle? How fast does it become charged and to what final value? Whipple and Chalmers[2] were interested in a quantitative model of Wilson's theory of thunderstorm electrification, which centered around how a particle could acquire charge while falling through essentially equal densities of positive and negative ions.[2]

In the following discussion, the particle being charged will be called the "drop," while the impacting particles will be termed ions. In fact, the "ions" might be fine macroscopic particulate being collected (scrubbed) by charged drops.[3]

At the outset, two useful parameters are identified. Regimes of charging are demarked by the critical charge

$$q_c \equiv 12\pi\varepsilon_o R^2 E \qquad (1)$$

which can be positive or negative, depending on the sign of E. Rates of charging will be characterized by the currents

$$I_{\pm} = \pi R^2 b_{\pm}\rho_{\pm} E = q_c \frac{b_{\pm}\rho_{\pm}}{\varepsilon_o 12} \qquad (2)$$

which are also determined in sign by E. The magnitudes of the positive and negative ion charge densities are $\rho_{\pm}$ respectively, uniformly distributed at "infinity," where the ions enter the volume neighboring the drops.

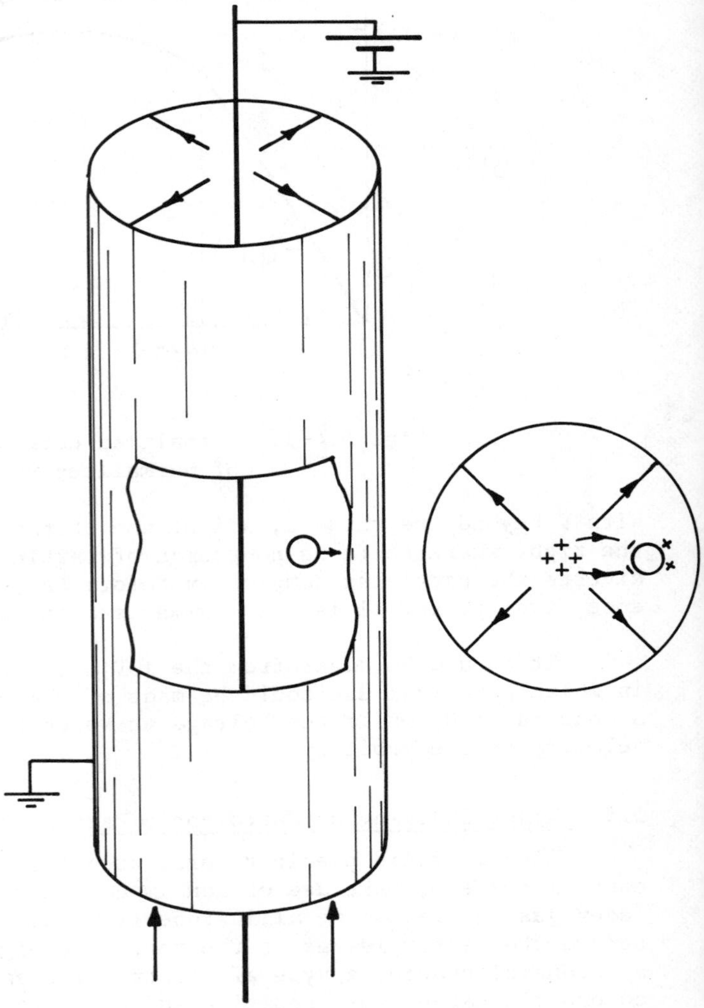

Fig. 5.5.1. Single-stage tube-type electrostatic precipitator.

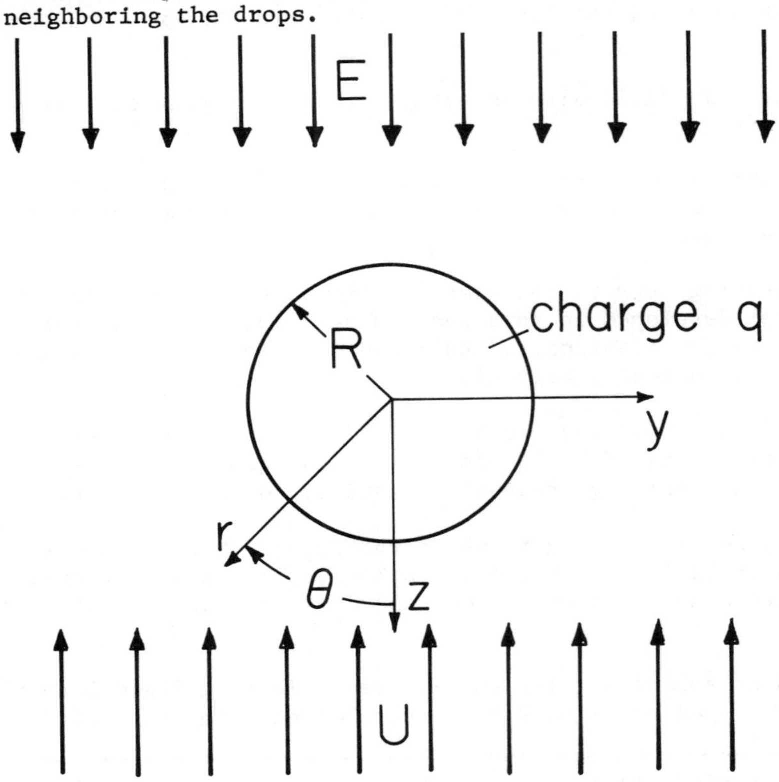

Fig. 5.5.2.

Spherical conducting drop in imposed electric field E and relative flow U that are uniform at infinity. In general, the electric field intensity E can be either positive or negative with E and U positive if directed as shown.

2. F. J. W. Whipple and J. A. Chalmers, "On Wilson's Theory of the Collection of Charge by Falling Drops," Quart. J. Roy. Meteorol. Soc. 70, 103 (1944).

3. J. R. Melcher, K. S. Sachar and E. P. Warren, "Overview of Electrostatic Devices for Control of Submicrometer Particles," Proc. IEEE 65, 1659 (1977).

That the charging rate is to be calculated implies that the electric field and hence the ion motions are not in the steady state. However, as discussed in Sec. 5.3, it is assumed that ion transit times through several particle radii R are short compared to charging times of interest. Hence, at any instant the particle charge is taken as a known constant, which then makes a contribution to the instantaneous electric field intensity.

The particle is taken as perfectly conducting. The electric potential is therefore constant at r = R, becomes $-E_o r\cos\theta$ far from the particle and is consistent with there being a net charge q on the particle. The appropriate combination of the potentials (satisfying Laplace's equation, as discussed in Sec. 2.16) $r\cos\theta$, $\cos\theta/r^2$ and $q/4\pi\varepsilon_o r$ therefore is the "imposed" field:

$$\vec{E} = -\nabla\phi = \{E(\frac{2R^3}{r^3}+1)\cos\theta + \frac{q}{4\pi\varepsilon_o r^2}\}\vec{i}_r + \{E(\frac{R^3}{r^3}-1)\sin\theta\}\vec{i}_\theta \qquad (3)$$

It follows from this result and Eq. 5.3.8a that the "stream" function for the electric field intensity is

$$\Lambda_E = ER^2[\frac{R}{r}+\frac{1}{2}(\frac{r}{R})^2]\sin^2\theta - \frac{q\cos\theta}{4\pi\varepsilon_o} \qquad (4)$$

The velocity distribution in the neighborhood of the particle must have both tangential and normal components that vanish on the particle surface, and must approach the uniform flow at infinity. Written in terms of a stream function, in accordance with Eq. 5.3.8b, the velocity distribution automatically is solenoidal (the flow is incompressible). Conservation of momentum supplies the additional law to determine the velocity distribution, but there is no exact analytical solution valid for all velocities. As is shown in Sec. 7.20, if forces due to viscosity dominate those due to inertia, Stokes's flow around a sphere applies, and the associated stream function is (from Eqs. 7.20.13 and 7.20.17)

$$\Lambda_v = \frac{-UR^2}{2}[(\frac{r}{R})^2 - \frac{3}{2}(\frac{r}{R}) + \frac{1}{2}\frac{R}{r}]\sin^2\theta \qquad (5)$$

The flow field found by using Eq. 5.3.8b is valid, provided the Reynolds number (Sec. 7.20) $R_y=\rho RU/\eta<1$, where η is the fluid viscosity. A fifty micron radius water droplet in free fall through air has $R_y \simeq 0.7$.

Given Eqs. 4 and 5, the characteristic lines are determined by substituting into Eq. 5.3.13b to obtain

$$\frac{1}{2}\frac{U}{b_\pm E}[(\frac{r}{R})^2 + \frac{1}{2}(\frac{R}{r}) - \frac{3}{2}(\frac{r}{R})]\sin^2\theta \mp [\frac{R}{r}+\frac{1}{2}(\frac{r}{R})^2]\sin^2\theta \pm \frac{3q}{q_c}\cos\theta = C \qquad (6)$$

The upper and lower signs, respectively, refer to positive and negative migrating particles and C is a constant which identifies the particular characteristic line.

Just what constant charge density should be associated with each of these lines is determined by a single boundary condition imposed wherever the line "enters" the volume of interest.

In terms of parameters now introduced, the object is to obtain the net instantaneous electrical current to the particle, $i_\pm(q,E,U,\rho_\pm)$. With the imposed field, velocity and charge densities held fixed, this expression then serves to evaluate the drop rate of charging

$$\frac{dq}{dt} = i_\pm(q) \qquad (7)$$

Permutations and combinations of flow velocity, imposed field, instantaneous drop charge, and sign of the incident particles are large, so an orderly approach is required to sort out the possible collection regimes. These are conveniently pictured in the (q,bE) plane: for positive particles Fig. 5.5.3, for negative ones Fig. 5.5.4.

First, recognize the surfaces which satisfy the condition of Eq. 5.3.6, and hence at which boundary conditions on the charge density are imposed. For positive particles (upper sign) and $b_+E > U$ the distribution of particle densities for particles entering at $z\to-\infty$ is required. Otherwise, the charge density is imposed as $z\to+\infty$ because the positive particles enter from below. These conditions therefore respectively apply to the left and right of the line $b_+E = U$ in Fig. 5.5.3. Characteristic lines originating on the particle surface carry zero charge density. Also, at the particle surface the normal fluid velocity is zero; hence the characteristic lines degenerate to $\pm b_\pm\vec{E}$. This greatly simplifies the charging process, because the electric field intensity given by Eq. 3 can be used to decide whether or not a given point on the particle surface can accept charge. Evaluation shows that

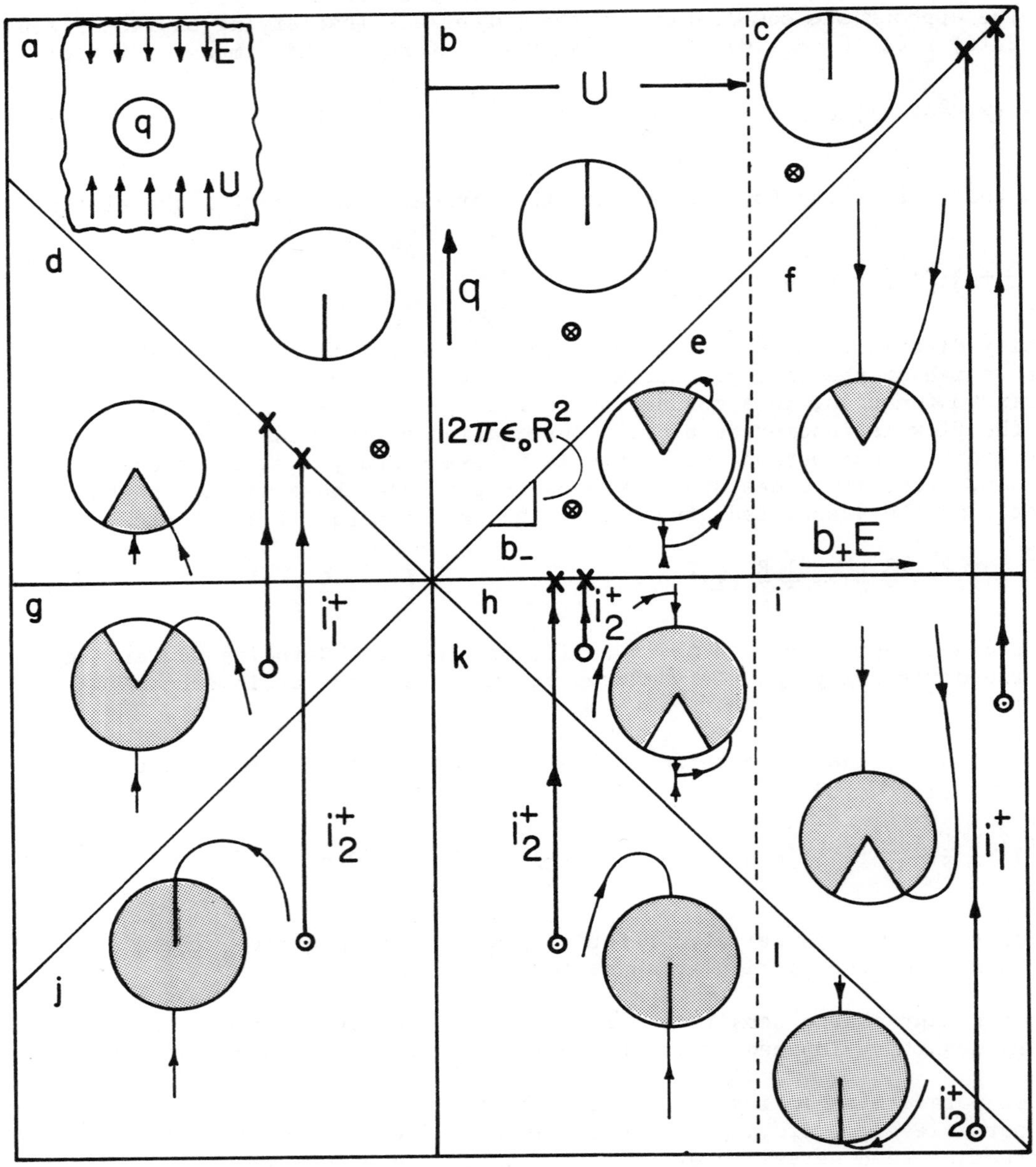

Fig. 5.5.3. Positive ion charging diagram. Charging regimes depicted in the plane of drop charge q and mobility-field product b_+E. With increasing fluid velocity, the vertical line of demarcation indicated by U moves to the right. Initial charges, indicated by $\odot$, follow the trajectories shown until they reach a final value given by $\otimes$. If there is no charging, the final and initial charges are identical, and indicated by $\otimes$. The inserted diagrams show the force lines $\vec{v} \pm b_{\pm}\vec{E}$.

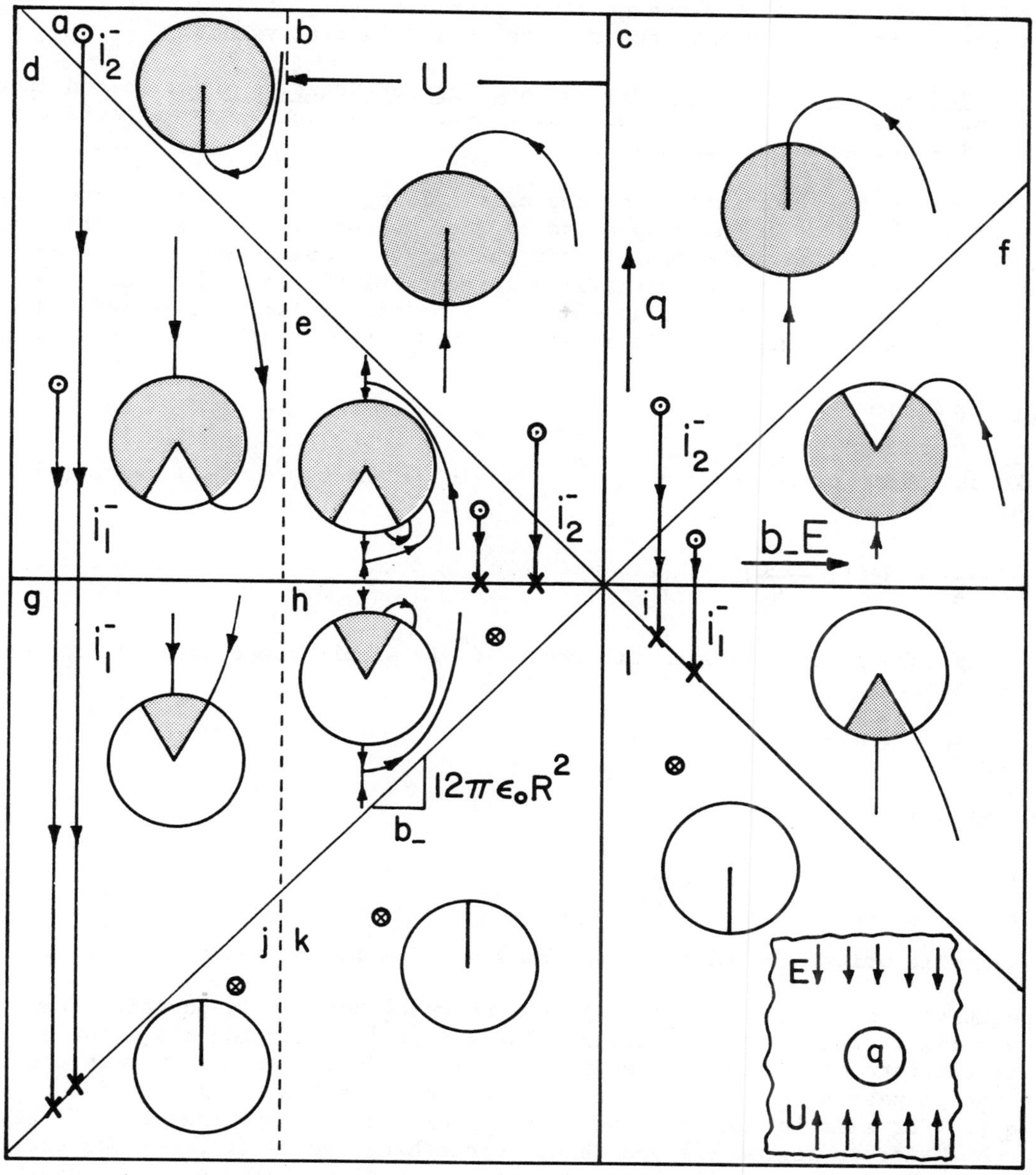

Fig. 5.5.4. Negative particle charging diagram. Conventions are as in Fig. 5.5.3. With increasing fluid velocity, the line of demarcation indicated by U moves to the left.

characteristic lines are directed into the particle surface wherever

$$\theta_c < \theta < \pi \quad \left| \begin{array}{l} \text{positive ions, } E > 0 \\ \text{negative ions, } E < 0 \end{array} \right. \tag{8}$$

$$0 < \theta < \theta_c; \quad \left| \begin{array}{l} \text{positive ions, } E < 0 \\ \text{negative ions, } E > 0 \end{array} \right. \tag{9}$$

where the critical angle, θ_c, demarking regions of inward and outward force lines, follows from the radial component of Eq. 3 evaluated at $r = R$:

$$\cos \theta_c = -\frac{q}{q_c} \tag{10}$$

A graphical representation of what has been determined is given by the direction of incident forc lines on the particle surfaces sketched in Figs. 5.5.3 and 5.5.4. Where directed inward, these force lines indicate a possible electric current density. Whether or not the current is finite depends on whether the given characteristic line originates elsewhere on the drop boundary or at infinity.

In any case, if a characteristic line is directed outward, there is no charging current density t the particle, and so without further derivations, regimes (a), (b) and (c) for the positive particles (Fig. 5.5.3) and (j), (k) and (ℓ) for the negative particles (Fig. 5.5.4) give no charging current. Fr Eq. 7, within these regimes the drop charge remains at its initial value.

Regimes (f) and (i) for Positive Ions; (d) and (g) for Negative Ions: To continue the characteriz tion of each regime shown in Figs. 5.5.3 and 5.5.4, upper and lower signs respectively will be used to refer to the positive and negative ion cases.

The characteristic line terminating at the critical angle on the drop surface reaches the $z \to -\infty$ surface at the radius y^* shown in the respective regimes in the Fig. 5.5.3. Particles entering within that radius strike the surface of the drop within the range of angles wherein the drop can accept ions. Hence, to compute the instantaneous drop charging current, simply find this radius y^* and compute the total current passing within that radius at $z \to -\infty$. The particular line is defined by Eq. 6 evaluated at the critical angle, and on the particle surface: $\theta = \theta_c$, $r = R$. Thus, the constant is evaluated as

$$C = \mp\frac{3}{2}[1 + (\frac{q}{q_c})^2] \tag{11}$$

To find y^*, take the limit of Eq. 6 ($r \to \infty$, $y^* = (r \sin\theta)$ and $\cos\theta \to -1$) using the constant of Eq. 11 to determine that

$$(y^*)^2(1 \mp \frac{U}{b_{\pm}E}) = 3R^2[1 - \frac{q}{q_c}]^2 \tag{12}$$

The current passing through the surface with radius y^* is simply the product of the current density and the circumscribed area:

$$i_1^{\pm} = \pm n_{\pm}q_{\pm}(\pm b_{\pm}E - U)\pi(y^*)^2 \tag{13}$$

The combination of Eqs. 12 and 13 is

$$i_1^{\pm} = 3I_{\pm}(1 - \frac{q}{q_c})^2 = \pm 3|I_{\pm}|\ (1 \mp \frac{q}{|q_c|})^2 \tag{14}$$

The second equality is written by recognizing the sign of E in the respective regimes.

In the positive ion regimes (f) and (i), the charging current is positive, tending to increase the drop charge until it reaches the limiting value $q = |q_c|$. Charging trajectories are shown in the figures, with i_1 the rate of charging, whether the initial drop charge is within the respective regimes or the charge passes from another regime into one of these regimes, and then passes on to its final value, $|q_c|$. For example, in the case of the positive ion charging, it will be shown that a drop charges at one rate in regime (ℓ) and then, on reaching regime (i), assumes the charging rate given by Eq. 14, which it obeys until the charge reaches a final value on the boundary between regimes (f) and (c).

Also sketched in Figs. 5.5.3 and 5.5.4 are the characteristic lines, and the critical angles defining those portions of the drop over which conduction can occur. As a drop charges and then passes from regime (i) to (f), and finally to the boundary between regimes (f) and (c) in the positive particl case, the angle over which the drop can accept particles decreases from a maximum of 2π to π at $q = 0$, and finally to zero when $q = |q_c|$. It is the closing of this "window" through which charge can be accepted to the particle surface which limits the drop charge to the critical or "saturation" value q_c.

Regimes (d) and (g) for Positive Ions; (f) and (i) for Negative Ions: These regimes are analogous to the four just discussed except that the particles enter at $z \to \infty$, rather than at $z \to -\infty$. The deriv tion is therefore as just described except that the limiting form of Eq. 6 is taken as $\theta \to 0$, with C again given by Eq. 11 to obtain

$$(y^*)^2(1 \mp \frac{U}{b_{\pm}E} = 3R^2(1 + \frac{q}{q_c})^2 \tag{15}$$

Then, the particle currents can be evaluated as

$$i_1^{\pm} = -3I_{\pm}(1 + \frac{q}{q_c})^2 = \pm 3|I_{\pm}|(1 \mp \frac{q}{|q_c|})^2 \quad (16)$$

As would be expected on physical grounds, the positive ion case gives charging currents and final drop charges in regimes (d) and (g) which are the same as those in (f) and (i).

Regimes (j) and (k) for Positive Ions; (b) and (c) for Negative Ions: For these regimes, the total surface of the drop can accept particles. The radius for the circular cross section of ions reaching the surface of the drop from $z \to \infty$ is determined by the line intersecting the drop surface at $\theta = \pi$. This line is defined by evaluating Eq. 6 at $r = R$, $\theta = \pi$ to obtain

$$C = \mp \frac{3q}{q_c} \quad (17)$$

Then, if the limit is taken $r \to \infty$, $\theta \to 0$ of Eq. 6, y^* is obtained and the current can be evaluated as

$$i_2^{\pm} = \pm n_{\pm} q_{\pm} (\pm b_{\pm} E + U)\pi (y^*)^2 = \frac{12|I_{\pm}|}{|q_c|} q \quad (18)$$

Note that in the positive ion regimes, q is negative, so the result indicates that the particle charges at this rate until it leaves the respective regimes when the charge $q = -|q_c|$.

Regime (ℓ) for Positive Ions; (a) for Negative Ions: The situation here is similar to that for the previous cases, except that ions enter at $z \to = -\infty$, so the appropriate constant for the critical characteristic lines given by Eq. 6 evaluated at $r = R$, $\theta = 0$, is the negative of Eq. 17. The limit of that equation given as $r \to \infty$, $\theta \to \pi$ gives y^* and evaluation of the current gives a value identical to that found with Eq. 18. In regime (ℓ), for positive particles, where the initial charge is negative, the charging current is positive, and tends to reduce the magnitude of the drop charge until it enters regime (i), where its rate of charging shifts to i_1 and it continues to acquire positive charge until it reaches the final value $|q_c|$ indicated on the diagram.

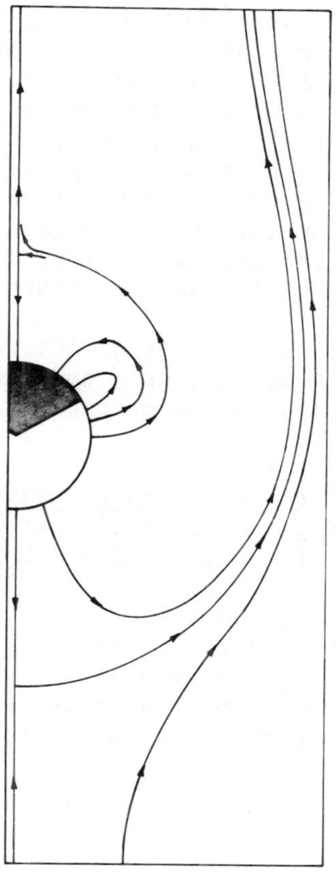

Fig. 5.5.5. Force lines in detail for regimes (e) for positive ions and (h) for negative ions. Here, all of the characteristic lines terminating on the particle also originate on the particle; hence, there is no charging. For the case shown, $U = \pm 2bE$, $q = \pm q_c/2$.

Regime (e), Positive Ions; Regime (h), Negative Ions: In regimes (e) and (h) for either sign of particles, the window through which the drop can accept a particle flux is on the opposite side from the incident particles. Typical force lines are drawn in Fig. 5.5.5. Force lines terminating within the window through which the drop can accept ions can originate on the drop itself. In that case, the charge dênsity on the characteristic line is zero, since the drop surface is incapable of providing particles.

To determine the particle charge that just prevents force lines originating at $z \to \infty$ from terminating on the particle surface, follow a line from the z axis where the drops enter at infinity back to the drop surface. That line has a constant determined by evaluating Eq. 6 with $\theta = 0$

$$C = \pm \frac{3q}{q_c} \quad (19)$$

Now, if Eq. 6 is evaluated using this constant, and $r = R$, an expression is found for the angular position at which that characteristic line meets the drop surface

$$\frac{3}{2} \sin\theta = \frac{3q}{q_c} (\cos\theta - 1) \quad (20)$$

Note that the quantity on the right is always negative if q/q_c is positive, as it is in regimes (e) for the positive particles and (h) for the negative. Thus, in regime (e) for the positive particles and (h) for the negative, the rate of charging vanishes and the drop remains at its initial charge.

Regime (h) for Positive Ions; (e) for Negative Ions: In these regimes, q/q_c is negative and Eq. 20 gives an angle at which the characteristic line along the z axis meets the drop surface. Typical force lines are shown in Fig. 5.5.6. To compute the rate of charging, the solution to this equation is not required because a circular area of incidence for ions at $z \to \infty$ is then determined by the characteristic line reaching the drop at $\theta = \pi$. Actually, no new calculation is necessary because that radius is the same as that found for regime (k) for the positive ions and (b) for the negative. The charging current is $i_2^{\pm}$, as given by Eq. 18. Drops in these regimes discharge until they reach zero charge. Moreover, if the initial drop charges place the drop in regimes (k) for the positive ions or (b) for the negative ions, the rate of discharge follows the same law through regimes (h) for the positive ions and (e) for the negative until the drop reaches zero charge.

As a matter of interest, in regimes (e) and (h) for both positive and negative ions a doughnut-shaped island of closed force lines is attached to the critical line if $0.5 < |bE|/U < 1$. An illustration is Fig. 5.5.7.

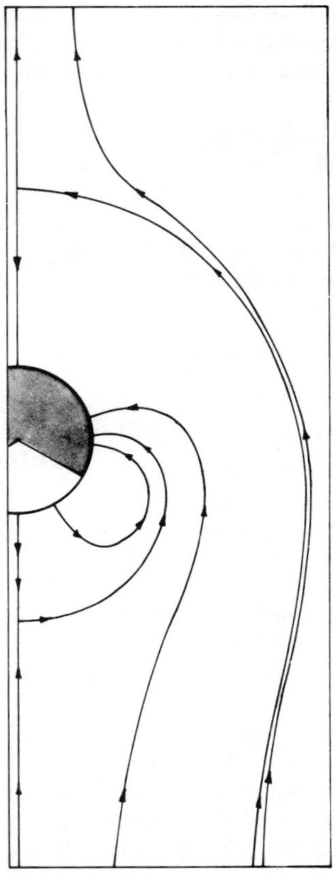

Fig. 5.5.6. Regimes (h) for positive ions and (e) for negative ions. Some of the characteristic lines extend to where the charge enters. $U = \pm 2b_{\pm}E$, $q = \pm q_c/2$.

Positive and Negative Particles Simultaneously: If both positive and negative particles are present simultaneously, the drop charging is characterized by simply superimposing the results summarized with Figs. 5.5.3 and 5.5.4. (The independence of species migration is discussed in Sec. 5.3.) The diagrams are superimposed with their origins (marked 0) coincident. A given point in either plane then specifies the drop charge and associated field experienced by both families of charges. This justifies superimposing the respective currents at the given point to find the total charging current:

$$\frac{dq}{dt} = i_+(q) + i_-(q) \tag{21}$$

Here, i_+ is i_1^+, i_2^+ or 0, in accordance with the charging regime and similarly, i_- is the appropriate current due to negative ions.

Drop Charging Transient: The quasistationary charging process is illustrated specifically by considering the fate of a drop starting out in regime (ℓ) of Fig. 5.5.3, in a field $b_+E > U$ and with a charge $q < -|q_c|$. Then, Eq. 7 with i_2^+ given by Eq. 18, becomes

$$\int_{q_o}^{q} \frac{dq}{q} = -\int_o^t \frac{12|I_+|}{|q_c|}\,dt = -\int_o^t \frac{\rho_+ b_+}{\varepsilon_o}\,dt \tag{22}$$

where q_o is the drop charge when $t = 0$. Thus, so long as the charge remains in regime (ℓ), the charging transient is

$$q = q_o e^{-t/\tau} \quad ; \quad \tau \equiv \varepsilon_o/\rho_+ b_+ \tag{23}$$

When the drop has been discharged to $q = -|q_c|$, the rate of discharge switches to i_1^+, given by Eq. 16. Thus, the charging equation is

$$\int_{-q_c}^{q} \frac{dq}{\left(1 \mp \frac{q}{|q_c|}\right)^2} = \pm 3|I_\pm| \int_o^{t'} dt' \tag{24}$$

where t' is the time measured relative to when the drop switches into regime (i). Integration gives a charging transient

$$q = |q_c|\left(\frac{t'/4\tau - \frac{1}{2}}{t'/4\tau + \frac{1}{2}}\right) \tag{25}$$

which completes the discharging of the drop and goes on in regime (f) to charge the drop positively until it approaches the saturation charge q_o.

Note that although the detailed temporal dependence of Eqs. 23 and 25 is quite different, the same charge relaxation time $\varepsilon_o/\rho_+ b_+$ characterizes the charging dynamics. It is this time that must be long compared to the particle transient time to justify the quasistationary model. The same time constant has a second complementary significance, brought out in the next section. There, it is possible to appreciate the relation of space-charge effects to the quasistationary model used in this section.

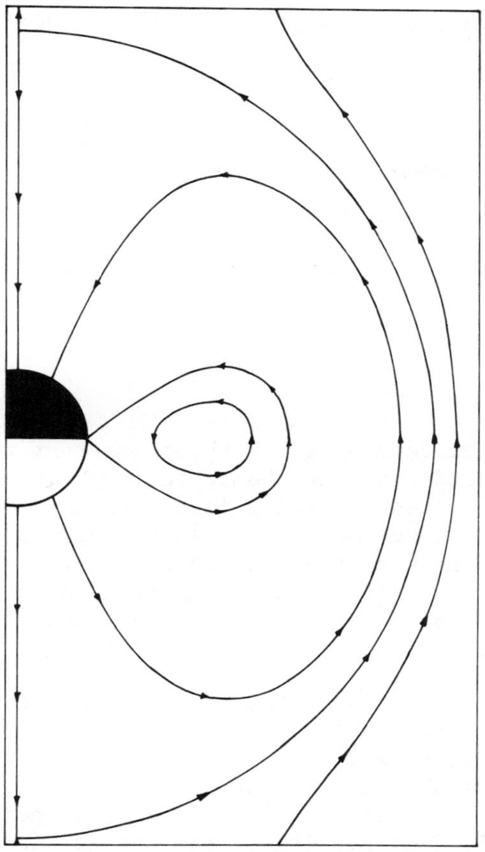

Fig. 5.5.7

In regimes (e) and (h) with $0.5<|bE|/|U|$, a "doughnut" of closed force lines is attached to critical line around drop. Positive particles are illustrated with $q = 0$ and $b_+E/U = 0.75$.

5.6 Unipolar Space Charge Dynamics: Self-Precipitation

Complementary assumptions in Secs. 5.3 - 5.5 are that the effect of the electric field on the flow can be ignored, and that the volume space charge density makes a negligible contribution to the imposed field. Although often good approximations in predicting the trajectories of dilute ions and charged macroscopic particles in moving gases and liquids, these are usually not good assumptions if there is to be an appreciable coupling between the electric field and the neutral fluid through which the charged particles migrate.

What are the effects of "self-fields" (space-charge contributions) left out in the imposed field approximation used in Secs 5.3 - 5.5? In this and the next section, this question is addressed while again considering a single species of either positively or negatively charged particles.

The pertinent laws are Eqs. 5.2.9 - 5.2.11 without source or recombination contributions (G-R=0) and with lengths of interest large enough to justify ignoring diffusion. In writing Eq. 5.2.9, the creation of a field divergence by the space charge is recognized by substituting on the right with Gauss' law, Eq. 5.2.10:

$$\frac{\partial \rho_{\pm}}{\partial t} + (\vec{v} \pm b_{\pm}\vec{E})\cdot\nabla\rho_{\pm} = -\frac{\rho_{\pm}^2 b_{\pm}}{\varepsilon} \qquad (1)$$

This expression is converted to one describing how the charge density changes with time for an observer moving along a characteristic (or force) line by following the procedure developed in Sec. 5.3.2. Instead of Eq. 5.3.3, Eq. 1 becomes

$$\frac{d\rho_{\pm}}{dt} = -\frac{\rho_{\pm}^2 b_{\pm}}{\varepsilon} \qquad (2)$$

along the characteristic lines

$$\frac{d\vec{r}}{dt} = \vec{v} \pm b_{\pm}\vec{E} \qquad (3)$$

Although the velocity $\vec{v}$ is still considered to be imposed, $\vec{E}$ in Eq. 3 has contributions from not only charges on the boundaries, but from those within the volume of interest as well. So it is that the time dependence of the charge density can be determined from an integration of Eq. 2. For a characteristic line originating when t = 0 where $\rho_\pm = \rho_o$

$$\int_{\rho_o}^{\rho_\pm} \frac{d\rho_\pm}{\rho_\pm^2} = -\frac{b_\pm}{\varepsilon}\int_o^t dt \qquad (4)$$

Thus,

$$\frac{\rho_\pm}{\rho_o} = \frac{1}{1 + t/\tau_e}; \quad \tau_e = \frac{\varepsilon}{\rho_o b_\pm} \qquad (5)$$

This result is both remarkably general and somewhat deceiving. Without apparent regard for the particulars of a physical situation, for the boundary conditions and hence for any imposed component of the field and for the locations of charges that image those evolving in the volume, the decay law of Eq. 5 is deduced. But, the law applies for an observer measuring time as he follows a given particle along a characteristic line, defined by Eq. 3, originating when t = 0 where $\rho_\pm = \rho_o$. At each step in the evolution, all of the charge (in the volume and on the boundaries) instantaneously contributes to $\vec{E}$. This contribution is embodied in Gauss' law. The characteristic viewpoint is now used to make some general deductions, and then by way of illustration, to make specific predictions.

General Properties: Suppose that when t = 0 the charged particles are uniformly distributed over some confined region V within the total volume, as suggested by Fig. 5.6.1. When t = 0, the volume of interest consists of regions either occupied by no charge density or by the uniform density ρ_u. At a later time, the cloud of charged particles has changed its shape and general location. Particles initially at the locations A, B and C are respectively found at A', B' and C'. At a point like A', with a characteristic line originating within the initial cloud of particles, the charge density is given by Eq. 5 as

$$\rho_\pm = \frac{\rho_u}{1 + t/\tau_e}; \quad \tau_e \equiv \frac{\varepsilon}{\rho_u b_\pm} \qquad (6)$$

Note that τ_e is itself dependent on the initial charge density.

Now, consider the time dependence of $\rho_\pm$ at a fixed location A'. So long as A' is within the region occupied by the charge cloud, this time dependence is also given by Eq. 6. At each instant, the point in question can be traced backward in time to a location in the cloud when t = 0 where the charge density is the same number, ρ_u. As time progresses, different locations originate the characteristic A', but because ρ_u is the same throughout the initial cloud, each of these has the same charge density ρ_u or no charge density at all. That is, at a position like B', the characteristic originates on no charge density, and there is no charge density at the instant in question.

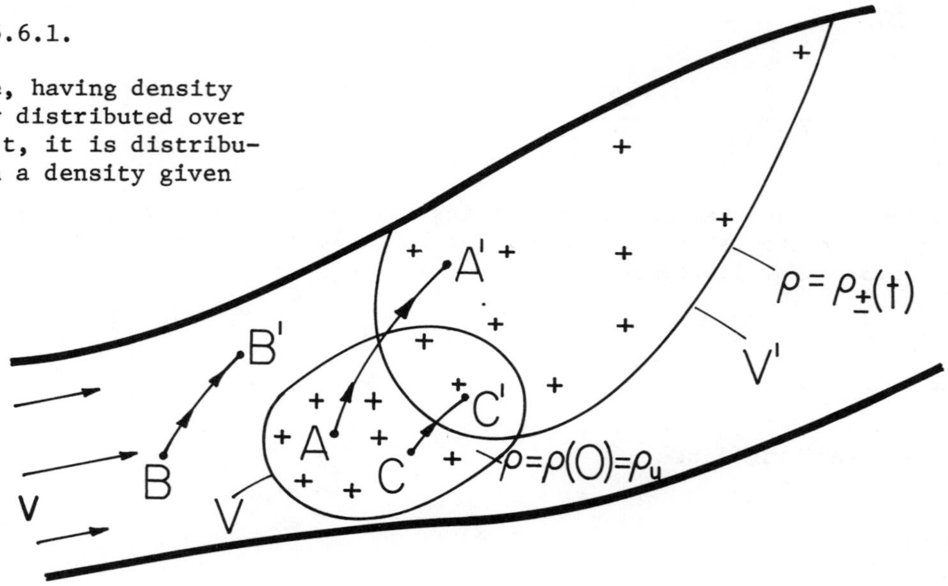

Fig. 5.6.1.

When t=0, charge, having density ρ_u, is uniformly distributed over V. By the time t, it is distributed over V' with a density given by Eq. 6.

So it is that the charge transient at any fixed location consists of either a charge density decaying according to Eq. 6 or no charge density at all. Generally, at a position like A', the charge density is zero until the "front" arrives. Then, the position A' is enveloped by the particle cloud which is expanding under its self-field so that the density decays in accordance with Eq. 6. At a position like C', there is no delay in the arrival of this front so that the decay is given by Eq. 6 from time t = 0. But, the time may come when the cloud passes beyond the point in question and the decay in charge density is then abruptly terminated by the density going to zero. When the front arrives and when the cloud has passed by is a matter that must be resolved by integrating to find the characteristic lines.

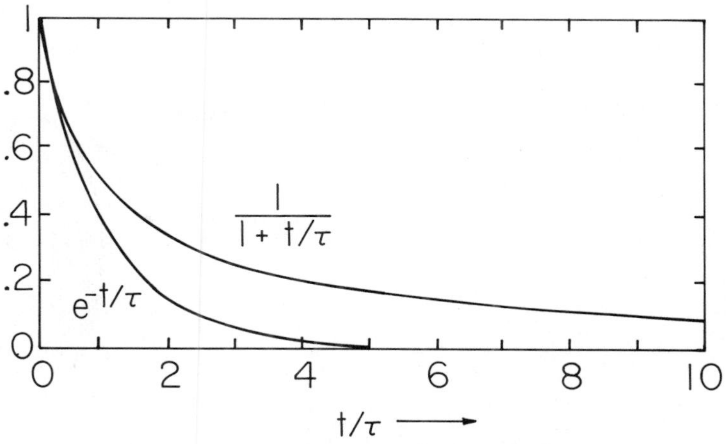

Fig. 5.6.2. Comparison of self-presipitation transient to exponential decay.

The tendency of the cloud to expand or self-precipitate, as the cloud as a whole is carried by the deforming medium and the total field, is described by a decay that is relatively slow. Figure 5.6.2 emphasizes this point by comparing Eq. 7 to an exponential decay.

The rate of decay along a characteristic line represented by τ_e is the same as the charging time constant for the "drop" in Sec. 5.5. An important observation can now be made relevant to taking into account space-charge effects on the collection of charged particles by isolated drops. If space-charge effects are really important, then processes of interest must occur on the time scale of τ_e. This implies that the drop described in Sec. 5.5 must change its charge in a time on this same scale. But, the drop charge contributes to the electric field, and in the analysis of Sec. 5.5 the electric field is assumed to be constant during the time that a particle migrates several drop radii. Thus it is apparent that if space-charge contributions to the field are to be taken into account, the quasi-steady approximation is not valid.

<u>A Space-Charge Transient</u>: As a simple illustration of the fate of a cloud of charged particles that is initially of uniform charge density, consider the radially symmetric configuration of Fig. 5.6.3. When t=0, the particles occupy the annular region $R_i < r < R_o$. Image charges are presumed sufficiently remote that the field can be regarded as radially symmetric. There is a source of fluid inside the region $r < R_i$ giving rise to a volume rate of flow Φ_v (m^3/sec). Because the flow is incompressible, the resulting velocity distribution is determined by the requirement that the material flux at any radius r be the same: $4\pi r^2 v_r = \Phi_v$. The characteristic lines are then found from the one nontrivial component of Eq. 3:

$$\frac{dr}{dt} = \frac{\Phi_v}{4\pi r^2} \pm b_{\pm} E_r \tag{7}$$

Because the initial charge distribution is uniform, any region within the cloud is known to have a density that decays according to Eq. 6. In this simple example it is easy to find the position of the outward propagating front, and hence locate the region where this decay applies. By the integral form of Gauss' law, Eq. 2.7.1a, the electric field at r, the leading edge of the cloud, is

$$E_r = \pm \frac{1}{3} \frac{\left(R_o^3 - R_i^3\right)\rho_u}{\varepsilon_o r^2} \tag{8}$$

Substitution of this expression into Eq. 7 and integration gives

$$\frac{r}{R_o} = \left\{1 + \left[\frac{3\Phi_v \tau_e}{4\pi R_o^2} + 1 - \left(\frac{R_i}{R_o}\right)^3\right]\frac{t}{\tau_e}\right\}^{1/3} \tag{9}$$

Similar arguments apply to the trailing edge, where $E_r = 0$ and hence

$$\frac{r}{R_o} = \left[\left(\frac{R_i}{R_o}\right)^3 + \left(\frac{3\Phi_v \tau_e}{4\pi R_o^3}\right)\frac{t}{\tau_e}\right]^{1/3} \tag{10}$$

These last two expressions define the region occupied by the charged particles. During the time that

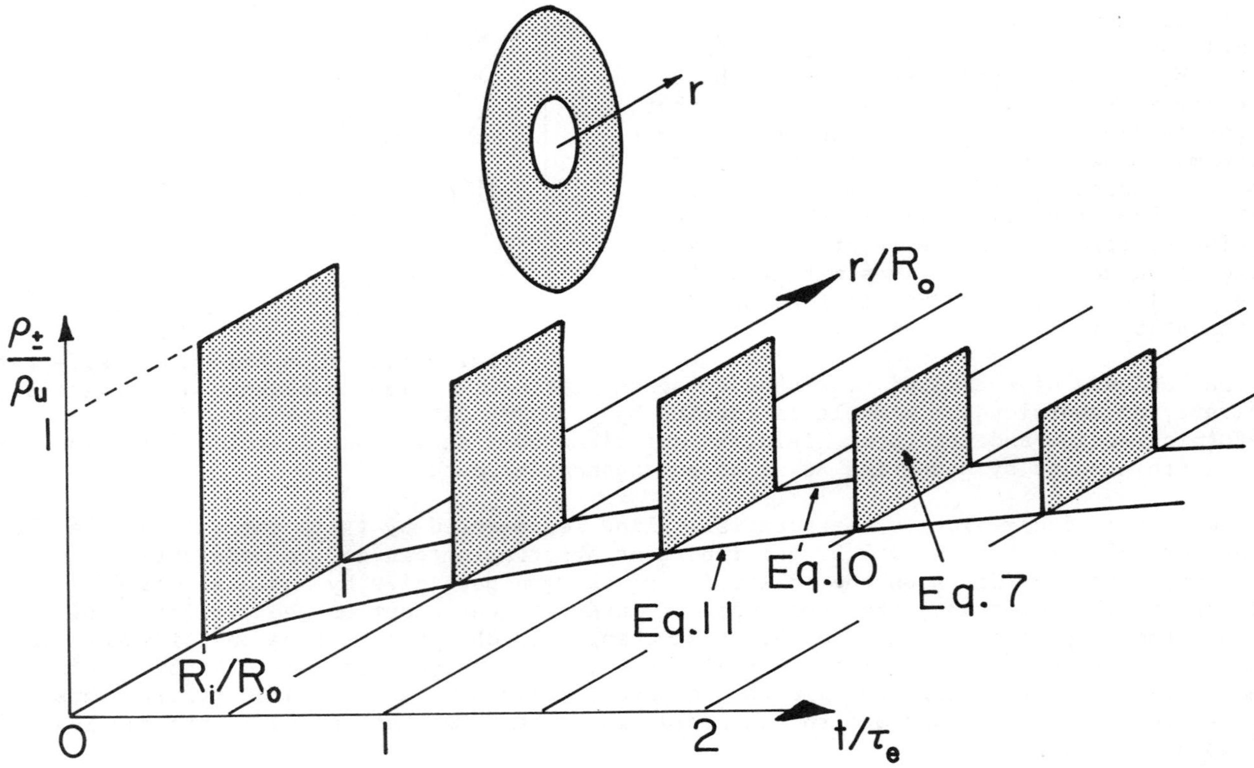

Fig. 5.6.3. When $t = 0$, a uniform density ρ_u of charged particles fills the spherical shell $R_i < r < R_o$. A source of gas at the origin imparts a radial velocity. For this plot, $R_o/R_i = 0.5$, $3\Phi_v\tau_e/4\pi R_o^3 = 1$. Remember that the charge is self-precipitating in three dimensions. At any time, the product of the charge density and the volume of the region filled by the charge is constant.

the particles surround a given fixed radial location, the temporal decay at that radius is given by Eq. 6. The evolution of the cloud is illustrated in Fig. 5.6.3. Fig.

Steady-State Space-Charge Precipitator: What from the laboratory frame of reference appears to be steady or stationary phenomenon is from the particle frame of reference still a transient. The characteristic time is typically a transport time ℓ/U, and the ratio

$$R_e = \frac{\tau_e}{\ell/U} = \frac{U\varepsilon_o}{\ell\rho_o b_\pm} \tag{11}$$

represents the degree to which convection competes with self-field migration in determining the distribution of the charged particles. R_e is defined as the electric Reynolds number.

As a specific illustration, consider the circular cylindrical duct shown in Fig. 5.6.4. Gas enters at the left with a uniform velocity profile carrying a uniform distribution of charged particles. The channel wall is at zero potential, and hence only the self-fields contribute to the migration. With the assumption that variations in the z direction of the particle density occur relatively slowly goes the quasi-one-dimensional model of an electric field that is dominantly in the radial direction. Hence, the characteristic lines are determined from Eq. 3 approximated as

$$\frac{d\vec{r}}{dt} = U\vec{i}_z \pm b_\pm E_r(r)\vec{i}_r \tag{12}$$

The z component of this expression can be integrated to describe the characteristic line associated with the solution given by Eq. 6, i.e., for the particles entering at $z = 0$, where $\rho_+ = \rho_o = \rho_u$ when $t = 0$:

$$z = Ut \tag{13}$$

Hence, the distribution of charge density with z is obtained directly by substituting Eq. 13 into Eq. 6.

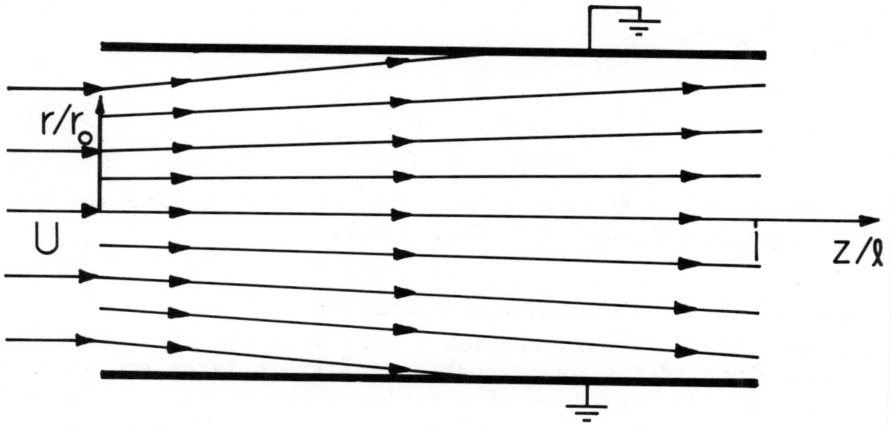

Fig. 5.6.4

Space-charge precipitator having circular cylindrical cross section and $R_e = 1$ showing characteristic lines.

$$\rho_{\pm} = \frac{\rho_u}{1 + \frac{z}{\ell}\left(\frac{\ell}{U\tau_e}\right)} \qquad (14)$$

The fact that the axial component of $\vec{E}$ is neglected has made it possible to find the spatial distribution of $\rho_{\pm}$ without solving the self-consistent characteristic equations.

A length ℓ of the channel might be used as a precipitator, for the removal of pollutant particles which are charged upstream. The cleaning efficiency of such a device follows from Eq. 14 integrated over the channel cross section A at $z = \ell$ and $z = 0$,

$$\eta \equiv \frac{\int_A \rho_u da - \int_A \rho_{\pm}(\ell) da}{\int_A \rho_u da} = \frac{R_e^{-1}}{1 + R_e^{-1}} \qquad (15)$$

and is determined by the ratio of transport time to τ_e. The dependence of η on R_e^{-1} is shown in Fig. 5.6.5. The relatively poor efficiency even with a residence time several times τ_e has its origins in the relatively slow decay depicted by Fig. 5.6.2.

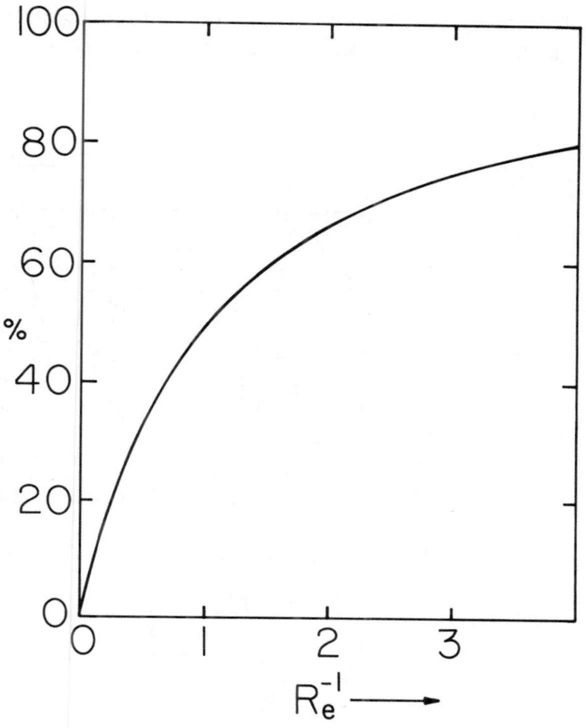

Fig. 5.6.5. Efficiency of space-charge precipitator as function of reciprocal electric Reynolds number, the ratio of residence time to τ_e.

The trajectories of the particles are determined from both the radial and axial components of Eq. 12. Gauss' law relates the charge density at a given cross section along the z axis to E_r:

$$\frac{1}{r}\frac{\partial}{\partial r}(rE_r) = \pm \frac{\rho_{\pm}}{\varepsilon_o} \qquad (16)$$

This expression can be integrated in the radial direction to give

$$E_r = \pm \frac{1}{r}\int_o^r \frac{r}{\varepsilon_o} \cdot \frac{\rho_u dr}{(1 + \frac{z}{\ell} R_e^{-1})} = \pm \frac{1}{2}\frac{r}{\varepsilon_o}\frac{\rho_u}{(1 + \frac{z}{\ell} R_e^{-1})} \qquad (17)$$

Thus, the radial component of Eq. 12 becomes

$$\frac{dr}{dt} = \frac{1}{2} b_{\pm} \frac{r}{\varepsilon_o} \frac{\rho_u}{1 + \frac{z}{\ell} R_e^{-1}} \qquad (18)$$

But, in view of the axial component of this same equation,

$$\frac{dr}{dt} = \frac{dr}{dz}\frac{dz}{dt} = U\frac{dr}{dz} \qquad (19)$$

Thus, the time is eliminated as a parameter to obtain an equation for the characteristic line in the (r,z) plane. Integration of Eqs. 18 and 19 gives

$$\frac{r}{r_o} = \sqrt{1 + \frac{z}{\ell} R_e^{-1}} \tag{20}$$

Which trajectory is considered is determined by r_o, the radial position at which a particle enters where $z = 0$. A sketch of the characteristic lines is included with Fig. 5.6.4.

5.7 Collinear Unipolar Conduction and Convection: Steady D-C Interactions

The electrohydrodynamic coupling undertaken in this section illustrates the electromechanical energy conversion processes that can take place if the space charge density is large enough to provide a significant contribution to the electric field. In the configuration shown in Fig. 5.7.1, a pair of electrically conducting grids at $z = 0$ and $z = \ell$ provide electrical "terminals" through which the fluid can pass and by which entrained charge particles are either injected or collected. The grids have the potential difference v. Charged particles are injected with zero potential at $z = 0$ and collected at potential v on the grid at $z = \ell$. Hence, with a load attached to the terminals, the charge carried by the fluid results in a current through the load, so that the configuration converts mechanical energy to electrical form. In this case, the fluid plays the role of the belt in a Van de Graaff generator. In fact, as for the Van de Graaff machine described in Sec. 4.14, it will be seen that generator, pump (motor) and brake operation are all possible.

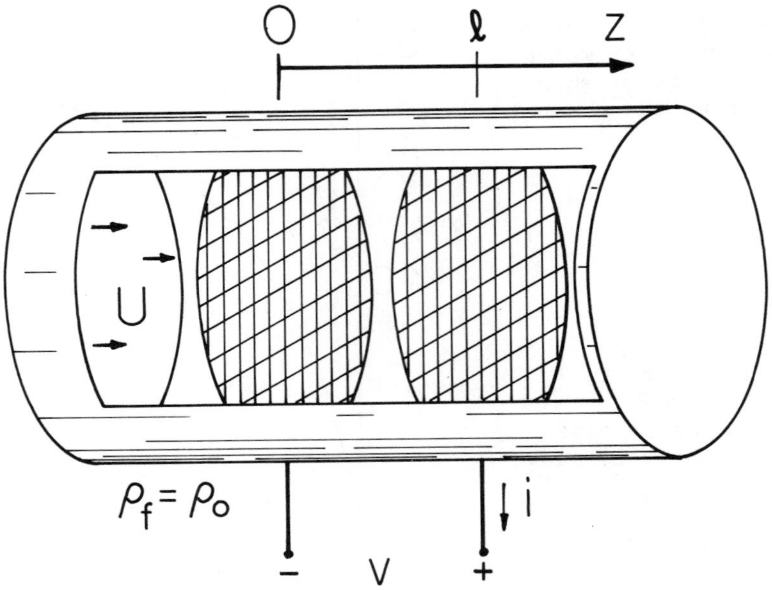

Fig. 5.7.1

One-dimensional unipolar d-c pump, generator or brake.

There is an important difference between the collinear configuration considered here and the Van de Graaff machine. In the latter, the generated field is orthogonal to the field associated with the charge carried by the belt. Here, transverse dimensions are very large and the charge entrained in the fluid produces a field that is collinear with the "generated" or "imposed" field associated with charges on the grids. Thus, $\vec{E} = \vec{i}_z E(z)$. As a result, the electromechanical energy conversion is through normal stresses, rather than shear stresses. The volume between the grids can be identified with the volume shown in the abstract by Fig. 4.15.1, or specifically by Fig. 5.7.2.

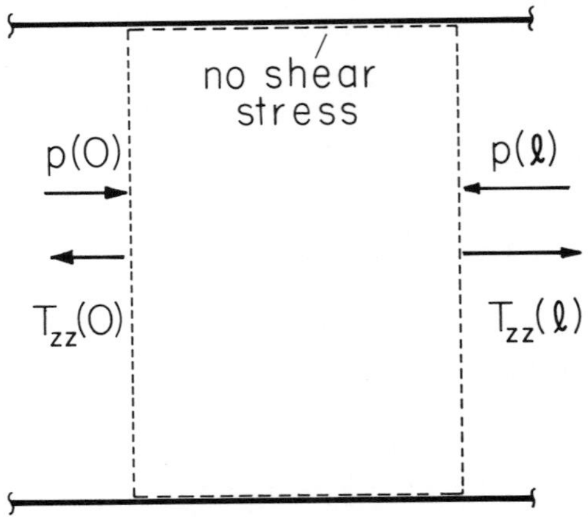

Fig. 5.7.2

The fluid volume between the grids is subject to the mechanical normal stresses (pressure) p defined as positive if acting inward.

Interest is confined to steady-state conditions, so conservation of charge, represented by Eq. 5.2.2 with G-R = 0 and $(\partial\rho/\partial t) = 0$, requires that the current density be solenoidal. The fluid is incompressible, so that $\vec{v}$ is also solenoidal. It follows from the one-dimensional model that the fluid velocity $\vec{v} = U\vec{i}_z$ and the current density $\vec{J} = \vec{i}_z J$, where U and J are independent of z:

$$J = \frac{i}{A} = \rho(bE + U) \tag{1}$$

The scale of interest is presumed large enough that effects of diffusion are negligible.

In addition to Eq. 1, Gauss' law relates ρ to $\vec{E} = \vec{i}_z E(z)$. Elimination of E between these equations gives

$$\rho^{-1}d\rho^{-1} = \frac{b}{\varepsilon J}\,dz \tag{2}$$

If the charge is injected with density $\rho(0) = \rho_o$, Eq. 2 is integrated to give

$$\frac{\rho}{\rho_o} = \left(1 + \frac{2}{R_e}\frac{i_o}{i}\frac{z}{\ell}\right)^{-1/2} \tag{3}$$

where $i_o \equiv \rho_o UA$ and the electric Reynolds number $R_e \equiv U\varepsilon/b\rho_o\ell$. Note that R_e is the ratio of the charge relaxation time $\varepsilon/b\rho_o$ (based on the charge density at the entrance) to the fluid transport time ℓ/U. Hence, if R_e is large, convection plays a dominant role in determining the charge distribution.

Now, if Eq. 3 is used with Eq. 1, the electric field intensity is known:

$$E \equiv \frac{U}{b}\left\{\frac{i}{i_o}\left[1 + \frac{2}{R_e}\frac{i_o}{i}\frac{z}{\ell}\right]^{1/2} - 1\right\} \tag{4}$$

and integration of E in turn gives the potential distribution

$$\Phi = \frac{U\ell}{b}\left\{\frac{z}{\ell} - \frac{R_e}{3}\left(\frac{i}{i_o}\right)^2\left[\left(1 + \frac{2}{R_e}\frac{i_o}{i}\frac{z}{\ell}\right)^{3/2} - 1\right]\right\} \tag{5}$$

Thus, because $\Phi(\ell)$ is the terminal voltage v, the "volt-ampere" characteristic of the device has been obtained.

The pressure rise $\Delta p \equiv p(\ell) - p(0)$ is balanced by the net electrical force on the fluid. Hence it is simply the difference in the normal electric stresses evaluated at the outlet and inlet. From Eq. 4,

$$\Delta p = \frac{\varepsilon}{2}[E^2(\ell) - E^2(0)] = \frac{\varepsilon U^2}{2b^2}\left\{\left[\frac{i}{i_o}\left(1 + \frac{2i_o}{R_e i}\right)^{1/2} - 1\right]^2 - \left[\frac{i}{i_o} - 1\right]^2\right\} \tag{6}$$

The pump, brake and generator energy conversion regimes can be identified by considering the dependence of the electrical power out, $P_e = vi$, and of the mechanical power in, $P_m = -\Delta pUA$, on the normalized current i/i_o. These dependences are shown in Fig. 5.7.3 with the electric Reynolds number, R_e, as a parameter. Note that as R_e is raised, the v-i relationship approaches that of a current source with $i = i_o$ (a vertical line through $i/i_o = 1$ on the plot). The short-circuit current i_{sc}, normalized to i_o, is determined by R_e. From Eq. 5 evaluated at $z = \ell$ and with $v = 0$,

$$R_e = \frac{3(4r^2 - 3) + [9(4r^2 - 3)^2 - 192r^3(r - 1)]^{1/2}}{12r^2(1 - r)} \quad ; \quad r \equiv \frac{i_{sc}}{i_o} \tag{7}$$

For convenience, R_e is expressed here as a function of i_{sc}/i_o. The current i_{bp} (at which the pressure rise is zero) follows from Eq. 6:

$$i_{bp}/i_o = 2R_e/(2R_e + 1) \tag{8}$$

These currents i_{sc} and i_{bp} are sketched as a function of R_e in Fig. 5.7.4. They demark the extremes of the brake regime of operation, and hence also define the upper and lower currents, respectively, of

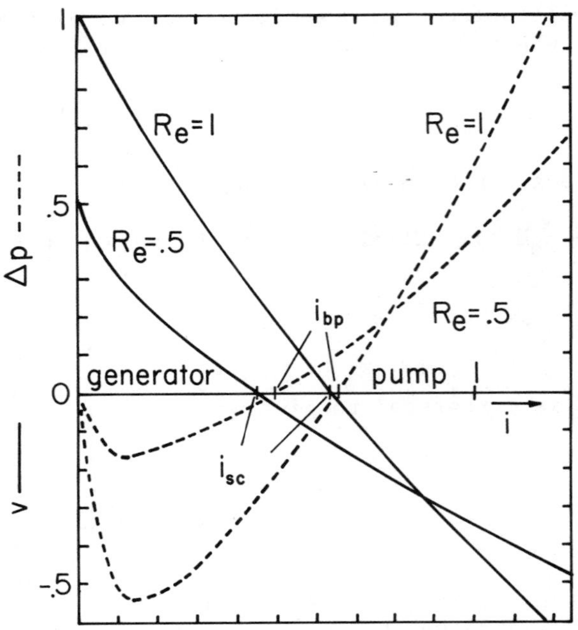

Fig. 5.7.3. Electrical and mechanical terminal characteristics as a function of the normalized terminal current with the electric Reynolds number R_e as a parameter. v is normalized to $\rho_o \ell^2/\varepsilon$, Δp is normalized to $(\rho_o \ell)^2/2\varepsilon$, and i to i_o.

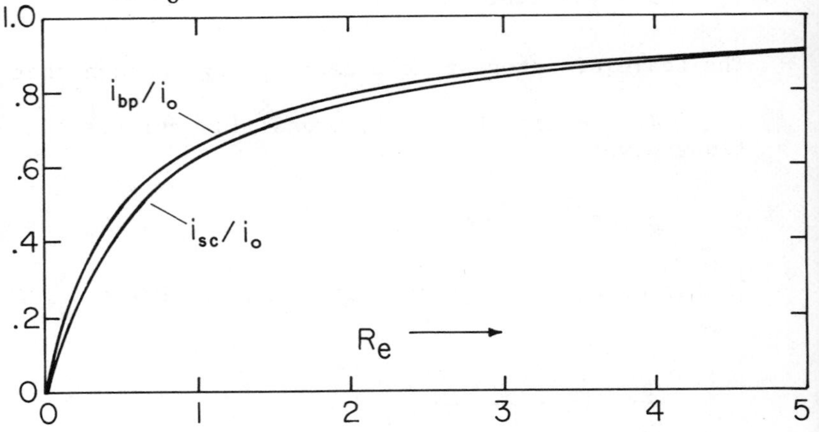

Fig. 5.7.4. Dependence of the normalized short-circuit and zero-pressure-drop currents on the electric Reynolds number R_e.

the generator and pump regimes.

The Generator Interaction: The optimum generator performance, from the point of view of electrical breakdown, is obtained by making $E(\ell) = 0$, so that the maximum pressure change is obtained for a given maximum E. In this case, it follows from Eq. 4 that i/i_o should be adjusted to make

$$\frac{i}{i_o} = -\frac{1}{R_e} + [(1/R_e)^2 + 1]^{1/2} \tag{9}$$

In any case, the electrical power output is given by Eq. 5 as

$$P_e = vi = \frac{\rho_o A U^2 \ell}{b} \left(\frac{i}{i_o}\right)\left\{1 - \frac{R_e}{3}\left(\frac{i}{i_o}\right)^2\left[\left(1 + \frac{2}{R_e}\frac{i_o}{i}\right)^{3/2} - 1\right]\right\} \tag{10}$$

For this particular case, the mechanical power input follows from Eqs. 6 and 9 as

$$P_m = -\Delta p U A = \frac{U^3 A \varepsilon}{2b^2}\left(\frac{i}{i_o} - 1\right)^2 \tag{11}$$

From these last two expressions, an electromechanical energy conversion efficiency is determined as a function of R_e or i/i_o:

$$\frac{P_e}{P_m} = \frac{1}{3}\left(1 + 2\frac{i}{i_o}\right) = \frac{1}{3}\left\{1 - \frac{2}{R_e} + 2[(1/R_e)^2 + 1]^{1/2}\right\} \tag{12}$$

The dependences of the energy conversion efficiency and i/i_o on R_e are summarized in Fig. 5.7.5.

The Pump Interaction: Consider now the distribution of fields that gives rise to the greatest pressure rise for a given maximum electric field intensity within the flow. From Eq. 6, in this case $E(0) = 0$: a condition obtained by making $i = i_o$. That is, at the entrance current is entirely carried by the convection, there being no slip velocity between the charge carriers and the neutral fluid. The electrical power P_e is again given by Eq. 10, but now $i/i_o = 1$. The mechanical power P_m follows from Eq. 6 and the current condition as

$$P_m = -\frac{U^3 A \varepsilon}{2b^2}[(1 + 2/R_e)^{1/2} - 1]^2 \tag{13}$$

The efficiency of the electrical to mechanical energy conversion is then fully determined by the

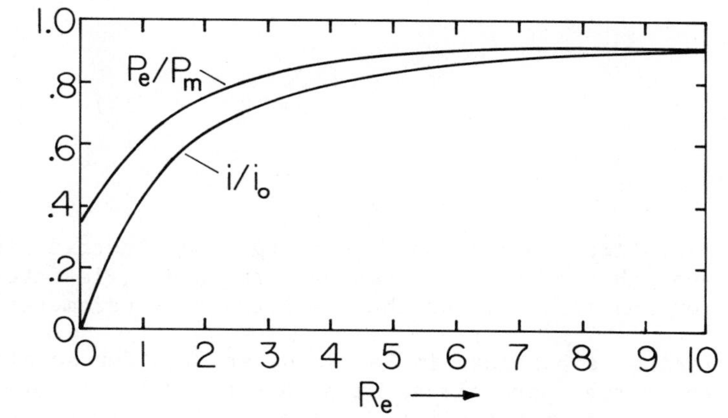

Fig. 5.7.5

Dependence of generator electromechanical energy conversion efficiency P_e/P_m and normalized terminal current i/i_o on the electric Reynolds number R_e.

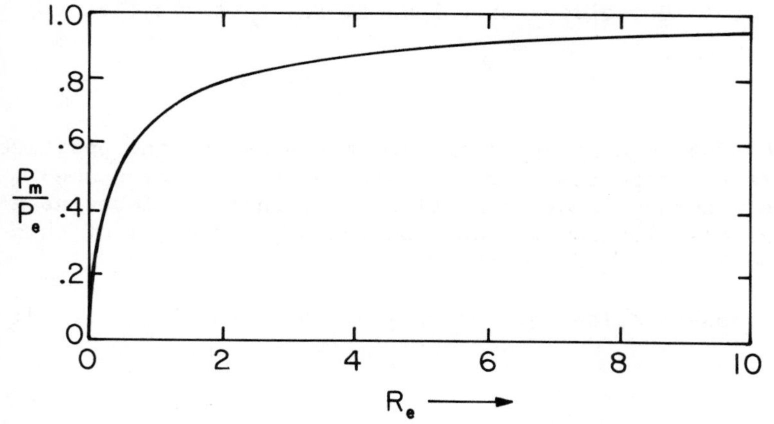

Fig. 5.7.6

Efficiency of electrical to mechanical energy conversion for unipolar one-dimensional interaction with current $i=i_o$ so that the entrance electric field intensity is zero.

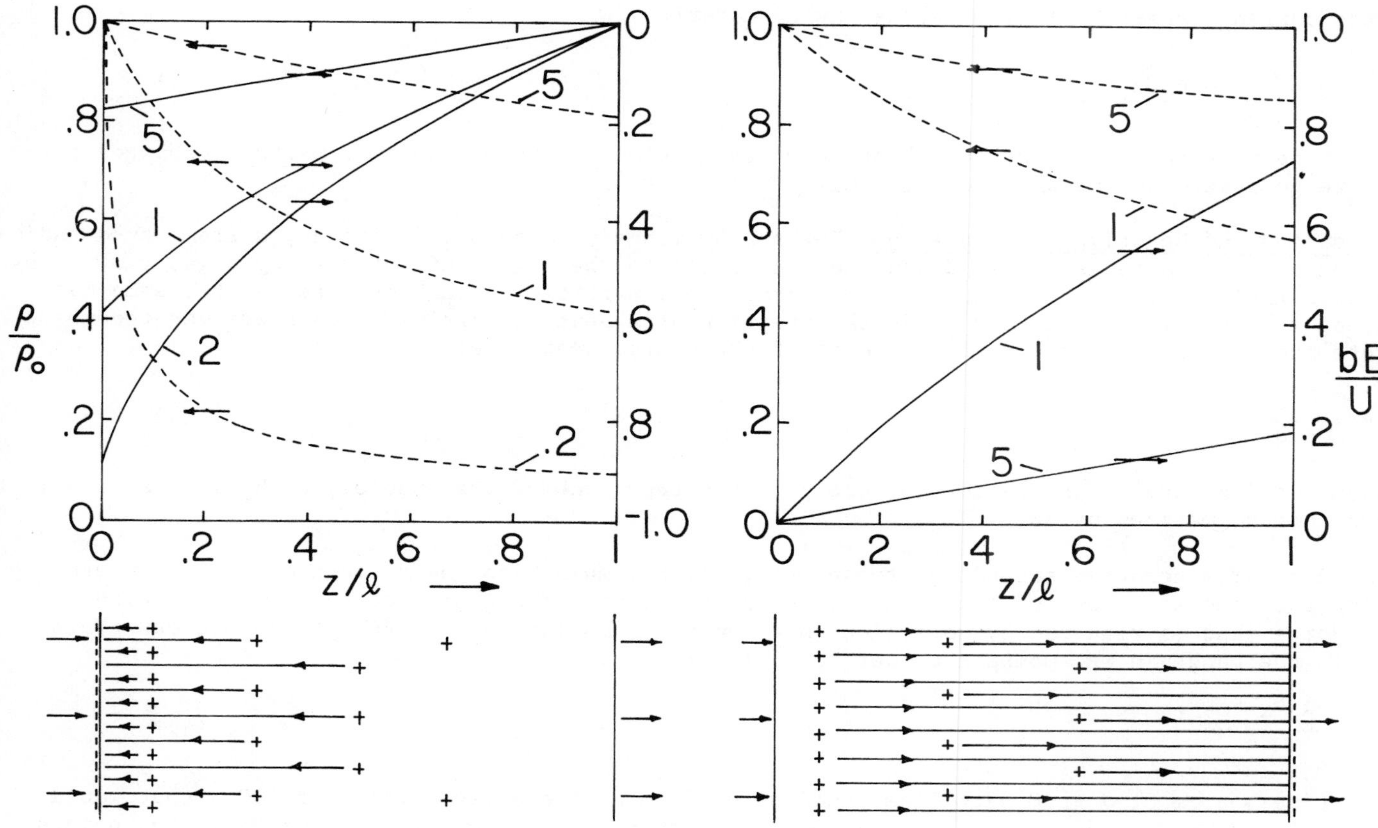

Fig. 5.7.7. Distribution of charge density and electric field intensity for generator and pump. The parameter $R_e \equiv (\varepsilon/b\rho_o)/(\ell/U)$ can be regarded as a normalized velocity.

electric Reynolds number

$$\frac{P_m}{P_e} = 3[2(1 + 2/R_e)^{1/2} + 1]^{-1} \tag{14}$$

The dependence on R_e is summarized in Fig. 5.7.6.

For both the generator and pump under these idealized conditions, the charge and electric field distributions are illustrated in Fig. 5.7.7. Because the relationship between i/i_o and R_e is determined by the operating conditions (Eq. 9 for the generator and $i/i_o = 1$ for the pump) the only parameter is R_e.

In this steady-state interaction, characteristics, emphasized in Sec. 5.6, still offer an alternative point of view. In the neighborhood of a given charged particle as it passes through the interaction region, the charge density must decay in accordance with Eq. 5.6.2. The spatial rate of decay shown in Fig. 5.7.7 decreases with increasing electric Reynolds number because the particle then spends less time in the interaction region.

5.8 Bipolar Migration with Space Charge

Common conduction phenomena involve more than one charge species. Media supporting one positive and one negative species are used here to illustrate interactions between carriers caused by space-charge fields, recombination and generation. The method of characteristics is further developed as a means of understanding the evolution of the charge distributions. Based on the bipolar model of this section, the limit of ohmic conduction is examined in the next section.

Each of the charge species is governed by a conservation equation taking the form of Eq. 5.2.3:

$$\frac{\partial \rho_\pm}{\partial t} + \nabla\cdot(\rho_\pm \vec{v} \pm \vec{J}'_\pm) = G_\pm - R_\pm \tag{1}$$

where the current density relative to the moving material is

$$\vec{J}'_\pm = b_\pm \rho_\pm \vec{E} \mp K_\pm \nabla\rho_\pm \tag{2}$$

Consider some physical situations to which these expressions pertain. Because pairs of charged particles are generated and recombined, $G_+ = G_- \equiv G$ and $R_+ = R_- \equiv R$.

Positive and Negative Ions in a Gas: Perhaps by means of a corona discharge, a flame or a radioactive source, ion pairs are created and then carried into the region of interest by a gas flow or by an electric field. With the proviso that the charge per particle of each species has the same magnitude, $q_\pm = q$, recombination results in the creation of a neutral particle. Carriers can recombine at a rate that is proportional to the product of the charge densities:[1]

$$R = \frac{\alpha \rho_+ \rho_-}{q} \tag{3}$$

One recombination results in the loss of one particle from each of the species, so $R_\pm$ is the same in the two equations summarized by Eq. 1.

At pressures somewhat exceeding atmospheric, the recombination coefficient α can be computed by picturing the process as one of oppositely charged particles being attracted to each other with a Coulomb force that is retarded by collisions between the ions and the neutral gas molecules. This results in the Langevin recombination coefficient:

$$\alpha = \frac{q(b_+ + b_-)}{\varepsilon_o} \tag{4}$$

A radioactive source of α or β particles could be used to create a generation term, $G_\pm$, that would then be dependent on the density of neutral particles at not only the point in question, but points in the gas between the radioactive source and the point of interest, since these could contribute to the slowing and hence final absorption of the ionizing particle.

---

[1] S. C. Brown, "Conduction of Electricity in Gases," in Handbook of Physics, E. U. Condon and H. Odishaw, eds., McGraw-Hill Book Company, New York, Toronto, London, 1958, pp. 4-166.

Aerosol Particles: Submicron particulate products of combustion are an example of macroscopic particles that often carry a natural charge of both signs. Self-agglomeration of overtly charged particles is also of interest in air pollution control.[2] In these cases, the charge per particle can be many electronic charges, and so electrically induced agglomeration of oppositely charged particles does not necessarily result in a neutral particle. Rather, with the assumption that the agglomeration is stable (the particles stick), yet another species of charged particles is created and the situation is generally much more complicated than can be described by the bipolar model. But, for a mixture of uniformly charged particles, the model applies with $G_{\pm} = 0$ and the self-agglomeration represented by the recombination term of Eq. 3.

Intrinsically Ionized Liquid: In liquids, thermal processes result in dissociation (ionization) of consituent molecules. For example, in pure water, a small fraction of the H_2O molecules disassociate into H^+ and OH^- ions. With these constituting the positive and negative species, there is a local thermal generation of ion pairs that is proportional to the number density, n, of neutral molecules:

$$G = \beta n \tag{5}$$

and a recombination rate given by Eq. 3 with $\varepsilon_o \rightarrow \varepsilon$. In the terminology of chemical kinetics, the recombination process would be regarded as a second order rate process.[3]

Partially Dissociated Salt in Solvent: When dissolved, materials such as NaCl or KCl tend to disassociate into positive and negative ions, Na^+Cl^- and K^+Cl^-. These then contribute to the conduction and, in this regard, can dominate over the intrinsic ionization. In that case, the conduction is represented in terms of just the two species, but it is also important to recognize that the unionized neutral molecules represent a third species. The number density, n, of this species is now, like the ion number densities n_+ and n_-, a function of space and time.

To describe the evolution of the neutral particles, a conservation equation is written much as for the ions, Eq. 1. However, because these particles are not charged, the only particle current density is due to diffusion. The migration term in Eq. 2 is absent. Also, generation of ion pairs now means that neutral particles are lost, and recombination means that neutrals are gained. Hence, terms on the right-hand side of the conservation equation are the negatives of those on the right in Eq. 1:

$$\frac{\partial n}{\partial t} + \nabla\cdot(n\vec{v} - K_D\nabla n) = -\frac{G}{q} + \frac{R}{q} \tag{6}$$

Summary of Governing Laws: Each of the illustrative situations that have been outlined can be described by deleting the inappropriate terms from the laws now summarized. The two charge densities contribute to Gauss' law:

$$\nabla\cdot\varepsilon\vec{E} = \rho_+ - \rho_- \tag{7}$$

where polarization is modeled as being linear and hence represented by the permittivity ε. In the following discussions, ε is taken as being uniform. The electric field is irrotational, and so

$$\vec{E} = -\nabla\Phi \tag{8}$$

With the understanding that the given material deformations are incompressible (that $\nabla\cdot\vec{v} = 0$), the carrier evolutions are represented by Eqs. 1 and 2, which in view of Gauss' law, Eq. 7, combine to become the two equations

$$\frac{\partial\rho_\pm}{\partial t} + (\vec{v} \pm b_\pm\vec{E})\cdot\nabla\rho_\pm = \mp\rho_\pm b_\pm\frac{(\rho_+ - \rho_-)}{\varepsilon} + \beta n - \frac{\alpha\rho_+\rho_-}{q} + K_\pm\nabla^2\rho_\pm \tag{9}$$

Here, Eqs. 3 and 5 are used to represent the recombination and generation. If n is a constant, or the generation term is absent, then the law governing the neutrals is not required; but if the neutral evolution is also part of the story, then Eq. 6 is added to the list:

$$\frac{\partial n}{\partial t} + \vec{v}\cdot\nabla n = -\frac{\beta n}{q} + \frac{\alpha}{q^2}\rho_+\rho_- + K_D\nabla^2 n \tag{10}$$

Equations 7 - 10 constitute one vector and 4 scalar equations in the unknowns $\vec{E}$, Φ, ρ_+, ρ_- and n.

---

2. J. R. Melcher, K. S. Sachar and E. P. Warren, "Overview of Electrostatic Devices for Control of Submicrometer Particles," Proc. IEEE 65, 1659 (1977).

3. K. J. Laidler, Chemical Kinetics, McGraw-Hill Book Company, New York, 1965, p. 535.

Characteristic Equations: With the understanding that lengths of interest are large enough to justify ignoring the diffusion contributions to Eqs. 9 and 10 (typically, the ratio given by Eq. 5.2.12 is small), Eqs. 9 can be written in the characteristic form introduced in Sec. 5.3:

$$\frac{d\rho_\pm}{dt} = \mp\rho_\pm b_\pm \frac{(\rho_+ - \rho_-)}{\varepsilon} + \beta n - \frac{\alpha}{q}\rho_+\rho_- \tag{11}$$

Here, the time rate of change is measured by an observer moving respectively with the $\pm$ ions, on the characteristic lines

$$\frac{d\vec{r}}{dt} = \vec{v} \pm b_\pm \vec{E} \tag{12}$$

Similarly, Eq. 10 becomes

$$\frac{dn}{dt} = -\frac{\beta n}{q} + \frac{\alpha}{q^2}\rho_+\rho_- \tag{13}$$

on the characteristic lines that are physically the particle lines for the neutrals

$$\frac{d\vec{r}}{dt} = \vec{v} \tag{14}$$

Following a particle of the neutral material, the neutral number density changes with time in accordance with the local balance between generation and recombination. What makes the bipolar situation more complex than for unipolar migration is that not only are the positive and negative species described by Eqs. 11 along different characteristic lines, but the space-charge term on the right has an effect that is proportional to the net charge, generally with contributions from both species.

One-Dimensional Characteristic Equations: Consider the one-dimensional configuration, illustrated by Fig. 5.8.1, in which densities and fields are independent of (y,z), with $\vec{E} = E(x,t)\vec{i}_x$ and $\vec{v} = U(t)\vec{i}_x$. Because $\vec{v}$ is solenoidal, U is at most a function of time only. Then, Eqs. 11-14 reduce to the first six ordinary differential equations summarized by Eq. 15:

$$\frac{d}{dt}\begin{bmatrix} \rho_+ \\ \rho_- \\ x_+ \\ x_- \\ n \\ x_n \\ E_+ \\ E_- \\ E_n \end{bmatrix} = \begin{bmatrix} -\frac{b_+}{\varepsilon}\rho_+(\rho_+ - \rho_-) + \beta n - \frac{\alpha}{q}\rho_+\rho_- \\ \frac{b_-}{\varepsilon}\rho_-(\rho_+ - \rho_-) + \beta n - \frac{\alpha}{q}\rho_+\rho_- \\ U + b_+E_+ \\ U - b_-E_- \\ -\frac{\beta}{q}n + \frac{\alpha}{q^2}\rho_+\rho_- \\ U \\ -\frac{(b_+ + b_-)}{\varepsilon}\rho_-E_+ + \frac{C(t)}{\varepsilon} \\ -\frac{(b_+ + b_-)}{\varepsilon}\rho_+E_- + \frac{C(t)}{\varepsilon} \\ -\frac{1}{\varepsilon}(b_+\rho_+ + b_-\rho_-)E_n + \frac{C(t)}{\varepsilon} \end{bmatrix} \tag{15}$$

where

$$\frac{C(t)}{\varepsilon} \equiv \frac{1}{d}\left\{\frac{dv}{dt} + \frac{1}{\varepsilon}\int_0^d [\rho_+(U + b_+E) - \rho_-(U - b_-E)]dx\right\}$$

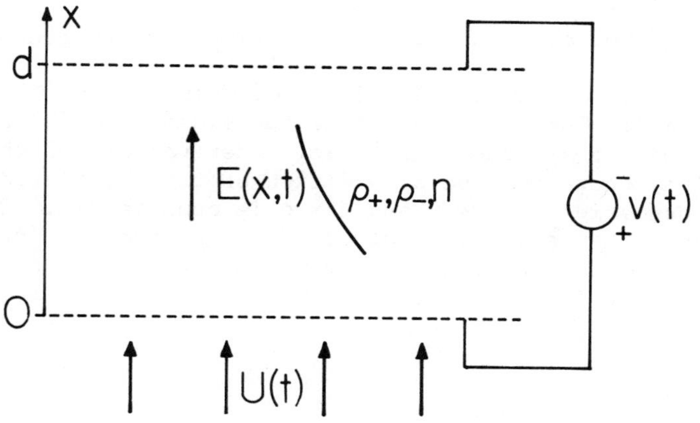

Fig. 5.8.1. One-dimensional bipolar migration configuration.

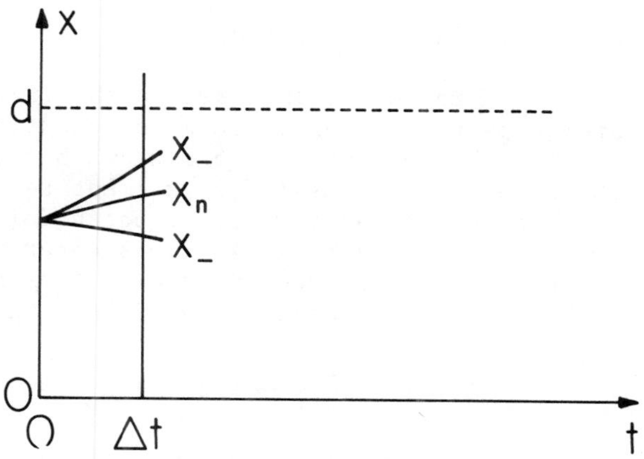

Fig. 5.8.2. Characteristic lines in (x-t) plane.

Here, subscripts are used to distinguish the characteristic lines. Thus the first two equations respectively apply along lines in the (x-t) plane represented respectively by the third and fourth expressions. Similarly, the fifth equation applies along the lines defined by the sixth expression.

In numerically integrating these equations it is convenient to take account of Gauss' law, Eq. 7, by having equations for the time rates of change of the electric field for an observer moving along each of the respective characteristic lines.[4] To this end, the time rate of change of Eq. 7 is written as

$$\frac{\partial}{\partial x}\left(\frac{\partial E}{\partial t}\right) = \frac{1}{\varepsilon}\frac{\partial}{\partial t}(\rho_+ - \rho_-) \tag{16}$$

The difference between Eqs. 9 becomes

$$\frac{\partial}{\partial t}(\rho_+ - \rho_-) + \frac{\partial}{\partial x}[\rho_+(U + b_+E) - \rho_-(U - b_-E)] = 0 \tag{17}$$

Elimination of the term in $\rho_+ - \rho_-$ between these equations leads to the conclusion that

$$\frac{\partial}{\partial x}\left\{\varepsilon\frac{\partial E}{\partial t} + [\rho_+(U + b_+E) - \rho_-(U - b_-E)]\right\} = 0 \tag{18}$$

The quantity in brackets, the sum of the displacement current and the migration currents, is defined as C(t). Integration of C(t) from x = 0 to x = d results in the expression given with Eq. 15. The voltage v, defined as the integral of E between the planes x = 0 and x = ℓ, brings in the remaining field law, Eq. 8.

Gauss' law can be used to eliminate the net charge $\rho_+ - \rho_-$ from C(t), the quantity in brackets in Eq. 18, to obtain

$$\frac{\partial E}{\partial t} + U\frac{\partial E}{\partial x} = \frac{C(t)}{\varepsilon} - \frac{1}{\varepsilon}(b_+\rho_+ - b_-\rho_-)E \tag{19}$$

What is on the left is the time rate of change of E for an observer moving on the neutral characteristic lines. Thus, Eq. 19 is the last of Eqs. 15. To obtain the time rates of change of E on the charged particle characteristic lines, add to both sides of Eq. 19 $\pm b_+E\partial E/\partial x$. On the left is then the time rate of change of E for an observer moving on the respective characteristic lines $x_\pm$. With Gauss' law used to replace $\partial E/\partial x$ on the right with $(\rho_+ - \rho_-)/\varepsilon$, these equations become the seventh and eighth expressions of Eq. 15.

The functions $E_+(t)$, $E_-(t)$ and $E_n(t)$ are numerically the same as E(x,t). Each is now regarded as solely a function of time because it is understood that the respective functions are measured by an observer moving along the lines $x_+(t)$, $x_-(t)$, and $x_n(t)$, respectively.

Numerical Solution: A beauty of the method of characteristics is that it reduces partial differential equations to a system of ordinary differential equations, Eqs. 15. Many numerical techniques

4. M. Zahn, "Transient Electric Field and Space Charge Behavior for Drift Dominated Bipolar Conduction," in Conduction and Breakdown in Dielectric Liquids, J. M. Goldschvartz, ed., Delft University Press, 1975, pp. 61-64.

exist for integrating nonlinear equations in this form, e.g., Runge-Kutta or predictor-corrector.[5]

The region of interest in the (x-t) plane is bounded by x = 0 and x = d, where screen electrodes are respectively constrained to potential v(t) and 0. These planes define two sides of a "U" shaped region, sketched in Fig. 5.8.2, with the initial line t = 0 the third side. Wherever one of the characteristic lines (x_+, x_-, x_n) enters this region, there must be a condition on the associated density ρ_+, ρ_- or n. In addition, the potential of the boundaries at x = 0 and x = d is constrained. Thus, when t = 0, characteristic lines enter the region with the initial values of ρ_+, ρ_- and n. Taken with the constraint on the potential difference between the screens, this determines the initial distribution of E(x,0), because at any time, Gauss' law can be integrated to obtain

$$E(x) = \int_o^x \frac{(\rho_+ - \rho_-)}{\varepsilon} dx' + E_o \tag{20}$$

The constant of integration, E_o, is determined by integrating E from x = 0 to x = d and requiring that the result be v. If the resulting value of E_o is substituted back into Eq. 20, an expression is obtained for E(x,t) in terms of ρ_+ and ρ_- when t = t:

$$E(x,t) = \int_o^x \frac{(\rho_+ - \rho_-)}{\varepsilon} dx' - \frac{1}{d}\int_o^d dx \int_o^x \frac{(\rho_+ - \rho_-)}{\varepsilon} dx' + \frac{v}{d} \tag{21}$$

With the initial values of all quantities on the right in Eq. 15 established, it is now possible to begin marching forward in time.

In the integration scheme used to generate the distributions shown, a predictor-corrector subroutine is used which calls a user-written subroutine for evaluation of derivatives (Eqs. 15) after each prediction or correction step. Because Eqs. 15 are a set of coupled ordinary nonlinear differential equations, there are readily available routines for carrying out the main integration (compiled subroutines for predictor-corrector integration are available, for example, in the International Mathematical & Statistical Library).

Note that the derivatives are not entirely determined by quantities naturally evaluated on the same characteristic line. For example, $d\rho_+/dt$ is determined by not only ρ_+, but by ρ_- and n as well, and these quantities are naturally found along their respective characteristic lines. If the distance d is broken into (i - 1) segments, there are (i) characteristic lines of each family emanating from the t=0 line into the region of interest. Equations 15 comprise (9i) coupled ordinary differential equations. The equations for values on a (+) characteristic line are coupled to those on neighboring characteristic lines by Eqs. 15b and 15e, and coupled to all the other characteristic lines thru C(t). Thus, at each step in time, values of ρ_- and n on the x_+ characteristic must be interpolated from values on the neighboring characteristics x_- and x_n. Similarly, values of ρ_+ and n must be interpolated from their respective characteristic lines onto the x_- lines for use in the equation for ρ_-, and values of ρ_+ and ρ_- must be interpolated onto the neutral characteristics in order to compute dn/dt. The interpolation for the examples illustrated here are done with a four-point Lagrangian formula. This fits a cubic equation to the nearest two data points on both sides of the interpolation point.

The charge and neutral density profiles are conveniently initiated with step singularities. In order to prevent the smearing out of these step edges, a two-point (linear) interpolation is used when near these edges, so that the data on one side of the edge does not influence the interpolated values on the other side.

The integration in C(t) is carried out in two parts: the $\rho_+(U_+ + b_+E)$ term is integrated over the (irregular) set of x_+ points using the readily available values of ρ_+ and E_+ on these points, and the $\rho_-(U - b_-E)$ term is similarly integrated over the x_- points.

Numerical Example: (The numerical analysis of this section was carried out by R. S. Withers) A situation which is the basis for gaining physical insights in this and the next section is sketched in Fig. 5.8.3. When t = 0, equal amounts of positive and negative charge uniformly occupy the region next to the lower screen, with the region above initially free of charge. Initially, neutral particles are absent throughout and there is no generation at any time. Because the effect of the convection in one dimension is to translate the material in the x direction, the material velocity U is taken as zero. Hence, the model is appropriate to describing what might be considered a "conducting layer" adjacent to an insulating layer of material sandwiched between plane-parallel electrodes. It is assumed that charged particles leaving the region by arriving at one or the other of the electrodes are neutralized and removed from the volume. Further, charged particles cannot be generated at the electrode

---

5. F. S. Acton, Numerical Methods that Work, Harper & Row, Publishers, New York, 1970.

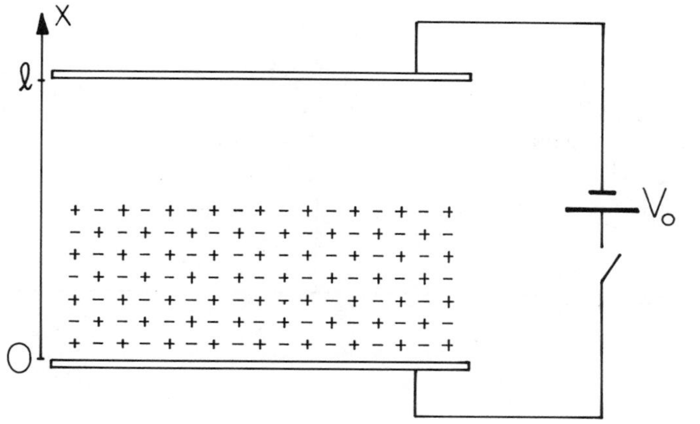

Fig. 5.8.3

When t=0, voltage is applied to plates. Initially, lower half of region between is filled with equal densities of positive and negative charges. Figure 5.8.4 shows evolution with time if there is no generation. Figure 5.9.3 illustrates what happens with generation.

surfaces, and hence characteristic lines emanating from the electrodes carry no associated particle density.

The evolution of electric field and net charge are displayed in Fig. 5.8.4a, where the x-t plane forms the "floor." Similarly, the x-t dependence of the particle densities is shown in Fig. 5.8.4b. The critical characteristic lines, x_+ and x_-, are also shown in these plots. (The neutral characteristics, x_n, are simply lines running parallel to the t axis.)

Considerable insight can be extracted from this example by identifying the dominant processes in each of the regions demarked in Fig. 5.8.5 by critical characteristic lines. In region I, bounded by x_+ originating at the lower electrode and x_- originating at the initial interface between the charged layer and the region above, the initial charge densities at A, A', and A" are the same. It follows that these initial points can be chosen such that B and B' occur at the same time (on the same line t = constant). Also, the initial conditions are the same so the values of ρ_+ and of ρ_- at the points B and B' are the same. In turn, the value of the charge densities at C are the same as at other positions in region I at this same time. It is concluded that Eqs. 15a and 15b describe the time dependence at any given fixed location x in region I. In the example, initial conditions set $\rho_+ = \rho_-$ and these equations reduce to the same equation for subsequent times. Thus, the net charge density $\rho_+ - \rho_-$ is zero in region I, and, through recombination alone, the individual charge species decay according to the law

$$\rho_+ = \frac{\rho_o}{1 + t/\tau}\ ;\ \tau = \frac{\varepsilon_o}{\rho_o(b_+ + b_-)} \tag{22}$$

where Eq. 5.8.4, the Langevin recombination coefficient, has been used.

Because there is no generation, the recombination simply feeds the neutral equation, and Eq. 15e shows that in region I

$$n = \int_o^t \frac{\rho_o}{q}\,\frac{d(t'/\tau)}{(1 + t'/\tau)^2} = \frac{\rho_o}{q}\,\frac{t/\tau}{(1 + t/\tau)} \tag{23}$$

Region II, like region I, has uniform initial conditions, so the same arguments apply. But, the initial conditions on (ρ_+, ρ_-, n) are all zero, and so these quantities remain zero throughout. It follows from the characteristic electric field equations, Eqs. 15g-15j, that E is uniform in this region.

In region III, the x_+ characteristics enter from the lower electrode carrying no ρ_+. At a point like D, Eq. 15a establishes $\rho_+ = 0$, and a step-by-step march into this region shows that at each point $\rho_+ = 0$. Hence, Eq. 15b applies with $\rho_+ = 0$ and it is concluded that along x_- in this region the charge evolution is as though the process were the unipolar self-precipitation process discussed in Sec. 5.6. Because there are only negative charges, there is no recombination.

Region IV, where the positive charges are moving upward along the E lines but the negative charges have been swept downward, is of essentially the unipolar character of region III. Because charges do not originate on the upper electrode, region VII is also unipolar.

Finally, it can be argued that in regions V, VI and VIII, $\rho_+ = 0$ and $\rho_- = 0$.

The neutral characteristics, x_n, do not enter into the classification of regimes because the coupling to n is "one-way." But, neutrals created by recombination remain behind the x_- wavefront defining the demarcation of regions IV and I. As a result, the distribution of n at a given time is uniform in region I (with amplitude given by Eq. 23) and makes a smooth transition to zero at the

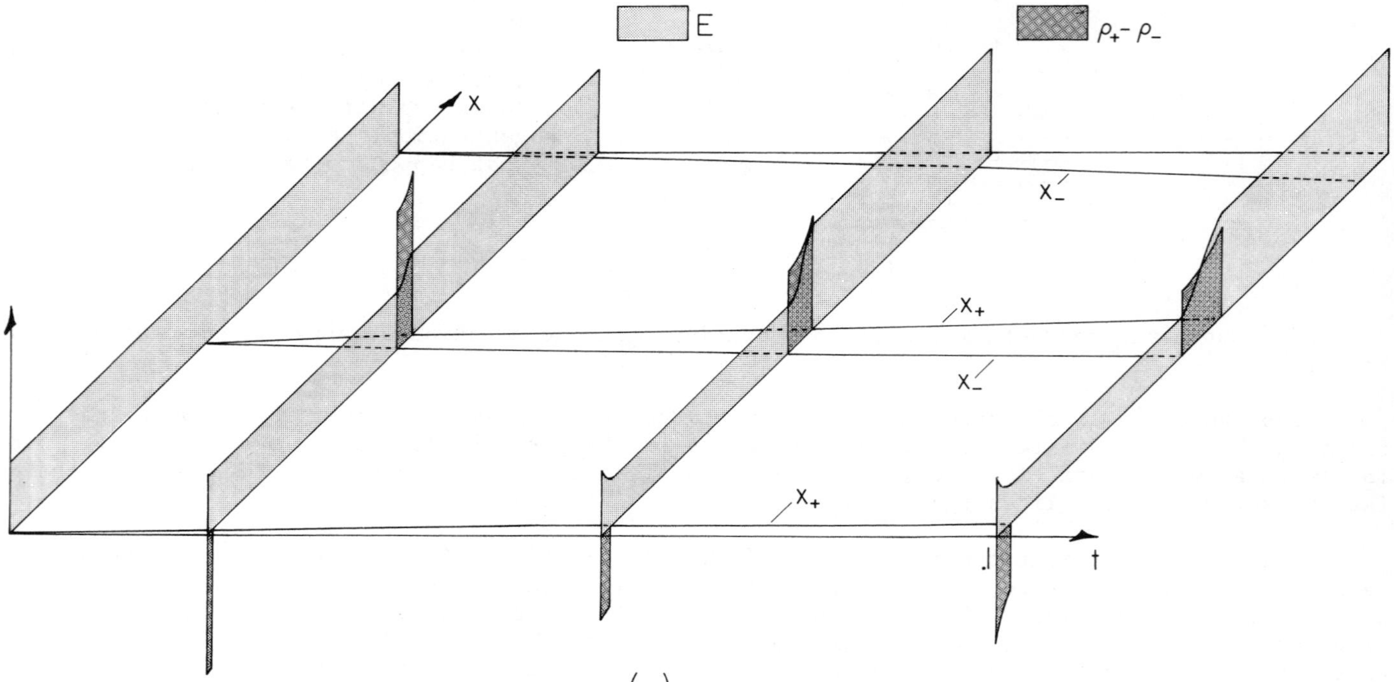

(a)

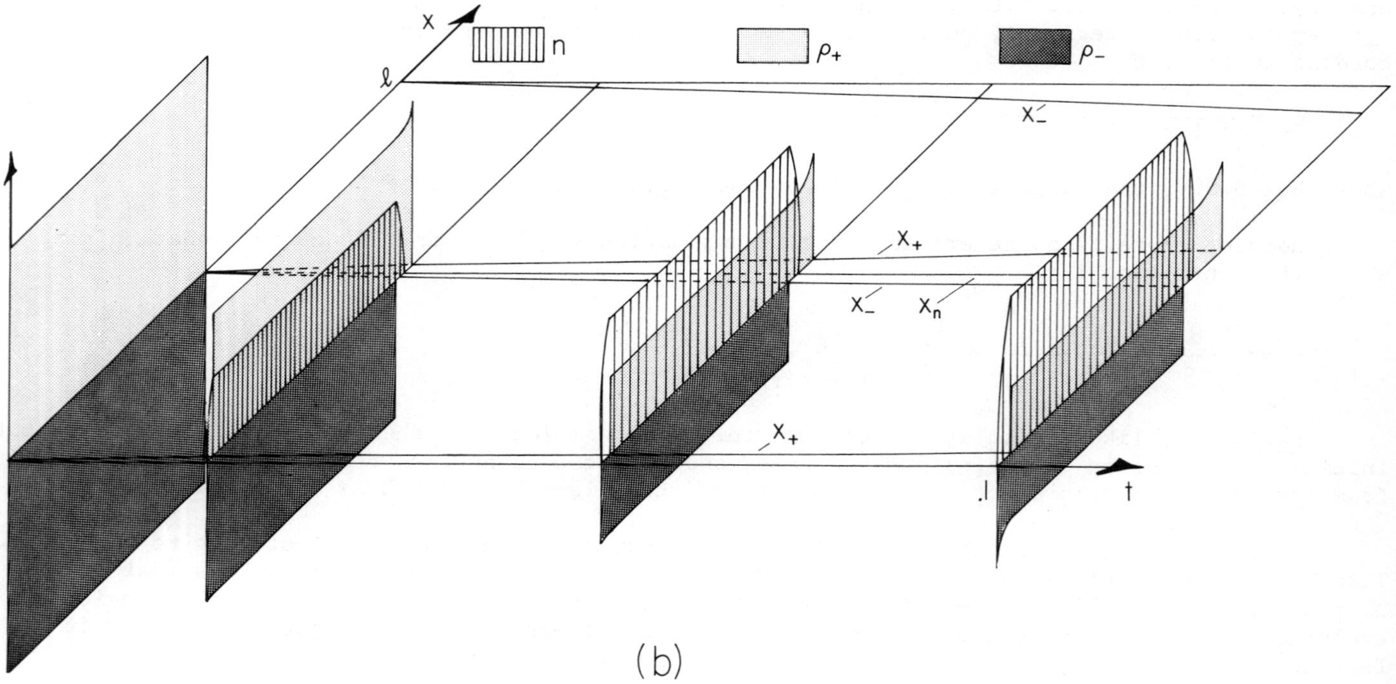

(b)

Fig. 5.8.4. Evolution of layer composed of equal densities of positive and negative carriers occupying lower half of region between capacitor plates. Initially there are no neutrals. Generation is absent ($\beta = 0$) so recombination results in neutrals. For the case shown, $t = \underline{t}[\ell/\mathscr{E}(b_+ + b_-)]$, where $\mathscr{E} \equiv V_o/\ell$. Also, $b_+ = b_-$ and the initial charge densities are such that $\rho_+ = \rho_- = 30(\varepsilon V_o/\ell^2)$. (a) Electric field and net charge density; (b) neutral density and positive and negative charge densities.

initial location of the region IV-I interface.

At a given instant, the net charge in region IV increases with x. The reason for this is apparent from following the x_+ characteristic originating at A in Fig. 5.8.5. At first, ρ_+ decays by recombination, with a time constant $\tau = \varepsilon_o/\rho_o(b_+ + b_-)$, until x_+ passes from regime I to region IV. The subsequent decay is due to self-precipitation and occurs with the larger (slower) unipolar time constant $\varepsilon_o/\rho_o b_+$. Thus, at G in Fig. 5.8.5, particles have spent more time in the unipolar regime and less time in the recombination regime than those at G'. This is why, at a given instant in Fig. 5.8.4a, the net charge at the x_+ wavefront of the regime IV (which is decaying at the unipolar rate) is greater than behind the front. (Note, however, that the ρ_o used in evaluating this latter time constant is the value of ρ_+ when the characteristic line enters region IV, which is not the same for points G and G'.)

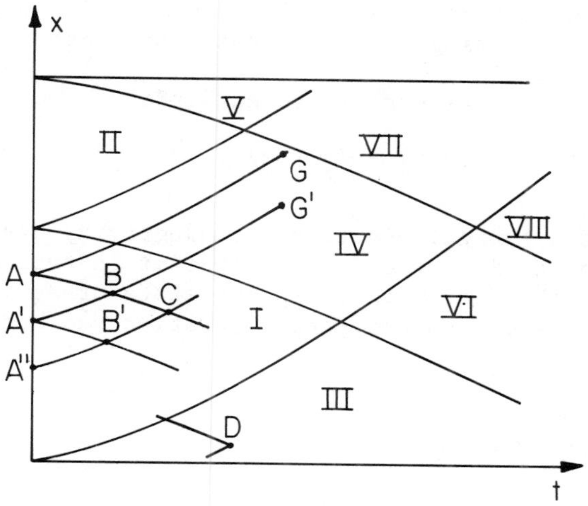

Fig. 5.8.5. Regions are delimited by x_+ and x_- characteristics emanating from interface and electrodes.

5.9 Conductivity and Net Charge Evolution with Generation and Recombination: Ohmic Limit

The net free charge density and conductivity for the bipolar systems treated in Sec. 5.8, defined as

$$\rho_f = \rho_+ - \rho_-;\ \sigma = b_+\rho_+ + b_-\rho_- \tag{1}$$

are natural variables for understanding the relationship between charge migration and relaxation. In terms of (ρ_f,σ), the charge densities ρ_+ and ρ_- are found by inverting Eqs. 1:

$$\rho_\pm = \frac{\sigma \pm b_\mp\rho_f}{b_+ + b_-} \tag{2}$$

With the objective of casting the charge evolution in terms of ρ_f and σ, the difference is taken between the conservation equations for + and - species, Eqs. 5.8.9, and ρ_+ and ρ_- are replaced on the right using Eqs. 2:

$$\frac{D\rho_f}{Dt} = -\vec{E}\cdot\nabla\sigma - \frac{\sigma\rho_f}{\varepsilon} + \left(\frac{K_+ - K_-}{b_+ + b_-}\right)\nabla^2\sigma + \left(\frac{K_+b_- + K_-b_+}{b_+ + b_-}\right)\nabla^2\rho_f \tag{3}$$

To similarly obtain an expression for σ, Eqs. 5.8.9 are respectively multiplied by $b_\pm$ and summed to obtain

$$\frac{D\sigma}{Dt} = -\vec{E}\cdot\nabla[(b_+-b_-)\sigma + b_+b_-\rho_f]-[(b_+-b_-)\sigma + b_+b_-\rho_f]\frac{\rho_f}{\varepsilon} + (b_++b_-)\beta n$$
$$- \frac{\alpha}{q(b_++b_-)}[\sigma^2-(b_+-b_-)\sigma\rho_f - b_+b_-\rho_f^2] + \left(\frac{K_+b_++K_-b_-}{b_++b_-}\right)\nabla^2\sigma + \frac{b_+b_-}{b_++b_-}(K_+-K_-)\nabla^2\rho_f \tag{4}$$

To complete the description, Eqs. 2 are used to write Eq. 10 as

$$\frac{Dn}{Dt} = \frac{-\beta}{q}n + \frac{\alpha}{q^2(b_++b_-)^2}[\sigma^2 - (b_+-b_-)\sigma\rho_f - b_+b_-\rho_f^2] + K_D\nabla^2 n \tag{5}$$

These last three expressions are an alternative to Eqs. 5.8.9 and 5.8.10 in describing the migration and diffusion of the carriers in a deforming material. The method of characteristics could be used to solve these expressions, much as illustrated in Sec. 5.8. But the objective in this section is to identify the rate processes encapsulated by these laws and hence to discern the dominant contributions to the equations. Limiting forms of the equations, for example the ohmic model emphasized here and used in the remainder of this chapter, are necessary if the conduction laws are to be embodied in models that bring in still other dynamical processes.

The approach now used is similar to that introduced in Sec. 2.3, where the quasistatic limits of the electrodynamic laws are recognized by using a normalization of the laws to discern the critical characteristic times. Given that dynamical times of interest are characterized by τ, what are the times characterizing the processes represented by Eqs. 3-5?

Variables are normalized such that

$$t = \underline{t}\tau, \quad (x,y,z) = (\underline{x},\underline{y},\underline{z})\ell, \quad \vec{v} = \underline{\vec{v}}\ell/\tau, \quad \sigma = \underline{\sigma}\Sigma, \quad \vec{E} = \underline{\vec{E}}\mathscr{E}, \quad \rho_f = \underline{\rho}_f \varepsilon \mathscr{E}/\ell \tag{6}$$

Thus, Σ is a typical electrical conductivity and $\mathscr{E}$ is a typical electric field intensity. The free charge density is normalized so that it is typically the charge density that would "shield out" the field $\mathscr{E}$ in the distance ℓ. In the state of equilibrium where the charge density is zero, while σ and n are uniform and constant, the generation and recombination terms balance. Thus, at each point

$$\frac{\beta}{q} = \frac{\alpha\sigma^2}{q^2(b_+ + b_-)^2 n} \tag{7}$$

This expression makes it possible to use equilibriun data to evaluate the generation coefficient, given the parameters on the right. It also suggests that the neutral number density be normalized such that

$$n = \underline{n}\,\frac{\alpha\Sigma^2}{q(b_+ + b_-)^2\beta} \tag{8}$$

Introduction of these normalizations into Eqs. 3-5 results in the expressions

$$\frac{D\rho_f}{Dt} = \frac{\tau}{\tau_e}(-\vec{E}\cdot\nabla\sigma - \sigma\rho_f) + \frac{\tau}{\tau_e}\frac{\tau_{mig}}{\tau_D}\frac{(K_+-K_-)(b_++b_-)}{(K_+b_-+K_-b_+)}\nabla^2\sigma + \frac{\tau}{\tau_D}\nabla^2\rho_f \tag{\underline{9}}$$

$$\begin{aligned}\frac{D\sigma}{Dt} = &-\frac{\tau}{\tau_{mig}}\frac{(b_+-b_-)}{(b_++b_-)}(\vec{E}\cdot\nabla\sigma + \sigma\rho_f) - \frac{\tau}{\tau_{mig}}\frac{\tau_e}{\tau_{mig}}\frac{b_+b_-}{(b_++b_-)^2}(\vec{E}\cdot\nabla\rho_f + \rho_f^2)\\
&+ \frac{\tau}{\tau_e}\left[\frac{\varepsilon\alpha}{(b_++b_-)q}\right]n - \frac{\tau}{\tau_e}\left[\frac{\varepsilon\alpha}{(b_++b_-)q}\right]\sigma^2 + \frac{\tau}{\tau_{mig}}\left[\frac{\varepsilon\alpha}{(b_++b_-)q}\right]\left(\frac{b_+-b_-}{b_++b_-}\right)\sigma\rho_f\\
&+ \frac{\tau}{\tau_{mig}}\frac{\tau_e}{\tau_{mig}}\left[\frac{\varepsilon\alpha}{(b_++b_-)q}\right]\frac{b_+b_-}{(b_++b_-)^2}\rho_f^2 + \frac{(K_+b_++K_-b_-)}{(K_+b_-+K_-b_+)}\nabla^2\sigma\\
&+ \frac{\tau}{\tau_{mig}}\frac{\tau_e}{\tau_D}\frac{b_+b_-}{(b_++b_-)^2}\frac{(K_+-K_-)(b_++b_-)}{(K_+b_-+K_-b_+)}\nabla^2\rho_f\end{aligned} \tag{\underline{10}}$$

$$\frac{Dn}{Dt} = -\frac{\tau}{\tau_{th}}n + \frac{\tau}{\tau_{th}}\left[\underline{\sigma}^2 - \frac{(b_+-b_-)}{(b_++b_-)}\frac{\tau_e}{\tau_{mig}}\sigma\rho_f - \frac{b_+b_-}{(b_++b_-)^2}\left(\frac{\tau_e}{\tau_{mig}}\right)^2\rho_f^2\right] + \frac{\tau}{\tau_D}\frac{K_D(b_++b_-)}{(K_+b_-+K_-b_+)}\nabla^2 n \tag{\underline{11}}$$

where the following characteristic times have been identified

$$\tau_e \equiv \frac{\varepsilon}{\Sigma}, \quad \tau_{mig} \equiv \frac{\ell}{\mathscr{E}(b_++b_-)}, \quad \tau_D \equiv \frac{\ell^2}{\left[\frac{K_+b_- + K_-b_+}{b_++b_-}\right]}, \quad \tau_{th} = \frac{q}{\beta} \tag{12}$$

The other dimensionless coefficients in Eqs. 9-11 are typically of the order of unity. (Note that at least for Langevin recombination, where α is given by Eq. 5.8.4, the coefficient $\varepsilon\alpha/(b_++b_-)q$ is unity).

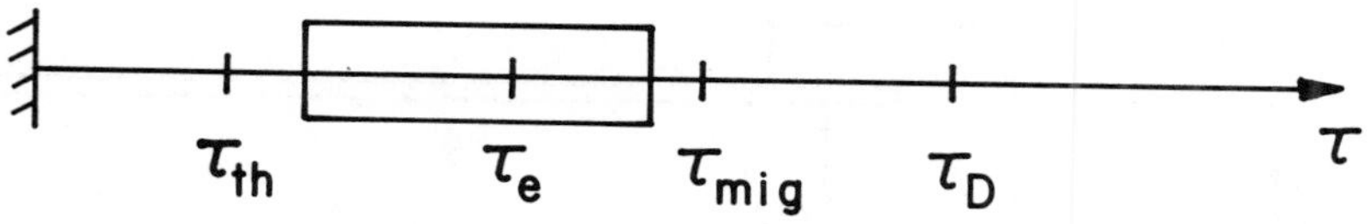

Fig. 5.9.1. Hierarchy of characteristic times and range of dynamical times appropriate to the use of an ohmic model.

With the objective of ordering the time constants of Eq. 13, τ_{th} is estimated by substituting the equilibrium values given by Eq. 7, α from Eq. 5.8.4 and $\sigma = (n_+b_+ + n_-b_-)q$:

$$\tau_{th} \equiv \frac{q}{\beta} \simeq \frac{q^2(b_+ + b_-)^2 n}{\alpha\sigma^2} = \tau_e \frac{n}{(n_+b_+ + n_-b_-)/(b_+ + b_-)} \tag{13}$$

Thus, τ_{th} is essentially τ_e multiplied by the ratio of the neutral to the charged particles. If β is large enough that essentially all of the particles available are ionized, then τ_{th} is a small fraction of the charge relaxation time τ_e.

The ordering of characteristic times shown in Fig. 5.9.1 is typical if a configuration is to be appropriately modeled as "ohmic." Because lengths of interest are relatively large, the diffusion time is extremely long. That the migration time τ_{mig} is also long compared to times of interest is also a matter of the length scale of interest, and is justified if the typical electric field intensities are not too large. Times of interest in the ohmic model are arbitrary relative to τ_e. They can be long or short compared to the charge relaxation time.

With the understanding that the equations are valid for processes in this dynamic range, Eqs. 9-11 are approximated by

$$\frac{D\rho_f}{Dt} = \frac{\tau}{\tau_e}(-\vec{E}\cdot\nabla\sigma - \rho_f\sigma) \tag{14}$$

$$\frac{D\sigma}{Dt} = \frac{\tau}{\tau_e}\left[\frac{\varepsilon\alpha}{(b_+ + b_-)q}\right](n - \sigma^2) \tag{15}$$

$$\frac{Dn}{Dt} = -\frac{\tau}{\tau_{th}}(n - \sigma^2) \tag{16}$$

By multiplying Eq. 16 by $(\tau_{th}/\tau_e)[\varepsilon\alpha/(b_+ + b_-)q]$ and adding it to Eq. 15, it follows that

$$\frac{D}{Dt}\left[\sigma + \frac{\tau_{th}}{\tau_e}\frac{\varepsilon\alpha}{(b_+ + b_-)q}n\right] = 0 \tag{17}$$

Now, if τ_{th} is short compared to times of interest, as depicted by Fig. 5.9.1, this expression becomes (with variables written in dimensional form),

$$\frac{D\sigma}{Dt} = 0 \tag{18}$$

For an observer attached to a given particle of the material, the conductivity is constant. In this limit, the conductivity can be regarded as a property of the material.

In unnormalized form, Eq. 14 is

$$\frac{D\rho_f}{Dt} = -\frac{\rho_f}{(\varepsilon/\sigma)} - \vec{E}\cdot\nabla\sigma \tag{19}$$

In this charge relaxation expression, σ can now be regarded as a given parameter. These last two expressions constitute the "ohmic" model.

Maxwell's Capacitor: In terms of an ohmic model, the bipolar migration with generation and recombination is the two-region lossy capacitor of Fig. 5.9.2. The lower region is a fixed material, which according to Eq. 18 conserves its initially uniform conductivity. The upper region is of the same permittivity, but is insulating.

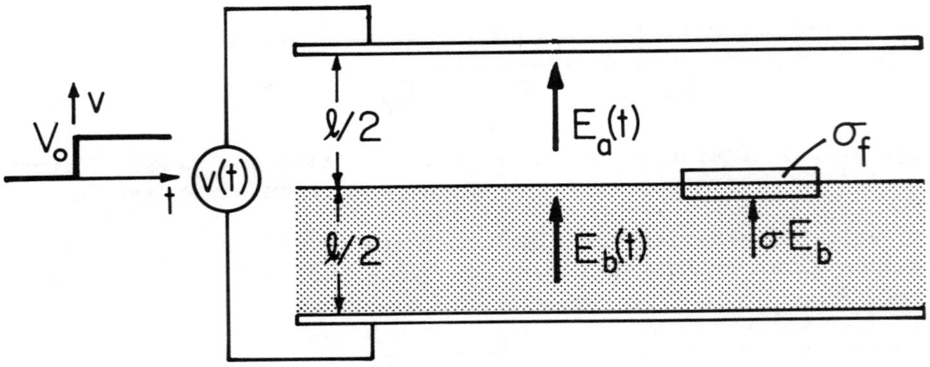

Fig. 5.9.2. Maxwell capacitor model for bipolar migration of Fig. 5.8.6.

As will be shown in the next section, with the application of a constant voltage V_o to the electrodes, there is never a net free charge density in the material. Hence, fields in each region are uniform, $E_a(t)$ and $E_b(t)$. Because of the voltage constraint,

$$E_a \frac{\ell}{2} + E_b \frac{\ell}{2} = v \tag{20}$$

Accumulation of surface charge $\sigma_f = \varepsilon E_a - \varepsilon E_b$ at the interface between the lossy material and the insulating upper region is caused by the conduction current σE_b feeding the interface. (This boundary condition is considered in general terms in Sec. 5.11). Thus,

$$\frac{d}{dt}(\varepsilon E_a - \varepsilon E_b) = \sigma E_b \tag{21}$$

These two expressions combine to give a differential equation for the field inside the lossy material with the applied voltage as a drive:

$$\frac{dE_b}{dt} + \frac{\sigma}{2\varepsilon} E_b = \frac{1}{\ell}\frac{dv}{dt}; \quad E_a = \frac{2v}{\ell} - E_b \tag{22}$$

It follows that the transient resulting from the application of a step in voltage to the amplitude V_o is

$$E_b = \frac{V_o}{\ell} e^{-t/\tau_e}; \quad \sigma_f = \frac{2\varepsilon V_o}{\ell}[1 - e^{-t/\tau_e}]; \quad \tau_e \equiv 2\varepsilon/\sigma \tag{23}$$

Numerical Example: (The numerical analysis of this section was carried out by R. S. Withers) Now, by comparing the predictions of the ohmic model to the "exact" solution afforded by the numerical scheme described in Sec. 5.8, consider the response of the Maxwell capacitor to a step in applied voltage. The configuration, shown in Fig. 5.8.3, is initially with the lower half of the region between the electrodes uniformly filled with positive and negative charge densities. In this lower region, generation and recombination are initially in equilibrium, as represented by Eq. 7. Thus, there is also an initial uniform distribution of n in the lower region.

With parameters arranged so that the characteristic times have the ordering shown in Fig. 5.9.1, the response to a step in applied voltage is displayed by Fig. 5.9.3. As would be expected from the ohmic Maxwell capacitor model, the electric field in the conducting region, shown by Fig. 5.9.3a, decays exponentially with the time constant τ_e, while the surface charge "density" builds up with a similar time constant (Eqs. 23).

Figure 5.9.4 identifies some of the regions demarked by the three families of characteristics, particularly those emanating from the initial position of the interface. Regions I and IV are described by the Maxwell capacitor model. This means that the electric field on the demarking characteristics x_+ and x_- is known. For example, on x_-, $\vec{E}$ is given by Eq. 23. Thus, the characteristic equation, Eq. 5.8.15c, can be integrated to delimit region I. In region I, charge neutrality prevails and generation is in equilibrium with recombination.

To further refine the picture, the role of the neutrals in determining the generation of new charged particle pairs must be recognized. Because region III is "ahead" of the neutral characteristic originating at the interface, this region is one where neutrals can only be created by recombination.

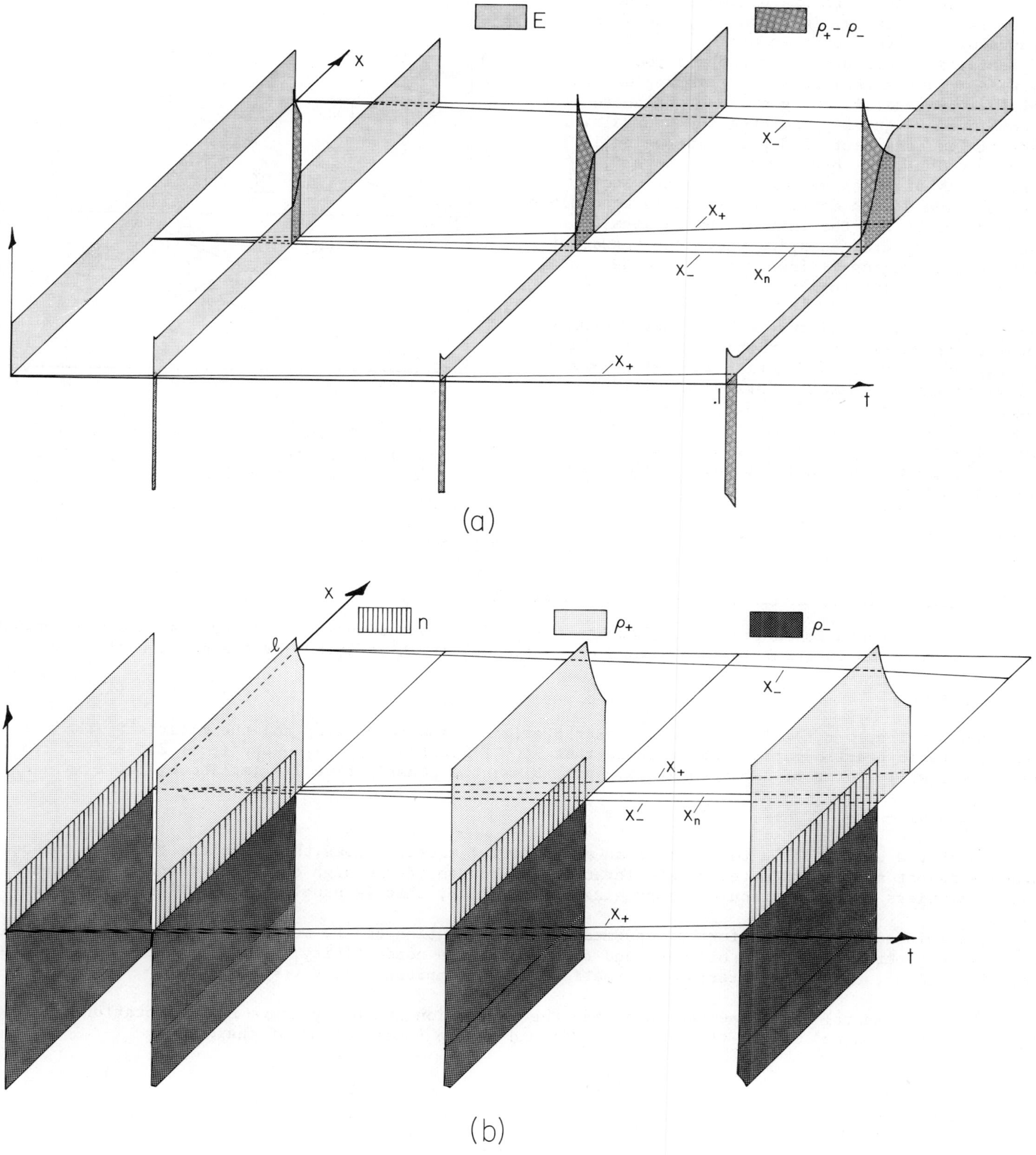

Fig. 5.9.3. Evolution of (a) field and free charge density and (b) charged particles and neutrals, with recombination and generation in the Maxwell capacitor configuration of Fig. 5.9.2. When t = 0, voltage is turned on. Characteristics x_+ and x_- are in the x-t plane. Neutral characteristics are x_n = constant. For the case shown, $b_+ = b_-$, $t = \underline{t}\tau_{mig}$, where τ_{mig} is given by Eq. 12 with $\mathscr{E} = V_o/\ell$. Also, initially $\rho_+ = \rho_- = 30\varepsilon_o V_o/\ell^2$, $n = \varepsilon V_o/q\ell^2$ (i.e., 30 ion pairs for each neutral so that according to Eq. 13, $\tau_{th} = \tau_e/30$ and β is equilibrium value given by Eq. 7. Recombination is Langevin (Eq. 5.8.4).

Because there are no negative charges in this region, there is no recombination and hence no neutrals. Initially, in region II, there are neutrals. However, because of the high degree of ionization intrinsic to the ohmic model ($\tau_{th} << \tau_e$), the generation in this region (which for lack of negative charges is not balanced by recombination) quickly depletes the neutrals. Essentially, the neutral density in region II is zero. Thus, in both regions II and III, essentially unipolar dynamics prevail, with the positive charge density decaying in accordance with Eq. 5.6.6 and the initial charge density, essentially determined by ρ_+ where the characteristic enters region II from region I, equal to ρ_+ in the equilibrium region.

This unipolar picture of the charge density decay along an x_+ characteristic in regions II and III explains why ρ_+ decays with increasing x at any given time. Characteristics x_+ entering region II at A and A' (Fig. 5.9.4) carry the same equilibrium charge density. Thus there is more time for decay of ρ_+ at point B than there is at B', even though B and B' are at the same instant in time.

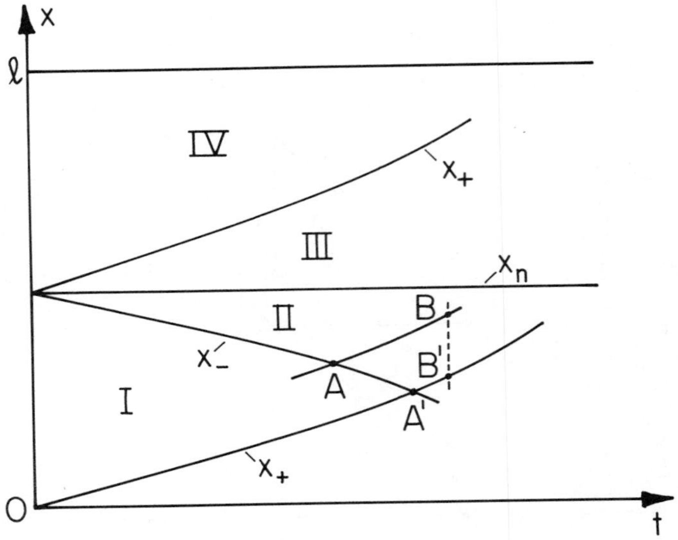

Fig. 5.9.4. Regions in x-t plane delimited by characteristic lines emanating from initial interface position.

DYNAMICS OF OHMIC CONDUCTORS

5.10 Charge Relaxation in Deforming Ohmic Conductors

If it is taken as an empirically substantiated fact that a material at rest is an ohmic conductor, then, moving in an inertial (primed) frame of reference, it is described by the constitutive law

$$\vec{J}_f' = \sigma\vec{E}' \tag{1}$$

The conductivity, $\sigma(\vec{r},t)$, is a parameter characterizing (and hence tied to) the material. The electroquasistatic transformation laws require that $\vec{E}' = \vec{E}$ but that $\vec{J}_f' = \vec{J}_f - \rho_f\vec{v}$ (Eqs. 2.5.9a and 2.5.12a) and show that in terms of laboratory-frame variables, the constitutive law implied by Eq. 1 is

$$\vec{J}_f = \sigma\vec{E} + \rho_f\vec{v} \tag{2}$$

With the use of Eq. 2 to describe an accelerating material goes the postulate that the conduction process is not altered by material accelerations. Because of the high collision frequency between charge carriers and the molecules comprising the material, this is usually an excellent assumption.

In this section, it is further assumed that polarization can be modeled in terms of a permittivity $\varepsilon(\vec{r},t)$, in general a function of space and time. Like the conductivity, ε is a property tied to the material. Also, the given material deformations are incompressible: $\nabla\cdot\vec{v} = 0$.

The fundamental laws required to define the relaxation process picture $\vec{E}$ as irrotational, relate ρ_f to $\vec{E}$ through Gauss' law ($\nabla\cdot\varepsilon\vec{E} = \varepsilon\nabla\cdot\vec{E} + \vec{E}\cdot\nabla\varepsilon$) and envoke conservation of charge:

$$\vec{E} = -\nabla\Phi \tag{3}$$

$$\nabla\cdot\vec{E} = \frac{\rho_f}{\varepsilon} - \frac{\vec{E}\cdot\nabla\varepsilon}{\varepsilon} \tag{4}$$

$$\nabla\cdot\vec{J}_f + \frac{\partial\rho_f}{\partial t} = 0 \tag{5}$$

The charge relaxation equation is obtained by entering $\vec{J}_f$ from Eq. 2 into Eq. 5, using Eq. 4 to replace the divergence of $\vec{E}$ and remembering that $\vec{v}$ is solenoidal,

$$\frac{\partial\rho_f}{\partial t} + \vec{v}\cdot\nabla\rho_f = -\frac{\sigma}{\varepsilon}\rho_f - \vec{E}\cdot\nabla\sigma + \frac{\sigma}{\varepsilon}\vec{E}\cdot\nabla\varepsilon \tag{6}$$

For a material of uniform permittivity, this is the same expression as Eq. 5.9.19, a fact that emphasizes the multispecies contribution to the conduction process necessary to justify the use of the ohmic model.

If characteristic lines are defined as the trajectories of fluid elements, then

$$\frac{d\vec{r}}{dt} = \vec{v} \tag{7}$$

and time is measured for an observer moving along a line satisfying Eq. 7, the charge relaxation equation, Eq. 6, becomes

$$\frac{d\rho_f}{dt} = -\frac{\sigma}{\varepsilon}\rho_f - \vec{E}\cdot\nabla\sigma + \frac{\sigma}{\varepsilon}\vec{E}\cdot\nabla\varepsilon \tag{8}$$

For an observer moving with the material, the three terms on the right are the possible contributors to a time rate of change of the charge density. Respectively, they represent the relaxation of the charge due to its self-field, the possible accumulation of charge where the electrical conductivity varies, and where the permittivity is inhomogeneous. Typically, these latter two terms are at interfaces, and hence are singular.

Region of Uniform Properties: In this case, the last two terms in Eq. 8 are zero, and the equation can be integrated without regard for details of geometry and boundary conditions:

$$\rho_f = \rho_o(\vec{r})e^{-t/\tau_e}; \quad \tau_e \equiv \varepsilon/\sigma \tag{9}$$

For the neighborhood of a given material particle, ρ_o is the charge density when $t = 0$. With Eq. 9, it has been deduced that at a given location within a deforming material having uniform conductivity and permittivity, the free charge density is zero unless that point can be traced backward in time along a particle line to a source of free charge density.

The general solution summarized by Eq. 9 has a physical significance which is best emphasized by considering two typical situations, one where the initial charge distribution is known, and the other involving a condition on the charge density where characteristic lines enter the volume of interest.

Suppose that the charge distribution is to be determined in an ohmic fluid as it passes between plane-parallel walls in the planes $x = 0$ and $x = d$. The flow is in the steady state with a velocity profile that is consistent with fully developed laminar flow:

$$\vec{v} = \frac{4x}{d}\left(1 - \frac{x}{d}\right)U\vec{i}_z \tag{10}$$

Initial Value Problem: When $t = 0$, the charge distribution throughout the flow is known to be

$$\rho_f(x,0) = \rho_t \sin(kz) \tag{11}$$

This distribution is sketched in Fig. 5.10.1a. For the given steady velocity distribution, it is simple to integrate Eq. 7 to find the characteristic lines $x = x_o$, $y = y_o$ and

$$z = \frac{4x}{d}\left(1 - \frac{x}{d}\right)Ut + z_o \tag{12}$$

The integration constant, z_o, is the z intercept of the characteristic line with the $t = 0$ plane. Figure 5.10.1b represents these characteristic lines in the x-z-t space. In the channel center, the characteristic line has its greatest slope (U) in the z-t plane, while at the channel edges the slope is zero. The lines take the same geometric shape regardless of z_o, and therefore other families of lines are generated by simply translating the picture shown along the z axis.

Now according to Eq. 9, the charge density at any time $t > 0$ is found by evaluating the initial charge density at the root of a characteristic line, when $t = 0$, and following that line to the point in question. The charge decays along this line by an amount predicted by the exponential equation using the elapsed time. If (x,z,t) represent the coordinates where the solution is required at some later time, then these coordinates are related to z_o through Eq. 12, and the initial charge density appropriate to the point in question is given by Eq. 11 with $z \to z_o$. Thus, the required solution is

$$\rho_f(x,z,t) = \rho_t \sin k\left[z - \frac{4x}{d}\left(1 - \frac{x}{d}\right)Ut\right]e^{-t/\tau_e} \tag{13}$$

This distribution is the one sketched in Fig. 5.10.1c.

The consequences of a boundary-value transient serve to provide further background for establishing the point of this section.

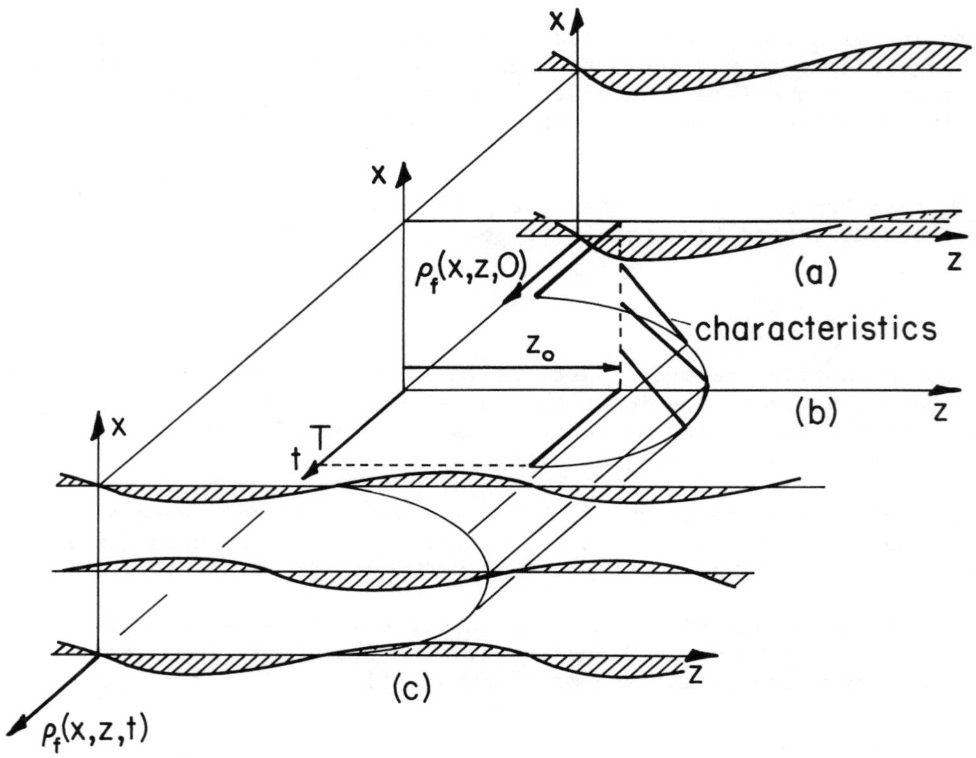

Fig. 5.10.1. (a) The initial distribution of charge density as a function of (x,z). (b) Characteristic lines in (x,z,t) space. Those lines originating along the cross section $z = z_o$ when $t = 0$ are shown. (c) Distribution of charge density by the time $t = T$. Charge is transported downstream in proportion to the stream velocity, and decays as $\exp(-t/\tau_e)$.

Injection from a Boundary: It is possible to inject charge into the bulk of an ohmic fluid so that a steady-state condition can be established with a space charge in the material volume. However, the position of interest in the material bulk must then be joined by a characteristic line to a source of charge. As an illustration, consider the case where, initially, there is no charge in the material. Again, the fluid flow of Eq. 10 is considered. However, now charge is introduced by a source in the plane $z = 0$. When $t = 0$, this source is turned on and provides a volume charge density ρ_s henceforth at $z = 0$. The problem is then one of finding the resulting downstream charge distribution. The boundary condition is shown graphically in Fig. 5.10.2a.

For this type of problem, the characteristic lines of Eq. 12 are more conveniently used if written in terms of the time $t = t_a$ when a given characteristic intercepts the $z = 0$ plane, where the source of charge is located, and it is known that for $t > 0$, the charge density is ρ_s. Then

$$z = \frac{4x}{d}\left(1 - \frac{x}{d}\right)U(t - t_a) \tag{14}$$

The family of characteristics having roots in the $z = 0$ plane when $t = t_a$ is sketched in Fig. 5.10.2b.

From the characteristic lines of the sketch, Fig. 5.10.2b, it follows that the distribution of charge can be divided into two regions, the surface of demarcation between the two being the surface formed by the characteristic lines with $t_a = 0$. For z greater than the envelope of these characteristic

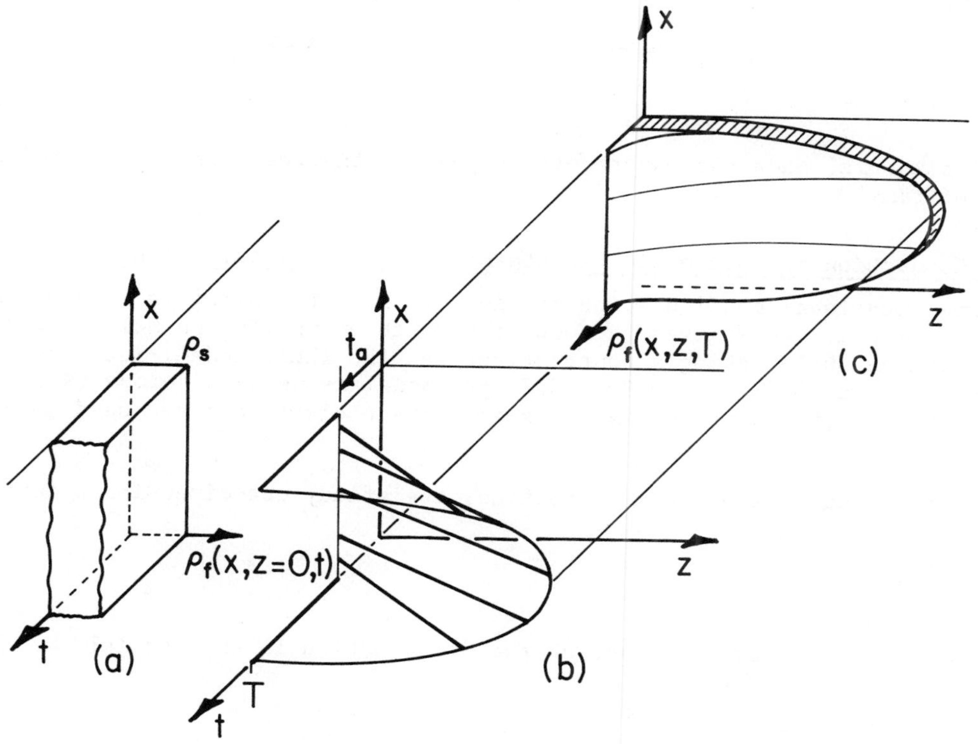

Fig. 5.10.2. (a) When t = 0, a uniform and henceforth constant source of charge is turned on at z = 0. (b) Characteristic lines. (c) Later distribution of charge density.

lines there is no response, because the characteristic lines originate from the z = 0 plane at a time when the charge density is constrained to be zero. For z less than the envelope, the initial charge distribution at z = 0 is the constant ρ_s. Thus, there is a wavefront between the two regions, as sketched in Fig. 5.10.2c. The charge density at any point behind the wavefront is determined by multiplying $\exp[(t-t_a/\tau_e)]$ times the charge density at z=0. That is, the appropriate evaluation of Eq. 9 is

$$\rho_f = \rho_s e^{-(t-t_a)/\tau_e} \tag{15}$$

and in view of the relation between a point in question (x,z,t) and the time of origination from the z = 0 plane, t_a (given by Eq. 14), the charge distribution of Eq. 15 can be written in terms of (x,z,t) as

$$\rho_f = \rho_s e^{-z/[v(x)\tau_e]}; \quad v(x) = \frac{4x}{d}\left(1 - \frac{x}{d}\right)U \tag{16}$$

This stationary distribution of charge is shown in Fig. 5.10.2c.

Because of the dependence of the velocity on x, the spatial rate of decay behind the front depends on the transverse position x. At the center of the channel, where the velocity is U, the spatial rate of decay is determined by the ratio of the relaxation time to the time required for the material to transport the charge to the given z position in question. This ratio is a measure of the influence of the material motion on the charge distribution: for a characteristic length ℓ in the z direction, it is convenient to define the electric Reynolds number of an ohmic conductor as

$$R_e \equiv (\varepsilon/\sigma)/(\ell/U) = \frac{\varepsilon U}{\sigma \ell} \tag{17}$$

and Eq. 16, written for the channel center where $x = d/2$, becomes

$$\rho_f(\frac{d}{2},z,t) = \rho_s e^{-\frac{z}{\ell}(1/R_e)} \tag{18}$$

At a given location z, once the wavefront has passed, the response represented in general by Eq. 9 is independent of time.

5.11 Ohmic Conduction and Convection in Steady State: D-C Interactions

The one-dimensional configuration of Fig. 5.7.1 is revisited in this section using an ohmic rather than a unipolar model. This gives the opportunity to exemplify the role of the electric field and boundary conditions while making a contrast between the ohmic model, introduced in Sec. 5.10 and the unipolar model of Sec. 5.7. As in Sec. 5.7, the model is used to demonstrate a type of "d-c" pump or generator exploiting longitudinal stresses. Again, screen electrodes are used to charge a uniform z-directed flow: $\vec{v} = U\vec{i}_z$.

Because the fluid has uniform properties, the steady one-dimensional form of Eq. 5.10.6 is

$$\frac{d\rho_f}{dz} + \frac{\sigma}{U\varepsilon}\rho_f = 0 \tag{1}$$

and it follows directly that the space-charge distribution is exponential:

$$\rho_f = \rho_o e^{-z/R_e \ell}; \quad R_e \equiv \frac{\varepsilon v}{\sigma \ell} \tag{2}$$

The electric Reynolds number R_e is introduced at this point because it reflects such attributes of the flow as the efficiency of energy conversion.

Conservation of charge requires that in the steady state $\vec{J}_f = J\vec{i}_z$ is a constant: the total current I divided by the area A. Thus the constitutive law, Eq. 5.10.2, can be solved for $\vec{E} = E(z)\vec{i}_z$ with ρ_f substituted from Eq. 2:

$$E = \frac{i}{\sigma A} - \frac{\rho_o U}{\sigma} e^{-z/R_e \ell} \tag{3}$$

In turn, the terminal potential is determined,

$$v = -\int_o^\ell E dz = -\frac{i\ell}{\sigma A} + \frac{\rho_o \ell U}{\sigma} R_e(1 - e^{-1/R_e}) \tag{4}$$

This is the electrical terminal relation for the interaction: a "volt-ampere" characteristic sketched in Fig. 5.11.1.

The electrical force on the charged particles is fully transmitted to the vehicle fluid, and hence the pressure rise between inlet and outlet is simply the difference in electric stresses at $z = \ell$ and $z = 0$, evaluated using Eq. 3:

$$\Delta p = p_o - p_i = \frac{1}{2}\varepsilon[E^2(\ell) - E^2(0)] = \frac{1}{2}\varepsilon[(\frac{i}{\sigma A} - \frac{\rho_o U}{\sigma} e^{-1/R_e})^2 - (\frac{i}{\sigma A} - \frac{\rho_o U}{\sigma})^2] \tag{5}$$

This mechanical "terminal relation" has a dependence on the terminal current i summarized by Fig. 5.11.1. Observe that $i_{sc} < i_{bp}$, where the short-circuit and zero pressure-rise currents follow from setting Eqs. 4 and 5 to zero:

$$I_{sc} = A\rho_o v R_e(1 - e^{-1/R_e}) \tag{6}$$

$$I_{bp} = A\rho_o v(1 + e^{-1/R_e})/2 \tag{7}$$

Three energy conversion regimes are defined by recognizing that the electrical power out is $P_e = VI$, while the mechanical power in is $P_m = -\Delta pUA$. Each of these quantities must be positive to give a generator function. Similarly, if both P_e and P_m are negative, energy is converted from electrical to mechanical form and the device is a pump. There is a midregion, which tends to vanish as R_e is increased, wherein both electrical and mechanical energy are absorbed. This region gives a braking effect at the expense of electrical energy. These three regimes are summarized by Fig. 5.11.1.

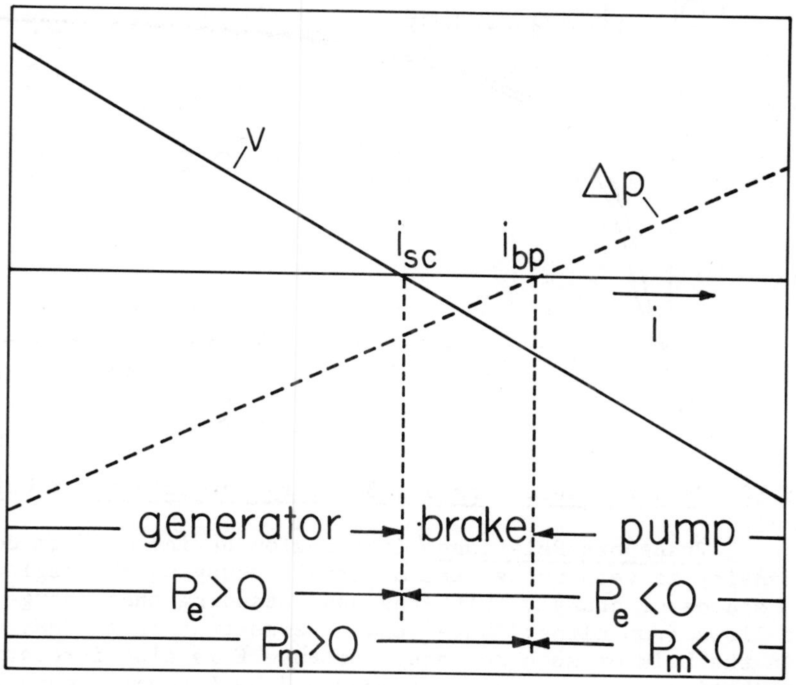

Fig. 5.11.1. Dependence of terminal voltage and pressure rise on terminal current i. Energy conversion regimes are as indicated.

The Generator Interaction: A primary limitation on electrohydrodynamic energy conversion devices is the relatively small electric pressure that can be obtained without incurring electrical breakdown. Difficulties in making an efficient converter are amplified by the extremely small fraction of the available mechanical energy that is altered by the electric coupling. It is clear from Eq. 5 that any electric stress at the outlet detracts from the total pressure change. To take the greatest advantage of the available electric stress, $E(\ell)$ should be adjusted to vanish. This can be done, according to Eq. 3, by operating with the space-charge density

$$\rho_o v = \frac{I}{A} e^{1/R_e} \tag{8}$$

It follows from Eq. 4 that (use upper sign):

$$P_e = -\frac{I^2 \ell}{\sigma A}[1 \pm R_e(1 - e^{\pm 1/R_e})] \tag{9}$$

while Eq. 5 shows that (upper sign)

$$P_m = \pm \frac{UA\varepsilon}{2}\left(\frac{i}{\sigma A}\right)^2 (1 - e^{\pm 1/R_e})^2 \tag{10}$$

The efficiency of energy conversion from mechanical to electrical form is then only a function of the electric Reynolds number (upper sign)

$$P_e/P_m = 2[\pm R_e(e^{\pm 1/R_e} - 1) - 1]/[\pm R_e(1 - e^{\pm 1/R_e})^2] \tag{11}$$

Of course, the conversion becomes perfectly efficient as $R_e \to \infty$. The detailed dependence is shown in Fig. 5.11.2.

The Pump Interaction: If it is a pumping function that is desired, Eq. 5 makes it clear that the space charge should be adjusted so that $E(0) = 0$, and it follows from Eq. 3 that

$$\frac{i}{A} = \rho_o v \tag{12}$$

The electrical and mechanical powers are now given by Eqs. 9 and 10 using the lower signs. In turn, the efficiency of electrical to mechanical conversion is the reciprocal of Eq. 11, using the lower signs. This pumping efficiency is summarized also in Fig. 5.11.2.

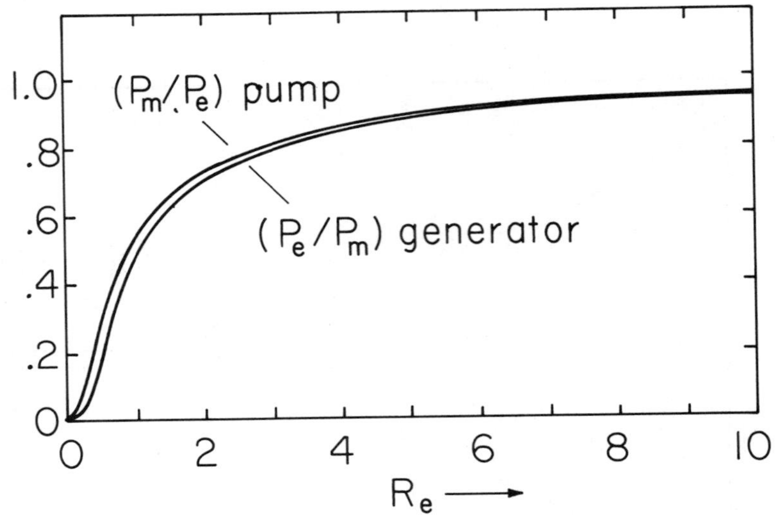

Fig. 5.11.2

Energy conversion efficiency of one-dimensional flow with ohmic fluid and immobile charged particles.

5.12 Transfer Relations and Boundary Conditions for Uniform Ohmic Layers

Transport Relations: In a region having uniform conductivity and permittivity, the free charge density is zero unless the material occupying the region can be traced back along a particle line to a source of charge. With the understanding that charge-free bulk regions are being described, it follows from either Gauss' Law or conservation of charge (Eqs. 5.10.4 or 5.10.5) that $\vec{E}$ is solenoidal in the bulk of such regions. Because $\vec{E}$ is also irrotational (Eq. 5.10.3), it follows that the distribution of potential Φ is governed by Laplace's equation. To describe the volume field distributions, the same relations are applicable as used to derive the flux-diplacement relations of Sec. 2.16. The transfer relations for planar layers, cylindrical annuli and spherical shells summarized in Sec. 2.16 are also applicable to regions having uniform conductivity. Because the effect of material motion on the fields comes from the convection of the free charge density, and ρ_f is zero in the material, these relations hold even if the material is moving. For example, the planar layer of Table 2.16.1 could be moving in the z direction with an arbitrary velocity profile.

In conjunction with the transfer relations, the conduction currents normal to the bounding surfaces (α,β) are of interest, and these are simply

$$\begin{bmatrix} \tilde{J}_n^\alpha \\ \tilde{J}_n^\beta \end{bmatrix} = \sigma \begin{bmatrix} \tilde{E}_n^\alpha \\ \tilde{E}_n^\beta \end{bmatrix} \tag{1}$$

where n signifies a coordinate normal to the (α,β) surfaces and σ has the value appropriate to the region between.

Conservation of Charge Boundary Condition: A typical model involves two or more materials having uniform properties and separated by interfaces. The boundary condition implied by the requirement that charge be conserved is given with some generality by Eq. 2.10.16. With the proviso that the regions neighboring the interface have the nature described in the previous paragraph, the volume current densities are simply $\vec{J}_f = \sigma\vec{E}$. In certain situations, the interface is itself comprised of a thin region over which the conductivity is appreciably greater than in the bulk. Then, a surface conductivity σ_s is used to model a surface conduction and the surface current density is

$$\vec{K}_f' = \sigma_s\vec{E}_t \Rightarrow \vec{K}_f = \sigma_s\vec{E}_t + \vec{v}_t\sigma_f \tag{2}$$

where the subscript t means that only components of the vector tangential to the interface contribute and σ_f is the surface charge density. Incorporating the appropriate values of $\vec{J}_f$ and $\vec{K}_f$, the required boundary condition, Eq. 2.10.16, becomes

$$\frac{\partial\sigma_f}{\partial t} + \nabla_\Sigma\cdot(\sigma_s\vec{E}_t + \vec{v}\sigma_f) + \vec{n}\cdot[\![\sigma\vec{E}]\!] = 0 \tag{3}$$

The tangential component of $\vec{E}$ is continuous at the interface, and so $\vec{E}_t$ or the potential can be evaluated on either side of the interface.

As an example used in subsequent sections, suppose that the interface is planar (in the y-z plane) and moves with the uniform velocity U in the z direction. Then, for $\vec{n} = \vec{i}_x$, Eq. 3 becomes

$$-\left(\frac{\partial \sigma_f}{\partial t} + U\frac{\partial \sigma_f}{\partial z}\right) = \sigma_s\left(\frac{\partial E_y}{\partial y} + \frac{\partial E_z}{\partial z}\right) + [\![\sigma E_x]\!] \tag{4}$$

Physically, this expression states that, for an observer moving with the material, the rate of decrease of σ_f with respect to time is proportional to the conduction current flowing out of the interfacial region in the plane of the interface and to the disparity between volume conduction currents leaving and entering from the bulk regions to either side of the interface.

5.13 Electroquasistatic Induction Motor and Tachometer

A configuration for establishing basic notions concerned with electric induction interactions is shown in Fig. 5.13.1, where a thin sheet having surface conductivity σ_s moves uniformly in the z-direction with the velocity U.[1] At a distance d above the sheet, a traveling wave of potential is imposed by means of electrodes, while the potential a distance d below is constrained by a solid electrode to be constant. The objective in this section is to determine the dependence of the electrical shear force tending to carry the sheet in the z direction on the frequency ω, the relative material and wave velocities, and the electrical surface conductivity. Later, the configuration is used to make a tachometer. In actual construction, the sheet might be wrapped around on itself to form a rotating shell.

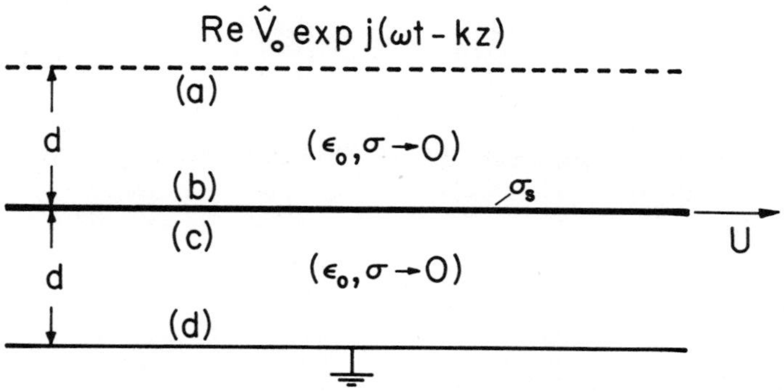

Fig. 5.13.1

A conducting sheet moves with velocity U and interacts with traveling waves of potential imposed on adjacent electrodes.

The active volume breaks into two regions joined by the conducting sheet. Thus, an analytical model simply involves the combination of transfer relations for the free space regions, and the boundary conditions for the sheet. The transfer relations of Table 2.16.1, Eqs. (a), become

$$\begin{bmatrix} \hat{D}_x^a \\ \hat{D}_x^b \end{bmatrix} = \begin{bmatrix} -\varepsilon_o k\coth(kd) & \dfrac{\varepsilon_o k}{\sinh(kd)} \\ \dfrac{-\varepsilon_o k}{\sinh(kd)} & \varepsilon_o k \coth(kd) \end{bmatrix} \begin{bmatrix} \hat{V}_o \\ \hat{\Phi}^b \end{bmatrix} \tag{1}$$

where the surface potentials have been identified as those of the electrodes and sheet, and the variables refer to the upper region with superscripts as defined in Fig. 5.13.1. From Eq. 5.12.4 with $\partial(\ )/\partial y = 0$ and $[\![J_x]\!] = 0$ (the regions adjacent to the sheet are insulating),

$$\sigma_s k^2 \hat{\Phi}^b + j(\omega - kU)(\hat{D}_x^b - \hat{D}_x^c) = 0 \tag{2}$$

where it is recognized that the net surface charge density on the sheet is $(\hat{D}_x^b - \hat{D}_x^c)$. Finally, the description is completed by the transfer relations for the lower region, again provided by Table 2.16.1:

$$\begin{bmatrix} \hat{D}_x^c \\ \hat{D}_x^d \end{bmatrix} = \begin{bmatrix} -\varepsilon_o k\coth(kd) & \dfrac{\varepsilon_o k}{\sinh(kd)} \\ \dfrac{-\varepsilon_o k}{\sinh(kd)} & \varepsilon_o k \coth(kd) \end{bmatrix} \begin{bmatrix} \hat{\Phi}^b \\ 0 \end{bmatrix} \tag{3}$$

---

1. For description of a somewhat similar device, see S. D. Choi and D. A. Dunn, "A Surface-Charge Induction Motor," Proc. IEEE 59, No. 5, 737-748 (1971).

Incorporated in the potentials on the right are the boundary conditions that $\Phi^b = \Phi^c$ and $\Phi^d = 0$. These three expressions can be viewed as five equations for the unknowns Φ^b and $(D_x^a, D_x^b, D_x^c, D_x^d)$. Before further manipulation is undertaken, it is advisable to look forward to the required variables.

Induction Motor: Summation of shear stresses on the sheet (see Eq. 4.2.2) shows that the space-average force density in the z direction is

$$\langle T_z \rangle_z = \frac{1}{2} \mathrm{Re} jk\hat{\Phi}^b [\hat{D}_x^b - \hat{D}_x^c]^* \tag{4}$$

The total complex surface charge density required in Eq. 4 follows from the subtraction of Eqs. 1b and 3a:

$$\hat{D}_x^b - \hat{D}_x^c = \frac{-\varepsilon_o k}{\sinh(kd)} \hat{V}_o + 2\varepsilon_o k \coth(kd)\hat{\Phi}^b \tag{5}$$

and substitution of this expression into Eq. 4 further reduces the surface force density to

$$\langle T_z \rangle_z = \frac{-\varepsilon_o k^2}{2\sinh(kd)} \mathrm{Re} j \hat{\Phi}^b \hat{V}_o^* \tag{6}$$

The complex sheet potential is found by again using Eq. 5, but this time to eliminate $\hat{D}_x^b - \hat{D}_x^c$ from Eq. 2:

$$\hat{\Phi}^b = \frac{jS_e \hat{V}_o}{2 \sinh kd(1 + jS_e \coth kd)} \tag{7}$$

where S_e is product of the angular frequency $(\omega - kU)$ measured by an observer moving with the material velocity U and the relaxation time constant $2\varepsilon_o/k\sigma_s$:

$$S_e \equiv \frac{2\varepsilon_o(\omega - kU)}{k\sigma_s} \tag{8}$$

The surface force density follows by substituting Eq. 7 into 6:

$$\langle T_z \rangle_z = \frac{\varepsilon_o k^2 \hat{V}_o \hat{V}_o^*}{4 \sinh^2(kd)} \frac{S_e}{(1 + S_e^2 \coth^2 kd)} \tag{9}$$

This result is analogous to one obtained for a magnetic induction machine in Sec. 6.4. It exhibits a maximum which is determined by the frequency in the frame of the moving sheet relative to the effective relaxation time. That is, the optimum or largest electric surface force density is

$$\langle T_z \rangle_z = \frac{\varepsilon_o k^2 \hat{V}_o \hat{V}_o^* \tanh(kd)}{8 \sinh^2(kd)} \; ; \; S_e = \tanh(kd) \tag{10}$$

Again, this result fits the general description of a "shearing" type of electromechanical energy converter given in Sec. 4.15. The surface force density takes the form of an electric stress $\varepsilon_o(kV_o)^2/2$ multiplied by factors reflecting the geometry and charge relaxation phenomena. The factor $(\sinh kd)^{-2}$ represents the Laplacian decay of the fields from the excitation to the sheet and then back again.

A sketch of the dependence of $\langle T_z \rangle_z$ on S_e is shown in Fig. 5.13.2. The physical origins of this curve are understood by interpreting Eq. 7. At very low material-frame frequencies, $S_e \to 0$ and $\Phi^b \to 0$. The sheet behaves as a perfect conductor, supports no tangential electric field intensity, and hence no electrical force in the z direction.

In the opposite extreme, the frequency is large compared to the reciprocal relaxation time for the system of sheet and adjacent regions of free space, and the amount of surface charge induced on the sheet becomes small. This follows from Eqs. 5 and 7. The optimum of Fig. 5.13.2 represents the compromise between the extremes of S_e small, and hence the wrong lag angle, and S_e large and hence reduced sheet surface charge.

Electroquasistatic Tachometer: It is the induced force upon the moving, semi-insulating sheet that is emphasized so far. The reverse effect of the motion on the field is emphasized by the slightly

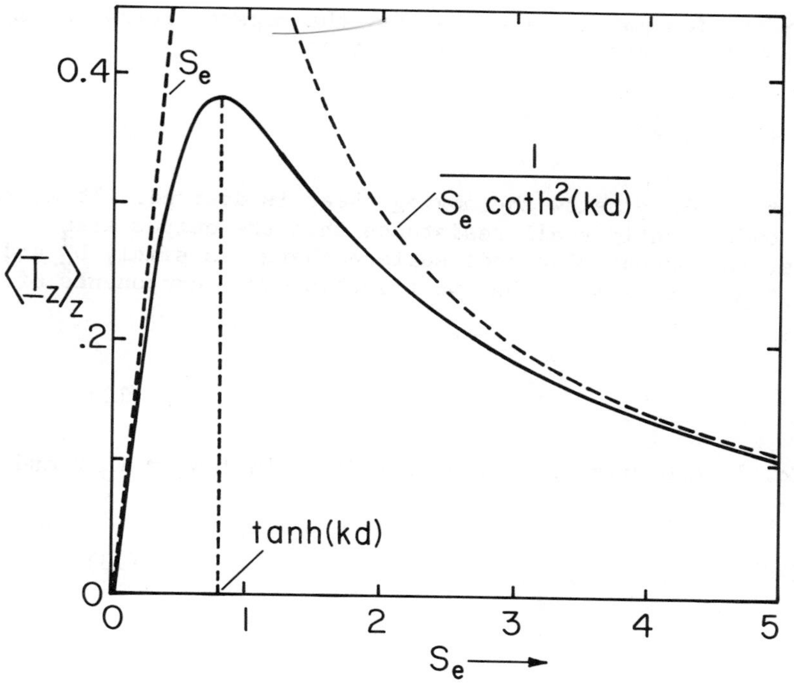

Fig. 5.13.2

Dependence of time-average surface force density normalized to $(\varepsilon_o|kV_o|^2/4 \sinh^2 kd)$ as a function of frequency in moving frame of reference, normalized to relaxation time. S_e is defined by Eq. 8 (kd = 1).

revised configuration of Fig. 5.13.3. Instead of a traveling wave, the imposed potential is now a standing wave. Points of zero amplitude retain fixed positions along the z axis. For the purpose of detecting the material velocity U, a pair of electrode segments is positioned in the grounded wall just below the moving sheet. The time variation of charge induced on these segments gives rise to a current, i, which is measured by means of external circuitry. Each segment is one half-wavelength, and positioned so that, in the absence of material motion, there is as much positive as negative surface charge induced on a segment surface. Thus, the electrodes are designed so that there is no output current in the absence of a material motion. But, with motion, the fields are skewed so that there is a net charge induced on each output segment. The result, an output signal v_o reflecting the material velocity U, is now going to be computed.

There is considerable analogy between the interaction studied here in the context of charge relaxation, and the magnetic diffusion example of Sec. 6.4. To make a practical device for measuring the rotational velocity of a shaft, the sheet pictured in Fig. 5.13.3 would be closed on itself, with the standing wave of imposed potential and the output segments perhaps arranged as in Fig. 5.13.4. By contrast with the conventional drag-cup tachometer, the sheet material in the device studied in this section would be made from semi-insulating material, rather than a metal.

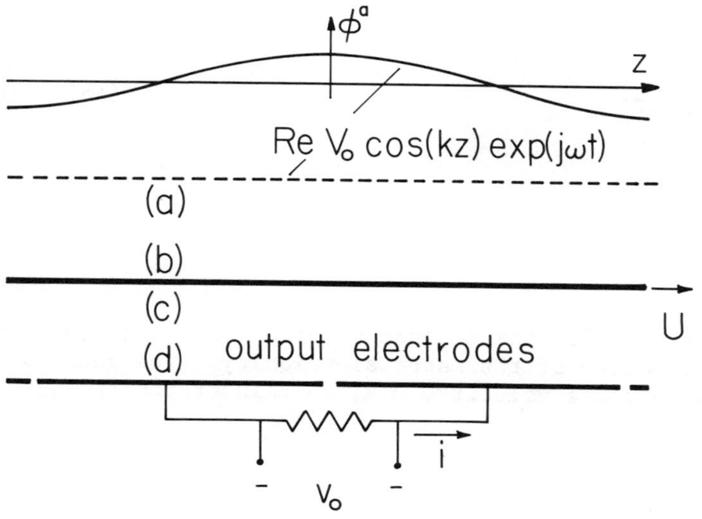

Fig. 5.13.3. A device for measuring the velocity U is made by exciting from above with a standing wave of potential and measuring the induced current on an electrode pair below the sheet.

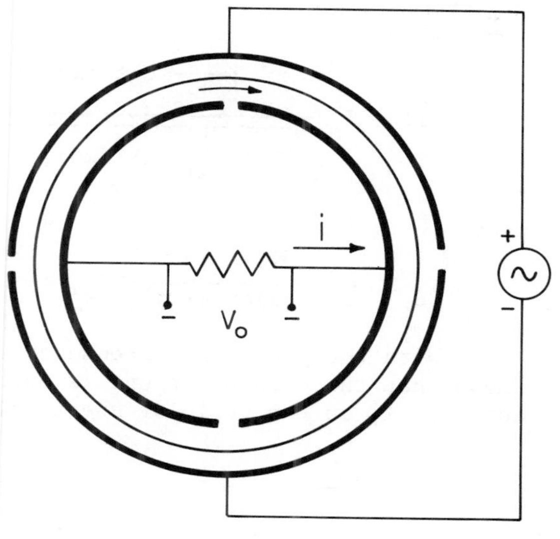

Fig. 5.13.4. Adaptation of the planar configuration of Fig. 5.13.3 to measure rotational velocity of shell of slightly conducting material.

The fields from a standing wave of excitation potential are simply the superposition of two of the traveling waves analyzed already. That is, the excitation can be written as

$$\Phi^a = \text{Re}\hat{V}_o \cos(kz)e^{j\omega t} = \text{Re}\frac{\hat{V}_o}{2}(e^{-jkz} + e^{jkz})e^{j\omega t} \tag{11}$$

The surface charge induced on the equipotential plane below the moving sheet is desired. It is assumed that the current, i, is measured through a sufficiently small resistance that the output electrodes remain at essentially zero potential. Thus, the output electrode surface charge is simply D_x^d and is found from Eq. 3b, as the superposition of the responses to the two traveling-wave components of the drive identified by Eq. 11:

$$\hat{D}_x^d = \frac{-\varepsilon_o k}{\sinh(kd)}(\hat{\Phi}_+^b + \hat{\Phi}_-^b) \tag{12}$$

The potential amplitudes called for with Eq. 12 are given by evaluating Eq. 7 with $\hat{V}_o \to \hat{V}_o/2$ and k first positive and then negative:

$$\hat{\Phi}_{\pm} = jS_{e\pm}\hat{V}_o/4\sinh(kd)[1 + jS_{e\pm}\coth(kd)] \tag{13}$$

$$S_{e\pm} \equiv 2\varepsilon_o(\omega \mp kU)/k\sigma_s$$

The combination of Eqs. 12 and 13 give the space-time dependence of the charge induced on the lower surface:

$$D_x^d = -\text{Re}j\frac{\varepsilon_o k\hat{V}_o}{4\sinh^2(kd)}\left\{\frac{S_{e+}e^{-jkz}}{1 + jS_{e+}\coth(kd)} + \frac{S_{e-}e^{jkz}}{1 + jS_{e-}\coth(kd)}\right\}e^{j\omega t} \tag{14}$$

The net charge on the right electrode is now computed by integrating the surface charge over its area, from z = 0 to z = π/k and over the width w of the electrode in the y direction. The required current is the time rate of change of the net charge on the electrode, and therefore given by

$$\hat{i} = j\omega\hat{q} = -\frac{j\omega w\varepsilon V_o}{2\sinh^2(kd)}\left[\frac{S_{e+}}{1 + jS_{e+}\coth(kd)} - \frac{S_{e-}}{1 + jS_{e-}\coth(kd)}\right] \tag{15}$$

As required, the net charge on the electrode vanishes in the absence of a material motion. To see the dependence of the output current on the material velocity, Eq. 15 is expanded, using the definition of $S_{e\pm}$ from Eq. 13:

$$|\hat{i}| = I_o\frac{\left[\frac{2\varepsilon_o\omega}{k\sigma_s}\coth(kd)\right](kU/\omega)}{\sqrt{[1 + S_{e+}^2\coth^2(kd)][1 + S_{e-}^2\coth^2(kd)]}} \tag{16}$$

where

$$I_o \equiv \frac{\omega\varepsilon_o|\hat{V}_o|w}{\sinh(kd)\cosh(kd)};\quad S_{e\pm} \equiv \frac{2\varepsilon_o\omega}{k\sigma_s}(1 \mp \frac{kU}{\omega})$$

With the excitation frequency large compared to kU, the dependence of $S_{e\pm}$ on U is weak, and Eq. 16 shows that the output current is then a linear function of the material velocity. The general dependence of $|\hat{i}|$ on the ratio of sheet velocity to wave phase velocity, ω/k, is illustrated in Fig. 5.13.5.[2]

---

2. For a similar approach to measuring fluid velocity, see J. R. Melcher, "Charge Relaxation on a Moving Liquid Interface," Phys. Fluids 10, 325-331 (1967).

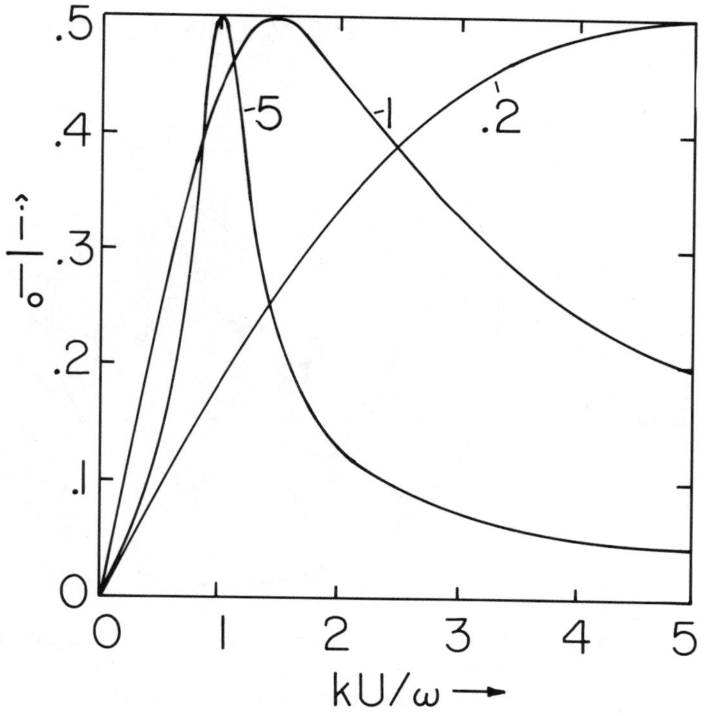

Fig. 5.13.5.

Dependence of output signal on material velocity U relative to phase velocity (ω/k) for tachometer of Fig. 5.13.3. Parameter is $2\varepsilon_o\omega\coth(kd)/k\sigma_s$.

5.14 An Electroquasistatic Induction Motor; Von Quincke's Rotor

The configuration of Fig. 5.14.1 gives the opportunity to study charge relaxation for finite-thickness conductors. Regions (a) and (b) are each composed of homogeneous materials having uniform conductivity and permittivity. The b-c interface moves to the right with a uniform velocity, U. The materials may move as rigid bodies with this same velocity, or might be composed of fluids which have some unspecified velocity profile $\vec{v} = v_z(x)\vec{i}_z$. They are bounded from below by a constant-potential plane, and from above by a system of electrodes used to impose a traveling wave of potential.

An objective is to determine the fields and hence the electrical surface force density acting on the interface in the direction of motion. From Sec. 5.10 it is known in advance that the only charges within the moving materials exist where the conductivity and permittivity have a spatial variation, at the interface. The planar configuration could be a developed model for a system "closed on itself" so that the interaction considered would be between a system of rotating, semi-insulating materials and an imposed rotating electric field. Except for geometric factors, the torque on the semi-insulating rotor sketched in Fig. 5.14.2 depends on the physical parameters and the imposed fields in essentially the same way as for the planar case study (see Problem 5.14.1).

The potential is the given traveling wave at the boundary denoted by (a), is continuous at the interface, and must vanish at the lower boundary ($\hat{\Phi}^a = \hat{V}_o$, $\hat{\Phi}^b = \hat{\Phi}^c$, $\hat{\Phi}^d = 0$). Thus, the transfer relations representing the field distributions in the bulk of each region, Eqs. (a) of Table 2.16.1, are

$$\begin{bmatrix} \hat{D}_x^a \\ \hat{D}_x^b \end{bmatrix} = \begin{bmatrix} -\varepsilon_a k\coth(ka) & \dfrac{\varepsilon_a k}{\sinh(ka)} \\ \dfrac{-\varepsilon_a k}{\sinh(ka)} & \varepsilon_a k\coth(ka) \end{bmatrix} \begin{bmatrix} \hat{V}_o \\ \hat{\Phi}^b \end{bmatrix} \tag{1}$$

$$\begin{bmatrix} \hat{D}_x^c \\ \hat{D}_x^d \end{bmatrix} = \begin{bmatrix} -\varepsilon_b k\coth(kd) & \dfrac{\varepsilon_b k}{\sinh(kb)} \\ \dfrac{-\varepsilon_b k}{\sinh(kb)} & \varepsilon_b k\coth(kb) \end{bmatrix} \begin{bmatrix} \hat{\Phi}^b \\ 0 \end{bmatrix} \tag{2}$$

By contrast with the model used in Sec. 5.13, there is no surface conduction, but rather a volume conduction, so that the boundary condition implied by conservation of charge for the interface,

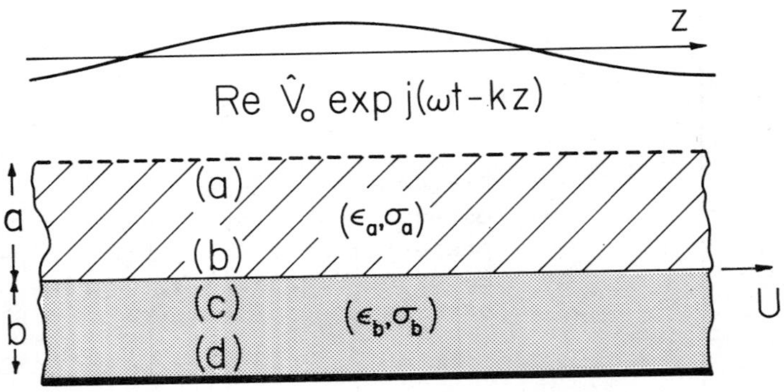

Fig. 5.14.1. Cross-sectional view of two planar layers of material having thicknesses a and b, respectively. The potential is constrained to be a traveling wave just above, and to be constant just below.

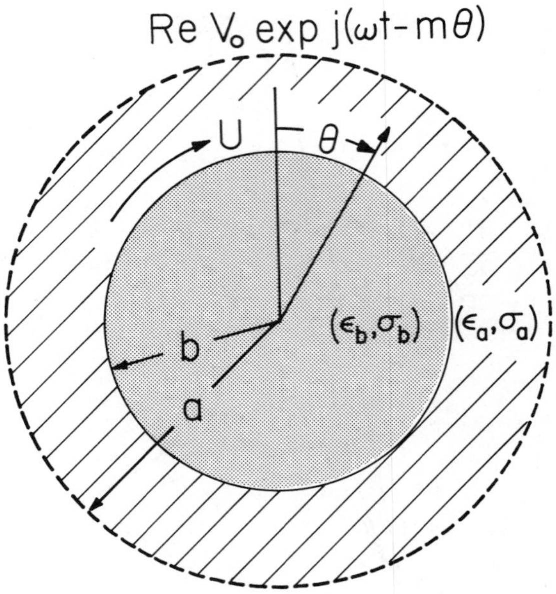

Fig. 5.14.2. Circular analog of the planar system of Fig. 5.14.1. Qualitatively, fields in the planar problem are the same as those in this circular configuration if the z axis of Fig. 5.14.1 is wrapped around on itself in an integral number of wavelengths.

and represented by Eq. 5.12.4, becomes

$$\left(\frac{\sigma_a}{\varepsilon_a}\hat{D}_x^b - \frac{\sigma_b}{\varepsilon_b}\hat{D}_x^c\right) + j(\omega - kU)(\hat{D}_x^b - \hat{D}_x^c) = 0 \tag{3}$$

The space average of the surface force density acting on the only charges within the volume of interest, those at the interface, is given by integrating the Maxwell stresses over an incremental volume enclosing the interface and having one wavelength in the z direction (see Sec. 4.2 for a similar calculation). Thus, the space-average force per unit area is

$$\langle T_z\rangle_z = \frac{1}{2}\text{Re}[(\hat{D}_x^b)^*\hat{E}_z^b - (\hat{D}_x^c)^*\hat{E}_z^c] = \frac{1}{2}\text{Re}jk\hat{\Phi}^b(\hat{D}_x^b - \hat{D}_x^c)^* \tag{4}$$

The jump in D_x called for with Eq. 4 is the surface charge density given by subtracting Eq. 2a from 1b:

$$\hat{D}_x^b - \hat{D}_x^c = \frac{-\varepsilon_a k\hat{V}_o}{\sinh(ka)} + \hat{\Phi}^b k[\varepsilon_a\coth(ka) + \varepsilon_b\coth(kb)] \tag{5}$$

Then, substitution into Eq. 4 shows that it is the interfacial potential which determines the space-average of the surface force density

$$\langle T_z\rangle_z = -\frac{\varepsilon_a k^2}{2\sinh(ka)}\text{Re}j\hat{\Phi}^b\hat{V}_o^* \tag{6}$$

With the objective of finding $\hat{\Phi}^b$, the first quantity in brackets in Eq. 3 is found in terms of the potential $\hat{\Phi}^b$ by multiplying Eq. 1b by σ_a/ε_a and subtracting Eq. 2a multiplied by σ_b/ε_b:

$$\frac{\sigma_a\hat{D}_x^b}{\varepsilon_a} - \frac{\sigma_b\hat{D}_x^c}{\varepsilon_b} = \frac{-\sigma_a k}{\sinh(ka)}\hat{V}_o + \hat{\Phi}^b k[\sigma_a\coth(ka) + \sigma_b\coth(kb)] \tag{7}$$

Then, substitution of Eqs. 5 and 7 into 3 gives the required surface potential in terms of the driving potential

$$\hat{\Phi}^b = \frac{[j(\omega - kU)\varepsilon_a + \sigma_a]\hat{V}_o}{\sinh(ka)\{[\sigma_a\coth(ka) + \sigma_b\coth(kb)] + j(\omega - kU)[\varepsilon_a\coth(ka) + \varepsilon_b\coth(kb)]\}} \tag{8}$$

For purposes of physical interpretation, it is helpful also to have the surface charge density given in terms of the driving potential by substituting Eq. 8 into Eq. 5:

$$\hat{D}_x^b - \hat{D}_x^c = \frac{-k\hat{V}_o\coth(kb)(\varepsilon_a\sigma_b - \varepsilon_b\sigma_a)}{\sinh(ka)[\sigma_a\coth(ka) + \sigma_b\coth(kb)](1 + jS_E)} \tag{9}$$

where

$$S_E = \omega\tau_E(1 - \frac{kU}{\omega}), \quad \tau_E = \frac{[\varepsilon_a\coth(ka) + \varepsilon_b\coth(kb)]}{[\sigma_a\coth(ka) + \sigma_b\coth(kb)]} \tag{10}$$

Finally, the electric surface force density is found by substituting Eq. 8 into Eq. 6:

$$\langle T_z\rangle_z = \frac{1}{2}\varepsilon_a(k\hat{V}_o)(k\hat{V}_o)^*K(\varepsilon_a\sigma_b - \varepsilon_b\sigma_a)\frac{S_E}{1 + S_E^2} \tag{11}$$

where

$$K = \coth(kb)/\sinh^2(ka)[\varepsilon_a\coth(ka) + \varepsilon_b\coth(kb)][\sigma_a\coth(ka) + \sigma_b\coth(kb)]$$

What has been computed relates to a number of different physical situations. If the material layers are solid, then Eq. 11 represents the force per unit x-y area acting on the layers. Even though Eq. 11 came from an integration of the stresses over a volume enclosing only the interface, because there is no free charge density anywhere else in the volume of the materials, it includes all of the force on the material. It is possible that one or the other, or both, of the materials could be fluids, in which case Eq. 11 is the surface force density acting at the interface and U is the interfacial velocity. The calculation remains correct, even if the material to either side of the interface moves with some velocity other than U.

To examine the physical implications of Eq. 11, suppose that the traveling-wave frequency is fixed, and interest is in the dependence of the electrical surface force density on the material velocity. First, note that for a given kU/ω, the sign of the surface stress depends on the relative permittivities and conductivities. If the lower material is sufficiently more conducting than the upper one, so that $\sigma_b\varepsilon_a > \sigma_a\varepsilon_b$, then for $kU/\omega < 1$ the force is in the same direction as the wave velocity. As a function of S_E, this stress first rises linearly, reaches a peak, and then falls off, in a manner familiar from Sec. 5.13. The dependence on U has the same nature except that the point where S_E vanishes is at the synchronous velocity $U = \omega/k$, and increasing U is equivalent to decreasing S_E. Hence, a plot of $\langle T_z\rangle_z$ as a function of the normalized velocity kU/ω is as shown in Fig. 5.14.3. If the lower material is a conducting solid or fluid, and the intervening material an insulator, such as air, and the interface moves at a velocity less than synchronous, there is an induced electrical force tending to pull the material in the direction of wave propagation.

If the electrical force is retarded by one proportional to the velocity, as would be the case with viscous damping, then the velocity at which there would be an equilibrium between the electrical force and the retarding viscous force is the intersection (i) of Fig. 5.14.3a. The material tends to follow the traveling wave at a somewhat lesser velocity than the phase velocity ω/k. Note that, if a perturbing force makes the velocity decrease in magnitude slightly, the electrical force dominates the viscous force and tends to return the material to its steady equilibrium position. An experiment illustrating the force as it pumps a liquid is shown in Fig. 5.14.4a.

So far, there is little qualitative difference between what has been found for the finite-thickness slab and the results of Sec. 5.13 for the sheet conductor. But now, suppose that the material adjacent to the traveling-wave structure conducts sufficiently more than that below so that $\sigma_a\varepsilon_b > \sigma_b\varepsilon_a$. From Eq. 11, it is clear that the electrical force now acts in a direction which opposes the direction of relative propagation for the field. Even more, there are now three velocities at which the electrical force can be equilibrated by a viscous retarding force. At position (ii), the material is moving in a direction opposite to that of the wave.

Arguments similar to those given for equilibrium (i) can be used to see that (ii) is also stable. Two equilibria are possible with the material moving faster than the traveling wave. Of these, (iii) is unstable and (iv) is stable.

The example illustrates that there are exceptions to the intuitive notion that in an induction type of interaction, the material always tends to follow the traveling wave, and that under conditions of "motor" operation, the material velocity is less than the phase velocity of the wave.

The seemingly mysterious finding, with $\sigma_a\varepsilon_b > \varepsilon_a\sigma_b$, is explained first by considering why the material follows the traveling wave in the case of Fig. 5.14.3a. Equation 9 gives the surface charge

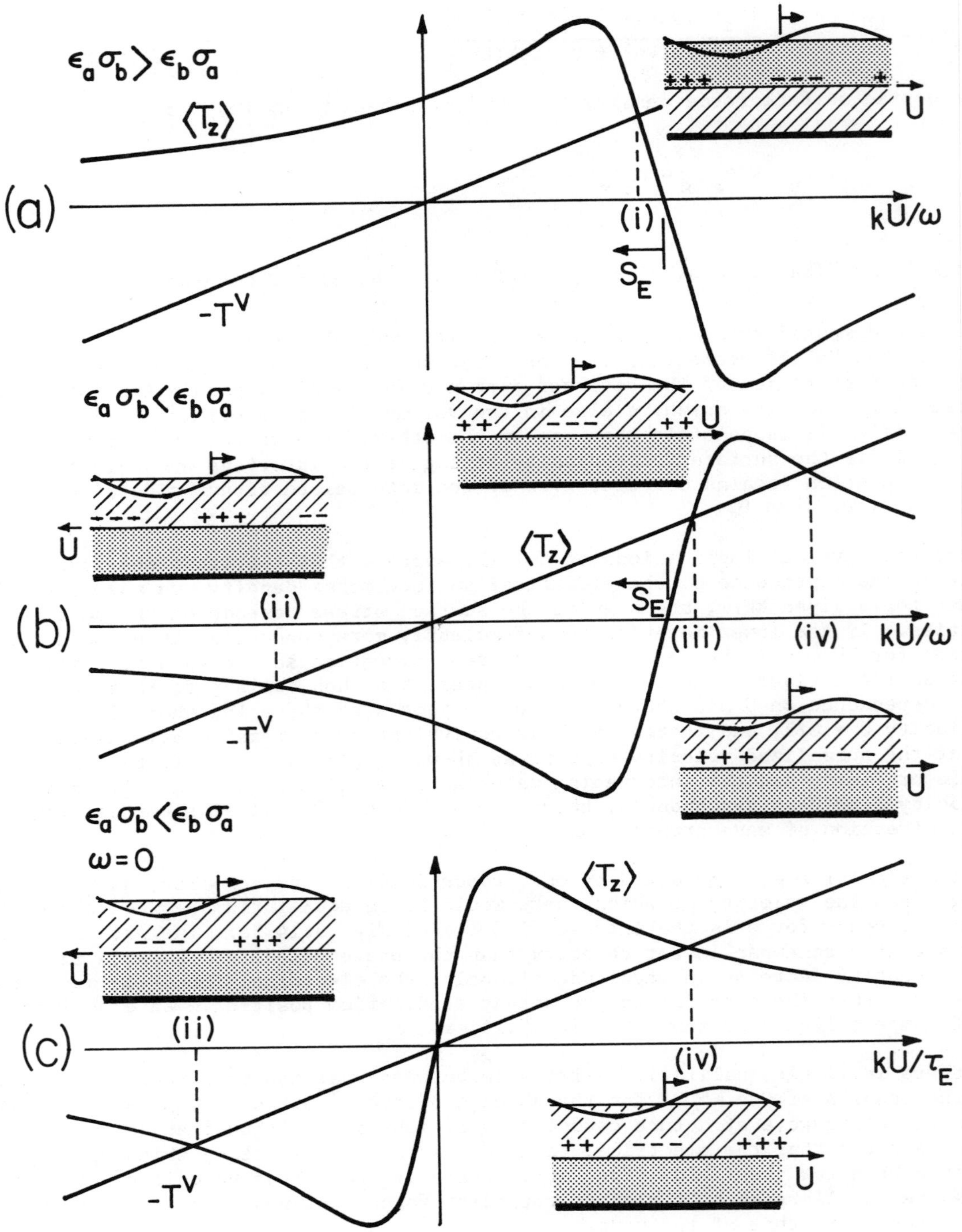

Fig. 5.14.3. Dependence of electric space-average surface force density on material velocity. (a) With the lower material more conducting, the material tends to follow the traveling wave. (b) With the conducting material next to the electrodes, the material can travel in a direction opposite to that of the imposed wave, or move faster than the traveling wave. (c) Applied frequency zero. Motion results from raising the applied field so that the slope of the curve exceeds that of the viscous force curve.

(a)

(b)

(c)

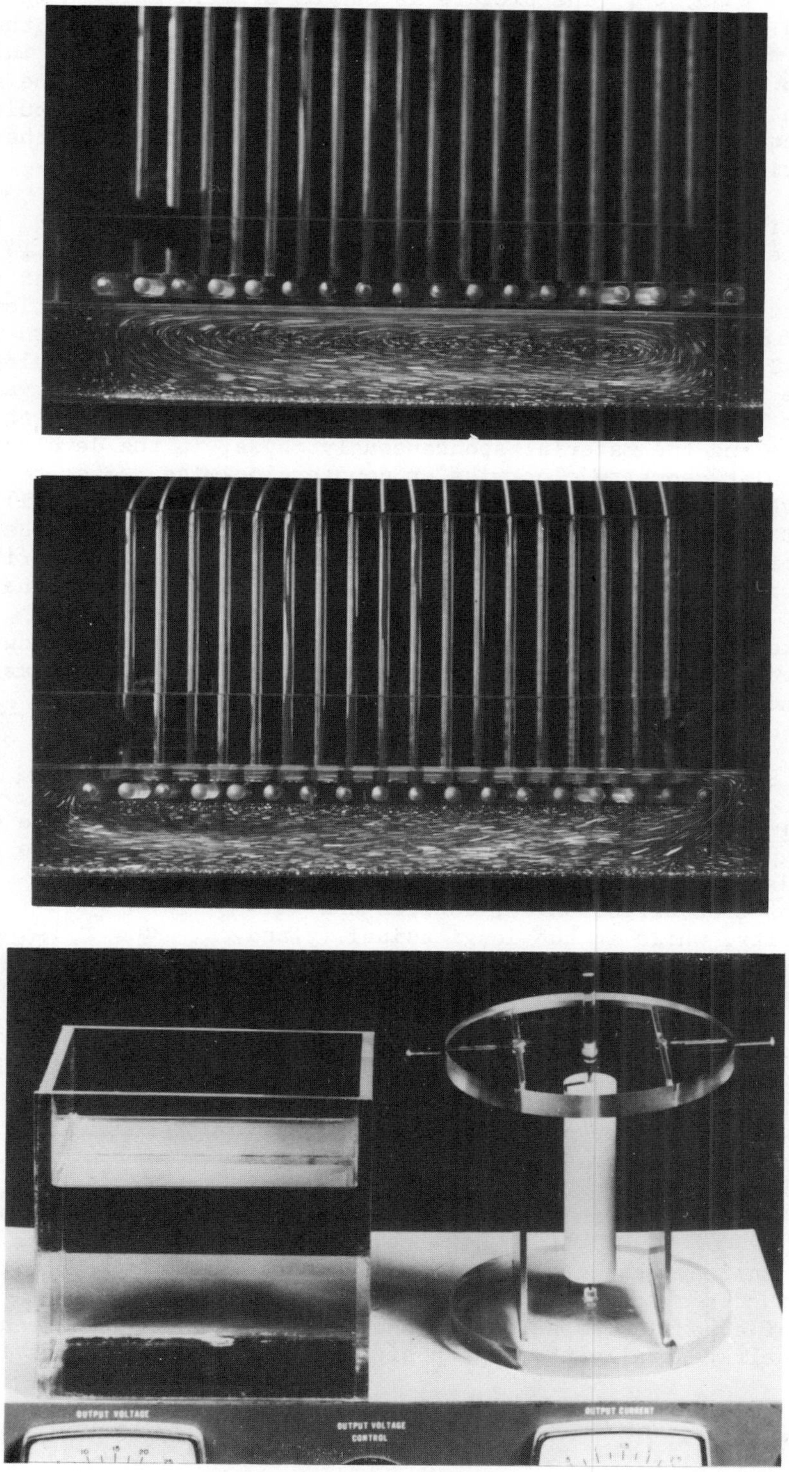

Fig. 5.14.4. (a) Electrodes embedded in a plastic sheet are driven by a 60-Hz 6-phase source so as to approximate a wave of potential traveling to the right. Separated from the electrode structure by an air gap, corn oil (doped to make $\omega\tau_E = 1$) has an interface that is pumped to the right, illustrating equilibrium (i) of Fig. 5.14.3a. To conserve mass, the liquid recirculates below the interface. (From film "Electric Fields and Moving Media," Reference 12, Appendix C.) (b) The traveling wave still propagates to the right but the electrode is immersed in the corn oil. The interface, which is now above, moves in the opposite direction of the wave. The configuration is Fig. 5.14.1 turned upside down, and the pumping illustrates equilibrium (ii) in Fig. 5.14.3b. (c) Von Quincke's rotor, consisting of a Teflon rotor immersed in a semi-insulating liquid. As a d-c potential applied between the electrodes is raised to about 20 kV, the rotor begins to rotate in either direction.

density, and shows that there is negative surface charge lagging the peak in potential on the electrode above by an angle less than 90°. The picture is one of a field axis on the fixed structure pulling along charges induced in the material. But, if the material adjacent to the electrodes is the conductor, so that $\sigma_a\varepsilon_b > \varepsilon_a\sigma_b$, then Eq. 9 shows that the sign of the charge at the interface is reversed. Regions of positive charge on the electrodes induce positive surface charge on the adjacent interface. What was a force of attraction in the case of Fig. 5.14.3a, becomes a force of repulsion in Fig. 5.14.3b. This is why the material can actually be repelled in a direction opposite to that of the traveling wave. An illustrative experiment is shown in Fig. 5.14.4b.

Equilibrium (iii) is best illustrated by considering the limit where the applied frequency vanishes. Thus, the applied potential is static. In the circular analog of Fig. 5.14.2 the applied field might be produced by a pair of parallel plates used to impose a field perpendicular to the z axis that, in the absence of the conducting materials, would be uniform. Such a configuration is Von Quincke's rotor, illustrated in Fig. 5.14.4c. The rotor is insulating relative to the corn oil in which it is immersed; hence $\varepsilon_a\sigma_b < \varepsilon_b\sigma_a$. The electrical force then depends on the material velocity, as sketched in Fig. 5.14.3c. If the applied field is raised, then there is a threshold value of field at which the slope of the electric force curve exceeds that of the viscous force. At that condition, equilibrium (iii) becomes unstable and the material spontaneously moves, in the developed model either to the right or left, in the circular geometry clockwise or counterclockwise, so as to establish a new equilibrium with a steady-state velocity either at (ii) or (iv). At the position (iii), the static field induces positive charges on the interface directly opposite positive charges on the electrodes. As a result, any small excursion of the material which tends to carry that charge distribution to the right or left is accompanied by a proportionate electric stress that tends to further the original deflection.

Spontaneous rotation of insulating objects immersed in somewhat conducting media and stressed by d-c fields are observed in seemingly unrelated situations. Examples are macroscopic particles in semi-insulating liquids and objects in ionized gases.

5.15 Temporal Modes of Charge Relaxation

Temporal Transients Initiated from State of Spatial Periodicity: The configurations of the two previous sections are typical of linear systems that are inhomogeneous in one direction only and excited from transverse boundaries. Pictured in the abstract by Fig. 5.15.1, the transverse direction, x, denotes the direction of inhomogeneity, while in the longitudinal (y and z) directions the system is uniform. In Secs. 5.13 and 5.14, it is at transverse boundaries (having x as the perpendicular) that driving conditions are imposed. In the picture, Φ_d imposes a driving frequency ω and a spatial dependence on the longitudinal coordinates that is periodic, either a pure traveling wave with known wavenumbers (k_y,k_z) or a Fourier superposition of these waves. The most common configuration in which spatial periodicity is demanded is one in which y or z "closes on itself," for example becomes the θ coordinate in a cylindrical system.

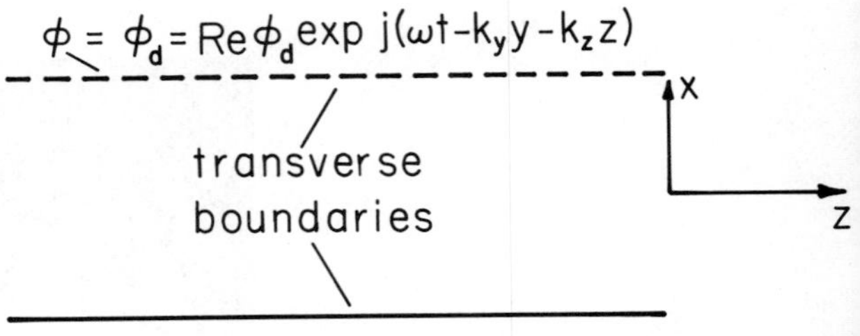

Fig. 5.15.1. Abstract view of systems that are inhomogeneous in a transverse direction, x, and uniform in longitudinal directions (y,z).

The temporal transient resulting from turning on the excitation when t = 0 with the system initially at rest can be represented as the sum of a particular solution (the sinusoidal steady-state driven response) and a homogeneous solution (itself generally the superposition of temporal modes having the natural frequencies s_n):

$$\Phi(x,y,z,t) = \mathrm{Re}\hat{\Phi}(x)e^{j(\omega t-k_y y-k_z z)} + \sum_n \mathrm{Re}\hat{\Phi}_n(x)e^{s_n t-j(k_y y+k_z z)} \tag{1}$$

Turning off the excitation results in a response composed of only the temporal modes. The coefficients $\Phi_n(x)$ are adjusted to guarantee that the total response satisfy the proper initial conditions for all values of x. In some situations this may require only one mode, whereas in others an infinite set of modes is entailed.

Identification of the eigenfunctions and their associated eigenfrequencies is accomplished in one of two ways. First, if the driven response is known, its complex amplitude takes the form

$$\hat{\Phi}(x) = \frac{\hat{\Phi}_d}{D(\omega,k_y,k_z)} \tag{2}$$

By definition, the natural modes are those that can exist with finite amplitude even in the limit of zero drive. This follows from the fact that the particular solution in Eq. 1 satisfies the driving conditions, so the natural modes must vanish at the driven boundaries. Thus, for given wavenumbers

(k_y,k_z) of the drive, the frequencies s_n must satisfy the dispersion relation

$$D(-js_n,k_y,k_z) = 0 \qquad (3)$$

Alternatively, if it is only the natural modes that are of interest, then the amplitudes are required to satisfy all boundary conditions, including those implied by setting the excitations to zero. In the abstract system of Fig. 5.15.1, $\Phi_d = 0$.

The natural modes identified in this way are only those that can be excited by means of the structure on the transverse excitation boundary. Thus, the implied distributions of sources within the volume are not arbitrary. The functions $\hat{\Phi}_n(x)$ are complete only in the sense that they can be used to represent arbitrary initial conditions on sources induced in this way. They are not sufficient to represent any initial distribution of the fields set up by some other means within the volume. The remainder of this section exemplifies this subject in specific terms. Magnetic diffusion transients, considered in Chap. 6, broaden the class of example.

Transient Charge Relaxation on a Thin Sheet: The build-up or decay of charge on a moving conducting sheet excited by a sinusoidal drive can be described by revisiting the example treated in Sec. 5.13. In terms of the complex amplitude of the sheet potential, $\hat{\Phi}^b$, and with x=0 at the sheet surface, the potential distributions above and below the sheet are (for a discussion of translating coordinate references to fit eigenfunctions to specific coordinates, see Sec. 2.20 in conjunction with Eq. 2.16.15)

$$\hat{\Phi}(x) = \begin{cases} \hat{V}_o \dfrac{\sinh(kx)}{\sinh(kd)} - \hat{\Phi}^b \dfrac{\sinh k(x-d)}{\sinh(kd)}; & x > 0 \\ \hat{\Phi}^b \dfrac{\sinh k(x+d)}{\sinh(kd)}; & x < 0 \end{cases} \qquad (4)$$

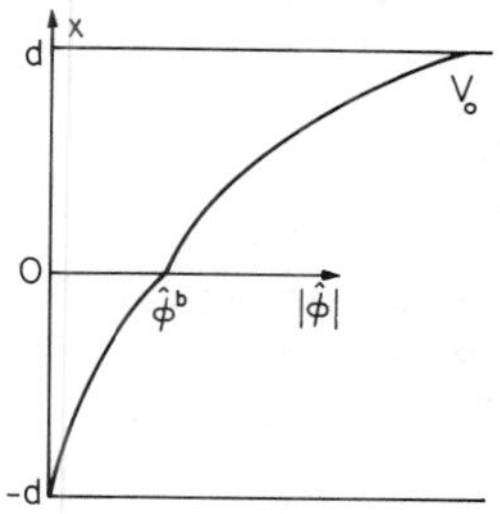

Fig. 5.15.2
Driven response

The eigenfrequency equation is the denominator of Eq. 5.13.7 set equal to zero and evaluated with $j\omega = s_n$:

$$\sinh(kd) + j\frac{2\varepsilon_o}{k\sigma_s}(-js_n - kU)\cosh(kd) = 0 \qquad (5)$$

This expression has only one root,

$$s_1 = jkU - \frac{k\sigma_s}{2\varepsilon_o}\tanh(kd) \qquad (6)$$

The one eigenfunction is determined by using the complex amplitudes of Sec. 5.13 with $j\omega = s_1$ and $\hat{V}_o=0$. In this example, the eigenfunction has the distribution with x of Eq. 4 with $\hat{V}_o=0$, and a complex amplitude $\hat{\Phi}_1$ determined by the initial conditions:

$$\hat{\Phi}_1(x) = \begin{cases} -\hat{\Phi}_1^b \dfrac{\sinh k(x-d)}{\sinh(kd)}; & x > 0 \\ \hat{\Phi}_1^b \dfrac{\sinh k(x+d)}{\sinh(kd)}; & x < 0 \end{cases} \qquad (7)$$

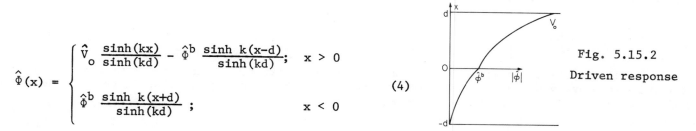

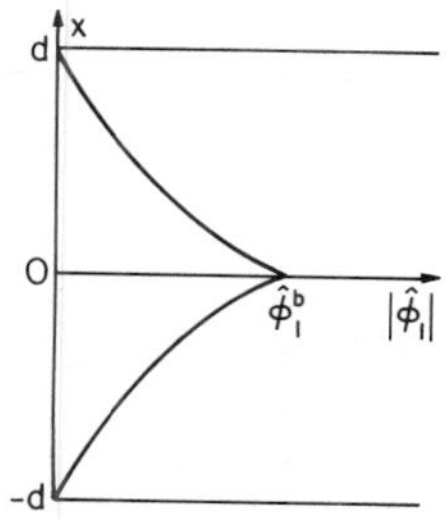

Fig. 5.15.3
Eigenfunction.

In general, the initial condition is on the charge distribution in the region $-d < x < d$. In this example, the charge is confined to the sheet and only the one eigenmode is needed to meet the initial condition.

Suppose that when t = 0 there is no sheet charge and the excitation is suddenly turned on. The total potential is given by Eq. 1 with $\hat{\Phi}(x)$ and $\hat{\Phi}_1(x)$ given by Eqs. 4 and 7. In terms of this potential, the surface charge is in general

$$\sigma_f(z,t) = D_x^b - D_x^c = -\varepsilon_o k \mathrm{Re}\left\{\left[\frac{\hat{V}_o}{\sinh(kd)} - 2\hat{\Phi}^b\coth(kd)\right]e^{j(\omega t-kz)} - 2\hat{\Phi}_1^b\coth(kd)e^{s_1t-jkz}\right\} \qquad (8)$$

To make $\sigma_f(z,0) = 0$, the eigenfunction amplitude must be such that when $t = 0$, Eq. 8 vanishes for all z:

$$\hat{\Phi}_1^b = -\hat{\Phi}^b + \frac{\hat{V}_o}{2\cosh(kd)} \tag{9}$$

When $t = 0^+$, the surface charge density is still zero, but the potential is finite over the entire region $-d < x < d$. It can be shown by using Eq. 9 in Eq. 1 (evaluated when $t = 0$ using Eqs. 4 and 7) that at this instant the potential is what it would be in the absence of the conducting sheet.

The surface charge builds up at a rate determined by s_1, which expresses the natural frequency as seen from a laboratory frame of reference. The oscillatory part is what is observed in the fixed frame as a spatially periodic distribution moves with the velocity of the material. If the driving voltage were suddenly turned off, the fields would decay in a way characterized by the same natural frequencies, with an oscillatory part reflecting the spatial periodicity of the initial charge distribution as it decays with a relaxation time $2\varepsilon_o/k\sigma_s \tanh kd$. Because the electric energy storage is in the free-space region, while the energy dissipation is within the sheet, this damping rate is not simply the bulk relaxation time of the conducting sheet.

In the long-wave limit, $kd \ll 1$, the relaxation in this inhomogeneous system can be largely attributed to energy storage in the transverse electric field and dissipation due to the longitudinal electric field. On a scale of the system as a whole, the charge actually diffuses rather than relaxes. This can be seen by taking the limit $kd \ll 1$ of Eq. 6:

$$s_1 + (-jk)U = \frac{\sigma_s d}{2\varepsilon_o}(-jk)^2 \tag{10}$$

to obtain the dispersion equation for diffusion with convection. By infering time and z derivatives from the complex frequency s and $-jk$ respectively, it can be seen from Eq. 10 that in the long-wave limit the surface charge density is governed by the equation

$$\left(\frac{\partial}{\partial t} + U\frac{\partial}{\partial z}\right)\sigma_f = \frac{\sigma_s d}{2\varepsilon_o}\frac{\partial^2 \sigma_f}{\partial z^2}$$

This model is consistent with the distributed network shown in Fig. 5.15.4. The rigorous deduction of Eq. 11 would exploit the space-rate expansion introduced in Sec. 4.12. The dominant electric fields are $\vec{E} = E_x(z,t)\vec{i}_x$ in the air gaps and $\vec{E} = E_z(z,t)\vec{i}_z$ in the sheet. This model is embedded in the discussion of the Van de Graaff machine given in Sec. 4.14, Eqs. 4.14.9 and 4.14.10.

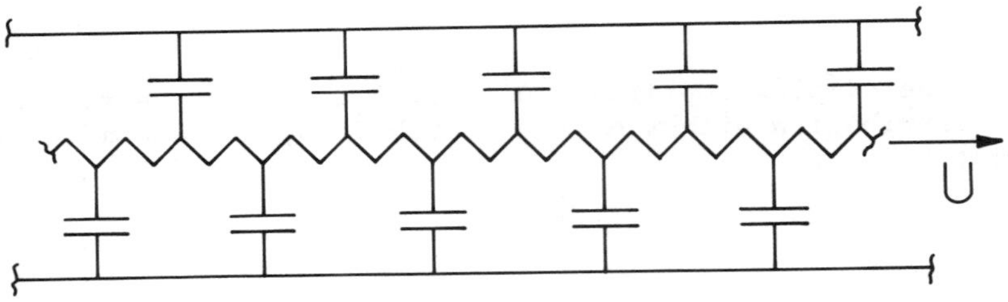

Fig. 5.15.4

Distributed network in the long-wave limit, equivalent to system of Fig. 5.13.1.

<u>Heterogeneous Systems of Uniform Conductors</u>: A generalization of the system of two uniformly conducting regions (the theme of Sec. 5.14) is shown in Fig. 5.15.5. Layers of material, each having the thickness d, have different conductivities and move to the right with the velocity profile $\vec{v} = U(x)\vec{i}_x$. Charge is confined to the interfaces, which have a negligible surface conductivity. Thus, the nth interface moves to the right with the velocity U_n and is bounded from above and below by regions having the uniform properties $(\varepsilon_n, \sigma_n)$ and $(\varepsilon_{n+1}, \sigma_{n+1})$ respectively. Variables evaluated just above and below the nth interface are denoted by n and n' respectively.

In the limit where the number of interfaces, N, becomes large, the "stair-step" conductivity distribution approaches that of a continuous distribution. The following illustrates the second method of determining the natural frequencies, while giving insight as to why an infinite number of natural modes exists in systems having a distributed conductivity.

The regions just above and just below the nth interface are described by the planar transfer relations representing Laplace's equation, Eq. (a) of Table 2.16.1:

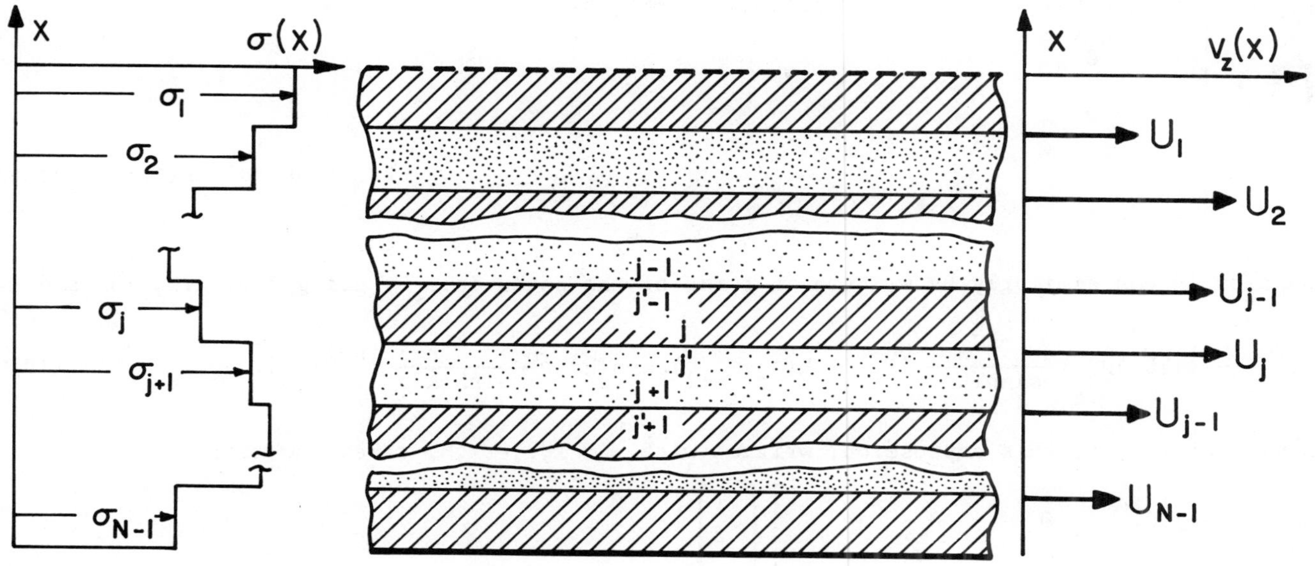

Fig. 5.15.5. A material having a conductivity that depends on x moves to the right with the velocity distribution $v_x = U(x)$.

$$\begin{bmatrix} \hat{D}_x^{n'-1} \\ \hat{D}_x^{n} \end{bmatrix} = \varepsilon_n k \begin{bmatrix} -\coth(kd) & \dfrac{1}{\sinh(kd)} \\ -\dfrac{1}{\sinh(kd)} & \coth(kd) \end{bmatrix} \begin{bmatrix} \hat{\Phi}^{n'-1} \\ \hat{\Phi}^{n} \end{bmatrix} \quad (12)$$

$$\begin{bmatrix} \hat{D}_x^{n'} \\ \hat{D}_x^{n+1} \end{bmatrix} = \varepsilon_{n+1} k \begin{bmatrix} -\coth(kd) & \dfrac{1}{\sinh(kd)} \\ -\dfrac{1}{\sinh(kd)} & \coth(kd) \end{bmatrix} \begin{bmatrix} \hat{\Phi}^{n'} \\ \hat{\Phi}^{n+1} \end{bmatrix} \quad (13)$$

At each interface, the potential is continuous:

$$\hat{\Phi}^{n'-1} = \hat{\Phi}^{n-1}; \quad \hat{\Phi}^{n'} = \hat{\Phi}^{n} \quad (14)$$

With the understanding that the natural modes now identified are associated with the response to potential constraints at the transverse boundaries, potentials at the upper and lower surfaces must vanish:

$$\hat{\Phi}^{o} = 0; \quad \hat{\Phi}^{N+1} = 0 \quad (15)$$

On the nth interface, conservation of charge (Eq. 5.12.4) requires the additional boundary condition:

$$(-s + jkU_n)\sigma_f^n = \frac{\sigma_n}{\varepsilon_n}\hat{D}_x^n - \frac{\sigma_{n+1}}{\varepsilon_{n+1}}\hat{D}_x^{n'} \quad (16)$$

At each interface, the surface charge is related to potentials at that and the adjacent interfaces, as can be seen by using Eqs. 12b, 13a and 14 to write

$$\hat{\sigma}_f^n = \hat{D}_x^n - \hat{D}_x^{n'} = k\left[\frac{-\varepsilon_n\hat{\Phi}^{n-1}}{\sinh(kd)} + (\varepsilon_n + \varepsilon_{n+1})\coth(kd)\hat{\Phi}^n - \frac{\varepsilon_{n+1}\hat{\Phi}^{n+1}}{\sinh(kd)}\right] \quad (17)$$

This expression holds at each of the N interfaces. In view of the boundary conditions at the transverse boundaries, Eqs. 15, Eqs. 17 are N equations for the N σ_f^n's in terms of the interfacial potentials $\hat{\Phi}^n$:

$$\left[\hat{\sigma}_f\right] = \left[A\right]\left[\hat{\Phi}\right] \quad (18)$$

where $[\sigma_f]$ and $[\hat{\Phi}]$ are Nth order column matrices and

$$[A] = \begin{bmatrix} k(\varepsilon_1+\varepsilon_2)\coth(kd) & -k\varepsilon_2/\sinh(kd) & 0 & & & \\ -k\varepsilon_2/\sinh(kd) & k(\varepsilon_2+\varepsilon_3)\coth(kd) & -k\varepsilon_3/\sinh(kd) & 0 & \cdot & \cdot \\ 0 & & & & & \\ 0 & & & & & \\ \cdot & & & 0 & -k\varepsilon_N/\sinh(kd) & k(\varepsilon_N+\varepsilon_{N+1})\coth(kd) \\ \cdot & & & & & \end{bmatrix}$$

Equation 16 can similarly be written in terms of the potentials by using Eqs. 12b, 13a and 14:

$$(-s+jkU_n)\hat{\sigma}_f^n = \frac{-k\sigma_n}{\sinh(kd)}\hat{\Phi}^{n-1} + k(\sigma_n+\sigma_{n+1})\coth(kd)\hat{\Phi}^n - \frac{k\sigma_{n+1}}{\sinh(kd)}\hat{\Phi}^{n+1} \tag{19}$$

In view of Eq. 15, this expression, written with n = 1,2,···,N, takes the matrix form

$$\begin{bmatrix} -s+jkU_1 & 0 & 0 & \\ 0 & -s+jkU_2 & 0 & \\ 0 & & & \\ \cdot & & & \\ \cdot & & & -s+jkU_N \end{bmatrix} \begin{bmatrix} \hat{\sigma}_f^1 \\ \\ \\ \\ \hat{\sigma}_f^N \end{bmatrix} = [B]\,[\hat{\Phi}] \tag{20}$$

where

$$[B] = \begin{bmatrix} k(\sigma_1+\sigma_2)\coth(kd) & \frac{-k\sigma_2}{\sinh(kd)} & 0 & & \\ \frac{-k\sigma_2}{\sinh(kd)} & k(\sigma_2+\sigma_3)\coth(kd) & \frac{-k\sigma_3}{\sinh(kd)} & 0 & \\ 0 & & & 0 & \\ \cdot & & & & \\ \cdot & & 0 & \frac{-k\sigma_N}{\sinh(kd)} & k(\sigma_N+\sigma_{N+1})\coth(kd) \end{bmatrix}$$

Now, if Eq. 18 is inverted, so that $[\hat{\Phi}] = [A]^{-1}[\hat{\sigma}_f]$ and the column matrix $[\hat{\Phi}]$ substituted on the right in Eq. 20, a set of equations are obtained which are homogeneous in the amplitudes $\hat{\sigma}_f^n$,

$$\begin{bmatrix} jkU_1-C_{11}-s & -C_{12} & -C_{13} & \\ -C_{21} & jkU_2-C_{22}-s & -C_{23} & \\ & & & \\ & & & \\ & & & jkU_N-C_{NN}-s \end{bmatrix} \begin{bmatrix} \hat{\sigma}_f^1 \\ \hat{\sigma}_f^2 \\ \cdot \\ \cdot \\ \cdot \\ \hat{\sigma}_f^N \end{bmatrix} = 0 \tag{21}$$

where $[C] = [B]\,[A]^{-1}$.

For the amplitudes to be finite, the determinant of the coefficients must vanish, and this constitutes the eigenfrequency equation $D(s,k_x,k_y) = 0$. The determinant takes the standard matrix form for a characteristic value problem.[1] Expanded, it is an Nth order polynomial in s, and hence has N roots which are the natural frequencies.

As an example, suppose that there is a single interface, N=1. Then, from Eqs. 18 and 20,

$$A^{-1} = \frac{1}{k(\varepsilon_1 + \varepsilon_2)\coth(kd)} \;;\quad B = k(\sigma_1 + \sigma_2)\coth(kd) \tag{22}$$

and it follows that $C_{11} = (\sigma_1 + \sigma_2)/(\varepsilon_1 + \varepsilon_2)$ so that Eq. 21 gives the single eigenfrequency

$$s_1 = jkU_1 - \left(\frac{\sigma_1 + \sigma_2}{\varepsilon_1 + \varepsilon_2}\right) \tag{23}$$

With a = b = d, this result is consistent with setting the denominator of Eq. 5.14.8 equal to zero and solving for $j\omega$.

With two interfaces, there are two eigenmodes, with frequencies determined from Eq. 21:

$$\begin{bmatrix} (jkU_1 - C_{11} - s) & -C_{12} \\ -C_{21} & (jkU_2 - C_{22} - s) \end{bmatrix} = 0 \tag{24}$$

The entries C_{ij} follow from $[C] = [B][A]^{-1}$

$$[C] = \begin{bmatrix} k(\sigma_1+\sigma_2)\coth(kd) & \dfrac{-k\sigma_2}{\sinh(kd)} \\ \dfrac{-k\sigma_2}{\sinh(kd)} & k(\sigma_2+\sigma_3)\coth(kd) \end{bmatrix} \begin{bmatrix} \dfrac{k(\varepsilon_2+\varepsilon_3)\coth(kd)}{DET} & \dfrac{k\varepsilon_2}{DET\ \sinh(kd)} \\ \dfrac{k\varepsilon_2}{DET\ \sinh(kd)} & \dfrac{k(\varepsilon_1+\varepsilon_2)\coth(kd)}{DET} \end{bmatrix}$$

$$= \frac{k^2}{DET}\begin{bmatrix} [(\sigma_1+\sigma_2)(\varepsilon_2+\varepsilon_3)\coth^2(kd) - \dfrac{\varepsilon_2\sigma_2}{\sinh^2(kd)}] & [(\sigma_1\varepsilon_2-\sigma_2\varepsilon_1)\dfrac{\coth(kd)}{\sinh(kd)} \\ [(\sigma_3\varepsilon_2 - \sigma_2\varepsilon_3)\dfrac{\coth(kd)}{\sinh(kd)} & [(\sigma_2+\sigma_3)(\varepsilon_1+\varepsilon_2)\coth^2(kd) - \dfrac{\sigma_2\varepsilon_2}{\sinh^2(kd)}] \end{bmatrix} \tag{25}$$

where

$$DET \equiv k^2[(\varepsilon_1 + \varepsilon_2)(\varepsilon_2 + \varepsilon_3)\coth^2(kd) - \varepsilon_2^2/\sinh^2(kd)]$$

The eigenfrequency equation, Eq. 24, is quadratic in s, and can be solved to obtain the two eigenfrequencies

$$\begin{pmatrix} s_1 \\ s_2 \end{pmatrix} = \tfrac{1}{2}[jk(U_1+U_2)-C_{11}-C_{22}] \pm \sqrt{\tfrac{1}{4}[jk(U_1+U_2)-C_{11}-C_{22}]^2 - (jkU_1-C_{11})(jkU_2-C_{22})-C_{12}C_{21}} \tag{26}$$

where the C_{ij} are given by Eq. 25.

The N eigenmodes can be used to represent the temporal transient resulting from turning on or turning off a spatially periodic drive. Although more complicated, the procedure is in principle much as illustrated in the sheet conductor example. As expressed by Eq. 1, the transient is in general a superposition of the driven response (for the turn on) and the natural modes. The N eigenmodes make it possible to satisfy N initial conditions specifying the surface charges on the N interfaces.

In the limit where N becomes infinite, the number of modes becomes infinite and the physical system is one having a smooth distribution of conductivity, $\sigma(x)$, and permittivity, $\varepsilon(x)$. This infinite

---

1. F. E. Hohn, Elementary Matrix Algebra, 2nd ed., Macmillan Company, New York, 1964, p. 273.

set of internal modes can also be used to account for initial conditions. Such modes are encountered again in Sec. 6.10, in connection with magnetic diffusion, where an infinite number of modes are possible even with systems having uniform properties. What has been touched on here is the behavior of smoothly inhomogeneous systems, described by linear differential equations with space-varying coefficients. The finite mode model, implicit to approximating $\sigma(x)$ and $\varepsilon(x)$ by the stair-step distributions, is one way to take into account the terms $\vec{E}\cdot\nabla\sigma$ and $\vec{E}\cdot\nabla\varepsilon$ in the charge relaxation law, Eq. 5.10.6.

5.16 Time Average of Total Forces and Torques in the Sinusoidal Steady State

Two descriptions are used to generalize the complex amplitude representations describing the steady state response to a sinusoidal drive having the angular frequency ω. If the system is spatially periodic or can be modeled by a portion of a periodic system, a Fourier series generalization of the complex amplitude description is appropriate. If it extends to "infinity," a Fourier transform is conveniently made the complex amplitude. The conventions and formulas for computing the time-average of field products, for example of forces, are summarized in this section.

Fourier Series Complex Amplitudes: With a periodicity length ℓ in the z direction, the Fourier series becomes one of complex amplitudes:

$$A(z,t) = \text{Re}\hat{A}(z,\omega)e^{j\omega t}; \quad \hat{A} = \sum_{n=-\infty}^{+\infty} \hat{A}_n(k_n,\omega)e^{-jk_n z} \tag{1}$$

where $k_n \equiv 2n\pi/\ell$. The series, which determines the phase as well as amplitude of the field at any given point, is in general complex. Thus, $\hat{A}_n$ is not necessarily equal to $\hat{A}^*_{-n}$. Each term in the series can be regarded as a traveling wave with phase velocity ω/k_n. The Fourier amplitudes are determined by multiplying both sides of Eq. 1b by $\exp(jk_m z)$, integrating both sides over the length ℓ and exploiting the orthogonality to solve for $\hat{A}_m$. With $m \to n$,

$$\hat{A}_n = \frac{1}{\ell}\int_0^{\ell} \hat{A}\, e^{jk_n z} dz \tag{2}$$

The time-average of a product of fields A and B, written in this form, is obtained by regarding each series as the complex amplitude (Eq. 2.15.14, with $k \to \omega$ and $z \to t$) to obtain

$$\langle AB\rangle_t = \frac{1}{2}\text{Re}\left[\sum_{n=-\infty}^{+\infty} \hat{A}_n e^{-jk_n z} \sum_{m=-\infty}^{+\infty} \hat{B}^*_m e^{jk_m z}\right] \tag{3}$$

The total time-average force (or some other physical quantity involving the product AB) is the space average of Eq. 3 multiplied by the length. To compute the space-average of the time average, think of writing out the first series in Eq. 3, and then successively multiplying it by each term from the second series and averaging over the length. Each term from the second series forms only one product having a finite integral over the length ℓ, the term with $m = n$. Thus, Eq. 3 becomes

$$\frac{1}{\ell}\int_z^{z+\ell} \langle AB\rangle_t dz = \frac{1}{2}\text{Re}\sum_{n=-\infty}^{+\infty} \hat{A}_n \hat{B}^*_n \tag{4}$$

Application of this expression is illustrated in Sec. 6.4. Its role with respect to Fourier series complex amplitudes is analogous to that of the formula developed next in connection with Fourier transform complex amplitudes.

Fourier Transform Complex Amplitudes: In a spatial transient situation, such as illustrated in Sec. 5.17, the complex amplitude takes the form of a Fourier superposition integral:

$$A(z,t) = \text{Re}\hat{A}(z,\omega)e^{j\omega t}; \quad \hat{A} = \frac{1}{2\pi}\int_{-\infty}^{+\infty} \hat{A}e^{-jkz}dk \tag{5}$$

The Fourier transform is found from the complementary integral

$$\hat{A} = \int_{-\infty}^{+\infty} \hat{A}e^{jkz}dz \tag{6}$$

and is not necessarily real. Hence, $\hat{A}(k)$ is not necessarily equal to $\hat{A}^*(-k)$.

To compute the total time-average force acting over the interval $-\infty < z < \infty$, use is first made of the complex amplitude theorem, Eq. 2.15.14, with $z \to t$ and $k \to \omega$:

$$\langle AB\rangle_t = \frac{1}{2}\text{Re}\,\hat{A}\,\hat{B}^* \tag{7}$$

The integral of this time average over z, perhaps the total time-average force, is

$$\int_{-\infty}^{+\infty} \langle AB \rangle_t \, dz = \frac{1}{2} \text{Re} \int_{-\infty}^{+\infty} \hat{A}(z)\hat{B}^*(z)dz \tag{8}$$

With the objective a Fourier transform analogue of Eq. 4, a convolution integral is defined such that f(0) is the integral required to evaluate Eq. 8:

$$f(\xi) \equiv \int_{-\infty}^{+\infty} \hat{A}(z)\hat{B}^*(z - \xi)dz \tag{9}$$

This function can be written as an integral on k (the equivalent of a summation on n in Eq. 4) by taking its Fourier transform. Then the inverse integral, Eq. 5, is the desired integration on k. Thus, the Fourier transform (defined by Eq. 6) is taken of Eq. 9, to obtain

$$\hat{f}(k) = \int_{-\infty}^{+\infty}\int_{-\infty}^{+\infty} \hat{A}(z)\hat{B}^*(z - \xi)e^{jk\xi}dz\, d\xi \tag{10}$$

Now, the substitution $z - \xi \rightarrow z'$ is made, so that, for an integration holding z fixed, $d\xi = -dz'$ and the limits of integration on ξ are reversed:

$$\hat{f}(k) = \int_{-\infty}^{+\infty}\int_{-\infty}^{+\infty} \hat{A}(z)\hat{B}^*(z')e^{jk(z-z')}\, dz\, dz' \tag{11}$$

Finally, this expression can be factored to make it clear that the transform of the integral defined with Eq. 9 is in fact the product of the individual transforms:

$$\hat{f}(k) = \left[\int_{-\infty}^{+\infty} \hat{A}(z)e^{jkz}\, dz\right]\left[\int_{-\infty}^{+\infty} \hat{B}^*(z')e^{-jkz'}dz'\right] = \hat{A}(k)\hat{B}^*(k) \tag{12}$$

Hence, by using the inverse integral, Eq. 5, it follows that

$$f(\xi) = \frac{1}{2\pi}\int_{-\infty}^{+\infty} \hat{A}(k)\hat{B}^*(k)e^{-jk\xi}\, dk \tag{13}$$

In summary, it has been found that the integration over z called for in Eq. 8 can alternatively be made an integration on k, because substitution of f(0) from Eq. 13 into Eq. 8 gives

$$\int_{-\infty}^{+\infty} \langle AB \rangle_t \, dz = \frac{1}{4\pi} \text{Re} \int_{-\infty}^{+\infty} \hat{A}(k)\hat{B}^*(k)dk \tag{14}$$

Application of this theorem is illustrated in Sec. 5.17.

5.17 Spatial Modes and Transients in the Sinusoidal Steady State

An abstract view of systems that are uniform in a longitudinal direction and inhomogeneous in a transverse direction is shown in Fig. 5.17.1. The thin sheet and finite conductor configurations of Secs. 5.13 and 5.14 are specific examples. In those sections, it is the spatially periodic sinusoidal steady-state response that is emphasized. In any real system, the excitation must be turned on, and so there is a temporal transient before this sinusoidal steady state is established. For spatially periodic systems, Section 5.15 introduced the temporal modes representing this turn-on transient. But, except for systems that are reentrant (for example rotating machines), the spatial extent of the excitation is also limited. In terms of Fig. 5.17.1, where the "system" extends over the length L, the excitation is applied to transverse boundaries of region II. Within this region, the excitation is spatially periodic. It might consist of a pure traveling wave having an "imposed" wavenumber β and frequency ω, or (by superposition) have an arbitrary periodic z-t dependence.

In terms of the longitudinal coordinate z and time t, the general response of the fields in some transverse plane can be pictured as shown in Fig. 5.17.2. When t = 0, the sinusoidal steady state excitation is turned on over region II. At any position along the z-axis, the response of a stable system consists of a transient beginning at the earliest when t = 0 and, as $t \rightarrow \infty$, approaching a temporal sinusoidal steady state with the same frequency ω as the drive. But at any given time t > 0, there is the possibility of a response outside the region ℓ as well as within. In the limit where the driven region is long (or the system is reentrant so that the extremes of region II are in fact the same location), the response in the middle of region II can be expected to have the same spatial periodicity as the drive. This is the limit in which the temporal transient and sinusoidal steady state of Secs. 5.13 or 5.14 and Sec. 5.15 pertain.

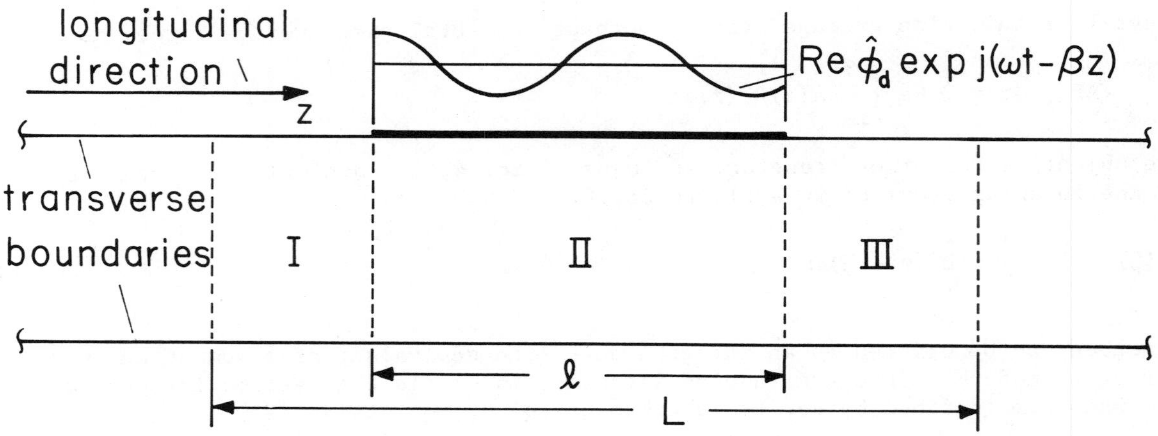

Fig. 5.17.1. Abstract view of systems having excitation on transverse boundaries which are in the temporal sinusoidal steady state but confined to the region ℓ.

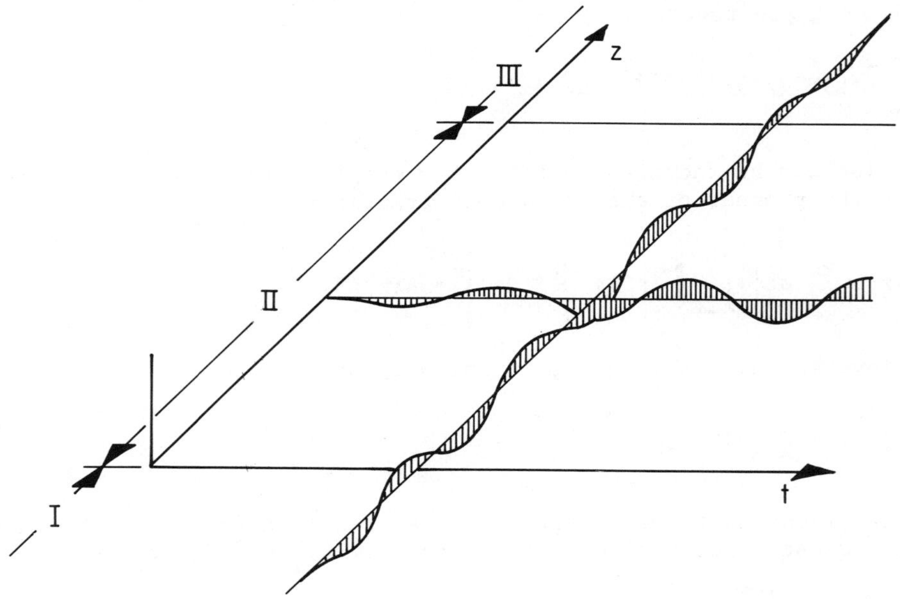

Fig. 5.17.2. Response in a given transverse plane of the system of Fig. 5.17.1 to a pure traveling wave turned on when t=0 and confined to the range $0 < z < \ell$.

In this section, a long enough time has elapsed that the temporal steady state has been established but the spatial extent of the excitation is not large enough to justify ignoring the end effects. A significant portion of region II is not in the spatial sinusoidal steady state. However, time has progressed to the point where the fields at any given location have the same temporal sinusoidal variation as the drive. Implicit to this section is the presumption of stability. If the turn-on transient gives rise to components that grow in time, then these will dominate the temporal sinusoidal steady state presumed to prevail as $t \to \infty$. A related question asks if the spatial transient in Regions I and III actually approaches zero far from the excitation. In this section, it is assumed that this is the case. It will be found in Chap.10 that to identify those systems where this assumption is not well founded it is necessary to consider the entire z-t transient.

<u>Spatial Modes for a Moving Thin Sheet:</u> The configuration shown in Fig. 5.17.3 is the same as that considered in Sec. 5.13, except that the excitation is confined to region II. A thin semi-insulating sheet, moving with velocity U, passes between electrodes constrained in potential as shown. In the range $0 < z < \ell$, the upper wall is excited with the traveling wave of potential. Elsewhere on the walls, both above and below, the potential is zero.

At every position in the system, fields have the same temporal frequency ω as the drive. Thus, at any location the temporal dependence is recovered by the operation

$$\Phi(x,z,t) = \text{Re}\ \hat{\Phi}(x,z,\omega)e^{j\omega t} \tag{1}$$

But then, the spatial Fourier transform of this complex amplitude is

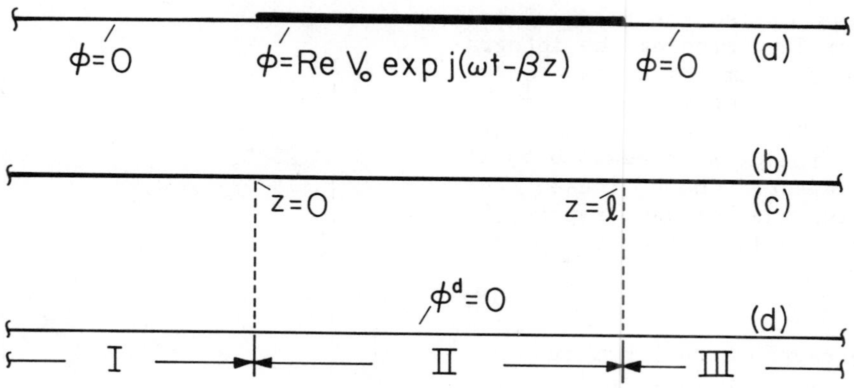

Fig. 5.17.3. A thin sheet moving with velocity U in the z direction enters the excitation region at z = 0 and leaves at z = ℓ.

$$\hat{\Phi}(x,k,\omega) = \int_{-\infty}^{+\infty} \overset{\blacktriangle}{\Phi}(x,z,\omega)e^{jkz} \, dz \qquad (2)$$

with an inverse

$$\overset{\blacktriangle}{\Phi}(x,z,\omega) = \frac{1}{2\pi} \int_{-\infty}^{+\infty} \hat{\Phi}(x,k,\omega)e^{-jkz} \, dk \qquad (3)$$

Because the rule for taking the transform of a derivative with respect to z is the same as if a substitution of the form Φe^{-jkz} is made, relations among complex amplitudes can now be regarded as relations among the Fourier transforms. For the specific problem at hand, these relations are developed in Sec. 5.13 where the Fourier transform of the sheet potential is given by Eq. 5.13.7:

$$\hat{\Phi}^b = \frac{jS_e\hat{\Phi}^a}{2 \sinh(kd) \; [1 + jS_e\coth(kd)]} \qquad (4)$$

$$S_e \equiv \frac{2\varepsilon(\omega - kU)}{k\sigma_s}$$

The fields are completely determined if the driving potential Φ^a is specified. For the traveling-wave driving potential of Fig. 5.17.3,

$$\Phi^a = \mathrm{Re}\,\hat{V}_o e^{-j\beta z} [u_{-1}(z) - u_{-1}(z - \ell)]e^{j\omega t} \qquad (5)$$

where $u_{-1}(z)$ is the function; unity for z > 0, 0 for z < 0. From Eq. 2, the transform follows as

$$\hat{\Phi}^a = \frac{\hat{V}_o}{j(k - \beta)} [e^{j(k-\beta)\ell} - 1] \qquad (6)$$

Thus, Eqs. 3 and 4 give the complex amplitude of the sheet potential as

$$\overset{\blacktriangle}{\Phi}^b = \frac{1}{2\pi} \int_{-\infty}^{+\infty} \frac{\hat{V}_o d\varepsilon(\omega - kU)(e^{j(\ell-z)k} e^{-j\beta\ell} - e^{-jkz})}{\sigma_s(k - \beta)D(\omega,k)} \, dk \qquad (7)$$

where $D(\omega,k)$ is the dispersion equation, Eq. 5.15.5, familiar from discussions of the temporal natural modes. In terms of normalized variables,

$$D(\omega,k) = k \sinh k + jU(\omega - k)\cosh k \qquad (8)$$

where

$$\underline{k} \equiv kd, \quad \underline{\omega} \equiv \frac{\omega d}{U}, \quad \underline{U} \equiv (\frac{2\varepsilon_o}{\sigma_s})U$$

The integration called for in Eq. 7 is conveniently performed by closing the integration at infinity in the complex k plane and evaluating by Cauchy's integral theorem.[1] The contributions to the integration are then seen to be a sum of residues determined by the zeros of the denominator. One of these, $k = \beta$, is associated with the "driven response," while the others are residues from the poles:

$$D(\omega,k) = 0 \tag{9}$$

Remember, ω is a prescribed real number. The roots of Eq. 9, k_n, are in general complex and are each associated with an eigenfunction $\hat{\Phi}_n(x)e^{-jk_nz}$ that satisfies all of the bulk conditions and boundary conditions in the interval $-d < x < d$ with the drive set equal to zero. Over the cross section, the eigenfunctions associated with a given root of Eq. 9 are

$$\hat{\Phi}_n(x) = \begin{cases} -\hat{\Phi}^b_n \dfrac{\sinh k_n(x-d)}{\sinh k_n d} & x > 0 \\ \hat{\Phi}^b_n \dfrac{\sinh k_n(x+d)}{\sinh k_n d} & x < 0 \end{cases} \tag{10}$$

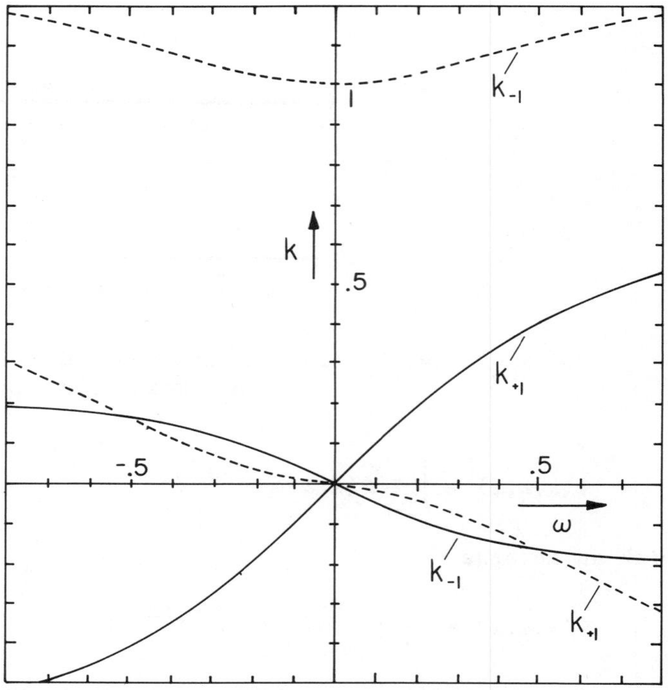

Fig. 5.17.4. Normalized complex wavenumber of lowest modes as function of normalized frequency.

The complex roots of Eq. 9 must be found numerically. However, the dominant roots are easily identified in the long-wave limit $|kd| << 1$ because then Eq. 9 is quadratic in k and can be solved for the two roots,

$$k_{\underline{+}1} = \left[j\frac{U}{2} \pm \sqrt{(\omega^2 - \frac{1}{2})\frac{U^2}{2} - j\omega U}\right]/(1 + \frac{j\omega U}{2}) \tag{11}$$

(Note that these are the same roots that would be determined from Eq. 5.15.10, with $s_1 \to j\omega$, and are therefore the only ones retained by a quasi-one-dimensional model.) Typical roots of Eq. 11, as a function of real ω, are shown in Fig. 5.17.4.

For kd not small compared to unity these roots retain the same qualitative nature. Thus k_{-1} and k_1 are respectively waves that have phase velocities in the - and + directions with the first decaying rapidly in the -z direction and the second decaying slowly in the +z direction. Although it is not in general possible to attribute certain of the modes to one aspect of the system or another, these two dominant modes are associated closely with the spectral build-up and decay of surface charge on the sheet.

The higher order modes are more closely connected with the fields that would exist in the free space regions in the absence of the sheet. In the limit where U is large enough that $U >> \omega/k$, the term in ω in Eq. 8 is ignorable, so that approximately

$$1 = jU\coth k \tag{12}$$

This expression has an infinite number of purely imaginary solutions $k = jk_i$, as can be seen by substituting into Eq. 12 to obtain

$$\tan k_i = U \tag{13}$$

which can be solved graphically. In the limit $\underline{U} >> 1$, roots $\underline{k} \to \underline{+}j(2n' - 1)\pi/2$, where n' is an integer. Note that these are the eigenmodes that would be obtained if the sheet were absent. The x distribution of potential associated with these approximate eigenvalues is given by Eq. 10, and is sinusoidal. The associated pure decay in the $\underline{+}z$ directions is typical of solutions to Laplace's equation that are periodic in x.

---

1. F. B. Hildebrand, Advanced Calculus for Applications, Prentice Hall, Englewood Cliffs, N.J., 1962, p. 548.

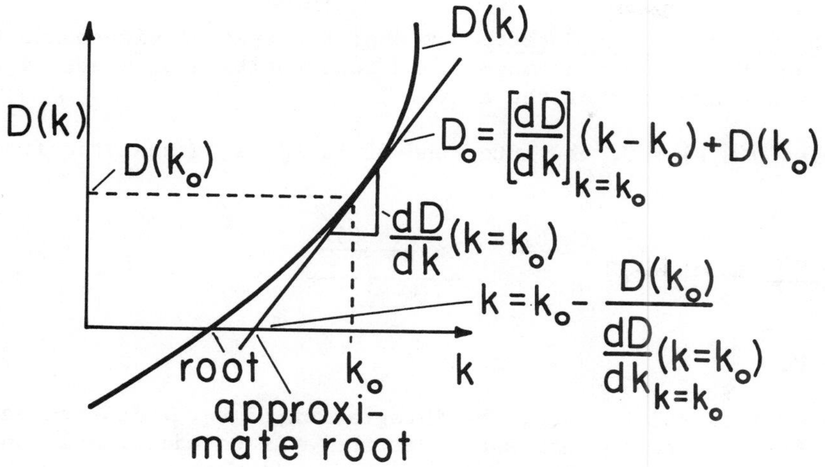

Fig. 5.17.5. Numerical solution illustrated graphically. The zero of the complex function D(k) of the complex variable is approximated at the trial value k_o by a straight line. The approximate root follows by setting $D_o(k) = 0$. This root can then be used as k_o in refining the approximation and the process repeated until the desired accuracy is obtained. To obtain roots of Fig. 5.17.6, k_o is first approximated by Eqs. 11 and 13.

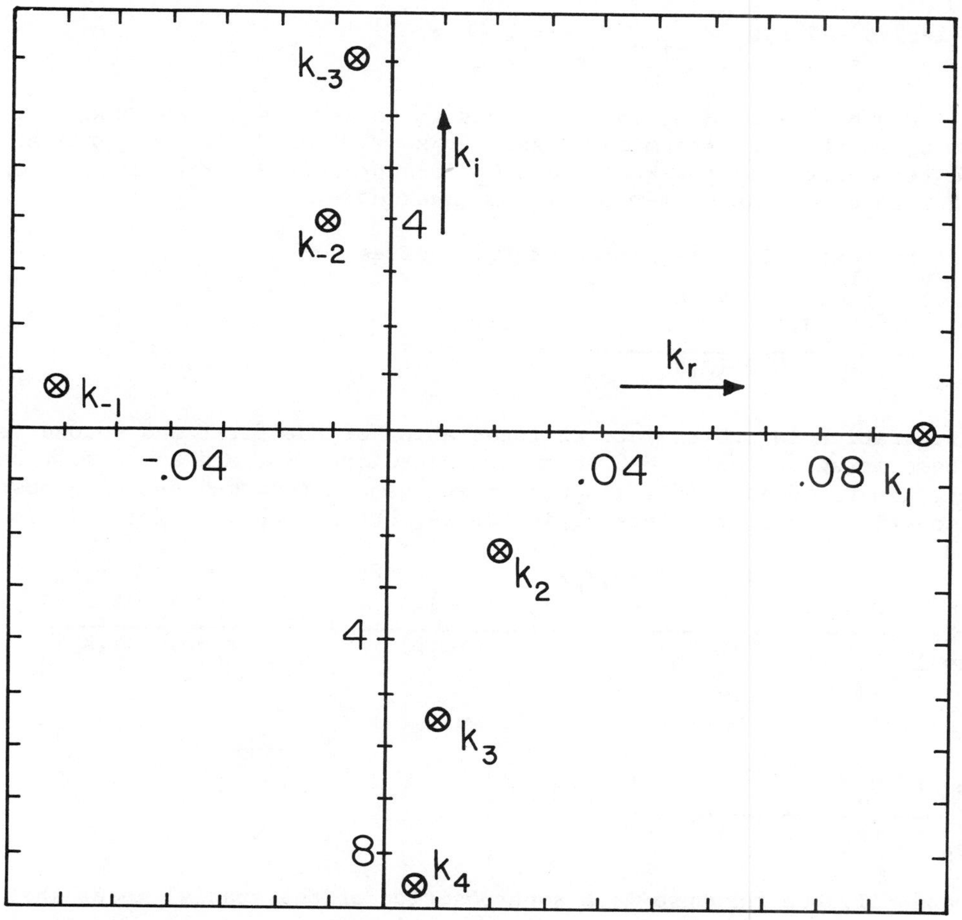

Fig. 5.17.6. Wavenumber eigenvalues given by Eqs. 8 and 9 for case $\underline{\underline{\omega U}} = 0.1$, $\underline{U} = 1$.

The numerical solution of $D(\omega,k)$ is described in Fig. 5.17.5. Given in Fig. 5.17.6 are specific roots conveniently found by using the approximate roots given from Eqs. 11 and 13 as a first approximation. Roots are denoted by the integer n, which ranges from $-\infty$ to $+\infty$, with $n = 0$ omitted.

Spatial Transient on Moving Thin Sheet: Now that the spatial eigenmodes have been found, consider how the integral solution, Eq. 7, is tantamount to a superposition of these eigenmodes and, in region II, a "driven" response with wavenumber β of the drive.

Except at the poles $D(\omega,k) = 0$, the integrand of Eq. 7 is an analytic function. This is even true at $k = \beta$, because

$$\frac{e^{j(\ell-z)k}e^{-j\beta\ell}-e^{-jkz}}{k-\beta} = je^{-jkz}e^{j\frac{(k-\beta)\ell}{2}}\frac{\sin\left[\frac{(k-\beta)\ell}{2}\right]}{\left[\frac{(k-\beta)}{2}\right]} \tag{14}$$

is not singular at $k = \beta$.

To apply the Cauchy integral theorem, the integration of Eq. 7 is extended to an integration around a closed contour, with the closure defined such that there is no additional contribution to the integral. For integration around a contour C in the counterclockwise direction,

$$\oint_C \frac{N(k)}{D(k)}\,dk = 2\pi j[K_1 + K_2 + \cdots] \tag{15}$$

where the residues K_n (at isolated singular points $k = k_n$) of a function $N(k)/D(k)$ are $N(k_n)/D'(k_n)$. Which of the contours shown in Fig. 5.17.7 is appropriate depends on the range of z of interest. With the three regions defined in Fig. 5.17.3, the appropriate contours are identified as follows. First, observe that with $k = k_r + jk_i$, the two z-dependent terms in Eq. 7 can be written as

$$e^{j(\ell-z)k} = e^{j(\ell-z)k_r}e^{-(\ell-z)k_i};\quad e^{-jkz} = e^{-jk_rz}e^{k_iz} \tag{16}$$

Thus, in region I, $z < 0$ and $(\ell - z) > 0$, so both terms go to zero as $k_i \to \infty$ and C_1 is appropriate. In region II, $(\ell - z) > 0$, so the first term converges for $k_i \to \infty$ and C_1 is appropriate. Also in region II, $\ell < z$, so the second term converges for $k_i \to -\infty$ and C_2 is appropriate. Finally, in region III, $\ell < z$ and $(\ell - z) < 0$, so each term decays as $k_i \to -\infty$ and C_2 is appropriate.

It follows that in region I, integration of Eq. 7 gives

$$\hat{\Phi}^b = \frac{j\hat{V}_o d\varepsilon}{\sigma_s}\sum_{n=-1}^{-\infty}\frac{(\omega - k_nU)[e^{j(k_n-\beta)\ell} - 1]}{(k_n - \beta)D'(\omega,k_n)}e^{-jk_nz} \tag{17}$$

In region II, the integration is broken into an integration of the first and second terms individually. Thus, for each of these integrations, $k = \beta$ becomes a singular point and $k - \beta$ must be included with $D(\omega,k)$ in determining the residues. This singular point can be regarded as being just below the axis, and hence as contributing to the integration on C_2, but not on C_1. Then, it follows that in region II

$$\hat{\Phi}^b = \frac{j\hat{V}_o d\varepsilon}{\sigma_s}\left\{\sum_{n=-1}^{-\infty}\frac{(\omega-k_nU)[e^{j(k_n-\beta)\ell}]e^{-jk_nz}}{(k_n-\beta)D'(\omega,k_n)} + \frac{(\omega-\beta U)e^{-j\beta z}}{D(\omega,\beta)} + \sum_{n=1}^{\infty}\frac{(\omega-k_nU)e^{-jk_nz}}{(k_n-\beta)D'(\omega,k_n)}\right\} \tag{18}$$

Finally, in region III

$$\hat{\Phi}^b = -\frac{j\hat{V}_o d\varepsilon}{\sigma_s}\left\{\sum_{n=1}^{\infty}\frac{(\omega-k_nU)[e^{j(k_n-\beta)\ell} - 1]}{(k_n-\beta)D'(\omega,k_n)}e^{-jk_nz}\right\} \tag{19}$$

In regions I and III, the response is a superposition of the spatial modes that decay in the -z and +z directions, respectively. These are the bow and stern waves. In region II, all of the spatial modes are involved in accounting for the finite length of the traveling-wave excitation. In addition there is the "driven" response at the same wavenumber as the excitation, the second term in Eq. 18. Note that for positions z well away from both ends, for example at $z = \ell/2$, the sums over the natural modes in Eq. 18 approach zero while the driven response that remains is the spatial sinusoidal steady-state response found in Sec. 5.13.

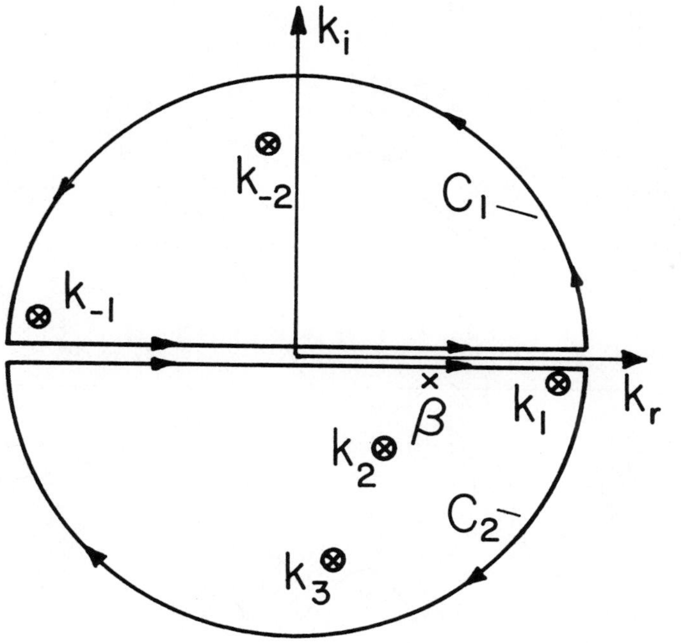

Fig. 5.17.7
Contours used to evaluate the integral of Eq. 7.

As a useful longwave approximation, only the two lowest spatial modes are used with k_1 and k_{-1} given by Eq. 11 and

$$D(\omega,k) \rightarrow d^2(k - k_1)(k - k_{-1}) \tag{20}$$

Then, Eqs. 17-19 reduce to:

$$\hat{\Phi}^b = \frac{j\hat{V}_o\varepsilon}{\sigma_s d}\left\{\frac{(\omega-k_{-1}U)(e^{j(k_{-1}-\beta)\ell} - 1)}{(k_{-1}-\beta)(k_{-1}-k_1)} e^{-jk_{-1}z}\right\};\text{ region I} \tag{21}$$

$$\hat{\Phi}^b = \frac{j\hat{V}_o\varepsilon}{\sigma_s d}\left\{\frac{(\omega-k_{-1}U)e^{jk_{-1}(\ell-z)}e^{-j\beta\ell}}{(k_{-1}-\beta)(k_{-1}-k_1)} + \frac{(\omega-\beta U)e^{-j\beta z}}{(\beta-k_1)(\beta-k_{-1})} + \frac{(\omega-k_1U)e^{-jk_1z}}{(k_1-\beta)(k_1-k_{-1})}\right\};\text{ region II} \tag{22}$$

$$\hat{\Phi}^b = \frac{-j\hat{V}_o\varepsilon}{\sigma_s d}\left\{\frac{(\omega-k_1U)(e^{j(k_1-\beta)\ell} - 1)e^{-jk_1z}}{(k_1-\beta)(k_1-k_{-1})}\right\};\text{ region III} \tag{23}$$

The z-t dependence of the sheet potential, recovered by using these equations in Eq. 1, is illustrated in Fig. 5.17.8.

Time-Average Force: To compute the total time-average force acting on the sheet, the steps are the Fourier transform extension of those leading from Eq. 5.13.5 to Eq. 5.13.11. The total force is the integral over the length of the sheet oι the time-average surface force density. This is in turn written as an integration over the wavenumbers, in accordance with Eq. 5.16.14:

$$\langle f_z\rangle_t = \frac{w}{2}\text{Re}\int_{-\infty}^{+\infty}\hat{E}_z^b(\hat{D}_x^b - \hat{D}_x^c)^* dz = w\text{Re}\frac{1}{2\pi}\int_{-\infty}^{+\infty}\frac{1}{2}\hat{E}_z^b(\hat{D}_x^b - \hat{D}_x^c)^* dk \tag{24}$$

Because $\hat{E}_z^b = jk\hat{\Phi}^b$, the integrand of Eq. 24 is the same as Eq. 5.13.4. Thus, with $\hat{V}_o \rightarrow \hat{\Phi}^a$, steps paralleling those of Eqs. 5.13.5 and 5.13.6 give the total time-average force as simply

$$\langle f_z\rangle_t = \frac{w}{2\pi}\int_{-\infty}^{+\infty}\frac{\varepsilon_o k^2 S_e \hat{\Phi}^a\hat{\Phi}^{a*}dk}{4\sinh^2(kd)[1 + S_e^2\coth^2(kd)]} \tag{25}$$

For the excitation represented by Eq. 6, this expression becomes

$$\langle f_z\rangle_t = \frac{w\varepsilon_o}{2\pi}\left|\hat{V}_o\right|^2\int_{-\infty}^{+\infty}\frac{k^2S_e\sin^2[\frac{\ell}{2}(k - \beta)]dk}{(k-\beta)^2\sinh^2(kd)[1 + S_e^2\coth^2(kd)]} \tag{26}$$

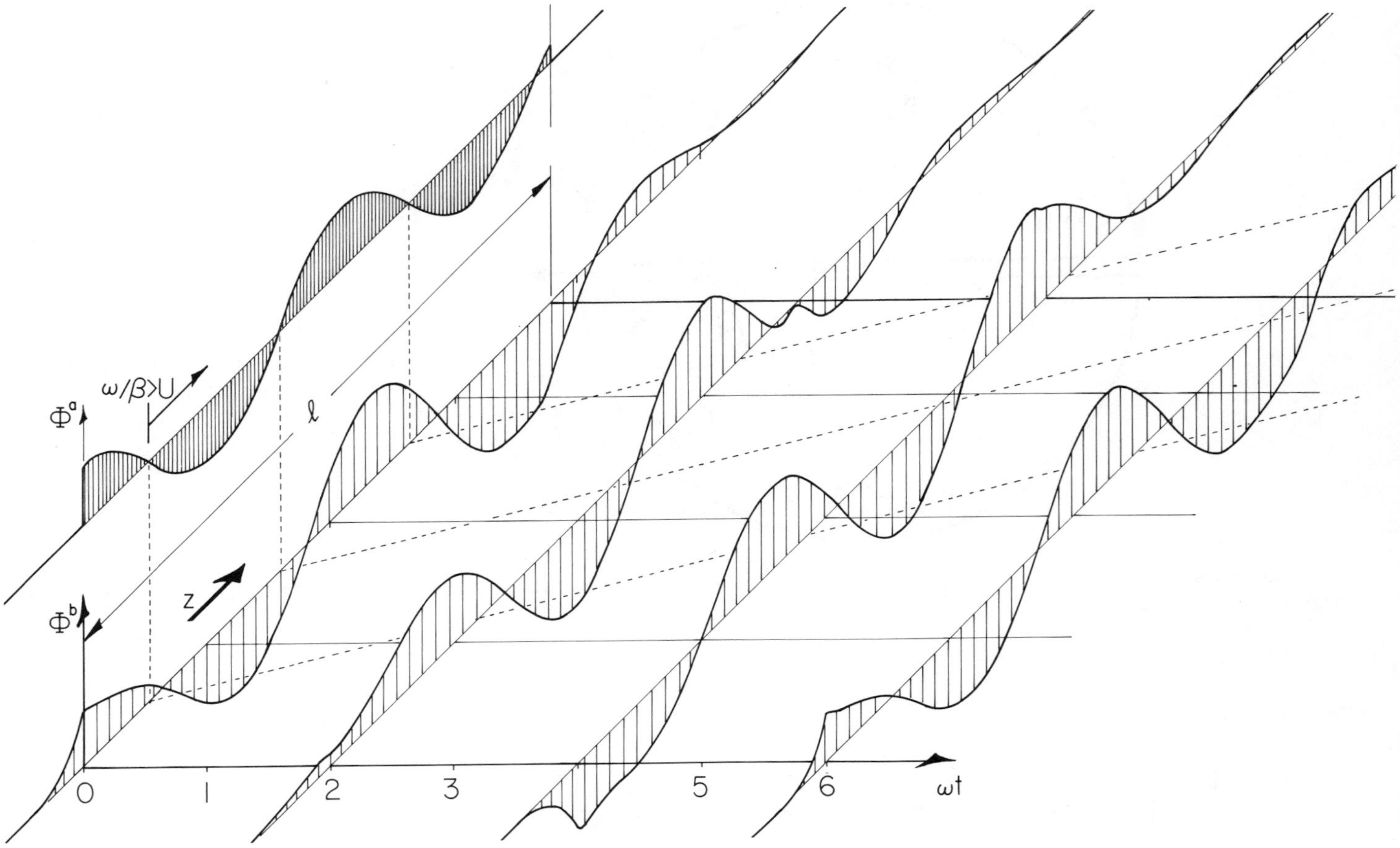

Fig. 5.17.8. Sheet potential Φ^b given by Eqs. 20-23 as a function of z and the normalized time ωt. The excitation potential Φ^a is also shown when t = 0. It takes the form of a traveling wave confined to the structure length ℓ with phases following the broken lines in the z-ωt plane. Note that, in the region under the excitation electrodes, the sheet potential, Φ^b, tends to a spatially periodic response. At any given location z, the fields are temporally periodic with the frequency ω. For the case shown, $\underline{\omega} = 0.5$, $\underline{U} = 0.5$ and $\underline{\beta} = 0.3$ so that $\underline{\omega}/\underline{\beta} = (\omega/\beta)/U > 1$, and the stator-wave phase velocity exceeds the sheet velocity.

Because the integrand of Eq. 26 is positive definite, and has a denominator that increases exponentially for large kd, numerical integration is straightforward. Typical results are illustrated by Fig. 5.17.9. For motor operation, the peak force per unit area and general frequency dependence is diminished by the end effects.

The integration over the Fourier components used to compute the total force in this section is one of two alternative approaches that can be used. In the second approach, the fields (expressed as functions of z) can be used to represent the stress, and this integrated on z to find the total force. The most convenient control volume is one that encloses the sheet, but extends across the air gap so that it has surfaces contiguous with the (a) and (d) surfaces of Fig. 5.17.3. Because the electric shear stress on the (a) and (d) surfaces is confined to the region between z = 0 and z = ℓ on the (a) surface, the integration reduces to one over that interval only. Care must be taken to include the singularities in E_z that appear at end points of the interval.

Fig. 5.17.9

Normalized force per unit length, Eq. 26, as a function of normalized frequency showing "end effect." The number of poles, p ≡ ($\underline{\beta}/\pi$)(ℓ/d) (the number of half-wave-lengths), is the parameter and $\underline{U}$ and $\underline{\beta}$ are 0.5 and 0.3 respectively. Note that the phase velocity of the drive exceeds that of the material velocity for $\underline{\omega} > \underline{\beta} = 0.3$.

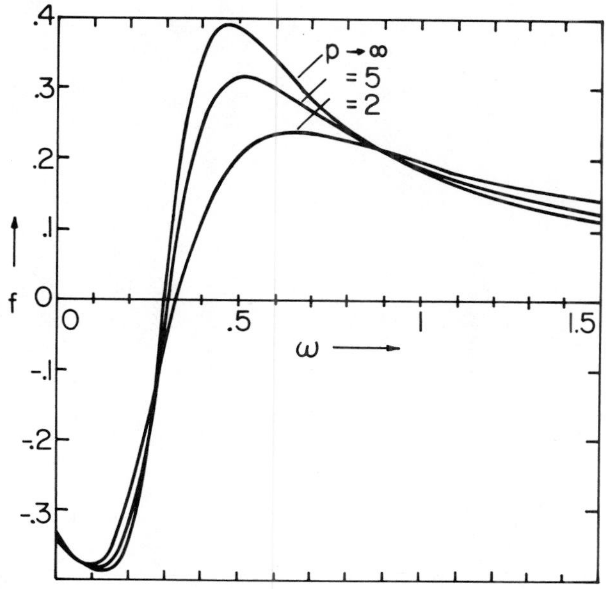

For Section 5.3:

Prob. 5.3.1 For flow and field that are two-dimensional and represented in Cartesian coordinates using the definitions suggested by Table 2.18.1, show that lines along which the charge density is constant are represented by Eq. 5.3.13a.

Prob. 5.3.2 For flow and field that are axisymmetric in cylindrical coordinates, as represented by that case in Table 2.18.1, show that lines along which the charge density is constant are given by Eq. 5.3.13b.

For Section 5.4:

Prob. 5.4.1 Gas passes through the planar channel shown in Fig. P5.4.1 with the velocity $4U(x/d)[1 - x/d]\vec{i}_y$. An electric field is imposed by placing the lower plane at potential V relative to the upper one. Between x = 0 and x = a on this lower plane, positively charged particles having mobility b are injected through a metallic grid. A goal is to determine the current i collected by an electrode imbedded opposite the injection grid. It is presumed that the potential of this electrode remains essentially zero.

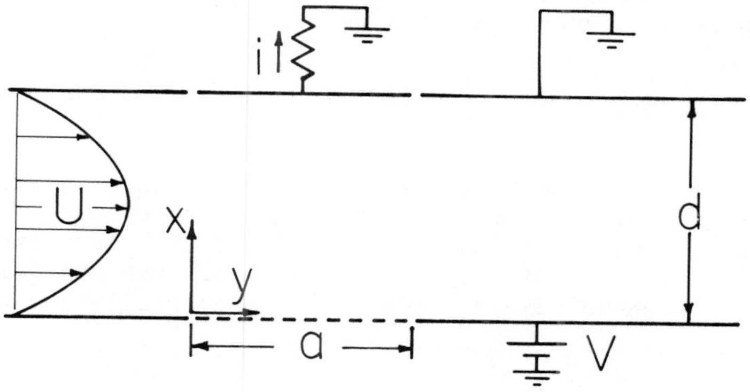

Fig. P5.4.1

(a) Use the result of Prob. 5.3.1 to show that the injected particles follow the characteristic lines

$$-2\frac{U}{d}x^2\left(1 - \frac{2x}{3d}\right) + \frac{bV}{d}y = \text{constant}$$

(b) Show that the current-voltage relation is

$$i = \begin{cases} \frac{bV}{d} nq\left[a - \frac{2}{3}\frac{Ud}{(bV/d)}\right], & V > \frac{2}{3}Ud^2/ba \\ 0, & V < \frac{2}{3}Ud^2/ba \end{cases}$$

Prob. 5.4.2 The potential of a spherical particle having radius R is constrained to be

$$\Phi(r=R) = V\cos\theta$$

(This could be accomplished by making the surface from electrode segments, properly constrained in potential.) The sphere is surrounded by fluid generally moving in the z direction. The flow is solenoidal and irrotational, consistent with its being inviscid and entering at $z \to -\infty$ without rotation. (See Fig. P5.4.1. Such flows are taken up in Chap. 7.) The fluid flow velocity is given as

$$\vec{v} = -\nabla\Phi_v \quad ; \quad \Phi_v = -UR\left[\frac{r}{R} + \frac{1}{2}\frac{R^2}{r^2}\right]\cos\theta$$

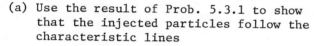

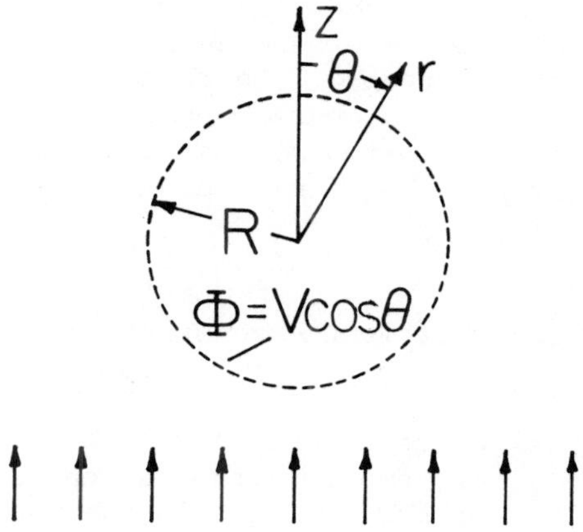

Fig. P5.4.2

There are no other sources of field than those on the sphere itself. The following steps establish the electrical current on the sphere created by ions entering uniformly with the fluid at $z \to -\infty$.

(a) Assume that the contribution of the ion space charge to the field is negligible, and represent $\vec{E}$ and $\vec{v}$ in terms of Λ_E and Λ_v.

Prob. 5.4.2 (continued)

(b) Find the expression for the particle trajectories in the form

$$f(\frac{r}{R}, \theta, \frac{Vb}{UR}) = \text{constant}$$

(c) Assume that $V > 0$ and that the ions are positive. Find the critical points in the region outside the sphere.

(d) Plot the characteristic lines in two cases: for $bV/RU < \frac{3}{2}$ and for $bV/RU > \frac{3}{2}$. Identify the critical points in the case where they exist in the region outside the sphere.

(e) Find the current i to the particle as a function of bV/RU. (Be sure to identify any "break points" in this V-i relation.

Prob. 5.4.3 A circular cylindrical conductor having radius a has the potential V relative to a surrounding coaxial cage having radius R_o (Fig. P5.4.3). Hence it imposes an electric field $\vec{E} = (V/r)/\ln(R_o/a)$ on the air in the region $a < r < R_o$. The wind passing perpendicular to this conductor has the velocity

$$\vec{v} = -U(1 - \frac{a^2}{r^2}) \cos\theta\, \vec{i}_r + U(1 + \frac{a^2}{r^2}) \sin\theta\, \vec{i}_\theta$$

consistent with an inviscid model. (Thus, there is a finite tangential wind velocity at the surface of the conductor.) Charged particles enter uniformly at the appropriate "infinity." This might be a model for the contamination of a high-voltage d-c conductor by naturally charged dust.

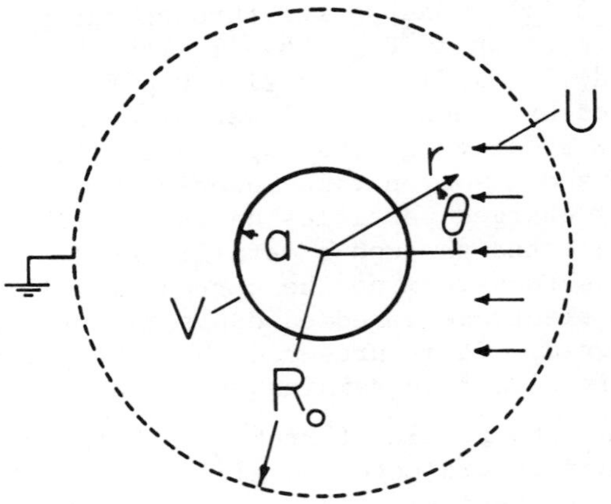

Fig. P5.4.3

(a) Consider two cases: (i) conductor and particles of the same polarity and, (ii) conductor and particles of opposite polarity. This is equivalent to taking the particles as positive and V as positive or negative. Find the critical points (lines).

(b) Find the characteristic lines and sketch them for the two cases.

(c) Determine the electrical current to the conductor as a function of V.

Prob. 5.4.4 Fluid enters the region between the electrodes shown in Fig. P5.4.4 through a slit at the top (where $x = c$). The system extends a length ℓ into the paper and the volume rate of flow through the slit is Q_V m^3/sec. The electrodes to left and right respectively are located at $xy = -a^2$ and $xy = a^2$ and have the constant potentials $-V_o$ and V_o. The electrodes in the plane $x = 0$ are essentially grounded, with the one between $x = -a$ and $x = a$ used to collect the current i. Entrained in the gas as it enters at $x = c$ is a charge density that is uniform over the cross section at that location. The charge density is ρ_o. The fluid velocity is

$$\vec{v} = 2C(x\vec{i}_x - y\vec{i}_y)$$

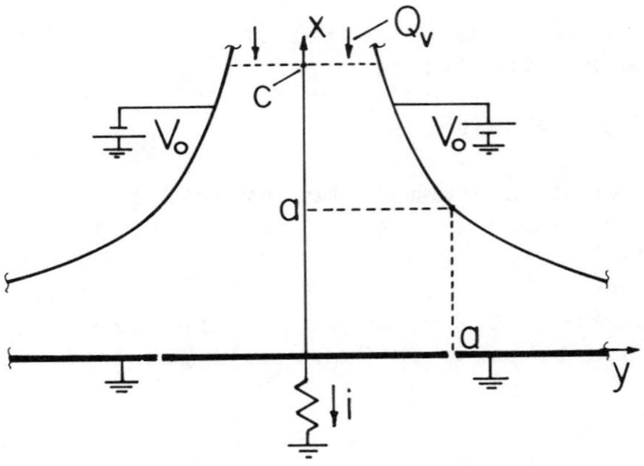

Fig. P5.4.4

(a) What is the constant C?

(b) Find the critical lines, if any.

(c) Given a certain volume rate of flow Q_V, find the current i to the center electrode as a function of bV_o, where b is the mobility of the charged particles. Present $i(bV_o)$ as a dimensioned sketch. (Assume that Q_V and V_o, as well as the charge density ρ_o, are positive.)

For Section 5.5:

Prob. 5.5.1 For a "drop" in an ambient electric field and flow as discussed in this section, both positive and negative "ions" are present simultaneously. The objective here is to make a charging diagram patterned after those of Figs. 5.5.3 and 5.5.4. Because there are now two different mobilities, b_+ and b_-, it is best to make the abscissa the imposed electric field $\vec{E}$. Construct the charging diagram, including charging trajectories, showing final values of charge. (With bipolar charging, the final charge can be less than q_c in magnitude. Expressions should be derived for these limiting values of charge.)

Prob. 5.5.2 The objective is to determine the charging diagrams, Figs. 5.5.3 and 5.5.4, with the low Reynolds number flow represented by Eq. 5.5.5 replaced by an inviscid flow. (See Sec. 7.8 for discussion of this class of flows.) Important here is the fact that such a flow can have a finite tangential velocity on a rigid boundary. The fluid velocity is given here as

$$\vec{v} = -U[1 - (\frac{R}{r})^3] \cos\theta\vec{i}_r + U[1 + \frac{R^3}{2r^3}] \sin\theta\vec{i}_\theta$$

(a) Find Λ_v and the general characteristic equation that replaces Eq. 5.5.6.

(b) Because both tangential and normal velocity are zero on the surface of the "drop" for the low Reynolds number flow, the points on the surface described by Eq. 5.5.10 are critical points. With an inviscid flow, matters are not so simple. Show that, as before, there are now two types of critical points, one type lying on the z axis and the other not. Find analytical expressions for the (r,θ) locations of these latter critical lines.

(c) Construct the charging diagrams for positive and negative "ions."

For Section 5.6:

Prob. 5.6.1 Unless some of an initial charge distribution reaches a boundary, self-precipitating charge of one polarity must conserve its total value. With the charge density given as a function of time by Eq. 5.6.6 and the volume filled by this density described by Eqs. 5.6.9 and 5.6.10, show that for the example of Fig. 5.6.3 this is indeed the case.

Prob. 5.6.2 Fig. P5.6.2 shows a one-dimensional configuration involving a unipolar conduction transient. Gas flows through a duct with the uniform velocity $U\vec{i}_z$. Screen electrodes at $z = 0$ and $z = \ell$ have the constant potential difference v. When $t = 0$, there is a uniform distribution of charged particles having charge density ρ_o and mobility b in the region between $z = z_B$ and $z = z_F$. The regions in front of this layer and behind it have no initial charge density. Assume that the charge is positive. In the following the evolution of the layer is to be described during the time that it has not encountered the screen electrodes.

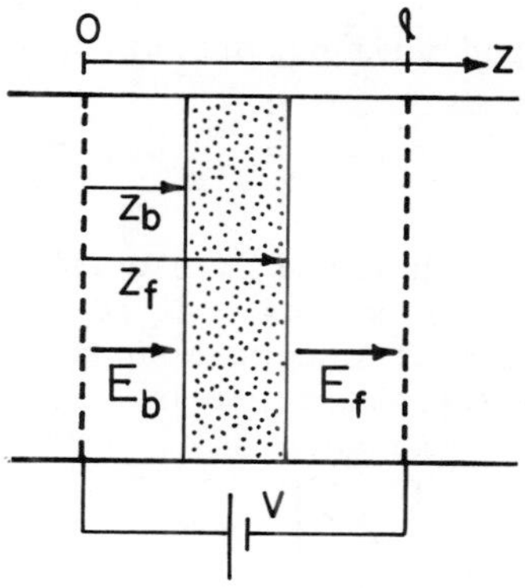

Fig. P5.6.2

(a) Show that the charge density within the layer remains uniform and find its dependence on time.

(b) Use Gauss' law to deduce that $(z_f - z_b) = (1 + t/\tau)(z_F - z_B)$; $\tau \equiv \varepsilon_o/\rho_o b$.

(c) Use Gauss' law and the potential constraint to relate $E_b(t)$, $E_f(t)$, $z_b(t)$ and $z_f(t)$.

(d) Use the second characteristic equations to also relate these four quantities.

(e) Find $z_f(t)$ and $z_b(t)$ and sketch the charge evolution in the z-t plane (as in Fig. 5.6.3).

For Section 5.7:

Prob. 5.7.1 The steady-state charge distribution of Eq. 5.7.3 is time-varying from the particle frame of reference. Hence, in accordance with Eqs. 5.6.2 and 5.6.3, the charge density decays from the frame of reference of a given particle. Start with these characteristic equations and deduce Eq. 5.7.3.

For Section 5.9:

Prob. 5.9.1 When $t = 0$, a region of fluid described by the bipolar laws, Eqs. 5.8.9 and 5.8.10, has uniform neutral density n_o and species charge densities $\rho_+ = \rho_- = 0$. A self-consistent picture of the ensuing dynamics has these densities evolving uniformly. This is possible because there is no applied

Prob. 5.9.1 (continued)

electric field and because $\rho_+ = \rho_-$, so there is no self-field either.

(a) Use the conservation laws to show that $\rho_+ = \rho_-$ is consistent with E = 0.

(b) Write an ordinary differential equation for n(t) and one for $\rho_\pm$ (t).

(c) Argue that the stationary equilibrium state is one having $\beta n = \frac{\alpha}{q}\rho_+ \rho_-$.

(d) Show that the time characterizing the early stages of the system's approach to this equilibrium is $\tau_{th} = q/\beta$.

For Section 5.10:

Prob. 5.10.1 (conductivity model) In the region 0 < x < d, the fluid velocity is $\vec{v} = U(x/d)\vec{i}_z$. When t = 0, the volume charge density is zero for z < 0 and is a constant ρ_o for 0 < z. Describe $\rho_f(x,z,t)$ for t > 0. Represent the distribution in the (x-z) plane, giving analytical expressions for wavefronts and decay rates.

Prob. 5.10.2 (conductivity model) The fluid velocity is as in Prob. 5.10.1. When t = 0, $\rho_f(x,z) = 0$ for z > 0. A source of charge is used to constrain the charge density to be a step function in the z = 0 plane. That is, $\rho_f(x,0,t) = \rho_s u_{-1}(t)$. Describe the charge evolution, including sketches in the x-z plane and analytical expressions for wavefronts and decay rates. What is the steady state condition and at a given position (x,z) when is it established?

Prob. 5.10.3 A particle initially has a net charge $q = q_o$ and is immersed in an electrolyte that has uniform conductivity and permittivity. Write integral statements of Gauss' law and the conservation of charge for a volume enclosing the particle. Show that $q(t) = q_o\exp(-t/\tau)$, where τ is the charge relaxation time ε/σ.

For Section 5.12:

Prob. 5.12.1 The planar layer of Table 2.16.1 is composed of a material having uniform permittivity ε and uniform anisotropic conductivity σ_{ij}, such that

$$\vec{J} = \sigma_x E_x \vec{i}_x + \sigma_y E_y \vec{i}_y + \sigma_z E_z \vec{i}_z$$

(a) Show that for variables taking the form $\Phi = \mathrm{Re}\hat{\Phi}(x)\exp j(\omega t - k_y y - k_z z)$, the current density $\mathcal{J}_x \equiv (j\omega\varepsilon + \sigma_x)\hat{E}_x$ (the sum of the displacement and conduction currents needed to write the conservation of charge boundary condition at an interface) evaluated at the (α,β) surfaces is related to the potentials there by

$$\begin{bmatrix} \mathcal{J}_x^\alpha \\ \mathcal{J}_x^\beta \end{bmatrix} = (j\omega\varepsilon + \sigma_x)\gamma \begin{bmatrix} -\coth\gamma\Delta & \dfrac{1}{\sinh\gamma\Delta} \\ \dfrac{-1}{\sinh\gamma\Delta} & \coth\gamma\Delta \end{bmatrix} \begin{bmatrix} \hat{\Phi}^\alpha \\ \hat{\Phi}^\beta \end{bmatrix}$$

where $\gamma^2 \equiv [k_y^2(\sigma_y + j\omega\varepsilon) + k_z^2(\sigma_z + j\omega\varepsilon)]/(\sigma_x + j\omega\varepsilon)$

(b) Consider as a special case $\sigma_y = \sigma_z = 0$, so that conduction is only in the x direction. Discuss implications of γ for penetration of the field in the x direction as function of frequency and of $k^2 \equiv k_y^2 + k_z^2$. In particular, what is the nature of field distribution in the limit $\omega \to 0$?

(c) Consider $\sigma_x = 0$ and $\sigma_y = \sigma_z = \sigma_o$, so that conduction is confined to y-z planes. Discuss the field distribution as in (b) and draw contrasts.

For Section 5.13:

Prob. 5.13.1 A circular analogue of the case study considered in this section is shown in Fig. P5.13.1. A rotating shell has radius R and angular velocity Ω. A traveling wave of potential is applied to electrodes around the shell at a radius a, while an equipotential electrode is at the center with radius b.

(a) Find the surface potential of the rotating shell.

(b) Determine the electrical torque acting on the shell.

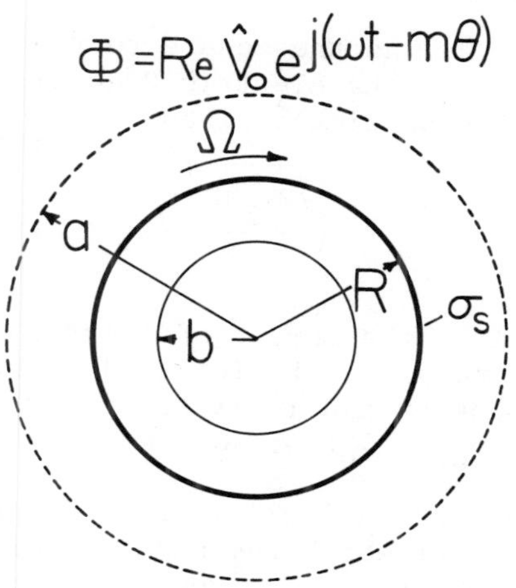

Fig. P5.13.1

Prob. 5.13.2 As a continuation of Prob. 5.13.1, a tachometer is constructed as shown in Fig. 5.13.4. Determine the output current in forms analogous to Eqs. 5.13.15 and 5.13.16.

For Section 5.14:

Prob. 5.14.1 The circular analogue of the planar configuration considered in this section is shown in Fig. 5.14.2. The following steps are intended to parallel those of the text for this configuration. Define the angular velocity of the rotor as $\Omega = U/R$.

(a) Write the electrical torque in a form analogous to Eq. 5.14.6.

(b) Find the surface potential of the rotor in a form analogous to that of Eq. 5.14.8.

(c) Write the electrical torque in a form like that of Eq. 5.14.11, identifying S_e and τ_E.

Prob. 5.14.2 Motions of Von Quincke's rotor, shown in Fig. 5.14.4c, can be of far greater complexity than the steady rotations considered here. To study these motions, it is appropriate to develop a "lumped parameter" model which exploits the fact that the dynamics enter only through the boundary conditions at the rotor interface. Plane parallel electrodes are used to impose an electric field $-E(t)\vec{i}_x$ perpendicular to the cylinder. The region surrounding the rotor is electrically taken as extending to "infinity," where the electric field is this imposed field. In the region immediately surrounding the rotor, the potential takes the form

$$\Phi = E(t)r\cos\theta + P_x(t)\frac{\cos\theta}{r} + P_y(t)\frac{\sin\theta}{r}$$

Permittivities of the surrounding fluid and the cylinder are respectively ε_a and ε_b. The cylinder is insulating while the fluid has conductivity σ. The rotor has radius b, moment of inertia per unit axial length I and a viscous damping torque per unit length $-B\Omega$, where $\Omega(t)$ is the rotor angular velocity.

(a) Show that motions of the rotor are in general described by the nonlinear equations

$$p_e^{-1}\dot{\underline{\Omega}} + \underline{\Omega} = \underline{E}\,\underline{P}_y$$

$$\dot{\underline{P}}_x + \underline{\Omega}\underline{P}_y + \underline{P}_x = H_e^2(-f\dot{\underline{E}} + \underline{E})$$

$$\dot{\underline{P}}_y - \underline{\Omega}\underline{P}_x + \underline{P}_y = fH_e^2\underline{E}\,\underline{\Omega}$$

where variables have been normalized such that

$$t = \underline{t}\tau_e \quad ; \quad \tau_e \equiv (\varepsilon_a + \varepsilon_b)/\sigma$$

$$\underline{\Omega} = \underline{\Omega}\tau_e \quad ; \quad \underline{E}(t) = E(t)/\mathscr{E}$$

$$\underline{P}_{\binom{x}{y}} = \frac{2\varepsilon_a\pi\tau_e}{B}P_{\binom{x}{y}}$$

so that $\mathscr{E}$ is a typical electric field intensity. For example, if E(t) is a constant, $\mathscr{E}$ is that constant and $\underline{E} = 1$. Other dimensionless parameters are the electric Hartmann number H_e (given in

Prob. 5.14.2 (continued)

Sec. 8.7 as the square root of the ratio of the charge relaxation time to the electro-viscous time τ_{EV}) and the electric Prandtl number p_e (the ratio of the charge relaxation time to the viscous diffusion time). Thus

$$H_e \equiv \sqrt{\frac{2\varepsilon_a \pi R^2 \tau_e}{B}} \quad ; \quad p_e = \tau_e / I/B \quad ; \quad f = \frac{\varepsilon_b - \varepsilon_a}{\varepsilon_b + \varepsilon_a}$$

If I is the moment of inertia of the rotor alone (ignoring inertial effects of the fluid), $I = \pi b^4 \rho/2$. If viscous diffusion in the liquid is complete, $B = 4\pi b^2 \eta$, where η is the fluid viscosity and ρ is the rotor mass density. (See Sec. 9.3). Then $H_e^2 = \tau_e/\tau_{EV}$; $\tau_{EV} \equiv 2\eta/\varepsilon_a \mathscr{E}^2$ and $p_e \equiv \tau_e/\tau_V$; $\tau_V \equiv \rho b^2/8\eta$.

(b) The imposed field is raised very slowly. Use the results of (a) to deduce the threshold value of H_e at which the static equilibrium of the rotor is unstable. What steady values of Ω result from raising H_e beyond this critical value?[1]

For Section 5.15:

Prob. 5.15.1 Identify the temporal modes for the rotor of Prob. 5.13.1.

Prob. 5.15.2 Identify the temporal modes for the rotor of Prob. 5.14.1.

Prob. 5.15.3 An insulating spherical particle having radius R and permittivity ε_b has angular velocity Ω about the z axis. It is surrounded by insulating material of infinite extent having permittivity ε_a. On its surface is a conducting coating having surface conductivity σ_s. Find the natural modes of decay for charge distributed on the surface. Modes included should represent the ϕ dependence exp $(jm\phi)$ by the mode number m, and the θ dependence by the mode number n of the function P_n^m. From these modes, pick the one that represents the rate of decay of a spherical particle initially in a uniform electric field, which is then suddenly turned on or off. Your result should be $\tau = R(2\varepsilon_a + \varepsilon_b)/2\sigma_s$.

Prob. 5.15.4 A particle has the properties given in Prob. 5.15.3. In addition, it has a bulk conductivity σ_b and the surrounding material has a bulk conductivity σ_a. Show that the relaxation time of the nth mode is

$$\tau_n = \frac{\varepsilon_a(n+1) + \varepsilon_b n}{\sigma_a(n+1) + \sigma_b n + \frac{\sigma_s}{R} n(n+1)}$$

Prob. 5.15.5 The planar layer described in terms of transfer relations in Prob. 5.12.1 is bounded in the planes $x = \Delta$ and $x = 0$ by equipotentials.

(a) Find an expression for the eigenfrequencies of the temporal modes.

(b) Show that as the material becomes isotropic in conductivity, so that $\sigma_x = \sigma_y = \sigma_z$, the infinite set of temporal modes all degenerate to the same eigenfrequency.

(c) Identify the eigenfrequencies for conduction confined to the x direction ($\sigma_y = \sigma_z = 0$) and plot as a function of $k \equiv \sqrt{k_y^2 + k_z^2}$ with the mode number n as a parameter.

(d) Proceed as in (c) for the case $\sigma_x = 0$, $\sigma_y = \sigma_z = \sigma_o$.

For Section 5.17:

Prob. 5.17.1 For the same configuration as developed in this section, define the sheet position as being at $x = 0$. Find the potential distribution for the regions above $(0 < x < d)$ and below $(-d < x < 0)$ the sheet. The expressions should reduce to Eqs. 5.17.17, 5.17.18 and 5.17.19 on the sheet surface $(x = 0)$.

---

1. Aperiodic motions such as these have been studied in connection with mathematically analogous models for thermal convection. See W.V.R. Malkus, "Nonperiodic Convection at High and Low Prandtl Number," Memoires Societe Royale des Sciences de Liege, 6 serie, tome IV, (1972), pp. 125-128.

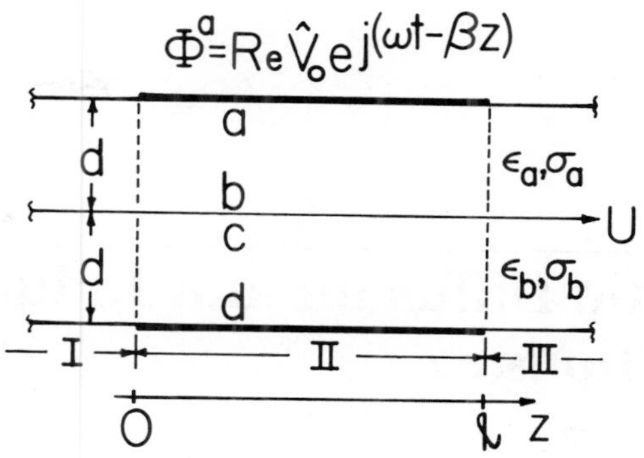

Fig. P5.17.2

Prob. 5.17.2 The system shown in Fig. P5.17.2 is the same as considered in Sec. 5.14, except that the excitation on the upper boundary starts at z = 0 and ends at z = ℓ. The potential upstream and downstream on this surface is zero. Also, the interface is midway between the transverse boundaries, so a and b from Sec. 5.14 are equal to d.

(a) The potential at the interface in the sinusoidal steady state is $\Phi^b(z,t) = \text{Re}\hat{\Phi}^b(z,\omega)e^{j\omega t}$. Show that

$$\hat{\Phi}^b = \frac{1}{2\pi}\int_{-\infty}^{+\infty} \frac{\hat{V}_o[(\omega-kU)\varepsilon_a - j\sigma_a][e^{j(\ell-z)k}\,e^{-j\beta\ell} - e^{-jkz}]dk}{(k-\beta)\,D(\omega,k)}$$

where

$$D(\omega,k) = \cosh kd[(\sigma_a+\sigma_b) + j(\omega-kU)(\varepsilon_a+\varepsilon_b)]$$

(b) Show that the wavenumbers of the spatial modes are

$$k_n = \begin{cases} \dfrac{\omega}{U} - j\dfrac{(\sigma_a+\sigma_b)}{U(\varepsilon_a+\varepsilon_b)}, & n = 0 \\[2ex] \mp\dfrac{(|2n|-1)}{d}\,j\dfrac{\pi}{2}, & n = \pm\infty \ldots \pm 1 \end{cases}$$

Sketch the transverse and longitudinal dependences of these modes. Why do modes $n \neq 0$ have no dependence on material properties, ω, or U?

(c) Use the Cauchy integral theorem to find $\Phi^b(z,t)$ from the result of part (a) and the modes of part (b).

(d) Find the total time average electrical force exerted in the z direction on the material. The expression can be left as an integral on k.

6

Magnetic Diffusion and Induction Interactions

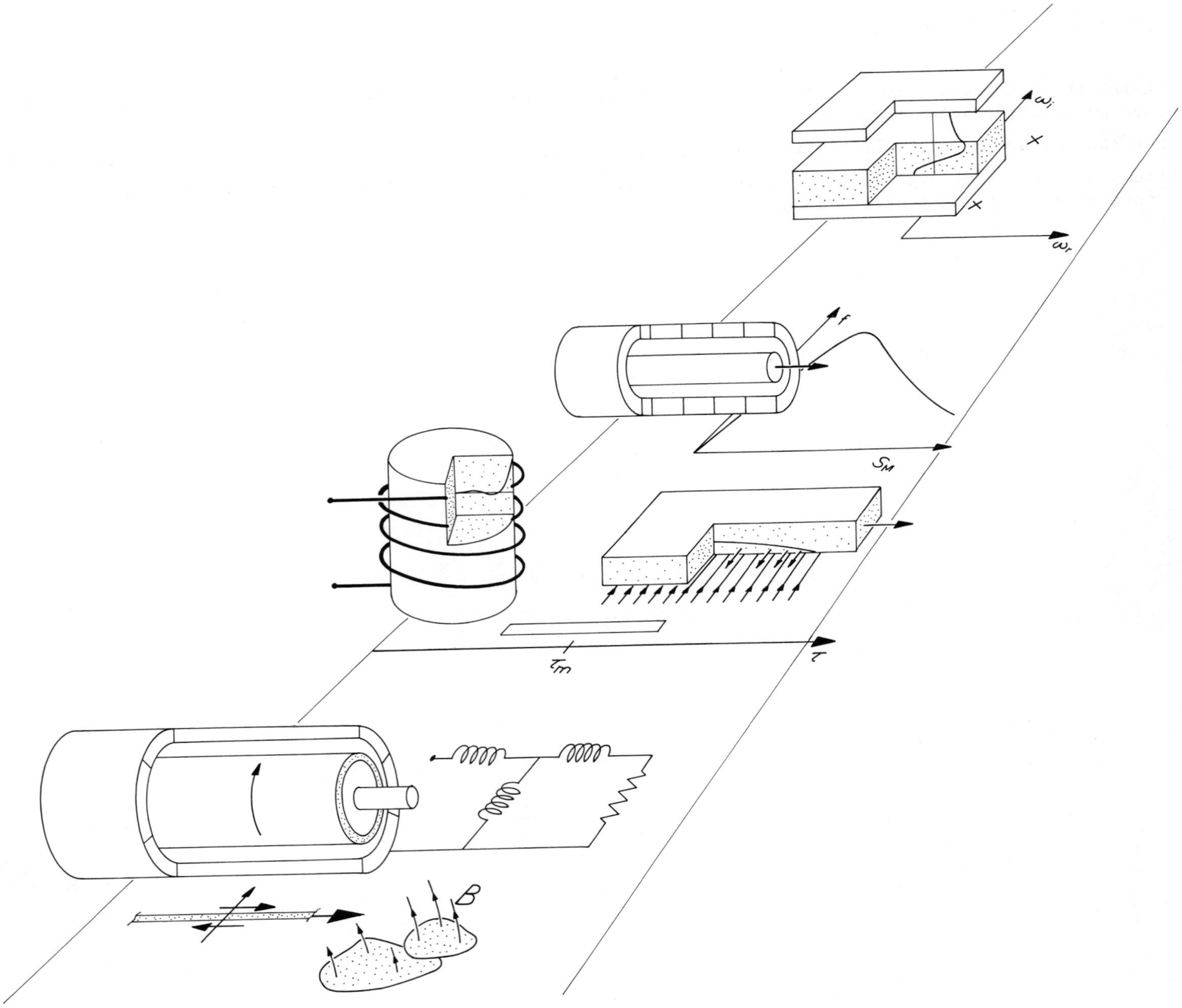

6.1 Introduction

Except that magnetoquasistatic rather than electroquasistatic systems are considered, in this chapter electromechanical phenomena are studied from the same viewpoint as in Chap. 5. Material deformations are again prescribed (kinematic) while the magnetic field sources, the distributions of current or magnetization density, evolve in a dynamical manner that is self-consistently described throughout the volume of interest. Most of the discussion in this chapter relates to magnetic diffusion with material convection.

In practical terms, this chapter takes leave of the windings and associated slip rings or commutators used in Chap. 4 to constrain current distributions in moving elements and takes up conductors in which the currents seek a distribution consistent with the magnetoquasistatic field laws and the imposed motion. The magnetic induction machine is an important example. Most often encountered as a rotating machine, it might also have as a moving member a "linear" sheet of metal or even a liquid. The study of temporal and spatial transients and of boundary layer models in Secs. 6.9-6.11 is pertinent to the linear induction machines, whether they be applied to train propulsion or manufacture of sheet metal. The "deep conductor" interactions considered in Secs. 6.6 and 6.7 give insights concerning liquid-metal induction pumping, a topic continued in Chap. 9.

The boundary conditions and transfer relations summarized in Secs. 6.3 and 6.5 are a basic resource for developing analytical models representing systems suggested by the case studies of Secs. 6.4 and 6.6. Similarly, the dissipation and skin-effect relations developed in Secs. 6.7 and 6.8 are designed to be of general applicability.

Much of the magnetic diffusion phenomena developed in this chapter, the mathematical relations as well as the physical insights, pertain as well to the diffusion of molecules or of heat. Hence, dividends from an investment in this chapter are in part collectable in Chap. 9. In addition, what in Sec. 6.2 is a theorem concerning the conservation of flux for material surfaces of fixed identity, in Chaps. 7 and 9 relates to fluid mechanics and becomes Kelvin's vorticity theorem. Diffusion of vorticity, a momentum transfer process in fluids taken up in Chap. 7, has much in common with magnetic diffusion.

The conduction model in this chapter is exclusively ohmic. The model is especially appropriate in the relatively highly conducting materials of interest if magnetic diffusion effects are an issue. Typically, conductors are solid or liquid metals, or perhaps highly ionized gases. The development is purposely one that parallels the sections on ohmic conductors in electroquasistatic systems, Secs. 5.10-5.16. A comparative study of electroquasistatic and magnetoquasistatic rate processes, models and examples results in the recognition of both analogies and contrasts.

Although resistive types of induction interactions are by far the most common, time-average forces can be developed through phase shifts created by other types of loss mechanisms. The important example of magnetization hysteresis interactions is used in Sec. 6.12 to exemplify not only how time-average magnetization forces can be developed, but by analogy, how polarization interactions can be created in an electroquasistatic context.

6.2 Magnetic Diffusion in Moving Media

For a material at rest in the primed frame of reference, Ohm's law is

$$\vec{J}_f' = \sigma\vec{E}' \tag{1}$$

where the conductivity σ is in general a function of position and time. This law, introduced in Sec. 3.3, implies at least two charge-carrier species and a Hall parameter (Eq. 3.3.4) that is small compared to unity. Use of the field transformations $\vec{J}_f = \vec{J}_f'$ (Eq. 2.5.11b) and $\vec{E}' = \vec{E} + \vec{v} \times \mu_o\vec{H}$ (Eq. 2.5.12b) expresses Eq. 1 in the laboratory frame of reference,

$$\vec{J}_f = \sigma(\vec{E} + \vec{v} \times \mu_o\vec{H}) \tag{2}$$

where $\vec{v}$ is the velocity of the material having the conductivity σ. This generalization of Ohm's law to represent conduction in a moving material is clearly valid provided that the material is moving with a constant velocity. But the law will be used throughout this chapter for materials that are accelerating. The assumption is made that accelerations have a negligible effect on the processes responsible for the conduction, for example, in a metallic conductor, that the acceleration of the ponderable material has a negligible effect on electronic motions.

Solution of Eq. 2 for $\vec{E}$ gives an expression that can be substituted into Faraday's law, Eq. 2.3.25b, to obtain

$$\nabla \times \left(\frac{\vec{J}_f}{\sigma}\right) = -\frac{\partial \vec{B}}{\partial t} + \nabla \times (\vec{v} \times \vec{B}) \tag{3}$$

where the definition $\vec{B} \equiv \mu_o(\vec{H} + \vec{M})$ has been used.

The embodiment of Ohm's and Faraday's laws, represented by Eq. 3, has a simple physical significance best seen by considering the integral form of these same laws. With $\vec{E}'$ replaced using Eq. 1, Faraday's integral law, Eq. 2.7.3b, becomes

$$\oint_C \frac{\vec{J}_f}{\sigma}\cdot d\vec{\ell} = -\frac{d}{dt}\int_S \vec{B}\cdot\vec{n}da \tag{4}$$

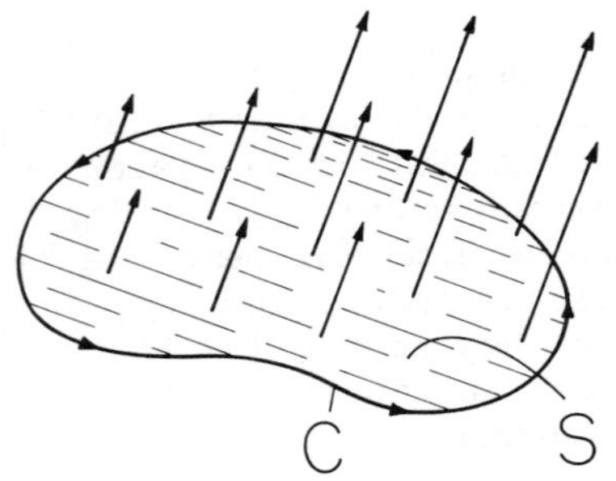

Fig. 6.2.1

Surface of fixed identity.

In writing this equation, the surface S enclosed by the contour C, Fig. 6.2.1, is one of fixed identity (one attached to the deforming material), so $\vec{v} = \vec{v}_s$. (The same expression would be obtained by integrating Eq. 3 over a surface of fixed identity and applying the generalized Leibnitz rule, Eq. 2.6.4.)

According to Eq. 4, the dissipation of total flux linked by a surface of fixed identity is proportional to the "iR" drop around the contour of fixed identity enclosing the surface. The statement is a generalization of one representing an ideal deforming inductor having the terminal variables (λ,i) shorted by a resistance R:

$$iR = -\frac{d\lambda}{dt} \tag{5}$$

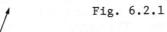

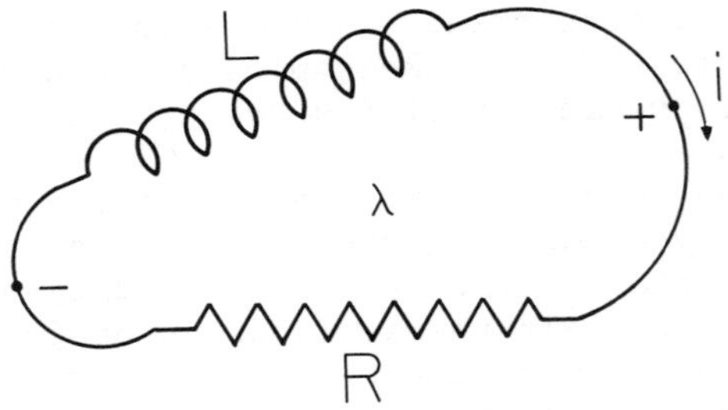

Fig. 6.2.2

Circuit equivalent to C in Fig. 6.2.1.

In the limit of "infinite" conductivity, the flux intercepted by a surface of fixed identity is invariant Equations 3 and 4 represent the same laws, so if the left side of Eq. 3 is negligible, it too implies that the flux linking a contour of fixed identity is conserved. The circuit helps to emphasize that in most of the chapter the subject is distributed "resistors" and "inductors" typified in their dynamics by "L/R" time constants.

Ampère's law, Eq. 2.3.23b, eliminates $\vec{J}_f$ from Eq. 3 in favor of $\vec{H}$. With magnetization described $\vec{B} = \mu\vec{H}$, where μ can be a function of space and time but not of $\vec{H}$, Eq. 3 then becomes

$$\nabla \times \frac{1}{\sigma}\left(\nabla \times \frac{1}{\mu}\vec{B}\right) = -\frac{\partial \vec{B}}{\partial t} + \nabla \times (\vec{v} \times \vec{B}) \tag{6}$$

In regions where the properties (σ,μ) and material velocity $\vec{v}$ are uniform, Eq. 6 becomes the convective diffusion equation*

$$\frac{1}{\mu\sigma}\nabla^2\vec{B} = \left(\frac{\partial}{\partial t} + \vec{v}\cdot\nabla\right)\vec{B} \tag{7}$$

On the right is the rate of change with respect to time for an observer moving with the velocity $\vec{v}$ of the material (Sec. 2.4). This convective derivative represents two ways in which time rates of change are experienced by a given element of material. Perhaps created by a time-varying field source, at a given fixed location there is a magnetic induction $\partial\vec{B}/\partial t$ with a rate characterized by a time τ. Motion of the material through a spatially varying field gives rise to a second magnetic induction contribution generally represented by the "speed" term $\nabla \times (\vec{v} \times \vec{B})$ (Eq. 6) and particularly reduced to $\vec{v}\cdot\nabla\vec{B}$ in Eq. 7. This contribution is characterized by a transport time ℓ/u, where ℓ and u are respectively a typical length and velocity. Parameters representing the competition between these two rates of change and the diffusion process are identified by writing Eq. 7 in terms of the dimensionless variables

$$t = \underline{t}\tau; \quad \vec{v} = \underline{\vec{v}}u; \quad (x,y,z) = (\underline{x},\underline{y},\underline{z})\ell \tag{8}$$

---

*$\nabla \times (\vec{v} \times \vec{B}) \equiv \vec{v}(\nabla\cdot\vec{B}) - \vec{B}(\nabla\cdot\vec{v}) + \vec{B}\cdot\nabla\vec{v} - \vec{v}\cdot\nabla\vec{B}$; $\nabla\cdot\vec{v} = 0, \nabla\cdot\vec{B} = 0$, $\nabla\vec{v} = 0$
$\nabla \times (\nabla \times \vec{B}) \equiv \nabla(\nabla\cdot\vec{B}) - \nabla^2\vec{B}$

Note that either Eq. 6 or Eq. 7 is linear in $\vec{B}$, so that the flux density need not be normalized. In terms of these variables, Eq. 7 becomes

$$\nabla^2\vec{B} = \frac{\tau_m}{\tau}\frac{\partial\vec{B}}{\partial t} + R_m\vec{v}\cdot\nabla\vec{B} \tag{9}$$

where

$\tau_m = \mu\sigma\ell^2$: magnetic diffusion time

$R_m = \mu\sigma u\ell$: magnetic Reynolds number

For μ,σ and $\vec{v}$ not uniform, τ_m and R_m are defined using typical magnitudes of these quantities.

If the diffusion term on the left in Eq. 9 (in Eq. 4) is negligible, the dynamics tend to be flux conserving. Thus, τ_m/τ and R_m are dimensionless numbers, really representing the same physical process, that are an index to the degree of flux conservation. If a process is steady so that $\partial\vec{B}/\partial t = 0$, then R_m, which is the ratio of the magnetic diffusion time to a typical transport time ℓ/u, is the appropriate index. If R_m is large, material convection tends to dominate in determining the field distribution.

Few physical situations involve only one dimension. Usually, practical systems are heterogeneous, in that they are made up of materials having different electrical properties each with its own dimensions. As a result, a model may involve several different τ_m's and R_m's. Identifying the most critical diffusion times and magnetic Reynolds numbers is an art developed by having as background examples such as those in the following sections.

Skin effect, a magnetic diffusion phenomenon, is conventionally characterized by the skin depth δ_m. As a parameter representing the extent to which a sinusoidal steady-state magnetic field diffuses into a conductor, it embodies the magnetic diffusion time τ_m. The extent to which the field diffuses into an "infinite" conductor is itself the characteristic length ℓ, while the characteristic time is the reciprocal of the imposed field frequency. In fact, setting $\tau_m/\tau = \mu\sigma\delta^2\omega = 2$ results in what will be identified in Sec. 6.6 as the magnetic skin depth:

$$\delta = \sqrt{\frac{2}{\omega\mu\sigma}} \tag{10}$$

The skin depth is the length that makes the magnetic diffusion time equal to twice the reciprocal angular frequency of a sinusoidal driving field.

Typical electrical conductivities for materials in which magnetic diffusion is of interest are given in Table 6.2.1. For these materials, the magnetic diffusion time, τ_m, is given as a function of the characteristic length ℓ in Fig. 6.2.3 and the skin depth, δ, is given as a function of frequency $f = \omega/2\pi$ in Fig. 6.2.4.

Table 6.2.1. Typical electrical conductivities of materials in which magnetic diffusion is of interest. Permeability is essentially μ_o unless otherwise stated.

Material	Conductivity σ (mhos/m)
Solids	
Copper	5.80×10^7
4% silicon-iron	1.7×10^6 ($\mu \sim 5000\ \mu_o$)
Silver	6.17×10^7
Aluminum	3.72×10^7
Graphite	7.27×10^4
Liquids	
Mercury	1.06×10^6
Sodium	1.04×10^7
Sodium potassium 22%-78%	2.66×10^6
Cerrelow-117 (tin-bismuth-lead-antimony alloy)	1.9×10^6
Seawater	4
Deionized pure water	4×10^{-6}
Alumium	4.31×10^6 (870°C)
Tin	2.1×10^6 (231.9°C)
Zinc	2.83×10^6 (419°C)
Gases	
Typical seeded combustion gases	~ 40

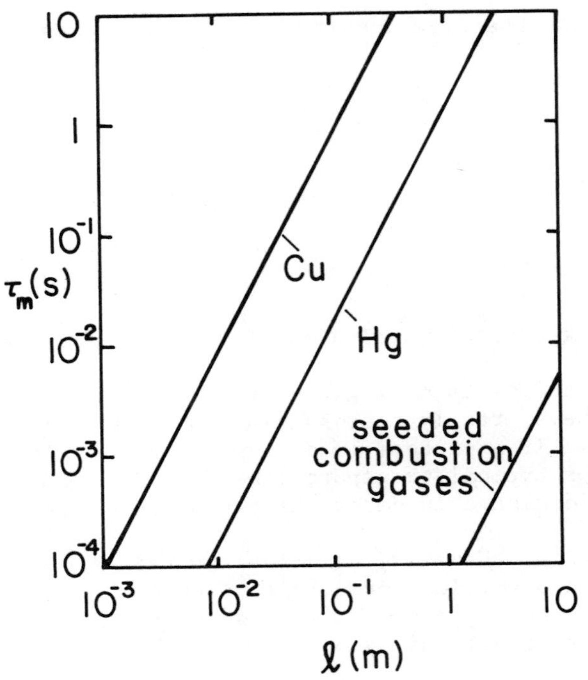

Fig. 6.2.3. Magnetic diffusion time as a function of characteristic length for solid copper, liquid mercury and gas typical of that used in MHD generator.

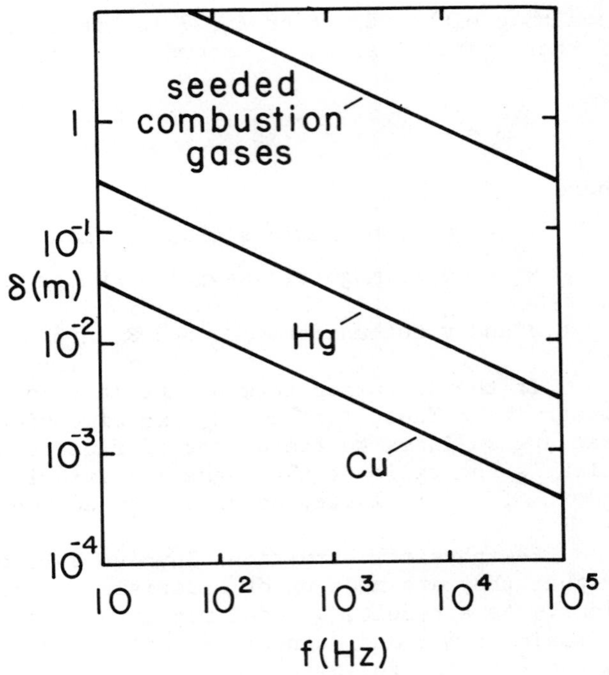

Fig. 6.2.4. Skin depth as function of frequency for materials of Fig. 6.2.3.

6.3 Boundary Conditions for Thin Sheets and Shells

Currents induced in a sufficiently thin conductor can be regarded as essentially uniform over its cross section. Some of the most important models for magnetic diffusion exploit the resulting simplification of the field representation. The magnetic diffusion process is condensed into a boundary condition at the surface occupied by the conductor, in Fig. 6.3.1, the surface separating regions (a) and (b).

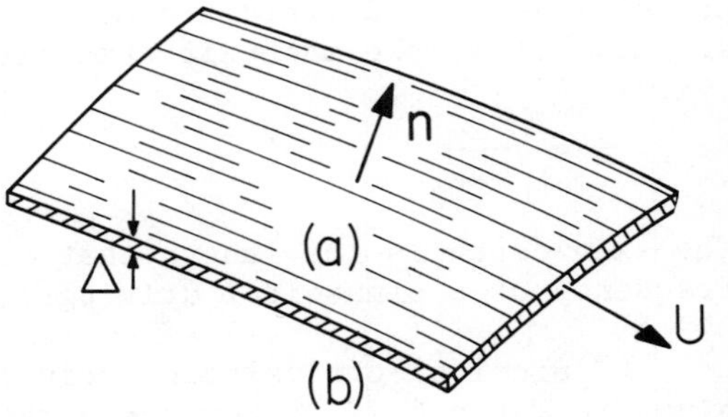

Fig. 6.3.1. Conducting sheet having normal flux density B_n, thickness Δ and hence surface conductivity $\sigma_s = \Delta\sigma$.

Because the conducting sheet is bounded from either side by insulators, the current distribution is essentially that of a surface current

$$\vec{K}_f = \Delta\sigma\vec{E}_t \equiv \sigma_s\vec{E}_t \tag{1}$$

The normal flux density is continuous, so it is denoted by B_n without distinguishing between regions (a) and (b).

Ohm s law and Faraday's law are embodied in Eq. 6.2.3. For the present purposes, the normal component of this equation is the essential one. Multiplied by the sheet thickness, Δ, it becomes

$$(\nabla \times \vec{K}_f)_n = -\sigma_s\frac{\partial B_n}{\partial t} + \sigma_s[\nabla \times (\vec{v} \times \vec{B})]_n \tag{2}$$

Continuity of current (Eq. 2.3.26) requires that

$$\nabla_\Sigma\cdot\vec{K}_f = 0 \tag{3}$$

where $\nabla_\Sigma\cdot$ is the two-dimensional divergence, the usual divergence with the vector component normal to the surface omitted.

Table 6.3.1. Boundary conditions on conducting moving sheets and shells. Normal flux density, B_n, is continuous and $\sigma_s \equiv \Delta\sigma$.

Configuration	Boundary condition
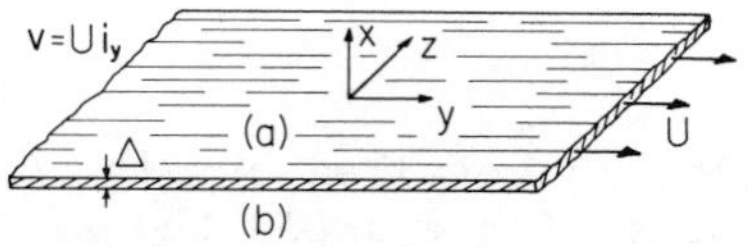 translating planar sheet	$\left(\frac{\partial^2}{\partial y^2} + \frac{\partial^2}{\partial z^2}\right) [\![H_y]\!] = -\sigma_s \frac{\partial}{\partial y}\left(\frac{\partial}{\partial t} + U\frac{\partial}{\partial y}\right)B_x$ (a)
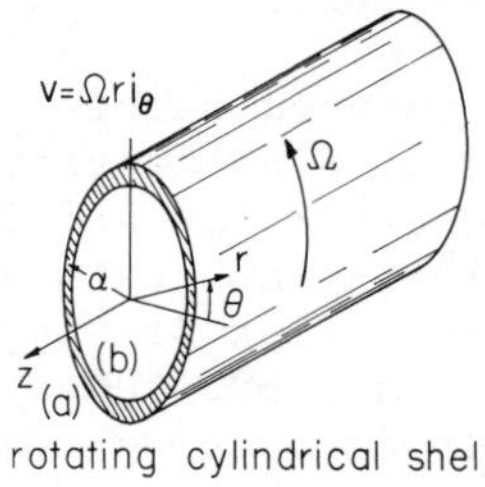 rotating cylindrical shell	$\left(\frac{1}{\alpha^2}\frac{\partial^2}{\partial \theta^2} + \frac{\partial^2}{\partial z^2}\right) [\![H_\theta]\!] = -\frac{\sigma_s}{\alpha}\frac{\partial}{\partial \theta}\left(\frac{\partial}{\partial t} + \Omega\frac{\partial}{\partial \theta}\right)B_r$ (b)
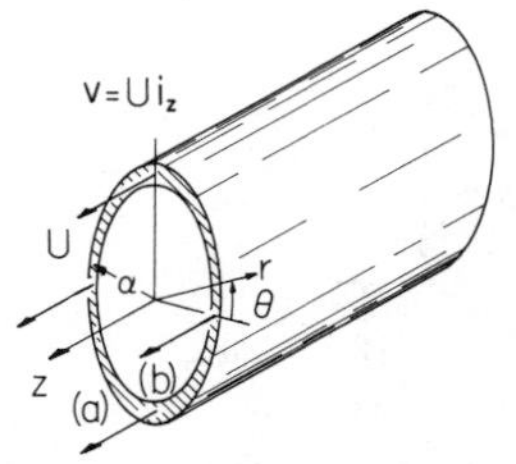 translating cylindrical shell	$\left(\frac{1}{\alpha^2}\frac{\partial^2}{\partial \theta^2} + \frac{\partial^2}{\partial z^2}\right) [\![H_z]\!] = -\sigma_s \frac{\partial}{\partial z}\left(\frac{\partial}{\partial t} + U\frac{\partial}{\partial z}\right)B_r$ (c)
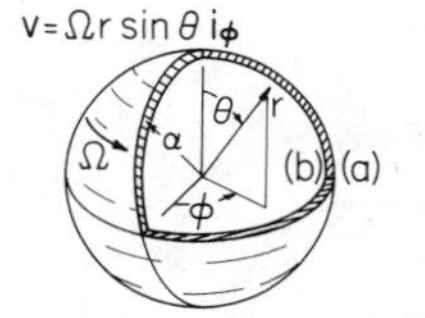 rotating spherical shell	Either $\left(\frac{\partial}{\partial \theta}\sin\theta\frac{\partial}{\partial \theta}\sin\theta + \frac{\partial^2}{\partial \phi^2}\right) [\![H_\phi]\!] = -\sigma_s\alpha\sin\theta\frac{\partial}{\partial \phi}\left(\frac{\partial}{\partial t} + \Omega\frac{\partial}{\partial \phi}\right)B_r$ (d) or $\left(\frac{\partial}{\partial \theta}\sin\theta\frac{\partial}{\partial \theta}\sin\theta + \frac{\partial^2}{\partial \phi^2}\right) [\![H_\theta]\!] = -\sigma_s\alpha\frac{\partial}{\partial \theta}\left[\sin^2\theta\left(\frac{\partial}{\partial t} + \Omega\frac{\partial}{\partial \phi}\right)B_r\right]$ (e)

Finally, the jump condition implied by Ampère's law relates $\vec{K}_f$ to the fields (Eq. 2.10.21):

$$\vec{n} \times [\![\vec{H}]\!] = \vec{K}_f \qquad (4)$$

These last three equations combine to provide a description of how the magnetic field diffuses through conductors of arbitrary geometry. Four typical geometries and associated boundary conditions are summarized in Table 6.3.1. The derivation of each of these conditions follows the steps now carried out in Cartesian coordinates.

Translating Planar Sheet: In this case, $B_n = B_x$ and $\vec{v} = U\vec{i}_y$. Then, Eqs. 2 and 3 become

$$\frac{\partial K_z}{\partial y} - \frac{\partial K_y}{\partial z} = -\sigma_s\left(\frac{\partial B_x}{\partial t} + U\frac{\partial B_x}{\partial y}\right) \qquad (5)$$

$$\frac{\partial K_y}{\partial y} + \frac{\partial K_z}{\partial z} = 0 \qquad (6)$$

Of the variables K_y and K_z, the latter is the more convenient. Hence, with the objective of eliminating K_y between these questions, $\partial/\partial y$ is taken of Eq. 5 and $\partial/\partial z$ is taken of Eq. 6 to generate a cross derivative that can be eliminated between these equations. Thus, Eqs. 5 and 6 become

$$\left(\frac{\partial^2}{\partial y^2} + \frac{\partial^2}{\partial z^2}\right) K_z = -\sigma_s \frac{\partial}{\partial y}\left(\frac{\partial}{\partial t} + U\frac{\partial}{\partial y}\right) B_x \tag{7}$$

Finally, Eq. 4 is used to write Eq. 7 as the required boundary condition, Eq. (a) of Table 6.3.1.

6.4 Magnetic Induction Motors and a Tachometer

A developed model for the rotating machine, shown in Fig. 6.4.1a, is detailed by Fig. 6.4.1b. It incorporates a stator structure much like that for a synchronous machine, for example the smooth air gap machine of Sec. 4.7. The windings shown here however have two rather than three phases, backed by a highly permeable magnetic material. The rotor also consists of a highly permeable material, but having its windings replaced by a sheet of conducting material wrapped on its periphery.

What makes this an induction machine is that the rotor currents are induced rather than imposed by means of windings and terminal pairs. The stator currents produce a magnetic flux density that has a component normal to the conducting sheet. Application of the integral induction law and Ohm's law (Eq. 6.2.4) to a surface lying in the plane of the sheet shows that circulating currents are induced in the sheet. These tend to produce their own fields and hence limit the normal flux density in the sheet. With the rotor assumed very long in the z direction compared to the rotor diameter, these sheet currents are modeled as mainly z-directed, closing in circulating paths on perfectly conducting "shorts" at $z=\pm\infty$.

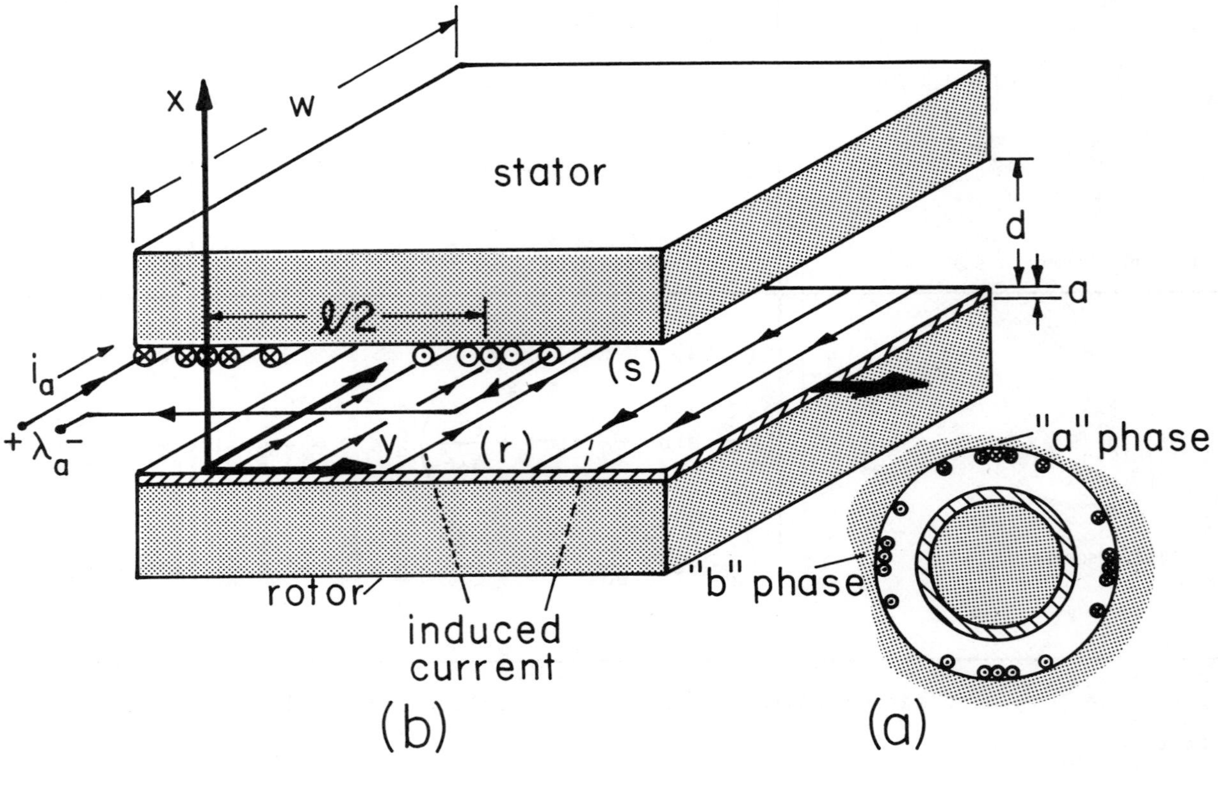

Fig. 6.4.1. (a) Cross section of rotating induction machine with thin-sheet conductor on rotor. (b) Developed model for (a) with air gap d and sheet conductor of thickness a. One of two phases on the stator is shown.

Viewed in terms of lumped parameter models, the induction machine is often represented by a conservative electromechanical coupling, such as developed for the synchronous machine in Sec. 4.7, with the rotor terminals shorted by resistors. In fact, some induction machines are constructed with wound rotors that can be connected to variable resistances through slip rings. However, most induction machines are made inherently more rugged by letting the currents flow through solid conductors, not windings. An important point made by the field representation used in this section is that the thin sheet model is in fact equivalent to the lumped parameter model, provided that the rotor is modeled by a properly distributed polyphase winding with equal resistances connected to each winding. But, if the sheet has finite thickness, the circuit model is not equivalent, as will be evident in Sec. 6.6.

Two-Phase Stator Currents: There are two windings on the stator, each with a sinusoidal distribution of turns density. The "b" phase is displaced by 90° relative to the "a" phase. Because magnetic induction depends explicitly on time rates of change, the description is one in terms of temporal complex amplitudes. Hence, the stator current is modeled as being the surface current,

$$K_z^s = \mathrm{Re}[\hat{i}_a e^{j\omega t} N_a \cos ky + \hat{i}_b e^{j\omega t} N_b \cos k(y - \tfrac{\ell}{4})] \tag{1}$$

where N_a and N_b are the peak turns per unit length on the respective phases and ℓ is the wavelength in the y-direction. For a two-pole rotating machine, ℓ is the rotor circumference. The complex amplitudes of the electrical terminal currents are $(\hat{i}_a, \hat{i}_b)$.

By using Euler's formula, $\cos\theta = (e^{j\theta} + e^{-j\theta})/2$, the cosines in Eq. 1 are written in terms of exponentials so that the surface current takes the alternative form

$$K_z^s = \mathrm{Re}[\hat{K}_+^s e^{j(\omega t - ky)} + \hat{K}_-^s e^{j(\omega t + ky)}] \tag{2}$$

where

$$K_\pm^s = \frac{1}{2}(\hat{i}_a N_a + \hat{i}_b N_b e^{\pm \frac{jk\ell}{4}})$$

Thus, the excitation is written in the form of a complex amplitude Fourier series. This type of representation is discussed in Sec. 5.16. In general, the series takes the form of Eq. 5.16.1. In the case at hand, there are only two terms, n = 1 ($k_1 = k$) and n = -1 ($k_{-1} = -k$), corresponding physically to waves propagating in the + and -z directions.

The fields satisfy linear bulk and boundary equations. Hence, the response to Eq. 2 is the superposition of the response to the first term and a response to the second, found from the first by simply replacing $\hat{K}_+^s \to \hat{K}_-^s$ and $k \to -k$.

Fields: Because the flux linkages are to be computed, it is convenient to describe the air-gap fields in terms of the vector potential, $\vec{A} = A\vec{i}_z$, the Cartesian coordinate case of Table 2.18.1; thus, $\hat{B}_x = -jk\hat{A}$. In view of the "infinitely" permeable stator and rotor materials, boundary conditions on single complex amplitudes of the fields at the stator and rotor surfaces follow from Ampère's law (Eq. 2.10.21). The boundary condition at the stator is thus

$$\hat{H}_y^s = -\hat{K}^s \tag{3}$$

and the composite boundary condition for the thin sheet, Eq. (a) of Table 6.3.1 with $\partial/\partial z = 0$, is

$$\hat{H}_y^r = \frac{\sigma_s}{k}(\omega - kU)\hat{B}_x^r = \frac{\sigma_s}{k}(\omega - kU)(-jk\hat{A}^r) \tag{4}$$

Fields at the stator and rotor surfaces are related by the transfer relations (b) of Table 2.19.1:

$$\begin{bmatrix} \hat{A}^s \\ \hat{A}^r \end{bmatrix} = \frac{\mu_o}{k}\begin{bmatrix} -\coth(kd) & \dfrac{1}{\sinh(kd)} \\ \dfrac{-1}{\sinh(kd)} & \coth(kd) \end{bmatrix}\begin{bmatrix} \hat{H}_y^s \\ \hat{H}_y^r \end{bmatrix} \tag{5}$$

With the objective of finding $\hat{H}_y^r$, which by Ampere's law is the rotor surface current, these last three equations are now combined. Equation 4 (solved for $\hat{A}^r$) and Eq. 3 are substituted into Eq. 5b. This expression is then solved for $\hat{H}_y^r$:

$$\hat{H}_{y\pm}^r = -\frac{\hat{K}_\pm^s S_{m\pm}[j + S_{m\pm}\coth(kd)]}{\sinh(kd)[1 + S_{m\pm}^2\coth^2(kd)]} \tag{6}$$

The dimensionless number S_m combines the ratio of a magnetic diffusion time $\tau_m = \mu_o\sigma_s/k$ to the characteristic time $1/\omega$ and a magnetic Reynolds number $\mu_o\sigma_s U$:

$$S_{m\pm} \equiv \frac{\mu_o\sigma_s}{k}(\omega \mp kU) \tag{7}$$

In writing Eq. 6, the components induced by the respective traveling waves of Eq. 2 are identified by replacing $\hat{K}^s \to \hat{K}_\pm^s$ and $k \to \pm k$. Note that coth(kd) and sinh(kd) are odd functions.

Time-Average Force: To determine the force of magnetic origin acting in the y direction on the rotor, the appropriate volume of integration is as shown in Fig. 4.2.1a. The only contribution to the integration of the stress over the enclosing surface comes from S_1, here taken as a surface adjacent to the rotor. It then follows from Eq. 5.16.4 that the time-average rotor force is simply

$$\langle f_y \rangle_t = (\frac{p}{2}\ell w)\frac{1}{2}\text{Re}[\hat{B}^r_{x+}(\hat{H}^r_{y+})^* + \hat{B}^r_{x-}(\hat{H}^r_{y-})^*] \tag{8}$$

where w is the rotor length in the z direction and p is the number of poles (the number of half-wave-lengths). Hence, $p\ell/2$ is the total rotor length in the y direction.

In Eq. 8, $\hat{B}^r_{x\pm} = \mp jk\hat{A}^r_{\pm}$, where $\hat{A}^r_{\pm}$ follow from Eqs. 3 and 5b:

$$\hat{A}^r_{\pm} = \frac{\mu_o}{k}\left[\frac{\hat{K}^s_{\pm}}{\sinh(kd)} + \coth(kd)\hat{H}^r_{y\pm}\right] \tag{9}$$

Thus, substitution for $\hat{B}^r_{x\pm}$ in Eq. 8 exploits the fact that self-fields can make no contribution to the total force to express the force as an interaction between stator and rotor surface currents:

$$\langle f_y \rangle_t = -\frac{p\ell w}{4}\frac{\mu_o}{\sinh(kd)}\text{Re}[j\hat{K}^s_+(\hat{H}^r_{y+})^* - j\hat{K}^s_-(\hat{H}^r_{y-})^*] \tag{10}$$

In terms of stator currents, Eq. 6 serves to evaluate this time-average force:

$$\langle f_y \rangle_t = \frac{p\ell w\mu_o}{4\sinh^2(kd)}\left[\frac{S_{m+}|\hat{K}^s_+|^2}{1 + S^2_{m+}\coth^2(kd)} - \frac{S_{m-}|\hat{K}^s_-|^2}{1 + S^2_{m-}\coth^2(kd)}\right] \tag{11}$$

Balanced Two-Phase Fields and Time-Average Force: The stator currents become a pure traveling wave if the (b) phase is made to temporally lag the (a) phase by 90°, and the windings have the same peak turns densities. Formally, this is seen from the definitions of $\hat{K}^s_{\pm}$ given with Eq. 2:

$$\begin{matrix}\hat{i}_b = \hat{i}_a e^{-j\pi/2} \\ N_a = N_b\end{matrix} \Longrightarrow \begin{matrix}\hat{K}^s_+ = \hat{i}_a N_a \\ \hat{K}^s_- = 0\end{matrix} \tag{12}$$

Only the first term in Eq. 11 contributes to the force. The dependence of this force on S_{m+} is familiar from the electroquasistatic analogue developed in Sec. 5.13. In Fig. 6.4.2a, the force is shown as a function of the material velocity divided by the traveling-wave phase velocity ω/k. Given the dependence of the force on S_{m+}, this plot is the result of first shifting the origin so that $S_{m+} = 0$ where $\omega = kU$ and then "flipping" the plot about the vertical axis passing through this origin.

The parameter S_{m+} is the effective magnetic diffusion time multiplied by the angular frequency $(\omega - kU)$ for an observer moving with the conducting sheet. The force is in the same direction as the traveling wave, provided S_{m+} is positive so that the traveling wave has a speed greater than that of the material. To understand the force-speed diagram, consider the phase relationship between stator and rotor surface currents, implied by Eq. 6 ($H_y = K_z$). For near synchronism between traveling wave and material, (i) typifies the operating point. In Eq. 6, small S_{m+} implies the complex amplitude (i) shown in the phase diagram of Fig. 6.4.2b. At a given instant, the rotor current spatially lags that on the stator by slightly more than 90°, as sketched in inset (i) of Fig. 6.4.2a. This current has just the right distribution for producing a force to the right, but because S_m is small (the time rate of change in a frame moving with the materials is small) the induced current is small. The magnetic field is distributed essentially as if there were no rotor current. Increasing S_m improves the magnitude of the current but at the price of compromising the relative spatial phase. The ultimate compromise between phase and magnitude comes at (ii) where $S_m = \tanh(kd)$. As S_m becomes large, currents in the rotor completely shield out the normal magnetic field. The rotor current becomes as large as is possible, but the spatial phase relation is wrong for producing a force in the y direction. Operating point (iii) is approaching this condition, with the magnetic field approximating that for a perfectly conducting sheet.

Electrical Terminal Relations: To compute the voltages (v_a, v_b) required to produce the terminal currents (i_a, i_b), the flux linkages (λ_a, λ_b) must be determined. For example, consider the (a) phase of a two-pole machine. The windings carrying current in the z direction at y' and returning the current at $y' + \ell/2$ each link a magnetic flux (Eq. (f) of Table 2.18.1):

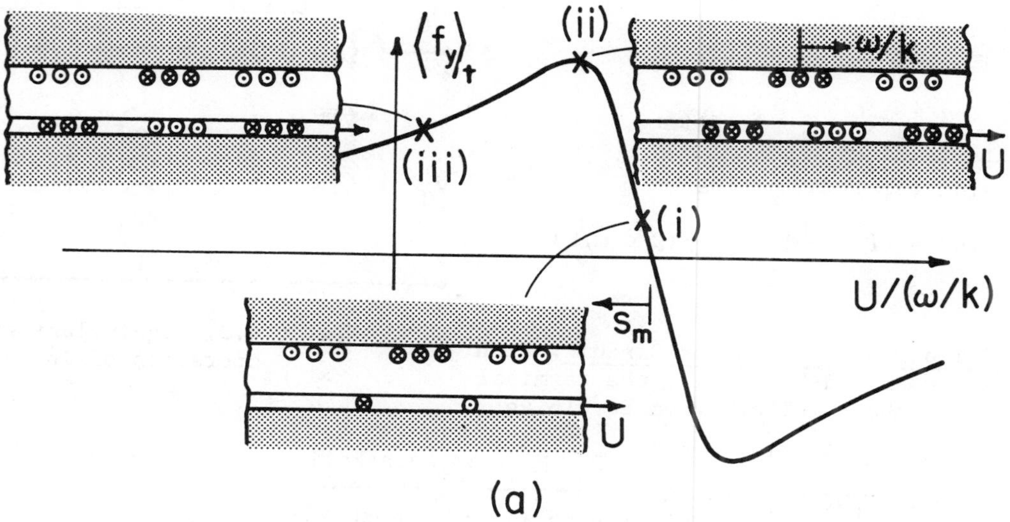

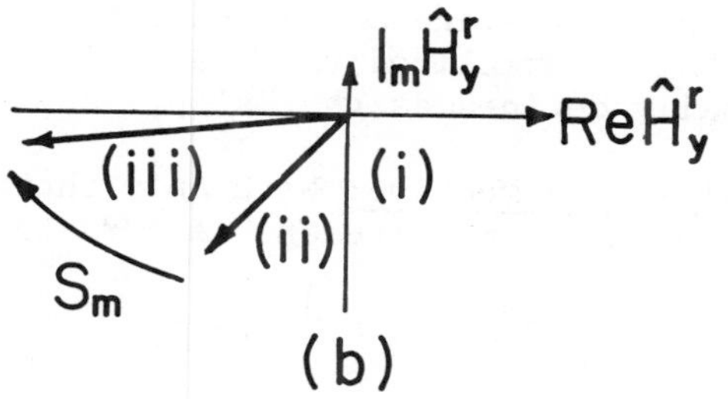

Fig. 6.4.2. (a) Time-average force for induction machine of Fig. 6.4.1 with balanced two-phase excitation. Abscissa is material velocity relative to wave phase velocity. The slip is $s_m \equiv S_{m+}/(\mu_o\sigma_s\omega/k)$. Insets show spatial phase of stator and rotor currents at a given instant. (b) Phasor $\hat{H}_y^r$, showing effect of increasing S_m on the phase and amplitude. Operating points (i) → (iii) are shown in (a). In nomenclature of lumped parameter induction machines, (i) is resistance dominated operation while (iii) is reactance dominated.

$$\Phi_\lambda = w[A^s(y') - A^s(y' + \ell/2)] \tag{13}$$

Written as the superposition of the two field components, so that the dependence on y' is explicit, this expression becomes ($k \equiv 2\pi/\ell$)

$$\begin{aligned}\Phi_\lambda &= w\mathrm{Re}[\hat{A}_+^s e^{j(\omega t-ky')} + \hat{A}_-^s e^{j(\omega t+ky')} - \hat{A}_+^s e^{-j\pi} e^{j(\omega t-ky')} - \hat{A}_-^s e^{j\pi} e^{j(\omega t+ky')} \\ &= w\mathrm{Re}2(\hat{A}_+^s e^{-jky'} + \hat{A}_-^s e^{jky'})e^{j\omega t}\end{aligned} \tag{14}$$

In the interval dy' in the neighborhood of y = y' there are N_a cos ky' dy' turns, so the flux linked by the (a) phase is altogether

$$\lambda_a = \int_{-\ell/4}^{\ell/4} \Phi_\lambda(y')N_a \cos ky'\, dy' = wN_a\mathrm{Re}\int_{-\ell/4}^{\ell/4}(\hat{A}_+^s e^{-jky'} + \hat{A}_-^s e^{jky'})(e^{jky'} + e^{-jky'})e^{j\omega t}dy' \tag{15}$$

Only the constant terms contribute to the integration, and substitution for $\hat{A}_\pm^s$ from Eq. 5a gives

$$\lambda_a = \frac{w\ell}{2}\frac{N_a}{k}\mu_o\mathrm{Re}\left[\coth(kd)(\hat{K}_+^s + \hat{K}_-^s) + \frac{(\hat{H}_{y+}^r + \hat{H}_{y-}^r)}{\sinh(kd)}\right]e^{j\omega t} \tag{16}$$

Remember that $K_\pm^s$ are given functions of the terminal currents, Eq. 2. Thus, $H_{y\pm}^r$ are also given as a function of the terminal currents by Eq. 6 and Eq. 16 is the required (a) phase terminal relation $\lambda_a(i_a,i_b)$. The same line of reasoning shows that λ_b is given by Eq. 16 with $N_a \to -jN_b$, $\hat{K}_+^s + \hat{K}_-^s \to \hat{K}_+^s - \hat{K}_-^s$ and $\hat{H}_{y+}^r + \hat{H}_{y-}^r \to \hat{H}_{y+}^r - \hat{H}_{y-}^r$:

$$L_1 \equiv L_2 \equiv \frac{wN_a^2\ell^2\mu_o}{4\pi}\tanh\left(\frac{\pi d}{\ell}\right)$$

$$M \equiv \frac{wN_a^2\ell^2\mu_o}{4\pi\sinh\left(\frac{2\pi d}{\ell}\right)};\quad R \equiv \frac{\ell wN_a^2}{2\sigma_s}$$

$$s_m \equiv \text{"slip"} = \left(1 - \frac{kU}{\omega}\right) = S_{m+}/(\mu_o\sigma_s\omega/k)$$

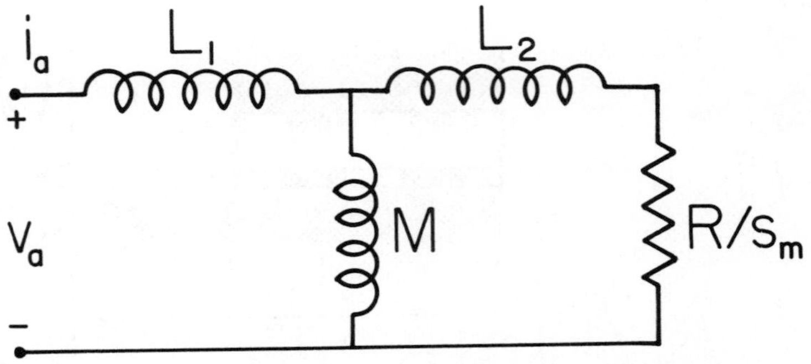

Fig. 6.4.3. Equivalent circuit for balanced operation of induction machine.

Balanced Two-Phase Equivalent Circuit: With excitations as summarized by Eq. 12, the terminal voltage on the (a) phase follows from Eq. 16 as

$$\hat{v}_a = j\omega\hat{\lambda}_a = j\omega\frac{w\ell N_a^2\mu_o}{2k}\left[\coth kd - \frac{S_{m+}[j + S_{m+}\coth(kd)]}{\sinh^2 kd(1+S_{m+}^2\coth^2 kd)}\right]\hat{i}_a \tag{17}$$

This relation of voltage and current is the same as is obtained for the circuit of Fig. 6.4.3. The parameter S_{m+}, normalized to a magnetic Reynolds number based on the wave velocity ω/k, is what is conventionally defined as the "slip," s_m.

Single-Phase Machine: With only the (a) phase excited, positive and negative traveling waves result having equal magnitudes. According to Eq. 2,

$$i_b = 0 \Rightarrow \hat{K}_+^s = \hat{K}_-^s = \frac{1}{2}N_a\hat{i}_a \tag{18}$$

The time-average force, Eq. 11, is the superposition of the forces that would be induced by purely forward and backward traveling waves. The resulting force-speed characteristic, sketched in Fig. 6.4.4, is rigorously the sum of the time-average forces from the traveling-wave components. At zero speed, these forces cancel. Provided the slope of the characteristic at zero speed is positive, once started in either direction, the rotor experiences a force tending to further increase the velocity. It follows from Eq. 11 that at U = 0 the slope is

$$\frac{d\langle f_y\rangle_t}{dU} = \frac{p\ell wN_a^2|i_a|^2}{8\sinh^2 kd}\frac{k}{\omega}R_M\frac{(R_M^2\coth^2 kd - 1)}{(1 + R_M^2\coth^2 kd)^2};\quad R_M \equiv \mu_o\sigma_s\omega/k \tag{19}$$

so the slope is positive, provided the frequency is high enough to make $R_M > \tanh(kd)$.

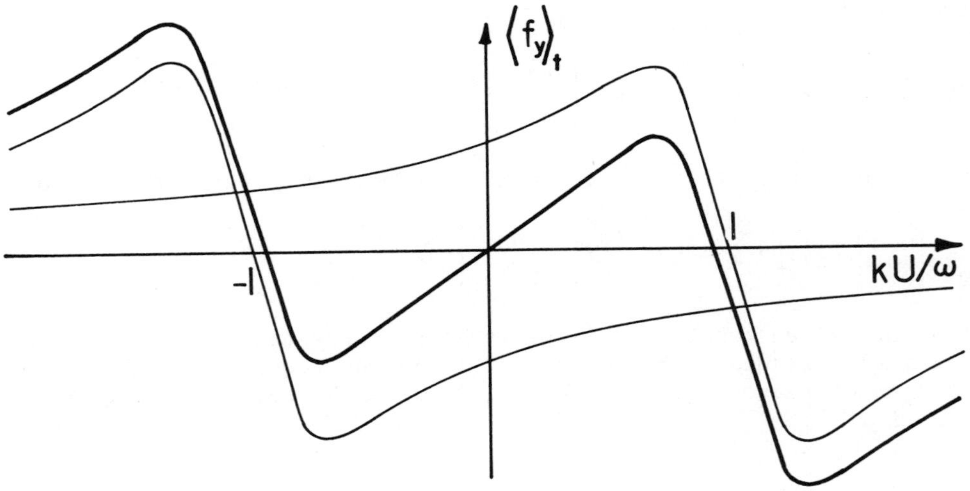

Fig. 6.4.4. Time-average force for single-phase induction machine as function of material velocity normalized to wave velocity. Total force is superposition of forces due to forward and backward wave components.

In practice, single-phase induction machines are started by pole shading or by using a (b) winding connected to the excitation in such a way that a temporal phase shift takes place, perhaps by a capacitor. Under start conditions, one force component then dominates the other.[1]

Fig. 6.4.5

Drag-cup tachometer with end cap and attached magnetic core removed so that thin-walled rotating cup is visible. The core is the lower highly permeable rotor material in Fig. 6.4.1, the cup is the moving conducting sheet. Coils adjacent to the cup rim are the stator windings. In this example the core is actually fixed and there is an appreciable air gap between core and cup.

Tachometer: One common way in which the induction machine sees application as a generator is for speed measurement. As a rotating machine, the model pertains to the drag-cup tachometer shown in Fig. 6.4.5. In linear geometry, the induction interaction might be used to measure the velocity of a moving conducting sheet. Single phase excitation, say of the (b) phase, is equivalent to a standing-wave excitation. With no motion, currents induced in the sheet also form a standing wave in spatial phase with the excitation. Material motion induces an imbalance in the forward and backward wave components. Thus, with no motion no signal is detected on the (a) phase, but with motion there is a sinusoidal signal at the frequency ω. The magnetic interaction exploited here is the analogue of that discussed for an electroquasistatic interaction in connection with Figs. 5.13.3 and 5.13.4.

With single phase excitation of the (b) phase, $\hat{i}_a = 0$ and according to Eq. 2, $\hat{K}^s_{\pm} = \mp \frac{j}{2}\hat{i}_b N_b$. The voltage on the (a) phase follows from Eq. 16:

$$\hat{v}_a = j\omega\hat{\lambda}_a = -j\frac{\omega w \ell N_a N_b \mu_o \hat{i}_b}{4k\sinh^2 kd}\left(\frac{S_{m+}}{1 + jS_{m+}\coth kd} - \frac{S_{m-}}{1 + jS_{m-}\coth kd}\right) \tag{20}$$

As expected, the output voltage is zero if $U = 0$ ($S_{m+} = S_{m-}$). The dependence of v_a on the velocity can be used to measure U. For example, the amplitude of the output follows from Eq. 20 as

$$|\hat{v}_a| = \frac{v_o\left[\frac{\mu_o \sigma s\omega}{k}\right]\left[kU/\omega\right]}{\sqrt{(1 + S^2_{m+}\coth^2 kd)(1 + S^2_{m-}\coth^2 kd)}} \tag{21}$$

where

$$v_o \equiv \frac{\omega w \ell N_a N_b \mu_o}{2k\sinh kd\coth kd}\ ;\ S_{m\pm} \equiv \frac{\mu_o \sigma_s \omega}{k}(1 \mp kU)$$

The analogy to the electroquasistatic tachometer of Sec. 5.13 is emphasized by the direct correspondence between Eqs. 20 and 5.13.15, and between Eqs. 21 and 5.13.16. The dependence on U given by Eq. 21 is illustrated by Fig. 5.13.5.

6.5 Diffusion Transfer Relations for Materials in Uniform Translation or Rotation

In terms of the vector potential $\vec{A}$, discussed in Sec. 2.18, magnetic diffusion in regions having uniform permeability and conductivity is described by Eq. 6.2.6 with $\vec{B} = \nabla \times \vec{A}$ and $\nabla\cdot\vec{A} = 0$:

---

1. For a description of induction machines in lumped-parameter terms, see H. H. Woodson and J. R. Melcher, Electromechanical Dynamics, Pt. I, John Wiley & Sons, New York, 1968, pp. 127-140; also, A. E. Fitzgerald, C. Kingsley, and A. Kusco, Electrical Machinery, McGraw-Hill Book Company, New York, 1971, pp. 525-531; S. A. Nasar and I. Boldea, Linear Motion Electric Machines, John Wiley and Sons, New York, 1976.

Table 6.5.1. Magnetic diffusion transfer relations with translation or rotation.

Planar layer	Rotating cylinder	Translating cylinder
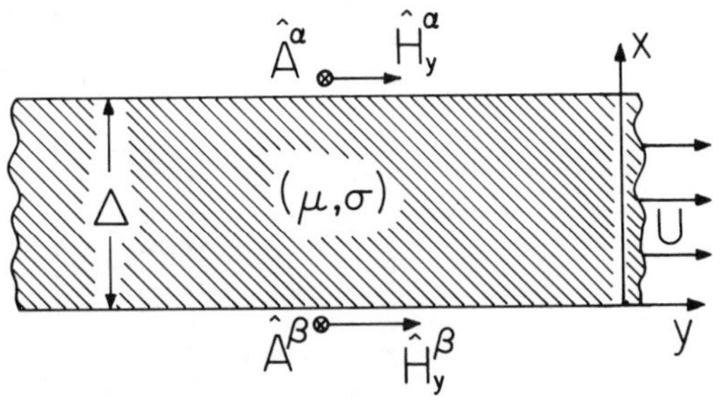	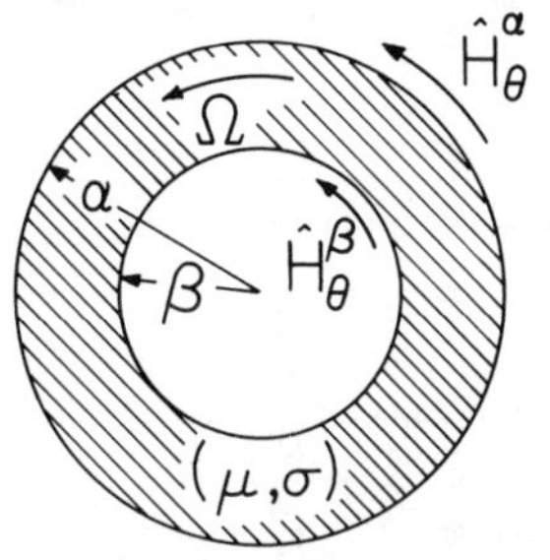	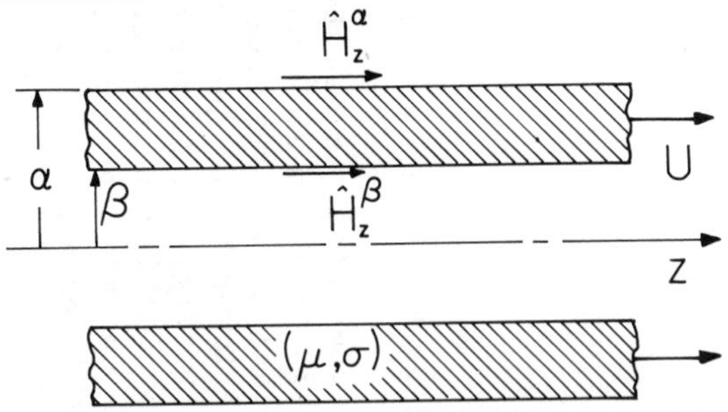
$\vec{A} = \vec{i}_z \text{Re}\hat{A}(x)\exp j(\omega t - ky)$	$\vec{A} = \vec{i}_z \text{Re}\hat{A}(r)\exp j(\omega t - m\theta)$	$\vec{A} = \vec{i}_\theta \hat{A}_\theta(r) \exp j(\omega t - kz)$
$\begin{bmatrix} \hat{H}_y^\alpha \\ \hat{H}_y^\beta \end{bmatrix} = \frac{\gamma}{\mu}\begin{bmatrix} -\coth\gamma\Delta & \frac{1}{\sinh\gamma\Delta} \\ \frac{-1}{\sinh\gamma\Delta} & \coth\gamma\Delta \end{bmatrix}\begin{bmatrix} \hat{A}^\alpha \\ \hat{A}^\beta \end{bmatrix}$ (a) $\begin{bmatrix} \hat{A}^\alpha \\ \hat{A}^\beta \end{bmatrix} = \frac{\mu}{\gamma}\begin{bmatrix} -\coth\gamma\Delta & \frac{1}{\sinh\gamma\Delta} \\ \frac{-1}{\sinh\gamma\Delta} & \coth\gamma\Delta \end{bmatrix}\begin{bmatrix} \hat{H}_y^\alpha \\ \hat{H}_y^\beta \end{bmatrix}$ (b) $\gamma \equiv \sqrt{k^2 + j\mu\sigma(\omega - kU)}$	$\begin{bmatrix} \hat{H}_\theta^\alpha \\ \hat{H}_\theta^\beta \end{bmatrix} = \frac{1}{\mu}\begin{bmatrix} f_m(\beta,\alpha,\gamma) & g_m(\alpha,\beta,\gamma) \\ g_m(\beta,\alpha,\gamma) & f_m(\alpha,\beta,\gamma) \end{bmatrix}\begin{bmatrix} \hat{A}^\alpha \\ \hat{A}^\beta \end{bmatrix}$ (c) $\begin{bmatrix} \hat{A}^\alpha \\ \hat{A}^\beta \end{bmatrix} = \mu\begin{bmatrix} F_m(\beta,\alpha,\gamma) & G_m(\alpha,\beta,\gamma) \\ G_m(\beta,\alpha,\gamma) & F_m(\alpha,\beta,\gamma) \end{bmatrix}\begin{bmatrix} \hat{H}_\theta^\alpha \\ \hat{H}_\theta^\beta \end{bmatrix}$ (d) f_m, g_m, F_m and G_m as defined in Table 2.16.2 with $k \rightarrow \gamma$ where $\gamma \equiv \sqrt{j\mu\sigma(\omega - m\Omega)}$	$\begin{bmatrix} \hat{H}_z^\alpha \\ \hat{H}_z^\beta \end{bmatrix} = -\frac{\gamma^2}{\mu}\begin{bmatrix} F_o(\beta,\alpha,\gamma) & G_o(\alpha,\beta,\gamma) \\ G_o(\beta,\alpha,\gamma) & F_o(\alpha,\beta,\gamma) \end{bmatrix}\begin{bmatrix} \frac{\hat{\Lambda}^\alpha}{\alpha} \\ \frac{\hat{\Lambda}^\beta}{\beta} \end{bmatrix}$ (e) $\begin{bmatrix} \frac{\hat{\Lambda}^\alpha}{\alpha} \\ \frac{\hat{\Lambda}^\beta}{\beta} \end{bmatrix} = -\frac{\mu}{\gamma^2}\begin{bmatrix} f_o(\beta,\alpha,\gamma) & g_o(\alpha,\beta,\gamma) \\ g_o(\beta,\alpha,\gamma) & f_o(\alpha,\beta,\gamma) \end{bmatrix}\begin{bmatrix} \hat{H}_z^\alpha \\ \hat{H}_z^\beta \end{bmatrix}$ (f) F_o, G_o, f_o and g_o as defined in Table 2.16.2 with $k \rightarrow \gamma$ where $\gamma \equiv \sqrt{k^2 + j\mu\sigma(\omega - kU)}$
$\begin{bmatrix} \hat{A}^\alpha \\ \hat{A}^\beta \end{bmatrix} = \frac{j}{k}\begin{bmatrix} \hat{B}_x^\alpha \\ \hat{B}_x^\beta \end{bmatrix}$	$\begin{bmatrix} \hat{A}^\alpha \\ \hat{A}^\beta \end{bmatrix} = \frac{j}{m}\begin{bmatrix} \alpha\hat{B}_r^\alpha \\ \beta\hat{B}_r^\beta \end{bmatrix}$	$\begin{bmatrix} \frac{\hat{\Lambda}^\alpha}{\alpha} \\ \frac{\hat{\Lambda}^\beta}{\beta} \end{bmatrix} = -\frac{j}{k}\begin{bmatrix} \hat{B}_r^\alpha \\ \hat{B}_r^\beta \end{bmatrix}$

$$\nabla \times [\frac{1}{\mu\sigma} \nabla \times (\nabla \times \vec{A}) + \frac{\partial \vec{A}}{\partial t} - \vec{v} \times (\nabla \times \vec{A})] = 0 \tag{1}$$

Because the curl of the gradient of any scalar, say $-\Phi$, is zero, a solution to this equation is

$$\frac{1}{\mu\sigma} \nabla \times (\nabla \times \vec{A}) + \frac{\partial \vec{A}}{\partial t} - \vec{v} \times (\nabla \times \vec{A}) = -\nabla\Phi \tag{2}$$

For a given material motion, this equation is linear in $\vec{A}$ so that solutions can be superimposed. The inhomogeneous solutions resulting from the "drive" on the right can be added to the homogeneous solutions satisfying Eq. 2 with $\Phi = 0$. A vector identity* converts the latter equation to

$$\frac{1}{\mu\sigma} \nabla^2\vec{A} = \frac{\partial \vec{A}}{\partial t} - \vec{v} \times (\nabla \times \vec{A}) \tag{3}$$

where the vector Laplacian must be distinguished from its scalar counterpart (Appendix A). This section is devoted to developing certain useful solutions to Eq. 3 in such a form that they can be used in problem solving. The geometries to be treated, summarized in Table 6.5.1, are extensions of those identified in Sec. 2.19, two-dimensional or symmetric configurations where the vector potential has a single component.

Planar Layer in Translation: In Cartesian coordinates, with $\vec{A} = A(x,y)\vec{i}_z$ and the material moving uniformly in the y direction, so that $\vec{v} = U\vec{i}_y$, Eq. 3 reduces to its z component, which is

$$\frac{1}{\mu\sigma} \nabla^2 A = \frac{\partial A}{\partial t} + U \frac{\partial A}{\partial y} \tag{4}$$

With solutions taking the complex-amplitude form $A(x,y,t) = \text{Re}\hat{A}(x)\exp j(\omega t - ky)$, this equation reduces to

$$\frac{d^2\hat{A}}{dx^2} - \gamma^2\hat{A} = 0; \quad \gamma^2 \equiv k^2 + j\mu\sigma(\omega - kU) \tag{5}$$

Transfer relations can now be deduced following the same line of reasoning used in preceding from Eq. 2.16.13 to the relations of Table 2.16.1, or from Eq. 2.19.3 to the Cartesian relations of Table 2.19.1. With $(\hat{A}^\alpha, \hat{A}^\beta)$ the complex amplitudes at $x = \Delta$ and $x = 0$, respectively, the solution to Eq. 5 is:

$$\hat{A} = \hat{A}^\alpha \frac{\sinh \gamma x}{\sinh \gamma\Delta} - \hat{A}^\beta \frac{\sinh \gamma(x-\Delta)}{\sinh \gamma\Delta} \tag{6}$$

Evaluation of $\hat{H}_y = -(1/\mu)d\hat{A}/dx$ (Table 2.18.1) at $x = \Delta$ and $x = 0$ then gives the transfer relations, Eqs. (a) of Table 6.5.1. Inversion of these relations gives Eqs. (b). Note that $\gamma = \gamma_r + j\gamma_i$ in Eq. 6. Thus,

$$\sinh \gamma x = \sinh(\gamma_r x + j\gamma_i x) = \sinh \gamma_r x \cos \gamma_i x + j \cosh \gamma_r x \sin \gamma_i x \tag{7}$$

is a complex function. In computer libraries it is usually the circular rather than the hyberbolic functions that are provided with the capability of having complex arguments. Then, evaluation is accomplished by replacing $\sinh \gamma x \to -j \sin j\gamma x$ in Eq. 6.

The diffusion transfer relations are the same as those for a nonconducting region (Table 2.19.1), except that k is replaced by γ. The transverse wavenumber governs the manner and degree of penetration of the field into the conductor, and is examined in Sec. 6.6. The transfer relations for a planar region are applied in Secs. 6.6-6.8 and 6.10.

Rotating Cylinder: In a material suffering rigid-body rotation with the angular velocity Ω, the velocity is $\vec{v} = \Omega r\vec{i}_\theta$. For field dynamics not depending on z, the appropriate form is $\vec{A} = A(r,\theta,t)\vec{i}_z$, the polar coordinate case of Table 2.18.1. Then, Eq. 3 reduces to its z component:

$$\frac{1}{\mu\sigma} \nabla^2 A = \frac{\partial A}{\partial t} + \Omega \frac{\partial A}{\partial \theta} \tag{8}$$

Substitution of $A = \text{Re}\hat{A}(r) \exp j(\omega t - m\theta)$ reduces this expression to

---

* $\nabla \times (\nabla \times \vec{A}) = \nabla(\nabla\cdot\vec{A}) - \nabla^2\vec{A}$

$$\frac{d^2\hat{A}}{dr^2} + \frac{1}{r}\frac{d\hat{A}}{dr} - (\gamma^2 + \frac{m^2}{r^2})\hat{A} = 0;\ \gamma^2 \equiv j\mu\sigma(\omega - m\Omega) \tag{9}$$

With the identification $k^2 \to \gamma^2$, Eq. 9 is Eq. 2.16.19, Bessel's equation. The appropriate solution for the cylindrical annulus shown in Table 6.5.1, with outer and inner radii at $r = \alpha$ and $r = \beta$, respectively, takes the same form as Eq. 2.16.25:

$$\hat{A} = \hat{A}^\alpha \frac{[H_m(j\gamma\beta)J_m(j\gamma r) - J_m(j\gamma\beta)H_m(j\gamma r)]}{[H_m(j\gamma\beta)J_m(j\gamma\alpha) - J_m(j\gamma\beta)H_m(j\gamma\alpha)]} + \hat{A}^\beta \frac{[J_m(j\gamma\alpha)H_m(j\gamma r) - H_m(j\gamma\alpha)J_m(j\gamma r)]}{[J_m(j\gamma\alpha)H_m(j\gamma\beta) - H_m(j\gamma\alpha)J_m(j\gamma\beta)]} \tag{10}$$

Evaluation of $H_\theta = -(1/\mu)d\hat{A}/dr$ (see Table 2.18.1) at the respective surfaces then gives the transfer relations (c) of Table 6.5.1. Inversion of these relations results in Eqs. (d).

The entries appearing in these transfer relations are those used to represent Laplacian fields, defined in Table 2.16.2, except that k is replaced by γ. In modeling a configuration composed of two or more regions having differing values of $\mu\sigma$, it is necessary to distinguish among two or more values of γ. By agreement, if the third argument is simply k, it is suppressed. For example,

$$f_m(x,y,k) \equiv f_m(x,y) \tag{11}$$

so that the transfer relation entries introduced in this section are natural generalizations of those introduced in Sec. 2.16.

Bessel and Hankel functions of complex argument bear much the same relationship to the real-argument limiting cases as do the circular functions in Cartesian coordinates. Computer library functions that allow complex arguments may be in terms of the Bessel function of second kind, N_m, in which case the definition of the Hankel function, Eq. 2.16.29, is used to evaluate H_m. For the rotating cylinder, the real and imaginary parts of the arguments are equal and, in this case, the Bessel and Hankel functions are tabulated as the Kelvin functions:[1]

$$\begin{aligned} \mathrm{ber}_m x + j\,\mathrm{bei}_m x &\equiv J_m(e^{j3\pi/4}x) \\ \mathrm{ker}_m x + j\,\mathrm{kei}_m x &\equiv \frac{j\pi}{2} H_m(e^{j3\pi/4}x) \end{aligned} \tag{12}$$

<u>Axisymmetric Translating Cylinder</u>: To complete Table 6.5.1, consider the annular shaped material moving with a uniform velocity $\vec{v} = U\vec{i}_z$ in the axial direction under axisymmetric conditions. Then, the appropriate vector potential is $\vec{A} = \vec{i}_\theta A(r,z,t)$ and Eq. 3 becomes

$$\frac{1}{\mu\sigma}\{\frac{\partial}{\partial r}[\frac{1}{r}\frac{\partial}{\partial r}(rA)] + \frac{\partial^2 A}{\partial z^2}\} = \frac{\partial A}{\partial t} + U\frac{\partial A}{\partial z} \tag{13}$$

Substitution of $A = \mathrm{Re}\hat{A}(r)\exp j(\omega t - kz)$ results in an equation of the same form as the homogeneous part of Eq. 2.19.9,

$$\frac{d^2\hat{A}}{dr^2} + \frac{1}{r}\frac{d\hat{A}}{dr} - (\gamma^2 + \frac{1}{r^2})\hat{A} = 0;\ \gamma^2 \equiv k^2 + j\mu\sigma(\omega - kU) \tag{14}$$

where k^2 has been replaced by γ^2. Thus, the solution is Eq. 2.19.10 with $k \to \gamma$:

$$\hat{\Lambda} \equiv r\hat{A} = \hat{A}^\alpha \frac{[H_1(j\gamma\beta)rJ_1(j\gamma r) - J_1(j\gamma\beta)rH_1(j\gamma r)]}{[H_1(j\gamma\beta)J_1(j\gamma\alpha) - J_1(j\gamma\beta)H_1(j\gamma\alpha)]} + \hat{A}^\beta \frac{[J_1(j\gamma\alpha)rH_1(j\gamma r) - H_1(j\gamma\alpha)rJ_1(j\gamma r)]}{[J_1(j\gamma\alpha)H_1(j\gamma\beta) - H_1(j\gamma\alpha)J_1(j\gamma\beta)]} \tag{15}$$

---

1. M. Abramowitz and I. A. Stegun, <u>Handbook of Mathematical Functions with Formulas, Graphs, and Mathematical Tables</u>, U.S. Government Printing Office, Washington D.C., 1964, p. 379 and pp. 430-433.

The transfer Eqs. (e) of Table 6.5.1 follow by evaluating $H_z = (1/\mu r)\partial\Lambda/\partial r$ at the outer and inner radii. Identities, Eqs. 2.19.12 and 2.16.26c, are used to write the entries in terms of previously defined functions. The inverse relations are Eqs. (f).

6.6 Induction Motor with Deep Conductor: A Magnetic Diffusion Study

While also being of practical significance, the induction interaction considered in this section is chosen to give insights concerning sinusoidal steady-state magnetic diffusion into the bulk of uniform conductors. The model is similar to the thin-sheet developed model shown in Fig. 6.4, except that the rotor conductor now has a finite thickness, a, that can in general be comparable to the effective skin depth δ', to the wavelength $2\pi/k$ of the imposed traveling wave of surface current on the stator and to the air gap d. The revised cross section is shown in Fig. 6.6.1. With the understanding that various stator configurations could be represented as in Sec. 6.4, the stator current is taken as a pure traveling wave.

The configuration allows for an examination of the thin-sheet model of Sec. 6.3 while also placing in perspective the opposite extreme, the short skin-depth model introduced in Sec. 6.8. The sinusoidal steady-state driven response emphasized in this section is also related to the temporal modes of the system in Sec. 6.10.

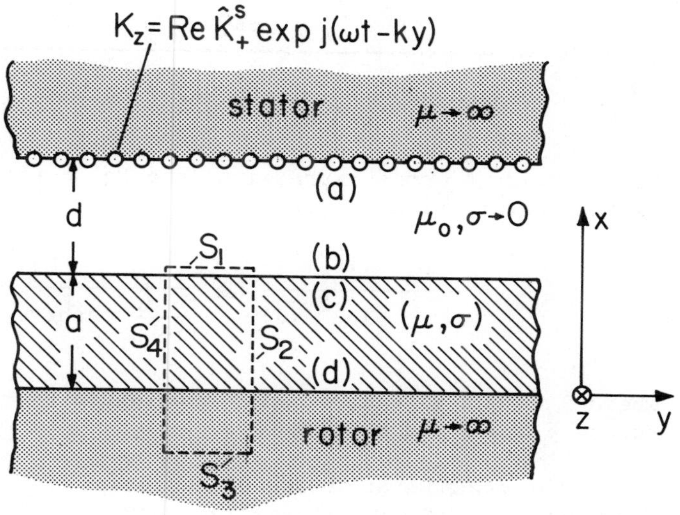

Fig. 6.6.1. Induction machine with rotor conductor having finite thickness a.

In terms of the locations defined in Fig. 6.6.1, boundary and jump conditions represent Ampère's law (Eq. 2.10.21):

$$H_y^a = -\mathrm{Re}\hat{K}_+^s e^{j(\omega t-ky)} \tag{1}$$

$$H_y^b = H_y^c \tag{2}$$

$$H_y^d = 0 \tag{3}$$

and continuity of magnetic flux density (Eq. 2.10.22),

$$B_x^b = B_x^c;\ A^b = A^c \tag{4}$$

Identification of the bulk relations (b) of Table 6.5.1, first with the air gap and then with the conducting layer, gives

$$\begin{bmatrix}\hat{A}^a\\ \hat{A}^b\end{bmatrix} = \frac{j}{k}\begin{bmatrix}\hat{B}_x^a\\ \hat{B}_x^b\end{bmatrix} = \frac{\mu_o}{k}\begin{bmatrix}-\coth(kd) & \dfrac{1}{\sinh(kd)}\\ \dfrac{-1}{\sinh(kd)} & \coth(kd)\end{bmatrix}\begin{bmatrix}-\hat{K}_+^s\\ \hat{H}_y^b\end{bmatrix} \tag{5}$$

$$\begin{bmatrix}\hat{A}^c\\ \hat{A}^d\end{bmatrix} = \frac{j}{k}\begin{bmatrix}\hat{B}_x^b\\ \hat{B}_x^d\end{bmatrix} = \frac{\mu}{\gamma}\begin{bmatrix}-\coth(\gamma a) & \dfrac{1}{\sinh(\gamma a)}\\ \dfrac{-1}{\sinh(\gamma a)} & \coth(\gamma a)\end{bmatrix}\begin{bmatrix}\hat{H}_y^b\\ 0\end{bmatrix} \tag{6}$$

In writing these expressions, the jump and boundary conditions have been inserted. These four equations determine the complex amplitudes ($\hat{B}_x^a$, $\hat{B}_x^b$, $\hat{B}_x^d$, $\hat{H}_y^b$) in terms of the stator surface current density. Before proceeding, it is prudent to determine which amplitude is required.

Time-Average Force: With a pure traveling-wave excitation, the time-average force per unit y-z area is independent of y. This is true because, except for a temporal phase shift, each "slice" of the material, shown in Fig. 6.6.1, is stressed by the same fields. Formally, this force per unit area is found by integrating the stress tensor over the surfaces $S_1 \ldots S_4$ shown in the figure. The sum of these surfaces is like that of Fig. 4.2.1a, except that its extent in the y direction is arbitrary; it is the time-average rather than the space-average that is being taken. With the understanding that $z \to t$, the complex-amplitude averaging theorem, Eq. 2.15.14, is applicable. The time-average stress integrated over surface S_2 cancels that integrated over surface S_4. The fields on S_3 are negligible,

$$\langle T_y \rangle_t = \frac{1}{2} \text{Re}[(\hat{B}_x^b e^{-jky})^* \hat{H}_y^b e^{-jky}] \tag{7}$$

and, as expected, the y dependence is eliminated. To eliminate the self-field term from this equation, Eq. 5b is substituted for $\hat{B}_x^b$:

$$\langle T_y \rangle_t = \frac{\mu_o}{2} \text{Re}\left[\frac{j(\hat{K}_+^s)^* \hat{H}_y^b}{\sinh(kd)}\right] \tag{8}$$

Thus, it is $\hat{H}_y^b$ that is required and so Eqs. 5b and 6a are equated and solved for $\hat{H}_y^b$:

$$\hat{H}_y^b = \frac{-\hat{K}_+^s}{\sinh(kd)\left[\frac{k}{\gamma}\frac{\mu}{\mu_o}\coth(\gamma a) + \coth(kd)\right]} \tag{9}$$

Substitution of this expression into Eq. 8 then gives the time-average force per unit area as a function of the stator surface current:

$$\langle T_y \rangle_t = -\frac{\mu_o}{2}\frac{|\hat{K}_+^s|^2}{\sinh^2(kd)} \text{Re}\frac{j}{\left[\frac{ka\mu}{\gamma a\mu_o}\coth(\gamma a) + \coth(kd)\right]}\ ; \ \gamma a \equiv \sqrt{(ka)^2 + jS_M} \tag{10}$$

where $S_M \equiv \mu\sigma a^2(\omega - kU)$. With a balanced two-phase excitation, $\hat{K}_+^s$ would be related to the terminal currents by 6.4.2 and 6.4.12.

The dependence of the time-average force on S_M, the normalized frequency as measured from the rotor frame of reference, is illustrated in Fig. 6.6.2. The function is odd in S_M. If the material velocity U exceeds the wave-phase-velocity ω/k, so that S_M is negative, the sign of the force is negative.

The dependence is somewhat similar to that for the thin-sheet interaction of Sec. 6.4 (see Fig. 5.13.2). A qualitative difference is that the deep-conductor force falls off less rapidly with increasing rotor-frame frequency than does the thin-sheet force. Two observations point to the origins of this difference. First, for S_M exceeding 2, the skin depth based on the rotor frame frequency, $\delta' \equiv \sqrt{2/|\omega-kU|\mu\sigma} = a\sqrt{2/|S_M|}$, is shorter than the conductor thickness. In the thick-conductor model, currents redistribute themselves in such a way that the effective L/R time constant remains on the order of the rotor-frame frequency (see discussion accompanying Eq. 6.2.10). Second, it is shown in Sec. 6.10 that whereas the thin-sheet model embodies a single natural temporal mode, the deep-conductor model retains an infinite number of such modes. At high frequencies, a spectrum of these contribute to the sinusoidal driven response, and tend to broaden the frequency dependence of the force.

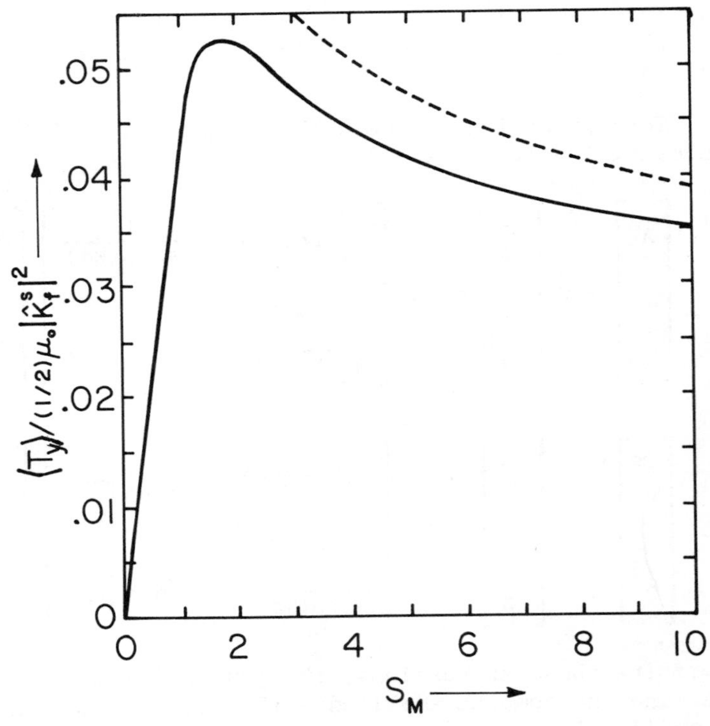

Fig. 6.6.2. Time-average force/unit area acting on deep conductor in direction of traveling wave. ka=kd=1, $\mu=\mu_o$. The force is an odd function of $S_M \equiv \mu\sigma a^2(\omega - kU)$. The broken curve is the high-frequency asymptote given by Eq. 11.

The high-frequency limit of Eq. 10 is taken by recognizing that if $|S_M| >> (ka)^2$, then $\gamma a \to (1 \pm j)\sqrt{|S_M|/2}$, where the upper and lower signs pertain for S_M positive and negative, respectively. As the magnitude of γa becomes large, $\coth(\gamma a) \to 1$.

Thus, Eq. 10 becomes approximately

$$\langle T_y\rangle_t = \pm \frac{\mu_o}{2} \frac{|\hat{K}_+^s|^2}{\sinh^2(kd)} \frac{(ak)\frac{\mu}{\mu_o}\sqrt{|S_M|/2}}{[(ak)\frac{\mu}{\mu_o} + \sqrt{|S_M|/2}\coth(kd)]^2 + [\sqrt{|S_M|/2}\coth(kd)]^2} \qquad (11)$$

This high frequency approximation is represented by the broken line curve of Fig. 6.6.2. Because of the skin effect, in this high-frequency limit, the force is inversely proportional to the square root of the rotor frequency. By contrast, in this limit with the thin-sheet model (represented by the first term in Eq. 6.4.11), the force varies inversely with the frequency.

Thin-Sheet Limit: What approximations are implicit to the thin-sheet model of Sec. 6.4? This is tantamount to asking what approximations are necessary if the thin sheet force for a pure traveling wave (the first term in Eq. 6.4.11) is to adequately approximate Eq. 10. It is clear from Eq. 10 that there are two measures of the conductor thickness a, one the quantity (ka) which is small compared to unity if $a < \lambda/2\pi$, where λ is the wavelength of the spatially periodic excitation. The other is γa (Eq. 10), which can alternatively be written in terms of a skin depth δ' based on the rotor frequency,

$$\gamma a = \sqrt{(ka)^2 + jS_M} = \sqrt{(ka)^2 \pm 2j(\frac{a}{\delta'})^2}; \quad \delta' \equiv \sqrt{\frac{2}{\mu\sigma|\omega - kU|}} = a/\sqrt{|S_M|/2} \qquad (12)$$

In order for $|\gamma a| \ll 1$, there are therefore two requirements, and these are the fundamental approximations validating the thin-sheet model:

$$ka \ll 1; \quad \frac{\delta'}{a} \gg 1 \qquad (13)$$

With these approximations, $\coth\gamma a \rightarrow 1/\gamma a$ and Eq. 10 can be written in the form of the first term in Eq. 6.4.11. Note that $S_M = (ka)S_m$. In the limit $(ka) \ll 1$, these expressions are in fact identical.

Conceptualization of Diffusing Fields: With the objective of picturing the space-time evolution of the fields in the conducting layer as a function of δ'/a and ka, remember that all fields have been represented in terms of

$$A = \mathrm{Re}\hat{A}(x)e^{j(\omega t - ky)} = \mathrm{Re}|\hat{A}(x)|e^{j[\omega t - ky + \theta(x)]} \qquad (14)$$

where $\hat{A}(x)$ in general is given by Eq. 6.5.6, and in particular for the configuration considered in this section (where $H_y(0) = 0$ and hence $dA/dx(0) = 0$) is

$$\hat{A} = \hat{A}^c \frac{\cosh(\gamma x)}{\cosh(\gamma a)} \equiv |\hat{A}(x)|e^{j\theta(x)} \qquad (15)$$

This expression can be deduced formally by manipulating the complex amplitudes, but is just as well found by inspection. From Eq. 15, the field intensity in the conductor follows from $\vec{H} = \nabla\times\vec{A}/\mu$ (Table 2.18.1), and the current density is

$$\hat{J} = \frac{1}{\mu}\nabla\times\nabla\times\vec{A} = -\frac{\nabla^2\vec{A}}{\mu} \Rightarrow \hat{J}_z = -\frac{1}{\mu}\left(\frac{d^2\hat{A}}{dx^2} - k^2\hat{A}\right) \qquad (16)$$

or in particular, because $-j = \exp(-j\pi/2)$

$$\hat{J}_z = -\frac{\hat{A}^c}{\mu} j\mu\sigma(\omega - kU)\frac{\cosh(\gamma x)}{\cosh(\gamma a)} = -j\frac{S_M}{\mu a^2}\hat{A}(x) = \frac{S_M}{\mu a^2}|\hat{A}(x)|e^{j(\theta - \pi/2)} \qquad (17)$$

Of course, $\hat{A}^c$ is determined from Eqs. 5 and 6 by the stator surface current density, but for the present purposes it is just as well to think of $\hat{A}^c$ as imposed at the air-gap surface of the conducting layer (at x = a). The amplitude and phase of $\hat{A}(x)$, defined by Eq. 15, are then typified by the distributions over the conductor cross section shown in Fig. 6.6.3. At any given plane x = constant in the conductor, the fields take the form of a sinusoid traveling in the y direction with the phase velocity ω/k. The amplitude of this wave, $|\hat{A}(x)|$, varies with distance into the conductor as shown in Fig. 6.6.3a. (Note that there is decay of the field in the -x direction even if $\delta'/a \rightarrow \infty$. This is simply the decay characterizing Laplace's equation in free space. For the plots, ka = 1.) Points of the same phase on the traveling sinusoidal wave, say of phase θ_o, have the space-time relationship

$$ky = \omega t + \theta(x) - \theta_o \qquad (18)$$

That is, for values of y given by Eq. 18, the exponential in Eq. 14 becomes $\exp j\theta_o$, a complex constant. Thus, at any instant, the plot of $\theta(x)$ shown in Fig. 6.6.3b is equivalent to the (x-y) distribution of

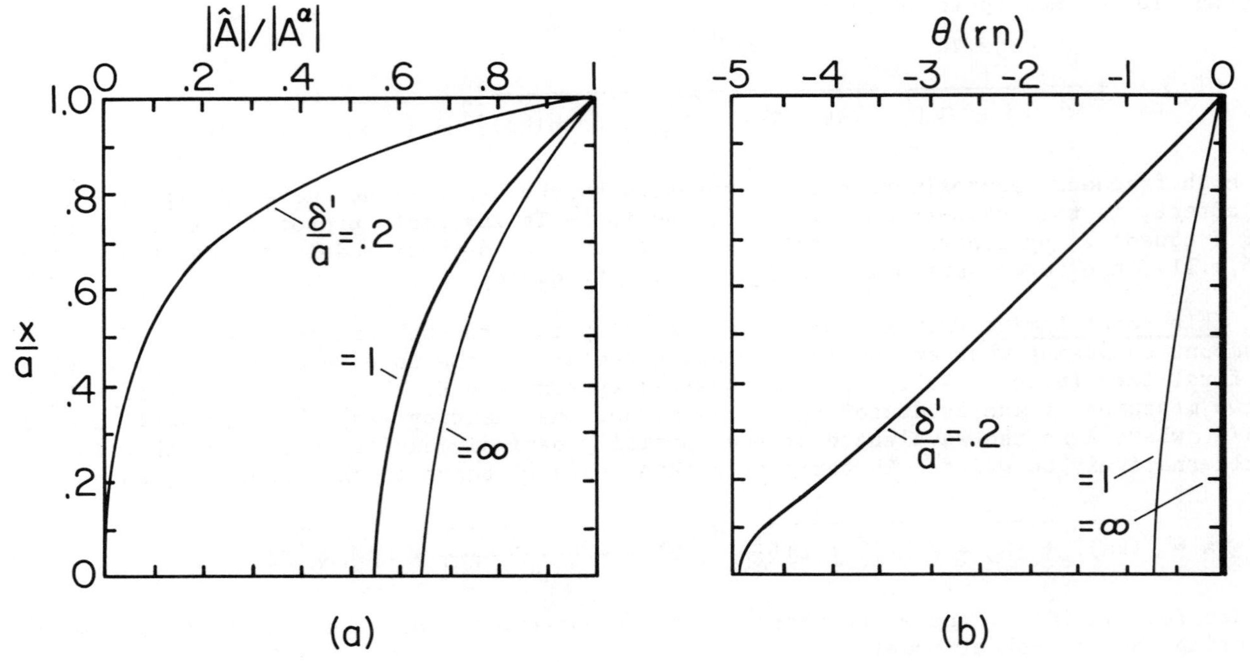

Fig. 6.6.3. Amplitude and phase of $A = \text{Re}|\hat{A}(x)|\exp j(\omega t - ky + \theta(x)|$ for fields diffusing through conductors of Fig. 6.6.1. The parameter is the skin depth, based on the material frame frequency, normalized to the conductor thickness, δ'/a; $ka = 1$.

the points of a given phase on the sinusoidal traveling waves. For example, when $t = 0$, an $x - y$ plot of the zero crossing for a co-sinusoid is given by Eq. 18 with $\theta_o = \pi/2$.

The distribution when $t = 0$ is now readily visualized in terms of the amplitude and phase plots of Fig. 6.6.3. As an example, Fig. 6.6.4 shows the distribution of A when $t = 0$ for $\delta'/a = 0.2$. As Eq. 18 shows, the time dependence is seen by simply letting this picture propagate to the right with the phase velocity ω/k.

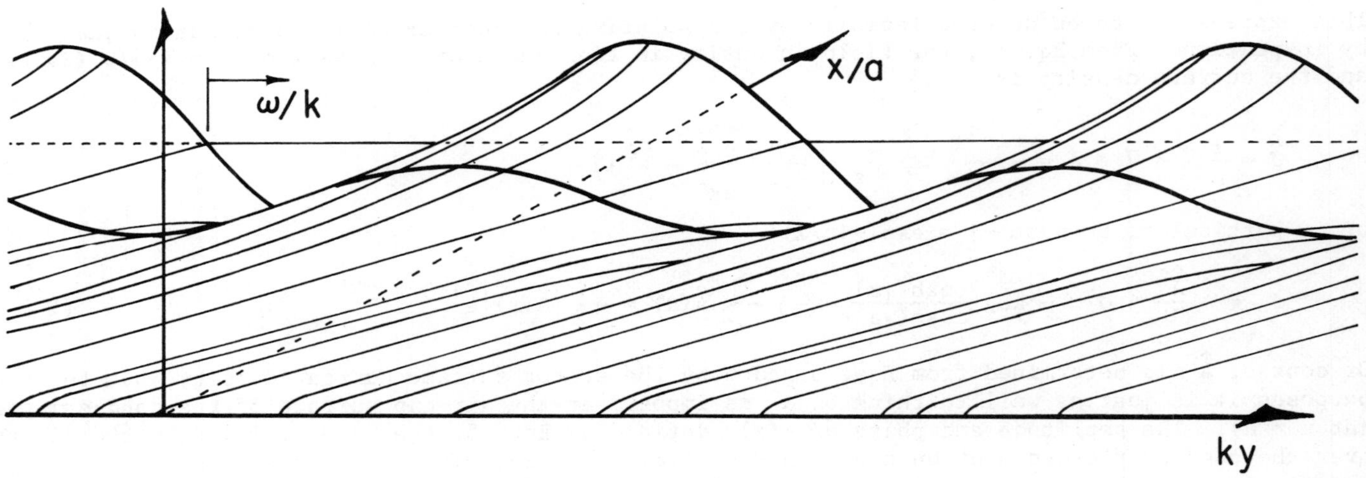

Fig. 6.6.4. Magnetic diffusion wave distribution A across conducting layer of Fig. 6.6.1 with pure traveling wave of excitation. For fields shown, phase velocity ω/k of wave exceeds material velocity U. As time proceeds, picture translates to the right with phase velocity ω/k.

The decay is of course least when the skin depth is largest, assuming a limiting value consistent with Laplace's equation. In this limit, the contours of constant phase in the x-y plane become parallel to the x-axis.

In the thin-sheet model, the Laplacian decay is negligible ($ka \ll 1$) and the skin depth is large enough compared to (a) that the amplitude and phase are essentially uniform over the cross section.

In the opposite extreme where the skin depth is short compared to the conductor thickness, there is little effect reflected back into the layer by the highly permeable backing material at $x = 0$. In this limit, the traveling wave leaves a trail of magnetic field in the conductor that appears at any given y plane as a rapidly attenuating wave with phases advancing in the -x direction.

For the picture shown, $S_M > 0$, meaning that the wave velocity exceeds that of the material. If the material moves faster than the wave, $S_M < 0$, and the sign of the imaginary part of γ is reversed. This reverses the sign of the phase shift. The lines of constant phase in a field picture like Fig. 6.6.4 now run to the right with increasing y rather than to the left. This is true even though the wave velocity ω/k is still to the right. To make this observation consistent with intuition, note that the material is moving even more rapidly to the right than the wave.

To emphasize the effect of the material motion, consider a thought experiment in which all parameters are fixed while the material velocity U is increased, starting at zero. At zero velocity, the picture is as in Fig. 6.6.4, with the skin depth determined by the imposed frequency ω alone. As the velocity is increased, the skin depth δ' increases. Hence, the decay and phase shift are reduced. At synchronism, the skin depth δ' is infinite, the decay is Laplacian and there is no phase shift. Further increase of the velocity results in a positive phase shift and a decreasing skin depth. The picture returns to that typified by Fig. 6.6.4, except that the constant phase lines "stream ahead" of the traveling wave.

The short skin depth approximation is the basis for a far-reaching boundary layer model, discussed in Sec. 6.8.

6.7 Electrical Dissipation

Induction interactions of the type exemplified in Sec. 6.6 involve electromechanical energy conversion at some price of electrical power converted to heat. In fact, one of the most common applications of induced currents is to the efficient electrodeless production of heat in the volume of a conducting material. But, even where the objective is electromechanical energy conversion, the heating is likely to be a significant consideration. In this section, general relations are derived that can be applied to any situation in which the canonical conducting layer of Sec. 6.5 is embedded.

Some preliminaries are required to have a way of representing power dissipated in terms of quantities evaluated at the surfaces of the layer. The magnetoquasistatic form of Poynting's theorem, Eq. 2.13.16 with terms given by Eq. 2.14.16, is written in the inertial (primed) frame moving with the material:

$$-\nabla'\cdot(\vec{E}'\times\vec{H}') - \frac{\partial}{\partial t'}\left(\frac{1}{2}\mu\vec{H}'\cdot\vec{H}'\right) = \vec{E}'\cdot\vec{J}_f' \tag{1}$$

Magnetization has been taken as linear, $\mu_o(\vec{H}' + \vec{M}') = \mu\vec{H}'$. For purposes of physical interpretation, note that the integral of this expression over a volume V' enclosing material of fixed identity takes the form of Eq. 2.13.12. This expression states that the total flux of power across the surface and into the volume either goes into increasing the total energy within the volume or it leaves the magnetoquasistatic subsystem in a way represented by the term on the right. In general, power can either leave as mechanical work done through the action of the magnetic force on the moving material, or it leaves as electrical dissipation. Because there is no velocity of the material in the frame for which Eq. 1 is written, the term on the right cannot include power flow into the mechanical subsystem. It must be the electrical dissipation density P_d.

For the present purposes, what is required is an integration of Eq. 1 over a volume that is fixed in the laboratory frame. Thus, Eq. 1 is rewritten in terms of fixed frame variables. That is, in accordance with Eq. 2.5.2, $\nabla' \rightarrow \nabla$ and $\partial(\;)/\partial t' \rightarrow \partial(\;)/\partial t + \vec{v}\cdot\nabla(\;)$. Also, because the system is magnetoquasistatic, $\vec{H}' = \vec{H}$ (Eq. 2.5.9b). Thus, Eq. 1 is equivalent to

$$P_d = -\left[\nabla\cdot(\vec{E}'\times\vec{H}) + \frac{\partial}{\partial t}\left(\frac{1}{2}\mu\vec{H}\cdot\vec{H}\right) + \vec{v}\cdot\nabla\left(\frac{1}{2}\mu\vec{H}\cdot\vec{H}\right)\right] \tag{2}$$

Because $\vec{v}$ is uniform, and hence $\nabla\cdot\vec{v} = 0$, the energy convection term can be taken inside the divergence:

$$P_d = -\nabla\cdot\left(\vec{E}'\times\vec{H} + \vec{v}\,\frac{1}{2}\mu\vec{H}\cdot\vec{H}\right) - \frac{\partial}{\partial t}\left(\frac{1}{2}\mu\vec{H}\cdot\vec{H}\right) \tag{3}$$

In the sinusoidal steady state, the time rate of change makes no time-average contribution. This expression therefore makes it possible to evaluate the electrical power dissipated by evaluating fields on the enclosing surface. Consider again the planar layer of material described in Sec. 6.5. It is embedded in a system that is periodic in the y direction and is in the sinusoidal steady state. The volume over which the electrical dissipation is to be found has the fundamental length of periodicity in the y direction and has y-z surfaces denoted by α and β adjacent to the upper and lower surfaces of the layer (Fig. 6.7.1).

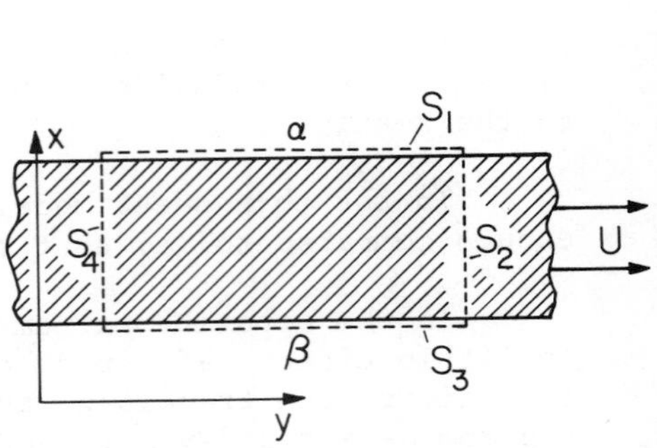

Fig. 6.7.1. Control volume fixed in laboratory frame with fundamental periodicity length in y direction.

The fields are presumed to be generally represented in terms of a Fourier complex-amplitude series, in the form of Eq. 5.16.1. Integration of the time average of Eq. 3 over the volume is converted by Gauss' theorem to an integration of the quantity inside the divergence over the enclosing surface. Because of the periodicity, contributions to surfaces cutting through the layer, surfaces S_2 and S_4 and those in the x-y plane, cancel or are zero. It follows from the averaging theorem, Eq. 5.16.4, that the integration over the surfaces S_1 and S_2 is evaluated by multiplying the area A of S_1 or S_3 by the spatial average

$$\langle S_d\rangle_{yt} \equiv \frac{1}{A}\int_V \langle P_d\rangle_t\, dV = \frac{1}{2}\,\mathrm{Re}\sum_{n=-\infty}^{+\infty}[(\hat{E}'_{zn}\hat{H}^*_{yn})^\alpha - (\hat{E}'_{zn}\hat{H}^*_{yn})^\beta] \tag{4}$$

Thus, the power dissipated over the cross section of the material within a volume having unit area in the y-z plane is evaluated in terms of complex amplitudes at the bounding surfaces. It is convenient to replace E'_z with variables already used in the transfer relations. By Ohm's law, Eq. 6.2.1 and Eq. 6.6.16,

$$\hat{E}'_z = \frac{\hat{J}_z}{\sigma} = -\frac{1}{\mu\sigma}\left(\frac{d^2\hat{A}}{dx^2} - k^2\hat{A}\right) \tag{5}$$

This expression is expressed in terms of the surface variables using Eq. 6.5.6 and the result evaluated at the respective surfaces. In view of the definition of γ^2, Eq. 6.5.5,

$$\begin{bmatrix}\hat{E}'^\alpha_z\\ \\ \hat{E}'^\beta_z\end{bmatrix} = -j(\omega - kU)\begin{bmatrix}\hat{A}^\alpha\\ \\ \hat{A}^\beta\end{bmatrix} \tag{6}$$

Thus, Eq. 4 can also be expressed as

$$\langle S_d\rangle_{yt} = -\,\mathrm{Re}\sum_{n=-\infty}^{+\infty} j\,\frac{(\omega - k_nU)}{2}\,[\hat{A}^\alpha_n(\hat{H}^\alpha_{yn})^* - \hat{A}^\beta_n(\hat{H}^\beta_{yn})^*] \tag{7}$$

This is the required time- and space-average power dissipation per unit y-z area in the layer. Similar relations can be derived for the other configurations of Table 6.5.1. Application of Eq. 7 is made in Sec. 6.8.

6.8 Skin-Effect Fields, Relations, Stress and Dissipation

In the short skin-depth limit, the planar layer of Table 6.5.1 becomes representative of all of the configurations in that table. The skin depth δ' is identified by writing γ (defined with Eq. 6.5.5) as

$$\gamma = \sqrt{k^2 \pm j2/(\delta')^2};\quad \frac{\omega}{k} \gtrless U;\quad \delta' \equiv \sqrt{\frac{2}{|\omega - kU|\mu\sigma}} \tag{1}$$

Note that the frequency that determines δ' is that experienced by the material; hence the appendage of a prime.

There are two approximations inherent to the model. First, the induced fields dominate over the "reactive" fields in determining the decay into the conductor:

$$k\delta' \ll 1 \Longrightarrow \gamma \rightarrow \frac{(1 \pm j)}{\delta'} \tag{2}$$

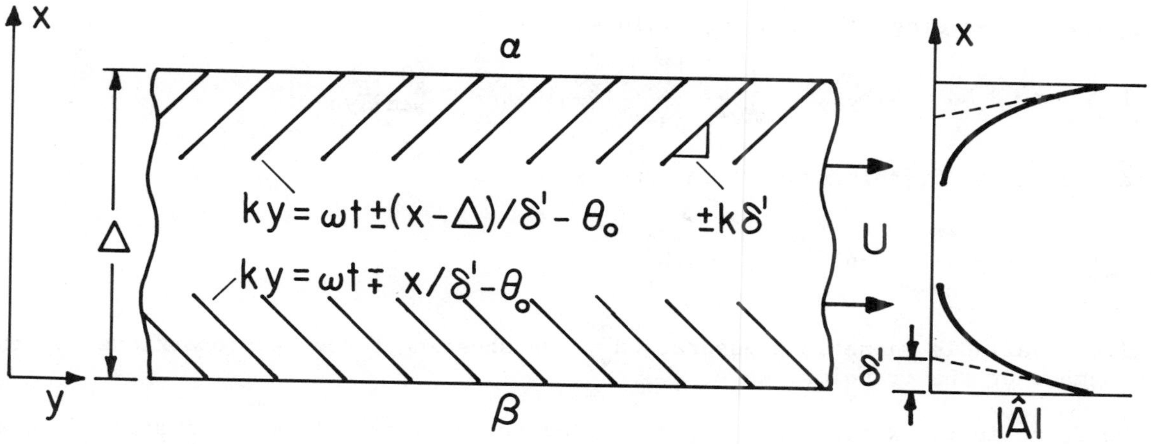

Fig. 6.8.1. Lines of constant phase move to right with velocity ω/k. For ω/k exceeding U, these lines form a "wake" to the left, as shown. If material velocity were to exceed ω/k, lines would slant to the right. Amplitude decays into material as shown, with depth for attenuation by e^{-1} equal to δ'.

Essentially, the skin depth is short compared to the wavelength of periodicity (divided by 2π).

Second, the skin depth is short compared to the thickness of the conductor:

$$|\gamma\Delta| >> 1 \tag{3}$$

Then, the fields represented by the vector potential, Eq. 6.5.6, become two independent rapidly decaying waves confined to the respective surfaces:

$$A \simeq \mathrm{Re}\left[\hat{A}^\alpha e^{(x-\Delta)/\delta'} e^{j(\omega t - ky \pm \frac{(x-\Delta)}{\delta'})} + \hat{A}^\beta e^{-(x/\delta')} e^{j(\omega t - ky \pm \frac{x}{\delta'})}\right] \tag{4}$$

These fields are of course a limiting case of the example depicted by Figs. 6.6.3 and 6.6.4. In the short skin-depth limit, the lines of constant phase, sketched in Fig. 6.8.1, are exactly straight lines. It is assumed in the sketch that the wave phase velocity ω/k exceeds the material velocity U.

Transfer Relations: In the short skin-depth limit summarized by Eqs. 2 and 3, the planar layer transfer relations take a form representative of all of the configurations of Table 6.5.1. The mutual coefficients tend to zero as the thickness becomes large compared to δ', so that the short skin-depth transfer relations are

$$\begin{bmatrix}\hat{B}_x^\alpha \\ \hat{B}_x^\beta\end{bmatrix} = -jk\begin{bmatrix}\hat{A}^\alpha \\ \hat{A}^\beta\end{bmatrix} \simeq \frac{1}{2}(\mp 1 - j)k\mu\delta'\begin{bmatrix}-1 & 0 \\ 0 & 1\end{bmatrix}\begin{bmatrix}\hat{H}_y^\alpha \\ \hat{H}_y^\beta\end{bmatrix} \;;\; \left|\frac{\omega}{k}\right| \gtrless U \tag{5}$$

According to these relations, in a frame of reference moving with the material, the fields diffuse into the conductor as though they were independent of y. That is, if the y component of the magnetic diffusion equation is written (Eq. 6.2.7), the contribution of the y derivative to the diffusion term is negligible compared to that from the x derivative. Thus, consistent with Eq. 5 is the approximation that

$$\frac{1}{\mu\sigma}\frac{\partial^2 H_y}{\partial x^2} = \left(\frac{\partial}{\partial t} + U\frac{\partial}{\partial y}\right)H_y \tag{6}$$

where the convective derivative on the right is the time rate of change for an observer moving with the material. If there were actually no y dependence, there would be no B_x. This is evident from the limit $k \to 0$ of Eq. 5. But, once having solved Eq. 6 to obtain H_y, the normal flux density can be found from the fact that $\hat{B}$ is solenoidal. The result would be Eq. 5. From a frame of reference moving with the conductor, short-skin-depth magnetic diffusion is as though the fields were independent of y.

Stress: But, without some y dependence there is no B_x and hence no magnetic stress. To compute the stress, the layer is enclosed by a control volume with surfaces as shown in Fig. 6.7.1. The force follows from an integration of the stress over this surface (as described in Sec. 4.2). The time-average force per unit y-z area tending to propel the slab in its direction of motion is found by applying the time-average theorem, Eq. 5.16.4:

$$\langle T_y\rangle_{yt} = \langle B_x^\alpha H_y^\alpha - B_x^\beta H_y^\beta\rangle_{yt} = \frac{1}{2}\,\text{Re}\sum_{n=-\infty}^{+\infty}[\hat{B}_{xn}^\alpha(\hat{H}_{yn}^\alpha)^* - \hat{B}_{xn}^\beta(\hat{H}_{yn}^\beta)^*] \tag{7}$$

Evaluated using Eq. 5, this expression becomes

$$\langle T_y\rangle_{yt} = \pm\frac{1}{4}\sum_{n=-\infty}^{+\infty}\mu k\delta'(|\hat{H}_{yn}^\alpha|^2 + |\hat{H}_{yn}^\beta|^2) \tag{8}$$

For a given available magnetic pressure, μH_y^2, the shearing force is proportional to the skin depth and the wavenumber of the traveling wave.

The force in the x direction might be used to levitate a layer or system of layers. Suppose that the layer is surrounded by free space, where $\mu = \mu_o$. In general the space-time average is then written in terms of quantities that are continuous across the surface as

$$\langle T_x\rangle_{yt} = \left\langle\frac{\mu_o}{2}\left[\frac{1}{\mu^2}(B_x^\alpha)^2 - (H_y^\alpha)^2\right] - \frac{\mu_o}{2}\left[\frac{1}{\mu^2}(B_x^\beta)^2 - (H_y^\beta)^2\right]\right\rangle_{yt} \tag{9}$$

Because the x component of $\vec{H}$ is of order $(k\delta)$ smaller than the y component, this expression is consistently approximated by $B_x \approx 0$ and hence

$$\langle T_x\rangle_{yt} \simeq \frac{\mu_o}{4}\sum_{n=-\infty}^{+\infty}\left[-|\hat{H}_{yn}^\alpha|^2 + |\hat{H}_{yn}^\beta|^2\right] \tag{10}$$

In the short skin-depth approximation, the normal force is simply the available magnetic pressure as it would exert itself on a layer of perfectly conducting material. In spite of the fact that the layer can be highly permeable, in the short skin-depth limit, the magnetic field "pushes" on the layer.

Dissipation: The power going into heating of the layer is computed in terms of the same surface variables as used to express the stress by applying Eq. 6.7.7. Evaluated using the short skin-depth transfer relations, Eqs. 5, it becomes

$$\langle S_d\rangle_{yt} = \frac{1}{2\sigma\delta'}(|\hat{H}_y^\alpha|^2 + |\hat{H}_y^\beta|^2) \tag{11}$$

For a given magnetic pressure, the power dissipation is inversely proportional to the skin depth. Hence, as the skin depth decreases, the heating increases and (from Eq. 8) the propulsion force decreases.

6.9 Magnetic Boundary Layers

An alternative title for Sec. 6.8 might be "magnetic boundary layers in the sinusoidal state." In essence, the skin-effect model is based on the same boundary layer approximation used in this section. Transverse magnetic diffusion dominates over that in the longitudinal (y) direction. Thus, in the magnetic diffusion equation, Eq. 6.2.7, the diffusion term is approximated by the second derivative with respect to the direction of field penetration, the x direction. With the conductor moving uniformly in the y direction, diffusion is therefore again governed by Eq. 6.8.3:

$$\frac{1}{\mu\sigma}\frac{\partial^2 H_y}{\partial x^2} = \left(\frac{\partial}{\partial t} + U\frac{\partial}{\partial y}\right)H_y \tag{1}$$

where it is presumed that $\partial(\ )/\partial z \approx 0$. Once the longitudinal field, H_y, is determined, the transverse field is determined by the rate-independent condition that the field be solenoidal:

$$\frac{\partial B_x}{\partial x} = -\mu\frac{\partial H_y}{\partial y} \tag{2}$$

The configuration of Fig. 6.9.1a is used in this section to illustrate the implications of the model. A relatively thick conductor moves to the right with velocity U. Just above the conductor, a fixed structure (perhaps windings driven by a current source) imposes a uniform current density $K_z = -H_o$ to the right of $y = 0$. This sheet is backed by an infinitely permeable material which extends over all

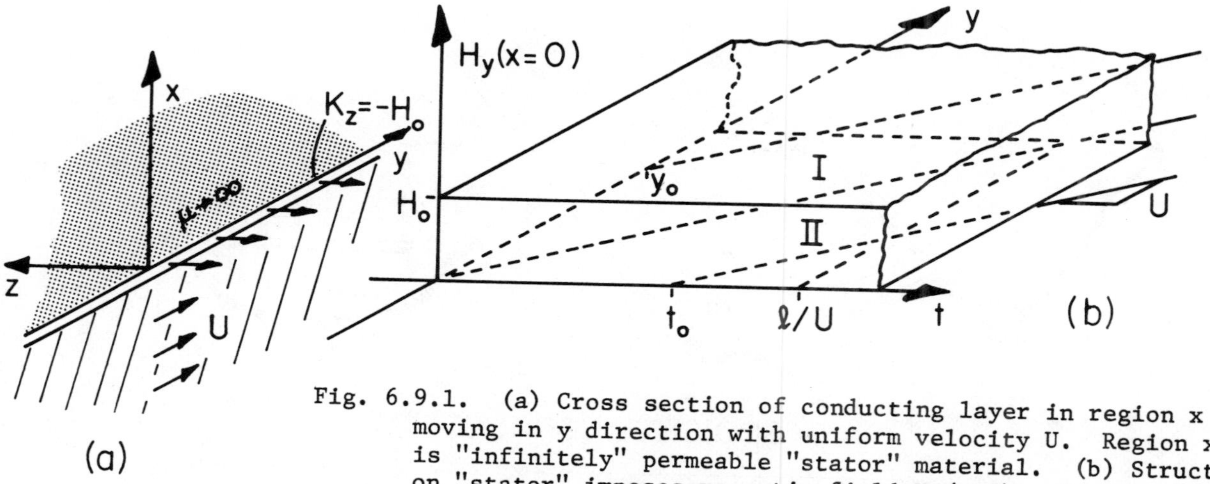

Fig. 6.9.1. (a) Cross section of conducting layer in region $x < 0$ moving in y direction with uniform velocity U. Region $x > 0$ is "infinitely" permeable "stator" material. (b) Structure on "stator" imposes magnetic field $H_y(x=0) = H_o$, which is turned on over region $y > 0$ when $t = 0$.

of the region $x > 0$. Because the distance between sheet and conductor is small compared to other dimensions of interest, the boundary condition imposed on H_y at the conductor surface is that it be H_o to the right of $y = 0$ and that it vanish to the left. When $t = 0$, the current excitation is turned on. A summary of the space-time dependence imposed on H_y at the conductor surface is given in Fig. 6.9.1b. What are the implications of the boundary layer approximation for the evolution of H_y in the moving conductor? How can the boundary layer model be used to compute the drag and lift on the excitation structure?

One of the more dramatic of many practical and proposed applications involving a magnetic diffusion process having the nature of that considered here is shown in Fig. 6.9.2.[1] The structure is in that case a magnetically levitated train and the conducting material the "rail." The y coordinate measures distance relative to the vehicle. From this frame of reference, the turn-on transient settles into a steady state in which the current imaging that on the structure in a given conductor element penetrates into the conductor to a depth determined by the time elapsed since the element passed the leading edge of the structure.

The convective derivative on the right in Eq. 1, the time rate of change for an observer moving with the velocity U of the conductor, can be written in terms of time t' measured from the reference frame of a material element (see Secs. 2.4 and 2.5):

$$\frac{1}{\mu\sigma}\frac{\partial^2 H_y}{\partial x^2} = \frac{\partial H_y}{\partial t'} \tag{3}$$

with $\partial(\ )/\partial t'$ defined as the partial derivative holding $y' = y - Ut$ constant. The lines of constant y', shown in the y-t plane of Fig. 6.9.1b, have intercepts (y_o, t_o) respectively with the positive y and t axes. These parameters both denote the constant y' and distinguish between those lines in regions I and II of the y-t plane separated by the line $y = Ut$:

$$y = \begin{cases} Ut + y_o; & y > Ut, \text{ region I} \\ U(t - t_o); & y < Ut, \text{ region II} \end{cases} \tag{4}$$

From the material frame of reference, the magnetic diffusion represented by the boundary layer equation, Eq. 3, is one-dimensional. Only the time dependence of the boundary condition on H_y at $x=0$ reflects the temporal transient. For the particular excitation shown graphically by Fig. 6.9.1b, H_y is a step function that turns on when $t = 0$ so long as $y > Ut$. Physically, material elements having a distance from the leading edge greater than the transit time Ut, see a uniform magnetic field applied when $t = 0$. But, for $y < Ut$, an element experiences a step that turns on when $t = t_o$. This is the time when the element passes the leading edge of the structure at $y = 0$. These general remarks pertain regardless of the details of the field excitation, once it is turned on. For example, the excitation might be a traveling wave confined to $y < 0$ and turned on when $t = 0$.

Here, discusson is confined to an excitation that is constant for $t > 0$ and $y > 0$.

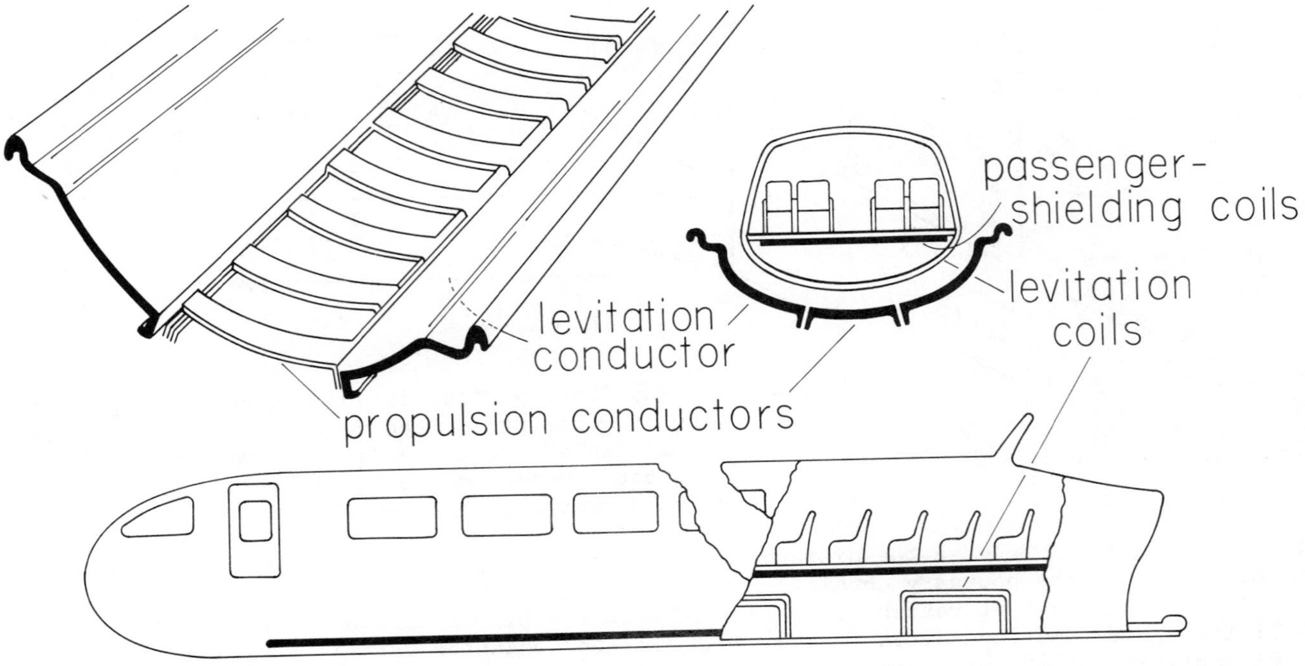

Fig. 6.9.2. Magneplane vehicle and "rail." Levitation results from interaction between conducting rail and magnetic fields from d-c excited superconducting coils mounted on vehicle. Currents are induced in rail by relative motion. The same d-c fields interact in synchronous fashion with traveling wave of magnetic field on center section of "rail" to provide propulsion.[1]

Similarity Solution: Can x and t' be related so that Eq. 3 becomes an ordinary differential equation? With t' understood to be the elapsed time since the field was turned on, it is expected that the field would have penetrated in the x-direction to a depth Δ_m typified by setting the magnetic diffusion time (defined with Eq. 6.2.9) equal to t' and solving for the length

$$\Delta_m = \sqrt{\frac{t'}{\mu\sigma}} \tag{5}$$

Thus, it is reasonable to scale the actual distance x to this length with a factor of 2 introduced to make the resulting equation assume a standard form

$$\xi \equiv \frac{x}{2}\sqrt{\frac{\mu\sigma}{t'}} \tag{6}$$

The conjecture is that the field intensity found at $x = x_1$ when the elapsed time from turn-on is $t' = t_1$ will be the same at time t where $x = x_1\sqrt{t/t_1}$. Evaluation of the derivatives in Eq. 3 justifies the supposition by converting the equation to

$$\frac{d^2H_y}{d\xi^2} + 2\xi\frac{dH_y}{d\xi} = 0 \tag{7}$$

In spite of the coefficient that depends on ξ, this equation has a simple solution satisfying the boundary condition $H_y(\xi = 0) = H_o$,

$$H_y = H_o[1 + \text{erf}(\xi)]; \quad \text{erf} \equiv \frac{2}{\sqrt{\pi}}\int_o^{\xi} e^{-\xi^2}d\xi \tag{8}$$

as can be seen by direct substitution. The error function,[2] erf(ξ), is normalized so that erf(ξ) $\rightarrow -1$ as $\xi \rightarrow -\infty$.

In applying Eq. 8, it is necessary to distinguish between regions I and II of the y-t plane, Fig. 6.9.1b. In region I, the elapsed time since turn-on of the field is simply t'=t. Hence, Eq. 8

1. See H. H. Kolm and R. D. Thornton, "Electromagnetic Flight," Sci. American 229, 17-25 (1973).

2. Jahnke-Emde-Losch, Tables of Higher Functions, McGraw-Hill Book Company, New York, 1960, pp. 26-31.

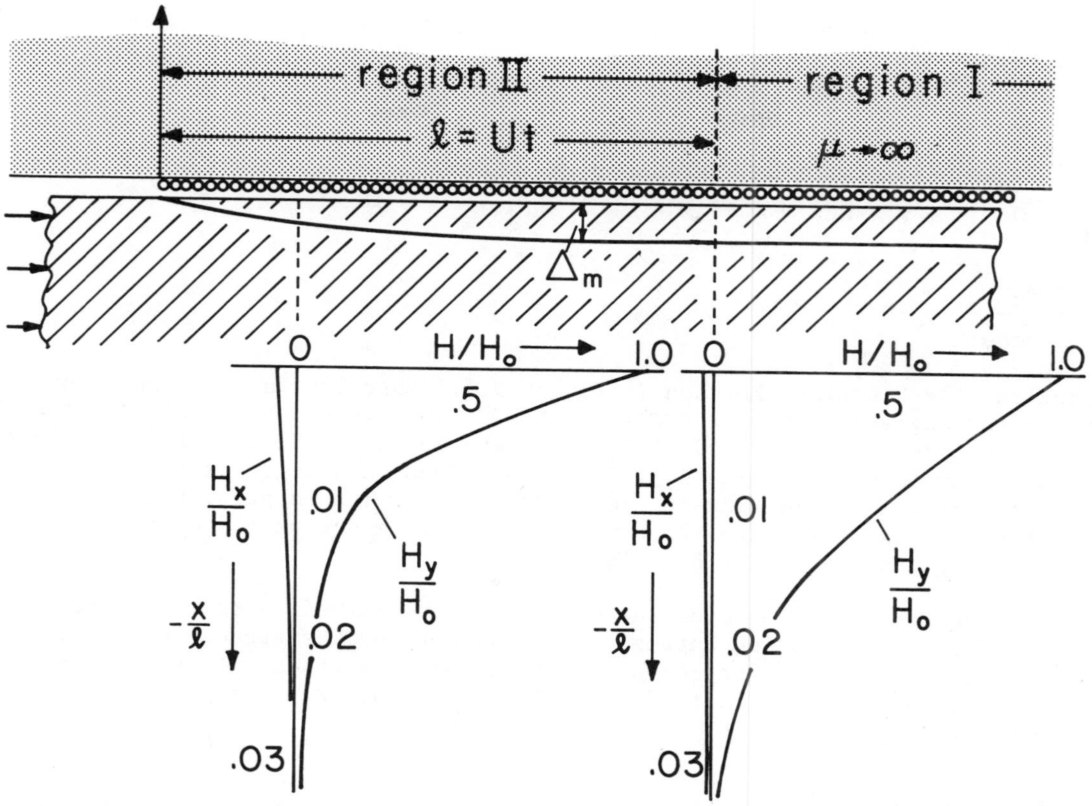

Fig. 6.9.3. Diffusion into moving conductor of magnetic field generated by current sheet $K_z=-H_o$ to right of y = 0 backed by highly permeable material and turned on when t = 0. Field has stationary profile to left of y = ℓ = Ut and profile that is independent of y but increasing its penetration with time to right of y = Ut. The plots show penetration of field at y = 0.25ℓ and y = ℓ with $R_m \equiv \mu\sigma U\ell = 100$. Note that magnitude of H_x is much less than H_y.

with ξ defined by Eq. 6 with t' → t gives the x-t dependence of H_y. The plot of H_y for y = ℓ = Ut in Fig. 6.9.3 illustrates the x-t dependence implied by the similarity solution. Region I is to the right of this location, so to the right the field is independent of y and increasing its depth of penetration with time.

In region II, between the leading edge and y = Ut, the elapsed time $t - t_o$ follows from Eq. 4b as t' = y/U and hence from Eq. 6

$$\xi = \frac{x}{2}\sqrt{\frac{\mu\sigma U}{y}} \tag{9}$$

With this parameter used in Eq. 8, it is clear that the field in region II is stationary, with the role of t replaced by y/U. Thus, in region II, the boundary layer grows in thickness with increasing y but remains constant in thickness at a given y. As time progresses, the front between the stationary field of region II and the temporally evolving field of region I moves to the right so that finally the stationary condition prevails. Of course, at some distance ℓ, the depth of penetration may be large enough to bring the finite thickness of the conductor into play. Alternatively, the length ℓ may reach the length L of the structure used to impose the field. In this latter case, a second boundary layer could be used to describe the field decay for y > L. The simple causal relation between excitation and downstream response can be traced to there being no longitudinal diffusion included in the boundary-layer model. There is no bow-wave in front of the leading edge and conditions downstream from the region of interest have no influence.

Normal Flux Density: To find the drag force on the conducting layer, the distribution of B_x is required. With H_y given by Eq. 8, it follows from Eq. 2 that

$$\frac{\partial B_x}{\partial x} = \begin{cases} 0 & ;\ y > Ut \\ \dfrac{\mu H_o}{2\sqrt{\pi y}}\sqrt{\dfrac{\mu\sigma U}{y}}\, x e^{-\frac{x^2}{4}\frac{\mu\sigma U}{y}} & ;\ 0 < y < Ut \end{cases} \tag{10}$$

Holding y constant, this expression is integrated from $x = -\infty$ (where B_x must vanish) to x to obtain $\left(\int xe^{-ax^2}dx = \frac{1}{2}\int e^{-ax^2}dx^2\right)$.

$$B_x = \begin{cases} 0 & ; \ y > Ut \\ \dfrac{-\mu H_o e^{-\xi^2}}{\sqrt{\pi\mu\sigma Uy}} & ; \ 0 < y < Ut \end{cases} \tag{11}$$

This distribution of B_x is also sketched in Fig. 6.9.3. Note that at the conductor surface, $B_x = \mu H_o/\sqrt{\pi R_m}$, where the magnetic Reynolds number $R_m = \mu\sigma Uy$ is based on the distance from the leading edge. The boundary layer model is only valid if $\Delta_m << y$, and in Region II this is equivalent to $R_m >> 1$. Thus, in the boundary layer approximation, B_x is much less than μH_o. As the boundary layer thickens in region II, the total magnetic flux in the y direction (which is proportional to $\mu H_o \Delta_m$) increases. Thus there must be a flux of B_x into the boundary layer from across the conductor surface and this is why a positive H_o implies a negative B_x.

Force: To find the total force on the conductor, the Maxwell stress is integrated over a surface enclosing the conductor and passing between the conductor and the structure in the $x = 0$ plane. The only contribution to the integration comes from this latter surface. Thus, the x-directed force on the conductor (the negative of the force tending to levitate the structure) due to a structure of length L and width w in the z direction is

$$f_x = w\int_0^L \frac{1}{2}\mu_o(H_x^2 - H_y^2)_{x=0}dy \tag{12}$$

In the boundary layer approximation, $H_x << H_y$. Therefore, consistent with this approximation is a normal force that is simply the product of the area of the conductor exposed to the magnetic stress and $(\frac{1}{2}\mu_o H_o^2)$.

Because region I has $B_x \approx 0$ and hence no shear stress, the force in the direction of motion is simply

$$f_y = w\int_0^{\ell}[B_x H_y]_{x=0}dy = -2\mu H_o^2 w\sqrt{\frac{\ell}{\pi\mu\sigma U}} \tag{13}$$

During the turn-on transient this drag force increases in proportion to $\ell = \sqrt{Ut}$ until ℓ reaches the full length L of the structure. Thereafter, the force is constant, given by Eq. 13 with $\ell = L$. With R_m again defined as $R_m = \mu\sigma UL$, this steady-state force can also be written as

$$f_y = -2\mu H_o^2 Lw/\sqrt{\pi R_m} \tag{14}$$

to make it clear that the final drag force is inversely proportional to the square root of the magnetic Reynolds number based on the length of the interaction region. From Eq. 14 it is clear that in the boundary layer limit, only a small fraction of the available magnetic stress, $\mu_o H_o^2$, contributes to the drag force.

6.10 Temporal Modes of Magnetic Diffusion

Temporal transients initiated from a state of spatial periodicity are introduced in Sec. 5.15. Just as that section revisited charge relaxation examples treated under sinusoidal steady-state conditions earlier in Chap. 5, this section returns to the configurations considered in Secs. 6.4 and 6.6. Analogies and contrasts between natural temporal modes of magnetic diffusion and charge relaxation are drawn by comparing the two magnetic configurations of this section to the corresponding electric pair from Sec. 5.15. It will be seen that there is a rather complete analogy between the thin sheet models. However, whereas a smoothly inhomogeneous conductor is required to give rise to an infinite set of natural modes in the charge relaxation bulk conduction model, here a uniform conductor is found to involve an infinite set of natural modes of magnetic diffusion.

Thin-Sheet Model: The natural frequencies for the system shown in Fig. 6.4.1 are given by setting the denominator of Eq. 6.4.6 equal to zero with $j\omega \rightarrow s_n$:

$$D(-js_n,k) = \sinh kd(-j + S_m \coth kd) = 0 \tag{1}$$

Only $S_{m+} \to S_m$ is considered in this expression because, in Eq. 1, k can be negative as well as positive. Solved for s_n, Eq. 1 becomes

$$s_n = +jkU - \left(\frac{k}{\mu_o \sigma_s}\right) \tanh(kd) \tag{2}$$

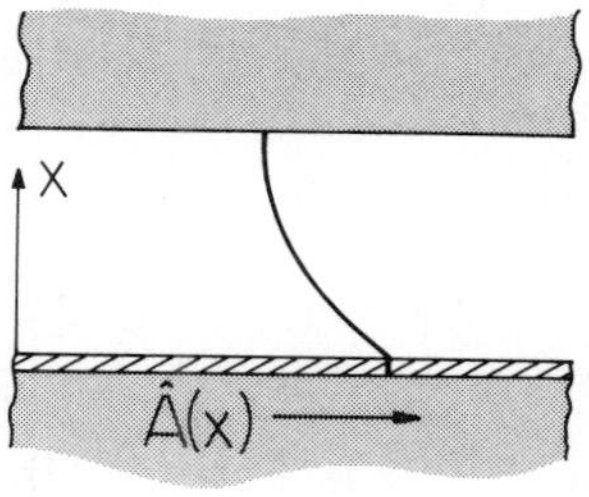

Fig. 6.10.1. Sheet model eigenmode.

The thin-sheet model implies a single natural mode having a damping part determined by the effective "L/R" time constant $[\mu_o\sigma_s/k \tanh(kd)]$ and an oscillatory part caused by the relative motion of the conductor through the spatially periodic fields. Note the complete analogy between Eqs. 2 and 5.15.6. In the air gap, the single eigenmode, $\hat{A}(x)$, associated with the eigenfrequency given by Eq. 2 is of the form of Eq. 2.19.3 with the coefficients adjusted to make the slope, $d\hat{A}/dx$, zero at the stator surface ($H_y = 0$) and to make $\hat{A}$ continuous at the sheet surface. Because the normal flux density is continuous through the sheet, $\hat{A}(x)$ is essentially uniform over the sheet cross section. This is consistent with H_y (which is proportional to dA/dx) being zero on the surface of the highly permeable rotor next to the conducting sheet. The distribution of A(x) is therefore given by

$$\hat{A} = \begin{cases} \hat{A}^c \cosh k(x-d); & 0 < x < d \\ \hat{A}^c & ; \text{ inside sheet} \end{cases} \tag{3}$$

which is sketched in Fig. 6.10.1. The significance of the thin-sheet model is further appreciated by considering the higher order modes which it does not embody.

<u>Modes in a Conductor of Finite Thickness</u>: For the same conductor air-gap configuration, but with account taken of the conductor thickness, consider now the temporal modes implied by Eq. 6.6.9:

$$D(-js_n,k) = \sinh kd\left(\frac{k}{\gamma}\frac{\mu}{\mu_o} \coth \gamma a + \coth kd\right) = 0 \tag{4}$$

The frequency enters in this expression through the parameter γ_n, defined according to Eq. 6.5.5 by

$$\gamma_n = \sqrt{k^2 + j\mu\sigma(-js_n - kU)} \tag{5}$$

In general, solution of Eq. 4 involves finding the complex roots s_n that make the real and imaginary parts of $D(-js_n,k) = 0$. Because s_n enters only through γ_n, it is convenient to find the roots, γ_n, and then use Eq. 5 to find the implied roots s_n. Fortunately, an infinite number of roots, γ_n, are purely imaginary, as can be seen by recognizing that coth u = jcot ju so that Eq. 4 becomes

$$\frac{\cot(j\gamma_n a)}{j\gamma_n a} = \frac{\mu_o}{\mu}\frac{\coth kd}{ka} \tag{6}$$

What is on the right in this expression is independent of $(j\gamma_n a)$ (and hence the frequency) and is real. Provided that $(j\gamma_n a)$ is real, what is on the left is also real. Hence, a graphical solution for the roots appears as shown in Fig. 6.10.2, where three of the roots $j\gamma_n a = \beta_n a$ (n = 0,1,2) are shown. Given the geometry and the layer permeability, which determine the right-hand side of Eq. 6, these roots are a set of numbers which can be inserted into Eq. 5 (solved for s_n) to determine the associated eigenfrequencies:

$$s_n = jkU - \frac{1}{\mu_o\sigma a^2}[(\beta_n a)^2 + (ka)^2] \tag{7}$$

Thus, there are an infinite number of modes, each having its own characteristic dependence on the transverse coordinate x. In terms of the vector potential $\hat{A}(x)$, Eq. 6.6.15 gives this dependence in the air gap, but this distribution is best found by simply adjusting the origin of the x coordinate so that a single hyperbolic function suffices to assure dA/dx = 0 at x = a + d and $\hat{A} = \hat{A}^c$ at x = a:

$$\hat{A}_n = \begin{cases} \hat{A}_n^c \dfrac{\cosh k[x-(a+d)]}{\cosh kd}; & a+d > x > a \\ \hat{A}_n^c \dfrac{\cosh \gamma_n x}{\cosh \gamma a} = \hat{A}_n^c \dfrac{\cos[\beta_n a(\frac{x}{a})]}{\cos \beta_n a}; & a > x > 0 \end{cases} \tag{8}$$

The three eigenvalues found graphically in Fig. 6.10.2 are used to plot the eigenfunctions of Eq. 8 in

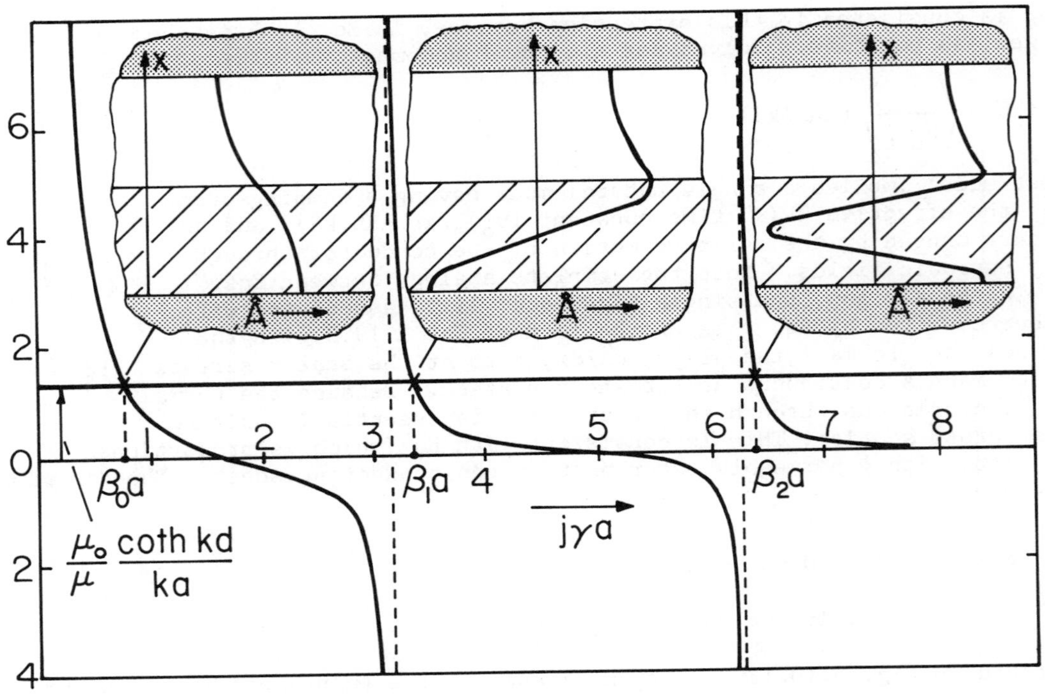

Fig. 6.10.2. Graphical solution for eigenvalues ($j\gamma_n a$) satisfying Eq. 6. Inserts show associated eigenfunctions, $\hat{A}(x)$, with ka = kd = 1 and μ/μ_o = 1. Roots shown are $\beta_o a$ = 0.776, $\beta_1 a$ = 3.364 and $\beta_2 a$ = 6.401.

the inserts to Fig. 6.10.2. Note that the n = 0 eigenmode approximates the sheet mode, Fig. 6.10.1.

Formally, the n = 0 mode becomes the thin-sheet mode in the limit of "small a." First, this means that $|\gamma a| \ll 1$, so that cot u → 1/u, and Eq. 6 can be solved approximately to obtain

$$(\gamma_o a)^2 = -\frac{\mu}{\mu_o} ka \tanh kd \tag{9}$$

Thus, the n = 0 eigenfrequency follows from Eq. 7 as

$$s_n \rightarrow jkU - \frac{ka}{\mu\sigma a^2}\left(\frac{\mu}{\mu_o}\tanh kd + ka\right) \tag{10}$$

If the second term in brackets can be dropped compared to the first, Eq. 10 indeed reduces to the eigenfrequency for the thin-sheet model, Eq. 2. Provided that $(\mu/\mu_o)\tanh kd$ is of the order of unity or more, this condition is met if ka << 1. This is the second condition to validate the thin-sheet model. Note that the two conditions for the thin-sheet model to approximate the lowest mode are just those given by Eq. 6.6.13.

An important proviso on the use of the thin-sheet model is apparent from these deductions. Unless the air gap is large compared to the sheet thickness, Eq. 10 does not follow from Eq. 9 and the thin-sheet model is not meaningful. In physical terms this is true because, in the model, magnetic energy storage within the sheet is ignored. To be meaningful, the sheet model must be incorporated into a system that allows for energy storage outside the sheet volume. In this example, that region is the air gap.

The general effect of decreasing the air gap can be seen from Fig. 6.10.2. As d is reduced, coth kd → ∞ and the horizontal curve moves upward. Thus, decreasing the gap decreases the values of $\beta_o \cdots \beta_\infty$ to the asymptotic roots $n\pi$, n = 0,1,.... It follows from Eq. 7 that reducing d results in a decrease in the damping, in an increase in the time constant for decay of the sheet currents. This is reasonable, because the reduction in gap width results in an increased inductance for current loops in the y-z plane. Note that the n = 0 mode has an eigenvalue β_o that approaches zero as the gap is reduced. Hence, in Eq. 7, the term ka (which represents the energy storage within the sheet) must be retained. In the n=o mode, electrical dissipation is in the sheet while magnetic energy storage is largely in the gap. In the higher order modes, energy storage in the conducting layer is appreciable.

Orthogonality of Modes: Given an initial distribution of currents in the conducting layer, the eigenmodes can be used to represent the resulting transient. More generally, the modes play the role of the homogeneous solution in describing the response of a system to spatially periodic excitations, as described in Sec. 5.15. This homogeneous solution is the superposition of the eigenmodes

$$J_z = \sum_{n=-\infty}^{+\infty} \hat{J}_n(x)e^{-jky}e^{s_n t} \tag{11}$$

The process by which the amplitudes $J_n(x)$ are determined, given the initial conditions, is similar to that for a Fourier series. But, because the eigenmodes do not satisfy simple boundary conditions, it is not clear that these modes are orthogonal, in the sense that

$$\int_o^a \hat{J}_n\hat{J}_m dx = 0, \; n \neq m \tag{12}$$

A proof that Eq. 12 is in fact valid follows from the differential properties of $\hat{J}_n$. The equation governing the current density modes follows from Eq. 6.6.16:

$$\hat{J}_z = -\frac{1}{\mu}\left(\frac{d^2}{dx^2} - k^2\right)\hat{A} \tag{13}$$

which is applied to Eq. 6.5.5 to see that

$$\frac{d^2\hat{J}_n}{dx^2} - \gamma_n^2\hat{J}_n = 0 \tag{14}$$

Now, Eq. 14 is multiplied by another eigenmode, $\hat{J}_m$, and the result integrated over the cross section of the conducting layer. The first term can be integrated by parts to generate terms evaluated at the conductor surfaces and an integral that is symmetric in m and n:

$$J_m\frac{d\hat{J}_n}{dx}\Bigg|_o^a - \int_o^a\left(\frac{d\hat{J}_m}{dx}\frac{d\hat{J}_n}{dx} + \gamma_n^2\hat{J}_n\hat{J}_m\right)dx = 0 \tag{15}$$

These same steps can be carried out with the roles of m and n reversed, and if the resulting expression is subtracted from Eq. 15, an expression is obtained that begins to look like Eq. 13:

$$\left[\hat{J}_m\frac{d\hat{J}_n}{dx} - J_n\frac{d\hat{J}_m}{dx}\right]_o^a = (\gamma_n^2 - \gamma_m^2)\int_o^a \hat{J}_n\hat{J}_m dx \tag{16}$$

In the usual orthogonality condition (for example Eq. 4.5.28) homogeneous boundary conditions apply at the extremes of the interval. Here, the nature of the fields in the air gap must be considered to see that the left-hand side of Eq. 16 is zero. To express this in terms of A, observe from Eqs. 6.5.5 and 6.6.16 that

$$\hat{J}_z = -\frac{1}{\mu}(\gamma^2 - k^2)\hat{A} = -j\sigma(-js_n - kU)\hat{A} \tag{17}$$

$$\frac{d\hat{J}_z}{dx} = -j\sigma(-js_n - kU)\frac{d\hat{A}}{dx} \tag{18}$$

It follows from this last expression that because $\hat{H}_y = -(1/\mu)d\hat{A}/dx = 0$ at $x = 0$, the left-hand side of Eq. 16 evaluated at the lower limit is zero. Using Eqs. 17 and 18, what remains on the left can be written as

$$\left[\hat{J}_m\frac{d\hat{J}_n}{dx} - J_n\frac{d\hat{J}_m}{dx}\right]^a = \sigma^2(-js_n - kU)(-js_m - kU)\left[\hat{A}_m\frac{d\hat{A}_n}{dx} - \hat{A}_n\frac{d\hat{A}_m}{dx}\right]^a \tag{19}$$

That this quantity also vanishes follows from the properties of the gap fields. In the gap, where $\gamma^2 \to k^2$, Eq. 6.5.2 becomes

$$\frac{d^2\hat{A}_n}{dx^2} - k^2\hat{A}_n = 0 \tag{20}$$

Following steps analogous to those leading from Eq. 14 to Eq. 16, the field properties represented by this expression are exploited to show that

$$\left[A_m \frac{d\hat{A}_n}{dx} - \hat{A}_n \frac{d\hat{A}_m}{dx}\right]_a^{a+d} = 0 \tag{21}$$

Because $d\hat{A}/dx = 0$ at $x = a+d$ (the highly permeable stator surface), it follows that Eq. 19 vanishes. So long as $s_n \neq s_m$, Eq. 12 is valid.

6.11 Magnetization Hysteresis Coupling: Hysteresis Motors

Although induction devices of the type discussed in Secs. 6.4 and 6.6 are of the most common variety, they are particular examples from a class of machines in which sources are induced in the moving material. A somewhat less common member of the family is the hysteresis motor, known for its relatively constant torque over speeds ranging from "start" to synchronism.

It is the magnetization that is induced in the rotor of the hysteresis motor, rather than free current, as in the induction motor. Basic to the advantages of a hysteresis motor is the magnetization characteristic of the moving member. The currents in the induction machine depend on a time rate of change for their existence. They are rate-dependent, and so the magnitude and spatial phase of the currents in the moving member, and hence the ponderomotive force, depend on the relative velocity of material and traveling wave. By contrast, the spatial phase and magnitude of the magnetization induced in the moving material through a hysteresis interaction tends to be state-dependent.

The quasi-one-dimensional model pictured in Fig. 6.11.1a gives the opportunity to explore the physical basis for the hysteresis interaction in a quantitative way, but still avoid the extreme complexity inherent to the complete understanding of a practical device. The model harks back to ones developed in Secs. 4.12 and 4.13 for the variable capacitance machine. The stator surface current density, $K_z(y,t)$, is a wave traveling in the y direction. Windings backed by a highly permeable "stator" structure are perhaps as described in Sec. 6.4. Across the air gap, a, the moving material consists of a highly magnetized "core" covered by a layer of magnetic material having thickness b, and the magnetization characteristic shown in Fig. 6.11.1b.

As suggested by the permanent polarization interactions of Sec. 4.4, all that is required to obtain a net force in the y direction is a spatial phase lag between the induced magnetization and the magnetic axis of the current sheet. This phase delay is provided by the hysteresis, which insures that the driving current must provide a certain coercive magnetic field intensity before the magnetization can be reversed.

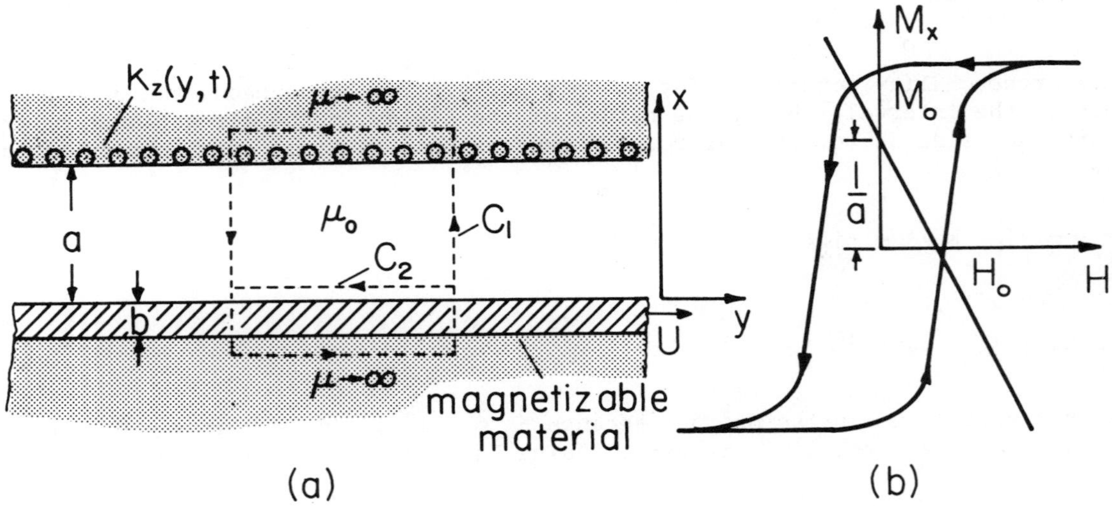

Fig. 6.11.1. (a) Cross-sectional view of quasi-one-dimensional model. (b) Magnetization characteristic approximated by hysteresis loop of Fig. 6.11.2.

At the risk of oversimplification, it is helpful to have a specific model in mind when dealing with the magnetization characteristic. Typically, magnetic materials used in electromechanical devices are polycrystalline, and can be thought of as composed of randomly oriented magnetlike domains. Application of a magnetic field intensity tends to align these domains, but because of what might be termed a "sticking friction," there is a threshold value of $\vec{H}$ at which the domains tend to flip into alignment with the imposed field. In some materials, complete orientation of the domains is very nearly achieved, once this threshold has been exceeded. For that reason, and because it is then possible to make a relatively simple analytical model, the hysteresis loop is now approximated by the rectangular loop shown in Fig. 6.11.2. (To some degree, the characteristic depends on the rapidity with which the fields vary, but for present purposes the curve is shown, regardless of time rates of change.) The loop is double-valued, so the manner of arrival at a given point must be stipulated. That is, the magnetization induced by the applied field depends on the state of the fields, and not on their rate of change. But also, it depends in an essential way on the history of the magnetization.

Because of the highly permeable surfaces backing the current sheet and the magnetizable layer, the dominant magnetic field in the gap is x-directed. Ampère's law in integral form for the contour C_1 of Fig. 6.11.1 shows that

$$-K_z\Delta y = [H_x^a(y+\Delta y) - H_x^a(y)]a + [H_x^b(y+\Delta y) - H_x^b(y)]b \qquad (1)$$

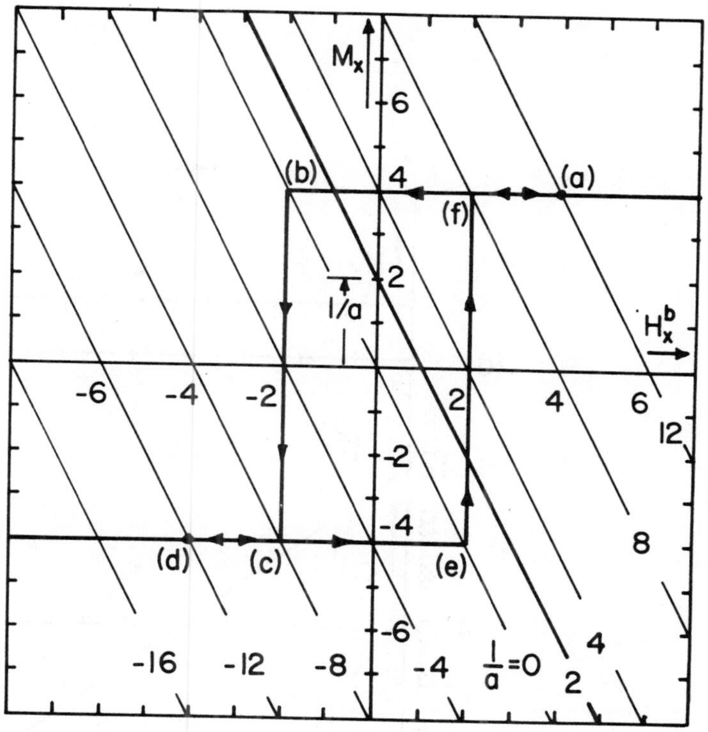

Fig. 6.11.2. Idealization of magnetization characteristic showing graphical solution $(a + b)/a = 2$.

In the limit $\Delta y \to 0$, this expression becomes

$$-K_z = a\frac{\partial H_x^a}{\partial y} + b\frac{\partial H_x^b}{\partial y} \qquad (2)$$

The flux density in the x direction is continuous at the air-gap/magnetic-layer interface, so

$$H_x^a = H_x^b + M_x \qquad (3)$$

These last two expressions combine to relate the magnetization and field intensity in the magnetized layer,

$$-K_z = (a+b)\frac{\partial H_x^b}{\partial y} + a\frac{\partial M_x}{\partial y} \qquad (4)$$

For the present purposes, the surface current density is a given function of y, and so Eq. 4 can be integrated:

$$M_x = \frac{I}{a} - \frac{(a+b)}{a}H_x^b; \quad I \equiv -\int K_z dy \qquad (5)$$

Under the assumption that steady-state operation implies that neither M_x or H_x^b have space-average values, it follows that if I(y,t) is defined as having no space-average value, the integration constant is zero. Because I is then a given function of y, Eq. 5 is a "load line" which can be used with the magnetization characteristic of Fig. 6.11.2 to graphically solve for (M_x,H_x). For illustrative purposes, the surface current is taken as a square wave, traveling to the right as sketched in Fig. 6.11.3a. Although there are no rate processes, it is essential to recognize that, if the moving member has a velocity less than that of the wave, the current distribution travels from left to right with respect to the material. The magnetic axis associated with the stator wave is indicated on Fig. 6.11.3a.

In the graphical solution of Eq. 5 and the magnetization characteristic depicted by Fig. 6.11.2, begin at point (a), where I/a has its peak amplitude. Because the wave travels from left to right, the magnetic material experiences a local evolution of I/a that proceeds from right to left on part (b) of Fig. 6.11.3. Thus, the points (a) - (f) denote the history of the (M_x,H_x) function in Fig. 6.11.2, and these points correspond to those indicated in Fig. 6.11.3. The graphical solutions for the magnetization and field intensity H_x are thus determined to be those shown in Fig. 6.11.3c. The induced magnetization lags the magnetic axis on the stator. The hysteresis has created the conditions for a force to the right.

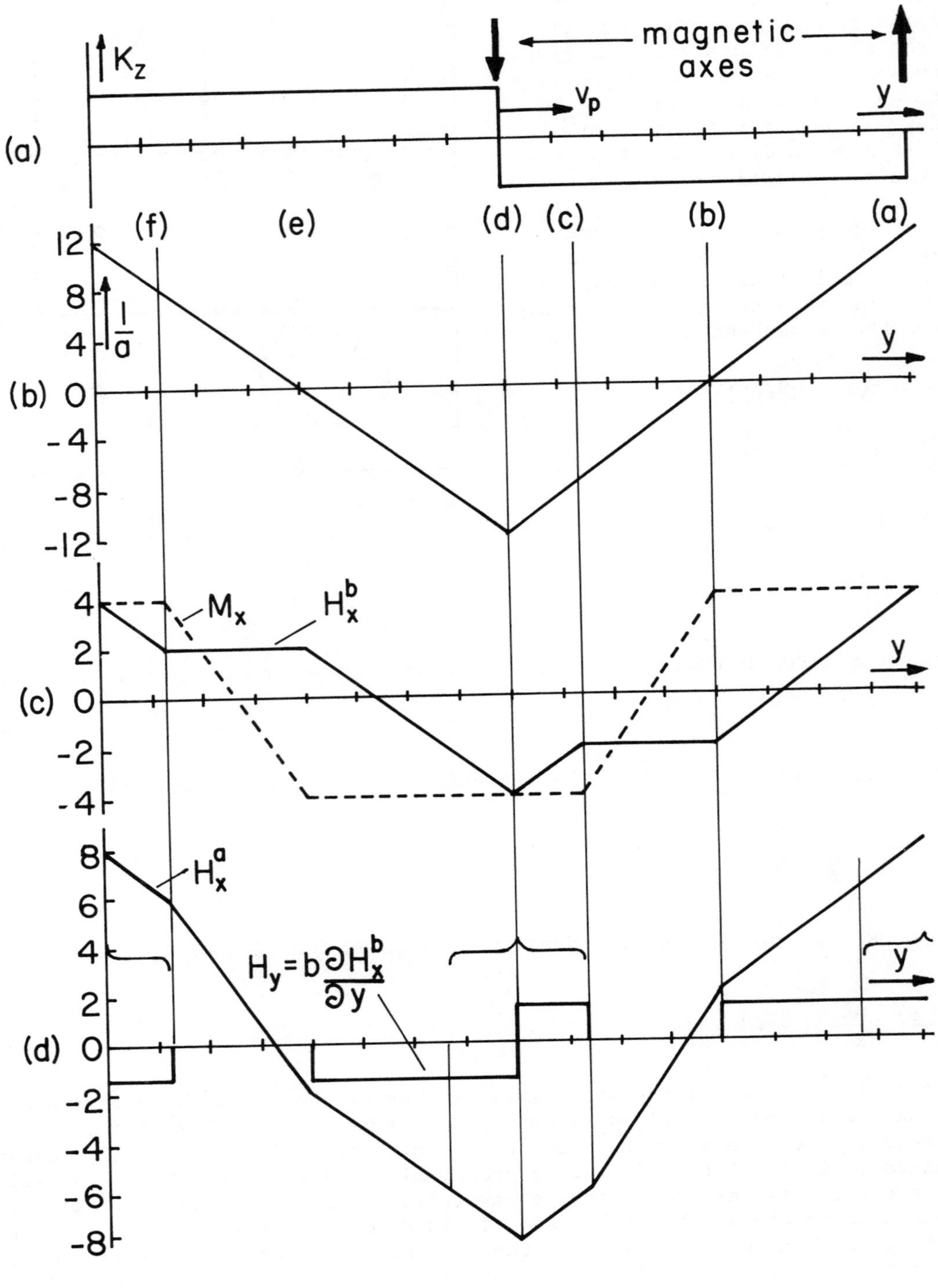

Fig. 6.4.3. (a) Distribution of surface current on "stator." (b) I/a; (c) distribution of magnetization and perpendicular magnetic field intensity in moving member; and (d) field components in adjacent air gap.

To determine the average force/unit area acting on one wavelength of the moving member, use is made of the free-space stress tensor. The force density is due entirely to magnetization, and might be taken as the Kelvin force density, Eq. 3.5.12 of Table 3.10.1, with $\vec{J}_f = 0$. However, from Table 3.10.1, the stress tensor evaluated in free space is the same regardless of the model for the force density. This stress tensor is now integrated over a volume one wavelength long in the y direction, with its upper surface at x = 0 and its lower surface adjacent to the perfectly permeable substrate. Because there is no shear stress on the bottom surface, the average force/unit y-z area is

$$\langle T_y \rangle_y = \langle T_{yx} \rangle_y = \langle \mu_o H_x^a H_y^a \rangle_y = \langle \mu_o H_x^a H_y^b \rangle_y \tag{6}$$

Note that Eq. 6 cannot be completed unless the y component of the magnetic field intensity is known. Ampère's law in integral form, written for the contour C_2 of Fig. 6.11.1a, relates H_y to fields already determined,

$$-\Delta y H_y^b(x = 0) + b[H_x^b(y + \Delta y) - H_x^b(y)] = 0 \tag{7}$$

In the limit $\Delta y \to 0$,

$$H_y^b = b \frac{\partial H_x^b}{\partial y} \tag{8}$$

and so Eq. 6 can be written as

$$\langle T_y \rangle_y = b \mu_o H_x^a \frac{\partial H_x^b}{\partial y} \tag{9}$$

The components of $\vec{H}$ required to evaluate Eq. 9 are sketched in Fig. 6.11.3d with $\partial H_x^b/\partial y$ determined by taking the derivative of H_x^b from (c) of that figure, and H_x^a following from Eq. 3.

It is easy to take the spatial average indicated by Eq. 9, because the net contributions of those segments indicated in brackets in Fig. 6.11.3d will cancel, and the remaining segments clearly give a positive contribution. Thus, a space-average surface force density is deduced. It is independent of the material velocity U, so that the force-velocity curve is as shown in Fig. 6.11.4. Once the material velocity exceeds that of the wave, the relative direction of the current excitation is from right to left, and the arguments already outlined lead to an oppositely directed magnetic force.

The simple quasi-one-dimensional model illustrates why a hysteresis "torque-speed" characteristic gives a torque that tends to be independent of speed. The induced magnetization has an effect similar to that of permanent magnets, with the desired phase relationship between imposed magnetic axis and material magnetization determined by the history of the rotor as it is magnetized by the stator current.

For design purposes, a more complete representation of the rotor material would be desirable, although attempts to make use of analytical models in dealing with hysteresis motors are not numerous.[1]

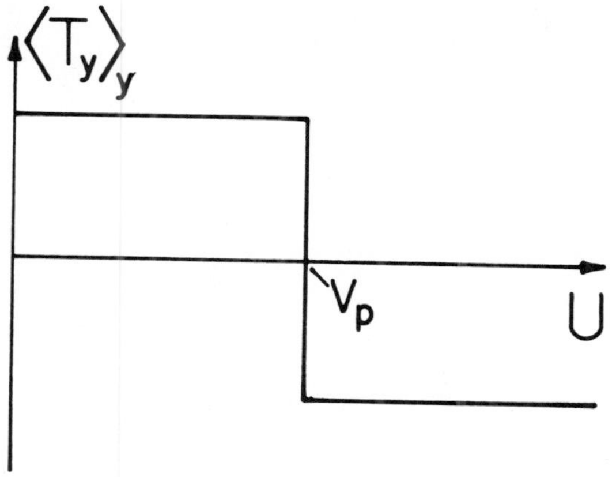

Fig. 6.11.4. Dependence of magnetic surface force density on speed for a hysteresis-type device.

1. M. A. Copeland and G. R. Slemon, "An Analysis of the Hysteresis Motor: I - Analysis of the Idealized Machine," IEEE Trans. on Power Apparatus and Systems, Vol. 82, April 1963, pp. 34-42, and II - "The Circumferential Flux Machine," ibid., Vol. 83, June 1964, pp. 619-625.

Problems for Chapter 6

For Section 6.2:

Prob. 6.2.1 Consider the configuration described in Prob. 2.3.3. In the MQS approximation and at low frequencies the configuration can be represented by an inductance in series with a resistance. Because the current is distributed, and in fact essentially uniform and x-directed, how should the inductance be computed?

(a) One method uses the field in the zero frequency limit to determine the magnetic energy density, and hence by integration the total stored energy. This is then equated to $\frac{1}{2}Li^2$ to obtain L. Use this method to find L and show that it is 1/3 of the value for electrodes without the conducting material but shorted at z = 0.

(b) Now, consider an alternative approach which considers the fields as quasistatic with respect to the magnetic diffusion time $\mu\sigma\ell^2$. In terms of the driving current, find the zero order fields as if they were static. Then, from Eq. 6.2.7 find the first order fields that result from time variations of the zero order field. Evaluate the voltage at the terminals and show that it has the form taken for a series inductance and resistance.

For Section 6.3:

Prob. 6.3.1 Show that Eq. (b) of Table 6.3.1 describes the rotating cylindrical shell shown in that table.

Prob. 6.3.2 Show that Eq. (c) of Table 6.3.1 describes the translating cylindrical shell shown in that table.

Prob. 6.3.3 Show that Eqs. (d) and (e) of Table 6.3.1 describe the rotating spherical shell shown in that table.

Prob. 6.3.4 If a sheet is of extremely high permeability, the normal flux density B_n is not continuous. Consider the sheets of Table 6.3.1 in the limit of zero conductivity but with a very high permeability and show that boundary conditions are

$$\vec{n} \times [\![\vec{H}]\!] = 0 \; ; \qquad \Delta\mu(\nabla_\Sigma \cdot \vec{H}) + [\![B_n]\!] = 0$$

These boundary conditions are appropriate if wavelengths in the plane of the sheet are long compared to the sheet thickness. Thus the boundary condition can be used to represent a thin region that would otherwise be represented by the flux-potential transfer relations of Sec. 2.16. To see this connection, show that for a planar sheet, the above boundary condition can be written as

$$\Delta\mu k^2 \tilde{\psi} + [\![\tilde{B}_x]\!] = 0$$

Take the long-wave limit of the transfer relations from Table 2.16.1 to obtain this same result.

Prob. 6.3.5 In the boundary conditions of Table 6.3.1 representing a thin conducting sheet, B_n is continuous while the tangential $\vec{H}$ is not. By contrast, for the condition found in Prob. 6.3.4 for a highly permeable sheet, B_n is discontinuous and tangential $\vec{H}$ is continuous. What boundary conditions should be used if the sheet is both highly permeable and conducting? To answer this question it is necessary to give the fields in the sheet some dependence on the normal coordinate. Consider the planar sheet and assume that the fields within take the form

$$B_x = B_x^b + \frac{x}{\Delta}(B_x^a - B_x^b) \; ; \qquad H_y = H_y^b + \frac{x}{\Delta}(H_y^a - H_y^b)$$

Define $\langle A\rangle = (A^a + A^b)/2$ and show that the boundary conditions are

$$\mu\Delta\nabla_\Sigma \cdot \langle\bar{H}\rangle + [\![B_n]\!] = 0$$

and Eq. (a) of Table 6.3.1 with $B_x \rightarrow \langle B_x\rangle$.

For Section 6.4:

Prob. 6.4.1 A type of tachometer employing a permanent magnet is shown in Fig. P6.4.1a. In the developed model, Fig. P6.4.1b, the magnetized material moves to the right with velocity U so that the magnetization is the given function of (y,t). M_o is a given constant. The thickness, a, of the conducting sheet is small compared to the skin depth. Find the time average force per unit y-z area acting on the conducting sheet in the y direction. How would you design the device so that the induced force is proportional to U?

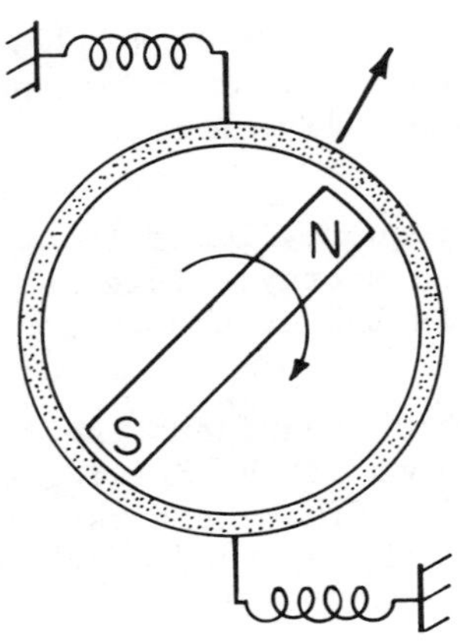

Fig. P6.4.1a

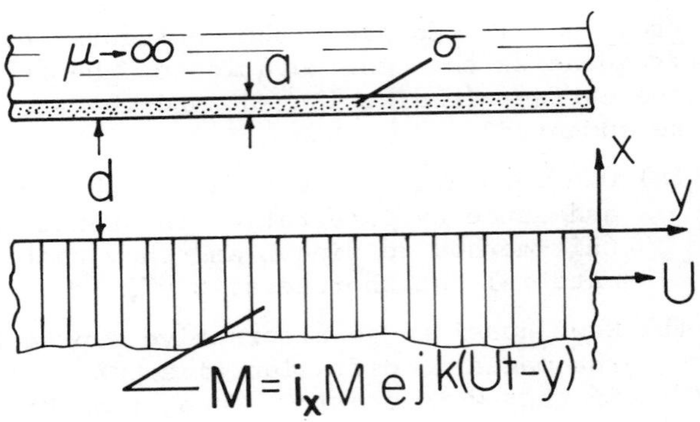

Fig. P6.4.1b

Prob. 6.4.2 Use the electrical terminal relations derived from the model, Eq. 6.4.17, to show that the equivalent circuit of Fig. 6.4.3 is valid.

Prob. 6.4.3 For the developed induction motor model shown in Fig. 6.4.1b, the time average force in the direction of motion is calculated. In certain applications, such as the magnetic levitation of vehicles (see Fig. 6.9.2), the lift force is also of importance. Find the time average lift force on the stator, $\langle f_x \rangle_t$, with two phase excitation. With single phase excitation, sketch this time average lift force as a function of S_m and explain in physical terms the asymptotic behavior.

Prob. 6.4.4 The cross section of a rotating induction machine is shown in Fig. 6.4.1a. The stator inner radius is (a), while the rotor has radius (b) and angular velocity Ω. The windings on the stator have p poles and two phases, as in the planar model developed in the section. For two phase excitation, find the time average torque on the rotor, an expression analogous to Eq. 6.4.11. Define θ as the clockwise angle from the vertical axis in Fig. 6.4.1a.

Prob. 6.4.5 For the rotating machine described in Prob. 6.4.4, find the two phase electrical terminal relations analogous to Eq. 6.4.17. Determine the parameters in the equivalent circuit, Fig. 6.4.3.

Prob. 6.4.6 This problem is intended to illustrate the application of the boundary conditions for a thin sheet that is both conducting and highly permeable, as in Prob. 6.3.5. In the plane x=0 there is a surface current density $\vec{K}_f = \vec{i}_z \text{Re}\, \hat{K}_o \exp j(\omega t - ky)$. The region x < 0 is infinitely permeable. In the plane x=d, a sheet of thickness Δ, permeability μ and conductivity σ moves in the y direction with velocity U. This sheet can shield the magnetic field from the region x > d either by virtue of its conductivity or its magnetizability. Find the magnetic potential just above the sheet ($x=d^+$). Consider $\mu \to \mu_o$ and show that for $\mu_o\sigma\Delta(\omega - kU)/k$ large, the field is excluded from the region x>d. Similarly, take $\sigma \to 0$ and show that if $k\Delta(\mu/\mu_o) >> 1$, shielding is obtained. Show that the effect of the permeability is to reduce the effectiveness of conduction shielding. In qualitative physical terms, why is there this conflict between the two types of shielding?

Prob. 6.4.7 A linear induction machine has the configuration of Fig. 6.4.1. However, the stator winding has a finite length ℓ in the y direction. Thus the stator surface current is

$$K_z^s = [u_{-1}(y) - u_{-1}(y-\ell)]\ \text{Re}\ \hat{K}_o \exp j(\omega t - \beta y)$$

Thus, the "stator" might be attached to a vehicle (such as that shown in Fig. 6.9.2) and the conducting sheet and magnetic backing might be the "rail." Using the approach of Sec. 5.17, show that the time average force exerted on the rail is

$$\langle f_z \rangle_t = \frac{\mu_o |\hat{K}_o|^2 w}{\pi} \int_{-\infty}^{+\infty} \frac{S_m \sin^2 [\frac{(k-\beta)\ell}{2}] dk}{(k-\beta)^2 \sinh^2 kd(1 + S_m^2 \coth^2 kd)}$$

where $S_m = \mu_o \sigma_s (\omega - kU)/k$.

Prob. 6.4.8 The induction machine rotor is a useful model for understanding phenomena observed if liquid metals are stressed by a-c magnetic fields. Motions of the liquid result from a competition of viscous and inertial forces with those from the magnetic field. Instability can result from the effect of the motion on the field. To illustrate, consider the single phase excitation of the configuration shown in Fig. 6.4.1. The "air gap" is filled with a liquid having viscosity η. Under the assumption that the flow in the gap resulting from the relative motion of the rotor and stator is fully developed and laminar, the viscous stress acting to retard the motion of the rotor is given by Eq. 7.13.1. As the magnetic field intensity $\hat{H}_o \equiv N_a\hat{i}_a$ is raised, there is a threshold at which the rotor spontaneously moves in one direction or the other. Write the condition for this instability in terms of the dimensionless numbers kd, R_M (product of frequency and magnetic diffusion time) and $\omega\tau_{MV}$ ($\tau_{MV} \equiv \eta/\mu_o H_o^2$, the magneto-viscous time as defined in Sec. 8.6).

For Section 6.5:

Prob. 6.5.1 Carry through the steps of Eqs. 6.5.8 - 6.5.10 leading to the transfer relations for rotating cylinders. Check relations (c) and (d) of Table 6.5.1.

Prob. 6.5.2 Carry through the steps beginning with Eq. 6.5.13 and leading to the transfer relations (e) and (f) of Table 6.5.1.

For Section 6.6:

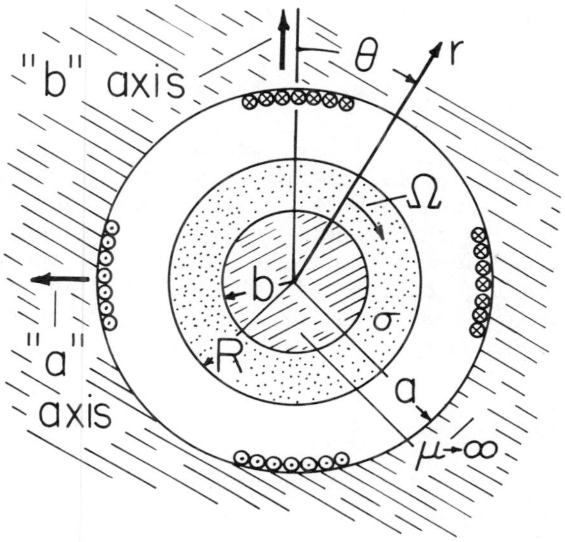

Fig. P6.6.1

Prob. 6.6.1 The rotor of an induction motor has finite thickness. Dimensions are defined in Fig. P6.6.1. The stator windings have p poles and two phases, the circular analogue of the windings for the developed model of Sec. 6.4. Hence the stator surface current distribution is the circular analogue of Eq. 6.4.1. Find the time average torque on the rotor.

Prob. 6.6.2 An induction machine is used to propel a circular cylindrical conductor in the longitudinal direction z. The "stator" consists of circumferential windings at the radius (a) surrounded by an infinitely permeable magnetic material in the region r > a. The material being propelled is coaxial with this structure and is of radius R, conductivity σ and permeability μ. Thus, there is an annular air gap of thickness a-R. The conducting rod has a velocity U in the z direction.

(a) The stator windings are in a three phase configuration driven by the three phase currents (i_a, i_b, i_c). Thus the surface current on the stator structure is

$$K_\theta = \text{Re}[\hat{i}_a e^{j\omega t} N_a \cos(kz) + \hat{i}_b e^{j\omega t} N_b \cos(kz-\tfrac{2\pi}{3}) + \hat{i}_c e^{j\omega t} N_c \cos(kz-\tfrac{4\pi}{3})]$$

Represent this driving surface current in the form

$$K_\theta = \text{Re}[\hat{K}_+^s e^{j(\omega t-kz)} + \hat{K}_-^s e^{j(\omega t+kz)}]$$

and identify $\hat{K}_+^s$ and $\hat{K}_-^s$ in terms of the terminal currents, turns per unit length N_a, N_b, N_c, etc.

(b) Find the time average longitudinal force $\langle f_z \rangle_t$ acting on a length ℓ of the rod.

Prob. 6.6.3 A linear induction machine has the configuration of Fig. 6.6.1, except that the stator surface current spans a limited length ℓ in the y direction. The driving current is

$$K_z^s = [u_{-1}(y)-u_{-1}(y-\ell)]\ \text{Re}\hat{K}_o \exp j(\omega t-\beta y)$$

Use the approach illustrated in Sec. 5.17 to show that the total force on the conducting slab and its highly permeable backing is

Prob. 6.6.3 (continued)

$$\langle f_y \rangle_t = -\frac{w\mu_o|K_o|^2}{\pi}\,\mathrm{Re}\int_{-\infty}^{+\infty}\frac{j\sin[\frac{(k-\beta)\ell}{2}]dk}{(k-\beta)^2\sinh^2 kd[\frac{k}{\gamma}\frac{\mu}{\mu_o}\coth\gamma a+\coth kd]}$$

where $a\gamma=\sqrt{(ak)^2+j\,S_M}$, $S_M\equiv\mu\sigma a^2(\omega-kU)$

For Section 6.7:

Prob. 6.7.1 The conducting layer of Fig. 6.7.1 represents the only lossy element in a linear induction machine. Arrangement of air gaps and magnetic materials is arbitrary. Special cases are the configurations of Fig. 6.4.1 and 6.6.1. Stator windings impose a pure traveling wave having phase velocity ω/k in the y direction. With P_m and P_d defined as the time average mechanical power output and electrical dissipation, respectively, the electrical power input is P_m+P_d. Show that the efficiency, $E_{ff}\equiv P_m/(P_m+P_d)$, is $U/(\omega/k)$. Define the "slip" by $s\equiv[(\omega/k)-U)]/(\omega/k)$, and show that $E_{ff}=1-s$.

Prob. 6.7.2 In terms of the same variables as used to express the time average force (Eq. 6.6.10), determine the time average electrical dissipation for the induction machine of Fig. 6.6.1.

For Section 6.8:

Prob. 6.8.1 A high frequency magnetic field is used to raise a liquid metal against gravity, as shown in Fig. P6.8.1. The skin depth is short compared to other dimensions of interest. Express the magnetic surface force density acting on the interface at the right in terms of the power dissipated in the liquid. What is the height ξ as a function of the power dissipated? (See Section 7.8 for the modicum of fluid statics needed here.)

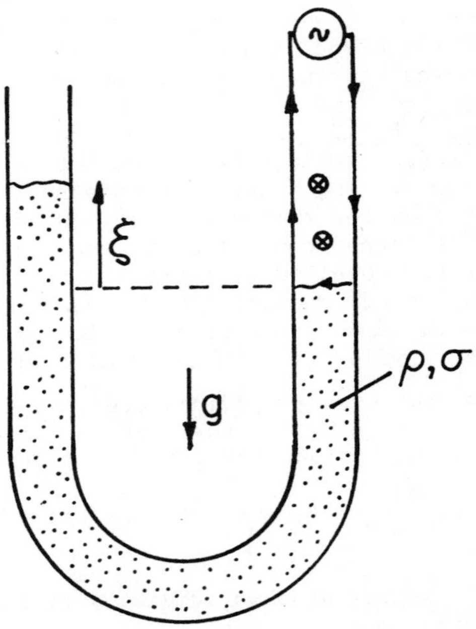

Fig. P6.8.1

For Section 6.9:

Prob. 6.9.1 Carry out the similarity transformation converting Eq. 6.9.3 to Eq. 6.9.7.

Prob. 6.9.2 A container holds a layer of liquid metal having depth b and length ℓ, as shown in Fig. P6.9.2. The system extends far enough in the z direction that it can be regarded as two-dimensional. At a distance h(y) above the interface is a bus-bar. Alternating current passes through this bar in the z direction and is returned through the liquid metal in the opposite z direction. Because the skin depth in both conductors is short compared to h(y) and b, magnetic flux is essentially ducted between the bus and the liquid metal, as sketched. The field throughout the air gap therefore has the same temporal phase. In the spirit of a quasi-one-dimensional model, in the air gap $\vec{H}$ has the zero order dependence $H_y=H_o a/h$, where $H_o=\mathrm{Re}\,\hat{H}_o\exp(j\omega t)$ is the field intensity at the left where $y=0$. The slope of the bus, dh/dy, at $y=0$ is given as S.

(a) Find H_y in the skin region of the liquid using the boundary layer model, Eq. 6.9.1. Assume that the fluid velocity has a negligible effect.

(b) Use the divergence law, Eq. 6.9.2, to approximate the normal flux density at the interface.

(c) Find the time average magnetic shearing surface force density acting over the thin skin layer.

(d) Show that if this quantity is to be independent of y, the bus geometry must be $h=a[1-2S(y/a)]^{-\frac{1}{2}}$.

(e) Show that this uniform surface force density is

$$\langle T_y\rangle_t=\frac{\mu_o}{4a}|\hat{H}_o|^2\,S\,\delta$$

Prob. 6.9.3 For the configuration described in Prob. 6.9.2, find the total power dissipation in the lower metal.

Prob. 6.9.4 For the configuration considered in this section, the magnetic structure has a total length L. As a function of time and y, compute the power dissipation in the conductor. What is the total power dissipation?

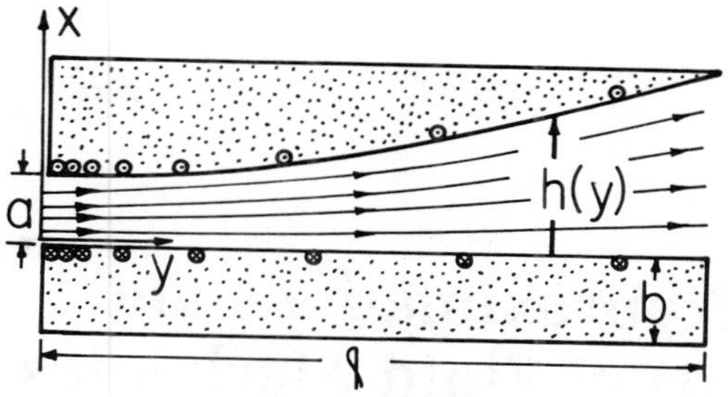

Fig. P6.9.2

For Section 6.10:

Fig. 6.10.1 A uniformly conducting slab of thickness 2a and permeability μ moves in the z direction with velocity U. To either side of the slab are air gaps of thickness d backed by infinitely permeable materials. Thus, half of the system is like that of Fig. 6.6.1 for $x > 0$, with $x=0$ a plane of symmetry. Because of the symmetry, temporal modes can be divided into those that are even and odd in H_y. Show that the odd modes are represented by Eq. 6.10.6. Find the analogous expression for the even modes, representing the graphical solution by a sketch similar to that of Fig. 6.10.2.

Prob. 6.10.2 A uniformly conducting circular cylindrical shell has outer radius a and inner radius b and spins about the z axis with angular velocity Ω. The regions outside and inside the shell are filled by infinitely permeable material. The system is long in the z direction compared to the outer radius a. However, the distance a-b is not small compared to the outer radius a.

(a) Find eigenfrequency equations from which the frequencies of the temporal modes can be determined. (The expression can be factored into two somewhat simpler expressions that define two classes of modes.)

(b) Define as a parameter the ratio b/a, and $\gamma a \equiv \sqrt{j\mu\sigma a^2(\omega-\Omega)}$ as another parameter representing the frequency. Describe how you would solve for the eigenfrequencies.

Prob. 6.10.3 A spherical shell has radius R and spins about the z axis with angular velocity Ω. It has a surface conductivity σ_s and is filled with an insulating material having permeability μ.

(a) Starting with the boundary condition, Eq. (d) of Table 6.3.1, find the temporal modes.

(b) Find the decay time resulting if a uniform external field directed along the z axis is suddenly turned off.

(c) What is the frequency of the temporal transient if a uniform field perpendicular to the z axis is suddenly turned off?

Prob. 6.10.4 For the configuration described in Prob. 6.6.2, the excitation is suddenly turned on or off. The resulting transient is initiated with the same k as imposed by the excitation.

(a) Find the transcendental equation that determined the eigenfrequencies of the temporal modes.

(b) Outline a procedure for numerically determining the eigenfrequencies. (Hint: Is it plausible that an infinite number of roots exist where the frequency measured in the frame of reference of the rod is purely imaginary?)

Prob. 6.10.5 In a configuration that generalizes that of Fig. 6.6.1, the entire region $0 < x < a+d$ is filled by a nonuniform conductor having conductivity $\sigma(x)$ and velocity $\vec{v}=U(x)\vec{i}_y$. Note that the uniformly conducting material partially filling the air gap and suffering rigid-body motion is a special case. Start with Eq. 6.2.6, keeping the x dependence of σ and U so that the expression is valid over the entire range of x. Show that the amplitudes $\hat{A}_n$ of the vector potential modes satisfy an orthogonality condition which is Eq. 6.10.12 with $\hat{J}_n \rightarrow \sqrt{\sigma(x)}\,\hat{A}_n$.

7

Laws, Approximations and Relations of Fluid Mechanics

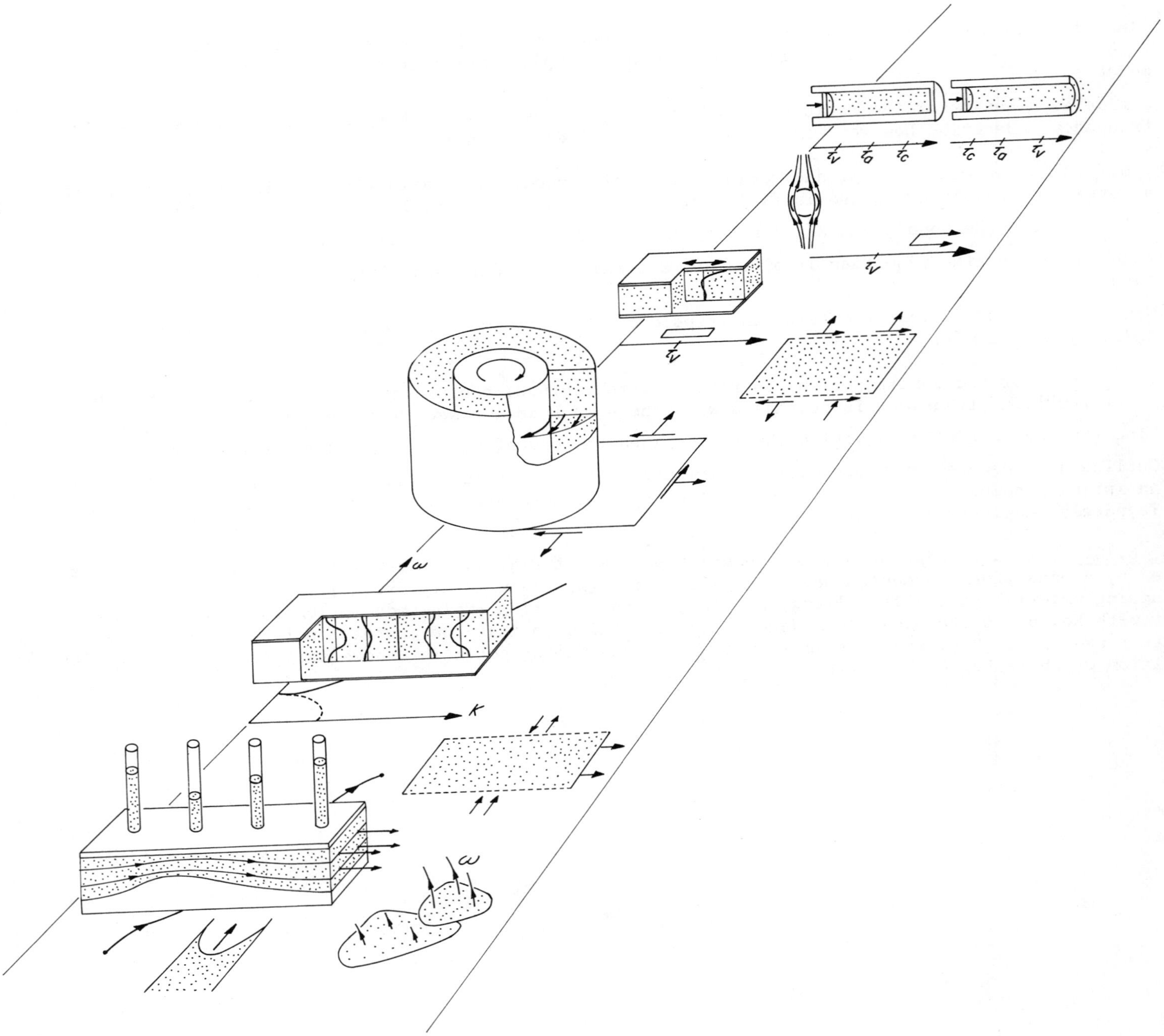

7.1 Introduction

The following chapters carry the subject of continuum electromechanics to its third level. Not only do the field sources assume distributions consistent with deformations of the support medium, the medium is itself free to respond to the associated electromagnetic forces. For gases and liquids, as well as fluid-like continua such as certain plasma models and electron beams, this response must be consistent with the mechanical laws and relations now derived. The role of this chapter is the mechanical analogue of the electromagnetic one played by Chap. 2.

The chapter is organized so that Secs. 7.2-7.9 are sufficient background in incompressible inviscid fluid mechanics to proceed directly with related electromechanical studies. An even wider range of electromechanical coupling mechanisms than might be imagined at this point are tied to fluid interfaces. This makes fluid interfaces (Sec. 7.5), surface tension (Sec. 7.6) and jump conditions (Sec. 7.7) appropriate for early discussion.

Compressibility and related acoustic phenomena are taken up in Secs. 7.10-7.12. Then, contributions of fluid friction, the consequence of fluid viscosity, are taken up in Secs. 7.13-7.17. The resulting Navier-Stokes's equations are summarized in Sec. 7.16.

Overlaying the derivation of the laws of fluid mechanics is the development of relations that play a role in the following chapters for describing the continuum mechanics that is analogous to that for the electric and magnetic transfer relations in the preceding chapters. Transfer relations describing an incompressible and inviscid inertial continuum (Sec. 7.9) will be used many times. Also for future reference are the relations of Sec. 7.11, which embody the acoustic phenomena associated with compressibility, those of Sec. 7.19, which establish the interplay between viscous and inertial effects, and of Sec. 7.20, which describe "creep flow," in which fluid friction overwhelms inertia.

Viscous diffusion, the diffusion of vorticity, has considerable analogy to magnetic diffusion. Thus, the studies of Chap. 6 are a useful background for understanding the interplay of inertia and fluid friction.

This chapter is largely concerned with general laws and relations. The chapters which follow make extensive use of these results in specific case studies.

Chapter 2 begins with a discussion of the two quasistatic limits of the general laws of electrodynamics, identifying rate processes brought in by electrical dissipation in each of these approximations. This chapter ends with a similar discussion.

7.2 Conservation of Mass

With the mass per unit volume of a continuous medium defined as ρ, a statement of mass conservation for a volume V of fixed identity is

$$\frac{d}{dt}\int_V \rho dV = 0 \qquad (1)$$

Here, the volume V is defined such that it always encloses the same material. The surface S enclosing the materials therefore moves with the material, and the velocity $\vec{v}$ is the velocity of surface and material alike.

With the integral theorem of Eq. 2.6.5, it is possible to express Eq. 1 as the integral form of mass conservation:

$$\int_V \frac{\partial\rho}{\partial t} dV + \oint_S \rho\vec{v}\cdot\vec{n}da = 0 \qquad (2)$$

Written in this form, the law applies for V and S either fixed or enclosing material of fixed identity. Using Gauss' theorem, the surface integral can again be expressed as a volume integral, so that the equation involves one integral over the volume, V. Because V is arbitrary, it follows that the integrand must vanish:

$$\frac{D\rho}{Dt} + \rho\nabla\cdot\vec{v} = 0 \qquad (3)$$

This is the required differential law of mass conservation.

<u>Incompressibility</u>: If fluid motions are typified by times that are long compared to the transit time of an acoustic wave through a length typifying the system, for important classes of flows the mass density in the vicinity of a given fluid particle remains constant. In view of the definition of the

convection derivative, Sec. 2.4, this means that

$$\frac{D\rho}{Dt} = 0 \tag{4}$$

For incompressible motions, the mass density evolves much as the free charge density in an insulating fluid (Sec. 5.10). If fluid particles of interest originate where the mass density is uniform, it follows that the mass density in the region occupied by this same fluid at a later time is also uniform. Thus, the solution to Eq. 4, ρ = constant, is a special "homogeneous" or "uniform" density case.

From Eqs. 3 and 4, it follows from conservation of mass that for an incompressible fluid

$$\nabla\cdot\vec{v} = 0 \tag{5}$$

whether the fluid is homogeneous or not.

The quasistatic nature of the incompressible model is investigated in Secs. 7.12 and 7.22.

7.3 Conservation of Momentum

Because momentum is a vector field, rather than a scalar one, it is convenient to deal with its individual components in Cartesian coordinates. Of course, this in no way restricts the validity of the resulting equation of motion.

Again, with the understanding that the volume V always encloses the same material, and hence that its surface deforms with the local velocity of the material, conservation of momentum for the ith component is

$$\frac{d}{dt}\int_V \rho v_i dV = \int_V F_i dV \tag{1}$$

The integral on the right represents contributions to the total force acting on the volume that come from the surrounding material (viscous and pressure forces) and from "external" sources, such as gravity and electromagnetic fields.

Use of the integral theorem, Eq. 2.6.5, gives the integral law for conservation of momentum:

$$\int_V \frac{\partial \rho v_i}{\partial t} dV + \oint_S \rho v_i \vec{v}\cdot\vec{n}da = \int F_i dV \tag{2}$$

Gauss' theorem, Eq. 2.6.2, makes possible a conversion of the surface integral to a volume integral:

$$\int_V \left(\frac{\partial \rho v_i}{\partial t} + \nabla\cdot\rho v_i\vec{v}\right)dV = \int_V F_i dV \tag{3}$$

Expansion of terms on the left gives

$$\int\left\{v_i\left[\frac{\partial\rho}{\partial t} + \nabla\cdot\rho\vec{v}\right] + \rho\left[\frac{\partial v_i}{\partial t} + \vec{v}\cdot\nabla v_i\right]\right\}dV = \int_V F_i dV \tag{4}$$

Again, the integrand of the volume integrations collected together must vanish, but note that conservatio of mass, Eq. 7.2.3, requires that the first term in brackets vanish. Thus, the differential law representing conservation of momentum is

$$\rho\left[\frac{\partial\vec{v}}{\partial t} + \vec{v}\cdot\nabla\vec{v}\right] = \vec{F} \tag{5}$$

On the left is the time-rate of change of $\vec{v}$ for an observer moving with the fluid, the convective derivative as discussed in Sec. 2.4. Even though the mass density appears "outside" the convective derivative this equation is valid even if ρ is a function of space and time.

7.4 Equations of Motion for an Inviscid Fluid

To complete the integral or differential force laws, Eqs. 7.3.2 and 7.3.5, it is necessary to take account of how the surrounding fluid exerts a force on the element of interest. This is naturally done by considering the associated traction exerted on the surface S that encloses the fluid volume V.

In an inviscid (frictionless) fluid, this traction acts normal to the surface and is of the same magnitude regardless of the local surface orientation. With $\vec{n}$ the local normal vector to the surface, the traction due to the surrounding fluid is

$$\vec{T} = -p\vec{n} \tag{1}$$

where the minus sign is introduced so that the pressure, p, will be a positive quantity. That this traction is consistent with a stress

$$T_{ij} = -p\delta_{ij} \tag{2}$$

can be seen by substituting Eq. 2 into the relation between stress and traction, Eq. 3.9.5.

The force density associated with this stress is found by taking the tensor divergence of Eq. 2 (Eq. 3.15.1),

$$F_i^p = -\frac{\partial}{\partial x_j}(p\delta_{ij}) = -\frac{\partial p}{\partial x_i} \tag{3}$$

With forces such as due to gravity and of electric or magnetic origin represented by the external force density $\vec{F}_{ex}$, the force equation, Eq. 7.3.5, becomes

$$\rho[\frac{\partial\vec{v}}{\partial t} + \vec{v}\cdot\nabla\vec{v}] + \nabla p = \vec{F}_{ex} \tag{4}$$

By way of discussing what is required to complete the formulation of the fluid mechanics, suppose that $\vec{F}_{ex}$ is a given driving function. Then, the dependent variables are $\vec{v}$, ρ and p. For incompressible fluid, Eqs. 7.2.4 and 7.2.5 are the two additional scalar laws required to describe the fluid mechanics. Constitutive laws for compressible flows are introduced in Sec. 7.10. Contributions of viscosity to the stress are taken up in Secs. 7.13–7.16.

7.5 Eulerian Description of the Fluid Interface

In electromagnetic theory, the boundary and the field are easily distinguished. In fluid mechanics, the boundary of a given fluid region may be the interface between two fluids. Then, the boundary is in fact a part of the fluid and flow is intrinsically linked to a deformation of the interface.

An interface can be represented analytically by

$$F(x,y,z,t) = 0 \tag{1}$$

That is, of all possible spatial coordinates (x,y,z), at some time, t, only those that make F = 0 comprise an interface. Figure 7.5.1 illustrates a particular case where it is convenient to denote the surface elevation above the y-z plane as $\xi(y,z,t)$, and

$$F = \xi - x = 0 \tag{2}$$

In the language of electrostatics, F could be regarded as a surface of zero potential. This observation is useful, because it is a reminder that the normal vector $\vec{n}$ to the interface is given by the geometry of the interface alone, and is

$$\vec{n} = \frac{\nabla F}{|\nabla F|} \tag{3}$$

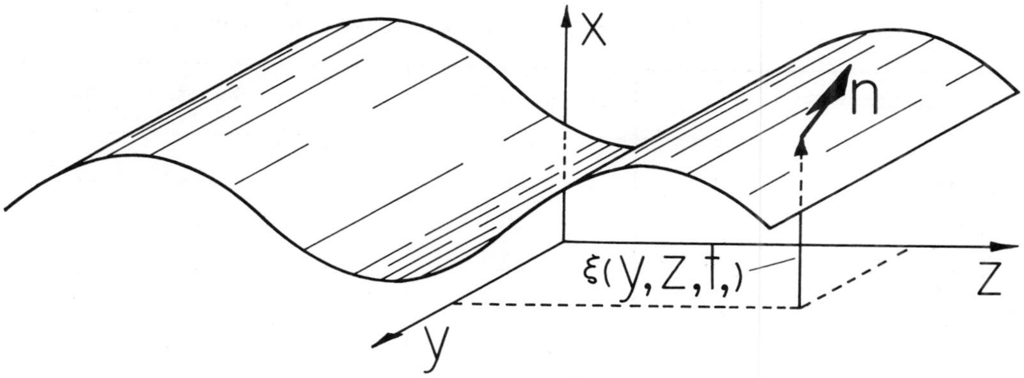

Fig. 7.5.1. Fluid interface.

The geometric relation between F and $\vec{n}$ is the same as that between the electric potential Φ and the electric field intensity $\vec{E}$. The normal to the interface is the gradient of F normalized to ensure unit magnitude.

What is the relationship between the interface geometry and the velocity $\vec{v}$ of the fluid adjacent to the interface? The interface is presumed to be a surface cut from the total fluid volume and always composed of the same material particles. Thus, the interface could be distinguished from the remainder of the fluid by dye markers. As the fluid deforms, it is presumed that the surface remains contiguous. Dyed particles always have adjacent dyed neighbors within the plane of the interface, and undyed neighbors in the adjacent regions of fluid bulk.

By definition, the convective derivative of Sec. 2.4 is the rate of change with respect to time for an observer moving with a particle of fluid. By the definition of what is meant by the "interface," the rate of change of F for an observer on the interface must be zero. Hence, the required relationship between the surface geometry and the fluid velocity is

$$\frac{DF}{Dt} = \frac{\partial F}{\partial t} + \vec{v}\cdot\nabla F = 0 \tag{4}$$

on F = 0.

For the particular case illustrated by Eq. 2 and Fig. 7.5.1, this condition requires that on the surface,

$$v_x = \frac{\partial \xi}{\partial t} + v_y \frac{\partial \xi}{\partial y} + v_z \frac{\partial \xi}{\partial z} \tag{5}$$

The relation is seen to be physically reasonable by considering limiting situations such as: (a) a flat interface that moves in the x direction in a time-varying fashion, $\xi = \xi(t)$; (b) an interface that is stationary but deformed, $\xi = \xi(y,z)$.

7.6 Surface Tension Surface Force Density

If viewed on a millimeter scale, a liquid can take on many of the appearances of an elastic solid. As if enclosed by an elastic "skin," drops of water suffer oscillations and capillary ripples have the appearance of a liquid surface covered by an elastic membrane. Although similar in effect to a membrane under tension, these attributes of the interface are a consequence of the difference between forces on a molecule deep within the bulk of a fluid and near an interface. Because of this difference, energy is required to make an interface between two fluids.

Energy Constitutive Law for a Clean Interface: A clean interface is one made up of molecules from one or the other of the bulk phases. Thus, there are no molecules attributable to the interface itself (as for example there are when an interface between water and air is covered by a film of oil). Because the nature of the interface is therefore completely determined by the bulk phases, it follows that increasing the interfacial area by the increment δA results in a proportionate increase in the energy W_s associated with the interface,

$$\delta W_s = \gamma \delta A \tag{1}$$

For a given pair of fluids, the surface tension is a constant physical property having the same units as for the tension of a membrane, newton/m. Typical values are given in Table 7.6.1.

Table 7.6.1. Illustrative values of surface tension.[1]

Substances	Temperature (°C)	Surface tension (newton/m)
Water/air	18	7.30×10^{-2}
Acetone/air	20	2.37×10^{-2}
Nitrobenzene/air	20	4.39×10^{-2}
Water/Carbon tetrachloride	20	4.5×10^{-2}
Water/mercury	20	3.75×10^{-1}

1. Values taken from Handbook of Chemistry and Physics, College Edition, 49th ed., Robert C. West, ed., The Chemical Rubber Co., Cleveland, Ohio, pp. F-30-32.

Surface Energy Conservation: With the objective a relationship between the geometry of an interface and an effective force per unit area $\vec{T}_s$ acting on the interface, the procedure is now analogous to that followed in Chap. 3. Instead of an electric or magnetic energy subsystem, energy conservation is now written for the "surface subsystem." Some external agent used to put an increment of energy into this system will either increase its stored energy by δW_s, or do work on the external mechanical subsystem through the agent of a force per unit area T_s displacing an area A of the interface by an amount $\delta\xi$. Thus,

$$\text{incremental input of energy} = \delta W_s + T_s A\delta\xi \tag{2}$$

Inputs on the left might come from changing the chemical nature of the bulk fluids. For interfaces of interest here, there are no such inputs of energy, and Eq. 2 is set equal to zero. The only way in which W_s can be altered is through the mechanical work done by displacing the interface. Thus for a clean interface, δW_s is given by Eq. 1,

$$\gamma\delta A + T_s A\delta\xi = 0 \tag{3}$$

To deduce T_s from this expression, δA must be related to the surface geometry and hence to $\delta\xi$.

Surface Force Density Related to Interfacial Curvature: In the geometric construction of Fig. 7.6.1, the local curvature of the elemental area A is represented by radii of curvature R_1 and R_2, defined for orthogonal directions within the local plane of the interface. To find the change in area δA, caused by the displacement $\delta\xi$, note that

$$A + \delta A = (x + \delta x)(y + \delta y) \approx xy + y\delta x + x\delta y \tag{4}$$

In addition, the similarity of triangles requires that

$$\frac{x + \delta x}{R_1 + \delta\xi} = \frac{x}{R_1}; \quad \frac{y + \delta y}{R_2 + \delta\xi} = \frac{y}{R_2} \tag{5}$$

which shows that

$$\delta x = \frac{x}{R_1}\delta\xi; \quad \delta y = \frac{y}{R_2}\delta\xi \tag{6}$$

From Eqs. 4 and 6, it follows that because $xy = A$

$$\delta A = y\delta x + x\delta y = A(\delta\xi)\left[\frac{1}{R_1} + \frac{1}{R_2}\right] \tag{7}$$

In turn, this result can be substituted into Eq. 3 to give

$$\left[\gamma\left(\frac{1}{R_1} + \frac{1}{R_2}\right) + T_s\right]A\delta\xi = 0 \tag{8}$$

Because $\delta\xi$ is arbitrary

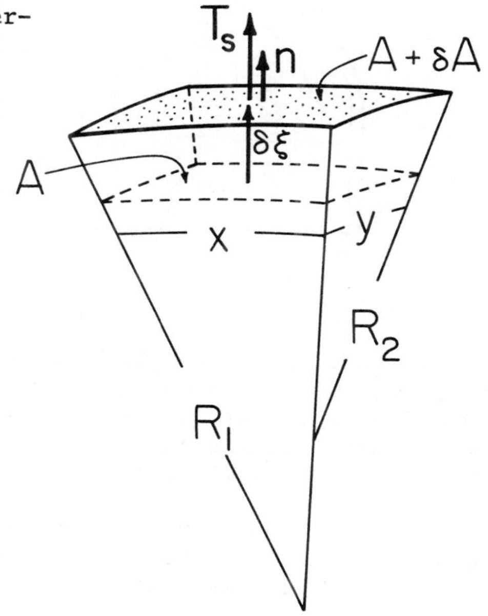

Fig. 7.6.1. Section of interface that suffers perpendicular displacement $\delta\xi$ to make new surface δA.

$$\vec{T}_s = -\gamma\left(\frac{1}{R_1} + \frac{1}{R_2}\right)\vec{n} \tag{9}$$

This surface force density of Young and Laplace[2] has been written as a vector which, if positive, acts in the direction of the normal $\vec{n}$. A radius of curvature has a sign that is positive if the associated center of curvature is in the region from which $\vec{n}$ is directed. If the center of curvature is in the region into which $\vec{n}$ is directed, the associated radius is taken as negative.

The implications of Eq. 9 for the static equilibrium of a liquid are illustrated in Fig. 7.6.2. The pair of glass plates are wetted by the liquid so that the radius of curvature of the interface is essentially equal to half the local distance between the plates. Thus, where the plates are closest together the radius of curvature is least and the surface force density is accordingly largest. Note that the radius of curvature is also negative, so that $\vec{T}_s$ acts from liquid to air with a net effect of making the interface rise between the plates. The height of rise is greatest to the right, where the plates are closest together. The height of rise, $\xi(r)$, is found in Sec. 7.8.

2. A. W. Adamson, Physical Chemistry of Surfaces, Interscience, New York, 1960, pp. 4-6.

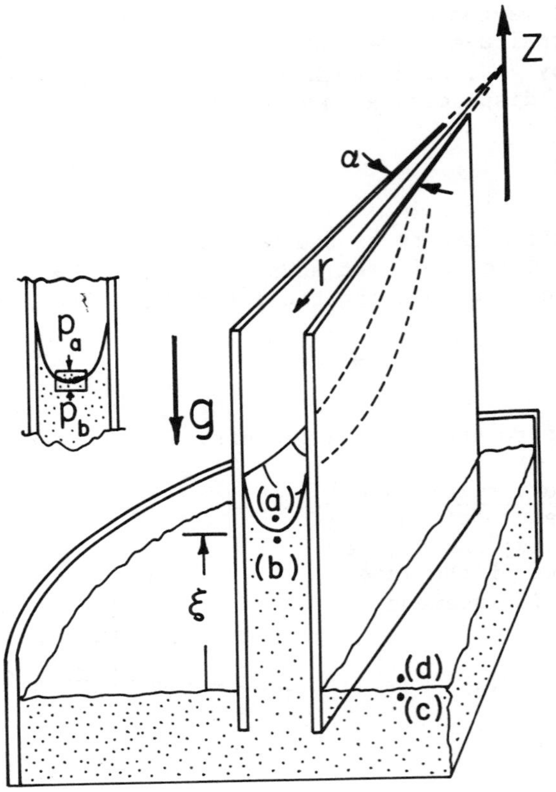

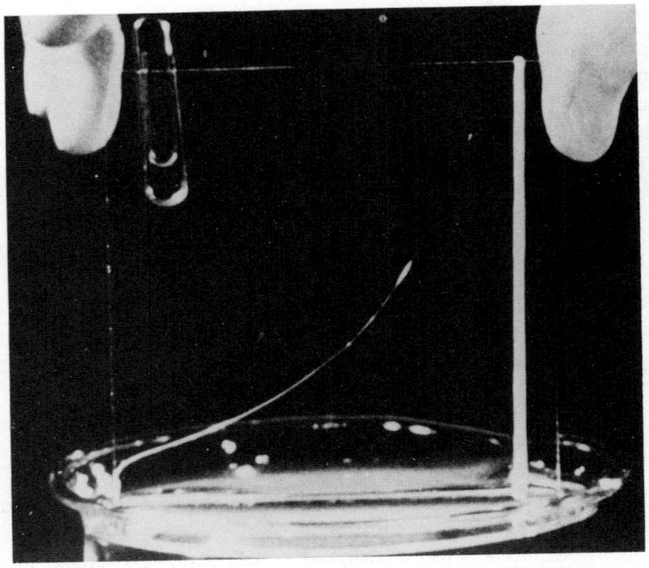

Fig. 7.6.2. Because of surface tension, fluid wetting pair of glass plates rises to a height $\xi(r)$ determined by the surface tension γ and local distance between plates. Experiment from film "Surface Tension in Fluid Mechanics" (Reference 9, Appendix C).

Surface Force Density Related to Interfacial Deformation: Three commonly encountered interfacial configurations are shown in Table 7.6.2. In "equilibrium," these are respectively planar, circular cylindrical and spherical in shape. To describe the dynamics of the interface, the surface force density due to surface tension must be expressed in terms of the perturbation ξ from these equilibria. This could be done by evaluating Eq. 9, but is more easily accomplished by returning to Eq. 3.

Consider the volume, shown in Fig. 7.6.3, that is "cut out" by the surface segment A as it displaces an amount $\delta\xi$. For this volume V, enclosed by the surface S having the outward normal vector $\vec{n}_s$, Gauss' theorem states that

$$\int_V \nabla\cdot\vec{C}dV = \oint_S \vec{C}\cdot\vec{n}_s da \qquad (10)$$

The vector $\vec{C}$ is arbitrary, and now chosen to be the vector $\vec{n}$ normal to the interface (not to the surface S enclosing the volume element). Thus, $\vec{n} = \vec{n}_s$ on the upper surface but $\vec{n} = -\vec{n}_s$ on the lower surface. On the remaining sides, $\vec{n}$ is perpendicular to $\vec{n}_s$. It follows that the right-hand side of Eq. 10 is the required change in area, δA. Because the area A is itself elemental, the left-hand side of Eq. 10 is $\nabla\cdot\vec{n}A\delta\xi$ and Eq. 10 becomes

$$\delta A = \nabla\cdot\vec{n}A\delta\xi \qquad (11)$$

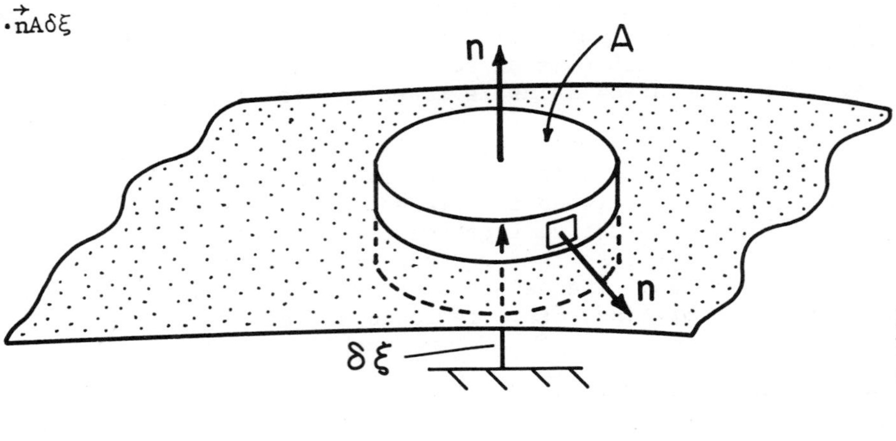

Fig. 7.6.3

Elemental volume V enclosed by surface S intersecting interface between fluids.

Table 7.6.2. Summary of normal vector and surface tension surface force density for small perturbations from planar, circular cylindrical and spherical equilibria.

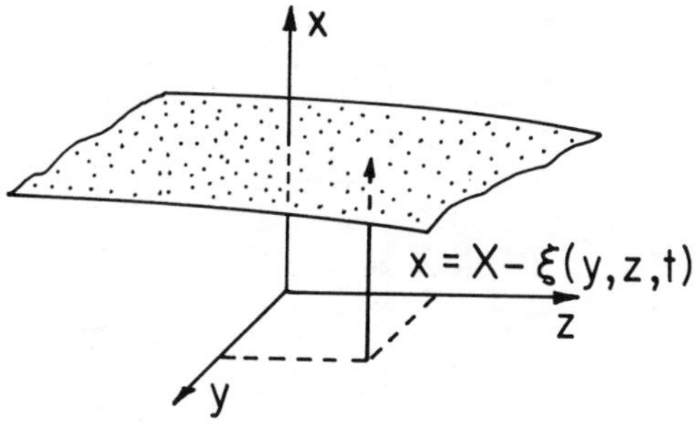	$\vec{n} = \vec{i}_x - \frac{\partial\xi}{\partial y}\vec{i}_y - \frac{\partial\xi}{\partial z}\vec{i}_z$	(a)
	$(\vec{T}_s)_x = \gamma\left(\frac{\partial^2\xi}{\partial y^2} + \frac{\partial^2\xi}{\partial z^2}\right)$	(b)
	$\xi = \mathrm{Re}\,\tilde{\xi}\,\exp -j(k_y y + k_z z)$	(c)
	$\tilde{T}_s = -\gamma(k_y^2 + k_z^2)\tilde{\xi}$	(d)
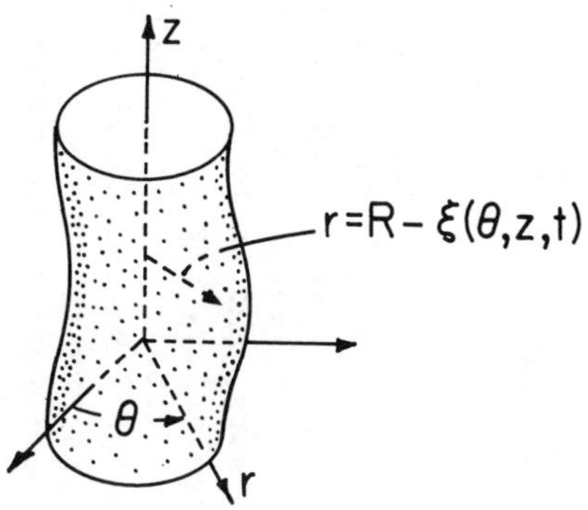	$\vec{n} = \vec{i}_r - \frac{1}{R}\frac{\partial\xi}{\partial\theta}\vec{i}_\theta - \frac{\partial\xi}{\partial z}\vec{i}_z$	(e)
	$(\vec{T}_s)_r = \gamma\left[-\frac{1}{R} + \frac{\xi}{R^2} + \frac{1}{R^2}\frac{\partial^2\xi}{\partial\theta^2} + \frac{\partial^2\xi}{\partial z^2}\right]$	(f)
	$\xi = \mathrm{Re}\,\tilde{\xi}\,\exp -j(m\theta + kz)$	(g)
	$\tilde{T}_s = \frac{\gamma}{R^2}\left[(1 - m^2) - (kR)^2\right]\tilde{\xi}$	(h)
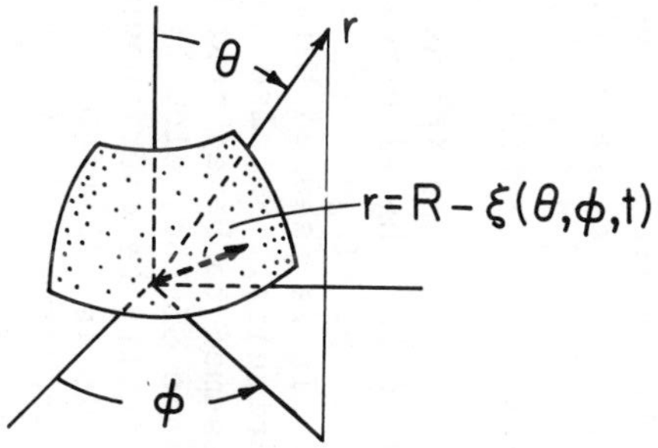	$\vec{n} = \vec{i}_r - \frac{1}{R}\frac{\partial\xi}{\partial\theta}\vec{i}_\theta - \frac{1}{R\sin\theta}\frac{\partial\xi}{\partial\phi}\vec{i}_\phi$	(i)
	$(\vec{T}_s)_r = \gamma\left[-\frac{2}{R} + \frac{2\xi}{R^2} + \frac{1}{R^2\sin\theta}\frac{\partial}{\partial\theta}\left(\sin\theta\frac{\partial\xi}{\partial\theta}\right) + \frac{1}{R^2\sin^2\theta}\frac{\partial^2\xi}{\partial\phi^2}\right]$	(j)
	$\xi = \mathrm{Re}\,\tilde{\xi}\,P_n^m(\cos\theta)e^{-jm\phi}$	(k)
	$\tilde{T}_s = -\frac{\gamma}{R^2}(n-1)(n+2)\tilde{\xi}$	(ℓ)

Substitution of Eq. 11 into Eq. 3 gives an alternative expression for the surface tension surface force density:

$$\vec{T}_s = -\gamma(\nabla\cdot\vec{n})\vec{n} \tag{12}$$

The use of this expression for relating $\vec{T}_s$ to interfacial deformations, as summarized in Table 7.6.2, is now illustrated for the cylindrical coordinate configuration. The interface is then described by

$$F = r - R - \xi(\theta,z,t) = 0$$

If terms that are quadratic in the perturbation amplitude ξ are ignored, it follows from Eq. 7.5.3 that $\vec{n}$ is given by Eq. (e) of Table 7.6.2. In turn,

$$\nabla\cdot\vec{n} = \frac{1}{r} - \frac{1}{r^2}\frac{\partial^2\xi}{\partial\theta^2} - \frac{\partial^2\xi}{\partial z^2} \tag{13}$$

Consistent with the small amplitude is the approximation $r^{-1} \simeq R^{-1} - \xi/R^2$. Thus, $\vec{T}_s$ is as given by Eq. (f) of Table 7.6.2. For $\xi = 0$, there is an equilibrium surface force density acting radially inward, tending to compress what is inside the surface much as if it were enclosed by a membrane under tension.

Also summarized in Table 7.6.2 are the complex amplitudes of $\vec{T}_s$. In the Cartesian and circular cylindrical geometries these are found by straightforward substitution. However, in the spherical case, Eq. (ℓ) is obtained by using the fact that P_n^m is a solution to Eq. 2.16.31a.

7.7 Boundary and Jump Conditions

It can be taken as phenomenologically based fact that there is neither tangential nor normal velocity of a fluid adjacent to a fixed rigid impermeable wall. Thus, boundary conditions for such a wall are

$$\vec{n}\cdot\vec{v} = 0 \tag{1}$$

$$\vec{n}\times\vec{v} = 0 \tag{2}$$

where $\vec{n}$ is the normal to the boundary.

The condition on the tangential component of $\vec{v}$ results because of the friction between wall and fluid, i.e., because of the fluid viscosity. If the fluid is modeled as inviscid, it is consistent to ignore the tangential velocity boundary condition. An inviscid model pictures the fluid as slipping adjacent to a fixed boundary. The extent of the error is investigated in Sec. 7.18.

The jump conditions at an interface between fluids are deduced from the integral laws, much as in Sec. 2.10 for the electromagnetic fields. But, before this can be done, it is necessary to specify the order of the singularity in mass density, pressure and velocity that is included in the interfacial model. It is assumed here that there is no surface mass density, that the density takes at most a step discontinuity. So also does the pressure, and in fact mechanical stresses including viscosity (Sec. 7.15) are assumed to be at most a step singularity. Because the viscous stresses depend on the spatial rates of change of the velocity (the strain rates), a self-consistent model for the interface requires that the velocity be continuous. But, in the inviscid limit, only the normal velocity must be continuous. That this is all required if the fluids are to have a common surface of demarcation can be seen from the relation between fluid velocity at the interface and interfacial geometry, Eq. 7.5.4. At a given location on the interface, ∇F has a normal direction. Hence, Eq. 7.5.4 involves only the velocity normal to the interface. Because the expression must hold whether $\vec{v}$ is evaluated on one or the other side of the interface, it is clear that the normal component of $\vec{v}$ must be continuous:

$$\vec{n}\cdot[\![\vec{v}]\!] = 0 \tag{3}$$

Conditions implied by the integral laws follow by using the same incremental volume of fixed identity used for some of the jump conditions in Sec. 2.10 and shown in Fig. 2.10.1. Because there is no surface mass density, mass conservation, Eq. 7.2.1, is automatically satisfied. Formally, this is seen from Eqs. 2.10.14 and 2.10.15 by replacing the free charge density with the mass density.

It is perhaps tempting to require that the mass flux $\rho\vec{v}$ normal to the interface be continuous. But, the interface considered here is composed of given fluid particles and deforms with the fluid.

The integral momentum-conservation law, expressed as Eq. 7.3.1, makes it clear that for similar reasons there is no contribution of the inertia (represented by the left-hand side) to the interfacial boundary condition. On the right, those force densities that are spatial impulses (surface force densities) make contributions in the limit $\Delta \to 0$. It is convenient to represent the mechanical and electrical surface force densities by writing them as the divergence of stress tensors, T^m_{ij} and T^e_{ij}. For an inviscid fluid, T^m_{ij} is $-p\delta_{ij}$ given by Eq. 7.4.2 while T^e_{ij} is one of the tensors summarized in Table 3.10.1. The contribution of surface tension has already been written as a surface force density, Eq. 7.6.9 or Eq. 7.6.12. With the use of the tensor form of Gauss' theorem, Eq. 3.9.4, the integral momentum law therefore becomes

$$\oint_S (T^m_{ij} + T^e_{ij})(i_n)_j\, da + \int_A (T_s)_i da = 0 \tag{4}$$

In the limit where A is incremental, the force (or stress) jump condition results:

$$[\![T^m_{ij} + T^e_{ij}]\!] n_j + (T_s)_i = 0 \tag{5}$$

This expression will be used with viscous fluids as well, but consider its special form for inviscid fluids and a clean interface so that T^m_{ij} is given by Eq. 7.4.2 and $\vec{T}_s$ is given by Eq. 7.6.12:

$$[\![p]\!] n_i = [\![T^e_{ij}]\!] n_j - \gamma(\nabla\cdot\vec{n})n_i \tag{6}$$

This vector jump condition has three components. Note that the pressure and surface tension contributions are normal to the interface. This makes it clear that to be consistent with the inviscid and clean interface model, the first term on the right, the surface force density of electric or magnetic origin, must also have no shearing components. Electromagnetic properties of interfaces meeting this requirement are taken up in Sec. 8.2.

7.8 Bernoulli's Equation and Irrotational Flow of Homogeneous Inviscid Fluids

In this section, external force densities take the form of the gradient of a scalar. Examples include the gravitational force density on a fluid having uniform density ρ. With $\vec{g}$ defined as the directed gravitational acceleration and $\vec{r} \equiv x\vec{i}_x + y\vec{i}_y + z\vec{i}_z$, this force density is

$$\vec{F}^g = \rho\vec{g} = \nabla(\rho\vec{g}\cdot\vec{r}) \tag{1}$$

Note that ρ must be uniform, or the last equality does not hold.

In general, electric and magnetic force densities do not take the form of the gradient of a scalar However, in many important situations, they are approximated by such a form. In fact, as illustrated in Chap. 8, it is often desirable to design a system so that this is the case. Thus, looking forward to such examples, the force densities of electric and magnetic origin are written as

$$\vec{F}^e = -\nabla\mathcal{E} \tag{2}$$

With these contributions to F_{ex}, the force equation, Eq. 7.4.4, becomes

$$\rho(\frac{\partial v}{\partial t} + \vec{v}\cdot\nabla v) + \nabla p = \nabla(\rho\vec{g}\cdot\vec{r} - \mathcal{E}) \tag{3}$$

A vector identity* makes it possible to rewrite Eq. 3 in a form that makes evident the contribution of vorticity $\vec{\omega} \equiv \nabla \times \vec{v}$, to the dynamics:

$$\rho(\frac{\partial\vec{v}}{\partial t} + \vec{\omega}\times\vec{v}) + \nabla(p + \frac{1}{2}\rho\vec{v}\cdot\vec{v} - \rho\vec{g}\cdot\vec{r} + \mathcal{E}) = 0 \tag{4}$$

Bernoulli's equation is a statement of invariance for a combination of dynamical quantities that represent the total energy. It is important to recognize that there are two essentially different circumstances under which similar equations apply.

First, consider points (a) and (b) in the flow, as sketched in Fig. 7.8.1, that can be joined by a streamline (not a particle line but rather a line always tangent to the instantaneous velocity vector $\vec{v}$). Then, integration of Eq. 4 along the line C gives no contribution from the second term, which must be perpendicular to the velocity $\vec{v}$, and hence the direction of integration. Further, in view of Eq. 2.6.1, the remaining terms integrate to

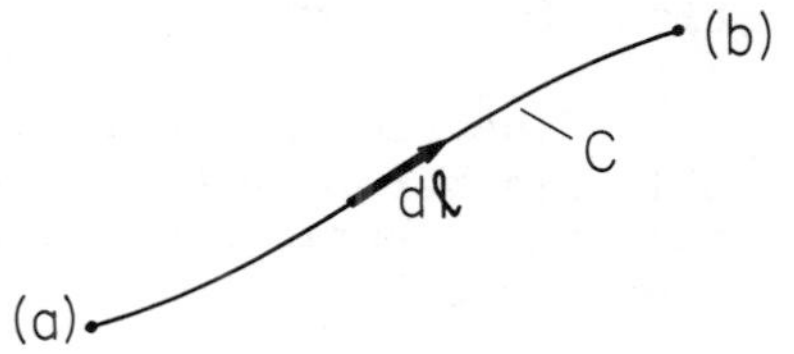

Fig. 7.8.1. Points (a) and (b) are joined by a streamline.

*$(\vec{v}\cdot\nabla)\vec{v} = (\nabla \times \vec{v}) \times \vec{v} + \frac{1}{2}\nabla(\vec{v}\cdot\vec{v})$.

$$\rho\int_a^b \frac{\partial\vec{v}}{\partial t}\cdot d\vec{\ell} + [p + \frac{1}{2}\rho\vec{v}\cdot\vec{v} - \rho\vec{g}\cdot\vec{r} + \mathcal{E}]_a^b = 0 \qquad (5)$$

This form of Bernoulli's equation applies to any two points joined by a streamline, regardless of the flow. Reference 8 of Appendix C gives experimental demonstrations of Bernoulli's law.

Second, consider irrotational flows, defined as having no vorticity, $\vec{\omega} = 0$. Then, it is appropriate to define a velocity potential Θ

$$\vec{v} = -\nabla\Theta \qquad (6)$$

and integration of Eq. 4 between fixed points a and b gives

$$[-\rho\frac{\partial\Theta}{\partial t} + p + \frac{1}{2}\rho\vec{v}\cdot\vec{v} - \rho\vec{g}\cdot\vec{r} + \mathcal{E}]_a^b = 0 \qquad (7)$$

This expression is restricted to irrotational flows, but applies to arbitrary fixed points a and b.

The importance of irrotational flows stems from the theorem on vorticity of Helmholtz and Kelvin. If at some instant fluid of fixed identity sustains an irrotational flow, then for this same material the irrotational condition prevails at a later instant. For example, if the flow was initiated from a static (and hence irrotational) condition, it must be irrotational.

Proof of this theorem follows by taking the curl of Eq. 4 and observing that the curl of a gradient is identically zero:

$$\frac{\partial\vec{\omega}}{\partial t} + \nabla\times(\vec{\omega}\times\vec{v}) = 0 \qquad (8)$$

If the vorticity, $\vec{\omega}$, is replaced by the magnetic flux density, $\vec{B}$, this expression is the same as that governing the magnetic field in a deforming perfect conductor, Eq. 6.2.3 in the limit $\sigma\rightarrow\infty$. Thus, the theorem on flux conservation for a perfectly conducting surface of fixed identity, Eq. 6.2.4, with $\sigma\rightarrow\infty$, becomes the theorem

$$\frac{d}{dt}\int_S \vec{\omega}\cdot\vec{n}da = 0 \qquad (9)$$

The vorticity linking a material surface S as it deforms with the flow is conserved. If there is no initial vorticity in a given region, the same material will have no vorticity in whatever region it occupies at a later time.

Conservation of mass requires that the flow be solenoidal (Eq. 7.2.5); this combines with the condition for irrotational flow (Eq. 6) to show that the velocity potential is governed by Laplace's equation

$$\nabla^2\Theta = 0 \qquad (10)$$

If boundary conditions involve only $\vec{v}$ (and hence Θ), this equation defines the flow distribution. With Π defined as a function of time alone set by flow conditions at a reference point, the associated pressure distribution follows from Eq. 7,

$$p = \rho\frac{\partial\Theta}{\partial t} - \frac{1}{2}\rho\vec{v}\cdot\vec{v} + \rho\vec{g}\cdot\vec{r} - \mathcal{E} + \Pi \qquad (11)$$

Although p is a nonlinear function of the velocity, it can be determined in such a problem "after the fact," once $\vec{v}$ has been found by solving a linear problem. That is, Laplace's equation is linear, in that superimposed solutions are also solutions. But, note that the pressure must be evaluated using the total velocity. Because Eq. 11 is a nonlinear function of $\vec{v}$, the pressure does not satisfy the conditions for superposition.

The flux potential relations derived in Sec. 2.16 for electric and magnetic cases are equally applicable here. With the identification $D_n/\varepsilon\rightarrow v_n$ and $\Phi\rightarrow\Theta$, the transfer relations and associated bulk distributions of Sec. 2.16 summarize solutions to Eq. 10 in Cartesian, cylindrical and spherical coordinates.

<u>A Capillary Static Equilibrium</u>: The static equilibrium illustrated in Fig. 7.6.2 is described by combining Bernoulli's equation with the capillary surface force density discussed in Sec. 7.6. The object is to find the interfacial profile, $\xi(r)$, of the water-air interface. Points (b) and (c) are related by Eq. 7, evaluated with $\partial/\partial t = 0$, $\vec{v} = 0$, $\vec{g} = -g\vec{i}_z$ and $\mathcal{E} = 0$:

$$p_c = p_b + \rho g\xi \qquad (12)$$

where ρ is the mass density of water. The mass density of the air is 10^3 times less than that of the water, so its contribution is ignored in connecting points (a) and (d) via Eq. 7 through the air:

$$p_a = p_d \qquad (13)$$

These two bulk relations are augmented by boundary conditions that relate the pressures on opposite sides of the interface. At the bottom of the meniscus, the z component of Eq. 7.7.6 is evaluated. It is assumed that the glass plates are perfectly wetted by the water and that the meniscus curvature is dominated by variations of the interface in the azimuthal direction. With the shape of the meniscus over the gap between plates approximated as being essentially circular, the local radius of curvature is approximately $\alpha r/2$ and Eq. 7.7.6 becomes

$$-(p_a - p_b) = -\gamma(\frac{2}{\alpha r}) \qquad (14)$$

The balance of surface force densities at (c-d), where the interface is flat, shows that

$$p_d - p_c = 0 \qquad (15)$$

The pressures can be eliminated by adding Eqs. 12-15 and the result solved for ξ:

$$\xi = (\frac{2\gamma}{\alpha\rho g})\frac{1}{r} \qquad (16)$$

This is essentially the interfacial radial profile shown in Fig. 7.6.2.

7.9 Pressure-Velocity Relations for Inviscid, Incompressible Fluid

Just as the electrical transfer relations introduced in Sec. 2.16 are a convenient building block for modeling complex systems, the mechanical relations derived in this section are useful in a variety of mechanical and electromechanical situations. They are restricted to perturbations described by the inviscid model of Sec. 7.8. The fluid is homogeneous and incompressible so that ρ is a constant. The transfer relations relate dynamical perturbations from a stationary equilibrium. In making use of the relations in a specific problem, it is important to first establish that the stationary (in special cases, static) conditions are satisfied.

Streaming Planar Layer: Consider first the planar layer of fluid shown in Table 7.9.1, having as a stationary state a uniform velocity in the z direction. Gravity acts in the -x direction, so $\vec{g} = -g\vec{i}_x$. The velocity takes the form

$$\vec{v} = U\vec{i}_z - \nabla\Theta' \qquad (1)$$

The equilibrium part has the velocity potential -Uz, which satisfies Laplace's equation, Eq. 7.8.10. By superposition, the perturbation Θ' must also satisfy this equation. Thus Θ' is described by the same derivation given in Sec. 2.16, Eqs. 2.16.11-2.16.16. With the identification $\hat{D}_x/\varepsilon \rightarrow \hat{v}_x$ and $\Phi \rightarrow \Theta$, the transfer relations of Table 2.16.1, Eq. (b), become

$$\begin{bmatrix} \hat{\Theta}^\alpha \\ \hat{\Theta}^\beta \end{bmatrix} = \frac{1}{\gamma}\begin{bmatrix} -\coth\gamma\Delta & \dfrac{1}{\sinh\gamma\Delta} \\ \dfrac{-1}{\sinh\gamma\Delta} & \coth\gamma\Delta \end{bmatrix}\begin{bmatrix} \hat{v}_x^\alpha \\ \hat{v}_x^\beta \end{bmatrix} \qquad (2)$$

Here it is understood that the complex amplitudes represent the perturbation. Because the next step brings in a time-rate of change, the time dependence has been specified in Eq. 2, as indicated by replacing ~ with ^. That is,

$$\Theta' = \text{Re}\,\hat{\Theta}(x)e^{j(\omega t-k_y y-k_z z)}\;;\quad \vec{v} = \text{Re}\,\hat{\vec{v}}(x)e^{j(\omega t-k_y y-k_z z)} + U\vec{i}_z \qquad (3)$$

To linear terms in the perturbations, Bernoulli's equation (Eq. 7.8.11) gives the pressure

$$p = -\frac{1}{2}\rho U^2 - \mathcal{E} + \Pi - \rho g x + \rho(\frac{\partial}{\partial t} + U\frac{\partial}{\partial z})\Theta' \qquad (4)$$

In terms of complex amplitudes, this expression becomes

$$p = -\frac{1}{2}\rho U^2 - \mathcal{E} + \Pi - \rho g x + \text{Re}\,\hat{p}(x)e^{j(\omega t-k_y y-k_z z)} \qquad (5)$$

Table 7.9.1. Pressure-velocity relations for perturbations of inviscid fluid.

Cartesian	Cylindrical	Spherical
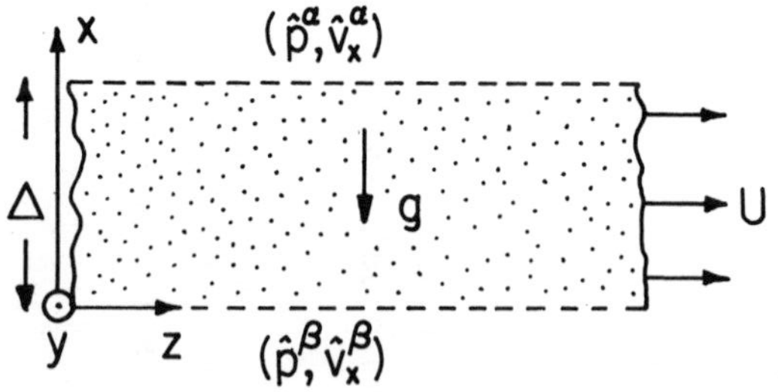	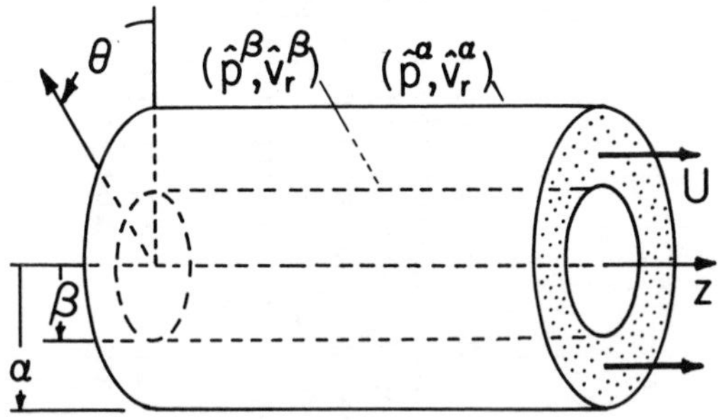	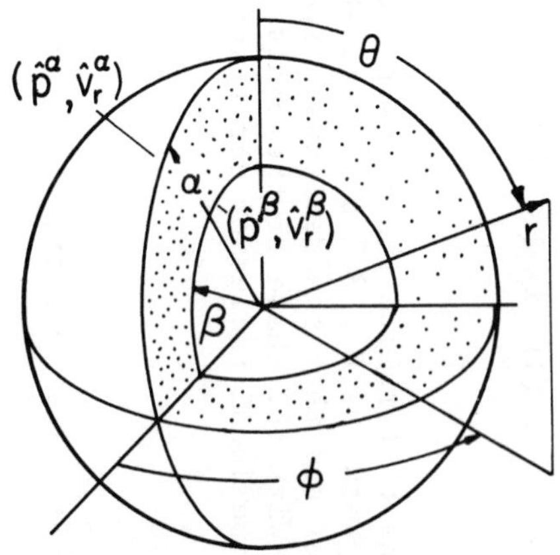
$p = \Pi - \frac{1}{2}\rho U^2 - \mathcal{E} - \rho g x + \text{Re}\,\hat{p}(x)e^{j(\omega t - k_y y - k_z z)}$ (a)	$p = \Pi - \frac{1}{2}\rho U^2 - \mathcal{E} + \text{Re}\,\hat{p}(r)e^{j(\omega t - m\theta - kz)}$ (d)	$p = \Pi - \mathcal{E} + \text{Re}\,\hat{p}(r)P_n^m(\cos\theta)e^{j(\omega t - m\phi)}$ (g)
$\hat{p}(x) = j(\omega - k_z U)\rho\hat{\Phi}(r)$ (b)	$\hat{p}(r) = j(\omega - kU)\rho\hat{\Phi}(r)$ (e)	$\hat{p}(r) = j\omega\rho\hat{\Phi}(r)$ (h)
$\begin{bmatrix}\hat{p}^\alpha \\ \hat{p}^\beta\end{bmatrix} = \frac{j(\omega - k_z U)\rho}{\gamma}\begin{bmatrix}-\coth\gamma\Delta & \frac{1}{\sinh\gamma\Delta} \\ \frac{-1}{\sinh\gamma\Delta} & \coth\gamma\Delta\end{bmatrix}\begin{bmatrix}\hat{v}_x^\alpha \\ \hat{v}_x^\beta\end{bmatrix}$ (c) $\gamma \equiv \sqrt{k_y^2 + k_z^2}$	$\begin{bmatrix}\hat{p}^\alpha \\ \hat{p}^\beta\end{bmatrix} = j(\omega - kU)\rho\begin{bmatrix}F_m(\beta,\alpha) & G_m(\alpha,\beta) \\ G_m(\beta,\alpha) & F_m(\alpha,\beta)\end{bmatrix}\begin{bmatrix}\hat{v}_r^\alpha \\ \hat{v}_r^\beta\end{bmatrix}$ (f) (See Table 2.16.2 for F_m and G_m)	$\begin{bmatrix}\hat{p}^\alpha \\ \hat{p}^\beta\end{bmatrix} = j\omega\rho\begin{bmatrix}F_n(\beta,\alpha) & G_n(\alpha,\beta) \\ G_n(\beta,\alpha) & F_n(\alpha,\beta)\end{bmatrix}\begin{bmatrix}\hat{v}_r^\alpha \\ \hat{v}_r^\beta\end{bmatrix}$ (i) (See Table 2.16.3 for F_n and G_n)
Compressible: $\gamma \equiv \sqrt{k_y^2 + k_z^2 - \frac{(\omega - k_z U)^2}{a^2}}$	Compressible; replace $k \to \gamma$ in F_m and G_m: $\gamma \equiv \sqrt{k^2 - \frac{(\omega - kU)^2}{a^2}}$	

where

$$\hat{p}(x) = j(\omega - k_z U)\rho\hat{\Theta}(x) \tag{6}$$

Note that the first four terms in Eq. 5, the "equilibrium" pressure, are independent of time; but, because of the gravitational force, this pressure is a linearly decreasing function of altitude, x.

With the understanding that it is only the part of the pressure that is a function of time at a fixed location (x,y,z) that is being described (the last term in Eq. 5), Eq. 6 is used to write Eq. 2 as the pressure-velocity relations summarized in Table 7.9.1.

Streaming Cylindrical Annulus: In the cylindrical configuration of Table 7.9.1, the fluid again assumes a stationary state of streaming in the z direction with the uniform velocity U. However, it is assumed that the effects of gravity are negligible. The relations summarized in Table 7.9.1 follow by exploiting the flux-potential relations of Table 2.16.2. The reasoning is identical to that for the planar relations.

Static Spherical Shell: In the spherical configuration, it is assumed that the fluid equilibrium is static, so that the perturbation velocity is the total velocity. Also, the effects of gravity are ignored. Then, the relations summerized in Table 7.9.1 follow from those of Table 2.16.3 using the reasoning already described.

7.10 Weak Compressibility

To specify the relationship between mass density and the other dynamical variables, it is helpful to distinguish between those tied to the material and to a given position in space. Thus, a constitutive law relating mass density to extensive variables, α_i, and pressure, p, takes the form

$$\rho = \rho(\alpha_1, \cdots, \alpha_m, p) \tag{1}$$

One of the α's might be a concentration (perhaps of salt in water) or might be the entropy density. In general, these variables are themselves described by still other laws that bring in additional rate processes. For example, the molecular diffusion in the face of material convection governs the concentrations, while heat conduction and convection determines the distribution of entropy. Coupling to additional subsystems is avoided (and hence closure of the laws needed to describe the dynamics obtained) by taking the α_i's as being conserved by fluid of fixed identity. Just as Eq. 7.2.1 then implies Eq. 3, it follows that

$$\frac{\partial \alpha_i}{\partial t} + \nabla \cdot \alpha_i \vec{v} = 0 \tag{2}$$

The pressure is not carried in this fashion by the material. Its role is simplified by confining the discussion to excursions of pressure that can be described as linear perturbations from a reference pressure p_r. Thus, Eq. 1 is specialized to

$$\rho = \rho(\alpha_1, \cdots, \alpha_m, p_r) + \frac{1}{a^2}(p - p_r) \tag{3}$$

where a, defined by

$$a^{-1} = \sqrt{\left(\frac{\partial \rho}{\partial p}\right)_{\alpha_i\text{'s}}} = \text{constant} \tag{4}$$

is taken as being independent of p, and is identified in the next section as the velocity of an acoustic wave.

If coupling to the thermodynamic subsystem were self-consistently included in the model (Sec. 7.23), it would be found that for processes having rates typical of acoustic applications, it is the entropy density that is held fixed (possibly along with other α_i's) in Eq. 4.

7.11 Acoustic Waves and Transfer Relations

Compressibility gives rise to time delays associated with the propagation of acoustic waves. For many purposes, acoustic phenomena can be represented in terms of small perturbations from an equilibrium of uniform density ρ_o and pressure p_o. In most acoustic applications, the equilibrium is also static, but to be able to represent doppler-related phenomena, included in this section is the possibility that the fluid streams with a uniform z-directed velocity, U.

The equations of motion that relate perturbations v', ρ', and p' in the velocity, density, and pressure, respectively, are conservation of mass and momentum, Eqs. 7.2.3 and 7.4.4 with $\vec{F}_{ex} = 0$. Written to linear terms in the perturbation quantities, these are

$$\rho_o\left(\frac{\partial}{\partial t} + U\frac{\partial}{\partial z}\right)\vec{v}' + \nabla p' = 0 \tag{1}$$

$$\left(\frac{\partial}{\partial t} + U\frac{\partial}{\partial z}\right)\rho' + \rho_o \nabla\cdot\vec{v}' = 0 \tag{2}$$

The equation of state, Eq. 7.10.3, provides the third relation. It follows that

$$p' = a^2\rho' \tag{3}$$

Typical values of the acoustic velocity, a, as well as the mass density and the acoustic impedance (to be defined in Sec. 7.13) are given in Table 7.11.1.

Table 7.11.1. Sound velocity, mass density and acoustic impedance for common fluids.[1]

Fluid	Temperature T^oC	Acoustic velocity a (m/sec)	Mass density ρ (kg/m^3)	Acoustic impedance Z_o $\left(\frac{n\text{-sec}}{m^3}\right)$
Gases				
Air	0	331.45	1.293	429
He	0	970	0.1785	173
CO_2	0	258	1.977	509
H_2	0	1269	0.08988	114
O_2	0	317	1.429	452
N_2	0	337	1.250	421
Liquids				
Water	17	1.43×10^3	0.999×10^3	1.43×10^6
Benzene	20	1.32×10^3	0.879×10^3	1.16×10^6
Glycerine	20	1.92 x 103	1.26×10^3	2.42×10^6
Mercury	20	1.45×10^3	1.35×10^4	1.96×10^7

The operators in Eqs. 1 and 2 are linear, and have constant coefficients. Thus, the velocity can be eliminated as a variable between the divergence of Eq. 1 and the convective derivative of Eq. 2, to obtain

$$\left(\frac{\partial}{\partial t} + U\frac{\partial}{\partial z}\right)^2 p' = a^2\nabla^2 p' \tag{4}$$

The second convective derivative on the left is the second derivative with respect to time for an observer moving with the velocity U in the z direction. Hence, in that moving frame, Eq. 4 is the wave equation and shows that waves have the velocity, a, relative to the fluid.

<u>Pressure-Velocity Relations for Planar Layer</u>: In the prototype configuration of Fig. 7.11.1, a layer of compressible but inviscid fluid fills the planar region between the α and β planes.

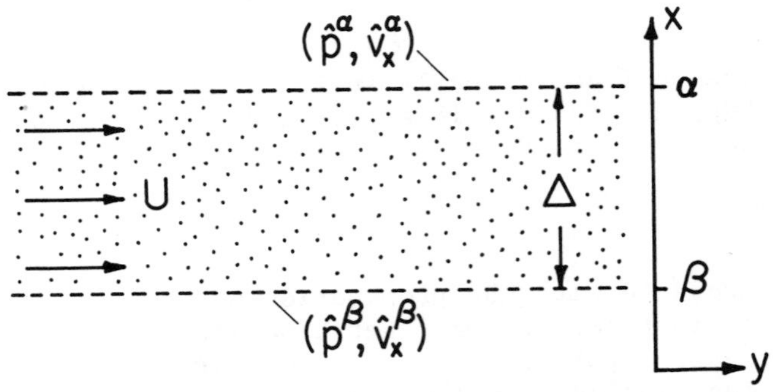

Fig. 7.11.1.

A layer of compressible fluid is bounded from above and below by surfaces having the perturbation deflections ξ^α and ξ^β. The pressures just inside the fluids adjacent to these surfaces are p^α and p^β, respectively.

1. L. L. Beranek, <u>Acoustic Measurements</u>, John Wiley & Sons, New York, 1949, pp. 40-46.

Solutions to Eqs. 1-4 take the form $p' = \text{Re}\,\hat{p}(x)e^{j(\omega t - k_y y - k_z z)}$. From Eq. 4, it follows that

$$\frac{d^2\hat{p}}{dx^2} - \gamma^2\hat{p} = 0 \tag{5}$$

where

$$\gamma \equiv \sqrt{k_y^2 + k_z^2 - \frac{(\omega - k_z U)^2}{a^2}} \tag{6}$$

The program is now the same as in Sec. 2.16. With perturbation pressures at $x = \alpha$ and $x = \beta$ denoted by (p^α, p^β), the solution to Eq. 5 is

$$\hat{p}(x) = \frac{1}{\sinh \gamma\Delta}[\hat{p}^\alpha \sinh \gamma(x - \beta) - \hat{p}^\beta \sinh \gamma(x - \alpha)] \tag{7}$$

The x component of Eq. 1 then gives $\hat{v}_x$ as

$$\hat{v}_x = \frac{j}{\rho_o(\omega - k_z U)}\frac{d\hat{p}}{dx}$$

$$= \frac{j\gamma}{\rho_o(\omega - k_z U)\sinh \gamma\Delta}\{\hat{p}^\alpha \cosh \gamma(x - \beta) - \hat{p}^\beta \cosh \gamma(x - \alpha)\} \tag{8}$$

Evaluation of this expression at $x = \alpha$ gives $\hat{v}_x^\alpha(\hat{p}^\alpha, \hat{p}^\beta)$ and at $x = \beta$ gives $\hat{v}_x^\beta(\hat{p}^\alpha, \hat{p}^\beta)$. This pair of equations is then inverted to give transfer relations (c) of Table 7.9.1, but with γ as defined by Eq. 6.

Pressure-Velocity Relations for Cylindrical Annulus: The same arguments as just outlined extend the cylindrical relations of Table 7.9.1 to include acoustic phenomena. With the substitution $p' = \text{Re}\,\hat{p}(r)\exp j(\omega t - m\theta - kz)$, Eq. 4 reduces to Bessel's equation, Eq. 2.16.19, with $\Phi \to \hat{p}$ and $k^2 \to \gamma^2$ where

$$\gamma^2 \equiv k^2 - \frac{(\omega - k_z U)^2}{a^2} \tag{9}$$

Thus, solutions for $\hat{p}(r)$ take the form of Eq. 2.16.25. From the radial component of Eq. 1, $\hat{v}_r$ is then evaluated at the α and β surfaces. The resulting transfer relations are the same as Eq. (f) of Table 7.9.1 if the functions F_m and G_m are evaluated replacing $k \to \gamma$. Because γ depends on the layer properties, these functions are now designated by three arguments. For example $F_m(x,y,\gamma)$ is F_m as summarized in Table 2.16.2 with $k \to \gamma$.

7.12 Acoustic Waves, Guides and Transmission Lines

In the configuration shown in Fig. 7.12.1, fluid having a static equilibrium is confined between a rigid wall at $x = 0$ and a deformable one at $x = d + \xi$. In addition to this transverse drive, a longitudinal excitation can be imposed at $z = 0$ and an acoustic load attached at $z = \ell$. In this section it is assumed that all excitations have the same real frequency ω and that sinusoidal steady-state conditions are established.

In specific terms, the acoustic response to the transverse drive demonstrates effects of compressibility on interactions across a layer of fluid. The compressible and inertial quasistatic limits discussed in general terms in Sec. 7.22, are exemplified by this response.

The eigenmodes of the response to the transverse drive represent fluid motions between rigid plates. The structure is then a planar acoustic waveguide. In a typical guide, a source having the frequency ω excites the system at one longitudinal boundary ($z = 0$) and a load exists at another ($z = \ell$). Both source and load are often electromechanical. If the frequency is lower than cutoff frequency determined in the following, interactions between longitudinal boundaries

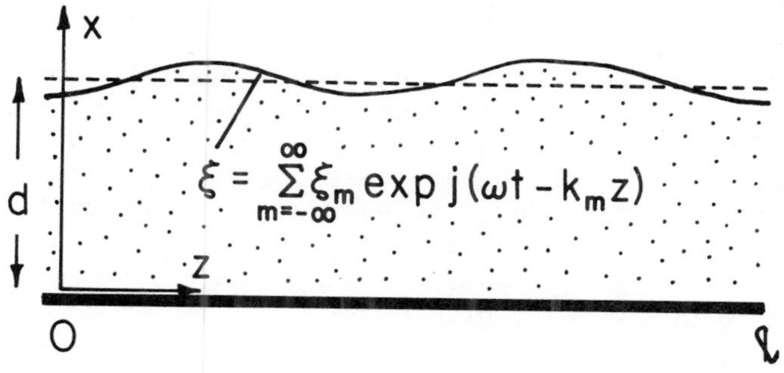

Fig. 7.12.1. Planar region is excited from transverse boundary at $x = d + \xi$. Longitudinal boundary conditions typically represent a load at $z = \ell$ and a source at $z = 0$.

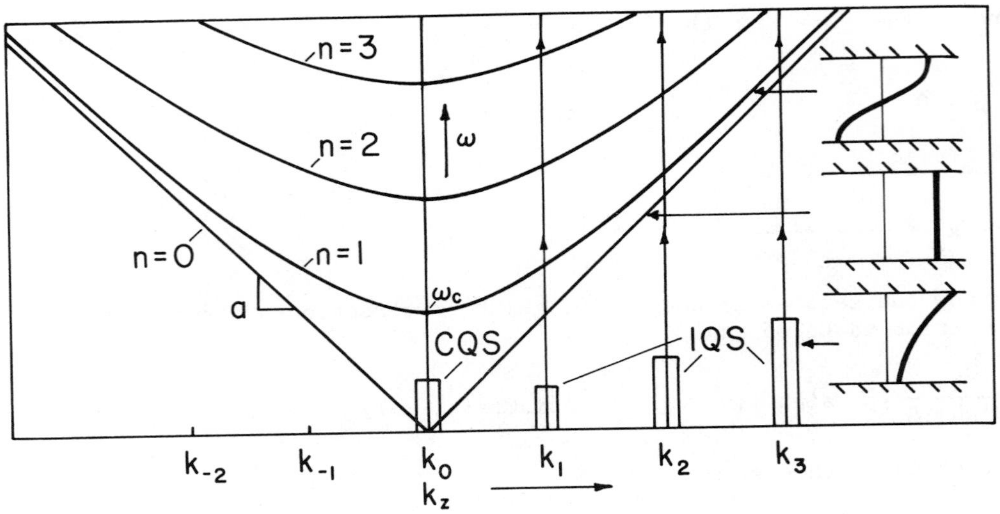

Fig. 7.12.2. Regions of ω-k_z plane characterize x dependence of response to transverse drive of each Fourier mode as driving frequency is raised.

can be represented in terms of the principal mode. This section carries the associated subject of acoustic transmission lines far enough to make clear the analogy with electromagnetic transmission lines.

Response to Transverse Drive: It follows from Eq. 7.5.5 that to linear terms the deformation of the upper boundary stipulates the velocity in the plane x = d. So, transverse boundary conditions are

$$\hat{v}_x^a = j\omega\hat{\xi}, \quad \hat{v}_x^b = 0 \tag{1}$$

Here, $\hat{\xi}$ is any one of the Fourier amplitudes, $\hat{\xi}_m$, specified in Fig. 7.12.1. It follows from Eq. (c) of Table 7.9.1 (with γ defined by Eq. 7.11.6) that the pressure amplitudes at the upper and lower boundaries are

$$\begin{bmatrix} p^a \\ p^b \end{bmatrix} = \frac{\omega^2\rho_o\hat{\xi}}{\gamma \sinh \gamma d}\begin{bmatrix} \cosh \gamma d \\ 1 \end{bmatrix} \tag{2}$$

These in turn are substituted into Eq. 7.11.7 to show that the pressure distribution over the duct cross section is

$$p = \text{Re} \sum_{m=-\infty}^{+\infty} \omega^2\rho_o \frac{\cosh \gamma_m x}{\gamma_m \sinh \gamma_m d} \hat{\xi}_m e^{j(\omega t - k_m z)} \tag{3}$$

where

$$\gamma_m = \sqrt{k_m^2 - \frac{\omega^2}{a^2}}$$

For the moment, consider that the system extends to "infinity" in the z direction, or alternatively that it closes on itself, so that the additional response from the longitudinal boundary conditions is absent. With the expression for γ_m given with Eq. 3 in view, the x dependence of each Fourier component can be pictured with the help of Fig. 7.12.2. At very low frequency, and for Fourier components other than m = 0, $\gamma_m \to k$. Thus, the x distribution is the decaying function familiar from the incompressible case. These low-frequency $m \neq 0$ components are termed the inertial (or incompressible) quasistatic (IQS) response. Note that they are the result of the part of the excitation that automatically conserves volume. The m = 0 part results from the "d-c" component of the surface displacement and so does not conserve volume. Nevertheless, at low frequencies the m = 0 component has a quasistatic nature. For this component, Eq. 3 takes the limiting form

$$\hat{p}_o \to -a^2\rho_o\left(\frac{\hat{\xi}_o}{d}\right) \tag{4}$$

At low frequencies, this compressible quasistatic (CQS) response has a pressure that is uniformly distributed over the layer cross section. It is just what would be expected as the pressure distribution

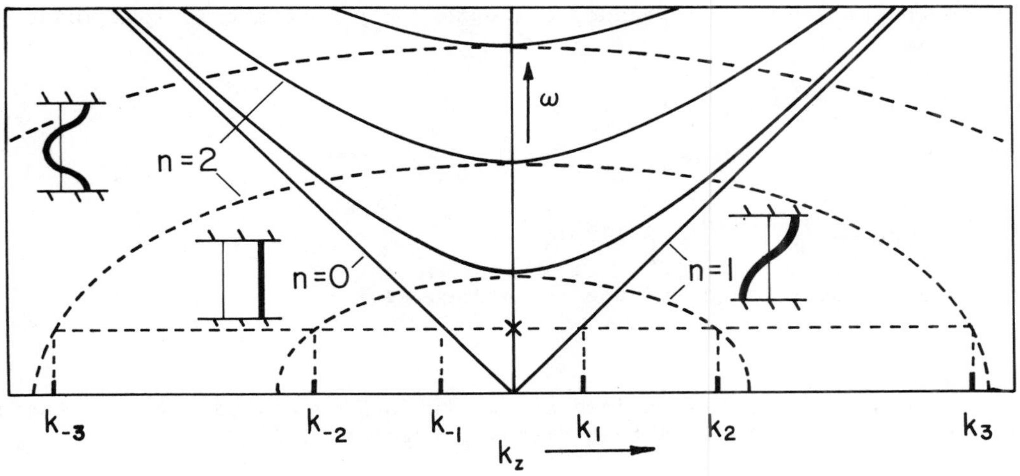

Fig. 7.12.3. Dispersion relation showing complex k_z for real ω. At the frequency shown, all but the n=0 modes are evanescent (cutoff).

in a fluid region slowly driven by vertical displacement of a horizontal piston.

As the frequency is raised, each $m \neq 0$ component takes on a uniform distribution at the frequency $|\omega| = a|k_m|$ (and hence $\gamma_m = 0$). For higher frequencies, γ_m is purely imaginary and the distribution becomes oscillatory. The curves shown in Fig. 7.12.2 are for $\gamma_m d = jn\pi$, where the frequency follows from Eq. 3 as

$$\omega = a\sqrt{k_m^2 + (\frac{n\pi}{d})^2} \tag{5}$$

and the transverse pressure distribution is n half-wavelengths. These curves also denote resonances in the driven response, as is evident from the fact that the denominator of Eq. 2 vanishes as the frequency meets the condition of Eq. 5, so that $\gamma_m d = jn\pi$.

Spatial Eigenmodes: Longitudinal conditions are satisfied by adding to the transverse driven response the eigenmodes consistent with both transverse boundaries being rigid (with $\xi = 0$). From Eq. 2,

$$\gamma d = jn\pi \tag{6}$$

where now k_z is a complex eigenvalue determined by combining Eq. 6 with the definition of γ_i

$$k_n = \pm\sqrt{\frac{\omega^2}{a^2} - (\frac{n\pi}{d})^2} \tag{7}$$

Thus, the spatial transient response to the longitudinal boundary conditions is composed of two or more propagating modes (real longitudinal wavenumbers) and an infinite number of evanescent modes. These wavenumbers are shown graphically in Fig. 7.12.3, where complex values of k_z are drawn for real values of ω. The nth mode is evanescent or cut off below the frequency.

$$\omega_c = a(\frac{n\pi}{d}) \tag{8}$$

These spatial evanescent plus propagating eigenmodes form an orthogonal set that can be used to satisfy longitudinal boundary conditions having an arbitrary dependence on x.

Acoustic Transmission Lines: The n = 0 mode has no cutoff frequency and propagates without dispersion at the velocity a, regardless of frequency. Such a mode is termed the "principal" mode. It is distinguished by having a pressure and velocity independent of x and y, and hence no transverse components of velocity anywhere. The principal mode is independent of the tube cross section. It exists in tubes of arbitrary geometry and is comprised of the same fluid motion as for a plane wave in free space. These principal modes are the most common in acoustic systems, and are conveniently pictured in terms of transmission line theory analogous to that used for TEM waves on electromagnetic transmission lines.[1]

1. P. C. Magnusson, Transmission Lines and Wave Propagation, Allyn and Bacon, Boston, Mass., 1970, pp. 57-111.

A few further steps show how impedance concepts apply to the principal mode. With the understanding that $k_1 = k$,

$$p = \mathrm{Re}\hat{p}e^{j\omega t} = \mathrm{Re}[\hat{p}^{+}e^{-jkz} + \hat{p}^{-}e^{jkz}]e^{j\omega t} \tag{9}$$

From Eq. 7.11.1 it follows that

$$v_z = \mathrm{Re}\hat{v}e^{j\omega t} = \mathrm{Re}\frac{1}{Z_o}[\hat{p}^{+}e^{-jkz} - \hat{p}^{-}e^{jkz}]e^{j\omega t} \tag{10}$$

where the characteristic acoustic impedance is defined as

$$Z_o \equiv a\rho_o \tag{11}$$

The (specific) acoustic impedance is defined as the ratio $\hat{P}/\hat{V}$, and is given by taking the ratio of complex amplitudes given by Eqs. 9 and 10, and then dividing through by $\hat{p}^{+}$:

$$Z = \frac{\hat{P}}{\hat{V}} = Z_o\left[\frac{1 + \hat{\Gamma}e^{2jkz}}{1 - \hat{\Gamma}e^{2jkz}}\right] \tag{12}$$

The reflection coefficient $\hat{\Gamma}$ has been defined as the ratio of reflected to forward wave amplitudes

$$\hat{\Gamma} = \frac{\hat{p}^{+}}{\hat{p}^{-}} \tag{13}$$

In terms of the impedance function, the analysis of a system proceeds by specifying the load impedance at $z = \ell$. For example, if there is a rigid wall at $z = \ell$, $v_z = 0$ and the impedance is infinite. Or, if the load is an absorber, then $\hat{P}/\hat{V}$ is a real number. Given the load impedance at $z = \ell$, Eq. 12 can be inverted to find the reflection coefficient Γ. Then, the impedance at any other point on the line can be determined by using Eq. 12 evaluated using the appropriate values of z and the previously determined value of reflection coefficient. The Smith chart, familiar in the theory of electromagnetic transmission lines, is a graphical representation of the calculation outlined here.

From Eq. 12, it is clear that if the reflection coefficient is to vanish, so that there is only a forward wave, then the load impedance must be Z_o. This line is then "matched." If there is no reflected wave, Z_o has the physical significance of being $\hat{P}/\hat{V}$ at any position z. Typical values of the characteristic (specific) acoustic impedance are given in Table 7.11.1. For a given velocity response, Z_o typifies the required pressure excursion. Values of Z_o in liquids are typically 3000 times greater than in gases.

7.13 Experimental Motivation for Viscous Stress Dependence on Strain Rate

Shear stress is exhibited by common fluids in motion, but not at rest. For most static fluids, the isotropic pressure of Sec. 7.4 is all that remains of the mechanical stress exerted on an element of fluid by its surroundings.

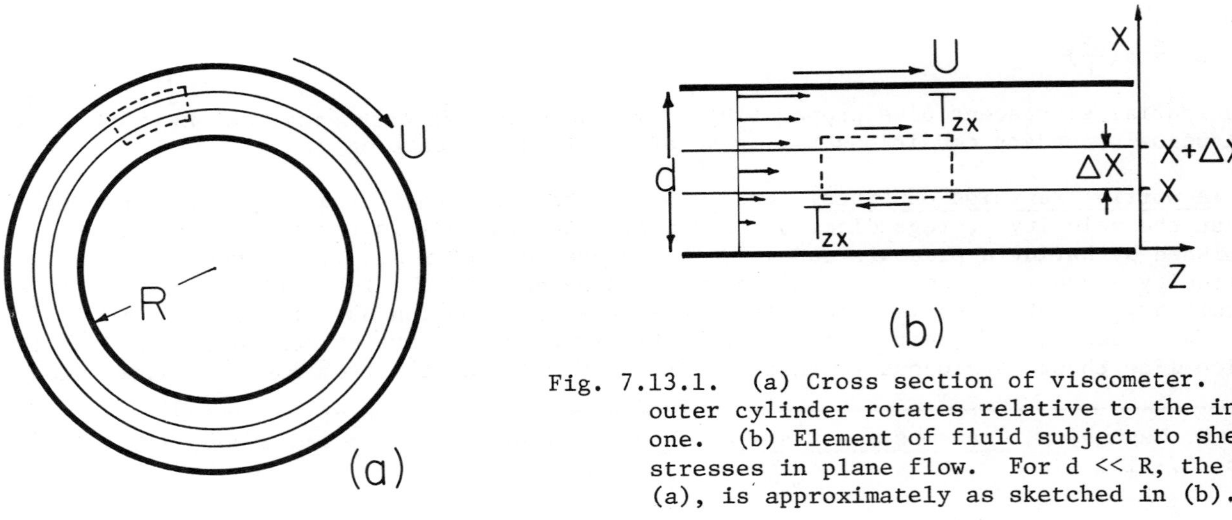

Fig. 7.13.1. (a) Cross section of viscometer. The outer cylinder rotates relative to the inner one. (b) Element of fluid subject to shear stresses in plane flow. For d << R, the flow, (a), is approximately as sketched in (b).

Typical of experiments that establish how a shear stress is transmitted across fluid layers suffering finite rate deformations is the Couette viscometer shown in Fig. 7.13.1a. The pair of concentric cylinders is arranged with the inside cylinder fixed and the outside one rotating at a constant peripheral velocity U. With the inner cylinder mounted on a torsion spring, static azimuthal deflections are a measure of the torque, and hence the shear stress, exerted by the surrounding fluid.

If the spacing d is small compared to the radius, a section of the annular region filled by fluid and bounded by the cylinders assumes the planar appearance of Fig. 7.13.1b. For common fluids, it is experimentally observed that the force per unit area, T_z, transmitted to the fixed inner plate by the moving outer one has the dependence on U and d,

$$T_z = \eta \left(\frac{U}{d}\right) \tag{1}$$

with η a constant defined as the <u>absolute viscosity</u> or the first coefficient of viscosity. Typical values of η and the <u>kinematic viscosity</u> $\nu \equiv \eta/\rho$ are given in Table 7.13.1.

Table 7.13.1. Typical viscosities of liquids and gases at 20°C and atmospheric pressure.*

	Absolute viscosity η (kg/sec m)	Mass density ρ (kg/m^3)	Kinematic viscosity ν (m^2/sec)
Water	1.002×10^{-3}	1.00×10^{3}	1.002×10^{-6}
Mercury	1.55×10^{-3}	13.6×10^{3}	1.14×10^{-7}
Heptane	0.409×10^{-3}	0.684×10^{3}	5.99×10^{-7}
Glycerin	1.49	1.26×10^{3}	1.18×10^{-3}
Carbon tetrachloride	0.969×10^{-3}	1.59×10^{3}	6.09×10^{-7}
Corn oil	5.5×10^{-2}	0.914×10^{3}	6.02×10^{-5}
Cerelow-117 alloy	$\sim 5 \times 10^{-4}$	8.8×10^{3}	$\sim 6 \times 10^{-8}$
Olive oil	0.138	0.918×10^{3}	1.51×10^{-4}
Turpentine	1.487×10^{-3}	0.87×10^{3}	1.71×10^{-6}
Air	1.83×10^{-5}	1.20	1.53×10^{-5}
Carbon dioxide	1.48×10^{-5}	1.98	7.47×10^{-6}
Hydrogen	0.87×10^{-5}	0.09	9.67×10^{-5}
Oxygen	2.02×10^{-5}	1.43	1.41×10^{-5}

*<u>Conversion</u>: η_{mks} (kg/sec m) = 0.1 η_{cgs} (Poise); Poise ≡ gm/sec cm

ν_{mks} (m^2/sec) = 10^{-4} ν_{cgs} (Stoke); Stoke ≡ cm^2/sec

Even with common fluids, at sufficiently large rotational velocities, Eq. 1 no longer holds. The planar motions are replaced by two- and three-dimensional ones, and eventually turbulence (motions that are never steady). The postulated viscometer flow is unstable at high velocities. The result is a complex flow, not the one postulated here.

The inverse dependence of T_z on d in Eq. 1 suggests that any pair of planes in the fluid are equivalent to the plates. Instead of d, the spacing is Δx, and instead of U, the relative velocity is the difference $v_z(x + \Delta x) - v_z(x)$. With T_{zx} the shear stress transmitted to the layer from the fluid above, Eq. 1 suggests that

$$T_{zx} = \eta\left[\frac{v_z(x + \Delta x) - v_z(x)}{\Delta x}\right] \tag{2}$$

The incremental layer must itself be in force equilibrium. For the incremental volume shown in Fig. 7.13.1b this means that the shear stress exerted on the layer by the fluid below is equal in magnitude to that given by Eq. 2 and that normal stresses acting in the z direction on the right and left surfaces cancel. In the viscometer, this is assured by the rotational symmetry of the flow, which excludes variations in the z direction.

In the limit $\Delta x \to 0$, Eq. 2 becomes

$$T_{zx} = \eta \frac{\partial v_z}{\partial x} \tag{3}$$

This simple but important example supports the postulate that viscous stresses are linear functions of spatial velocity derivatives.

It also illustrates the steps involved in finding the stresses on an arbitrary volume of fluid. First, the particular spatial derivatives that can reasonably give rise to mechanical stresses are defined as the components of the strain-rate tensor. Then, appeal is made to conditions of isotropy and experiments like the Couette viscometer to relate the strain-rate tensor to the stress. To carry the derivation one step further, the divergence of the viscous stress tensor finally gives the required viscous force density. These three steps are carried out in the next three sections.

7.14 Strain-Rate Tensor

Consider the difference in fluid velocity at two points separated by the incremental distance $\Delta\vec{r}$, as shown in Fig. 7.14.1. The ith component, expanded in a Taylor expansion about the position $\vec{r}$, is

$$v_i(\vec{r}+\Delta\vec{r},t)-v_i(\vec{r},t) = v_i(\vec{r},t) + \frac{\partial v_i}{\partial x_j}(\vec{r},t)\Delta x_j - v_i(\vec{r},t) \tag{1}$$

As $\Delta x_j \to 0$, all that remains in this expression is the second term, which can be written identically as

$$v_i(\vec{r}+\Delta\vec{r},t) - v_i(\vec{r},t) = \frac{1}{2}\left[\frac{\partial v_i}{\partial x_j} - \frac{\partial v_j}{\partial x_i}\right]\Delta x_j + \overset{o}{e}_{ij}\Delta x_j \tag{2}$$

where $\overset{o}{e}_{ij}$ is the strain-rate tensor, defined as

$$\overset{o}{e}_{ij} \equiv \frac{1}{2}\left(\frac{\partial v_i}{\partial x_j} + \frac{\partial v_j}{\partial x_i}\right) \tag{3}$$

Just as translational fluid motions cannot give rise to a viscous stress, neither can combinations of the spatial velocity derivatives that represent a pure rotation. Note that the first term in Eq. 2 is composed of a sum on products of Δx_j and components of the curl $\vec{v}$. Thus, it represents relative fluid motion in the neighborhood of $\vec{r}$ that is circulating about the point. This combination of spatial derivatives is not expected to be proportional to the viscous stress. Thus the strain rate, Eq. 3, is identified as that combination of the spatial derivatives that should be proportional to the stress components.

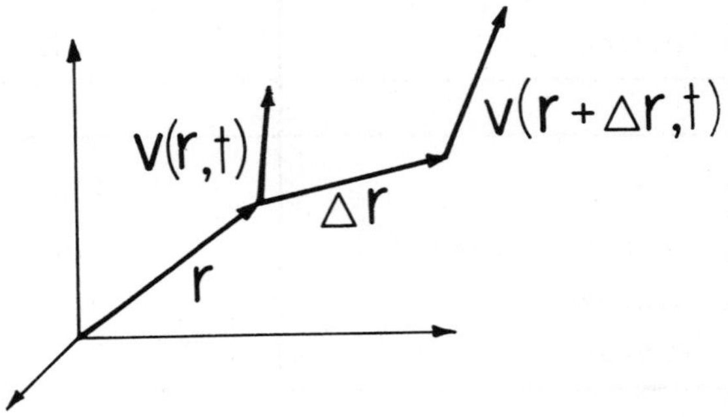

Fig. 7.14.1. Positions in the flow separated by an incremental distance $\Delta\vec{r}$.

The components of the strain rate take on physical significance if associated with the types of flow shown in Fig. 7.14.2. The diagonal components i = j represent dilatational motion, while the components i ≠ j stand for relative motions such that fluid particles located on initially perpendicular lines are found an instant later on lines at an acute angle.

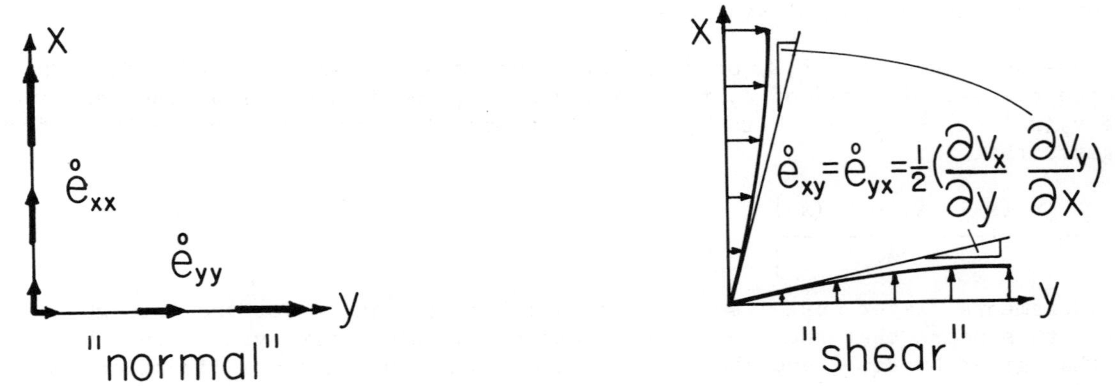

Fig. 7.14.2. Illustration of the geometric significance of "normal" and "shear" strain rates.

The viscous force density is a mechanism for introducing vorticity and hence local circulation to a flow. This point is developed in Sec. 7.18. That the viscous stress is here postulated to be independent of local rotation is a seeming contradiction. The stress tensor must be distinguished from its tensor divergence, the force density. Even though the vorticity is not linked to the local viscous stress by linear constitutive laws, its spatial rates of change are an essential part of the force density.

Fluid Deformation Example: The plane flow shown in Fig. 7.13.1b is $\vec{v} = U(x/d)\vec{i}_z$. That the flow has translational, rotational and strain-rate parts is illustrated by following the same procedure of adding and subtracting equal parts used in going from Eq. 1 to Eq. 2:

$$U\frac{x}{d}\vec{i}_z = \frac{U}{2}\vec{i}_z + \frac{U}{4}\left[\left(\frac{2x}{d} - 1\right)\vec{i}_z - \left(\frac{2z}{d}\right)\vec{i}_x\right] + \frac{U}{4}\left[\left(\frac{2x}{d} - 1\right)\vec{i}_z + \left(\frac{2z}{d}\right)\vec{i}_x\right] \quad (4)$$

The respective terms have the physical significance shown in Fig. 7.14.3.

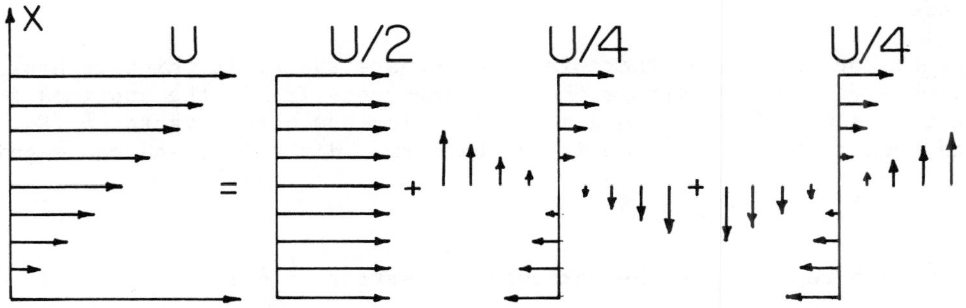

Fig. 7.14.3. Plane shear flow divided into translation, rotation and strain-rate flow.

Strain Rate as a Tensor: A discussion of the tensor character of the stress is given in Sec. 3.9. To similarly prove that e_{ij} transforms from one coordinate system to another in accordance with

$$\overset{o}{e}'_{ij} = a_{ik}a_{j\ell}\overset{o}{e}_{k\ell} \quad (5)$$

the vector nature of $\vec{v}$ is exploited:

$$v'_i = a_{ij}v_j \quad (6)$$

It follows from this relation and the definition of the direction cosines a_{ij} (Eq. 3.9.7 and discussion following Eq. 3.9.11) that

$$\frac{\partial v'_i}{\partial x'_j} = a_{ik}\frac{\partial v_k}{\partial x'_j} = a_{ik}\frac{\partial v_k}{\partial x_\ell}\frac{\partial x_\ell}{\partial x'_j} = a_{ik}a_{j\ell}\frac{\partial v_k}{\partial x_\ell} \quad (7)$$

From this expression, the definition of e'_{ij}, Eq. 3 written in the primed frame of reference, becomes

$$e'_{ij} = \frac{1}{2}\left(\frac{\partial v'_i}{\partial x'_j} + \frac{\partial v'_j}{\partial x'_i}\right) = a_{ik}a_{j\ell}\frac{1}{2}\left(\frac{\partial v_k}{\partial x_\ell} + \frac{\partial v_\ell}{\partial x_k}\right) \quad (8)$$

and Eq. 5 follows. The tensor nature of $\overset{o}{e}_{ij}$ is exploited in the next section.

7.15 Stress-Strain-Rate Relations

It is a postulate that the fluids of interest can be described by a linear relationship between viscous stress and strain rate. With $c_{ijk\ell}$ coefficients defined as properties of the fluid, the most general linear constitutive relation is

$$T_{ij} = c_{ijk\ell}\overset{o}{e}_{k\ell} \quad (1)$$

Even though these properties must be deduced in the laboratory, the number that must actually be measured can be greatly reduced by exploiting the isotropy of the fluid.

All arguments in this section pertain to relations at a given fixed location in the fluid. The coordinate systems (primed and unprimed) have a common origin at this point, as suggested by Fig. 3.9.3. The fluid is in general not necessarily homogeneous. The properties $c_{ijk\ell}$ can be functions of position.

At any given point in an isotropic material, the properties do not depend on the coordinates. Hence, in a primed frame of reference, the constitutive law of Eq. 1 is

$$T'_{ij} = c'_{ijk\ell}\overset{o}{e}{}'_{k\ell} \tag{2}$$

and isotropy requires that the properties are the same:

$$c_{ijk\ell} = c'_{ijk\ell} \tag{3}$$

For example, if shear stress and shear strain rate $(T_{ij},\overset{o}{e}_{ij})$ are related by a viscosity coefficient in one coordinate system, the same components $(T'_{ij},\overset{o}{e}{}'_{ij})$ will be related by the same coefficient in the primed frame of reference.

<u>Principal Axes</u>: For any tensor there is a coordinate system in which it has only normal components. To see this first observe that the stress, having components T_{ij} in the unprimed frame of reference, gives rise to the traction $T_i = T_{ij}n_j$ on a surface having the normal vector $\vec{n}$ (Eq. 3.9.5). Suppose that a plane is defined such that the traction is in the normal direction, and has magnitude T. Then

$$T_{ij}n_j = Tn_i = Tn_j\delta_{ij} \tag{4}$$

With the components of $\vec{n}$ regarded as the unknowns, by setting i = 1,2 and 3, this expression is three equations:

$$\begin{bmatrix} T_{11} - T & T_{12} & T_{13} \\ T_{21} & T_{22} - T & T_{23} \\ T_{31} & T_{32} & T_{33} - T \end{bmatrix}\begin{bmatrix} n_1 \\ n_2 \\ n_3 \end{bmatrix} = 0 \tag{5}$$

These homogeneous relations have a solution if the determinant of the coefficients vanishes. This condition gives three eigenvalues, $T = T_1,T_2,T_3$, which are the normal components of stress in three directions.

To actually find one of these directions, the associated eigenvalue T is inserted into Eqs. 5, a value of n_1 is assumed and any pair of the expressions then solved for n_2 and n_3. The magnitudes of these components of $\vec{n}$ are then adjusted so that $|\vec{n}| = 1$.

That the three directions found in this way are orthogonal follows from Eq. 4, which gives the traction associated with each of the eigenvalues. Suppose that the eigenvalues T_a and T_b, respectively give the normal vectors $\vec{n} = \vec{a}$ and $\vec{n} = \vec{b}$. Then, from Eq. 4

$$T_{ij}a_j = T_a a_i \tag{6}$$

$$T_{ij}b_j = T_b b_i \tag{7}$$

Multiplication of Eq. 6 by b_i and of Eq. 7 by a_i and subtraction gives

$$b_iT_{ij}a_j - a_iT_{ij}b_j = (T_a - T_b)a_ib_i \tag{8}$$

Each of the indices is summed, so they are dummy variables which can be relabeled. In the first term on the left, i and j can be interchanged. Then, so long as T_{ij} is symmetric, it is clear that the terms on the left cancel. Provided that the eigenvalues T_a and T_b are distinct, it follows that $a_ib_i = \vec{a}\cdot\vec{b} = 0$. These axes, shown here to be orthogonal, are called the principal axes.

<u>Strain-Rate Principal Axes the Same as for Stress:</u> The strain rate, like the stress, is a symmetric tensor. This is shown in Sec. 7.14. Suppose that the unprimed coordinates are the principal axes for the strain rate. Then, according to Eq. 1, the shear stress T_{yz} is

$$T_{yz} = c_{yzxx}\overset{o}{e}_{xx} + c_{yzyy}\overset{o}{e}_{yy} + c_{yzzz}\overset{o}{e}_{zz} \tag{9}$$

The axes in a primed coordinate system gotten from this one by rotating it 180° about the z axis must also be principal axes. Hence, hence Eq. 2 becomes

$$T'_{yz} = c'_{yzxx}\overset{o}{e}'_{xx} + c'_{yzyy}\overset{o}{e}'_{yy} + c'_{yzzz}\overset{o}{e}'_{zz} \tag{10}$$

Formally, the transformation of the stress and strain rate tensors between these coordinates is $T'_{ij} = a_{ik}a_{j\ell}T_{k\ell}$ (Eq. 3.9.11) and $\overset{o}{e}'_{ij} = a_{ik}a_{j\ell}\overset{o}{e}_{k\ell}$ (Eq. 7.14.6), where

$$a_{ij} = \begin{bmatrix} -1 & 0 & 0 \\ 0 & -1 & 0 \\ 0 & 0 & 1 \end{bmatrix} \tag{11}$$

so with the use of the isotropy condition, Eq. 3, Eq. 10 becomes

$$-T_{yz} = c_{yzxx}\overset{o}{e}_{xx} + c_{yzyy}\overset{o}{e}_{yy} + c_{yzzz}\overset{o}{e}_{zz} \tag{12}$$

Comparison of this expression with Eq. 9 shows that $T_{yz} = 0$. Similar arguments show that the other shear stress components are zero.

It is concluded that in a coordinate system where the strain rate has only normal components, the stress must also be normal.

Principal Coordinate Relations: That the stress and strain rate have the same principal axes effectively reduces the number of independent coefficients to nine, because in such a coordinate system (now the primed system) Eq. 1 reduces to

$$\begin{bmatrix} T'_{xx} \\ T'_{yy} \\ T'_{zz} \end{bmatrix} = \begin{bmatrix} c_{xxxx} & c_{xxyy} & c_{xxzz} \\ c_{yyxx} & c_{yyyy} & c_{yyzz} \\ c_{zzxx} & c_{zzyy} & c_{zzzz} \end{bmatrix} \begin{bmatrix} \overset{o}{e}'_{xx} \\ \overset{o}{e}'_{yy} \\ \overset{o}{e}'_{zz} \end{bmatrix} \tag{13}$$

But, the isotropy requires a further reduction in this number. For the x axis, it is clear that either e_{yy} or e_{zz} must have the same effect on T_{xx}. Hence, the first of Eqs. 13 reduces to the first of the following relations

$$\begin{bmatrix} T'_{xx} \\ T'_{yy} \\ T'_{zz} \end{bmatrix} = \begin{bmatrix} k_1 & k_2 & k_2 \\ k_2 & k_1 & k_2 \\ k_2 & k_2 & k_1 \end{bmatrix} \begin{bmatrix} \overset{o}{e}'_{xx} \\ \overset{o}{e}'_{yy} \\ \overset{o}{e}'_{zz} \end{bmatrix} \tag{14}$$

Because of the isotropy there is no distinction between the x axis and the other two. The same coefficient relates T'_{yy} to $\overset{o}{e}'_{yy}$ as relates T'_{xx} to $\overset{o}{e}'_{xx}$, for example. To complete the last step in the deduction of the stress-strain rate relations, observe that Eq. 14 can also be written as

$$\begin{aligned} T'_{xx} &= k_2\overset{o}{e}'_{nn} + (k_1 - k_2)\overset{o}{e}'_{xx} \\ T'_{yy} &= k_2\overset{o}{e}'_{nn} + (k_1 - k_2)\overset{o}{e}'_{yy} \\ T'_{zz} &= k_2\overset{o}{e}'_{nn} + (k_1 - k_2)\overset{o}{e}'_{zz} \end{aligned} \tag{15}$$

where $\overset{o}{e}_{nn} = \nabla\cdot\vec{v}$ is the same number regardless of the coordinates used in the evaluation.

Isotropic Relations: The constitutive laws expressed in the form of Eq. 15 are now transformed to the arbitrary unprimed frame by using the transformation law $T_{ij} = a_{ki}a_{\ell j}T'_{k\ell}$ (Eq. 3.9.11 and subsequent discussion):

$$\begin{aligned} T_{ij} = {} & a_{xi}a_{xj}[k_2\overset{o}{e}'_{nn} + (k_1 - k_2)\overset{o}{e}'_{xx}] \\ & a_{yi}a_{yj}[k_2\overset{o}{e}'_{nn} + (k_1 - k_2)\overset{o}{e}'_{yy}] \\ & a_{zi}a_{zj}[k_2\overset{o}{e}'_{nn} + (k_1 - k_2)\overset{o}{e}'_{zz}] \end{aligned} \tag{16}$$

Because $a_{ki}a_{kj} = \delta_{ij}$ (Eq. 3.9.14 and discussion following Eq. 3.9.11) and $\overset{o}{e}_{ij} = a_{ki}a_{\ell j}\overset{o}{e}_{k\ell}$, it follows from Eq. 16 that

$$T_{ij} = k_2\overset{o}{e}_{nn}\delta_{ij} + (k_1 - k_2)\overset{o}{e}_{ij} \tag{17}$$

To be consistent with the coefficient of viscosity defined with Eq. 7.13.3, it is observed that for that plane flow situation, all components of $\overset{o}{e}_{ij} = 0$, except $\overset{o}{e}_{zx} = \overset{o}{e}_{xz} = (\partial v_z/\partial x)/2$. Thus, Eq. 7.13.3 is $T_{zx} = 2\eta\overset{o}{e}_{zx}$, and Eq. 17 reduces to this expression if

$$k_1 - k_2 = 2\eta \tag{18}$$

By convention, a second coefficient of viscosity, λ, is defined such that

$$k_2 = \lambda - \frac{2}{3}\eta \tag{19}$$

Thus, the viscous stress-strain-rate relations for an isotropic fluid are

$$T_{ij} = (\lambda - \frac{2}{3}\eta)\delta_{ij}\overset{o}{e}_{kk} + 2\eta\overset{o}{e}_{ij} \tag{20}$$

In general, the viscosities η and λ are functions of position.

7.16 Viscous Force Density and the Navier-Stokes Equation

The total mechanical stress, S_{ij}, is the sum of the viscous stress given by Eq. 7.15.20 and the isotropic pressure stress remaining with strain rate absent (Eq. 7.4.2). In terms of the strain rate

$$S_{ij} = -p\delta_{ij} + 2\eta\overset{o}{e}_{ij} + (\lambda - \frac{2}{3}\eta)\delta_{ij}\overset{o}{e}_{kk} \tag{1}$$

while substitution for $\overset{o}{e}_{ij}$ from Eq. 7.14.3 gives

$$S_{ij} = -p\delta_{ij} + \eta(\frac{\partial v_i}{\partial x_j} + \frac{\partial v_j}{\partial x_i}) + (\lambda - \frac{2}{3}\eta)\frac{\partial v_k}{\partial x_k}\delta_{ij} \tag{2}$$

The tensor divergence of this expression (Eq. 3.9.1) is the force density required for writing the force balance equation. In taking this divergence, η and λ are for the first time taken as constants. The ith component is

$$F_i^v = \frac{\partial S_{ij}}{\partial x_j} = -\frac{\partial p}{\partial x_i} + \eta\frac{\partial^2 v_i}{\partial x_j \partial x_j} + (\lambda + \frac{1}{3}\eta)\frac{\partial}{\partial x_i}(\frac{\partial v_k}{\partial x_k}) \tag{3}$$

and translated into vector notation

$$\vec{F}^v = -\nabla p + \eta\nabla^2\vec{v} + (\lambda + \frac{1}{3}\eta)\nabla(\nabla\cdot\vec{v}) \tag{4}$$

With the use of a vector identity ($\nabla^2\vec{v} = \nabla(\nabla\cdot\vec{v}) - \nabla \times \nabla \times \vec{v}$), the essential role of vorticity becomes apparent:

$$\vec{F}^v = -\nabla p - \eta\nabla \times (\nabla \times \vec{v}) + (\lambda + \frac{4}{3}\eta)\nabla(\nabla\cdot\vec{v}) \tag{5}$$

Note that in an incompressible fluid, the last term in both Eqs. 4 and 5 vanishes.

With $\vec{F}_{ex}$ denoting the sum of all force densities other than the internal ones due to pressure and viscosity, the force equation, Eq. 7.4.4, becomes

$$\rho\frac{D\vec{v}}{Dt} + \nabla p = \vec{F}_{ex} + \eta\nabla^2\vec{v} + (\lambda + \frac{1}{3}\eta)\nabla(\nabla\cdot\vec{v}) \tag{6}$$

This form of the momentum conservation law is termed the Navier-Stokes equation.

7.17 Kinetic Energy Storage, Power Flow and Viscous Dissipation

A statement of kinetic energy conservation is made by starting with the ith component of the force equation, written using a vector identity*

$$\rho\{\frac{\partial v_i}{\partial t} + [(\nabla\times\vec{v})\times\vec{v}]_i + \frac{\partial}{\partial x_i}\frac{1}{2}v_j v_j\} = (F_{ex})_i + \frac{\partial}{\partial x_j}(S_{ij}) \tag{1}$$

Dot multiplication of this expression by $\vec{v}$ eliminates the second term on the left, and mass conservation, Eq. 7.2.3, makes it possible to manipulate the remaining inertial terms so that they take the form required for a conservation statement (for example, the form of Eq. 3.13.13):

$$\frac{\partial}{\partial t}(\frac{1}{2}\rho v_i v_i) + \frac{\partial}{\partial x_i}[(\frac{1}{2}\rho v_j v_j)v_i] = (F_{ex})_i v_i + \frac{\partial}{\partial x_j}(v_i S_{ij}) - S_{ij}\frac{\partial v_i}{\partial x_j} \tag{2}$$

The viscous stress and pressure term on the right has also been written as a perfect divergence minus what is required to make it agree with the original expression. Integration of Eq. 2 over an arbitrary volume V then results in perfect divergence terms on the left and right that, by virtue of the tensor form of Gauss' theorem, Eq. 9.6.2, can be converted to surface integrals:

$$\int_V \frac{\partial}{\partial t}\frac{1}{2}\rho\vec{v}\cdot\vec{v}dV + \oint_S(\frac{1}{2}\rho\vec{v}\cdot\vec{v})\vec{v}\cdot\vec{n}da = \int_V \vec{F}_{ex}\cdot\vec{v}dV + \oint_S v_i S_{ij}n_j da - \int_V S_{ij}\frac{\partial v_i}{\partial x_j}dV \tag{3}$$

The volume V can either be fixed in space, or be one of fixed identity. In the latter case, where the surface S moves with the material itself, what is on the left in Eq. 3 will be recognized as the rate of change with respect to time of the volume integral of the kinetic energy density $\rho\vec{v}\cdot\vec{v}/2$ (see the scalar form of the generalized Leibnitz rule, Eq. 2.6.5).

According to Eq. 3, the rate of increase of the total kinetic energy in V is equal to the rate at which the external force density does work through the volume, plus the rate at which stresses (that balance the viscous and pressure stresses) do work on the volume through the surface S, minus the last term. That this last term apparently represents a part of the input power that does not go into kinetic energy suggests that it is power leaving the kinetic energy subsystem in the form of heat (viscous dissipation) to be stored in the internal energy of the fluid. To support this interpretation, note that reindexing and then exploiting the symmetry of S_{ij} gives

$$S_{ij}\frac{\partial v_i}{\partial x_j} \equiv S_{ji}\frac{\partial v_j}{\partial x_i} = S_{ij}\frac{\partial v_j}{\partial x_i} \tag{4}$$

Thus,

$$S_{ij}\frac{\partial v_i}{\partial x_j} = S_{ij}\frac{1}{2}(\frac{\partial v_i}{\partial x_j} + \frac{\partial v_j}{\partial x_i}) = S_{ij}\mathring{e}_{ij} \tag{5}$$

With use made of Eq. 7.16.1 to write S_{ij} in terms of the strain rate, it follows from some algebraic manipulation that

$$S_{ij}\mathring{e}_{ij} = -p\nabla\cdot\vec{v} + \phi_v \tag{6}$$

where the positive definite quantity

$$\phi_v \equiv \lambda(\mathring{e}_{kk})^2 + 4\eta(\mathring{e}_{xy}^2 + \mathring{e}_{yz}^2 + \mathring{e}_{zx}^2) + \frac{2}{3}\eta[(\mathring{e}_{xx} - \mathring{e}_{yy})^2 + (\mathring{e}_{yy} - \mathring{e}_{zz})^2 + (\mathring{e}_{zz} - \mathring{e}_{xx})^2] \tag{7}$$

is identified as the viscous dissipation density. In terms of this density, the integral statement of kinetic power flow (Eq. 3) becomes the statement that the rate of doing work on the fluid is equal to the rate of increase of kinetic energy (the first two terms on the right), plus the rate of increase of energy stored internally by compressing the fluid (the third term on the right), plus the viscous dissipation:

$$\int_V \vec{F}_{ex}\cdot\vec{v}dV + \oint_S v_i S_{ij}n_j da = \int_V \frac{\partial}{\partial t}(\frac{1}{2}\rho\vec{v}\cdot\vec{v})dV + \oint_S(\frac{1}{2}\rho\vec{v}\cdot\vec{v})\vec{v}\cdot\vec{n}da - \int_V p\nabla\cdot\vec{v}dV + \int_V \phi_v dV \tag{8}$$

In general, by mechanisms such as heat conduction, some of the internal energy can be dissipated. But according to the "weak compressibility" model introduced in Sec. 7.10, dilatations result in energy

* $\vec{v}\cdot\nabla\vec{v} = (\nabla\times\vec{v})\times\vec{v} + \frac{1}{2}\nabla(\vec{v}\cdot\vec{v})$

storage. This is clarified by first using mass conservation, Eq. 7.2.3, and then using Eqs. 7.10.2 and 7.10.3 to write the compressibility energy storage term as

$$-p\nabla\cdot v = \frac{p}{\rho}\frac{D\rho}{Dt} = \frac{p}{\rho}\frac{\partial\rho}{\partial p}\frac{Dp}{Dt} \tag{9}$$

Given the constitutive law of Eq. 7.10.3, an energy density W_c can be defined:

$$W_c = \int_{p_r}^{p}\frac{p}{\rho}\frac{\partial\rho}{\partial p}dp = (p-p_r) - [a^2\rho(\alpha_1,\cdots\alpha_m,p_r)-p_r]\ln\left[1+\frac{(p-p_r)}{a^2\rho(\alpha_1,\cdots\alpha_m,p_r)}\right] \tag{10}$$

such that Eq. 9 is

$$-p\nabla\cdot\vec{v} = \frac{DW_c}{Dt} \tag{11}$$

Hence, what is added up by the volume integration of Eq. 11, called for in Eq. 8, is the time-rate of change of an energy density W_c as measured by a fluid particle of fixed identity.

7.18 Viscous Diffusion

The theme of this section is the interplay between inertial and viscous forces. Approximations underlying relations derived in Secs. 7.19 - 7.21 are established here.

Throughout, the fluid is presumed incompressible, so that

$$\nabla\cdot\vec{v} = 0 \tag{1}$$

Even more, the mass density is uniform, as is also the viscosity.

External forces are represented by scalar and vector potentials:

$$\vec{F}_{ex} = -\nabla\mathcal{E} + \nabla\times\vec{G} \tag{2}$$

and the Navier-Stokes's equation, Eq. 7.6.6 (written using 7.6.5 rather than 7.6.4), becomes

$$\rho[\frac{\partial\vec{v}}{\partial t} + (\nabla\times\vec{v})\times\vec{v}] + \nabla(\frac{1}{2}\rho\vec{v}\cdot\vec{v} + p + \mathcal{E}) = -\eta\nabla\times(\nabla\times\vec{v}) + \nabla\times\vec{G} \tag{3}$$

Convective Diffusion of Vorticity: It is shown in Sec. 7.8 that in an inviscid fluid, the net vorticity linking a surface of fixed identity is conserved. The basis for proving that this is so, the force equation written in terms of the vorticity $\vec{\omega} \equiv \nabla\times\vec{v}$ (Eq. 7.8.3), is now examined to identify viscous stresses and other rotational forces (represented by $\vec{G}$) as generators of vorticity. The curl of Eq. 3 is

$$\frac{\partial\vec{\omega}}{\partial t} + \nabla\times(\vec{\omega}\times\vec{v}) = -\frac{\eta}{\rho}\nabla\times(\nabla\times\vec{\omega}) + \frac{1}{\rho}\nabla\times(\nabla\times\vec{G}) \tag{4}$$

Without the external force, comparison of this expression to that governing magnetic diffusion in a deforming conductor (Eq. 6.2.6) shows a complete analogy. The role of the vorticity, $\vec{\omega}$, is played by the magnetic flux density. Just as the magnetic flux linking a surface of fixed identity is dissipated by joule heating, viscous losses tend to dissipate the net vorticity. This is stated formally by integrating Eq. 4 over a surface of fixed identity and exploiting the generalized Leibnitz rule for surface integrals, Eq. 2.6.4:

$$\frac{d}{dt}\int_S \vec{\omega}\cdot\vec{n}da = -\frac{\eta}{\rho}\oint_C(\nabla\times\vec{\omega})\cdot\vec{d\ell} + \frac{1}{\rho}\oint_C\nabla\times\vec{G}\cdot\vec{d\ell} \tag{5}$$

In the neighborhood of a fixed wall, for example, an inviscid fluid can slip. In a real fluid, the tangential velocity must vanish. The modification of velocity in the neighborhood of the boundary enters through the viscosity term on the right in Eq. 5 to generate vorticity.

In Chap. 6, the material deformation represented by $\vec{v}$ is given, and so the magnetic analogue to Eq. 4 is linear. In the vorticity equation, $\vec{\omega}$ really represents the unknown $\vec{v}$, and so Eq. 4 is not linear. But, two important approximations are now identified in which linear differential equations do describe flows. Because $\vec{v}$ is solenoidal, it is first convenient to represent it in terms of a vector potential, familiar from Sec. 2.18,

$$\vec{v} = \nabla\times\vec{A}_v;\quad \nabla\cdot\vec{A}_v = 0 \tag{6}$$

Substitution into Eq. 3 then gives

$$\nabla\pi + \nabla \times \vec{C} + \rho\nabla \times (\nabla \times \vec{A}_v) \times (\nabla \times \vec{A}_v) = 0 \tag{7}$$

where

$$\pi = p + \frac{1}{2}\rho\vec{v}\cdot\vec{v} + \mathcal{E}$$

$$\vec{C} = \rho\frac{\partial\vec{A}_v}{\partial t} - \eta\nabla^2\vec{A}_v - \vec{G}$$

Perturbations from Static Equilibria: In the equilibrium state, $\vec{A}_v = 0$. For incremental flows, the third term in Eq. 7, which is proportional to the product of perturbation quantities, can be ignored. The curl of the remaining terms gives a fourth order expression for $\vec{A}_v$:

$$\nabla \times \nabla \times \left[\rho\frac{\partial\vec{A}_v}{\partial t} - \eta\nabla^2\vec{A}_v - \vec{G}\right] = 0 \tag{8}$$

Given $\vec{A}_v$, and hence $\vec{C}$, π is determined by integrating the first two terms of Eq. 7 between some reference point $\vec{r}_o$ and the position $\vec{r}$ of interest,

$$\int_{\vec{r}_o}^{\vec{r}} \nabla\pi\cdot\vec{d\ell} = \pi(\vec{r}) - \pi(\vec{r}_o) = -\int_{\vec{r}_o}^{\vec{r}} \nabla \times \vec{C}\cdot\vec{d\ell} \tag{9}$$

Thus, the relation between pressure and the vector potential is

$$p = p(\vec{r}_o) + \mathcal{E}(\vec{r}_o) - \mathcal{E}(\vec{r}) - \int_{\vec{r}_o}^{\vec{r}} \nabla \times \vec{C}\cdot\vec{d\ell} \tag{10}$$

where the dynamic pressure term, $\rho\vec{v}\cdot\vec{v}/2$, is dropped from π because it is the square of a perturbation.

Equations 8 and 10 are used in Sec. 7.19 to derive general relations that are used extensively in the following chapters. Further physical insights are the objective of Sec. 7.20.

Low Reynolds Number Flows: The terms that make Eq. 7 nonlinear arise because of the inertial force density. For flows that are slow enough that viscous diffusion is complete, this force density has a negligible effect. The third term in Eq. 7 is then ignorable for a reason other than its nonlinearity. Indeed, the terms in π and $\vec{C}$ involving the mass density are also negligible.

To clarify what is meant by this "creep-flow" approximation, external forces are not considered. The Navier-Stokes's equation, Eq. 7.16.6, is written in terms of normalized variables:

$$(x,y,z) = (\underline{x},\underline{y},\underline{z})\ell,\ t = \underline{t}\tau,\ \vec{v} = \underline{\vec{v}}u,\ p = \underline{p}\frac{\eta u}{\ell} \tag{11}$$

$$\frac{\tau_v}{\tau}\frac{\partial v}{\partial t} + R_y\vec{v}\cdot\nabla\vec{v} = -\nabla p + \nabla^2\vec{v} \tag{12}$$

where

$$\tau_v \equiv \frac{\rho\ell^2}{\eta} = \text{viscous diffusion time}$$

$$R_y \equiv \frac{\rho u\ell}{\eta} = \text{Reynolds number}$$

Shear stresses set a fluid into motion in spite of its inertia at a rate typified by the viscous diffusion time. If processes of interest occur on a time scale τ that is long compared to this time, then the effect of the first inertial term in Eq. 12 is ignorable. The Reynolds number, which is the ratio of τ_v to a residence time ℓ/u, represents the importance of inertia relative to viscosity for processes that are typified by a velocity rather than a time. Examples are flows in the steady state. Alternatively, R_y typifies the ratio of an inertial force density to a viscous force density.

In the "low Reynolds number approximation," the terms on the left in Eq. 12 are neglected. This expression is equivalent to the curl of Eq. 7 without its inertial terms:

$$\nabla \times \nabla \times (\eta\nabla^2\vec{A}_v + \vec{G}) = 0 \tag{13}$$

The pressure then follows from Eq. 10 with the inertial terms omitted:

$$p = p(\vec{r}_o) + \mathcal{E}(\vec{r}_o) - \mathcal{E}(\vec{r}) + \int_{\vec{r}_o}^{\vec{r}} \nabla \times (\eta\nabla^2\vec{A}_v + \vec{G})\cdot\vec{d\ell} \tag{14}$$

Without compromise concerning the amplitude of the flow, these linear expressions are used to predict flows that are extremely viscous, that involve extremely small dimensions or that occur over long periods of time. They are applied in Secs. 7.20 and 7.21.

7.19 Perturbation Viscous Diffusion Transfer Relations

Consider small-amplitude motions in the x-y plane of a viscous fluid with no external rotational forces ($\vec{G} = 0$). Then, in Cartesian coordinates, the vector potential reduces to just the z component, with amplitude A_v, and Eq. 7.18.8 reduces to the single scalar equation

$$\nabla^2(\rho \frac{\partial A_v}{\partial t} - \eta\nabla^2 A_v) = 0 \tag{1}$$

Here, a vector identity* and the solenoidal character of A_v have been used (Eq. 7.18.6). This is the first of the four symmetric configurations summarized by Table 2.18.1 that are represented by a single component of the vector potential. The others are handled as illustrated by the Cartesian case considered now.

With the objective of obtaining relations that can be adapted to a variety of physical situations, consider the motions within a planar region having thickness Δ, as shown in Fig. 7.19.1.

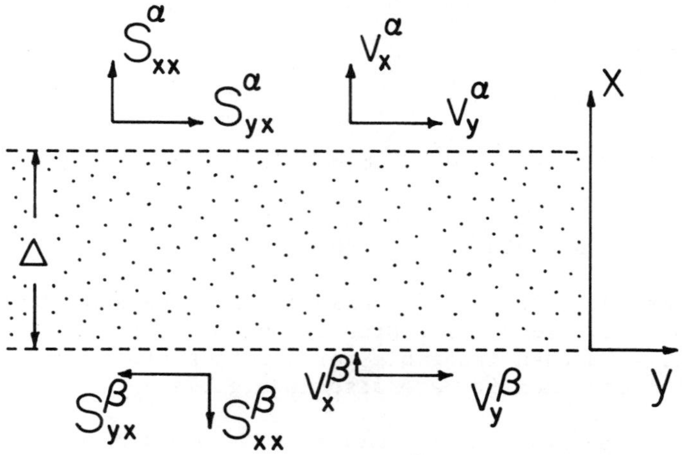

Fig. 7.19.1

Planar region filled by viscous fluid with stress components (S_{xx}, S_{yx}) and velocity components (v_x, v_y) in the α and β planes related by Eq. 13.

For perturbations having the form Re $\hat{A}_v(x)$ exp $j(\omega t - ky)$, Eq. 1 requires that the complex amplitude, $A_v(x)$, satisfy a fourth-order differential equation that has two solutions familiar from the incompressible inviscid fluid model of Sec. 7.9,

$$(\frac{d^2}{dx^2} - k^2)(\frac{d^2}{dx^2} - \gamma^2)\hat{A}_v = 0 \tag{2}$$

where

$$\gamma^2 = k^2 + j\frac{\omega\rho}{\eta}$$

The other two are solutions to the diffusion equation, familiar from magnetic diffusion as discussed in Sec. 6.5. Thus,

$$\hat{A}_v = \hat{A}_1 \sinh kx + \hat{A}_2 \sinh k(x - \Delta) + \hat{A}_3 \sinh \gamma x + \hat{A}_4 \sinh \gamma(x - \Delta) \tag{3}$$

The two lengths that typify the interactions between α and β surfaces are evident in this equation. For the first two solutions, which represent pressure attenuation across the layer, the length is $2\pi/k$. Identification of these components in Eq. 3 with the pressure follows from taking the gradient and then the divergence of Eq. 7.18.10 to show that p satisfies Laplace's equation. The last two terms bring in the second length scale, $2\pi/|\gamma|$, which is at most $2\pi/k$ and at least the viscous skin depth defined (analogous to the magnetic skin depth, Eq. 6.2.10) as

$$\delta = \sqrt{\frac{2\eta}{\omega\rho}} \tag{4}$$

*$\nabla \times \nabla \times \vec{F} = \nabla(\nabla\cdot\vec{F}) - \nabla^2 F;\ \nabla\cdot\vec{F} \equiv 0$

This length, which represents the transmission of shear stress across the layer through the action of viscosity inhibited by the fluid inertia, is shown as a function of frequency for some typical fluids in Fig. 7.19.2. The viscosity and mass density are taken from Table 7.13.1. Even with relatively modest frequencies, the viscous skin depth can be quite short.

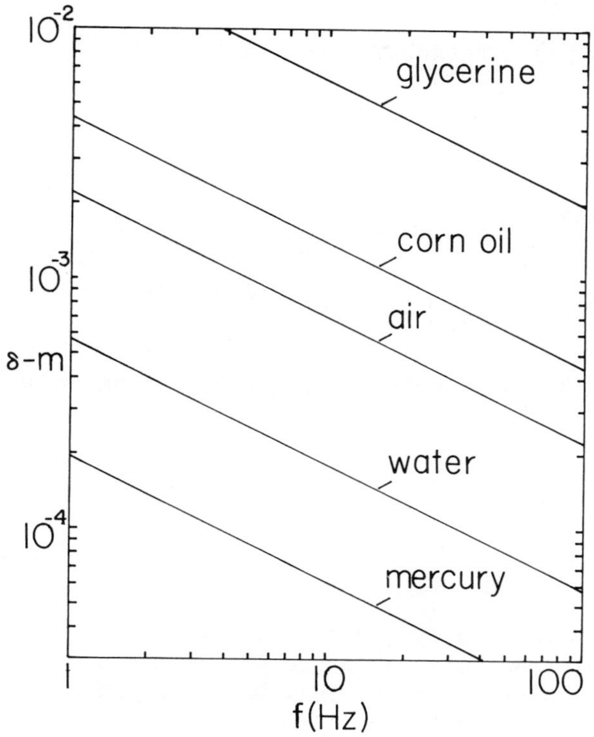

Fig. 7.19.2

Viscous skin depth as function of frequency.

In the remainder of this section, the relationships between the velocities in the α and β planes and the stress components in these planes are determined. First, this is done without further approximations. Then, the interaction between boundary layers is illustrated by taking the limit $\delta \ll \Delta$, so that the transmission of stresses across the layer is through the pressure modes alone. Finally, useful relations are derived between stress and velocity with not only $\delta \ll \Delta$, but $k\Delta \ll 1$, so that the surfaces are uncoupled.

Layer of Arbitrary Thickness: The velocity components are written in terms of the coefficients A_i by taking the curl of $\vec{A}$, Eq. 3 (Eq. (b), Table 2.18.1). Evaluated at the respective planes $x = \Delta$ and $x = 0$, these are

$$\begin{bmatrix} \hat{v}_x^\alpha \\ \hat{v}_x^\beta \\ \hat{v}_y^\alpha \\ \hat{v}_y^\beta \end{bmatrix} = \begin{bmatrix} -jk \sinh k\Delta & 0 & -jk \sinh \gamma\Delta & 0 \\ 0 & jk \sinh k\Delta & 0 & jk \sinh \gamma\Delta \\ -k \cosh k\Delta & -k & -\cosh \gamma\Delta & -\gamma \\ -k & -k \cosh k\Delta & -\gamma & -\gamma \cosh \gamma\Delta \end{bmatrix} \begin{bmatrix} \hat{A}_1 \\ \hat{A}_2 \\ \hat{A}_3 \\ \hat{A}_4 \end{bmatrix} \tag{5}$$

Inversion of these equations is the first chore in determining the transfer relations. Cramer's rule gives

$$[\hat{A}] = [M][\hat{v}] \tag{6}$$

where $[\hat{A}]$ and $[\hat{v}]$ are the column matrices and $[M]$ is the inverse of the 4 x 4 matrix appearing in Eq. 5. Even though it is the velocity and stress amplitudes that are usually used when the transfer functions represent a piece of a more complex system, the entries in M are worth saving so that the distribution with x can be reconstructed from the velocity amplitudes:

$$M_{11} = -M_{22} = jk\gamma \sinh \gamma\Delta[\gamma \sinh k\Delta \sinh \gamma\Delta + k(1 - \cosh k\Delta \cosh \gamma\Delta)]/D$$

$$M_{12} = -M_{21} = jk^2\gamma \sinh \gamma\Delta(\cosh k\Delta - \cosh \gamma\Delta)/D$$

$$M_{13} = M_{24} = k^2 \sinh \gamma\Delta(\gamma \sinh k\Delta \cosh \gamma\Delta - k \cosh k\Delta \sinh \gamma\Delta)/D$$

$$M_{14} = M_{23} = k^2 \sinh \gamma\Delta(k \sinh \gamma\Delta - \gamma \sinh k\Delta)/D \qquad (7)$$

$$M_{31} = -M_{42} = jk^2 \sinh k\Delta[\gamma(1 - \cosh \gamma\Delta \cosh k\Delta) + k \sinh \gamma\Delta \sinh k\Delta]/D$$

$$M_{32} = -M_{41} = jk^2\gamma \sinh k\Delta(\cosh \gamma\Delta - \cosh k\Delta)/D$$

$$M_{33} = M_{44} = k^2 \sinh k\Delta(k \cosh k\Delta \sinh \gamma\Delta - \gamma \sinh k\Delta \cosh \gamma\Delta)/D$$

$$M_{43} = M_{34} = k^2 \sinh k\Delta(k \sinh \gamma\Delta - \gamma \sinh k\Delta)/D$$

where

$$D \equiv k^4 \sinh k\Delta \sinh \gamma\Delta\left[\frac{2\gamma}{k}(1 - \cosh \gamma\Delta \cosh k\Delta) + \sinh k\Delta \sinh \gamma\Delta\left(\frac{\gamma^2}{k^2} + 1\right)\right]$$

The stress components are written in terms of the velocity components and the pressure using Eq. 2 (with the last term omitted because $\nabla\cdot\vec{v} = 0$):

$$\hat{S}_{xx} = -\hat{p} + 2\eta\left(\frac{d\hat{v}_x}{dx}\right) \qquad (8)$$

$$\hat{S}_{yx} = \eta\left(\frac{d\hat{v}_y}{dx} - jk\hat{v}_x\right) \qquad (9)$$

With the objective of evaluating these in terms of the A's, $(\hat{v}_x,\hat{v}_y)$, found earlier from Eq. 3, are substituted into these expressions. But, the pressure must also be expressed in terms of the A's by using Eq. 3 to evaluate Eq. 7.18.10. With p defined as Π_o at (x,y) = (0,0), the line integration results in

$$p = \Pi_o + \mathrm{Re}\ \omega\rho(\hat{A}_1 + \hat{A}_2 \cosh k\Delta)e^{j\omega t} + \mathrm{Re}\ \hat{p}(x)e^{j(\omega t-ky)} \qquad (10)$$

where the complex amplitude representing the part of p that depends on (x,y,t) is

$$\hat{p} = -\omega\rho[\hat{A}_1 \cosh kx + \hat{A}_2 \cosh k(x - \Delta)] \qquad (11)$$

Note that the definition of $\vec{C}$, Eq. 7.18.7, insures that the Laplacian solutions contribute to p, to the exclusion of the diffusion solutions.

With the stress components expressed as functions of x in terms of the A's, they are evaluated at the respective planes, to obtain

$$\begin{bmatrix} \hat{S}^\alpha_{xx} \\ \hat{S}^\beta_{xx} \\ \hat{S}^\alpha_{yx} \\ \hat{S}^\beta_{yx} \end{bmatrix} = \eta \begin{bmatrix} -j(k^2+\gamma^2)\cosh k\Delta & -j(k^2+\gamma^2) & -2j\gamma k \cosh \gamma\Delta & -2j\,\gamma k \\ -j(k^2+\gamma^2) & -j(k^2+\gamma^2)\cosh k\Delta & -2j\,\gamma k & -2j\,\gamma k \cosh \gamma\Delta \\ -2k^2 \sinh k\Delta & 0 & -(\gamma^2+k^2)\sinh \gamma\Delta & 0 \\ 0 & 2k^2 \sinh k\Delta & 0 & (\gamma^2+k^2)\sinh \gamma\Delta \end{bmatrix} \begin{bmatrix} \hat{A}_1 \\ \hat{A}_2 \\ \hat{A}_3 \\ \hat{A}_4 \end{bmatrix} \qquad (12)$$

In compact notation, this expression is equivalent to $[\hat{S}] = [N]\,[\hat{A}]$. Finally, the transfer relations are obtained by substituting Eq. 6 for the column matrix $[A]$ in Eq. 12 and performing the multiplication $[N]\,[M] \equiv \eta\,[P]$:

$$\begin{bmatrix} \hat{S}^\alpha_{xx} \\ \hat{S}^\beta_{xx} \\ \hat{S}^\alpha_{yx} \\ \hat{S}^\beta_{yx} \end{bmatrix} = \eta[P_{ij}] \begin{bmatrix} \hat{v}^\alpha_x \\ \hat{v}^\beta_x \\ \hat{v}^\alpha_y \\ \hat{v}^\beta_y \end{bmatrix} \qquad (13)$$

where

$$F = \frac{2\gamma}{k}(1 - \cosh\gamma\Delta\cosh k\Delta) + \sinh\gamma\Delta\sinh k\Delta\left[\left(\frac{\gamma}{k}\right)^2 + 1\right]$$

$$P_{11} = -P_{22} = k\left[1 - \left(\frac{\gamma}{k}\right)^2\right]\left[\frac{\gamma}{k}\cosh\gamma\Delta\sinh k\Delta - \left(\frac{\gamma}{k}\right)^2\cosh k\Delta\sinh\gamma\Delta\right]/F$$

$$P_{12} = -P_{21} = \gamma\left[1 - \left(\frac{\gamma}{k}\right)^2\right]\left[\frac{\gamma}{k}\sinh\gamma\Delta - \sinh k\Delta\right]/F$$

$$P_{13} = -P_{31} = P_{24} = -P_{42} = jk\left\{\frac{\gamma}{k}\left[3 + \left(\frac{\gamma}{k}\right)^2\right]\left[1 - \cosh k\Delta\cosh\gamma\Delta\right] + \left[1 + 3\left(\frac{\gamma}{k}\right)^2\right]\sinh k\Delta\sinh\gamma\Delta\right\}/F$$

$$P_{14} = P_{23} = P_{32} = P_{41} = jk\left(\frac{\gamma}{k}\right)\left[1 - \left(\frac{\gamma}{k}\right)^2\right](\cosh\gamma\Delta - \cosh k\Delta)/F$$

$$P_{33} = -P_{44} = k\left[\left(\frac{\gamma}{k}\right)^2 - 1\right]\left(\frac{\gamma}{k}\sinh k\Delta\cosh\gamma\Delta - \sinh\gamma\Delta\cosh k\Delta\right)/F$$

$$P_{34} = -P_{43} = k\left[\left(\frac{\gamma}{k}\right)^2 - 1\right]\left(\sinh\gamma\Delta - \frac{\gamma}{k}\sinh k\Delta\right)/F$$

These transfer relations are used to describe a variety of problems, not only of a fluid-mechanical nature, but involving electromechanical coupling that can be relegated to deformable interfaces. Examples are given in Chaps. 8 and 9.

Short Skin-Depth Limit: By way of illustrating the two lengths typifying the dynamics of the viscous layer, suppose that the viscous skin depth is small so that $\delta \ll \Delta$ and hence $|\gamma\Delta| \gg 1$, but that $k\Delta$ is arbitrary. Then, viscous diffusion is confined to boundary layers adjacent to the α and β planes. Instead of Eq. 3, solutions exploiting the approximation would conveniently take the form

$$\hat{A}_v = \hat{A}_5\sinh kx + \hat{A}_6\sinh k(x - \Delta) + \hat{A}_7 e^{-\gamma x} + \hat{A}_8 e^{\gamma(x-\Delta)} \tag{14}$$

where it is understood that γ is defined such that $\mathrm{Re}\gamma > 0$. The diffusion terms are respectively negligible when evaluated in the α and β planes. This could be exploited in simplifying a derivation of the transfer relations for this limiting case, one that parallels that begun with Eq. 3. Because the result is easily found by taking the appropriate limit of Eq. 13, it suffices to draw attention to the apparent role of the pressure in coupling one viscous boundary layer to the other. Even though the viscous skin depth is short compared to the layer thickness, the coupling between planes afforded by the pressure results in diffusion motions at one plane caused by excitations at the other. For example, the shear stress $\hat{S}_{yx}^{\alpha}$ in the α plane caused by a shearing velocity $\hat{v}_y^{\beta}$ in the β plane is proportional to ηP_{34}. From Eq. 13, even in the limit $|\gamma\Delta| \gg 1$, but $k\Delta \sim 1$,

$$P_{34} = \eta k/\sinh k\Delta \tag{15}$$

It is only in the limit $k\Delta \gg 1$, so that the pressure perturbations cannot penetrate the layer, that the shearing interactions across the layer cease.

Infinite Half-Space of Fluid: With both $|\gamma\Delta| \gg 1$ and $k\Delta \gg 1$, motions in one plane are uncoupled from those in the other. With the understanding that $\mathrm{Re}\gamma > 0$, and that upper signs refer to a lower half space bounded from above by the α plane while lower signs are for an upper half space bounded from below by the β plane, appropriate solutions to Eq. 2 are

$$A_v = \hat{A}_1 e^{\pm kx} + \hat{A}_2 e^{\pm\gamma x} \tag{16}$$

Transfer relations are determined following the same steps just outlined. First, the velocity amplitudes are written in terms of $(\hat{A}_1, \hat{A}_2)$ and then these relations are inverted to obtain

$$\begin{bmatrix}\hat{A}_1\\ \hat{A}_2\end{bmatrix} = \frac{1}{k-\gamma}\begin{bmatrix}-j\frac{\gamma}{k} & \mp 1\\ j & \pm 1\end{bmatrix}\begin{bmatrix}\hat{v}_x^{\alpha\beta}\\ \hat{v}_y^{\alpha\beta}\end{bmatrix} \tag{17}$$

In terms of the potential amplitudes, the respective stress components are

$$\begin{bmatrix} \hat{S}_{xx}^{\alpha\beta} \\ \hat{S}_{yx}^{\alpha\beta} \end{bmatrix} = \begin{bmatrix} \mp j\eta(k^2+\gamma^2) & \mp 2j\eta\gamma k \\ -2\eta k^2 & -\eta(\gamma^2+k^2) \end{bmatrix} \begin{bmatrix} \hat{A}_1 \\ \hat{A}_2 \end{bmatrix} \tag{18}$$

Finally, the transfer relations follow by combining Eqs. 17 and 18:

$$\begin{bmatrix} \hat{S}_{xx}^{\alpha\beta} \\ \hat{S}_{yx}^{\alpha\beta} \end{bmatrix} = \begin{bmatrix} \pm\eta\frac{\gamma}{k}(\gamma+k) & -j\eta(\gamma-k) \\ j\eta(\gamma-k) & \pm\eta(\gamma+k) \end{bmatrix} \begin{bmatrix} \hat{v}_x^{\alpha\beta} \\ \hat{v}_y^{\alpha\beta} \end{bmatrix} \tag{19}$$

Remember that k has been assumed positive. If a wave propagating in the -y direction is to be represented, the derivation can be repeated with $k \rightarrow -k$. For this negative traveling wave, Eq. 19 is altered by a sign reversal of the two off-diagonal terms.

7.20 Low Reynolds Number Transfer Relations

In terms defined with Eq. 7.18.12, the inertial force density is negligible compared to that due to viscosity if the viscous diffusion time is short compared to times of interest, or equivalently, if the Reynolds number is small:

$$\tau_v = \rho\ell^2/\eta \ll 1; \quad R_y = \rho u\ell/\eta \ll 1 \tag{1}$$

In this extreme, the dynamic response is a sequence of stationary states. The governing volume equation Eq. 7.18.13, is written as the biharmonic equation using a vector identity,*

$$\nabla^2(\eta\nabla^2\vec{A}_v + \vec{G}) = 0 \tag{2}$$

It involves no time rates of change. The flow is therefore an arbitrary function of time determined by boundary conditions and the external rotational force density. The flow at any instant can adjust itsel throughout the volume without the time delays associated with viscous diffusion.† A consequence is flow reversibility. For a graphic demonstration, see Reference 6, Appendix C. Moreover, so long as the conditions of Eq. 1 prevail, the amplitude of the response is also arbitrary. There is no implied linearization. Finally, because Eq. 2 is linear, a superposition of solutions is also a solution.

The vector potential reduces to a single scalar component for the configurations of Table 2.18.1. In the following subsections, two of these are considered. First, the dynamics of a planar layer is revisited and then the transfer relations for axisymmetric flows in spherical geometry are derived.

Planar Layer: With $\vec{G} = 0$ and $\vec{A}_v = \mathrm{Re}\tilde{A}_v(x,t)e^{-jky}\vec{i}_y$, Eq. 2 requires that the x dependence satisfy

$$\left(\frac{d^2}{dx^2} - k^2\right)^2 \tilde{A}_v = 0 \tag{3}$$

Formally, this is the limit $\omega\rho/\eta \ll k^2$ and hence $\gamma \rightarrow k$ of Eq. 7.19.2 (but of course the underlying approximations do not limit the solution to small amplitudes). Because the viscous and Laplacian roots of Eq. 3 have now degenerated into the same roots, two solutions are linear combinations of $\exp(\pm kx)$ and the other two are combinations of $x\exp(\pm kx)$:

$$\tilde{A}_v = \tilde{A}_1 \sinh kx + \tilde{A}_2\left(\frac{x}{\Delta}\right)\sinh kx + \tilde{A}_3 \sinh k(x-\Delta) + \tilde{A}_4\left(\frac{x}{\Delta}\right)\sinh k(x-\Delta) \tag{4}$$

By contrast with the amplitudes of Sec. 7.19, the $\tilde{A}_i$'s are arbitrary functions of time.

The outline for finding the transfer relations for the planar layer shown in Fig. 7.19.1 is now the same as illustrated in Sec. 7.19. With the caveat that the result does not have the same limitations as the viscous diffusion relations, it is possible to obtain the transfer relations as a limit of

---

*$\nabla\times\nabla\times\vec{F} = \nabla(\nabla\cdot\vec{F}) - \nabla^2\vec{F};\ \nabla\cdot F \equiv 0$

the results of Sec. 7.19 in which $\omega\rho/\eta \to 0$. As a practical matter, it is perhaps easier to repeat the derivation.

For reference, the potential amplitudes of Eq. 4 are related to the velocity amplitudes by

$$\begin{bmatrix} \tilde{A}_1 \\ \tilde{A}_2 \\ \tilde{A}_3 \\ \tilde{A}_4 \end{bmatrix} = [Q_{ij}] \begin{bmatrix} \tilde{v}_x^\alpha \\ \tilde{v}_x^\beta \\ \tilde{v}_y^\alpha \\ \tilde{v}_y^\beta \end{bmatrix} \qquad (5)$$

where

$$H = [\sinh^2(k\Delta) - (k\Delta)^2]/4k$$

$$Q_{11} = j[\sinh(k\Delta) + k\Delta\cosh(k\Delta)]/4k^2H$$

$$Q_{21} = Q_{42} = -j\Delta[\cosh(k\Delta)\sinh(k\Delta) + k\Delta]/4k\sinh(k\Delta)H$$

$$Q_{41} = -Q_{12} = Q_{22} = j\Delta[\sinh(k\Delta) + k\Delta\cosh(k\Delta)]/4k\sinh(k\Delta)H$$

$$Q_{32} = j[(k\Delta)^2 - \sinh^2(k\Delta)]/4k^2\sinh(k\Delta)H$$

$$Q_{13} = -Q_{23} = Q_{44} = \Delta\sinh k\Delta/4kH$$

$$Q_{14} = -Q_{24} = Q_{43} = \Delta^2/4H$$

$$Q_{31} = Q_{33} = Q_{34} = 0$$

The stress-velocity transfer relations are then

$$\begin{bmatrix} \tilde{S}_{xx}^\alpha \\ \tilde{S}_{xx}^\beta \\ \tilde{S}_{yx}^\alpha \\ \tilde{S}_{yx}^\beta \end{bmatrix} = \eta[P_{ij}] \begin{bmatrix} \tilde{v}_x^\alpha \\ \tilde{v}_x^\beta \\ \tilde{v}_y^\alpha \\ \tilde{v}_y^\beta \end{bmatrix} \qquad (6)$$

where

$$P_{11} = -P_{22} = [\tfrac{1}{4}\sinh(2k\Delta) + \tfrac{k\Delta}{2}]/H$$

$$P_{33} = -P_{44} = [\tfrac{1}{4}\sinh(2k\Delta) - \tfrac{k\Delta}{2}]/H$$

$$P_{21} = -P_{12} = [\tfrac{k\Delta}{2}\cosh(k\Delta) + \tfrac{1}{2}\sinh(k\Delta)]/H$$

$$P_{31} = -P_{13} = -P_{24} = P_{42} = j(k\Delta)^2/2H$$

$$P_{14} = P_{23} = P_{32} = P_{41} = -j(\tfrac{k\Delta}{2})\sinh k\Delta/H$$

$$P_{43} = -P_{34} = \tfrac{1}{2}[\sinh(k\Delta) - k\Delta\cosh(k\Delta)]/H$$

Application of these relations is illustrated in Chap. 9

<u>Axisymmetric Spherical Flows</u>: To describe motions around small particles, bubbles and the like, creep flows are now considered in spherical coordinates. The relations developed are limiting forms of those for a spherical shell. First, stress-velocity relations are obtained relating variables on

a spherical surface of radius α enclosing the region of interest. (The shell's inner radius $\beta \to 0$.) Then, they are found for an infinite region exterior to a surface of radius β ($\alpha \to \infty$).

In spherical coordinates, flows with no azimuthal dependence are described by the vector potential of Table 2.18.1:

$$\vec{A}_v = \frac{\Lambda(r,\theta)}{r \sin\theta}\vec{i}_\phi \tag{7}$$

In substituting this form into Eq. 2 (with $\vec{G} = 0$), observe that

$$\nabla^2\left(\frac{\Lambda\vec{i}_\phi}{r\sin\theta}\right) = \frac{1}{r\sin\theta}\left[\frac{\partial^2\Lambda}{\partial r^2} + \frac{\sin\theta}{r^2}\frac{\partial}{\partial\theta}\left(\frac{1}{\sin\theta}\frac{\partial\Lambda}{\partial\theta}\right)\right]\vec{i}_\phi \tag{8}$$

To evaluate Eq. 2, the vector Laplacian is now taken of this expression. Because it takes the same form as Eq. 7, with the quantity in [$\cdot$] playing the role of Λ, it follows that Eq. 2 reduces to

$$\frac{1}{r\sin\theta}\left[\frac{\partial^2}{\partial r^2} + \frac{\sin\theta}{r^2}\frac{\partial}{\partial\theta}\left(\frac{1}{\sin\theta}\frac{\partial}{\partial\theta}\right)\right]^2\Lambda = 0 \tag{9}$$

That variable separable solutions to Eq. 9 take the form

$$\Lambda = \sin\theta\, P_n^1(\cos\theta)\tilde{\Lambda}(r,t) \tag{10}$$

can be seen by observing from Eqs. 2.16.31a and 2.16.34 that the Legendre polynomial P_n^1 satisfies

$$\frac{d}{d\theta}\left[\frac{1}{\sin\theta}\frac{d}{d\theta}(P_n^1\sin\theta)\right] = \frac{1}{\sin\theta}\frac{d}{d\theta}\left(\sin\theta\frac{dP_n^1}{d\theta}\right) - \frac{P_n^1}{\sin^2\theta} = -n(n+1)P_n^1 \tag{11}$$

Hence, substitution of Eq. 10 into Eq. 9 results in a fourth-order differential equation determining the radial dependence:

$$\left[\frac{d^2}{dr^2} - \frac{n(n+1)}{r^2}\right]^2\tilde{\Lambda} = 0 \tag{12}$$

Further substitution shows that two solutions to Eq. 12 are of the form r^q, where $q = 1 + n$ and $-n$. Two more solutions follow as r^2r^q, so that the radial dependence is expressed in terms of four time-dependent amplitudes, $\tilde{\Lambda}_j$,

$$\tilde{\Lambda} = \tilde{\Lambda}_1\left(\frac{r}{R}\right)^{n+1} + \tilde{\Lambda}_2\left(\frac{r}{R}\right)^{-n} + \tilde{\Lambda}_3\left(\frac{r}{R}\right)^{2-n} + \tilde{\Lambda}_4\left(\frac{r}{R}\right)^{n+3} \tag{13}$$

The radius R will be identified with either α or β.

The velocity components are evaluated from Eq. 13 by using Eq. (k) of Table 2.18.1:

$$v_r = \tilde{v}_r\left\{\frac{1}{\sin\theta}\frac{d}{d\theta}[\sin\theta\, P_n'(\cos\theta)]\right\};\ \tilde{v}_r \equiv \frac{\tilde{\Lambda}}{r^2} \tag{14}$$

$$v_\theta = \tilde{v}_\theta\, P_n'(\cos\theta);\ \tilde{v}_\theta \equiv -\frac{1}{r}\frac{d\tilde{\Lambda}}{dr} \tag{15}$$

The θ dependences of the two components differ. For convenience, these are summarized in Table 7.20.1. The amplitudes $\tilde{v}_\theta$ and $\tilde{v}_r$ are multiplied by the respective functions from Table 7.20.1 to recover the θ dependence.

Flow within a volume enclosed by a spherical surface having radius α includes the origin. Because the velocity implied by the second and third terms in Eq. 13 is singular at the origin, these terms are excluded. Evaluation of Eqs. 14 and 15 at $r = \alpha$ then gives a pair of expressions in $(\tilde{\Lambda}_1, \tilde{\Lambda}_4)$ which can be inverted to obtain

$$\begin{bmatrix}\tilde{\Lambda}_1\\ \tilde{\Lambda}_4\end{bmatrix} = \begin{bmatrix}\frac{(n+3)\alpha^2}{2} & \frac{\alpha^2}{2}\\ \frac{-(n+1)\alpha^2}{2} & \frac{-\alpha^2}{2}\end{bmatrix}\begin{bmatrix}\tilde{v}_r^\alpha\\ \tilde{v}_\theta^\alpha\end{bmatrix} \tag{16}$$

Table 7.20.1. Angular dependence of velocity and stress functions.

n	v_θ and $S_{\theta r}$ $P_n^1(\cos\theta)$	v_r and S_{rr} $\frac{1}{\sin\theta}\frac{d}{d\theta}[\sin\theta\, P_n^1(\cos\theta)]$
1	$\sin\theta$	$2\cos\theta$
2	$3\sin\theta\cos\theta$	$3(3\cos^2\theta - 1)$
3	$\frac{3}{2}\sin\theta(5\cos^2\theta - 1)$	$15\cos\theta(1 - 2\sin^2\theta) - 3\cos\theta$
4	$\frac{5}{2}\sin\theta(7\cos^3\theta - 3\cos\theta)$	$5\cos\theta(7\cos^3\theta - 3\cos\theta) + \frac{5}{2}\sin^2\theta(3 - 21\cos^2\theta)$

For the flow in the region exterior to the surface having radius $r = \beta$, contributions to Eq. 13 that are singular as $r \to \infty$ are excluded. The n=1 mode is special, in that it represents flow that is uniform in the z direction far from $r = \beta$. Thus, the second and third terms in Eq. 13 contribute for all values of n, but the first term also contributes when n=1. For a uniform parallel flow, $\vec{v} = U\vec{i}_z$ at infinity, and it follows that for n=1, $\tilde{\Lambda}_1 = U\beta^2/2$. Two equations for $(\tilde{\Lambda}_2, \tilde{\Lambda}_3)$ are then written by evaluating Eqs. 14 and 15 at $r = \beta$. These are inverted to obtain

$$\begin{bmatrix} \tilde{\Lambda}_2 \\ \tilde{\Lambda}_3 \end{bmatrix} = \begin{bmatrix} \frac{(2-n)\beta^2}{2} & \frac{\beta^2}{2} \\ \frac{n\beta^2}{2} & \frac{-\beta^2}{2} \end{bmatrix} \begin{bmatrix} \tilde{v}_r^\beta - \frac{1}{2}U\delta_{1n} \\ \tilde{v}_\theta^\beta + U\delta_{1n} \end{bmatrix} \tag{17}$$

In spherical coordinates, the stress components are

$$S_{rr} = -p + 2\eta \frac{\partial v_r}{\partial r} \tag{18}$$

$$S_{\theta r} = \eta \left[r \frac{\partial}{\partial r}\left(\frac{v_\theta}{r}\right) + \frac{1}{r}\frac{\partial v_r}{\partial \theta} \right] \tag{19}$$

To evaluate the pressure in terms of the $\tilde{\Lambda}_i$'s, Eq. 7.18.14 is evaluated using Eq. 8. The line integration can be carried out along the ϕ,θ and finally r directions. Because the integration is only a function of the end points, it is clear that the θ-dependent part comes from the last integration. Thus,

$$\tilde{p} = \eta \int \frac{1}{r^2}\left[\frac{d^2\tilde{\Lambda}}{dr^2} - \frac{n(n+1)}{r^2}\tilde{\Lambda}\right] dr \tag{20}$$

with the θ dependence the same as for v_r. Equations 18 and 19 are now evaluated, first at $r = R = \alpha$ (the region $r < R$, where $\tilde{\Lambda}_2$ and $\tilde{\Lambda}_3 = 0$):

$$\begin{bmatrix} \tilde{S}_{rr}^\alpha \\ \tilde{S}_{\theta r}^\alpha \end{bmatrix} = \frac{\eta}{\alpha^3} \begin{bmatrix} 2(n-1) & \frac{2}{n}(n^2-n-3) \\ 2(n+1)(1-n) & -2n(n+2) \end{bmatrix} \begin{bmatrix} \tilde{\Lambda}_1 \\ \tilde{\Lambda}_4 \end{bmatrix} \tag{21}$$

and then at $r = R = \beta$ (the region $r > R$, where $\tilde{\Lambda}_1 = U\beta^2/2$, $\tilde{\Lambda}_4 = 0$):

$$\begin{bmatrix} \tilde{S}_{rr}^\beta \\ \tilde{S}_{\theta r}^\beta \end{bmatrix} = \frac{\eta}{\beta^3} \begin{bmatrix} -2(n+2) & \frac{-2(n^2+3n-1)}{n+1} \\ -2n(n+2) & 2(1-n^2) \end{bmatrix} \begin{bmatrix} \tilde{\Lambda}_2 \\ \tilde{\Lambda}_3 \end{bmatrix} \tag{22}$$

Note that the term in $\tilde{\Lambda}_1$ does not contribute to Eq. 22 because its coefficient is zero for n = 1.

To recover the θ dependence, the amplitudes $\tilde{S}_{rr}$ and $\tilde{S}_{\theta r}$ are respectively multiplied by the functions summarized in Table 7.20.1. That the θ dependence of $S_{\theta r}$ is indeed simply $P_n'(\cos\theta)$ is shown by making use of Eq. 11.

The stress-velocity transfer relations now follow by substituting Eqs. 16 and 17 respectively into Eqs. 21 and 22:

$$\begin{bmatrix} \tilde{S}^{\alpha}_{rr} \\ \\ \tilde{S}^{\alpha}_{\theta r} \end{bmatrix} = \frac{\eta}{\alpha} \begin{bmatrix} \frac{1}{n}(2n^2+n+3) & \frac{3}{n} \\ \\ 3(n+1) & (2n+1) \end{bmatrix} \begin{bmatrix} \tilde{v}^{\alpha}_{r} \\ \\ \tilde{v}^{\alpha}_{\theta} \end{bmatrix} \tag{23}$$

$$\begin{bmatrix} \tilde{S}^{\beta}_{rr} \\ \\ \tilde{S}^{\beta}_{\theta r} \end{bmatrix} = \frac{-\eta}{\beta} \begin{bmatrix} \frac{(2n^2+3n+4)}{n+1} & \frac{3}{n+1} \\ \\ 3n & (2n+1) \end{bmatrix} \begin{bmatrix} \tilde{v}^{\beta}_{r} - \frac{1}{2}U\delta_{1n} \\ \\ \tilde{v}^{\beta}_{\theta} + U\delta_{1n} \end{bmatrix} \tag{24}$$

The stress and velocity components in these relations are multiplied by the functions of θ given in Table 7.20.1 to recover the θ dependence. Application of these relations is illustrated in Sec. 7.21.

7.21 Stokes's Drag on a Rigid Sphere

Certainly the most celebrated low Reynolds number flow is that around a rigid sphere placed in what would otherwise be a uniform flow. Of particular interest is the total drag force on the sphere, found by integrating the z component of the traction, $S_{rr}\cos\theta - S_{\theta r}\sin\theta$, over its surface,

$$f_z = \int_o^{\pi} [S_{rr}\cos\theta - S_{\theta r}\sin\theta]2\pi R^2 \sin\theta d\theta \tag{1}$$

The exterior n=1 flow of Sec. 7.20 is now identified with that around the sphere. The uniform z-directed velocity far from the sphere is U. Because the sphere surface at r = R is rigid, both velocity components vanish there. In Eq. 7.20.24,

$$\tilde{v}^{\beta}_{r} = 0, \; \tilde{v}^{\beta}_{\theta} = 0 \tag{2}$$

and the stress components are

$$\begin{bmatrix} \tilde{S}^{\beta}_{rr} \\ \\ \tilde{S}^{\beta}_{\theta r} \end{bmatrix} = \frac{-\eta}{2R} \begin{bmatrix} -\frac{3U}{2} \\ \\ 3U \end{bmatrix} \tag{3}$$

Using these amplitudes, as well as the θ dependence given in Table 7.20.1, Eq. 1 becomes

$$f_z = 6\pi\eta RU \tag{4}$$

For a particle falling through a static fluid, U is the particle velocity. This "Stokes's drag" force is a good approximation, provided the Reynolds number based on the particle diameter is small compared to unity.

7.22 Lumped Parameter Thermodynamics of Highly Compressible Fluids

That additional laws are required to model highly compressible fluids is evident from the appearance of additional dependent variables in the constitutive law for the mass density. In this section, certain constitutive laws and thermodynamic relations are introduced. In Sec. 7.23 these are used to formulate integral and differential statements of energy conservation for the internal energy subsystem.

These laws are used extensively in Secs. 9.15-9.19.

Mechanical Equations of State: For a weakly compressible fluid, as defined in Sec. 7.10, the mass density is a function of pressure and parameters reflecting either the fluid composition or state. That air is buoyant when heated at constant pressure makes it evident that the mass density also depends on temperature. A commonly used mechanical constitutive law, representing a single-component perfect gas, is

$$p = \rho RT \tag{1}$$

The temperature, T, is measured in degrees Kelvin ($T_{Kelvin} = T_{centigrade} + 273.15$). The gas constant, R, is $R = R_g/M$, where $R_g = 8.31 \times 10^3$ is the universal gas constant and M is the molecular weight of the fluid. Using N_2 as an example, the molecular weight is 28, R = 297 and it follows from Eq. 1 that at atmospheric pressure ($p=1.013\times10^5$ n/m^2) and 20°C the mass density in mks units is $\rho=p/RT=1.16$ kg/m^3.

Energy Equation of State for a Perfect Gas: The specific internal energy, W_t, is defined as the energy per unit mass stored in the thermal motions of the molecules. In a perfect gas, it depends only on the temperature. Incremental changes in internal energy and temperature are related by

$$\delta W_T = c_v \delta T \tag{2}$$

and a simple constitutive law takes the specific heat at constant volume, c_v, as being constant over the temperature range of interest.

Conservation of Internal Energy in CQS Systems: There is a formal correspondence between conservation of energy statements exploited in describing lumped-parameter electromechanical coupling in Sec. 3.5 and used now for thermal-mechanical coupling in a fluid. As a reminder, suppose an EQS electromechanical subsystem having single electrical and mechanical degrees of freedom is represented electrically by a charge q at the potential v and mechanically by the displacement ξ of material subject to the force of electrical origin f. Energy conservation for a subsystem defined as being free of dissipation is expressed by

$$v\delta q = \delta w + f\delta\xi \tag{3}$$

where w is the electrical energy stored.

Now, consider the thermal lumped-parameter system exemplified by Fig. 7.22.1. The first law of thermodynamics, conservation of energy for this subsystem, states that an increment of heat, δq_T (measured in joules) goes either into increasing the energy stored, or into doing mechanical work on an external system

$$\delta q_T = \delta w_T + p\delta\upsilon \tag{4}$$

Here, w_T plays the role of w and is energy stored in kinetic (thermal) motions at the molecular level. The mechanical work done is expressed in terms of the change in the total volume, υ, and the pressure, p. That this term plays the role of the last term in Eq. 3 is seen by considering the work done by the displacement of p pistons in Fig. 7.22.1. With A_i the area of the ith piston, the net change in volume is

$$\delta\upsilon = \sum_{i=1}^{p} A_i\delta\xi_i \tag{5}$$

Because the gas is quasistatic (in the CQS sense of Sec. 7.25) the pressure exerted on each of the pistons is the same. Thus,

$$\sum_{i=1}^{p} p_i A_i \delta\xi_i = p\sum_{i=1}^{p} A_i\delta\xi_i = p\delta\upsilon \tag{6}$$

so that $p\delta\upsilon$ is indeed the mechanical work resulting from the net motions of the pistons.

Comparison of Eqs. 3 and 4 makes it natural to represent the incremental heat addition in terms of two variables. One of these, the potential, v, in the electrical analogue represented the intensity through which the heat addition is made and is the temperature, T. The other variable, defined as the entropy s, is analogous to the charge. It expresses the quantity or extent of the heat addition in units of joules/°K. With the understanding that the incremental heat addition is indeed to a "conservative system" (that the thermal input can be recovered), the statement of energy conservation, Eq. 4, becomes

$$T\delta s = \delta w_T + p\delta\upsilon \tag{7}$$

In working with a continuum, it is convenient to use extensive variables that are normalized to the mass density. This is accomplished in the lumped-parameter system now being considered by dividing Eq. 7 by the (constant) total mass of the system. Thus, Eq. 7 becomes

$$T\delta S = \delta W_T + p\delta V \tag{8}$$

where the entropy per unit mass or specific entropy is S, and the specific volume $V = 1/\rho$ will be recognized as the reciprocal mass density.

Just as it is natural to think of (q,ξ) as independent variables in Eq. 1, (S,V) are independent variables in Eq. 8. Thus, the specific thermal energy is a state function $W_T(S,V)$. The coenergy, $W'(v,\xi)$, is introduced in the electromechanical system if it is more convenient to use the potential rather than the charge as an independent variable. With a similar motivation, energy-function alternatives to W_T are often introduced.

Where p is a natural independent variable, the identity $p\delta V = \delta(pV) - V\delta p$ converts Eq. 8 to

$$T\delta S = \delta H_T - V\delta p \tag{9}$$

where the specific enthalpy, $H_T \equiv W_T + pV$, is the convenient energy function. The specific enthalpy, like W_t, is a state function. But even more, for a perfect gas it is a function only of T. This is clear from the definition of H_T, the fact that for a perfect gas, $W_T = W_T(T)$ and because (from Eq. 1) $pV = (p/\rho) = RT$.

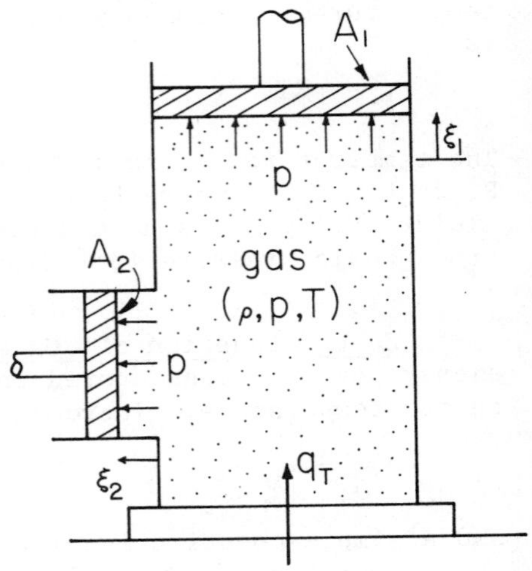

Fig. 7.22.1. Schematic view of lumped-parameter thermodynamic subsystem.

An energy equation of state equivalent to Eq. 2 can be stated in terms of the specific enthalpy

$$\delta H_T = c_p \delta T \tag{10}$$

and since the specific enthalpy is a defined function, it is not surprising that specific heat at constant pressure, c_p, is related to c_v and R. To determine this relationship, write Eq. 9 using Eq. 10:

$$T\delta S = c_p\delta T - V\delta p \tag{11}$$

Subtract Eq. 8 evaluated using Eq. 2 from this relation and it follows that

$$(c_p - c_v)\delta T = \delta(pV) = R\delta T \tag{12}$$

where the second equality comes from Eq. 1. Thus,

$$c_p - c_v = R \tag{13}$$

7.23 Internal Energy Conservation in a Highly Compressible Fluid

In a moving fluid, the thermodynamic variables are generally functions of position and time. Strictly, neither the equations of state nor the thermodynamic statement of energy conservation from Sec. 7.22 applies to media in motion. The approach now taken in regard to the state equations is simila[r] to that used in the latter part of Sec. 3.3 to broaden the application of Ohm's law to conductors in motion.

First, the laws must hold with the thermodynamic variables evaluated in the primed or moving fram[e] of reference, at least for a fluid element undergoing uniform and constant translation. Equations of state are expressed in the laboratory frame of reference by transforming variables from the primed to the unprimed frame. The thermodynamic variables of temperature, specific entropy, etc., are scalars. They are the same in both reference frames, and hence the mechanical and energy equations of state, Eqs. 7.22.1 and 7.22.2, are used even if the fluid they describe is in motion.

The seeming ubiquity of these state equations should not obscure the underlying assumption that accelerations and relative deformations of the material have negligible effect on the mechanical and energy equations of state. The notion that the fluid can be described in terms of state functions rests on there being a local equilibrium condition for the internal energy subsystem. Because processes occur at a finite rate and in an accelerating frame of reference, extension of the first law to continuum systems rests on the assumption that each element of the medium reaches this equilibrium state at each stage of the process.

A further assumption in what follows is that it is meaningful to separate the thermal and electric or magnetic subsystems. If the constitutive laws, Eqs. 7.22.1 and 7.22.2 for example, are modified by the electromagnetic fields, then this is not possible.

In this section, three subsystems are distinguished, each including dissipations. Two are the electric or magnetic and the mechanical subsystems. Each of these couples to the third, the internal energy subsystem. Given fluid of fixed identity filling a volume V enclosed by a surface S, the objective now is to write a continuum statement of internal energy conservation that makes the same physical statement as Eq. 7.22.8.

Power Conversion from Electromagnetic to Internal Form: To begin with, consider the inputs of heat to the volume. Whether the system is EQS or MQS, the electrical input of heat per unit volume is $\vec{J}'\cdot\vec{E}'$. To see this, observe from the conservation of energy statement for the electric (Eq. 2.13.16) or magnetic (Eq. 2.14.16) subsystem that the power density leaving that system is $\vec{J}_f\cdot\vec{E}$. This density either goes into the mechanical subsystem (into moving the fluid) or into the internal subsystem (into heating the fluid). Given the force densities, it is now possible to isolate the dissipation density. For an EQS system where polarization effects are negligible, the electrical dissipation must therefore be

$$\vec{J}_f\cdot\vec{E} - \rho_f\vec{E}\cdot\vec{v} = (\vec{J}_f' + \rho_f\vec{v})\cdot\vec{E}' - \rho_f\vec{E}'\cdot\vec{v} = \vec{J}_f'\cdot\vec{E}' \tag{1}$$

Here, the EQS transformation laws (Eqs. 2.5.9a, 2.5.11a and 2.5.12a) have been used. For an MQS system without magnetization, the electrical dissipation density is

$$\vec{J}_f\cdot\vec{E} - \vec{J}_f\times\mu_o\vec{H}\cdot\vec{v} = \vec{J}_f'\cdot(\vec{E}' - \vec{v}\times\mu_o\vec{H}) - \vec{J}_f'\times\mu_o\vec{H}\cdot\vec{v} = \vec{J}_f'\cdot\vec{E}' \tag{2}$$

where the MQS transformation laws (Eqs. 2.5.9b, 2.5.11b and 2.5.12b) and an identity* have been used. Hence, the electrical dissipation density makes the same appearance for EQS and MQS systems.

Power Flow Between Mechanical and Internal Subsystems: Just as the statement of energy conservation is the basis for identifying the electrical dissipation density (Eqs. 1 and 2), the kinetic energy conservation statement, Eq. 7.17.8, makes it possible to identify the last two terms in that expression as power flowing from the mechanical system into the internal energy subsystem. Because the first of these two terms has been interpreted as heat generated by mechanical dissipation, it is now written on the left of the internal energy equation. However, the second of these terms represents mechanical power input in the form of a compression of the gas, and is therefore moved to the other side of the expression (with its sign of course reversed).

Integral Internal Energy Law: The continuum version of Eq. 7.22.8 is now written as

$$\int_V \vec{E}'\cdot\vec{J}_f'dV + \int_V \phi_v dV - \oint_S \vec{\Gamma}_T'\cdot\vec{n}da = \frac{d}{dt}\int_V \rho W_T dV + \int_V p\nabla\cdot\vec{v}dV \tag{3}$$

In addition to the first two heat input terms on the left, there has been added one representing the conduction of heat across the surface S and into the volume V. Typically, the thermal heat flux, $\vec{\Gamma}_T$, is represented by a thermal conduction constitutive law, to be introduced in Sec. 10.2. On the right is the time rate of change of energy stored within the volume (which is one of fixed identity) plus the work done on the mechanical system through the expansion of the fluid.

Differential Internal Energy Law: To convert Eq. 3 to a differential statement, Gauss' theorem, Eq. 2.6.2, is used to write the surface integral as a volume integral. In addition, the generalized Leibnitz rule, Eq. 2.6.5, is used to take the time derivative inside the integral on the right. Then, conservation of mass, Eq. 7.2.3, is used to simplify that integrand. Because the volume V is arbitrary, it follows that

$$\vec{E}'\cdot\vec{J}_f' + \phi_v - \nabla\cdot\vec{\Gamma}_T' = \rho\frac{DW_T}{Dt} + p\nabla\cdot\vec{v} \tag{4}$$

Combined Internal and Mechanical Energy Laws: Especially in dealing with steady flows, it is often convenient to add the mechanical energy equation, Eq. 7.17.8, to the internal energy equation, Eq. 3:

$$\int_V \vec{E}\cdot\vec{J}_f dV + \oint_S v_i S_{ij} n_j da - \oint_S \vec{\Gamma}_T'\cdot\vec{n}da = \frac{d}{dt}\int_V \rho(W_T + \frac{1}{2}\vec{v}\cdot\vec{v})dV \tag{5}$$

Here, Eqs. 1 or 2 have been used in reverse, with $\vec{F}_{ex}$ taken as being of electric or magnetic origin. The surface integral is converted to a volume integral and the Leibnitz rule used on the right. Then, the

---

* $\vec{A}\cdot\vec{B}\times\vec{C} = \vec{A}\times\vec{B}\cdot\vec{C}$

integrands are equated to give

$$\vec{E}\cdot\vec{J}_f + \frac{\partial}{\partial x_j}(v_i S_{ij}) - \nabla\cdot\vec{I}_T' = \frac{\partial}{\partial t}\rho(W_T + \frac{1}{2}\vec{v}\cdot\vec{v}) + \nabla\cdot\rho\vec{v}(W_T + \frac{1}{2}\vec{v}\cdot\vec{v}) \tag{6}$$

Now, if the flow is steady so that $\partial(\ )/\partial t = 0$, substitution of $S_{ij} = -p\delta_{ij} + T_{ij}^v$ gives

$$\vec{E}\cdot\vec{J}_f + \frac{\partial}{\partial x_j}(v_i T_{ij}^v) - \nabla\cdot\vec{I}_T' = \nabla\cdot[\rho v(H_T + \frac{1}{2}\vec{v}\cdot\vec{v})] \tag{7}$$

where the pressure part of S_{ij} has been moved to the right and absorbed in the specific enthalpy, $H_T \equiv W_T + p/\rho$ (Eq. 7.22.9).

Entropy Flow: That the energy equation, Eq. 4, is the continuum version of Eq. 7.22.8 is made evident if it is recognized from mass conservation, Eq. 7.2.3, that

$$\nabla\cdot v = -\frac{1}{\rho}\frac{D\rho}{Dt} = \rho\frac{D}{Dt}(\frac{1}{\rho}) \tag{8}$$

Remember that the specific volume $V \equiv 1/\rho$. Thus, the right-hand side of Eq. 7.22.8 multiplied by ρ is the same as Eq. 4, provided that the variations δW_T and δV are replaced by convective derivatives of these functions. This suggests that the left side of Eq. 5 can be identified with $T\delta S$, so that Eq. 4 becomes

$$\rho T\frac{DS}{Dt} = \rho\frac{DW_T}{DT} + \rho p\frac{D}{Dt}(\frac{1}{\rho}) \tag{9}$$

For an ideal gas, it follows from the mechanical and energy equations of state, Eqs. 7.22.1 and 7.22.2, that

$$\frac{DS}{Dt} = \frac{c_v}{T}\frac{DT}{Dt} + \rho R\frac{D}{Dt}(\frac{1}{\rho}) = c_v(\frac{\rho}{p})\frac{D}{Dt}(\frac{p}{\rho}) + \rho R\frac{D}{Dt}(\frac{1}{\rho}) \tag{10}$$

Because $R = c_p - c_v$ (Eq. 7.22.13), with $\gamma \equiv c_p/c_v$, this expression becomes

$$\frac{DS}{Dt} = c_v\left(\frac{1}{p}\frac{Dp}{Dt} - \frac{\gamma}{\rho}\frac{D\rho}{Dt}\right) = \frac{D}{Dt}[c_v \ln(p\rho^{-\gamma})] \tag{11}$$

It follows that along a particle line passing through a point where the properties are $S = S_o$, $p = p_o$ and $\rho = \rho_o$, the specific entropy of a perfect gas is

$$S = S_o + c_v \ln[\frac{p}{p_o}(\frac{\rho_o}{\rho})^\gamma] \tag{12}$$

If in particular there are no heat additions to the element of fluid, so that the left side of Eq. 4 is zero, then the element of fluid sustains isentropic dynamics: $S = S_o$ and the pressure and density are related by the isentropic equation of state,

$$\frac{p}{p_o}(\frac{\rho_o}{\rho})^\gamma = 1 \Rightarrow \frac{p}{p_o} = (\frac{\rho}{\rho_o})^\gamma \tag{13}$$

For isentropic flow, Eq. 13 represents an invariant along the trajectory of a given fluid element. If the volume of gas of interest originates where the properties are uniform, then Eq. 13 is equivalent to a constitutive law relating pressure and density throughout that volume. Thus, isentropic dynamics fall within the framework of the weakly compressible dynamics considered in Secs. 7.10 - 7.12. With the understanding that it is the specific entropy that is being held constant, the acoustic velocity follows from Eqs. 7.10.4 and Eq. 13 as

$$a = \sqrt{\left(\frac{\partial p}{\partial \rho}\right)_S} = \sqrt{\gamma\frac{p_o}{\rho_o^\gamma}\rho^{\gamma-1}} = \sqrt{\gamma\frac{p}{\rho}} = \sqrt{\gamma RT} \tag{14}$$

For a perfect gas, the acoustic velocity depends only on the temperature and ratio of specific heats. Note that if the dynamics were isothermal (constant temperature) rather than isentropic, the acoustic velocity would be $a = \sqrt{RT}$. Because γ ranges between unity and two, such a velocity would always be less than that for an isentropic process.

7.24 Overview

The fluid continuum developed in this chapter is capable of storing energy in two forms, the kinetic energy associated with the fluids having inertia and internal energy associated with its compressibility. Dissipation has been represented by the Newtonian model, in which stress is linearly related to strain rate. In summary, the differential laws are the equation of state, Eq. 7.10.3,

$$\rho = \rho(\alpha_1,\cdots,\alpha_m,p_s) + \frac{p - p_s}{a^2} \tag{1}$$

conservation statements for the properties α_i, Eq. 7.10.4,

$$\frac{\partial \alpha_i}{\partial t} + \nabla\cdot(\alpha_i\vec{v}) = 0 \tag{2}$$

conservation of mass, Eq. 7.2.3, which can be combined with Eqs. 1 and 2 to give

$$\frac{1}{a^2}\left(\frac{\partial p}{\partial t} + \vec{v}\cdot\nabla p\right) = -\left(\rho - \alpha_i\frac{\partial \rho}{\partial \alpha_i}\right)\nabla\cdot\vec{v} \tag{3}$$

and conservation of momentum, Eq. 7.16.6,

$$\rho\left(\frac{\partial \vec{v}}{\partial t} + \vec{v}\cdot\nabla\vec{v}\right) + \nabla p = \vec{F}_{ex} + \eta\nabla^2\vec{v} + (\lambda + \tfrac{1}{3}\eta)\nabla(\nabla\cdot\vec{v}) \tag{4}$$

The relations and approximations which have been developed are now placed in perspective by identifying the characteristic times underlying these laws and recognizing the hierarchy of these times implicit to the various models. The discussion is to the laws of fluid mechanics what that of Sec. 2.3 is to the laws of electrodynamics.

The laws are normalized by introducing dimensionless variables,

$$(x,y,z) = \ell(\underline{x},\underline{y},\underline{z}),\quad t = \tau\underline{t},\quad \vec{v} = \vec{\underline{v}}(\ell/\tau),\quad \rho = R\underline{\rho} \tag{5}$$

With the objective a time-rate parameter expansion for the dependent variables, the pressure is given two different normalizations designed to make the zero order approximation all that is required in a wide range of physical situations. Thus, p is normalized to

reflect the dependence of density on pressure as represented by Eq. 1

$$p = a^2R\underline{p} \tag{6a}$$

and Eqs. 1-4 become

$$\underline{\rho} = \underline{\rho}(\alpha_1,\cdots,\alpha_m,p_s) + \underline{p} - \underline{p}_s \tag{7a}$$

reflect the dynamic pressure (inertia) appearing in Bernoulli's equation

$$p = \frac{R\ell^2}{\tau^2}\underline{p} \tag{6b}$$

$$\underline{\rho} = \underline{\rho}(\alpha_1,\cdots,\alpha_m,p_s) + (\underline{p} - \underline{p}_s) \tag{7b}$$

$$\frac{\partial \alpha_i}{\partial \underline{t}} + \underline{\nabla}\cdot(\alpha_i\underline{\vec{v}}) = 0 \tag{8}$$

$$\left(\frac{\partial \underline{p}}{\partial \underline{t}} + \underline{\vec{v}}\cdot\underline{\nabla}\underline{p}\right) = -\left(\underline{\rho} - \alpha_i\frac{\partial \underline{\rho}}{\partial \alpha_i}\right)\underline{\nabla}\cdot\underline{\vec{v}} \tag{9a}$$

$$\beta\underline{\rho}\left(\frac{\partial \underline{\vec{v}}}{\partial \underline{t}} + \underline{\vec{v}}\cdot\underline{\nabla}\underline{\vec{v}}\right) + \underline{\nabla}\underline{p} = \underline{\vec{F}}_{ex} + \frac{\tau_c}{\tau}\left[\underline{\nabla}^2\underline{\vec{v}} + \left(\frac{\lambda}{\eta} + \frac{1}{3}\right)\underline{\nabla}(\underline{\nabla}\cdot\underline{\vec{v}})\right] \tag{10a}$$

where

$$\underline{\vec{F}}_{ex} = \frac{\ell}{Ra^2}\vec{F}_{ex}$$

$$\tau_c = \frac{\eta}{Ra^2} \tag{11a}$$

$$\beta\left(\frac{\partial \underline{p}}{\partial \underline{t}} + \underline{\vec{v}}\cdot\underline{\nabla}\underline{p}\right) = -\left(\underline{\rho} - \alpha_i\frac{\partial \underline{\rho}}{\partial \alpha_i}\right)\underline{\nabla}\cdot\underline{\vec{v}} \tag{9b}$$

$$\underline{\rho}\left(\frac{\partial \underline{\vec{v}}}{\partial \underline{t}} + \underline{\vec{v}}\cdot\underline{\nabla}\underline{\vec{v}}\right) + \underline{\nabla}\underline{p} = \underline{\vec{F}}_{ex} + \frac{\tau}{\tau_v}\left[\underline{\nabla}^2\underline{\vec{v}} + \left(\frac{\lambda}{\eta} + \frac{1}{3}\right)\underline{\nabla}(\underline{\nabla}\cdot\underline{\vec{v}})\right] \tag{10b}$$

$$\underline{\vec{F}}_{ex} = \frac{\tau^2}{R\ell}\vec{F}_{ex}$$

$$\tau_v = \frac{R\ell^2}{\eta} \tag{11b}$$

and

$$\beta = \left(\frac{\tau_a}{\tau}\right)^2; \; \tau_a \equiv \ell/a \tag{12}$$

The time-rate parameter β is the ratio of an acoustic wave transit time, τ_a, to characteristic times of interest. The viscous dissipation brings in a second characteristic time, either τ_c or τ_v. The viscous diffusion time, τ_v, is familiar from Sec. 7.18 (Eq. 7.18.12), where its analogy to the magnetic diffusion time is discussed. The viscous relaxation time, τ_c, is analogous to the charge relaxation time. For example, both τ_c and τ_e are independent of the characteristic length. Moreover, as can be seen by substitution from Eqs. 11 and 12, the geometric mean of τ_c and τ_v is the acoustic transit time

$$\sqrt{\tau_v \tau_c} = \tau_a \tag{13}$$

The analogy to the electrodynamic relation between τ_m, τ_e and τ_{em}, Eq. 2.3.11, points to there being two quasistatic limits, each resulting because $\beta << 1$.

These can be identified by expanding the normalized dependent variables in power series in β. For example,

$$p = p_o + \beta p_1 + \beta^2 p_2 + \cdots \tag{14}$$

To zero order in β, Eqs. 7-10 become the quasistatic laws. In un-normalized form these are

Compressible quasistatic (CQS)

$$\rho = \rho(\alpha_1, \cdots, \alpha_m, p_s) + \frac{p - p_s}{a^2} \tag{15a}$$

Incompressible (inertial) quasistatic (IQS)

$$\rho = \rho(\alpha_1, \cdots, \alpha_m, p_s) \tag{15b}$$

$$\frac{\partial \alpha_i}{\partial t} + \nabla \cdot (\alpha_i \vec{v}) = 0 \tag{16}$$

CQS:

$$\frac{\partial \rho}{\partial t} + \nabla \cdot (\rho \vec{v}) = 0 \tag{17a}$$

$$\nabla p = \vec{F}_{ex} + \eta \nabla^2 \vec{v} + (\lambda + \tfrac{1}{3}\eta) \nabla (\nabla \cdot \vec{v}) \tag{18a}$$

IQS:

$$\nabla \cdot \vec{v} = 0 \tag{17b}$$

$$\rho\left(\frac{\partial \vec{v}}{\partial t} + \vec{v} \cdot \nabla \vec{v}\right) + \nabla p = \vec{F}_{ex} + \eta \nabla^2 \vec{v} \tag{18b}$$

where the ordering of characteristic times is respectively as indicated in Fig. 7.22.1.

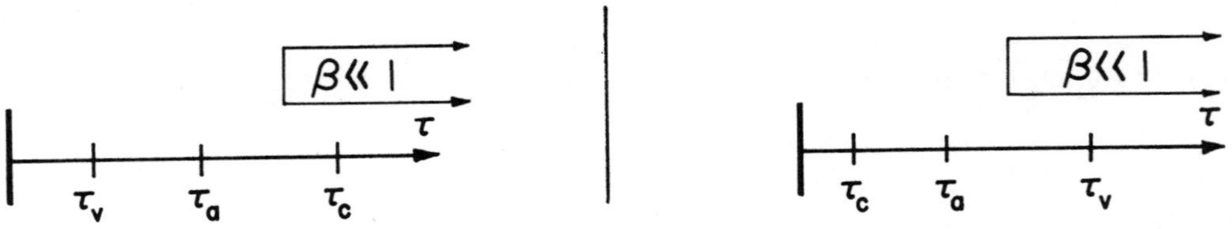

Fig. 7.22.1. Ordering of τ_a, τ_v and τ_c and domain of mechanical quasistatics.

Which of the normalized laws, Eqs. 7-10, is used is arbitrary. However, if for example the left normalization were used for a configuration in which the quasistatic motions were incompressible, the zero-order approximation would be zero, and the appropriate solution would be first-order in β. Examples in which boundary conditions clearly require the CQS limit are those where the total volume of the fluid must change, as in the slow compression of a gas in a rigid-walled vessel by a piston. The IQS and CQS limits are identified for a specific problem, without viscous dissipation, in Sec. 7.12.

Usually, it is the IQS limit that is considered when $\beta << 1$. Note that with the exception of Secs. 7.10-7.12, Eqs. 15b - 18b have received most of the attention in this chapter. The inviscid incompressible model pertains to $\tau_a << \tau << \tau_v$. The low Reynolds number limit is one in which not only is $\tau_a << \tau$, but $\tau_v << \tau$ as well.

Nature makes unlikely the CQS ordering of characteristic times. For $\tau_v/\tau_c < 1$, it is necessary that the characteristic length $\ell < \eta/Ra$. In air under standard conditions this length is a fraction of a micrometer. Because this is about the molecular mean free path, the continuum fluid model is of doubtful validity on a length scale small enough to make viscous relaxation important.

For Section 7.2:

Prob. 7.2.1 In Sec. 3.7, α_i is defined such that in the conservative subsystem, Eq. 3.7.3 holds. Show that α_i satisfies Eq. 7.2.3 with $\rho \rightarrow \alpha_i$. Further, show that if a "specific" property β_i is defined such that $\beta_i \equiv \rho\alpha_i$, then by virture of conservation of mass, the convective derivative of β_i is zero.

For Section 7.6:

Prob. 7.6.1 Show that Eq. (b) of Table 7.6.2 is correct.

Prob. 7.6.2 Show that Eqs. (j) and (ℓ) from Table 7.6.2 are correct.

Prob. 7.6.3 A pair of bubbles are formed with the tube-valve system shown in the figure. Bubble 1 is blown by closing valve V_2 and opening V_1. Then, V_1 is closed and V_2 opened so that the second bubble is filled. Each bubble can be regarded as having a constant surface tension γ. With the bubbles having the same initial radius ξ_o, when $t = 0$, both valves are opened (with the upper inlet closed off). The object of the following steps is to describe the resulting dynamics.

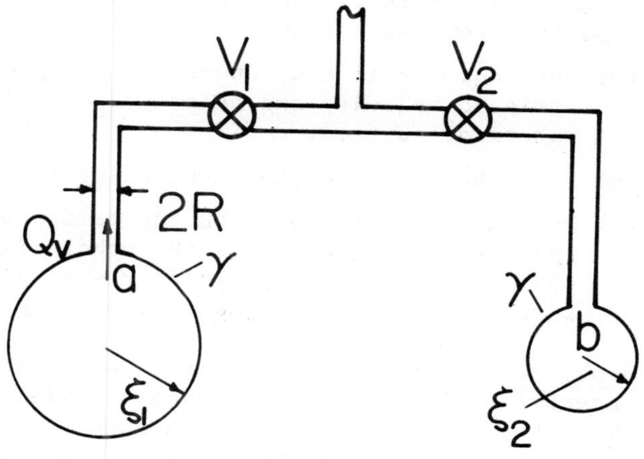

Fig. P7.6.3

(a) Flow through the tube that connects the bubbles is modeled as being fully developed and viscous dominated. Hence, for a length of tube ℓ having inner radius R and with a viscosity of the gas η, the volume rate of flow is related to the pressure difference by

$$Q_v = \frac{\pi R^4}{8\eta}\frac{(p_a - p_b)}{\ell} \qquad m^3/sec$$

The inertia of the gas and bubble is ignored, as is that of the surrounding air. Find an equation of motion for the bubble radius ξ_1.

(b) With the bubbles initially of equal radius ξ_o, there is a slight departure of the radius of one of the bubbles from equilibrium. What happens?

(c) In physical terms, explain the result of (b).

For Section 7.8:

Prob. 7.8.1 A conduit forming a closed loop consists of a pair of tubes having cross-sections with areas A_r and A_ℓ. These are arranged as shown with a fluid having density ρ_b filling the lower half and a second fluid having density ρ_a filling the upper half. The object of the following steps is to determine the dynamics of the fluid, specifically the time dependence of the interfacial positions ξ_r and ξ_ℓ.

(a) Use mass conservation to relate the displacements (ξ_r, ξ_ℓ) to each other and to the fluid velocities (v_r, v_ℓ) on the right and left respectively. Assume that the fluid is inviscid and has a uniform profile over the cross-section of a tube.

(b) Use Bernoulli's equation, Eq. 7.8.5 to relate quantities evaluated at the interfaces in the lower fluid, and in the upper fluid.

(c) Write the boundary conditions that relate quantities across the interfaces.

(d) Show that these laws combine to give an equation of motion for the right interface having the form

$$m\frac{d^2\xi_r}{dt^2} + \frac{1}{2}(\rho_b - \rho_a)\left[\left(\frac{d\xi_r}{dt}\right)^2 - \left(\frac{d\xi_\ell}{dt}\right)^2\right] + K\xi_r = 0$$

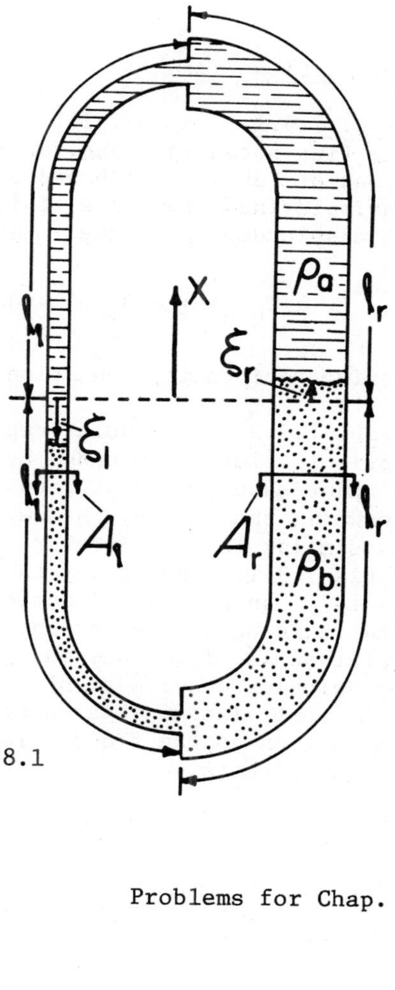

Fig. P7.8.1

Prob. 7.8.1 (continued)

What are the effective mass per unit length, m, and "spring-constant" K?

(e) Now, assume that the departures from equilibrium are small (linearize) and determine the natural frequencies of the system. Under what conditions will the system be unstable?

(f) A U tube is filled with water and open to the air. With a length of water in the tube (of uniform cross-section), ℓ, what are the natural frequencies?

Prob. 7.8.2 A hemispherical object rests on a flat plate. Fluid passes over and around the sphere with a velocity that is to be determined. The flow is uniform but a function of time far from the hemisphere.

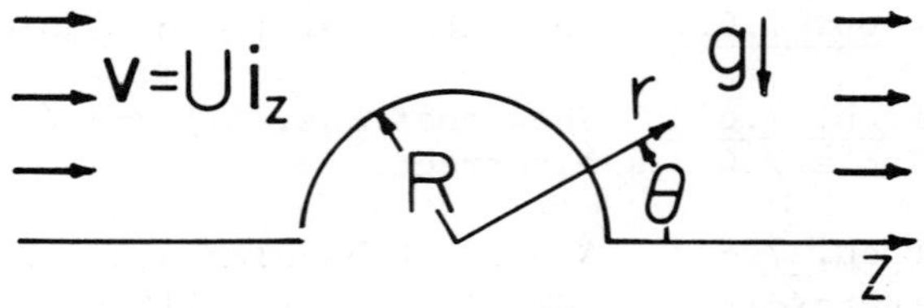

Fig. P7.8.2

(a) Note Eq. 7.8.11 and subsequent discussion. Find the inviscid velocity and velocity potential on the hemispherical surface.

(b) Find the pressure distribution on the hemisphere.

(c) What is the lift force on the hemisphere? (Assume that the pressure inside the sphere is the same as that at the stagnation point $r = R$, $\theta = \pi$ just outside the sphere, as would be the case if there were a small hole through the shell at this point.)

Prob. 7.8.3 An electromagnetic rocket constrained by a test stand is shown in the figure. In the interior region there is a space occupied by an apparatus that produces a normal surface force density T_n on the surface S_i. A tube connects this space to the outside, and hence equalizes the pressures inside S_i and outside the rocket. The fluid inside S_i and outside the rocket has negligible mass density. There are no external forces in the fluid bulk. Thus the pressure in the surrounding homogeneous fluid is $p=T_n$. The volume is large enough that the fluid inside the rocket has negligible velocity and an essentially steady flow condition prevails. It is expelled through the throat and reaches a point where its velocity U is essentially uniform and x directed; the pressure is equal to that of the surroundings (say p=0) and the cross-sectional area is A. Gravitational effects are negligible. Use Eqs. 7.8.5, 7.3.2 and 7.4.3 to find the total force on the rocket in terms of T_n and A.

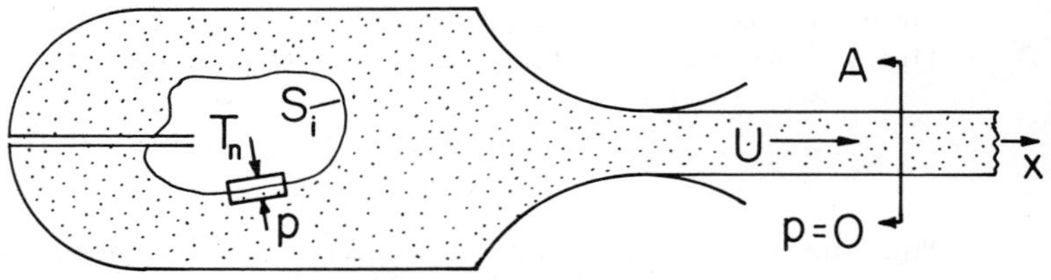

Fig. 7.8.3

For Section 7.9:

Prob. 7.9.1 In Sec. 2.17, conservation of electric energy is used to derive reciprocity conditions for the flux-potential transfer relations. The object here is a similar derivation for the transfer relations of Table 7.9.1 based on conservation of kinetic energy. Start with the assumption that for an inviscid incompressible fluid having uniform mass density, the change in kinetic energy is the result of displacements at the α and β planes.

$$\delta(W_{kin}) = -\oint_S p\delta\vec{\xi}\cdot\vec{n}da$$

Derive reciprocity conditions similar to Eq. 2.17.10.

Prob. 7.9.2 An annular region of incompressible inviscid fluid is bounded by outer and inner coaxial boundaries of radius α and β respectively, as shown in the figure. Hence, the configuration is similar to the circular cylindrical case of Table 7.9.1. However, rather than being in a state of uniform axial motion when in equilibrium, the fluid here is rotating. This equilibrium rotation is rigid body and could be established by spinning a cylinder of fluid for a long enough time that viscous shear stresses could transmit the motion to the fluid volume. Ignore gravitational effects.

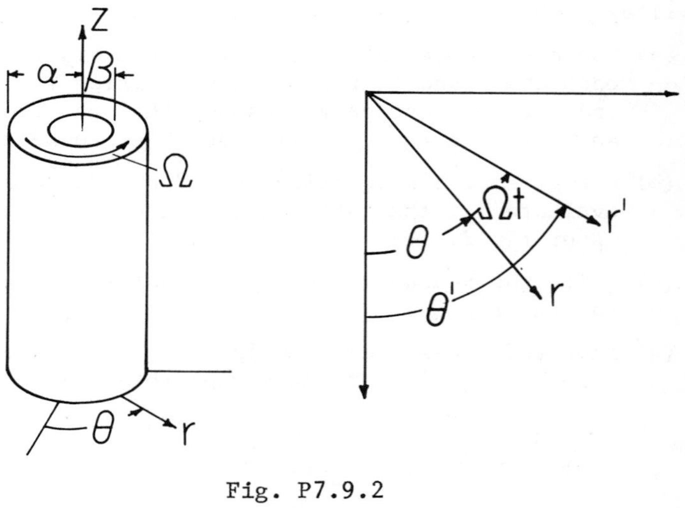

Fig. P7.9.2

Prob. 7.9.2 (continued)

(a) What is the vorticity of the equilibrium motion?

(b) Show that the equilibrium pressure distribution is

$$P_o = \frac{1}{2}\rho^2 r^2 + \Pi$$

(c) Write the perturbation continuity and force equations. Transform these expressions from the laboratory frame (primed) to a rotating frame (unprimed) where

$$r = r' \qquad v_r = v_r'$$

$$\theta = \theta' - \Omega t \qquad v_\theta = v_\theta' - \Omega r$$

$$z = z' \qquad p = p'$$

$$t = t'$$

and show that the perturbation equations are

$$\frac{1}{r}\frac{\partial(rv_r)}{\partial r} + \frac{1}{r}\frac{\partial v_\theta}{\partial \theta} + \frac{\partial v_v}{\partial z} = 0$$

$$\rho\left(\frac{\partial v_r}{\partial t} - 2\Omega v_\theta\right) + \frac{\partial p}{\partial r} = 0$$

$$\rho\left(\frac{\partial v_\theta}{\partial t} + 2\Omega v_r\right) + \frac{1}{r}\frac{\partial p}{\partial \theta} = 0$$

$$\rho\left(\frac{\partial v_z}{\partial t}\right) + \frac{\partial p}{\partial z} = 0$$

(d) Show that the pressure complex amplitude satisfies the equation

$$r^2\frac{d^2\hat{p}}{dr^2} + r\frac{d\hat{p}}{dr} - \hat{p}\left[m^2 + r^2k^2\left(1 - \frac{4\Omega^2}{\omega^2}\right)\right] = 0$$

where ω is the frequency in the rotating frame of reference.

(e) Show that transfer relations are

$$\begin{bmatrix}\hat{p}^\alpha \\ \hat{p}^\beta\end{bmatrix} = \frac{\rho(4\Omega^2-\omega^2)}{j\omega D}\begin{bmatrix}\left[f_m(\alpha,\beta,\gamma) + \frac{2\Omega m}{\omega\beta}\right] & -g_m(\alpha,\beta,\gamma) \\ -g_m(\beta,\alpha,\gamma) & \left[f_m(\beta,\alpha,\gamma) + \frac{2\Omega m}{\omega\alpha}\right]\end{bmatrix}\begin{bmatrix}\hat{v}_r^\alpha \\ \hat{v}_r^\beta\end{bmatrix}$$

where

$$\gamma^2 \equiv k^2\left(1 - \frac{4\Omega^2}{\omega^2}\right);$$

$$D \equiv \left[f_m(\beta,\alpha,\gamma) - \frac{m\omega}{2\Omega\alpha}\right]\left[f_m(\alpha,\beta,\gamma) - \frac{m\omega}{2\Omega\beta}\right] - g_m(\beta,\alpha,\gamma)\,g_m(\alpha,\beta,\gamma)$$

For Section 7.11:

Prob. 7.11.1 Determine the transfer relations for the spherical shell of Table 7.9.1. Define functions F_n and G_n such that these relations take the form of Eq. (i) of that table. Describe the temporal modes of a gas surrounded at a radius r = R by a rigid boundary.

For Section 7.12:

Prob. 7.12.1 Gas flows with a uniform velocity U in the z direction through a rigid tube having inner radius R.

(a) Determine the dispersion equation for acoustic waves propagating in the z direction with pressure dependence of the form $p = \text{Re}\hat{p}(r)\exp j(\omega t - m\theta - kz)$.

(b) What are the wavenumbers of the spatial modes? Sketch the dispersion equation for a < U and for a > U.

Prob. 7.12.2 An acoustic waveguide consists of rigid plane parallel walls in y-z planes having the spacing (a+b) bounding planar layers of fluid respectively having the thicknesses (a) and (b), mass densities ρ_a and ρ_b and acoustic velocities a_a and a_b. Ignore gravity and surface tension effects at the interface. Determine the dispersion equation for waves propagating in the z direction between the walls. This expression is transcendental, and hence requires numerical solution. Consider two limits in which explicit expressions can be derived.

(a) The waves are very long, so that $\gamma_a a \ll 1$ and $\gamma_b b \ll 1$. What is the wave velocity for the resulting "principal" mode?

(b) Here, $a_a \ll a_b$ (for example, air and water) and $k^2 \gg \omega^2/a_b^2$. Use a graphical solution to find the wavenumbers of the approximate spatial modes.

For Section 7.13:

Prob. 7.13.1 The equations describing the incremental motions of a perfectly elastic isotropic solid can be developed in steps that follow those for a Newtonian fluid. The first problem following each of Secs. 7.13.1 through 7.16.1 is a step in developing these equations, which are summarized for reference in the table of Prob. 7.16.1.

(a) It is natural to use the displacement $\vec{\xi}(\vec{r}_o,t)$, rather than the velocity $\vec{v}(\vec{r},t)$, as a variable. By contrast with the Eulerian variable, the displacement is a Lagrangian variable in that a material particle originally at $\vec{r}_o$ is found at the position $r_o + \vec{\xi}(\vec{r}_o,t)$ (see Sec. 2.4). Show that to linear terms $\vec{\xi}(\vec{r}_o,t) \approx \vec{\xi}(\vec{r}_o+\vec{\xi},t)$, so that for incremental displacements ξ can be regarded as either an Eulerian or Lagrangian variable.

(b) The annulus of Fig. 7.13.1 is filled with an elastic solid rigidly attached to the walls. Instead of being given a steady velocity U, the boundary is given a steady displacement Ξ_z. It is found that $T_z = G_s(\Xi_z/d)$, where the coefficient G_s is the shear modulus. What is the elasticity version of Eq. 7.13.3?

(c) A thin rod of initial length ℓ is fixed at x=0 and subjected to a surface force density T_x at its end (where x=ℓ originally). Because the rod is "thin," the transverse stresses are negligible compared to T_{xx}. It is found that $T_x = E_s \Xi_x/\ell$, where the coefficient E_s is the elastic modulus. Write an equation expressing force equilibrium for an incremental length Δx of the rod, and obtain the analogue of Eq. 7.13.3 for dilatational deformations.

Typical values of G_s and E_s are given in Table P7.13.1.

Table P7.13.1. Elastic properties of various materials.

Material	Shear Modulus G_s (N/m^2)	Elastic Modulus E_s (N/m^2)	Poisson's Ratio ν_s	Mass Density kg/m^3
Aluminum	$2.6\text{x}10^{10}$	$7.3\text{x}10^{10}$	0.33	$2.8\text{x}10^3$
Steel	$7.8\text{x}10^{10}$	$2.1\text{x}10^{11}$	0.27	$7.8\text{x}10^3$
Rubber	$\sim .6\text{x}10^6$	$\sim 2\text{x}10^6$	~ 0.50	$1.1\text{x}10^3$

For Section 7.14:

Prob. 7.14.1 Define the strain tensor e_{ij} for incremental deformations, using arguments paralleling those from Eq. 7.14.1 to 7.14.3. The result should be Eq. (a) of Table P7.16.1. Geometrically interpret the shear and normal components of e_{ij}.

For Section 7.15:

Prob. 7.15.1 Starting with the assumption that the stress-strain constitutive laws for an isotropic perfectly elastic solid take the form

$$T_{ij} = c_{ijk\ell}e_{k\ell}$$

show that Eq. (b) of Table P7.16.1 is the desired relation with G_s as defined in Prob. 7.13.1 and λ_s, a second property of the material. Remember that in the thin-rod experiment of Prob. 7.13.1, the transverse stress components were essentially zero. Use this fact to show that G_s and E_s are related to λ_s by Eq. f of Table P7.16.1. In terms of the thin-rod experiment, Poisson's ratio ν_s is defined as the negative of the strain ratio e_{yy}/e_{xx} or e_{zz}/e_{xx}. Show that this property is related to G_s and E_s by Eq. (g) of the table.

Prob. 7.15.2 Following Eq. 7.15.15, it is argued that $\overset{o}{e}_{nn}$ is invariant under a transformation between coordinate systems. Confirm this by using the transformation properties of $\overset{o}{e}_{ij}$ (note Eq. 3.9.14).

Prob. 7.15.3 For the velocity distribution of Eq. 7.14.4

(a) What is S_{ij}?

(b) Use Eq. 7.15.5 to find the principal axes and the associated normal stresses.

For Section 7.16:

Prob. 7.16.1 The relations required to write the force equation representing an isotropic perfectly elastic solid, as derived in Probs. 7.13.1, 7.14.1 and 7.15.1 are summarized in Table P7.16.1. Show that the force equation can be written as Eq. (d) and, hence, as Eq. (e). (Note the discussion in Sec. 2.4.)

Table P7.16.1. Definitions, relations and equations of motion for an isotropic perfectly elastic solid.

Strain-displacement	$e_{ij} = \frac{1}{2}\left(\frac{\partial\xi_i}{\partial x_j} + \frac{\partial\xi_j}{\partial x_i}\right)$	(a)
Stress-strain	$T_{ij} = 2G_s e_{ij} + \lambda_s\delta_{ij}e_{kk}$	(b)
	$e_{ij} = \frac{1}{2G_s}T_{ij} - \frac{\nu_s}{E_s}\delta_{ij}T_{kk}$	(c)
Force equations	$\rho\frac{\partial^2\xi_i}{\partial t^2} = \frac{\partial T_{ij}}{\partial x_j} + (F_{ex})_i$	(d)
	$\rho\frac{\partial^2\bar{\xi}}{\partial t^2} = (2G_s+\lambda_s)\nabla(\nabla\cdot\bar{\xi}) - G_s\nabla\times(\nabla\times\bar{\xi})+\bar{F}_{ex}$	(e)
Constitutive relations	$\lambda_s = (E_s-2G_s)/[3-(E_s/G_s)]$	(f)
	$\nu_s = (E_s/2G_s) - 1$	(g)

For Section 7.18:

Prob. 7.18.1 The equation of motion for a perfectly elastic isotropic solid is Eq. (e) of Table P7.16.1. The external force density is represented by Eq. 7.18.2, while the displacement is represented in terms of vector and scaler potentials

$$\vec{\xi} = \nabla \times \vec{A}_s - \nabla \psi_s \; ; \; \nabla \cdot \vec{A}_s = 0$$

Show that $\vec{A}_s$ and ψ_s respectively represent rotational and dilational deformations. Show that A_s and ψ_s respectively satisfy wave equations with the wave velocities $v_s = \sqrt{G_s/\rho}$ and $v_c = \sqrt{(2G_s+\lambda_s)/\rho}$.

Prob. 7.18.2 In Sec. 7.21, it is shown that the viscous drag force on a rigid sphere having radius R moving through a fluid with velocity U is $6\pi\eta RU$. A spherical particle has mass density ρ_p much greater than that of the surrounding fluid so that the mass of the fluid can be ignored (the Reynolds number is low, as it must be for the Stokes drag force to be correct). Write the force equation for the slowing of the particle from some initial velocity. Show that the velocity decreases exponentially, with a time constant 2/9 of the viscous "diffusion" time based on the particle radius and density and the fluid viscosity.

For Section 7.19:

Prob. 7.19.1 An incompressible elastic solid is one in which deformations are solenoidal. It can be pictured as having $\nabla \cdot \vec{\xi} \to 0$ and $2G_s+\lambda_s \to \infty$ in such a way that the product $(2G_s+\lambda_s)\nabla \cdot \vec{\xi} \to -p$, where the pressure p is finite. It is an appropriate model if the transit time of compressional waves having velocity v_c as found in Prob. 7.18.1 is very short compared to times of interest but that of shear waves is arbitrary. Equations (f) and (g) of Table P7.16.1 combine to show that $(2G_s+\lambda_s) \to \infty$ as $\nu_s \to 0.5$, so the incompressible model is especially appropriate in working with materials such as Jello or rubber. The force equation, Eq. e of Table 7.16.1, becomes

$$\rho \frac{\partial^2 \vec{\xi}}{\partial t^2} = -\nabla p + G_s \nabla^2 \vec{\xi} + \vec{F}_{ex}$$

(a) Show that the associated stress tensor is

$$S_{ij} = -p + G_s\left(\frac{\partial \xi_i}{\partial x_j} + \frac{\partial \xi_j}{\partial x_i}\right)$$

(b) Show that the transfer relations derived in this section, Eqs. 7.19.13 and 7.19.19, can be adapted to an incompressible solid by making the identification of variables

$$\hat{\vec{v}} \to j\omega\hat{\vec{\xi}} \; ; \; j\omega\eta \to G_s \; ; \; \hat{p} \to \hat{p}, \; \hat{S}_{ij} \to \hat{S}_{ij}$$

Prob. 7.19.2 Show that an infinite half-space of elastic material is described by the transfer relations

$$\begin{bmatrix} \hat{S}^{\alpha}_{xx}{}^{\beta} \\ \hat{S}^{\alpha}_{yx}{}^{\beta} \end{bmatrix} = \frac{\rho}{\gamma_c\gamma_s - k^2} \begin{bmatrix} \pm\gamma_s v_c^2(\gamma_c^2-k^2) & jk[k^2(v_c^2-2v_s^2) - v_c^2\gamma_c^2 + 2\gamma_c\gamma_s v_s{}^2] \\ jkv_s^2(\gamma_s^2+k^2-2\gamma_c\gamma_s) & \pm v_s^2\gamma_c(\gamma_s^2-k^2) \end{bmatrix} \begin{bmatrix} \hat{\xi}^{\alpha}_{x}{}^{\beta} \\ \hat{\xi}^{\alpha}_{y}{}^{\beta} \end{bmatrix}$$

where $\gamma_c^2 \equiv k^2 - \omega^2/v_c^2$ and $\gamma_s^2 \equiv k^2 - \omega^2/v_s^2$ and v_s and v_c are the velocities of shear and compressional waves as defined in Prob. 7.18.1.

Prob. 7.19.3 Rayleigh waves propagate on the free surface of an elastic material without dispersion. Because the associated deformations are readily accessible from the adjacent free space, these waves have been made the basis for surface acoustic wave (SAW) devices. The Rayleigh wave is neither a shear wave nor a dilatational wave, but rather a combination of these.

Prob. 7.19.3 (continued)

(a) Using the transfer relations for the lower half space derived in Prob. 7.19.2, show that the dispersion equation for this wave is

$$(\gamma_s^2+k^2)^2 = 4\gamma_s\gamma_c k^2$$

(b) Use the definition of γ_s and γ_c to show that the dispersion equation can be expressed as

$$\underline{\omega}^6 - 8\underline{\omega}^4 + 8\underline{\omega}^2\left(3 - \frac{2v_s^2}{v_c^2}\right) - 16\left(1 - \frac{v_s^2}{v_c^2}\right) = 0$$

where $\underline{\omega} \equiv \omega/kv_s$.

(c) Note that the coefficients of this expression do not depend on k. Extraneous roots have been generated in deriving this expression. One root represents the Rayleigh wave. Argue that the surface wave propagates without dispersion.

(d) Show that $v_s^2/v_c^2 = (1-2\nu_s)/2(1-\nu_s)$, so that $\underline{\omega}$ is determined by ν_s.

Prob. 7.19.4 The transfer relations, Eq. 7.19.3, are to be extended to describe the fluid response not only because of external interactions which have their effect on the layer through the surfaces, but also because of an imposed force density

$$\vec{F}_{ex} = \nabla \times \vec{G};\quad \vec{G} = \text{Re}\, \hat{G}(x)\, \exp j(\omega t - ky)\vec{i}_z \tag{1}$$

The extension follows lines similar to those taken in Sec. 4.5 for the flux-potential relations.

(a) Write Eq. 7.19.1 including the effect of the force density.

(b) Given any particular solution to this equation, $\hat{A}_P(x)$ with associated velocity and stress functions denoted by subscripts P, show that the transfer functions are:

$$\begin{bmatrix} \hat{S}_{xx}^{\alpha} \\ \hat{S}_{xx}^{\beta} \\ \hat{S}_{yx}^{\alpha} \\ \hat{S}_{yx}^{\beta} \end{bmatrix} = \eta[P_{ij}] \begin{bmatrix} \hat{v}_x^{\alpha} \\ \hat{v}_x^{\beta} \\ \hat{v}_y^{\alpha} \\ \hat{v}_y^{\beta} \end{bmatrix} + \begin{bmatrix} (\hat{S}_{xx}^{\alpha})_P \\ (\hat{S}_{xx}^{\beta})_P \\ (\hat{S}_{yx}^{\alpha})_P \\ (\hat{S}_{yx}^{\beta})_P \end{bmatrix} - \eta[P_{ij}] \begin{bmatrix} (\hat{v}_x^{\alpha})_P \\ (\hat{v}_x^{\beta})_P \\ (\hat{v}_y^{\alpha})_P \\ (\hat{v}_y^{\beta})_P \end{bmatrix} \tag{2}$$

(c) For a y directed force density F_o that is independent of x, $\hat{G}(x) = F_o x$. Evaluate Eq. 2 in this case.

Prob. 7.19.5 The fluid layer shown in Fig. 7.19.1 is bounded in the x=0 and x=d planes by rigid walls. Find the frequencies of the temporal modes. To do this use γ/k as a parameter representing the frequency ω, and write a transcendental equation of the form $D(\gamma/k,kd)=0$ which (given kd) can be solved for γ/k and hence ω. Illustrate how a graphical construction can be used to find roots of this expression, wherein γ/k is imaginary.

For Section 7.20:

Prob. 7.20.1 The equations of motion for an elastic solid are summarized in Prob. 7.19.1.

(a) Show that the transfer relations developed in this section can be used to describe an incompressible "inertia-less" elastic material by making the substitution

$$\hat{\vec{v}} \to \hat{\vec{\xi}},\ \hat{S}_{ij} \to \hat{S}_{ij},\ \eta \to G_s.$$

(b) Argue that the relations hold for deformation that are quasistatic with respect to the transit times of both the compressional and shear waves.

Prob. 7.21.1 Use Eqs. 7.21.17 and 7.21.13 to show that the Stoke's flow around a sphere is represented by Eq. 5.5.5.

Prob. 7.21.2 A rigid sphere having radius R is subject to an externally applied slowly varying z directed force f_z. Show that the resulting displacement Ξ in the z direction is related to this force by $f_z = 6\pi G_s R\Xi$. (See Prob. 7.20.1.)

8

Statics and Dynamics of Systems Having a Static Equilibrium

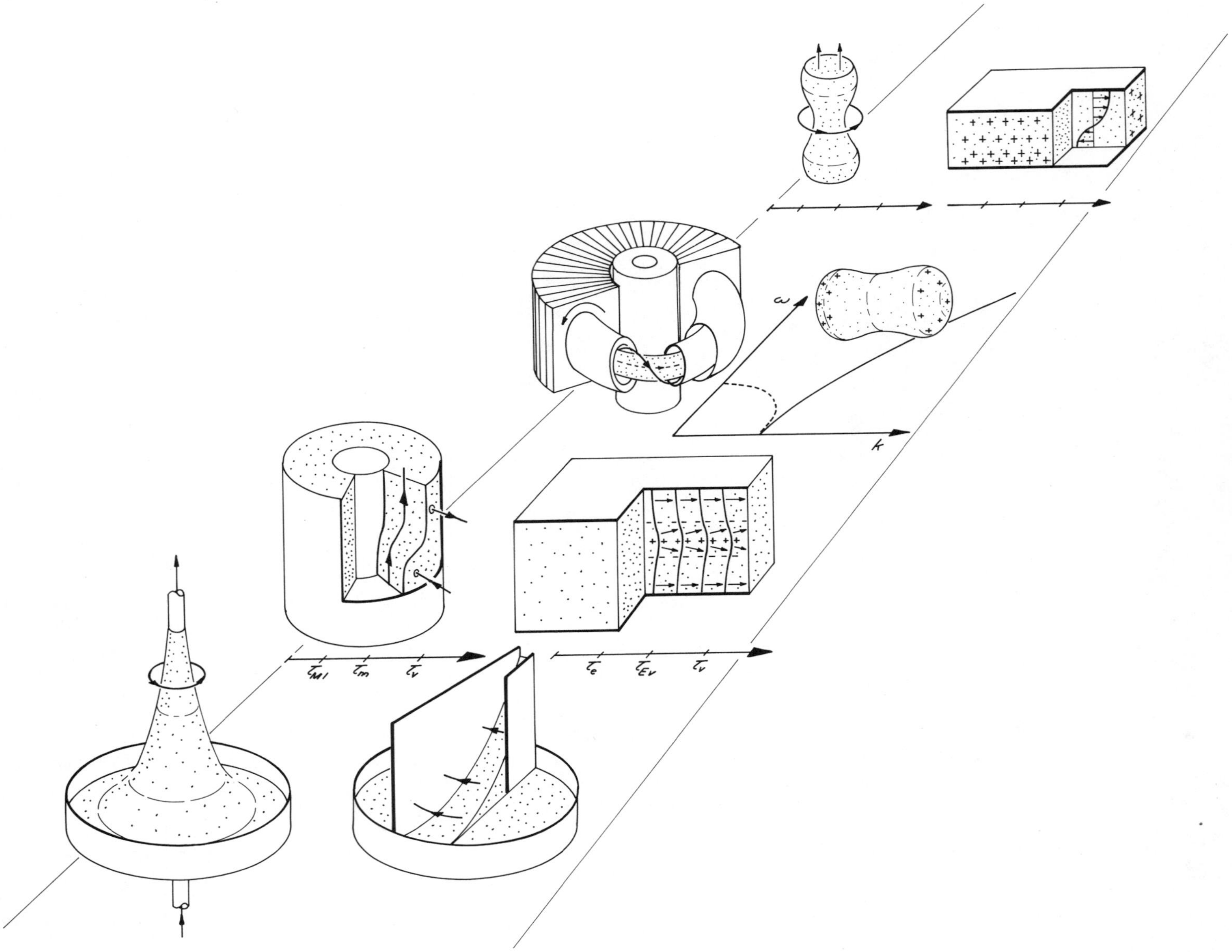

8.1 Introduction

In general, it is not possible for a fluid to be at rest while subject to an electric or magnetic force density. Yet, when a field is used to levitate, shape or confine a fluid, it is a static equilibrium that is often desired. The next section begins by identifying the electromechanical conditions required if a state of static equilibrium is to be achieved. Then, the following three sections exemplify typical ways in which these conditions are met. From the mathematical viewpoint, the subject becomes more demanding if the material deformations have a significant effect on the field. These sections begin with certain cases where the fields are not influenced by the fluid, and end with models that require numerical solution.

The magnetization and polarization static equilibria of Sec. 8.3 also offer the opportunity to explore the attributes of the various force densities from Chap. 3, to exemplify how entirely different distributions of force density can result in the same incompressible fluid response and to emphasize the necessity for using a consistent force density and stress tensor.

Given a static equilibrium, is it stable? This is one of the questions addressed by the remaining sections, which concern themselves with the dynamics that result if an equilibrium is disturbed. Some types of electromechanical coupling take place in regions having uniform properties. These are exemplified in Secs. 8.6-8.8. However, most involve inhomogeneities. The piecewise homogeneous models developed in Secs. 8.9-8.16 are chosen to exemplify the range of electromechanical models that can be pictured in this way.

The last sections, on smoothly inhomogeneous systems, serve as an introduction to a viewpoint that could equally well be exemplified by a range of electromechanical models. Once it is realized that the smoothly inhomogeneous systems can be regarded as a limit of the piecewise inhomogeneous systems, it becomes clear that all of the models developed in this chapter have counterparts in this domain.

The five electromechanical models that are a recurring theme throughout this chapter are summarized in Table 8.1.1.

Table 8.1.1. Electromechanical models.

Model	Approximation
Magnetization (MQS) or polarization (EQS)	No free current or charge Instantaneous magnetization or polarization
Flux conserving (MQS)	$\tau \ll \tau_m$
Charge conserving (EQS)	$\tau \ll \tau_e$ or τ_{mig}
Instantaneous magnetic diffusion (MQS)	$\tau \gg \tau_m$
Instantaneous charge relaxation (EQS)	$\tau \gg \tau_e$ or τ_{mig}

Magnetization and polarization models for incompressible motions require an inhomogeneity in magnetic or electric properties. The remaining interactions involve free currents or charges which generally bring in some form of magnetic diffusion or charge relaxation (or migration). How such rate processes come into the electromechanics is explicitly illustrated in the sections on homogeneous systems, Secs. 8.6 and 8.7. However, in the more complex inhomogeneous systems, the last four models of Table 8.1.1 not only result in analytical simplifications, but give insights that would be difficult to glean from a more general but complicated description. "Constant potential" continua fall in the category of instantaneous charge relaxation models.

STATIC EQUILIBRIA

8.2 Conditions for Static Equilibria

Often overlooked as an essential part of fluid mechanics is the subject of fluid statics. A reminder of the significance of the subject is the equilibrium between the gravitational force density and the hydrostatic fluid pressure involved in the design of a large dam. On the scale of the earth's surface, where g is essentially constant, the gravitational force acting on a homogeneous fluid obviously is of a type that can result in a static equilibrium.

Except for scale, electric and magnetic forces might well have been the basis for Moses' parting of the Red Sea. Fields offer alternatives to gravity in the orientation, levitation, shaping or

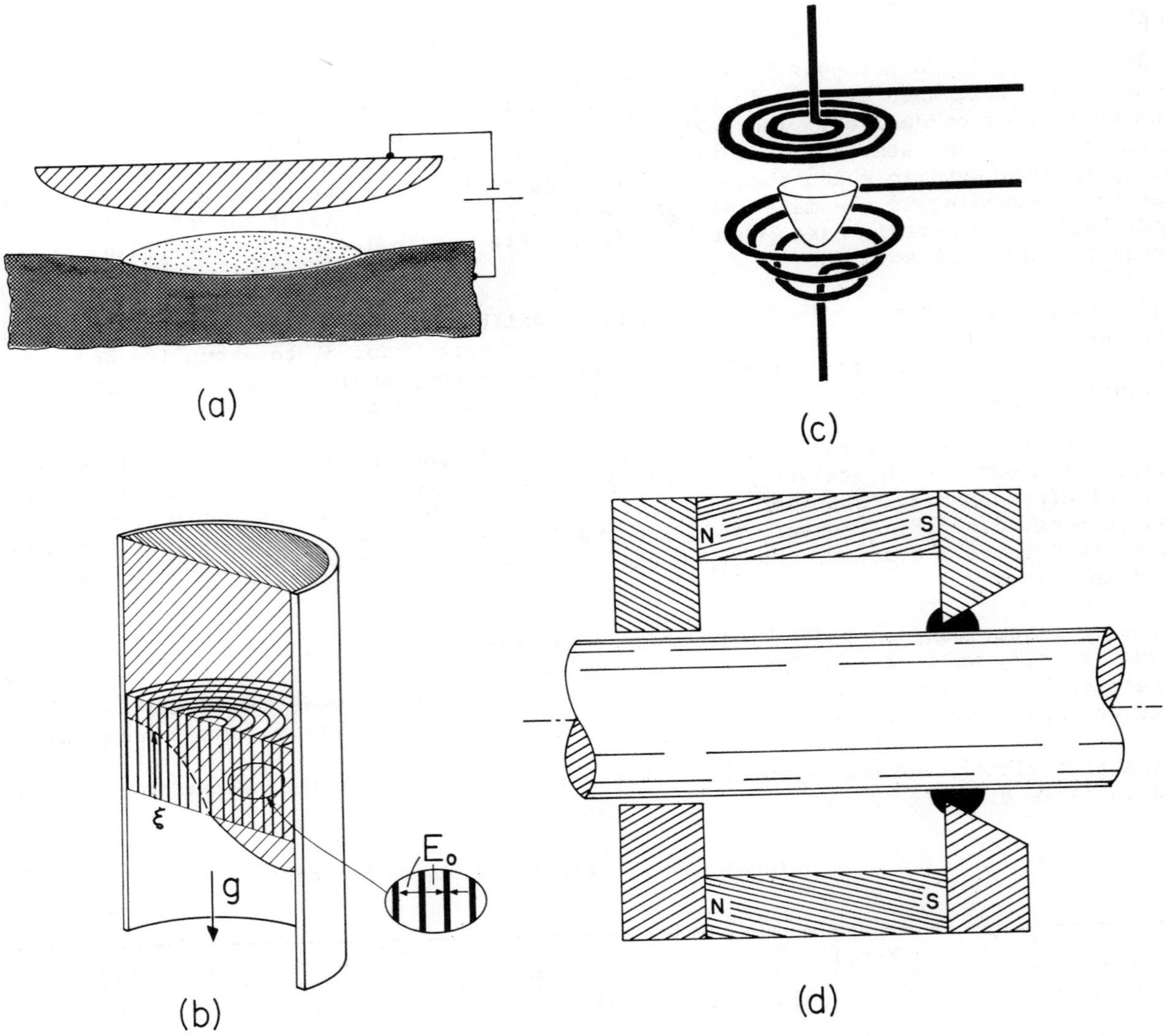

Fig. 8.2.1. (a) Electric field used to shape a "lens" of conducting liquid resting on a pool of liquid metal. Molten plastics and glass are sufficiently conducting that they can be regarded as "perfect" conductors. (b) Polarization forces used to orient a highly insulating liquid in the top of a tank regardless of gravity. The scheme might be used for providing an artificial bottom in cryogenic fuel storage tanks under the zero-gravity conditions of space. (c) Liquid metal levitator that makes used of forces induced by a time-varying magnetic field. At high frequencies, the flux is excluded from the metal, and hence the fields tend toward a condition of zero shearing surface force density. (d) Cross-sectional view of axisymmetric magnetic circuit and magnetizable shaft with magnetizable fluid used to seal penetration of rotating shaft through vacuum containment.

otherwise controlling of static fluid configurations. Examples are shown in Fig. 8.2.1.[1-3]

For what force distributions can each element of a fluid be in static equilibrium? If the external electric or magnetic force density is $\vec{F}^e$, then the force equation reduces to

$$-\nabla(p - \rho\vec{g}\cdot\vec{r}) = \vec{F}^e \tag{1}$$

This expression is a limiting form of Eq. 7.4.4 with the velocity zero. Even if effects of viscosity

---

1. J. R. Melcher, D. S. Guttman and M. Hurwitz, "Dielectrophoretic Orientation," J. Spacecraft and Rockets 6, 25 (1969).
2. E. C. Okress et al., "Electromagnetic Levitation of Solid and Molten Metals," J. Appl. Phys. 23, 545 (1952).
3. R. E. Rosensweig, G. Miskolczy and F. D. Ezekiel, "Magnetic-Fluid Seals," Machine Design, March 28, 1968.

are included in the model, because $\vec{v} = 0$, Eq. 1 still represents the static equilibrium. Thus, it is also the static limit of Eq. 7.4.4. The curl of a gradient is zero. So, the curl of Eq. 1 gives a necessary condition on $\vec{F}^e$ for static equilibrium:

$$\nabla \times \vec{F}^e = 0 \tag{2}$$

To achieve a static equilibrium, the force density must be the gradient of a scalar, $-\nabla\mathcal{E}$. Then Eq. 1 becomes

$$\nabla(p - \rho\vec{g}\cdot\vec{r} + \mathcal{E}) = 0 \tag{3}$$

which will be recognized as Eq. 7.8.4 in the limit $\vec{v} = 0$.

More often than not, in an electromagnetic field a fluid does not reach a static equilibrium. Electromagnetic forces do not generally satisfy Eq. 2. Fields designed to achieve an irrotational force density are exemplified by Secs. 8.3-8.5.

These sections also illustrate that stress balance at interfaces is similarly restricted. A clean static interface is incapable of sustaining a net electrical shearing surface force density. Formally, this is seen from the interfacial stress balance, Eq. 7.7.6, which states that the normally directed pressure jump and surface tension surface force density must be balanced by the electrical force density. The last, $[\![T^e_{ij}]\!] n_j$, is in general not normal to the interface.

To be specific about what types of interfaces do satisfy this requirement, consider an interface having a normal vector in the x direction. Then, $n_j = \delta_{jx}$ and for the directions $i \neq x$ the surface force density is

$$\begin{aligned} [\![T_{ix}]\!] &= [\![E_i D_x]\!] = E_i [\![D_x]\!] \quad \text{(EQS)} \\ [\![T_{ix}]\!] &= [\![H_i B_x]\!] = B_x [\![H_i]\!] \quad \text{(MQS)} \end{aligned} \tag{4}$$

In writing the second equalities, advantage is taken of the continuity of tangential $\vec{E}$ (EQS) and normal $\vec{B}$ (MQS). From Eq. 4a, two EQS idealizations are distinguished for having no electrical shearing surface force density at the interface. First, the tangential electric field intensity can vanish, in which case (4a) is satisfied. The interface is "perfectly" conducting. Secondly, the jump in electric displacement at the interface can vanish, and again, there is no shear stress at the interface. The interface then supports no free surface charge density. Two MQS circumstances exist for achieving no shearing surface force density. First, the normal flux density can vanish at the interface. Physically, this is realized if the interface is perfectly conducting. Alternatively, the jump in tangential $\vec{H}$ can vanish, and this means that there is no surface current density on the interface.

The four static equilibria of Fig. 8.2.1 exemplify the four limiting situations in which there is no electrical shearing force density at an interface. In Fig. 8.2.1a, the lens is pictured as sufficiently highly conducting that it excludes the electric field, and hence behaves as a perfect conductor. Molten glass is more than conducting enough to satisfy this condition. Polarization forces are used to orient highly insulating fluids with no free charge density either on the interface or in the bulk, as illustrated in Fig. 8.2.1b. Metallurgists use high-frequency magnetic fields to make a crucible with magnetic walls, as shown in Fig. 8.2.1c. Here, because of the high frequency used, the magnetic field penetrates the liquid metal only slightly, and tends to the limit of no normal flux density. Thus, a static configuration with the melt levitated in mid-air is in principle possible. Magnetic fluids are being exploited as the basis for making vacuum seals for shaft penetrations as sketched in Fig. 8.2.1d. Here, the magnetic field is used to orient the liquid in the region between shaft and walls. Generally, the magnetizable fluids are highly insulating and so there is not only no surface current to produce a surface shearing force density, but also no volume force density due to $\vec{J} \times \vec{B}$.

In all of the examples in Fig. 8.2.1, the electromechanical forces can be regarded as confined to interfaces. This is clear for the free charge and free current interactions of parts (a) and (c) of that figure, because there are no fields inside the material. In the polarization and magnetization interactions, the properties are essentially uniform in the bulk. Thus, the force density expressed as Eq. 3.7.19 or 3.8.14 is concentrated at the interfaces.

Some common static configurations involving volume forces are evident from symmetry. For example, if the force density is in one direction and only depends on that direction, i.e., if

$$\vec{F}^e = F_x(x)\vec{i}_x \tag{5}$$

then it is clear that the force density is the gradient of $(-\mathcal{E})$:

$$\mathcal{E} = -\int F_x(x)dx \tag{6}$$

Similar arguments can be used if the force density is purely in a radial direction.

Other approaches to securing a static equilibrium using bulk force densities are illustrated in Sec. 8.4.

8.3 Polarization and Magnetization Equilibria: Force Density and Stress Tensor Representations

For an incompressible fluid, the pressure is a dangling variable. It only appears in the force equation. Its role is to be whatever it must be to insure that the velocity is solenoidal. As a consequence, those external forces which are gradients of "pressures" have no influence on the observable incompressible dynamics. Any "pressure" can be lumped with p and a new pressure defined. Although true for dynamic as well as static situations, this observation is now illustrated by two static equilibria.

The first of these illustrates polarization forces, and is depicted by Fig. 8.3.1. A pair of diverging conducting electrodes are dipped into a liquid having permittivity ε. A potential difference V_o applied between these plates results in the electric field

$$\vec{E} = \frac{V_o}{\alpha r}\vec{i}_\theta \tag{1}$$

in the interior region well away from the edges. At any given radius r, the situation is essentially the dielectric of Fig. 3.6.1, drawn into the region between parallel capacitor plates. Because the field increases to the left, so also does the liquid height. What is this height of rise, $\xi(r)$?

There are two reasons that this experiment is a classic one. The first stems from the lack of coupling between the fluid geometry and the electric field. The interface tends to remain parallel with the θ-direction, and as a result the electric field given by Eq. 1 remains valid regardless of the height of rise. As a result, the description is greatly simplified. The second reason pertains to its use as a counterexample against any contention that the polarization force density is $\rho_p\vec{E}$, where ρ_p is the polarization charge density. In this example, there is neither polarization charge in the liquid bulk (in the region between the electrodes and even in the fringing field near the lower edges of the electrodes in the liquid) nor is there surface polarization charge at the interface (where $\vec{E}$ is tangential). If $\rho_p\vec{E}$ were the force density, the liquid would not rise!

Illustrated now are two self-consistent approaches to determining the height of rise, the first using Kelvin's force density and the second exploiting the Korteweg-Helmholtz force density.

Kelvin Polarization Force Density: The force density and associated stress tensor are in this case (Table 3.10.1)

$$\vec{F} = \vec{P}\cdot\nabla\vec{E} \tag{2}$$

$$T_{ij} = E_iD_j - \frac{1}{2}\delta_{ij}\varepsilon_oE_kE_k \tag{3}$$

The liquid is modeled as electrically linear with $\vec{P}$ and $\vec{E}$ collinear,

$$\vec{P} = (\varepsilon - \varepsilon_o)\vec{E} \tag{4}$$

Throughout the liquid, ε is uniform. Hence, Eqs. 2 and 3 and the fact that $\vec{E}$ is irrotational combine to show that the force density is

$$(\vec{P}\cdot\nabla\vec{E})_i = (\varepsilon - \varepsilon_o)E_j\frac{\partial E_i}{\partial x_j} = (\varepsilon - \varepsilon_o)E_j\frac{\partial E_j}{\partial x_i} = (\varepsilon - \varepsilon_o)\frac{\partial}{\partial x_i}(\frac{1}{2}E_jE_j) \tag{5}$$

So long as the force density is only used where ε is constant (in the bulk of the liquid or of the air) Eq. 6 is in the form of the gradient of a pressure,

$$\vec{F} = -\nabla\mathcal{E}; \quad \mathcal{E} \equiv -\frac{1}{2}(\varepsilon - \varepsilon_o)\vec{E}\cdot\vec{E} \tag{6}$$

This makes it clear that the polarization force density is irrotational throughout the bulk. In the bulk, Eq. 8.2.3 applies. With $\mathcal{E}$ evaluated using Eq. 1, it follows that in the bulk regions

$$p + \rho gz - \frac{(\varepsilon - \varepsilon_o)V_o^2}{2\alpha^2r^2} = \text{constant} \tag{7}$$

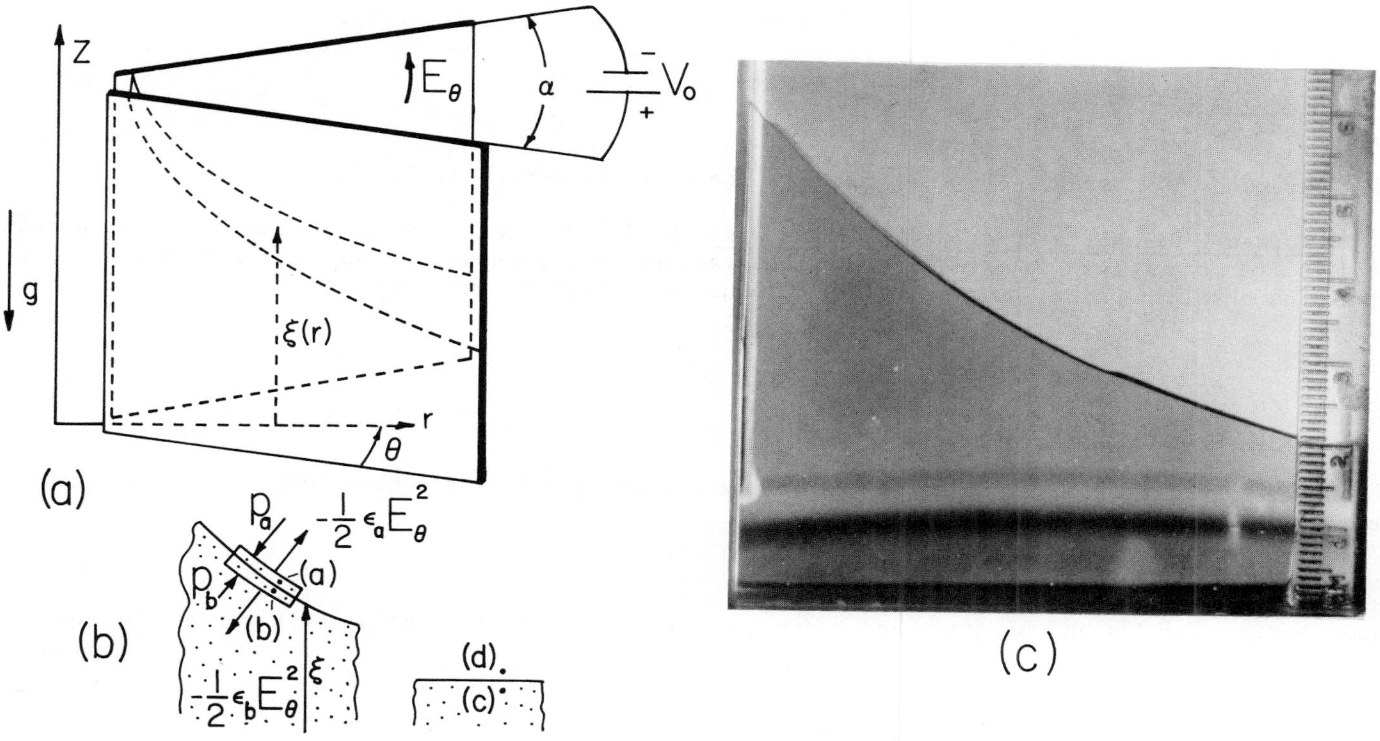

Fig. 8.3.1. (a) Diverging conducting plates with potential difference V_o are immersed in dielectric liquid. (b) Interfacial stress balance. (c) From Reference 12, Appendix C; corn oil ($\varepsilon = 3.7\ \varepsilon_o$) rises in proportion to local E^2. Upper fluid is compressed nitrogen gas ($\varepsilon \simeq \varepsilon_o$) so that E can approach 10^7 V/m required to raise liquid several cm. To avoid free charge effects, fields are 400 Hz a-c. The fluid responds to the time-average stress. The interface position is predicted by Eq. 12.

Thus, with the interface elevation, ξ, measured relative to the liquid level well removed from the electrodes, positions a and d in the air (where $\varepsilon = \varepsilon_o$ and $\rho \simeq 0$) and positions b and c (in the liquid) are joined by Eq. 7:

$$p_a = p_d \tag{8}$$

$$p_b + \rho g\xi - \frac{(\varepsilon - \varepsilon_o)V_o^2}{2\alpha^2 r^2} = p_c \tag{9}$$

To complete the formulation, account must be taken of any surface force densities at the interface that would make the pressure discontinuous at the interface. In general, the boundary condition is Eq. 7.7.6. As discussed in Sec. 8.2, there is no free surface charge, so there is no shearing component of the surface force density. If the electrodes are very close together, capillarity will contribute to the height of rise, as described by the example in Sec. 7.8. Here the electrodes are sufficiently far apart that the meniscus has a negligible effect.

If the local normal to the interface is in the x direction, the surface force density is $[\![T_{xx}]\!]$. Because the electric field is entirely perpendicular to x and is continuous at the interface, it follows from Eq. 3 that $[\![T_{xx}]\!] = [\![-\frac{1}{2}\varepsilon_o E_\theta^2]\!] = 0$, so that there is no surface force density. Hence, the stress equilibrium for the interface at locations a-b and c-d is simply represented by

$$p_a - p_b = 0 \tag{10}$$

$$p_c - p_d = 0 \tag{11}$$

The pressures are eliminated between the last four relations by multiplying Eq. 8 by (-1) and adding the four equations. The resulting expression can then be solved for $\xi(r)$:

$$\xi = \frac{(\varepsilon - \varepsilon_o)V_o^2}{2\alpha^2 \rho g r^2} \tag{12}$$

This dependence is essentially that shown in the photograph of Fig. 8.3.1.

<u>Korteweg-Helmholtz Polarization Force Density</u>: It is shown in Sec. 3.7 that this force density differs from the Kelvin force density by the gradient of a pressure. Thus, the same height of rise should be obtained using (from Table 3.10.1) the force density and stress tensor pair

$$\vec{F} = -\frac{1}{2} E^2 \nabla\varepsilon \tag{13}$$

$$T_{ij} = \varepsilon E_i E_j - \frac{1}{2}\delta_{ij}\varepsilon E_k E_k \tag{14}$$

Now, there is no electrical force in the volume and the static force equation, Eq. 8.2.3, simply requires that

$$p + \rho g z = \text{constant} \tag{15}$$

Thus, points a and d and points b and c are joined through the respective bulk regions by Eq. 15 to obtain

$$p_a = p_d \tag{16}$$

$$p_b + \rho g \xi = p_c \tag{17}$$

By contrast with Eqs. 8 and 9 there is no bulk effect of the field. Now, the electromechanical coupling comes in at the interface where ε suffers a step discontinuity and hence a surface force density exists. At the interface, $[\![T_{xx}]\!] = -\frac{1}{2}(\varepsilon_o - \varepsilon)E_\theta^2$, so that the stress balances at the interface locations a-b and c-d are respectively

$$p_a - p_b = -\frac{(\varepsilon_o - \varepsilon)V_o^2}{2\alpha^2 r^2} \tag{18}$$

$$p_c - p_d = 0 \tag{19}$$

Multiplication of Eq. 16 by (-1) and addition of these last four equations eliminates the pressure and leads to the same deflection as obtained before, Eq. 12.

<u>Korteweg-Helmholtz Magnetization Force Density:</u> The force density and stress tensor pair appropriate if the fluid has a nonlinear magnetization are (from Table 3.10.1)

$$\vec{F} = \sum_{k=1}^{m} \frac{\partial W}{\partial \alpha_k} \nabla \alpha_k \tag{20}$$

$$T_{ij} = H_i B_j - \delta_{ij} W' \tag{21}$$

where $\vec{B}$ and $\vec{H}$ are collinear:

$$\vec{B} = \mu(\alpha_1, \alpha_2, \cdots, \alpha_m, H^2)\vec{H} \tag{22}$$

In the experiment of Fig. 8.3.2, the magnetic field

$$\vec{H} = \frac{I}{2\pi r}\vec{i}_\theta \tag{23}$$

is imposed by means of the vertical rod, which carries the current I. The ferrofluid in the dish has essentially uniform properties α_i throughout its bulk, but tends to saturate as the field exceeds about 100 gauss.

The Korteweg-Helmholtz force density has the advantage of concentrating the electromechanical coupling where the properties vary. In this example, this is at the liquid-air interface. Because

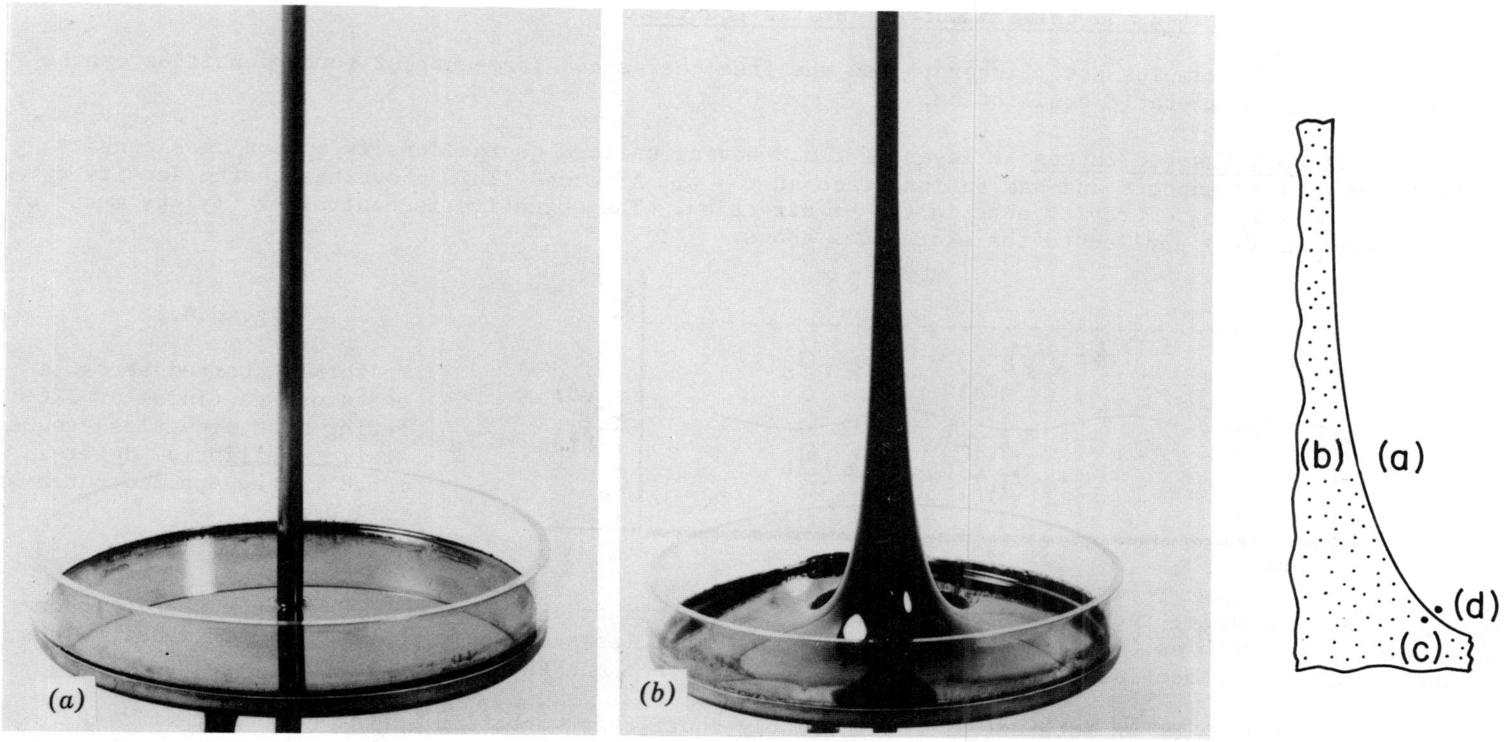

Fig. 8.3.2. A magnetizable liquid is drawn upward around a current-carrying wire in accordance with Eq. 29. (Courtesy of AVCO Corporation, Space Systems Division.)

Eq. 20 is zero throughout the bulk regions, Eqs. 16 and 17 respectively pertain to these regions.

Stress balance at the interface is represented by evaluating the surface force density acting normal to the interface, to write

$$p_a - p_b = -[\![W']\!] \tag{24}$$

$$p_c - p_d = 0 \tag{25}$$

for locations a-b and c-d, respectively. The pressures are eliminated between Eqs. 16, 17, 24 and 25 to obtain

$$\xi = -\frac{[\![W']\!]}{\rho g} \tag{26}$$

To complete the evaluation of $\xi(r)$, the magnetization characteristic of the liquid must be specified. As an example, suppose that

$$\vec{B} = \frac{\vec{H}}{\alpha_1\sqrt{\alpha_2^2 + H^2}} + \mu_o\vec{H} \tag{27}$$

where α_1 and α_2 are properties of the liquid. Then, the coenergy density (Eq. 2.14.13) is

$$W' = \int_o^{\vec{H}} \vec{B}\cdot\delta\vec{H} = \frac{1}{\alpha_1}\sqrt{\alpha_2^2 + H^2} - \frac{\alpha_2}{\alpha_1} + \frac{1}{2}\mu_o H^2 \tag{28}$$

and, in view of Eq. 23, Eq. 26 becomes

$$\xi = \frac{1}{\rho g}\left[\frac{1}{\alpha_1}\sqrt{\alpha_2^2 + \left(\frac{I}{2\pi r}\right)^2} - \frac{\alpha_2}{\alpha_1}\right] \tag{29}$$

As for the electric-field example considered previously, the relative simplicity of Eq. 26 originates in the independence of $\vec{H}$ and the liquid deformation. If there were a normal component of $\vec{H}$ at the interface, the field would in turn depend on the liquid geometry and a self-consistent solution would be more complicated.

8.4 Charge Conserving and Uniform Current Static Equilibria

A pair of examples now illustrate how the free-charge and free-current force densities can be arranged to give a static equilibrium.

Uniformly Charged Layers: A layer of fluid having uniform charge density q_b and mass density ρ_b rests on a rigid support and has an interface at $x = \xi$. A second fluid above has charge density q_a and mass density ρ_a. Gravity acts in the -x direction. The objective is control of $\xi(y)$ by means of the potential V(z) applied to the electrodes above.

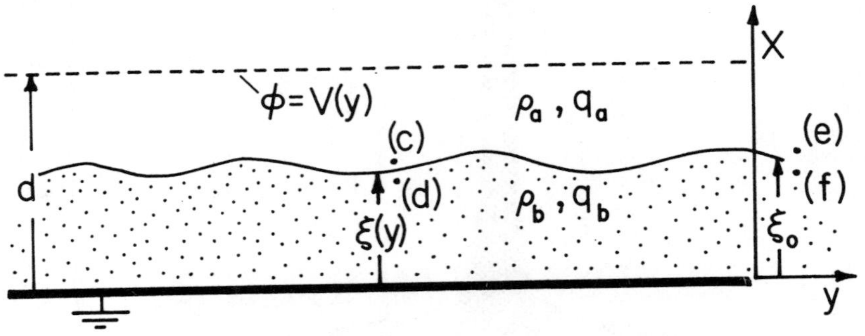

Fig. 8.4.1

Uniformly charged aerosols entrained in fluids of differing mass densities assume static equilibrium determined by the applied potential V(y).

As an example, the upper fluid might be air which is free of charge ($q_a = 0$) and the lower one a heavier gas such as CO_2 with entrained submicron particles previously charged by ion impact. Thus, the fluids have essentially the permittivity of free space and there is no surface tension.

The time-scales of interest are sufficiently short that migration of the charged particles relative to the fluids is inconsequential. Thus, the charge is frozen to the gas. Because the gas is incompressible ($\nabla\cdot\vec{v} = 0$), the charge density of a gas element is conserved. Regardless of the particular shape of the interface, the charge densities above and below remain uniform, q_a and q_b respectively. It is for this reason and because $\vec{E}$ is irrotational that the force density in each fluid is irrotational:

$$\vec{F} = q\vec{E} = -q\nabla\Phi = -\nabla(q\Phi) \tag{1}$$

Thus, Eq. 8.2.3 shows that within a given fluid region

$$p + \rho g x + q\Phi = \text{constant} \tag{2}$$

Evaluation of the constant at the points (e) and (f) adjacent to the interface where $\xi = \xi_o$ gives

$$\begin{aligned} p + \rho_a g x + q_a\Phi &= p^e + \rho_a g\xi_o + q_a\Phi(\xi_o);\ x > \xi \\ p + \rho_b g x + q_b\Phi &= p^f + \rho_b g\xi_o + q_b\Phi(\xi_o);\ x < \xi \end{aligned} \tag{3}$$

The force density suffers a step discontinuity at the interface. This means that there is no surface force density, so that the pressure is continuous at the interface. Continuity of p also follows formally from the stress jump condition, Eq. 7.7.6 with the surface tension $\gamma = 0$.

So that stability arguments can be made, an external surface force density $T_{ext}(y)$ is pictured as also acting on the interface. By definition $T_{ext} = 0$ at location (e-f):

$$p^c - p^d = T_{ext};\quad p^e - p^f = 0 \tag{4}$$

Subtraction of Eqs. 3a and 3b then gives

$$g(\xi - \xi_o)(\rho_b - \rho_a) + (q_b - q_a)[\Phi(\xi) - \Phi(\xi_o)] = T_{ext} \tag{5}$$

where $\Phi(\xi)$ is the potential evaluated at the interface.

Of course, the potential distribution is determined by the presently unknown geometry of the interface and the field equations. Here, the relation of field and geometry is simplified by considering long-wave distributions of the interface. The electric field is approximated as being dominantly in the x direction. Thus, Poisson's equation reduces to simply

$$\frac{\partial^2\Phi}{\partial x^2} = \frac{-q}{\varepsilon_o};\quad q = \begin{cases} q_a: & x > \xi \\ q_b: & x < \xi \end{cases} \tag{6}$$

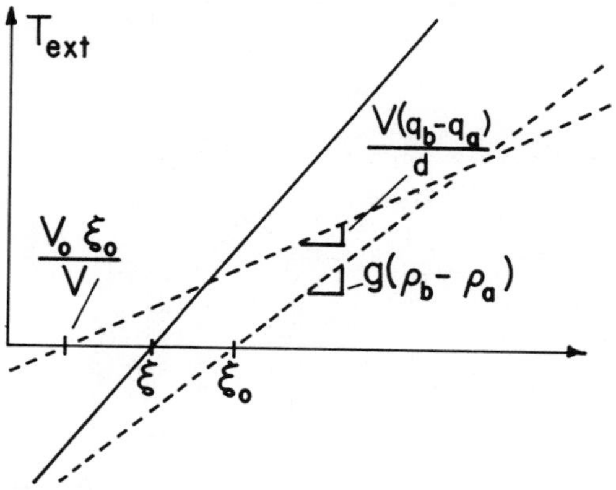

Fig. 8.4.2

Graphical representation of Eq. 9.

With the boundary conditions that $\Phi(d) = V(y)$, that $[\![\Phi]\!] = 0$ and $[\![\partial\Phi/\partial x]\!] = 0$ at the interface and that $\Phi(0) = 0$, it follows that

$$\Phi(\xi) = \frac{V\xi}{d} + \frac{q_a}{2\varepsilon_o d}\xi(d-\xi)^2 + \frac{q_b}{2\varepsilon_o d}\xi^2(d-\xi) \tag{7}$$

Thus, with $T_{ext} = 0$, Eq. 5 becomes a cubic expression that can be solved for $\xi(y)$ given $V(y)$

$$g(\xi-\xi_o)(\rho_b-\rho_a) + (q_b-q_a)\left(\frac{V\xi}{d} - \frac{V\xi_o}{d}\right)$$
$$+ (q_b-q_a)\left\{\frac{q_a}{2\varepsilon_o d}[\xi(d-\xi)^2 - \xi_o(d-\xi_o)^2] + \frac{q_b}{2\varepsilon_o d}[\xi^2(d-\xi)-\xi_o^2(d-\xi_o)]\right\} = T_{ext} \tag{8}$$

Given a desired $\xi(y)$, Eq. 8 can also be solved for the required $V(y)$. If the field imposed by the electrode potential $V(y)$ is large compared to the space charge field, the last term in Eq. 8 can be ignored. Then, the equilibrium is represented by

$$g(\xi-\xi_o)(\rho_b-\rho_a) + (q_b-q_a)\left(\frac{V\xi}{d} - \frac{V_o\xi_o}{d}\right) = T_{ext} \tag{9}$$

To picture how the interface responds to $V(y)$, it is helpful to use the graphical solution of Fig. 8.4.2. The interfacial deflection is given by $T_{ext} = 0$. Increasing V has the effect of decreasing the intercept and increasing the slope of the electrical "force" curve.

In this imposed field limit, Eq. 9 can be solved for the layer thickness as a function of the imposed potential:

$$\frac{\xi}{\xi_o} = \frac{1+\underline{V}_o}{1\quad\underline{V}}; \quad \underline{V} \equiv \frac{(q_b-q_a)}{gd(\rho_b-\rho_a)}V(y) \tag{10}$$

Illustrated in Fig. 8.4.3 is an example which represents what would happen if the potential shown were imposed on a light layer over a heavier layer, with the upper one uncharged and the lower one negatively charged.

Stability of the equilibrium can be argued from the dependence of T_{ext} on ξ. If

$$g(\rho_b-\rho_a) + \frac{V}{d}(q_b-q_a) > 0 \tag{11}$$

a positive force is required to produce a positive deflection, much as if the interface were equivalent to a spring with a positive spring constant. Thus, the condition of Eq. 11 is required for stability. In terms of the normalized voltage used in expressing Eq. 10, the interface is stable where $\underline{V} > -1$.

A more complete stability argument that includes the effects of space charge is given in Sec. 8.14.

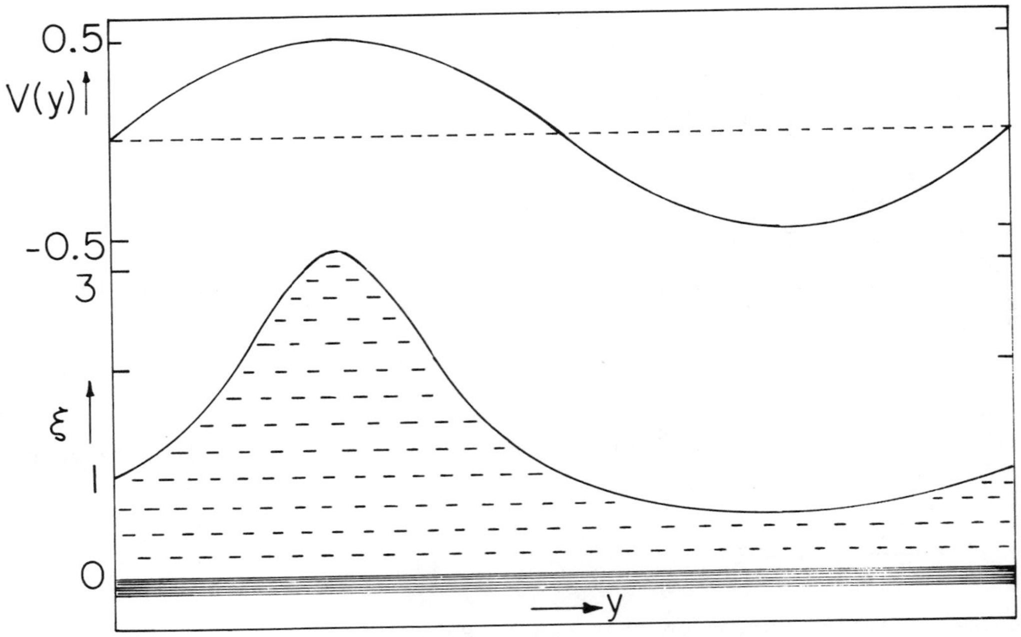

Fig. 8.4.3

Imposed field equilibrium with $\underline{V} = -0.7 \sin(\underline{y})$. Shape of charge layer is given by Eq. 10.

Uniform Current Density: Static equilibrium with the free-current force density $\vec{J}_f \times \mu_o \vec{H}$ distributed throughout the volume of a fluid is now illustrated. In the MQS system of Fig. 8.4.4, a layer of liquid metal rests on a rigid plane at $x = 0$ and has a depth $\xi(y)$. The system, including the fields and currents, is assumed to have a uniform distribution with the z direction, so that the view shown is any cross section.

The magnetic field is to be used in deforming the liquid interface. A d-c electromagnet produces a magnetic flux density with components in the x-y plane. In addition, a voltage source drives a uniform current density J_o in the z direction throughout the fluid volume. This current density interacts with the imposed flux density to produce a vertical component of magnetic force in the liquid, and a resultant deformation of the interface. Note that because the fields are static, there are no surface currents. Also, the liquid metal is not magnetizable, so there are no magnetization forces to consider. Finally, effects of surface tension are ignored. Therefore, the interface is in stress equilibrium, provided the pressure there is continuous.

The essential approximation in obtaining the irrotational force density throughout the volume is that the imposed magnetic flux density is very large compared to the flux density induced by the imposed current density J_o. Thus, the force density takes the approximate form

$$\vec{F} = J_o \vec{i}_z \times [B_x \vec{i}_x + B_y \vec{i}_y] \tag{12}$$

The vector potential is convenient for dealing with $\vec{B}$, because if the substitution is made $\vec{B} = \nabla \times \vec{A}$, then Eq. 12 becomes $\vec{F} = -\nabla \mathcal{E}$, wherein

$$\mathcal{E} = -J_o A(x,y) \tag{13}$$

The imposed field approximation and the uniform imposed current result in the irrotational force density required for static equilibrium. Given the particular field structure and the magnitude of the field excitation, $A(x,y)$ is known.

In an engineering application, the liquid metal might serve as a base for the casting of plastic or glass products.[1] The magnetic field can be controlled so that there is a ready means of altering the shape of the mold without a need for replacing the casting material. If a quiescent fluid state is desirable, conditions for a static equilibrium are essential. From Eq. 8.2.3 and Eq. 13

$$p + \rho g x - J_o A = \text{constant} \tag{14}$$

There is no current density in the gas above the interface, and hence no force density. The depth as $y \to -\infty$ is defined as ξ_∞, and $A\ (x = \xi,\ y \to -\infty)$ is defined as A_∞. Then, Eq. 14 shows that for points

1. See U.S. Patent #3,496,736, "Sheet Glass Thickness Control Method and Apparatus," February 24, 1970, M. Hurwitz and J. R. Melcher.

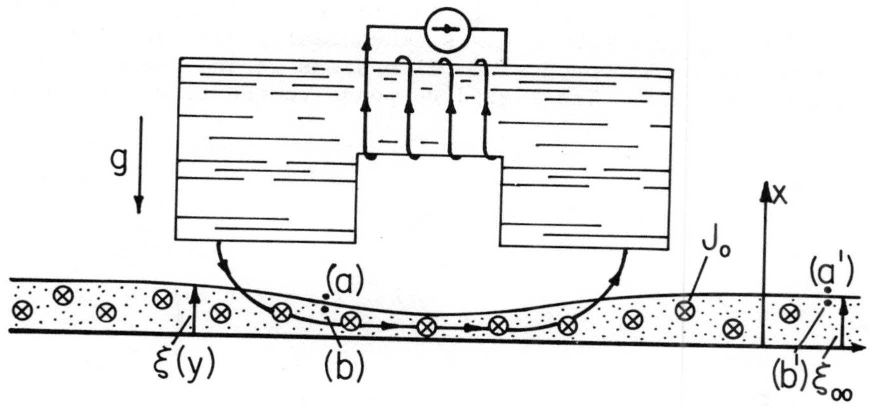

Fig. 8.4.4. Layer of liquid metal has the depth $\xi(y)$ which is controlled by the interaction of a uniform z-directed current density J_o and a magnetic flux density induced by means of the magnetic structure.

(a) and (a') of Fig. 8.4.4

$$p_{a'} + \rho_a g\xi_\infty = p_a + \rho_a g\xi \tag{15}$$

and for points (b) and (b')

$$p_{b'} + \rho_b g\xi_\infty - J_o A_\infty = p_b + \rho_b g\xi - J_o A \tag{16}$$

Because the hydrostatic pressures are the same at the primed and unprimed positions, subtraction of Eq. 15 from Eq. 16 gives a relation that can be solved for the height $\xi(y)$:

$$\xi = \xi_\infty - J_o(A_\infty - A)/g(\rho_b - \rho_a) \tag{17}$$

The vector potential has the physical significance of being a flux linkage per unit length in the z direction. To see this, define $\lambda(y)$ as the flux linked by a loop having one edge outside the field region to the right, the other edge at the position y and height ξ of the interface and unit depth in the z direction. Then the flux linked per unit length is

$$\lambda = \int_S \vec{B}\cdot\vec{n}da = \int_C \vec{A}\cdot d\ell = A_\infty - A(\xi,y) \tag{18}$$

and in terms of this flux, Eq. 18 becomes

$$\xi = \xi_\infty - \frac{J_o\lambda}{g(\rho_b - \rho_a)} \tag{19}$$

The flux passing through the interface to the right of a given point determines the depression at that point. Proceeding from right to left, the flux is at first increasing, and hence the depression is increasing. But near the middle, additions to the total flux reverse, and the net flux tends toward zero. Hence, ξ returns to ξ_∞, as sketched in Fig. 8.4.4. Even if used only qualitatively, Eq. 19 gives a picture of the interfacial deformation that is useful for engineering design. Measurements can be used to determine $\lambda(x,y)$.

8.5 Potential and Flux Conserving Equilibria

Typical of EQS systems in which an electric pressure is used to shape the interface of a somewhat conducting liquid is that shown in Fig. 8.5.1a. Provided that the region between the cylindrical electrode and the liquid is highly insulating compared to the liquid, the interface is an equipotential. Because the applied voltage is constant and the equilibrium is static, this is true even for what might be regarded as relatively insulating liquids. Certainly water, molten glass, plasticizers and even used transformer oil will behave as equipotentials with air insulation between electrodes and interface. The liquid is in a reservoir. By virtue of its surface tension, the interface attaches to the reservoir's edges at $y = \pm\ell$. Thus, continuity requires that the upward deflection of the interface under the electrode be compensated by a downward deflection to either side. To be considered in this section is how the static laws make it possible to account for such requirements of mass conservation.

In the MQS system of Fig. 8.5.1b, the liquid is probably a metal. To achieve the conditions for a static equilibrium, the driving flux source F_o is sinusoidally varying with a sufficiently high frequency that the skin depth is small compared to dimensions of interest. Thus, the normal flux density at the interface approaches zero. The liquid responds to the time average of the normal magnetic stress.

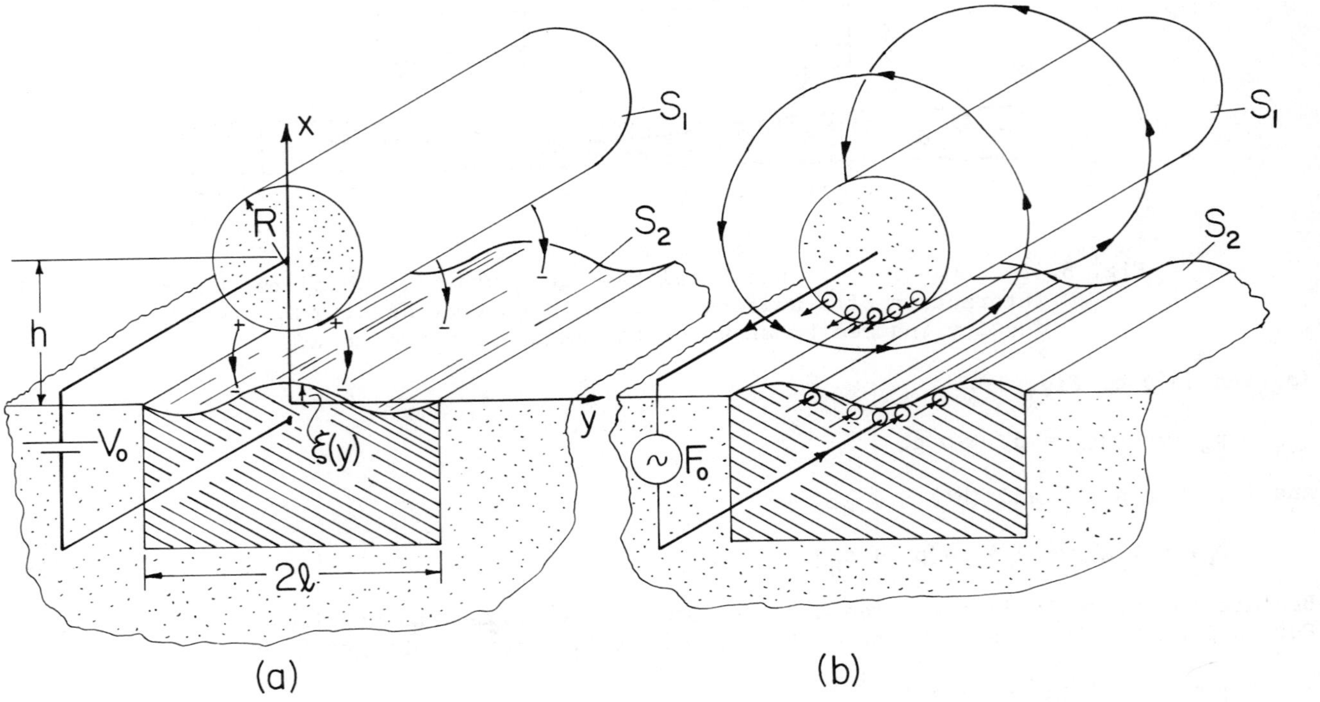

Fig. 8.5.1. (a) EQS system; liquid interface stressed by d-c field is equipotential. (b) MQS system; driving current has sufficiently high frequency that currents are on surfaces of liquid and electrode. Liquid responds to time average of magnetic pressure.

This pair of case studies exemplifies the free charge and free current static equilibria, from Sec. 8.2, involving electromagnetic surface force densities. The EQS static equilibrium is possible because there is no electric field tangential to the interface, while the MQS equilibrium results because there is essentially no normal magnetic flux density.

Antiduals: The two-dimensional fields in the two systems have an interesting relationship. For the moment, suppose that the geometry of the interfaces is known. Then, the electric field is represented by the potential, while the magnetic flux density is represented in terms of the z component of the vector potential, as summarized by Eqs. (a)-(c) of Table 2.18.1. Thus, in the regions between electrodes and interfaces,

$$\nabla^2\Phi = 0 \quad\Big|\quad \nabla^2 A = 0 \tag{1}$$

Boundary conditions on the respective systems are

$$\Phi = V_o \text{ on } S_1 \quad\Big|\quad A = F_o \text{ on } S_1 \tag{2}$$

$$\Phi = 0 \text{ on } S_2 \quad\Big|\quad A = 0 \text{ on } S_2 \tag{3}$$

where S_1 is the surface of the electrode or bus above the interface and S_2 is the interface and adjacent surface of the container. By definition, F_o is the flux per unit length (in the z direction) passing between the bus and the interface. Note that to make the magnetic field tangential to these surfaces, A is constant on the interface and on the surface of the bus.

With the understanding that n denotes the direction normal to the local interface, the electric and magnetic stresses on the interfaces are

$$T_{nn} = \frac{1}{2}\varepsilon_o E_n^2 = \frac{1}{2}\varepsilon_o\left(\frac{\partial\Phi}{\partial n}\right)^2 \quad\Big|\quad T_{nn} = -\frac{1}{2}\mu_o H_t^2 = -\frac{1}{2}\mu_o\left(\frac{1}{\mu_o}\frac{\partial A}{\partial n}\right)^2 \tag{4}$$

Thus, if the interface had the same geometry in the two configurations, the magnetic stress would "push" on the interface to the same degree that the electric stress would "pull." The magnetic stress is the negative of the electric stress and can be formally found by replacing $\varepsilon_o \rightarrow \mu_o$ and $\partial\Phi/\partial n \rightarrow (\partial A/\partial n)/\mu_o$.

Although limited to two-dimensional fields, the antiduality makes it possible to extend the electromechanical description of one class of configurations to another by simply changing the sign of the electromechanical coupling term. Provided that charge can relax sufficiently rapidly on the EQS interface to render it an equipotential even under dynamic conditions, and provided that motions remain slow compared to the period of the sinusoidal excitation for the MQS system (so that the interface responds primarily to the time-average magnetic stress), the antiduality is valid for dynamic as well as static interactions.

Bulk Relations: Bernoulli's equation, Eq. 7.8.7, applied to the air and liquid bulk regions, shows that

$$p = \begin{cases} \Pi_a & ; \; x > \xi \\ \Pi_b - \rho g x & ; \; x < \xi \end{cases} \tag{5}$$

where Π_a and Π_b are constants. The mass density of the air is ignored compared to that of the liquid.

Stress Equilibrium: The normal component of the stress balance, Eq. 7.7.6, requires that

$$[\![p]\!] = T_{nn} - \gamma \nabla\cdot\vec{n} \tag{6}$$

Evaluation of the pressure jump using Eqs. 5 and of $\nabla\cdot\vec{n}$ with $\vec{n}$ given by Eq. 7.5.3 gives

$$(\Pi_a - \Pi_b) + \rho g \xi = T_{nn} + \gamma \frac{d}{dy}\left\{\frac{d\xi}{dy}\left[1 + \left(\frac{d\xi}{dy}\right)^2\right]^{-\frac{1}{2}}\right\} \tag{7}$$

Evaluation of Surface Deflection: Suppose that in the absence of a field, the interface is flat. Then, as the excitation V_o or F_o is raised, $\xi(y)$ increasingly departs from this initial state, $\xi = 0$. One way to compute $\xi(y)$ at a given excitation is to find the deflections as the excitation is raised, in stages, to this final value. Thus, $T_{nn}(y)$ in Eq. 7 is approximated by solving Eq. 1 with $\xi(y)$ approximated by its shape at the previous somewhat lower level of excitation. Thus, T_{nn} is a known function of y and the new $\xi(y)$ is approximated by integrating Eq. 7. Once this is done, the new $\xi(y)$ can be used to refine the determination of the fields. This interaction can be repeated until a desired accuracy is achieved. Then, the excitation can be incrementally raised and the process repeated.

For a system that is symmetric about the x axis boundary conditions appropriate to the solution of the second-order differential equation, Eq. 7, are

$$\frac{d\xi}{dy}(0) = 0 \tag{8}$$

$$\xi(-\ell) = 0 \tag{9}$$

In addition, mass conservation requires that

$$\int_{-\ell}^{o} \xi dy = 0 \tag{10}$$

This condition translates into a determination of the pressure jump. In view of Eqs. 8 and 10, integration of Eq. 7 between $y = -\ell$ and $y = 0$ shows that

$$\underline{\Pi}_a - \underline{\Pi}_b = \int_{-1}^{o} \underline{T}_{nn} d\underline{y} - \frac{1}{W}\left(\frac{u}{\sqrt{1+u^2}}\right)_{\underline{y}=-1} \tag{11}$$

where normalized variables and dimensionless parameters are

$$y = \ell \underline{y}; \quad \Pi_a - \Pi_b = (\underline{\Pi}_a - \underline{\Pi}_b)(\tfrac{1}{2}\varepsilon_o V_o^2/\ell^2); \quad \xi = \ell\underline{\xi}$$
$$T_{nn} = (\tfrac{1}{2}\varepsilon_o V_o^2/\ell^2)\underline{T}_{nn}; \quad W \equiv \tfrac{1}{2}\varepsilon_o V_o^2/\gamma\ell; \quad G \equiv \rho g \ell^2/\gamma \tag{12}$$

and u is the slope of the interface, defined as

$$\frac{d\underline{\xi}}{d\underline{y}} = u \tag{13}$$

In terms of u, Eq. 7 is normalized and written as a first-order differential equation

$$\frac{du}{dv} = (1 + u^2)^{3/2}[(\Pi_a - \Pi_b)W + G\xi - WT_{nn}] \tag{14}$$

This last pair of relations, equivalent to Eq. 7, take a form that is convenient for numerical integration. (The integration of systems of first-order nonlinear equations, given "initial conditions," is carried out using standard computer library subroutines. For example, in Fortran IV, see IMSL Integration Package DEVREK.) With $T_{nn}(y)$ given from the solution of Eqs. 1-3 (to be discussed shortly), the integration begins at $\underline{y} = -1$ where Eq. 9 provides one boundary condition. To make a trial integration of Eqs. 12 and 13, a trial value of u(-1) is assumed. Thus, from Eq. 11, the value of $\Pi_a - \Pi_b$ that insures conservation of mass is determined. Integration of Eqs. 12 and 13 is then carried out and evaluated at y=0. Using u(-1) as a parameter, this process is repeated until the condition u(0) = 0 (boundary condition, Eq. 8) is satisfied. One way to close in on the appropriate value of u(-1) is by halving the separation of two u(-1)'s yielding opposite-signed slopes at y = 0.

Evaluation of Stress Distribution: To provide $T_{nn}(y)$ at each step in the determination of the surface deflection which has just been described, it is necessary to solve Eq. 1 using the boundary conditions of Eqs. 2 and 3. A numerical technique that is well suited to this task results in the direct evaluation of the surface charge density σ_f on the interface. Because $T_{nn} = \varepsilon_o E_n^2/2 = \sigma_f^2/2\varepsilon_o$, this is tantamount to a direct determination of the desired stress distribution.

In the two-dimensional configuration of Fig. 8.5.2, the solution of Laplace's equation can be represented by a potential (at the location $\vec{r}$) that is the superposition of potentials due to incremental line charges per unit length $\sigma_f ds'$:

$$\Phi(\vec{r}) = \frac{-1}{2\pi}\int \sigma_f(\vec{r}\,')\ln|\vec{r} - \vec{r}\,'|ds' \tag{15}$$

This expression is normalized such that

$$\Phi = V_o\underline{\Phi}; \quad \sigma_f = \frac{\varepsilon_o V_o}{\ell}\underline{\sigma}_f; \quad s = \ell\underline{s} \tag{16}$$

Although $\ln|\vec{r} - \vec{r}\,'| = \ln|\underline{\vec{r}} - \underline{\vec{r}}\,'| + \ln \ell$, so long as the net charge in the system is zero, integration of the $\ln \ell$ term gives no contribution and so is omitted from Eq. 15. The desired (normalized) surface charge is $\underline{\sigma}_f$ and $d\underline{s}'$ is the (normalized) incremental segment of boundary.

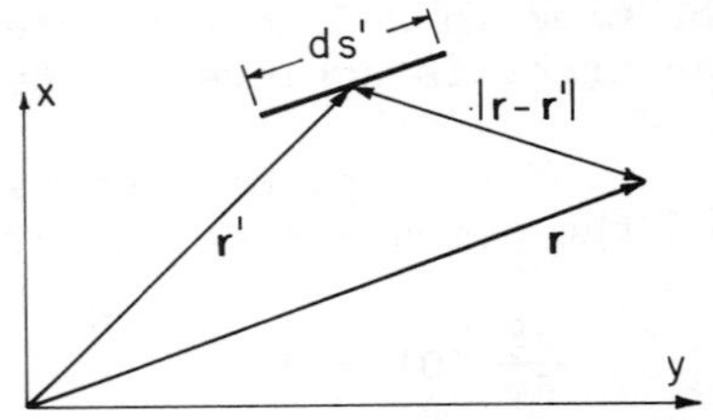

Fig. 8.5.2. Potential given by Eq. 15 at $\vec{r}$ is superposition of potentials due to line charges at $\vec{r}\,'$.

The integral equation is solved numerically by approximating the integral by a sum over segments of the boundaries. These are denoted by the index n, as shown in Fig. 8.5.3. The first N segments are on the zero potential interface, the next 2M are on the surrounding zero potential plane and the remaining P segments are on the cylindrical electrode, and hence have the potential $\underline{\Phi} = 1$. Thus, the potential at the mth segment is the superposition of integrations over each of the charge segments. Because the latter have a length Δs that is small, the surface charge on each segment can be approximated as constant and the integration carried out analytically. For example, the contribution to the potential of the mth segment from the surface charge σ_n on the nth segment is (see Fig. 8.5.4),

$$\Phi_m = -\frac{\sigma_n}{2\pi}\int_{s_n}^{s_n+\Delta s} \ln\sqrt{d_n^2 + s^2}\, ds \tag{17}$$

Thus, Eq. 15 becomes

$$\Phi_m = \sum_{n=1}^{N+2M+P} a_{mn}\sigma_n \tag{18}$$

where

$$a_{mn} = -\frac{1}{2\pi}\{\tfrac{1}{2}(\Delta s + s_n)\ln[(\Delta s + s_n)^2 + d_n^2] - \Delta s - \tfrac{1}{2}s_n \ln[s_n^2 + d_n^2] + d_n \tan^{-1}\left(\frac{\Delta s + s_n}{d}\right) - d_n \tan^{-1}\frac{s_n}{d}\} \tag{19}$$

Now, Eq. 18 can be written for each of the N+2M+P segments. Thus, it represents a set of N+2M+P equations, linear in as many unknowns σ_n. These equations are then inverted to obtain the desired σ_n's. (Matrix inversion is carried out using standard computer library subroutines. For example, in Fortran I see IMSL Matrix Inversion Routine LINV1F.)

Because $\underline{T}_{nn} = \underline{\sigma}_n^2/2$, the normalized stress distribution on each segment follows. So that the numerical integration of the surface equations, Eqs. 13 and 14, can be carried out with an arbitrary step size, the discrete representation of T_{nn} on the interface is conveniently converted to a smooth function by fitting a polynomial to the values of T_{nn}. (Polynomial fit can be carried out using a Least Square Polynomial Fit Routine such as the Math Library Routine LSFIT.)

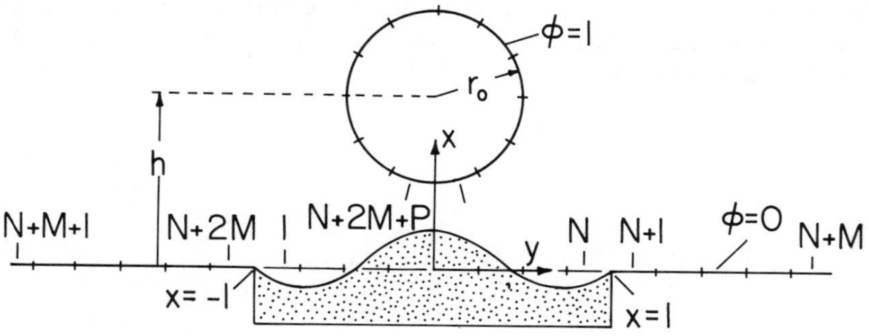

Fig. 8.5.3. Definition of segments and geometry for numerical solution.

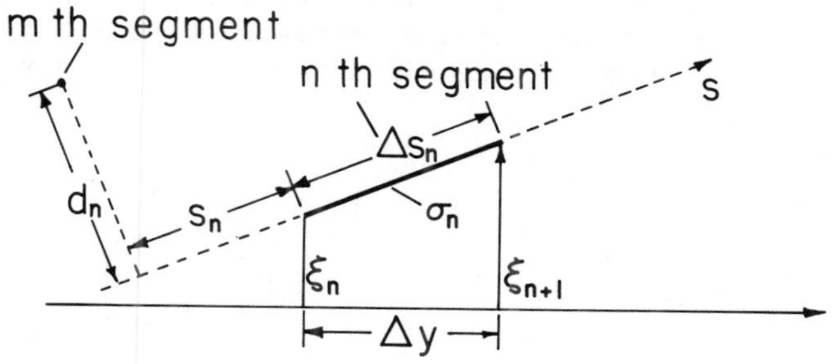

Fig. 8.5.4. Typical segment on interface.

Typical results of the combined numerical integration to determine $T_{nn}(y)$ and the interfacial deformation are shown in Fig. 8.5.5. (These computations were carried out by Mr. Kent R. Davey.) The procedure begins with a modest value of W and a flat interface and starts with a determination of T_{nn}. Then, Eqs. 13 and 14 are integrated and this integration repeated until the boundary condition $u(0) = 0$ is satisfied. Using this revised distribution of $\xi(y)$, the distribution of T_{nn} is recalculated, followed by a recalculation of the interface shape. This process is repeated until a desired accuracy is achieved.

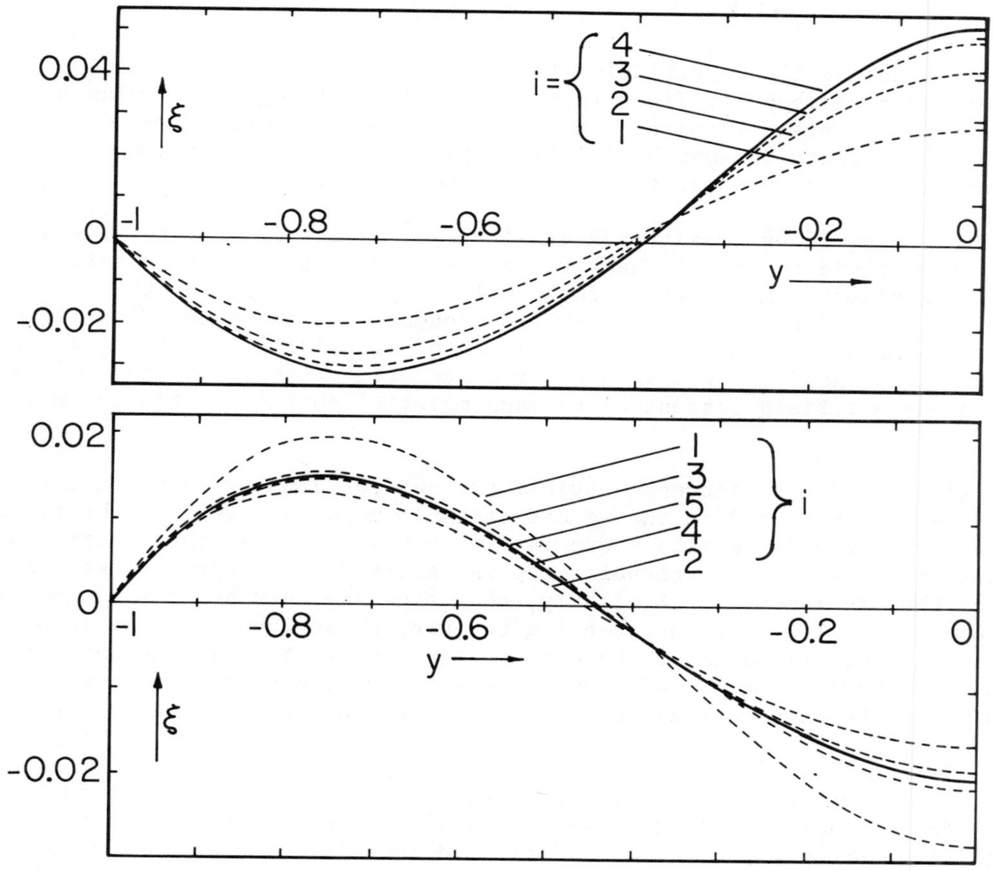

Fig. 8.5.5

Shape of interface with $G = 3$, $\underline{r}_o = 0.5$ and $\underline{h} = 1$. Broken curves are for successive iterations (i) with W fixed. (a) EQS system with $W = 0.5$. (b) MQS system with $W = -0.5$. Note that electric case converges monotonically, while magnetic one oscillates.

With W raised to a somewhat higher value, the previously determined shape is used as a starting point in repeating the iteration described.

8.6 Flux Conserving Continua and Propagation of Magnetic Shear Stress

Alfvén waves that propagate along magnetic field lines in the bulk of a highly conducting fluid result from the tendency for arbitrary fluid surfaces of fixed identity to conserve their flux linkage. The physical mechanisms involved are apparent in the one-dimensional motions of a uniformly conducting incompressible fluid permeated by an initially uniform magnetic field intensity $H_o\vec{i}_x$, as in Fig. 8.6.1a. By assumption, each fluid particle in a y-z plane executes the same motion.

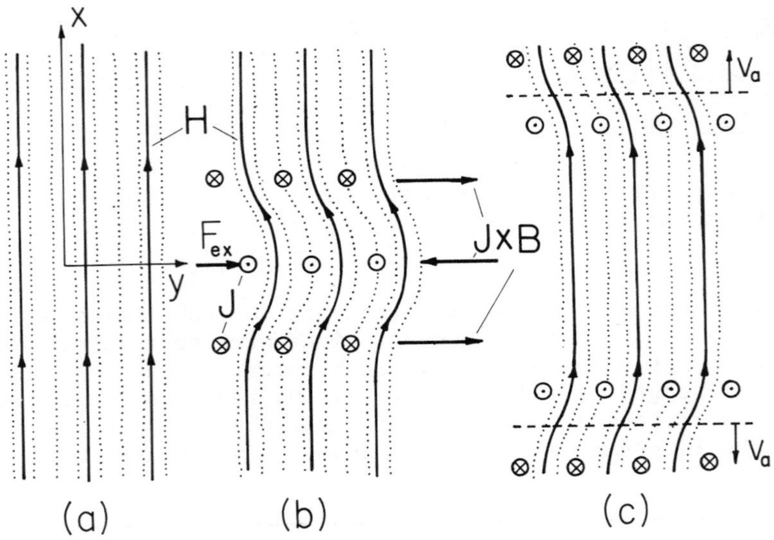

Fig. 8.6.1. (a)Perfectly conducting fluid initially at rest in uniform magnetic field. (b) For flux conservation of loops of fixed identity initially lying in x-z planes, translation of layer in y-z plane requires induced currents shown. (c) Force densities associated with currents induced by initial motion. (d) Translation of layers resolves into wave fronts propagating along magnetic field lines.

Consider the consequences of using an external force density $F_{ex}\vec{i}_y$ (Fig. 8.6.1b) to give a y-directed translation to a layer of fluid in one of these y-z planes. Because of the translation, fluid elements initially in any x-z plane form a surface that would be pierced twice by the initial field H_o. It is shown in Sec. 6.2 that if the fluid is perfectly conducting, the total flux linked by such a surface of fixed identity must be conserved. As a result of material deformation, a current density (sketched in Fig. 8.6.1b) is induced in just such a way as to create the y component of magnetic field required to maintain the <u>net</u> field tangential to each material surface initially in an x-y plane.

Note that because charge accumulation is inconsequential, the current density is solenoidal, so that current in the z direction must be returned in the -z direction in adjacent planes. The force density associated with these return currents is also shown in Fig. 8.6.1b. Because these currents are proportional to the displacement of a layer, the external force is retarded by a "spring-like" force proportional to the magnitude of the displacement. Similarly, the returning currents in adjacent y-z layers cause magnetic forces above and below, but here tending to carry these layers in the same direction as the original displacement. Thus, fluid layers to either side tend to move in the same direction as the layer subjected to the external force. Adjacent layers in the y-z planes are coupled by a magnetic shear stress representing the force associated with currents induced to preserve the constant flux condition.

In the absence of viscosity, the magnetic shear stress on adjacent layers is only retarded by inertia. There is some analogy to the viscous diffusion (Sec. 7.19), with the interplay between viscosity and inertia now replaced by one between magnetic field and inertia. The viscous shear stress of Sec. 7.19 is proportional to the shear-strain rate. By contrast, the magnetic shear stress in the perfect conductor is proportional to the shear strain (the spatial rate of change of the material displacement rather than velocity). Thus, rather than being diffusive in nature, the motion resulting from the magnetic shear stress in a perfect conductor is wave-like. As suggested by Fig. 8.6.1c, the motion propagates along the lines of magnetic field intensity as a transverse electromechanical wave. Just how perfectly the fluid must conduct and how free of viscosity it must be to observe these waves is now determined by a model that includes magnetic and viscous diffusion.

A layer of fluid having conductivity σ, viscosity η and thickness Δ is shown in Fig. 8.6.2. In static equilibrium, it is permeated by a uniform x directed magnetic field intensity H_o. Because the magnetic flux density is solenoidal, it is written in the form $\vec{B} = \mu H_o \vec{i}_x + \nabla \times \vec{A}$, where $\vec{A}$ is governed by the magnetic diffusion equation, Eq. 6.5.3. Fluid deformations that are now considered are independent of z and confined to x-y planes, and so only the z component of $\vec{A}$ exists; $\vec{A} = A\vec{i}_z$. Moreover, motions are taken as independent of y, so $\vec{v} = v_y(x,t)\vec{i}_y$ and $A = A(x,t)$. Thus,

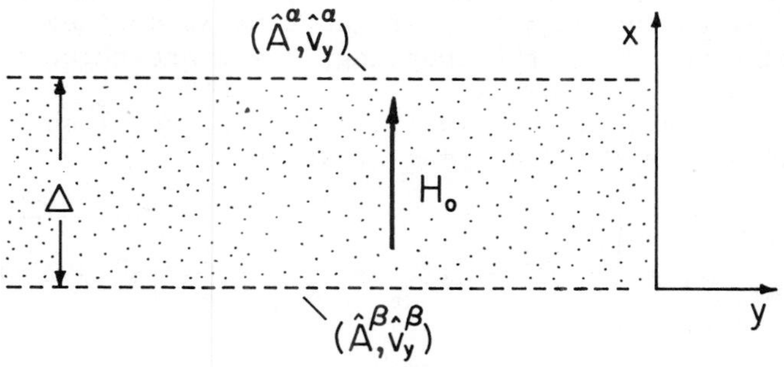

Fig. 8.6.2. Layer of liquid metal or plasma with ambient magnetic field H_o.

$$\frac{1}{\mu\sigma}\frac{\partial^2 A}{\partial x^2} = \frac{\partial A}{\partial t} + \mu H_o v_y \tag{1}$$

where [Eq. (b) of Table 2.18.1]

$$H_y = -\frac{1}{\mu}\frac{\partial A}{\partial x} \tag{2}$$

The fact that motions are independent of y and that $\vec{B}$ is solenoidal combine to show that B_x is independent of x, and hence $B_x = \mu H_o$ even as the motion occurs. There is no linearization implied by the last term of Eq. 1.

For the one-dimensional incompressible motions, conservation of mass is identically satisfied and only the y component of the force equation is pertinent. With the magnetic stress substituted into Eq. 7.16.1, it follows from Eq. 2 that

$$\rho\frac{\partial v_y}{\partial t} = -H_o\frac{\partial^2 A}{\partial x^2} + \eta\frac{\partial^2 v_y}{\partial x^2} \tag{3}$$

where the magnetic shear stress is $T_{yx} = \mu H_o H_y$ and the viscous shear stress is

$$S_{yx} = \eta\frac{\partial v_y}{\partial x} \tag{4}$$

The self-consistent coupling between field and fluid is expressed by Eqs. 1 and 3. These represent the one-dimensional response of the layer shown in Fig. 8.6.2. Given the amplitudes $[\hat{A}^\alpha, \hat{A}^\beta, \hat{v}_y^\alpha, \hat{v}_y^\beta]$ at the boundaries, what are the transfer relations for the amplitudes $[\hat{H}_y^\alpha, \hat{H}_y^\beta, \hat{S}_{yx}^\alpha, \hat{S}_{yx}^\beta]$ in these same planes? (Note that these relations are the limit $k \to 0$ of more general transfer relations for traveling wave dependences on y. For the two-dimensional motions implied by such a dependence, v_x becomes an additional variable, and the normal stress S_{xx} is its complement. Thus, the more general two-dimensional transfer relations relate two potentials and four velocity components to two tangential fields and four stress components, evaluated at the α and β surfaces.)

For complex amplitude solutions of the form $A = \text{Re}\,\hat{A}(x)\exp(j\omega t)$, Eqs. 1 and 3 become differential laws for the x dependence:

$$\frac{1}{\mu\sigma}\frac{d^2\hat{A}}{dx^2} - j\omega\hat{A} - \mu H_o\hat{v}_y = 0 \tag{5}$$

$$\eta\frac{d^2\hat{v}_y}{dx^2} - j\omega\rho\hat{v}_y - H_o\frac{d^2\hat{A}}{dx^2} = 0 \tag{6}$$

These constant coefficient expressions admit solutions $\hat{A} \propto \exp(\gamma x)$ and $\hat{v}_y \propto \exp(\gamma x)$. Substitution shows that γ must satisfy the relation ($\gamma\Delta = \underline{\gamma}$):

$$(\underline{\gamma}^2 - j\omega\tau_m)(\underline{\gamma}^2 - j\omega\tau_v) - \left(\frac{\tau_m\tau_v}{\tau_{MI}^2}\right)\underline{\gamma}^2 = 0 \tag{7}$$

Thus, the spatial distribution with x is determined by the magnetic diffusion time, τ_m, the viscous diffusion time, τ_v, and the magneto-inertial time, τ_{MI}:

$$\tau_m \equiv \mu\sigma\Delta^2; \quad \tau_v \equiv \rho\Delta^2/\eta; \quad \tau_{MI} = \Delta\sqrt{\rho/\mu H_o^2} \tag{8}$$

In the absence of the equilibrium magnetic field ($H_o = 0$), Eq. 7 shows that what remains is viscous diffusion (Secs. 7.18 and 7.19) and magnetic diffusion (Secs. 6.5 and 6.6). The parameter expressing the coupling in Eq. 7, the ratio of the geometric mean of the magnetic and viscous diffusion times to the magneto-inertial time is defined as the Magnetic Hartmann number $H_m = \sqrt{\tau_m\tau_v}/\tau_{MI} = \Delta\mu H_o\sqrt{\sigma/\eta}$. With the coupling, there are three characteristic times that determine the dynamics.

Even so, the biquartic form of Eq. 7 shows that there are still only four solutions to Eqs. 5 and 6, $\gamma = \pm\gamma_1$ and $\gamma = \pm\gamma_2$, where

$$\gamma_{\substack{1\\2}} = \left\{\frac{1}{2}\left[H_m^2 + j\omega(\tau_m + \tau_v)\right] \pm \frac{1}{2}\sqrt{\left[H_m^4 - \omega^2(\tau_m - \tau_v)^2\right] + 2j\omega(\tau_m + \tau_v)H_m^2}\right\}^{1/2} \tag{9}$$

Thus, in terms of coefficients $\hat{A}_1\cdots\hat{A}_4$, the solution is

$$\hat{A} = \hat{A}_1 \sinh \gamma_1 x + \hat{A}_2 \sinh \gamma_1(x - \Delta) + \hat{A}_3 \sinh \gamma_2 x + \hat{A}_4 \sinh \gamma_2(x - \Delta) \tag{10}$$

Equation 5 shows how to find $\hat{v}_y$ in terms of these same four coefficients:

$$\hat{v}_y = \frac{1}{\mu^2 H_o \sigma}\left(\frac{d^2\hat{A}}{dx^2} - j\omega\mu\sigma\hat{A}\right) \tag{11}$$

Given the potential and velocity in the α and β planes, Eqs. 10 and 11 become four expressions that can be inverted to determine $\hat{A}_1\cdots\hat{A}_4$. Fortunately, $\hat{A}_1$ and $\hat{A}_3$ are determined by the α variables alone, and A_2 and A_4 by the β variables alone, so this task is not all that difficult. In fact, with a bit of hindsight, the desired linear combination of solutions can be written by inspection:

$$\begin{aligned}\hat{A} = \Bigg\{&\left[-(\gamma_2^2 - j\omega\mu\sigma)\hat{A}^\alpha + \mu^2 H_o\sigma\hat{v}_y^\alpha\right]\frac{\sinh \gamma_1 x}{\sinh \gamma_1\Delta} + \left[(\gamma_2^2 - j\omega\mu\sigma)\hat{A}^\beta - \mu^2 H_o\sigma\hat{v}_y^\beta\right]\frac{\sinh \gamma_1(x-\Delta)}{\sinh \gamma_1\Delta}\\ &+ \left[(\gamma_1^2 - j\omega\mu\sigma)\hat{A}^\alpha - \mu^2 H_o\sigma\hat{v}_y^\alpha\right]\frac{\sinh \gamma_2 x}{\sinh \gamma_2\Delta} + \left[-(\gamma_1^2 - j\omega\mu\sigma)\hat{A}^\beta + \mu^2 H_o\sigma\hat{v}_y^\beta\right]\frac{\sinh \gamma_2(x-\Delta)}{\sinh \gamma_2\Delta}\Bigg\}\Big/(\gamma_1^2 - \gamma_2^2)\end{aligned} \tag{12}$$

Now, by use of Eqs. 11 and 12 in 2 and 4, the transfer relations follow:

$$\begin{bmatrix}\hat{H}_y^\alpha\\ \hat{H}_y^\beta\\ \hat{S}_{yx}^\alpha\\ \hat{S}_{yx}^\beta\end{bmatrix} = \left[M_{ij}\right]\begin{bmatrix}\hat{A}^\alpha\\ \hat{A}^\beta\\ \hat{v}_y^\alpha\\ \hat{v}_y^\beta\end{bmatrix} \tag{13}$$

where with $\underline{\gamma}_k \equiv \gamma_k\Delta$ and $q_k^2 \equiv \underline{\gamma}_k^2 - j\omega\mu\sigma\Delta^2$, $k = 1$ or 2:

$$M_{1\left(\substack{1\\2}\right)} = -M_{2\left(\substack{2\\1}\right)} = \frac{1}{\mu}\left[\underline{\gamma}_1 q_2^2\begin{pmatrix}\cosh \gamma_1\\ -1\end{pmatrix}\sinh \gamma_2 - \gamma_2 q_1^2\begin{pmatrix}\cosh \gamma_2\\ -1\end{pmatrix}\sinh \gamma_1\right]\Big/F$$

$$M_{1\left(\substack{3\\4}\right)} = -M_{2\left(\substack{4\\3}\right)} = \mu H_o\sigma\Delta^2\left[\gamma_1\begin{pmatrix}\cosh \gamma_1\\ -1\end{pmatrix}\sinh \gamma_2 - \gamma_2\begin{pmatrix}\cosh \gamma_2\\ -1\end{pmatrix}\sinh \gamma_1\right]\Big/F$$

$$M_{3\left(\substack{1\\2}\right)} = -M_{4\left(\substack{2\\1}\right)} = \frac{\eta}{\mu^2 H_o^2\sigma\Delta^2}q_1^2 q_2^2\left[\gamma_1 \sinh \gamma_2\begin{pmatrix}-\cosh \gamma_1\\ 1\end{pmatrix} - \underline{\gamma}_2 \sinh \gamma_1\begin{pmatrix}-\cosh \gamma_2\\ 1\end{pmatrix}\right]\Big/F$$

$$M_{3\left(\substack{3\\4}\right)} = -M_{4\left(\substack{4\\3}\right)} = \eta\left[\gamma_1 q_2^2\begin{pmatrix}\cosh \gamma_2\\ -1\end{pmatrix}\sinh \gamma_1 - \gamma_1 q_1^2\begin{pmatrix}\cosh \gamma_1\\ -1\end{pmatrix}\sinh \gamma_2\right]\Big/F$$

$$F = \Delta(\gamma_1^2 - \underline{\gamma}_2^2)\sinh \gamma_1 \sinh \gamma_2$$

Temporal Modes: Suppose that the layer is excited in the α and β planes by perfectly conducting rigid boundaries that (perhaps by dint of a displacement in the y direction) provide excitations $(\hat{v}_y^\alpha, \hat{v}_y^\beta)$. The perfect conductivity assures $\hat{A}^\alpha = 0$ and $\hat{A}^\beta = 0$ (Eq. 6.7.6). Thus, the electrical and mechanical variables on the right in Eq. 13 are determined. The temporal modes for this system (that represent the homogeneous response to initial conditions and underlie the driven response) are then given by F = 0. The roots of this equation are simply

$$\gamma_1 = jn\pi; \ \gamma_2 = jn\pi, \quad n = 1,2,\cdots \tag{14}$$

With these values of γ, Eq. 7 can be solved for the eigenfrequencies

$$\omega_n = j\frac{(n\pi)^2}{2}\left[\frac{1}{\tau_m} + \frac{1}{\tau_v}\right] \pm \sqrt{(n\pi)^2\left[\frac{1}{\tau_{MI}^2} - \frac{(n\pi)^2}{4}\left(\frac{1}{\tau_m} - \frac{1}{\tau_v}\right)^2\right]} \tag{15}$$

In the extreme where τ_m and τ_v are long compared to τ_{MI},

$$\omega_n = \pm\frac{n\pi}{\tau_{MI}} \tag{16}$$

This oscillatory natural frequency is the result of an Alfvén wave resonating between the boundaries. The wave transit time is $\tau_{MI} = \Delta/v_a$, so $v_a = \sqrt{\mu H_o^2/\rho}$ is the velocity of this Alfvén wave.

Typical of an experiment using a sodium-based liquid metal are the parameters

$$\left.\begin{array}{ll} \sigma = 10^6 \text{ mhos/m} & \Delta = 0.1 \text{ m} \\ \rho = 10^3 \text{ kg/m}^3 & \mu H_o = 1 \text{ tesla} \\ \eta = 10^{-3} \text{ newton-sec/m}^2 & \end{array}\right\} \begin{array}{l} \tau_v = 10^4 \text{ sec} \\ \tau_m = 1.25 \times 10^{-2} \text{ sec} \\ \tau_{MI} = 3.53 \times 10^{-3} \text{ sec} \end{array} \tag{17}$$

Thus, the characteristic times have the ordering $\tau_{MI} < \tau_m < \tau_v$ with the magnetic diffusion time far shorter than the viscous diffusion time. (The ratio of these times is sometimes defined as the magnetic Prandtl number $P_m = \tau_m/\tau_v = \eta\mu\sigma/\rho$. For the numbers given by Eq. 17, $P_m = 1.25 \times 10^{-6}$.) Thus, in Eq. 15, $1/\tau_v$ can be neglected compared to $1/\tau_m$ and it is seen that the natural frequency will display an oscillatory part if

$$\frac{\tau_m}{\tau_{MI}} > \frac{n\pi}{2} \tag{18}$$

That the transit time for the Alfvén wave be short compared to the time for appreciable magnetic diffusion underscores the flux-conserving nature of the wave dynamics. For the numbers of Eq. 17, $\tau_m/\tau_{MI} = 3.54$. As a practical matter, Alfvén waves observed in the laboratory are relatively damped. Note that as Δ increases, the inequality of Eq. 18 is better satisfied. The dependence of the natural frequency on the mode number n reflects how damping increases with the wave number $j\gamma$ in the x direction. Near the origin in Fig. 8.6.3, the linear relation of frequency and mode number is typical of nondispersive wave phenomena. As the mode number increases, magnetic (and possibly viscous) diffusion damps the oscillations, which then give way to totally damped modes. The oscillatory modes would of course appear as resonances in the sinusoidal steady-state driven response.

Spatial Structure of Sinusoidal Steady-State Response: The penetration of a sinusoidal excitation from the surfaces into the bulk is determined by γ_1 and γ_2, Eq. 9. As the magnetic field is raised, the viscous and magnetic skin effect are taken over by the electromechanical coupling. In Fig. 8.6.4, the transition of these complex wave numbers is shown, with the magnetic Hartmann number H_m representing the magnetic field. In terms of characteristic times, H_m is increased until the magneto-inertial time becomes sufficiently short that the Alfvén wave can penetrate the layer before the flux diffuses to its original uniform distribution. The magnetic shear stress is then able to penetrate the layer (tending to set the whole of it into motion) to a greater extent than would be possible via the magnetic or viscous diffusion alone. This is indicated by the lower of the roots shown, which has an imaginary part $\gamma \rightarrow \pm\sqrt{\tau_m\tau_v}/\Delta H_m = \pm\omega\tau_{MI}/\Delta$ as H_m becomes large. In this same limit of large H_m, the other branch becomes strongly decaying, with value $\gamma = \pm H_m/\Delta$. The physical nature of the dynamics represented by this mode is recognized by observing that $H_m = \sqrt{\tau_m/\tau_{MV}}$, where τ_{MV} is the magneto-viscous time. The electrical analogue of this time, which expresses the rate at which a process occurs involving a competition of viscous and magnetic stresses, will play an essential role in the next section. An experiment demonstrating Alfvén waves is sketched in Fig. 8.6.5.[1]

1. See also J. R. Melcher and E. P. Warren, "Demonstration of Magnetic Flux Constraints and a Lumped Parameter Alfvén Wave," IEEE Transactions on Education, Vol. E-8, Nos. 2 and 3, June-September, 1965, pp. 41-47.

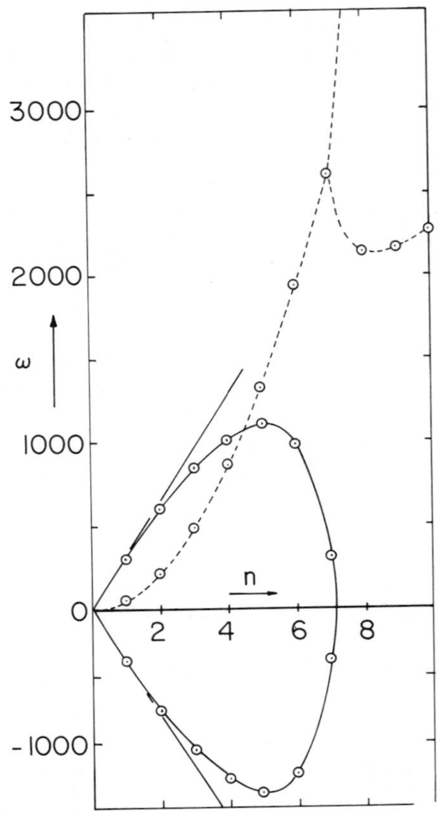

Fig. 8.6.3. Eigenfrequencies of temporal modes as a function of mode number for $\tau_{MI} = 0.01$, $\tau_m = 0.1$, and $\tau_v = 1$. ω_r ———, ω_i -----. H_m=31.6.

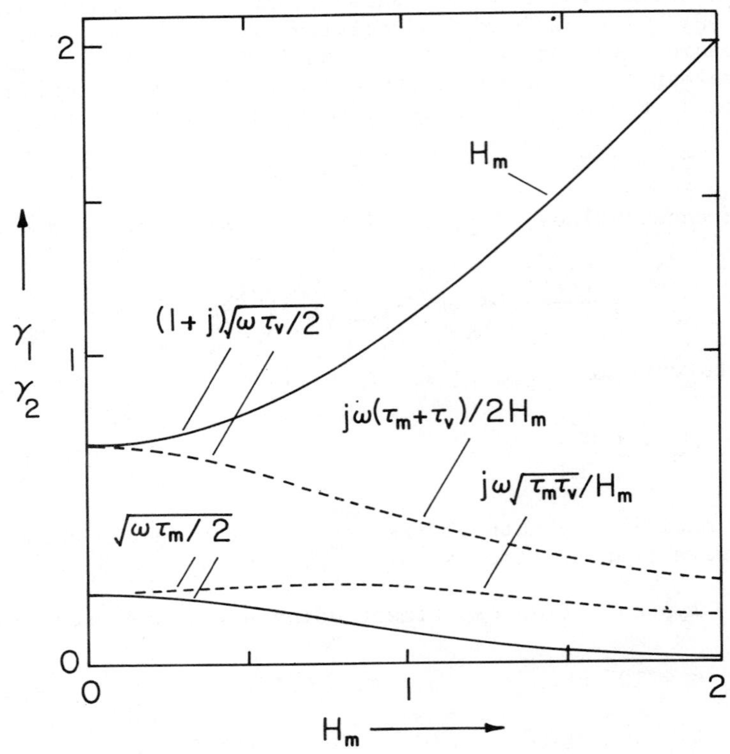

Fig. 8.6.4. Real (——) and imaginary (---) parts of γ_1 and γ_2 (Eq. 9) as functions of $H_m \equiv \Delta\mu H_o\sqrt{\sigma/\eta}$. Low- and high-$H_m$ approximations are shown. Note that the Alfvén wave branch is represented by $j\omega\sqrt{\tau_m\tau_v}/H_m = j\omega\tau_{MI}$.

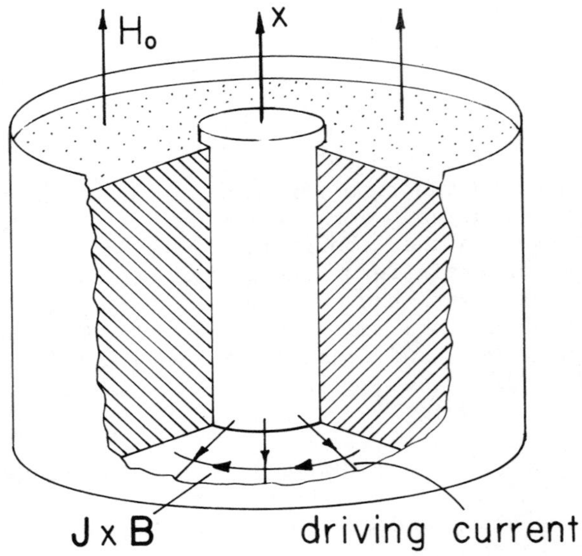

Fig. 8.6.5

Alfvén wave, as demonstrated by Shercliff in film "Magnetohydrodynamics" (Reference 7, Appendix C). Liquid NaK (sodium-potassium eutectic) fills conducting circular metal container having coaxial inner and outer walls. Wave is excited at bottom by radial driving current and detected at middle by coil that senses the change in magnetic field accompanying the passage of the upward-propagating electromechanical wave. As viewed radially inward, layers of liquid metal undergo shearing motions depicted by Fig. 8.6.1.

8.7 Potential Conserving Continua and Electric Shear Stress Instability

In an electric counterpart to the magnetic flux conserving fluid introduced in Sec. 8.6, a fluid element having fixed identity tends to retain its potential even as it moves. Under what physical circumstances could a homogeneous continuum tend to conserve its potential in this way? Figure 8.7.1 gives a schematic illustration (see Prob. 5.12.1 for charge relaxation in anisotropic conductors).

Initially, the volume is filled with static layers of miscible fluid having the same mechanical properties. Alternate layers are rendered conducting, perhaps by doping the same fluid as used for the other layers. At the upper and lower extremities, the conducting layers make electrical contact with

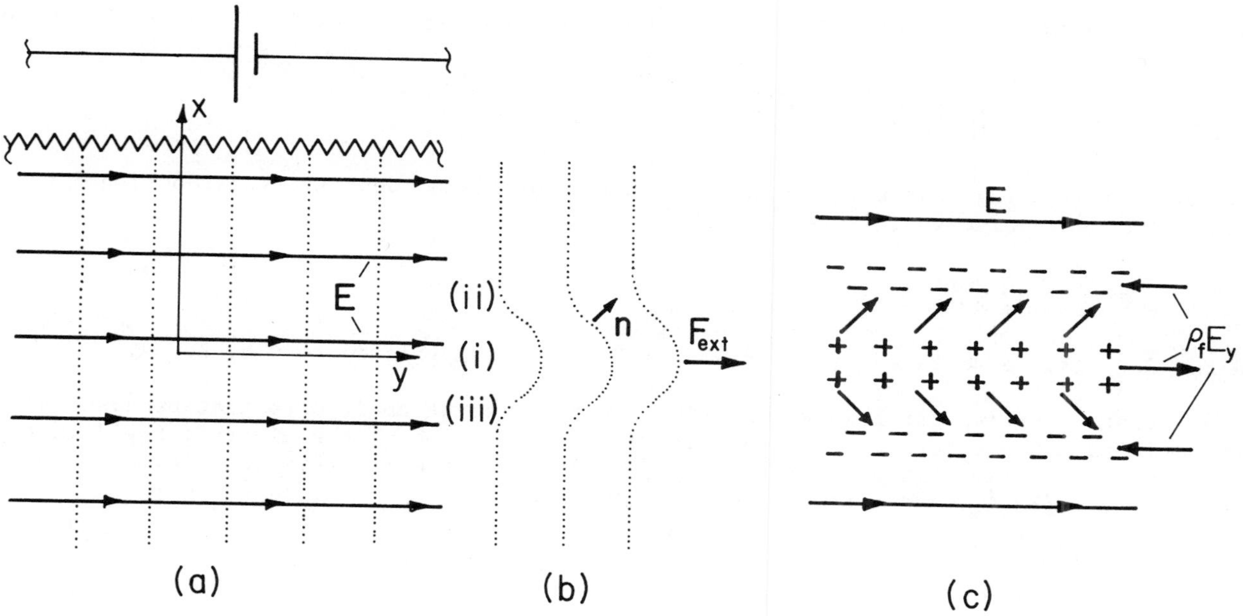

Fig. 8.7.1. (a) Example of potential conserving fluid made from numerous conducting layers buffered by relatively insulating layers. On a macroscale, a given fluid region tends to retain its potential as it deforms. (b) Shearing displacement causing elevation of potential in plane (i) relative to that at the same position y in planes (ii) and (iii). (c) Charge density implied by potential conservation, showing electrical force induced by the motion in adjacent layers.

surfaces having a linear potential distribution in the y direction. Thus, there is an initial ambient electric field $\vec{E} = E_o\vec{i}_y$ throughout the volume. What would be termed an isotropic inhomogeneous system on a microscale typified by the interlayer dimensions, is an anisotropic homogeneous system on the macroscale considered here. On this macroscale, a material element tends to retain its initial potential. In the model considered here, the conducting layers are of finite conductivity, but the layers between are considered perfect insulators. Just how faithfully the potential is conserved therefore depends on the electrical relaxation time of the composite.

By way of forming an intuitive impression of why the electric field induces instability, consider motions that are purely y-directed but depend on x. Suppose that the external force density $F_{ext}\vec{i}_y$ is used to translate a fluid layer in the y-z plane, denoted by (i) in Fig. 8.7.1b. To begin with, the potential of this and the adjacent layers decreases linearly in the y direction. So, at a given position along the y axis, the translation results in the potential in the plane (i) becoming elevated with respect to that of the adjacent layers (ii) and (iii). The adjacent layers form capacitor plates with the (i) layer which, in accordance with the relative potentials, are charged as sketched in Fig. 8.7.1c.

The field- and deformation-induced charge of the initially displaced layer, (i), are such that it is subject to an electrical force tending to further encourage the deformation. Thus, with the adjacent layer fixed, the external force would act against a negative spring constant. However, the adjacent layers are not fixed and experience electrical forces tending to carry them in a direction opposite that of the original displacement. There is an electrical shear stress acting between adjacent layers that is proportional to the negative of the strain. By contrast with the magnetic shear stress that gives rise to Alfvén waves, the electric stress tends to cause instability.

The laws needed to formulate a model begin with a constitutive law for the conduction. With $\vec{n}$ defined as a unit normal to a material surface of fixed identity that is initially in an x-z plane, as shown in Fig. 8.7.1b, the component of the electric field that is tangential to this surface is $-\vec{n} \times \vec{n} \times \vec{E}$. Thus, if the average conductivity in the plane of the conducting layer is σ, the current density in a stationary sample of the anisotropic material is

$$\vec{J}_f' = -\sigma\vec{n} \times \vec{n} \times \vec{E} \tag{1}$$

Because $\vec{J}_f = \vec{J}_f' + \rho_f\vec{v}$, it follows that the statement of charge conservation, Eq. 2.3.25a, is

$$\nabla\cdot[-\sigma(\vec{n} \times \vec{n} \times \vec{E}) + \rho_f\vec{v}] + \frac{\partial\rho_f}{\partial t} = 0 \tag{2}$$

The normal vector can be eliminated from this expression by first expressing it in terms of the surface $y = \xi(x,t)$

$$\vec{n} = [\vec{i}_y - \frac{\partial \xi}{\partial x}\vec{i}_x][1 + (\frac{\partial \xi}{\partial x})^2]^{-\frac{1}{2}} \tag{3}$$

and then recognizing that because this surface is of fixed identity, the function $F = y - \xi$ must have a convective derivative that is zero (Sec. 7.5):

$$v_y = \frac{\partial \xi}{\partial t} + v_x \frac{\partial \xi}{\partial x} \tag{4}$$

In Eq. 2, $\vec{n}$ can be replaced by Eq. 3, where ξ is in turn related to $\vec{v}$ by Eq. 4.

Before carrying out this elimination for the case at hand, note that because the electric field is irrotational and the perturbation quantities only depend on x, the electric field in the y direction is not a function of x. Pinned at E_o in any y-z plane, E_y remains this value even as the fluid deforms: $\vec{E} = E_o\vec{i}_y - (\partial\Phi/\partial x)\vec{i}_x$. As a result, Gauss' Law becomes

$$\frac{\partial^2 \Phi}{\partial x^2} = -\frac{\rho_f}{\varepsilon} \tag{5}$$

The motions considered are only in the y direction: $\vec{v} = v_y(x,t)\vec{i}_y$. With this understanding, Eqs. 2, 3 and 4 are linearized and combined to eliminate ξ, and Eq. 5 is substituted for ρ_f, to obtain

$$\frac{\partial^2}{\partial x^2}[E_o v_y - \frac{\partial}{\partial t}(\Phi + \frac{\varepsilon}{\sigma}\frac{\partial \Phi}{\partial t})] = 0 \tag{6}$$

This statement of the effect of the motion on the fields reduces to the linearized version of $D\Phi/Dt = 0$ in the limit where the charge relaxation time, ε/σ, is short compared to times of interest. If the charge can relax instantaneously, the potential of an element of fluid is conserved even as it deforms.

The y component of the force equation, Eq. 7.16.6 with $\nabla\cdot\vec{v} = 0$ and $\vec{F}_{ex}$ represented by the divergence of the stress tensor (given with Eq. 3.7.22 of Table 3.10.1), is

$$\rho\frac{\partial v_y}{\partial t} = -\varepsilon E_o \frac{\partial^2 \Phi}{\partial x^2} + \eta\frac{\partial^2 v_y}{\partial x^2} \tag{7}$$

The x-component simply determines the pressure distribution required to equilibrate the x component of the electrical force density. Equations 6 and 7 represent the electromechanical coupling.

The quantity in brackets in Eq. 6 is zero throughout the volume when the fluid is in static equilibrium. Hence, the two constants resulting from integrating Eq. 6 twice on x are zero. Then, with the substitutions $v_y = \text{Re}\hat{v}_y(x)e^{j\omega t}$ and $\Phi = \text{Re}\hat{\Phi}(x)e^{j\omega t}$, Eqs. 6 and 7 become

$$E_o\hat{v}_y = j\omega[1 + \frac{j\omega\varepsilon}{\sigma}]\hat{\Phi} = 0 \tag{8}$$

$$(j\omega\rho - \eta\frac{d^2}{dx^2})\hat{v}_y + \varepsilon E_o\frac{d^2\hat{\Phi}}{dx^2} = 0 \tag{9}$$

By contrast with the magnetohydrodynamic system represented by Eqs. 8.6.5 and 8.6.6, the system is only second order in x, so that there are only two boundary conditions that can be imposed on a layer having the thickness Δ (Fig. 8.6.2). Imposing a boundary condition on $\hat{\Phi}$ is (through Eq. 8) tantamount to a condition on v_y. Substitution into Eqs. 8 and 9 of solutions having the form $\hat{v}_y = \exp(\gamma x)$ and $\hat{\Phi} = \exp(\gamma$ gives a pair of homogeneous relations

$$\begin{bmatrix} E_o & -j\omega\left(1 + \frac{j\omega\varepsilon}{\sigma}\right) \\ j\omega\rho - \eta\gamma^2 & \varepsilon E_o\gamma^2 \end{bmatrix}\begin{bmatrix} v_y \\ \hat{\Phi} \end{bmatrix} = 0 \tag{10}$$

and the requirement that the determinant of the coefficients vanish gives an expression for the allowed values of γ:

$$\gamma = \pm\gamma_1; \quad \gamma_1 \equiv \sqrt{\frac{j\omega\rho}{\eta + \dfrac{j\varepsilon E_o^2}{\omega(1 + \dfrac{j\omega\varepsilon}{\sigma})}}} \tag{11}$$

The situation is now no different than in dealing with Laplace's equation, where solutions take the form of Eq. 2.16.15 with $\gamma \to \gamma_1$. Thus, the transfer relation for the layer is (Table 2.16.1):

$$\begin{bmatrix} \hat{D}_x^\alpha \\ \hat{D}_x^\beta \end{bmatrix} = \frac{\gamma_1\varepsilon}{\sinh(\gamma_1\Delta)} \begin{bmatrix} -\cosh(\gamma_1\Delta) & 1 \\ -1 & \cosh(\gamma_1\Delta) \end{bmatrix} \begin{bmatrix} \hat{\Phi}^\alpha \\ \hat{\Phi}^\beta \end{bmatrix} \tag{12}$$

In terms of these variables, the mechanical variables follow from Eq. 8 as

$$\hat{v}_y = \frac{j\omega}{E_o}\left[1 + \frac{j\omega\varepsilon}{\sigma}\right]\hat{\Phi} \tag{13}$$

$$\hat{S}_{yx} = \eta\frac{d\hat{v}_y}{dx} = \frac{j\omega\eta}{E_o}\left[1 + \frac{j\omega\varepsilon}{\sigma}\right]\frac{d\hat{\Phi}}{dx} \tag{14}$$

Temporal Modes: Because the system is unstable, the temporal modes are of most interest. For a system bounded by planes maintaining the linear equilibrium distribution in potential (constrained to zero perturbation potential), the condition on ω resulting from there being a finite solution $(\hat{D}_x^\alpha, \hat{D}_x^\beta)$ with $(\hat{\Phi}^\alpha, \hat{\Phi}^\beta) = 0$ is $\sinh(\gamma_1\Delta) = 0$. Thus, the eigenvalues are

$$\gamma_1\Delta = jn\pi, \quad n = 1,2,3\cdots \tag{15}$$

The eigenfrequencies follow by substituting γ_1 from this expression into Eq. 11. The result is a cubic equation which determines the allowed frequencies ω:

$$\omega^3 - \omega^2 j\left[\frac{(n\pi)^2}{\tau_v} + \frac{1}{\tau_e}\right] - \omega\frac{(n\pi)^2}{\tau_v\tau_e} - \frac{j(n\pi)^2}{\tau_e\tau_v\tau_{EV}} = 0 \tag{16}$$

$$\tau_v \equiv \frac{\rho\Delta^2}{\eta}; \quad \tau_e \equiv \frac{\varepsilon}{\sigma}; \quad \tau_{EV} = \frac{\eta}{\varepsilon E_o^2}$$

As a function of the mode number $n\pi$, the solutions $s_n = j\omega$ of this expression are illustrated in Fig. 8.7.2. For each sinusoidal distribution represented by a given n, there are three temporal modes, one unstable and two decaying.

Typical of a 2-cm liquid layer having 50 times the viscosity of water, the density of water, an electrical relaxation time of 10^{-2} sec and $E_o = 2 \times 10^{+5}$ V/m are the times given in the caption. Note that $\tau_e < \tau_{EV} < \tau_v$.

The roots to Eq. 16 in the limit $\tau_e \to 0$ give a good idea of what is happening on time scales long compared to τ_e. The quadratic limit of Eq. 16 can then be solved to give

$$s = \frac{(n\pi)^2}{2\tau_v}\left[-1 \pm \sqrt{1 + \frac{4\tau_v}{\tau_{EV}(n\pi)^2}}\right] \tag{17}$$

Thus, there are roots $s_n > 0$ representing an exponentially growing instability. The fastest growing modes are those having the largest number of wavelengths in the x direction. In the limit $n\pi \to \infty$, this mode has a growth rate τ_{EV}. (In fact, there would be a finite mode exhibiting the maximum rate of growth, since wavelengths in the x direction shorter than the distance between layers are not described by the model.) By contrast with the electro-viscous nature of the short-wavelength instability, the long wavelengths (small mode numbers) are electro-inertial in nature. In the limit $n\pi \to 0$, Eq. 17 reduces to $s_n = 1/\tau_{EI}$, where $\tau_{EI} = \sqrt{\tau_v\tau_{EV}} = \Delta\sqrt{\rho\varepsilon E^2}$. Until its rate of decay becomes comparable to τ_e, the decaying mode can also be approximated using Eq. 17. At short wavelengths, the basically viscous diffusion mode and charge relaxation mode couple to produce a pair of modes that are damped in a sinusoidal fashion.

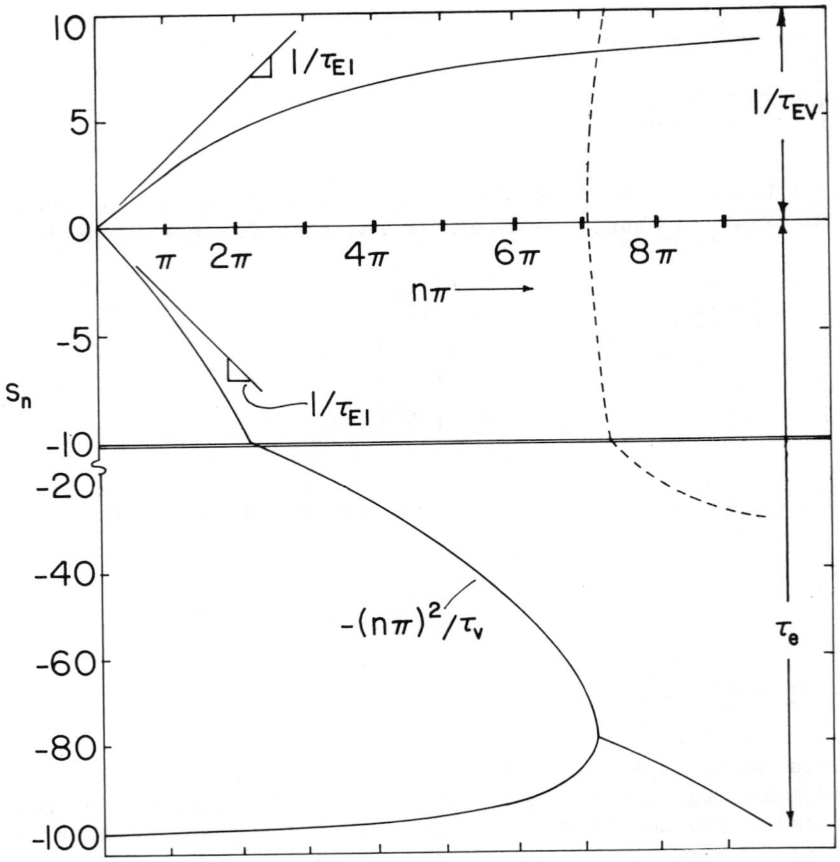

Fig. 8.7.2. Frequencies of temporal eigenmodes, $s_n = j\omega$; --- $(s_n)_r$, —— $(s_n)_i$. For each n there are three modes. $\tau_e = 10^{-2}$ sec, $\tau_{EV} = 0.1$ sec, $\tau_v = 10$ sec.

The instability is fundamental to many situations where electric fields are used to augment mass, heat and momentum transfer. Usually a more complicated model is required even to recognize the linear stages of instability. Shown in Fig. 8.7.3 is an example for which the illustration given in this section is itself a useful model. The Couette mixer exploits a rotating inner cylinder to promote large scale mixing. Two liquids entering at the bottom are typically the highly viscous components of a polymer. Because of the rotation, these form laminae of relatively insulating and conducting liquids that work their way upward to the exit. With the application of a radial electric field, instability leads to mixing. The electrohydrodynamic instability provides mixing on a length scale that bridges the gap between what can be efficiently produced by the mechanical stirring and what is required to insure

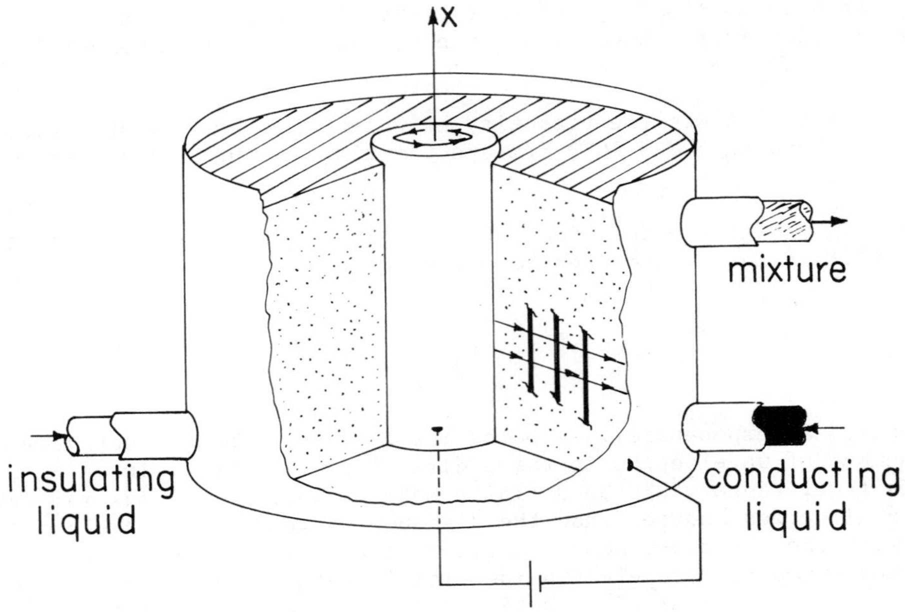

Fig. 8.7.3

Couette mixer exploiting instability of components stressed by electric field.

genuine molecular scale mixing.[1] For successful operation the residence time of the liquids must at least exceed $\tau_{EV} = \eta/\varepsilon E_o^2$. Even in its nonlinear stages and on length scales shorter than the distance between layers, τ_{EV} is found to scale the rate at which mixing processes occur.[2,3] In practical applications, the "insulating" component actually is itself semi-insulating so the growth rate for instability is reduced by a factor reflecting the ratio of the component conductivities.

8.8 Magneto-Acoustic and Electro-Acoustic Waves

Electromechanical coupling through dilatational deformation is illustrated in this section. First considered as one-dimensional examples are perfectly conducting limits of the MQS and EQS continua of Secs. 8.6 and 8.7, respectively. Then, the incremental motions of a system of magnetizable particles randomly suspended in a uniform magnetic field are modeled.

Both the MQS and EQS configurations are shown in Fig. 8.8.1. Also shown in each case are the distributed elements that embody the same physical phenomena as represented by the continuum models. Without electromechanical coupling, the one-dimensional acoustic wave propagates through a continuum of masses (represented by the perfectly conducting plates) interconnected by layers of fluid comprising the springs.

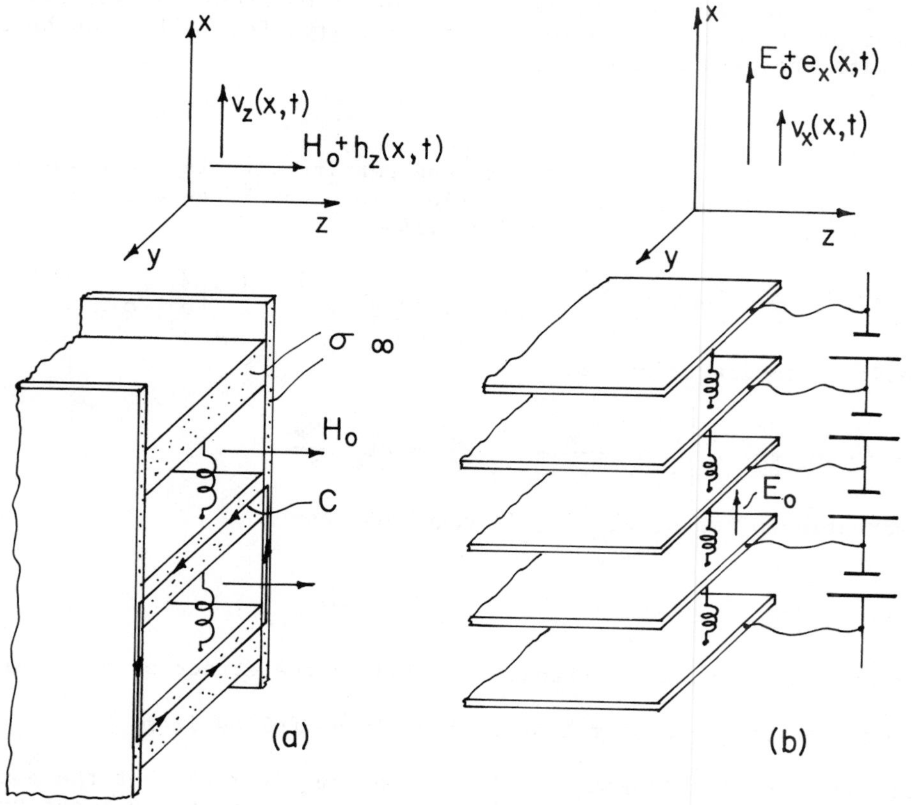

Fig. 8.8.1. One-dimensional compressional motions. (a) Magneto-acoustic waves in perfectly conducting liquid across uniform magnetic field. (b) electro-acoustic waves in potential conserving continuum along uniform electric field. Lumped models emphasize salient features of dynamics.

In the magnetohydrodynamic case, the fluid is uniform and perfectly conducting. When at rest, it is permeated by a uniform magnetic field H_o directed transverse to the direction of propagation. Compression of the fluid results in a decrease in enclosed area for a contour such as C which is attached to the fluid. To retain the same flux linkage, a current is induced around this contour. The associated force density tends to counteract the dilatation, thus having the effect of a magnetic spring between elements. It is not surprising that the magnetic field tends to increase the velocity of propagation of waves.

---

1. C. A. Rotz, "A Generalized Approach to Increased Mixing Efficiency for Viscous Liquids," S.M. Thesis, Department of Mechanical Engineering, Massachusetts Institute of Technology, Cambridge, Mass., 1976.

2. J. H. Lang, J. F. Hoburg and J. R. Melcher, "Field Induced Mixing Across a Diaphragm," Phys. Fluids 19, 917 (1976).

3. J. F. Hoburg and J. R. Melcher, "Electrohydrodynamic Mixing and Instability Induced by Collinear Fields and Conductivity Gradients," Phys. Fluids 20, 903 (1977).

In the electrohydrodynamic case, a given element of fluid conserves its potential, as described in Sec. 8.7. Either the fluid is a stratification of insulating and conducting components, or it actually consists of thin conducting sheets dispersed through the fluid. Because the motions are compressional, such sheets would not inhibit the motions. The equivalent distributed lumped parameter system, shown in Fig. 8.7.1b, consists of perfectly conducting layers constrained to have the same potential difference even as their relative spacing changes. As a "plate" approaches one of its neighbors, the intervening electric field increases. So also does the electric force associated with the charge on that side of the plate. Thus, the electric field is equivalent in its effect to a spring with a negative spring constant. It has the effect of diminishing the stiffness of the "spring" separating a pair of plates. The field is expected to reduce the velocity of a wave propagating in the x direction.

Now, consider the interactions in analytical terms. In both cases, the linearized longitudinal force equation is simply

$$\rho_o \frac{\partial v_x}{\partial t} + \frac{\partial p'}{\partial x} = \frac{\partial T_{xx}}{\partial x} \tag{1}$$

where ρ_o is the equilibrium mass density, p' is the perturbation pressure, and T_{xx} is the Maxwell stress. With the assumption that pressure is only a function of density, Eq. 7.11.3 can be used to replace the perturbation pressure with the perturbation density,

$$p' = a^2\rho' \tag{2}$$

where a is the acoustic velocity. The permeability and permittivity in the respective situations are taken as constant. Thus, with $\vec{h}$ and $\vec{e}$ the perturbations in $\vec{H}$ and $\vec{E}$ respectively, to linear terms, T_{xx} becomes simply (Table 3.10.1, Eqs. 3.7.22 and 3.8.14)

$$T_{xx} \simeq -\frac{\mu}{2}(H_o+h_z)^2 \simeq -\frac{1}{2}\mu H_o^2 - \mu H_o h_z \quad \Bigg| \quad T_{xx} \simeq \frac{1}{2}\varepsilon(E_o+e_x)^2 \simeq \frac{1}{2}\varepsilon E_o^2 + E_o e_x \tag{3}$$

These last three equations combine to become

$$\rho_o \frac{\partial v_x}{\partial t} + a^2\frac{\partial \rho'}{\partial x} = -\mu H_o \frac{\partial h_z}{\partial x} \quad \Bigg| \quad \rho_o \frac{\partial v_x}{\partial t} + a^2\frac{\partial \rho'}{\partial x} = \varepsilon E_o \frac{\partial e_x}{\partial x} \tag{4}$$

To linear terms, conservation of mass, Eq. 7.2.3, requires that

$$\frac{\partial \rho'}{\partial t} + \rho_o \frac{\partial v_x}{\partial x} = 0 \tag{5}$$

These last two statements represent the mechanics, including the effect of the fields.

The reciprocal effects of the deformation on the fields follow from

the requirement that the flux linked by a surface of fixed identity be constant, Eq. 8.6.1. To linear terms	the requirement that the potential, Φ, of an element of fixed identity be constant, Eq. 8.7.1. To linear terms

$$H_o \frac{\partial v_x}{\partial x} = -\frac{\partial h_z}{\partial t} \quad \Bigg| \quad \frac{\partial \Phi'}{\partial t} - E_o v_x = 0 \tag{6}$$

where $e_x = -\nabla\Phi'$

To combine these last three statements, take the time derivative of Eq. 4 and the space derivative of Eqs. 5 and 6 and eliminate p and h_z or e_x:

$$\frac{\partial^2 v_x}{\partial t^2} = a_m^2 \frac{\partial^2 v_x}{\partial x^2} \quad \Bigg| \quad \frac{\partial^2 v_x}{\partial t^2} = a_e^2 \frac{\partial^2 v_x}{\partial x^2} \tag{7}$$

These wave equations make it clear that the effect of the fields is to replace the acoustic velocity with a magneto-acoustic velocity:

$$a_m = \sqrt{a^2 + \frac{\mu H_o^2}{\rho_o}} \quad \Bigg| \quad a_e = \sqrt{a^2 - \frac{\varepsilon E_o^2}{\rho_o}} \tag{8}$$

Acoustic velocities, given in Table 7.11.1, are typically 300 m/sec in gases and 1500 m/sec in liquids. In gases, the Alfvén velocity, $\sqrt{\mu H_o^2/\rho_o}$, can be made to dominate in its contribution to the magneto-acoustic velocity. In liquid metals the magnetic contribution to a_m is greatly reduced by the increased mass density, although it is still possible for it to be significant. But in the electro-acoustic wave, electrical breakdown limits the effect of the electric field to a level that would make it difficult to even measure the effect.

Magnetization Dilatational Waves: Although electromechanical effects on dilatational motions in natural materials are likely to be small, continua formed from "molecules" that are actually macroscopic in their dimensions can give rise to significant electromechanical effects. As an example, magnetizable spheres are suspended in a random array, with the voidage a gas or even vacuum. Interest is confined to deformations characterized by lengths that are large compared to the distance between particles. Unperturbed, the system is uniform on the macroscopic scale, and is subjected to a uniform z-directed magnetic field intensity H_o. Because the spheres can interact with each other only through the magnetic field, the pressure is taken as zero.

Perhaps determined experimentally, the effective permeability of the continuum has been related to the mass density through a constitutive law, $\mu = \mu(\rho)$. Thus, the force density of Eq. 3.8.17 from Table 3.10.1 is applicable. With perturbations from the equilibrium mass density and magnetic field, ρ_o and $H_o\vec{i}_z$, denoted by ρ' and $\vec{h}$, respectively, this force density is linearized to become

$$\vec{F} = \rho_o\nabla[H_o(\frac{\partial\mu}{\partial\rho})_o h_z + \frac{1}{2}H_o^2(\frac{\partial^2\mu}{\partial\rho^2})_o\rho'] \tag{9}$$

Because there are no free currents, $\vec{H}$ is irrotational and hence $\vec{H} = H_o\vec{i}_z - \nabla\psi$. Thus, the force equation, Eq. 7.4.4 written with $p = 0$, is

$$\rho_o\frac{\partial\vec{v}}{\partial t} = -\rho_o H_o(\frac{\partial\mu}{\partial\rho})_o\nabla(\frac{\partial\psi}{\partial z}) + \frac{\rho_o}{2}H_o^2(\frac{\partial^2\mu}{\partial\rho^2})_o\nabla\rho' \tag{10}$$

Mass conservation is represented by a linearized version of Eq. 7.2.3:

$$\frac{\partial\rho'}{\partial t} + \rho_o\nabla\cdot\vec{v} = 0 \tag{11}$$

In terms of the scalar potential, ψ, the linearized statement that $\mu\vec{H}$ is solenoidal is

$$-\mu(\rho_o)\nabla^2\psi + H_o(\frac{\partial\mu}{\partial\rho})_o\frac{\partial\rho'}{\partial z} = 0 \tag{12}$$

To obtain an expression for ρ' alone, the divergence of Eq. 10 is taken. Then Eq. 11 eliminates $\nabla\cdot\vec{v}$, while the $\partial(\ )/\partial z$ of Eq. 12 can be used to eliminate ψ. Thus, the expressions combine to give

$$\frac{\partial^2\rho'}{\partial t^2} = \frac{\rho_o H_o^2}{\mu(\rho_o)}(\frac{\partial\mu}{\partial\rho})_o^2\frac{\partial^2\rho'}{\partial z^2} - \frac{\rho_o}{2}H_o^2(\frac{\partial^2\mu}{\partial\rho^2})_o\nabla^2\rho' \tag{13}$$

A possible relation between permeability and mass density is the Clausius-Mossotti law:[1]

$$\frac{(\frac{\mu}{\mu_o}-1)}{(\frac{\mu}{\mu_o}+2)} = C\rho \Rightarrow \frac{\partial\mu}{\partial\rho} = \frac{\mu_o}{3}(\frac{\mu}{\mu_o}+2)(\frac{\mu}{\mu_o}-1)\rho^{-1} \Rightarrow \frac{\partial^2\mu}{\partial\rho^2} = \frac{2\mu_o}{9}(\frac{\mu}{\mu_o}-1)^2(\frac{\mu}{\mu_o}+2)\rho^{-2} \tag{14}$$

where C is determined by the nature of the spheres.

It follows from Eqs. 13 and 14 that compressional motions across the field lines (in the x direction) are unstable, while those in the direction of the field propagate with the velocity

$$a_M = \sqrt{\frac{\mu_o H_o^2}{\rho_o}\frac{2}{9}(\frac{\mu}{\mu_o}+2)(\frac{\mu}{\mu_o}-1)^2\frac{\mu_o}{\mu}} \tag{15}$$

---

1. J. A. Stratton, Electromagnetic Theory, McGraw-Hill Book Company, New York, 1941, p. 140.

8.9 Gravity-Capillary Dynamics

The incompressible dynamics of fluids that are inhomogeneous in mass density are as commonplace as wave motions in a teacup or at the interface between sea and atmosphere. At the interface, the mass density suffers a step discontinuity. Fundamentally, the pertinent laws express the fact that the mass density in the neighborhood of a particle of fixed identity remains constant, Eq. 7.2.4, that mass is conserved, Eq. 7.2.5, and that inertial and pressure forces balance. For the present purposes the fluid is represented as being inviscid, and hence the pertinent force law is Eq. 7.4.4 with the external force density that due to gravity, $\vec{F}_{ex} = \rho\vec{g}$.

Because inhomogeneities in electrical properties are often accompanied by variations in mass density, electromechanical interactions with inhomogeneous systems are commonly interwoven with the fluid mechanics resulting from effects of gravity. In this section, the mechanics of a fluid interface illustrates effects of gravity in systems that are inhomogeneous in mass density. If the interface is between immiscible fluids, effects of capillarity are also important.

In the configuration shown in Fig. 8.9.1, planar layers of fluid each have uniform properties designated by the subscripts "a" (above) and "b" (below), respectively, and a common interface at $x = \xi(y,z,t)$. The lower fluid rests on a rigid boundary while the upper one consists of a deformable structure. The system is driven from this structure by the traveling-wave excitation shown in the figure. What is the response of the fluids, and in particular of their interface?

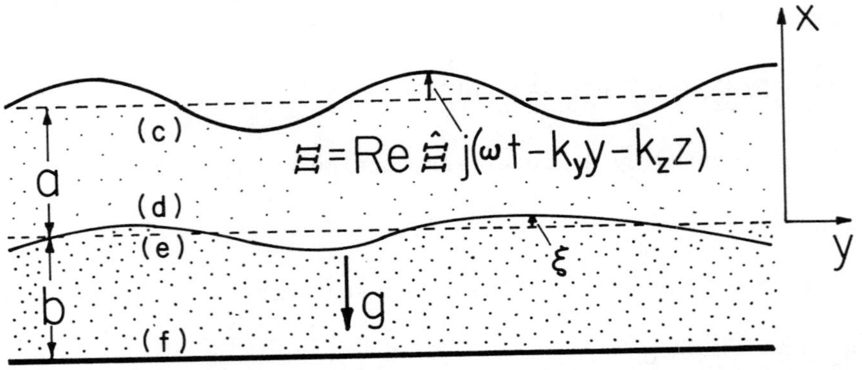

Fig. 8.9.1. Fluids of differing mass densities have interface at ξ and are driven by structure at Ξ.

In the absence of the excitation, the fluids are in static equilibrium with the gravitational force density. Thus, the fluid velocity $\vec{v} = 0$ and the pressure balances the gravitational force density. From the force equation, Eq. 7.8.3, applied to each region:

$$p = \begin{cases} -\rho_a g x + \Pi_a; & x > 0 \\ -\rho_b g x + \Pi_b; & x < 0 \end{cases} \tag{1}$$

Perturbations from this static equilibrium are represented in terms of complex amplitudes. To linear terms the pressure and velocity are

$$p = -\rho g x + \Pi + p'(x,y,z,t);\ p' = \mathrm{Re}\,\hat{p}(x)\exp j(\omega t - k_y y - k_z z) \tag{2}$$

$$\vec{v} = \mathrm{Re}\,\hat{\vec{v}}(x)\exp j(\omega t - k_y y - k_z z) \tag{3}$$

Within a given fluid region the mass density is uniform. Thus, the complex amplitudes in the respective planes designated in Fig. 8.9.1 are related by the transfer relations for an inviscid fluid given by Eq. (c) of Table 7.9.1:

$$\begin{bmatrix}\hat{p}^c \\ \hat{p}^d\end{bmatrix} = \frac{j\omega\rho_a}{k}\begin{bmatrix}-\coth(ka) & \dfrac{1}{\sinh(ka)} \\ \dfrac{-1}{\sinh(ka)} & \coth(ka)\end{bmatrix}\begin{bmatrix}\hat{v}_x^c \\ \hat{v}_x^d\end{bmatrix};\quad \begin{bmatrix}\hat{p}^e \\ \hat{p}^f\end{bmatrix} = \frac{j\omega\rho_b}{k}\begin{bmatrix}-\coth(kb) & \dfrac{1}{\sinh(kb)} \\ \dfrac{-1}{\sinh(kb)} & \coth(kb)\end{bmatrix}\begin{bmatrix}\hat{v}_x^e \\ \hat{v}_x^f\end{bmatrix} \tag{4}$$

Complex amplitudes are evaluated in the equilibrium planes. But, the jump conditions apply wherever the interface is actually located and that location is in fact yet to be determined! This difficulty is sidestepped by linearizing the jump conditions in such a way that they are expressed in terms of perturbation variables evaluated at the equilibrium positions of the boundaries.

Taking boundary and jump conditions from top to bottom, observe first that the position of the deformable upper structure is related to the velocity of the adjacent fluid by Eq. 7.5.5, which to linear terms is

$$\hat{v}_x^c = j\omega\hat{\Xi} \tag{5}$$

where it is appropriate to use the complex amplitude evaluated at the equilibrium position because the difference between that and $\hat{v}_x$ $(x = a + \Xi)$ is second order in the perturbation amplitude, Ξ.

Similarly, at the interface the velocities are related to the interfacial deformation by

$$\hat{v}_x^d = j\omega\hat{\xi}; \quad \hat{v}_x^e = j\omega\hat{\xi} \tag{6}$$

Again, this jump condition, which expresses mass conservation for the interface, has been written in terms of amplitudes evaluated at the equilibrium interfacial position. Stress balance for the interface is represented by Eq. 7.7.6, which has only a normal component. To linear terms, this is represented by the i = x component

$$[-\rho_a g\xi + \Pi_a + p'^d(x=\xi)] - [-\rho_b g\xi + \Pi_b + p'^e(x=\xi)] = \gamma\left(\frac{\partial^2\xi}{\partial y^2} + \frac{\partial^2\xi}{\partial z^2}\right) \tag{7}$$

where the surface tension force density is given by Eq. (c) of Table 7.6.1. For static equilibrium, $\Pi_a - \Pi_b = 0$. Also, to linear terms the perturbation pressures evaluated at the perturbed position ξ are equal to these pressures evaluated at the equilibrium position of the interface. Thus, Eq. 7 reduces to

$$\hat{p}^d - \hat{p}^e = g\hat{\xi}(\rho_a - \rho_b) - \gamma k^2\hat{\xi} \tag{8}$$

It is because the fluid is inviscid that the other two components of the interfacial stress balance equation are, to linear terms, identically satisfied. Finally, on the rigid lower boundary

$$\hat{v}_x^f = 0 \tag{9}$$

The boundary and jump conditions, Eqs. 5, 6, 8 and 9, are now used to "splice" together the bulk solutions represented by Eqs. 4. Of the four equations summarized by these relations, the expressions for $\hat{p}^c$ and $\hat{p}^f$ simply serve to determine these pressures once the fluid motions have been determined. The other two, Eqs. 4b and 4d, are evaluated using the boundary conditions, Eqs. 6, 7 and 10, and substituted into the stress balance condition, Eq. 9, to obtain

$$-\frac{\omega^2}{k}[\rho_a \coth(ka) + \rho_b \coth(kb)]\hat{\xi} + [\gamma k^2 + g(\rho_b - \rho_a)]\hat{\xi} = -\frac{\omega^2\rho_a}{k\sinh(ka)}\hat{\Xi} \tag{10}$$

This relation has the same form as would be used to describe the deflections of a spring attached to a mass at one end and to a displacement source at the other. The "mass" reflects the inertia of the fluids to either side of the interface while the "spring" results from the combined gravitational and capillary forces.

From Eq. 10, it follows that the complex amplitude of the interfacial response is

$$\hat{\xi} = -\frac{\omega^2\rho_a}{k\sinh(ka)}\,\frac{\hat{\Xi}}{D(\omega,k)} \tag{11}$$

where the dispersion equation, $D(\omega,k)$, is

$$D(\omega,k) = -\frac{\omega^2}{k}[\rho_a \coth(ka) + \rho_b \coth(kb)] + [\gamma k^2 + g(\rho_b - \rho_a)] \tag{12}$$

Driven Response: The response having the same wave number and frequency as the drive would represent all of the motions if the system were reentrant in the direction of the traveling wave and sufficient time had elapsed for a temporal sinusoidal state to be established. (This presumes that the temporal natural modes are stable.) Under the assumption that $\gamma k^2 + g(\rho_b - \rho_a) > 0$ (which is assured regardless of wavelength if the lower fluid is the heavier), the frequency response of the interface is as shown in Fig. 8.9.2. Because there are no dissipation mechanisms included in the model, the interface is either in phase or 180° out of phase with the excitation.

Gravity-Capillary Waves: The resonance comes at that frequency that gives synchronism between the phase velocity ω/k of the drive and phase velocity of a gravity-capillary wave propagating on the interface. Solution for ω/k of Eq. 13 set equal to zero identifies the phase velocity of these waves as

$$v_p = \sqrt{\frac{\gamma k + g(\rho_b - \rho_a)/k}{\rho_a \coth(ka) + \rho_b \coth(kb)}} \tag{13}$$

Long waves are dominated by gravity while short ones are of a capillary nature. Often, the waves are short enough that effects of the transverse boundaries are not significant, $|ak| >> |bk| >> 1$. Then, Eq. 13 reduces to

$$v_p = \sqrt{\frac{\gamma k}{\rho_a + \rho_b} + \frac{g(\rho_b - \rho_a)}{k(\rho_b + \rho_a)}} \tag{14}$$

This makes it evident that there is a wave number for minimum phase velocity, found by setting the derivative with respect to k of Eq. 14 equal to zero. The wavelength, $2\pi/k$, of this minimum will be termed the Taylor wavelength, λ_T:

$$\lambda_T = 2\pi \sqrt{\frac{\gamma}{g(\rho_b - \rho_a)}} \tag{15}$$

At wavelengths longer than λ_T, gravity waves prevail, while shorter wavelengths represent capillary ripples. For an air-water interface, $\lambda_T = 1.7$ cm.

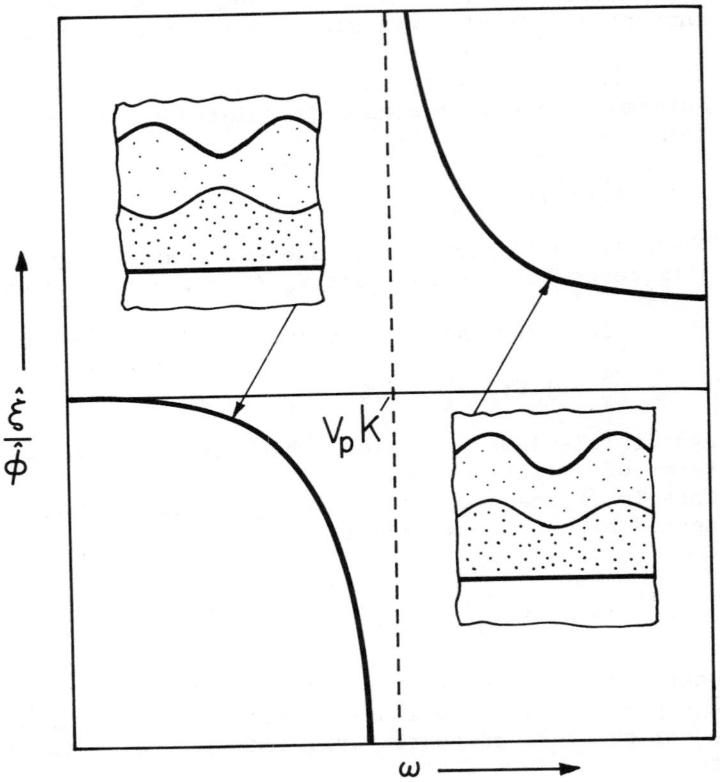

Fig. 8.9.2. Driven response of gravity-capillary wave system.

In the opposite limit of long waves, $|ka| << 1$ and $|kb| << 1$, the phase velocity becomes

$$v_p = \sqrt{\frac{\gamma k^2 + g(\rho_b - \rho_a)}{[(\rho_a/a) + (\rho_b/b)]}} \tag{16}$$

and the gravity wave (which is likely to dominate in a long-wave situation) propagates without dispersion. A quasi-one-dimensional model for long gravity waves results in the wave equation with a velocity given by Eq. 16 without the capillary term.

Temporal Eigenmodes and Rayleigh-Taylor Instability: Temporal transients, initiated from conditions that are periodic in the horizontal plane, are described by $D(\omega,k) = 0$ with k real and $j\omega$ the eigenfrequencies s_n. The role of the temporal modes in this chapter is very much as introduced in Sec. 5.15. The roots of $D(s_n,k) = 0$ are either purely real or imaginary. Resonance in the driven response results from the coincidence of the natural frequency and the driving frequency. Of most interest is the instability resulting from having the heavier fluid on top and sufficiently long wavelengths that

$$\gamma k^2 < g(\rho_a - \rho_b) \tag{17}$$

Note that this condition prevails for wavelengths longer than the Taylor wavelength defined with Eq. 15. The eigenfrequencies can be pictured as poles in the complex s plane, with the density difference $\rho_b - \rho_a$ a variable parameter. For $\rho_b > \rho_a$, the poles are conjugates on the imaginary axis. With decreasing density difference and long enough wavelength, the poles migrate to the origin, and as the condition of Eq. 17 prevails, the poles separate on the real axis. The instability is incipient at zero frequency. In general, there might be an infinite set of eigenfrequencies. If all pass into the right-half s plane through the origin, the principle of exchange of stabilities applies. That is, the incipient condition could be identified by setting $\omega = 0$ at the outset and asking for the condition on $\rho_b - \rho_a$ that makes it possible for all of the fluid mechanics laws to be satisfied. Here, as in Sec. 5.15 where the charge relaxation eigenfrequencies for a step discontinuity in electrical properties is considered, there are a finite number of eigenfrequencies (two). There it is shown that a smooth distribution of electrical properties leads to an infinite set of temporal modes. It should come as no surprise that a smoothly distributed density distribution similarly leads to an infinite set of eigenmodes. In that case taken up in Sec. 8.18, the principle of exchange of stabilities also applies.

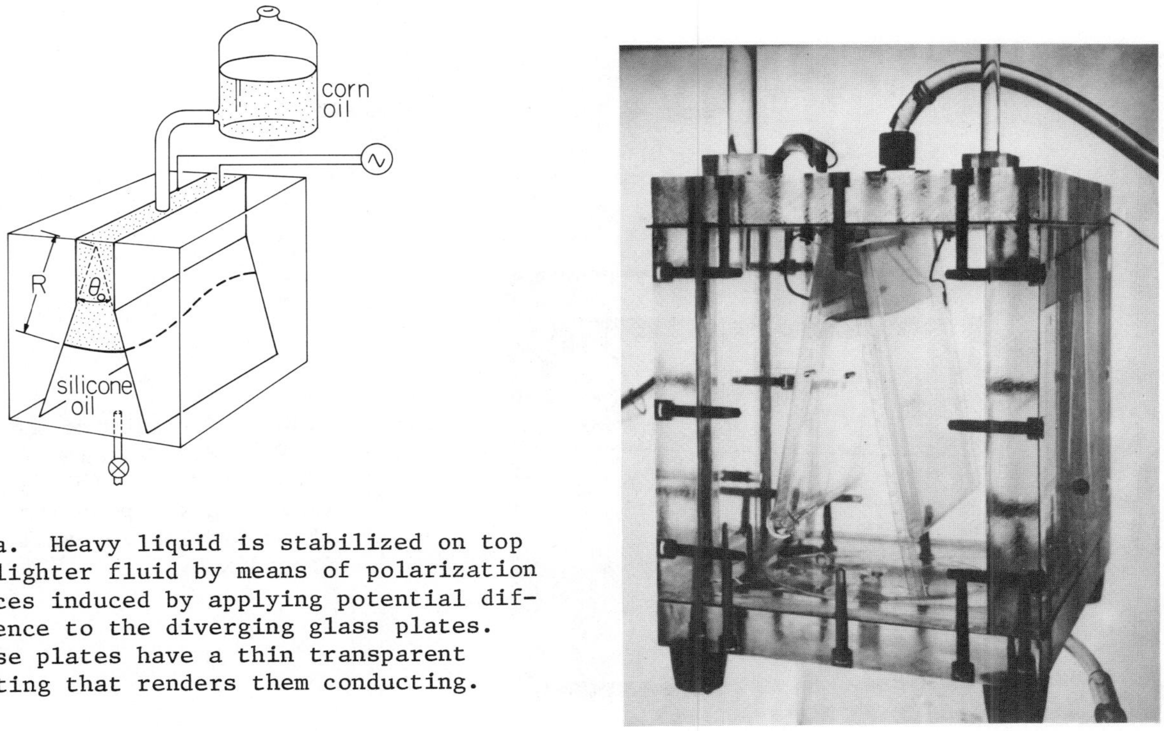

Fig. 8.9.3a. Heavy liquid is stabilized on top of lighter fluid by means of polarization forces induced by applying potential difference to the diverging glass plates. These plates have a thin transparent coating that renders them conducting.

The inviscid model is especially justified for predicting the incipience, because there are then no temporal rates involved. Thus the effects of viscosity vanish.

In the example of this Rayleigh-Taylor instability shown in Fig. 8.9.3,[1] polarization forces are used to stabilize a static equilibrium with a heavy liquid on top of a lighter one. (The electromechanics is developed in Sec. 8.11.) When the field is removed, the unstable temporal eigenmode is evident. Some fluid rises so that some can fall. The sinusoidal deflection predicted by the linear theory gives way to a plume extending into the lighter liquid. It is characteristic of this purely mechanical instability that the nonlinear "process" initiated by the instability becomes blunted in its advanced stages. The bulbous plume can itself be unstable if the viscosity is low. This characteristic appearance, which is commonly seen "upside down" as warm air rises into the atmosphere, is in sharp contrast with the electromechanical forms of Rayleigh-Taylor instability considered in the following sections.

Spatial Eigenmodes: Spatial modes are introduced in Sec. 5.17. With longitudinal boundary conditions, the sinusoidal steady-state response consists not only of a part having the same wave number as the transverse drive, but an infinite set of eigenmodes having the same frequency as the drive, each with its own wave number. These are in general complex, $k = k_r + jk_i$, and found by solving the dispersion equation $D(\omega,k) = 0$ for k, given that ω is the same as for the drive. In general this expression is transcendental, so that it must be solved numerically. Here, an infinite set of eigenvalues can be identified by a simple graphical solution. First, there are the two propagating modes in which $k = k_r$ and the dispersion equation becomes

$$\omega^2 = \frac{[\gamma k_r^2 + g(\rho_b - \rho_a)]k_r}{\rho_a \coth(k_r a) + \rho_b \coth(k_r b)} \tag{18}$$

A graphical solution is obtained by finding the intersection of curves representing the right and left sides of this expression as a function of (ak_r). This is shown in Fig. 8.9.4a. An infinite set of modes are evanescent, $k = jk_i$. With k purely imaginary, the dispersion equation is again purely real ($\coth jx = -j \cot x$):

$$\omega^2 = \frac{[\gamma k_i^2 - g(\rho_b - \rho_a)]k_i}{\rho_a \cot(k_i a) + \rho_b \cot(k_i b)} \tag{19}$$

so that graphical solution gives rise to an infinite set of k_i's, as illustrated in Fig. 8.9.4b. The functions on the right in these last two expressions are even in the wave number, so for each positive root there is a negative one as well. The two propagating modes have an exponential dependence on depth, while the evanescent modes are sinusoidal in their depth dependence, with a number of zero crossings in the x direction that increases with the mode number.

1. See J. R. Melcher and M. Hurwitz, "Gradient Stabilization of Electrohydrodynamically Oriented Liquids," J. Spacecraft and Rockets 4, 864-881 (1967).

(Figure 8.9.3 continued)

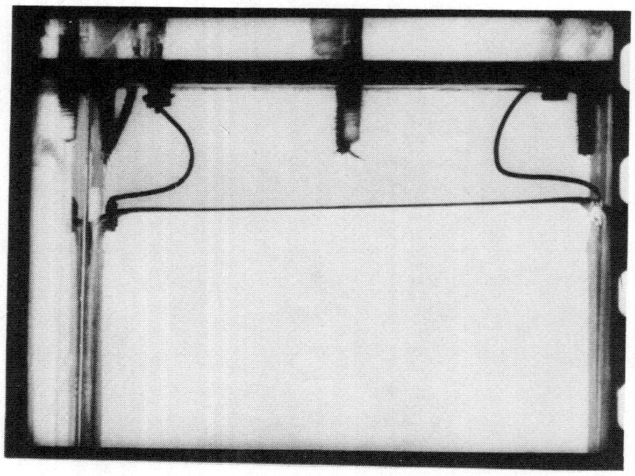

(a)

Fig. 8.9.3b. Side view of apparatus shown in Fig. 8.9.3a. (a) Equilibrium with field on. (b)-(e) Sequential view of developing instability. (From Complex Waves II, Reference 11, Appendix C.)

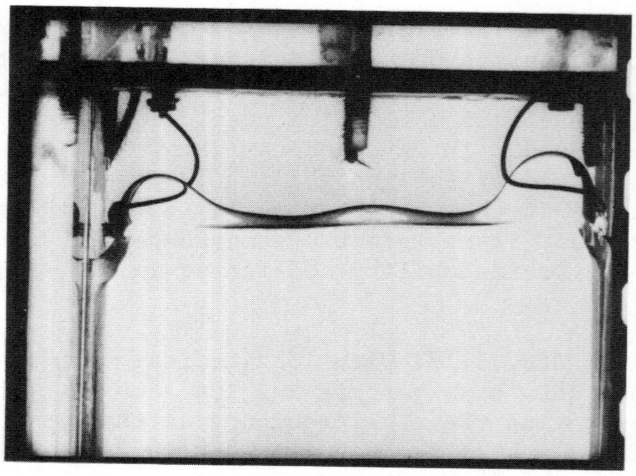

(b)

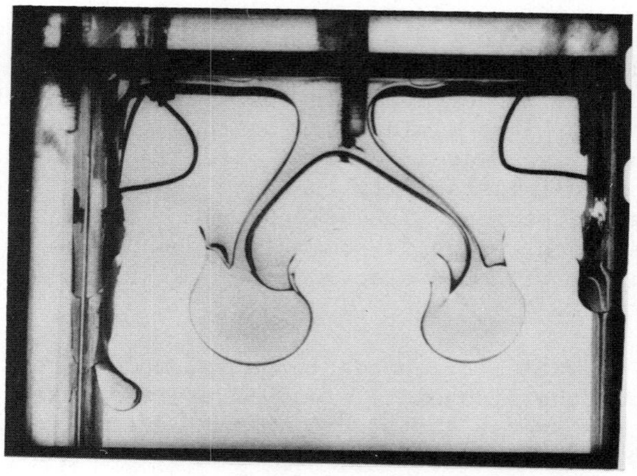

(d)

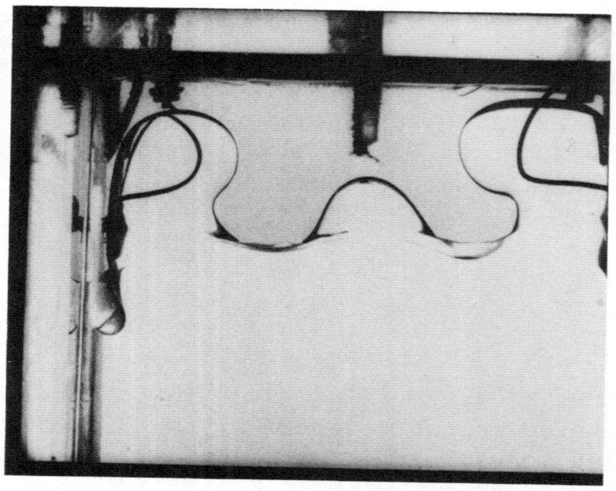

(c)

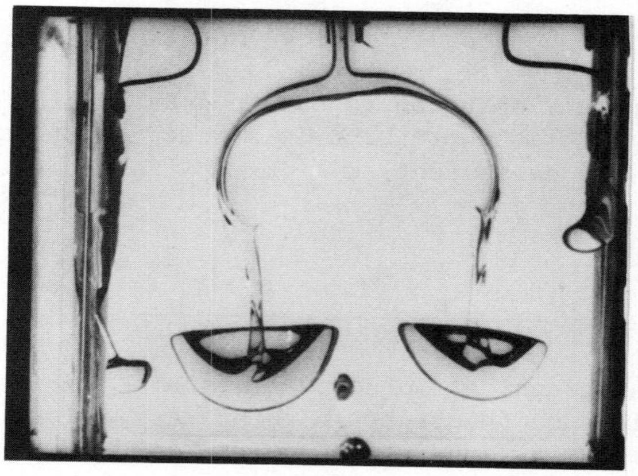

(e)

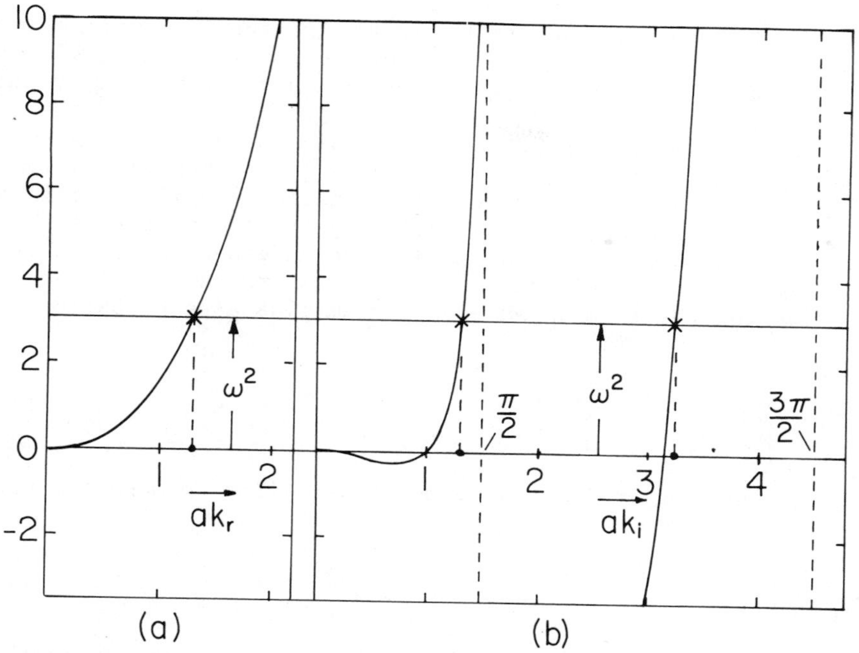

Fig. 8.9.4

Graphical solution for spatial eigenmodes. (a) Equation 18 for propagating modes. (b) Equation 19 for evanescent modes. For case shown, $\rho_a = 0$ and $\gamma/\rho_b g a^2 = 1$.

As an example, a gravity-capillary resonator might be constructed with rigid walls in the planes $y = 0$ and $y = \ell_y$ and $z = 0$ and $z = \ell_z$. These propagating and evanescent modes would in general also be excited by the transverse drive. In general, the evanescent modes are required to insure there being no normal velocity on the longitudinal boundaries. With the surface tension comes still another boundary condition. For example, by virtue of the surface tension, the interface can cling to a sharp edge. Note that for $\rho_b > \rho_a$ the lowest evanescent mode in fact exists because of the surface tension. It represents the effect of the surface tension reaching out into the interfacial region from the longitudinal boundary. The higher order modes are more closely connected with the inertia and mass conservation represented by Laplace's equation in the fluid bulk.

8.10 Self-Field Interfacial Instabilities

If a magnet is held over or under the free surface of a ferrofluid so that the field is normal to the interface, sprouts of liquid will be seen to extend into the air. With the magnet fixed, the sprouts are fixed. Even if stressed by an initially perfectly uniform magnetic field (so that hydrostatic pressure can balance the magnetic forces to maintain a static equilibrium with the interface flat), the sprouts represent a new static equilibrium preferred by the fluid. The electromechanical form of Rayleigh-Taylor instability that takes place as the planar interface, stressed by a uniform magnetic field, gives way to the new configuration, is one of the results from the model now developed. The configuration, shown in Fig. 8.10.1a, consists of planar layers having different permeabilities (μ_a, μ_b), mass densities (ρ_a, ρ_b) and equilibrium thicknesses (a, b). The common interface is at $x = \xi$, while rigid boundaries (infinitely permeable pole faces) bound the layers from above and below. The liquids are water based or even hydrocarbon based ferrofluids. Hence, in MQS terms, the materials are essentially insulating. Only the magnetization force density, Eq. 3.8.14 with $\vec{J}_f = 0$, is responsible for the electromechanical coupling.

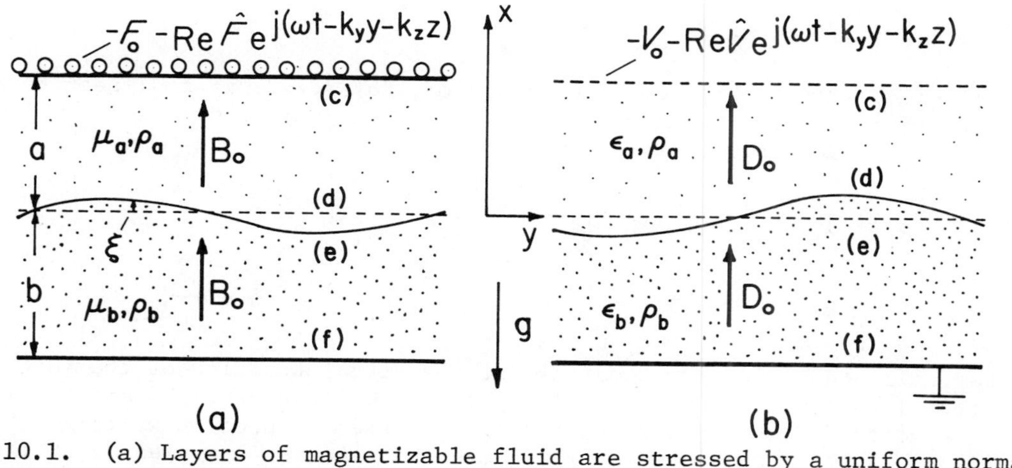

Fig. 8.10.1. (a) Layers of magnetizable fluid are stressed by a uniform normal magnetic flux density, B_o. Polarizable liquid layers are stressed by a normal electric displacement, D_o.

The time and space-varying drive is taken as imposed on the upper transverse boundary by means of a coil structure. Thus, the magnetic potential in this surface is an equilibrium value $\mathcal{F}_o$ (relative to the lower surface) representing the magnet field plus a traveling wave having the complex amplitude $\hat{\mathcal{F}}$.

The EQS system, consisting of layers of insulating polarizable fluid as shown in Fig. 8.10.1b, is described with the same model by simply identifying $\mu \to \varepsilon$, $B_o \to D_o$ and $\mathcal{F} \to \mathcal{V}$. There is an important physical difference between the two systems. To obtain a purely polarization coupling, it is necessary to use an alternating electric field having a high enough frequency to guarantee that free charge does not enter into the electromechanics. This field can be considered as being essentially static provided the frequency is also high enough to insure that the fluid responds to its rms value. In the respective regions the magnetic field is taken as having the form of an equilibrium plus a perturbation:

$$\vec{H} = \begin{Bmatrix} H_a \\ \\ H_b \end{Bmatrix} \vec{i}_x + \vec{h}; \quad \vec{h} = -\nabla\Psi \tag{1}$$

The equilibrium magnetic flux density in each region is related to the equilibrium magnetic potential difference between the pole faces by

$$B_o = \mu_a H_a = \mu_b H_b = \frac{\mathcal{F}_o}{[(a/\mu_a) + (b/\mu_b)]} \tag{2}$$

The magnetization force density is confined to the interface, where it acts on the equilibrium interface as a normal surface force density. The equilibrium pressure difference $\Pi_a - \Pi_b$ then holds the interface in static equilibrium. In the bulk regions, the magnetic field is uncoupled from the fluid nechanics. Thus, the perturbation mechanics of each layer is described by the inviscid pressure-velocity relations from Table 7.9.1, Eqs. 8.9.4. Similarly, the perturbation magnetic field is described by the flux-potential transfer relations, Eqs. (a) of Table 2.16.1 ($k \equiv \sqrt{k_y^2 + k_z^2}$)

$$\begin{bmatrix} \hat{h}_x^c \\ \\ \hat{h}_x^d \end{bmatrix} = k \begin{bmatrix} -\coth(ka) & \dfrac{1}{\sinh(ka)} \\ \\ \dfrac{-1}{\sinh(ka)} & \coth(ka) \end{bmatrix} \begin{bmatrix} \hat{\Psi}^c \\ \\ \hat{\Psi}^d \end{bmatrix}; \quad \begin{bmatrix} \hat{h}_x^e \\ \\ \hat{h}_x^f \end{bmatrix} = k \begin{bmatrix} -\coth(kb) & \dfrac{1}{\sinh(kb)} \\ \\ \dfrac{-1}{\sinh(kb)} & \coth(kb) \end{bmatrix} \begin{bmatrix} \hat{\Psi}^e \\ \\ \hat{\Psi}^f \end{bmatrix} \tag{3}$$

The essence of the electromechanics is in the boundary conditions, which must be consistent with the electromagnetic and mechanical laws used in the model. Proceeding from top to bottom in Fig. 8.10.1a, the magnetic potential must be that of the drive at the upper boundary. The boundary is rigid, so

$$\hat{\Psi}^c = \hat{\mathcal{F}} \tag{4}$$

$$\hat{v}_x^c = 0 \tag{5}$$

At the interface, continuity requires that

$$\hat{v}_x^d = \hat{v}_x^e = j\omega\hat{\xi} \tag{6}$$

The x component of the stress balance jump condition, Eq. 7.7.3, is to linear terms equivalent to the normal component of the stress balance. With i = x, that jump condition is evaluated using the stress tensor with Eq. 3.8.14 in Table 3.10.1:

$$[-\rho_a gx + \Pi_a + p'^d]_{x=\xi} - [-\rho_b gx + \Pi_b + p'^e]_{x=\xi}$$
$$\simeq [\tfrac{1}{2}\mu_a (H_a + h_x^d) - \tfrac{1}{2}\mu_b (H_b + h_x^e)^2]_{x=\xi} + \gamma\left(\frac{\partial^2\xi}{\partial y^2} + \frac{\partial^2\xi}{\partial z^2}\right) \tag{7}$$

where, remember, all quantities are evaluated at the actual position of the interface. The normal vector is written in terms of ξ by means of Eq. (a) from Table 7.6.1. Terms from the stress that are nonlinear in the perturbation amplitudes have already been dropped in writing Eq. 7. To linear terms, the perturbation quantities evaluated at $x = \xi$ are the same as if evaluated at the equilibrium interfacial position $x = 0$. Also, the equilibrium magnetic field is uniform (not a function of x like the equilibrium pressure), so these terms are the same at $x = 0$ as at $x = \xi$. The equilibrium part of Eq. 7 expresses the condition for static equilibrium,

$$\Pi_a - \Pi_b = \frac{1}{2}(\mu_a H_a^2 - \mu_b H_b^2) = \frac{1}{2} B_o (H_a - H_b) \tag{8}$$

and the perturbation part becomes the required jump condition representing stress balance at the interface:

$$(\hat{p}^d - \hat{p}^e) + \hat{\xi}g(\rho_b - \rho_a) = B_o(\hat{h}_x^d - \hat{h}_x^e) - k^2\gamma\hat{\xi} \tag{9}$$

The conditions of Eqs. 6 and 9 guarantee that the mechanical laws are satisfied through the interface. Similarly, on the magnetic side, $\vec{H}$ is irrotational and $\vec{B}$ is solenoidal, so $\vec{n} \times [\![\vec{H}]\!] = 0$ and $\vec{n} \cdot [\![\mu\vec{H}]\!] = 0$ (Eqs. 21 and 22 of Table 2.10.1). With $\vec{n}$ again given by Eq. (a) of Table 7.6.1, either the y or z components of the condition that tangential $\vec{H}$ be continuous reduces to

$$\hat{\psi}^d - \hat{\psi}^e = \hat{\xi}(H_a - H_b) \tag{10}$$

while the continuity of normal flux density is to linear terms given by

$$\mu_a \hat{h}_x^d = \mu_b \hat{h}_x^e \tag{11}$$

Finally, there are the mechanical and magnetic conditions at the lower rigid and infinitely permeable boundary:

$$\hat{v}_x^f = 0 \tag{12}$$

$$\hat{\psi}^f = 0 \tag{13}$$

With the objective of finding the driven response and in the process deducing the dispersion equation, the stress and field continuity conditions, Eqs. 9, 10 and 11, are now written with the p's and h_x's substituted from the bulk equations, Eqs. 8.9.4 and 3. These latter relations are themselves first written using the remaining simple boundary conditions. Thus, Eqs. 9, 10 and 11 respectively become

$$\begin{bmatrix} \frac{\omega^2}{k}[\rho_a\coth(ka) + \rho_b\coth(kb)] - g(\rho_b-\rho_a) - k^2\gamma & kB_o\coth(ka) & kB_o\coth(kb) \\ H_a - H_b & -1 & +1 \\ 0 & \mu_a k\coth(ka) & \mu_b k\coth(kb) \end{bmatrix} \begin{bmatrix} \hat{\xi} \\ \hat{\psi}^d \\ \hat{\psi}^e \end{bmatrix} = \begin{bmatrix} \frac{B_o k\hat{\mathscr{F}}}{\sinh(ka)} \\ 0 \\ \frac{\mu_a k\hat{\mathscr{F}}}{\sinh(ka)} \end{bmatrix} \tag{14}$$

Solution for $\hat{\xi}$ gives

$$\hat{\xi} = -\frac{kB_o(\mu_b - \mu_a)\coth(kb)\hat{\mathscr{F}}}{\sinh(ka)[\mu_b\coth(kb) + \mu_a\coth(ka)]}\frac{1}{D(\omega,k)} \tag{15}$$

where

$$D(\omega,k) = -\frac{\omega^2}{k}[\rho_a\coth(ka) + \rho_b\coth(kb)] + [\gamma k^2 + g(\rho_b - \rho_a)] - \frac{kB_o^2(\mu_b - \mu_a)^2}{\mu_a\mu_b[\mu_b\tanh(ka) + \mu_a\tanh(kb)]} \tag{16}$$

The many types of information that can be gleened from Eq. 15 are illustrated in Sec. 8.9. Concerning the driven response, it is here simply observed that its frequency dependence is similar to that illustrated by Fig. 8.9.2, with the frequency of the resonance occurring as the excitation phase velocity coincides with that of a field coupled surface wave having the phase velocity

$$v_p = \sqrt{\frac{\gamma k + g(\rho_b-\rho_a)/k - B_o^2(\mu_b-\mu_a)^2/\mu_a\mu_b[\mu_b\tanh(ka) + \mu_a\tanh(kb)]}{\rho_a\coth(ka) + \rho_b\coth(kb)}} \tag{17}$$

The effect of the field is to reduce the gravity-capillary phase velocity and hence the frequency. This phenomenon is a "self-field" effect, in the sense that a deformation of the interface distorts the magnetic field and this in turn creates a magnetization perturbation surface force density that tends to further increase the deflection.[1]

---

1. For experimental documentation of resonance frequency shift with magnetic field, see R. E. Zelazo and J. R. Melcher, "Dynamics and Stability of Ferrofluids: Surface Interactions," J. Fluid Mech. 39, 1 (1969).

The tendency for this self-field coupling to precipitate instability makes the temporal modes of particular interest. In the short-wave limit ka << 1 and kb << 1, solution of the dispersion equation $D(\omega,k) = 0$ for ω^2 results in

$$\omega^2 = \frac{gk(\rho_b - \rho_a)}{\rho_a + \rho_b} + \frac{k^3\gamma}{\rho_a + \rho_b} - \frac{k^2B_o^2(\mu_b - \mu_a)^2}{\mu_a\mu_b(\mu_a + \mu_b)(\rho_a + \rho_b)} \tag{18}$$

Even with the lighter fluid on top (say air over a ferroliquid) so $\rho_b > \rho_a$, the magnetic field can make $\omega^2 \to 0$ and hence one of the eigenmodes unstable. Figure 8.10.2 shows ω^2 as given by Eq. 18 as a function of k. As B_o is raised, there is a critical value at which the curve just kisses the $\omega^2 = 0$ axis. Under this condition, instability impends at the wave number k^*. For greater values of B_o, wave numbers between the roots of Eq. 18 with $\omega^2 = 0$, k_u and k_ℓ, are unstable. These roots coalesce as the discriminant of the quadratic formula vanishes. Thus, the incipient condition is

$$\left[\frac{B_o^2(\mu_b - \mu_a)^2}{\mu_a\mu_b(\mu_a+\mu_b)\gamma}\right]^2 = \frac{4g(\rho_b - \rho_a)}{\gamma} \tag{19}$$

The critical wave number is what remains from the quadratic formula, which in view of Eq. 19 is

$$k^* = \sqrt{\frac{g(\rho_b - \rho_a)}{\gamma}} \tag{20}$$

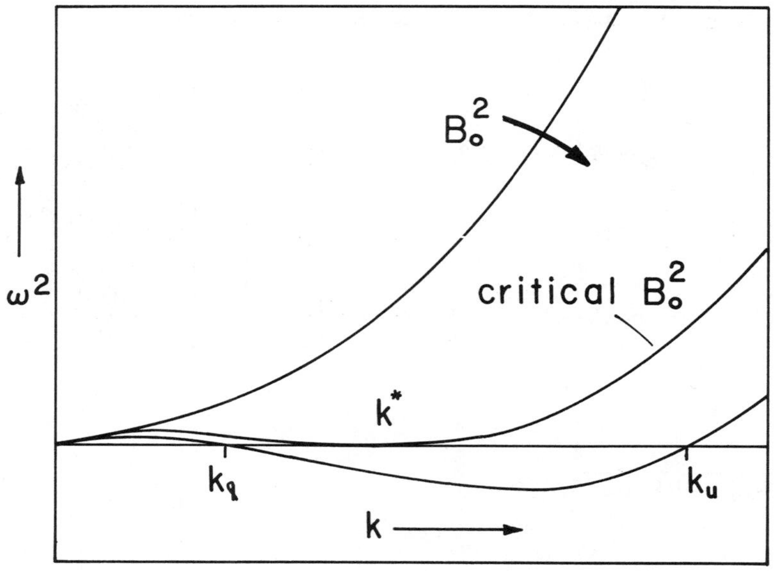

Fig. 8.10.2

Dependence of $\omega^2(k)$ as given by Eq. 18 with B_o^2 as a parameter.

Note that the first perturbations to become unstable as the field reaches the level predicted by Eq. 19 have the Taylor wavelength given by Eq. 15.[2]

What happens if the field is raised above the value consistent with Eq. 19? The initial rate of growth is given by the linear theory, although because a rate process is now involved, this may be strongly influenced by the viscosity. But, the ultimate state will depend on the nature of the electromechanical coupling. In the magnetization example at hand, the interface typically reaches a new state of static equilibrium. The protrusions shown in Fig. 8.10.3 are typical. Consistent with the fact that the interface is always free of a shearing surface force density, they are perfectly static.

As discussed in the introduction to this section, to obtain a similar instability in the EQS polarization configuration of Fig. 8.10.1b, it is usually necessary to use an alternating field.[3] If the frequency of this field is low enough that the natural modes can interact with its pulsating component, parametric instabilities can also result. By contrast with the coupling described here, these instabilities

2. Conditions for instability are studied by M. D. Cowley and R. E. Rosensweig, J. Fluid Mech. 30, 721 (1969).

3. E. B. Devitt and J. R. Melcher, "Surface Electrohydrodynamics with High-Frequency Fields," Phys. Fluids 8, 1193 (1965).

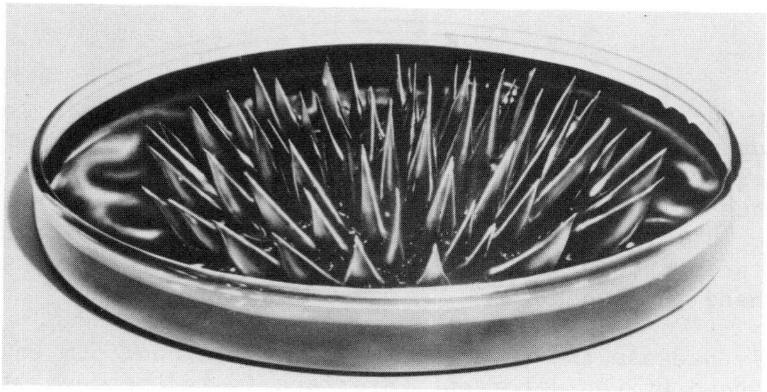

Fig. 8.10.3. System of static fluid sprouts represents a new static equilibrium formed once planar interface in perpendicular field becomes unstable. (Courtesy of Ferrofluidics Corp., Burlington, Mass.)

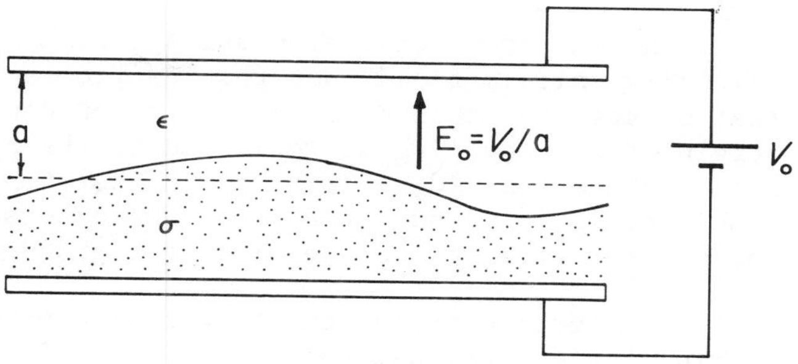

Fig. 8.10.4. Rigid plane-parallel electrodes bound liquids having common interface. The upper liquid is insulating relative to the lower one.

are dynamic in character and can result in splattering or atomization of the interface.[4]

To appreciate the perfectly static equilibrium of the polarization sprouts resulting from the instability of the flat interface, consider by contrast some of the possibilities resulting when the interface of a conducting fluid bounded by a relatively insulating one is stressed by a normal electric field E_o. The configuration is shown in Fig. 8.10.4. For example, the upper fluid might be air and the lower one water (or any other liquid having a charge relaxation time ε/σ short compared to times of interest).[5]

The boundary condition at the interface is that it sustains no tangential electric field. This is formally equivalent to the (analogous) magnetic field situation in the limit where the lower fluid is infinitely permeable. That is, in the limit $\mu_b \to \infty$, the interfacial tangential magnetic field just above the interface of Fig. 8.10.1a must vanish. The magnetic field above this infinitely permeable fluid then satisfies the same boundary conditions as the electric field does in the physically very different situation of Fig. 8.10.4.

It follows from Eq. 19 with the substitution $\mu_a \to \varepsilon$, $\mu_b \to \infty$ and $B_o \to E_o/\varepsilon = \mathscr{V}_o/a$ that the voltage required to just induce instability of the interface is

$$\mathscr{V}_o = a\left[\frac{4g(\rho_b - \rho_a)\gamma}{\varepsilon^2}\right]^{1/4} \qquad (21)$$

The danger in exploiting the formal equivalence of the infinitely permeable and the "infinitely" conducting lower fluid is that the physics of the two situations will be confused. In the case now considered, the surface force density acting upward on the interface is due to free surface charges. That these are free to conduct accounts for the diverse processes that can be triggered by the instability.

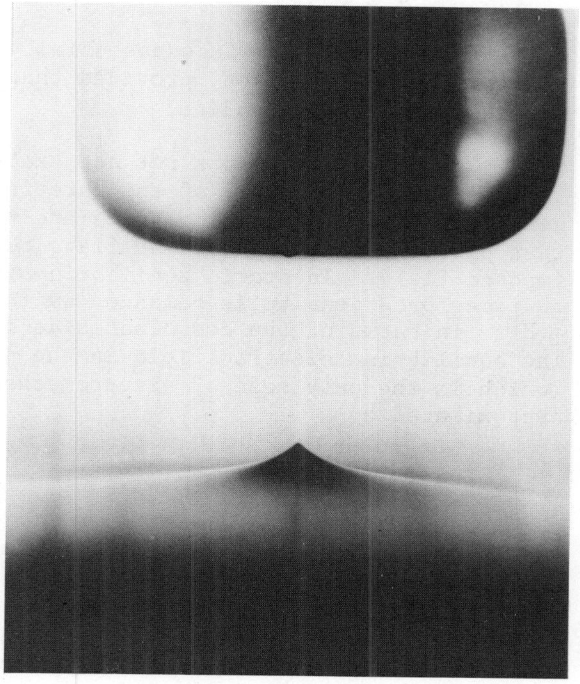

Fig. 8.10.5. Nonlinear stages of surface instability caused by applying 30 kV d-c between electrode above and glycerine interface below. Insulation is mixture of air and gaseous Freon.

A typical appearance shortly after incipience is shown in Fig. 8.10.5. An extremely sharp spike has formed. In the neighborhood of this point, the nonlinear stages of instability are generally dynamic, and often involve dielectric breakdown in some region of the insulating fluid. Depending on properties and breakdown strength, it is very likely that simultaneous spraying and corona discharge will be observed.

---

4. T. B. Jones, "Interfacial Parametric Electrohydrodynamics of Insulating Dielectric Liquids," J. Appl. Phys. 43, 4400 (1972).

5. For experiments and a more general treatment of stability conditions, see J. R. Melcher, Field-Coupled Surface Waves, The M.I.T. Press, Cambridge, Mass., 1963, Chaps. 3 and 4.

8.11 Surface Waves with Imposed Gradients

The electromechanical coupling exemplified in Sec. 8.10 is entirely caused by the distortion of the initially uniform field that results from a deformation of the interface. It is this perturbation field that creates the change in surface force density tending to destabilize the interface. The "self-field" origin of the coupling is reflected in the dependence of the coupling on the square of the jump in electrical properties $[(\mu_a - \mu_b)^2$ in the last term of Eq. 8.10.16]. The perturbation self-field is proportional to $\mu_a - \mu_b$ and the surface force density is proportional to this field multiplied by $(\mu_a - \mu_b)$. The net effect is proportional to the product of these and hence to $(\mu_a - \mu_b)^2$.

The surface force density can also vary simply because the interface moves in a nonuniform equilibrium field. Because the change in field experienced by the deforming interface is independent of the jump in property, it can be expected that this imposed field type of coupling is linearly proportional to the property jump.

To exemplify imposed field effects and at the same time highlight electromechanical surface waves that propagate along field lines, the electromechanics of the configuration shown in Fig. 8.9.3 is now considered. Both fluids can be regarded as perfectly insulating so that the relevant force density is given by Eq. 3.7.22 of Table 3.10.1. How is it that the polarization interaction can stabilize the initial equilibrium with the heavier liquid on top? What is the role of self-field effects when the equilibrium electric field is tangential to the interface?

The cross section of the system is shown in Fig. 8.11.1. Diverging transparent electrodes (which are tin oxide coated glass in Fig. 8.9.3a) are used to impose the field

$$\vec{E} = \vec{i}_\theta \frac{\mathcal{V}_o}{\theta_o r} \tag{1}$$

on fluids with an interface essentially at $r = R$. Note that Eq. 1 gives the exact solution, provided that the interface approximately has this equilibrium radius.

Because gravity does not act exactly in the radial direction, the equilibrium geometry of the interface is in fact somewhat field dependent. The essential physics are retained in a Cartesian model that pictures the interface as flat, but subject to a nonuniform imposed field. In static equilibrium the x-directed polarization surface force density is balanced by the jump in equilibrium pressure $[\![\Pi]\!]$. In terms of the coordinates defined in Fig. 8.11.1, $r = R - x$. The equilibrium electric field in the neighborhood of the interface (which is the only seat of electromechanical coupling) is therefore approximated by

$$\vec{E} = \vec{i}_y E_o (1 + \frac{x}{R}); \quad E_o \equiv \frac{\mathcal{V}_o}{\theta_o R} \tag{2}$$

Because of the quasi-Cartesian approximation, this equilibrium field is not irrotational.

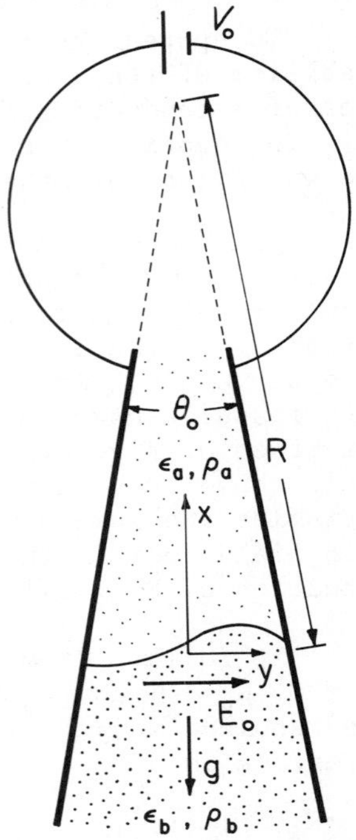

Fig. 8.11.1. Cross section of experiment shown in Fig. 8.9.3a with Cartesian coordinates for planar model.

Bulk Relations: Perturbations in the electric field are both irrotational and solenoidal in the uniform bulk of the fluids. In applying the flux-potential transfer relations representing Laplace's equation above and below the interface (Eqs. (a) of Table 2.16.1), perturbations on the interface having wave number $k \equiv \sqrt{k_y^2 + k_z^2}$ are assumed short enough that boundaries above and below the interface can be considered as being at $x = \pm\infty$. Thus, with the understanding that Re$k > 0$, perturbation fields evaluated at the equilibrium interfacial position are related by

$$\hat{e}_x^a = k\hat{\Phi}^a \tag{3}$$

$$\hat{e}_x^b = -k\hat{\Phi}^b \tag{4}$$

In the bulk regions, the pressure balances the gravitational force density. Hence, in each region the pressure takes the form

$$p = \Pi - \rho g x + p'(x,y,z,t) \tag{5}$$

From the inviscid pressure-velocity transfer relations (Eqs. (c) of Table 7.9.1) the perturbation part of Eq. 5 evaluated at the equilibrium interfacial position is related to the velocity there by

$$\hat{p}^a = \frac{j\omega\rho_a}{k}\hat{v}_x^a \tag{6}$$

$$\hat{p}^b = -\frac{j\omega\rho_b}{k}\hat{v}_x^b \tag{7}$$

Jump Conditions: To assure that the laws defining the model prevail through the interface, there are two electrical boundary conditions. First, $\vec{n} \times [\![\vec{E}]\!] = 0$ is evaluated at the interface using $\vec{n}$ expressed in terms of ξ (Eq. (a) of Table 7.6.1) and $\hat{e}_y = jk_y\hat{\Phi}$ or $\hat{e}_z = jk_z\hat{\Phi}$ to obtain

$$\hat{\Phi}^a - \hat{\Phi}^b = 0 \tag{8}$$

Second, by assumption there is no free surface charge so $\vec{n}\cdot[\![\varepsilon\vec{E}]\!] = 0$, which to linear terms requires that

$$\varepsilon_a\hat{e}_x^a - \varepsilon_b\hat{e}_x^b + jk_y\hat{\xi}E_o(\varepsilon_a - \varepsilon_b) = 0 \tag{9}$$

In addition, two mechanical conditions are required, the first representing continuity

$$\hat{v}_x^a = j\omega\hat{\xi} = \hat{v}_x^b \tag{10}$$

and the second force equilibrium. To linear terms, the normal force balance is the x component of Eq. 7.7.6 with the surface tension contribution given by Eq. (b) of Table 7.6.1,

$$[\Pi_a - \rho_a g\xi + p'^a(x=0)] - [\Pi_b - \rho_b g\xi + p'^b(x=0)]$$
$$= -\frac{1}{2}\varepsilon_a[E_o(1+\frac{\xi}{R}) + e_y^a(x=0)]^2 + \frac{1}{2}\varepsilon_b[E_o(1+\frac{\xi}{R}) + e_y^b(x=0)]^2 - \gamma\left[\frac{\partial^2\xi}{\partial y^2} + \frac{\partial^2\xi}{\partial z^2}\right] \tag{11}$$

The balance of the equilibrium surface force density by the equilibrium pressure is represented by the equilibrium part of Eq. 11:

$$\Pi_a - \Pi_b = -\frac{1}{2}(\varepsilon_a - \varepsilon_b)E_o^2 \tag{12}$$

so that in terms of complex amplitudes evaluated at the equilibrium position of the interface, the perturbation stress balance requires that

$$\hat{p}^a - \hat{p}^b + g(\rho_b - \rho_a)\hat{\xi} = (\varepsilon_b - \varepsilon_a)\frac{E_o^2}{R}\hat{\xi} - jk_yE_o(\varepsilon_a\hat{\Phi}^a - \varepsilon_b\hat{\Phi}^b) - \gamma k^2\hat{\xi} \tag{13}$$

Dispersion Equation: Of the possible types of information about the dynamics that can be gleaned from this model, it is the temporal modes that are of interest here. One way that they can be identified is to find the response to a transverse drive in the form of Eq. 8.9.11 for example. Then the condition is $D(\omega,k) = 0$. Here, there is no drive and the temporal modes are identified by asking for the relation between ω and k that makes it possible for surface distortions to exist, consistent with all the laws, but with homogeneous boundary conditions. To this end, Eqs. 3 and 4, 6 and 7 and 9 are substituted into Eq. 13 using Eqs. 8 and 10 in the process. The resulting expression is of the form $D(\omega,k)\hat{\xi} = 0$. If $\hat{\xi}$ is to be finite, it follows that $D(\omega,k) = 0$. This relation,

$$\omega^2(\rho_a+\rho_b) = gk(\rho_b-\rho_a) + \gamma k^3 + (\varepsilon_a-\varepsilon_b)\frac{kE_o^2}{R} + k_y^2E_o^2\frac{(\varepsilon_a - \varepsilon_b)^2}{(\varepsilon_a + \varepsilon_b)} \tag{14}$$

is an expression of the fact that the inertia of the fluid above and below the interface is equilibrated by forces due to gravity, surface tension, imposed fields and self-fields.

Temporal Modes: In addition to the now familiar gravity and capillary contributions to the phase velocity, ω/k, there are now the polarization contributions. In the absence of an imposed gradient the effect of the field is to stabilize perturbations with peaks and valleys running perpendicular to the electric field. To see why, consider the perturbation fields resulting from the deformation of the interface shown in Fig. 8.11.2a. With $\varepsilon_a < \varepsilon_b$, the equilibrium field, E_o induces polarization surface charges. As shown, these in turn give rise to the perturbation fields. Remember that the polarization surface force density on an interface stressed by a tangential field acts in the direction of decreasing permittvity. Thus, at the downward peaks where the perturbation field reinforces the applied field there is an increase in the upward directed surface force density, and this tends to restore the inter-

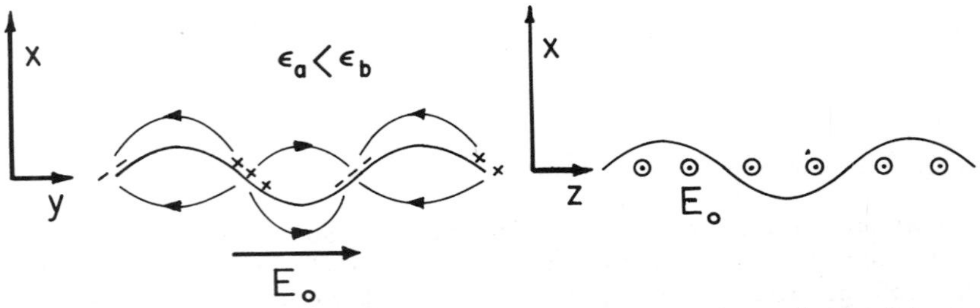

Fig. 8.11.2. (a) Perturbation fields for waves propagating along lines of electric field. (b) Perturbation fields are absent for waves propagating across $\vec{E}$ lines.

face to its equilibrium position. That perturbations propagating in the z direction are not influenced by the self-fields is evident from the fact that the equilibrium field remains unaltered by such deformations of the interface.

Note that the self-field stiffening cannot stabilize the interface with the heavy fluid on top; modes appearing as in Fig. 8.11.2b, sometimes called exchange modes because the fluid can be displaced without an associated change in stored electric energy, are unstable despite a uniform imposed field.

However, the imposed gradient can be used to stabilize all wavelengths. Regardless of wave number, the interface is stable provided that

$$(\varepsilon_a - \varepsilon_b)\frac{E_o^2}{R} > g(\rho_a - \rho_b) \tag{15}$$

So, by making the upper fluid have the greater permittivity, the equilibrium can be made stable even with the heavier fluid on top.

In the experiment of Fig. 8.9.3, the region between the electrodes is sealed. Thus, hydrostatic pressure maintains the equilibrium, while the electric field stabilizes it. If too much of the upper fluid is run into the region between the electrodes, it simply breaks through the interface until enough is lost to satisfy Eq. 15.[1]

Considerations of stability are essential to the design of systems for orienting liquids. An example is the use of polarization forces for orienting liquid fuels in the zero gravity environments of space.[2] Magnetization interactions with ferrofluids are analogous to those described here.[3]

8.12 Flux Conserving Dynamics of the Surface Coupled z-θ Pinch

The magnetic field levitation of a liquid metal, sketched in Fig. 8.2.1c, is based on time-average forces caused by currents induced because the field is oscillating with a period short compared to a magnetic diffusion time. Transient, rather than steady-state forces, are similarly induced if the field is abruptly switched on. The confinement of a highly ionized gas in many fusion experiments[1] is based on this tendency for the plasma to behave as a "perfect conductor" over several magnetic diffusion times. Not only does the magnetic field "bottle up" the plasma, but it can also be the means of compressing the gas. The stability of the pinch configuration shown in Fig. 8.12.1 is examined in this section.

An axial current on the surface of the cylindrical conductor gives an azimuthal magnetic field, H_a, and hence a surface force density that compresses the conductor radially inward. An example is shown in Fig. 8.12.2.[2] If the conductor is an ionized gas, this pressure will evidence itself in the constriction of the conducting volume, thereby producing an increase in the plasma density and local conductivity. In turn, because the magnetic field intensity in the neighborhood of the conducting path is inversely proportional to the radius of the conductor, the magnetic pressure is itself increased. As a scheme for heating of plasmas for thermonuclear experiments, the magnetic field serves the dual purpose of compressing and confining the plasma column.

---

1. J. R. Melcher and M. Hurwitz, "Gradient Stabilization of Electrohydrodynamically Oriented Liquids," J. Spacecraft and Rockets 4, 864 (1967).
2. J. R. Melcher, D. S. Guttman and M. Hurwitz, "Dielectrophoretic Orientation," ibid., 6, 25 (1969).
3. R. E. Zelazo and J. R. Melcher, "Dynamics and stability of ferrofluids: surface interactions," J. Fluid Mech. 39, 1-24 (1969).

1. See, for example, D. J. Rose and M. Clarke, Jr., Plasmas and Controlled Fusion, The MIT Press and John Wiley & Sons, New York, 1961, p. 336.
2. See F. C. Jahoda. E. M. Little, W. E. Quinn, F. L. Ribe and G. A. Sawyer,"Plasma Experiments with a 570-kJ Theta-Pinch," J. Appl. Phys. 35, 2351-2363 (1964).

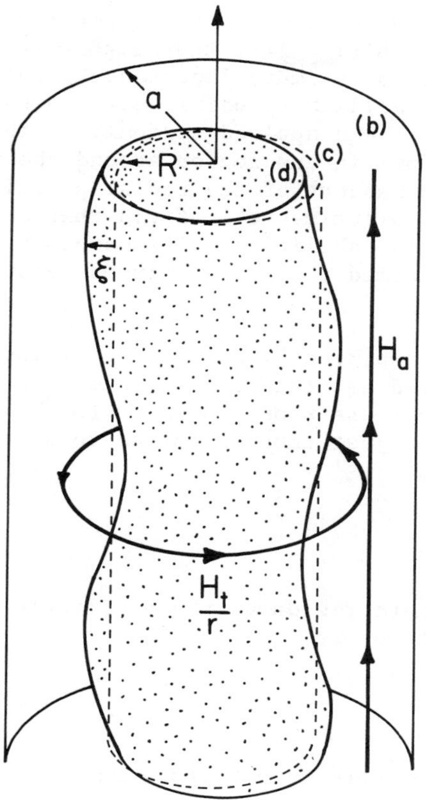

Fig. 8.12.1. Plasma column showing equilibrium radius R and equilibrium magnetic fields.

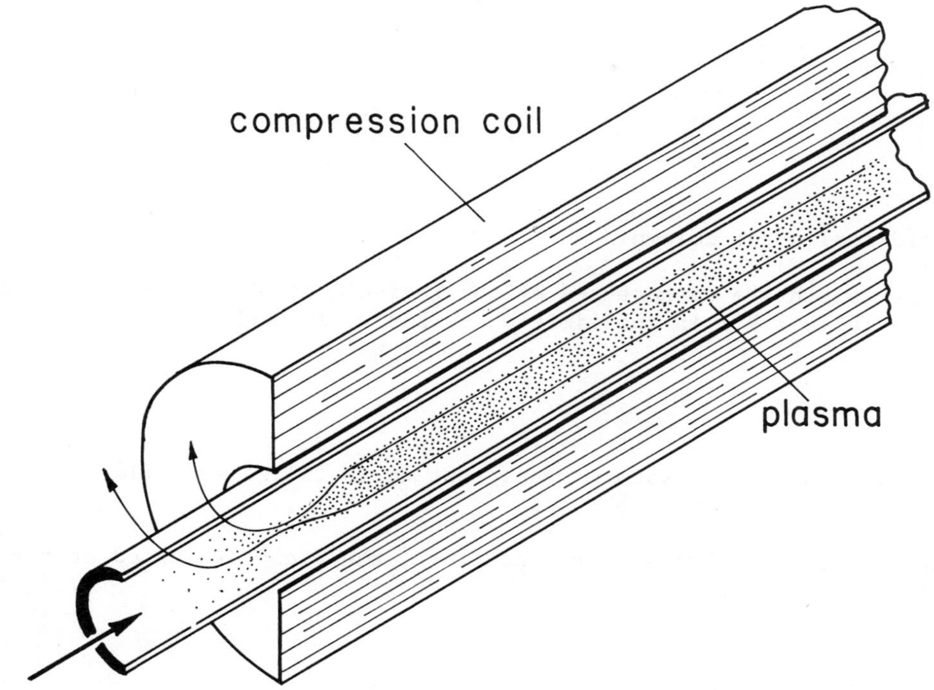

2.4 sec.

3.6 sec.

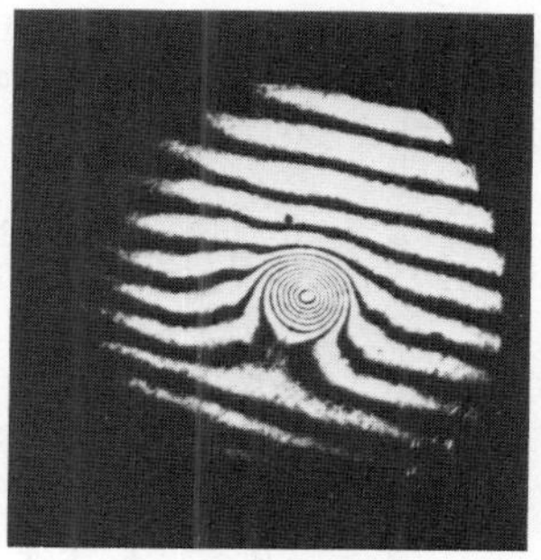

4.9 sec.

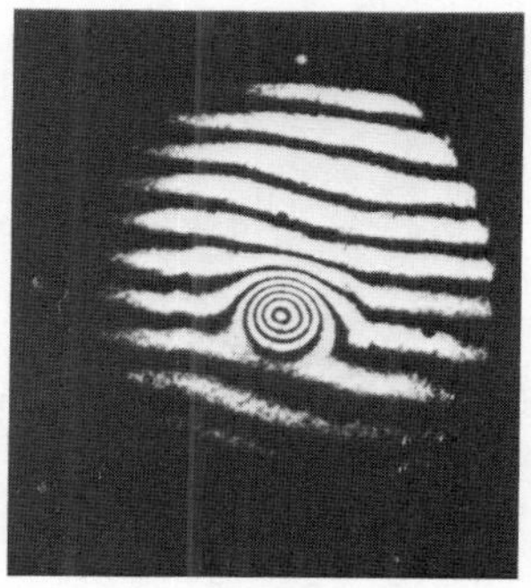

6.1 sec.

Fig. 8.12.2

Theta-pinch experiment showing magnetic compression of plasma cross section as viewed by means of interferometer. Peak magnetic field is about 100 kgauss. (Courtesy of Los Alamos Scientific Laboratory.)

The axial or z pinch, with the current in the direction of the columnar axis and the induced magnetic field azimuthally directed, is inherently unstable: a fact that emphasized early in the fusion effort that the stability of confinement schemes was of primary importance. The theta pinch of Fig. 8.12.2 avoids the inherent tendency toward instability by using currents that flow azimuthally around the column. These are induced by a magnetic field applied suddenly in the axial direction. The applied magnetic field has the virtue of being uniform in the region around the plasma, and thus the magnetic stress at the surface of the column is independent of the radial position of the interface. As will be seen, it is the 1/r dependence of the equilibrium magnetic field that makes the axial pinch naturally unstable. The imposed field gradient is destabilizing. The combined axial and theta pinch configuration, shown in Fig. 8.12.1, is sometimes termed the "screw pinch" because of the helical shape of the magnetic field lines.

Equilibrium: The plasma column is modeled as a perfectly conducting cylinder of incompressible and inviscid fluid. Although the equilibrium is pictured as static, the fields are nevertheless applied and the column motion of interest completed in times that are short compared to the time for the field to diffuse into the column. Thus, surface currents are just those required to shield the applied fields from the column:

$$\vec{H} = \frac{R}{r} H_t \vec{i}_\theta + H_a \vec{i}_z \tag{1}$$

where H_a and H_t are, respectively, the axial and theta fields at the equilibrium surface of the column. The equilibrium surface current on the column is therefore

$$\vec{K} = -H_a \vec{i}_\theta + H_t \vec{i}_z \tag{2}$$

Stress equilibrium requires that the equilibrium pressure jump balance the magnetic surface force density:

$$\Pi_c - \Pi_d = -\frac{1}{2}\mu_o (H_t^2 + H_a^2) \tag{3}$$

Bulk Relations: With the column surface represented in the complex amplitude form $\xi = \text{Re}\hat{\xi}\exp j(\omega t - m\theta - kz)$, perturbations in the magnetic field around the column, $\vec{h} = -\nabla\Psi$, where Ψ satisfies Laplace's equation. Thus, the flux potential relations, Eq. (c) of Table 2.16.2, pertain to the region between column and wall:

$$\begin{bmatrix} \hat{\Psi}^b \\ \\ \hat{\Psi}^c \end{bmatrix} = \begin{bmatrix} F_m(R,a) & G_m(a,R) \\ \\ G_m(R,a) & F_m(a,R) \end{bmatrix} \begin{bmatrix} \hat{h}_r^b \\ \\ \hat{h}_r^c \end{bmatrix} \tag{4}$$

There is no perturbation magnetic field inside the column.

The perturbation mechanics of the column are represented by the inviscid model of Sec. 7.9. The pressure-velocity relations, Eq. (f) of Table 7.9.1 in the limit where $\beta \to 0$, show that

$$\hat{p}^d = j\omega\rho F_m(0,R)\hat{v}_r^d \tag{5}$$

That the region surrounding the column is essentially vacuum means that it is filled with fluid of negligible density and hence zero perturbation pressure: $\hat{p}^c \simeq 0$.

Boundary and Jump Conditions: Because the equilibrium $\vec{H}$ is nonuniform, the field evaluated at the perturbed position of the interface is to linear terms

$$\begin{aligned} \vec{H} &= \frac{R}{R+\xi} H_t \vec{i}_\theta + H_a \vec{i}_z + \vec{h}(r = R + \xi) \\ &\simeq H_t \vec{i}_\theta + H_a \vec{i}_z - \frac{\xi}{R} H_t \vec{i}_\theta + \vec{h}(r = R) \end{aligned} \tag{6}$$

The effect of the mechanics on the magnetic field is represented by the condition that there be no magnetic flux linked by contours lying in the deforming perfectly conducting interface. With the normal vector related to ξ by Eq. (e) of Table 7.6.1, it follows that to linear terms

$$\hat{h}_r^c = -j\left(\frac{mH_t}{R} + kH_a\right)\hat{\xi} \qquad (7)$$

where $\hat{h}_r^c$ is evaluated at the unperturbed position of the interface.

The physical nature of the outer wall will be left open. For now, it is presumed that there is some normal magnetic field at the outer wall having the complex amplitude $\hat{\mathscr{H}}$:

$$\hat{h}_r^b = \hat{\mathscr{H}} \qquad (8)$$

To express the effect of the fields on the mechanics, continuity requires that

$$\hat{v}_x^d = j\omega\hat{\xi} \qquad (9)$$

Then, stress equilibrium is represented by Eq. 7.7.6. As applied to plasmas, the model need not include the surface tension. Of the three components of the stress condition, only the normal component is appropriate. Fundamentally, this is because a perfectly conducting interface sustains no magnetic shear stress (see Sec. 8.2). To linear terms, it is the radial component of the stress condition that represents the normal stresses. Thus, in view of Eq. 6 ($\hat{h}_z = jk\hat{\Psi}$, $\hat{h}_\theta = jm\hat{\Psi}/R$)

$$-\hat{p}^d = \frac{\mu_o H_t^2}{R}\hat{\xi} - j\mu_o\left(\frac{m}{R}H_t + kH_a\right)\hat{\Psi}^c \qquad (10)$$

where $\hat{p}^c = 0$.

Dispersion Equation: Equations 4b and 5 are evaluated using Eqs. 7, 8 and 9 and substituted into Eq. 10 to obtain

$$\omega^2\rho F_m(0,R)\hat{\xi} = \frac{\mu_o H_t^2}{R}\hat{\xi} - \mu_o\left(\frac{m}{R}H_t + kH_a\right)^2 F_m(a,R)\hat{\xi} - j\mu_o\left(\frac{m}{R}H_t + kH_a\right)G_m(R,a)\hat{\mathscr{H}} \qquad (11)$$

In particular, if the outer wall is perfectly conducting, Eq. 11 shows that the appropriate dispersion equation is

$$-\omega^2\rho F_m(0,R) = -\frac{\mu_o H_t^2}{R} + \mu_o\left(\frac{m}{R}H_t + kH_a\right)^2 F_m(a,R) \qquad (12)$$

It is shown in Sec. 2.17 that $F_m(0,R) = 1/f_m(0,R) < 0$ (see Fig. 2.16.2b for typical behavior) and $F_m(a,R) > 0$.

The first term on the right in Eq. 12 arises from the imposed gradient in azimuthal magnetic field. That it tends to make the equilibrium unstable is not surprising because the inward directed magnetic surface force density associated with the imposed θ field decreases as the interface moves outward. The question of stability hinges on whether or not the self-field coupling represented by the last term in Eq. 12 "saves the day."

Certainly, the self-fields stiffen the interface. However, for deformations having azimuthal and axial wave numbers related by $(m/R)/k = -H_a/H_t$, this stiffening is absent. To appreciate the origins of this result, observe that a vector perpendicular to crests and valleys of the surface perturbation is $\vec{p} = (m/R)\vec{i}_\theta + k\vec{i}_z$, as shown in Fig. 8.12.3. Also, as a vector in the $(\theta R,z)$ plane, the equilibrium magnetic field is given by Eq. 1. The perturbations that produce no self-field effect have $\vec{p}\cdot\vec{H} = 0$ in the surface of the column. Thus the modes that cause no perturbation in $\vec{H}$ propagate across the lines of equilibrium field. If the equilibrium field circles the z axis in the clockwise direction shown in Fig. 8.12.3, the perturbations that produce no self-fields have crests and valleys that also follow these helical lines, as shown in Fig. 8.12.3b. Note that for the z pinch, where $H_a = 0$, these are the sausage modes $m = 0$. These modes that have no self-fields, sometimes called exchange modes, are similar to the polarization and magnetization modes of Sec. 8.11.

From another point of view, it is Alfvén surface waves propagating along the lines of magnetic field intensity that are described by Eq. 12. The flux conserving dynamics is similar to that for the bulk interactions. However, the phase velocity of waves is now dependent on k, the surface waves are dispersive.

The theta pinch ($H_t = 0$) is at worst neutrally stable. Only the self-field remains on the right in Eq. 12. However, for "exchange" perturbations with crests running in the axial direction, this term is zero, so that the frequency is zero, and the system is on the verge of instability. In fact, the theta pinch has been found to be a useful approach to obtaining confinement for extremely short periods

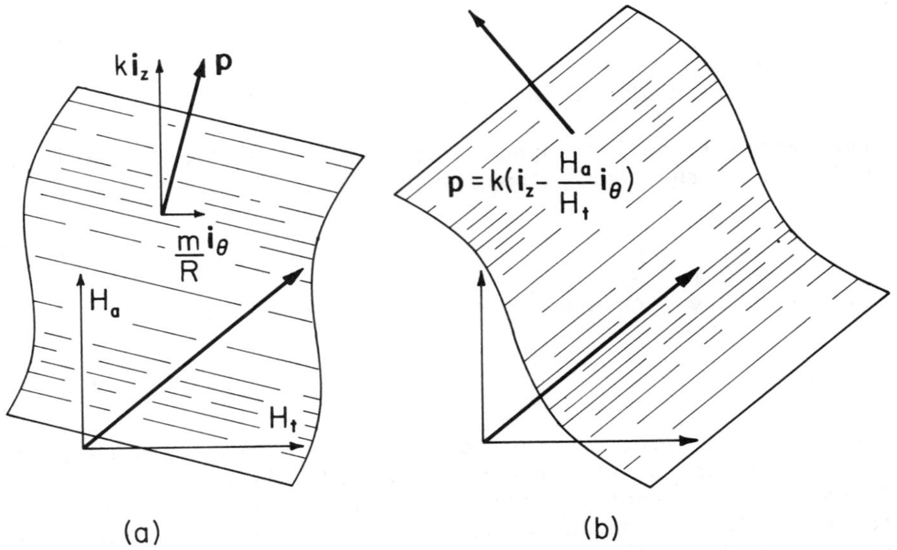

Fig. 8.12.3

(a) Equilibrium $\vec{H}$ and propagation vector in $(R\theta,z)$ plane at $r = R$. (b) Exchange modes showing $\vec{p}\cdot\vec{n} = 0$ and hence lines of constant phase parallel to equilibrium $\vec{H}$.

of time. Experiments are illustrated by Fig. 8.12.2. From the hydromagnetic viewpoint, the stability of the theta pinch depends on effects not included here, such as the necessary curvature of the imposed fields if the column is closed on itself. Internal modes associated with volume distributions of current are thought to come into play in pinch devices and especially in the tokamaks. Such modes are taken up in Secs. 8.17-8.18. In any case, there are many other forms of instability associated with a highly ionized gas that are not described by a hydromagnetic theory.

One approach to stabilizing the equilibrium is to sense the position of the interface and feedback fields to a structure located on the outer wall. For example, in the limit of a continuum of samples and feedback stations, the normal magnetic field at the wall might be made proportional to the deflection of the interface at the same (θ,z) location, $\hat{\mathscr{H}} = A\hat{\xi}$. With this expression introduced into Eq. 11, the revised dispersion equation follows. But, note that no matter what the nature of the feedback scheme, the last term in Eq. 11 has a factor $[(m/R)H_t + kH_a]$. No matter what the feedback, in the framework of this linear model, it will not couple to the exchange modes. The origins of this difficulty are clear from the stress balance, Eq. 10, which shows that field perturbations perpendicular to the imposed field result in no perturbation stress. This is true whether Ψ^c (Eq. 4b) is the result of the self-field (Eq. 7) or caused by the feedback at the outer wall.

8.13 Potential Conserving Stability of a Charged Drop: Rayleigh's Limit

Charged drops and droplets are exploited in devices such as ink jet printers that use electric fields to deflect and direct the ink, charged droplet scrubbers for air pollution control and electrostatic paint sprayers. Of possible importance in these applications is the limiting amount of charge that can be placed on a drop without producing mechanical rupture. It is this Rayleigh's limit,[1] determined as it is by considerations of stability, that is an objective in this section. The example gives the opportunity to put to work relations derived in Chaps. 2 and 7 in spherical coordinates.

The drop, perhaps of water, is assumed to be perfectly conducting and to have the equilibrium radius R and surface tension γ. Its interface has the radial position $r = R + \xi(\theta,\phi,t)$, as sketched in Fig. 8.13.1. The drop is initially in static equilibrium with a total charge, q, evenly distributed over its surface. Thus, an equilibrium electric field

$$\vec{E} = E_o\left(\frac{R}{r}\right)^2; \quad q = 4\pi\varepsilon_o R^2 E_o \tag{1}$$

surrounds the drop with the radial electric surface force density $\varepsilon E_o^2/2$ balanced by the jump in equilibrium pressure $\Pi_c - \Pi_d$ and the surface tension force density $-2\gamma/R$.

Surface deformations take the form

$$\xi = \text{Re}\,\hat{\xi}P_n^m(\cos\theta)e^{j(\omega t - m\phi)} \tag{2}$$

1. Lord Rayleigh, "On the Equilibrium of Liquid Conducting Masses Charged with Electricity," Phil. Mag. 14, 184-186 (1882).

with normal vector and surface tension force density summarized in Table 7.6.2.

Bulk Relations: With the perturbation in electric field from that given by Eq. 1 represented by $\vec{e} = -\nabla\Phi$, the Laplacian nature of the fields surrounding the drop is represented by the flux-potential transfer relation, Eq. (d) of Table 2.16.3:

$$\hat{e}_r^c = \frac{(n+1)}{R}\hat{\Phi}^c \tag{3}$$

Similarly, the inviscid fluid within is represented by the pressure-velocity relation, Eq. (i) of Table 7.9.1 in the limit $\beta \to 0$,

$$\hat{p}^d = j\omega\rho F_n(0,R)\hat{v}_r^d = -j\omega\rho\frac{R}{n}\hat{v}_r^d \tag{4}$$

Boundary Conditions: The electrical boundary condition at the drop interface requires that there be no tangential electric field: $\vec{n} \times \vec{E} = 0$. This condition prevails if frequencies of interest are low compared to the reciprocal charge relaxation time of the drop. With the objective of evaluating the electric field at the perturbed position of the interface, note that to linear terms Eq. 1 is evaluated at the interface as

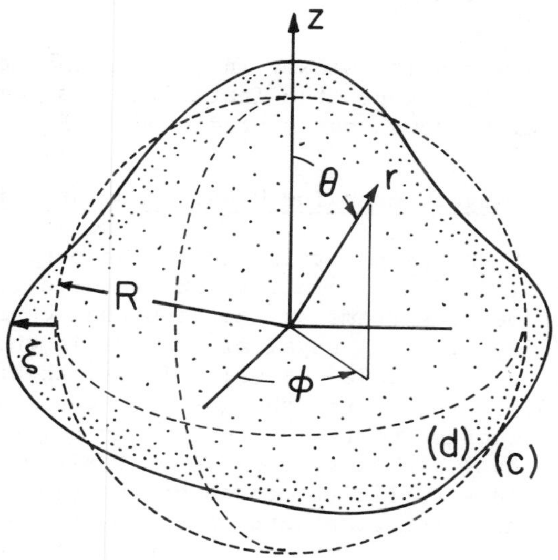

Fig. 8.13.1. Spherically symmetric equilibrium for a drop having total charge q uniformly distributed over its surface.

$$E_o\left(\frac{R}{r}\right)^2\Bigg|_{r=R+\xi} \simeq E_o\left(1 - \frac{2\xi}{R}\right) \tag{5}$$

Then, Eq. (e) of Table 7.6.1 is used to represent $\vec{n}$ and, to linear terms in ξ and hence $\vec{e}$, the boundary condition is written in terms of amplitudes evaluated at the unperturbed interfaces

$$\hat{\Phi}^c = E_o\hat{\xi} \tag{6}$$

Continuity requires that (Eq. 7.5.5 to linear terms)

$$\hat{v}_r^d = j\omega\hat{\xi} \tag{7}$$

Stress equilibrium for the interface, in general given by Eq. 7.7.3, is written with the perturbation pressure outside the drop ignored because the density there is negligible compared to that of the drop. Thus,

$$\Pi_c - \Pi_d - (p')^d = T_{rj}n_j + (T_s)_r \simeq \frac{1}{2}\varepsilon_o E_o^2 - \frac{2\varepsilon_o E_o^2}{R}\xi + \varepsilon_o E_o e_r^c + (T_s)_r \tag{8}$$

The equilibrium terms balance out, so that with the complex amplitude of $(T_s)_r$ given by Eq. (ℓ) of Table 7.6.2,

$$-\hat{p}^d = -\frac{2\varepsilon_o E_o^2}{R}\hat{\xi} + \varepsilon_o E_o\hat{e}_r^c - \frac{\gamma}{R^2}(n-1)(n+2)\hat{\xi} \tag{9}$$

Dispersion Relation and Rayleigh's Limit: All terms in the stress balance, Eq. 9, are written in terms of ξ by using Eq. 6 in Eq. 3 for $\hat{e}_r^c$, and Eq. 7 in Eq. 4 for $\hat{p}^d$. The factor multiplying $\hat{\xi}$ in the resulting homogeneous equation is the dispersion equation:

$$\omega^2\rho R^2 = (n-1)n\left[\frac{\gamma}{R}(n+2) - \varepsilon_o E_o^2\right] \tag{10}$$

The surface deflections are pictured with the help of Table 2.16.3. Conservation of mass excludes the n = 0 mode. From Eq. 10, the two n = 1 modes are neutrally stable. These are pure translations, either along or transverse to the z axis.

The first modes to become unstable as E_o is increased are the three n = 2 modes. This is seen by solving Eq. 10 for the E_o that makes the term in brackets vanish and recognizing that this is first true for the lowest allowed value of n, n = 2. Thus, because $E_o = q/4\pi\varepsilon R^2$, it follows that Rayleigh's limit on the total drop charge consistent with a stable equilibrium is

$$q = 8\pi\sqrt{\varepsilon\gamma R^3} \tag{11}$$

From this result, slowly increasing the net charge causes the drop to burst by fissioning into two drops. In most situations, the instability is dominated by the most rapidly growing of a spectrum of unstable modes with growth rates predicted by Eq. 10.

8.14 Charge Conserving Dynamics of Stratified Aerosols

If charge can relax instantaneously on the time scale of interest, an interface and even bulk material of fixed identity can preserve its potential. Examples are given in Secs. 8.13 and 8.7. In the opposite extreme are motions that conserve the charge density in the neighborhood of material of fixed identity. A physical example is the transport of submicron charged particles entrained in air. By virtue of applied or self-fields, these particles migrate according to the laws investigated in Sec. 5.6. But there, the gas flow was assumed to be known. What if the force transmitted to the gas by the charged particles results in a gas motion that dominates the migration of the particles relative to the gas? In fact, because of their extremely low mobilities, fine particles of high density can result in a sufficient force on the gas that the resulting fluid motions dominate over migration in determining the transport of the particles. Typically, what is observed is transport of particles by turbulent mixing with its origins in the electrohydrodynamic instability examplified in this section.

If fluid convection dominates over migration (or relaxation) in the transport of charged particles by an incompressible fluid, then the charge density is related to the fluid flow by

$$\frac{D\rho_f}{Dt} = 0 \tag{1}$$

In Sec. 7.2, this same statement was made for the mass density of an incompressible fluid. The general laws and relations subsequently developed in Secs. 7.8 and 7.9 bear on the motions of a mass density stratified fluid in a gravitational field much as does this section on motions of a charged fluid in an electric field. The discussion of gravity-capillary dynamics, Sec. 8.9, exemplifies the dynamics of fluids stratified in mass density, and is an example of how piecewise continuous models represent systems that are inhomogeneous in mass density.

At least as discussed here, where effects of self-gravitation are ignored, $\vec{g}$ in the gravitational force density is constant, whereas the electric field $\vec{E}$ in the electric force density is a function of the distribution of the field source, in this case ρ_f. But, in regions where the charge density is constant, say $\rho_f = q$, the force density transmitted to the fluid by the charged particles nevertheless takes the form of the gradient of a pressure:

$$\vec{F} = \rho_f\vec{E} = -\rho_f\nabla\Phi = -\nabla\mathcal{E};\quad \mathcal{E} \equiv q\Phi \tag{2}$$

Note that this statement prevails only where ρ_f is constant. It cannot be used to deduce a stress tensor at a boundary where ρ_f is discontinuous, for example.

That the force density in regions of uniform charge density is the gradient of a pressure effectively uncouples the bulk fluid mechanics from the electromagnetics. The inviscid equations of motion are as given in Sec. 7.8, with $\mathcal{E}$ as defined by Eq. 2. Thus, in the bulk, vorticity is conserved by a surface of fixed identity, and Eqs. 7.8.10 and 7.8.11 determine the velocity and pressure of motions initiated from a state of zero vorticity.[1]

Planar Layer: Suppose that a planar layer is embedded in a system in such a way that the equilibrium fields generated by the space charge are x-directed, as shown in Fig. 8.14.1. Because the following comments are general, for the moment consider the layer to have an equilibrium uniform translation in the z direction with velocity U. With $\mathcal{E}$ defined by Eq. 2, the pressure follows from Bernoulli's equation, Eq. 7.9.4, as

$$p = p_o(x) + p'(x,y,z,t);\quad \begin{cases} p_o(x) = -\frac{1}{2}\rho U^2 - q\Phi_o + \Pi - \rho g x \\ p' = \rho\left(\frac{\partial}{\partial t} + U\frac{\partial}{\partial z}\right)\Theta' - q\Phi' \end{cases} \tag{3}$$

---

1. The piecewise uniform approximation used here is developed in various geometries by M. Zahn, "Space Charge Coupled Interfacial Waves," Phys. Fluids 17, 343 (1974).

where primes indicate the time varying perturbation. A hybrid perturbation pressure is now defined,

$$\pi' \equiv p' + q\Phi' \tag{4}$$

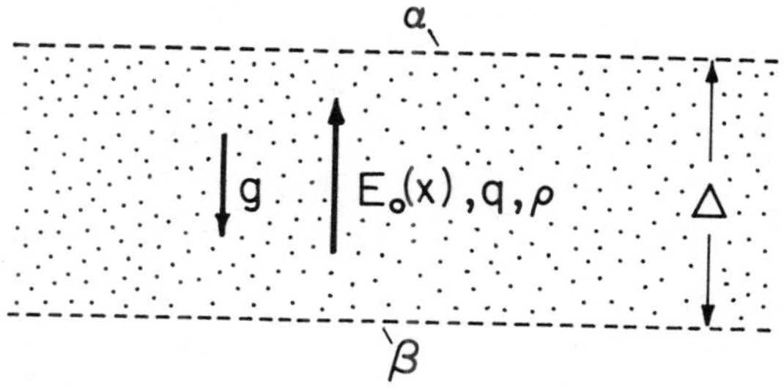

Fig. 8.14.1. Uniformly charged planar layer of charge conserving fluid.

It follows from Eq. 3 that π is related to the velocity potential by

$$\hat{\pi} = j\rho(\omega - k_z U)\hat{\Theta} \tag{5}$$

Thus, $\hat{\pi}$ now has the same relationship to the velocity potential as did $\hat{p}$ in Sec. 7.9 (Eq. 7.9.6). Here, as in Sec. 7.9, Θ satisfies Laplace's equation. Thus, the pressure-velocity relations of Table 7.9.1 apply with $\hat{p} \rightarrow \hat{\pi}$.

On the electrical side, Poisson's equation must be satisfied at every point in the bulk. However, because ρ_f is constant, the equilibrium field equilibrates the charge density in Poisson's equation and perturbations in the potential must satisfy Laplace's equation. Thus, fields take the form

$$\Phi = \Phi_o(x) + \text{Re}\hat{\Phi}\, e^{j(\omega t - k_y y - k_z z)}; \quad \frac{dE_o}{dx} = \frac{q}{\varepsilon_o}; \quad E_o = -\frac{d\phi_o}{dx} \tag{6}$$

where the flux-potential transfer relations of Table 2.16.1 apply to the perturbation, $\hat{\Phi}$.

<u>Boundary Conditions</u>: The electromechanical coupling occurs in the regions of singularity between layers of uniformly charged fluid. Interfacial boundary conditions representing the mechanical equations come from continuity, which requires that

$$\hat{v}_x = j\omega\hat{\xi} \tag{7}$$

and stress equilibrium. The charge density has a step discontinuity at the interface, but there is no surface charge. Further, there is no discontinuity in the permittivity at the interface. Thus, the surface force density, represented by the first term on the right side of Eq. 7.9.6, is zero. For layers of charged aerosol, it is appropriate to ignore the surface tension, so the boundary condition is simply

$$\vec{n}[\![p]\!] = 0 \tag{8}$$

In view of Eq. 3, this condition is represented by its x-component evaluated to linear terms on the interface at $x = \xi$ (say) to give

$$[\![qE_o - \rho g]\!]\hat{\xi} + [\![\hat{\pi}]\!] - [\![q\hat{\Phi}]\!] = 0 \tag{9}$$

where E_o is now the equilibrium electrical field evaluated at the unperturbed interface.

The potential must be continuous at the perturbed interface. Because there is no surface charge and no discontinuity in permittivity, it is also true that $[\![E_o]\!] = 0$, so this condition requires that

$$[\![\hat{\Phi}]\!] = 0 \tag{10}$$

Because there is no surface charge even on the perturbed interface, a further boundary condition reflecting Poisson's equations is that $\vec{n}\cdot[\![\varepsilon_o\vec{E}]\!] = 0$, so this condition requires that

$$[\![q]\!]\hat{\xi} + \varepsilon_o[\![e_x]\!] = 0 \tag{11}$$

where Eq. 6 is used to replace $d\varepsilon_o E_o/dx$ by (q). The four boundary conditions, Eqs. 7, 9, 10 and 11, are evaluated at the unperturbed position of the interface.

<u>Stability of Two Charge Layers</u>: As a specific example, consider the motions of the layers shown in Fig. 8.14.2. In the bulk, the mechanics in each layer is represented by Eqs. (c) of Table 7.9.1 with $\hat{p} \rightarrow \hat{\pi}$:

$$\begin{bmatrix} \hat{\pi}^c \\ \hat{\pi}^d \end{bmatrix} = \frac{j\omega\rho_a}{k}\begin{bmatrix} -\coth(ka) & \dfrac{1}{\sinh(ka)} \\ \dfrac{-1}{\sinh(ka)} & \coth(ka) \end{bmatrix}\begin{bmatrix} \hat{v}_x^c \\ \hat{v}_x^d \end{bmatrix} \tag{12}$$

$$\begin{bmatrix} \hat{\Pi}^e \\ \hat{\Pi}^f \end{bmatrix} = \frac{j\omega\rho_b}{k}\begin{bmatrix} -\coth(kb) & \frac{1}{\sinh(kb)} \\ \frac{-1}{\sinh(kb)} & \coth(kb) \end{bmatrix}\begin{bmatrix} \hat{v}_x^e \\ \hat{v}_x^f \end{bmatrix} \tag{13}$$

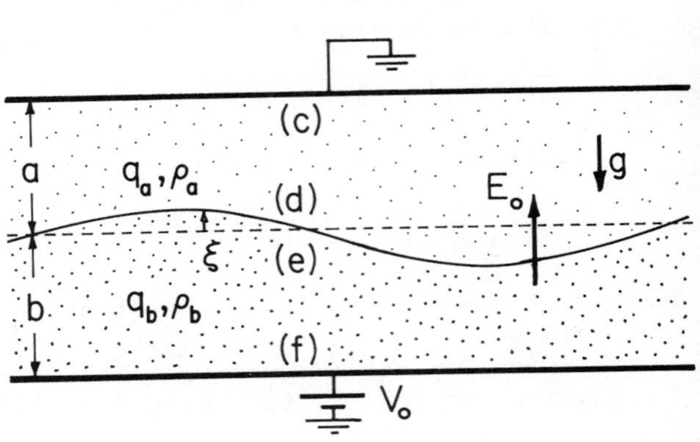

Fig. 8.14.2. Fluid layers of different uniform charge and mass densities have an interface, d-e, and are bounded by rigid electrodes.

Similarly, the fields follow from Eqs. (a) of Table. 2.16.1:

$$\begin{bmatrix} \hat{e}_x^c \\ \hat{e}_x^d \end{bmatrix} = k\begin{bmatrix} -\coth(ka) & \frac{1}{\sinh(ka)} \\ \frac{-1}{\sinh(ka)} & \coth(ka) \end{bmatrix}\begin{bmatrix} \hat{\Phi}^c \\ \hat{\Phi}^d \end{bmatrix} \tag{14}$$

$$\begin{bmatrix} \hat{e}_x^e \\ \hat{e}_x^f \end{bmatrix} = k\begin{bmatrix} -\coth(kb) & \frac{1}{\sinh(kb)} \\ \frac{-1}{\sinh(kb)} & \coth(kb) \end{bmatrix}\begin{bmatrix} \hat{\Phi}^e \\ \hat{\Phi}^f \end{bmatrix} \tag{15}$$

Boundary conditions at the top electrode are

$$\hat{v}_x^c = 0 \tag{16}$$

$$\hat{\Phi}^c = 0 \tag{17}$$

at the interface are Eqs. 5, 7, 8 and 9:

$$\hat{v}_x^d = \hat{v}_x^e = j\omega\hat{\xi} \tag{18}$$

$$[E_o(q_a-q_b) - g(\rho_a-\rho_b)]\hat{\xi} + (\hat{\Pi}^d-\hat{\Pi}^e) - (q_a\hat{\Phi}^d-q_b\hat{\Phi}^e) = 0 \tag{19}$$

$$\hat{\Phi}^d - \hat{\Phi}^e = 0 \tag{20}$$

$$(q_a - q_b)\hat{\xi} + \varepsilon_o(\hat{e}_x^d - \hat{e}_x^e) = 0 \tag{21}$$

and at the bottom electrodes are

$$\hat{v}_x^f = 0 \tag{22}$$

$$\hat{\Phi}^f = 0 \tag{23}$$

It is a simple matter to substitute Eqs. 16-18,20,22, and 23 into the bulk relations. Substitution of the resulting Eqs. 14b and 15a into Eq. 21 then shows that

$$\hat{\Phi}^d = \hat{\Phi}^e = \frac{-(q_a-q_b)\hat{\xi}}{\varepsilon_o k[\coth(ka)+\coth(kb)]} \tag{24}$$

The force-equilibrium boundary condition, Eq. 19, is finally evaluated using Eqs. 12b and 13a and Eq. 24 to obtain the dispersion equation

$$\frac{\omega^2}{k}[\rho_a\coth(ka) + \rho_b\coth(kb)] = g(\rho_b-\rho_a)+E_o(q_a-q_b) + \frac{(q_a - q_b)^2}{\varepsilon_o k[\coth(ka) + \coth(kb)]} \tag{25}$$

Remember that E_o is the equilibrium electric field evaluated at the unperturbed position of the interface. The equilibrium fields imply that the voltage V_o is related to E_o and the charge densities by

$$E_o = \frac{V_o}{b+a} + \frac{q_b b^2 - q_a a^2}{2\varepsilon_o(a+b)} \tag{26}$$

The last "self-field" term in Eq. 25 is positive regardless of the relative charge densities, and hence tends to stabilize all wavelengths. However, for short waves ($ka \gg 1$ and $kb \gg 1$) its contribution is negligible compared to the gravitational and "imposed field" term. Thus, a necessary and sufficient condition for all wavelengths to be stable is that the first two terms on the right in Eq. 25 be positive,

$$g(\rho_b - \rho_a) + E_o(q_a - q_b) > 0 \tag{27}$$

The static arguments used in Sec. 8.4 lead to a similar condition, Eq. 8.4.11, because instability is incipient at zero frequency.

If the inequality of Eq. 27 is not satisfied, Eq. 25 shows that the growth rate of instabilities increases linearly with the wave number. Actually, there is a wavelength for maximum rate of growth that would be predicted if the model included effects of viscosity (which come into play at short wavelengths) or recognized the finite structure of the discontinuity in charge density.

The model of a charge density that is frozen to the fluid is of course relevant only if the processes described take place on a time scale short compared to the migration time of the charged particles. To what physical situations might the model apply?

Suppose that the electromechanical waves are of interest and V_o is adjusted to make $E_o = 0$. For a fluid of uniform mass density ($\rho_a = \rho_b = \rho$), according to Eq. 25, short waves have the frequency

$$\omega = \frac{|q_a - q_b|}{\sqrt{4\varepsilon_o\rho}} \tag{28}$$

(Note that this is a reciprocal electro-inertial time.) For particles having charge q, number density n and mobility b, the self-precipitation time due to migration is $\tau_e = \varepsilon_o/nqb$) (Eq. 5.6.6). The frozen charge model is valid if the electro-inertial frequency given by Eq. 28 is high compared to the reciprocal of the self-precipitation time. That is, for $|q_a - q_b| \simeq nq$, it is valid if

$$\omega\tau_e = \frac{1}{2}\frac{\sqrt{\varepsilon_o/\rho}}{b} \gg 1 \tag{29}$$

The summary of typical mobilities given by Table 5.2.1 makes it clear that the model does not apply to ions in a gas. However, it could apply to charged macroscopic particles in air[2] and to ions in liquids.[3,4] In fact, as a consequence of the electrohydrodynamic instability that prevails when Eq. 27 is not satisfied, the electrically induced convection can be a dominant charge transport mechanism.

The effect of the instability on transport of an aerosol is demonstrated by the experiment shown in Fig. 8.14.3.[5] Generated by dry ice immersed in water, the aerosol passes from left to right as a layer, bounded from below by an electrode and from above by clear air. Thus, the configuration is essentially that of Fig. 8.14.2 with the upper region uncharged. The aerosol is negatively charged by ion impact at the left. From the picture center to the right, the layer is subjected to a vertically applied electric field. In Fig. 8.14.3a, the applied field is upward and hence the configuration is stable. Some migration is observed, but little convection. In Fig. 8.14.3b, the field is reversed. Electrohydrodynamic instability is apparent in its contribution to the transport of charge out of the gas stream. For this experiment, $\sqrt{\varepsilon_o/\rho}/2b > 10$, so effects of convection are expected to be important.

---

2. R. S. Withers, J. R. Melcher and J. W. Richmann, "Charging, Migration and Electrohydrodynamic Transport of Aerosols," J. Electrostatics 5, 225-239 (1978)
3. P. K. Watson, J. M. Schneider and H. R. Till, "Electrohydrodynamic Stability of Space-Charge-Limited Currents in Dielectric Liquids," Phys. Fluids 13, 1955 (1970).
4. E. J. Hopfinger and J. P. Gosse, "Charge Transport by Self-Generated Turbulence in Insulating Liquids Submitted to Unipolar Injection," Phys. Fluids 14, 1671 (1971).
5. R. S. Colby, "Electrohydrodynamics of Charged Aerosol Flows," B.S. Thesis, Department of Electrical Engineering and Computer Sciences, Massachusetts Institute of Technology, Cambridge, Mass., 1978.

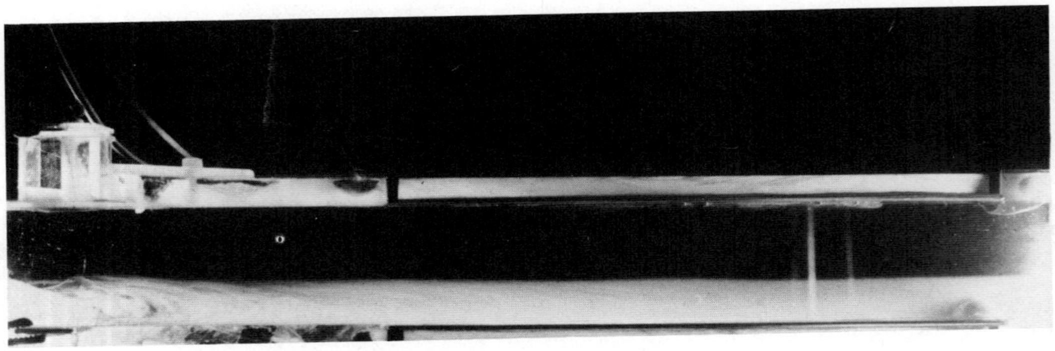

(a)

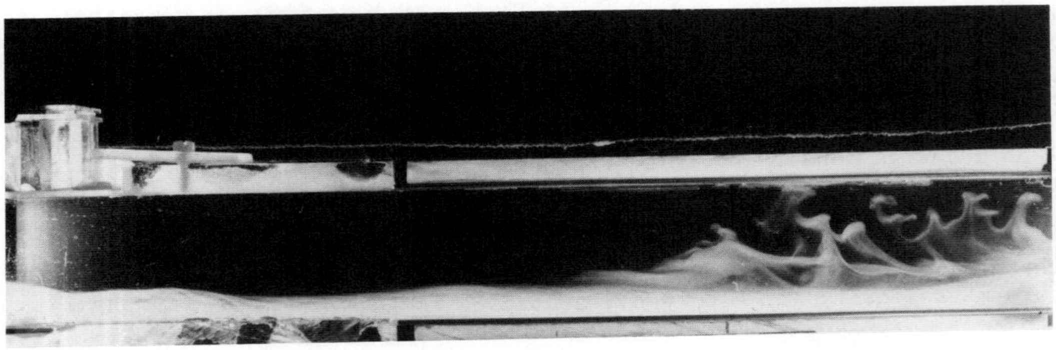

(b)

Fig. 8.14.3. Aerosol passed through ion-impact charging region at left and into region of applied electric field from the center to the right. The aerosol is charged negatived. (a) Stable configuration with applied field directed upward. (b) Unstable configuration with applied field reversed.

8.15. The z Pinch with Instantaneous Magnetic Diffusion

The model exemplified in this section pertains to the MQS dynamics of electrical conduction in the opposite extreme of that considered in Sec. 8.12. There time scales of interest were short compared to the magnetic diffusion time, so that the magnetic flux linked by a surface of fixed identity was conserved. In the opposite extreme considered here, the diffusion of magnetic field on the time scales of interest is instantaneous. In the magnetic diffusion equation, Eq. 6.2.2, the induction and "speed-voltage" terms are now negligible. That is, the magnetic diffusion time $\tau_m = \mu\sigma\ell^2$ is short compared to times of interest and the magnetic Reynolds number $R_m = \mu\sigma\ell v$ is small (Eq. 6.3.9).

In this limit of instantaneous magnetic diffusion, the effect of the material deformation on the magnetic field comes from the heterogeneity of the conductor. The distribution of $\vec{J}$ and hence $\vec{H}$ is determined by the geometry of the conductors. This is best emphasized by dealing with the current density rather than the magnetic field. Because $R_m \ll 1$, the effect of motion on the current density is ignorable. Thus

$$\vec{J} = \sigma\vec{E} \tag{1}$$

It follows from the law of induction, Eq. 6.2.3 with $\tau_m/\tau \ll 1$ and $R_m \ll 1$, that

$$\nabla \times \left(\frac{\vec{J}}{\sigma}\right) \simeq 0 \tag{2}$$

In the MQS approximation, the current density is also solenoidal:

$$\nabla\cdot\vec{J} = 0 \tag{3}$$

This is insured by Ampère's law, which represents $\vec{J}$ in terms of a "vector potential" which happens to be the magnetic field intensity:

$$\vec{J} = \nabla \times \vec{H} \tag{4}$$

In regions where σ and μ are uniform, it follows from Eqs. 2 and 4 and the solenoidal nature of $\vec{B}$ that

$$\nabla^2\vec{H} = 0 \tag{5}$$

which is of course the limit $\tau_m/\tau \ll 1$ and $R_m \ll 1$ of Eq. 6.2.6.

Liquid Metal z Pinch: The column of liquid metal shown in Fig. 8.15.1 initially has a uniform circular cross section and carries a longitudinal current density, J_o, that is uniform over this cross section,

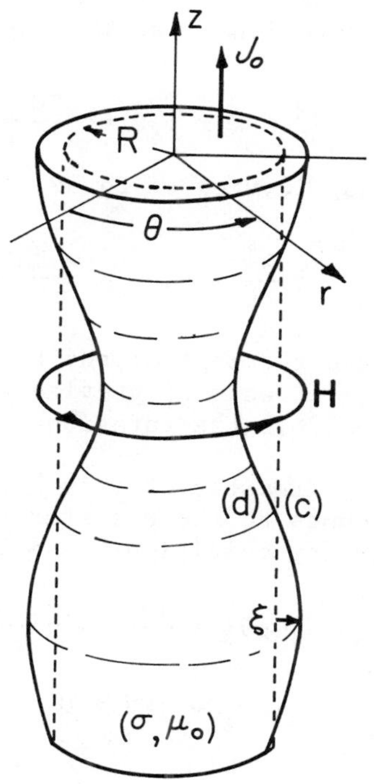

Fig. 8.15.1. Column of liquid metal has static equilibrium with $\xi = 0$ and uniform axial current density.

$$\vec{J} = J_o\vec{i}_z \tag{6}$$

Thus, by contrast with the perfectly conducting pinch of Sec. 8.12 where the current is on the surface, the equilibrium magnetic field has completely diffused into the conductor. It assumes the linear distribution consistent with Ampère's law and Eq. 6:

$$\vec{H} = \begin{cases} \dfrac{J_o r}{2}\vec{i}_\theta & r < R \\ \dfrac{J_o R^2}{2r}\vec{i}_\theta & r > R \end{cases} \tag{7}$$

Static equilibrium prevails because the radial pressure distribution, p(r), just balances the associated radial magnetic force density and surface tension surface force density. With p defined as zero in the air surrounding the column,

$$p = -\frac{1}{4}\mu_o J_o^2(r^2 - R^2) + \frac{\gamma}{R} \tag{8}$$

An experiment demonstrating the dynamics to be described (Ref. 2, Appendix C) makes use of a liquid jet of mercury. In the model now developed, the longitudinal streaming of the jet is ignored. Instabilities exhibiting a temporal growth here can be displayed as a spatial growth as a result of the streaming. Such effects of streaming are taken up in Chap. 11.

Bulk Relations: With the vector potential $\vec{A} \to \vec{H}$ and $\vec{B} \to \vec{J}$, the situation is formally the same as described by Table 2.18.1. Axisymmetric perturbations from this static equilibrium now considered can be described in terms of one component of the magnetic field, $\vec{H} = \vec{H}_\theta(r,z,t)\vec{i}_\theta$. Here, $H_\theta = \Lambda/r$ and in terms of $\Lambda(r,z,t)$, the perturbation current density is

$$\vec{J} = -\frac{1}{r}\frac{\partial\Lambda}{\partial z}\vec{i}_r + \frac{1}{r}\frac{\partial\Lambda}{\partial r}\vec{i}_z \tag{9}$$

The axisymmetric solutions of Eq. 5 in cylindrical coordinates are discussed in Sec. 2.19. Solutions are of the form of Eq. 2.19.10 with $\beta \to 0$:

$$\hat{\Lambda} = \hat{H}_\theta^d \frac{rJ_1(jkr)}{J_1(jkR)} \tag{10}$$

That is, the perturbation current density in the bulk is uncoupled from the mechanics and determined by the geometry of the interface, which will determine the coefficient $\hat{H}_\theta^d$.

By contrast, the mechanics is bulk coupled to the field distribution. The strategy in Sec. 8.14 was to represent the electromechanical bulk coupling in terms of a force density that was the gradient of a pressure. Essentially, this attributes the coupling to interfaces. Here, part of the force density is rotational, so that matters are not so simple. It follows from Eqs. 6 and 7 that

$$\vec{F} = \vec{J} \times \mu_o \vec{H} = -\frac{\mu_o \mathcal{J}_o^2 r}{2} \vec{i}_r + \vec{J}' \times \frac{\mu_o \mathcal{J}_o r}{2} \vec{i}_\theta + \mathcal{J}_o \vec{i}_z \times \mu_o \vec{H}' \qquad (11)$$

Thus, in view of Eq. 9, the force equation for the fluid becomes

$$\rho \frac{\partial \vec{v}}{\partial t} + \nabla \pi = -\frac{\mathcal{J}_o \mu_o \Lambda}{r} \vec{i}_r; \quad \pi \equiv p' + \frac{\mathcal{J}_o \Lambda}{2} \qquad (12)$$

where the part of the force density that is the gradient of a pressure is lumped with the perturbation p'. Effects of gravity and viscosity are not included in Eq. 12. What is on the right in the force of Eq. 12 is the rotational part of the magnetic force density.

Because of this "one-way" coupling of the field to the fluid, it is necessary to rederive what amounts to the transfer relations for the fluid. The r and z components of Eq. 12, as well as the continuity condition $\nabla \cdot \vec{v} = 0$, give three relations for the mechanical perturbations:

$$j\omega\rho\hat{v}_r + \frac{d\hat{\pi}}{dr} = -\frac{\mathcal{J}_o \mu_o \hat{\Lambda}}{r} \qquad (13)$$

$$j\omega\rho\hat{v}_z - jk\hat{\pi} = 0 \qquad (14)$$

$$\frac{1}{r}\frac{d}{dr}(r\hat{v}_r) - jk\hat{v}_z = 0 \qquad (15)$$

Elimination of $\hat{v}_z$ between Eqs. 14 and 15 gives an expression that can be solved for $\hat{\pi}$,

$$\hat{\pi} = \frac{\omega\rho}{jk^2}\frac{1}{r}\frac{d}{dr}(r\hat{v}_r) \qquad (16)$$

Substitution of this expression into Eq. 13 gives

$$\frac{d^2\hat{v}_r}{dr^2} + \frac{1}{r}\frac{d\hat{v}_r}{dr} - \frac{\hat{v}_r}{r^2} - k^2\hat{v}_r = -\frac{jk^2 \mathcal{J}_o \mu_o}{\omega\rho}\frac{\hat{\Lambda}}{r} \qquad (17)$$

In the absence of the bulk coupling, these last two expressions could be used to derive the pressure-velocity transfer relations of Table 7.9.1. Added to the homogeneous solutions of Eq. 17 (that comprise these transfer relations) is now a particular solution satisfying the equation with Eq. 10 substituted on the right. Substitution and recognition that $J_o(jkr)$ satisfies Eq. 2.16.19 with m = 0 shows that a particular solution is

$$\frac{\mathcal{J}_o \mu_o k \hat{H}_\theta^d}{2\omega\rho J_1(jkR)} r J_o(jkr) \qquad (18)$$

where Eq. 2.16.26c has been used. Of the two homogeneous solutions, the one that is not singular at the origin is $J_1(jkr)$. The linear combination of particular and homogeneous solutions that makes $\hat{v}_r(R) = \hat{v}_r^d$ is

$$\hat{v}_r = \hat{v}_r^d \frac{J_1(jkr)}{J_1(jkR)} - \frac{\mathcal{J}_o \mu_o k \hat{H}_\theta^d}{2\omega\rho J_1(jkR)}\left[\frac{R J_o(jkR) J_1(jkr)}{J_1(jkR)} - r J_o(jkr)\right] \qquad (19)$$

Thus, in view of Eqs. 12 and 16, the amplitude of the perturbation pressure is

$$\hat{p} = \frac{\omega\rho}{k}\frac{J_o(jkr)}{J_1(jkR)}\hat{v}_r^d - \frac{\mathcal{J}_o \mu_o \hat{H}_\theta^d}{2jkJ_1(jkR)}\left[\frac{jkRJ_o(jkR)}{J_1(jkR)} J_o(jkr) - 2J_o(jkr) + 2jkrJ_1(jkr)\right] \qquad (20)$$

<u>Boundary Conditions</u>: The effect of the boundary condition on the distribution of current density, and hence magnetic field, is represented by the condition that at the interface, $\vec{n} \cdot \vec{J} = 0$. To linear terms, with $\vec{n}$ written in terms of $\xi(z,t)$ (Eq. (e) of Table 7.6.2),

$$\hat{J}_r^d + jk\mathcal{J}_o\hat{\xi} = 0 \qquad (21)$$

The radial current density, J_r, is substituted into this expression using Eqs. 9 and 10 to show that

$$\hat{H}_\theta^d = -\mathcal{J}_o\hat{\xi}$$

This condition represents the effect of the mechanics (geometry) on the field.

The return effect of the field on the fluid is taken into account in writing stress equilibrium for the interface. Note that there is no singularity in the magnetic force density at the interface. That is, there is no surface current and no discontinuity in magnetizability of the material. Hence, the magnetic surface force density, $[\![T_{ij}^e]\!]n_j$, makes no contribution to the stress equilibrium, Eq. 7.7.6. Because the fluid surrounding the column is of considerably lesser density than the column, the perturbation pressure, $\hat{p}^c$, is ignored. Thus, the jump in total pressure evaluated at the perturbed position of the interface is balanced by the surface tension surface force density, Eq. (f) of Table 7.6.2:

$$-\{-\frac{1}{4}\mu_o\mathcal{J}_o^2[(R+\xi)^2-R^2]+\frac{\gamma}{R}+p^d\} = \gamma[-\frac{1}{R}+\frac{\xi}{R^2}+\frac{\partial^2\xi}{\partial z^2}] \qquad (23)$$

By design, the equilibrium part of this balance cancels out. In terms of complex amplitudes, the perturbation part is

$$\frac{1}{2}\mu_o\mathcal{J}_o^2R\hat{\xi} - \hat{p}^d = \frac{\gamma}{R^2}[1-(kR)^2]\hat{\xi} \qquad (24)$$

Evaluated using Eqs. 20, 22 and the continuity condition $\hat{v}_r^d = j\omega\hat{\xi}$, this expression becomes the dispersion equation

$$\omega^2\frac{\rho R^3}{\gamma} = \frac{kRI_1(kR)}{I_o(kR)}\left\{\left[\frac{\mathcal{J}_o^2\mu_o R^3}{2}\right]\left[\frac{I_o^2(kR)}{I_1^2(kR)} - \frac{2I_o(kR)}{kRI_1(kR)} - 1\right] + [(kR)^2-1]\right\} \qquad (25)$$

Rayleigh-Plateau Instability: The normalized frequency given by Eq. 25 is shown as a function of wave number by Fig. 8.15.2 with the magnetic pressure $\mu_o(\mathcal{J}_oR)^2/2$ normalized to the surface tension pressure γ/R as a parameter. Negatives of the quantities shown are also solutions. Note that even in the absence of an axial current, perturbations $kR < 1$ (wavelengths longer than $2\pi R$) are unstable. Any perturbation results in major radii of curvature that differ in sign. For a region that is necking off, the curvature associated with the axial dependence tends to restore the equilibrium whereas that caused by the circular cross section of the column tends to further neck off the column. For perturbations having wavelength $\lambda > 2\pi R$, the latter wins and the equilibrium is unstable. The wavelength for maximum rate of growth, given by $kR \simeq 0.7$, can be used to give a rough prediction of the size of drops formed from a liquid jet. According to the linear theory, a drop having radius r_o would have a volume equal to that of one wavelength of the jet, $\pi R^2 = 4/3(\pi r_o^3)$.

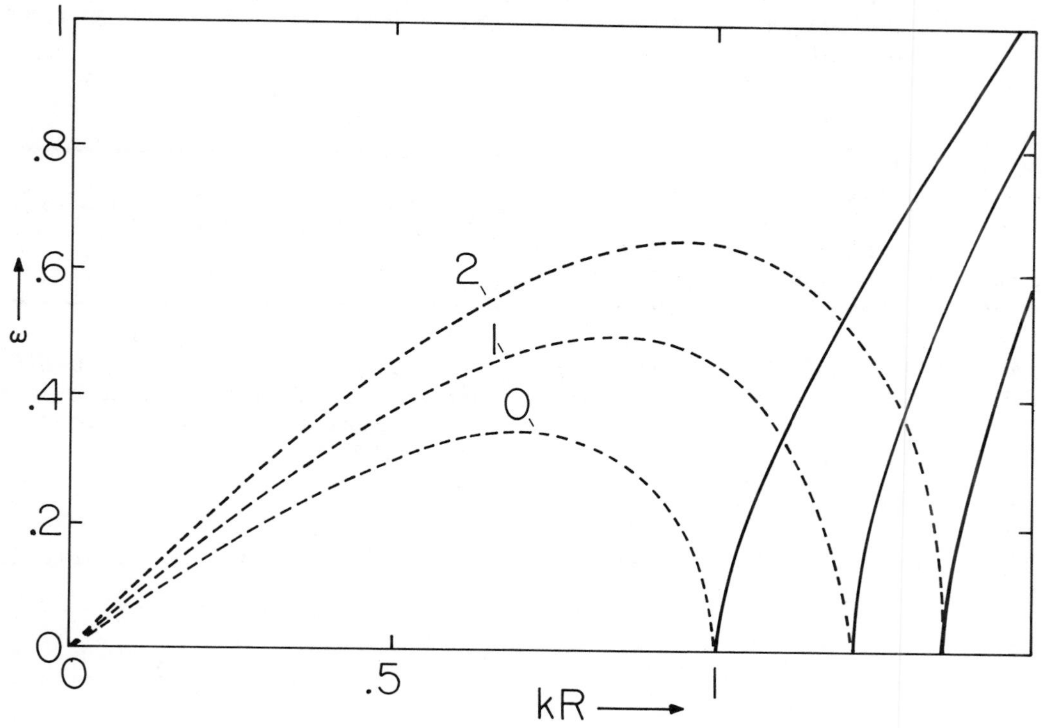

Fig. 8.15.2

Normalized frequency $\underline{\omega} \equiv \omega\sqrt{\rho R^3/\gamma}$ as a function of wave number. --- ω_i, —— ω_r. The parameter is $\mathcal{J}_o^2\mu_o R^3/2\gamma$.

z-Pinch Instability: The general nature of the pinch instability is qualitatively similar to that found with the flux conserving pinch of Sec. 8.12. Because the current through the column must be conserved, both the current density and the magnetic field intensity in the fluid adjacent to the interface go up wherever the column tends to neck off. The result is an inward magnetic force density that tends to further encourage the necking off. Unless wavelengths are sufficiently short to be stabilized by surface tension, they are unstable. According to the model, it is only the inertia of the column that limits the rate of growth of the instability.

Finally, is the instantaneous magnetic diffusion model appropriate for the description of a mercury column having a radius of 1 cm or less? From Eq. 25 and Fig. 8.15.2 the frequency can be taken as of the order of $\sqrt{\gamma/\rho R^3}$. For the approximation to be justified, the product of this frequency (or growth rate) and the magnetic diffusion time (based here on the column radius) must be small:

$$\omega\tau_m = \sqrt{\frac{\mu^2\sigma^2\gamma R}{\rho}} \tag{26}$$

Typically, this number is less than 10^{-3}.

The major electromechanical effect that would be experimentally observed but not accounted for by this model is magnetic damping.

8.16 Dynamic Shear Stress Surface Coupling

It is a straightforward process to include the effects of viscosity in the piecewise homogeneous models developed in Secs. 8.9–8.15. The fluid mechanics is represented by the viscous diffusion transfer relations of Sec. 7.19 rather than the inviscid pressure-velocity relations of Sec. 7.11. With the viscosity come additional boundary conditions. At an interface, not only is the normal velocity continuous, but so also is the tangential velocity (Eq. 7.7.3). Also, the shearing stresses acting at an interface, Eq. 7.7.6, are not automatically balanced. In Secs. 8.9–8.15, the interfacial stress balance is for interfaces free of shearing surface force densities. Thus, any of these examples have stress balance equations in directions tangential to the equilibrium interface that are identically satisfied.

In this section, the example treated not only illustrates how viscosity is taken into account in piecewise homogeneous systems, but also involves an electric shearing surface force density. Hence, the viscous shear stresses are necessary for the formulation of a self-consistent model.

A highly insulating liquid, such as hexane, has a free surface which is bounded from above by a gas, as shown in Fig. 8.16.1. Perhaps by means of a very small radioactive source, some ion pairs are provided in the bulk of the liquid. By means of a potential applied between the planar electrodes, half of this charge is swept to the interface where it forms a monolayer of surface charge that shields the electric field from the liquid; thus, $\sigma_o = \varepsilon_o E_o$. Subjected to a tangential electric field, common interfacial ions migrate relative to the liquid at a rate that is negligible compared to that due to convection. A good model pictures the charge as frozen to the liquid interface. What are the modes of motion characterizing the adjustment of the interface to a perturbation field?

Because the fluids to either side of the interface have uniform permittivities and no free charge density, the electromechanical coupling is confined to the interface. In the following, it is assumed that the depth of the liquid and the distance to the upper electrode from the interface are large compared to typical perturbation wavelengths on the interface.

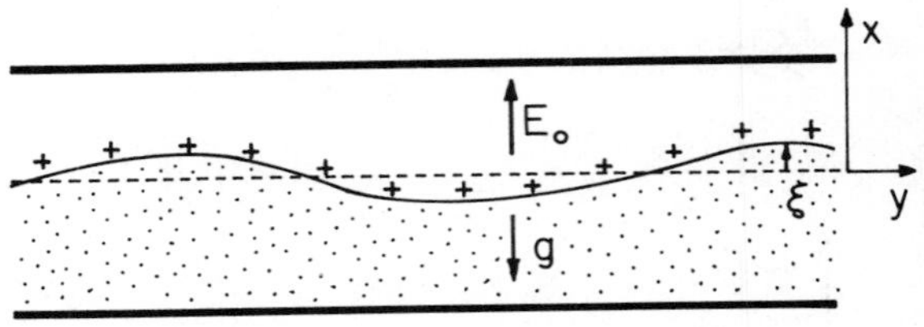

Fig. 8.16.1. Cross section of liquid-air interface supporting surface charge density σ_o. Charges are modeled as frozen to the liquid.

Static Equilibrium: With the interface flat and $\vec{v} = 0$, the electric field is

$$\vec{E} = \begin{cases} E_o\vec{i}_x = \dfrac{\sigma_o}{\varepsilon_o}\vec{i}_x; & x > 0 \\ 0 \qquad\qquad\quad ; & x < 0 \end{cases} \tag{1}$$

and the pressure balances the gravitational force density in the liquid with a jump at the interface to equilibrate the surface force density $\varepsilon_o E_o^2/2$:

$$S_{xx} = -p = \begin{cases} -\Pi & ; \quad x > 0 \\ \rho g x + \frac{1}{2}\varepsilon_o E_o^2 - \Pi; & x < 0 \end{cases} \tag{2}$$

Bulk Perturbations: With the perturbation electric field represented by $\vec{e} = -\nabla\Phi$, the flux-potential relations describe the fields in the bulk regions. Application of Eqs. (a) from Table 2.16.1 in the limit $(k\Delta) \to \infty$ gives

$$\hat{e}_x^d = k\hat{\Phi}^d; \quad \hat{e}_x^e = -k\hat{\Phi}^e \tag{3}$$

for the regions above and below the interface respectively. Because the system is invariant to rotation about the x axis, there is no loss in generality if perturbations are taken as independent of z; $\Phi = \text{Re}\exp j(\omega t - ky)$.

For the half-space of liquid, the mechanical perturbation stress-velocity relations are given by Eq. 7.19.19, where $\gamma_V \equiv \sqrt{k^2 + j\omega\rho/\eta}$,

$$\begin{bmatrix} \hat{S}_{xx}^e \\ \\ \hat{S}_{yx}^e \end{bmatrix} = \begin{bmatrix} \eta\dfrac{\gamma_V}{k}(\gamma_V + k) & -j\eta(\gamma_V - k) \\ \\ j\eta(\gamma_V - k) & \eta(\gamma_V + k) \end{bmatrix} \begin{bmatrix} \hat{v}_x^e \\ \\ \hat{v}_y^e \end{bmatrix} \tag{4}$$

Jump Conditions: Each of the laws prevailing in the bulk must be consistently represented in the highly singular neighborhood of the interface. The charge forms a monolayer, but not a double layer, and hence consistent with the irrotational nature of $\vec{E}$ is the condition that its tangential component is continuous. In writing this condition, note that $\hat{v}_x = j\omega\xi$ where $\vec{n}$ is given in terms of ξ by Eq. (a) of Table 7.6.2:

$$\hat{\Phi}^d - \hat{\Phi}^e = -\frac{jE_o}{\omega}\hat{v}_x^e \tag{5}$$

The remaining electrical laws are charge conservation, Eq. 23 of Table 2.10.1, and Gauss' law. Together, these require that

$$\frac{\partial\sigma_f}{\partial t} + \nabla_\Sigma\cdot(\sigma_f\vec{v}) = 0; \quad \sigma_f = \vec{n}\cdot[\![\varepsilon\vec{E}]\!] \tag{6}$$

To linear terms, Gauss' law and conservation of charge are then represented by

$$\omega(\varepsilon_o\hat{e}_x^d - \varepsilon\hat{e}_x^e) - k\sigma_o\hat{v}_y^e = 0 \tag{7}$$

For the mechanical jump conditions, continuity of the velocity components does not enter because the contributions of the upper fluid to the stress equilibrium is negligible. Stress equilibrium, represented by Eq. 7.7.5, includes the normal surface force density due to surface tension, γ (given by Eq. (d) of Table 7.6.2):

$$[\![S_{ij}]\!]n_j + [\![T_{ij}^e]\!]n_j - \gamma(\nabla\cdot\vec{n})n_i = 0 \tag{8}$$

Physically, the x component of this expression represents (to linear terms) the balance of stresses normal to the distorted interface. Note that the total normal stress, S_{xx}, is the sum of an equilibrium part and the perturbation:

$$S_{xx} = \Pi + \frac{1}{2}\varepsilon_o E_o^2 + \rho g x + \text{Re}\,\hat{S}_{xx}(x)\exp j(\omega t - ky) \tag{9}$$

Thus, because $\hat{\xi} = \hat{v}_x^e/j\omega$, the x component of Eq. 8 is

$$-\hat{S}_{xx}^e - \rho g\frac{\hat{v}_x^e}{j\omega} + \varepsilon_o E_o\hat{e}_x^d - k^2\gamma\frac{\hat{v}_x^e}{j\omega} = 0 \tag{10}$$

What is new is the shearing component, the y component, of Eq. 8. In linearizing this expression, remember that S_{yy} also has an equilibrium part. Above the interface, it is $-\Pi$, while below the interface it is $-\Pi + \frac{1}{2}\varepsilon_o E_o^2 + \rho g x$ (Eq. 2). Thus, to linear terms, $[\![S_{yy}]\!]n_y \simeq [-\Pi - (-\Pi + \frac{1}{2}\varepsilon_o E_o^2)](-\partial\xi/\partial y)$ and this adds to one of the two terms resulting from the electric stress contribution. Also, $\hat{e}_y = jk\hat{\Phi}$, so the shearing component of the stress equilibrium reduces to

$$-S_{yx}^e - \varepsilon_o E_o^2\frac{k\hat{v}_x^e}{\omega} + j\varepsilon_o E_o k\hat{\Phi}^d = 0 \tag{11}$$

Dispersion Equation: By using Eqs. 3, the components $(\hat{e}_x^d, \hat{e}_x^e)$ can be eliminated from the electric jump conditions, Eqs. 5 and 7, and these solved for $\hat{\Phi}^d$,

$$\hat{\Phi}^d = \frac{-j\varepsilon E_o \hat{v}_x^e + \sigma_o \hat{v}_y^e}{\omega(\varepsilon_o + \varepsilon)} \tag{12}$$

This expresses the effect of the mechanics on the fields.

The self-consistent electromechanics is now represented by the two stress conditions, Eqs. 10 and 11, written in terms of the velocity amplitudes $(\hat{v}_x^e, \hat{v}_y^e)$. The stress amplitudes are eliminated in favor of these variables using Eqs. 4, while $\hat{e}_x^d$ is written in terms of $\hat{\Phi}^d$ by using Eq. 3a, and $\hat{\Phi}^d$ in turn eliminated using Eq. 12. Thus, the two expressions are

$$\begin{bmatrix} \left[-\frac{\eta\gamma_v}{k}(\gamma_v+k) - (\rho g + k^2\gamma)\frac{1}{j\omega} - \frac{j\varepsilon k E_o^2}{\omega(1+\frac{\varepsilon}{\varepsilon_o})}\right] & \left[j\eta(\gamma_v - k) + \frac{\sigma_o E_o k}{\omega(1+\frac{\varepsilon}{\varepsilon_o})}\right] \\ \left[-j\eta(\gamma_v - k) - \frac{\varepsilon_o k E_o^2}{\omega(1+\frac{\varepsilon}{\varepsilon_o})}\right] & j\left[j\eta(\gamma_v + k) + \frac{\sigma_o k E_o}{\omega(1+\frac{\varepsilon}{\varepsilon_o})}\right] \end{bmatrix} \begin{bmatrix} \hat{v}_x^e \\ \hat{v}_y^e \end{bmatrix} = \begin{bmatrix} 0 \\ 0 \end{bmatrix} \tag{13}$$

Physical insights are more easily obtained by adding j times Eq. 13b to Eq. 13a, and writing both equations (each multiplied by ω) in terms of the variables $\hat{v}_x^e$ and $(-\hat{v}_x^e + j\hat{v}_z^e)$. Use is made of the definitions $\sigma_o = \varepsilon_o E_o$ and $\gamma_v^2 \equiv k^2 + j\omega\rho/\eta$:

$$\begin{bmatrix} -\left[\frac{j\omega^2\rho}{k} + 4\eta k\omega - j(\rho g + k^2\gamma) + j\varepsilon_o k E_o^2\right] & [-2\eta k\omega] \\ [2j\eta k\omega] & \left[j\eta(\gamma_v + k)\omega + \frac{k\varepsilon_o E_o^2}{1+\frac{\varepsilon}{\varepsilon_o}}\right] \end{bmatrix} \begin{bmatrix} \hat{v}_x^e \\ -\hat{v}_x^e + j\hat{v}_y^e \end{bmatrix} = \begin{bmatrix} 0 \\ 0 \end{bmatrix} \tag{14}$$

The dispersion equation is obtained by setting either the determinant of the coefficients from Eq. 13 or from Eq. 14 equal to zero. But, written in the second form, it is clear that as the viscosity approaches zero, two modes can be distinguished. These have limiting dispersion equations given by setting the diagonal terms to zero. The frequencies resulting from the upper left and lower right terms, respectively, are then

$$\omega = j\left(\frac{2\eta k^2}{\rho}\right) \pm \omega_o; \quad \omega_o \equiv \left[\frac{k}{\rho}(\rho g + k^2\gamma - k\varepsilon_o E_o^2) - \left(\frac{2\eta k^2}{\rho}\right)^2\right]^{1/2} \tag{15}$$

$$\omega = \omega_c\left[\left(\frac{\sqrt{3}}{2} + \frac{j}{2}\right)\right]; \quad \omega_c \equiv \left[\frac{\sigma_o^2}{\sqrt{\eta\rho}}\frac{k}{(\varepsilon_o + \varepsilon)}\right]^{2/3} \tag{16}$$

The modes can be distinguished in this way only if the frequencies given by Eqs. 15 and 16 are disparate. In general, the higher order dispersion equation must be solved.

When Eq. 15 is satisfied, Eq. 14 shows that $\hat{v}_x^e \simeq j\hat{v}_z^e$, and similarly, if Eq. 16 holds, then the vertical motions are dominated by the horizontal ones, $\hat{v}_x^e \simeq 0$. Thus, the dispersion equation (Eq. 15) is identified with gravity-capillary like waves coupled to an electric field in much the same way as discussed in the latter part of Sec. 8.10.

The main effect of viscosity on the gravity-capillary modes is damping, represented by the imaginary term in Eq. 15. Perhaps a surprising feature of these modes in this low-viscosity limit is that the electric field has the same destabilizing effect as if the interface were perfectly conducting. For example, the condition for incipient instability is the same as given by Eq. 8.10.21, even though that result was derived for an equipotential interface. In this low-viscosity limit, the surface charge on the insulating interface is convected sufficiently rapidly to maintain the interfacial potential constant.

The electromechanical oscillations or shear waves, represented by Eq. 16, involve interfacial dilatations. If an interfacial region is horizontally compressed, self-fields give rise to horizontal electric repulsion forces, much as if there were an elastic film on the interface. Because this electrical "film" is coupled to the inertia of the liquid below through the viscous shear stress, an initial horizontal dilatation of the interface results in oscillations. The oscillations are highly damped because the electrical "spring" is coupled to the "mass" only through the viscous "damper." The frequency ω_c typifies how rapidly nonuniformities in a charged interface can adjust, so that the interface is free of electrical shear stress. The motion stops when the interface is an equipotential.

SMOOTHLY INHOMOGENEOUS SYSTEMS AND THEIR INTERNAL MODES

8.17 Frozen Mass and Charge Density Transfer Relations

A static EQS equilibrium with mass density $\rho_o(x)$ and charge density $q_o(x)$ continuously varying with vertical position is shown in Fig. 8.17.1. The equilibrium vertical gravitational and electrical force densities are balanced by a vertical gradient in pressure. It is the objective in this section to describe small amplitude perturbations from this equilibrium.

The mass density and charge density are conserved by a fluid element of fixed identity,

$$\frac{D\rho}{Dt} = 0; \quad \frac{D\rho_f}{Dt} = 0 \tag{1}$$

The fluid has uniform permittivity ε and it is inviscid.

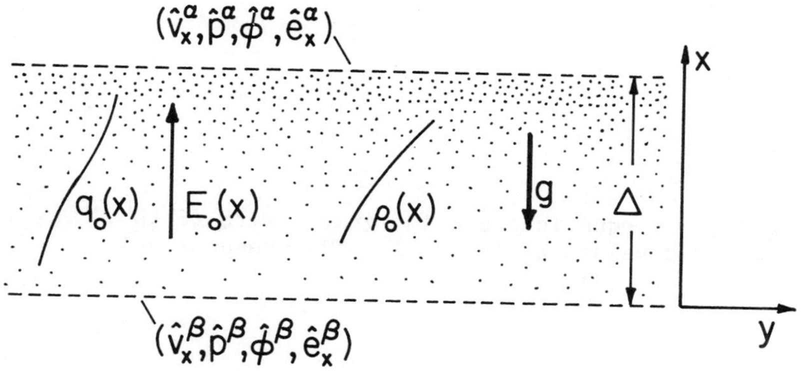

Fig. 8.17.1

Planar layer of fluid with vertical inhomogeneities in mass and charge densities.

It will be recognized that this system is a generalization of the piecewise homogeneous systems considered in Secs. 8.9 and 8.14. In principle, any distribution of $\rho_o(x)$ and $q_o(x)$ could be approximated by "stair-steps" representing stratified layers, with uniform densities, as illustrated in Fig. 8.17.2. The transfer relations for the homogeneous layers might then be used to represent the approximated system. With each interface goes a pair of modes, so that the piecewise homogeneous approximation represents the dynamics in terms of twice as many modes as interfaces. In the limit of a smooth distribution, an infinite number of modes are brought into play. Hence, it should come as no surprise that associated with the smoothly distributed inhomogeneities are an infinite number of "internal" modes. The objective in this and the next sections is to explore an approach that is an alternative to the piecewise homogeneous models.

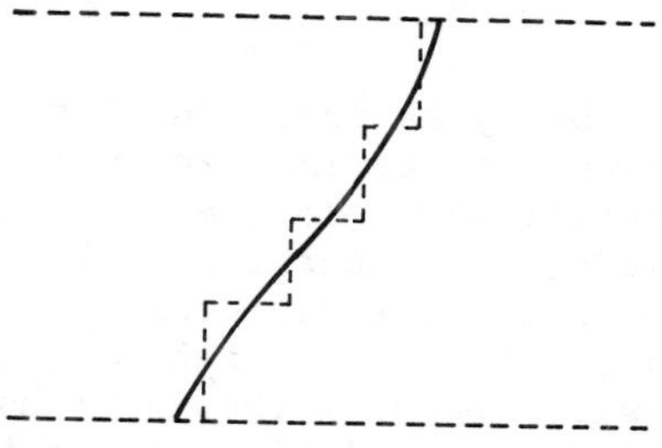

Fig. 8.17.2. Stair-step approximation to smooth inhomogeneity in $\rho_o(x)$ or $q_o(x)$.

In manipulations that follow, remember that ρ_o, q_o and E_o are functions of x. By Gauss' law, $DE_o = q_o/\varepsilon$, where $d(\ )/dx \equiv D(\ )$. Thus, in terms of complex amplitudes and $\hat{\xi} \equiv \hat{v}_x/j\omega$, Eqs. 1 relate perturbations in mass and charge density to the deformation

$$\hat{\rho} = -(D\rho_o)\hat{\xi}; \quad \hat{q} = -(Dq_o)\hat{\xi} \tag{2}$$

The additional statements represent force balance, mass conservation, that the electric field is irrotational and Gauss' law. These are unraveled so as to obtain four first-order differential equations in $(\hat{v}_x, \hat{p}, \hat{\Phi}, \varepsilon\hat{e}_x)$.

The z-component of the force equation can be solved for $\hat{v}_z$ to obtain

$$\hat{v}_z = \frac{k}{\omega\rho_o}\hat{\pi} \tag{3}$$

where the perturbation electric field $\vec{e} = -\nabla\Phi$ and $\hat{\pi} \equiv \hat{p} + q_o\hat{\Phi}$. Thus, the continuity equation, $\nabla\cdot\vec{v} = 0$, requires that

$$D\hat{\xi} = \frac{k}{\omega}\hat{v}_z = \frac{k^2}{\omega^2\rho_o}\hat{\pi} \tag{4}$$

In view of Eqs. 2, the x component of the force equation requires that

$$D\hat{\pi} = (\omega^2\rho_o - E_oDq_o + gD\rho_o)\hat{\xi} + (Dq_o)\hat{\Phi} \tag{5}$$

where $\hat{e}_x$ is replaced by $-D\hat{\Phi}$ and p represented in terms of $\hat{\pi}$. That $\vec{e}$ is irrotational is also explicitly stated,

$$D\hat{\Phi} = -\frac{\varepsilon\hat{e}_x}{\varepsilon} \tag{6}$$

Finally, Gauss' law, together with Eq. 2b, gives

$$D(\varepsilon\hat{e}_x) = \hat{q} + jk\varepsilon\hat{e}_z = -Dq_o\hat{\xi} - k^2\varepsilon\hat{\Phi} \tag{7}$$

Given the amplitudes $(\hat{\xi}^\beta,\hat{p}^\beta,\hat{\Phi}^\beta,\varepsilon\hat{e}_x^\beta)$ at the lower extremity of the layer (say x = 0), these last four equations can be numerically integrated and the amplitudes evaluated at the upper extremity. Thus the relations

$$\begin{bmatrix}\hat{\xi}^\alpha\\ \hat{\pi}^\alpha\\ \hat{\Phi}^\alpha\\ \varepsilon\hat{e}_x^\alpha\end{bmatrix} = [B_{ij}]\begin{bmatrix}\hat{\xi}^\beta\\ \hat{\pi}^\beta\\ \hat{\Phi}^\beta\\ \varepsilon\hat{e}_x^\beta\end{bmatrix} \tag{8}$$

are obtained. For example, to compute the B_{ij}'s, the equations are integrated with $(\hat{\xi}^\beta,\hat{p}^\beta,\hat{\Phi}^\beta,\varepsilon\hat{e}_x^\beta) = (1,0,0,0)$. Then, $(B_{11},B_{12},B_{13},B_{14})$ are the computed values of $(\hat{\xi}^\alpha,\hat{p}^\alpha,\hat{\Phi}^\alpha,\varepsilon\hat{e}^\alpha)$, respectively.

Transfer relations in the form

$$\begin{bmatrix}\hat{\pi}^\alpha\\ \hat{\pi}^\beta\\ \varepsilon\hat{e}_x^\alpha\\ \varepsilon\hat{e}_x^\beta\end{bmatrix} = [C_{ij}]\begin{bmatrix}\hat{\xi}^\alpha\\ \hat{\xi}^\beta\\ \hat{\Phi}^\alpha\\ \hat{\Phi}^\beta\end{bmatrix} \tag{9}$$

follow by manipulating Eqs. 8. With the 4x4 matrix C_{ij} divided into four 2x2 submatrices, transduction between electrical and mechanical surface variables is represented by the upper right and lower left submatrices. In the absence of coupling (say, with $q_o = 0$), these entries should vanish. In this same limit, the upper left submatrix relates the pressure to the velocity amplitudes and these relations play the role of those derived in Sec. 7.9. Of course, here the layer has a nonuniform equilibrium mass density. Also in this limit, the lower right matrix relates the electric perturbation flux to the potentials. Because the layer has uniform electrical properties, these should become the same as the 2x2 entries in relations given by Eq. (a) of Table 2.16.1.

An alternative way of expressing Eqs. 4-7 results from combining the first three of these expressions to obtain

$$D(\rho_oD\hat{\xi}) + k^2\left(\frac{\mathcal{N}}{\omega^2} - \rho_o\right)\hat{\xi} = \frac{k^2Dq_o}{\omega^2}\hat{\Phi} \tag{10}$$

where $\mathcal{N} \equiv E_oDq_o - gD\rho_o$ and the last two to obtain

$$(D^2 - k^2)\hat{\Phi} = \frac{Dq_o\hat{\xi}}{\varepsilon} \tag{11}$$

This pair of second-order expressions can be used to determine $(\hat{\xi},\hat{\Phi})$ and the remaining pair of variables $(\hat{\pi},\varepsilon\hat{e}_x)$ then evaluated using Eqs. 4 and 6. The first of these expressions represents force equilibrium between the inertial force density and the gravitational and electric force densities. The "imposed-field" electric force density is on the right. The second expression is Poisson's equation.

On the right is the perturbation space charge generated by the convection of the nonuniform equilibrium charge density.

The driven response, spatial modes and temporal modes are illustrated in Sec. 8.18.

Weak-Gradient Imposed Field Model: Two approximations make it possible to obtain analytical expressions for the C_{ij}. First, the mass and charge densities are taken as being linear functions of x. Hence,

$$\rho_o = \rho_m + (D\rho_m)x; \quad q_o = q_e + (Dq_e)x \tag{12}$$

where $\rho_m, D\rho_m, q_e$ and Dq_e are constants and neither ρ_o nor q_o departs greatly from a mean value. Then, Eqs. 10 and 11 are approximated by

$$D^2\hat{\xi} + \gamma^2\hat{\xi} = \frac{k^2 Dq_e}{\omega^2\rho_m}\hat{\Phi}; \quad \gamma^2 \equiv \left(\frac{\mathcal{N}}{\rho_m\,\omega^2} - 1\right)k^2 \tag{13}$$

$$(D^2 - k^2)\hat{\Phi} = \frac{Dq_e}{\varepsilon}\hat{\xi} \tag{14}$$

Secondly, the field E_o is regarded as largely imposed by means of external sources. Then, not only is E_o approximated by a constant, but the coupling between fluid and field, represented by the terms on the right in Eqs. 13 and 14, is relatively weak. This breaks the electromechanical feedback loop.

First, to determine the mechanical response, the effect of the motion on the charge distribution is ignored in determining the potential distribution. With the term on the right in Eq. 14 set to zero,

$$\hat{\Phi} = \hat{\Phi}^\alpha \frac{\sinh(kx)}{\sinh(k\Delta)} - \hat{\Phi}^\beta \frac{\sinh k(x-\Delta)}{\sinh(k\Delta)} \tag{15}$$

This potential is used as a "drive" to evaluate the right-hand side of Eq. 13. By inspection, the solution satisfying the boundary conditions that $\hat{\xi}$ is $\hat{\xi}^\alpha$ and $\hat{\xi}^\beta$ at the respective planes is

$$\hat{\xi} = \left[\hat{\xi}^\alpha - \frac{k^2 Dq_e\hat{\Phi}^\alpha}{\omega^2\rho_m(k^2+\gamma^2)}\right]\frac{\sin(\gamma x)}{\sin(\gamma\Delta)} + \left[-\hat{\xi}^\beta + \frac{k^2 Dq_e\hat{\Phi}^\beta}{\omega^2\rho_m(k^2+\gamma^2)}\right]\frac{\sin\gamma(x-\Delta)}{\sin(\gamma\Delta)}$$
$$+ \frac{k^2 Dq_e}{\omega^2\rho_m(k^2+\gamma^2)}\left[\hat{\Phi}^\alpha\frac{\sinh(kx)}{\sinh(k\Delta)} - \hat{\Phi}^\beta\frac{\sinh k(x-\Delta)}{\sinh(k\Delta)}\right] \tag{16}$$

To find the approximate electrical response, the procedure is reversed. Given that $\hat{\xi}$ is $\hat{\xi}^\alpha$ and $\hat{\xi}^\beta$ at the respective planes, solution of Eq. 13 with the term on the right ignored gives

$$\hat{\xi} = \hat{\xi}^\alpha\frac{\sin(\gamma x)}{\sin(\gamma\Delta)} - \hat{\xi}^\beta\frac{\sin\gamma(x-\Delta)}{\sin(\gamma\Delta)} \tag{17}$$

In turn, the solution of Eq. 14 is

$$\hat{\Phi} = \left[\hat{\Phi}^\alpha + \frac{Dq_e\hat{\xi}^\alpha}{\varepsilon(\gamma^2+k^2)}\right]\frac{\sinh(kx)}{\sinh(k\Delta)} - \left[\hat{\Phi}^\beta + \frac{Dq_e\hat{\xi}^\beta}{\varepsilon(\gamma^2+k^2)}\right]\frac{\sin k(x-\Delta)}{\sin(\gamma\Delta)}$$
$$- \frac{Dq_e}{\varepsilon(\gamma^2+k^2)}\left[\hat{\xi}^\alpha\frac{\sin(\gamma x)}{\sin(\gamma\Delta)} - \hat{\xi}^\beta\frac{\sin\gamma(x-\Delta)}{\sin(\gamma\Delta)}\right] \tag{18}$$

where coefficients are determined by inspection so that the boundary conditions on $\hat{\Phi}$ are satisfied at the respective planes. The covariables $(\hat{\Pi}, \varepsilon\hat{e}_x)$ follow from Eqs. 4 and 6 and are evaluated at the respective boundaries to give the transfer relations, Eqs. 9, with

$$
\begin{aligned}
C_{11} &= -C_{22} = \frac{\omega^2 \rho_m \gamma}{k^2}\cot(\gamma\Delta) \\
C_{21} &= -C_{12} = \frac{\omega^2 \rho_m \gamma}{k^2 \sin(\gamma\Delta)} \\
C_{13} &= -C_{24} = -C_{31} = C_{42} = \frac{Dq_e}{k^2+\gamma^2}\,[k\coth(k\Delta) - \gamma\cot(\gamma\Delta)] \\
C_{14} &= -C_{23} = -C_{32} = C_{41} = \frac{Dq_e}{k^2+\gamma^2}\left[\frac{\gamma}{\sin(\gamma\Delta)} - \frac{k}{\sinh(k\Delta)}\right] \\
C_{33} &= -C_{44} = -\varepsilon k\coth(k\Delta)' \\
C_{34} &= -C_{43} = \frac{\varepsilon k}{\sinh(k\Delta)}
\end{aligned}
\tag{19}
$$

Although the weak coupling approximation is sufficient to give the mechanical response to an electrical drive or the electrical response to a mechanical drive, the electrical-to-electrical response, represented by C_{33}, C_{34}, C_{43} and C_{44} is devoid of any of the electromechanics. Electromechanical effects on the transfer between electrical signals depend on there being a "two-way" interaction.

Reciprocity and Energy Conservation: That some coefficients, C_{ij}, in the transfer matrix have equal magnitudes suggests that basic relations exist between off-diagonal coefficients even with arbitrary gradients and fields. The frozen charge model is free of dissipation and allows for energy storage in electrical, kinetic and gravitational forms. With variables as defined in Eq. 9, this requires that the submatrix representing the hybrid pressure responses to electrical excitations is the negative of that representing the electrical flux responses to mechanical deformations. It also requires that mutual electrical and mutual mechanical coefficients are respectively negatives. The proof is a generalization of that developed in Sec. 2.17 for a region storing only electric energy.

Incremental changes in the total electrical, kinetic and gravitational energy stored by a system having volume V enclosed by a surface S are respectively

$$
\delta w_e = \int_V \Phi\delta\rho_f dV - \oint_S \Phi\delta\vec{D}\cdot\vec{n}da \tag{20}
$$

$$
\delta w_k = \int_V \rho_f \vec{E}\cdot\delta\vec{\xi}dV - \oint_S p\delta\vec{\xi}\cdot\vec{n}da + \int_V \rho\vec{g}\cdot\delta\vec{\xi}dV \tag{21}
$$

$$
\delta w_g = \int_V (-\vec{g}\cdot\vec{r})\delta\rho dV \tag{22}
$$

The electrical contribution is familiar from Sec. 2.13 (Eq. 2.13.4). The kinetic statement exploits Newton's law and the incompressibility condition to state that all work done by the electrical, mechanical and gravitational subsystems goes into the creation of kinetic energy (Eq. 7.17.3). The gravitational energy storage is familiar as a specialized analogue of the electric one. The scale is small enough that gravitational self-fields are neglected and g is constant. Thus, by contrast with the potential ϕ for the electrical system, the gravitational potential is "imposed" and is simply $-\hat{g}\cdot\hat{r}$.

Charge migration is negligible, so the charge carried by fluid of fixed identity is conserved. Because $\nabla\cdot\delta\vec{\xi} = 0$, it follows (from Eq. 3.7.5 with $\alpha_i \rightarrow q$) that

$$
\delta q = -\nabla q\cdot\delta\vec{\xi} \tag{23}
$$

Similarly, the mass density of fluid of fixed identity is conserved,

$$
\delta\rho = -\nabla\rho\cdot\delta\vec{\xi} \tag{24}
$$

These expressions are now used in writing the sum of Eqs. 20-22 as

$$
\begin{aligned}
\delta(w_e + w_k + w_g) = &-\oint_S \Phi\delta\vec{D}\cdot\vec{n}da - \oint_S p\delta\vec{\xi}\cdot\vec{n}da - \int_V (\Phi\nabla q + q\nabla\Phi)\cdot\delta\vec{\xi}dV \\
&- \int_V [\rho\nabla(-\vec{g}\cdot\vec{r}) + (-\vec{g}\cdot\vec{r})\nabla\rho]\cdot\delta\vec{\xi}dV
\end{aligned}
\tag{25}
$$

where use has also been made of the relations $\vec{E} = -\nabla\Phi$ and $\vec{g} = \nabla(\vec{g}\cdot\vec{r})$. The volume integrals are converted to surface integrals by first using a vector identity to contract the integrands $[\Phi\nabla\Psi+\Psi\nabla\Phi=\nabla(\Psi\Phi)]$ and then exploiting the fact that $\nabla\cdot\delta\vec{\xi} = 0$ to make the integrands take the form of perfect divergences $(\nabla\Psi\cdot\vec{A} = \nabla\cdot\Psi\vec{A} - \vec{A}\cdot\nabla\Psi)$. From Gauss' theorem, it follows that

$$\delta(w_e + w_k + w_g) = -\oint_S \Phi\delta\vec{D}\cdot\vec{n}da - \oint_S [p + \Phi q + (-\vec{g}\cdot\vec{r})\rho]\delta\vec{\xi}\cdot\vec{n}da \tag{26}$$

The desired reciprocity relations are between perturbation quantities, now designated by primes to distinguish them from the zero-subscripted equilibrium variables. Thus, incremental changes $\delta\vec{D}\cdot\vec{n}$ and $\delta\vec{\xi}\cdot\vec{n}$ on S lead to changes in the total energy given by Eq. 26 expressed up to quadratic terms in the perturbations as

$$\begin{aligned}\delta(w_e + w_k + w_g) = &-\oint_S \Phi_o\delta\vec{D}_o\cdot\vec{n}da - \oint(\Phi_o\delta\vec{D}' + \Phi'\delta\vec{D}_o + \Phi'\delta\vec{D}')\cdot\vec{n}da \\ &- \oint_S [p_o + \Phi_o q_o + (-\vec{g}\cdot\vec{r})\rho_o]\delta\vec{\xi}'\cdot\vec{n}da - \oint_S [p' + \Phi_o q' + q_o\Phi' + (-\vec{g}\cdot\vec{r})\rho']\delta\vec{\xi}'\cdot\vec{n}da\end{aligned} \tag{27}$$

The surface S is now made one enclosing a section of the planar layer shown in Fig. 8.17.1 that has the wavelengths $2\pi/k_y$ and $2\pi/k_z$ in the y and z-directions, respectively. Because $\vec{D}_o$ is x-directed, the first term makes contributions only on the α and β surfaces. Perturbations are assumed to take the complex-amplitude form $\xi = \text{Re}\tilde{\xi}\exp(-jk_y - jk_z z)$, where k_y and k_z are real. The spatial periodicity in the y and z directions insures that contributions to the surface integrations from the second and third terms only come from the α and β surfaces. Moreover, because the integrands of these terms are linear in the perturbation quantities, they "average out" and make no contribution. The quadratic perturbation terms from the last intergral, which are also periodic and hence make contributions only on the α and β surfaces, can be represented using the space-average theorem, Eq. 2.15.14:

$$\begin{aligned}\delta(w_e + w_k + w_g) = &-(\Phi_o^\alpha\delta D_{xo}^\alpha - \Phi_o^\beta\delta D_{xo}^\beta) - \tfrac{1}{2}\text{Re}[\tilde{\Phi}^\alpha\delta(\tilde{D}_x^\alpha)^* - \tilde{\Phi}^\beta\delta(\tilde{D}_x^\beta)^*] \\ &- \tfrac{1}{2}\text{Re}[\tilde{p}^\alpha\delta(\tilde{\xi}^\alpha)^* - \tilde{p}^\beta\delta(\tilde{\xi}^\beta)^*] - \tfrac{1}{2}\text{Re}[\Phi_o^\alpha\tilde{q}^\alpha\delta(\tilde{\xi}^\alpha)^* - \Phi_o^\beta\tilde{q}^\beta\delta(\tilde{\xi}^\beta)^*] \\ &- \tfrac{1}{2}\text{Re}[q_o^\alpha\tilde{\Phi}^\alpha\delta(\tilde{\xi}^\alpha)^* - q_o^\beta\tilde{\Phi}^\beta\delta(\tilde{\xi}^\beta)^*] - \tfrac{1}{2}\text{Re}(-\vec{g}\cdot\vec{r})[\tilde{\rho}^\alpha\delta(\tilde{\xi}^\alpha)^* - \tilde{\rho}^\beta\delta(\tilde{\xi}^\beta)^*]\end{aligned} \tag{28}$$

With the understanding that the incremental variations are made with the equilibrium potentials Φ_o held fixed on the transverse boundaries, the first terms on the right become perfect differentials, $\Phi_o\delta D_{xo} \rightarrow \delta(\Phi_o D_{xo})$, so these equilibrium terms are moved to the left side of the equation.

In the remaining terms, it is now assumed that all complex amplitudes are real. It is entirely possible to proceed without making this assumption by treating the real and imaginary parts of the variables $(\tilde{\xi}^\alpha,\tilde{\xi}^\beta,\tilde{\Phi}^\alpha,\tilde{\Phi}^\beta)$ as independent. However, there is little to be learned from this generalization because it is obvious from Eqs. 4-7 (which, provided ω^2 is real, have real coefficients) that the coefficients C_{ij} are real. Hence, given that the amplitudes $(\tilde{\xi}^\alpha,\tilde{\xi}^\beta,\tilde{\Phi}^\alpha,\tilde{\Phi}^\beta)$ are real, the amplitudes of the conjugate variables are clearly real.

In the forth and sixth terms of Eq. 28, Eqs. 2 are used to substitute

$$\Phi_o\tilde{q}\delta\tilde{\xi} = -\Phi_o Dq_o\tilde{\xi}\delta\tilde{\xi} = -\tfrac{1}{2}\delta(\Phi_o Dq_o\tilde{\xi}^2) \tag{29}$$

$$(-\vec{g}\cdot\vec{r})\tilde{\rho}\delta\tilde{\xi} = -(-\vec{g}\cdot\vec{r})D\rho_o\tilde{\xi}\delta\tilde{\xi} = -\tfrac{1}{2}\delta[(-\vec{g}\cdot\vec{r})D\rho_o\tilde{\xi}^2] \tag{30}$$

respectively. The second equalities are based on recognition that if variations in the $\tilde{\xi}$'s and $\tilde{D}_x$'s result in variations of $D q_o$ or $D\rho_o$, the latter can be neglected, because the terms in which they appear are already quadratic in the perturbations. With the substitution of Eqs. 29 and 30, the fourth and sixth terms also become perfect differentials and are therefore moved to the left side of Eq. 28. Finally, in the second term on the right the transformation $\tilde{\Phi}\delta\tilde{D}_x = \delta(\tilde{\Phi}\tilde{D}_x) - \tilde{D}_x\delta\tilde{\Phi}$ is made and the perfect differential moved to the left-hand side. Thus, the energy statement becomes

$$\delta w' = -\frac{1}{2}(\tilde{\pi}^\alpha\delta\tilde{\xi}^\alpha - \tilde{\pi}^\beta\delta\tilde{\xi}^\beta) + \frac{1}{2}(\tilde{D}_x^\alpha\delta\tilde{\Phi}^\alpha - \tilde{D}_x^\beta\delta\tilde{\Phi}^\beta) \tag{31}$$

where

$$w' \equiv w_e + w_k + w_g + \frac{1}{2}(\tilde{\Phi}^\alpha\tilde{D}_x^\alpha - \tilde{\Phi}^\beta\tilde{D}_x^\beta) - \frac{1}{4}[\Phi_o^\alpha Dq_o^\alpha(\tilde{\xi}^\alpha)^2 - \Phi_o^\beta Dq_o^\beta(\tilde{\xi}^\beta)^2]$$

$$- \frac{1}{4}(-\vec{g}\cdot\vec{r}\,[D\rho_o^\alpha(\tilde{\xi}^\alpha)^2 - D\rho_o^\beta(\tilde{\xi}^\beta)^2] + (\Phi_o^\alpha\delta D_{xo}^\alpha - \Phi_o^\beta\delta D_{xo}^\beta)$$

and

$$\tilde{\pi} \equiv \tilde{p} + q_o\tilde{\Phi}$$

Now, with the assumption that w' is a state function $w'(\tilde{\xi}^\alpha,\tilde{\xi}^\beta,\tilde{\Phi}^\alpha,\tilde{\Phi}^\beta)$, the incremental change $\delta w'$ can also be written as

$$\delta w' = \frac{\partial w'}{\partial\tilde{\xi}^\alpha}\delta\tilde{\xi}^\alpha + \frac{\partial w'}{\partial\tilde{\xi}^\beta}\delta\tilde{\xi}^\beta + \frac{\partial w'}{\partial\tilde{\Phi}^\alpha}\delta\tilde{\Phi}^\alpha + \frac{\partial w'}{\partial\tilde{\Phi}^\beta}\delta\tilde{\Phi}^\beta \tag{32}$$

Because the variables $(\tilde{\xi}^\alpha,\tilde{\xi}^\beta,\tilde{\Phi}^\alpha,\tilde{\Phi}^\beta)$ are independent, it follows from Eqs. 31 and 32 that corresponding coefficients must be equal:

$$\tilde{\pi}^\alpha = -2\frac{\partial w'}{\partial\tilde{\xi}^\alpha}\;;\quad \tilde{\pi}^\beta = 2\frac{\partial w'}{\partial\tilde{\xi}^\beta} \tag{33}$$

$$\tilde{D}_x^\alpha = 2\frac{\partial w'}{\partial\tilde{\Phi}^\alpha}\;;\quad \tilde{D}_x^\beta = -2\frac{\partial w'}{\partial\tilde{\Phi}^\beta} \tag{34}$$

The reciprocity relations follow by taking cross-derivatives of these relations. For example, in view of Eqs. 33a and 34b together with Eq. 9,

$$\frac{\partial\tilde{\pi}^\alpha}{\partial\tilde{\Phi}^\beta} = \frac{\partial\tilde{D}_x^\beta}{\partial\tilde{\xi}^\alpha} \Rightarrow C_{14} = C_{41} \tag{35}$$

Thus, if C_{ij} is broken into four 2x2 matrices K, L, M and N such that

$$C_{ij} = \begin{bmatrix} K & L \\ M & N \end{bmatrix} \tag{36}$$

where K and N are each antisymmetric and L is the negative of M.

The next section exemplifies the implications of the transfer relations, both found by numerical integration and approximated by the weak-gradient imposed-field model.

8.18 Internal Waves and Instabilities

The frozen charge and mass density transfer relations derived in Sec. 8.17 are now applied to the study of space-charge gravity waves excited in the sinusoidal steady state from transverse boundaries. Also discussed are the temporal and spatial modes. Instability conditions are exemplified and a general proof given that the principle of exchange of stabilities is satisfied. With the objective of both gaining physical insight for this type of dynamics and for ways in which it can be represented, two models are developed and compared. First, the weak-gradient imposed-field approximation of Sec. 8.17 is used to obtain an analytical representation of the response. Then, as a recourse that is applicable for an arbitrary distribution of charge and mass density, numerical integration is used to determine the response. Because one of these representations depends on numerical procedures, it is convenient to normalize variables at the outset.

Configuration: The stratified layer shown in Fig. 8.18.1 is bounded from above by fixed excitation electrodes upon which a spatially and temporally periodic potential is imposed. From below, it is bounded by a conducting rigid electrode, essentially constrained in potential to the constant equilibrium value V_o.

Normalization: To be specific about the distributions in charge and mass density, they are taken as linear and written in terms of the constants defined in Fig. 8.18.1:

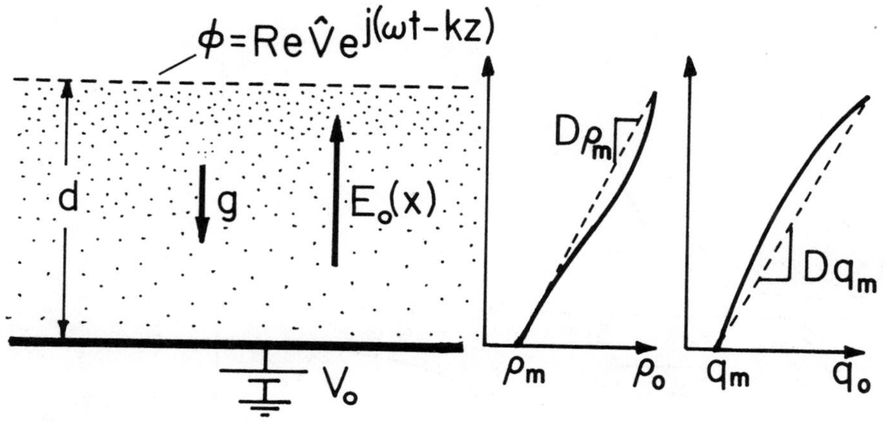

Fig. 8.18.1

Cross section of system in which internal space-charge gravity waves are excited.

$$\underline{q}_o \equiv \frac{q_o(x)}{q_e} = 1 + \frac{Dq_e}{q_e}\underline{x}; \quad \underline{\rho}_o \equiv \frac{\rho_o}{\rho_m} = 1 + \frac{D\rho_m}{\rho_m}\underline{x} \tag{1}$$

In terms of these quantities, variables are normalized such that

$$\begin{aligned} x &= \underline{x}d & \hat{\xi} &= \hat{\underline{\xi}}d \\ k &= \underline{k}/d & \hat{\pi} &= \hat{\underline{\pi}}|Dq_e||V_o|d \\ \omega^2 &= \frac{\underline{\omega}^2|V_o||Dq_e|}{\rho_m d} & \hat{\Phi} &= \hat{\underline{\Phi}}|V_o| \\ \hat{V} &= \hat{\underline{V}}|V_o| & \hat{d}_x &= \varepsilon\hat{e}_x = \hat{\underline{d}}_x|Dq_e|d^2 \end{aligned} \tag{2}$$

For other equilibrium distributions, the same normalization could be used with the quantities ρ_m and $|Dq_e|$ defined as mean values.

From the one-dimensional form of Gauss' law and the equilibrium potential boundary conditions, the equilibrium distribution of electric field is written in terms of the normalized variables as

$$E_o = \frac{|V_o|}{d}\left\{\frac{V_o}{|V_o|} + S\left[\frac{q_e}{|Dq_e|}(\underline{x} - \tfrac{1}{2}) + \frac{1}{2}\frac{Dq_e}{|Dq_e|}(\underline{x}^2 - \tfrac{1}{3})\right]\right\} \tag{3}$$

where $S \equiv |Dq_e|d^2/\varepsilon|V_o|$ represents the influence of the space charge on the imposed field.

<u>Driven Response</u>: Boundary conditions reflect electrode constraints on the normal motion of the fluid and on the potential:

$$[\hat{\xi}^a,\hat{\xi}^b,\hat{\Phi}^a,\hat{\Phi}^b] = [0,0,\hat{V},0] \tag{4}$$

Given the electrical excitation at the upper boundary, what is the mechanical and electrical response of the fluid, and in particular, what perturbation pressure and normal electric field would be expected on instruments embedded in the lower electrode? These follow from Eq. 8.17.9 as

$$\frac{\hat{\pi}^b}{\hat{V}} = C_{23}; \quad \frac{\hat{d}_x^b}{\hat{V}} = C_{43} \tag{5}$$

In the weak-gradient imposed-field approximation, it is possible to evaluate the C_{ij}'s by using Eqs. 8.17.19. Thus normalized, Eq. 3 becomes

$$\frac{\hat{\underline{\pi}}^b}{\hat{\underline{V}}} = \underline{C}_{23} \equiv \frac{C_{23}}{d|Dq_e|} = \frac{-1}{\underline{k}^2+\gamma^2}\left(\frac{\gamma}{\sin\gamma} - \frac{\underline{k}}{\sinh\underline{k}}\right) \tag{6}$$

$$\frac{\hat{\underline{d}}_x^b}{\hat{\underline{V}}} = \underline{C}_{43} \equiv -\frac{C_{43}|V_o|}{|Dq_e|d^2} = -S^{-1}\frac{\underline{k}}{\sinh\underline{k}} \tag{7}$$

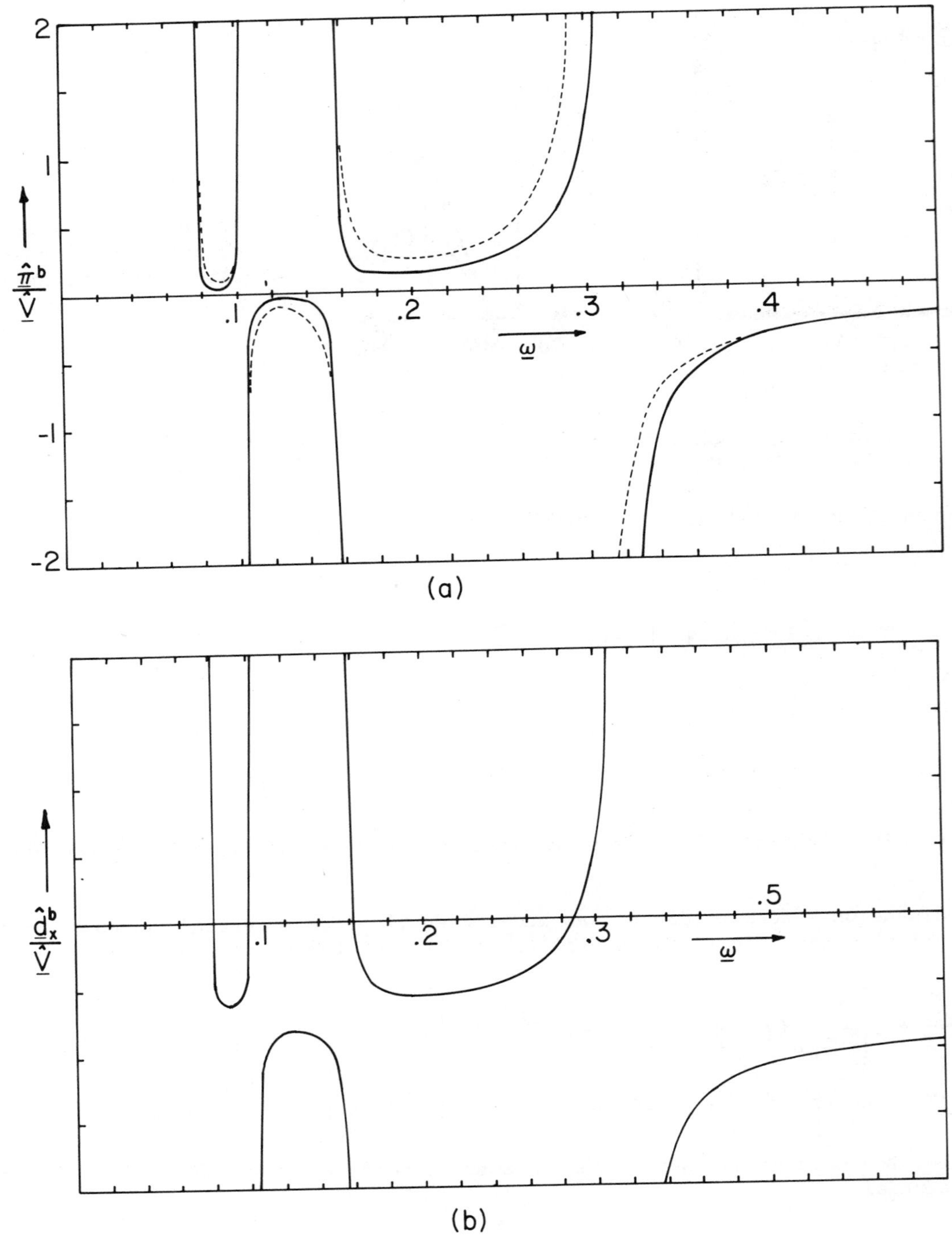

Fig. 8.18.2. Driven response of charged layer showing prediction of weak-gradient imposed-field model (broken line) for comparison with numerically determined response (solid line). The response below $\underline{\omega} = 0.08$ is not shown because it displays an infinite number of resonances crowded toward the origin. In both cases, $\underline{k} = 1$ and $\underline{V}_o$ and $\underline{D}q_e$ are both positive or both negative so that equilibrium is stable. The solid numerically predicted curves are for $\underline{D}q_e/q_e = 1$ and $S = 1$. (a) Hybrid pressure response at lower electrode as a function of frequency for electrical excitation at upper electrode. (b) Electric flux at lower electrode.

where

$$\gamma \equiv k\sqrt{\pm\frac{1}{\omega^2}\left(1-\frac{gD\rho_m d}{V_o Dq_e}\right)-1} \qquad (8)$$

The upper sign applies if V_o and Dq_e are both positive or both negative. The lower sign is to be used if V_o and Dq_e have opposite signs.

The weak-gradient imposed-field driven responses are illustrated as a function of frequency in Fig. 8.18.2. Because of approximations inherent to this model, the electrical-to-electrical response is no more than that of the layer without the charged fluid. This result will be refined to include the electromechanical effects shortly. The resonances in the hybred pressure response that dominate the picture reflect the electromechanical coupling. In this loss-free system, they serve notice that the natural frequencies of the stable temporal modes are real and that there are an infinite number of spatial modes having real wave numbers. The conditions for the resonances follow from Eq. 6:

$$\sin\gamma = 0 \Rightarrow \gamma = n\pi, n = 1,2,\cdots \qquad (9)$$

Thus, the resonance frequencies are found by evaluating γ in Eq. 8 and solving for ω,

$$\omega^2 = \frac{k^2\mathcal{N}}{k^2+(n\pi)^2}\ ;\ \mathcal{N} = \frac{\frac{V_o}{d}Dq_e - gD\rho_m}{\frac{|V_o|}{d}|Dq_e|} \qquad (10)$$

The associated distributions, $\hat{\xi}(x)$, in the neighborhood of a resonance follow from Eq. 8.17.17 as being $\sin(n\pi\underline{x})$. These are pictured by the broken curves of Fig. 8.18.3. Implicit to the discussion thus far is the presumption that $\mathcal{N} > a$.

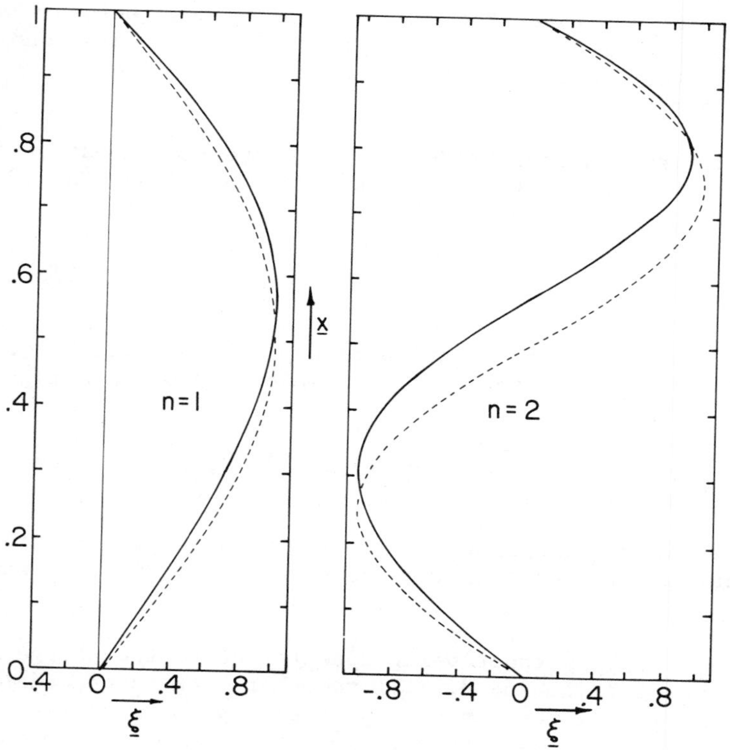

Fig. 8.18.3

Vertical displacement of fluid as a function of vertical position. Response is shown in the neighborhood of first and second resonances, and hence represents first and second temporal eigenmodes. Solid curves are predicted numerically using parameters of Fig. 8.18.2, while broken curves are weak-gradient imposed-field approximation.

Consider now the more general approach of numerically integrating Eqs. 8.17.4-7 to find the transfer relations. Normalized, these equations are

$$D\hat{\xi} = \frac{k^2}{\omega^2\rho_o}\pi \qquad (11)$$

$$D\pi = (\omega^2\rho_o - \mathcal{N})\hat{\xi} + \frac{Dq_o}{|Dq_e|}\hat{\Phi} \qquad (12)$$

$$D\Phi = -Sd_x \tag{13}$$

$$Dd_x = -\frac{Dq_o}{|Dq_e|}\hat{\xi} - \frac{1}{S}k^2\hat{\Phi} \tag{14}$$

These expressions are applicable with arbitrary charge and mass distributions. For the specific linear distributions, ρ_o is given by Eq. 2, $Dq_o = Dq_e$ and

$$\zeta = \left\{\frac{V_o}{|V_o|} + \frac{Dq_e}{|Dq_e|} S\left[\frac{q_e}{Dq_e}(x - \tfrac{1}{2}) + \tfrac{1}{2}(x^2 - \tfrac{1}{3})\right]\right\}\frac{Dq_e}{|Dq_e|} - \frac{gdD\rho_m}{|V_o||Dq_e|} \tag{15}$$

The coefficients required to evaluate the responses, Eqs. 3, follow by converting the transfer relations of Eq. 8.17.8 to those of Eq. 8.17.9. The coefficients needed here are

$$C_{23} = -B_{14}/D; \quad C_{43} = B_{12}/D; \quad D \equiv B_{12}B_{34} - B_{14}B_{32} \tag{16}$$

Coefficients in the transfer relations have been normalized so that $\underline{C}_{ij}$ and $\underline{B}_{ij}$ relate normalized veriables. The $\underline{B}_{ij}$'s are determined by numerical integration of Eqs. 11-14 following the procedure indicated following Eq. 8.17.8. (Numerical integration of systems of first-order differential equations written in the form of Eqs. 11-14 is conveniently carried out using standard library subroutines. Used here was the IMSLIB Routine DVERK.)

For purposes of comparison, the numerically determined frequency responses are shown with those predicted by the weak-gradient imposed-field model in Fig. 8.18.2. For the numerical case shown, $Dq_e/q_e = 1$ and $S = 1$, so both the weak-gradient and the imposed-field approximations are somewhat invalid. Note that the electrical-to-electrical response now displays the characteristic resonances of the internal waves. The numerically determined mechanical displacement and potential distributions with the frequency in the neighborhood of the first and of the second resonances are shown in Fig. 8.18.3.

Spatial Modes: Still in the sinusoidal steady state, these modes satisfy homogeneous transverse boundary conditions and are needed to make the total solution obey longitudinal boundary conditions. (Spatial modes are introduced in Sec. 5.17.) For example, what is the response to a drive at some z plane with the duct walls free of excitations?

From the weak-gradient imposed-field driven response of Eq. 6, the dispersion equation is $D(\omega,k) = \sin\gamma = 0$. This has roots that are the same as for the resonance conditions, Eq. 10. Here, however, interest is in complex k for a real driving frequency ω,

$$k = \pm\frac{n\pi\omega}{\sqrt{\mathcal{N} - \omega^2}} \tag{17}$$

Under the assumption once again that $\mathcal{N} > 0$, the dispersion equation is typified by Fig. 8.18.4. Note that all modes have the same cut-off frequency $\underline{\omega} = 1$. With $\underline{\omega} < 1$, all modes are propagating, whereas with $\underline{\omega} > 1$, all modes are evanescent.

The resonances below $\underline{\omega} = 1$ in the driven frequency response, Fig. 8.18.2, result from a coincidence of the imposed wave number and the purely real wave number of the propagating spatial modes.

Temporal Modes: When t = 0, initial conditions are spatially periodic in the z direction, with wave number k. What modes are to be superimposed in representing the ensuing transient? (Temporal modes are introduced in Sec. 5.15.)

A mode $\hat{\xi}_n(x)$ has the eigenfrequency $j\omega_n \equiv s_n$. Without being specific as to the charge and density distributions, it can be deduced from Eqs. 8.17.10 and 8.17.11 together with the boundary conditions that these eigenfrequencies are either purely real or purely imaginary so s_n^2 is real. Equation 8.17.10 is multiplied by $\hat{\xi}_n^*$ and integrated over the cross section. The first term is then integrated by parts to obtain

$$\rho_o\hat{\xi}_n^* D\hat{\xi}_n\Big]_o^d - \int_o^d \rho_o D\hat{\xi}_n D\hat{\xi}_n^* dx - k^2\int_o^d\left(\rho_o + \frac{\mathcal{N}}{s_n^2}\right)\hat{\xi}_n\hat{\xi}_n^* dx = -k^2\int_o^d \frac{Dq_o\hat{\Phi}_n\hat{\xi}_n^*}{s_n^2}dx \tag{18}$$

Similarly, the complex conjugate of Eq. 8.17.11 is multiplied by $k^2\varepsilon\hat{\Phi}_n$ and integrated over the cross section. Again, the first term is integrated by parts to obtain

$$k^2\varepsilon\hat{\Phi}_n D\hat{\Phi}_n^*\Big]_o^d - k^2\varepsilon\int_o^d D\hat{\Phi}_n (D\hat{\Phi}_n)^* dx - k^4\varepsilon\int_o^d \hat{\Phi}_n\hat{\Phi}_n^* dx = k^2\int_o^d Dq_o\hat{\xi}_n^*\hat{\Phi}_n dx \tag{19}$$

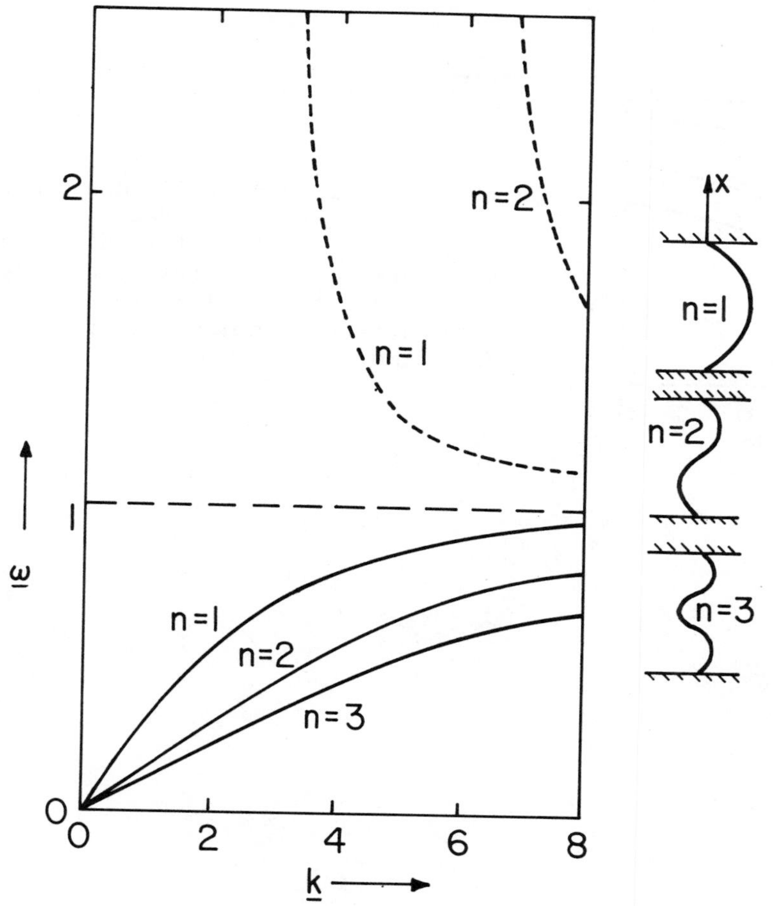

Fig. 8.18.4

Complex normalized frequency as function of real longitudinal wave number for spatial modes in weak-gradient imposed-field approximation. $\underline{\omega}_r$ ——, $\underline{\omega}_i$ ----. All modes have common asymptotic frequency at $\underline{\omega} = 1$, above which they are evanescent.

The point of these manipulations is to obtain positive definite integrands and to make the right-hand sides of these expressions negatives. Because of the boundary conditions on $\hat{\xi}_n$ and $\hat{\Phi}_n$, the terms evaluated on the boundaries vanish. Thus, the last two expressions give

$$\int_0^d (\rho_o|D\hat{\xi}_n|^2 + \rho_o k^2|\hat{\xi}_n|^2)\,dx = -\frac{k^2}{s_n^2}\int_0^d [\mathcal{N}|\hat{\xi}_n|^2 + \varepsilon(|D\hat{\Phi}|^2 + k^2|\hat{\Phi}|^2)]dx \tag{20}$$

This expression can be solved for the square of the eigenfrequency, s_n^2,

$$s_n^2 = \frac{-k^2\int_0^d [\mathcal{N}|\hat{\xi}_n|^2 + \varepsilon(D\hat{\Phi}D\hat{\Phi}^* - k^2\hat{\Phi}\hat{\Phi}^*)]dx}{\int_0^d [\rho_o(|D\hat{\xi}_n|^2 + k^2|\hat{\xi}_n|^2)\,dx} \tag{21}$$

Terms on the right are real, and it therefore follows that s_n^2 is real. Moreover, because terms in the denominator are positive definite, as is $k^2|\hat{\xi}_n|^2$ in the numerator, it is clear that if $\mathcal{N}$ is everywhere positive, the eigenmodes are all stable:

$$\mathcal{N} \equiv E_o Dq_o - gD\rho_o > 0 \tag{22}$$

Similarly, if $\mathcal{N}$ is everywhere negative, the eigenmodes have an exponential dependence, half of them decaying and half of them growing in time.

Using the weak inhomogeneity imposed-field approximations, the eigenfrequencies follow from Eq. 10 where this time k is a given real number. These are shown as a function of $\underline{k}$ in Fig. 8.18.5. According to this model, in the unstable configuration ($\mathcal{N} < 0$) the n = 1 mode is the most rapidly growing.

It is worthwhile to make a comparative study of the discretely and smoothly stratified charge layers. The condition of Eq. 22 plays a role relative to the smoothly inhomogeneous system that is played by Eq. 8.14.25 for the piecewise homogeneous system of Sec. 8.14.

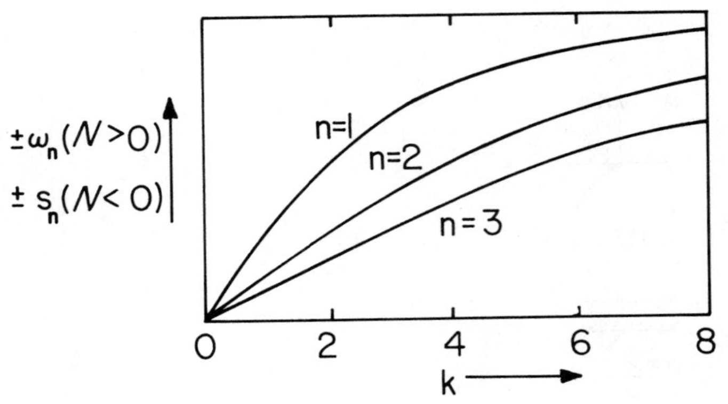

Fig. 8.18.5

Weak-gradient imposed-field eigenfrequencies of temporal modes as a function of wave number. For $\mathcal{N} > 0$, all modes are stable and purely oscillatory. For $\mathcal{N} < 0$, they are either exponentially growing or decaying with time.

For Section 8.3:

Prob. 8.3.1 A pair of electrodes is constructed from thin sheets separated by a thin sheet of insulator. This dielectric "sandwich" is dipped into an insulating liquid having the polarization constitutive law

$$\vec{D} = \frac{\vec{E}}{\alpha_1\sqrt{\alpha_2^2 + E^2}} + \varepsilon_o\vec{E}$$

where α_1 and α_2 are constant parameters. The objective here is to describe the rise of the dielectric liquid around the outside edges of the electrodes, where there is a strong surrounding fringing field. Assume that the applied voltage is alternating at a sufficiently high frequency so that free charge effects are absent and effects of the time-varying part of the electric stress are "ironed out" by the fluid viscosity and inertia.

(a) Determine the electric field in the neighborhood of one of the edges under the assumption that the dielectric rises in an axisymmetric fashion ($\xi = \xi(r)$, with r as defined in Fig. P8.3.1). The right and left edges of the electrodes (see the side view in the figure) are sufficiently far apart so that they can be considered not to influence each other.

(b) Find $\xi(r)$.

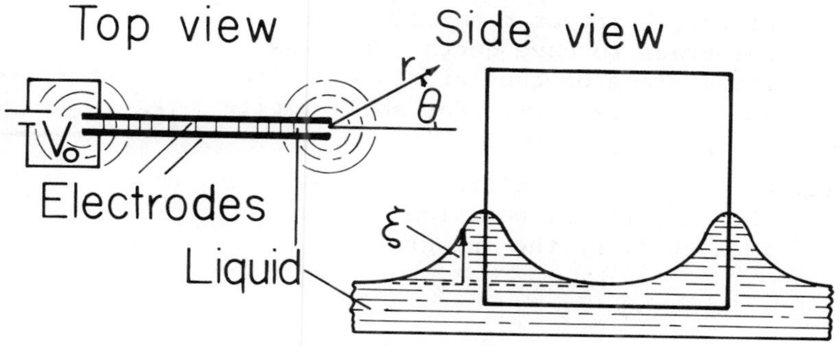

Fig. P8.3.1

Prob. 8.3.2 An insulating liquid is represented by the constitutive law

$$|\vec{D}| = \varepsilon_o|E| + \alpha_1 \tanh \alpha_2|E|$$

where $\vec{D}$ and $\vec{E}$ are collinear and α_1 and α_2 are properties of the fluid. The liquid is placed in a dish as shown in Fig. P8.3.2. Shaped electrodes are dipped into the liquid and held at a potential difference V_o. The variable spacing s(z) between the electrodes is small compared to the electrode dimensions in the x and z directions, so the electric field can be taken as essentially in the y direction. With the application of the field, the liquid reaches a static equilibrium of profile $\xi(z)$. Find an expression for $\xi(z) - \xi_o$.

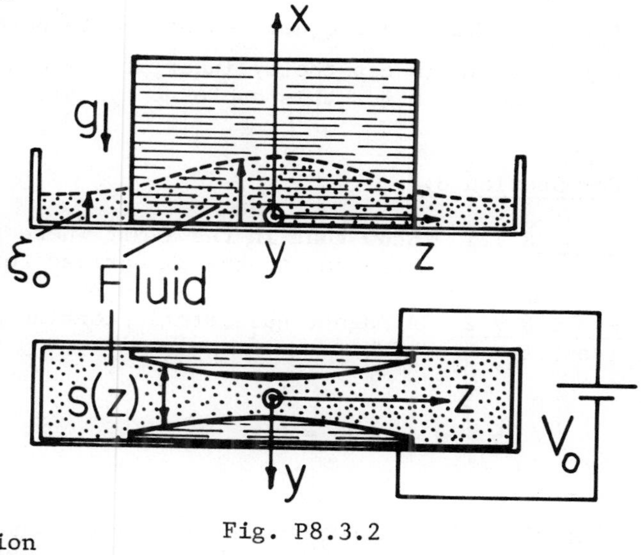

Fig. P8.3.2

For Section 8.4:

Prob. 8.4.1 The configuration of Fig. 8.4.4 is altered by replacing the magnet with a periodic distribution of magnets. These constrain the normal magnetic flux density in the plane x = d to be $B_o \cos ky$. As in the example treated, ignore effects of the self fields and of surface tension. Assume that $\xi = \xi_1$ at y = 0.

(a) Show that an implicit expression for $\xi(y)$ is

$$k(\xi-\xi_o)\, e^{-k(\xi-\xi_o)} = e^{k(d-\xi_o)} \frac{J_oB_o}{g(\rho_b-\rho_a)} \sin ky$$

(b) Make sketches of the left side of this expression (as a function of $(\xi=\xi_o)$ and the right side of the expression (as a function of ky) and describe in graphical terms how you would find $(\xi-\xi_o)$ as a function of y. What is the significance of there being two solutions for $\xi-\xi_o$ or none at all? For what value of J_oB_o would you expect the static equilibrium to be unstable?

Prob. 8.4.2 In the configuration of Fig. 8.4.1, the lower fluid is a perfectly conducting liquid while the upper one is an insulating gas ($\rho_a << \rho_b$). Surface deformations have a very long characteristic length in the y direction compared to $d-\xi$, so that the electric field normal to the interface in

Prob. 8.4.2 (continued)

the gas can be approximated as the voltage divided by the spacing $d-\xi$.

(a) Show that for a given V(y) static deformations of the interface are described by

$$\gamma \frac{d}{dy}\left\{[1+(\frac{d\xi}{dy})^2]^{-\frac{1}{2}}\frac{d\xi}{dy}\right\} + \tfrac{1}{2}\varepsilon_o \frac{V^2(y)}{(d-\xi)^2} - \rho g(\xi-b) = 0$$

where $\xi = b$ at a location where V=0.

(b) Now consider the application of this equation to the special case shown in Fig. P8.4.2. The plane horizontal electrode is of uniform potential V. An infinite pool of liquid to the left communicates liquid to the region below the electrode. In the fringing region, the interface is covered by a flat electrode. At y=0 the sharp edge of the electrode constrains the interface to have depth ξ=b. The field elevates the interface to the height ξ_o as $y \to \infty$. For small amplitudes $\xi-b$, determine $\xi(y)$.

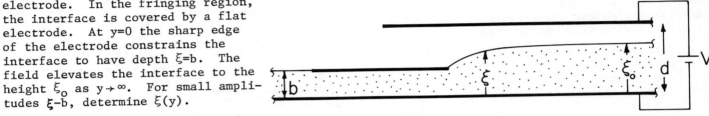

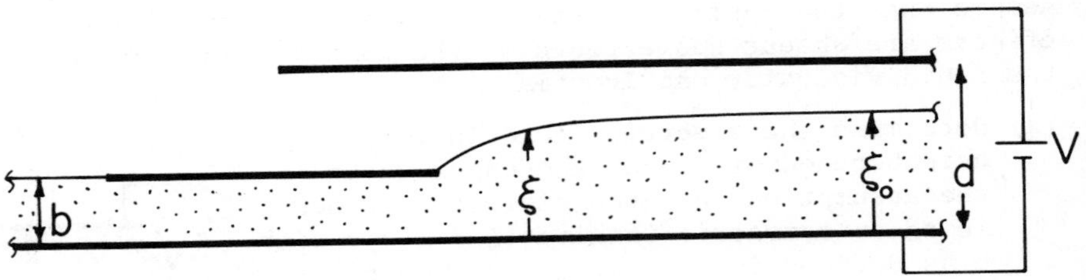

Fig. P8.4.2

(c) Show that for arbitrary deformations, the interfacial position is given implicitly by the integral

$$y = \int_b^{\xi} \frac{d\xi}{\sqrt{[1+P(\xi_o)-P(\xi)]^2-1}} \quad ; \quad P(\xi) \equiv \tfrac{1}{2}\frac{\varepsilon_o V^2}{\gamma(d-\xi)} - \rho\frac{g}{\gamma}(\xi-b)^2$$

For Section 8.6:

Prob. 8.6.1 In Prob. 7.9.2, the transfer relations are found for an annular region of fluid that is perturbed from an equilibrium in which it suffers a rigid-body rotation of angular velocity Ω about the z axis. Based upon those results, consider now the dynamics of fluid completely filling a container having radius R (there is no inner cylindrical region).

(a) Find the eigenfrequencies of the temporal modes having wavenumber k but m = 0.

(b) Rigid walls cap the cylinder at z = 0 and z = ℓ. What are the natural frequencies of the temporal modes m = 0 for this enclosed system?

For Section 8.7:

Prob. 8.7.1 Show that in the limit where times of interest are long compared to the relaxation time ε/σ, Eq. 8.7.6 reduces to the linearized form of $D\Phi/Dt = 0$.

Prob. 8.7.2 A magnetoquasistatic continuum conserves the free current linking any surface of fixed identity

$$\frac{d}{dt}\int_S \vec{J}_f \cdot \vec{n}da = 0$$

Show that the appropriate equations for an incompressible fluid are

$$\nabla \cdot \vec{v} = 0$$

$$\frac{\partial \vec{J}_f}{\partial t} - \nabla \times (\vec{v} \times \vec{J}_f) = 0$$

$$\rho \frac{D\vec{v}}{Dt} + \nabla p = \vec{J}_f \times \vec{B} + \eta \nabla^2 \vec{v}$$

$$\nabla \times \vec{H} = \vec{J}_f \quad ; \quad \nabla \cdot \vec{J}_f = 0$$

where Faraday's law is used only if the electric field is required.

Prob. 8.7.3 As a particular example of the current-conserving continua from Prob. 8.7.2, the configuration shown in Fig. P8.7.3 consists of a layer of fluid having essentially zero conductivity in the y and z directions compared to that in the x direction. The walls are composed of segments, each constrained to constant current. Thus, in static equilibrium, there is a uniform current density $J_o\vec{i}_x$ throughout and an imposed magnetic field $B_o\vec{i}_x$. Assume that the magnetic field induced by $\vec{J}_f$ is negligible compared to B_o. As the fluid moves, the current through any given open surface of fixed identity remains constant.

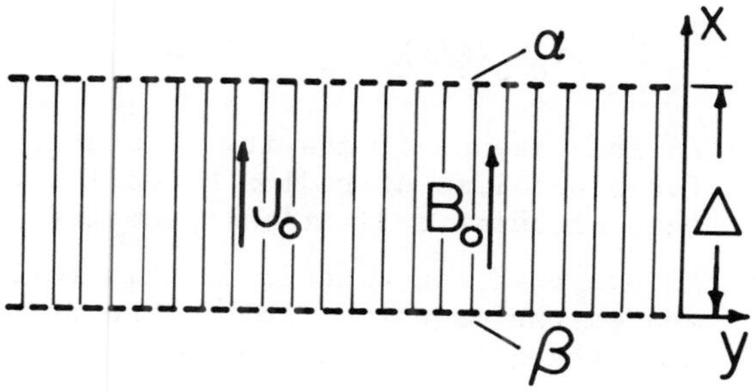

Fig. P8.7.3

The fluid has the electrical nature of conducting "wires" insulated from each other and stretched in the x direction. The "wires" deform with the fluid, and might actually consist of conducting fluid columns in an insulating fluid having the same mechanical properties.[1]

(a) Assume that motions and field depend only on (x,t) and show that the equations formed in Prob. 8.7.2 are satisfied by solutions of the form

$$\vec{v} = v_y(x,t)\vec{i}_y + v_z(x,t)\vec{i}_z \text{ and } \vec{J} = J_o\vec{i}_x + J_y(x,t)\vec{i}_y + J_z(x,t)\vec{i}_z$$

where

$$\frac{\partial J_y}{\partial t} - J_o \frac{\partial v_y}{\partial x} = 0$$

$$\frac{\partial J_z}{\partial t} - J_o \frac{\partial v_z}{\partial x} = 0$$

$$\rho \frac{\partial v_y}{\partial t} = B_o J_z + \eta \frac{\partial^2 v_y}{\partial x^2}$$

$$\rho \frac{\partial v_z}{\partial t} = -B_o J_y + \eta \frac{\partial^2 v_z}{\partial x^2}$$

(b) Describe how you would establish transfer relations for the layer, given that the surface variables are the velocities and the shear stresses. Show that in the limit where there is no electromechanical coupling, $B_o = 0$, there is no coupling between the y directed motions and the z directed motions.

(c) As a specific example, rigid boundaries are imposed at $x = 0$ and $x = \ell$. Find the eigenfrequencies of the resulting temporal modes.

Prob. 8.7.4 A spherical particle is impact-charged to saturation so that its mobility is given by Eq. (a) of Table 5.2.1. It is pulled through a fluid by the same electric field used to achieve this saturation charging. Show that the electroviscous time based on this field and the fluid viscosity is the time required for the particle to move a distance equal to its own diameter.

---

1. For discussion of the related dynamics of a current conserving "string" in a similar configuration, see H. H. Woodson and J. R. Melcher, Electromechanical Dynamics, Part II, John Wiley & Sons, New York, 1968, p. 627.

For Section 8.10:

Prob. 8.10.1 A planar layer of insulating liquid having a mass density ρ_s has the equilibrium thickness d. The layer separates infinite half-spaces of perfectly conducting liquid, each half-space having the same mass density ρ. The interfaces between insulating and conducting liquids each have a surface tension γ, but ρ_s is sufficiently close to ρ so that gravity effects can be ignored. Voltage applied between the conducting fluids results in an electric field in the insulating layer. In static equilibrium, this field is E_o. Determine the dispersion equations for kinking and sausage modes on the interfaces. Show that in the long-wave limit $kd \ll 1$, the effect of the field on the kinking motions is described by a voltage-dependent surface tension. In this long-wave limit, what is the condition for incipient instability?

For Section 8.11:

Prob. 8.11.1 A vertical wire carries a current I so that there is a surface magnetic field

$$\vec{H} = \vec{i}_\theta H_o(R/r), \quad H_o \equiv I/2\pi R$$

(a) In the absence of gravity, a static equilibrium exists in which a ferrofluid having permeability μ forms a column of radius R coaxial with the wire. (The equilibrium shown in Fig. 8.3.2b approaches this circular cylindrical geometry.) Show that conditions for a static equilibrium are satisfied.

(b) Assume that the wire is so thin that its presence has a negligible effect on the fluid mechanics and on the magnetic field. The ferrofluid has a surface tension γ and a mass density much greater than that of the surrounding medium. Find the dispersion equation for perturbations from this equilibrium.

(c) Show that the equilibrium is stable provided the magnetic field is large enough to prevent capillary instability. How large must H_o be made for the equilibrium to be stable?

(d) To generate a significant magnetic field using an isolated wire requires a substantial current. A configuration that makes it easy to demonstrate the electromechanics takes advantage of the magnet from a conventional loudspeaker. A cross section of such a magnet is shown in Fig. P8.11.1. In the region above the magnet, the fringing field has the form H_oR/r. Ferrofluid placed over the gap will form an equilibrium figure that is roughly hemispherical with radius R. Viewed from the top, each half-cylindrical segment of the hemisphere closes on itself with a total length ℓ. For present purposes, the curvature introduced by this closure is ignored so that the axial distance is approximated by z with the understanding that z = 0 and z = ℓ are the same position. Effects of surface tension and gravity are ignored. Argue that the m = 0 mode represented by the dispersion equation from (b) is mechanically and magnetically consistent with this revised configuration.

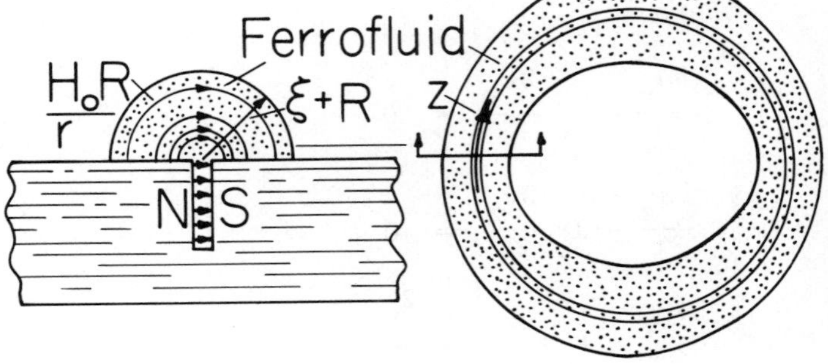

Fig. P8.11.1

(e) Show that, in the long-wave limit $kR \ll 1$, the m = 0 waves that propagate in the z direction (around the closed loop of ferrofluid) do so without dispersion. What is the dispersion equation?

(f) One way to observe these waves exploits the fact that the fluid is closed in the z direction, and therefore displays resonances. Again using the long-wave approximations, what are the resonant frequencies? How would you excite these modes?

For Section 8.12:

Prob. 8.12.1 The planar analog of the axial pinch is the sheet pinch shown in Fig. P8.12.1. A layer of perfectly conducting fluid (which models a plasma as an incompressible inviscid fluid), is in equilibrium with planar interfaces at x = $\pm$ d/2. At distances a to the left and right of the interfaces are perfectly conducting electrodes that provide a return path for surface currents which pass vertically through the fluid interfaces. The equilibrium magnetic field intensity to right and left is H_o, directed as shown. Regions a and b are occupied by fluids having negligible density.

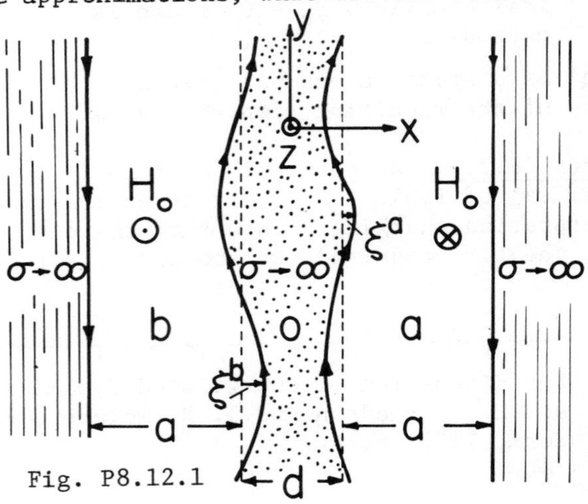

Fig. P8.12.1

Prob. 8.12.1 (continued)

(a) Determine the equilibrium difference in pressure between the regions a and b and the fluid o.

(b) Show that deflections of the interfaces can be divided into kink modes $[\xi^a(y,z,t) = \xi^b(y,z,t)]$, and sausage modes $[\xi^a(y,z,t) = -\xi^b(y,z,t)]$.

(c) Show that the dispersion equation for the kink modes is*, with $k \equiv \sqrt{k_y^2 + k_z^2}$,

$$\frac{\rho\omega^2}{k}\tanh\left(\frac{kd}{2}\right) = \mu_o H_o^2 \frac{k_z^2}{k}\coth(ka)$$

while the dispersion equation for the sausage modes is

$$\frac{\rho\omega^2}{k}\coth\left(\frac{kd}{2}\right) = \mu_o H_o^2 \frac{k_z^2}{k}\coth(ka)$$

(d) Is the equilibrium, as modeled, stable? The same conclusion should follow from both the analytical results and intuitive arguments.

Prob. 8.12.2 At equilibrium, a perfectly conducting fluid (plasma) occupies the annular region $R < r < a$ (Fig. P8.12.2.) It is bounded on the outside by a rigid wall at $r = a$ and on the inside by free space. Coaxial with the annulus is a "perfectly" conducting rod of radius b. Current passing in the z direction on this inner rod is returned on the plasma interface in the -z direction. Hence, so long as the interface is in equilibrium, the magnetic field in the free-space annulus $b < r < R$ is

$$\vec{H} = H_o \frac{R}{r}\vec{i}_\theta$$

(a) Define the pressure in the region occupied by the magnetic field as zero. What is the equilibrium pressure Π in the plasma?

(b) Find the dispersion equation for small-amplitude perturbations of the fluid interface. (Write the equation in terms of the functions $F(\alpha,\beta)$ and $G(\alpha,\beta)$.)

(c) Show that the equilibrium is stable.

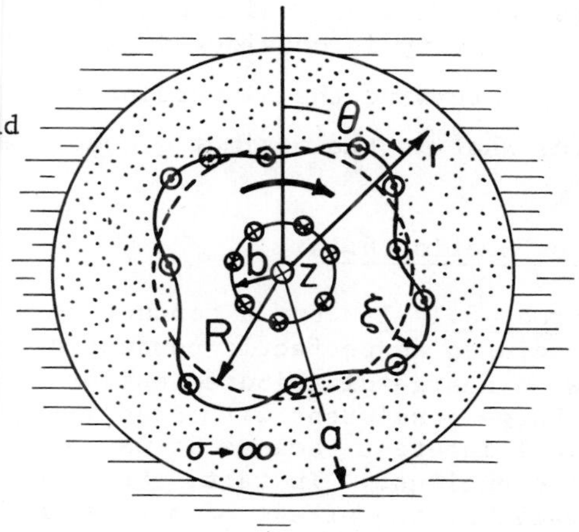

Fig. P8.12.2

Prob. 8.12.3 A "perfectly" conducting incompressible inviscid liquid layer rests on a rigid support at $x = -b$ and has a free surface at $x = \xi$. At a distance a above the equilibrium interface $\xi=0$ is a thin conducting sheet having surface conductivity σ_s. This sheet is backed by "infinitely" permeable material. The sheet and backing move in the y direction with the imposed velocity U. With the liquid in static equilibrium, there is a surface current $K_z = -H_o$ in the conducting sheet that is returned on the interface of the liquid. Thus, there is an equilibrium magnetic field intensity $\vec{H} = H_o\vec{i}_y$ in the gap between liquid and sheet. Include in the model gravity acting in the -x direction and surface tension. Determine the dispersion equation for temporal or spatial modes.

Prob. 8.12.4 In the pinch configuration of Fig. 8.12.1, the wall at r=a consists of a thin conducting shell of surface conductivity σ_s (as described in Sec. 6.3) surrounded by free space.

(a) Find the dispersion equation for the plasma column coupled to this lossy wall.

(b) Suppose that the frequencies of modes have been found under the assumption that the wall is perfectly conducting. Under what condition would these frequencies be valid for the wall of finite conductivity?

(c) Now suppose that the wall is very lossy. Show that the dispersion equation reduces to a quadratic expression in $(j\omega)$ and show that the wall tends to induce damping.

For Section 8.13:

Prob. 8.13.1 A cylindrical column of liquid, perhaps water, of equilibrium radius R, moves with uniform equilibrium velocity U in the z direction, as shown in Fig. P8.13.1. A coaxial cylindrical electrode is used to impose a radially symmetric electric field intensity

---

* $\coth kd - \dfrac{1}{\sinh kd} \equiv \tanh\left(\dfrac{kd}{2}\right)$; $\coth kd + \dfrac{1}{\sinh kd} \equiv \coth\left(\dfrac{kd}{2}\right)$

Prob. 8.13.1 (continued)

$$\vec{E} = E_o \frac{R}{r} \vec{i}_r$$

in the region between the electrode and liquid.

Assume that the density of the liquid is large compared to that of the surrounding gas. Moreover, consider the liquid to have a relaxation time short compared to any other times of interest, and assume that the cylindrical electrode is well removed from the surface of the liquid.

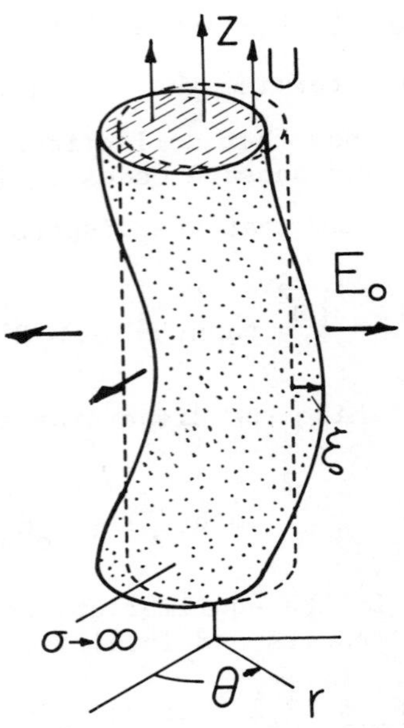

Fig. P8.13.1

(a) Determine the equilibrium pressure jump at the interface.

(b) Show that the dispersion equation is

$$(\omega-kU)^2 = \frac{\gamma}{\rho R^3}[-Rf_m(0,R)]\left\{m^2-1+(kR)^2 + \frac{\varepsilon_o E_o^2 R}{\gamma}[1-Rf(\infty,R)]\right\}$$

by using the transfer relations of Tables 2.16.2 and 7.9.1.

Prob. 8.13.2 A spherical drop of insulating liquid is of radius R and permittivity ε. At its center is a metallic, spherical particle of radius $b < R$ supporting the charge q. Hence, in equilibrium, the drop is stressed by a radial electric field.

(a) What is the equilibrium $\vec{E}$ in the drop $(b < r < R)$ and in the surrounding gas, where the mass density is considered negligible and $\varepsilon \simeq \varepsilon_o$?

(b) Determine the dispersion equation for perturbations from the equilibrium.

(c) What is the maximum q consistent with stability for $b \ll R$?

For Section 8.14:

Prob. 8.14.1 For a conducting drop, such as water in air, the model of Sec. 8.13, where the drop is pictured as perfectly conducting, is appropriate. Here, the drop is pictured as perfectly insulating with charge distributed uniformly over its volume. The goal is to find the limit on the net drop charge consistent with stability; i.e., the analogue of Rayleigh's limit. This model is of historical interest because it was used as a starting point in the formulation of the liquid drop model of the nucleus.[2] In fact, the term in that model from nuclear physics that accounts for fission is motivated by the effect of a uniform charge density. Assume that the drop is uniformly charged, has a net charge Q but has permittivity equal to that of free space. Find the maximum charge consistent with stability.

Prob. 8.14.2 Consider the same configuration as developed in this section with the following generalization. The fluids in the upper and lower regions have permittivities ε_a and ε_b respectively.

(a) Write the equilibrium and perturbation bulk and boundary conditions.

(b) Find the dispersion equation and discuss the implications of the terms.

For Section 8.15:

Prob. 8.15.1 This problem is similar to that treated in the section. However, the magnetic field is imposed and the motions are two-dimensional, so that it is possible to represent the magnetic force density as the gradient of a scalar. This makes the analysis much simpler. A column of liquid-metal carries the uniform current density J_o in the z direction but suffers deformations that are independent of z. A wire at the center of the column also carries a net current I along the z axis. The field associated with this current is presumed much greater than that due to J_o. Thus, self fields due to J_o are ignored. Assume that the wire provides a negligible mechanical constraint on the motion and that the mass density of the gas surrounding the column is much less than that of the column.

(a) Show that the magnetic force density is of the form $-\nabla\mathcal{E}$, where

$$\mathcal{E} = -\frac{\mu_o I}{2\pi} \ln\left(\frac{r}{R}\right)$$

2. I. Kaplan, Nuclear Physics, Addison-Wesley Publishing Company, Reading, Mass., 1955, p. 425.

(b) The column has an equilibrium radius R and surface tension γ. Find the dispersion equation for perturbations $\xi = \xi(\theta,t)$.

(c) Show that the column is unstable in the m = 1 mode if $J_oI < 0$, and is stable in all modes if $J_oI > 0$. Use physical arguments to explain this result.

For Section 8.16:

Prob. 8.16.1 The fluid of Fig. 8.16.1 is perfectly conducting rather than perfectly insulating. Show that the dispersion equation is

$$j\omega\eta \frac{[k(\gamma_v - k)^2 - \gamma_v(\gamma_v + k)^2]}{k(\gamma_v + k)} = \rho g + \gamma k^2 - \varepsilon_o k E_o^2$$

Show that in the limit of low viscosity the dispersion equation is Eq. 8.16.15, and that in the opposite extreme, where $\gamma_v \approx k + j\omega\rho/2\eta k$, the dispersion equation is

$$\frac{3}{2}\frac{\omega^2\rho}{k} = 2j\omega\eta k + \rho g + \gamma k^2 - \varepsilon_o k E_o^2$$

Discuss effects of viscosity on incipience and rates of growth of instability in these two limits.

Prob. 8.16.2 The magnetohydrodynamic counterpart of the interaction studied in this section might be taken as that shown in Fig. P8.16.2. The interface between a perfectly insulating liquid in the lower half space and the air above is covered by a layer of perfectly conducting liquid. In static equilibrium, a uniform magnetic field H_o is imposed in the x direction. Instead of space-charge electroviscous oscillations caused by conservation of charge and stress equilibrium, there are now magnetoviscous oscillations within the plane of the interface caused by conservation of flux for any loop of fixed identity in the conducting layer. Assume that the layer has the same mechanical properties as the fluid below.

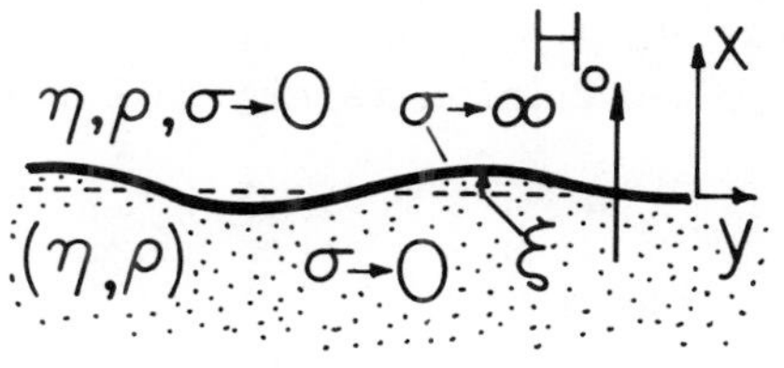

Fig. P8.16.2

Show that the thin perfectly conducting layer can be represented by the boundary condition

$$\frac{\partial H_x}{\partial t} = -H_o \frac{\partial v_y}{\partial y} \quad \text{at } x = \xi$$

Determine the dispersion equation for perturbations of the interface. Show that in the low-viscosity limit there are shearing modes of oscillation similar to those described by Eq. 8.16.16, except that

$$\omega_o = \left[\frac{2\mu_o H_o^2 k}{\sqrt{\eta\rho}}\right]^{2/3}$$

and that there are transverse modes of oscillation. Discuss the effect of viscosity on the latter in the limit where the transverse modes have a frequency that is high and that is low compared to ω_o.

Prob. 8.16.3 In the configuration of Fig. 8.16.1, the liquid layer has equilibrium thickness b, and uniform viscosity η, mass density ρ, permittivity ε and electrical conductivity σ. The upper electrode, at a distance a from the interface, has a potential -V relative to the rigid electrode at x = -b. Because the region between electrode and interface is highly insulating relative to the liquid, the equilibrium electric field is $V/a = E_o$ between the interface and the electrode and zero in the liquid layer. Effects due to the depth b and of the width a of the air gap are to be included.

(a) Write the perturbation boundary conditions and bulk conditions in terms of complex amplitudes.

(b) Show that the normalized dispersion equation is

$$M_{11}M_{22} - M_{12}M_{21} = 0$$

where in terms of normalized variables

$$M_{11} = -P_{11}j\omega - \rho - k^2 + \frac{kURS(j\omega r + 1)}{j\omega rC + R}$$

Prob. 8.16.3 (continued)

$$M_{12} = -P_{13} + j\frac{\varepsilon_o}{\varepsilon}\frac{rUkS}{j\omega r C + R}$$

$$M_{21} = -P_{31}j\omega - jUk + j\frac{kU(j\omega r+1)R}{j\omega rC + R}$$

$$M_{22} = -P_{33} - \frac{kU\frac{\varepsilon_o}{\varepsilon} r}{j\omega rC + R}$$

The normalizations are

$\underline{\omega} = \omega b\eta/\gamma$, $\underline{\rho} = \rho g b^2/\gamma$, $\underline{k} = kb$, $\underline{a} = a/b$, $U = b\varepsilon_o E_o^2/\gamma$, $r = (\gamma/b\eta)(\varepsilon/\sigma)$, $\underline{P}_{ij} = bP_{ij}$ (defined by Eq. 7.19.13 or 7.33.6), $C = (\varepsilon_o/\varepsilon)\coth \underline{ka} + \coth \underline{k}$, $S = \coth \underline{ka}$

(c) Interpret the characteristic time used to normalize ω and form the dimensionless numbers $\underline{\rho}$, r and U.

(d) In the limit of complete viscous diffusion ($\omega\rho b^2/\eta \ll 1$) and instantaneous charge relaxation ($\omega\varepsilon/\sigma \ll 1$), show that this expression reduces to simply

$$j\omega = (kUS-\rho-k^2)P_{33}/(P_{11}P_{33}+P_{13}^2)$$

(e) Again, viscous diffusion is complete but the liquid is sufficiently insulating that charge relaxation is negligible ($r \gg 1$). Show that the dispersion equation becomes

$$a(j\omega)^2 + b(j\omega) + c = 0$$

where

$$a \equiv P_{11}P_{33}+P_{13}^2 \;;\quad b = [(\rho+k^2)P_{33}+Uk(\frac{P_{11}}{C}\frac{\varepsilon_o}{\varepsilon} - \frac{RS}{C}P_{33})-2j\frac{kUP_{13}}{C}\frac{S\varepsilon_o}{\varepsilon}] \;;\quad c = \frac{kU}{C}\frac{\varepsilon_o}{\varepsilon}(\rho+k^2-UkS)$$

Prob. 8.16.4 In the configuration of Fig. 8.16.1, the liquid is replaced by a perfectly elastic incompressible solid that can be regarded as perfectly conducting (perhaps Jello). The interface, like that in the case of the viscous fluid, must be described by a balance of both normal and shear stresses. Directly applicable transfer relations are deduced in Prob. 7.19, and in the limit $\omega \to 0$ in Prob. 7.20. The solid layer, which has a thickness b, is rigidly attached to the lower solid plate. The mass density and viscosity of the gas make negligible contributions to the dynamics.

(a) Determine the dispersion equation for deformations of the solid.

(b) Under the assumption that the principle of exchange of stabilities holds (that instability is incipient with $\omega=0$) and that perturbation wavelengths are very short compared to b, determine the voltage threshold for instability.

For Section 8.18:

Prob. 8.18.1 An important connection between smoothly inhomogeneous systems and the piece-wise uniform ones considered in Sec. 8.14 is made by considering the temporal modes from another point of view. As shown in Fig. P8.18.1, the distribution of charge and mass density is approximated by two layers, each uniform in its properties.

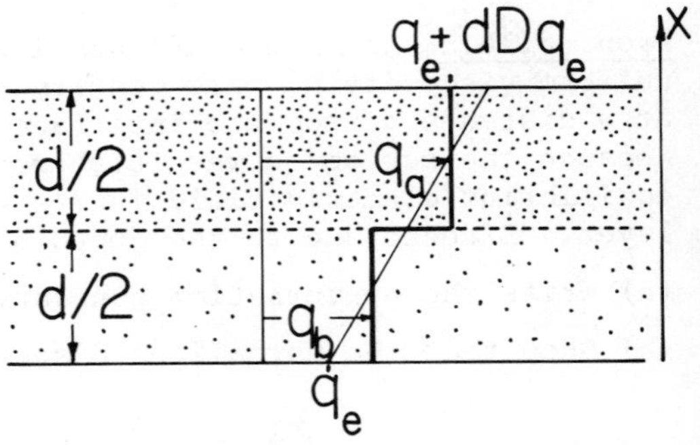

Fig. P8.18.1

(a) Show that for layers of equal thickness,

$$q_a = q_e + \frac{3}{4}Dq_e d \;:\quad q_b = q_e + \frac{1}{4}Dq_e d \;:\quad E_o = \frac{V_o}{d} - \frac{d^2 Dq_e}{16\varepsilon_o}$$

where, consistent with the usage in Section 8.14, E_o is the equilibrium electric field evaluated at the interface between layers.

(b) Show that the dispersion equation for the layer model, based on the results of Section 8.14, takes the normalized form

$$\frac{\underline{\omega}^2 \coth(\frac{\underline{k}}{2})}{\underline{k}}\left(2 + \frac{D\rho_m}{\rho_m}\right) = \frac{1}{2}\left[\frac{\frac{V_o}{d} Dq_e - gD\rho_m}{\frac{|V_o|}{d}|Dq_e|}\right] + S\left(\frac{1}{8\underline{k}} - \frac{1}{16}\right)$$

(c) Using $\underline{k} = 1$, $D\rho_m = 0$, $V_o/|V_o| = 1$, $Dq_e/|Dq_e| = 1$ and $S = 1$, compare the prediction of the first eigenfrequency to the first resonance frequency predicted in the weak-gradient approximation and to the "exact" result shown in Fig. 8.18.2a. Compare the analytical expression to that for the weak-gradient imposed field approximation in the long-wave limit. Should it be expected that the layer approximation would agree with numerical results for very short wavelengths?

(d) How should the model be refined to include the second mode in the prediction?

Prob. 8.18.2 A layer of magnetizable liquid is in static equilibrium, with mass density and permeability having vertical distributions $\rho_s(x)$ and $\mu_s(x)$ (Fig. P18.8.2). The equilibrium magnetic field $H_s(x)$ is assumed to also have a weak gradient in the x direction, even though such a field is not irrotational. (For example, this gradient represents fields in the cylindrical annulus between concentric pole faces, where the poles have radii large compared to the annulus depth ℓ. The gradient in H_s is a quasi-one-dimensional model for the circular geometry.) Assume that the fluid is perfectly insulating and inviscid.

(a) Show that the perturbation equations can be reduced to

$$D(\mu_s D\hat{h}_z) - k^2\mu_s\hat{h}_z - j\frac{k_z^2}{\omega}H_s D\mu_s\hat{v}_x = 0$$

$$D(\rho_s D\hat{v}_x) - k^2\left(\rho_s - \frac{\mathcal{N}}{\omega^2}\right)\hat{v}_x + j\frac{k^2 H_s D\mu_s}{\omega}\hat{h}_z = 0$$

where $k^2 = k_y^2 + k_z^2$, $\vec{H} = H_s\vec{i}_z + \vec{h}$ and $\mathcal{N} = -g\,D\rho_s + \frac{1}{2}D\mu_s DH_s^2$

(b) As an example, assume that the profiles are $\rho_s = \rho_m \exp\beta x$, $\mu_s = \mu_m \exp\beta x$, H_s = constant. Show that solutions are a linear combination of $\exp\gamma x$, where

$$\gamma = \frac{\beta}{2} \pm c\ ; \quad c_\pm = \left[\left(\frac{\beta}{2}\right)^2 + k^2 + a \pm b\right]^{1/2}\ ; \quad b = \left[\left(\frac{g\beta k^2}{2\omega^2}\right)^2 + \frac{k^2k_z^2}{\omega^2}\frac{H_s^2\mu_m^2\beta^2}{\rho_m}\right]^{1/2}$$

$$a = g\beta k^2/2\omega^2$$

(c) Assume that boundary conditions are $\hat{v}_x(^{\ell}_{0}) = 0$, $\hat{h}_z(^{\ell}_{0})$, and show that the eigenvalue equation is

$$\frac{2b}{a^2 - b^2}\sinh c_+\ell \,\sinh c_-\ell = 0$$

and that eigenfrequencies are

$$\omega_n^2 = \frac{k^2k_z^2H_s^2\mu_m^2\beta^2}{K_n^4\rho_m} - \frac{g\beta k^2}{K_n^2}\ ; \quad K_n^2 = \left(\frac{n\pi}{\ell}\right)^2 + \left(\frac{\beta}{2}\right)^2 + k^2$$

(d) Discuss the stabilizing effect of the magnetic field on the bulk Rayleigh-Taylor instability.

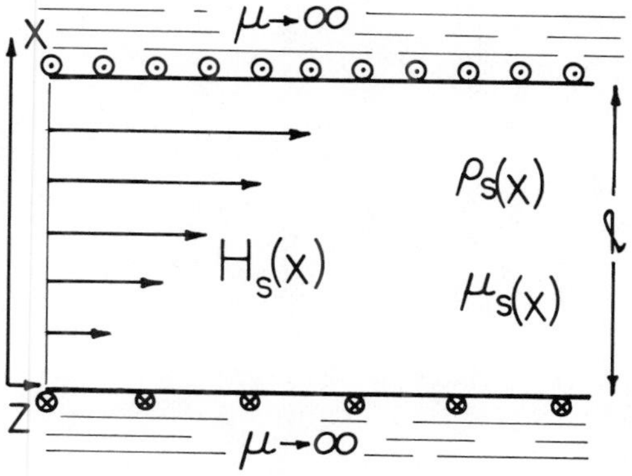

Fig. P8.18.2

Fig. 8.18.2 (continued)

(e) Discuss the analogous electric coupling with $\mu_s \to \varepsilon_s$ and $H_s \to E_s$ and describe the analogous physical configuration.

Prob. 8.18.3 As a continuation of Problem 8.18.2, prove that the principle of exchange of stabilities holds, and specifically that the eigenfrequencies are given by

$$\omega^2 = \frac{k^2 k_z^2 |I_4|^2 + I_1 I_2}{I_1 I_2}$$

where

$$I_1 = \int_0^\ell (\mu_s |D\hat{h}_z|^2 + k^2 \mu_s |\hat{h}_z|^2) dx \; ; \qquad I_2 = \int_0^\ell (\rho_s |D\hat{v}_x|^2 + k^2 \rho_s |\hat{v}_x|^2) dx$$

$$I_3 = \int_0^\ell k^2 N |\hat{v}_x|^2 dx \; ; \qquad I_4 = \int_0^\ell H_s D\mu_s \hat{v}_x^* \hat{h}_z dx$$

9

Electromechanical Flows

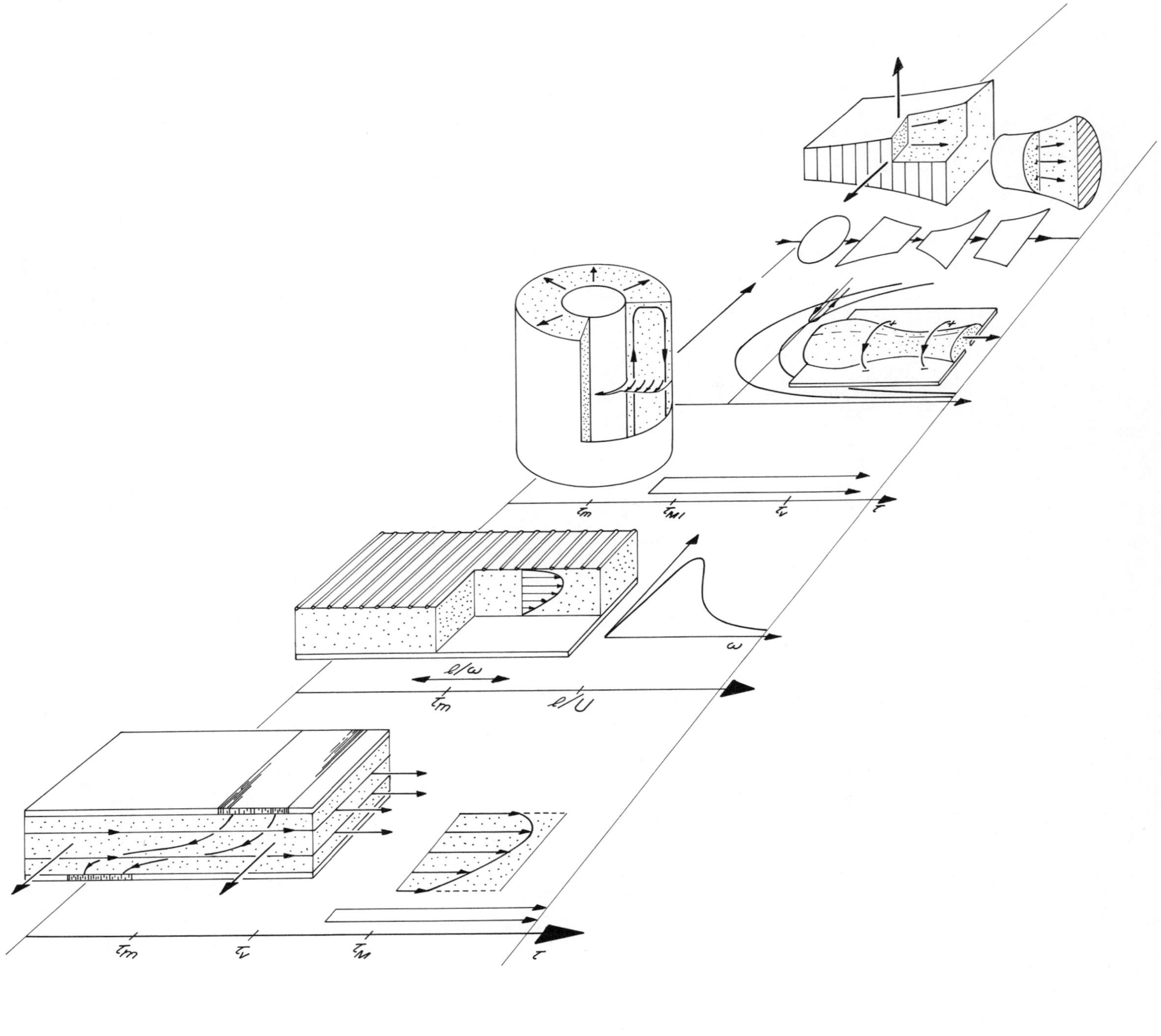

9.1 Introduction

The dynamics of fluids perturbed from static equilibria, considered in Chap. 8, illustrate mechanical and electromechanical rate processes. Identified with these processes are characteristic times. Approximations are then motivated by recognizing the hierarchy of these times and the temporal range of interest. For example, if the response to a sinusoidal steady-state drive having the frequency ω is of interest, the range is in the neighborhood of $\tau = 1/\omega$. Even for a temporal transient, where the natural frequencies are at the outset unknown, approximations are eventually justified by seeing where the reciprocal of a given natural frequency fits into the hierarchy of characteristic times.

In this chapter, it is steady flows and the establishment of such flows that is of interest. Typically, the characteristic times from Chap. 8 are now to be compared to a transport time, ℓ/u. The recognition of simplifying approximations becomes even more important, because nonlinear equations are likely to be an essential part of a model.

The requirement for a static equilibrium that force densities be irrotational is emphasized in Sec. 8.2. Taken up in Sec. 9.2 is the question, what types of flow can result from application of such force densities? This is the first of 11 sections devoted to homogeneous flows, where such properties as the mass density and electrical conductivity are uniform throughout the flow region.

Some of the most practical interactions between fields and fluids can be represented by a force density or surface force density that is determined without regard for the fluid motion or geometry. Such imposed surface and volume force density flows are the subject of Secs. 9.3-9.8. In Sec. 9.3, fully developed flows are described in such a way that their application to a wide range of problems should be evident. By way of illustration, surface coupled and volume coupled electric and magnetic flows are then discussed in Secs. 9.4 and 9.5. Liquid metal magnetohydrodynamic induction pumps usually fit the model of Sec. 9.5.

To appreciate a fully developed flow, it is necessary to consider the flow development. In Sec. 9.6, this is done by examining the temporal transient that results as a closed system is suddenly turned on and the steady flow allowed to establish itself. Then, in terms of boundary layers, the spatial transient is discussed. In addition to its application to surface coupled flows, illustrated in Sec. 9.7, the boundary layer model is applied to a self-consistent bulk coupled flow in Sec. 9.12. In Secs. 9.6 and 9.7, viscous diffusion is of interest, both fluid inertia and viscosity are important and times of interest are, by definition, on the order of the viscous diffusion time.

Illustrated in Sec. 9.8 are an important class of electromechanical models in which the bulk flow is described by linear equations. Here, transport times are long compared to the viscous diffusion time and "creep flow" prevails.

The self-consistent imposed field flows of Secs. 9.9-9.12 give the opportunity to broaden the range of dynamical processes. In the first two of these sections, magnetohydrodynamic processes are taken up. The magnetic diffusion time is short compared to the other times of interest, the viscous diffusion time and the magneto-inertial time. These sections first illustrate how the field alters fully developed flows and then considers how the electromechanics contributes to temporal flow development. The electrohydrodynamic approximation discussed and illustrated in the last two sections of this part is based on having a self-precipitation time for unipolar charges that is long compared to other times of interest, for example, an electroviscous time.

With the introduction of inhomogeneity come more characteristic dynamical times. These are illustrated for systems having a static equilibrium and abrupt discontinuities in properties in Secs. 8.9-8.16. Typically, the associated characteristic times represent propagation of surface waves. Smoothly distributed inhomogeneities, Secs. 8.17-8.18, give rise to related internal waves with their characteristic times. The flow models developed in Sec. 9.13 and illustrated in Sec. 9.14 incorporate wave phenomena similar to those from Chap. 8. The wave phenomena show up in steady flow situations through critical conditions, often expressed in terms of the ratio of a convective velocity to a wave velocity, i.e., as a Mach number. In essence these numbers are the ratio of transport times to wave transit times. Times of interest in these sections, which reflect the existence of waves, are a capillary time $\tau_\gamma = \sqrt{\gamma/\rho\ell^2}$ (Sec. 8.9), a gravity time $\tau g = \sqrt{g\,(\rho_b - \rho_a)/(\rho_b + \rho_a)\ell}$ (Sec. 8.9) and various magneto- and electro-inertial times (Secs. 8.10-8.15).

In view of Sec. 8.8 on magneto-acoustic and electro-acoustic waves, it should be expected that additional times introduced in the remaining sections on compressible flow are the transit times for acoustic and acoustic related waves. Sections 9.15 and 9.17 bring into the discussion the additional physical laws required to represent interactions with the internal energy subsystem of a gas. Here, the energy equation is derived and thermodynamic variables needed in subsequent sections defined. These laws are not only necessary for the description of thermal-to-electrical energy conversion (to be taken up in Secs. 9.21-9.23), but also can be used to describe convective heat transfer.

The quasi-one-dimensional model introduced in Sec. 9.19 is the basis for the various energy conversion systems discussed in the remaining sections. Once again, even in steady flows, the role of wave propagation is unavoidable. As in Sec. 9.14, flow through energy conversion devices is dependent on the fluid velocity relative to a wave velocity. This time, the waves are acoustic related.

In Secs. 9.21 and 9.23, the energy conversion process is again highlighted. These models, which include the thermodynamics as well as the electromechanics, hark back to the prototype magnetic and electric d-c machines introduced in Secs. 4.10 and 4.14. The MHD and EHD energy converters combine the functions of the turbine and a generator in a conventional power generating plant. Thus, they give the opportunity to understand the overall thermodynamic limitations of the energy conversion process.

To fully appreciate the steady flows of inhomogeneous fluids, Sec. 9.14, and of compressible fluids, Secs. 9.20-9.22, flow transients predicted by the same quasi-one-dimensional models should be studied. These are taken up in Chap. 12, where the method of characteristics is applied to nonlinear flows involving propagating wave phenomena. In this chapter, nonlinear processes represented by quasi-one-dimensional models are represented by systems of ordinary differential equations. Similarity solutions, introduced in Sec. 6.9, are now extended to nonlinear equations.

9.2 Homogeneous Flows with Irrotational Force Densities

The static equilibria of Secs. 8.1-8.5 illustrate electrical to mechanical coupling approximated by irrotational magnetic and electric force densities. In this section, yet another field configuration that can be represented by a force density of the form $\vec{F} = -\nabla\mathcal{E}$ is introduced. But more important, steady flows are to be illustrated. The point in this section is that now, given boundary conditions stipulating the fluid velocity, an irrotational force density interacts with the flow of a homogeneous incompressible fluid to alter the pressure distribution, but not the flow pattern.

Inviscid Flow: Recall that for an irrotational inviscid flow, the velocity potential satisfies Laplace's equation (Eq. 7.8.10)

$$\nabla^2\Theta = 0; \quad \vec{v} = -\nabla\Theta \tag{1}$$

With boundary conditions on $\vec{v}$ specified over the surface enclosing the volume of interest, the flow is therefore uniquely determined without regard for the force densities. However, through $\mathcal{E}$, the force density does contribute to the pressure distribution. From Eq. 7.8.11,

$$p = -\frac{1}{2}\rho\vec{v}\cdot\vec{v} + \rho\vec{g}\cdot\vec{r} - \mathcal{E} + \Pi \tag{2}$$

As an example of an irrotational force density, consider the MQS low magnetic Reynolds number flow in two dimensions (x,y) through a region where a perpendicular uniform magnetic field, $\vec{H} = H_o\vec{i}_z$, is imposed. The current density is solenoidal with components in the x-y plane only. It is therefore represented in terms of the z component of a vector potential (Cartesian coordinates, Table 2.18.1)

$$\vec{J} = \frac{\partial A}{\partial y}\vec{i}_x - \frac{\partial A}{\partial x}\vec{i}_y \tag{3}$$

Because current induced by the motion is negligible compared to that imposed, throughout a region of uniform electrical conductivity, $\vec{J}$ is also approximately irrotational. Hence, within such a region

$$\nabla^2 A = 0 \tag{4}$$

This expression is justified provided that $R_m \equiv \mu\sigma\ell u \ll 1$, as is evident from a normalization of Eq. 6.5.3 in the fashion of Eq. 6.2.9. The force density is expressed in terms of A by using Eq. 3 for $\vec{J}$ and approximating the field intensity as the imposed field:

$$\vec{F} = \vec{J} \times \mu_o H_o \vec{i}_z = -\nabla\mathcal{E}; \quad \mathcal{E} = \mu_o H_o A \tag{5}$$

Remember that H_o is by assumption much greater than the field induced by $\vec{J}$. From Ampere's law, this requires that $|J|\ell \ll H_o$, where ℓ is a typical length.

Uniform Inviscid Flow: The channel flow sketched in Fig. 9.2.1a has fluid entering at the left with a uniform velocity profile and leaving at the right with the same profile. A flow satisfying Eq. 1 and the additional boundary conditions that there be no normal velocity on the rigid upper and lower boundaries is simply a uniform velocity everywhere, $\Theta = -Uy$.

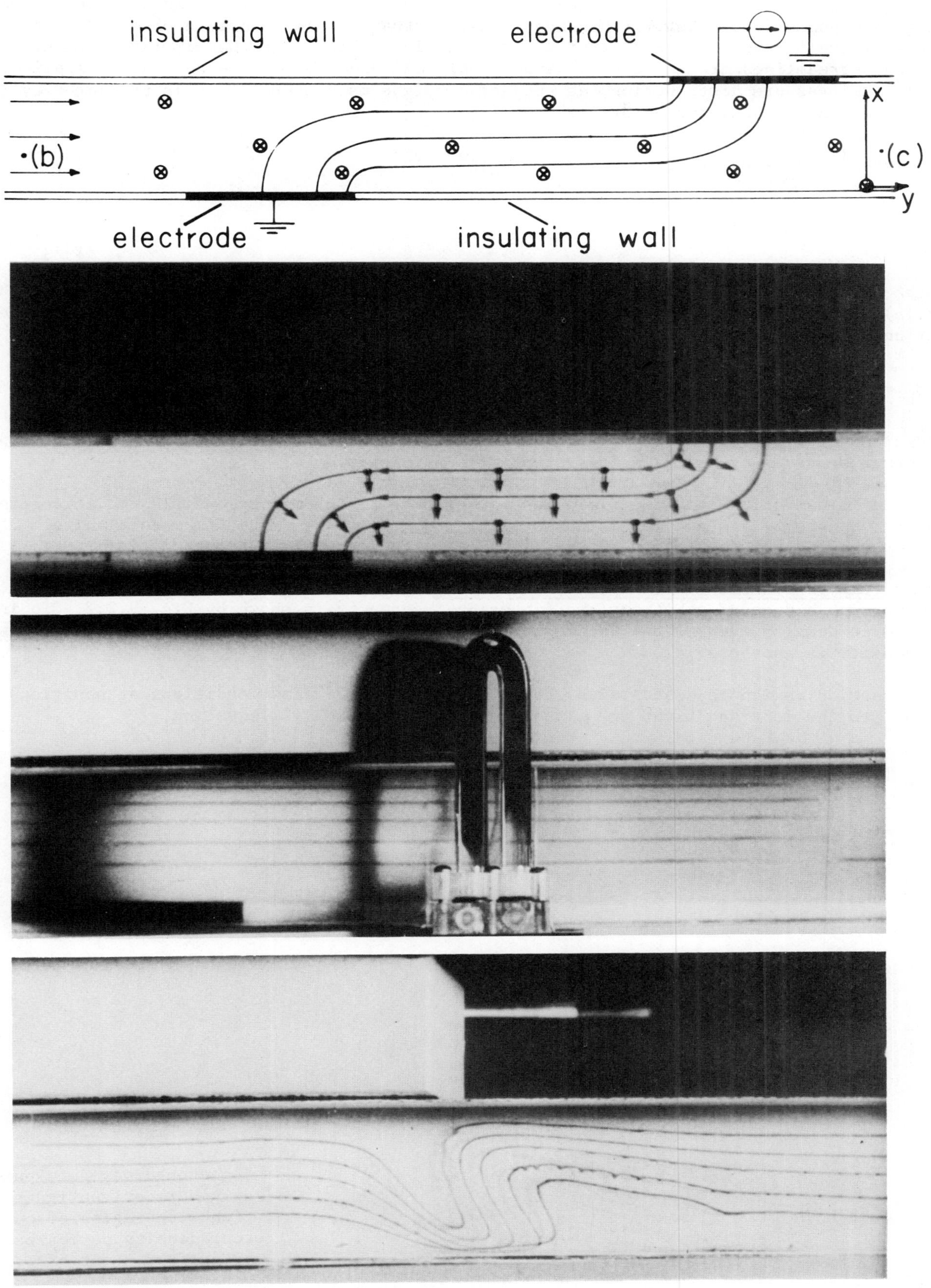

Fig. 9.2.1. (a) Electrolyte is channeled by insulating walls through region of uniform magnetic field perpendicular to flow (positive in the y direction). (b) From Reference 7, Appendix C, sketch of current and $\vec{J} \times \vec{B}$ densities in experiment with H_o positive and I negative. (c) With H_o uniform, extremely nonuniform but irrotational force distribution of (b) leaves plane flow undisturbed (shown by streamlines) but results in pressure rise (shown by manometer). (d) With H_o nonuniform, strong acceleration caused by rotational force density is evident in the stirring of the flow.

Electrodes embedded in the lower and upper walls are used to pass a current through the flow. As sketched in Fig. 9.2.1b, the resulting force density, which has both vertical and horizontal components, is complicated and nonuniform. Yet, it has been asserted that the flow pattern observed in the absence of a current would be the same as seen after the current is applied. In the experiment shown in Fig. 9.2.1c, the streamlines are in fact not appreciably different after the current is applied. What does change is the pressure distribution, as suggested by the manometer. This is predicted by Eqs. 2 and 5, which show that for any two points (α) and (β),

$$p_\alpha - p_\beta = -\frac{1}{2}\rho(v_\alpha^2 - v_\beta^2) - \rho g(x_\alpha - x_\beta) - \mu_o H_o(A^\alpha - A^\beta) \tag{7}$$

Note that

$$A^\alpha - A^\beta = i \tag{8}$$

where i is the current linked by a surface having unit length in the z direction and edges at (α) and (β) (see Sec. 2.18). Thus, with α and β the locations c and b respectively in Fig. 9.2.1a, i is the total current (per unit length perpendicular to the paper) I, and Eq. 7 becomes

$$p_c - p_b = -\frac{1}{2}\rho(v_c^2 - v_b^2) - \rho g(x_c - x_b) - \mu_o H_o I \tag{9}$$

For the conditions of Fig. 9.2.1c, $v_c = v_b$, H_o is positive and I is negative (as sketched in Fig. 9.2.1b) so the pressure rise given by Eq. 9 is consistent with intuition.

A dramatic illustration that the fluid must accelerate if the magnetic field conditions for an irrotational force density are not met is shown in Fig. 9.2.1d. The magnet imposes a uniform H_o over the region to the left, but the region to the right is in the nonuniform fringing field.

<u>Inviscid Pump or Generator with Arbitrary Geometry</u>: The generalization of the configuration to a channel flow through a duct of arbitrary two-dimensional geometry is shown by Fig. 9.2.2. To insure an irrotational force density everywhere, the magnetic field need only be uniform over the region where the current density is appreciable.

The interaction region is described by Eq. 9. To relate the flow conditions at positions (d) and (a), the "legs" to the left and right are also described by Bernoulli's equation,

$$p_d - p_c = -\frac{1}{2}\rho(v_d^2 - v_c^2) - \rho g(x_d - x_c) \tag{10}$$

$$p_d - p_a = -\frac{1}{2}\rho(v_b^2 - v_a^2) - \rho g(x_b - x_a) \tag{11}$$

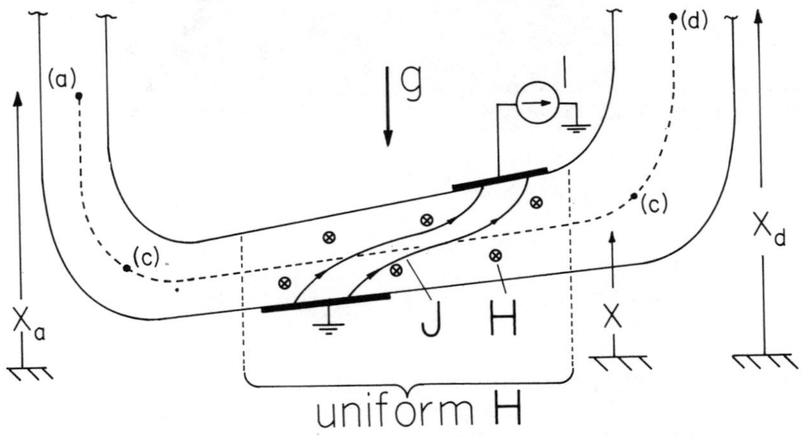

Fig. 9.2.2

Magnetohydrodynamic pump or generator configuration with region of current density permeated by uniform $\vec{H}$ out of paper.

Addition of these last three expressions gives the desired pressure-velocity relation for the entire system:

$$p_d - p_a = -\frac{1}{2}\rho(v_d^2 - v_a^2) - \rho g(x_d - x_a) - \mu_o H_o I \quad (12)$$

Again, note that this simple relation applies regardless of electrode geometry.

Viscous Flow: Finally, observe that if the force density is irrotational, and hence takes the form $\vec{F} = -\nabla\mathcal{E}$, it can be lumped with the pressure gradient. For an incompressible homogeneous flow, the pressure appears only in the force equation.

With the redefinition of the pressure, $p \to p + \mathcal{E}$, the equations of motion are no different than in the absence of the field. Thus, if the boundary conditions do not involve the pressure, it is clear that the flow pattern must be the same with and without the field. In the experiment of Fig. 9.2.1, the flow is probably more nearly fully developed (as defined in Sec.9.3) than inviscid and hence has vorticity. Yet, the only effect of the irrotational magnetic force density is to revise the pressure distribution.

FLOWS WITH IMPOSED SURFACE AND VOLUME FORCE DENSITIES

9.3 Fully Developed Flows Driven by Imposed Surface and Volume Force Densities

Fully developed flows are stationary equilibria established after either a temporal or a spatial transient. Flow established by setting the coaxial wall of a Couette viscometer into steady rotation is an example of the former. Typical of a spatial transient is steady flow through a conduit of uniform cross section. As the fluid first enters a pipe, the velocity profile is determined by the entrance conditions. But, as an element progresses, the viscous shear stresses from the walls penetrate into the flow until they are effective over the entire cross section. At this point, the flow becomes independent of longitudinal position and is said to be fully developed.

For a region of rectangular cross section, with its x dimension much less than the y dimension, the fully developed flow is a special type of plane flow:

$$\vec{v} = v(x)\vec{i}_y \quad (1)$$

Note that continuity is automatically satisfied, i.e., $\vec{v}$ is solenoidal.

The objective of this and the next two sections is an illustration of how viscous forces can balance electric and magnetic forces imposed either at surfaces or throughout the fluid volume. By "imposed," it is meant that the fluid motion does not play a significant part in determining the electromagnetic force distribution. Sections 9.4 and 9.5 illustrate the flow itself.

Because $\partial\vec{v}/\partial t = 0$ and (from Eq. 1) $\vec{v}\cdot\nabla\vec{v} = 0$, there is no acceleration. The Navier-Stokes equation, Eq. 7.16.6, becomes

$$\nabla p = \nabla(\rho\vec{g}\cdot\vec{r}) + \vec{F} + \eta\nabla^2\vec{v} \quad (2)$$

The force density is only a function of x, so a scalar $\mathcal{E}$ can always be found such that $F_x = -\partial\mathcal{E}(x)/\partial x$. Thus, the x component of Eq. 2 becomes

$$\frac{\partial p'}{\partial x} = 0; \quad p' \equiv p - \rho\vec{g}\cdot\vec{r} + \mathcal{E}(x) \quad (3)$$

It follows that p' is uniform over the cross section. The x dependence of p is whatever it must be to balance the transverse gravitational and electromagnetic force components.

In terms of p', the longitudinal component of Eq. 2 becomes

$$\frac{\partial p'}{\partial y} = F_y(x) + \eta\frac{\partial^2 v}{\partial x^2} \quad (4)$$

Terms on the right are independent of z, so the longitudinal hybrid pressure gradient, $\partial p'/\partial z$, must also be independent of y.

Because the force density F_y is independent of y, it can be written in terms of a tensor divergence which reduces to simply $F_y = T_{yx}/\partial x$.

Integrated on x, Eq. 4 then represents the balance of electromagnetic and viscous shear stresses:

$$\frac{\partial p'}{\partial y}x = T_{yx}(x) - T_{yx}(0) + \eta\left[\frac{\partial v}{\partial x}(x) - \frac{\partial v}{\partial x}(0)\right] \tag{5}$$

It is instructive to note the physical origins of this expression. It can also be obtained by writing the y component of force balance for the fluid within a control volume of incremental length in the y direction, unit depth in the z direction and with transverse surfaces at $x = 0$ and $x = x$, respectively. In the absence of a hybrid pressure gradient, the fully developed flow is simply a balance of viscous and electromagnetic shear stresses.

To determine the velocity profile, Eq. 5 is once again integrated from $x = 0$ where $v = v^\beta$ to $x = x$, and solved for $v(x)$. The constant $\partial v/\partial x(0)$ is determined by evaluating $v(x)$ at $x = \Delta$ where it equals v^α. The resulting velocity profile is the first of those given in Table 9.3.1.

The circulating flow and axial flow through a circular cylindrical annulus, also shown in the table, are other examples where a fully developed flow is found by what amounts to the same stress balance as exploited in the planar case. Note that for the circulating flow, the pressure gradient in the flow direction is zero. Determination of the velocity profiles summarized in Table 9.3.1 is left for the problems.

Table 9.3.1. Three fully developed flows.

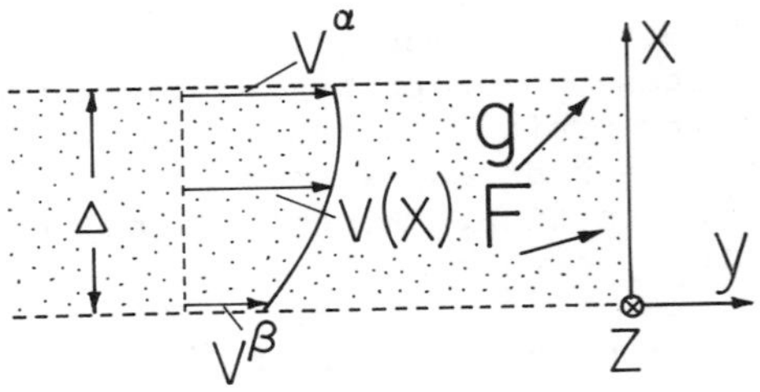	$v(x) = v^\beta\left(1 - \frac{x}{\Delta}\right) + v^\alpha \frac{x}{\Delta} + \frac{\Delta^2}{2\eta}\frac{\partial p'}{\partial y}\left[\left(\frac{x}{\Delta}\right)^2 - \frac{x}{\Delta}\right] - \frac{1}{\eta}\int_0^x T_{yx}dx + \frac{x}{\eta\Delta}\int_0^\Delta T_{yx}dx$	(a)
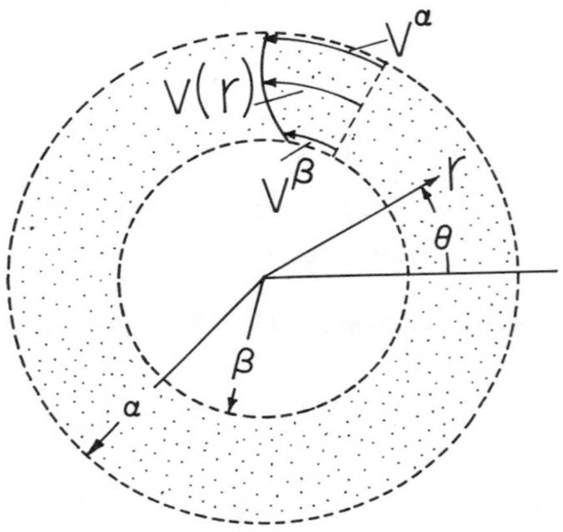	$v = \frac{1}{\left(\frac{\alpha}{\beta} - \frac{\beta}{\alpha}\right)}\left[v^\alpha\left(\frac{r}{\beta} - \frac{\beta}{r}\right) + v^\beta\left(\frac{\alpha}{r} - \frac{r}{\alpha}\right)\right] + \alpha\frac{\left(\frac{r}{\beta} - \frac{\beta}{r}\right)}{\left(\frac{\alpha}{\beta} - \frac{\beta}{\alpha}\right)\eta}\int_\beta^\alpha \frac{T_{\theta r}}{r}dr - \frac{r}{\eta}\int_\beta^r \frac{T_{\theta r}}{r}dr$	(b)
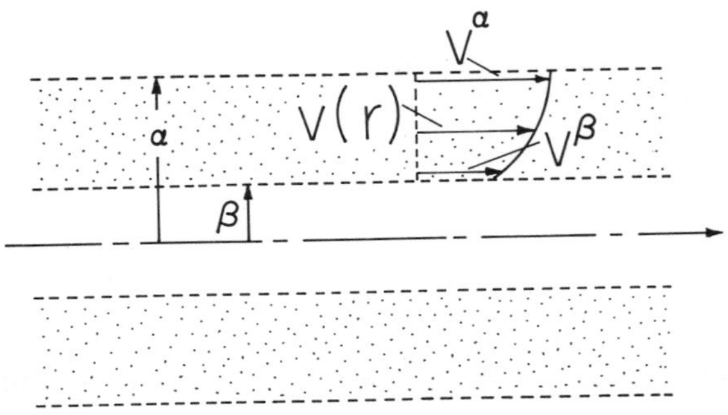	$v = v^\alpha\frac{\ln\left(\frac{r}{\beta}\right)}{\ln\left(\frac{\alpha}{\beta}\right)} + v^\beta\frac{\ln\left(\frac{\alpha}{r}\right)}{\ln\left(\frac{\alpha}{\beta}\right)} + \frac{1}{4\eta}\frac{\partial p'}{\partial z}\left[(r^2 - \beta^2) - (\alpha^2 - \beta^2)\frac{\ln\left(\frac{r}{\beta}\right)}{\ln\left(\frac{\alpha}{\beta}\right)}\right] - \frac{1}{\eta}\int_\beta^r T_{zr}dr + \frac{\ln\left(\frac{r}{\beta}\right)}{\eta\ln\left(\frac{\alpha}{\beta}\right)}\int_\beta^\alpha T_{zr}dr$ where $p' \equiv p - \rho(\vec{g}\cdot\vec{r}) + \mathcal{E}(r);\quad F_r = -\frac{d\mathcal{E}}{dr}$	(c)

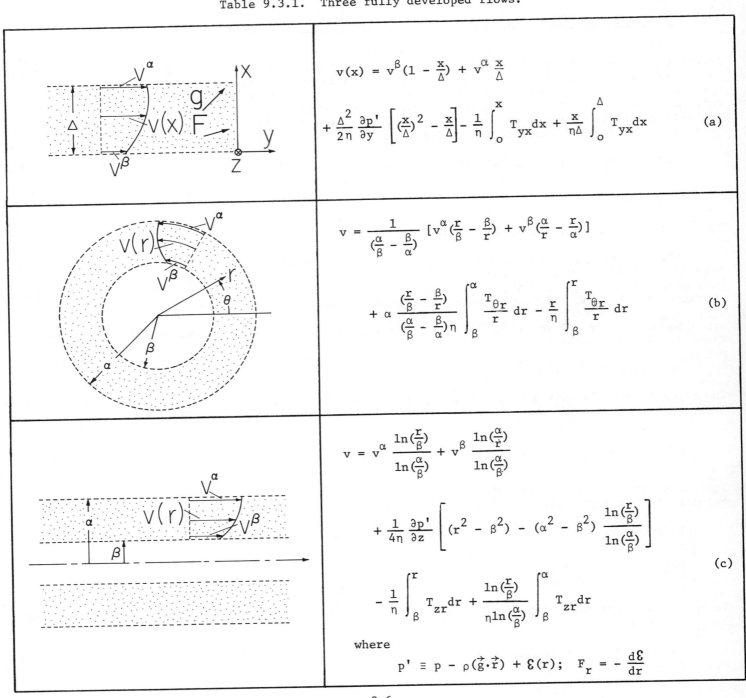

9.4 Surface-Coupled Fully Developed Flows

Fully developed flows are often used in quasi-one-dimensional models. Examples in this section illustrate by using the results of Sec. 9.3 to describe liquid circulations. They also illustrate how shearing surface force densities can act in consort with viscous shear stresses to give rise to volume fluid motions. The example treated in detail is EQS, with the surface force density resulting from the combination of a monolayer of charge and a tangential electric field at a "free" interterface. If the magnetic skin depth is short compared to the depth of the fluid, similar flows can result from subjecting the interface of a liquid metal to a magnetic shear stress, as suggested by Sec. 6.8. Consider first a specific case study, after which it is appropriate to identify the general nature of the interactions that it illustrates.

Charge-Monolayer Driven Convection: A semi-insulating liquid fills an insulating container to a depth b. Electrodes to the right and left have the potential difference V_o. Provided the charge convection at the interface is not appreciable, the resulting current density in the liquid is uniformly distributed throughout the volume of the liquid. As discussed in Sec. 5.10, at least insofar as it can be described by an ohmic conduction model, the liquid does not support a volume free charge density. It also has a uniform permittivity. Hence, there is no volumetric electrical force. However, an electrical force does exist at the interface, where the conductivity is discontinuous. In this "Taylor Pump" the electrode above the interface is canted in just such a way as to make the resulting electric shearing surface force density tend to be uniform over most of the interface. To see that this is so, observe that, if effects of convection can be ignored, in the liquid

$$\Phi = V_o \frac{(y-\ell)}{\ell}\ ; \quad \vec{E} = -\frac{V_o}{\ell}\vec{i}_y; \quad 0 < x < b \tag{1}$$

Because a << ℓ, the electric field between interface and slanted plate is essentially in the x direction and given by the plate-interface potential difference divided by the spacing:

$$\vec{E} = \frac{V_o y/\ell}{h(y)}\vec{i}_x = \frac{V_o}{a}\vec{i}_x; \quad b < x < \frac{ay}{\ell} + b \tag{2}$$

Note that, at the interface, the tangential electric field is continuous and there is no normal electric field on the liquid side. Thus, the interfacial surface force density is

$$\vec{T} = \frac{1}{2}[\![\varepsilon E_x^2 - \varepsilon E_y^2]\!]\vec{i}_x + [\![\varepsilon E_x E_y]\!]\vec{i}_y = \frac{1}{2}\left(\frac{V_o}{a}\right)^2\left[\varepsilon_o + \frac{a^2}{\ell^2}(\varepsilon - \varepsilon_o)\right]\vec{i}_x - \varepsilon_o \frac{V_o^2}{a\ell}\vec{i}_y \tag{3}$$

and, as required for a fully developed model, both the normal and shear components are independent of y.

The normal component of $\vec{T}$ is equilibrated by the liquid pressure. With the pressure of the air defined as zero, normal stress balance at the interface, where $x = b + \xi$, requires that

$$T_x = -p(x = b + \xi) \tag{4}$$

In the liquid bulk, where the flow is modeled as fully developed, Eq. 9.3.3 shows that p' is only a function of y. Here, p' is determined by substituting p' evaluated at $x = b + \xi$ into Eq. 4. It follows that

$$p' = \rho g(b + \xi) - \frac{1}{2}\left(\frac{V_o}{a}\right)^2\left[\varepsilon_o + \left(\frac{a}{\ell}\right)^2(\varepsilon - \varepsilon_o)\right] \tag{5}$$

Here, $\xi(y)$ is yet to be determined. If this vertical deflection of the interface is much less than the depth of the liquid layer, insofar as the flow is concerned, the fluid depth can be approximated as simply b.

Three conditions are required to determine the variables v^α, v^β and $\partial p/\partial y$ in Eq. (a) of Table 9.3.1. Two of these come from the facts that the velocity at the tank bottom is zero and that the net flow through any x-y plane is zero:

$$v(x = 0) = 0 \tag{6}$$

$$\int_o^b v\,dx = 0 \tag{7}$$

The third follows from the shear stress equilibrium at the interface, where the electrical shear stress is balanced by the viscous shear stress,

$$\frac{-\varepsilon_o V_o^2}{a\ell} = \eta \frac{\partial v}{\partial x}\ (x = b) \qquad (8)$$

It follows from Eq. 6 that $v^\beta = 0$. Then, substitution of Eq. (a) of Table 9.2.1 into Eq. 7 gives the longitudinal pressure gradient in terms of the surface velocity:

$$\frac{\partial p'}{\partial y} = \frac{6\eta}{b^2} v^\alpha \qquad (9)$$

This result and Eq. 8 (evaluated using Eq. (a) of Table 9.3.1) then make it possible to evaluate the surface velocity:

$$v^\alpha = -\frac{\varepsilon_o V_o^2 b}{4a\ell\eta} \qquad (10)$$

This velocity results from a competition between electric and viscous stresses, so it is no surprise that the transport time b/v^α is found to be on the order of the electro-viscous time $\tau_{EV} = \eta/\varepsilon_o(V_o^2/a\ell)$.

Evaluated using Eqs. 9 and 10, the velocity profile follows from Eq. (a) of Table 9.3.1 as

$$v = -\frac{\varepsilon_o V_o^2 b}{2\eta a\ell}\left[\frac{3}{2}\left(\frac{x}{b}\right)^2 - \frac{x}{b}\right] \qquad (11)$$

This is the profile shown in Fig. 9.4.1a.

As a reminder of the vertical pressure equilibrium implied by the model, it is now possible to evaluate the small variation in the liquid depth caused by the horizontal flow. Integration of Eq. 9 with v^α from Eq. 11 gives

$$p' = -\frac{3}{2}\frac{\varepsilon_o V_o^2}{a\ell}\left(\frac{y}{b}\right) + \text{constant} \qquad (12)$$

where p' is also given by Eq. 5. The constant is set by equating these expressions and defining the position where $\xi = 0$ as being $y = 0$. It follows that the depth varies as

$$\xi = -\frac{3}{2}\frac{\varepsilon_o V_o^2}{ab\rho g}\left(\frac{y}{\ell}\right) \qquad (13)$$

That the liquid depth is greatest at the left reflects the fact that the pressure is greatest at the left. Thus, in the lower 2/3's of the liquid (where there is no horizontal force density to propel the liquid) the pressure propels the liquid to the right.

The field and charge distributions have been computed under the assumption that the effects of material motion are not important. This is justified only if the fluid conductivity is large enough that the interfacial convection of charge does not compete appreciably with the volume conduction in determining the interfacial charge distribution. In retrospect, an estimate of the implied condition is obtained by considering conservation of charge for a section of the interface near the left end. Here, the surface velocity falls from its peak value to zero in a horizontal distance on the order of the depth b. If the current convected at the interface is to be small compared to that conducted to the electrode from the bulk, then

$$|\sigma_f v^\alpha| << \left|\frac{b\sigma V_o}{\ell}\right| \qquad (14)$$

According to the approximate theory, the surface charge is given from Eq. 2 and Gauss' law as $\sigma_f = \varepsilon_o V_o/a$, so that Eq. 14 is equivalent to

$$R_e \equiv \frac{\varepsilon_o v^\alpha \ell}{ab\sigma} << 1 \qquad (15)$$

Hence, the imposed stress model is valid in the low electric Reynolds number approximation. The physical significance of R_e, here the ratio of the charge relaxation time $(\varepsilon_o/\sigma)(\ell/a)$ to the transport time b/v^α, is discussed in Sec. 5.10. Too great a velocity or too small a conductivity results in an electric stress in part determined by the fluid response.

Of course, the velocity in Eq. 15 is actually determined by the fields themselves, so a more explicit statement can be made. From Eq. 10, v^α is related to the fields so that R_e becomes

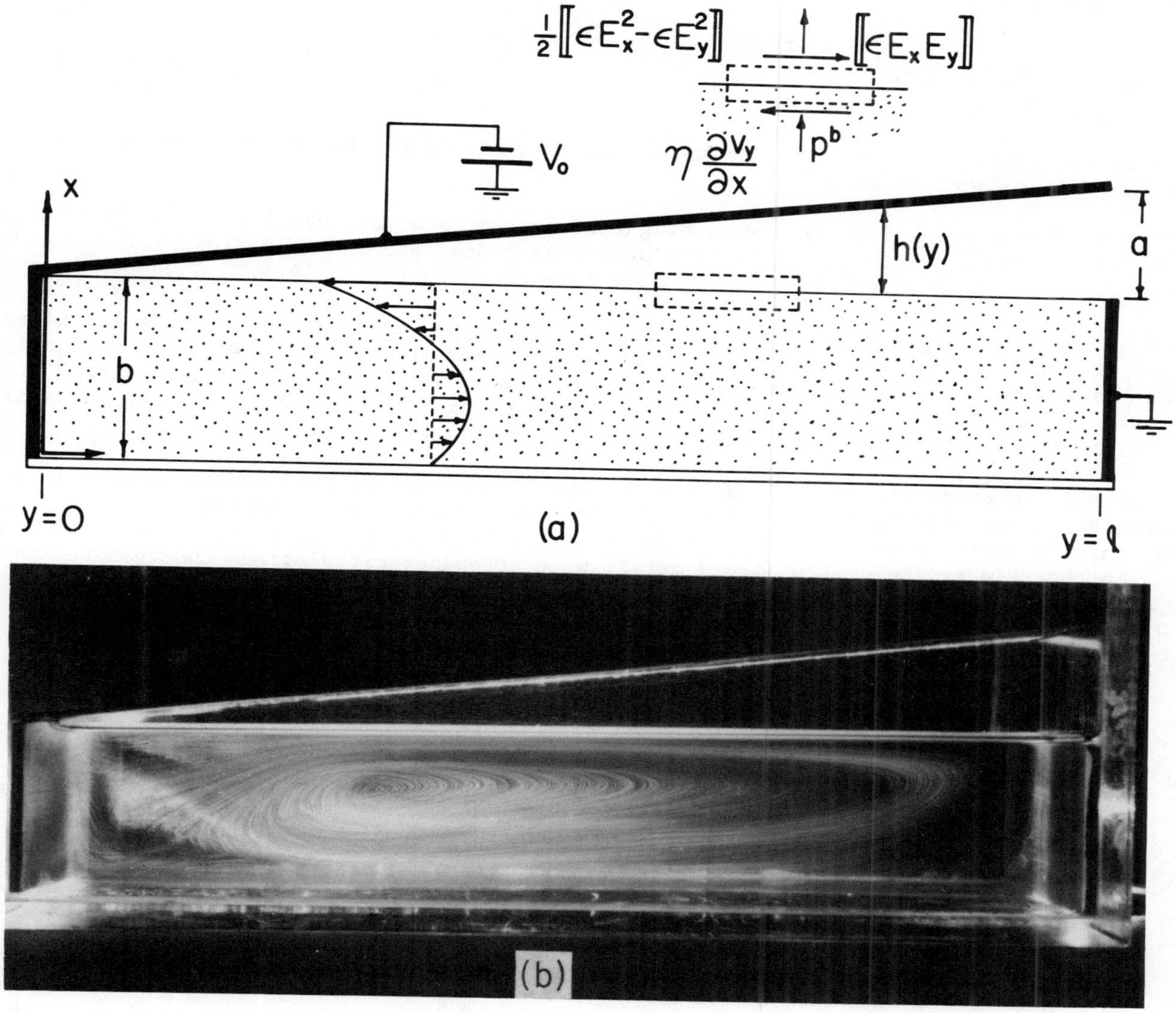

Fig. 9.4.1. (a) Cross section of liquid layer driven to the left at its interface by surface force density. Electrodes at the left and above have zero potential relative to the electrode at the right, which has potential V_o. The liquid is slightly conducting and contained by an insulating tank. (b) Time exposure of bubbles entrained in liquid show stream lines with experimental configuration essentially that of (a). The liquid is corn oil, with depth of a few cm and surface velocities at voltages in the range 10-20 kV on the order of 5 cm/sec. (For experimental correlation, see J. R. Melcher and G. I. Taylor, "Electrohydrodynamics: A Review of the Role of Interfacial Shear Stresses," in Annual Review of Fluid Mechanics, Vol. 1, W. R. Sears, Ed., Annual Reviews, Inc., Palo Alto, Calif.. 1969, pp. 111-146. The experiment is shown in Reference 12, Appendix C.

$$H_e^2 \equiv \frac{\varepsilon_o V_o^2}{4a^2\eta\sigma} << 1 \qquad (16)$$

This more useful expression of the approximation is in terms of what will be termed the electric Hartmann number, H_e. As the square root of the ratio of the charge relaxation time ε_o/σ, to the electro-viscous time, $\eta/\varepsilon(V_o/a)^2$, this number also appears in Sec. 8.6.

If the viscosity is too low, the fully developed flow is not observed. Rather, the shear force at the interface cannot entrain the fluid near the bottom before an element has passed from one end of the tank to the other. Then, only a boundary layer is set into motion. A suitable model is discussed in Sec. 9.7.

At the ends, the "turn around" also involves accelerations. Under what conditions does the resulting inertial force density, $\rho\vec{v}\cdot\nabla\vec{v}$, compete significantly with that from viscosity? With the spatial derivatives characterized by the reciprocal length b^{-1}, the ratio of acceleration to viscous force density is of the order

$$R_y \equiv \frac{[\rho(v^\alpha)^2/b]}{\eta v^\alpha/b^2} = \frac{\rho v^\alpha b}{\eta} << 1 \tag{17}$$

Defined as the ratio of the viscous diffusion time, $\rho b^2/\eta$, to the transport time, b/v^α, the Reynolds number R_y is introduced in Sec. 7.18.

EQS Surface Coupled Systems: Two configurations that are very similar to the "Taylor pump" with fully developed flows providing quasi-one-dimensional models are shown by Figs. 9.4.2a and 9.4.2b. Note that the experiments to which these models apply are shown in Fig. 5.14.4.

MQS Systems Coupled by Magnetic Shearing Surface Force Densities: In pumping liquid metals with alternating fields, if the magnetic skin depth is short compared to the depth of the liquid, the surface-coupled model exemplified in this section again applies. In the configuration of Fig. 9.4.2c, a traveling wave is used to induce circulations in a liquid metal. Such a pump is useful in handling liquid metals in open conduits, perhaps in metallurgical processing.

The MQS system of Fig. 9.4.2d is in a way the counterpart of the "Taylor pump." In the air gap, the alternating magnetic field has essentially the same temporal phase throughout the air gap. However, because this field is nonuniform in the y direction, a time-average shearing surface force density is induced in the skin region of the liquid metal, with attendant circulations that can be modeled by the fully developed flow.

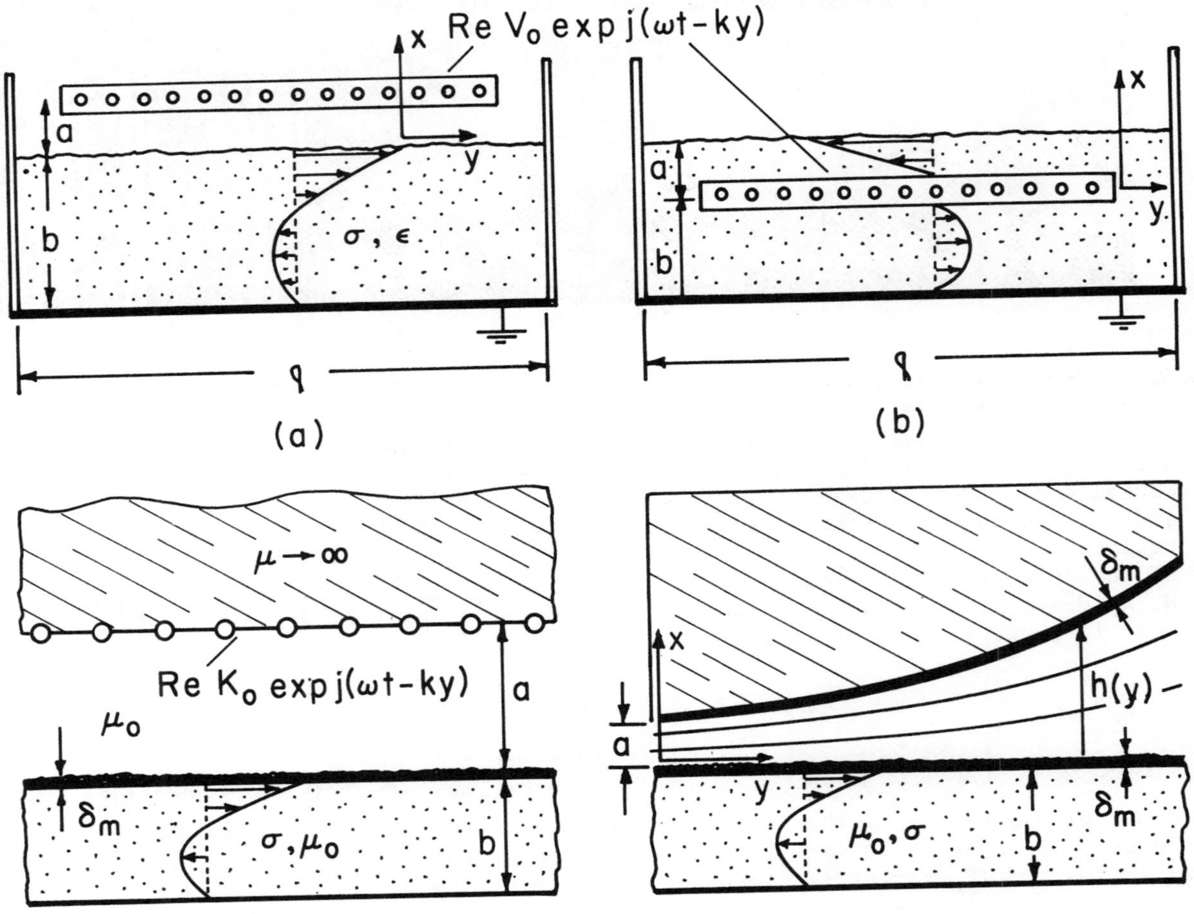

Fig. 9.4.2. (a) EQS traveling-wave-induced convection model for experiment shown in Fig. 5.14.4a. (b) Model for experiment of Fig. 5.14.4b. (c) MQS MHD surface pump. Traveling wave of current imposed above air gap induces currents in liquid metal with magnetic skin depth much less than b. (d) Surface current in skin layer has the same temporal phase as a function of y, but because the field is nonuniform there is a time-average surface force density driving liquid circulations.

9.5 Fully Developed Magnetic Induction Pumping

The magnetic induction motor discussed in Sec. 6.6 is readily adapted to pumping conducting liquids. The arrangement of driving current, magnetic material and fluid is typically like that of Fig. 9.5.1 in a class of pumps that has the advantage of not requiring mechanical moving parts or electrical contact with the liquid. The conduit can be insulating. A natural application is to the pumping of liquid metals such as sodium, which can react violently when exposed.[1]

In the liquid metal pump, each x-y layer of the fluid is analogous to the conducting sheet of Sec. 6.4. Induced currents result in both longitudinal and transverse traveling-wave forces. At a given position, these forces are composed of time-average and second-harmonic parts. With the traveling-wave frequency in the frame of the moving fluid ($\omega - kv_y$) sufficiently high, the liquid (limited as it is by its inertia and viscosity) usually can react only to the time-average part.

First, observe that the components of the magnetic stress have time averages that are independent of y. For example $\langle T_{yx}\rangle_t = \frac{1}{2}\mathrm{Re}\mu_o\hat{H}_x(x)\hat{H}_y^*(x)$. Hence, the time-average magnetic force density is simply

$$\langle\vec{F}\rangle_t = \frac{d}{dx}\langle T_{xx}\rangle_t\vec{i}_x + \frac{d}{dx}\left(\frac{1}{2}\mu_o\mathrm{Re}\hat{H}_x\hat{H}_y^*\right)\vec{i}_y \qquad (1)$$

and takes the form assumed in Sec. 9.3, where

$$\mathcal{E} \rightarrow -\langle T_{xx}\rangle_t \quad \text{and} \quad T_{yx} \rightarrow \frac{1}{2}\mu_o\mathrm{Re}\hat{H}_x\hat{H}_y^*$$

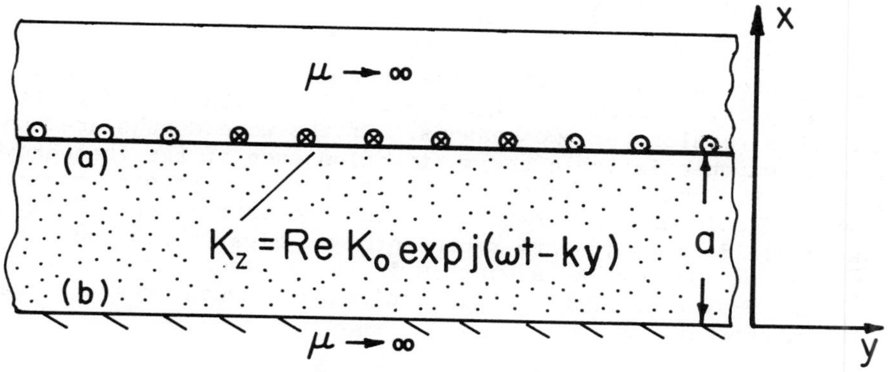

Fig. 9.5.1

Planar magnetohydrodynamic induction pump.

At the walls, where x = 0 and x = a, the no-slip condition requires that $v_y \equiv v$ vanish, and hence with the identification of $\Delta \rightarrow a$, the velocity profile of Eq. (a) from Table 9.3.1 becomes

$$\underline{v} = \frac{\partial \underline{p}'}{\partial \underline{y}}(\underline{x}^2 - \underline{x}) + V(\underline{x},\underline{\omega},\underline{k}) \qquad (2)$$

where

$$V(\underline{x},\underline{\omega},\underline{k}) \equiv -\int_o^{\underline{x}} \mathrm{Re}\hat{\underline{H}}_x\hat{\underline{H}}_y^* d\underline{x} + \underline{x}\int_o^1 \mathrm{Re}\hat{\underline{H}}_x\hat{\underline{H}}_y^* d\underline{x}$$

and variables are normalized such that

$$\vec{H} = \vec{\underline{H}}K_o \qquad k = \underline{k}/a$$

$$p = \underline{p}\mu_o K_o^2 \qquad \omega = \underline{\omega}/\mu\sigma a^2$$

$$v = (a\mu_o K_o^2/2\eta)\underline{v} \qquad (x,y) = (a\underline{x}, a\underline{y})$$

Implicit is the assumption that the magnetic field distribution is not altered by the liquid motion. In fact, to some extent, it must be. But, if the fluid velocity at all points is small compared to the wave velocity, ω/k, then the fields are not dependent on the motion. This is suggested by the example of the moving sheet in Sec. 6.4, where the sheet represents a liquid layer. The liquid velocity enters in determining the time average force of Eq. 6.4.11 through S_m, as expressed by Eq. 6.4.7. Currents responsible for the force are induced because a magnetic diffusion time $\tau_m = \mu_o\sigma\ell^2$ is on the order of ω^{-1}, whereas convective effects on this induction are ignorable because the magnetic Reynolds number based on

1. For extensive treatment, see E. S. Pierson and W. D. Jackson, "The MHD Induction Machine," Tech. Rep. AFAPL-TR-65-107, Air Force Aero Propulsion Laboratory, Research and Technology Division, Air Force Systems Command, Wright-Patterson Air Force Base, Dayton, Ohio, 1966.

The convection velocity, $R_m = \mu_o \sigma v \ell$, is still small compared to unity. Note that of the two possible lengths, a and $2\pi/k$, the latter is used here to represent rates of change in the y direction.

In this limit of small R_m, the transfer relations (b) from Table 6.5.1 with $U \to 0$ and $\Delta \to a$ give the magnetic field distribution. Identification of $\alpha \to a$ and $\beta \to b$ and use of the boundary conditions $\hat{H}_y^a = \hat{K}_o, \hat{H}_y^b = 0$ specializes the relations to

$$\begin{bmatrix} \hat{A}^a \\ \hat{A}^b \end{bmatrix} = -\frac{\mu_o \hat{K}_o}{\gamma} \begin{bmatrix} \coth \gamma a \\ \frac{1}{\sinh \gamma a} \end{bmatrix} \tag{3}$$

where $\gamma \equiv \sqrt{k^2 + j\omega\mu\sigma}$. Substituted into Eqs. 6.5.6, these coefficients give the distribution of $\hat{A}$ and hence of $\hat{H}_x$ and $\hat{H}_y$. In normalized form

$$\hat{H}_x = \frac{jk}{\gamma}\left[\coth \gamma \frac{\sinh \gamma x}{\sinh \gamma} - \frac{\sinh \gamma(x-1)}{\sinh^2 \gamma}\right] \tag{4}$$

$$\hat{H}_y = \coth \gamma \frac{\cosh \gamma x}{\sinh \gamma} - \frac{\cosh \gamma(x-1)}{\sinh^2 \gamma}; \quad \gamma \equiv \sqrt{k^2 + j\omega} \tag{5}$$

Substituted into Eq. 2, these expressions determine the velocity profile as a function of the pressure gradient and the driving current.

Although now reduced to a straightforward integration, the explicit evaluation of the x dependence is conveniently done numerically. The profiles shown in Fig. 9.5.2 reflect the tendency for the velocity to peak near the driving windings. This results for two reasons. If the wavelengths are short compared to the channel width, the fields decay exponentially in the x direction even if the frequency is sufficiently low to give no induced currents. But even more, as the frequency is raised, the induced currents shield the magnetic field out of the lower fluid regions to further enhance this decay of the force density. The details of the magnetic field diffusion are represented in Figs. 6.6.3 and 6.6.4. Note that δ'/a as defined there is $\sqrt{2/\underline{\omega}}$. For a pump having a width w in the z direction, the volume rate of flow, Q_v, is the integral of v over the x-y cross section. The relation between pressure gradient and volume rate of flow thus follows by integrating Eq. 2,

$$\underline{Q}_v = \int_0^1 \underline{v}dx = -\frac{1}{6}\frac{\partial \underline{p}'}{\partial \underline{y}} + Q(\underline{\omega},\underline{k}); \quad Q \equiv \int_0^1 V(\underline{x},\underline{\omega},\underline{k})d\underline{x} \tag{6}$$

where

$$Q_v = \underline{Q}_v w a^2 \mu_o K_o^2 / 2\eta$$

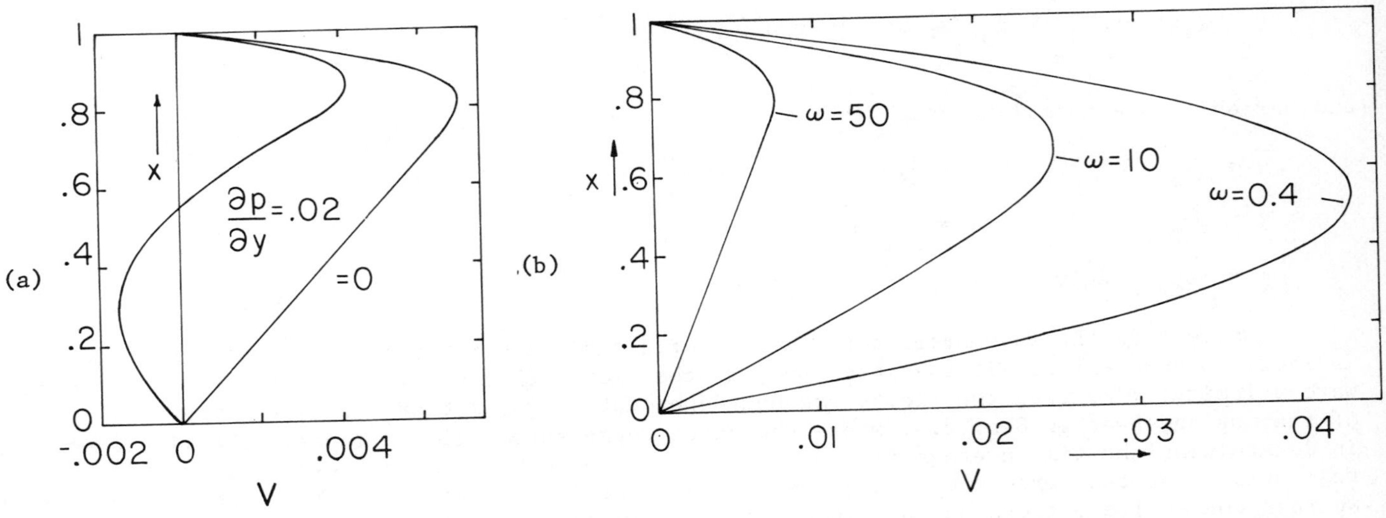

Fig. 9.5.2. (a) Normalized velocity profile with pressure gradient as parameter. $\underline{\omega} = 50$ $(\delta'/a = 0.2)$ and $\underline{k} = 1$. (b) Normalized velocity profile with zero pressure gradient showing effect of frequency $\underline{\omega}$. For $\underline{\omega} = 50$, the force density is confined to upper 20% of layer so that profile in the region below is the linear one typical of Couette flow.

The pump characteristic therefore has the general form illustrated by Fig. 9.5.3. The intercepts are functions of (ω,k). The dependence on $\underline{\omega}$ is typically that of the induction machine developed in Secs. 6.4 and 6.6 and is illustrated in Fig. 9.5.4.

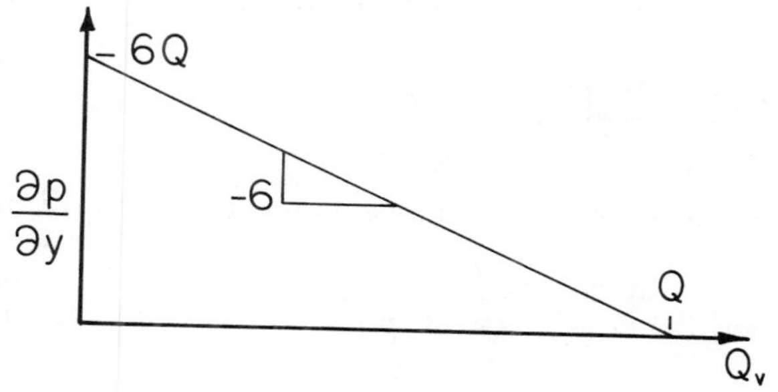

Fig. 9.5.3. Normalized pressure gradient as a function of normalized volume rate of flow.

To achieve pumping over the entire cross section, the design calls for making the wavelength and skin depth large compared to the channel depth a. Mathematically, this is the limit $\gamma a \ll 1$, and the limiting forms taken by Eqs. 4 and 5 show that $\hat{H}_x$ is then uniform over the cross section, while $\hat{H}_y$ decays in a linear fashion from the current sheet to the magnetic wall below:

$$\hat{H}_x \to j\frac{k}{\gamma^2}; \quad \hat{H}_y = x \tag{6}$$

The integrations in Eqs. 2 and 6 are now carried out to give

$$V(x,\omega,k) = \frac{\omega k}{2(k^4 + \omega^2)}(x - x^2) \tag{7}$$

$$Q(\omega,k) = \frac{\omega k}{12(k^4 + \omega^2)} \tag{8}$$

The approximate magnetic force density implied by the magnetic field of Eq. 6 is uniform over the channel cross section. This is why the approximate long-wavelength long-skin-depth velocity profile has the same parabolic x dependence as if the flow were driven by a negative pressure gradient.

In practice, "end effects" are likely to be important. Such effects result from the spatial transient needed to establish the spatial sinusoidal steady state described in this section. In the imposed force density approximation used here, this transient is akin to those illustrated in Sec. 9.7, superimposed on a magnetic diffusion spatial transient.

Windings that could be used to drive the system are illustrated in Sec. 4.7. The electrical terminal relations are then found following the same approach taken in Sec. 6.4.

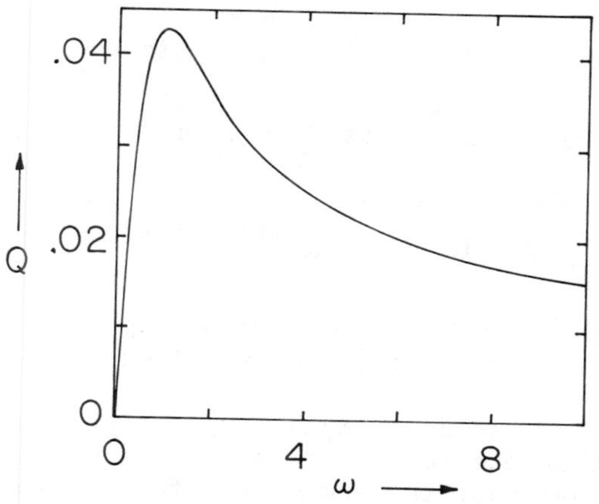

Fig. 9.5.4. Dependence of normalized Q (Fig. 9.5.3 and Eq. 6) on normalized frequency with $\underline{k} = 1$.

9.6 Temporal Flow Development with Imposed Surface and Volume Force Densities

Under what conditions is a flow fully developed? The answer to this question can either be one of "when?" or "where?" If the configuration is reentrant, as for example in the Couette geometry of Table 9.3.1, and volume and surface force densities which are uniformly distributed with respect to the longitudinal directions are suddenly turned on, the question is one of, when? On the other hand, if a steady state prevails in a system having a finite length and the fluid enters with some velocity profile other than the fully developed one, the question is one of, where? In either case, the development is governed by viscous diffusion.

In this section, the temporal transient is considered. The spatial transient is taken up in Sec. 9.7.

Turn-On Transient of Reentrant Flows: Suppose that the plane flow considered in Sec. 9.3 (first of the configurations in Table 9.3.1) is reentrant, so that there is no longitudinal pressure gradient, $\partial p'/\partial y = 0$. Boundaries (or surface stresses) and volume force densities are applied when $t = 0$. How long before the fully developed flow described by Eq. (a) of Table 9.3.1 pertains?

The incompressible mass conservation and momentum force equations can be satisfied by a time-varying plane flow: $\vec{v} = v(x,t)\vec{i}_y$. The longitudinal force equation is then

$$\rho\frac{\partial v}{\partial t} = F_y(x) + \eta\frac{\partial^2 v}{\partial x^2} \tag{1}$$

for $t > 0$. Fully developed flow can be regarded as a particular solution, $v_{fd}(x)$. This solution both balances the force distribution in the volume and satisfies the boundary conditions $v(\Delta) = v^\alpha$ and $v(0) = v^\beta$. With the understanding that the total solution is $v = v_{fd} + v_h(x,t)$, it follows that

$$\rho \frac{\partial v_h}{\partial t} = \eta \frac{\partial^2 v_h}{\partial x^2} \tag{2}$$

where v_h satisfies the boundary conditions $v = 0$ at $x = \Delta$ and $x = 0$.

In the terminology introduced in Sec. 5.15, the required solutions to Eq. 2 are the temporal modes $\mathrm{Re}\,v_n(x)\exp s_n t$ (with no longitudinal dependence and hence with $k_y = 0$). Substitution converts Eq. 2 to

$$\frac{d^2 v_n}{dx^2} + \gamma_n^2 v_n = 0; \quad \gamma_n^2 = -\frac{\rho s_n}{\eta} \tag{3}$$

Solutions to Eq. 3 that satisfy the homogeneous boundary conditions are

$$v_n = V_n \sin \gamma_n x \tag{4}$$

where because $\sin \gamma_n \Delta = 0$,

$$\gamma_n = \frac{n\pi}{\Delta}; \quad s_n = -\frac{\eta}{\rho}(\gamma_n)^2$$

Thus, the velocity distribution evolves at a rate determined by the sum of modes, each having a time constant $\tau_n = \rho(\Delta/n\pi)^2/\eta$, the viscous diffusion time based on a length $\Delta/n\pi$. The total solution is in general

$$v = v_{fd}(x) + \sum_{n=1}^{\infty} V_n \sin\left(\frac{n\pi}{\Delta}x\right)e^{s_n t} \tag{5}$$

The coefficients V_n are determined by the initial conditions on the flow, $v(x,0) = 0$,

$$-v_{fd} = \sum_{n=1}^{\infty} V_n \sin\left(\frac{n\pi}{\Delta}x\right) \tag{6}$$

The temporal modes are orthogonal, in this case simply Fourier modes, so the coefficients are determined from Eq. 5, much as explained in Sec. 2.15.

As an example, suppose that when $t = 0$, the upper boundary is set into motion with velocity U, that the lower one is fixed and that there is no volume force density. Then, $v_{fd} \doteq (x/\Delta)U$ and it follows that the sum of the fully developed and homogeneous solutions gives

$$\frac{v}{U} = \frac{x}{\Delta} + \sum_{n=1}^{\infty} \frac{2(-1)^n}{n\pi} \sin\left(\frac{n\pi}{\Delta}x\right)e^{s_n t} \tag{7}$$

This developing flow is shown in Fig. 9.6.1.

The boundary conditions satisfied by the temporal modes are determined by the way in which the transverse drive is applied. Suppose that the upper boundary is a "free" surface to which an electric stress is suddenly applied when $t = 0$. An example would be the electrically driven flow of Fig. 5.14.4a, but closed on itself in the longitudinal direction. (It is assumed that the traveling-wave velocity is much greater than that of the interface, and that, in terms of variables used in that section, the flow responds to the time-average surface force density $T_o \equiv \langle T_z \rangle_z$ which is suddenly turned on when $t = 0$.)

The fully developed flow is again simply $(x/\Delta)U$. However, the surface velocity is in general a function of time, $U = U(t)$, and for the fully developed flow is determined by the condition that the interfacial viscous shear stress balance the applied surface force density: $\eta\partial v/\partial x(x=\Delta) = T_o u_{-1}(t)$. Because the driving condition is balanced by the fully developed part, the homogeneous solution to Eq. 3 must now satisfy homogeneous boundary conditions: $\partial v_h/\partial x(x=\Delta) = 0$ and $v_h(0) = 0$. Thus, the temporal modes are determined. The resulting solution is

$$v/(\Delta T_o/\eta) = \frac{x}{\Delta} - \sum_{n=0}^{\infty} \frac{2(-1)^n}{[\pi(n+\frac{1}{2})]^2} \sin\left[\left(n+\frac{1}{2}\right)\frac{x}{\Delta}\right]e^{s_n t} \tag{8}$$

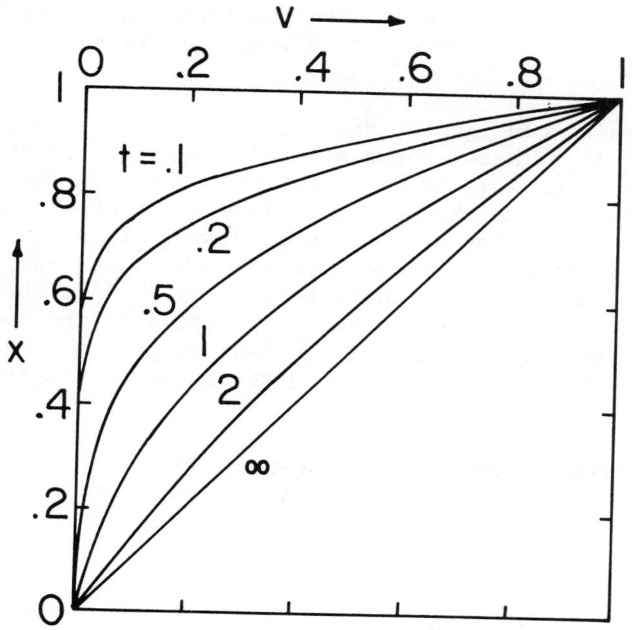

Fig. 9.6.1

Temporal transient leading to fully developed plane Couette flow as velocity in plane $x=\Delta$ is suddenly constrained to be U. v,x and t respectively, normalized to U, Δ, and $\rho\Delta^2/\eta\pi^2$.

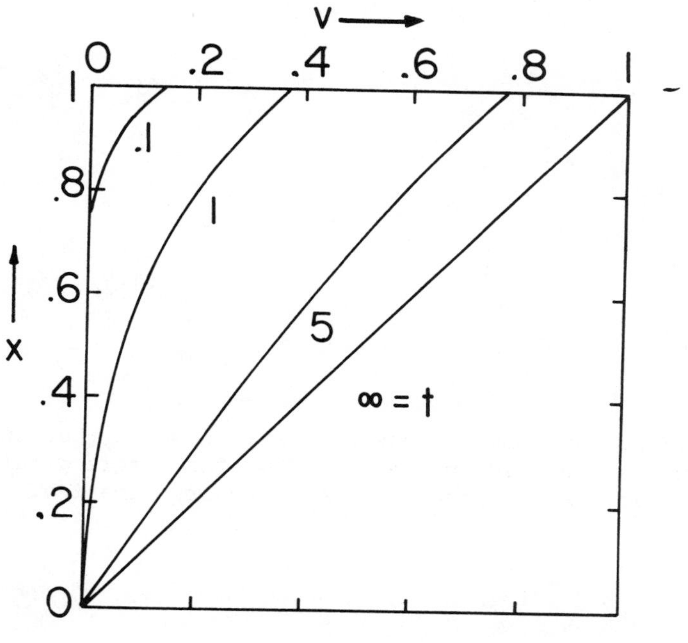

Fig. 9.6.2

Temporal transient leading to fully developed plane Couette flow initiated by application of constant surface force density, T_o, at free upper interface. v,x and t respectively, normalized to $\Delta T_o/\eta$, Δ, and $\rho\Delta^2/\eta\pi^2$.

where

$$s_n = -\frac{\eta\pi^2}{\rho\Delta^2}\left(n+\frac{1}{2}\right)^2$$

This transient, shown in Fig. 9.6.2, shows how both the interface and the fluid beneath approach the fully developed plane Couette flow.

9.7 Viscous Diffusion Boundary Layers

It is clear from the temporal viscous diffusion transients considered in Sec. 9.6 that in the early stages of development, motions imparted by a boundary are confined to the adjacent fluid. Examples are shown by Figs. 9.6.1 and 9.6.2. For times short compared to the viscous diffusion time based on the channel width Δ, the second boundary is of no influence and the diffusion phenomenon effectively picks out its own natural length. For the temporal transients considered, this length increases with time until the diffusion reaches another boundary.

With increasing time, the viscous process remains confined to the neighborhood of a boundary in two important situations. One is encountered in Sec. 7.19. There, boundary excitations are in the sinusoidal steady state and motions are confined to within a viscous skin depth of the boundary. In the second situation, there is a mean flow involved having a transport time through the volume of interest that is short compared to the viscous diffusion time based on a typical dimension of that volume. Thus, the distance into the flow that boundary effects can diffuse is limited to a viscous skin depth (based on the reciprocal transport time). Thus, there are two spatial scales. One, characterized by ℓ, describes variations in the longitudinal (dominant flow) direction y. The other scale is typified by the boundary layer thickness, which represents variations in the transverse direction. What makes the subject of boundary layers require some foresight is that this characteristic transverse length, d, is at the outset unknown.

The approach now taken is akin to that introduced in Sec. 4.12, a space-rate-parameter expansion is made in the ratio of lengths, $\gamma \equiv (d/\ell)^2$. The Navier-Stokes's equations (in two dimensions) and the continuity equation are written in normalized form as

$$\gamma\left(\frac{\partial v_x}{\partial t} + v_x\frac{\partial v_x}{\partial x} + v_y\frac{\partial v_x}{\partial y}\right) + \frac{\partial p}{\partial x} = \frac{\eta}{\rho\ell U}\left(\frac{\partial^2 v_x}{\partial x^2} + \gamma\frac{\partial^2 v_x}{\partial y^2}\right) + F_x \tag{1}$$

$$\frac{\partial v_y}{\partial t} + v_x\frac{\partial v_y}{\partial x} + v_y\frac{\partial v_y}{\partial y} + \frac{\partial p}{\partial y} = \frac{1}{\gamma}\left(\frac{\eta}{\rho\ell U}\right)\left(\frac{\partial^2 v_y}{\partial x^2} + \gamma\frac{\partial^2 v_y}{\partial y^2}\right) + F_y \tag{2}$$

$$\frac{\partial v_x}{\partial x} + \frac{\partial v_y}{\partial y} = 0 \tag{3}$$

where

$$\begin{aligned} x &= \underline{x}d & v_x &= \underline{v}_x U\frac{d}{\ell} & F_x &= \underline{F}_x\frac{\rho U^2}{d} \\ y &= \underline{y}\ell & v_y &= \underline{v}_y U & F_y &= \underline{F}_y\frac{\rho U^2}{\ell} \\ t &= \underline{t}(\ell/U) & p &= \underline{p}\rho U^2 & & \end{aligned} \tag{4}$$

Formally, an expansion is now made of the normalized variables in powers of γ. However, not only is this space-rate parameter small, so also is the reciprocal Reynolds number based on the longitudinal length: $\eta/\rho\ell U$. That is, the viscous diffusion time $\rho\ell^2/\eta$ is long compared to the transport time ℓ/U. Thus, to zero order, Eq. 1 is simply

$$\frac{\partial p}{\partial x} = F_x \tag{5}$$

This means that the transverse pressure distribution is determined without regard for the inertial and viscous force densities. The flow outside the boundary layer, which is essentially inviscid, determines the exterior pressure distribution. Because F_x is imposed, from Eq. 13 it is deduced that the pressure distribution, p(y,t), within the layer is therefore a <u>given</u> function. In ordinary fluid mechanics, p(y,t is usually determined by solving for the inviscid fluid motion in the volume subject to boundary conditions appropriate to an inviscid model.

In Eq. 2, it is clear that, to zero order in γ, the second term on the right can be dropped compared to the first. But, because both γ and $\eta/\rho\ell U$ are small, the parameter $(\eta/\rho\ell U)/\gamma$ is of the order of unity, so that the first term on the right is retained. The continuity equation contains no parameters. Hence, the boundary layer equations are Eq. 5 and

$$\frac{\partial v_y}{\partial t} + v_x\frac{\partial v_y}{\partial x} + v_y\frac{\partial v_y}{\partial y} + \frac{1}{\rho}\frac{\partial p}{\partial y} = \frac{\eta}{\rho}\frac{\partial^2 v_y}{\partial x^2} + \frac{1}{\rho}F_y \tag{6}$$

$$\frac{\partial v_x}{\partial x} + \frac{\partial v_y}{\partial y} = 0 \qquad (7)$$

In these last two expressions, p is to be regarded as a predetermined function. If $\vec{F}$ is known, they comprise two equations for determining v_y and v_x.

Linear Boundary Layer: Suppose that liquid fills the half-space $x < 0$, has a "free surface" in the plane $x = 0$ and, in the absence of electrical excitations, undergoes a uniform translation to the right with velocity U. An electric or magnetic structure, sketched in Fig. 9.7.1, is used to impose a surface force density that is turned on when $t = 0$ and extends from $y = 0$ to the right. There is no bulk imposed force density. What is the perturbation in velocity or viscous stress distribution induced in the liquid by this excitation? Effects of the gas above the liquid will be ignored.

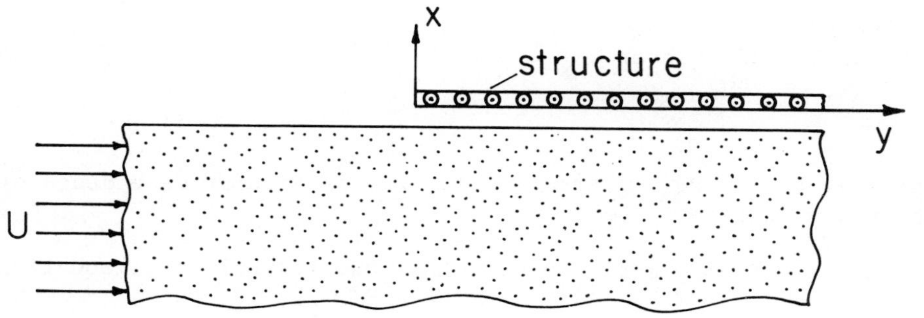

Fig. 9.7.1

Fluid moving uniformly to right encounters imposed surface force density where $y > 0$. Structure might induce electric or magnetic surface force density, as suggested in Secs. 5.14 and 6.8, respectively.

The imposed pressure is zero. The velocity can be written as $\vec{v} = v_x'\vec{i}_x + (U + v_y')\vec{i}_y$, where primes indicate perturbations. Thus, for small amplitudes, Eq. 14 reduces to a linear expression in v_y alone:

$$\left(\frac{\partial}{\partial t} + U\frac{\partial}{\partial y}\right)v_y' = \frac{\eta}{\rho}\frac{\partial^2 v_y'}{\partial x^2} \qquad (8)$$

and Eq. 15 determines v_x' once v_y' is known.

The boundary condition at $x = 0$ is that $\eta\partial v_y/\partial x \equiv S_{yx} = T_o u_{-1}(t)u_{-1}(y)$, so it is convenient to take the derivative of Eq. 8 and introduce the stress as the dependent variable:

$$\frac{\partial S_{yx}}{\partial t'} = \left(\frac{\partial}{\partial t} + U\frac{\partial}{\partial y}\right)S_{yx} = \frac{\eta}{\rho}\frac{\partial^2 S_{yx}}{\partial x^2} \qquad (9)$$

Here, t' is the rate of change with respect to time for an observer moving with the velocity U. This expression and the associated initial value and boundary value problem is the viscous analogue of the magnetic diffusion example treated in Sec. 6.9. Compare Eqs. 6.9.3 for example. Thus, the picture of temporal and finally spatial boundary layer evolution given there, for example by Fig. 6.9.3 with $H_y \rightarrow S_{yx}$, pertains equally well here.

The notion of an electric or magnetic surface force density implies that the coupling is confined to a region that is thin compared to that of the viscous boundary layer. In the case of a magnetic skin-effect coupling, the magnetic skin depth must be short compared to the viscous skin depth if the model suggested here is to be appropriate.

Stream-Function Form of Boundary Layer Equations: So that the continuity equation, Eq. 7, is automatically satisfied, it is convenient to introduce the stream function (from Table 2.18.1)

$$\vec{v} = \frac{\partial A_v}{\partial y}\vec{i}_x - \frac{\partial A_v}{\partial x}\vec{i}_y \qquad (10)$$

Substitution converts the longitudinal force equation, Eq. 6, to

$$\frac{\partial^2 A_v}{\partial t\partial x} + \frac{\partial A_v}{\partial y}\frac{\partial^2 A_v}{\partial x^2} - \frac{\partial A_v}{\partial x}\frac{\partial^2 A_v}{\partial x\partial y} - \frac{\eta}{\rho}\frac{\partial^3 A_v}{\partial x^3} = \frac{1}{\rho}\frac{\partial p}{\partial y} - \frac{1}{\rho}\vec{F}_y \qquad (11)$$

This expression is now applied to two examples in the remainder of this section.

Irrotational Force Density; Blasius Boundary Layer: Suppose that, as in Sec. 9.2, the imposed force density is irrotational; $\vec{F} = -\nabla\mathcal{E}$. Also, steady conditions prevail, so $\partial(\ )/\partial t = 0$. Thus, the boundary layer describes the first stages of steady-state flow development adjacent to a planar boundary. Perhaps the fluid makes an entrance with uniform velocity profile (at $y = 0$) to the region of interest, as shown in Fig. 9.7.2.

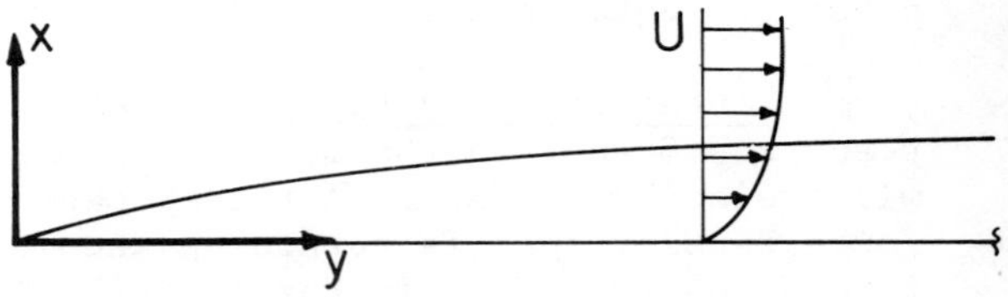

Fig. 9.7.2. Viscous diffusion boundary layer near entrance to channel.

Conditions in the core of the flow are determined from the inviscid laws. Given that the flow enters free of vorticity, Bernoulli's equation, Eq. 7.8.11, shows that

$$p + \mathcal{E} \equiv P = \Pi - \frac{1}{2}\rho v^2 \tag{12}$$

where Π is a constant.

The transverse component of the boundary layer equation, Eq. 9.7.5, requires that across the boundary layer

$$\frac{\partial}{\partial x}(P) = 0 \tag{13}$$

so it follows that within the boundary layer, $P = P(y)$. From Eq. 12, the particular dependence of $P(y)$ is determined by the bulk flow velocity distrubtion.

Because it follows from Eq. 10 that $p = P - \mathcal{E}$ and $F_y = -\nabla\mathcal{E}$, the longitudinal force equation, Eq. 11, reduces to

$$\frac{\partial A_v}{\partial y}\frac{\partial^2 A_v}{\partial x^2} - \frac{\partial A_v}{\partial x}\frac{\partial^2 A_v}{\partial x \partial y} - \frac{\eta}{\rho}\frac{\partial^3 A_v}{\partial x^3} = \frac{1}{\rho}\frac{dP}{dy} \tag{14}$$

Consider now a flow that enters at $y = 0$ in Fig. 9.7.2 with a uniform velocity profile $\vec{v} = U\vec{i}_y$. In the core, where the inviscid laws apply, the flow remains uniform with this same velocity. Thus, because v in Eq. 12 is independent of y, it follows that the pressure gradient on the right in Eq. 14 is zero. By introducing a similarity parameter, such as illustrated for magnetic diffusion in Sec. 6.9, it is then possible to reduce Eq. 14 to an ordinary differential equation.

By way of motivating the similarity parameter, observe that at a location y fluid has had the transit time $\tau = y/U$ for viscous diffusion. The rate of this process is typified by the viscous diffusion time, $\tau_v = \rho(x/2)^2/\eta$, based on half of the transverse position x of interest. Thus, it is plausible that viscous diffusion will have proceeded to the same degree at locations (x,y) preserving the ratio

$$\xi = \sqrt{\frac{\tau_v}{\tau}} = \frac{x}{2}\sqrt{\frac{\rho U}{\eta y}} \tag{15}$$

This similarity parameter is the analogue of the magnetic diffusion parameter given by Eq. 6.9.9.

With a function $f(\xi)$ defined such that $A_v = -f(\xi)\sqrt{\eta U y/\rho}$, Eq. 14 then reduces to the ordinary differential equation

$$\frac{d^3 f}{d\xi^3} + f\frac{d^2 f}{d\xi^2} = 0 \tag{16}$$

This third order expression is equivalent to the three first-order equations

$$\frac{d}{d\xi}\begin{bmatrix} f \\ g \\ h \end{bmatrix} = \begin{bmatrix} g \\ h \\ -fh \end{bmatrix} \tag{17}$$

Appropriate boundary conditions for flow over the flat plate are

$$v_x(0,y) = 0 \Rightarrow f(0) = 0, \quad v_y(0,y) = 0 \Rightarrow g(0) = 0, \quad v_y(\infty,y) \to U \Rightarrow g(\infty) \to 2 \tag{18}$$

Numerical integration of Eqs. 18 subject to these boundary conditions is conveniently carried out using standard library subroutines. (Used here was the IMSLIB routine DVERK.) To satisfy the conditic as $\xi \to \infty$, $h(0)$ is used as an iteration parameter and found to be 1.328.

The velocity profile $v_y = -(U/2)df/d\xi$ is shown in Fig. 9.7.3. Note that eighty-five percent of the free stream velocity is obtained at $\xi = 1.5$. (For demonstration of this boundary layer, as well as exposition of layers with free stream pressure gradients, their transition to turbulence and turbulent boundary layers, see Reference 5, Appendix C.)[1]

The viscous stress on the flat plate then follows as

$$S_{yx}(0,y) = \frac{1}{4}\, U\eta\sqrt{\frac{\rho U}{\eta y}}\, h(0) = 0.332U\eta\sqrt{\frac{\rho U}{\eta y}} \tag{19}$$

This y dependence is shown in Fig. 9.7.4a. Stream lines are illustrated by Fig. 9.7.4b. Even though the boundary layer approximation breaks down at the leading edge, the total viscous force, f_y, on a plate of width w and length L, found by integrating Eq. 19, is well behaved:

$$f_y = w\int_o^L S_{yx}(0,y)\,dy = 0.664wU\sqrt{\rho\eta UL} \tag{20}$$

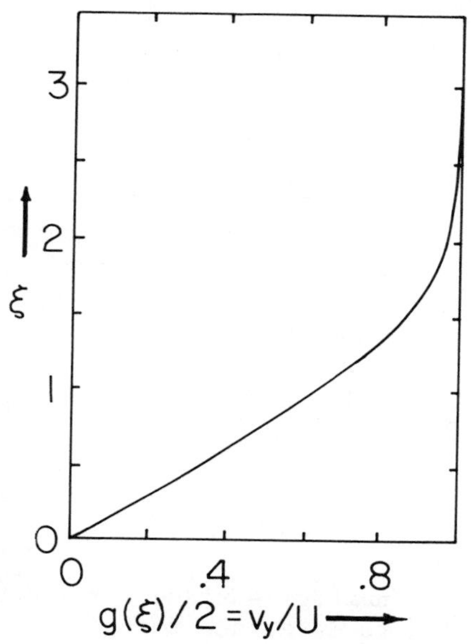

Fig. 9.7.3. Velocity profile of Blasius boundary layer as function of similarity parameter ξ, defined by Eq. 15.

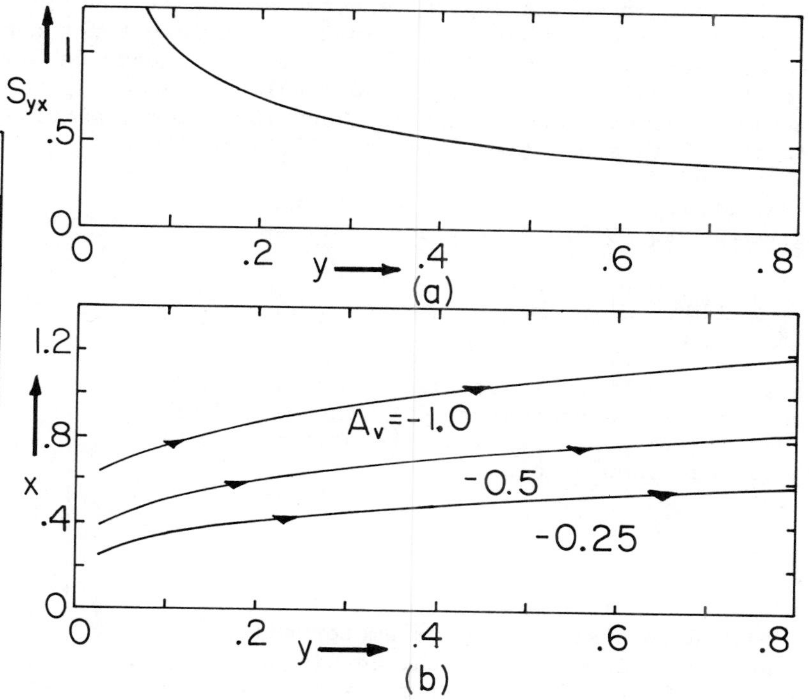

Fig. 9.7.4. (a) Distribution of viscous stress with longitudinal position $\underline{y} = y/L$. $\underline{S}_{yx} \equiv S_{yx}\, U\eta\sqrt{\rho U/\eta L}$. (b) Streamlines with $\underline{A}_v \equiv A_v/\sqrt{\eta UL/\rho}$, $\underline{x} \equiv (x/2)\sqrt{\rho U/L\eta}$.

What is there to be learned from this classical similarity solution that can serve as a guide in attacking the next example? First, observe that the similarity parameter can be thought of as an alternative coordinate. Lines of constant ξ form a family of parabolas in the x-y plane. One similarity coordinate is perpendicular to this family. In the x-y plane this similarity coordinate has the shape of an ellipse, as exemplified by Fig. 9.7.5. Not only does the boundary layer equation become an ordinary differential equation in this coordinate, but the boundary conditions are also a function of ξ alone.

---

1. Standard references on boundary layers are: A. Walz, Boundary Layers of Flow and Temperature, The MIT Press, Cambridge, Mass., 1969; and H. Schlichting, Boundary Layer Theory, McGraw-Hill Book Company, New York, 1960.

In Fig. 9.7.5, boundary conditions at A (the flat plate) and C (the free stream) are the same as at A' and C'. Otherwise, the solution found by integrating Eq. 17 along AC and A'C' would not give the same result at B as at B'. Because $f = f(\xi)$, the value of f must be the same at these two points.

A rational procedure for seeking a similarity parameter as well as the y dependence of A_v would begin by letting $\xi = c_1 x y^n$ and $A_v = c_2 f(\xi) y^m$, where n and m are to be determined. It follows from the assumed form for A_v that $v_y = -c_2 y^{m+n} df/d\xi$. If this velocity is to be the same constant, U, in the free stream regardless of the trajectory in the x-y plane, it follows that $m = -n$. To make the boundary layer equation reduce to an ordinary differential equation, it is then necessary that $m = -1/2$. Thus, the assumed forms for ξ and A_v are deduced.

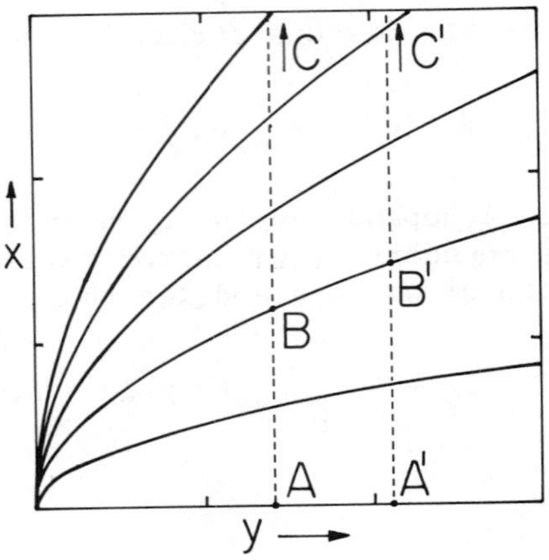

Fig. 9.7.5. Lines of constant similarity parameter, ξ, in (x-y) plane.

Stress-Constrained Boundary Layer: Typical of boundary layer development with an imposed surface force density is the system shown in Fig. 9.7.6. The electrode structure imposes a time-average surface force density T_o at the interface to the right of $y = 0$. Well below the interface, the fluid is essentially quiescent, and so the only motion is the result of the electromechanical drive. A typical electromechanical coupling is that of Fig. 5.14.4a, where a time-average surface force density acts on that part of the interface under the electrode structure. For the boundary layer model now developed to apply, the fluid should be doped Freon, which is about 100 times less viscous than the fluid shown (see Reference 12, Appendix C).

First observe that, in terms of the normalization given by Eq. 9.7.4, the viscous stress is

$$S_{yx} = \frac{\eta U}{d}\left(\gamma \frac{\partial \underline{v}_x}{\partial \underline{y}} + \frac{\partial \underline{v}_y}{\partial \underline{x}}\right) \tag{21}$$

Thus, in the boundary layer approximation (γ small), the viscous stress is approximated by the second of the two derivatives. In terms of the stream function, the stress then becomes

$$S_{yx} = -\eta \frac{\partial^2 A_v}{\partial x^2} \tag{22}$$

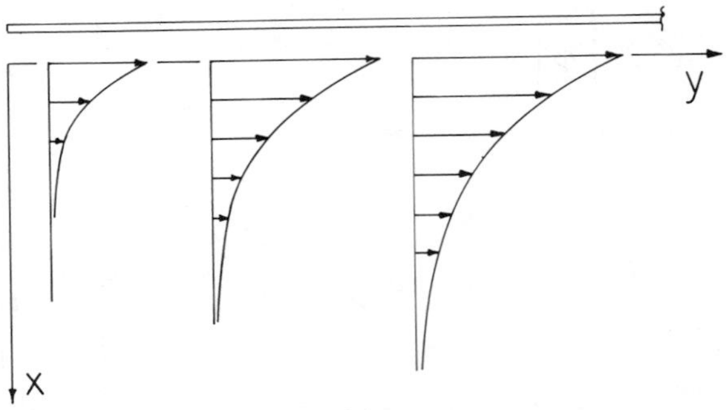

Fig. 9.7.6. A uniform surface force density is applied to interface for $0 < y$. Developing velocity profile is v_y.

What is desired is a similarity parameter and stream function defined so that the condition that S_{yx} be constant at $x = 0$ for all $y > 0$ is met by evaluating $f(\xi)$ at one value of ξ. Thus, S_{yx} must be a function of the similarity parameter alone. With m and n at the outset unknown and c_1 and c_2 normalizing parameters, trial forms are

$$\xi = c_1 x y^n; \quad A_v = c_2 f(\xi) y^m \tag{23}$$

It follows from Eq. 22 that if S_{yx} is to be a function of the similarity parameter alone, $m + 2n = 0$. Substitution into Eq. 14 then shows that $n = -1/3$. Thus, the boundary layer equation, Eq. 14 with $dP/dy = 0$, reduces to

$$\frac{d}{d\xi}\begin{bmatrix} f \\ g \\ h \end{bmatrix} = \begin{bmatrix} g \\ h \\ \frac{1}{3}g^2 - \frac{2}{3}fh \end{bmatrix} \tag{24}$$

where S_{ys} is normalized to T_o so that (this similarity solution was identified for the author by Mr. Richard M. Ehrlich while a graduate student)

$$\xi = \left(\frac{\rho T_o}{\eta^2}\right)^{1/3} x y^{-1/3} \tag{25}$$

$$A_v = -\left(\frac{\rho T_o}{\rho^2}\right)^{1/3} f(\xi) y^{2/3} \tag{26}$$

Boundary conditions consistent with having a constant surface force density T_o acting in the y direction, no vertical velocity at the interface and a stagnant-free stream are

$$v_x(0,y) = 0 \Rightarrow f(0) = 0, \; S_{yx}(0,y) = T_o \Rightarrow h(0) = -1, \; v_y(\infty,y) \to 0 \Rightarrow g(\infty) \to 0 \tag{27}$$

To match the boundary conditions as $\xi\to\infty$, g(0) is used as an iteration parameter which is adjusted to make $g\to 0$ as $\xi\to\infty$ with the other two conditions at $\xi=0$ satisfied. From this iteration it follows that g(0) = 1.296. The universal profiles $f(\xi)$ and $g(\xi)$ are shown in Fig. 9.7.7. The velocity profile, recovered by using the relation $v_y = (T_o^2/\rho\eta)^{1/3} g(\xi) y^{1/3}$, is as exemplified in Fig. 9.7.6. With increasing longitudinal position y, the interface has increasing velocity and the motion penetrates further into the bulk.

The velocity of the interface is simply

$$v_y = \left(\frac{T_o^2}{\rho\eta}\right)^{1/3} (1.296) y^{1/3} \tag{28}$$

and is shown in Fig. 9.7.8.

Streamlines help to emphasize that the fluid is being drawn into the boundary layer from below. These lines of constant A_v, given by Eq. 26, are illustrated in Fig. 9.7.9.

In retrospect, what is the physical origin of the difference between similarity parameters for the constant velocity and the imposed stress boundary layers? In fact ξ as defined by Eq. 25 is again the ratio of a time for viscous diffusion in the x direction to a transport time in the y direction. However, with the stress at the interface constrained, the transport velocity in fact varies as $y^{1/3}$. Based on a transport time consistent with this variation in velocity, it is again found that ξ is the square root of the ratio of the viscous diffusion time to the transport time.

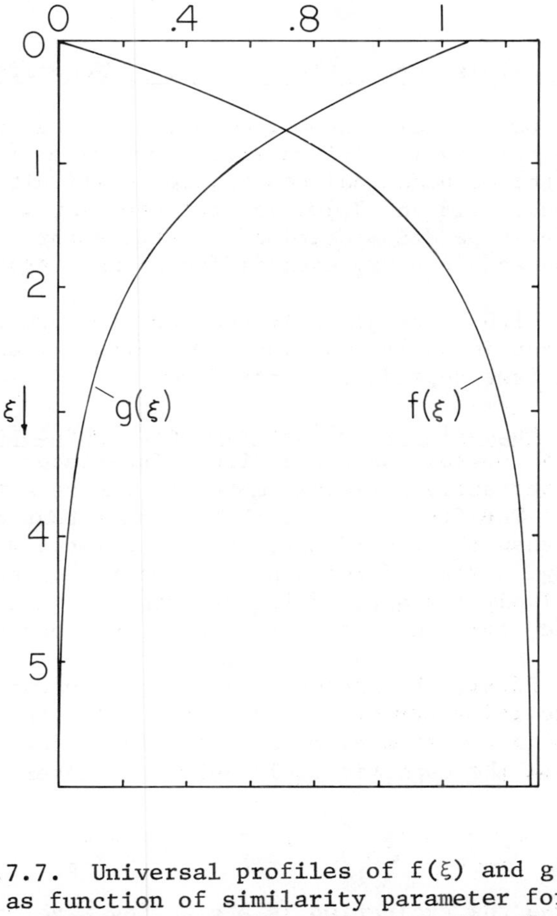

Fig. 9.7.7. Universal profiles of $f(\xi)$ and $g(\xi)$ as function of similarity parameter for boundary layer with uniform surface force density.

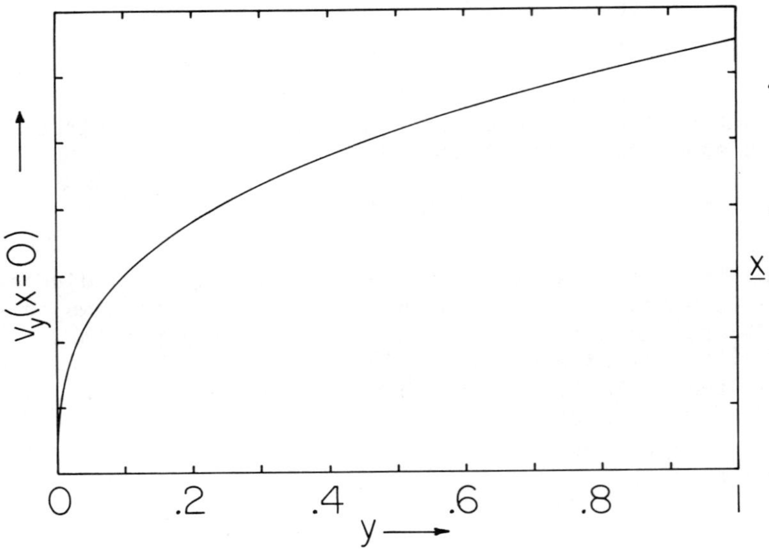

Fig. 9.7.8. Interfacial velocity of interface subject to uniform surface force density T_o. v_y and y normalized to $(T_o^2 L/\rho\eta)^{1/3}$ and L respectively.

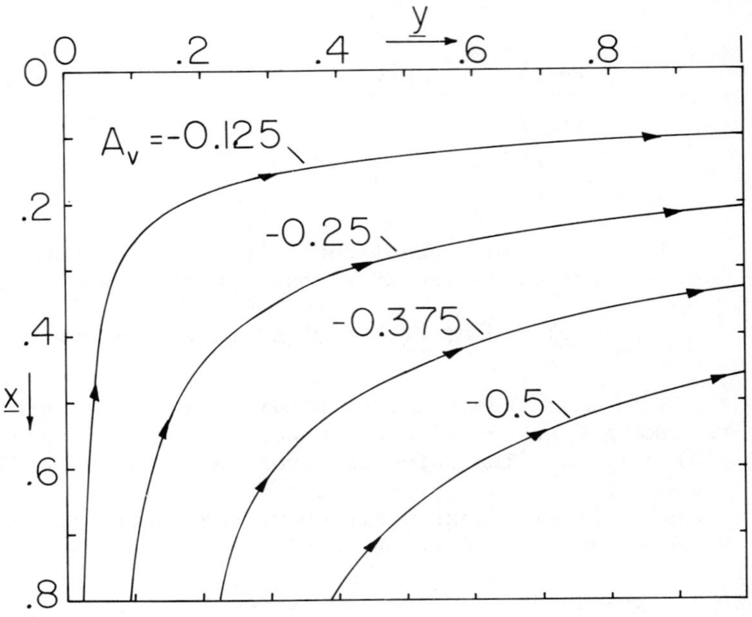

Fig. 9.7.9. Streamlines for stress-constrained boundary layer, as would result in configuration of Fig. 9.7.6. Variables are normalized.

9.8 Cellular Creep Flow Induced by Nonuniform Fields

Low-Reynolds-number models are often used to describe fluid circulations where, if it were not for a relatively high viscosity or for a relatively low velocity, the nonlinear acceleration term would make the mathematical description difficult. The main virtue of this approximation, which is discussed in Secs. 7.18 and 7.20, is that the flow is then described by linear differential equations. Thus, a Fourier-type decomposition of surface force densities results in a flow that can be represented by responses, in a way exemplified by many spatially periodic examples from previous chapters.

Illustrated in this section are such circulating imposed surface density flows. They are of interest in their own right, but also are useful in developing models where the surface force density is in fact dependent on the flow.

Magnetic Skin-Effect Induced Convection: The layer of liquid metal shown in Fig. 9.8.1 rests on a rigid bottom and has a "free" interface. Separated from the interface by an air gap, windings backed by a perfectly permeable material impose a tangential magnetic field that takes the form of a standing wave. The frequency ω is high enough that the magnetic skin depth δ in the liquid (Eq. 6.2.10) is much less than the liquid depth b. Associated with this skin region are both normal and shearing time-average surface force densities acting on the material within the layer (Eqs. 6.8.8 and 6.8.10). At relatively low applied fields, gravity maintains an essentially flat interface in spite of the normal surface force density. However, the shearing component establishes cellular motions, as now derived.

First, the imposed time-average magnetic shearing surface force density is computed. A region having thickness of the order of δ near the interface is pictured as subject to a force per unit area which is the time average of the force density $\vec{J} \times \mu_o \vec{H}$ integrated over the thickness of the layer. Because the magnetic field below the layer is zero, this shearing surface force density is

$$\langle T_y \rangle_t = \langle B_x^d H_y^d - B_x^e H_y^e \rangle_t = \langle B_x^d H_y^d \rangle_t \qquad (1)$$

Because the excitation is a standing wave, there is no net force on a section of the skin region one wavelength long in the y direction. Rather, there is a spatially periodic distribution of the time-average surface force density that has twice the periodicity of the imposed field.

To exploit the complex-amplitude transfer relations, observe that the excitation surface current can be written as the sum of two traveling waves:

$$K_z = \mathrm{Re}\hat{K}_o \cos \beta y e^{j\omega t} = \mathrm{Re}(\hat{K}_+ e^{-j\beta y} + \hat{K}_- e^{j\beta y})e^{j\omega t};\ \hat{K}_\pm \equiv \hat{K}_o/2 \quad (2)$$

The backward wave is gotten from the forward one by replacing $\beta \to -\beta$. Using this decomposition, Eq. 1 becomes

$$\langle T_y \rangle_t = \mathrm{Re}\,\frac{1}{2}(\hat{B}^d_{x+} e^{-j\beta y} + \hat{B}^d_{x-} e^{j\beta y})(\hat{H}^{d*}_{y+} e^{+j\beta y} + H^{d*}_{y-} e^{-j\beta y})$$

$$= \frac{1}{2}\,\mathrm{Re}[\hat{B}^d_{x+}\hat{H}^{d*}_{y+} + \hat{B}^d_{x-}\hat{H}^{d*}_{y-} + \hat{B}^d_{x-}\hat{H}^{d*}_{y+} e^{j2\beta y} + \hat{B}^d_{x+}\hat{H}^{d*}_{y-} e^{-j2\beta y}] \quad (3)$$

The normal and tangential fields above the interface are related by the skin-effect transfer relations, Eq. 6.8.5. The upper sign is appropriate because it is assumed that the peak interfacial velocity is still much less than ω/β. Thus, substitution for $\hat{B}^d_{x+}$ and $\hat{B}^d_{x-}$ shows that the space average part of Eq. 3 cancels out while the remaining terms give

$$\langle T_y \rangle_t = \frac{1}{4}\,\mathrm{Re}[(1 + j)\beta\mu_o \delta \hat{H}^d_{y-}\hat{H}^{d*}_{y+} e^{j2\beta y} - (1 + j)\beta\mu_o \delta \hat{H}^d_{y+}\hat{H}^{d*}_{y-} e^{-j2\beta y}] \quad (4)$$

Use can be made of the air-gap transfer relation to represent $\hat{H}^d_y$ in terms of the driving current $\hat{K}_o$. For simplicity it is assumed that $\beta a << 1$, so that the tangential field imposed by the surface current at (b) is essentially experienced at (c) as well:

$$\hat{H}^d_{y\pm} \simeq \hat{H}^c_{y\pm} = -\frac{\hat{K}_o}{2} \quad (5)$$

Thus, the time-average magnetic surface force density of Eq. 4 is simply

$$\langle T_y \rangle_t = \frac{\mu_o \beta \delta |\hat{K}_o|^2}{8} \sin 2\beta y \quad (6)$$

This is the distribution sketched at the interface of Fig. 9.8.1.

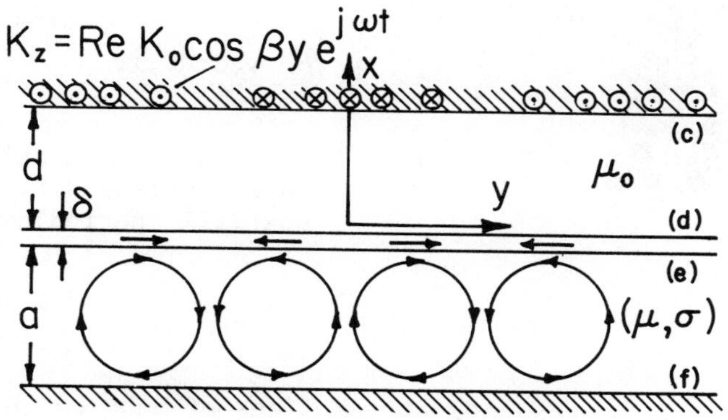

Fig. 9.8.1

Cross section of liquid metal layer set into cellular convection by spatially periodic a-c magnetic field inducing magnetic shear stress in skin layer at interface.

Now that the imposed magnetic surface force density has been determined, the flow response can be computed. In Sec. 7.20, this too is represented in terms of complex amplitudes, so the drive, Eq. 7, is again decomposed into traveling-wave parts:

$$\langle T_y \rangle_t = \mathrm{Re}(\tilde{T}_+ e^{-j2\beta y} + \tilde{T}_- e^{j2\beta y});\quad \tilde{T}_\pm \equiv \pm j\mu_o \beta\delta |K_o|^2/16 \quad (7)$$

That the interface, modeled here as having a thickness several times δ, be in shear stress equilibrium requires that

$$\tilde{S}^e_{yx\pm} = \tilde{T}_\pm \quad (8)$$

With the assumption that gravity holds the interface essentially flat in spite of the normal magnetic surface force density goes the boundary condition

$$\tilde{v}_x^e = 0 \tag{9}$$

At the rigid lower boundary, both velocity components are zero:

$$\tilde{v}_x^f = 0; \; \tilde{v}_y^f = 0 \tag{10}$$

The stress-velocity relations for the layer, Eq. 7.20.6, can now be used to represent the bulk fluid mechanics. In particular, Eq. 7.20.6c is evaluated using Eq. 8 on the left and Eqs. 9 and 10 on the right. Solved for the interfacial shear velocity, that expression becomes:

$$\tilde{v}_{y\pm}^e = \tilde{T}_\pm/\eta P_{33}; \; P_{33} \equiv \frac{[\frac{1}{4}\sinh(4\beta a) - \beta d](8\beta)}{[\sinh^2(2\beta a) - (2\beta a)^2]} \tag{11}$$

Note that P_{33} is an even function of k and hence the same number whether evaluated with $k\Delta = 2\beta a$ or $k\Delta = -2\beta a$.

The last three equations specify all of the velocity amplitudes, so that equations 7.20.4 and 7.20.5 can be used to reconstruct the x-y dependence of the flow field if that is required. At the interface, it follows from Eq. 11 that the y dependence is

$$\tilde{v}_y = \text{Re}\,\frac{1}{\eta P_{33}}(\tilde{T}_+ e^{-2j\beta y} + \tilde{T}_- e^{2j\beta y}) = \frac{\mu_o \beta \delta |\tilde{K}_o|^2}{8\eta P_{33}} \sin 2\beta y \tag{12}$$

Thus, the flow pattern is as sketched in Fig. 9.8.1.

The hydromagnetic convection modeled here is akin to that obtained in the quasi-one-dimensional configuration of Fig. 9.4.2d. There the field nonuniformity is obtained by using a shaped bus. Here, the windings are used to shape the field.

Charge-Monolayer Induced Convection: Surface charge induced convection, akin to that of Fig. 9.4.1, takes a cellular form in the EQS experiment of Figs. 9.8.2 and 9.8.3. In the model developed in Prob. 9.8.1, the flow is slow enough that it has negligible effect on the field.[1]

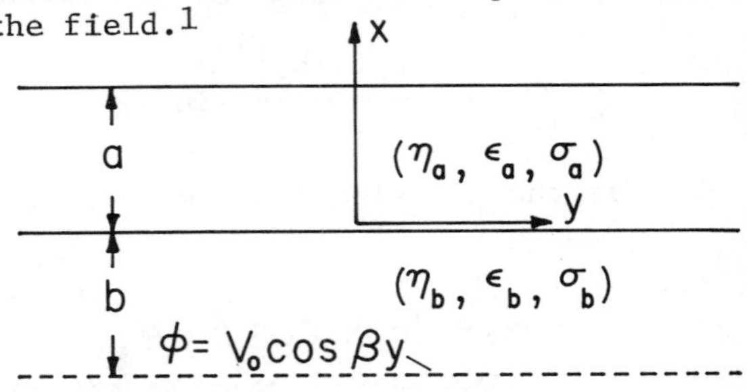

Fig. 9.8.2. Semi-insulating liquid layers stressed by static spatially periodic potential.

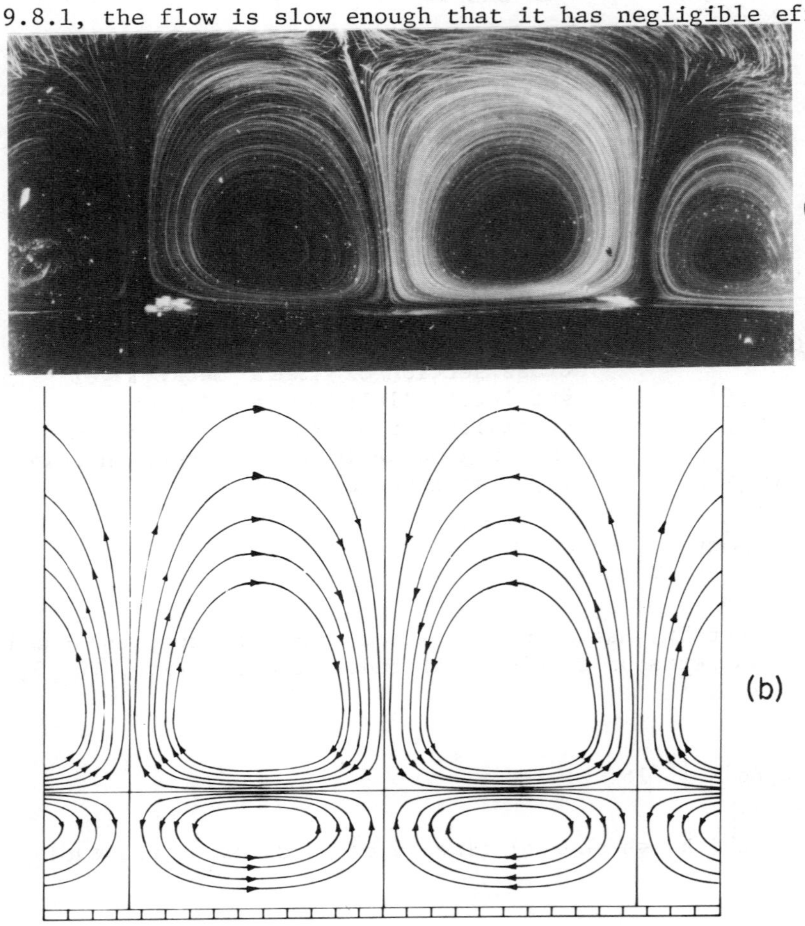

(a)

(b)

Fig. 9.8.3

(a) Streak lines of bubbles entrained in flow induced in configuration shown in Fig. 9.8.2. Upper fluid has properties $\varepsilon = 3.1\varepsilon_o$, $\sigma = 5\text{x}10^{-11}$ mhos/m while lower one has $\varepsilon = 6.9\varepsilon_o$, and $\sigma = 3\text{x}10^{-9}$. (b) Theoretical streamlines in limit where upper boundary is at infinity. In the experiment shown in (a), the cells in the upper region actually interact appreciably with the upper wall.

1. See C. V. Smith and J. R. Melcher, "Electrohydrodynamically Induced Spatially Periodic Cellular Stokes-Flow," Phys. Fluids 10, No. 11, 2315 (1967).

9.9 Magnetic Hartmann Type Approximation and Fully Developed Flows

Approximation: In typical laboratory situations involving the flow of electrolytes, liquid metals or even some plasmas through a magnetic field, magnetic diffusion times are short compared to times of interest. Nevertheless currents induced by the motion can make an appreciable contribution to the magnetic force density. The magnetic field associated with induced currents is then small compared to the imposed field.

The appropriate approximations to the magnetohydrodynamic equations are seen by writing those equations in normalized form:

$$\nabla \times \vec{E} = -\frac{\partial \vec{H}}{\partial t} \tag{1}$$

$$\nabla \times \vec{H} = \frac{\tau_m}{\tau}\sigma(\vec{E} \times \vec{v} \times \vec{H}) \tag{2}$$

$$\nabla \cdot \vec{H} = 0 \tag{3}$$

$$\frac{\partial \vec{v}}{\partial t} + \vec{v}\cdot\nabla\vec{v} + \nabla p = \frac{\tau_m}{\tau_{MI}}\frac{\tau}{\tau_{MI}}(\vec{E} + \vec{v} \times \vec{H}) \times \vec{H} + \frac{\tau}{\tau_V}\nabla^2\vec{v} \tag{4}$$

$$\nabla \cdot \vec{v} = 0 \tag{5}$$

It is assumed that the fluid is an ohmic conductor with characteristic conductivity σ_o and essentially the permeability of free space. The normalization used here, summarized by Eqs. 2.3.4b, takes the electric field as being of the order $\mu_o \ell \mathscr{H}/\tau$, as it would be if induced by the motion. The three characteristic times

$$\tau_V = \frac{\rho \ell^2}{\eta}; \quad \tau_m = \sigma_o \mu_o \ell^2; \quad \tau_{MI} = \sqrt{\rho \ell^2/\mu_o \mathscr{H}^2} \tag{6}$$

are the viscous diffusion time, magnetic diffusion time and the magneto-inertial time, respectively, familiar from Sec. 8.6.

In the imposed field approximation, these times have the order shown in Fig. 9.9.1, and times of interest, τ, are long compared to τ_m but arbitrary relative to τ_{MI} and τ_V. Of course, for steady flows the characteristic time is a transport time ℓ/u. Then, the approximation requires that the magnetic Reynolds number be small, but that the Reynolds number $\tau_V/\tau = \rho\ell u/\eta$ and the ratio of fluid velocity to Alfvén velocity $\tau_{MI}/\tau = u/\sqrt{\mu\mathscr{H}/\rho}$ be arbitrary.

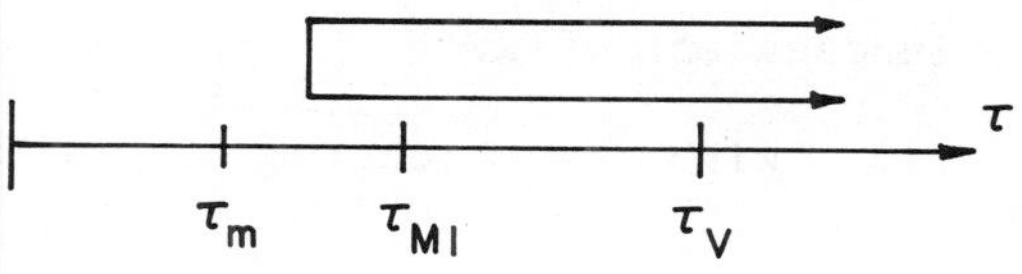

Fig. 9.9.1. Ordering of characteristic times in magnetic imposed field approximation.

Because τ_m/τ is small, the induced currents on the right in Eq. 2 are negligible. The magnetic field is imposed by means of currents in external windings. (More generally, there might be contributions from imposed volume currents which would arise from an electric field greater in order than $\mu_o \ell \mathscr{H}/\tau$, as presumed in the normalization of Eq. 2.)

Note that to zero order in τ_m/τ, the divergence of Eq. 2 still requires that the divergence of the induced current density vanish. Thus, Eqs. 2 and 3 reduce to expressions that determine $\vec{H}$,

$$\nabla \times \vec{H} = 0 \tag{7}$$

$$\nabla \cdot \mu_o \vec{H} = 0 \tag{8}$$

and with the understanding that $\vec{H}$ is the imposed field only,

$$\nabla \times \vec{E} = -\frac{\partial \vec{H}}{\partial t} \tag{9}$$

$$\nabla \cdot \vec{J} = 0; \quad \vec{J} = \sigma(\vec{E} + \vec{v} \times \mu_o \vec{H}) \tag{10}$$

$$\rho\left(\frac{\partial \vec{v}}{\partial t} + \vec{v}\cdot\nabla\vec{v}\right) + \nabla p = \vec{J} \times \mu_o \vec{H} + \eta\nabla^2\vec{v} \tag{11}$$

$$\nabla\cdot\vec{v} = 0 \tag{12}$$

The simple dimensional arguments given here presume that there is only one characteristic length ℓ. In general, more lengths and perhaps more than one characteristic time might be involved, and then approximations must hinge on a more detailed knowledge of the physical situation. The fully developed flow now considered involves one characteristic length, the transverse dimension d of the channel.

Fully Developed Flow: The magnetohydrodynamic pumping or generating configuration of Fig. 9.9.2 is an adaptation of the d-c kinematic (rotating machine) interaction from Sec. 4.10 and a refinement of the model introduced in Sec. 9.2. What is new is the internal redistribution of velocity caused by the magnetic force density.

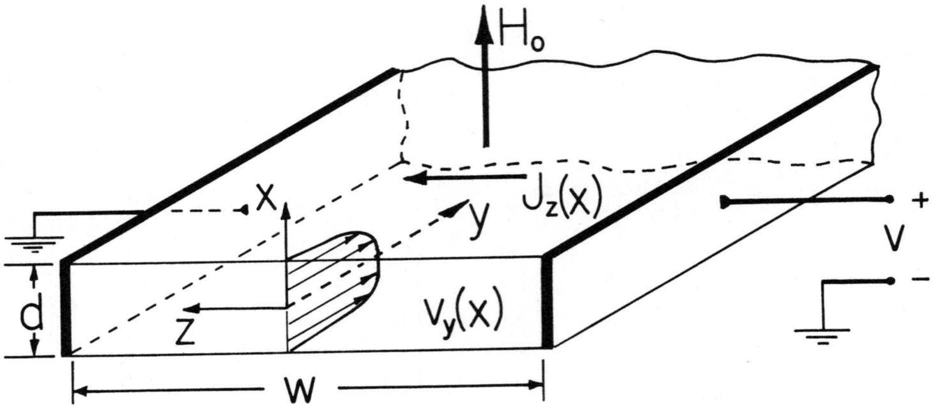

Fig. 9.9.2

Configuration for Hartmann flow. The aspect ratio $d/w >> 1$ so that the velocity is essentially a function only of x: thus so also is the current density J_z.

A conducting fluid moves in the y direction through the rectangular channel (Fig. 9.9.2) having a width w much greater than the depth d. Hence, the viscous shear from the upper and lower walls dominates that due to the side walls and the velocity profile can be considered a function of x alone.

The side walls are conducting electrodes that make electrical contact with the fluid and are connected to an external load or excitation. With the application of a transverse magnetic field H_o in the x direction, there is a magnetic force density $\vec{J}\times\vec{B}$ in the y direction tending to retard or accelerate the flow. Effects of gravity are absorbed in the pressure p. In this configuration, external currents generate the imposed $\vec{H}$ which is uniform and the constant H_o. Thus, Eqs. 7 and 8 are satisfied and the right-hand side of Eq. 9 is zero. Even if the flow is time-varying, the electric field is irrotational. If flow and field quantities are to be independent of z, it follows from the y component of Faraday's law that

$$E_z = E_z(t) = \frac{v}{w} \tag{13}$$

independent of x.

With the objective of finding a plane flow solution, $\vec{v} = v_y(x,t)\vec{i}_y$, note that the current density is

$$J_z = \sigma(E_z - \mu_o v_y H_o) \tag{14}$$

so that the y component of Eq. 11 reduces to

$$\rho\frac{\partial v_y}{\partial t} + \frac{\partial p}{\partial y} = \mu_o\sigma H_o E_z - (\mu_o H_o)^2\sigma v_y + \eta\frac{\partial^2 v_y}{\partial x^2} \tag{15}$$

Also, Eq. 12 is automatically satisfied. The x component of the force equation, Eq. 11, shows that p is independent of x. In fully developed flow, the longitudinal pressure gradient, $\partial p/\partial y$, is also independent of y.

Temporal flow development is considered in the next section. For the remainder of this section, consider the flow to be steady, so that Eq. 15 reduces to

$$\frac{d^2v_y}{dx^2} - \frac{\sigma(\mu_o H_o)^2 v_y}{\eta} v_y = -\frac{\mu_o\sigma H_o}{\eta}E_z + \frac{1}{\eta}\frac{\partial p}{\partial y} \tag{16}$$

where the terms on the right are independent of x.

Boundary conditions for the configuration of Fig. 9.10.2 require that the velocity vanish at $x = \pm d/2$. Solution of Eq. 16 then gives

$$v_y = \left[\frac{1}{H_m^2}\frac{d^2}{4\eta}\frac{\partial p}{\partial y} - \frac{E_z}{\mu_o H_o}\right]\left[\frac{\cosh(H_m 2x/d)}{\cosh H_m} - 1\right] \quad (17)$$

where the Hartmann[1] number, H_m, is defined as

$$H_m = \mu_o H_o \frac{d}{2}\sqrt{\frac{\sigma}{\eta}} \quad (18)$$

The velocity profile given by Eq. 17 is illustrated in Fig. 9.9.3. In the absence of a magnetic field, plane-Poiseuille flow prevails and the profile is a parabola. The tendency of the magnetic field to flatten the profile should have been expected. The current density has a direction determined by E_z', the term in brackets in Eq. 14. Wherever the velocity is so great that the "speed" field $\mu_o H_o v_y$ exceeds E_z, the force density tends to retard the motion. Thus, there is a tendency for the fluid bulk to suffer a rigid-body motion, with the strain rate confined to fully developed boundary layers. It follows from Eq. 16 that this "Hartmann layer" has an exponential profile with a thickness $\delta = d/2H_m$.

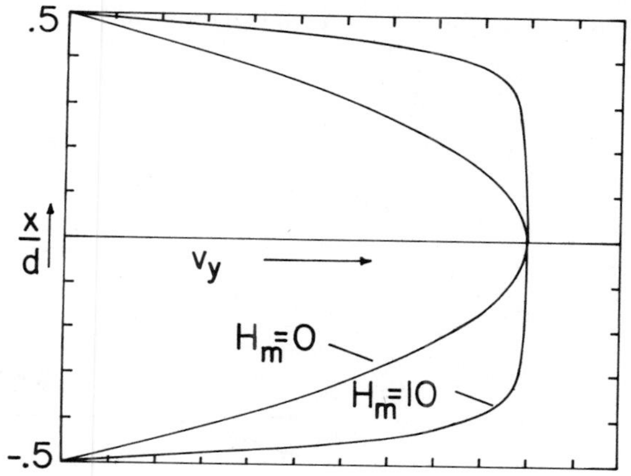

Fig. 9.9.3. Velocity profile of Hartmann flow ($H_m = 10$) and plane-Poiseuille flow ($H_m = 0$).

The Hartmann number indicates the degree to which the field competes with the viscosity in determining the fully developed profile. By one definition, the magnetic Hartmann number is the square root of the ratio of that part of the magnetic force density attributable to the material motion to the viscous force density. From Eq. 14, the motion-dependent part of $J \approx \sigma\mu_o v_y H_o$, so that the magnetic force density is of the order $(\sigma\mu_o v_y H_o)(\mu_o H_o)$. Using as a typical length $d/2$, the viscous force density is of the order $\eta v_y/(d/2)^2$. The square root of the ratio of these two quantities is H_m as defined by Eq. 18. This dimensionless number is alternatively defined in Sec. 8.6 as the square root of the ratio of a magnetic diffusion time to a magneto-viscous time.

The Hartmann flow was originally studied as a model for a liquid metal pump. The electromechanical terminal relations help to emphasize the energy conversion issues.

In practice, it is difficult to make an electrical contact between a liquid metal and a metallic electrode that does not have an appreciable contact resistance. However, with the understanding that v is the voltage across the fluid (the contact resistance might then be included in the external circuit equations), $E_z = v/w$. On the mechanical side, the pressure gradient is the pressure rise Δp divided by the length of the system in the flow direction, ℓ. Thus, Eq. 17 can be used to deduce the electromechanical terminal relations for the system by integrating over the x-z cross section to obtain the volume rate of flow Q_v:

$$Q_v \equiv w\int_{-\frac{d}{2}}^{\frac{d}{2}} v_y dx = \frac{d^3}{4\eta H_m^2}\left(\frac{\tanh H_m}{H_m} - 1\right)\frac{\Delta p}{\ell} - \frac{d}{\mu_o H_o}\left(\frac{\tanh H_m}{H_m} - 1\right)v \quad (19)$$

The electrical counterpart of this relation between the "terminals" of the system is obtained by using Eq. 17 in Eq. 14 to evaluate v_y and integrating the latter expression over the area of the input electrode:

$$i = \frac{\sigma\ell d}{w}v - \frac{\sigma\mu_o H_o \ell}{w}Q_v \quad (20)$$

With the volume rate of flow, Q_v, and voltage, v, constrained, it is convenient to solve Eq. 19 for the pressure rise and express the mechanical power output of the flow as

---

1. J. Hartmann and F. Lazarus, Kgl. Danske Videnskab. Selskab., Mat.-Fys. Medd. 15 , Nos. 6 & 7 (1937)

$$\Delta p Q_v = \frac{4\eta\ell}{wd^3} Q_v^2 H_m \left(\frac{H_m^2}{\tanh H_m - H_m} + \underline{V} \right) \tag{21}$$

where $\underline{V} \equiv (v/Q_v) d^2 \sqrt{\sigma/\eta}/2$.

The electrical power input is similarly expressed by using Eq. 20:

$$vi = \frac{2\sigma\ell}{wd} \sqrt{\frac{\eta}{\sigma}}\, v Q_v (\underline{V} - H_m) \tag{22}$$

In these last two expressions, H_m represents the magnetic field. It plays the role of the field current, i_f, in the d-c machine of Sec. 4.10. The modes of energy conversion obtained by varying the field are seen from the dependences given by Eqs. 21 and 22 and illustrated by Fig. 9.9.4. The energy conversion regimes are as would be expected from those for the prototype machine from Sec. 4.10 (Fig. 4.10.5). The new brake regíme to the left and the expanded one to the right reflect the new loss mechanism, the viscous dissipation.

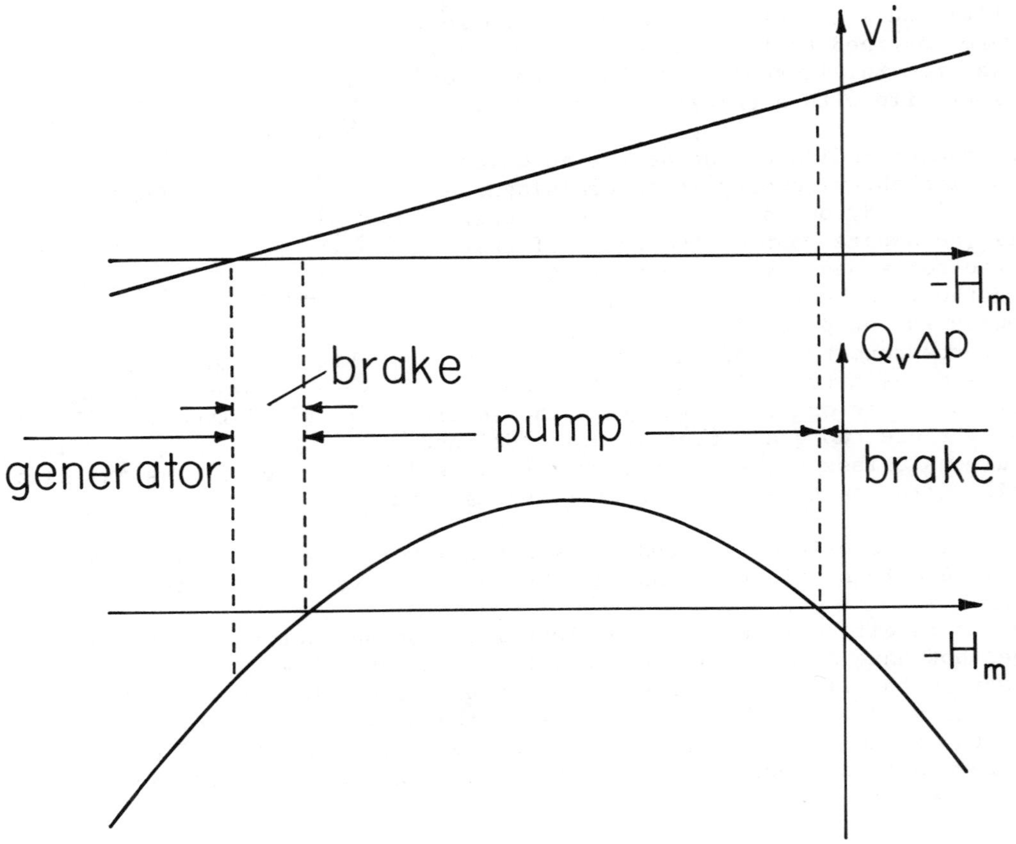

Fig. 9.9.4. Regimes of energy conversion for fully developed Hartmann flow with $\underline{V}$ = 10.

9.10 Flow Development in the Magnetic Hartmann Approximation

In the absence of electromechanical interactions, thè viscous diffusion time determines the time (or distance) for flow development. With the imposition of a magnetic field come processes characterized by the magneto-inertial time (Fig. 9.9.1). Because $\tau_{MI} < \tau_V$, there is now a stronger mechanism than viscous diffusion for establishing a fully developed flow.

To illustrate how induced currents can result in the establishment of fully developed flow at a rate that can be more rapid than would be expected on the basis of viscous diffusion alone, consider the configuration shown in Fig. 9.10.1. The system of Fig. 9.9.2 is essentially "wrapped around on itself" in the y direction. The annulus is thin enough compared to the radius ($a - b \equiv d \ll a$) that the planar model from Sec. 9.9 can be used. The annulus of what amounts to a Couette viscometer is filled with a liquid metal and subjected to a radial magnetic field, H_o. Motion is imparted by the rotation of the inner wall, which has a velocity U. Azimuthal fluid motion therefore induces currents in the z direction as shown in the figure.

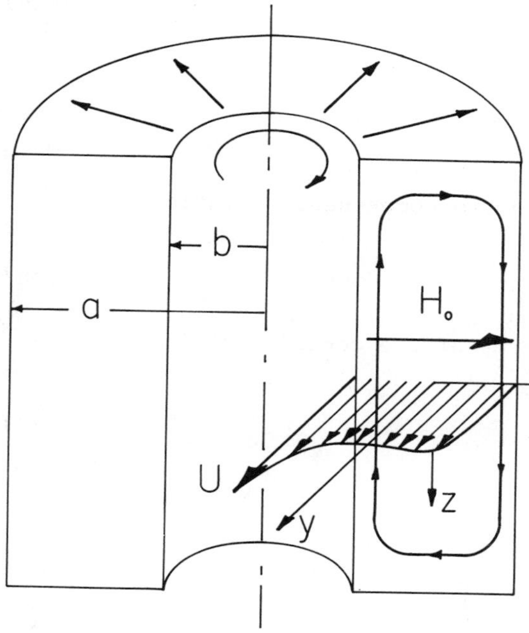

Fig. 9.10.1. Couette Hartmann flow. Inner wall rotates while outer one is fixed.

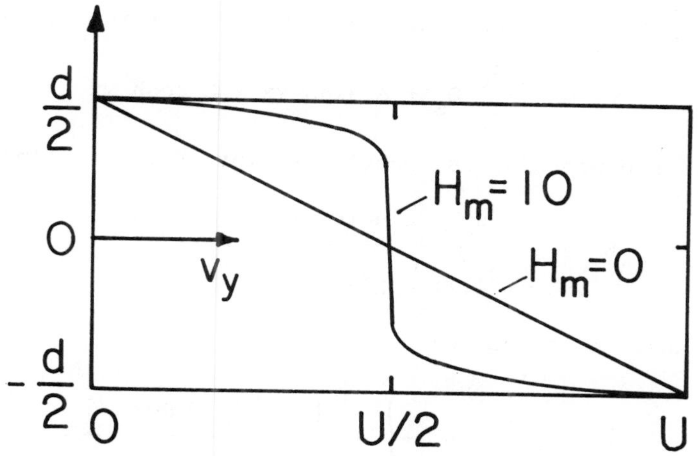

Fig. 9.10.2. Fully developed profile in Couette Hartmann flow of Fig. 9.10.1.

A filmed experiment (Reference 7, Appendix C) shows how the liquid responds as the inner wall is suddenly set into steady motion. Because the upper and lower surfaces of the annular region of liquid metal are bounded by insulators, the current that flows in the z direction over the region of the annulus well removed from the end circulates through the end regions. Thus, the net current in the z direction at any instant is zero. Questions to be answered here include, what is the fully developed velocity profile and what characteristic times govern in its establishment?[1]

The channel closes on itself in the azimuthal direction, and hence the pressure gradient in that direction is zero. This is the y direction in the planar model, and hence Eq. 9.9.15 reduces to

$$\rho \frac{\partial v_y}{\partial t} = \mu_o \sigma H_o E_z - (\mu_o H_o)^2 \sigma v_y + \eta \frac{\partial^2 v_y}{\partial x^2} \tag{1}$$

Because the net current in the z direction must be zero, the integral of J_z over the cross section must be zero. With J_z given by Eq. 9.9.14, it follows that E_z is related to v_y by the condition

$$E_z = \frac{\mu_o H_o}{d} \int_{-\frac{d}{2}}^{\frac{d}{2}} v_y dy \tag{2}$$

Representation of the temporal transient leading to the fully developed flow is carried out as in Sec. 9.6. The fully developed flow plays the role of a particular solution. It follows from Eq. 1 with $\partial v_y/\partial t \rightarrow 0$ and Eq. 2 together with the boundary conditions that $v_y(d/2) = 0$ and $v_y(-d/2) = U$, that

$$v_y = \frac{U}{2}\left[1 - \frac{\sinh(H_m 2x/d)}{\sinh H_m}\right] \tag{3}$$

where $H_m \equiv \mu_o H_o (d/2)\sqrt{\sigma/\eta}$. (This expression follows using the same steps as lead to Eq. 9.9.17.) The profile is shown in Fig. 9.10.2.

To satisfy the initial conditions, superimposed on this fully developed flow are the temporal modes. These are solutions to Eqs. 1 and 2 with the homogeneous boundary conditions $v_y(\pm d/2) = 0$.

---

1. For model in circular geometry, see W. H. Heiser and J. A. Shercliff, "A Simple Demonstration of the Hartmann Layer," J. Fluid Mech. 22, 701-707 (1965).

The temporal modes are assumed to take the form $v_y = \text{Re}\,\hat{v}_y(x)\exp(st)$. Thus, Eq. 1 becomes

$$\frac{d^2\hat{v}_y}{dx^2} - \gamma^2\hat{v}_y = -\frac{\mu_o\sigma H_o}{\eta}\hat{E}_z \tag{4}$$

where

$$\gamma^2 \equiv \frac{s\rho}{\eta} + \frac{\mu_o^2 H_o^2\sigma}{\eta}$$

Solutions to Eq. 4 are the sum of a particular solution and two homogeneous solutions

$$\hat{v}_y = \frac{\mu_o\sigma H_o}{\eta\gamma^2}\hat{E}_z + A\sinh\gamma x + C\cosh\gamma x \tag{5}$$

This expression is substituted into Eq. 2 to find E_z in terms of the coefficient C:

$$\hat{E}_z = C\frac{\mu_o H_o\sinh(d\gamma/2)}{(\frac{d\gamma}{2})\left[1 - H_m^2/(\frac{d\gamma}{2})^2\right]} \tag{6}$$

Thus, Eq. 5 becomes

$$\hat{v}_y = A\left\{\sinh\gamma x\right\} + C\left\{\frac{H_m^2\sinh(d\gamma/2)}{(d\gamma/2)^3\left[1 - H_m^2/(d\gamma/2)^2\right]} + \cosh\gamma x\right\} \tag{7}$$

The coefficients A and C are now adjusted to insure that $v_y = 0$ at $x = \pm d/2$. Because these coefficients can respectively be identified with the odd and even temporal modes, it is possible to determine the eigenvalues and associated eigenmodes by inspection. Odd modes can be made to satisfy the boundary conditions by having $C = 0$ and the coefficient of A vanish at either of the boundaries:

$$\sinh\left(\frac{d\gamma_o}{2}\right) = 0 \tag{8}$$

Here, o is used to denote the odd eigenvalues. Similarly, the even modes follow from Eq. 7 as resulting if $A = 0$ and the coefficient of C vanishes at either of the boundaries:

$$\frac{H_m^2\sinh(\frac{d\gamma_e}{2})}{(\frac{d\gamma_e}{2})^3\left[1 - H_m^2/(\frac{d\gamma_e}{2})^2\right]} + \cosh(\frac{\gamma_e d}{2}) = 0 \tag{9}$$

Thus, the total solution, the sum of the fully developed profile from Eq. 3 and the transient solution given by Eq. 7, is

$$v_y = \frac{U}{2}\left[1 - \frac{\sinh(H_m\frac{2x}{d})}{\sinh H_m}\right] + \text{Re}\sum_{o=1}^{\infty} A_o\sinh(\gamma_o x)e^{s_o t} + \text{Re}\sum_{e=1}^{\infty} C_e\left[\cosh(\gamma_e x) - \cosh(\frac{\gamma_e d}{2})\right]e^{s_e t} \tag{10}$$

Here, the coefficient of C has been simplified by using Eq. 9. The eigenfrequencies s_o and s_e of the even and odd modes follow from the definition of γ given with Eq. 4. Roots of Eq. 8 are simply $\gamma_o d/2 = jo\pi$, and hence the odd modes have the eigenfrequencies

$$\omega_o = \frac{j}{\tau_V}[H_m^2 + (o\pi)^2];\; o = 1,2,\cdots \tag{11}$$

where τ_V is the viscous diffusion time $(\rho/\eta)(d/2)^2$ based on the annulus half-width. These modes are so simply described because the condition on E_z is automatically satisfied by the odd modes with $E_z = 0$. Thus it is that temporal modes found here are a limiting case of those found in Sec. 8.6. That is, in the limit $\tau_m \ll \tau_{MI}$, Eq. 8.6.15 reduces to Eq. 11, where $H_m^2 \equiv \tau_m\tau_V/\tau_{MI}^2$.

To find the eigenvalues, γ_e, and eigenfrequencies, s_e, of the even modes, it is convenient to replace $\gamma_e \to j\beta_e$ and write Eq. 9 as

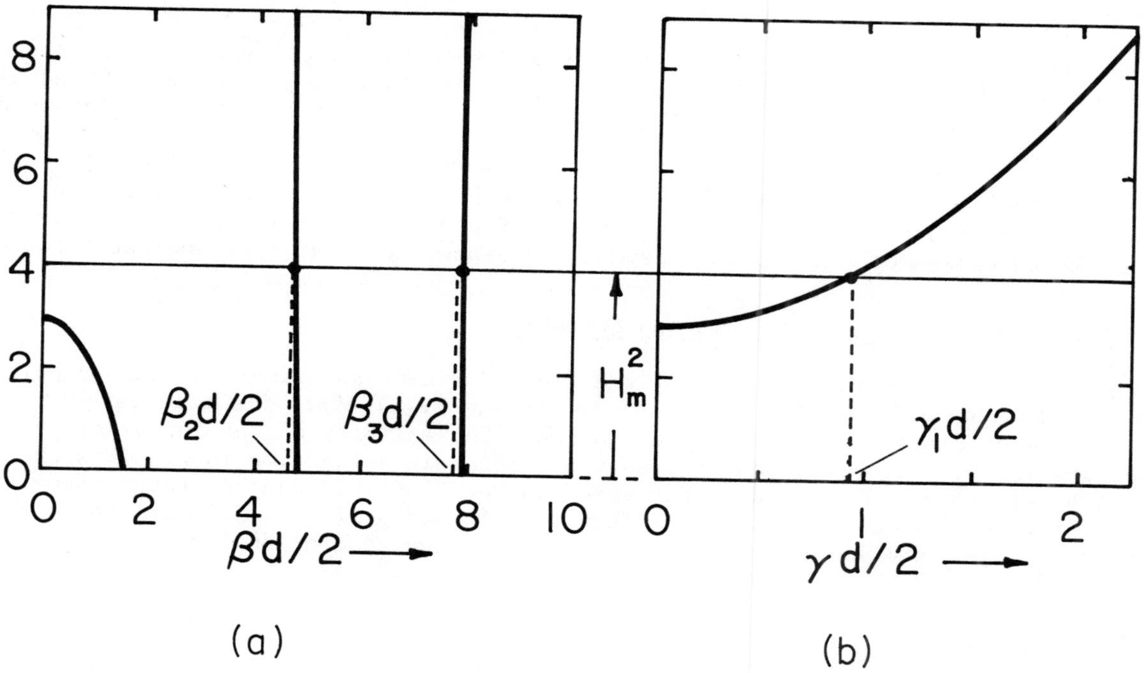

Fig. 9.10.3. Graphical solution of Eq. 12 for eigenvalues $\gamma_e d/2 \equiv j\beta_e d/2$ of even temporal modes.

$$H_m^2 = -\left(\frac{\beta_e d}{2}\right)^3 / \left[\left(\frac{\beta_e d}{2}\right) - \tan\left(\frac{\beta_e d}{2}\right)\right] \tag{12}$$

so that a graphical solution, Fig. 9.10.3a, gives the required modes. For these modes, the eigenvalues are themselves a function of H_m^2. From the definition of γ^2(Eq. 4) the even-mode eigenvalues thus determined then give the eigenfrequencies:

$$s_e = -\frac{1}{\tau_V}\left[H_m^2 - \left(\frac{\gamma_e d}{2}\right)^2\right] \tag{13}$$

The even modes, $e \neq 1$, have eigenvalues that are essentially independent of H_m: $\beta_e d/2 \simeq 3\pi/2$, $5\pi/2, \cdots$. These even modes therefore have characteristic times having much the same nature as for the odd modes. With the magnetic field raised to a level such that H_m exceeds several multiples of π, the lower order modes have decay rates that are of the order $\tau_V/H_m^2 = \tau_{MI}(\tau_{MI}/\tau_m)$. These modes represent the relative adjustment of the profile so that the core of the fluid suffers essentially rigid-body translation. One way to envision the magnetic damping represented by these eigenfrequencies is to select a contour of fixed identity as shown in Fig. 9.10.4. Any vorticity results in an increasing flux linkage for such a loop. The current induced in response to the resulting rate of change of flux linkage results in a force tending to flatten the profile.

Although the magnetic field has a strong effect on the rate at which rigid-body motion is seen in the fluid bulk, the fluid nevertheless comes up to speed at a much slower rate. This process is represented by the lowest even mode, e=1. As H_m^2 is raised, the eigenvalue decreases to zero (at $H_m^2 = 3$) and then becomes purely real with a graphical solution gotten by plotting Eq. 12 with $j\beta_e \to \gamma_e$. The graphical solution is illustrated in Fig. 9.10.3b. As H_m becomes large, this root can be approximated by $(d\gamma/2)^2 \to H_m^2 - H_m$. Then the associated eigenfrequency is

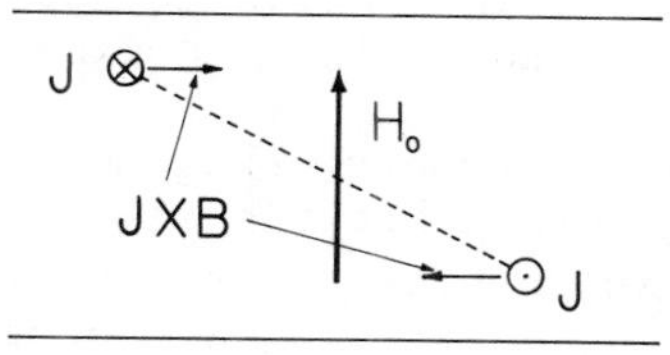

Fig. 9.10.4. Contour of fixed identity in fluid.

$$s = -\frac{H_m}{\tau_V} = -\frac{\mu_o H_o \sqrt{\sigma\eta}}{\rho(d/2)} \tag{14}$$

Thus, the time required to get the rigid translating core of the fluid up to its steady velocity U/2 is τ_V/H_m, which is longer than the dominant time for the relative motion to establish itself, τ_V/H_m^2.

There is a simple picture to go with the transient represented by this lowest even mode. With H_m large, the profile consists of Hartmann boundary layers connected by a uniform profile. In the neighborhood of the boundary, the steady profile is exponential with a decay length

$$\delta = d/2H_m \tag{15}$$

The viscous stress imparted to the fluid by the wall is of the order $\eta v_{wall}/\delta$. This stress must accelerate the core of the fluid to half of the velocity at the wall, and hence must be equal to $s\rho d v_{wall}/2$. Balancing of the inertial and viscous stresses results in a characteristic frequency consistent with Eq. 14.

Because currents circulate within the fluid, there is no net magnetic force on the fluid to contribute directly to its acceleration. The magnetic field plays a role in Eq. 14 only because it determines the thickness of the boundary layer, and hence the shear rate and the viscous stress.

9.11 Electrohydrodynamic Imposed Field Approximation

With the material motion prescribed, the imposed field approximation with unipolar conduction is as introduced in Sec. 5.3. In the region of interest, the electric field is largely due to external charges, perhaps on electrodes bounding the volume. The validity of the approximation hinges on the self-precipitation time τ_e being longer than the charge migration time τ_{mig}. This characteristic time interpretation of the approximation is discussed in Sec. 5.6. It can be stated formally by observing that the pertinent EQS equations of motion (Eqs. 11, 10 and 9 of Sec. 5.2, written for one species and no diffusion), together with the force and continuity equation for an incompressible fluid, take the normalized form

$$\nabla \times \vec{E} = 0; \quad \vec{E} = -\nabla\Phi \tag{1}$$

$$\nabla\cdot\vec{E} = \frac{\tau_{mig}}{\tau_e}\rho_f \tag{2}$$

$$\frac{\partial\rho_f}{\partial t} + \left(\frac{\tau}{\tau_{mig}}\vec{E} + \vec{v}\right)\cdot\nabla\rho_f + \frac{\tau}{\tau_e}\rho_f^2 = 0 \tag{3}$$

$$\frac{\partial v}{\partial t} + \vec{v}\cdot\nabla\vec{v} + \left(\frac{\tau}{\tau_{EI}}\right)^2\nabla p = \left(\frac{\tau}{\tau_{EI}}\right)^2\rho_f E + \frac{\tau}{\tau_V}\nabla^2\vec{v} \tag{4}$$

$$\nabla\cdot v = 0 \tag{5}$$

Here, the normalization is as used in connection with Eqs. 4a, ρ_o and $\mathscr{E}$ are typical of the free charge density and imposed electric field, and the times that have been identified are

$$\tau_e \equiv \frac{\varepsilon_o}{\rho_o b}; \quad \tau_{mig} = \frac{\ell}{b\mathscr{E}}; \quad \tau_V = \frac{\rho\ell^2}{\eta}; \quad \tau_{EI} = \sqrt{\frac{\rho\ell}{\rho_o\mathscr{E}}} \tag{6}$$

In the imposed field approximation, times of interest, τ, are short compared to the self-precipitation time τ_e. If processes involve viscous diffusion, particle migration and electromechanical coupling to the fluid, then for the imposed approximation to be appropriate, the associated characteristic times must all be shorter than τ_e. But, regardless of the ordering of times, τ_{mig} must be shorter than τ_e if the approximation is to apply (Fig. 9.11.1). This means that the volume charge density term on the right in Eq. 2 is also ignorable, as is also the last (self-precipitation) term in Eq. 3.

In summary, the electric field is approximated as being both irrotational and solenoidal. The charge density is governed by the same rules as outlined in Sec. 5.3. Thus, ρ_f is constant along characteristic lines (Eqs. 5.3.3 and 4.3.4). Unless processes represented by the viscous diffusion and electro-inertial times can be ignored, the mechanical laws are represented by the Navier-Stokes equation, with the force density $\rho_f\vec{E}$, and the condition that $\vec{v}$ be solenoidal.

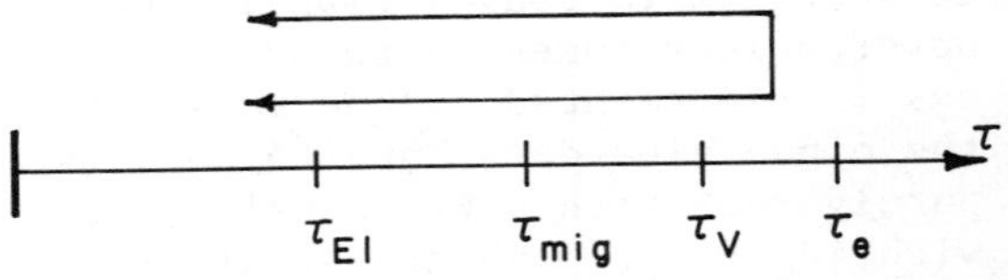

Fig. 9.11.1. Ordering of characteristic times in the EQS imposed field approximation.

9.12 Electrohydrodynamic "Hartmann" Flow

The competition between viscous and magnetic stresses that establishes the fully developed Hartmann flow illustrated in Sec. 9.11 has as an EQS analogue the fully developed flow in the "ion drag" configuration of Fig. 9.12.1. Charged particles, uniformly injected at the inlet where $z = 0$, might be ions generated upstream by a corona discharge, or might be charged macroscopic particles. They are collected by a screen electrode at $z = \ell$. Although it might be used as a pump in the conventional sense, practical interest in the interaction illustrated would more likely come from a need to account for fluid-mechanical effects on the transport of macroscopic particles. Without a self-consistent representation of the effect of the field on the material motion, the interaction is developed in Sec. 5.7. There, space-charge effects are included, whereas here the electric field is approximated by the imposed field. The objective here is to illustrate the reaction of the field on the flow.

The conduction law and force density for charge carriers that individually transmit the electrical force to a neutral medium are discussed in Secs. 3.2, 3.3 and 5.2. In terms of the mobility b, the current density in the z direction is

$$J_z = \rho_f(bE_z + v_z) \tag{1}$$

where bE_z is the particle velocity relative to the air, and the fluid is itself moving at the velocity v_z. There is only one species of particles, and effects of diffusion and generation are negligible.

Because the electric field induced by charges in the fluid is negligible compared to that imposed by means of the electrodes,

$$E \approx E_o\vec{i}_z = \frac{V}{\ell}\vec{i}_z \tag{2}$$

and Gauss' law is ignored in further developments. Note that Eq. 2 is consistent with there being no current density normal to the insulating walls. Fully developed solutions are of the form

$$\vec{v} = v_z(r)\vec{i}_z; \quad \rho_f = \rho_f(r)$$
$$\vec{J}_f = J_z(r)\vec{i}_z; \quad \frac{\partial p}{\partial z} = \text{constant} \tag{3}$$

and hence $\vec{v}$ and $\vec{J}_f$ are automatically solenoidal so that mass and charge conservation are insured. Effects of gravity are lumped with the pressure, and therefore only the z component of the Navier-Stokes equation remains to be satisfied:

$$\frac{\partial p}{\partial z} = \rho_f E_o + \frac{\eta}{r}\frac{\partial}{\partial r}\left(r\frac{\partial v_z}{\partial r}\right) \tag{4}$$

The current density, J_f, has a radial dependence determined at the inlet. Here, $J_f = J_o$ is taken as uniform over the cross section so that Eq. 1 is solved for the charge density and substituted into Eq. 4 to obtain a differential equation for the velocity profile:

$$\frac{\partial p}{\partial z} = \frac{J_oE_o}{(bE_o + v_z)} + \frac{\eta}{r}\frac{d}{dr}\left(r\frac{dv_z}{dr}\right) \tag{5}$$

This nonlinear expression is reduced to a linear one by restricting attention to circumstances when $bE_o >> v_z$ so that $(bE_o + v_z)^{-1} \approx (bE_o)^{-1} - v_z(bE_o)^{-2}$ and Eq. 5 can be written as a linear equation but with space varying coefficients:

$$\frac{1}{r}\frac{d}{dr}\left(r\frac{dv_z}{dr}\right) - \frac{J_o}{\eta b^2 E_o}v_z = \frac{1}{\eta}\left(\frac{\partial p}{\partial z} - \frac{J_o}{b}\right) \tag{6}$$

Homogeneous solutions to Eq. 6 are zero order modified Bessel's functions (introduced in

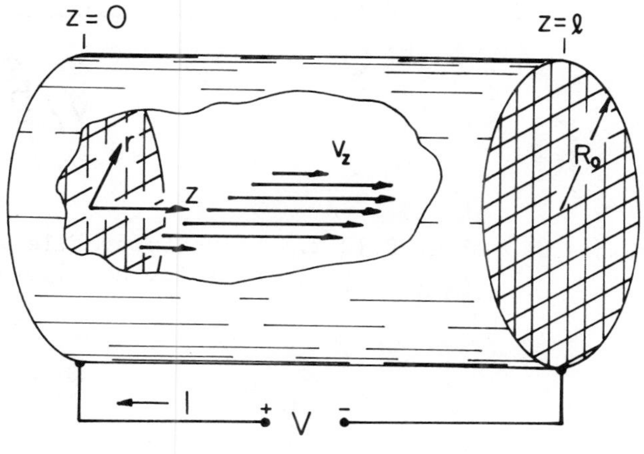

Fig. 9.12.1. Circular cylindrical conduit having insulating wall supporting screens at $z = 0$ and $z = \ell$. Charged particles are injected at left and pulled through the fluid to provide electrohydrodynamic pumping. Flow is electric analogue of Hartmann flow.

Sec. 2.16, Eqs. 2.16.19 and 2.16.25). Because $r = 0$ is included in the flow, the singular solution is excluded. The particular plus homogeneous solution to Eq. 6 that makes $v_z(R_o) = 0$ is then

$$v_z = \frac{R_o^2}{\eta H_e^2}\left[\frac{J_o}{b} - \frac{\partial p}{\partial z}\right]\left[1 - \frac{I_o\left(H_e \frac{r}{R_o}\right)}{I_o(H_e)}\right] \qquad (7)$$

where the electric Hartmann number is $H_e \equiv \sqrt{J_o R_o^2/\eta b^2 E_o}$. Note the analogy between this profile and that for the magnetic Hartmann flow represented by Eq. 9.10.17. Here, H_e^2 is the ratio of that part of the electric force density that is proportional to the fluid velocity to the viscous force density. An alternative interpretation comes from recognizing that $H_e = \sqrt{\tau_{mig}/\tau_{EV}}$ where $\tau_{mig} = R_o/bE_o$ (the time for a particle to migrate the radius R_o relative to the fluid) and $\tau_{EV} \equiv \eta b/J_o R_o$. The electro-viscous time, τ_{EV}, assumes the form $\eta/\varepsilon E^2$ familiar from Sec. 8.7, provided that $J_o \approx \rho_f b E_o$ and one of the E's is recognized from Gauss' law to be of the order $\rho_f R_o/\varepsilon$.

The pump characteristic is obtained by integrating Eq. 7 over the channel cross section, defining Q_v as the volume rate of flow and recognizing that the pressure rise Δp through a channel of length ℓ is $\ell(\partial p/\partial z)$ [from Eq. 2.16.26a, the integral of $xI_o(x)$ is $xI_1(x)$],

$$Q_v = \frac{2\pi R_o^4}{\eta H_e^2}\left[\frac{\Delta p}{\ell} - \frac{J_o}{b}\right]\left[\frac{I_1(H_e)}{H_e I_o(H_e)} - \frac{1}{2}\right] \qquad (8)$$

The velocity profile given by Eq. 7 has the dependence on the electric Hartmann number illustrated in Fig. 9.12.2. Because $\vec{E}$ is taken as constant throughout, the force density is proportional to the charge density. With a constant current density, it is seen from Eq. 1 that the charge density is least where the velocity is the most. In spite of the viscous retarding stresses, the tendency is for elements near the wall to catch up with those nearer the center.

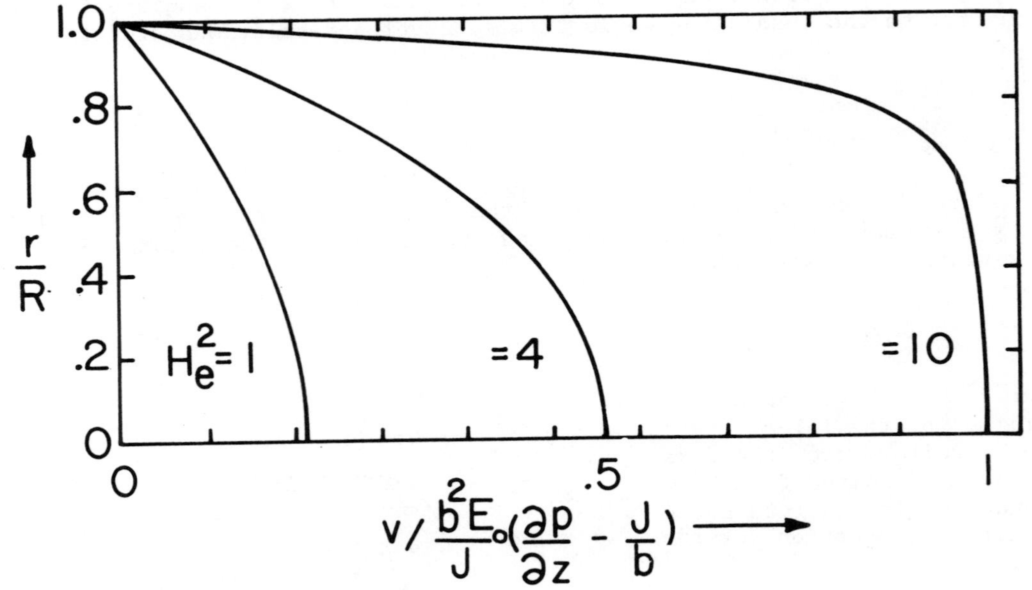

Fig. 9.12.2. Velocity profile with electric Hartmann number as a parameter for configuration of Fig. 9.12.1.

9.13 Quasi-One-Dimensional Free Surface Models

Channel flows, such as in rivers, canals and aqueducts, are a hydrodynamic example of the class of mechanical and electromechanical flow configurations considered in this and the next section. Thus, as an example, homogeneous incompressible fluid, typically water, rests on a bottom having the elevation b(z), as shown in Fig. 9.13.1. The interface at $x = \xi(z,t)$ forms the upper "wall" of a natural conduit for the flow. Gravity confines the fluid to the neighborhood of the bottom. The height of this upper channel boundary, $\xi(z,t)$, is itself determined by the fluid mechanics.

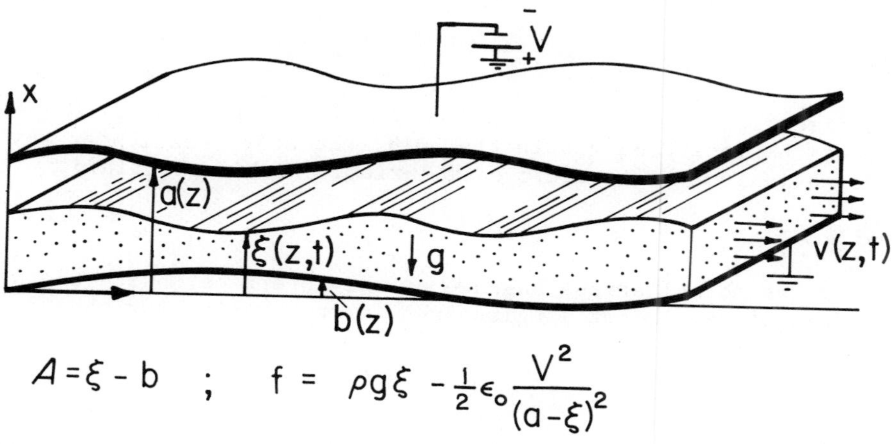

Fig. 9.13.1. Gravity flow with constant potential interface stressed by electric field. Quasi-one-dimensional model expressed by Eqs. 11 and 12 reduces to classic gravity-wave model if V = 0. The hydromagnetic flux conserving antidual of this potential conserving continuum is suggested by Sec. 8.5.

In the purely hydrodynamic context, a canoeist might ask of a long-wave model, given a downstream rock hidden at the bottom of the river [represented by b(z)], can he expect the surface he sees above the rock to be elevated or depressed? In the next section, it will be seen that the answer to such a question depends on the upstream flow conditions relative to the velocity of propagation of a gravity wave. Questions to be asked, where electric or magnetic forces alter or replace gravity, are similar. By way of illustration, an electrode is placed over the flow in Fig. 9.13.1 to impose an electric stress on the interface. The charge relaxation time in the liquid is presumed short enough that the interface can be regarded as retaining a constant potential. However, the electric surface force density is determined by not only V and a(z), but by the position of the interface as well. Of course, it is hardly on the scale of a canoe that the electric field could compete with effects of gravity. But on a scale somewhat larger than a Taylor wavelength, variations in a(z) can affect the flow in a way that depends on the upstream flow relative to a wave velocity altered by the electric field.

The electromechanical coupling due to the electric stress is typical of a wide range of electromechanical interactions that can be modeled using the prototype laws derived in this section. The configuration of Fig. 9.13.1 is typical because the fluid is subject to a volume force density (due to gravity) that can be represented as the gradient of a pressure and because the surface force density (due to the surface free-charge force density) acts normal to the interface.

Many seemingly different mechanical and electromechanical configurations have in common the following properties:

a) The dominant flow is in an axial direction, usually denoted here by z. The viscous skin depth is small compared to the transverse dimensions of the flow. Effects of viscosity are therefore ignored compared to inertial effects, and the longitudinal flow velocity is essentially independent of the transverse coordinates:

$$\vec{v} = \vec{v}_T(x,y,z,t) + v(z,t)\vec{i}_z \tag{1}$$

Because the interfaces are not subject to shear stresses, this approximation is especially appropriate for free surface flows.

b) In the absence of flow, the free surface can assume a shape such that the conditions for a static equilibrium as defined in Sec. 8.2 prevail. Thus, electrical force densities are of the form $\vec{F} = -\nabla\mathcal{E}$ and surface force densities act normal to the interface.

c) Variations of the configuration with respect to the longitudinal direction are sufficiently slow that a quasi-one-dimensional model is appropriate.

A formal derivation of the canonical equations of motion for this class of flows is based on the space-rate parameter expansion introduced in Sec. 4.12 and applied to the Navier-Stokes and continuity equations (in two dimensions) in Sec. 9.7. With ℓ and d respectively representing typical dimensions in the longitudinal and transverse directions, $(d/\ell)^2 \ll 1$. What approximations are appropriate in the laws of fluid mechanics follow from a review of Eqs. 9.7.1 - 9.7.4.

Longitudinal Force Equation: First, the transverse force equation is approximated by a balance between the pressure gradient and any volume force density that is present. To first order in $(d/\ell)^2$

$$\nabla_T(p + \mathcal{E} - \rho\vec{g}\cdot\vec{r}) = 0 \qquad (2)$$

as illustrated in two dimensions by Eq. 9.7.1. Here ∇_T is the gradient in the transverse directions (x,y). In the large, at the interface and in the bulk, the cross section at any given longitudinal position is in a state equivalent to a static equilibrium. Within the fluid,

$$p + \mathcal{E} - \rho\vec{g}\cdot\vec{r} = f(z,t) \qquad (3)$$

where f is determined by the normal stress balance at the interface. With the presumption that the velocity takes of the form of Eq. 1, the longitudinal force equation for the fluid becomes simply

$$\rho\left(\frac{\partial v}{\partial t} + v\frac{\partial v}{\partial z}\right) + \frac{\partial f}{\partial z} = 0 \qquad (4)$$

At each z-t plane, the pressure, p, and surface force density (if any) must balance. This uniquely specifies the cross-sectional geometry and f in terms of one scalar function, the transverse area A(z,t):

$$p + \mathcal{E} - \rho\vec{g}\cdot\vec{r} = f(A) \qquad (5)$$

This hybrid pressure function serves to evaluate Eq. 4, which then becomes one of two mechanical equations of motion in the variables (v,A). If electromechanical coupling is involved, the pressure of Eq. 5 will also be dependent on electric or magnetic variables.

Mass Conservation: All of the terms in the statement of mass conservation are of the same order in $(d/\ell)^2$. (For example, see Eq. 9.7.3.) Thus, all terms are retained. Because the fluid is homogeneous and incompressible, the integral statement of mass conservation for a section on the fluid having incremental lengths Δz, shown in Fig. 9.13.2, is

$$\oint_S \vec{v}\cdot\vec{n}da = \Delta z\int_C \vec{v}\cdot\vec{n}d\ell + A(z+\Delta z)v(z+\Delta z) - A(z)v(z) = 0 \qquad (6)$$

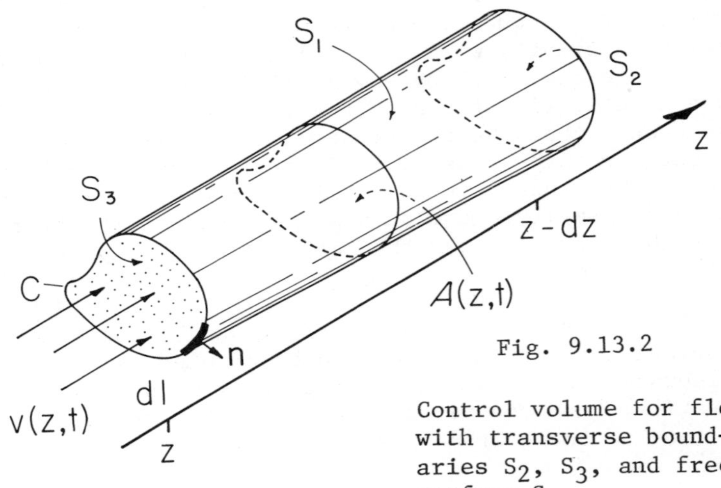

Fig. 9.13.2

Control volume for flow with transverse boundaries S_2, S_3, and free surface S_1.

Portions of the transverse surface S_1 are bounded by rigid walls, while others are the free surface. Integrations over the cross-sectional surfaces S_2 and S_3, which have fixed locations $z + \Delta z$ and z, account for the last two terms in Eq. 6. By definition, the surface S_1 deforms with the interface, so the velocity in the integrand of the first term on the right in Eq. 6 is the interfacial velocity. To first order in the incremental length Δz, the integration on S_1 is reduced to an integration around the contour C multiplied by the length Δz.

The simple geometric significance of the contour integral in Eq. 6 is seen by using the volume form of the generalized Leibnitz rule for differentiation of an integral over a time-varying volume. With $\zeta = 1$, and applied to a right cylinder having the cross section A (not the volume element of Fig. 9.13.2, but rather a right cylinder with fixed ends), Eq. 2.6.5 becomes

$$\frac{d}{dt}\int_V dV = \Delta z\int_C \vec{v}\cdot\vec{n}d\ell \qquad (7)$$

The longitudinal length of the volume in Eq. 7 is fixed, so it follows that this expression is equivalent to

$$\Delta z \left.\frac{dA}{dt}\right|_z = \Delta z \int_C \vec{v}\cdot\vec{n}d\ell \tag{8}$$

The desired quasi-one-dimensional statement of mass conservation follows by substituting the contour integral of Eq. 8 into Eq. 6 and taking the limit $\Delta z \to 0$,

$$\frac{\partial A}{\partial t} + \frac{\partial}{\partial z}(Av) = 0 \tag{9}$$

In the derivation, z has been considered a fixed quantity. Because A is not only a function of t, but of z as well, the temporal partial derivative (the time derivative holding z fixed) is now used in Eq. 9.

Gravity Flow with Electric Surface Stress: As a specific application of the long-wave model, consider the configuration of Fig. 9.13.1. In the long-wave approximation, the zero order electric field in the gap between interface and upper electrode is (see Sec. 4.12 for a formal space-rate expansion):

$$\vec{E} \simeq \vec{i}_x \frac{V}{a(z) - \xi(z,t)} \tag{10}$$

To zero order, this is also the electric field, E_n, normal to the interface. Balance of stresses at the interface requires that $p(\xi) = -\frac{1}{2}\varepsilon_o E_n^2$, where, because the mass density of the upper fluid is much less than that below, the pressure above is defined as zero. Gravity causes the only force density in the fluid volume, so $\mathcal{E} - \rho\vec{g}\cdot\vec{r} = \rho g x$. Thus, evaluation of Eq. 3 at the interface gives $f = p(\xi) + \rho g\xi$. This result makes it possible to express the longitudinal force equation, Eq. 4, in terms of (v,ξ):

$$\rho\left(\frac{\partial v}{\partial t} + v\frac{\partial v}{\partial z}\right) + \frac{\partial}{\partial z}\left[\rho g\xi - \frac{1}{2}\varepsilon_o\frac{V^2}{(a-\xi)^2}\right] = 0 \tag{11}$$

Because the flow is independent of y, the flow area is taken as an area per unit length in the y direction, $A \to \xi - b$. Thus, Eq. 9 becomes

$$\frac{\partial \xi}{\partial t} + \frac{\partial}{\partial z}[v(\xi - b)] = 0 \tag{12}$$

With a(z) and b(z) prescribed, these last two nonlinear expressions comprise the quasi-one-dimensional model. With the removal of the voltage, they become the classic equations for gravity waves and flows.

A second configuration having a small enough scale that capillary effects dominate those due to gravity is shown in Fig. 9.13.3.[1] Here, polarization forces augment and stabilize the tendency of the capillary forces to provide a flow having most of its surface "free." Such "wall-less" flow structures provide for a gravity-independent channeling of a flowing liquid while permitting the interface to be active in heat or mass transfer processes.

It is instructive to linearize Eqs. 11 and 12. With the electrode and bottom flat, so that a and b are constants, and for perturbations from a static equilibrium in which the fluid depth is constant, the dispersion equation must agree with what is obtained from a linear (small-amplitude) theory as would develop following the approach of Sec. 8.10. Illustrated once again is the equivalence between a linearized quasi-one-dimensional model and a long-wave limit of a linearized model (Fig. 4.12.2).

Steady flow phenomena predicted by the models developed in this section are illustrated in Sec. 9.14. Nonlinear temporal transients are taken up using the method of characteristics in Chap. 11. That even steady-state phenomena depend on the causal effect of wave propagation is already evident in Sec. 9.14.

9.14 Conservative Transitions in Piecewise Homogeneous Flows

Piecewise irrotational steady flows are illustrated in this section with a quasi-one-dimensional model that can be applied to a variety of interactions with fields. Typical is the configuration shown in Fig. 9.14.1. Liquid flows in the y direction with variations in the depth $\xi(y)$ slow enough that the velocity profile is essentially independent of depth: $\vec{v} = v\vec{i}_y$. (The longitudinal coordinate is taken as y rather than the z used in Sec. 9.13.)

---

1. See T. B. Jones, Jr., and J. R. Melcher, "Dynamics of Electromechanical Flow Structures," Phys. Fluids 16, 393-400 (1973).

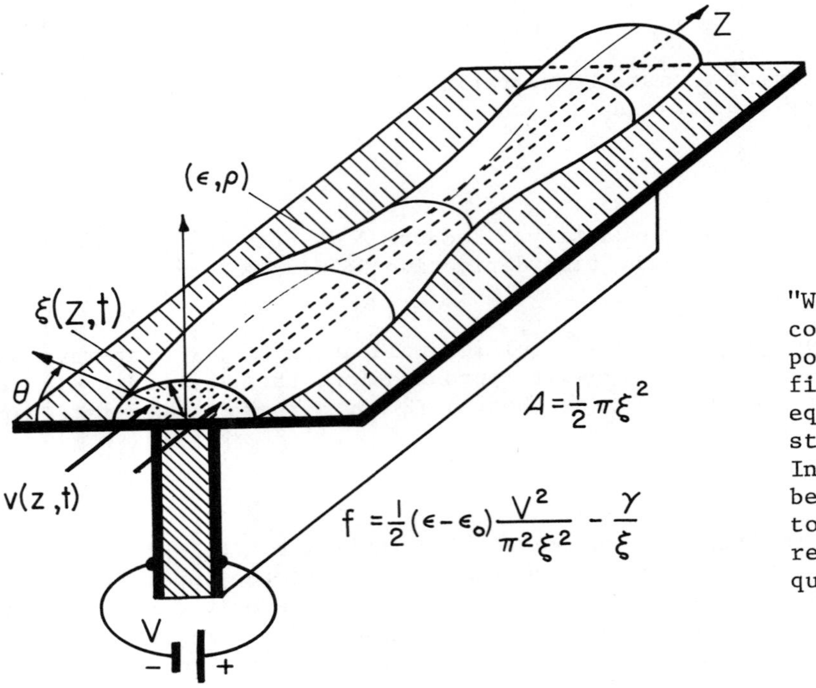

Fig. 9.13.3

"Wall-less" pipe in which fluid is confined by means of capillary and polarization forces. The electric field also stabilizes the transverse equilibrium against the pinch instability caused by surface tension. In practice the applied field should be a-c having a high enough frequency to avoid free charge and mechanical response at twice the applied frequency.

A magnetic field is imposed in the x-y plane. The fluid is an electrolyte or even a liquid metal, so that a uniform current density, J_o, can be imposed in the z direction. However, the flow velocity and conductivity are low enough that the magnetic Reynolds number is small. Currents induced by the motion through the imposed magnetic field can therefore be ignored. So also can the magnetic field generated by J_o.

Given the velocity v_∞ and depth ξ_∞ where the fluid enters at the left, what are these quantities as a function of y? For purposes of illustration the magnetic field is imposed by a two-dimensional magnetic dipole adjacent to the channel bottom (at the origin).

First, observe that the magnetic field and current configuration are the same as illustrated in the last part of Sec. 8.4. Thus, the magnetic force density takes the form $\vec{F} = -\nabla\mathcal{E}$ where, if J_o and A(x,y) are respectively the z-directed current density and vector potential for the imposed magnetic field, $\mathcal{E} = -J_oA$ (Eq. 8.4.13). For an N-turn coil with elements having the spacing s, as shown in Fig. 9.14.1, a driving current, i, results in the vector potential

$$A = \frac{sNi}{2\pi}\frac{\sin\theta}{r} \tag{1}$$

Transformed to Cartesian coordinates, this function becomes

$$A = \frac{sNi}{2\pi}\frac{x}{x^2+y^2} \tag{2}$$

Steady-state conservation of mass, as expressed by Eq. 9.13.9, requires that the volume rate of flow be the same over the cross section at any position y:

$$\xi v_y = \xi_\infty v_\infty \tag{3}$$

In the longitudinal force equation, Eq. 9.13.4, $-\rho\vec{g}\cdot\vec{r} = \rho g x$ and $\partial/\partial t = 0$; and, by recognizing that $v\partial v/\partial y = \partial(\frac{1}{2}v^2)/\partial y$, it follows that

$$\frac{\partial}{\partial y}[\frac{1}{2}\rho v^2 + f] = 0 \Rightarrow \frac{1}{2}\rho v^2 + f = \Pi \tag{4}$$

This is the same expression that is obtained from Bernnoulli's equation, Eq. 7.8.11, if $\vec{v} \approx v\vec{i}_y$.

At the interface there are no surface currents Also the fluid has negligible magnetizability, so there is no magnetic surface force density. Variations of the interface are on a scale long enough (compared to the Taylor wavelength, Eq. 8.9.15) that surface tension can be ignored. Thus, interfacial stress balance shows that the pressure is continuous at the interface. Because the mass density and current density above the layer are negligible, the pressure there is constant and can be defined as zero. Thus, evaluation of Eq. 9.13.3 at the interface, where p = 0, gives f, and Eq. 4 becomes

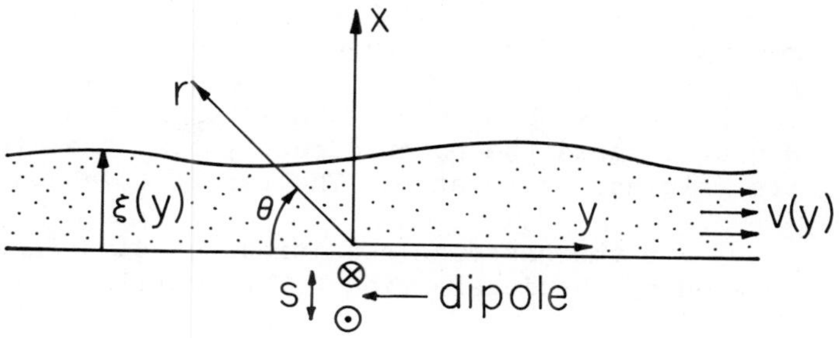

Fig. 9.14.1. Cross section of fluid flowing to right through imposed magnetic dipole. Uniform current density is imposed into paper.

$$\frac{1}{2}\rho v^2 + \rho g\xi + \mathcal{E}(\xi,y) = \Pi \tag{5}$$

For any given flow, the "head" Π is conserved. By using Eqs. 2 and 3, Eq. 5 is converted to an implicit expression for $\xi(y)$ as a function of y:

$$\frac{1}{2}\rho\left(\frac{\xi_\infty v_\infty}{\xi}\right)^2 + \rho g\xi - \frac{sNi}{2\pi} J_o \frac{\xi}{\xi^2 + y^2} = \Pi \tag{6}$$

The viewpoint now used to understand the implications of Eq. 6 would be familiar to a hydraulic engineer. But rather than being concerned with variations in the depth of a river, perhaps caused by an obstruction in the bottom, interest here is in the effect on the depth of the nonuniform magnetic field.

With the flow conditions, mass density, and currents i and J_o set, the left side of Eq. 6 can be plotted as a function of ξ with the longitudinal position y as a parameter. An example is shown in Fig. 9.14.2 where, because ξ measures a vertical distance, it is the ordinate. Flow conditions

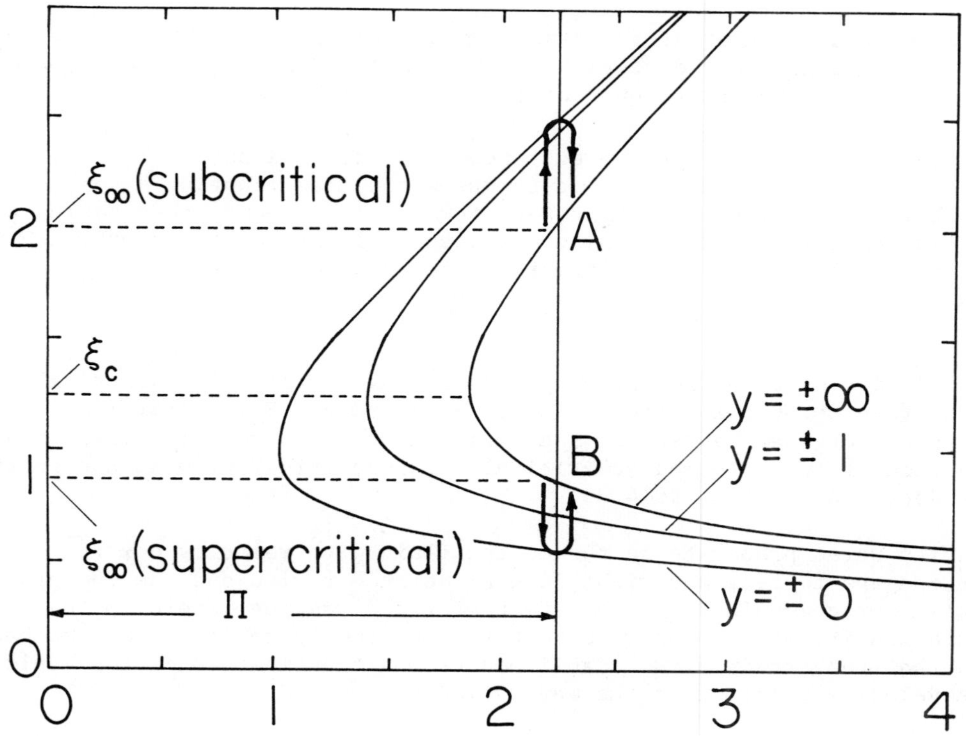

Fig. 9.14.2. Head diagram representing graphical solution of Eq. 6. $\rho(\xi_\infty v_\infty)^2/2 = 1$, $\rho g = 1$, and $sNiJ_o/2\pi = 1$.

to the left establish Π. For the value shown, the entrance depth is either at A or B. The same head is established by a relatively deep but slowly moving entrance as by a shallower but more rapidly moving flow. The dependence of ξ on y can now be sketched by observing that flow entering at A or at B must conserve Π. Thus, entrance at depth A leads to a depth that increases between y=1 and y=0 to the values obtained by the intersections of the appropriate curves with the constant head line. Having reached the point directly over the dipole at y = 0, the depth further downstream returns to its original value at A. The result is shown in Fig. 9.14.3a.

For the entrance conditions of B in Fig. 9.14.2, the fluid depth is decreased rather than increased by the interaction. The profile is illustrated in Fig. 9.14.3b.

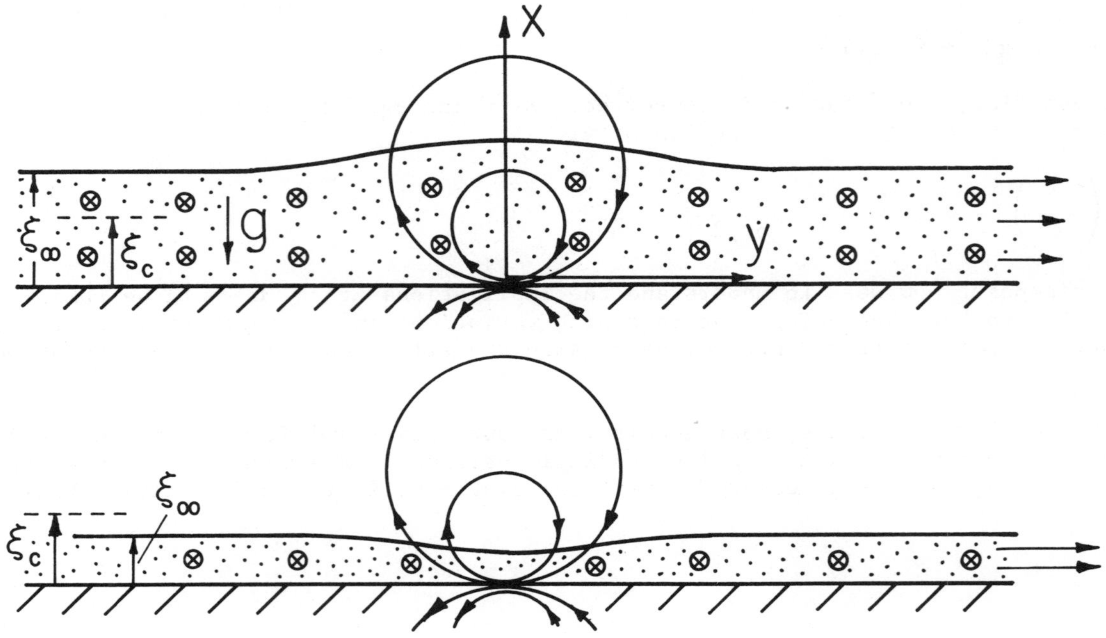

Fig. 9.14.3. Conservative transition of steady flow carrying uniform current density in z direction as it passes through field of magnetic dipole: (a) subcritical entrance; (b) supercritical entrance.

What evidently distinguishes the two entrance conditions, A and B, is their being above and below a critical depth, ξ_c, defined as the depth where the head function at $y \to -\infty$ is a minimum. This critical depth is found from Eq. 6 by taking the limit $y \to -\infty$, then taking the derivative with respect to ξ, setting that expression equal to zero and solving for the depth, $\xi \equiv \xi_c$. The important point is that the flow velocity obtained from Eq. 3 for this depth is

$$v = \sqrt{g\xi_c} \tag{7}$$

This is also the velocity of a shallow gravity wave on the surface of an initially stationary fluid having the depth ξ_c (see Eq. 8.9.16 with $\gamma \to 0$, $\rho_a \to 0$, and $b \to \xi_\infty$). It follows that the case of Fig. 9.14.3a is typical of what happens if the fluid enters at a velocity less than that of a gravity wave. Such an entrance flow is termed subcritical. Supercritical flow at the entrance results in a depression of the depth, as in Fig. 9.14.3b.

Shallow gravity waves propagate on the moving fluid with velocity $v \pm \sqrt{g\xi}$. If the flow is subcritical, waves propagate to left and right in the entrance region and the one propagating upstream provides a mechanism for communicating the effect of the downstream field to the entrance. With supercritical flow, both gravity waves propagate to the right and there is no such mechanism. Hence, it might be expected that the steady flow established from a transient condition would depend intimately on the convection velocity relative to the wave velocity.

Any of the configurations discussed in Secs. 8.3 - 8.5 which resulted in static equilibria suggest steady flows that can be represented by quasi-one-dimensional conservative flow transitions. Examples are shown in Figs. 9.13.1 and 9.13.3.

Any of the configurations discussed in Secs. 8.3 - 8.5 which resulted in static equilibria suggest flows that can similarly undergo conservative transitions. Examples are shown in Figs. 9.13.1 and 9.13.3. If fields exist in the entrance region, there are in general electromechanical contributions to the criticality condition, reflecting the effect of the field on the propagation velocity of surface waves.

Compressible flow transitions through ducts have much in common with those described in this section. Acoustic related waves play the role of the surface waves in this section for determining the criticality conditions.

GAS DYNAMIC FLOWS AND ENERGY CONVERTERS

9.15 Quasi-One-Dimensional Compressible Flow Model

Gas flow through ducts having slowly varying cross-sectional areas is not only of interest in regards to understanding the performance of nozzles and diffusers, but also basic to magnetohydrodynamic and electrohydrodynamic energy conversion configurations. The basic model is developed in this section with sufficient generality that it can be applied directly to these problems in the following sections.

The duct with its rigid walls is depicted schematically in Fig. 9.15.1. In the same spirit as in Sec. 9.13 on free surface flows, the formulation is to be reduced to one involving the single independent spatial variable z. The model hinges on having a cross-sectional area A(z) that varies slowly with z. Even though there is some motion transverse to the z axis, the dominant flow is in the z direction with the transverse flow of "higher order." Effects of viscosity are ignored, and hence the fluid is allowed to slip at the walls. Thus, it is assumed at the outset that the dominant velocity component, as well as the pressure and mass density, are independent of the cross-sectional position:

$$\vec{v} = v(z)\vec{i}_z;\ p = p(z),\ \rho = \rho(z) \qquad (1)$$

The integral laws of mass, momentum and energy conservation, used in conjunction with the incremental control volume of Fig. 9.15.1, are the basis for deriving the quasi-one-dimensional differential equations. Consider first the steady form of mass conservation, Eq. 7.2.2 with $\partial\rho/\partial t = 0$ and S the surface of the incremental volume. Because there is no velocity normal to the channel walls,

$$[\rho vA]_{z+\Delta z} - [\rho vA]_z = 0 \qquad (2)$$

In the limit of vanishing Δz, Eq. 2 becomes the first of the laws listed in Table 9.15.1.

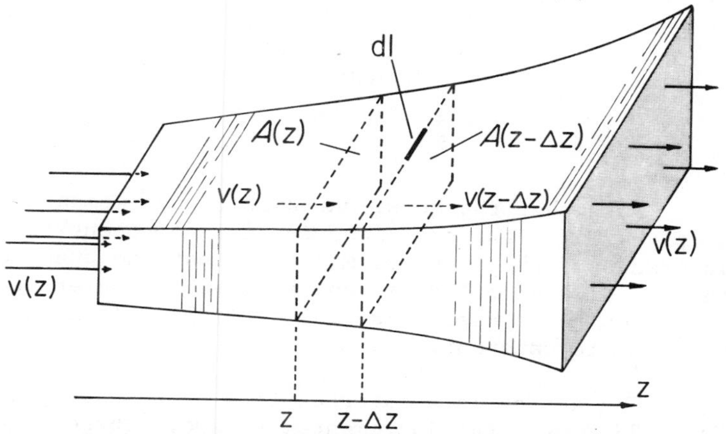

Fig. 9.15.1. Schematic view of duct having slowly varying cross section A(z).

The integral form of conservation of momentum, given by Eq. 7.3.3 with $\partial\vec{v}/\partial t = 0$, $\vec{F} \rightarrow -p + \vec{F}$ and $-\int_V \nabla p\, dV = \oint_S p\vec{n}\cdot d\vec{a}$, becomes

$$[\rho v^2 A]_{z+\Delta z} - [\rho v^2 A]_z + [pA]_{z+\Delta z} - [pA]_z + p\oint_C \Delta z n_z d\ell = AF_z\Delta z \qquad (3)$$

Note that included is an integration over the walls of that component of the normal force acting in the z direction. An incremental section of the wall is sketched in Fig. 9.15.2. For a slowly varying cross section,

$$p\oint_C \Delta z n_z d\ell \simeq -p[A(z+\Delta z) - A(z)] \qquad (4)$$

Now, substitution of Eq. 4 into Eq. 3 gives, in the limit $\Delta z \rightarrow 0$, the differential expression

$$\frac{d}{dz}(\rho v^2 A) + \frac{d(pA)}{dz} - p\frac{dA}{dz} = AF_z \qquad (5)$$

The momentum conservation equation of Table 9.15.1 follows if the conservation of mass statement, Eq. (a) of the table, is used to simplify the first term. Subscripts are dropped from both v_z and F_z for convenience.

Table 9.15.1. Summary of quasi-one-dimensional flow equations for ideal gas subject to force density F and power density JE.

Law	Equation	
Mass conservation	$\frac{d}{dz}(\rho v A) = 0$	(a)
Momentum conservation	$\rho v \frac{dv}{dz} + \frac{dp}{dz} = F$	(b)
Energy conservation	$\rho v \frac{d}{dz}(H_T + \frac{1}{2}v^2) = EJ$	(c)
Mechanical state equation	$p = \rho RT$	(d)
Thermal state equation	$\delta H_T = c_p \delta T$	(e)

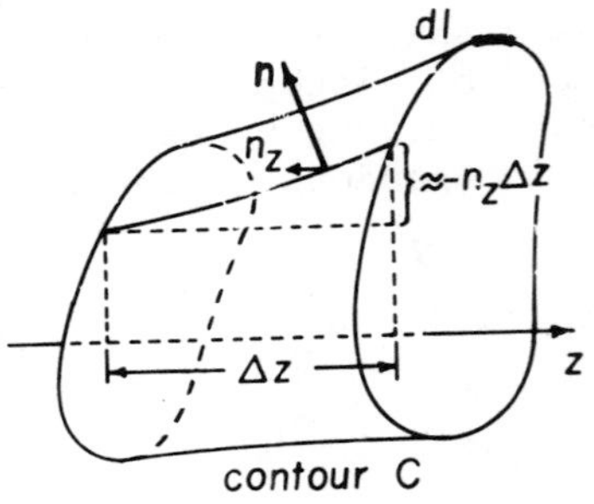

Fig. 9.15.2

Incremental control volume showing normal vector $\vec{n}$ at duct walls and cross section enclosed by contour C.

Note that the quasi-one-dimensional momentum conservation law would be correctly obtained by simply writing the one-dimensional z component of the differential equation of motion. The misleading inference of this finding might be that the quasi-one-dimensional model is obtained by simply writing the one-dimensional differential laws. However, the mass conservation law gives clear evidence that such is not the case: Eq. (a) of Table 9.15.1 is certainly not the one-dimensional form of $\nabla\cdot\rho\vec{v} = 0$ unless A is constant.

The appropriate integral form representing conservation of energy follows from integration of Eq. 7.23.7 over the incremental volume. There is no velocity normal to the walls of the incremental volume, and hence

$$[\rho(H_T + \frac{1}{2}v^2)vA]_{z+\Delta z} - [\rho(H_T + \frac{1}{2}v^2)vA]_z = AEJ\Delta z \qquad (6)$$

Here, E and J represent dominant components of $\vec{E}$ and $\vec{J}_f$. If the limit of vanishing Δz is taken first, and then Eq. (a) of Table 9.15.1 exploited, Eq. (c) of that table follows.

To have a summary of the model, the mechanical and thermal equations of state for an ideal gas are also listed in Table 9.15.1. Given the duct geometry A(z), and the field induced quantities F, E and J, the quasi-one-dimensional model is complete.

9.16 Isentropic Flow Through Nozzles and Diffusers

By definition, a duct shaped to accelerate a gas serves as a nozzle, while one that functions as a diffuser decelerates the flow. The actual variation of cross section depends on the gas velocity relative to the acoustic velocity, i.e., on whether the flow is subsonic or supersonic. An immediate objective in this section is an understanding of the relationship between duct geometry and the steady flow evolution in a purely aerodynamic situation.

But in a broader context, the study illustrates once again how propagation effects can influence steady flow phenomena. The analogy to free surface gravity channel flows from Secs. 9.13 to 9.14 is often cited. There, gravity waves replace acoustic waves in propagating disturbances and concern is with the variation of the fluid depth ξ rather than the mass density ρ. But just as a given variation in the duct cross-sectional area A can lead to either an increased or a decreased mass density (depending on whether the flow is subsonic or supersonic), a variation in the height of the channel bottom can lead to an increased or decreased liquid depth.[1]

The wide variety of free surface electromechanical flows from Secs. 9.13-9.14 are also analogous in their behavior to the flow of compressible gas. The role of acoustic waves is played by electromechanical waves.

For the purely aerodynamic situation considered in this section, F = 0 and J = 0 in the equations of Table 9.15.1. This makes it possible to find integrals of the flow. In any case, conservation of mass as expressed by Eq. (a) of that table shows that

$$\rho v A = \rho_o v_o A_o \tag{1}$$

where subscripts o denote variables evaluated at a given position s_o. The energy conservation and thermal state equations of Table 9.15.1 show that

$$c_p T + \frac{1}{2} v^2 = c_p T_o + \frac{1}{2} v_o^2 \tag{2}$$

As a representation of momentum conservation, Eqs. (b)-(e) of Table 9.15.1 combine to give the equation of state

$$p\rho^{-\gamma} = p_o \rho_o^{-\gamma} \tag{3}$$

This manipulation is carried out without making the quasi-one-dimensional approximation with Eqs. 7.23.8-7.23.13. Recall also that the acoustic velocity is related to the local temperature by Eq. 7.23.6:

$$a = \sqrt{\gamma R T} \tag{4}$$

This last relation should be regarded only as a definition of a. Its use in the following developments in no way implies that the equations have been linearized.

The subscripts used in defining the constants of the flow are now identified with a particular position along the duct. Given v_o, p_o, T_o and ρ_o, the flow velocity, pressure, temperature and density at points downstream in the flow are to be determined. It is convenient to define the Mach number of the flow at the point o as

$$M_o = v_o/a_o = v_o/\sqrt{\gamma R T_o} \tag{5}$$

The objective is a relationship between the velocity v and the area A, with the other flow variables eliminated. Thus, from Eq. 1 the density is eliminated by writing

$$\rho = \frac{\rho_o v_o A_o}{vA} \tag{6}$$

In turn, it follows that Eq. 3 can be used to find the pressure from v and A:

$$p = p_o \left(\frac{v}{v_o}\right)^{-\gamma} \left(\frac{A}{A_o}\right)^{-\gamma} \tag{7}$$

The temperature follows from the perfect gas law and Eqs. 6-7

$$T = T_o \left(\frac{v}{v_o}\right)^{1-\gamma} \left(\frac{A}{A_o}\right)^{1-\gamma} \tag{8}$$

Now, if this last expression is introduced into Eq. 2, it can be solved for the area ratio as a function of the velocity ratio with the Mach number at the point o as a parameter:

$$\left(\frac{A}{A_o}\right) = \frac{v_o}{v}\left\{1 + (\gamma - 1)\frac{M_o^2}{2}\left[1 - \left(\frac{v}{v_o}\right)^2\right]\right\}^{1/1-\gamma} \tag{9}$$

---

1. For a discussion in depth, see A. H. Shapiro, Compressible Fluid Flow, Vol. I, Ronald Press Company, New York, 1953, pp. 73-105.

Given an area ratio and Mach number, Eq. 9 defines v/v_o. The remaining flow variables follow from Eqs. 6-8.

For a given gas (given γ), Eq. 9 can be represented by curves in the $v/v_o - A/A_o$ plane with M_o as a parameter. Illustrated in Fig. 9.16.1 are curves typical of flows that are supersonic and subsonic at the point o. The point (1,1) represents the flow condition at point o. Consider the subsonic flow with the Mach number at o equal to $\sqrt{0.5}$. If the area decreases, the $M_o^2 = 0.5$ trajectory requires that the velocity must increase, because the trajectory is from o to a in Fig. 9.16.1. The section behaves as a nozzle in that it increases the flow velocity. Similar arguments for trajectories b-d motivate the appearance and function of the ducts shown in Fig. 9.16.1.

Note that a supersonic flow behaves in a fashion that is just the reverse of what would be expected from simple incompressible flow concepts. An increase in the local area gives rise to an increase in the flow velocity, the duct functions as a nozzle, while the diffuser function is obtained by making a converging channel.

It is the slope of the $v/v_o - A/A_o$ curve at (1,1) that determines whether the velocity increases or decreases with increasing cross-sectional area. That slope is found from Eq. 9 to be

$$\left.\frac{d(A/A_o)}{d(v/v_o)}\right|_{v/v_o=1} = M_o^2 - 1 \qquad (10)$$

an expression which makes it clear that the velocity-area relationship reverses as the Mach number is increased through unity.

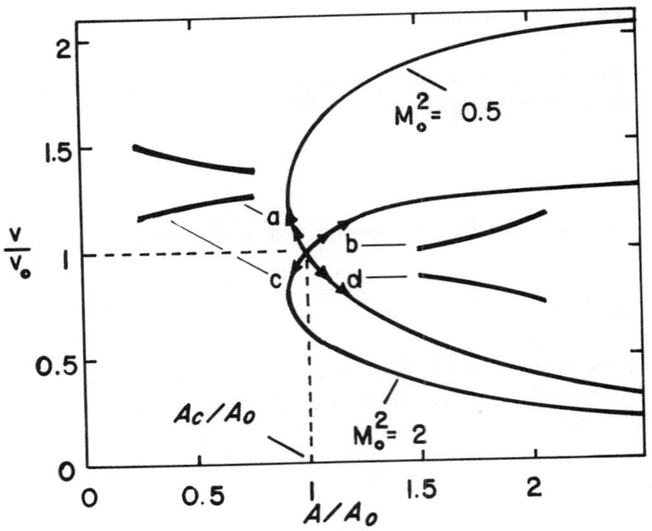

Fig. 9.16.1. Velocity-area relationship in flow transition from position "o": "a" subsonic Nozzle, "b" supersonic nozzle, "o" supersonic diffuser, and "d" subsonic diffuser. Trajectories indicated in the $v/v_o - A/A_a$ plane have the physical interpretation shown by the channel cross sections.

The trajectories of Fig. 9.16.1 make it clear that the laws used to describe the flow cannot pertain if the area is decreased by more than a critical ratio (A_c/A_o). It can be seen that as the area is reduced to this critical ratio, the flow approaches unity Mach number (see Prob. 9.16.1). The flow is then said to be choked. The existence of a greater area ratio negates the assumptions basic to the model.

The choking crisis can be responsible for generation of shocks, highly dissipative discontinuities in the flow. To understand transitions from subsonic to supersonic flow requires combining the conservative flow transitions of this section with the shock relations to be derived in the problems.

The Laval nozzle of Fig. 9.16.2 provides the means of accelerating a stationary gas to supersonic velocities and illustrates one consequence of choking. The channel converges to a smallest cross-sectional area A_c at the throat, and then diverges. Gas enters from a large room at the left and leaves under vacuum at the right. The manometer heights record the pressure. From Eq. (b) of Table 9.15.1, it is clear that a falling pressure implies an increasing velocity and vise versa. With the pressure at the left constant, the pressure at the right is decreased by opening a valve.

The conservative transition through the channel is understood in terms of the velocity-area curves of Fig. 9.16.3. A reduction in outlet pressure causes an increase in the Mach number at the upstream position o, and hence an alteration of the curves as shown. With low pressure drop, $M_o^2 = H_1^2 < 1$, say, and the trajectory is b in the figure. The velocity first increases until the throat is reached and then decreases until the original pressure is very nearly recovered. The transition is "conservative." However, as the outlet pressure is reduced, the flow at the throat becomes sonic, as in trajectory c. It is not possible to further increase the Mach number at o. Rather a further decrease in the outlet pressure results in supersonic flow beyond the throat. This is shown, experimentally in Fig. 9.16.2c because the velocity continues to increase beyond the throat. In the supersonic region, upstream boundary conditions prevail. Hence, the supersonic region just to the right of the throat isolates the flow upstream from that downstream, and upstream flow remains essentially the same even as the outlet pressure is further reduced. But, if the supersonic region is controlled by upstream conditions, how then does the gas adjust its flow so as to match the outlet flow conditions? The shock shown to the right of the throat in Fig. 9.16.2 solves this dilemma by making an abrupt transitio

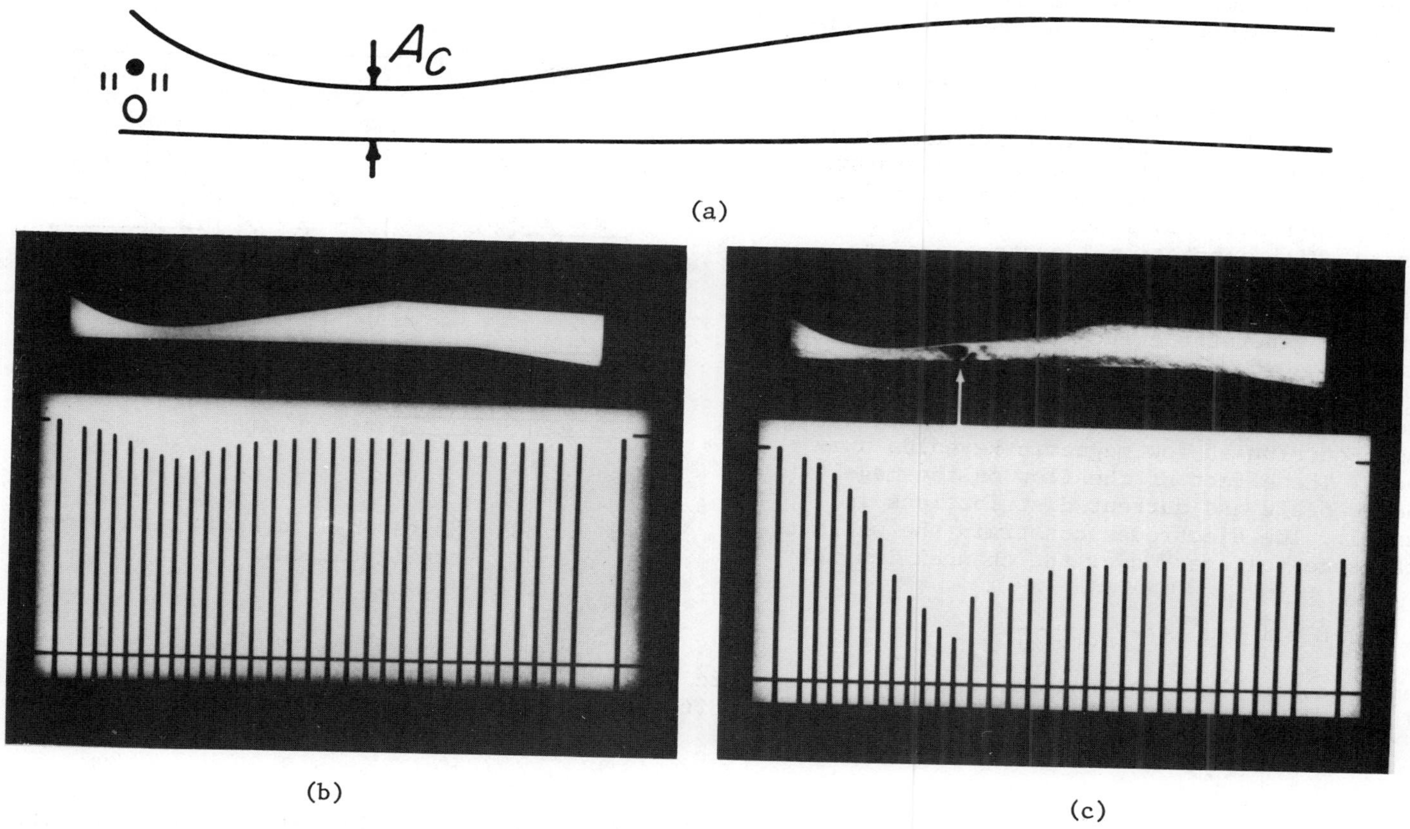

Fig. 9.16.2. Laval nozzle. (a) Cross section with flow from left to right; (b) subsonic transition; (c) subsonic-to-supersonic transition with shock discontinuity beyond throat seen by Schlieren optics. From film "Channel Flow of a Compressible Fluid." (Reference 1, Appendix C).

from supersonic-to-subsonic flow. After the shock, the velocity decreases rather than increases, just as is expected for a subsonic flow in a diverging section. After the shock, the channel behaves as a subsonic diffuser. An observation to be made from the Laval nozzle is that if the flow is to make a transition from subsonic to supersonic, then this must be done at the throat and the flow there must be sonic.

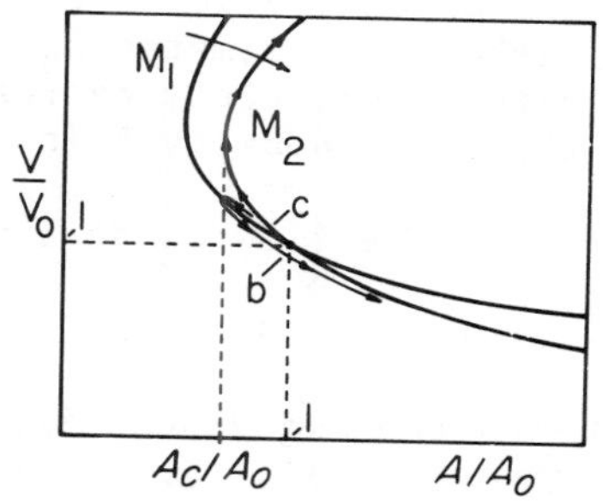

Fig. 9.16.3. Velocity-area diagram showing trajectories of flow corresponding to "b" and "c" of Fig. 9.16.2.

Phenomena illustrated in this section have analogues in the flows developed in Sec. 9.14. In the inhomogeneous incompressible flows, there are also "subcritical-to-supercritical" transitions and the analogue of the shock is a "jump," or sudden change in the flow accompanied by dissipation, usually through the agent of turbulence. Shocks are taken up in Sec. 9.20, and the analogies explored in the problems.

9.17 A Magnetohydrodynamic Energy Converter

The magnetohydrodynamic generator shown in Fig. 9.17.1 combines the magnetic d-c interactions of Sec. 4.10 with the compressible channel flows of Sec. 9.15. The gas is rendered electrically conducting by ionization in a combustion process, and the object is to convert the thermal.energy to electrical form. The interaction region serves as both the turbine and the generator in a conventional plant. From the combustion zone, the gas is accelerated to velocity v_o at the entrance to the conversion section by use of a nozzle, as discussed in Sec. 9.16. By virtue of its conductivity, the gas can play the role of the armature conductors of a d-c machine as it passes through a transverse magnetic field imposed by an external magnet. Electrical continuity through the moving gas and an external circuit connected to the load is provided by electrodes placed on the walls. These play the role of brushes in a rotating d-c machine.[1]

One of the most significant problems in making magnetohydrodynamic generators practical is the relatively low electrical conductivities that can be attained. The conductivity is relatively small

1. For an in-depth treatment, see G. W. Sutton and A. Sherman, Engineering Magnetohydrodynamics, McGraw-Hill Book Company, New York, 1965.

even at extremely high temperatures, and although seeding of the gas and other techniques are used to increase the degree of ionization, the gas is far too low in conductivity at reasonable outlet temperatures to make the generator a practical substitute for existing turbine-generator systems. As a result, such generators are being currently developed as "topping" units, with conventional systems used to convert some of the significant amount of energy remaining in the gas as it leaves the MHD generator exit.

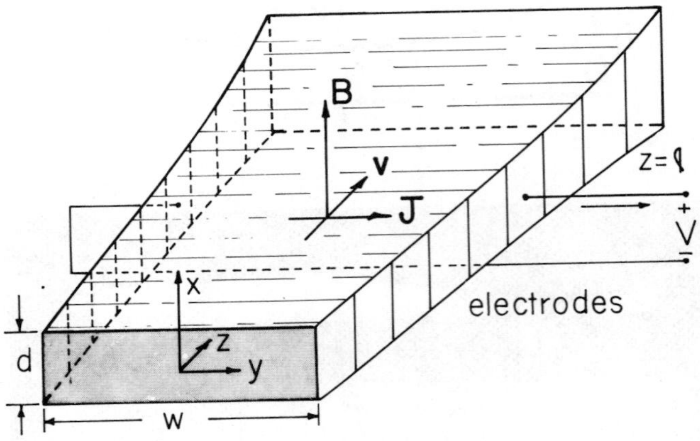

Fig. 9.17.1. Magnetohydrodynamic generator configuration.

MHD Model: Because the conductivity is relatively low, the flow can be regarded as occurring at low magnetic Reynolds number. The effect of the flow on the magnetic field and current distributions is small. The electrodes constrain the walls to the same voltage V over the channel length:

$$\vec{E} \simeq \vec{i}_y E(z) = -\vec{i}_y \frac{V}{w(z)} \tag{1}$$

and because the upper and lower walls are magnetic surfaces, with a constant magnetomotive force (the ampere turns driving the external magnetic circuit)

$$\vec{H} = \vec{i}_x \frac{\mathcal{F}}{d(z)} \tag{2}$$

The generator is constructed with a constant aspect ratio, so that if the width increases, so also does the height:

$$\frac{w}{d} = \text{constant} \tag{3}$$

The objective is to determine the electrical power output, given the inlet conditions of the gas, and either the geometry of the converter or the desired flow process. Because the power conversion density is correctly expressed as EJ, the quasi-one-dimensional model of Table 9.15.1 is applicable once the force density F is stipulated.

The flow is at low magnetic Reynolds number so the magnetic force is essentially imposed,

$$F = -JB \tag{4}$$

Thus the magnetoquasistatic laws for the fields do not come into the formulation. However, to relate J to E, Ohm's law for the moving fluid, as expressed by Eq. 6.2.2, is required. In terms of the model variables,

$$J = \sigma(E + vB) \tag{5}$$

Given the voltage (or a relationship between V and the current as imposed by the external load), E is known. Then, Eq. 5 provides the additional law needed because the additional unknown, J, is introduced by the MHD coupling.

The electrical load connected to each pair of segments is characterized by a "loading factor" K defined by

$$K \equiv -\frac{E}{vB} \tag{6}$$

If the object is as much electrical power output as possible, the resistance of the load on each segment should be adjusted to make K = 1/2. This can be argued by recognizing that in terms of K,

$$J = \sigma vB(1 - K) \tag{7}$$

and hence the output power from a section of the electrodes having unit length in the z direction is

$$JVd = -J(wE)d = wd\sigma v^2B^2(1 - K)K \tag{8}$$

With the imposed field and local flow velocity held fixed, the output power per unit length is maximum with K = 1/2. (Note that in fact the current in turn can alter the velocity, so that the actual optimum K could be somewhat different from 1/2. Nevertheless, it is useful to think of the loading as being in the range of K = 1/2.) It is now assumed that the load is adjusted over the generator length to make K a constant.

Constant Velocity Conversion: Most likely, the geometry is considered fixed and the flow variables are to be determined. However, the generator can be designed such that one of the flow variables assumes a desired distribution throughout the generator. Following this latter approach, consider now the particular case in which the flow velocity v is to be maintained constant throughout and A(z) determined accordingly.

Because $v = v_o$ = constant throughout, Eqs. (a) - (e) of Table 9.15.1 [with Eq. (b) augmented by Eqs. 4 and 6 and Eq. (c) supplemented by Eqs. 6 and 7] become

$$\rho A = \rho_o A_o \tag{9}$$

$$\frac{dp}{dz} = -\sigma v_o B^2 (1 - K) \tag{10}$$

$$\rho c_p \frac{dT}{dz} = -\sigma v_o B^2 (1 - K)K \tag{11}$$

$$p = \rho RT \tag{12}$$

The last three of these relations combine to show that ($\gamma \equiv c_p/c_v$ and $c_p - c_v = R$)

$$\frac{K dp}{p} = \frac{\gamma}{\gamma - 1}\frac{dT}{T} \tag{13}$$

Thus, integration relates the pressure and temperature:

$$\frac{p}{p_o} = \left(\frac{T}{T_o}\right)^{\gamma/[(\gamma-1)K]} \tag{14}$$

From Eq. 12, the density can be related to T:

$$\frac{\rho}{\rho_o} = \frac{p}{p_o}\frac{T_o}{T} = \left(\frac{T}{T_o}\right)^{[\gamma-(\gamma-1)K]/[(\gamma-1)K]} \tag{15}$$

In turn, the area follows from Eq. 9:

$$\frac{A}{A_o} = \frac{\rho_o}{\rho} = \left(\frac{T}{T_o}\right)^{-[\gamma-(\gamma-1)K]/[(\gamma-1)K]} \tag{16}$$

With these last three equations, it is clear that a determination of T(z) would lead to a specification of all flow variables. The temperature is simply obtained from Eq. 11 which can be written as

$$\frac{dT}{dz} = -\frac{\sigma v (AB^2)(1 - K)K}{(A\rho)c_p} \tag{17}$$

and since $A\rho$ and AB^2 are constants (see Eqs 9, 2 and 3), the term on the right is constant. Integration therefore gives a linear dependence of temperature on distance:

$$\frac{T}{T_o} = 1 - (\sigma v_o \mu_o \ell)\left(\frac{\mu_o H_o^2}{\rho_o c_p T_o}\right)(1 - K)K\left(\frac{z}{\ell}\right) \tag{18}$$

The pressure, density and area then follow from Eqs. 14 - 16. With a loading factor K = 0.5 and $\gamma = 1.5$, the exponential in Eq. 15 is 5. Thus, the gas density decreases while, from Eq. 16, the channel area must be made to increase. The electrical power out per unit length is given by Eq. 8. Because B varies inversely with d and the aspect ratio is constant, the power out per unit length is independent of z. Thus the total power output is obtained by evaluating Eq. 8 at the inlet and multiplying by the channel length ℓ:

$$V\int_0^\ell dJdz = (w_o d_o)(\ell\sigma v_o \mu_o) v_o \mu_o H_o^2 (1 - K)K \tag{19}$$

Here, the power output is written as the product of an "active" area w_od_o, a magnetic Reynolds number based on the channel length, a product of magnetic pressure and velocity and a dimensionless factor representing the degree of loading. Thus, the generator output takes the form of an area-velocity-magnetic-pressure product, familiar from Sec. 4.15. The modifying factor of the magnetic Reynolds number is present because it is the alteration in magnetic stress caused by the current that accounts for the interaction with the field. The magnetic Reynolds number is the ratio of the induced-to-the-imposed magnetic field. Thus, one component of H_o^2 in the magnetic pressure term represents the imposed field, and the product of the other value of H_o and R_m represents the spatial variations in $\vec{H}$ induced by the motion.

Given the temperature T_d at the generator outlet where $z = \ell$, the electrical power output is alternatively evaluated using Eq. (c) of Table 9.15.1. The negative of the right-hand side, integrated over the generator volume, is the total electric power output, while the integral of the left side is simply mass rate of flow multiplied by the drop in specific enthalpy. Hence, because ρvA is constant

$$V\int_0^{\ell} dJdz = \rho_o v_o A_o (H_T^o - H_T^d) \tag{20}$$

where $H_T(z = \ell) \equiv H_T^d$. For an ideal gas, $H_T = c_pT$, and with the use of Eq. 18 for the temperature, Eq. 20 is identical to Eq. 19. As seen from Eq. (c) of Table 9.15.1, if the generator operates with a variable velocity, then it is the stagnation specific enthalpies $H_T^* = H_T + \frac{1}{2}v^2$ that appear in Eq. 20. Why is it that even though there is ohmic heating accounted for by JE, all of the drop in enthalpy turns up as electrical power output? The answer comes from recognizing that the electrical heating is of the gas itself. Hence, heating at one position results in thermal energy storage which can be recovered downstream. Ohmic heating in the electrodes or external conductors that is removed from the system is another matter and subtracts from the right-hand side of Eq. 20.

There is of course a price paid even for the ohmic heating of the gas itself. This can best be appreciated by inserting the generator into a thermodynamic cycle and seeing how the increase in entropy caused by the ohmic dissipation dictates an increased heat rejection and hence a diminished overall efficiency. This is discussed in Sec. 9.19.

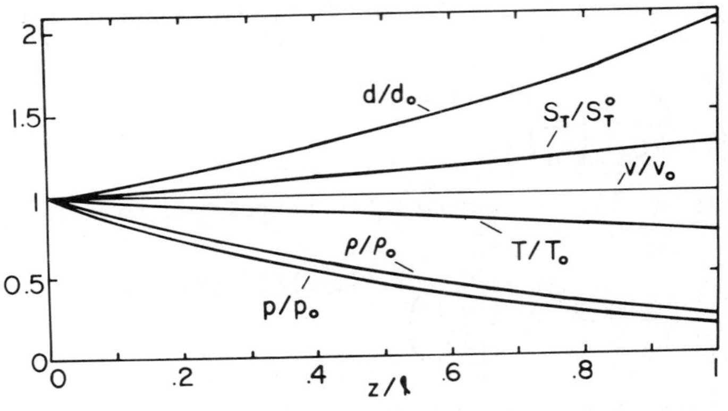

Fig. 9.17.2. Flow evolution through MHD generator of Fig. 9.17.3 with $A(z)$ and hence $a(z)$ designed to give constant velocity.

The increase in entropy through the generator is evaluated by using the pressure and density ratios found with Eqs. 14 and 15 in the entropy equation of state for a perfect gas, Eq. 7.23.12:

$$S_T = S_T^o - c_p \frac{(1-K)}{K} \ln\left(\frac{T}{T_o}\right) \tag{21}$$

Thus, the decrease in temperature predicted by Eq. 18 is accompanied by an increase in the specific entropy, S_T.

To summarize, the area distribution has been designed to make $v = v_o$ throughout, with the other flow conditions represented by Eqs. 14, 15, 18, and 21. Evolution of the flow is typified by Fig. 9.17.2. The temperature decreases with z in a linear fashion. For $\gamma = 1.5$ and $K = 1/2$, the area ratio A/A_o and specific volume are then proportional to $(T/T_o)^{-5}$ and hence increase with z. According to Eq. 3, this means that d/d_o is proportional to $(T/T_o)^{-5/2}$. The pressure varies as $(T/T_o)^6$ and hence drops even more rapidly than the temperature. Some of the implications of these characteristics for an energy conversion system are explored in Sec. 9.19.

Finally, observe that because the acoustic velocity is proportional to $T^{\frac{1}{2}}$ (Eq. 7.23.14) and hence increasing with z, while v is constant, the Mach number is increasing. This suggests the alternative mode of operation of Prob. 9.17.1.

9.18 An Electrogasdynamic Energy Converter

Just as the MHD convertor of Sec. 9.17 is a variation on the d-c magnetic machines of Sec. 4.10, the electrogasdynamic or EGD device of Fig. 9.18.1 is closely related to the Van de Graaff machine of Sec. 4.14.

Electromechanical coupling is through the free charge force density $\rho_f\vec{E}$. With the objective of obtaining a net space charge, and hence an electrical force density on the gas, charged particles are

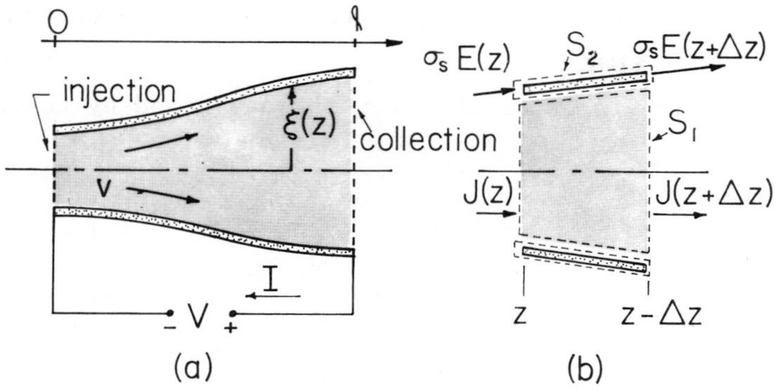

Fig. 9.18.1

Cross section of circular EGD conversion channel having walls with surface conductivity σ_s. (a) Charge is injected at the left and removed at the right. (b) Incremental volume element for deriving quasi-one-dimensional model

injected at the left and removed at the right, thus giving rise to the generator current I. Although ions can be used, charged solid particles or droplets are used to achieve small electrical losses. These particles are of sufficient size to insure that their slip velocity relative to the gas is small compared to the gas velocity. Once charged, the particles can be modeled as having a velocity proportional to the electric field intensity, with the constant of proportionality the mobility b. In the frame of the gas, the current density is $\vec{J}_f' = \rho_f b\vec{E}'$, and hence in the laboratory frame the electroquasistatic transformations give

$$\vec{J}_f = \rho_f(b\vec{E} + \vec{v}) \tag{1}$$

The first objective in this section is a substantive discussion of the electric field alternative to MHD energy conversion. A second is the illustration of how the quasi-one-dimensional modeling extends into the electrical side of the interaction when the effects of the motion on the field are dominant. Thus, by contrast to the low magnetic Reynolds number limit used in Sec. 9.17, considered here are interactions with entrained particles of sufficiently low mobility that the distribution of charge is strongly influenced by the motion. This necessitates a self-consistent electromechanical formulation and the augmentation of the quasi-one-dimensional mechanical equations formulated in Sec. 9.15. Because of limits on achievable electric pressure imposed by electrical breakdown, it is difficult to demonstrate much of a reaction on the flow from the electrical forces. Nevertheless no restrictions are made in that regard.

The EGD Model: The development of a model serves to further describe the nature of the interaction. It hinges on there being no interest in the distribution of the charge over the channel cross section. In fact, the flow is likely to be turbulent with an associated mixing that makes the charge density uniform over a given duct cross section. The generated field $E_z \equiv E(z)$ is also assumed to be constant over the cross section. The radial field E_r is defined as that evaluated adjacent and normal to the wall. The cross section is circular with radius ξ so that $A(z) = \pi\xi^2(z)$.

Conservation of charge for the control volume of incremental length Δz in Fig. 9.18.1b requires that

$$[\rho_f(bE + v)A]_{z+\Delta z} - [\rho_f(bE + v)A]_z + \rho_f bE_r 2\pi\xi\Delta z = 0 \tag{2}$$

The first two terms account for charged particles leaving and entering the volume in the z direction traveling with the velocities bE + v. With the last term, it is recognized that in the gas, unlike on the belt of a conventional Van de Graaff machine, a transverse electric field E_r can cause particle motion relative to the gas with as much ease as the axial field E. Thus, there is in general a current to the wall represented by the last term in Eq. 2.

To understand what determines the radial field E_r, it is necessary to specify the physical nature of the wall. Here it is modeled as having a surface conductivity σ_s, so that current carried to the wall is then carried along the walls to the electrical terminals. The conservation of charge equation, now applied to an annular volume with surface S_2 enclosing the section of wall having length Δz, requires that

$$\rho_f bE_r 2\pi\xi\Delta z = [2\pi\xi\sigma_s E]_{z+\Delta z} - [2\pi\xi\sigma_s E]_z \tag{3}$$

A further and important relation between E and the space-charge density is written using the integral form of Gauss' law for the surface S_1 of Fig. 9.18.1b:

$$[\varepsilon_o\pi\xi^2 E]_{z+\Delta z} - [\varepsilon_o\pi\xi^2 E]_z + \varepsilon_o 2\pi\xi E_r\Delta z = \rho_f\pi\xi^2\Delta z \tag{4}$$

The first two terms account for electric displacement through the surfaces with normals in the $\pm z$ directions, while the third is the net radial flux.

The differential equations describing the electrical side of the coupling are found by taking the limits $\Delta z \rightarrow 0$ of Eqs. 2-4, and are summarized now by

$$\frac{d}{dz}[\rho_f(bE + v)\xi^2\pi] + 2\rho_f bE_r\xi\pi = 0 \tag{5}$$

$$\sigma_s \frac{d}{dz}(\xi E) - \rho_f bE_r\xi = 0 \tag{6}$$

$$\frac{d}{dz}(\xi^2 E) + 2\xi E_r - \frac{\rho_f \xi^2}{\varepsilon_o} = 0 \tag{7}$$

Without raising the order of Eqs. 5 and 7, E_r is eliminated in these equations by using Eq. 6:

$$\frac{d}{dz}[\rho_f(bE + v)\pi\xi^2 + 2\pi\sigma_s\xi E] = 0 \tag{8}$$

$$\frac{d}{dz}(\xi^2 E) + \frac{2\sigma_s}{\rho_f b}\frac{d}{dz}(\xi E) = \frac{\rho_f \xi^2}{\varepsilon_o} \tag{9}$$

In addition to these two statements, representing the electrical side of the interaction, there are the mechanical relations from Table 9.15.1, with $F = \rho_f E$ and EJ reflecting the fact that the wall is thermally insulated and insulating, so that electrical heat losses in the wall are also available to the gas. Thus, the incremental volume used in deriving the quasi-one-dimensional model (Fig. 9.15.2 and Eq. 9.15.6) includes a section of wall having length Δz,

$$\frac{d}{dz}(\rho v\pi\xi^2) = 0 \tag{10}$$

$$\rho v \frac{dv}{dz} + \frac{dp}{dz} = \rho_f E \tag{11}$$

$$\rho v \frac{d}{dz}(c_p T + \frac{1}{2}v^2) = \rho_f E(bE + v) + \frac{2\sigma_s E^2}{\xi} \tag{12}$$

$$p = \rho RT \tag{13}$$

Given $\xi(z)$, these last six expressions describe the evolution of the flow in terms of the six independe variables ρ_f, E, v, ρ, p and T. The terminal variables are then given by evaluating

$$I = \rho_f(bE + v)\pi\xi^2 + 2\pi\xi\sigma_s E \tag{14}$$

$$V = -\int_o^{\ell} Edz \tag{15}$$

Note that according to Eq. 8, I is the same evaluated at any position, z.

In view of Eqs. 10, 14 and 15, the energy equation, Eq. 12, can be multiplied by the area $\pi\xi^2$ and integrated from the entrance to the exit to show that

$$\pi\xi_o^2\rho_o v_o[(H_T + \frac{1}{2}v^2)_{z=\ell} - (H_T + \frac{1}{2}v^2)_{z=0}] = I\int_o^{\ell} Edz = -VI \tag{16}$$

That is, the difference between entrance and exit enthalpy plus kinetic energy is equal to the electric power output. Electrical heating, due to particle slip in the gas and ohmic heating in the wall, is to some extent recovered downstream. However, the ohmic heating does show up as an increase in entropy at the outlet.

Problem 9.18.3 illustrates how the equations are written in a form convenient for numerical integration. The formulation is similar to that for the MHD generator in Sec. 9.17. However, because the field variables are as much a part of the coupling as are the flow variables, E and ρ_f play roles on a par with ρ, v, etc. The channel geometry can be regarded as given and the flow determined, or the dependence of one of the field or flow variables on z can be specified and the geometry determined along with the other variables.

Electrically Insulating Walls: Physically, Eq. 8 requires that the sum of the total convection and conduction currents in the gas and the conduction current in the wall passing any position z must be the same at any other z. Equation 9 is Gauss law, written to the exclusion of E_r so as to make possible a comparison between radial and longitudinal electric fields in accounting for the space charge. If the wall equivalent conductivity σ_s/ξ is large compared to the equivalent bulk conductivity $\rho_f b$, then most of the images for the space charge are at the same axial position on the duct walls. But in the opposite extreme where

$$\frac{2\sigma_s}{\xi\rho_f b} << 1 \tag{17}$$

space charge results mainly in the divergence of an axial field E and the radial field is negligible. If, in addition to Eq. 17, the wall current is negligible compared to that in the gas,

$$\rho_f(bE + v)\xi^2 >> 2\sigma_s\xi E \tag{18}$$

then the last term in Eq. 8 is ignorable. If the wall is sufficiently electrically insulating compared to the volume that both Eqs. 17 and 18 are satisfied, then the radial electric field can be ignored in Eqs. 5 and 7. Physically, this is because a surface charge of the same polarity as the space charge builds up on the walls. This surface charge is just that required to make the electric field be tangential to the wall.

Although the quasi-one-dimensional model presumes that the channel cross section is a slowly varying function of z, it does not presume that the channel is short. Geometrically, the channel would be made to look similar to a Van de Graaff generator. But what has been learned is that using a homogeneous substance such as the gas to replace the belt of a Van de Graaff machine results in a steady-state space-charge field that is of necessity in the same direction as the generated field. This is in contrast to the Van de Graaff machine. The only way to make the space-charge field predominantly perpendicular to z is to make the wall compete for an appreciable fraction of the generated current. This may be practical for the generation of high voltages, but because it implies an electrical loss in the walls on the same order or greater than that generated, it is impractical in making bulk power.

Note that the inequality of Eq. 15 also justifies ignoring the last term in the energy equation, Eq. 12, compared to the first term on the right. Thus, for an insulating wall the appropriate model is represented by Eqs. 8-13 with $\sigma_s \to 0$.

Zero Mobility Limit with Insulating Wall: For efficient generation, it is desirable that the mobility be sufficiently small that

$$|bE| << v \tag{19}$$

in which case bE can be ignored in Eqs. 8 and 12. Limitations on wall conductivity implied by Eq. 17 become even more stringent as it is again assumed that terms in Eqs. 8, 9 and 12 proportional to σ_s are negligible.

With zero mobility, conservation of charge and mass, Eqs. 8 and 10, show that

$$\frac{\rho_f}{\rho} = \frac{I}{\rho_d\xi_d^2\pi v_d} \tag{20}$$

where the subscript d denotes variables evaluated at the downstream end of the generator where $z = \ell$. Thus, the force equation, Eq. 11, becomes

$$\frac{d}{dz}\left[\frac{1}{2}v^2 + \left(\frac{I}{\rho_d\xi_d^2\pi v_d}\right)\Phi\right] + \frac{1}{\rho}\frac{dp}{dz} = 0 \tag{21}$$

where the potential, Φ, is defined by $E = -d\Phi/dz$. Similarly, the energy equation, Eq. 12, becomes

$$\frac{d}{dz}\left[c_pT + \frac{1}{2}v^2 + \left(\frac{I}{\rho_d\xi_d^2\pi v_d}\right)\Phi\right] = 0 \tag{22}$$

These last two expressions make it clear that the duct flow with no electrical coupling is equivalent to that with coupling if we replace $\frac{1}{2}v^2 \to (\frac{1}{2}v^2 + I\Phi/\rho_d\xi_d^2\pi v_d)$. Thus, the flow is isentropic, as can be seen by manipulating Eqs. 21 and 22, together with the mechanical equation of state to obtain Eq. 7.23.13:

$$\frac{p}{p_d} = \left(\frac{\rho}{\rho_d}\right)^{\gamma} = \frac{\rho}{\rho_d}\frac{T}{T_d} \tag{23}$$

Of course, this must be true because the rate of heat generation is zero in the limit $\sigma_s \to 0$ and then $b \to 0$.

Constant Velocity Conversion: Suppose that the channel is designed to implement a constant gas velocity $v = v_d$ throughout. Then Eqs. 8 and 10 show that

$$\rho_f \xi^2 = \rho_{fd} \xi_d^2; \; \rho \xi^2 = \rho_d \xi_d^2 \tag{24}$$

and the right-hand side of Eq. 9, representing Gauss' law, is constant, so that that expression can be integrated to obtain

$$\xi^2 E = -\xi^2 \frac{d\Phi}{dz} = \frac{\rho_{fd}}{\varepsilon_o} \xi_d^2 (z - \ell) \tag{25}$$

Here, the generator is designed (ξ prescribed) for constant velocity operation with E at the outlet adjusted to zero. This is motivated by an interest in generator operation and hence a desire to impose as large a net electric force in the -z direction as possible.

The temperature is related to the area variation by combining Eqs. 23 and 24b:

$$\frac{T}{T_d} = \left(\frac{\xi_d}{\xi}\right)^{2(\gamma-1)} \tag{26}$$

The quantity in brackets in Eq. 22 is constant, and hence relates the temperature of Eq. 26 to the potential. Thus, the potential is defined as V at $z = \ell$ and also written in terms of ξ^2.

$$\Phi = V + (T_d - T)\frac{c_p \rho_d \xi_d^2 \pi v_d}{I} = V + \frac{c_p \rho_d \xi_d^2 \pi v_d T_d}{I}\left[1 - \left(\frac{\xi_d}{\xi}\right)^{2(\gamma-1)}\right] \tag{27}$$

Now, by substituting Eq. 27 for the potential in Eq. 25, an expression is obtained for the cross section as a function of z. Integration, and the condition that $\xi(\ell) = \xi_d$, gives

$$\frac{\xi^2}{\xi_d^2} = \left\{1 - \left[\frac{(\rho_{fd}\ell)^2(2-\gamma)}{2\varepsilon_o \rho_d c_p T_d(\gamma-1)}\right](1 - \frac{z}{\ell})^2\right\}^{1/(2-\gamma)} \tag{28}$$

where the definition of $I \equiv \rho_{fd}\pi\xi_d^2 v_d$ has been used. With this result, Eqs. 24-27 give the dependence of ρ_f, ρ, E, T and Φ on z. The normalized distance upstream from the exit is $(1 - z/\ell)$. Thus, the duct radius is least at the inlet and increases to its maximum at the exit. The temperature therefore decreases in accordance with Eqs. 26 and 28, as it must if the velocity is to remain constant and yet electrical power is to be removed.

The major limitation on an electric field device is likely to be the maximum electric stress that can be developed without causing sparking. From Eq. 25 it is clear that the most critical point in this regard is at the inlet where $E = E_o$ is evaluated using Eqs. 25 and 28 with z = 0. From Eq. 25, it follows that $\rho_{fd} = -(\varepsilon_o E_o/\ell)(\xi_o/\xi_d)^2$. Then, if that result is used in Eq. 28 also evaluated at z = 0, and it is recognized from Eqs. 26 and 23 that $T_d = T_o(\xi_o/\xi_d)^{2(\gamma-1)}$ and hence $\rho_d = \rho_o(\xi_o/\xi_d)^2$, it follows that the area ratio and largest electric pressure (normalized to the entrance enthalpy) are related by

$$\frac{\varepsilon_o E_o^2/2}{\rho_o c_p T_o} = \left(\frac{\gamma-1}{2-\gamma}\right)\left[\left(\frac{\xi_d^2}{\xi_o^2}\right)^{2-\gamma} - 1\right] \tag{29}$$

Given the thermal entrance conditions and γ, the maximum electric field consistent with electrical breakdown serves to determine the area ratio. In turn, all of the other parameters are then determined. For example, the ratio of electrical power out to thermal power entering the duct is found from combining Eqs. 26, 27 (evaluated at z = 0 where $T = T_o$ and $\Phi = 0$) and 29:

$$\frac{VI}{\rho_o c_p T_o \xi_o^2 \pi v_o} = 1 - \left[\left(\frac{2-\gamma}{\gamma-1}\right)\frac{(\varepsilon_o E_o^2/2)}{\rho_o c_p T_o} + 1\right]^{-\left(\frac{\gamma-1}{2-\gamma}\right)} \tag{30}$$

Thus, the entrance ratio of electric pressure to thermal energy per unit volume determines the fraction of thermal energy that can be extracted in a single stage device. To see that the electrical power output is again approximated by an area-velocity-electric-pressure product, consider the particular case where $\gamma = 1.5$ (compared to 1.4 for air). Then, Eq. 30 becomes

$$VI = \pi\xi_o^2 v_o(\frac{1}{2}\varepsilon_o E_o^2)\left[1 + \frac{\frac{1}{2}\varepsilon_o E_o^2}{\rho_o c_p T_o}\right]^{-1} \tag{31}$$

The factor modifying the expected form $Av(\frac{1}{2}\varepsilon_o E_o^2)$ insures that as the electric pressure is increased, the output saturates and never exceeds the available thermal power.

In practice, even with the use of high pressures and electronegative gases to prevent sparking, the electric pressure is likely to be small compared to $\rho_o c_p T_o$. One way to scale the conversion magnitude upward in spite of this is to use many of the individual converters in series, so as to extract a reasonable fraction of the available energy. However, frictional losses (which are ignored here) are likely to give pressure drops on the order of $\frac{1}{2}\varepsilon_o E_o^2$, and create a source of entropy that cannot easily be made manageable by multiple staging. Frictional losses are reduced if the walls are essentially removed and the charged stream is allowed to expand in a "natural" fashion. Some developments are along these lines,[1] with momentum transferred from the expanding stream to a second recirculating flow.

9.19 Thermal-Electromechanical Energy Conversion Systems

To appreciate the limitations imposed on engines that convert heat into electrical power through an electromechanical process, the converter must be seen in the overall context of a steady-state cycle. Use made of thermal energy available in a fuel depends primarily on thermodynamic considerations, and cycle refinements such as reheat loops are essential to the achievement of efficiencies such as are found in modern power systems. Objectives in this section are served by considering a basic system, with refinements a subject in itself.

In the steady state, a process can be characterized by what happens to enthalpy, volume and entropy of a given mass as it passes through its cycle. Thus, the specific extensive variables H_T, ρ^{-1} and S_T are used along with pressure, temperature and velocity to represent the state of a system at a given position in the cycle. The understanding is that the enthalpy, for example, passing a given location in time Δt is $(Av\Delta t)\rho H_T$.

Either the MHD or EGD converter can be the generator in the cycle of Fig. 9.19.1. The state of a unit mass of the fluid as it passes a given station denoted by a-e in Fig. 9.19.1 is given by the state-space trajectories of Fig. 9.19.2.

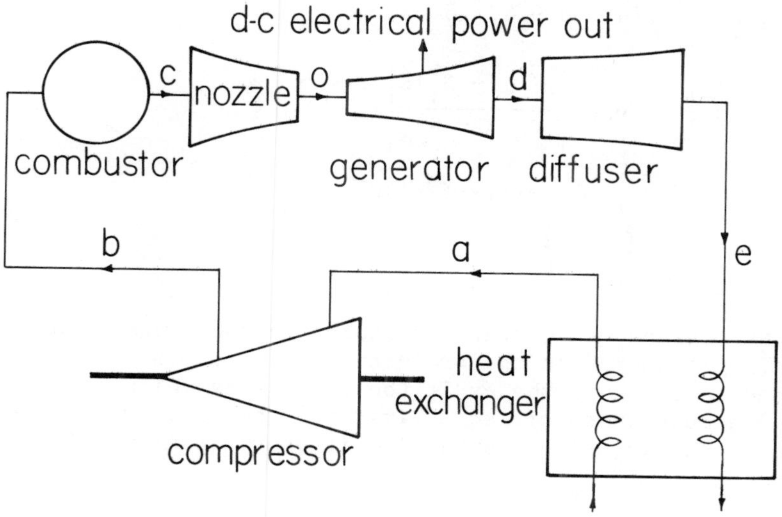

Fig. 9.19.1. Open cycle with either MHD or EGD generator.

Regardless of the cycle, it is important to first recognize the relationship between variables implied by the state equation for a perfect gas. The entropy and mechanical state equations, Eqs. 7.23.12 and 7.22.1, relate T to S_T with p as a parameter, as required for the T-S_T diagram:

$$\frac{T}{T_o} = (\frac{p}{p_o})^{(\gamma-1/\gamma)} \exp\left(\frac{S_T - S_T^o}{\gamma c_v}\right) \tag{1}$$

Hence, the lines of constant pressure shown in Fig. 9.19.2a. For Fig. 9.19.2b the same state equations are solved for the pressure as a function of the specific volume ρ^{-1} with S_T as a parameter:

$$\frac{p}{p_o} = (\frac{\rho_o}{\rho})^{-\gamma} \exp\left(\frac{S_T - S_T^o}{c_v}\right) \tag{2}$$

---

1. M. O. Lawson and J. A. Decaire, "Investigation on Power Generation Using Elecrofluid-Dynamic Processes," Intersociety Energy Conversion Engineering Conference, Miami Beach, Florida, August 13-17, 1967 (participating societies including ASME, IEEE and AIAA).

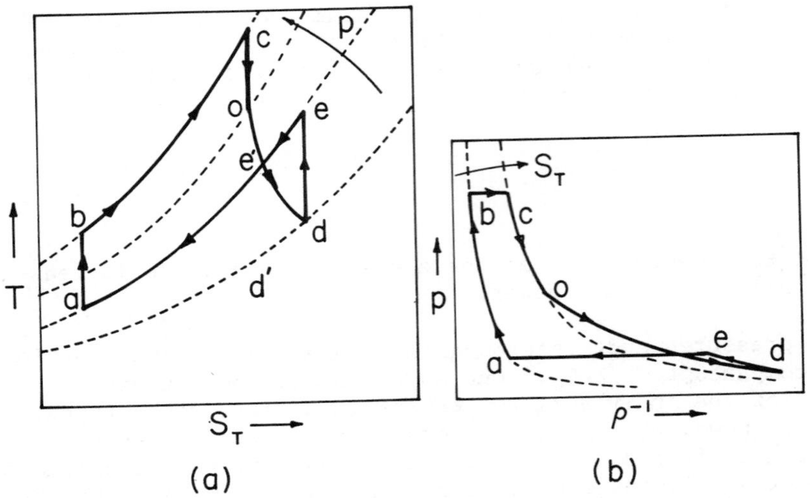

Fig. 9.19.2

(a) Temperature-specific entropy trajectory for cycle of Fig. 9.19.1. Broken lines are at constant pressure; (b) pressure-specific volume trajectory, broken lines are at constant entropy.

The physical position in the cycle of a unit mass of the gas and its state at each position are now described beginning with a. The cycle might be "open," in that ambient air is taken in at a with temperature and pressure thereof pinned to atmospheric conditions. Open cycle or not, it is desirable to make the compression from a to b essentially isentropic.

Combustion involves the working gas as a primary constituent and results in heat addition from b to c. At this point the gas is essentially at rest. From c to o the nozzle converts some of the thermal energy into kinetic form, ideally through an isentropic expansion. The developments of Sec. 9.16 therefore describe the nozzle and, subsequently, the diffuser. According to Eq. 9.16.2,

$$H_T^c = H_T^o + \frac{1}{2} v_o^2 \tag{3}$$

where h_c is the "stagnation enthalpy" in the combustor and o denotes entrance conditions to the generator (nomenclature consistent with Secs. 9.17 or 9.18).

From o to d is described in the previous two sections. For example, in the MHD interaction the T-S_T relation through the generator is Eq. 9.17.21, and the p - ρ^{-1} relation follows from Eqs. 9.17.14 and 9.17.15:

$$\frac{T}{T_o} = \exp\left[\frac{-(S_T - S_T^o)K}{c_p(1 - K)}\right] \tag{4}$$

$$\frac{p}{p_o} = \left(\frac{\rho^{-1}}{\rho_o^{-1}}\right)^{-\gamma/[\gamma-(\gamma-1)K]} \tag{5}$$

The thermodynamic state reflected in the plots is not the whole story. There is also a change in kinetic energy in the transitions from c → e. Upon reaching d, the gas has a residual kinetic energy and to complete the cycle the process by which it is brought to rest with the same ambient conditions as a must be specified. First the gas is brought to rest, d → e, in as nearly an isentropic manner as possible using a subsonic diffuser. Again, using Eq. 9.16.2,

$$H_T^d + \frac{1}{2} v_d^2 = H_T^e \tag{6}$$

Then, by means of a heat exchanger, or simply by expelling the gas to the atmosphere, the gas is returned to ambient temperature and pressure. For the latter, heat rejection from e → a represents a loss of energy and a major contribution to the overall inefficiency. A regeneration system recovering some of the rejected heat is described in Prob. 9.19.1.

An overview of the energy conversion cycle comes from representing the system by the specific enthalpy. To this end, note that in the steady state, the combined internal and kinetic energy conservation statement for a volume V enclosed by a surface S is the integral form of Eq. 7.23.7. Viscous and thermal losses are neglected, so that

$$\int_V \vec{E}\cdot\vec{J}_f dV = \oint_S \rho\vec{v}(H_T + \frac{1}{2}\vec{v}\cdot\vec{v})\cdot\vec{n}da \tag{7}$$

As derived, the left side of this expression is the sum of the total ohmic heating and work done by external forces (Eq. 7.23.1 or 7.23.2).

First, think of the ohmic heating as equivalent to the heat of combustion and apply Eq. 7 to the combustor. In the combustor, the kinetic energy is ignorable and therefore Eq. 7, divided by the mass rate of flow, becomes

$$\frac{\text{thermal energy input/unit time}}{\text{mass/unit time}} = \frac{A\rho v(H_T^c - H_T^b)}{A\rho v} = H_T^c - H_T^b \tag{8}$$

Second, Eq. 7 is applied to the converter section, where the left-hand side becomes the negative of the electrical power output:

$$\frac{\text{electrical power output}}{\text{mass/unit time}} = \frac{VI}{A\rho v} = A\rho v \frac{[(H_T^o + \frac{1}{2} v_o^2)-(H_T^d + \frac{1}{2} v_d^2)]}{A\rho v} = H_T^c - H_T^e \tag{9}$$

Here, the third equality brings in the nozzle and diffuser functions, represented by Eqs. 3 and 6.

Third, Eq. 7 is used to represent the compressor. This time, the left side represents mechanical work done by an external force density of mechanical origin:

$$\frac{\text{compressor energy input/unit time}}{\text{mass/unit time}} = H_T^b - H_T^a \tag{10}$$

Finally, with the neglect of electrical losses (other than in the MHD or EGD generator), heat transfer losses and frictional losses, the overall efficiency can be written as

$$\eta = \frac{\text{electrical power out - compressor power}}{\text{thermal power in}} = \frac{(H_T^c - H_T^e)-(H_T^b - H_T^a)}{H_T^c - H_T^b} \tag{11}$$

Written as it is in terms of the specific enthalpy, this relation is quite general. For an ideal gas $H_T = c_v T$ and Eq. 11 takes a form emphasizing the importance of having a high combustor temperature:

$$\eta = \frac{(T_c - T_e) - (T_b - T_a)}{T_c - T_b} \tag{12}$$

It is now possible to see why an entropy increase in the generator implies a loss of efficiency. If the generator operated isentropically, then points e and d in Fig. 9.19.2a would become points e and d; thus $T_c - T_e$ would be increased in Eq. 12 with the result an improvement in efficiency.

By writing Eq. 11 in the equivalent form

$$\eta = \frac{(H_T^c - H_T^b) - (H_T^e - H_T^a)}{H_T^c - H_T^b} = \frac{\text{heat in - heat rejected}}{\text{heat in}} \tag{13}$$

it is seen that the entropy increase requires a greater heat rejection and for that reason a decreased efficiency.

Note that the rejected heat could be put to useful purposes, for example in heating or refrigerating buildings. The high priority put on increasing the efficiency as defined by Eq. 13 reflects the presumption that the heat rejected is indeed wasted.

Problems for Chapter 9

For Section 9.3:

Prob. 9.3.1 Plane Couette flow exists in a planar channel if there are no electromagnetic stresses and no longitudinal pressure gradients. (This would be the case if the channel were a model for a reentrant flow.)

(a) What is the velocity profile?

(b) With both boundaries fixed and no electromagnetic stresses, the flow is driven by the pressure gradient and called plane Poiseuille flow. Describe the velocity profile and use it to relate the volume rate of flow, Q_v, through a channel of length ℓ and width w to the pressure drop.

Prob. 9.3.2 Carry out the derivation of Eq. 9.3.5 described in the paragraph following that equation.

Prob. 9.3.3 The circulating flow shown in Table 9.3.1 is reentrant, and hence has no azimuthal hybrid pressure gradient. Show that the radial dependence of the azimuthal velocity is given by Eq. (b) of that table.

Prob. 9.3.4 In the absence of electromagnetic forces, a rotor having radius b rotates with the angular velocity Ω_b. It is surrounded by a viscous fluid in an annulus with an outer wall at the radius a having angular velocity Ω_a. Hence, with $\Omega_b = 0$, the configuration is that of the Couette viscometer shown in Fig. 7.13.1.

(a) Find the viscous torque acting on the inner rotor.

(b) Show that in the limit where b >> (a-b), the flow reduces to plane Couette flow (Prob. 9.3.1).

Prob. 9.3.5 Axial flow through an annular region with circular cylindrical boundaries is depicted in Table 9.3.1. Show that the velocity profile is as summarized by Eq. (c) of the table.

Prob. 9.3.6 A pipe has radius R.

(a) Use Eq. (c) of Table 9.3.1 to deduce the velocity profile as a function of the pressure gradient. This is Couette flow in cylindrical geometry.

(b) Find the relationship between pressure drop and volume rate of flow Q_v for a pipe having length ℓ.

For Section 9.4:

Prob. 9.4.1 A tank, shown in Fig. 9.4.2a, is made of insulating material and holds a semi-insulating liquid so that it forms a layer of depth b with a free surface at x = 0. At a distance a above the interface, an electrode structure runs parallel to the interface and imposes the traveling wave of potential Re $\hat{V}_o$ expj(ωt-ky). Thus, the experiment shown in Fig. 5.14.4a is modeled. The time average surface force density is derived in Section 5.14. Using the fully developed flow model, find an expression for the velocity profile as a function of the system parameters and the imposed voltage amplitude.

Prob. 9.4.2 In the configuration of Fig. 9.4.2b, the electrodes are immersed in the liquid. The model is for the experiments shown in Fig. 5.14.4b. Thus there is a layer of liquid above the structure having a depth a; a free upper surface; and a layer of the returning liquid below having a depth b and bounded from below by a rigid equipotential surface. Take the lower surface of the box to be an equipotential surface, and the region of the free interface as extending to infinity. Use fully developed flow models for the regions above and below the electrodes to approximate the volume rate of flow for the circulation around the electrode structure.

Prob. 9.4.3 A layer of liquid metal has an interface carrying skin currents induced by means of a traveling wave of surface current backed by an infinitely permeable material, as shown in Fig. 9.4.2c. Use the sinusoidal steady-state skin-effect model of Section 6.8 and the fully developed flow model to find the surface velocity of the liquid in the tank.

Prob. 9.4.4 The configuration shown in Fig. 9.3.2d is a model for ciculation in liquid metals by nonuniformities in a high frequency imposed magnetic field. The magnetic skin depth is much less than b. Fluid motions are slow enough that they have little effect on the fields. The upper bus-electrode is designed in Prob. 6.9.2 to give a uniformly distributed surface force density. Using the stress derived in that problem, find the interfacial velocity induced by the nonuniformity in field.

For Section 9.5:

Prob. 9.5.1 The planar fluid layer shown in Fig. (a) of Table 9.3.1 is a liquid metal driven by a traveling magnetic-field wave that imposes a tangential field $H_y^\alpha = \text{Re}\,\hat{H}_y^\alpha \exp j(\omega t - ky)$ at the upper surface. The structure used to produce this field might be like that of Fig. 9.5.1, or the layer might be embedded in a heterogeneous system. The skin depth $\delta = \sqrt{2/\omega\mu\sigma}$ is much less than both the layer width Δ and the wavelength $2\pi/k$.

(a) With the velocities at the upper and lower surfaces and the pressure gradient left arbitrary, show that the velocity profile is approximately

$$v = v^\beta(1 - \frac{x}{\Delta}) + \frac{x}{\Delta} v^\alpha + \frac{\Delta^2}{2\eta}\frac{\partial p'}{\partial y}\left[\left(\frac{x}{\Delta}\right)^2 - \frac{x}{\Delta}\right] - \frac{\mu k|\hat{H}_y^\alpha|^2\delta^2}{8\eta}\left(e^{\frac{2(x-\Delta)}{\delta}} - \frac{x}{\Delta}\right)$$

(b) Sketch the magnetic contribution to this profile and compare it to the high frequency profile shown in Fig. 9.5.2.

Prob. 9.5.2 The cross section of a liquid metal induction pump is shown in Fig. P9.5.2. As the circular analogue of the planar configuration considered in this section, it consists of liquid metal in the annulus between highly permeable coaxial cylinders. The inner cylinder has outer radius b while the outer one has inner radius a. A winding, disposed essentially on the surface at r = a, imposes a surface current $\vec{K} = \vec{i}_z\,\text{Re}\,\hat{K}_o \exp j(\omega t - m\theta)$ so that the fluid is pumped azimuthally. Use the velocity profile of Table 9.3.1, Eq. b, and the magnetic diffusion relations summarized by Eq. 6.5.10 and Table 6.5.1 to determine the velocity of the fluid in the annulus. Set up the integrations so that they can be evaluated numerically, as in this section. Include an evaluation of the volume rate of flow.

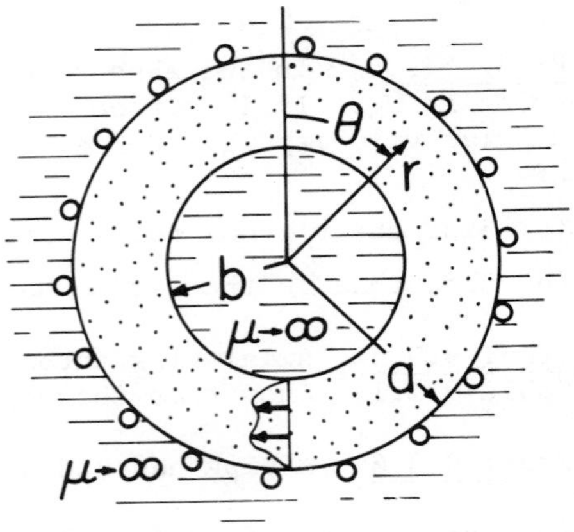

Fig. P9.5.2

Prob. 9.5.3 Table 9.3.1c shows the geometry of a circular induction pump. The liquid metal is in the annulus between coaxial walls at r = a and r = b. The region inside the inner wall can be taken as infinitely permeable while that outside the outer wall is a traveling wave structure backed by an infinitely permeable material. The winding is excited so that at r = a there is a surface current $\vec{K} = \text{Re}\,\hat{K}_o e^{j(\omega t - kz)}\vec{i}_\theta$. The fluid is pumped in the axial direction. Use the velocity profile of Table 9.3.1, Eq. c, and the magnetic diffusion relations summarized by Eqs. 6.5.15 and Table 6.5.1 to determine the fully developed velocity profile. Set up the integrations so that they may be conveniently evaluated numerically, including the relation between pressure gradient and volume rate of flow.

For Section 9.6:

Prob. 9.6.1 A reentrant flow is modeled as in this section by a plane flow. When t = 0, the fluid is static and a uniform force density $\vec{F} = F_o\vec{i}_y$ is suddenly applied. Walls at x = 0 and x = Δ are fixed. Find the fluid response.

Prob. 9.6.2 Find a force density profile $F_y(x)$ such that the fluid velocity profile has the same relative distribution as the fluid comes up to speed. Assume that this force density is suddenly applied when t = 0 and remains constant in time thereafter.

For Section 9.7:

Prob. 9.7.1 There are electromechanical situations where a fluid essentially "slips" relative to a fixed boundary. An example results when a double layer exists between an insulating boundary plate and an electrolyte and a tangential electric field is applied. The resulting flow, which is taken up in Chap. 10, is dominated by viscous stresses within the double layer. Insofar as the bulk flow is concerned, the fluid in the vicinity of the plate is moving with a uniform velocity $v_y = U$ (i.e., the velocity is independent of y). Suppose that $v_y(0,y) = U$ for y > 0, that the rigid plate requires that $v_x(0,y) = 0$ and that the fluid is stagnant as $x \to \infty$. Formulate the similarity problem. What is the viscous stress acting on the plate and what is the total force acting on a length L of the plate?

Prob. 9.7.2 For the stress-constrained boundary layer, what is the transit time between y = 0 where the stress begins and y = y for a particle on the surface? Show that the similarity parameter, Eq. 25, is the square root of the ratio of the viscous diffusion time to this transit time.

Prob. 9.7.3 The fluid interface shown in Fig. 9.7.6 is subject to the imposed surface force density $T(y) = T_o(y/a)^k$, where T_o, and k are constants.

(a) Show that appropriate similarity parameter and function are

$$\xi = \left(\frac{T_o \rho}{a^k \eta^2}\right)^{1/3} xy^{\left(\frac{k-1}{3}\right)} \;; \qquad A_v = -\left(\frac{T_o \eta}{\rho^2 a^k}\right)^{1/3} f(\xi) y^{\left(\frac{k+2}{3}\right)}$$

(b) Show that the boundary layer equations are

$$\frac{d}{d\xi}\begin{bmatrix} f \\ g \\ h \end{bmatrix} = \begin{bmatrix} g \\ h \\ \left(\frac{2k+1}{3}\right) g^2 - \left(\frac{k+2}{3}\right) fh \end{bmatrix}$$

(c) Argue that appropriate boundary conditions are given by Eq. 27.

Prob. 9.7.4 The configuration shown in Fig. P9.7.4 has a planar layer of relatively inviscid ohmic liquid having depth b and charge relaxation time short compared to transport times of interest. The liquid has a "free" surface at x = 0 which, because the mass density of the liquid is much greater than that of the air above, is held flat by gravity. Electrodes in the plane x = b constrain the potential of the liquid as shown in the figure, where V_b and b are constants.

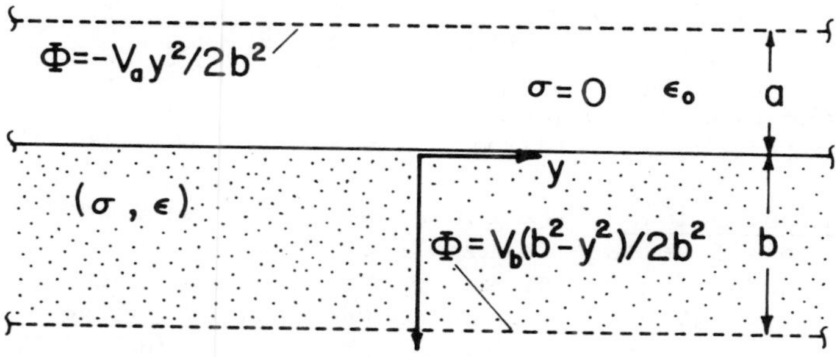

Fig. P9.7.4

(a) Show that in the liquid the electric potential is

$$\Phi = \frac{V_b}{2b^2}(x^2 - y^2)$$

(b) At a distance a above the interface, electrodes constrain the potential to be as shown in the figure. Assume that a is small enough so that E_x in the air can be approximated as the voltage divided by the spacing. What is the electric shearing surface force density acting on the interface?

(c) Show that the boundary layer resulting from this surface force density can be represented as in Prob. 9.7.3. Assume that b is so much greater than the boundary layer thickness that the fluid outside this layer can be regarded as stagnant. What is the value of k?

For Section 9.8:

Prob. 9.8.1 Two semi-insulating liquid layers having ohmic conductivities (σ_a, σ_b), permittivities $(\varepsilon_a, \varepsilon_b)$ and viscosities (η_a, η_b), respectively, are shown in Fig. 9.8.2. Assume that the flow has little effect on the distribution of fields, that gravity holds the interface flat and that the Reynolds number is small.

(a) What is the shearing surface force density $T_y(y)$ due to the field?

(b) Sketch the expected cellular flow pattern.

(c) What is the velocity of the interface $v_y(y)$ as a function of the driving voltage V_o?

(d) What conditions must prevail to insure that effects of motion on the field are negligible and that R_y is small?

For Section 9.9:

Prob. 9.9.1 Fully developed Hartmann flow exists in the half-space x > 0. In the plane x = 0 there is an insulating rigid flat plate. Throughout the fluid, there is a uniform electric field $\vec{E} = E_z \vec{i}_z$ and the pressure gradient in the y direction is constant. Determine the velocity profile $\vec{v} = v_y(x)\vec{i}_y$. What is the thickness of the Hartmann boundary layer?

For Section 9.14:

Prob. 9.14.1 Flow over an uneven bottom is shown in Fig. P9.14.1.

(a) Write the quasi-one-dimensional equations of motion in terms of $\xi(y,t)$ and $v_y(y,t)$.

(b) Draw a "head" diagram analogous to Fig. 9.14.2 and discuss the steady transition of the fluid depth as it passes over an elevation in the bottom. How does the profile depend on the entrance velocity relative to the velocity of a gravity wave?

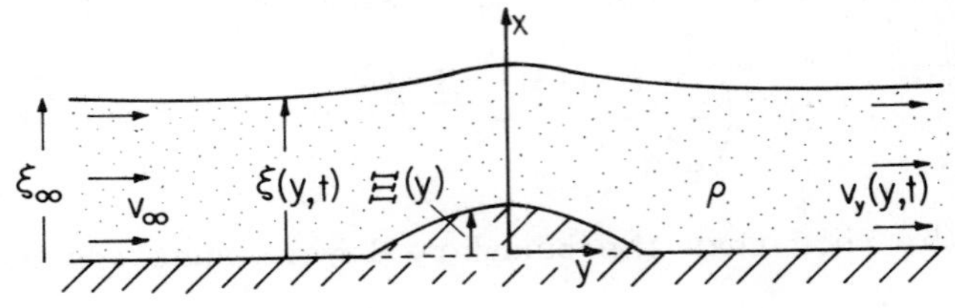

Fig. P9.14.1

Prob. 9.14.2 An alternative to the deduction of the quasi-one-dimensional derivation given here is to use the space-rate expansion illustrated in Sec. 4.12. For a gravity flow, where $\xi = 0$ and $\rho\vec{g}\cdot\vec{r} = -\rho g x$, normalize variables such that

$$x = d\underline{x},\quad y = \ell\underline{y},\quad t = \ell\underline{t}/\sqrt{dy},\quad p = dg\rho\underline{p}$$

$$v_y = \sqrt{dg}\,\underline{v}_y,\quad v_x = \frac{d}{\ell}\sqrt{dg}\,\underline{v}_x$$

and deduce the quasi-one-dimensional model by expanding the dependent variables in powers of the space-rate parameter $(d/\ell)^2$. See Fig. P9.14.1 for the configuration.

Prob. 9.14.3 The cross section of an electromechanical flow structure is shown in Fig. P9.14.3. The applied voltage is high frequency a-c, so that free charge cannot accumulate in the highly insulating liquid. Under the assumption that the mechanical response is only to the time average of the field, V_a is taken as the rms of the applied voltage and henceforth regarded as being d-c. The flow dynamics in the z direction is to be described under the assumption that as the fluid cross section varies the interfaces remain in the regions to right and left, respectively, well removed from the position of minimum spacing between electrodes.

(a) For static equilibrium in directions transverse to z, what are $p(\xi_a)$ and the cross-sectional area $A(\xi_a)$? (Assume p = 0 exterior to the liquid.)

(b) Show that the quasi-one-dimensional equations of motion are

$$\frac{\partial \xi_a^2}{\partial t} + \xi_a^2\frac{\partial v}{\partial z} + v\frac{\partial \xi_a^2}{\partial z} = 0$$

$$\rho\left(\frac{\partial v}{\partial t} + v\frac{\partial v}{\partial z}\right) + \frac{1}{2}\frac{(\varepsilon-\varepsilon_o)}{\alpha^2(\xi_a^2)^2}V_a^2\frac{\partial(\xi_a^2)}{\partial z} = 0$$

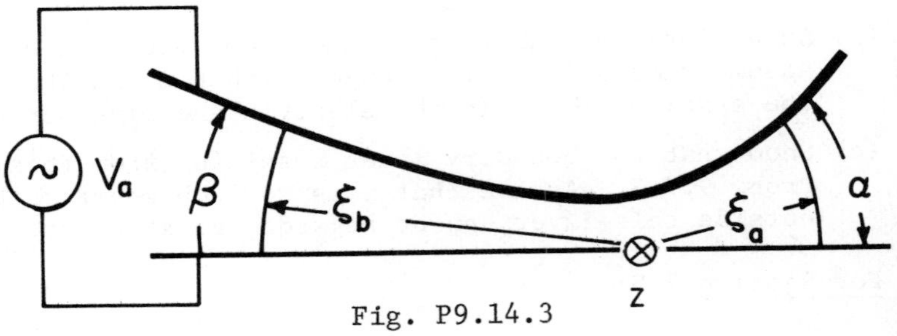

Fig. P9.14.3

For Section 9.16:

Prob. 9.16.1 Derive the area-velocity relation of Eq. 9.16.9.

Prob. 9.16.2 Along the trajectories a and o of Fig. 9.16.1, in both the subsonic nozzle and supersonic diffuser, the area ratio decreases in the direction of flow. Show that as the channel reaches the critical area ratio, defined as the minimum ratio consistent with isentropic steady flow, the Mach number is unity.

Prob. 9.16.3 Use Eqs. (b) - (e) of Table 9.15.1, with no external coupling (F = 0, J = 0) to show that

$$\frac{d}{dz}(p\rho^{-\gamma}) = 0$$

for the quasi-one-dimensional flow described in this section.

For Section 9.17:

Prob. 9.17.1 Because constant velocity implies an increasing Mach number, the flow discussed in this section approaches sonic velocity even if initially subsonic. To avoid the associated losses in the

Prob. 9.17.1 (continued)

subsequent diffuser, used to bring the gas to rest after passage through the MHD duct, it can be advantageous to make $A(z)$ such that the Mach number remains constant: $M^2 \equiv v^2/\gamma RT = M_o^2$. Assume that B is also constant with respect to z and observe that $v dv = \gamma R M_o^2\, dT/2$. Use conservation of energy and momentum to show that

$$\frac{p}{p_o} = \left(\frac{T}{T_o}\right)^{\alpha}; \qquad \alpha \equiv \frac{\gamma}{K(\gamma-1)}\left[1 - \frac{1}{2}(\gamma-1)M_o^2(K-1)\right]$$

In turn, find p, ρ and v in terms of T. Then find and integrate a differential equation for T. Finally, what is $A(z)$ and the specific entropy $S_T(z)$?

Prob. 9.17.2 In general, the z dependence of flow variables cannot be found in analytical form. However, numerical integration of the equations from given inlet conditions is relatively straightforward once the differential equations have been written as a system of first order equations.

The loading is allowed to be arbitrary so that E is now independent of vB. Write the quasi-one-dimensional laws in the systematic form ($\rho' \equiv d\rho/dz$)

$$\begin{bmatrix} \rho v & 0 & \rho v & 0 & 0 \\ 0 & p & \rho v^2 & 0 & 0 \\ 0 & 0 & \rho v^3 & \dfrac{\rho v^3}{M^2(\gamma-1)} & 0 \\ p & -p & 0 & p & 0 \\ 0 & 0 & -2 & 1 & 1 \end{bmatrix} \begin{bmatrix} \dfrac{\rho'}{\rho} \\ \dfrac{p'}{p} \\ \dfrac{v'}{v} \\ \dfrac{T'}{T} \\ \dfrac{(M^2)'}{M^2} \end{bmatrix} = \begin{bmatrix} -\rho v \dfrac{A'}{A} \\ -\sigma B(E+vB) \\ \sigma E(E+vB) \\ 0 \\ 0 \end{bmatrix}$$

where, from top to bottom, these equations represent mass, momentum and energy conservation, the differential forms of the mechanical equation of state and definition of M^2. Under the assumption that $A(z)$ is given, invert these equations and show that written in terms of "influence coefficients" they are

$$\begin{bmatrix} \dfrac{\rho'}{\rho} \\ \dfrac{p'}{p} \\ \dfrac{v'}{v} \\ \dfrac{T'}{T} \\ \dfrac{(M^2)'}{M^2} \end{bmatrix} = \frac{1}{1-M^2} \begin{bmatrix} M^2 & -\dfrac{1}{p} & -\dfrac{(\gamma-1)}{\gamma v p} \\ \gamma M^2 & -\dfrac{1}{p}\left[1+M^2(\gamma-1)\right] & -\dfrac{M^2(\gamma-1)}{pv} \\ -1 & \dfrac{1}{p} & \dfrac{(\gamma-1)}{\gamma v p} \\ M^2(\gamma-1) & -\dfrac{M^2(\gamma-1)}{p} & -\dfrac{(\gamma-1)(\gamma M^2-1)}{\gamma p v} \\ -M^2(\gamma-1)-2 & \dfrac{M^2(\gamma-1)+2}{p} & \dfrac{(\gamma M^2+1)(\gamma-1)}{\gamma p v} \end{bmatrix} \begin{bmatrix} \dfrac{A'}{A} \\ \sigma B(E+vB) \\ \sigma E(E+vB) \end{bmatrix}$$

Discuss how these equations would be integrated numerically. Describe a systematic approach to specifying $A(z)$ such that one of the flow variables has a prescribed evolution with z.

Prob. 9.17.3 The three modes of operation for a d-c machine with a rigid conductor are summarized in Fig. 4.10. To say whether the MHD duct as a whole gives generation, braking or pumping, the distribution of flow variables and load must be determined. In general, for a compressible conductor, solutions to the equations found in Prob. 9.17.2 are required. Consider here the local interaction, the effect of specifying E and B at a given location by means of segmented electrodes. Use the results of Prob. 9.17.2 specialized to a channel of uniform cross section to find the signs of the rates of change of flow parameters for the following cases:

(a) Generator operation with local electrical power out, $EJ < 0$, and a retarding magnetic force, $JB > 0$, for $M^2 \gtrless 1$.

(b) Pump or accelerator operation with $EJ > 0$ and $JB < 0$ for $M^2 \lesseqgtr 1$.

(c) In both of the above, identify those cases where acceleration is reversed from what would be expected for the assigned JB and explain.

For Section 9.18:

Prob. 9.18.1 Use the procedure outlined before Eq. 9.18.23 to show that the zero mobility flow is isentropic and hence satisfies Eq. 9.18.23.

Prob. 9.18.2 The zero mobility generator is to be designed for constant temperature throughout. Show that the pressure and mass density are then also constant. Given the outlet conditions denoted by subscripts d, find v and Φ in terms of the channel area $A = \pi\xi^2$. In turn, show that the area is governed by the equation

$$\frac{d^2}{dz^2}A^{-1} - \frac{\rho_{fd} I \rho_d^2}{\varepsilon_o (A_d \rho_d v_d)^3} A = 0$$

Show that this expression can be integrated, with boundary conditions $E(\ell) = 0$ and $A(\ell) = A_d$, to obtain the implicit dependence of A on z:

$$F(x)e^{x^2} = (\ell - z)\left[\frac{\rho_{fd}^2}{\varepsilon_o v_d^2 \rho_d}\right]^{1/2}$$

$$F(x) \equiv e^{-x^2}\int_0^x e^{x^2}dx \quad ; \quad x \equiv \left[\ln\left(\frac{A_d}{A}\right)\right]^{1/2}$$

where F(x) is tabulated as the Dawson integral.[1] Use subscripts to denote inlet variables and show that

$$\frac{\varepsilon_o E_o^2/2}{\rho_o v_o^2} = \ln\left[\left(\frac{A_o}{A_d}\right)\right]$$

Show that the electrical output power VI can be written in terms of the inlet electric pressure as

$$VI = \frac{v_o A_o}{r}\left(\frac{1}{2}\varepsilon_o E_o^2\right)[1 - \exp(-r)]$$

where

$$r \equiv \left(\frac{1}{2}\varepsilon_o E_o^2\right)/\left(\frac{1}{2}\rho_o v_o^2\right).$$

Prob. 9.18.3 A systematic approach to writing the quasi-one-dimensional equations in terms of influence coefficients is outlined in Prob. 9.17.2. Consider here the analogous electrohydrodynamic flow with

---

1. M. Abramowitz and I. A. Stegun, Handbook of Mathematical Functions, NBS Applied Math. Series 55, U.S. Printing Office, Washington, DC, 1964, p. 319.

finite mobility and wall conductivity. Write the appropriate flow equations in a form analogous to the first equations in Prob. 9.17.2. The geometry can be taken as given so that ξ'/ξ is known, and the unknowns are ρ'_f/ρ_f, E'/E, v'/v, ρ'/ρ, p'/p, T'/T and $(M^2)'/M^2$. Invert this system of seven equations to show that the influence-coefficient representation of the equations is

$$\begin{bmatrix} v'/v \\ p'/p \\ T'/T \\ (\frac{bE}{v}+1)\ \rho_f'/\rho_f \\ E'/E \\ \rho'/\rho \\ M^{2\prime}/M^2 \end{bmatrix} = \frac{1}{M^2-1}\left[A_{ij}\right] \begin{bmatrix} -2\xi'/\xi \\ \rho_f E/\rho \\ \frac{\rho_f E}{\rho v}(\frac{\gamma-1}{\gamma})(bE+v) \\ -2(bE+v+\frac{\sigma_s E}{\xi\sigma_f})\ (\frac{\xi'}{\xi}) \\ Q \end{bmatrix}$$

where

$$A_{14} = A_{15} = A_{24} = A_{25} = A_{34} = A_{35} = A_{51} = A_{52} = A_{53} = A_{54} = A_{64} = A_{65} = A_{74} = A_{75} = 0$$

$$A_{12} = -A_{13} = A_{41} = -A_{42} = A_{43} = -A_{62} = A_{63} = 1; \quad -A_{71} = A_{72} = M^2(\gamma-1) + 2$$

$$A_{11} = A_{61} = M^2; \qquad A_{21} = A_{23} = M^2$$

$$A_{22} = -M^2(\gamma-1) - 1; \qquad A_{31} = -A_{32} = M^2(\gamma-1)$$

$$A_{33} = -A_{73} \doteq M^2\gamma - 1; \qquad A_{44} = 1/\sigma_s$$

$$A_{45} = -(M^2-1)\ \frac{E}{v}(b + \frac{2\sigma_s}{\xi\rho_f}); \qquad A_{55} = M^2 - 1$$

and

$$Q \equiv -2(1 + \frac{\sigma_s}{\xi\rho_f b})\ (\frac{\xi'}{\xi}) + \frac{\rho_f/\varepsilon_o E}{1+(2\sigma_s/\xi\rho_f b)}$$

Prob. 9.18.4 In the configuration of Fig. 9.18.1, ions are injected at the left and collected at the right with no gas flow ($v = 0$). The total current is I and the inlet radius is ξ_o. Determine the radius $\xi(z)$ required to keep the electric field $E = E_o$ independent of z. What is the associated space-charge distribution?

For Section 9.19:

Prob. 9.19.1 In the diffuser, from d to e, it is assumed that the pressure rises. Show that if the flow at the generator outlet is subsonic, $p_e > p_d$.

Prob. 9.19.2 In a "conventional" thermal power plant, shaft power from a turbine is used to drive a synchronous alternator which generates electrical power. Thus the generator of Sec. 4.7 integrates into a system fundamentally like that of Fig. P9.19.2a. The turbine plays a role in this Rankine steam cycle analogous to that of the MHD generator, directly producing shaft rather than electrical power.[2] The steam cycle is summarized by the T-S_T plot, which shows the demarcation between liquid, wet vapor and

2. The analogy to a turbine extends to the manner in which frictional heat generated at one stage can be partially recovered downstream with the inefficiency showing up through the entropy production (ignored in this problem). See E. F. Church, Steam Turbines, McGraw-Hill Book Co., 1950, Chap. 14.

superheated gas phase. Clearly the perfect gas model is not appropriate. Use the enthalpy function defined at the marked stations, and assume that the turbine acts isentropically. Find the overall efficiency, defined as the electrical power output divided by thermal power input. Assume that the generator has an efficiency η_g for mechanical to electrical conversion and that the compressor is not used. Now the MHD generator has the disadvantage that relatively high outlet temperatures must be maintained in order that the thermal ionization responsible for the gas conductivity remains effective. Thus the cycle of Fig. 9.19.1 is operated as a topping unit with the rejected heat used to drive the steam cycle of Fig. P9.19.2a. Find the overall efficiency of the combined system in terms of the enthalpy function. Show that it can be written in the form of Eq. 9.19.13 where the heat rejected is that rejected by the steam cycle.

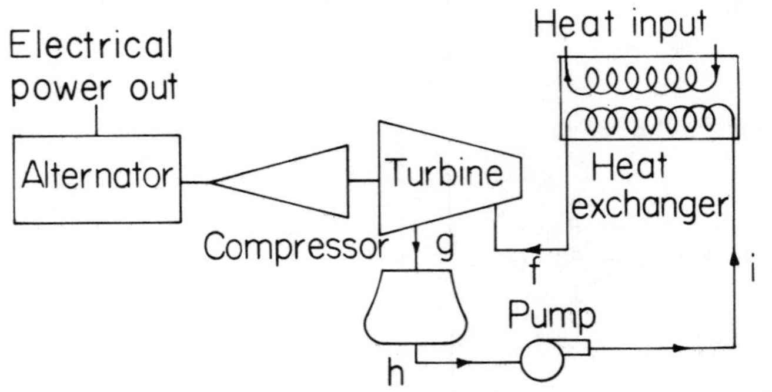

P9.19.2a

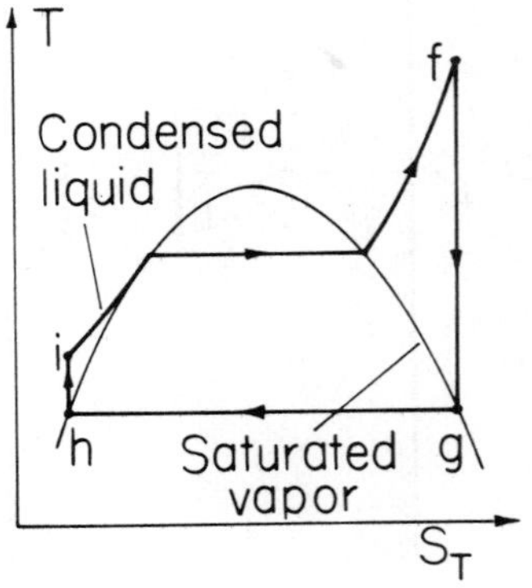

P9.19.2b

10

Electromechanics with Thermal and Molecular Diffusion

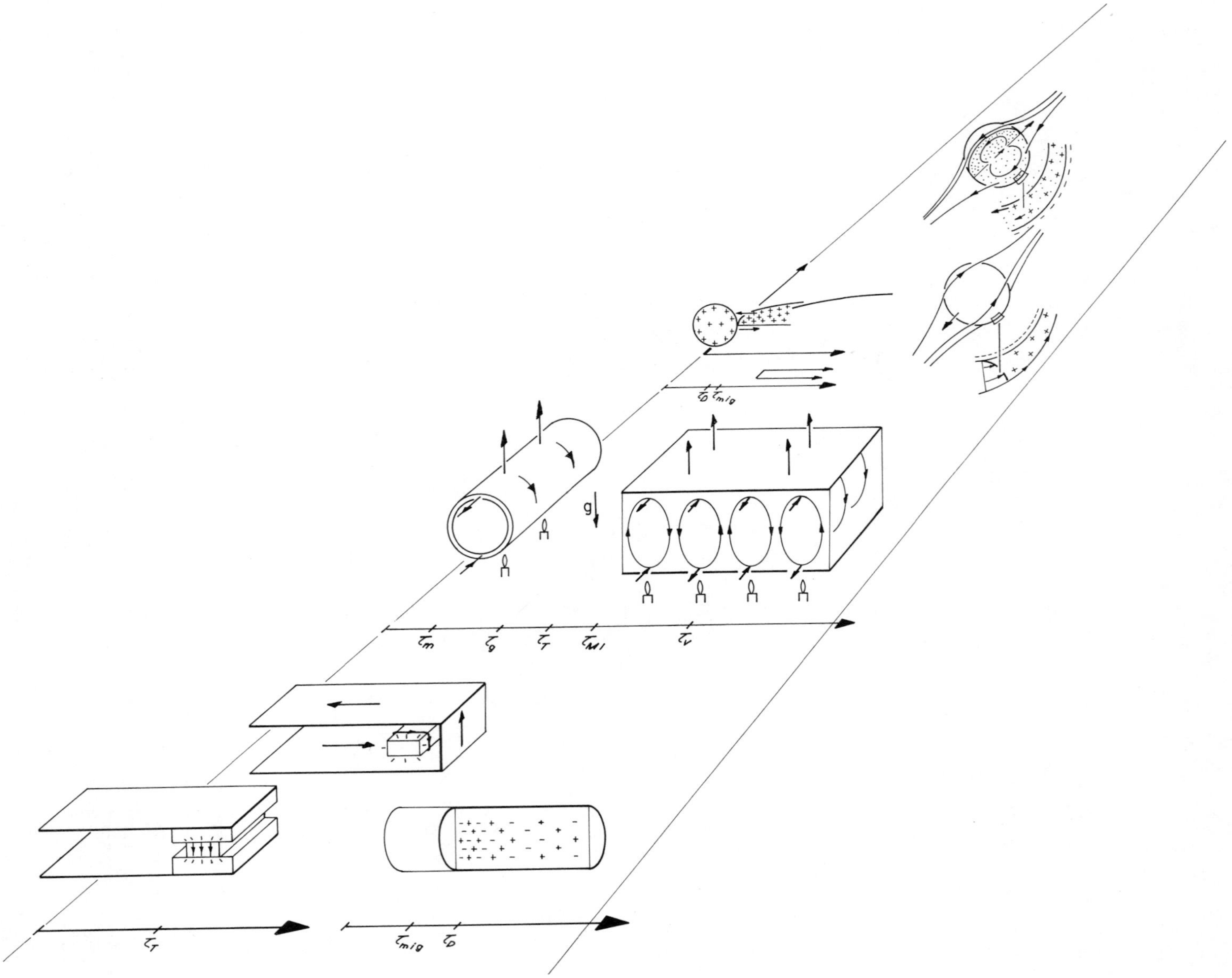

10.1 Introduction

The general three-way coupling between electromagnetic, mechanical and thermal or molecular subsystems might be pictured as in Fig. 10.1.1. Thermal interactions are the subject of the first half of this chapter while the second is concerned with the molecular subsystem.

Diffusion dynamics is familiar from the magnetic diffusion of Chap. 6 and the viscous diffusion of Chap. 7. For both thermal and neutral molecular diffusion processes, Sec. 10.2 builds on this background by identifying the characteristic times, lengths and dimensionless numbers with analogous parameters from these previous dynamical studies. Much of the sinusoidal steady-state and transient dynamics, boundary layer models and transfer relations are equally applicable here.

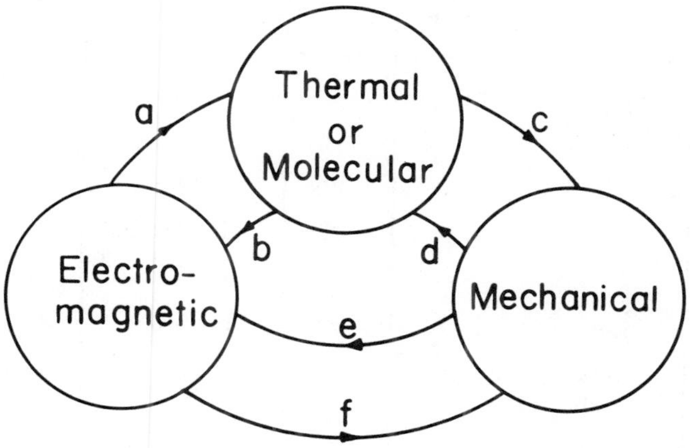

Fig. 10.1.1. Three-way coupling.

Electrical heating and the need for conduction and transport of that heat is often crucial in engineering problems. Section 10.3 is therefore devoted to this one-way coupling in which heat generated electrically in a volume is removed by thermal diffusion, (a) in Fig. 10.1.1. The three-way coupling illustrated in Sec. 10.4 involves an electrical conductivity that is a function of temperature, (b) in Fig. 10.1.1, an electric force created by the resulting property inhomogeneity, (f), and a convection that contributes to the heat transfer, (d).

The rotor model introduced in Sec. 10.5 should incite an awareness of analogies with dynamical phenomena encountered in Chaps. 5 and 6 on circulating fluids, but it should not be forgotten that the diffusion phenomena discussed in many of these sections also occur in solids. The magnetic-field-stabilized Bènard type of instability discussed in Sec. 10.6 is an example of a continuum phenomena that might be modeled by the rotor. This study gives an opportunity to illustrate how the Rayleigh-Taylor types of instability from Chap. 8 are modified if property gradients have their origins in thermal or molecular diffusion.

Because the effect of molecular diffusion of neutral species is similar to that of thermal convection, the sections on molecular diffusion are confined to the diffusion of charged species. Diffusional charging of small macroscopic particles subjected to unipolar ions is the subject of Sec. 10.7. Section 10.8 is aimed at picturing the standoff between diffusion and migration that makes a double layer possible. Based on this simple model, shear-flow electromechanics are modeled in Sec. 10.9 and used to introduce electro-osmosis and streaming potential as electrokinetic phenomena. Another electrokinetic phenomenon, electrophoresis of particles, is taken up in Sec. 10.10. Sections 10.11 and 10.12 introduce electrocapillary phenomena, where the double-layer surface force density from Sec. 3.11 comes into play. Sections 10.7 and 10.8 involve links (a) and (b) in Fig. 10.1.1, while Secs. 10.9, 10.10 and 10.12 involve all links. The sections on molecular diffusion suggest the scale and nature of electromechanical processes found in electrochemical, biological and physiological systems.

10.2 Laws, Relations and Parameters of Convective Diffusion

Thermal Diffusion: The most common thermal conduction constitutive relation between heat flux and temperature is Laplace's law:

$$\vec{\Gamma}'_T = -k_T \nabla T \tag{1}$$

where k_T is the coefficient of thermal conductivity. Not only in a perfect gas, but also for many purposes in a liquid, the internal energy is usefully taken as proportional to the temperature. Thus, the energy equation, Eq. 7.23.4, becomes

$$\frac{\partial T}{\partial t} + \vec{v}\cdot\nabla T = K_T \nabla^2 T + \frac{\phi_d}{\rho c_v}; \quad \phi_d \equiv \vec{E}'\cdot\vec{J}'_f + \phi_v - p\nabla\cdot\vec{v} \tag{2}$$

where the thermal diffusivity is defined as $K_T \equiv k_T/\rho c_v$. From left to right, terms in this expression represent the thermal capacity, convection and conduction. The last term is due to electrical and viscous dissipation and power entering the thermal system because of dilatations. Although c_v and k_T are in general functions of temperature, thermally induced variations of other parameters are usually more important and so c_v and k_T have been taken as constant in writing Eq. 2.

Table 10.2.1. Thermal diffusion parameters for representative materials.

Material	Temp. (^oC)	Mass density ρ (kg/m^3)	Specific heat (J/kg^oC)	Thermal conductivity k_T ($watts/m^oK$)	Thermal diffusivity K_T (m^2/s)	Prandtl number $P_T \equiv \frac{\eta}{\rho K_T}$
Liquid			c_p			
Water	10	1.000×10^3	4.19×10^3	0.58	1.38×10^{-7}	9.5
"	30	0.996×10^3	4.12×10^3	0.61	1.46×10^{-7}	5.5
"	70	0.978×10^3	3.96×10^3	0.66	1.61×10^{-7}	2.6
"	100	0.958×10^3	3.82×10^3	0.67	1.66×10^{-7}	1.8
Glycerine	10-70	1.26×10^3	2.5×10^3	0.28	0.89×10^{-7}	1.3×10^4
Carbon tetra-chloride	15	1.59×10^3	0.83×10^3	0.11	0.832×10^{-7}	7.3
Mercury	20	13.6×10^3	0.14×10^3	8.0	4.2×10^{-6}	2.7×10^{-2}
CErelow-117	50	8.8×10^3	0.15×10^3	16.5	1.25×10^{-5}	$\sim 5 \times 10^{-3}$
Gases			c_v			
Air	20	1.20	0.72×10^3	2.54×10^{-2}	2.1×10^{-5}	0.72
"	100	0.95	0.72×10^3	3.17×10^{-2}	3.3×10^{-5}	0.70
Solids			c_p			
Aluminum	25	2.7×10^3	0.90×10^3	240	9.4×10^{-7}	-
Copper	25	8.9×10^3	0.38×10^3	400	11×10^{-7}	-
Vitreous quartz	50	2.2×10^3	0.77×10^3	1.6	9.4×10^{-7}	-

With electrical and viscous heating given, and work done by dilatations negligible (as is usually the case in liquids), Eq. 2 becomes a convective diffusion equation analogous to magnetic diffusion equations in Chap. 6 and viscous diffusion equations in Secs. 7.18-7.20. Instead of the magnetic or viscous diffusion times, the thermal diffusion time

$$\tau_T = \ell^2/K_T \tag{3}$$

characterizes transients having ℓ as a typical length. For processes determined by convection, it is the ratio of this thermal diffusion time to the transport time, ℓ/u, that is relevant. With u a typical fluid velocity, this dimensionless number is defined as the thermal Peclet number,

$$R_T = \ell u/K_T \tag{4}$$

The response to sinusoidal steady-state thermal excitations with angular frequency ω is likely to have a spatial scale that is much shorter than other lengths of interest, in which case the thermal diffusion skin depth

$$\delta_T = \sqrt{\frac{2K_T}{\omega}} \tag{5}$$

is the length over which the thermal inertia of the bulk equilibrates the oscillatory conduction of heat. It is this length that makes $\omega\tau_T = 2$.

Typical thermal parameters are given in Table 10.2.1. In liquids, c_p and c_v are essentially equal. Even at relatively low frequencies the thermal skin depth is perhaps shorter than might be intuitively expected, as illustrated by Fig. 10.2.1.

Molecular Diffusion of Neutral Particles: The analogy between thermal and molecular diffusion is evident from a comparison of the equation for conservation of neutral particles (Eq. 5.2.9 with b = 0, G - R = 0 and $\rho_i \to n$),

$$\frac{\partial n}{\partial t} + \vec{v} \cdot \nabla n = K_D \nabla^2 n \tag{6}$$

to Eq. 2. Transient molecular diffusion, steady diffusion in a steady flow and periodic diffusion are respectively characterized by

$$\tau_D = \ell^2/K_D \qquad \text{molecular diffusion time} \qquad (7)$$

$$R_D = \ell U/K_D \qquad \text{molecular Peclet number} \qquad (8)$$

$$\delta_D = \sqrt{2K_D/\omega} \qquad \text{molecular diffusion skin depth} \qquad (9)$$

Typical parameters are given in Table 10.2.2. The molecular diffusion skin depth is presented as a function of frequency in Fig. 10.2.1, where it can be compared to the thermal skin depth for representative fluids and solids. Simple kinetic models support the observation that, in gases, molecular and thermal diffusion processes have comparable characteristic numbers.[1] Relatively long molecular diffusion times, high molecular Peclet numbers and short skin depths typify liquids on ordinary length scales. In liquids, the molecular diffusion processes occur much more slowly than for thermal diffusion.

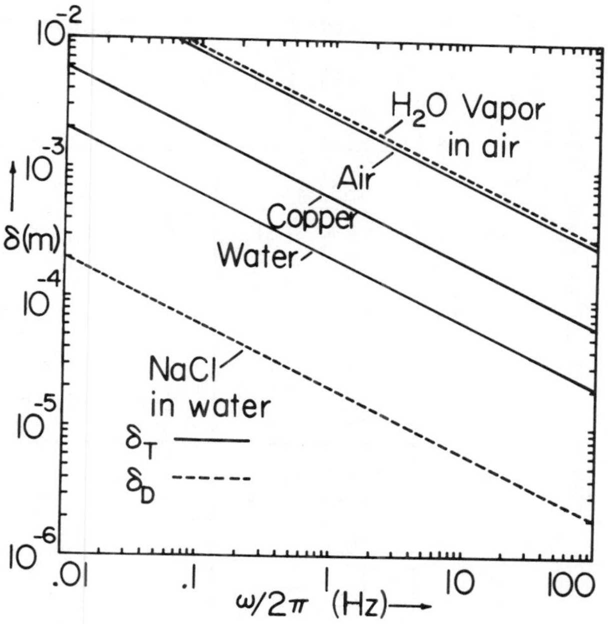

Fig. 10.2.1. Skin depth for sinusoidal steady-state diffusion of heat (solid lines) and molecular diffusion (broken lines) at frequency $f = \omega/2\pi$.

<u>Convection of Properties in the Face of Diffusion</u>: One of the most common ways in which coupling arises between the diffusion subsystem and either the electromagnetic or mechanical subsystem is through the dependence of properties on temperature or concentration. The electrical conductivity is an example. In liquids, it can be a strong function of temperature. If $\sigma = \sigma(T)$, it follows from Eq. 2 that

$$\frac{D\sigma}{Dt} = \frac{\partial\sigma}{\partial T}\frac{DT}{Dt} = \frac{\partial\sigma}{\partial t}\left[K_T\nabla^2 T + \frac{\phi_d}{\rho c_v}\right] \qquad (10)$$

so that, in the absence of diffusion and heat generation, the conductivity is a property carried by the material. That is, the right-hand side of Eq. 10 is zero. Subsequent to the transport of material having an enhanced conductivity into a region of lesser σ, the diffusion tends to return the temperature, and hence the conductivity, to the local value.

In a liquid, the electrical conductivity is linked to the molecular diffusion in a more complicated way. Suppose that an ionizable material is added to a fluid, which in the absence of the added material does not have an appreciable conductivity. Ionization is into bipolar species having charge densities $\rho_{\pm}$ with the unionized material having the number density, n.

The conservation equations for such a system were written in terms of the net charge density and conductivity in Sec. 5.9, Eqs. 9-11. Written in normalized form, the terms in these equations can be sorted out by establishing an ordering of the intrinsic times relative to times of interest, τ. Typical of relatively conducting, certainly aqueous electrolytes, is the ordering shown in Fig. 10.2.2. Because $\tau/\tau_{th} >> 1$, generation and recombination terms dominate all others in the conservation of neutrals expression, Eq. 5.9.11. It follows that

$$n = \sigma^2 - \frac{(b_+ - b_-)}{b_+ + b_-}\frac{\tau_e}{\tau_{mig}}\sigma\rho_f - \frac{b_+b_-}{(b_+ + b_-)^2}\left(\frac{\tau_e}{\tau_{mig}}\right)^2\rho_f^2 \qquad (\underline{11})$$

In the net charge density equation, Eq. 5.9.9, $\tau/\tau_e >> 1$, so that the convective derivative on the left and the last term on the right are negligible compared to the other terms. Hence, that expression becomes

$$\vec{E}\cdot\nabla\sigma + \sigma\rho_f = \frac{\tau_{mig}}{\tau_D}\frac{(K_+ - K_-)(b_+ + b_-)}{K_+b_- + K_-b_+}\nabla^2\sigma \qquad (\underline{12})$$

In Eq. 5.9.10, the first term on the right, multiplying τ/τ_{mig}, is expressed using Eq. 12, the second is negligible because $\tau_e/\tau_{mig} << 1$, the third through the sixth cancel by virtue of Eq. 11, while

1. J. O. Hirschfelder, C. F. Curtiss and R. B. Bird, <u>Molecular Theory of Gases and Liquids</u>, John Wiley & Sons, New York, London, 1954, pp. 9-16.

Table 10.2.2. Typical molecular diffusion parameters.[2]
(Prandtl number $P_D \equiv \eta/\rho K_D$).

Material	in Liquid	Temperature (^oC)	Diffusion coefficient K_D (m^2/s)	Molecular Prandtl number $P_D=\tau_D/\tau_v=\eta/\rho K_D$
NaCl	H_2O	18	1.3×10^{-9}	770
		5	0.9×10^{-9}	1700
KNO_3	H_2O	18	1.5×10^{-9}	670
HCl	H_2O	19	2.5×10^{-9}	400
KCl	H_2O	18	1.5×10^{-9}	670
I_2	Ethyl alcohol	18	1.1×10^{-9}	
Material	**in Gas**		K_D*	P_D†
O_2	Air		$K_o = 1.78 \times 10^{-5}$	0.8
H_2	N_2		$K_o = 6.74 \times 10^{-5}$	0.2
H_2O	Air		$K_o = 2.20 \times 10^{-5}$	0.7

*For these gases, $K_D = K_o(T/273)^2/p$; T in ^oK, p in atms.

†Evaluated at 0^oC.

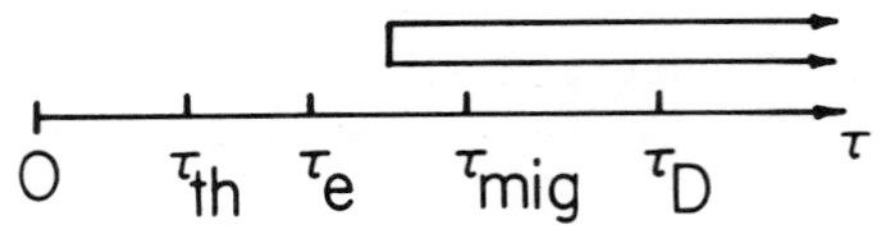

Fig. 10.2.2

Hierarchy of characteristic times for ambipolar diffusion of conductivity.

because $\tau_e \ll \tau_{mig}$, the last term is negligible compared to the next-to-last term. Hence, in dimensional form, the expression becomes

$$\frac{D\sigma}{Dt} = K_a\nabla^2\sigma; \quad K_a \equiv \frac{K_+b_- + K_-b_+}{b_+ + b_-} \tag{13}$$

Thus, the conductivity is subject to convective diffusion, but with the ambipolar diffusion coefficient, K_a. Although oppositely charged ions may have different mobilities and diffusion coefficients, the electric field generated by separation of species tends to make the species diffuse together. According to Eq. 12, the net charge can relax essentially instantaneously. Given the distribution of σ from Eq. 13, coupled through $\vec{v}$ to the mechanical subsystem, Eq. 12 can be used to find the distribution of net charge density and hence the force density.

2. For further data and indication of accuracy see E. W. Washburn, International Critical Tables, Vol. 5, McGraw-Hill Book Company, New York, 1929, p. 63.

10.3 Thermal Transfer Relations and an Imposed Dissipation Response

Fully developed flows responding to imposed force densities (Secs. 9.3-9.5) are similar in their description to the sinusoidal steady-state thermal diffusion exemplified in this section. Dissipation densities and material deformation are known, and therefore not influenced by the resulting distribution of temperature and heat flux.

A typical example, shown in Fig. 10.3.1, is an MQS induction system in which a conducting layer having thickness Δ is subject to currents induced by tangential magnetic fields at the upper and lower surfaces:

$$H_y^\alpha = \text{Re}\hat{H}_y^\alpha e^{j(\omega t-ky)}; \quad H_y^\beta = \text{Re}\hat{H}_y^\beta e^{j(\omega t-ky)} \tag{1}$$

The layer, which might be a developed model for the conductor in a rotating machine, translates in the y direction with the velocity U. Given the electrical dissipation density $\phi_d=\vec{J}_f'\cdot\vec{E}'$, what is the distribution of temperature in the layer? This density has a time-average part that depends only on x and a second harmonic traveling-wave part that depends on (x,y,t). Fortunately, for a given motion, the conduction equation, Eq. 10.2, is linear,

$$\frac{\partial T}{\partial t} + U\frac{\partial T}{\partial y} - K_T\nabla^2 T = \frac{\phi_d}{\rho c_v} \tag{2}$$

so that a transfer relation approach can be taken that combines ideas familiar from Secs. 2.16, 4.5 and 9.3. The system is in the temporal and spatial sinusoidal steady state.

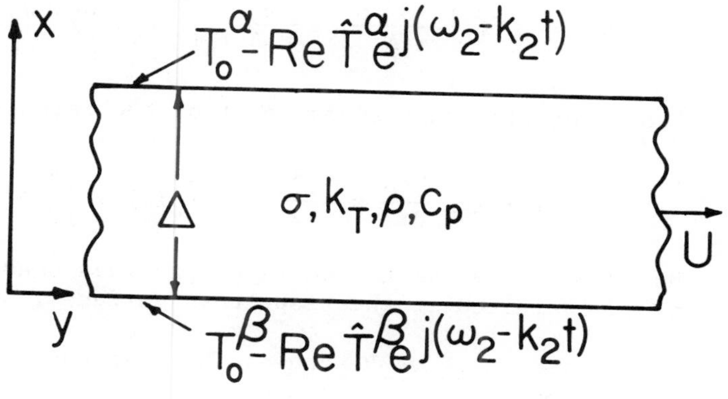

Fig. 10.3.1. Electrical dissipation due to currents induced in moving layer result in steady and second-harmonic temperature response.

Electrical Dissipation Density: The traveling-wave magnetic excitations at the (α,β) surfaces are in general determined by the structure outside the layer. If the layer is bounded by current sheets backed by infinitely permeable material, the amplitudes $(\hat{H}_y^\alpha,\hat{H}_y^\beta)$ are simply $(-\hat{K}_z^\alpha,\hat{K}_z^\beta)$. Regardless of the specific system, magnetic diffusion in the layer is described by the transfer relations (b) of Table 6.5.1. In terms of the resulting amplitudes $(\hat{A}^\alpha,\hat{A}^\beta)$, the distribution of the vector potential follows from Eq. 6.5.6:

$$\hat{A} = \hat{A}^\alpha\frac{\sinh\gamma_m x}{\sinh\Delta_m} - \hat{A}^\beta\frac{\sinh\gamma_m(x-\Delta)}{\sinh\gamma_m\Delta}; \quad \gamma_m = \sqrt{k^2 + j\mu\sigma(\omega - kv)} \tag{3}$$

The electrical dissipation follows by evaluating

$$\phi_d = \vec{E}'\cdot\vec{J}_f' = \frac{\vec{J}_f\cdot\vec{J}_f}{\sigma} = \frac{1}{2\pi}[\hat{J}_z\hat{J}_z^* + \text{Re}\hat{J}_z^2 e^{j(2\omega t-2ky)}] \tag{4}$$

with the current density related to $\hat{A}$ by Eq. 6.7.5,

$$\hat{J}_z = -\frac{1}{\mu}\left(\frac{d^2A}{dx^2} - k^2\hat{A}\right) = -j\sigma(\omega - kU)\hat{A}(x) \tag{5}$$

Thus, the dissipation density is determined, with a steady x-dependent part and a second-harmonic traveling-wave component,

$$\phi_d = \phi_o(x) + \text{Re}\hat{\phi}(x)e^{j(2\omega t-2ky)} \tag{6}$$

where

$$\phi_o = \frac{1}{2}\sigma(\omega - kU)^2\hat{A}\hat{A}^*; \quad \hat{\phi} = -\frac{1}{2}\sigma(\omega - kU)^2\hat{A}^2$$

The temperature response is now the superposition of parts that are respectively due to the steady and to the second-harmonic drives,

$$T = T_o(x) + \mathrm{Re}\hat{T}(x)e^{j(\omega_2 t - k_2 y)} \tag{7}$$

where $\omega_2 \equiv 2\omega$ and $k_2 \equiv 2k$.

Steady Response: Because the steady dissipation depends only on x and the system extends to infinity in the y directions, Eq. 1 reduces to

$$\frac{d^2 T_o}{dx^2} = -\frac{\phi(x)}{k_T} \tag{8}$$

This expression is integrated twice, using as boundary conditions that the steady part of the temperatures at $x = \Delta$ and $x = 0$ are respectively T_o^α and T_o^β:

$$T_o = T_o^\beta + \left[\frac{(T_o^\alpha - T_o^\beta)}{\Delta} + \frac{1}{k_T\Delta}\int_o^\Delta\int_o^{x'} \phi_o(x'')dx''dx'\right]x - \frac{1}{k_T}\int_o^x\int_o^{x'} \phi_o(x'')dx''dx' \tag{9}$$

Associated with this steady part of the response is the heat flux

$$\Gamma_o(x) = -k_T\frac{dT_o}{dx} = -\frac{k_T}{\Delta}(T_o^\alpha - T_o^\beta) + \int_o^x \phi_o(x')dx' - \frac{1}{\Delta}\int_o^\Delta\int_o^{x'} \phi_o(x'')dx''dx' \tag{10}$$

The system external to the layer provides constraints on (T_o^α, T_o^β) and $(\Gamma_o^\alpha, \Gamma_o^\beta)$ which, together with Eq. 10 evaluated at the respective surfaces, specialize these general relations.

Traveling-Wave Response: The response to the traveling wave of dissipation can itself be divided into a homogeneous and particular part. Each takes the complex-amplitude form $\mathrm{Re}\hat{T}\exp j(\omega_2 t - k_2 y)$, and so Eq. 2 requires that

$$\left(\frac{d^2}{dx^2} - \gamma_T^2\right)\begin{bmatrix}\hat{T}_H \\ \hat{T}_P\end{bmatrix} = \begin{bmatrix}0 \\ -\frac{\hat{\phi}}{k_T}\end{bmatrix}; \quad \gamma_T^2 \equiv k^2 + \frac{j(\omega_2 - k_2 U)}{K_T} \tag{11}$$

The homogeneous expression takes the same form, Eq. 2.16.13 with $\gamma \to \gamma_T$, as for the flux-potential relations from Table 2.16.1, so the heat-flux temperature transfer relations can be written by analogy:

$$\begin{bmatrix}\hat{\Gamma}_H^\alpha \\ \hat{\Gamma}_H^\beta\end{bmatrix} = k_T\gamma_T\begin{bmatrix}-\coth\gamma_T\Delta & \frac{1}{\sinh\gamma_T\Delta} \\ \frac{-1}{\sinh\gamma_T\Delta} & \coth\gamma_T\Delta\end{bmatrix}\begin{bmatrix}\hat{T}_H^\alpha \\ \hat{T}_H^\beta\end{bmatrix} \tag{12}$$

The total solution is $\hat{T} = \hat{T}_H + \hat{T}_P$ and it follows that $\hat{T}_H = \hat{T} - T_P$. Substitution of this and the associated heat flux $\hat{\Gamma}_H = \Gamma - \Gamma_P$ on the left and right in Eq. 12 results in transfer relations expressing the combined response of the layer to internal and external dissipations:

$$\begin{bmatrix}\hat{\Gamma}^\alpha \\ \hat{\Gamma}^\beta\end{bmatrix} = k_T\gamma_T\begin{bmatrix}-\coth\gamma_T\Delta & \frac{1}{\sinh\gamma_T\Delta} \\ \frac{-1}{\sinh\gamma_T\Delta} & \coth\gamma_T\Delta\end{bmatrix}\begin{bmatrix}\hat{T}^\alpha - \hat{T}_P^\alpha \\ \hat{T}^\beta - \hat{T}_P^\beta\end{bmatrix} + \begin{bmatrix}\hat{\Gamma}_P^\alpha \\ \hat{\Gamma}_P^\beta\end{bmatrix} \tag{13}$$

Any particular solution can be used to evaluate these expressions; but, following the approach used in Sec. 4.5, suppose that both the dissipation density and the particular solution are expressed as a summation of the same modes $\Pi_i(x)$:

$$\hat{\phi} = \sum_{i=0}^{\infty} \hat{\phi}_i\Pi_i(x); \quad \hat{T}_P = \sum_{i=0}^{\infty} \hat{T}_i\Pi_i(x) \tag{14}$$

Then, Eq. 11b shows that these modes satisfy the equation

$$\left(\frac{d^2}{dx^2} + \nu_i^2\right)\Pi_i = 0; \quad \nu_i^2 \equiv \frac{\hat{\phi}_i}{k_T\hat{T}_i} - k_2^2 - \frac{j(\omega_2 - k_2U)}{K_T} \tag{15}$$

Boundary conditions to be satisfied by these modes are a matter of convenience in writing Eq. 13 or expressing Π_i. Here, T_p, and hence Π_i, is taken as zero at the boundaries,

$$\Pi_i = \sin \nu_i x; \quad \nu_i = \frac{i\pi}{\Delta} \tag{16}$$

and it follows from the definition of ν_i, Eq. 15, that

$$\hat{T}_i = \hat{\phi}_i/k_T\left[(\frac{i\pi}{\Delta})^2 + k_2^2 + \frac{j(\omega_2 - k_2U)}{k_T}\right] \tag{17}$$

The amplitudes, $\hat{\phi}_i$, are in this case simply Fourier amplitudes evaluated exploiting the orthogonality of the modes, Π_i,

$$\hat{\phi}_i = \frac{2}{\Delta}\int_0^\Delta \hat{\phi}(x)\sin \nu_i x dx \tag{18}$$

Thus, because $T_p = 0$ on the α and β surfaces, the total temperature response to the traveling-wave part of the dissipation is

$$T = \text{Re}\left\{\left[\hat{T}^\alpha \frac{\sinh \gamma_T x}{\sinh \gamma_T\Delta} - \hat{T}^\beta \frac{\sinh \gamma_T(x-\Delta)}{\sinh \gamma_T\Delta}\right] + \sum_{i=1}^{\infty} \hat{T}_i \sin (\frac{i\pi}{\Delta} x)\right\} e^{j(\omega_2 t - k_2 y)} \tag{19}$$

In terms of the same temperature amplitudes, $(\hat{T}^\alpha,T^\beta)$, the heat flux at the boundaries follows by evaluating Eq. 13:

$$\begin{bmatrix}\hat{f}^\alpha \\ \hat{f}^\beta\end{bmatrix} = k_T\gamma_T\begin{bmatrix}-\coth \gamma_T\Delta & \frac{1}{\sinh \gamma_T\Delta} \\ \frac{-1}{\sinh \gamma_T\Delta} & \coth \gamma_T\Delta\end{bmatrix}\begin{bmatrix}\hat{T}^\alpha \\ \hat{T}^\beta\end{bmatrix} - \sum_{i=1}^{\infty} \frac{(\frac{i\pi}{\Delta})\hat{\phi}_i}{\left[(\frac{i\pi}{\Delta})^2 + k_2^2 + \frac{j(\omega_2 - k_2U)}{K_T}\right]}\begin{bmatrix}(-1)^i \\ 1\end{bmatrix} \tag{20}$$

These transfer relations between temperatures and heat fluxes at the (α,β) surfaces of the layer, are applicable to the description of different thermal conduction systems in which the layer might be embedded.

In practice, the thermal diffusion skin depth $\delta_T \equiv \sqrt{2k_T/(\omega_2 - k_2U)}$, based on the Doppler frequency $(\omega_2 - k_2U)$, is likely to be short compared either to the thickness of the layer or to half the wavelength of the magnetic field, $2\pi/k_2$. For example, from the curve for copper in Fig. 10.21, $\delta_T \approx .06$ mm at $\omega/2\pi = 100$ Hz. Thus for the lowest values of i in either Eq. 17 or Eq. 20, it is likely that

$$(\frac{i\pi}{\Delta})^2 + k_2^2 << \frac{|\omega_2 - k_2U|}{K_T} \equiv \frac{2}{\delta_T'^2} \tag{21}$$

The Fourier coefficients $\hat{T}_i$ are therefore proportional to $\hat{\phi}_i$ for the lowest terms in the series, and the driven response has essentially the same profile over the layer cross section as does the dissipation density. In this case, the thermal capacity absorbs the heat with a 90° time delay of the temperature relative to the dissipation density. There is insufficient time for the heat to diffuse appreciably. Also note that in this short thermal skin-depth limit these lowest order terms are proportional to $\delta_T'^2$, and so the thermal inertia represented by the heat capacity tends to suppress the oscillatory part of the temperature response.

10.4 Thermally Induced Pumping and Electrical Augmentation of Heat Transfer

By means of a simple one-dimensional flow, illustrated in this section is the three-way interaction between electric, kinetic and thermal subsystems. The flow is essentially incompressible. Shown in Fig. 10.4.1 is a section from a duct for fluid flowing in the y direction. Grids in the planes $y = 0$ and $y = \ell$ constrain the fluid temperatures in these planes to be T_a and T_b, respectively. Typical of many semi-insulators (such as doped hydrocarbons, plasticizers and even chocolate), this liquid has an electrical conductivity that is a function of temperature. For this example,

$$\sigma = \sigma_a[1 + \alpha_T(T - T_a)] \tag{1}$$

where σ and α_T are constant material properties.

With the application of a potential difference, V, between the grids, there is a current density J_o that flows in the y direction between the grids. Continuity requires that J_o be independent of y, and hence that the electric field between the grids be nonuniform. The charge density attending this nonuniformity conspires with the electric field itself to give an electric force density tending to pump the liquid. However, fluid motion implies the convection of heat and a field induced contribution to the temperature distribution and hence to the heat transferred between the grids.

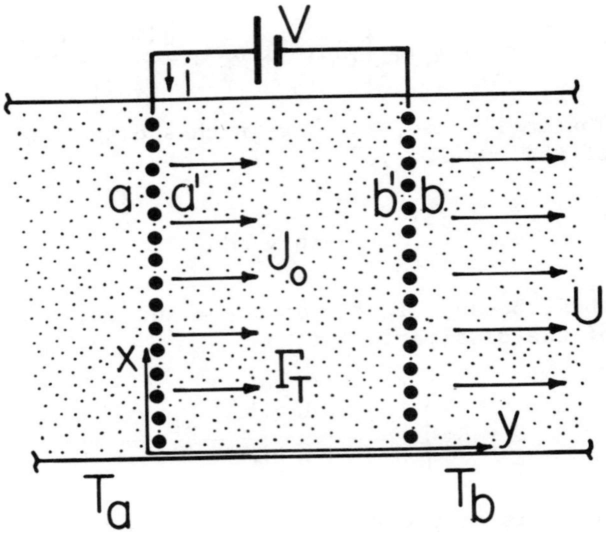

Fig. 10.4.1. Configuration for electrothermally induced pumping and electrically augmented heat transfer.

The width of the channel is large compared to ℓ. Hence, the velocity profile is uniform with respect to the transverse direction. Viscous effects are confined to the flow through the grids and reflected in a pressure drop through each of the grids. Because the flow is one-dimensional and essentially incompressible, $\vec{v} = U\vec{i}_y$. In terms of this velocity and the locations indicated in Fig. 10.4.1, the pressure drops through the grids are taken as

$$p_a - p_{a'} = \frac{c\eta U}{d}; \quad p_{b'} - p_b = \frac{c\eta U}{d} \tag{2}$$

where the dimensionless coefficient c is determined by the geometry of the grids. The dependence of the grid pressure drops on the viscosity and velocity is consistent with flow through the grids at a low Reynolds number based on a characteristic dimension, d, of the grid.

<u>Elecrical Relations</u>: Consistent with the geometry is an electric field having the form $\vec{E} = E(y)\vec{i}_y$. It is assumed that in this EQS system, the charge relaxation time, τ_e, is short compared to the thermal diffusion time, τ_T, and that the transport time, ℓ/U, is long compared to τ_e but arbitrary relative to τ_T. That $\tau_e << \ell/U$ means that the convection current density, $\rho_f U$, can be ignored compared to the electrical conduction current density, σE. Thus, even with the fluid motion, Ohm's law is simply $\vec{J} = \sigma\vec{E}$ with σ given by Eq. 1. Because $\vec{J}_f = J_o\vec{i}_y$ is independent of y, this makes it possible to specify E(y) in terms of the yet to be determined temperature distribution, T(y):

$$E = J_o\{\sigma_a[1 + \alpha_T(T - T_a)]\}^{-1} \tag{3}$$

With the terminal current, i, taken as the cross-sectional area, A, times J_o, the electrical terminal relation is then given by

$$V = \int_o^\ell E dx = \frac{i}{A\sigma_a}\int_o^\ell \frac{dx}{[1 + \alpha_T(T - T_o)]} \tag{4}$$

<u>Mechanical Relations</u>: Only the longitudinal component of the Navier-Stokes equation is relevant, and because the flow is one-dimensional, neither inertial nor viscous force densities make a contribution between the grids. With the electrical force density written as the divergence of the Maxwell stress ($\rho_f E = d(\frac{1}{2}\varepsilon E^2)/dy$), the force equation then becomes simply

$$\frac{\partial}{\partial y}(p - \frac{1}{2}\varepsilon E^2) = 0 \tag{5}$$

The quantity in brackets is independent of y and can be evaluated by letting $p(0) = p_{a'}$ and $E(0) = E(T = T_a)$. Thus, evaluation of p at $y = \ell$ where $p = p_{b'}$ gives

$$p_{a'} - p_{b'} = \frac{1}{2}\varepsilon(E_a^2 - E_b^2) \tag{6}$$

By means of Eqs. 2, this expression is expressed in terms of the pressures just outside the grids:

$$p_b - p_a = -\frac{2c\eta U}{d} + \frac{1}{2}\varepsilon\frac{J_o^2}{\sigma_a^2}\{[1 + \alpha_T(T_b - T_a)]^{-2} - 1\} \tag{7}$$

As with the electrical relations, Eqs. 3 and 4, the temperature distribution is required to evaluate this mechanical terminal relation.

<u>Thermal Relations</u>: With the electrical and viscous dissipations taken as negligible compared to thermal inputs from the grids, the energy equation, Eq. 10.2.2, reduces to

$$\rho c_v U\frac{dT}{dy} = k_T\frac{d^2T}{dy^2} \tag{8}$$

With one integration, this expression simply states that the heat flux, Γ_T, is independent of y:

$$\Gamma_T = -k_T\frac{dT}{dy} + \rho c_v UT \tag{9}$$

The temperature distribution is then determined by solving Eq. 9 subject to the condition that $T(0)=T_a$:

$$T = \frac{\ell\Gamma_T}{k_T R_T}(1 - e^{R_T(y/\ell)}) + T_a e^{R_T(y/\ell)} \tag{10}$$

Here, $R_T \equiv \rho c_v U\ell/k_T$ is the thermal Peclet number.

What might be termed the thermal terminal relation is found by evaluating Eq. 10 at $y = \ell$ where $T = T_b$ and solving for the heat flux, now determined by T_a and T_b and the velocity U (represented by R_T):

$$\Gamma_T = \frac{T_b - T_a e^{R_T}}{1 - e^{R_T}}\,\frac{k_T R_T}{\ell} \tag{11}$$

By way of emphasizing the degree to which convection contributes to the heat flux, the Nusselt number, N_u, is defined as the ratio of Γ_T to what the flux would be at the same temperature difference if only thermal conduction were present:

$$Nu \equiv \frac{\Gamma_T}{k_T(T_a - T_b)/\ell} = \frac{R_T}{1 - e^{R_T}}\left[\frac{\dfrac{T_b}{T_a} - e^{R_T}}{1 - \dfrac{T_b}{T_a}}\right] \tag{12}$$

Given terminal constraints on the external pressure difference, $p_b - p_a$, electrical current, $i = J_oA$ and temperatures T_a and T_b, the remaining variables are now known. The distribution of electric field intensity and the voltage are given by Eqs. 3 and 4. The flow velocity, U, follows from Eq. 7, and hence R_T is determined. Finally, the temperature distribution and heat flux (or Nusselt number) are given by Eqs. 10-12.

Illustrated in Fig. 10.4.2 is N_u as a function of R_T for the case where $T_a > T_b$. From Eq. 7, note that if the flow were re-entrant so that $p_a = p_b$, the fluid velocity and hence R_T would be proportional to J_o^2. Typical distributions of the temperature are shown in the insets to Fig. 10.4.2. Illustrated is the tendency of the convection to skew the temperature profile in the streamwise direction from the linear profile for conduction alone.

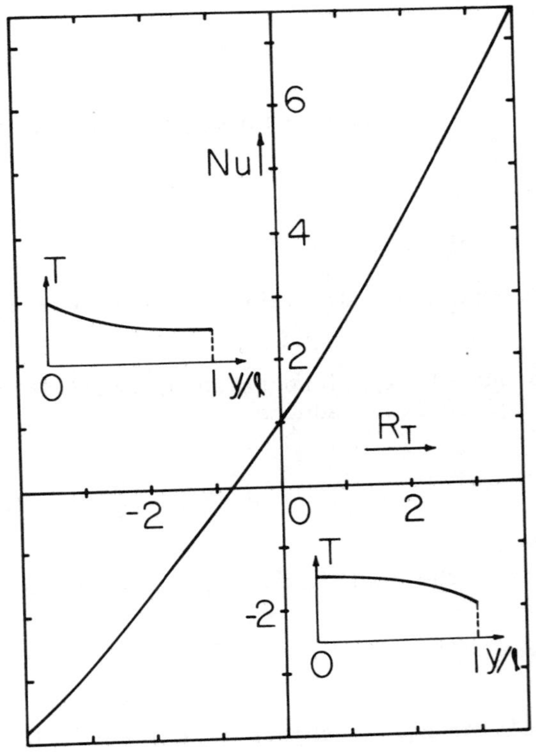

Fig. 10.4.2

Nusselt number as a function of thermal Peclet number for $T_b/T_a = 0.5$. Inserts show temperature distributions typical of positive ($R_T = 4$) and negative ($R_T = -4$) flows.

10.5. Rotor Model for Natural Convection in a Magnetic Field

When heated, most fluids decrease in mass density. In a gravitational field, the result is a tendency for hot fluid to rise and be replaced by falling cold fluid. Heating and cooling systems exploit the transport of heat through the agent of this "natural" convection.

The electrothermal pumping illustrated in Sec. 10.4 is an electromechanical analogue of this process. Gravity is replaced by the electric field and the role of the temperature-dependent mass density taken by the electrical conductivity.

The model developed in this section can be applied to understanding such aspects of thermally induced convection as the instability that starts the convection with the thermally stratified system satisfying conditions for a static equilibrium.

Thermally induced circulations are often undesirable. An example is in the growth of crystals, where convection is a source of imperfections in the product. Especially in liquid metals, it is possible to damp these circulations by applying a magnetic field. Such damping is included in the model.

In Sec. 10.6, the incipience of the instability and its magnetic stabilization are considered again in terms of the more general fluid mechanics, but for small-amplitude circulations. The model developed here retains nonlinear dynamical effects and is similar to models that have proved useful in gaining insights into magnetohydrodynamic circulations of the earth's core.[1]

The cylindrical rotor, shown in Fig. 10.5.1, is both a thermal and an electrical conductor, such as a metal. It is free to rotate with angular velocity Ω. Surrounding the rotor is a jacket, the exterior of which is constrained in temperature to $T_{ext}(\theta)$. Specifically, representing heating from below and cooling from above would be the exterior temperature distribution

$$T_{ext} = T_E - T_e \sin\theta \tag{1}$$

if T_E were positive.

Heat transferred across the thickness, d, of the layer to the shell has alternative mechanisms for reaching the top of the cylinder and being transferred back across the layer to the exterior. Along the shell periphery, the heat can be thermally conducted, or if the shell turns out to be moving,

1. W. V. R. Malkus, "Non-periodic Convection at High and Low Prandtl Number," Mem. Soc. Roy. Sci. Liège 4 [6], 125-128 (1972).

it can be convected. For conduction alone, the heat flux is symmetric and so also is the temperature distribution. Thus, if there is no motion, thermally induced changes in mass density on the right are the same as to the left, and the effect of gravity gives rise to no net torque. But, if there is motion, conduction of heat is augmented on one side but inhibited on the other, and there is a skewing of the temperature distribution. The result is an expansion of the material on one side that exceeds that on the other, and a net gravitional torque that tends to further encourage the motion. This tendency toward instability that depends on the rate of rotation is countered by two other rate processes. One results in viscous drag from the fluid surrounding the cylinder, modeled here by the thin layer of fluid. As an additional damping mechanism, a magnetic field $\vec{H} = H_o\vec{i}_y$ is imposed. Thus, in response to the motion, z-directed currents are induced in the cylinder in the neighborhoods of the north and the south poles, and these conspire with H_o to produce a rate-dependent damping torque on the rotor.

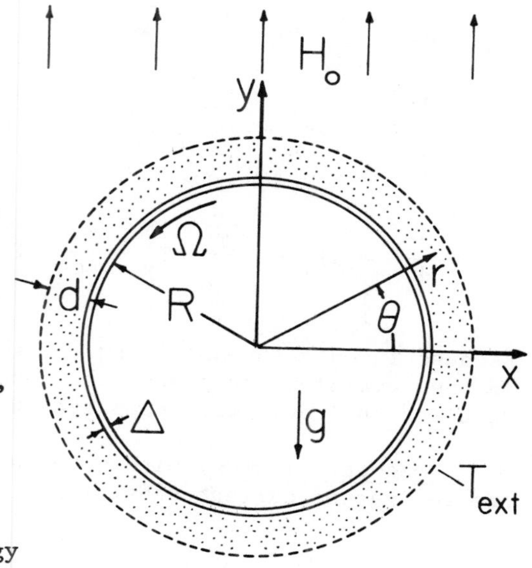

Fig. 10.5.1. Cross section of rotor used to model thermally induced convection.

Heat Balance for a Thin Rotating Shell: An incremental section of the shell, shown in Fig. 10.5.2, is described by the energy equation in integral form, Eq. 7.23.3. Consistent with the materials being only weakly compressible is the neglect of $p\nabla\cdot\vec{v}$. The objective here is a quasi-one-dimensional model playing a heat-transfer role that is analogous to that of the shell models introduced in Sec. 6.3 for magnetic diffusion.

The rate of increase of the thermal energy stored in the section of shell is accounted for by the net convection and conduction of heat into the section plus the volume dissipation,

$$\rho c_v[R(\Delta\theta)\Delta]\frac{\partial T}{\partial t} = -\rho c_v \Delta\Omega R[T(\theta+\Delta\theta) - T(\theta)]$$

$$+ k_T\Delta[\frac{1}{R}\frac{\partial T}{\partial\theta}(\theta+\Delta\theta) - \frac{1}{R}\frac{\partial T}{\partial\theta}(\theta)] - R(\Delta\theta)[\![\Gamma_r]\!] + \phi_d R(\Delta\theta)\Delta \qquad (2)$$

Divided by $\Delta\theta$ and in the limit $\Delta\theta \to 0$, Eq. 2 becomes

$$(\frac{\partial}{\partial t} + \Omega\frac{\partial}{\partial\theta})T = \frac{k_T}{\rho c_v R^2}\frac{\partial^2 T}{\partial\theta^2} - \frac{1}{\Delta\rho c_v}[\![\Gamma_r]\!] + \frac{\phi_d}{\rho c_v} \qquad (3)$$

In the following, it is assumed that the volume dissipation, ϕ_d, associated for example with ohmic heating, is negligible compared to heating from the exterior.

In Eq. 3, the angular velocity (like the temperature) is a dependent variable. The expression is nonlinear. Because the shell can only suffer rigid-body rotation, it is appropriate to reduce the thermal aspects of the problem to "lumped-parameter" terms as well. If the thermal excitation were more complicated than Eq. 1, it would be necessary to represent the temperature distribution in terms of a Fourier series. But for the given single harmonic external temperature distribution, only the first harmonic in the series is required:

$$T = T_o(t) + T_x\cos\theta + T_y\sin\theta \qquad (4)$$

The components (T_x, T_y) represent the components of a "thermal axis" for the cylinder.

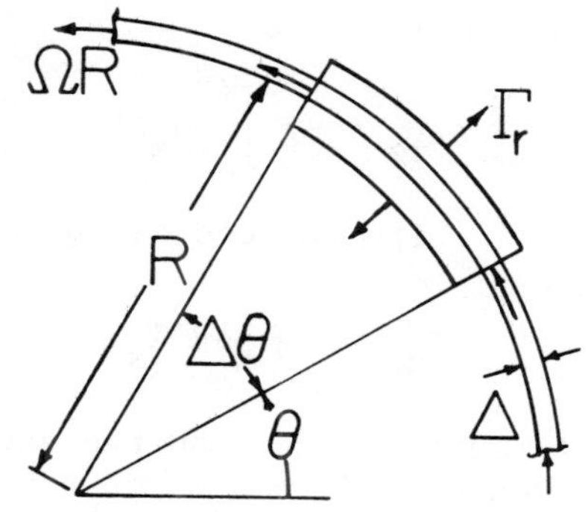

Fig. 10.5.2. Incremental section of thermally conducting moving shell.

Heat flux through the jacket is represented in terms of a surface coefficient of heat transfer, h, so that $\Gamma_r = h(T - T_e)$. For pure conduction through a fluid layer having thermal conductivity, k_{Tf}, and thickness d, $h = k_{Tf}/d$.

Substitution of Eq. 4 into Eq. 3 results in terms that are independent of θ and that multiply $\cos\theta$ and $\sin\theta$, respectively. The equation is satisfied by making each of these groups vanish. Hence the three expressions

$$\frac{dT_o}{dt} = -\frac{h}{\Delta\rho c_v}(T_o - T_E) \tag{5}$$

$$\frac{dT_x}{dt} = -\Omega T_y - \frac{k_T}{\rho c_v R^2} T_x - \frac{h}{\Delta\rho c_v} T_x \tag{6}$$

$$\frac{dT_y}{dt} = \Omega T_x - \frac{k_T}{\rho c_v R^2} T_y - \frac{h}{\Delta\rho c_r} T_y - \frac{hT_e}{\Delta\rho c_v} \tag{7}$$

Because T_o only appears in Eq. 5, that expression serves to determine the mean temperature distribution. In the remaining equations, the dependent variables are (T_x, T_y, Ω). Thus, a mechanical (torque) equation for the rotor will complete the description.

Magnetic Torque: Within the electrically conducting shell, Ohm's law (Eq. 6.2.2) requires that the z-directed current density be

$$J_z = \sigma(E_z - \mu_o H_o \Omega R \sin\theta) \tag{8}$$

Here, the magnetic field intensity due to the current in the rotor is ignored. The ends of the cylindrical shell are pictured as being shorted electrically by perfect conductors. Because the electric field in this imposed field approximation is irrotational, and there is no magnetization contribution to Faraday's law, the shorts require that $E_z = 0$ in Eq. 8. Thus, the magnetic torque per unit length in the z direction is

$$\tau_{zm} = -\Delta\int_o^{2\pi}(\vec{J} \times \mu_o H_o \vec{i}_y) R \sin\theta (Rd\theta) = -\Delta R^3 \mu_o^2 H_o^2 \sigma\Omega \int_o^{2\pi} \sin^2\theta = -\Delta R^3 \pi\sigma\mu_o^2 H_o^2 \Omega \tag{9}$$

Consistent with the low magnetic Reynolds number approximation used is a torque proportional to speed that tends to retard the rotation.

Buoyancy Torque: Typically, an increase in temperature results in a decrease in density, although there are exceptions. For small excursions in temperature the surface mass density (kg/m^2) is taken as

$$\sigma_m = \sigma_M[1 - \alpha(T - T_E)] \tag{10}$$

where α is typically positive. Of course, associated with an increase in surface mass density is a local extension of the shell. The resulting effect on the radius tends to be cancelled by contractions elsewhere, but in any case will be neglected. Thus, the net gravitational torque per unit length on the shell is

$$\tau_{zg} = -g\int_o^{2\pi} \sigma_M[1 - \alpha(T - T_E)]R \cos\theta \, Rd\theta \tag{11}$$

With the use of Eq. 4, this integral reduces to

$$\tau_{zg} = \pi g \sigma_M \alpha R^2 T_x \tag{12}$$

A positive T_x means the shell is hotter on the right than on the left, and for positive α, material should tend to rise on the right and fall on the left. As expressed, this buoyancy torque is indeed positive under such circumstances.

Viscous Torque: The fluid in the jacket surrounding the shell is presumed thin enough that its inertia is negligible compared to that of the shell. Also, viscous diffusion is complete in times of interest. Then, the flow can be pictured as plane Couette with a shear stress $-\eta\Omega R/d$. Thus, the viscous torque is

$$\tau_{zv} = -\frac{2\pi\eta R^3}{d}\Omega \tag{13}$$

Torque Equation: The shell has essentially a moment of inertia per unit length $2\pi R^3\sigma_M$. (Small changes due to the expansion are ignored.) Thus, the torques from Eqs. 9, 12 and 13 are set equal to the inertial torque:

$$2\pi R^3\sigma_M \frac{d\Omega}{dt} = -\Delta R^3\pi\sigma\mu_o^2H_o^2\Omega + \pi g\sigma_M R^2 T_x - \frac{2\pi\eta R^3}{d}\Omega \tag{14}$$

Along with Eqs. 6 and 7, this expression provides a relationship between T_x, T_y, and Ω.

Dimensionless Numbers and Characteristic Times: Normalization of the three equations of motion so that

$$T_x = \underline{T}_x/T_e,\ T_y = \underline{T}_y/T_e,\ t = \underline{t}\tau_T,\ \Omega = \underline{\Omega}/\tau_T \tag{15}$$

identifies characteristic times:

$$\tau_T \equiv \frac{\rho c_v R^2}{k_T};\quad \tau_t \equiv \frac{\Delta\rho c_v}{h};\quad \tau_g^2 = \frac{R\sigma_M}{g(\sigma_M\alpha T_e)};\quad \tau_m = \mu\Delta R\sigma;\quad \tau_v = \frac{d\sigma_M}{\eta};\quad \tau_{MI}^2 = \frac{R\sigma_M}{\mu_o H_o^2}$$

and dimensionless numbers

$$R_{av} \equiv \frac{Rdg\sigma_M\alpha T_e\rho c_v}{2\eta k_T} = \frac{\tau_v\tau_T}{2\tau_g^2};\quad R_{am} \equiv \frac{Rg\sigma_M\alpha T_e\rho c_v}{\Delta\sigma\mu_o^2H_o^2k_T} = \frac{\tau_{MI}^2\tau_T}{\tau_m\tau_g^2}$$

and leaves Eqs. 6,7 and 14 in the form

$$\frac{dT_x}{dt} = -\Omega T_y - T_x(1+f) \tag{16}$$

$$\frac{dT_y}{dt} = \Omega T_x - T_y(1+f) - f \tag{17}$$

$$\frac{1}{p_T}\frac{d\Omega}{dt} = -\Omega + R_a T_x \tag{18}$$

where $\tau_T/\tau_t \equiv f$. Thus, only three dimensionless numbers specify the physical situation, f,

$$R_a \equiv [R_{av}^{-1} + R_{am}^{-1}]^{-1} \text{ and } p_T \equiv \left(\frac{\tau_T}{\tau_v} + \frac{1}{2}\frac{\tau_m}{\tau_{MI}^2}\tau_T\right) \tag{19}$$

The thermal diffusion and relaxation times τ_T and τ_t, respectively, represent the dynamics of heat conduction in the azimuthal direction and radially through the jacket, in the face of the shell's thermal inertia. The period of a gravitational pendulum having differential surface mass density $\sigma_M\alpha T_e$ and total surface mass density σ_M is familiar from the gravity waves described in Sec. 8.9. The thin-shell magnetic diffusion time, τ_m, is the time for circulating currents to decay (Sec. 6.10), while τ_v is a viscous diffusion time based on the fluid viscosity but the mass density of the shell (Sec. 7.18).

The Rayleigh number, R_a, is large if the time for gravitational acceleration is short (τ_g is small) compared to the geometric mean of the time for viscous slowing of the shell, τ_v, and the time for the shell temperature to return to a uniform distribution, τ_T. Put another way, τ_g^2/τ_v is a gravity-viscous time representing the competition of gravitational and viscous forces. The Rayleigh number is then the ratio of the thermal diffusion time to this gravity-viscous time.

The magnetic Rayleigh number, R_{am}, is large if τ_g is short compared to the geometric mean of τ_T and τ_{MI}^2/τ_m, where the latter is the time required for the magnetic damping to slow the shell despite its inertia.

In the absence of the magnetic field, p_T plays the role of a thermal Prandtl number, the ratio of the thermal to the viscous diffusion time. With negligible viscosity but a magnetic field, the number becomes what might be termed a thermal-magnetic Prandtl number, where the viscous diffusion time is replaced by the time τ_{MI}^2/τ_m.

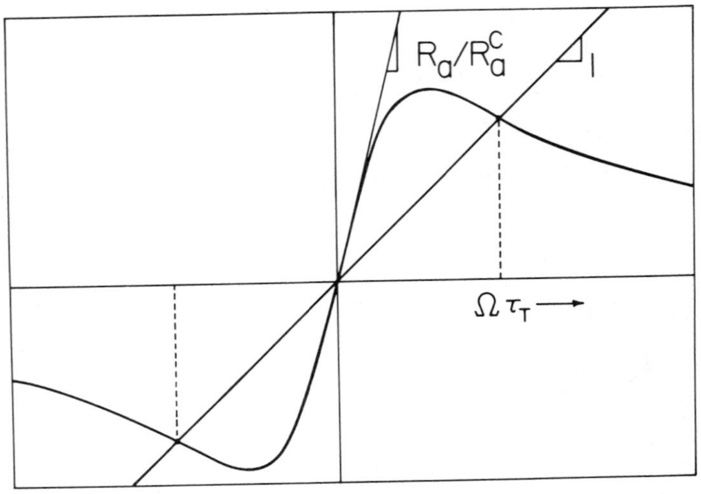

Fig. 10.5.3. Graphical solution of Eq. 21.

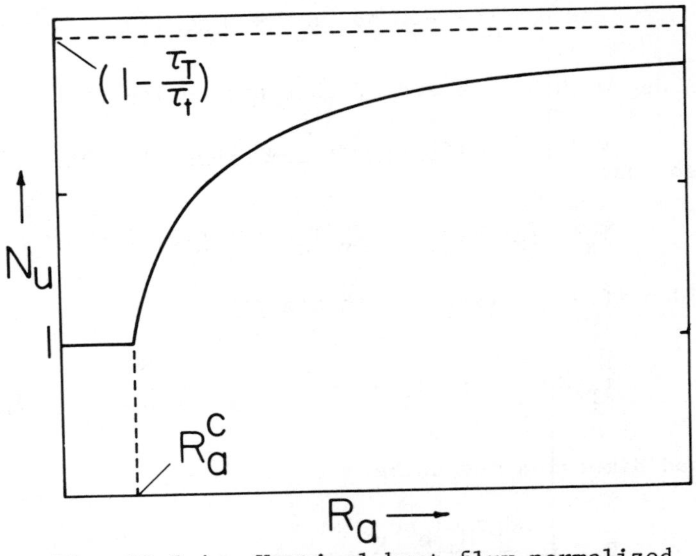

Fig. 10.5.4. Vertical heat flux normalized to flux in absence of rotation as a function of R_a.

Onset and Steady Convection: The similarity between the thermal rotor model and the model for electroconvection developed in Sec. 5.14 (see Prob. 5.14.2) suggests looking for a stationary state. In Eqs. 16-18, the time derivatives are taken as zero and from the first two equations it follows that

$$T_x = \frac{-\Omega T_y}{(1+f)} = \frac{f\Omega}{(1+f)^2+\Omega^2} \tag{20}$$

Hence, the torque equation is expressed in terms of the angular velocity:

$$\Omega = \frac{R_a f\Omega}{(1+f)^2+\Omega^2} \tag{21}$$

The graphical solution of this expression, pictured in Fig. 10.5.3, is familiar from the electric rotor of Sec. 5.14. If R_a is small, the only intersection of the two curves is at the origin and the rotor is stationary. A negative or positive velocity obtains if R_a exceeds R_a^c, where

$$R_a^c = (1+f)^2/f \tag{22}$$

so that the slope of the thermal torque curve at the origin exceeds that of the viscous-magnetic torque curve (which in normalized form is unity). Solution of Eq. 21 gives this velocity and Eqs. 20 give the associated components of the temperature:

$$\Omega = \sqrt{(R_a - R_a^c)f};\quad T_x = \Omega/R_a;\quad T_y = -(1+f)/R_a \tag{23}$$

These steady conditions are interpreted as the result of an instability having its threshold at $R_a = R_a^c$ and resulting in steady rotation in either direction. As R_a becomes large compared to its critical value, R_a^c, the reciprocal angular velocity is approximated by the product of the gravitational time and the square root of the ratio of the fluid thermal diffusion time to a time representing the combined damping effects of viscosity and magnetic diffusion.

The rotation is reflected in the vertical heat flux. Heat passing into the jacket over the lower half and leaving over the top half is augmented by the motion. From Eq. 5 for the steady motion, it follows that $T_o = T_E$. Using Eqs. 1 and 4, the heat flux is computed from

$$Q_T = \int_0^\pi \Gamma_r R d\theta = \int_0^\pi h(T - T_{ext})Rd\theta = 2hR(T_y + T_e) \tag{24}$$

The Nusselt number, N_u, is now defined as the ratio of this heat flux to what it would be in the absence of rotation (convection),

$$N_u \equiv \frac{Q_T(\Omega)}{Q_T(\Omega=0)} = \frac{1 + T_y(\Omega)}{1 + T_y(0)} \tag{25}$$

Through Eq. 23, it follows that

$$N_u = \begin{cases} (1 + f)[1 - \frac{(1 + f)}{R_a}]; & R_a > R_a^c \\ 1 \quad ; & R_a < R_a^c \end{cases} \tag{26}$$

The Nusselt number is shown as a function of R_a in Fig. 10.5.4. This type of dependence is typical of fluid layers heated from below. At most, the effect of the steady convection is to render the rotor isothermal, but even then conduction through the jacket limits the flux. Hence, the asymptote (1+f) for N_u as $R_a \to \infty$. Raising the magnetic field reduces R_a and hence suppresses the heat flux. Of course, if the magnetic field is large enough to prevent the convection altogether by making $R_a < R_a^c$, then heat transfer is solely due to conduction and $N_u \to 1$.

The dynamical model can be used to study transient behavior. A hint that the predicted phenomena are of great variety is given by considering the stability of the steady rotation just described. Perturbation of the steady rotation shows that oscillatory instability (overstability) can result at high R_a (see Prob. 10.5.1). Because the rotor inertia now comes into play, p_T is therefore a critical parameter.

If heated from the side, the rotor is not in a state of static indeterminancy. It can execute steady rotation in one direction without a threshold. This configuration is also useful for modeling practical natural convection systems. These observations are developed in the problems.

10.6 Hydromagnetic Bénard Type Instability

What is conventionally termed Bénard instability is commonly seen when a layer of cooking oil in the bottom of a pan is heated from below.[1] If heat were applied with perfect uniformity over the horizontal plane, density stratification would result because the lighter fluid is on the bottom. What is seen is cellular convection, as illustrated in Fig. 10.6.1, and it results because, in the gravitational field, the configuration of mass density is unstable, as might be expected from Sec. 8.18. Because material of fixed identity tends to lose its heat to its surroundings, and hence to take on the same mass density, thermal diffusion requires a finite vertical heat flux before the convection is observed.

The rotor of Sec. 10.5 is a finite-amplitude model for this cellular convection. Recognized now are the infinite number of degrees of freedom of the actual fluid, but the continuum model is restricted to perturbations from the static equilibrium.

The layer, shown in Fig. 10.6.2, is horizontal. Driven by a temperature difference $T_b - T_a$, the static layer sustains a uniform vertical heat flux Γ_o. The heat conduction through this static layer is in the steady state, so the temperature distribution is linear and the heat flux independent of x. With DT_s the stationary gradient in temperature, this flux is $\Gamma_o = -k_T DT_s$.

There is no equilibrium magnetic force density, so gravity alone is responsible for the vertical pressure gradient. Conditions for the magnetic Hartmann-type of approximation prevail, in that the magnetic diffusion time, τ_m, is much less than the magneto-inertial time τ_{MI}, while τ_{MI} is much less than the viscous diffusion time, τ_v (see Sec. 9.9). In fact, in this section, viscous effects will be ignored altogether.

The gravitational acceleration of the fluid has its origins in the dependence of the mass density on the temperature. For the relatively small changes in mass density typical of liquids,

$$\rho = \rho_o[1 + \alpha_\rho(T - T_E)] \tag{1}$$

where ρ_o and α are constants and T_E is the average equilibrium temperature. The coefficient of thermal expansion, α_ρ, is typified in Table 10.6.1.

---

1. S. Chandrasekhar, Hydrodynamic and Hydromagnetic Stability, Clarendon Press, Oxford, 1961, pp.9-75. For effect of magnetic field see pp. 177-186.

Fig. 10.6.1

Cellular convection subsequent to incipience of thermally induced Bénard instability. A layer of silicone oil is heated from below in a frying pan. (Reference 4, Appendix C).

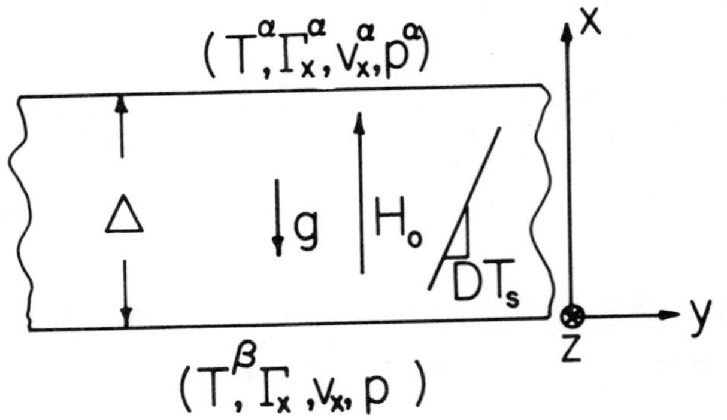

Fig. 10.6.2

Layer of conducting fluid such as liquid metal supporting uniform vertically directed heat flux and magnetic field intensity.

Table 10.6.1. Coefficient of thermal expansion $\alpha_\rho \equiv -(\partial\rho/\partial T)/\rho$ for representative fluids at 200^oC

Liquid	Coefficient of thermal expansion α_ρ $(^oC^{-1})$
Water	-2.1×10^{-4}
Glycerol	-4.7×10^{-4}
Mercury	-1.8×10^{-4}
n-Xylene	-9.9×10^{-4}
Gas (at constant pressure)	
Dry air	-3.4×10^{-3}

For small temperature excursions, mass conservation becomes

$$\nabla\cdot\vec{v} = -\alpha_\rho \frac{DT}{Dt} \tag{2}$$

so the flow is not exactly solenoidal. It is straightforward to include dilatational terms in the force and energy equations, but the additional analytical effort is not justified in the class of flows of interest here. Because α_ρ is small, $\nabla\cdot\vec{v} \simeq 0$. However, it does not follow that the mass density of a given element of fluid remains constant.

The perturbation part of the thermal equation, Eq. 10.2.2 with $\phi_d \simeq 0$, makes evident why the temperature (and hence the mass density) of fluid of fixed identity varies:

$$\frac{\partial T'}{\partial t} + (DT_s)v_x = K_T\nabla^2 T' \tag{3}$$

On the left is the time rate of change of T' for a given element of fluid, and on the right the thermal diffusion that accounts for this rate of change.

The force equation is written neglecting the viscous force density:

$$\rho_o \frac{\partial \vec{v}}{\partial t} + \nabla p = -\rho_o[1 - \alpha_\rho(T_s - T_E)]g\vec{i}_x + \alpha_\rho\rho_o gT'\vec{i}_x + \vec{J} \times \mu_o H_o \vec{i}_x \tag{4}$$

Consistent with the Hartmann type approximation considered in Sec. 9.9, H_o is imposed both in the force equation and in the constitutive law

$$\vec{J} = \sigma(\vec{E} + \vec{v} \times \mu_o H_o \vec{i}_x) \tag{5}$$

needed in Eq. 4. Also, because the imposed $\vec{H}$ is constant,

$$\nabla \times \vec{E} \approx 0 \tag{6}$$

With Eq. 5 substituted into Eq. 4, the pressure is eliminated from the latter by taking the curl. In fact the desired equation for v_x, devoid of $\vec{E}$, is obtained by taking the curl again and exploiting the identity $\nabla \times \nabla \times \vec{v} = \nabla(\nabla\cdot\vec{v}) - \nabla^2\vec{v}$. Then the x component is simply

$$\rho \frac{\rho}{\partial t} \nabla^2 v_x = -\alpha_\rho \rho_o g\left(\frac{\partial^2 T'}{\partial y^2} + \frac{\partial^2 T'}{\partial z^2}\right) - \sigma(\mu_o H_o)^2 \frac{\partial^2 v_x}{\partial x^2} \tag{7}$$

Here, the mass density has been approximated as uniform in the inertial term and $\nabla\cdot\vec{v} \approx 0$. This last approximation is valid provided $\alpha_\rho \ell DT_s \ll 1$, where ℓ is a typical length, perhaps the thickness Δ of the layer. (For a layer of mercury, 1 cm thick, subject to a 100^oC temperature difference, this number is 1.8×10^{-2}.) The electric field appears in the other components of the force equation operated on in this fashion, but not in the x component.

In normalized form, Eqs. 3 and 7 become

$$\left[\frac{j\omega}{p_{TM}}(D^2 - k^2) + D^2\right]\hat{v}_x - R_{am}k^2\hat{T} = 0 \tag{8}$$

$$\hat{v}_x + [j\omega - (D^2 - k^2)]\hat{T} = 0 \tag{9}$$

where

$$\omega = \underline{\omega}/(\Delta^2/k_T); \quad \hat{T} = \underline{\hat{T}}\Delta DT_s; \quad \hat{\Gamma}_x = \underline{\hat{\Gamma}}_x k_T DT_s$$

$$x = \underline{x}\Delta; \quad \hat{v}_x = \underline{\hat{v}}_x K_T/\Delta; \quad p = \underline{p}K_T^2\rho_o/\Delta^2$$

The variables that complement $(\hat{T}, \hat{v}_x)$ are the thermal flux,

$$\hat{\Gamma}_x = -D\hat{T} \tag{10}$$

and the pressure, found from the x component of the force equation, Eq. 4, in terms of $(\hat{T}, \hat{v}_x)$:

$$D\hat{p} = R_{am}p_{TM}\hat{T} - j\omega\hat{v}_x \tag{11}$$

The magnetic Rayleigh number and thermal-magnetic Prandtl number are familiar from Sec. 10.5, where they are written as ratios of characteristic times.

Because Eqs. 8 and 9 have constant coefficients, solutions take the form

$$\hat{T} = \sum_{m=1}^{4} \hat{T}_m \exp(\gamma_m x) \tag{12}$$

$$\hat{v}_x = -\sum_{m=1}^{4} [j\omega - (\gamma_m^2 - k^2)]\hat{T}_m \exp(\gamma_m x) \tag{13}$$

where the latter follows from Eq. 9. The characteristic equation gotten by substituting into Eqs. 8 and 9 is quadratic in γ^2. Thus, roots take the form $\gamma = \pm\gamma_a$ and $\gamma = \pm\gamma_b$. With $a \equiv [j\omega - (\gamma_a^2 - k^2)]$ and $b \equiv [j\omega - (\gamma_b^2 - k^2)]$, the conditions that Eqs. 12 and 13 assume the correct values at the α and β surfaces are

$$\begin{bmatrix} e^{\gamma_a} & e^{-\gamma_a} & e^{\gamma_b} & e^{-\gamma_b} \\ 1 & 1 & 1 & 1 \\ ae^{\gamma_a} & ae^{-\gamma_a} & be^{\gamma_b} & be^{-\gamma_b} \\ a & a & b & b \end{bmatrix} \begin{bmatrix} \hat{T}_1 \\ \hat{T}_2 \\ \hat{T}_3 \\ \hat{T}_4 \end{bmatrix} = \begin{bmatrix} \hat{T}^\alpha \\ \hat{T}^\beta \\ \hat{v}^\alpha \\ \hat{v}^\beta \end{bmatrix} \qquad (14)$$

The procedure for deducing transfer relations between $[\hat{T}^\alpha, \hat{T}^\beta, \hat{v}_x^\alpha, \hat{v}_x^\beta]$ and $[\hat{\Gamma}_x^\alpha, \hat{\Gamma}_x^\beta, \hat{p}^\alpha, \hat{p}^\beta]$ is now similar to that given in Sec. 7.19. Here attention is confined to the temporal modes and the critical conditions for instability.

Suppose that the boundaries are actually rigid walls, so that $(\vec{v}_x^\alpha, \vec{v}_x^\beta) = 0$, and are constrained to be isothermal, so that $(\hat{T}^\alpha, \hat{T}^\beta) = 0$. Then the determinant of the coefficients in Eq. 14 must vanish. The determinant is easily reduced by subtracting the second and fourth columns from the first and third, respectively. Thus

$$4(b - a)^2 \sinh \gamma_a \sinh \gamma_b = 0 \qquad (15)$$

Nontrivial roots to Eq. 15 are either $\gamma_a = jn\pi$ or $\gamma_b = jn\pi$, $n = 1,2,\cdots$. To determine the associated eigenfrequencies $j\omega \equiv s_n$ of the temporal modes, the characteristic equation, found by substituting $\exp(\gamma x)$ into Eqs. 8 and 9,

$$[s_n(\gamma_n^2 - k^2) + p_{TM}\gamma_n^2][s_n - (\gamma_n^2 - k^2)] + p_{TM}R_{am}k^2 = 0 \qquad (16)$$

is evaluated with $\gamma_n = jn\pi$. This expression can be solved for s_n to give

$$s_n = \frac{-B \pm \sqrt{B^2 - 4p\{(n\pi)^2[(n\pi)^2 + k^2] - R_{am}k^2\}[(n\pi)^2 + k^2]}}{2[(n\pi)^2 + k^2]} \qquad (17)$$

where $B \equiv p(n\pi)^2 + [(n\pi)^2 + k^2]^2$. Provided that the quantity in {} under the radical is greater than zero, all roots are negative, because then the radical has a magnitude less than B. However, if that term is negative, half of the roots represent growing exponentials. Thus, the critical condition for the onset of cellular convection of each mode, n, at a wavelength $2\pi/k$ is

$$R_{am} = \frac{(n\pi)^2}{k^2}[(n\pi)^2 + k^2] \qquad (18)$$

Note that R_{am} is indeed positive for the typical fluid heated from below, because α_ρ is typically negative and DT_s is also negative. In addition to the transverse modal structure represented by n, there is the longitudinal dependence represented by k. According to the model, the n = 1 mode with infinitely short wavelength (infinite k) is the most critical with incipience at

$$R_{am}^c = \pi^2 \qquad (19)$$

To have a better approximation as to the critical longitudinal wavelength of the most critical mode, it would be necessary to add further physical processes, such as viscous diffusion, to the model. The way in which viscosity plays the damping role of the magnetic field is illustrated in Sec. 10.5 and Prob. 10.6.3.

In the rotor model, there are two thermal time constants, with a ratio $\tau_T/\tau_t \equiv f$. In the fluid layer, there is no such dimensionless ratio, because azimuthal and radial conditions involve the same spatial scale and the same fluid properties. Hence, the critical Rayleigh number that is equivalent to Eq. 19 is given by Eq. 10.5.22. The steady convection and overstability of that convection predicted using the rotor model give some hint as to the nonlinear phenomena that ensue as R_{am} is raised beyond R_a^c. At first, due to cellular convection, there is an augmentation of the heat transfer, as typified by a Nusselt number that increases with R_a. The steady cellular motion is itself potentially unstable with an ultimate turbulent (nonsteady) state the result. The transition to turbulence should be expected to be a function not only of R_a but also of p_T.

10.7 Unipolar-Ion Diffusion Charging of Macroscopic Particles

Ions encountering the surface of a macroscopic particle tend to become attached. This is especially true in gases, where macroscopic particles are commonly charged in passage through an ion filled region. This is illustrated in Sec. 5.5, where an imposed electric field is responsible for the migration of ions to the surface of the particle. The result is "impact" or "field" charging. The model in Sec. 5.5 neglects the fact that, on a sufficiently small scale, there is also a diffusional contribution to the ion flux. Through diffusion, ions also reach the surface and hence charge the particle. This contribution can exceed that due to impact for sufficiently small particles.

As diffusion charging proceeds, it does so at a decreasing rate because the electric field generated by the charging tends to produce an ion migration that counters the ion diffusion. The determination of this charging rate and hence of the particle charge gives the opportunity to discuss some general features of the diffusion of a single charged species while obtaining a useful result.

The continuum conservation laws from Sec. 5.2 include contributions from molecular diffusion. What is now described is a continuum in which almost all particles are neutral and uniform. A relatively small fraction of the particles are charged. For a single charged species, taken for purposes of illustration as positive, the conservation of mass equation is Eq. 5.2.9, with G = 0 and R = 0. Combined with Gauss' law, it gives

$$\frac{\partial \rho}{\partial t} + (\vec{v} + b\vec{E})\cdot\nabla\rho = K_+\nabla^2\rho - \frac{\rho^2 b}{\varepsilon} \tag{1}$$

With a characteristic length ℓ and time τ, fluid velocity U and electric field E, the respective terms in Eq. 1 are of the order

$$(\frac{1}{\tau});\quad (\frac{U}{\ell} \equiv \frac{1}{\tau_{trans}});\quad (\frac{bE}{\ell} \equiv \frac{1}{\tau_{mig}});\quad (\frac{K_+}{\ell^2} \equiv \frac{1}{\tau_D});\quad (\frac{\rho b}{\varepsilon_o} \equiv \frac{1}{\tau_e}) \tag{2}$$

where the expression has been divided by a characteristic amplitude of ρ. For the charging of a particle having radius a, ℓ might be taken as a. The competition between diffusion and migration is represented by the terms in τ_{mig} and τ_D. These terms are equal if $\tau_{mig} = \tau_D$; and, because of the Einstein relation, Eq. 5.2.8, this is equivalent to

$$\ell E = kT/q \tag{3}$$

Thus, thermal diffusion and migration are of equal importance if the thermal voltage is equal to the voltage drop over a characteristic length. For $E = 10^5$ V/m (typical of fields in an electrostatic precipitator) the length that makes diffusion and migration equal is 2.5×10^{-7} m. The radius, a, of the macroscopic particle is taken as being of this order.

The diffusion time is estimated by taking as a typical ion mobility from Table 5.2.1, $b = 10^{-4}$, which (for an ion of one electronic charge) gives as a typical diffusion coefficient $K_+ = 2.5 \times 10^{-6}$. Thus, the diffusion time is only 2.5×10^{-8} sec.

By comparison, the self-precipitation time τ_e is long. Whether ions are present in a given volume by virtue of convection or migration, $\tau_e \approx 10^{-3}$ sec or longer is typical. After all, the ions are self-precipitating with this time and some other mechanism having an equally short characteristic time must be available to secure the required density.

The transport time is estimated by taking as typical the velocity of a charged submicron particle in a field of 10^5 V/m, say 10^{-2} m/sec. Thus $\tau_{trans} = 2.5 \times 10^{-5}$ sec, which is still 100 times longer than τ_D and τ_{mig}.

Consistent with ignoring the self-precipitation term is the neglect of contributions to $\vec{E}$ from the diffusing ions. Thus, $\vec{E}$ in Eq. 1 is taken as imposed, in general by the charge, Q, on the macroscopic particle and charges on external electrodes. With this understanding, and one more observation, Eq. 1 then reduces to

$$\nabla\cdot(b\vec{E}\rho - K_+\nabla\rho) = 0 \tag{4}$$

The first term in Eq. 1 has been neglected because the time scale for the charging process is very long compared to the diffusion and migration times. Looking ahead, it will be found that the charging time is of the order of τ_e. The charging process is quasi-stationary in the volume with the transient resulting only because of the field's dependence on the charge, Q, of the macroscopic particle.

In general, the solution of Eq. 4 with an externally applied electric field is difficult. Here, it will now be assumed that any ambient electric field is small compared to $(kT/q)/\ell$. Thus, in Eq. 4 the electric field is now taken as

$$\vec{E} = \frac{Q}{4\pi\varepsilon_o r^2}\vec{i}_r \tag{5}$$

With this field there is a radial symmetry, so Eq. 4 can be integrated once to obtain

$$4\pi r^2 K_+ \frac{d\rho}{dr} - \frac{bQ}{\varepsilon_o}\rho = i \tag{6}$$

Here, i(t) is the electrical current to the particle.

Superposition of particular and homogeneous solutions to Eq. 6 results in

$$\rho = (\rho_o + \frac{i\varepsilon_o}{bQ}) \exp\left[-\frac{bQ}{4\pi\varepsilon_o K_+ r}\right] - \frac{i\varepsilon_o}{bQ} \tag{7}$$

where the coefficient in front of the second term, the homogeneous solution, has been adjusted to make $\rho \to \rho_o$ far from the particle.

The diffusion model pictures ions in the neighborhood of a given point as having a random distribution of velocities. At the surface of the particle, those moving inward are absorbed and this forces the ion density there to zero. Thus, a second boundary condition is $\rho(a) = 0$ and Eq. 7 then becomes a relation between the particle charge and the rate of charging, i(t):

$$\frac{dQ}{dt} = i = \frac{b\rho_o}{\varepsilon_o}\frac{Qe^{-fQ}}{1 - e^{-fQ}} \tag{8}$$

where in view of the Einstein relation, Eq. 5.2.8, $f \equiv q/4\pi\varepsilon_o akT$.

Rewritten so as to be integrable, this Fuchs-Pluvinage equation[1] becomes

$$\int_o^Q \frac{e^{fQ} - 1}{Q}\, dQ = \int_o^t \frac{b\rho_o}{\varepsilon_o}\, dt \tag{9}$$

Integration then gives

$$\sum_{m=1}^{\infty} \frac{\underline{Q}^m}{m\, m!} = \underline{t} \tag{10}$$

where $\underline{Q} = Q/Q_D$ ($Q_D = 1/f = 4\pi\varepsilon_o akT/q$ is the charge needed to terminate the thermal "field", $(kT/q)/a$, on the surface of the particle) and where $\underline{t} = t/\tau_e$ ($\tau_e = \varepsilon_o/\rho_o b$, the self-precipitation time for the ions based on the ion density far from the particle). This charging characteristic is shown in Fig. 10.7.1.

Diffusion charging is expected to dominate over impact charging if the particle is sufficiently small that $aE < kT/q$, where E is the imposed or ambient electric field. Thus, in a field of 10^5 V/m, particles must be smaller than about 0.2 μm for diffusion charging to prevail. In fact, for the model to be valid, there is also a lower limit on size. The continuum picture of diffusion depends on the particle having a radius that is large compared to the mean free path of the ions and neutrals. In air at atmospheric pressure, this distance is 0.09 μm. For particles somewhat smaller than this, the continuum diffusion model is called into question. Models based on having a mean free path much greater than the particle radius give a charging law that is surprisingly similar to Eq. 10,[2] so the result is actually useful for particles smaller than the mean free path. Effects of the ambient field (impact charging) in combination with diffusion have been considered.[3]

---

1. N. A. Fuchs, Izv. Akad. Nauk USSR, Ser. Geogr. Geophys. 11, 341 (1947); P. Pluvinage, Ann. Geophys. 3, 2 (1947).
2. H. J. White, Industrial Electrostatic Precipitation, Addison-Wesley, Reading, Mass., 1963, pp. 137-141.
3. B. Y. H. Liu and H. C. Yeh, J. Appl. Phys. 39, 1396 (1968).

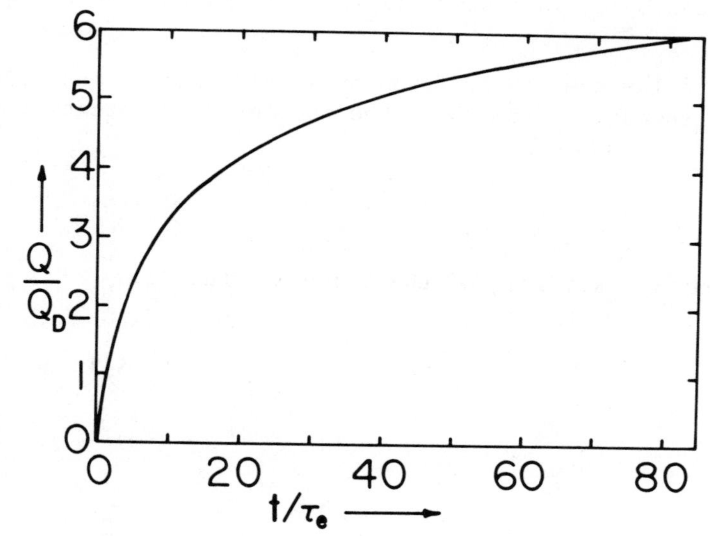

Fig. 10.7.1

Normalized charge on a macroscopic particle having radius a, as a function of normalized time, where charging is by diffusion alone, $Q_D \equiv 4\pi\varepsilon_o akT/q$ and $\tau_e \equiv \varepsilon_o/\rho_o b$.

10.8 Charge Double Layer

Considered in this section is the competition between migration and diffusion that creates a double layer at an interface between a bipolar conductor and an insulating boundary. The fluid is some form of electrolyte in which dissociation has created ion species having densities $\rho_\pm$ with a background of molecules having density n. The conservation laws for the charged and neutral species are Eqs. 5.8.9 and 5.8.10 and the system is EQS.

At the outset, the electrolyte is presumed to be highly ionized. As discussed in Sec. 5.9, this means that generation largely depletes the neutral density. In the neutral conservation equation, Eq. 5.8.10, terms on the left are essentially zero while recombination and generation on the right almost exactly balance. As a result, G-R is negligible in the charged particle equations, Eqs. 5.8.9, as well.

Consider the quasi-stationary distribution of ions in the vicinity of an insulating boundary. For now, there is no fluid convection, so $\vec{v} = 0$. The steady one-dimensional particle conservation statements then reduce to

$$\frac{d}{dx}\left[b_+E_x\rho_+ - K_+\frac{d\rho_+}{dx}\right] = 0 \tag{1}$$

$$\frac{d}{dx}\left[-b_-E_x\rho_- - K_-\frac{d\rho_-}{dx}\right] = 0 \tag{2}$$

with Gauss' law linking the electric field to the charge densities

$$\frac{d\varepsilon E_x}{dx} = \rho_+ - \rho_- \tag{3}$$

The polarizability of the fluid is assumed uniform, so ε is a constant.

The wall, at x = 0, is taken as insulating or "polarized," in that there is no current due to either species through its surface. Hence, the current densities in brackets in Eqs. 1 and 2 are each zero. These expressions are then solved for E_x. Because $\rho^{-1}d\rho/dx = d(\ell n\rho)/dx$ and $E_x = -d\Phi/dx$, it follows that

$$\Phi = -\frac{K_+}{b_+}\ln\frac{\rho_+}{\rho_o} = -\frac{kT}{q}\ln\frac{\rho_+}{\rho_o} \tag{4}$$

$$\Phi = \frac{K_-}{b_-}\ln\frac{\rho_-}{\rho_o} = \frac{kT}{q}\ln\frac{\rho_-}{\rho_o} \tag{5}$$

Here, as an integration constant, the charge densities have been taken as reaching the same uniform density, ρ_o, far from the boundary. Also, the Einstein relation, Eq. 5.2.8, has been used to express the ratio of diffusion coefficient to mobility in terms of the thermal voltage kT/q. Consistent with the positive and negative charges being generated by an ionization is the assumption that the q is the same for each ionized species.

The charge densities required to express Gauss' law can now be found by solving Eqs. 4 and 5 for $\rho_\pm$. Thus, Eq. 3 becomes the classic Debye-Hückel[1] expression from which the double-layer potential is determined:

$$\frac{d^2\Phi}{dx^2} = \frac{2\rho_o}{\varepsilon} \sinh\,[\Phi/(kT/q)] \tag{6}$$

Normalization of the potential and length makes clear the key role of the Debye length, δ_D:

$$x = \underline{x}\delta_D, \quad \Phi = \underline{\Phi}kT/q, \quad \delta_D \equiv \sqrt{\frac{\varepsilon kT}{2\rho_o q}} \tag{7}$$

because then Eq. 6 becomes simply

$$\frac{d^2\underline{\Phi}}{d\underline{x}^2} = \sinh \underline{\Phi} \tag{8}$$

The Debye length is that distance over which the potential developed by separating a charge density ρ_o from the background charge of the opposite polarity is equal to the thermal voltage kT/q. By substituting for $kT/q = K/b$, δ_D can also alternatively be considered the distance over which the molecular diffusion time δ_D^2/K is equal to the self-precipitation time $\varepsilon/\rho b$. Thus, δ_D varies from about 100 Å in aqueous electrolytes to microns in semi-insulating liquids.

To integrate Eq. 8, multiply by $D\underline{\Phi}$ and form the perfect differential

$$\frac{d}{d\underline{x}}\left[\frac{1}{2}\left(\frac{d\underline{\Phi}}{d\underline{x}}\right)^2 - \cosh \underline{\Phi}\right] = 0 \tag{9}$$

Far from the layer, the potential is defined as zero. Because there is no current flow there and the charge densities neutralize each other in this region, the electric field $-D\underline{\Phi}$ also goes to zero far from the boundary. Thus, the x-independent quantity in brackets in Eq. 9 is unity, and the expression can be solved for $D\underline{\Phi}$. The x- and Φ-dependence of that expression can be separated so that it can be integrated:

$$\int_o^{\underline{x}} d\underline{x} = \pm \int_{-\underline{\zeta}}^{\underline{\Phi}} \frac{d\underline{\Phi}}{\sqrt{2(\cosh \underline{\Phi} - 1)}} \tag{10}$$

As a function of the normalized zeta potential $\underline{\zeta}$, this result is illustrated in Fig. 10.8.1. The exponential character of the potential distribution is best seen directly from Eq. 6 by recognizing that if $\underline{\Phi} \ll 1$, sinh can be approximated by its argument. It follows that the solution is simply $\underline{\Phi} = -\underline{\zeta}\exp(-\underline{x})$. For $\underline{\zeta} > 1$, the rate of decay is faster than would be expected from low $\underline{\zeta}$ limit.

On the interface is a surface charge given by

$$\underline{\sigma}_f = -\frac{d\underline{\Phi}}{d\underline{x}}; \quad \sigma_f = \underline{\sigma}_f\sqrt{\frac{2\rho_o \varepsilon kT}{q}} \tag{11}$$

and this has image charge distributed throughout the diffuse half of the double layer. Found from the potential by inverting Eqs. 4 and 5, the charge densities $\rho_\pm$ and net charge density ρ_f are illustrated in Fig. 10.8.2.

Double layers can exist not only at interfaces between an insulating material and an electrolyte, but even at the interface between a liquid metal such as mercury and an electrolyte. What is required is an interface that, for lack of chemical reaction, largely prevents the transfer of charge. For potential differences under about a volt or so, even a mercury-electrolyte interface can prevent the passage of current. Double layers at such interfaces are taken up in Sec. 10.11. In the next two sections, the double layer abuts a material that is itself a rigid electrical insulator.

---

1. P. Delahay, Double Layer and Electrode Kinetics, Interscience Publishers, New York, 1966, pp. 33-52.

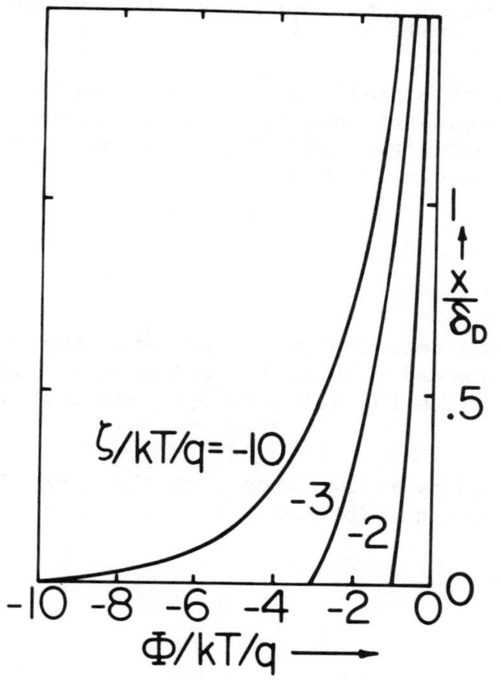

Fig. 10.8.1. Potential distribution in diffuse part of double layer with zeta potential as parameter.

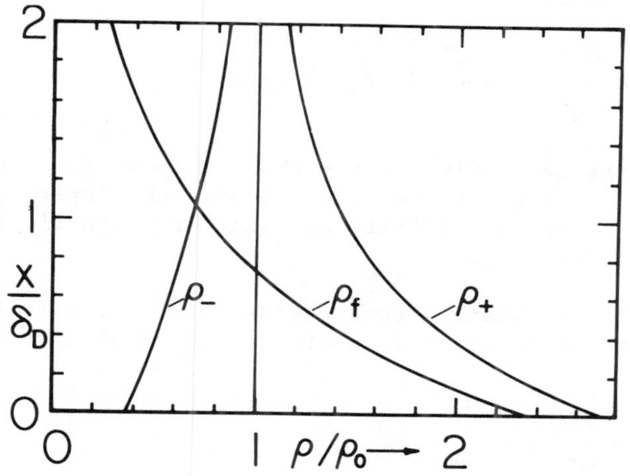

Fig. 10.8.2. Charge density distributions for $\zeta = kT/q$.

10.9 Electrokinetic Shear Flow Model

A double layer in an electrolyte abuting an insulating solid is sketched in Fig. 10.9.1. Even though this layer tends to be extremely thin, the application of an electric field tangential to the boundary can result in a significant relative motion between the solid and fluid. From the boundary frame of reference, the field E_y exerts a force density $\rho_f E_y$ on the fluid, and shear flow results. Because pressure forces prevent motion in the x direction, flow is essentially orthogonal to the double-layer diffusion and migration currents. Thus it can be superimposed on the static double-layer distribution discussed in Sec. 10.8. In layers that are "wrapped around" a particle, as taken up in Sec. 10.10, a component of the applied field tends to compete with the fields internal to the layer. The model now developed can only be applied to such situations if the x component of the applied field is small compared to the double-layer internal field.

The relative flow is inhibited by the viscous stresses associated with strain rates developed within the layer itself. These strain rates are inversely proportional to the layer thickness (of the order of the Debye length) so the relative velocity tends to be small. Nevertheless, such electrokinetic flows are important in fine capillaries and in the interstices of membranes. Electrophoretic motions of both macroscopic and microscopic particles in electrolytes also have their origins in this streaming.

The simple model developed now is used in this section to describe electro-osmosis through pores. It will be used to describe electrophoresis of particles in Sec. 10.10.

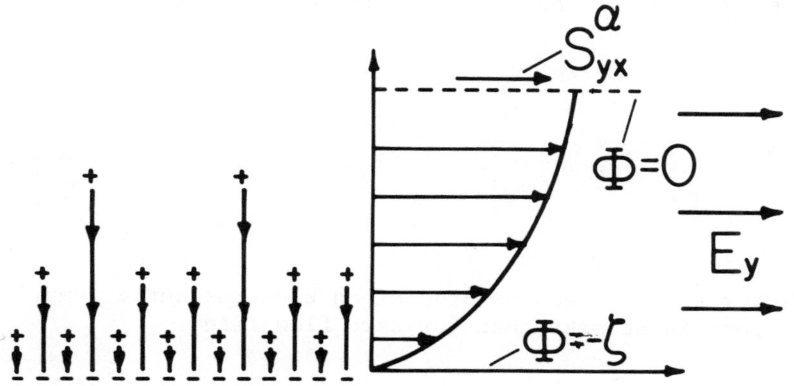

Fig. 10.9.1

Schematic view of double layer subject to imposed field in y direction resulting in shear flow.

Zeta Potential Boundary Slip Condition: For flows that have a scale that is large compared to the Debye thickness, the electromechanical coupling can be reduced to a quasi-one-dimensional model that amounts to a boundary condition for the flow.

On the scale of the double layer, the imposed electric field can be considered to be uniform. The velocity is fully developed in its distribution in the sense of Sec. 9.3. Also, because the double-layer region is so thin, the viscous force density far outweighs the pressure gradient in the y direction. Thus, the force equation, Eq. 9.3.4, takes the one-dimensional form

$$\eta \frac{d^2 v_y}{dx^2} = -\frac{dT_{yx}}{dx}; \quad T_{yx} = \varepsilon E_y E_x \tag{1}$$

where the derivative of the shear component of the stress tensor simply represents the force density $\rho_f E_y$. To prescribe the flow outside the layer, it is assumed that at the distance d from the slip plane, there is a fictitious plane at which fluid moves with the velocity $v_y^\alpha \vec{i}_y$ and sustains a viscous shear stress S_{yx}^α.

The constant from integration of Eq. 1 is evaluated by recognizing that the electric shear stress falls to zero at x = d, where the external viscous stress equilibrates the internal stress:

$$\eta \frac{dv_y}{dx} = -T_{yx} + S_{yx}^\alpha \tag{2}$$

A second integration is possible because E_y is constant and $E_x = -d\Phi/dx$. Also, S_{yx}^α is a constant, so that

$$\eta v_y = \int_o^x \varepsilon E_y \frac{d\Phi}{dx} dx + \int_o^x S_{yx}^\alpha dx$$
$$= \varepsilon E_y[\Phi(x) - \Phi(0)] + xS_{yx}^\alpha \tag{3}$$

In terms of the conventions used in Sec. 10.9, the potential of the slip plane is taken as $-\zeta$, while that at x = d is zero, so Eq. 3 becomes

$$v_y^\alpha = \frac{\varepsilon E_y \zeta}{\eta} + \frac{d}{\eta} S_{yx}^\alpha \tag{4}$$

If the external stress, S_{yx}^α, comes from shear rates determined by flow on a scale large compared to δ_D, the last term in Eq. 4 can be ignored. The mechanical boundary condition representing the double layer is then simply

$$v_y = \frac{\varepsilon E_y \zeta}{\eta} \tag{5}$$

In refining the simple model, a distinction is sometimes made between the potential evaluated in the slip plane and evaluated on the other side of a compact zone of charge that forms part of the double layer but is not in the fluid and hence cannot move.

Electro-Osmosis: Flow through a planar duct, such as shown in Table 9.3.1, illustrates the application of Eq. 4. Suppose that the duct width, Δ, is much greater than a Debye length. In the volume of the flow, there are no electrical stresses, so Eq. (a) of Table 9.3.1 gives the velocity as

$$v_y(x) = \frac{\varepsilon E_y \zeta}{\eta} + \frac{\Delta^2}{2\eta} \frac{\partial p'}{\partial y} \left[\left(\frac{x}{\Delta}\right)^2 - \frac{x}{\Delta}\right] \tag{6}$$

The volume rate of flow per unit z follows as

$$Q_v = -\frac{\Delta^3}{12\eta} \frac{dp}{dy} + \frac{\varepsilon E_y \zeta}{\eta} \Delta \tag{7}$$

This relation gives the trade-off between flow rate and pressure drop of an electrokinetic pump. The pressure rise developed in a length ℓ of the pore is at most that for zero flow rate,

$$\Delta p = \varepsilon E_y \zeta \ell / \Delta^2 \tag{8}$$

In situations where Δ is small (but, to validate the model, still larger than a Debye length), this pressure can be appreciable. For example, with $\zeta = 0.1$ V (about four times the thermal voltage kT/q), $\Delta = 1\ \mu$m, $E_y = 10^4$ V/m, $\ell = 0.1$ m and $\varepsilon \approx 5\varepsilon_o$, the pressure rise is about 5×10^3 n/m^2, which would raise water to a height of about 0.5 m.

In membranes composed of a matrix of materials, perhaps of a biological origin, surrounded by double layers, flow of fluid through the interstices is modeled as flow through a system of pores, each described by a relation such as Eq. 7.[1] The velocity profile for δ_D arbitrary relative to Δ is found in Prob. 10.9.1.

Electrical Relations; Streaming Potential: Associated with the electric field and flow in the y direction, there is a current density

$$J_y = (\rho_+ b_+ + \rho_- b_-)E_y + (\rho_+ - \rho_-)v_y \tag{9}$$

The charge densities in this expression are as found in Sec. 10.8 and illustrated by Fig. 10.8.2. The conductivity, $\rho_+b_+ + \rho_-b_-$, tends to remain uniform through the double layer, but, if $\underline{\zeta} > 1$, tends to be increased somewhat over the bulk value. The convection term is concentrated in the region of net charge, and hence (on the scale of an external flow having characteristic lengths large compared to δ_D) comprises a surface current. Because it results from motion of the fluid, it might be termed a convection current. However, it results from fluid motion within a Debye length or so of the boundary, and this motion is caused by the externally applied pressure difference and the field itself. For a small zeta potential $\sinh \underline{\Phi} \simeq \underline{\Phi}$ and $\Phi \simeq -\zeta \exp(-x/\delta_D)$, and so it follows from Eqs. 10.8.4 and 10.8.5 that

$$\rho_+ - \rho_- = -\frac{2\rho_o \Phi}{(kT/q)}\Phi = \frac{2\rho_o\zeta}{kT/q}\exp(-x/\delta_D) \tag{10}$$

and that the velocity of Eq. 3 is

$$v_y = \frac{\varepsilon E_y}{\eta}\zeta[1 - \exp(-x/\delta_D)] + \frac{x}{\eta}S^\alpha_{yx} \tag{11}$$

The current density of Eq. 9 can be divided into a volume density represented by the first term evaluated with $\rho_+ \simeq \rho_o$ and a surface current density represented by the second term

$$K_y = \int_o^\infty (\rho_+ - \rho_-)v_y dx = \int_o^\infty \frac{2\rho_o\zeta}{kT/q} e^{-x/\delta_D}\left[\frac{\varepsilon E_y \zeta}{\eta}(1 - e^{-x/\delta_D}) + \frac{xS^\alpha_{yx}}{\eta}\right]dx$$
$$= \frac{2\rho_o\zeta}{\eta(kT/q)}\left(\frac{\zeta\varepsilon\delta_D}{2}E_y + \delta_D^2 S_{yx}\right) \tag{12}$$

Both terms in this surface current density are due to convection, but the first reflects motion caused by the field itself. This contribution therefore appears much as if the material had a surface conductivity $\rho_o\zeta^2\varepsilon\delta_D/\eta(kT/q)$. Its origins are more apparent if it is recognized as the product of the surface charge $\rho_o\delta_D\zeta/(kT/q)$ and the slip velocity $\varepsilon\zeta E_y/\eta$.

The total current, i (per unit length in the z direction), flowing through a channel having width Δ is then the sum of the surface currents at each of the walls and the bulk current

$$i = \sigma\Delta E_y + 2K_y \tag{13}$$

where K_y is given by Eq. 12. For the case at hand where $\Delta >> \delta_D$, the wall stress, S_{yx}, can be approximated using Eq. 5 as a boundary condition, and so is determined by the pressure gradient. (See Prob. 10.9.2.)

10.10 Particle Electrophoresis and Sedimentation Potential

Electrophoretic motions account for the "migration" of a wide variety of particles in an applied electric field intensity. Particles may be as small as large molecules or as large as macroscopic particles (in the micron-diameter range). If these motions persist over times much longer than the charge relaxation time, it is clear that the particle and its immediate surroundings carry no net charge. The particle is not pulled through the fluid by the electric field, but rather by dint of the field "swims" through the fluid.

1. A. J. Grodzinsky and J. R. Melcher, "Elecromechanical Transduction with Charged Polyelectrolyte Membranes," IEEE Trans. on Biomedical Eng., BME-23, No. 6, 421-33 (1976).

Electrophoresis is used by chemists as a means of classifying particles. For example, protein molecules can be distinguised by electrophoretic techniques, and the electrophoretic motion of particles through a liquid absorbed in paper or comprising the main constituent of a gel is used for routine clinical tests (paper and gel electrophoresis). Electrophoretic motions are also used to control particles of pigment in liquids, for example in large-scale painting of metal surfaces.

Electrophoretic motions are now modeled under the assumption that the particle is much larger in its extreme dimensions than the thickness of the double layer. The particles are insulating, and approximated as spherical with a radius R, as shown in Fig. 10.10.1. The particle is taken as fixed, with the fluid having a uniform relative flow at $z \to \pm\infty$, as illustrated. External electrodes are used to apply the electric field intensity E_o, which is also uniform in the z direction. As $z \to \pm\infty$,

$$\Phi \to -E_o r \cos\theta \qquad \vec{v} \to U\vec{i}_z \tag{1}$$

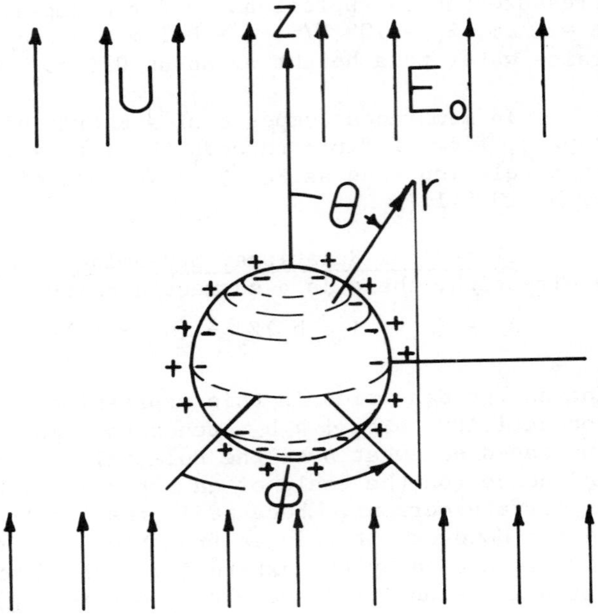

Fig. 10.10.1. Solid insulating particle supporting double layer.

Electric Field Distribution: For a control volume which cuts through the double layer, as shown in Fig. 10.10.2, conservation of charge requires that the conduction current from the bulk of the liquid be balanced by the divergence of convection surface current along the interface:

$$\vec{n}\cdot\vec{J}_f + \nabla_\Sigma\cdot\vec{K}_f = 0 \tag{2}$$

Here, $\vec{K}_f$ is the integral of the tangential current density $\rho_f\vec{v}$ over the mobile part of the double layer and takes the form of Eq. 10.9.11. It is assumed that, because the external viscous stress results from strain rates on the scale of R, and relative motions of the liquid are due to the field itself, the stress term in Eq. 10.9.12 is negligible compared to the first term. In terms of the spherical coordinates, Eq. 2 therefore requires that at r = R,

$$-\sigma\frac{\partial\Phi}{\partial r} + \frac{1}{R\sin\theta}\frac{\partial}{\partial\theta}(\sigma_s E_\theta \sin\theta) = 0 \tag{3}$$

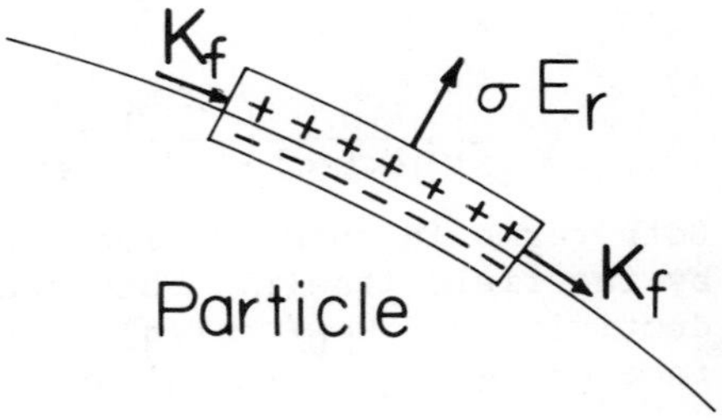

Fig. 10.10.2. Control volume enclosing double layer.

where $\sigma_s \equiv \rho_o \zeta^2 \varepsilon \delta_D/\eta(kT/q)$. To satisfy the condition on Φ at infinity, Eq. 1, Φ is taken as having the form

$$\Phi = -E_o r\cos\theta + A\frac{\cos\theta}{r^2} \tag{4}$$

It follows from Eq. 3 that

$$A = -\frac{E_o R^3}{2}\frac{[\sigma - \frac{2\sigma_s}{R}]}{[\sigma + \frac{\sigma_s}{R}]} \tag{5}$$

and hence that at r = R,

$$E_\theta = -\frac{3E_o\sigma}{2(\sigma + \frac{\sigma_s}{R})}\sin\theta \tag{6}$$

What has been solved has the appearance of being an electrical conduction problem. But, remember that the surface conductivity reflects the convection of net charge by the slip velocity of the fluid relative to the particle.

Fluid Flow and Stress Balance: The slip velocity follows from Eq. 10.9.3 evaluated using Eq. 6:

$$v_\theta = \frac{\varepsilon\zeta}{\eta} E_\theta = \tilde{v}_\theta \sin\theta; \quad \tilde{v}_\theta \equiv -\frac{3}{2}\frac{\varepsilon\zeta}{\eta} \frac{E_o}{1 + \frac{\sigma_s}{\sigma R}} \tag{7}$$

In addition to this boundary condition, the radial velocity is essentially zero at $r = R$ and the velocity approaches the uniform one of Eq. 1 far from the particle. Because of the small particle size and relatively low velocities, the conditions for low Reynolds number are likely to prevail.

The boundary conditions fit the exterior, n=1, high Reynolds number flows of Table 7.20.1. Thus, the stress components follow directly from Eq. 7.20.24 evaluated using Eq. 7 and $v_r = 0$:

$$\begin{bmatrix} \tilde{S}_{rr} \\ \\ \tilde{S}_{\theta r} \end{bmatrix} = -\frac{\eta}{R} \begin{bmatrix} \frac{9}{2} & \frac{3}{2} \\ \\ 3 & 3 \end{bmatrix} \begin{bmatrix} -\frac{U}{2} \\ \\ \tilde{v}_\theta + U \end{bmatrix} \tag{8}$$

Here the complex amplitudes represent the θ dependence summarized in Table 7.20.1.

The net force on the particle in the z direction can be computed from these stresses by integrating the appropriate components over the spherical surface, as in Eq. 7.21.1:

$$f_z = \frac{8}{3}\pi R^2(\tilde{S}_{rr} - \tilde{S}_{\theta r}) = -\pi R\eta(6U + 4\tilde{v}_\theta) \tag{9}$$

There are no external forces acting on the particle, so $f_z = 0$. It therefore follows from Eq. 9 that the particle "swims" at a velocity

$$U = -\frac{2}{3}\tilde{v}_\theta = \frac{\varepsilon\zeta}{\eta} \frac{E_o}{1 + \frac{\sigma_s}{\sigma R}} \tag{10}$$

where $\sigma_s \equiv \rho_o\zeta^2\varepsilon\delta_D/\eta(kT/q)$. This velocity is now interpreted as the velocity of the particle due to an applied field with the fluid stationary. Note that it is in a direction opposite to that of the applied field (assuming that the zeta potential is positive, or that the charge in the liquid is positive, as indicated in Fig. 10.10.1). The charges in the fluid surrounding the particle carry the fluid in the direction of the field. The resulting force on the particle is in an opposite direction. The particle moves as if it were subject to the net force QE, where Q is proportional to the net charge on the particle side of the double layer.

As would be expected, for small zeta potentials, the particle velocity increases with ζ. However, as ζ becomes "large," this velocity peaks and finally becomes inversely proportional to ζ. This finding might at first seem surprising, but relates to the fact that for large ζ, the motion is impeded by fields generated by the build-up of charge carried forward by convection. According to the model, convected charge must be carried back again by conduction through the surrounding liquid. Thus it is that the tendency of an increasing ζ to decrease the particle mobility is avoided by increasing the conductivity of the surrounding fluid.[1]

With external forces such as those due to gravity or centrifugal acceleration forcing a particle through the liquid, reciprocal coupling occurs. Convection of charge in the double layer results in a dipole of electric field intensity and current density around the particle. If many particles are present, these generated fields add, to induce a "macroscopic" field measurable by electrodes immersed in the liquid through which the collection of particle move. This sedimentation potential (or "Dorn effect") is the subject of Prob. 10.10.3.

10.11 Electrocapillarity

A simple experiment that would prove baffling without an appreciation for the action of double layers at interfaces between liquids is sketched in Fig. 10.11.1. Mercury drops fall from a pipette

1. For extensive discussion see V. G. Levich, Physicochemical Hydrodynamics, Prentice-Hall, Englewood Cliffs, N.J., 1963, pp. 472-93.

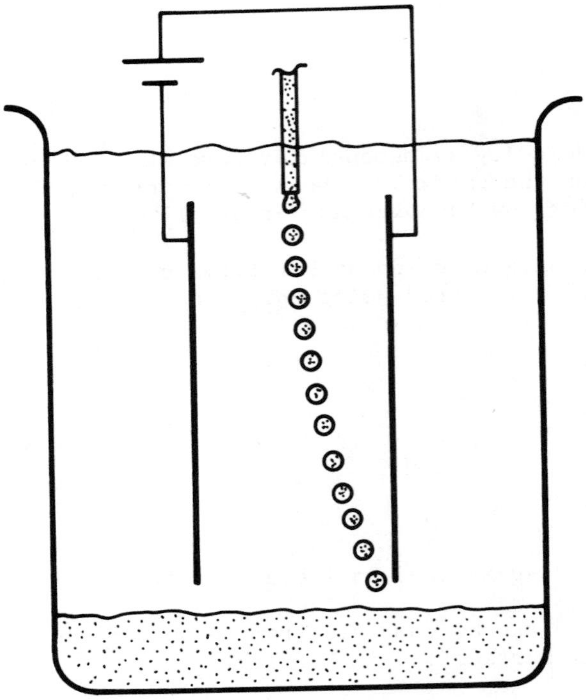

Fig. 10.11.1

Falling mercury drops surrounded by NaCl electrolyte are deflected as they pass through imposed field E_o. Typical for a drop having radius R = 1 mm passing through field of E_o = 100 V/m would be a horizontal velocity of 5 cm/sec.

through an electrolyte between electrodes to which a potential difference of a few volts has been applied. The drops are strongly deflected to one of the electrodes.

It is natural to simply attribute a net charge to each drop. However, the electrolyte is relatively conducting and this means that any net charge would leak away in a few relaxation times (Prob. 5.10.3). For the experiment of Fig. 10.11.1 this time is about 10^{-8} sec! Clearly, the drop and its immediate surroundings can carry no net charge on the time scale of the experiment. The drops must be "swimming," much as for the electrophoresing particles of Sec. 10.10. However, there are two important ways in which the drops do not fit the electrophoresis model. First, the drop is much more conducting than its surroundings. More important, it moves much too fast to be accounted for by the electrophoresis model and reasonable zeta potentials.

Up to potential differences on the order of a volt or so, the mercury-electrolyte interface can be polarized, in the sense that there are no chemical reactions to sustain a current flow so that the interface acts as an insulator. The result is an electric field within the double layer that is far larger than that in the electrolyte, on the order of 10^8 V/m compared to 10^2 V/m. The conditions are established for having a double-layer surface force density, as discussed in Sec. 3.11.

If the drops were rigid, the surface force density would have no effect. On a closed surface, there is no net force resulting from a surface force density (Prob. 3.11.2). However, the liquid surface can be set into motion. The shear rate is determined by the scale of the drop and not the scale of the double layer. This is why the drops move with such surprising speed relative to particles subject to electrophoresis.

The double layer also provides a mechanism for mechanical-to-electrical transduction. In the mercury drop experiment, electrical signals are generated in the electrolyte by the passing drops. Here again is cause for surprise, because generation of an appreciable electric field by the motion implies a significant electric Reynolds number. Based on the bulk properties of the electrolyte and the time for a drop to migrate one radius, this number is typically 10^{-7}. The lesson here is that the relevant relaxation time should reflect the heterogeneity of the system. The electric energy storage is in the double layer but the electrical loss is in the surrounding medium. Hence, the correct electric Reynolds number is modified by the ratio of the drop radius to the double layer thickness, a number that is of the order of $10^{-3}/10^{-8} = 10^5$. Drop motions are taken up in Sec. 10.12.

That the double layer electric surface force density of Sec. 3.11 takes a form similar to that found in Sec. 7.6 for surface tension, is a warning that in dealing with naturally occurring double layers it is not possible to make a clear distinction between electrical and mechanical surface force densities. The microstructure of the fields within the layer is in general not known. For example, through the electrochemical interaction of mercury and electrolyte, interior fields are generated which can be altered by an externally applied potential difference, but are not solely determined by external constraints.

Developments in this section make no distinction between electrical and mechanical surface forces. Rather, a surface tension γ_e is used to represent both electrical contributions and those ordinarily associated with the surface tension. The starting point is a statement of conservation of energy for an element of the interface. Such a statement defines the energy in terms of the local geometry and potential of the interface. If the exterior field contribution to the energy of the system is significant, then the energy stored in the electric field is a function of the geometry of the interface and of neighboring conductors and dielectrics. This contribution of the exterior fields is represented by the first term in Eq. 3.11.8. In what follows, it is assumed that exterior energy storage is negligible.

The surface tension γ_e is to the interface what the stress is to the volume. With the understanding that $\gamma_E \rightarrow \gamma_e$, the control volume of Fig. 3.11.1 is used, where γ_e is visualized as a force per unit length acting normal to the edges. Because the interface can be expected to have properties independent of rotations about the normal vector $\vec{n}$, it is assumed at the outset that the surface tension acting in the μ direction is the same as that acting in the ζ direction. Also, the edges are pictured as free of interfacial shear stresses. (A monomolecular interfacial film, residing on the interface as a distinguishable phase, can behave as two-dimensional fluid or solid. For the former, γ_e is replaced by a two-dimensional tensor γ_{ij}, with components departing from the diagonal form $\gamma_{ij} = \gamma_e\delta_{ij}$ used here because of relative motion (because of surface viscosity). The role played by the pressure in the mechanical three-dimensional force density is taken by γ_e on the surface. The scalar surface tension can be regarded as an inviscid model for the interface that is particularly appropriate if the interface is clean.[2]

Force equilibrium for the control volume requires that Eq. 3.11.8 relate the surface force density and the surface tension:

$$\vec{T} = -\vec{n}\gamma_e\left[\frac{1}{R_1} + \frac{1}{R_2}\right] + \nabla_\Sigma\gamma_e \tag{1}$$

where external stress contributions are dropped.

With the objective of relating γ_e to the double-layer charge, consider conservation of energy for a uniform section of the interface. An incremental increase in the energy W_s stored in the section of interface having area A can either be caused by doing work by means of the surface stress along the edges, or by increasing the total double layer charge q_d placed on the electrolyte side of the interface in the face of the potential difference v_d:

$$\delta W_s = \gamma_e\delta A + v_d\delta q_d \tag{2}$$

The mechanical and electrical work in this expression make it analogous to the conservation of energy statement for a lumped parameter electroquasistatic coupling system, for example Eq. 3.5.1. One difference is that in Chap. 3 the force is assumed to be of purely electrical origin.

A second useful connection is between Eq. 2 and similar thermodynamic relations used in Sec. 7.22 for compressible fluids. In the volumetric deformations of a gas, $p\delta V$ plays a role analogous to that of the term $\gamma_e\delta A$ in Eq. 2.

With the objective of using the double layer potential difference v_d as an independent variable, recognize that $v_d\delta q_d = \delta(v_d q_d) - q_d\delta v_d$ so that Eq. 2 becomes

$$\delta W_s' = -\gamma_e\delta A + q_d\delta v_d; \quad W_s' \equiv q_d v_d - W_s \tag{3}$$

where W_s' is an electrocapillary coenergy function. In a manner familiar from Sec. 3.5, the assumption that W_s' is a state function of A and v_d makes it possible to write

$$\delta W_s' = \frac{\partial W_s'}{\partial A}\delta A + \frac{\partial W_s'}{\partial v_d}\delta v_d \tag{4}$$

and to conclude by comparing Eqs. 3 and 4 that

$$\gamma_e = -\frac{\partial W_s'}{\partial A}; \quad q_d = \frac{\partial W_s'}{\partial v_d}; \quad \frac{\partial \gamma_e}{\partial v_d} = -\frac{\partial q_d}{\partial A} \tag{5}$$

2. For discussion, see for example G. L. Gaines, Jr., The Physical Chemistry of Surface Films, Reinhold Publishing Corp., New York, 1952.

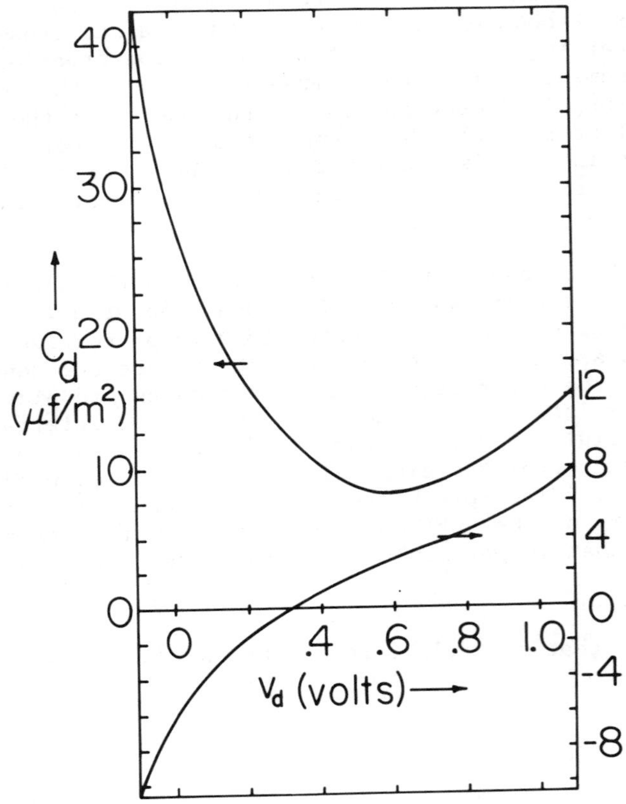

Fig. 10.11.2a. Incremental capacitance and charge per unit area as function of voltage for mercury-KNO_3. Here the electrolyte is 0.2 M KNO_3 in gel. This solid-liquid interface exhibits properties typical of liquid-liquid interfaces.[3]

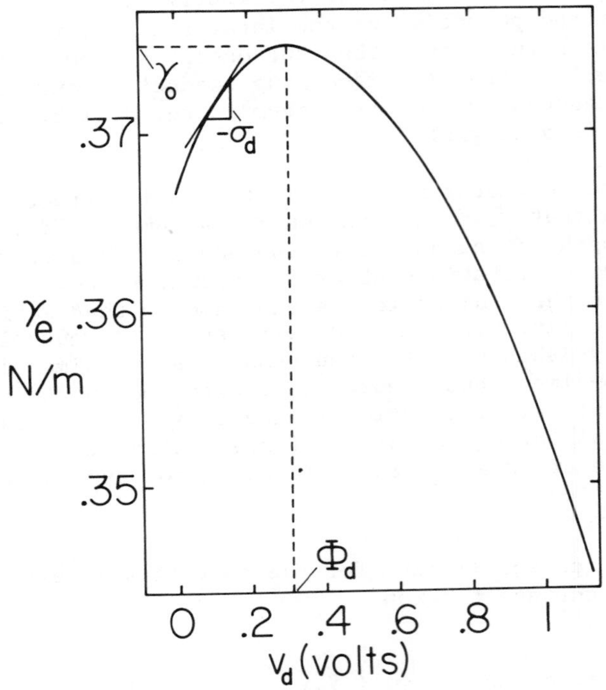

Fig. 10.11.2b. Surface tension as function of voltage for data of (a). γ_o has been defined as value for H_2O-Hg interface.

The third of these expressions follows by taking cross-derivatives of the previous two expressions.

An example of a constitutive law expressing the dependence of the charge on A is

$$q_d = A\sigma_d(v_d) \tag{6}$$

This expression pertains to a "clean" interface because it stipulates that provided the potential difference is held fixed, increasing the area of exposure between mercury and electrolyte proportionately increases the total charge. Such a law would not apply if, for example, the layer were a thin region of insulating liquid that conserved its mass and therefore thinned out as the area increased.

With the use of Eq. 6, Eq. 5c becomes the Lippmann equation:

$$-\sigma_d = \frac{\partial \gamma_e}{\partial v_d} \tag{7}$$

The graphical significance of Eq. 6 for an electrocapillary curve is depicted by Fig. 10.11.2.[3] double-layer charge/unit area determined from γ_e by Eq. 7 does not depend on a specific model.

That an alternative view has been taken of the same type of surface force density treated in Sec. 3.11 is illustrated by taking the coenergy stored in the area A as being proportional to that area and the integral of a coenergy density over the cross section of the layer,

$$W_s' = A\int_{o^-}^{o^+} W'(\nu)d\nu$$

3. A. J. Grodzinsky, "Elastic Electrocapillary Transduction," M.S. Thesis, Department of Electrical Engineering, Massachusetts Institute of Technology, Cambridge, Mass., 1971.

Then, with the use of Eq. 5a, an expression is obtained for γ_e comparable to that for γ_E given with Eq. 3.11.8. Of course, here W' can include contributions of a mechanical origin, whereas in Sec. 3.11 it does not. To preserve the generality inherent to Eq. 3, it is integrated along the state space contour of Fig. 10.11.3:

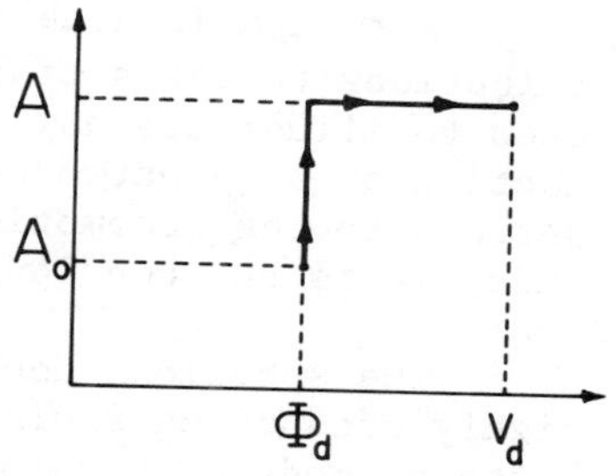

Fig. 10.11.3. Line integration in state space (A, v_d) to determine co-energy function, Eq. 9.

$$W_s' = -\int_{A_o}^{A} \gamma_o \delta A + A\int_{\Phi_d}^{v_d} \sigma(v_d)\delta v_d \tag{9}$$

where an electrical "clean-interface" constitutive law, Eq. 6, is assumed. The surface tension is defined as γ_o with the potential equal to Φ_d. Thus, measurement of σ_d and integration is one procedure for determining γ_e, which by virtue of Eqs. 5a and 9 is

$$\gamma_e = \gamma_o - \int_{\Phi_d}^{v_d} \sigma_d(v_d)\delta v_d \tag{10}$$

Conventionally, σ_d is determined by electrical measurements. With the area held fixed, a section of the interface is driven by a voltage composed of a constant part V_d and a small perturbation v_d'. The measured current is then to linear terms

$$i_d = AC_d \frac{dv_d}{dt}; \quad C_d \equiv \frac{\partial \sigma_d}{\partial v_d}(V_d) \tag{11}$$

so that the incremental capacitance $C_d(v_d)$ can be deduced. The surface charge then follows from the integration:

$$\sigma_d = \int_{\Phi_d}^{v_d} C_d(V_d)dV_d \tag{12}$$

The constant of integration must be independently determined, say by measuring the voltage at which there is no mechanical linear response to a tangential perturbation field. Thus, the electrocapillary curve can be determined by two successive integrations, the first Eq. 12 and the second Eq. 10. An independent measurement of the surface tension, say at the voltage for zero charge, Φ_d, is required for the second integration constant. The three curves for the differential capacitance, C_d, double layer charge density σ_d, and surface tension γ_e are illustrated in Fig. 10.11.2.

Finally, note that for a clean interface the double-layer shear force density can still be thought of as the product of σ_d and the tangential electric field on the electrolyte side. This is seen by combining the potential and tangential field boundary conditions of Eqs. 2.10.10 and 2.10.11 (with $E_t = 0$ and Φ = constant on the metal side of the interface) to write

$$\vec{E}_t = -\nabla_\Sigma v_d \tag{13}$$

Then, if γ_e varies <u>only</u> by virtue of v_d,

$$\nabla_\Sigma \gamma_e = \frac{\partial \gamma}{\partial v_d}\nabla_\Sigma v_d = \sigma_d E_t \tag{14}$$

This expression for the shear component of $\vec{T}$ applies if the layer is homogeneous in the sense that any section of the interface is characterized by the same constitutive law, Eq. 8.

Electrocapillary phenomena illustrate how double layers can impart a net electric surface force density to an interface. Although most studied and best understood for H_g-electrolyte interfaces,[4] electrocapillarity serves as a thought provoking example in developing models involving other more complex combinations of materials.

---

4. P. Delahay, Double Layer and Electrode Kinetics, Interscience Publishers, New York, 1966.

10.12 Motion of a Liquid Drop Driven by Internal Currents

Although incapable of causing a net electric force on a closed surface, the double-layer contributions to the surface force density can nevertheless induce net motion. The specific example used to illustrate how is depicted by Fig. 10.12.1, and intended as a primitive model for the transduction of an electrochemically generated current into net mechanical migration. Perhaps it might pertain to the locomotion of a biological entity. With the driving current outside rather than inside, it is the configuration of the dropping mercury electrode. (See Prob. 10.12.1.)

The spherical double-layer interface separates an electrolytic fluid inside from a relatively highly conducting fluid outside. At the center, there is a current source having the nature of a battery, modeled here as a dipole current source. A source of I amps is separated along the z axis by a distance $d \ll R$ from a sink of I amps (the positive and negative terminals of the battery). The objective is to determine the velocity of the drop relative to the surrounding fluid, which is stationary at infinity. So that the flow is steady, use is made of a frame of reference fixed to the center of the drop. The surrounding fluid then appears to have a uniform velocity $U\vec{i}_z$ far from the drop.

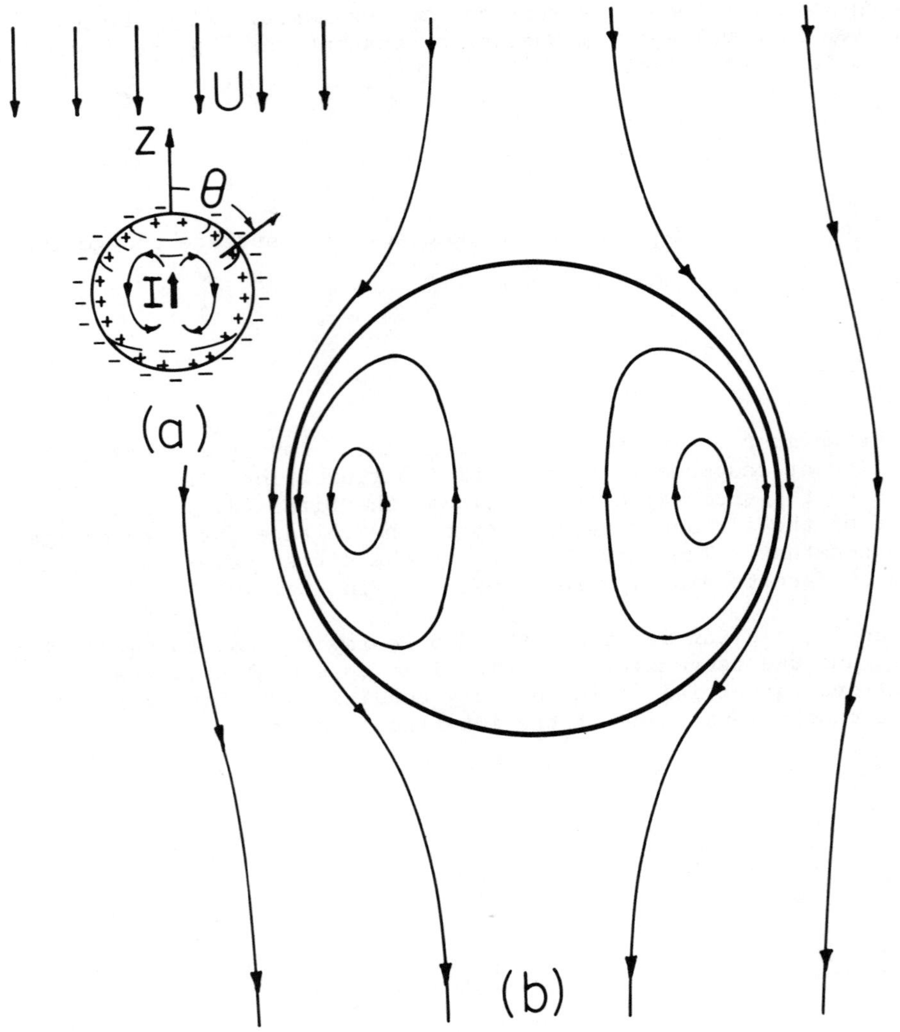

Fig. 10.12.1. (a) Liquid drop separated from surrounding liquid by ideally polarized double layer. Dipole current source is located at drop center. (b) Stream lines for fluid motion as viewed from frame fixed to drop.

With the double layer positive on the inside and I positive, U will be found to be negative, meaning that the drop is propelled in the z direction or in the direction of the dipole. Thus the magnitude and direction of migration is determined by the dipole. The physical mechanism is the double-layer shear surface force density tending to propel the interface from north to south. This density is largest at the equator. The consequent bulk flow is sketched in Fig. 10.12.1b. Inside, a doughnut-shaped cellular motion results, while outside fluid is pumped in the -z direction. Viscous shear stresses at the interface are typically determined by the interfacial velocity and a characteristic distance on the order of the drop radius R. The double-layer thickness is many times smaller than R, and hence the viscous shear stresses within the double layer (which are based on the thickness of the double layer) make the layer move essentially as a whole. Thus, for the present purposes, the fluid velocity is continuous through the double layer, and it is the net surface force density discussed in Sec. 3.11 that is the drive.

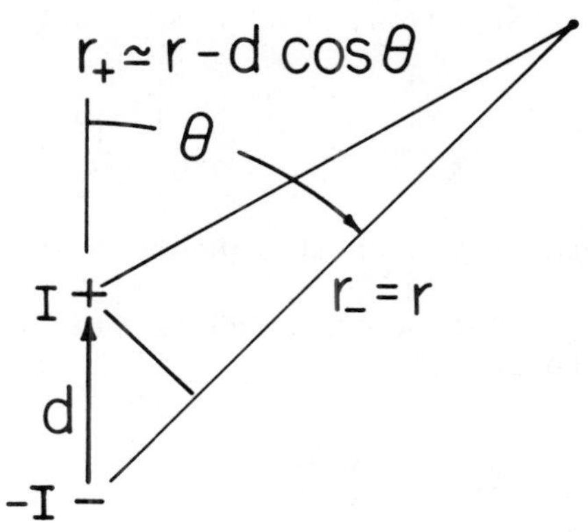

Fig. 10.12.2. Definition of dipole current source in terms of a source and sink of current disposed along the z axis a distance d apart.

The physical explanation for the drop motions applies (turned inside out) to the drop motions discussed in Prob. 10.12.1. It is because the interface can flow that the drops sustain a net electrically driven motion. Propulsion of a boat is in a way analogous. The double layer simply "rows" the drop through the surrounding fluid.

A self-consistent model for radial and tangential stress equilibrium, as well as conservation of double-layer charge, could in general be complicated. The remarkable fact is that a relatively simple model can be formulated combining electrical and mechanical distributions that have the θ dependence cos θ or sin θ. The drop is assumed to remain spherical. The assumption is subsequently shown to be valid.

First, the electrical current-dipole is represented. In the electrolyte, the current density is given by $\vec{J}_f = -\sigma_b \nabla\Phi$ and hence there is an electric potential associated with the point current source and sink, $\Phi_\pm = \pm I/4\pi\sigma_b r_\pm$. The distances $r_\pm$ are sketched in Fig. 10.12.2. By taking the limit d << r of the superimposed source and sink potential, it is seen that in the neighborhood of the origin, the potential must be

$$\Phi \rightarrow \frac{Id}{4\pi\sigma_b}\frac{\cos\theta}{r^2} \tag{1}$$

On a spherical surface radius c << R, the potential takes the form $\mathrm{Re}\tilde{\Phi}^c \cos\theta$, which is the n = 1 and m = 0 case from Table 2.16.3. The complex amplitude on the β → c surface surrounding the dipole at r = c is

$$\tilde{\Phi}^c = \frac{Id}{4\pi\sigma_b c^2} \tag{2}$$

The electric transfer relation, Eq. (a) from Table 2.16.3, again in the limit c << R, then gives the radial field at r = R in terms of the current drive and the potential at the interface:

$$\tilde{E}_r^b = -\frac{\tilde{\Phi}^b}{R} + \frac{3c^2}{R^3}\tilde{\Phi}^c = -\frac{\tilde{\Phi}^b}{R} + \frac{3Id}{4\pi\sigma_b R^3} \tag{3}$$

The θ dependence is recovered by multiplying by cos θ.

The region r > R is highly conducting, so for now the potential there is taken as uniform. The coupling at the interface is a two-way one.

Charge Conservation: Because the interface moves in a nonuniform fashion, charge carried by convection must be supplied by conduction at one pole and similarly removed at the other. The interface is presumed to be ideally polarized, so that charge conservation requires an equilibrium between the convection of the double-layer charge associated with the interior region and conduction normal to the interface from the interior:

$$\nabla_\Sigma(\sigma_d\vec{v}^b) = \frac{1}{R\sin\theta}\frac{\partial}{\partial\theta}\left[\sigma_d v_\theta^d \sin\theta\right] = \sigma_b E_r^b \tag{4}$$

A similar relation applies to the exterior side. In the absence of the electrical drive, the interface has a uniform charge σ_o consistent with a potential difference V_d. The surface potential variation caused by I is reflected in a charge variation. In the following it is assumed that the total departure of the potential from V_d is relatively small so that the double layer charge, σ_d, in Eq. 4 can be approximated by σ_o.

The transfer relations for the viscous flow, as developed in Sec. 7.20, suggest that $v_\theta = \tilde{v}_\theta \sin\theta$, so that Eq. 4, with cos θ factored out, becomes

$$\frac{2\sigma_o \tilde{v}_\theta}{R} = \sigma_b \tilde{E}_r^b = \sigma_b \left[-\frac{\tilde{\Phi}^b}{R} + \frac{3Id}{4\pi\sigma_b R^3} \right] \qquad (5)$$

Here, Eq. 3 establishes the second equality.

The combination of electric double-layer boundary conditions, Eqs. 2.10.10 and 2.10.11, reduces here to

$$E_\theta = -\frac{1}{R}\frac{\partial \Phi^b}{\partial \theta} \qquad (6)$$

serving as a reminder that just inside the interface there is a tangential electric field.

Stress Balance: The radial and tangential balance of mechanical stresses, with the surface force density given by Eq. 10.11.1 and with the shear term expressed as Eq. 10.11.14, are represented by

$$\Pi^a + S_{rr}^a - \Pi^b - S_{rr}^b - \frac{2\gamma_e}{R} = 0 \qquad (7)$$

$$S_{\theta r}^a - S_{\theta r}^b + \sigma_d E_\theta = 0 \qquad (8)$$

With the outside potential defined as zero, it is appropriate to let Φ be the departure from potential V_d in the interior. Then

$$\gamma_e \approx \gamma_c - \sigma_d \Phi; \quad \sigma_d \equiv -\left(\frac{\partial \gamma_e}{\partial v_d}\right)_{V_d} \qquad (9)$$

where γ_c is the surface tension at the equator and σ_d is in accordance with the Lippman Eq. 10.11.7. The θ-independent part of the surface tension radial force is balanced by a uniform pressure jump $\Pi^a - \Pi^b$ at the interface. With the assumption that $\sigma_d \approx \sigma_o$, Eq. 7 is satisfied for each value of θ if

$$2\tilde{S}_{rr}^a - 2\tilde{S}_{rr}^b + \frac{2\sigma_b}{R}\tilde{\Phi}^b = 0 \qquad (10)$$

where cos θ has been factored out and amplitudes are introduced consistent with Table 7.20.1.

Similarly, according to Eqs. 8 and 6, tangential force equilibrium results at each value of θ if

$$\tilde{S}_{\theta r}^a - \tilde{S}_{\theta r}^b + \frac{\sigma_o}{R}\tilde{\Phi}^b = 0 \qquad (11)$$

where sin θ is factored out.

That the double layer moves as a whole at a given interfacial location implies a tangential velocity at the interface that is continuous, while the assumption that spherical geometry is retained requires that the interior and exterior radial velocities vanish:

$$\tilde{v}_\theta^a = \tilde{v}_\theta^b; \quad \tilde{v}_r^a = 0; \quad \tilde{v}_r^b = 0 \qquad (12)$$

With velocity amplitudes so related, viscous stresses are given for the outside region by Eq. 7.20.24 with the radius $\beta \to R$ and for the interior by Eq. 7.20.23 with radius $\alpha \to R$. These are now substituted into Eqs. 10 and 11. Three conditions on the amplitudes, physically representing conservation of charge, Eq. 5, and these radial and tangential interfacial stress balances are

$$\begin{bmatrix} \frac{2\sigma_o}{R} & \frac{\sigma_b}{R} & 0 \\ -\frac{3}{R}(\eta_a+2\eta_b) & \frac{2\sigma_o}{R} & \frac{3\eta_a}{2R} \\ -\frac{3}{R}(\eta_a+\eta_b) & \frac{\sigma_o}{R} & -\frac{3\eta_a}{2R} \end{bmatrix} \begin{bmatrix} \tilde{v}_\theta^b \\ \tilde{\Phi}^\beta \\ U \end{bmatrix} = \begin{bmatrix} \frac{3Id}{4\pi R^3} \\ 0 \\ 0 \end{bmatrix} \tag{13}$$

The velocity of the drop relative to an exterior fluid at infinity is the negative of U, where from Eq. 13

$$U = \frac{-Id}{4\pi R^2\sqrt{\sigma_b(\eta_a + \frac{3}{2}\eta_b)}} \frac{H_e}{(1 + H_e^2)}; \quad H_e \equiv \frac{\sigma_o}{\sqrt{\sigma_b(\eta_a + \frac{3}{2}\eta_b)}} \tag{14}$$

The associated interfacial velocity follows by subtracting twice Eq. 13c from Eq. 13b,

$$\tilde{v}_\theta = -\frac{3U}{2} \tag{15}$$

With I and σ_o positive, the signs are consistent with Fig. 10.12.1 and the introductory discussion.

The normalized double layer charge density, H_e, also takes the form of an electric Hartmann number, $\sqrt{\tau_e/\tau_{EV}}$. This is seen by recognizing that $\sigma_o \to \varepsilon\mathcal{E}$ where $\mathcal{E}$ is typical of the electric field inside the layer. The dependence of U on H_e sketched in Fig. 10.12.3 makes it clear that an optimum charge density exists. With H_e small, the motion is mainly limited by viscosity and so increases in linear proportion to σ_o. But if $H_e \gg 1$, then the interfacial velocity, and hence U, is limited by the ability of the electrolyte to conduct away the convected charge.

To discover what limits the magnitude of U, suppose that σ_o is made the optimum value so that $H_e = 1$. Then, Eq. 14 becomes

$$U_{opt} = \frac{-Id}{8\pi R^2\sigma_b}\sqrt{\frac{\sigma_b}{(\eta_a + \frac{3}{2}\eta_b)}} \tag{16}$$

The magnitude of I is limited by the maximum excursion of the double-layer potential from V_d. From Eqs. 13a and 15, the interfacial potential variation has the amplitude

$$\tilde{\Phi}^b = \frac{3Id}{4\pi R^2\sigma_b} + \frac{3\sigma_o}{\sigma_b}U \tag{17}$$

These voltage contributions are respectively due to conduction and convection. With $H_e = 1$, the second term cancels half of the first, so that the pole-to-pole excursion in potential, $2\tilde{\Phi}^b$, can be used to write Eq. 16 as

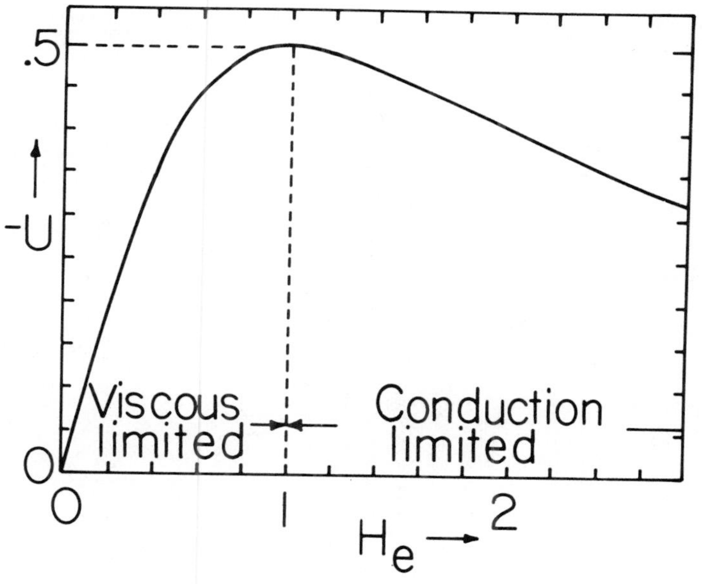

Fig. 10.12.3. Dependence of drop velocity on normalized double layer charge.

$$U_{opt} \to \frac{2\tilde{\Phi}^b}{6}\sqrt{\frac{\sigma_b}{(\eta_a + \frac{3}{2}\eta_b)}} \tag{18}$$

Typical values are $2\tilde{\Phi}^b = 0.1$ V, $\sigma_b = 10^{-2}$ mhos/m, $\eta_a = \eta_b = 10^{-3}$ for water based electrolytes, and hence velocities on the order of 3 cm/sec. Of course, the low Reynolds number condition may not be met with such a velocity. But clearly, the double layer mechanism can be the basis for significant motions.

The mechanical response to an electrical drive has been emphasized. That fields are generated by the motion is a reminder that the electrocapillary double layer can be the site of a reverse transduction.

Problems for Chapter 10

For Section 10.2:

Prob. 10.2.1 The region between two planes, at $x = \Delta$ and $x = 0$, is filled with a material having uniform thermal properties that sustains fully developed flow with velocity $\vec{v} = v(x)\vec{i}_y$. The surfaces are at the respective constant temperatures (T^α, T^β). In the volume, there is an arbitrary dissipation $\phi_d(x)$.

(a) Determine the temperature distribution $T(x)$.

(b) What is the thermal flux at the boundaries? Note that this is one of a group of "fully developed" heat conduction configurations, playing a role in heat transfer analogous to the fluid mechanics relations of Table 9.3.1.

For Section 10.3:

Prob. 10.3.1 The magnetically excited layer considered in this section is embedded in a system in which the surroundings are relatively thermally insulating. The temperature of the layer rises to a sufficient extent that the steady dissipation is accommodated by the steady heat flux. However, insofar as the time-varying part of the heat flux is concerned, the layer surfaces are bounded by thermal insulators. What are the temperatures at the layer surfaces?

Prob. 10.3.2 The moving slab of Fig. 10.3.1 is now a semi-insulating dielectric having uniform electrical conductivity σ and permittivity ε. Potential distributions at the α and β surfaces are respectively $\text{Re}\hat{\Phi}^\alpha \exp j(\omega t-ky)$ and $\text{Re}\hat{\Phi}^\beta \exp j(\omega t-ky)$.

(a) Write the electrical dissipation density in the form of Eq. 10.3.6.

(b) Find the temperature distribution throughout the slab and the heat fluxes at its surfaces. Assume that at the α and β surfaces the respective temperatures are

$$T_o^\alpha + \text{Re}\hat{T}^\alpha \exp j(\omega_2 t-k_2 y) \quad \text{and} \quad T_o^\beta + \text{Re}\hat{T}^\beta \exp j(\omega_2 t-k_2 y).$$

For Section 10.4:

Prob. 10.4.1 A ferrofluid has a permeability that has the temperature dependence $\mu = \mu_a[1-\alpha_\mu(T-T_a)]$, where μ_a and α_μ are constant parameters. In the channel of Fig. 10.4.1, the fluid is subjected to a uniform transverse magnetic field intensity H_o. The object is to pump the fluid by imposing the temperatures T_a and T_b on the grids, and hence producing a variation in the permeability in the direction of the heat flux. Assume that the boundary layer thickness is small compared to the channel cross section, so that the velocity is uniform across the channel. Determine the pressure-velocity relation that is analogous to Eq. 10.4.7 and the temperature distribution and heat flux.

For Section 10.5:

Prob. 10.5.1 The rotor described by Eqs. 10.5.16 - 10.5.18 is in the state of steady rotation described by Eqs. 10.5.23.

(a) Show that this stationary state is overstable if R_a exceeds

$$R_a = p_T \frac{(1+f)}{f} \frac{p_T+4(1+f)}{p_T-2(1+f)}$$

(b) Show that the frequency of oscillation at the onset of this instability is

$$\omega = \frac{1}{\tau_T}\sqrt{2p_T(1+f)\frac{[p_T+(1+f)]}{[p_T-2(1+f)]}}$$

Prob. 10.5.2 The rotor of Fig. 10.5.1 is heated from the side rather than from below. Thus the external temperature distribution is given by Eq. 10.5.1 with $\sin\theta \rightarrow \cos\theta$.

(a) Deduce the equations of motion, similar to Eqs. 10.5.16 - 10.5.18.

(b) Use a graphical solution similar to that pictured by Fig. 10.5.3 to determine the steady angular velocity. Explain qualitatively the direction of rotation.

For Section 10.6:

Prob. 10.6.1 Implicit to Eq. 10.6.17 is the principle of exchange of stabilities. That is, as R_{am} is raised, each temporal mode becomes unstable with $s_n = 0$. If it is only the condition for onset of instability that is of interest, it can be assumed at the outset that $s_n = 0$ and R_{am} can be treated as an eigenvalue. Thus $(R_{am})_n$ is the value of R_{am} that reduces the frequency of the nth mode to zero.

(a) Use Eqs. 10.6.8 and 10.6.9 with the boundary conditions that $\underline{T} = 0$ and $\underline{v}_x = 0$ on the boundaries $\underline{x} = 0$, $\underline{x} = 1$ to show that, provided $R_{am} > 0$, the principle of exchange of stabilities holds. (See the Temporal Modes subsection of Sec. 8.18.)

(b) Set $\underline{\omega} = 0$ in Eqs. 8 and 9 and solve the eigenvalue problem for R_{am}. The result should be Eq. 10.6.18 and hence 10.6.19.

Prob. 10.6.2 For the thermal-hydromagnetic layer between the planes α and β as treated in this section, determine the transfer relations

$$\begin{bmatrix} \hat{\underline{\Gamma}}^{\alpha} \\ \hat{\underline{\Gamma}}^{\beta} \\ \hat{\underline{p}}^{\alpha} \\ \hat{\underline{p}}^{\beta} \end{bmatrix} = \begin{bmatrix} C_{ij} \end{bmatrix} \begin{bmatrix} \hat{\underline{T}}^{\alpha} \\ \hat{\underline{T}}^{\beta} \\ \hat{\underline{v}}_x^{\alpha} \\ \hat{\underline{v}}_x^{\beta} \end{bmatrix}$$

Prob. 10.6.3 Consider the layer of Fig. 10.6.2, but with viscosity.

(a) Show that the normalized equations replacing Eqs. 10.6.8 and 10.6.9 are

$$\left[(D^2-k^2) - \frac{j\omega}{P_T}(D^2-k^2) - H_m^2 D^2\right]\hat{v}_x = -R_a \hat{T}$$

$$\left[j\omega - (D^2-k^2)\right]\hat{T} = -\hat{v}_x$$

where

$$\omega = \underline{\omega} K_T/\Delta^2 \qquad \hat{T} = \hat{\underline{T}}\Delta DT_s$$

$$x = \underline{x}\Delta \qquad \hat{v}_x = \hat{\underline{v}}_x K_T/\Delta$$

$$k = \underline{k}/\Delta \qquad \underline{p} = p\Delta^2/K_T^2\rho_o$$

and the conventional Rayleigh, Prandtl and Hartmann numbers are

$$R_a \equiv \frac{\alpha\rho_o g\Delta^4 DT_s}{K_T\eta} \quad ; \quad P_T \equiv \frac{\eta}{\rho K_T} = \frac{\tau_T}{\tau_v} \quad ; \quad H_m^2 \equiv \frac{\mu\sigma\Delta^2\mu_o H_o^2}{\eta}$$

(b) Outline a scheme to determine the transfer relations expressing the surface stresses and heat flux $(\hat{S}_x^{\alpha}, \hat{S}_x^{\beta}, \hat{S}_y^{\alpha}, \hat{S}_y^{\beta}, \hat{\Gamma}_x^{\alpha}, \hat{\Gamma}_x^{\beta})$ in terms of the surface velocities and temperatures $(\hat{v}_x^{\alpha}, \hat{v}_x^{\beta}, \hat{v}_y^{\alpha}, \hat{v}_y^{\beta}, \hat{T}^{\alpha}, \hat{T}^{\beta})$. The motions may be assumed to be independent of z, so $k_z = 0$.

For Section 10.7:

Prob. 10.7.1 A thin metal cylinder having radius R is charged by unipolar ions having the density ρ_o at the radius a from the cylinder's center. Assume that at a given instant the charge per unit length on the cylinder is λ and that the self fields of the ions in the volume are negligible compared to those due to the charge on the cylinder.

(a) Determine the ion charge density as a function of radial distance r.

(b) What is the current per unit length collected by the cylinder as a function of the voltage of the cylinder relative to that at r = a?

(c) If the cylinder is allowed to charge up, what is $\lambda(t)$ given that when t = 0, $\lambda = 0$?

Prob. 10.7.2 A pair of horizontal capacitor plates are used to impose a perpendicularly directed electric field on a homogeneous layer of liquid bounded from above (at $x = 0$) by air. Model the liquid as devoid of all but one positive species of electrical carriers with charge density ρ_+. Assume that charge in the neighborhood of the interface shields the field from the liquid bulk so that $\vec{E} = E_o\vec{i}_x$ at $x = 0$ and $\vec{E} \to 0$ as $x \to -\infty$. Hence, self fields of the ions are included.

(a) With negligible net current through the air, and hence in the liquid, show that the electric field and charge density comprising the monolayer of surface charge for $x < 0$ are

$$E_x = E_o/(1-x/\ell_d)\ ; \qquad \rho_+ = (\varepsilon E_o/\ell_d)/(1-x/\ell_d)^2\ ; \qquad \ell_d = 2K_+/bE_o$$

(b) For $E_o = 10^4$ v/m, what is a typical value of ℓ_d?

For Section 10.8:

Prob. 10.8.1 An electrolyte is bounded by plane parallel boundaries, each having the potential $-\zeta$. They are positioned at $x = 0$ and $x = \Delta$.

(a) Under the assumption that $\underline{\Phi} \ll 1$, what is the distribution of $\underline{\Phi}$? What is the potential $\underline{\Phi} \equiv \underline{\Phi}_c$ at the midplane?

(b) For arbitrary magnitude of $\underline{\Phi}$, show that in terms of normalized variables the potential distribution is

$$\underline{x} = \int_{-\underline{\zeta}}^{\underline{\Phi}} \frac{d\underline{\Phi}}{\sqrt{2\cosh\underline{\Phi} - \cosh\underline{\Phi}_c}}$$

where again $\underline{\Phi}_c$ is the potential at the midplane.

(c) Given the normalized spacing $\underline{\Delta} \equiv \Delta/\delta_o$, describe a numerical procedure for finding $\underline{\Phi}_c$ and hence determining the potential distribution.

(d) For $\underline{\Delta} = 2$ and $\underline{\zeta} = 3$, what is $\underline{\Phi}_c$? Plot the potential distribution.

For Section 10.9:

Prob. 10.9.1 The boundaries of a planar duct, such as pictured in Table 9.3.1, have a spacing Δ that is not necessarily large compared to δ_d.

(a) Used Eq. a from Table 9.3.1 to express the velocity distribution in terms of the potential distribution.

(b) Show that this expression reduces to Eq. 10.9.5 in the case where the Debye length is short compared to the channel width.

(c) In Prob. 10.8.1, a procedure is developed for finding the potential distribution with arbitrary wall spacing. Show that the velocity distribution can be written in the normalized form

$$\underline{v} = \frac{\Delta^2 q}{2\varepsilon E_y kT}\frac{\partial p'}{\partial y}\left[\underline{x}^2\left(\frac{\delta_D}{\Delta}\right)^2 - \underline{x}\frac{\delta_D}{\Delta}\right] + \underline{\Phi}(\underline{x}) + \underline{\zeta}$$

where $v = \underline{v}\varepsilon E_y kT/\eta q$ and $x = \underline{x}\delta_D$ and where $\underline{\Phi}(x)$ follows from Prob. 10.8.1.

Prob. 10.9.2 A two-dimensional channel having width Δ has walls with potentials $\Phi = -\zeta$. The current density in the y direction is "fully developed" and hence the total current through the channel is given by Eq. 10.9.13.

(a) Show that the current is related to the imposed E_y and the pressure gradient $\partial p/\partial y$ by

$$i = \sigma\Delta + \frac{2\rho_o\zeta^2\varepsilon\delta_D}{\eta(kT/q)}E_y - \frac{\zeta\Delta\varepsilon}{\eta}\frac{\partial p'}{\partial y}$$

(b) For an "open-circuit" channel ($i = 0$) having a length ℓ and pressure difference $\Delta p = -\ell\partial p/\partial y$, what is the streaming potential $v \equiv -E_y\ell$?

For Section 10.10:

Prob. 10.10.1 Following Eq. 10.10.2, it is argued that the shear stress induced surface current is ignorable compared to that driven by the imposed field. Approximate the shear stress contribution using the velocity U that was determined and justify this approximation.

Prob. 10.10.2 The particle considered in this section is fixed on a "stinger" which does not distort the field or impede the flow but does constrain the particle to a fixed position relative to the fluid at infinity. What is the force imparted by the electric field to the stinger?

Prob. 10.10.3 The particle is fixed on a stinger, as in Prob. 10.10.2, but both a uniform electric field and a uniform flow velocity are imposed at infinity. Because the flow is now forced, the contributions of the shear stress to the surface current can be significant. In view of Eq. 10.9.12, represent the surface current as

$$K_\theta = \sigma_s E_\theta + \beta S_{\theta r}$$

where for $\zeta < kT/q$, $\beta = 2\rho_o \zeta \delta_D^2/\eta(kT/q)$ and determine the potential distribution around the particle as a function of E_o and U. What is the potential if $E_o = 0$? What is f_z?

For Section 10.11:

Prob. 10.11.1 A clean interface is modeled as having a surface tension γ_o at the voltage $v_d = \Phi_d$, the tension being independent of the area A, and a Helmholtz double layer consisting of a plane parallel capacitor having spacing Δ, permittivity ε and zero double layer charge at $v_d = \Phi_d$. Determine C_d, σ_d and W_s', and compare to Fig. 10.11.1.

Prob. 10.11.2 A hemisphere of mercury submerged in an electrolyte is shown in cross section in Fig. P10.11.2. The interface between liquids forms a double layer of thickness Δ, pictured here as being a "Helmholtz" layer. (Prob. 10.11.1)

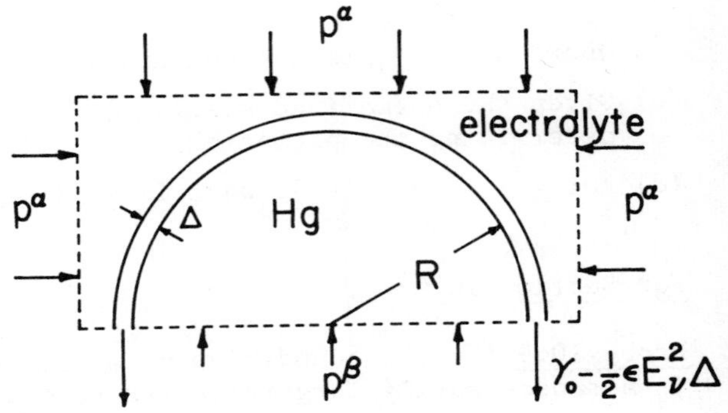

Fig. P10.11.2

(a) Write an expression for static equilibrium using the control volume shown to balance the pressure forces against those due to the combined surface tension and Maxwell stresses. Show that the resulting expression is as would be deduced from Eq. 10.11.1, where the electrocapillary surface tension is found in Prob. 10.11.1.

(b) Now suppose that, by means of an orifice at the center of the hemisphere, a small additional amount of mercury is introduced, so that the interface expands from R to R + $\delta\xi$. Use the result of (a) to compute the incremental change in pressure implied by the electrocapillary model.

(c) An alternative model might depict the double layer as composed of a film of insulating fluid. In that case, the equilibrium would take the same form as found in (a). But, suppose that with the addition of an increment of mercury the surface expands in such a way that the insulating layer of fluid preserves its volume. Find an expression for the change in pressure associated with an incremental change in radius $\delta\xi$. Compare the result to that found in (b) and explain the difference.

For Section 10.12:

Prob. 10.12.1 With the objective of determining the mobility $b = U/E_o$ of the mercury drop in an electrolyte, consider a drop that is highly conducting, with a surrounding electrolyte permeated by an electric field which is $E_o\vec{i}_z$ far from the drop. Following steps paralleling those in this section, show that the mobility is

$$b = \sigma_o R/(\frac{\sigma_o}{\sigma})^2 + (2\eta_b + 3\eta_a)$$

A mercury drop in an electrolyte is the configuration of a dropping mercury electrode, widely used to study electrochemical double layers because the surface is constantly renewed by continual generation of drops.[1] The dropping mercury electrode is used in analytical chemistry as a sensitive means of measuring trace constituents of the electrolyte.[2]

Prob. 10.12.2 A linear volume rate of flow is secured in the configuration of Fig. P10.12.2 by exploiting the double layer shearing surface force density. An electrolyte is bounded from above by insulating walls and from below by alternate sections of insulator and pools of mercury, each having length ℓ >> a or b.

Electrodes fixed adjacent to the pool edges are driven by an external current source and cause a "standing wave" of current with the distribution sketched. Hence, the ideally polarized double layer experiences a shearing surface force density tending to carry the liquid in one direction, while the insulating sections prevent backward motion where that force density would be reversed.

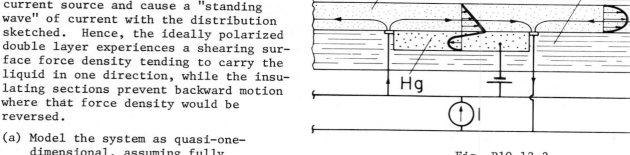

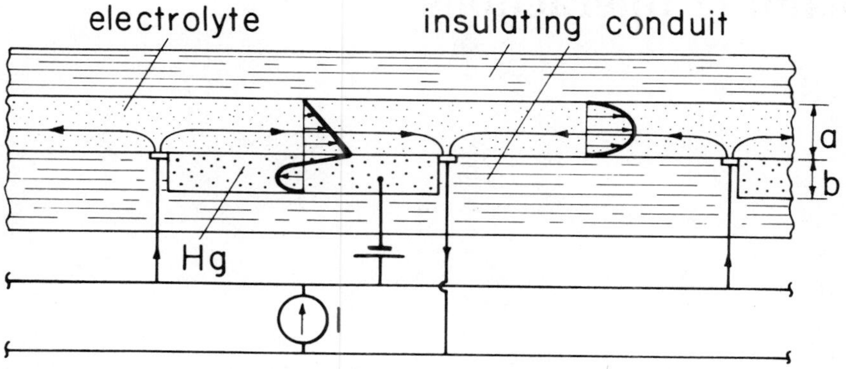

Fig. P10.12.2

(a) Model the system as quasi-one-dimensional, assuming fully developed plane flow in each of the sections and using mass and momentum conservation to piece these flows together at the pool edges. Assume that gravity holds the interface flat and that the system is closed on itself. Assume that the electrolyte is sufficiently highly conducting that charge convection at the interface can be ignored and the interface can be regarded as essentially uniformly polarized (even with the driving current producing a voltage drop in the interfacial plane).

(b) Find the volume rate of flow of the electrolyte as a function of the driving current.

---

1. An extensive treatment of the subject is given by V. G. Levich, Physicochemical Hydrodynamics, Prentice-Hall, Englewood Cliffs, N.J., 1965, pp. 493-551.

2. J. Heyrovský and K. Jaroslav, Principles of Polarography, Academic Press, New York, 1966.

11

Streaming Interactions

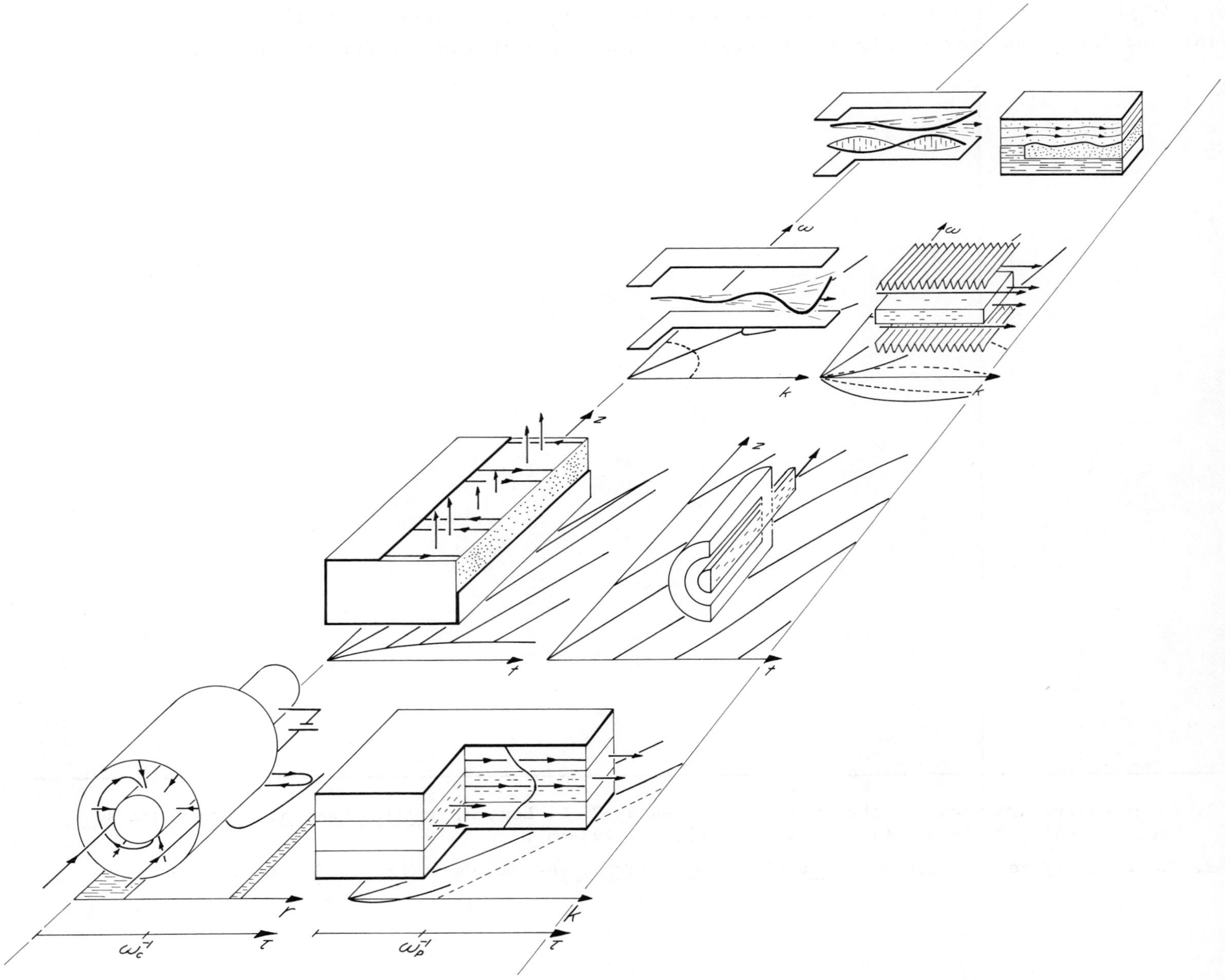

11.1 Introduction

Some of the most significant interactions between continua having large relative velocities involve charged particle beams accelerated under near-vacuum conditions. Thus, the first part of this chapter gives some background in electron beam dynamics. The charged particles of Chap. 5 become a continuum in their own right because their inertia is dominant. Section 11.2, on the laws and theorems for a charged particle gas, draws on the fluid mechanics of Chap. 7, and leads to the steady electron flows considered in Secs. 11.3 and 11.4. Flows illustrated in these latter sections are typical of those found in magnetrons and in electric and magnetic electron beam lenses. Pictured as they are in Lagrangian coordinates, the motions appear to be time varying. But, if viewed in Eulerian coordinates, the electron flows of these sections are steady and might be considered in Chap. 9. The remaining sections relate not only to electron beams, but to electromechanical continua introduced in previous chapters.

Sections 11.6-11.10 have as a common theme the use of the method of characteristics to understand dynamics in "real" space and time. The approach is restricted to two dimensions, here one space and the other time, but makes it possible to investigate such nonlinear phenomena as shock formation and nonlinear space-charge oscillations. Thus it is that these sections are concerned with quasi-one-dimensional models. As pointed out in Sec. 4.12, the small-amplitude limits of these models are identical with the long-wave limits of two- and three-dimensional models. Thus, the quasi-one-dimensional model represents what physical content there is to the dominant modes from the infinite number of spatial modes of a linear system. However, nonlinear phenomena can be incorporated into the quasi-one-dimensional model.

In addition to giving the opportunity to develop nonlinear phenomena, the method of characteristics gives the opportunity to explore the implications of causality for longitudinal boundary conditions and the general domain of dependence of a response pictured in the z-t plane. This gives an alternative to the complex-wave point of view, taken up in the remainder of the chapter, in appreciating the difference between absolute and convective instabilities and between evanescent and amplifying waves. The prototype configurations examined in Sec. 11.10 are analogous to traveling-wave electron beam or beam plasma systems taken up in later sections.

Sections 11.11-11.17 return to a theme of complex waves. Spatial transients in the sinusoidal steady state are considered in Sec. 5.17 with the tacit assumption that the response decays away from the excitation source. As illustrated in these sections, the response could just as well amplify from the region of excitation. How is an evanescent wave, which simply decays from the region of excitation, to be distinguished from one that amplifies? Temporal transients are first introduced in Sec. 5.15, and instability, defined as an unbounded response in time, illustrated in Sec. 8.9. In a system that is infinitely long in the longitudinal direction, a dispersion relation that gives "unstable" ω's for real k's can either imply that the response is unbounded in time at a given fixed location, or that there is unlimited growth for an observer moving with the response. In a given situation, how is an absolute instability to be distinguished from one that is convective? For special hyperbolic systems, these questions are answered in terms of the method of characteristics in Sec. 11.10. Sections 11.11 and 11.12 are devoted to the alternative of answering these questions in terms of complex waves. The remaining sections illustrate with classic examples.

BALLISTIC CONTINUA

11.2 Charged Particles in Vacuum; Electron Beams

Equations of Motion: In terms of the Eulerian coordinates of Sec. 2.4, Newton's law for a particle having mass m and charge q, subject to the Lorentz force (Eq. 3.2.1), is

$$m\left(\frac{\partial\vec{v}}{\partial t} + \vec{v}\cdot\nabla\vec{v}\right) = q(\vec{E} + \vec{v}\times\mu_o\vec{H}) \tag{1}$$

Multiplied by the particle number density, n, this expression is almost what would be written to describe a fluid. The pressure and viscous stress terms are absent from Eq. 1. To each point in space is ascribed the velocity, $\vec{v}$, of the particle that happens to be at that point at the given instant in time.

Because the pressure and viscous stresses are absent, much of the literature of electron beams pictures the motions in Lagrangian terms, as discussed in Sec. 2.4. Then, the initial coordinates of each particle are the independent variables as the partial derivative with respect to time is taken:

$$m\frac{\partial v}{\partial t} = q(\vec{E} + \vec{v}\times\mu_o\vec{H}) \tag{2}$$

Thus, for example, in cylindrical coordinates the equations of motion for a particle having the instantaneous position (r,θ,z) are

$$\frac{d^2r}{dt^2} - r(\frac{d\theta}{dt})^2 = \frac{q}{m} E_r + \frac{q}{m} \mu_o H_z r \frac{d\theta}{dt} \tag{3}$$

$$r \frac{d^2\theta}{dt^2} + 2 \frac{dr}{dt}\frac{d\theta}{dt} = \frac{1}{r}\frac{d}{dt}(r^2 \frac{d\theta}{dt}) = \frac{q}{m}(\mu_o H_r \frac{dz}{dt} - \mu_o H_z \frac{dr}{dt}) \tag{4}$$

$$\frac{d^2z}{dt^2} = \frac{q}{m} E_z - \frac{q}{m} \mu_o H_r r \frac{d\theta}{dt} \tag{5}$$

where the second terms on the left in Eqs. 3 and 4 are respectively the centripetal and Coriolis accelerations of rigid-body mechanics.

The dynamics of interest can be pictured as EQS with an imposed magnetic field. In Sec. 3.4 it is argued that in EQS systems, the magnetic force is negligible compared to the electric force. Now, the particles of interest include electrons or ions in vacuum. Their velocities can easily be large enough to make magnetic forces due to the imposed field important. The arguments of Sec. 3.4 show that the part of the force attributable to a magnetic field induced by the displacement current (or the current density associated with the accumulation of net charge) is still negligible provided that times of interest are long compared to the transit time of an electromagnetic wave.

The laws required to complete the description are usually written in Eulerian coordinates, much as in the description of charged migrating and diffusing particles in Sec. 5.2. With the charge density defined as nq, conservation of charge, Gauss' law and the condition that the electric field be irrotational are written as

$$\frac{\partial\rho}{\partial t} + \nabla\cdot\rho\vec{v} = 0 \tag{6}$$

$$\nabla\cdot\varepsilon_o\vec{E} = \rho \tag{7}$$

$$\vec{E} = -\nabla\Phi \tag{8}$$

Either Eq. 1 or 2 and these three expressions comprise two vector and two scalar equations in the dependent variables $\vec{v}$, $\vec{E}$ and ρ,Φ.

Energy Equation: The equivalent of Bernoulli's equation for charged ballistic particles is obtained following the same steps as in Sec. 7.8. With the use of a vector identity* and Eq. 8, Eq. 1 becomes

$$m(\frac{\partial\vec{v}}{\partial t} + \vec{\omega} \times \vec{v}) + \nabla(\frac{1}{2} m\vec{v}\cdot\vec{v} + q\Phi) = q\vec{v} \times \mu_o\vec{H} \tag{9}$$

where the vorticity, $\vec{\omega} \equiv \nabla \times \vec{v}$. Because both the vorticity and magnetic field terms in this expression are perpendicular to $\vec{v}$, it can be integrated along a stream line joining points a and b to obtain

$$\int_a^b m \frac{\partial\vec{v}}{\partial t}\cdot d\vec{\ell} + [\frac{1}{2} m\vec{v}\cdot\vec{v} + q\Phi]_a^b = 0 \tag{10}$$

In the steady state, the sum of the kinetic and electric potential energies of a particle are constant, regardless of the imposed magnetic field. Note that, in Eulerian coordinates, particle motions are steady provided Φ is constant.

Theorems of Kelvin and Busch: The curl of Eq. 9 describes the vorticity in Eulerian coordinates:

$$\frac{\partial\vec{\omega}}{\partial t} + \nabla \times (\vec{\omega} \times \vec{v}) = \frac{q}{m} \nabla \times (\vec{v} \times \mu_o\vec{H}) \tag{11}$$

This expression can be integrated over an open surface S enclosed by a contour C moving with the particles. The generalized Leibnitz rule, Eq. 2.6.4, then gives

---

*$\vec{v}\cdot\nabla\vec{v} = (\nabla \times \vec{v}) \times v + \frac{1}{2}\nabla(\vec{v}\cdot\vec{v})$

$$\frac{d}{dt}\int_S \vec{\omega}\cdot\vec{n}\, da = \frac{q}{m}\oint_C (\vec{v} \times \mu_o\vec{H})\cdot d\vec{\ell} \tag{12}$$

Provided there is no imposed magnetic field, the vorticity is conserved over a surface of fixed identity. The magnetic field generates vorticity.

Kelvin's theorem, represented in Eulerian terms by Eq. 12, is often exploited in Lagrangian terms in dealing with axisymmetric electron beams having no θ components of electric or magnetic field. Then, Eq. 4 describes the θ-directed particle motions. Because particle current contributions to $\vec{H}$ are ignored, $\vec{B}$ is solenoidal and can be represented in terms of a vector potential, $\vec{A} = [\Lambda(r,z)/r]\vec{i}_\theta$ (Eq. (g) of Table 2.18.1). Thus,

$$\frac{1}{r}\frac{d}{dt}(r^2\frac{d\theta}{dt}) = -\frac{q}{m}[\frac{1}{r}\frac{\partial\Lambda}{\partial z}\frac{dz}{dt} + \frac{1}{r}\frac{\partial\Lambda}{\partial r}\frac{dr}{dt}] \tag{13}$$

What is on the right in Eq. 13 is the rate of change of $\Lambda(r,z)$ for a given particle,

$$d(r^2\frac{d\theta}{dt}) = -\frac{q}{m}d\Lambda \tag{14}$$

From Sec. 2.18, $2\pi\Lambda$ is the total magnetic flux linking a circle of radius r. Thus, with $2\pi\Lambda_o$ defined as the flux linked by a surface on which the particles have no angular velocity, Eq. 14 can be integrated to obtain Busch's theorem:[1]

$$r^2\frac{d\theta}{dt} = -\frac{q}{m}(\Lambda - \Lambda_o) \tag{15}$$

This result is a useful integral of one of the equations of motion. It also lends immediate insight to the result of directing a beam of particles through a complex magnetic field, for if the beam enters from a field-free region with no angular velocity, so that $\Lambda_o = 0$, then it is clear that it leaves the magnetic field with no angular velocity.

11.3 Magnetron Electron Flow

Electron flow in a type of magnetron configuration illustrates the implications of the laws given in Sec. 11.2. A uniform magnetic field, B_z, is imposed collinear with the axis of a cylindrical cathode, surrounded by a coaxial anode, as shown in Fig. 11.3.1. The arrangement is essentially that of a cyclotron-frequency magnetron, an early type of device for converting d-c energy (supplied by the source constraining the anode to a potential V relative to the cathode) to microwave-frequency a-c.

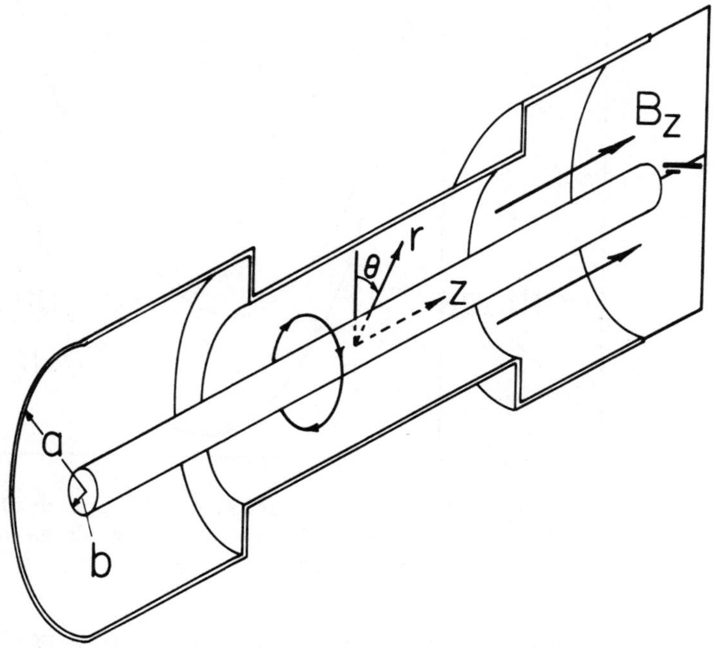

Fig. 11.3.1.

In configuration of cyclotron-frequency magnetron, electrons emitted from inner cathode execute cyclotron motions as they are accelerated toward anode across axial magnetic field.

1. J. R. Pierce, Theory and Design of Electron Beams, D. Van Nostrand Company, New York, 1949, p. 35.

Busch's theorem, Eq. 11.2.15, describes the tendency of the electrons to rotate about the magnetic field. Here, the flux density is uniform, so $2\pi\Lambda = \pi r^2 B_z$. Also, the electrons have no angular velocity at r = a, so $2\pi\Lambda_o = \pi b^2 B_z$. Thus, Eq. 11.2.15 becomes

$$\frac{d\theta}{dt} = \frac{1}{2}\omega_c\left(1 - \frac{b^2}{r^2}\right) \quad (1)$$

where q = -e, and the electron cyclotron frequency is defined as $\omega_c \equiv B_z e/m$.

An electron in the vicinity of the cathode is accelerated in the radial direction by the imposed electric field. As its radial position increases, Eq. 1 shows that it picks up an angular velocity, just as would be expected from the Lorentz force generated by the radial motion. The radial force equation is required to describe the trajectory. Because motions are in the steady state, the energy equation, Eq. 11.2.10, provides a convenient first integral of this equation. The potential and kinetic energies are both zero at the cathode, so that with the use of Eq. 1 for the angular velocity, it follows that

$$\left(\frac{dr}{dt}\right)^2 + \frac{r^2\omega_c^2}{4}\left(1 - \frac{b^2}{r^2}\right)^2 = \frac{2e}{m}\Phi \quad (2)$$

Thus, the electron executes radial motions in a potential well determined by the combination of the electric field tending to pull the electron outward and the magnetic field tending to divert it into an angular motion and eventually back toward the cathode. For the coaxial geometry, and in the absence of space-charge effects, the potential distribution is

$$\Phi = V \ln\left(\frac{r}{b}\right)/\ln\left(\frac{a}{b}\right) \quad (3)$$

and so, Eq. 2 becomes an expression for the velocity as a function of radial position,

$$\frac{d\underline{r}}{d\underline{t}} = \frac{1}{2}\sqrt{\underline{V}\frac{\ln \underline{r}}{\ln \underline{a}} - \underline{r}^2\left(1 - \frac{1}{\underline{r}^2}\right)^2} \quad (4)$$

where the normalization has been introduced,

$$\underline{r} = r/b, \quad \underline{a} = a/b, \quad \underline{t} = t\omega_c,$$

$$\underline{V} \equiv \frac{8\,eV}{\omega_c^2 mb^2} \qquad \omega_c \equiv \frac{B_z e}{m} \quad (5)$$

Typical potential wells are shown in Fig. 11.3.2. For $\underline{V}$ = 10, the electron is returned to the cathode while for $\underline{V}$ = 16 it collides with the anode. For a critical value, $\underline{V} = \underline{V}_c$, the electron just grazes the anode. This critical value is determined by setting Eq. 4 equal to zero with $\underline{r} = \underline{a}$,

$$\underline{V}_c = \left(1 - \frac{1}{\underline{a}^2}\right)\underline{a}^2 \quad (6)$$

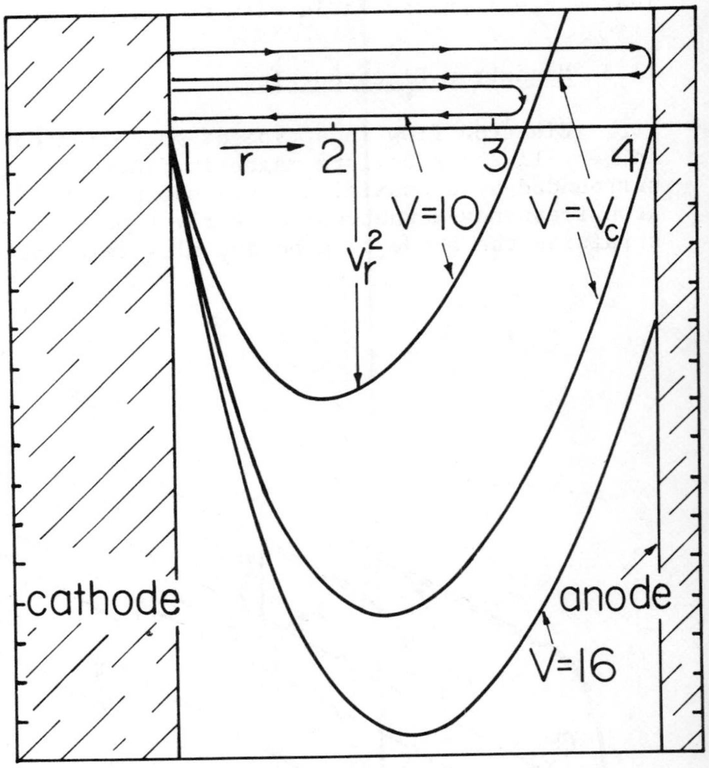

Fig. 11.3.2. Potential wells for cyclotron motions. All variables are normalized.

Integration of Eq. 4 gives

$$\underline{r}(\underline{t}) = 2\int_o^{\underline{r}} \frac{d\underline{r}}{\sqrt{\underline{V}\frac{\ln \underline{r}}{\ln \underline{a}} - \underline{r}^2\left(1 - \frac{1}{\underline{r}^2}\right)^2}} \quad (7)$$

Numerical integration gives the radial dependence shown in Fig. 11.3.3. In turn, the angular position follows from Eq. 1:

$$\theta = \frac{1}{2}\int_o^{\underline{t}}\left(1 - \frac{1}{\underline{r}^2}\right)d\underline{t} \quad (8)$$

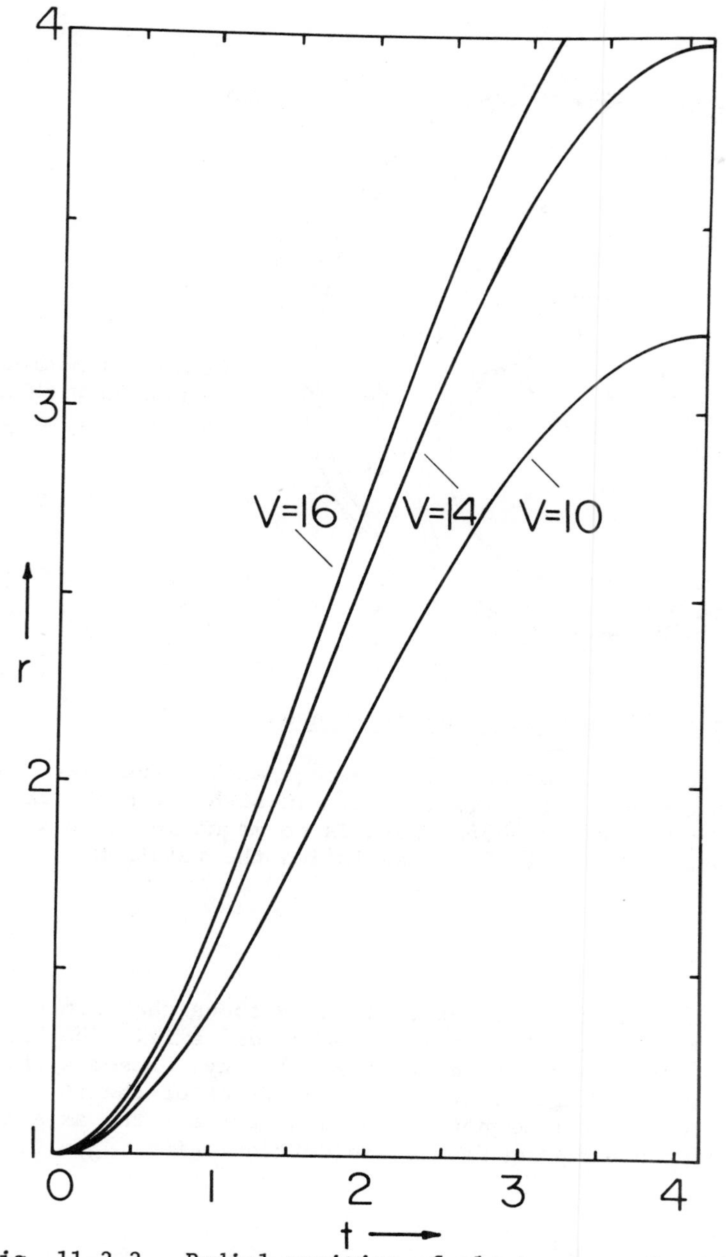

Fig. 11.3.3. Radial position of electrons as function of time. All variables are normalized.

The results from the radial integration can be used to numerically evaluate this expression and obtain the trajectories shown in Fig. 11.3.4.

For these trajectories, where the potential is held fixed, the electron kinetic plus potential energy is conserved. With the introduction of a potential component varying at a frequency on the order of ω_c, energy imparted to an electron by the d-c field can be removed in a-c form. With the d-c voltage adjusted to make $V = V_c$, the effect of a small increase in potential is dramatically different from that of a small decrease. Suppose that as a given electron departs from the cathode, the potential increases. The electron is accelerated by the potential and hence takes energy from the source. But, it also strikes the anode or cathode and is removed after only one orbit. By contrast, an electron that leaves the cathode as the potential is decreasing will be decelerated and hence give up energy to the a-c source. This electron does not strike the anode, and in fact tends to remain in the annulus for many cycles, contributing, along with electrons having a similar phase relation to the a-c field, to giving up energy to the a-c source.

If the a-c source is replaced by a low-loss resonator, the device can sustain self-oscillation. Hence, it can be used as a generator of energy having a frequency on the order of ω_c. For electrons with $B_z = 0.1$ tesla, $\omega_c/2\pi = 2.8$ GHz. Common magnetrons make use of resonators to provide for a traveling-wave interaction with the gyrating electrons.[1]

---

1. H. J. Reich, P. F. Ordnung, H. L. Krauss and J. G. Skalnik, Microwave Theory and Techniques, D. Van Nostrand Company, Princeton, N.J., pp. 708-735.

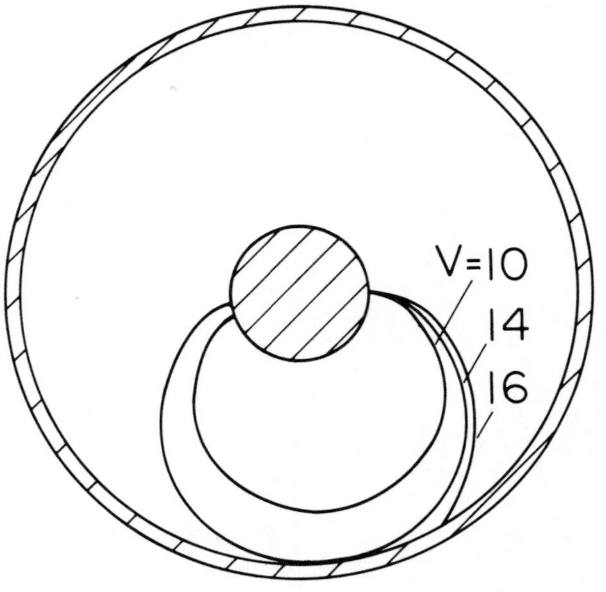

Fig. 11.3.2

Cyclotron motions in cylindrical magnetron with normalized voltage as a parameter.

11.4 Paraxial Ray Equation: Magnetic and Electric Lenses

Oscilloscopes and electron microscopes are devices exploiting electric and magnetic lenses. Simple lens configurations are shown in Fig. 11.4.1. In both the magnetic and electric configurations, the electron beam enters from a region where there is no magnetic field with an axial velocity which, according to the energy equation, Eq. 11.2.10, satisfies the relation

$$\frac{1}{2}m\left(\frac{dz}{dt}\right)^2 = e\Phi \tag{1}$$

The tendency of the axisymmetric magnetic field to focus the beam can be seen by considering the Lorentz force on an electron entering the field somewhat off axis. The longitudinal velocity crosses with the radial component of $\vec{B}$ to produce an angular velocity. Busch's theorem, Eq. 11.2.15, exploits the solenoidal character of $\vec{B}$ to represent this effect of the rotational field in terms of the axial field alone. Thus, even if the magnetic flux density near the axis is approximated as independent of r, for an electron entering from a field-free region, $\Lambda_o = 0$, and the angular velocity can be simply taken as

$$\frac{d\theta}{dt} = \frac{eB_z}{2m} \tag{2}$$

This component in turn crosses with the axial component of $\vec{B}$ to deflect the electron toward the axis. Thus, while in the magnetic field, the electron is deflected toward the z axis; but, in accordance with Eq. 2, once through the field it continues toward the axis without an angular velocity.

In the electric lens, the electron tends to be focused toward the axis as it enters the fringing field, but to be diverted toward the electrode as it leaves the field. The net focusing effect derives from the fringing field having a greater intensity as the electron enters than as it leaves.

Paraxial Ray Equation: An equation for the radial position, r, of an electron as a function of its longitudinal position, z, is the basis for designing both magnetic and electric lenses. It pertains to electrons traversing the fields near the axis where the magnetic flux density is essentially independent of radius, i.e., of the form $B_z(z)$. The radial component of $\vec{B}$ implied by the z dependence is already built into Eq. 2. The radial component of $\vec{E}$ near the axis has an r dependence that can be represented in terms of a given dependence of the potential $\Phi = \Phi(z)$ at $r \approx 0$ by exploiting Gauss' law (Prob. 4.12.1):

$$E_r \simeq -\frac{r}{2}\frac{d^2\Phi}{dz^2} \tag{3}$$

Given $B_z(z)$ and $\Phi(z)$, what is $r(z)$? With the use of Eqs. 2 and 3, the radial component of the force equation, Eq. 11.2.3, becomes

$$\frac{d^2r}{dt^2} + \frac{1}{4}\left(\frac{eB_z}{m}\right)^2 r = -\frac{e}{m}\frac{r}{2}\frac{d^2\Phi}{dz^2} \tag{4}$$

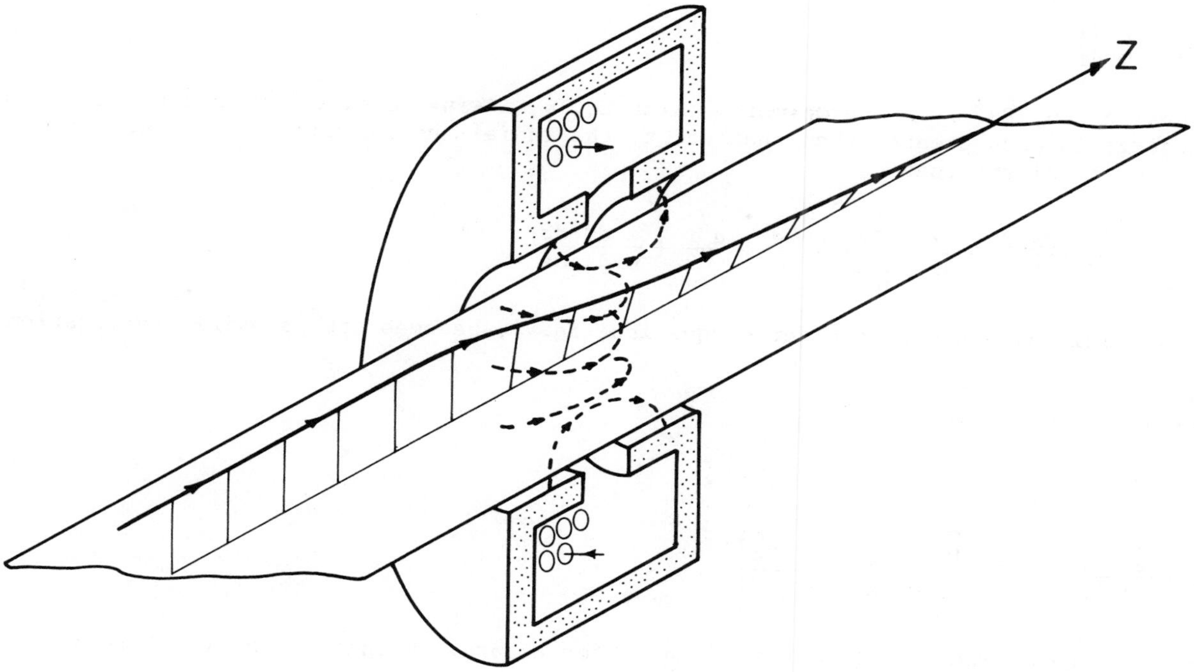

Fig. 11.4.1. (a) Magnetic electron beam lens approximated by fields and focal length of Fig. 11.4.2.

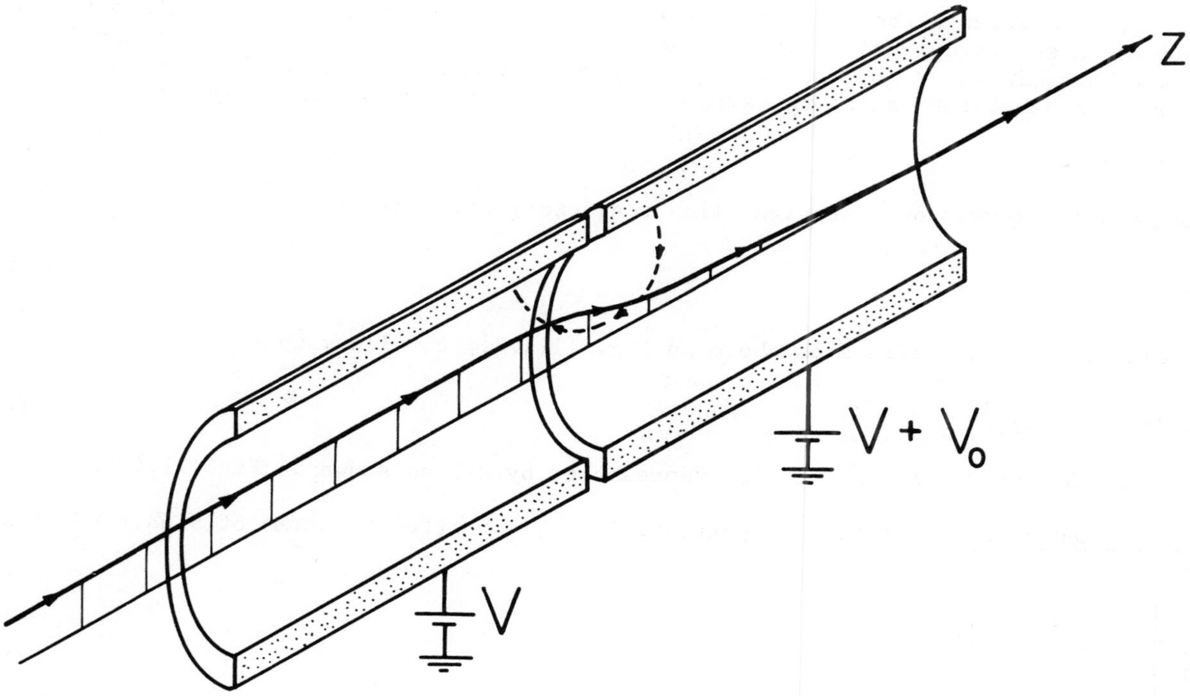

Fig. 11.4.1. (b) Electric electron beam lens with trajectory exemplified by Fig. 11.4.3.

Here, the electron position is pictured with time as the independent parameter. With Φ and B_z independent of time, the electron flow is steady so that time can be eliminated as a parameter and $r = r(z)$. With the objective of writing Eq. 4 with z as the independent variable, observe that

$$\frac{dr}{dt} = \frac{dr}{dz}\frac{dz}{dt} \tag{5}$$

and from the time derivative of Eq. 1 that

$$\frac{d^2z}{dt^2} \simeq \frac{e}{m}\frac{d\Phi}{dz} \tag{6}$$

In the lens region, Eq. 1 is approximate. With Eq. 6, it is therefore assumed that the longitudinal kinetic energy is much greater than that due to the radial and angular velocities. With the use of Eqs. 5 and 6 it follows that

$$\frac{d^2r}{dt^2} = \frac{d^2r}{dz^2}\left(\frac{dz}{dt}\right)^2 + \frac{dr}{dz}\left(\frac{d^2z}{dt^2}\right) = \frac{2e}{m}\Phi\frac{d^2r}{dz^2} + \frac{e}{m}\frac{d\Phi}{dz}\frac{dr}{dz} \tag{7}$$

Thus, the radial component of the force equation, Eq. 4, becomes the paraxial ray equation,[1]

$$\frac{d^2r}{dz^2} + A\frac{dr}{dz} + Cr = 0 \tag{8}$$

where

$$A = \frac{1}{2\Phi}\frac{d\Phi}{dz}; \quad C = \frac{1}{\Phi}\left[\frac{1}{8}\frac{e}{m}B_z^2 + \frac{1}{4}\frac{d^2\Phi}{dz^2}\right]$$

<u>Magnetic Lens</u>: The limiting form of Eq. 8 for a purely magnetic lens is misleading in its simplicity (Prob. 11.4.1):

$$\frac{d^2r}{dz^2} + \kappa^2 r = 0; \quad \kappa \equiv \sqrt{\frac{e}{m8\Phi}}\,B_z \tag{9}$$

Through Eq. 1, Φ represents the incident axial velocity. A reasonable approximation to the on-axis axial field for a solenoidal coil that is long compared to its radius is the distribution shown in Fig. 11.4.2. Thus, in Eq. 9, $B_z = 0$ for $z < 0$ and $z > \ell$, and $B_z = B_o$ over the length ℓ of the lens. An electron entering at the radius r_o, with $dr/dz = 0$, therefore has the trajectory

$$r = r_o \cos \kappa z \tag{10}$$

inside the lens and leaves on a straight-line trajectory with the slope

$$\frac{dr}{dz}(z = \ell) = -r_o\kappa \sin \kappa\ell \tag{11}$$

With the focal length, f, defined as shown in Fig. 11.4.2, it follows that

$$\frac{f}{\ell} = (\kappa\ell \sin \kappa\ell)^{-1} \tag{12}$$

Thus, the focal length decreases with B_z (represented by $\kappa\ell$) as shown in Fig. 11.4.2.

<u>Electric Lens</u>: For numerical integration, Eq. 8 is written in terms of a pair of first-order equations

$$\frac{d}{d\underline{z}}\begin{bmatrix}\underline{r}\\ \underline{u}\end{bmatrix} = \begin{bmatrix}\underline{u}\\ -\underline{A}\underline{u} - \underline{C}\underline{r}\end{bmatrix} \tag{13}$$

where $\underline{z} = z/a$, $\underline{r} = r/a$, $\underline{u} = u$, $\underline{A} = aA$, and $\underline{C} = a^2C$.

As an example, consider the electric lens of Fig. 11.4.1. The potential distribution inside the abutting cylindrical electrodes with radius a that comprise the lens is found in Prob. 5.17.3 to be an infinite series with radial dependences represented by Bessel functions. The z dependences of the dominant terms have an exponential decay away from the plane $z = 0$, $\exp \beta\underline{z}$, where $\beta = 2.4$ is the first root of the Besssel function $J_o(\beta)$. For the present purposes, this longitudinal distribution of

1. J. R. Pierce, <u>Theory and Design of Electron Beams</u>, D. Van Nostrand Company, New York, 1949, pp. 72-91.

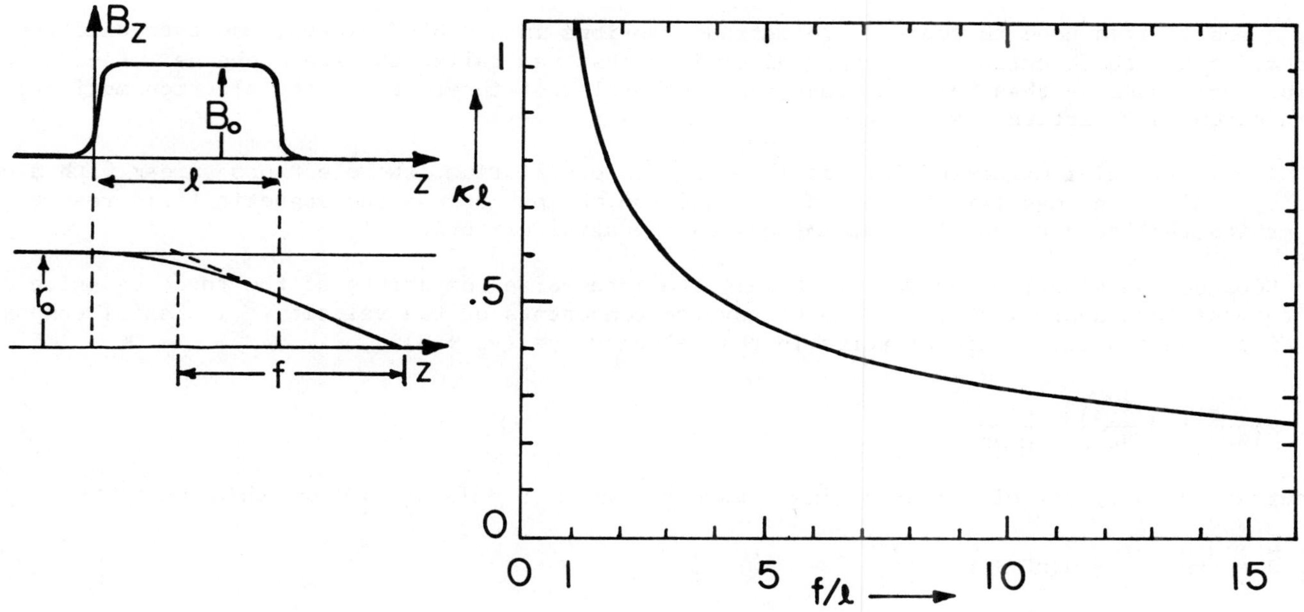

Fig. 11.4.2. For magnetic lens of Fig. 11.4.1a, having essentially uniform axial field B_o over length ℓ, focal length f normalized to ℓ is given as a function of $\kappa\ell$, defined with Eq. 10.

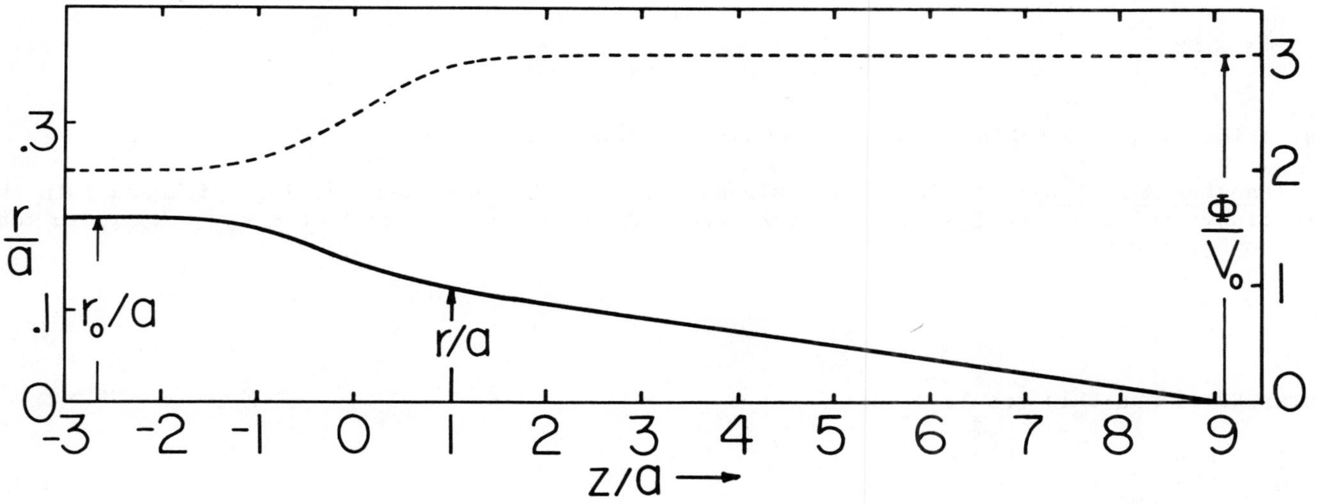

Fig. 11.4.3. For the electric lens of Fig. 11.4.1b, the axial potential distribution is represented by the broken curve. The solid curve is the electron trajectory predicted by Eqs. 13 and 14 with $V_o/V = 0.5$ and $\beta = 2$.

potential is approximated by

$$\Phi = V + \frac{V_o}{2}\left(1 + \tanh \frac{\beta z}{2}\right) \qquad (14)$$

This distribution of potential is shown in Fig. 11.4.3.

Using Eq. 14, A and C (given with Eq. 8) are given functions of z. Note that, although it has space-varying coefficients, the paraxial ray equation, Eq. 8, is linear. Thus, general properties of the electron flow can be deduced using superposition. Numerical integration, using Eqs. 13, is straightforward and results in electron trajectories typified in Fig. 11.4.3. Note that the electron velocity, deduced at any given point from Eq. 1, is increased in the conservative transition through the lens.

11.5 Plasma Electrons and Electron Beams

A model often used to represent electronic motions in a "cold" plasma, and even electron beams in "vacuum," gives the electrons a background of ions that neutralize the space charge. Because the ions are much more massive than the electrons, on time scales of interest for the electron motions, the ions remain essentially motionless.

A uniform axial magnetic field is imposed. In equilibrium, the electrons stream with a uniform velocity U along the magnetic field lines. Electron motions across the magnetic field result in cyclotron orbits that tend to confine the motions to the axial direction.

Because the electron motions are axial, the transverse components of the force equation only give an after-the-fact approximation to the transverse components of the velocity. The axial component of the force equation is, to linear terms in the velocity $\vec{v} = (v_z + U)\vec{i}_z$,

$$m\left(\frac{\partial v_z}{\partial t} + U\frac{\partial v_z}{\partial z}\right) = \frac{e}{m}\frac{\partial \Phi}{\partial z} \tag{1}$$

The current density for electrons having a number density $n_o+n(x,y,z,t)$ and this velocity is

$$\vec{J} = -en_oU\vec{i}_z - e(nU + n_ov_z)\vec{i}_z \tag{2}$$

so that to linear terms, conservation of charge requires that

$$-\frac{\partial}{\partial z}(enU + en_ov_z) - \frac{\partial(ne)}{\partial t} = 0 \tag{3}$$

Finally, because the equilibrium electronic space-charge density, $-n_oe$, is cancelled by that due to positive ions, Gauss' law requires that

$$\nabla^2\Phi = \frac{ne}{\varepsilon_o} \tag{4}$$

Perturbations v_z, n and Φ are described by Eqs. 1, 3 and 4.

Transfer Relations: Consider now a planar layer of plasma or beam having thickness Δ in the x direction. Solutions to Eqs. 1 and 3 take the form $\hat{v}_z = \text{Re}\hat{v}_z(x)\exp j(\omega t - k_y y - k_z z)$, so substitution into Eq. 1 gives

$$\hat{v}_z = -\frac{e}{m}\frac{k_z\hat{\Phi}}{(\omega - kU)} \tag{5}$$

In turn, Eq. 3 and this result give

$$\hat{n} = \frac{k_z n_o \hat{v}_z}{(\omega - kU)} = -\frac{k_z^2 n_o e}{(\omega - kU)^2 m}\hat{\Phi} \tag{6}$$

Finally, this relation combines with Eq. 4 to show that the potential distribution must satisfy

$$\frac{d^2\hat{\Phi}}{dx^2} - \gamma^2\hat{\Phi} = 0; \quad \gamma^2 = k_y^2 + k_z^2\left[1 - \frac{\omega_p^2}{(\omega - kU)^2}\right] \tag{7}$$

where the plasma frequency is defined as $\omega_p = \sqrt{n_oe^2/\varepsilon_o m}$. This relation is of the same form as for representing Laplace's equation in Sec. 2.16. Thus, the transfer relations are the same as in Table 2.16.1, provided that γ is defined as in Eq. 7. Note that a similar derivation leads to transfer relations in cylindrical geometry so that the transfer relations for an annular beam are as given by the relations of Table 2.16.2 with coefficients suitably defined. For example, $f_m(x,y) \rightarrow f_m(x,y,k \rightarrow \gamma)$, where γ may differ from one annular region to another.

Space-Charge Dynamics: The sheet beam shown in Fig. 11.5.1 exemplifies the dynamics of beams in uniform structures. The surrounding region is free space, with walls to either side constrained by a traveling wave of potential. The wall potential can be regarded as a given drive, although more generally it can be made consistent with external electromagnetic structures. The velocity is purely axial, so the boundaries do not deform and boundary conditions are simply

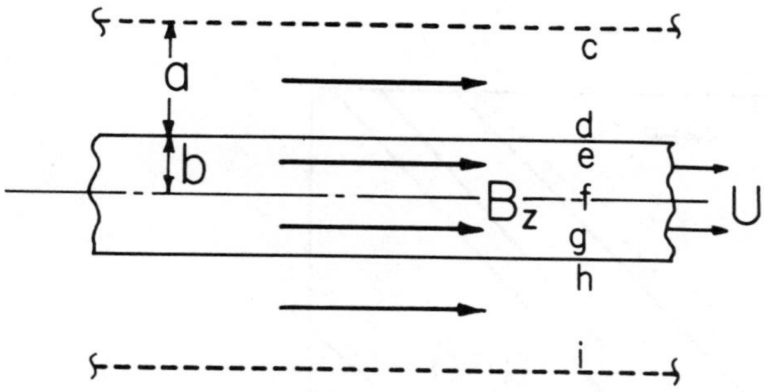

Fig. 11.5.1

Planar electron beam in uniform axial magnetic field.

$$\hat{\Phi}^c = \hat{V}_o, \quad \hat{\Phi}^d = \hat{\Phi}^e, \quad \hat{D}_x^d = \hat{D}_x^e, \quad \hat{D}_x^f = 0 \tag{8}$$

Here, interest is restricted to motions that are even in the potential, so with the last boundary condition it is presumed that $\hat{\Phi}^c = \hat{\Phi}^i$.

Transfer relations for the free-space region and for the beam follow from Eq. 7 and Table 2.16.1:

$$\begin{bmatrix} \hat{D}_x^c \\ \hat{D}_x^d \end{bmatrix} = \epsilon k \begin{bmatrix} -\coth ka & \dfrac{1}{\sinh ka} \\ -\dfrac{1}{\sinh ka} & \coth ka \end{bmatrix} \begin{bmatrix} \hat{\Phi}^c \\ \hat{\Phi}^d \end{bmatrix} \tag{9}$$

$$\begin{bmatrix} \hat{D}_x^e \\ 0 \end{bmatrix} = \epsilon \gamma \begin{bmatrix} -\coth \gamma b & \dfrac{1}{\sinh \gamma b} \\ -\dfrac{1}{\sinh \gamma b} & \coth \gamma b \end{bmatrix} \begin{bmatrix} \hat{\Phi}^d \\ \hat{\Phi}^f \end{bmatrix} \tag{10}$$

The last equation gives $\hat{\Phi}^f$ in terms of $\hat{\Phi}^d$. This is inserted into Eqs. 9b and 10a, set equal to each other. The resulting expression can be solved for $\hat{\Phi}^d$. Substituted into Eq. 9a, that gives

$$\hat{D}_x^c = -\epsilon k \frac{(k + \gamma \coth ka \tanh \gamma b)}{D(\omega,k)} \hat{\Phi}^c; \quad D \equiv k \coth ka + \gamma \tanh \gamma b \tag{11}$$

This result gives the driven response, but also embodies the temporal modes and spatial modes, as discussed in Secs. 5.15 and 5.17. These are described by the dispersion equation, $D(\omega,k) = 0$.

<u>Temporal Modes:</u> It is clear that there are no roots of this expression having γ purely real. Purely imaginary roots abound, as is evident by substituting $\gamma \rightarrow j\alpha$:

$$(\alpha b)\tan(\alpha b) = (kb)\coth[(kb)\tfrac{a}{b}] \tag{12}$$

Graphical solution of this expression, for a given a/b, results in roots, α_n. The frequencies of associated temporal modes follow from the definition of $\gamma^2 = -\alpha^2$ given with Eq. 7:

$$\frac{\omega}{\omega_p} = bk\left(\frac{U}{\omega_p b}\right) \pm \left[1 + \frac{(b\alpha_n)^2}{(bk)^2}\right]^{-1/2} \tag{13}$$

Here it has been assumed that $k_y = 0$. This dispersion equation is represented graphically by Fig. 11.5.2.

For each real wavenumber, k, there are two eigenfrequencies representing space-charge waves. In the absence of convection these have phase velocities, ω/k, in the positive and negative directions. With convection, these waves are respectively the fast and slow space-charge waves that are central to a variety of electron-beam devices and interactions. The transverse dependence of a temporal mode having a real wavenumber $k_z = k$ is sinusoidal within the beam and exponential in the surrounding regions of free space, as depicted by the inserts to Fig. 11.5.2.

Without the equilibrium streaming, the temporal modes are similar to those of the internal electrohydrodynamic space-charge waves of Sec. 8.18. There are an infinite number of modes, $n > p$, within the region of the ω-k plot bounded by the fast and slow wave branches for any given mode $n = p$. In the

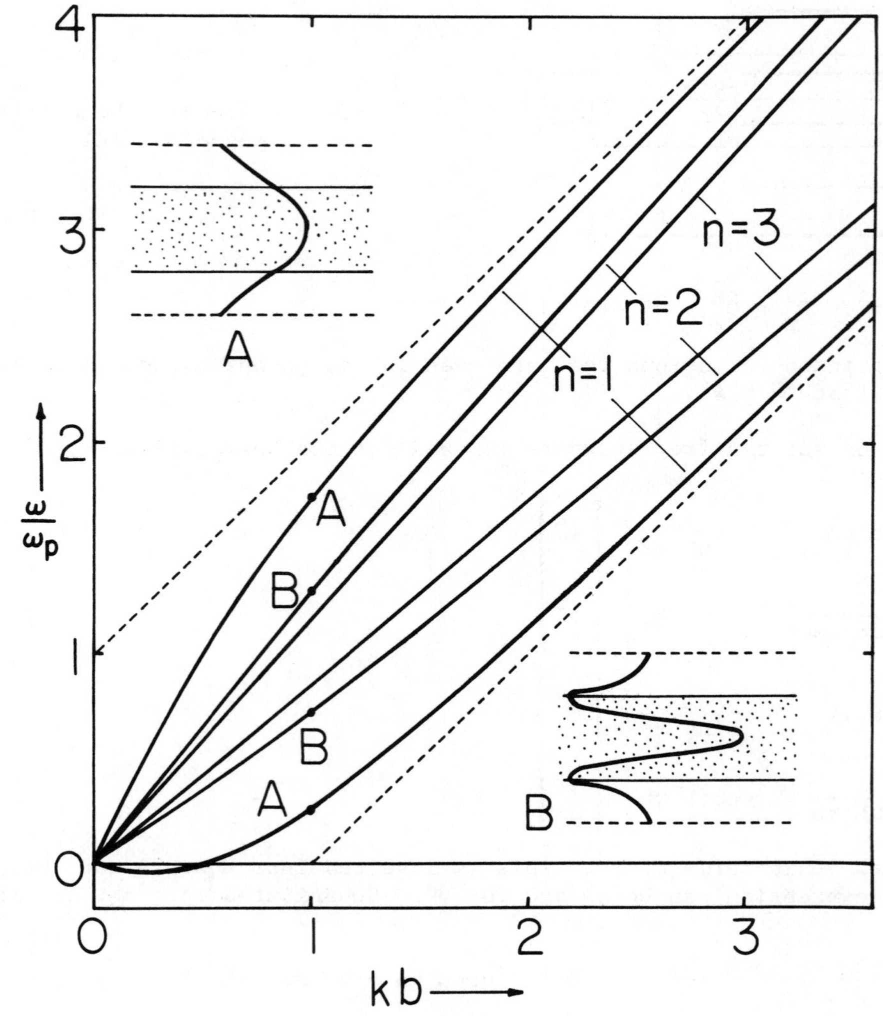

Fig. 11.5.2. Normalized angular frequency as a function of normalized wavenumber for space-charge waves on planar electron beam of Fig. 11.5.1. Inserts show transverse distribution of potential for two lowest eigenmodes at longitudinal wavenumber kb = 1.

frame of reference moving with the beam velocity U, the frequencies, ω-kU, of these modes approach zero as the mode number, and hence the number of oscillations over the transverse dimension of the beam, approaches infinity.

Spatial Modes: Typically, in electron beam devices, it is the response to a given driving frequency that is of interest. The homogeneous part of the response is made up of spatial modes having wavenumbers that are solutions to $D(\omega,k) = 0$, with ω the specified driving frequency. In general, these are complex roots of a complex equation. However, those modes having real wavenumbers for the real driving frequency can be identified from Fig. 11.5.2.

11.6 Method of Characteristics

In representing the evolution of a continuum of particles, it is natural to express the partial differential equations of motion as ordinary equations with time as the independent variable. As illustrated in Secs. 5.3, 5.6 and 5.10, what can result is a complete picture of the temporal evolution, but one viewed along a characteristic line in $(\vec{r},t)$ space. The price paid for the characteristic formulation is an implicit dependence on space. That the characteristic lines do not have to be identified with particles is illustrated in this and the next five sections. Physically, the characteristics now represent waves rather than particles. However, the objective is again to reduce partial differential equations to ordinary ones.

If the equations, written as a system of first-order expressions, have coefficients that are not functions of the independent variables, they are said to be quasi-linear. An example comes from the one-dimensional longitudinal motions of a highly compressible gas.

For now, there is no external force density and viscous effects are ignored. Thus, with the assumption that $\vec{v} = v(z,t)\vec{i}_z$, $p = p(z,t)$ and $\rho = \rho(z,t)$, the force equation, Eq. 7.16.6, becomes

$$\rho\left(\frac{\partial v}{\partial t} + v\frac{\partial v}{\partial z}\right) + \frac{\partial p}{\partial z} = 0 \tag{1}$$

The flow is not only assumed adiabatic, but initiated in such a way that every fluid particle can be traced backward in time along a particle line to a point in space and time when it had the same state $(p,\rho) = (p_o,\rho_o)$. The flow is initiated from a uniform state. Thus, Eq. 7.23.13 holds throughout the region of interest, and it follows that

$$\frac{\partial p}{\partial z} = a^2\frac{\partial \rho}{\partial z}; \quad a \equiv \sqrt{\gamma\frac{p_o}{\rho_o}}\left(\frac{\rho}{\rho_o}\right)^{(\gamma-1)/2} \tag{2}$$

It follows that Eq. 1 can be written as

$$(0)\frac{\partial \rho}{\partial t} + (a^2)\frac{\partial \rho}{\partial z} + (\rho)\frac{\partial v}{\partial t} + (\rho v)\frac{\partial v}{\partial z} = 0 \tag{3}$$

Conservation of mass, Eq. 7.2.3, provides the second equation in (v,ρ):

$$(1)\frac{\partial \rho}{\partial t} + (v)\frac{\partial \rho}{\partial z} + (0)\frac{\partial v}{\partial t} + (\rho)\frac{\partial v}{\partial z} = 0 \tag{4}$$

These last two expressions typify systems of first-order partial differential equations with two independent variables (z,t). They are not linear, but do have coefficients depending only on the dependent variables (v,ρ). The characteristic equations are now deduced following the reasoning of Courant and Friedrichs.[1]

First Characteristic Equations: Arbitrary incremental changes in the time and position result in changes in (ρ,v) given by

$$d\rho = \frac{\partial \rho}{\partial t}dt + \frac{\partial \rho}{\partial z}dz \tag{5}$$

$$dv = \frac{\partial v}{\partial t}dt + \frac{\partial v}{\partial z}dz \tag{6}$$

The objective now is to find a linear combination of Eqs. 3 and 4 that takes the form

$$f(\rho)d\rho + g(v)dv = 0 \tag{7}$$

because this equation can be integrated. To this end, note that a line in the z-t plane along which $d\rho$ and dv are to be evaluated has not yet been specified. It can be selected to guarantee the desired form of the equations of motion.

---

1. R. Courant and K. O. Friedrichs, Supersonic Flow and Shock Waves, Interscience Publishers, New York, 1948, pp. 40-45.

A linear combination of Eqs. 3 and 4 is written by multiplying Eq. 3 by the parameter λ_1 and Eq. 4 by λ_2 and taking the sum:

$$\frac{\partial \rho}{\partial t}(\lambda_1) + \frac{\partial \rho}{\partial z}(\lambda_1 v + \lambda_2 a^2) + \frac{\partial v}{\partial t}(\rho\lambda_2) + \frac{\partial v}{\partial z}(\lambda_1 \rho + \lambda_2 v\rho) = 0 \tag{8}$$

If this expression is to have the same form as Eq. 7, where $d\rho$ and dv are given by Eqs. 5 and 6, then

$$\frac{dz}{dt} = \frac{\lambda_1 v + \lambda_2 a^2}{\lambda_1}; \quad \frac{dz}{dt} = \frac{\lambda_1 \rho + \lambda_2 v\rho}{\rho\lambda_2} \tag{9}$$

These expressions are linear and homogeneous in the coefficients (ρ,v):

$$\begin{bmatrix} \frac{dz}{dt} - v & -a^2 \\ -\rho & \frac{dz}{dt} - v \end{bmatrix} \begin{bmatrix} \lambda_1 \\ \lambda_2 \end{bmatrix} = \begin{bmatrix} 0 \\ 0 \end{bmatrix} \tag{10}$$

It follows that if the coefficients are to be finite, the determinant of the coefficients must vanish. Thus,

$$\frac{dz}{dt} = v \pm a; \quad C^{\pm} \tag{11}$$

If v,ρ and hence $a(\rho)$ were known functions of (z,t), these expressions could be solved to give families of curves along which Eq. 8 would take the form of Eq. 7. Apparently two such families have been found. They are called the Ist characteristic equations and respectively designated by C^+ and C^-.

Differential equations, such as Eqs. 3 and 4, for which the Ist characteristic equations are real, are said to be hyperbolic. Elliptic equations, for which the Ist characteristics are not real, must be solved by some other method than now described.

Second Characteristic Lines: The goal of writing Eq. 8 in the form of Eq. 7 is achieved by factoring λ_1 from the first two terms and λ_2 from the third and fourth terms, and then substituting for the coefficients of the second and fourth terms, respectively, using Eqs. 5 and 6:

$$\lambda_1\left(\frac{\partial \rho}{\partial t} + \frac{\partial \rho}{\partial z}\frac{dz}{dt}\right) + \rho\lambda_2\left(\frac{\partial v}{\partial t} + \frac{\partial v}{\partial z}\frac{dz}{dt}\right) = 0 \tag{12}$$

If this expression is multiplied by dt and divided by λ_1, the desired form follows:

$$d\rho + \rho\frac{\lambda_2}{\lambda_1}dv = 0 \tag{13}$$

To establish the ratio λ_2/λ_1, either of Eqs. 10 can be used. For example, substituting λ_2/λ_1 as found from Eq. 10a gives

$$d\rho + \frac{\rho}{a^2}\left(\frac{dz}{dt} - v\right)dv = 0 \tag{14}$$

This expression is further simplified by using the Ist characteristic equations, Eqs. 11, to write

$$dv \pm \frac{a}{\rho}d\rho = 0 \quad \text{on } C^{\pm} \tag{15}$$

where the choice of signs is determined by which sign is being used in Eqs. 11.

With $a(\rho)$ specified by Eq. 2, the IInd characteristic equations can be integrated:

$$v \pm \frac{2a(\rho)}{\gamma - 1} = c_{\pm} \quad \text{on } C^{\pm} \tag{16}$$

Here, c_+ and c_- are respectively invariants along the C^+ and C^- characteristic lines.

Systems of First-Order Equations: The method used to determine the first and second characteristic equations makes their deduction a logical response to the objective. As long as the number of independent variables (z,t) remains only two, the same technique can be used with more complex problems.

But, it is convenient in dealing with several first order equations to use a more formal approach to finding the characteristics. Although the formalism now considered appears to be different, in fact the characteristic equations are the same.

Equations 3 and 4 are particular cases of the first two expressions in the set of four,

$$\begin{bmatrix} C \\ D \\ d\rho \\ dv \end{bmatrix} = \begin{bmatrix} A_1 & A_2 & A_3 & A_4 \\ B_1 & B_2 & B_3 & B_4 \\ dt & dz & 0 & 0 \\ 0 & 0 & dt & dz \end{bmatrix} \begin{bmatrix} \frac{\partial \rho}{\partial t} \\ \frac{\partial \rho}{\partial z} \\ \frac{\partial v}{\partial t} \\ \frac{\partial v}{\partial z} \end{bmatrix} \tag{17}$$

Here, the coefficients A_i and B_i are in general functions of (ρ,v,z,t). Also, for generality, $C = C(\rho,v,z,t)$ and $D = D(\rho,v,z,t)$ represent the possibility that the differential equations are inhomogeneous in the sense that they have terms which do not involve partial derivatives. The last two expressions will be recognized as the differential relations for $d\rho$ and dv, Eqs. 5 and 6.

Following the formalism leading to Eqs. 11 and 15, Eqs. 17a and 17b are multiplied respectively by λ_1 and λ_2, and added. Then the ratio of coefficients for the respective $\partial/\partial t$'s and $\partial/\partial z$'s are required to be dz/dt, and the result is two homogeneous equations in the λ's:

$$\begin{bmatrix} (A_2 - A_1 \frac{dz}{dt}) & (B_2 - B_1 \frac{dz}{dt}) \\ (A_4 - A_3 \frac{dz}{dt}) & (B_4 - B_3 \frac{dz}{dt}) \end{bmatrix} \begin{bmatrix} \lambda_1 \\ \lambda_2 \end{bmatrix} = 0 \tag{18}$$

The first characteristic equations are found by requiring that the determinant of the coefficients in Eq. 18 vanish.

This same condition is obtained by requiring that the determinant of the coefficients in Eq. 17 vanish. To see this, rows three and four are multiplied by $(dt)^{-1}$. Then rows three and four are multiplied respectively by $-A_1$ and $-A_3$ and added to row one. Similarly, rows three and four can be multiplied respectively by $-B_1$ and $-B_3$ and added to row two. The result is the determinant

$$\begin{bmatrix} 0 & A_2 - A_1 \frac{dz}{dt} & 0 & A_4 - A_3 \frac{dz}{dt} \\ 0 & B_2 - B_1 \frac{dz}{dt} & 0 & B_4 - B_3 \frac{dz}{dt} \\ dt & dz & 0 & 0 \\ 0 & 0 & dt & dz \end{bmatrix} = 0 \tag{19}$$

Now, by expanding about the dt's that appear in columns with all other entries zero, the same requirement as given by Eq. 18 is obtained. The first characteristic equations are obtained by writing the differential equations in the form of Eq. 17 and simply requiring that the determinant of the coefficients vanish. The same approach can be used with an arbitrary number of dependent variables.

To solve Eqs. 17 for any one of the four partial differentials would require substituting the column on the left for the column of the square matrix corresponding to the desired partial derivative, and to divide the determinant of the resulting matrix by the coefficient determinant. However, the coefficient determinant has already been required to vanish, since that is just the condition for obtaining the Ist characteristic equations. Thus, if the partial derivatives are not infinite, the numerator determinants must also vanish. Four equations result which are reducible to the IInd characteristic equations.

As an example, consider once again Eqs. 3 and 4. The coefficient determinant is

$$\begin{bmatrix} 1 & v & 0 & \rho \\ 0 & a^2 & \rho & \rho v \\ dt & dz & 0 & 0 \\ 0 & 0 & dt & dz \end{bmatrix} = 0 \tag{20}$$

and can be expanded, taking advantage of the zeros, to give Eqs. 11. Then the numerator determinant for finding $\partial\rho/\partial t$ is the coefficient matrix with the column matrix on the left in Eq. 17 substituted for the first column on the right, or

$$\begin{bmatrix} 0 & v & 0 & \rho \\ 0 & a^2 & \rho & \rho v \\ d\rho & dz & 0 & 0 \\ dv & 0 & dt & dz \end{bmatrix} = 0 \qquad (21)$$

With the use of the first characteristic equations, Eq. 21 reduces to the second characteristic equations given by Eqs. 15. A check of the other three equations obtained by substituting the column matrix in the second, third and fourth columns gives the same result.

11.7 Nonlinear Acoustic Dynamics: Shock Formation

The longitudinal motions of a gas under adiabatic conditions both serve as a vehicle for seeing how the characteristic equations are used, and provide insight into the nonlinear phenomena that are responsible for wave steepening and shock formation.[1] (See Reference 10, Appendix C.)

Initial Value Problem: The characteristic equations are given by Eqs. 11.6.11 and 11.6.16. Although there is no necessity for linearizing, it is helpful to realize how perturbations from a uniform flow with velocity U, density ρ_o and acoustic velocity a_o are represented by the characteristics. (The linearized versions of Eqs. 11.6.3 and 11.6.4 are the one-dimensional forms of Eqs. 7.11.1-7.11.3.) In that limit, the first characteristic equations have $U \pm a_o$ on the right, and are therefore integrable to give straight lines with slopes equal to the wave velocities $U \pm a_o$ in the $\pm z$ directions. These families of lines are illustrated in Fig. 11.7.1a.

In general, v and a can vary and the characteristic lines sketched in Fig. 11.7.1b in the z-t plane are not known. However, the functions v and $a(\rho)$ are known at an intersection of the C^+ and C^- characteristics, wherever that may be. That is, suppose the initial values of v and a at points A and B shown in Fig. 11.7.2 are given (at t = 0 but at different points along the z axis). Then, from Eq. 11.6.16a, c_+ is

$$c_+^a = [v]_A + [\frac{2a(\rho)}{\gamma - 1}]_A \qquad (1)$$

Similarly, from the initial conditions at B,

$$c_-^b = [v]_B - [\frac{2a(\rho)}{\gamma - 1}]_B \qquad (2)$$

Now, c_+^a is invariant along the C^+ characteristic, and c_-^b is invariant along the C^- characteristic. At point C, certainly at a later time and generally at a different point in space than either A or B, Eqs. 11.6.16 both hold, with $c_+ = c_+^a$ and $c_- = c_-^b$. Hence, they can be solved simultaneously for either v or $a(\rho)$. For the former, addition yields

$$[v]_C = \frac{c_+^a + c_-^b}{2} \qquad (3)$$

Given conditions at A and B, the solution at C is established. The solution is known, but where and when does it apply? The Ist characteristic equations must be integrated to determine the location of C in the z-t plane.

The characteristic lines have the physical interpretation of being wave fronts. Conditions at A and B propagate along the respective lines and combine at C to give the response. It is remarkable that what happens at C depends only on the conditions at A and B. But the "location" of C is determined by initial conditions everywhere between A and B, as can be seen by considering how a computer can be used to "march" from left to right in the z-t plane and determine the solution in a stepwise fashion.

---

1. R. Courant and K. O. Friedrichs, Supersonic Flow and Shock Waves, Interscience Publishers, New York, 1948, pp. 40-45.

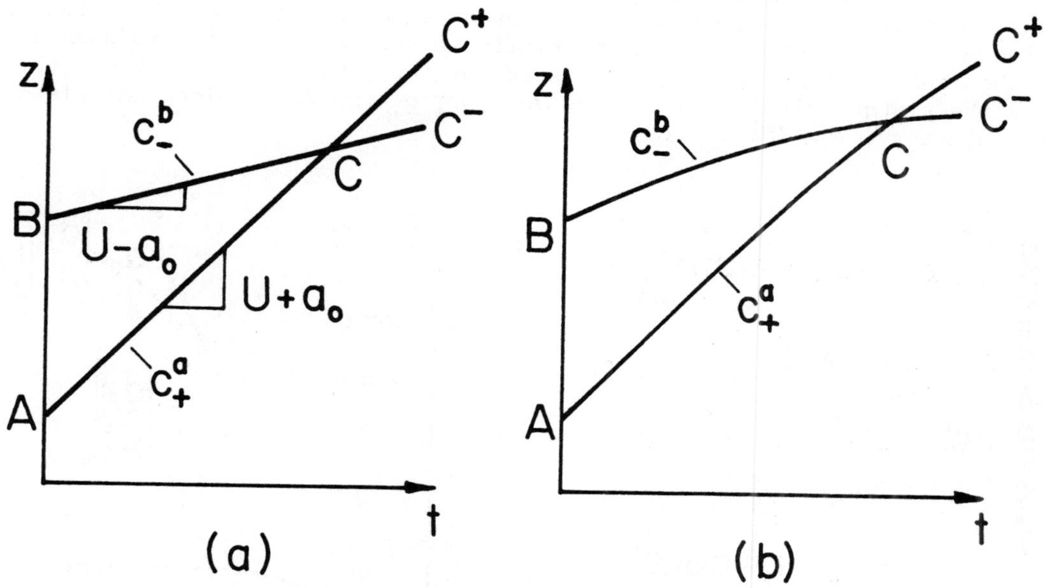

Fig. 11.7.1. (a) Linear case where v and a are known constants and the characteristics are straight lines. (b) c_+^a and c_-^b are established from the initial conditions at A and B. Because they are invariant along the C^+ and C^- characteristics respectively, the solution is established where the characteristics intersect at C.

<u>The Response to Initial Conditions</u>: In the linearized case, the characteristics are straight lines as shown in Fig. 11.7.1a. The point of intersection, C, is then determined because the z coordinates of A and B, as well as the characteristic slopes, are known. The effect of the nonlinearity is to bend the characteristic lines in the z-t plane. This is not surprising, because it would be expected that the velocity of propagation of wavefronts (the slope of dz/dt of a characteristic line) depends on the local speed of sound superimposed on the local velocity of the fluid.

In Fig. 11.7.2, a discrete representation of the characteristics is made. Initial values are given at the positions $z = z_i$. The C^+ line emanating from z_j and the C^- line from z_k cross at some point (j,k). To find the solution throughout the z-t plane,

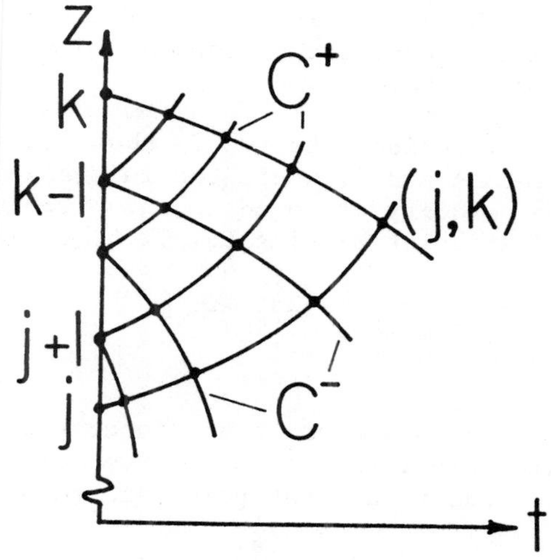

Fig. 11.7.2. The (z-t) intersection of jth C^+ characteristic and kth C^- characteristic is denoted by (j,k).

a) Evaluate the invariants using the initial values in Eqs. 1 and 2:

$$c_{\pm}^{i} = [v] \pm [\frac{2a}{\gamma-1}] \tag{4}$$

b) Tabulate solutions at all intersections (j,k) by solving simultaneously Eqs. 4:

$$[v]_{(j,k)} = \frac{1}{2}(c_+^j + c_-^k)$$
$$[a]_{(j,k)} = \frac{1}{2}(c_+^j - c_-^k)(\frac{\gamma-1}{2}) \tag{5}$$

c)Use the results of (b) to tabulate all characteristic slopes at the intersections (j,k):

$$[\frac{dz}{dt}]^{\pm}_{(j,k)} = [v]_{(j,k)} \pm [a]_{(j,k)} \tag{6}$$

d) Start when t = 0 and build up grid by approximating characteristic lines as being straight between points of intersection. Coordinates and slopes at neighboring points $(z_{j,k-1})$ and $(z_{j+1,k})$ determine z-t coordinates of point $(z_{j,k})$. Thus both the solution and the z-t coordinate at which it applies are determined.

Simple Waves: Initial and boundary conditions are illustrated in Fig. 11.7.3 in which the fluid is initially static [$v = 0$, $a(\rho) = a$] and is driven by a piston at one end. The piston, with position shown as a function of time in the figure, is initially at rest at $z = 0$ and is pushed into the gas until it reaches the final position $z = z_o$. The slope of the piston trajectory has the physical significance of being the piston velocity; hence the velocity of the fluid along the piston trajectory is the slope of the trajectory, $(dz/dt)_p$.

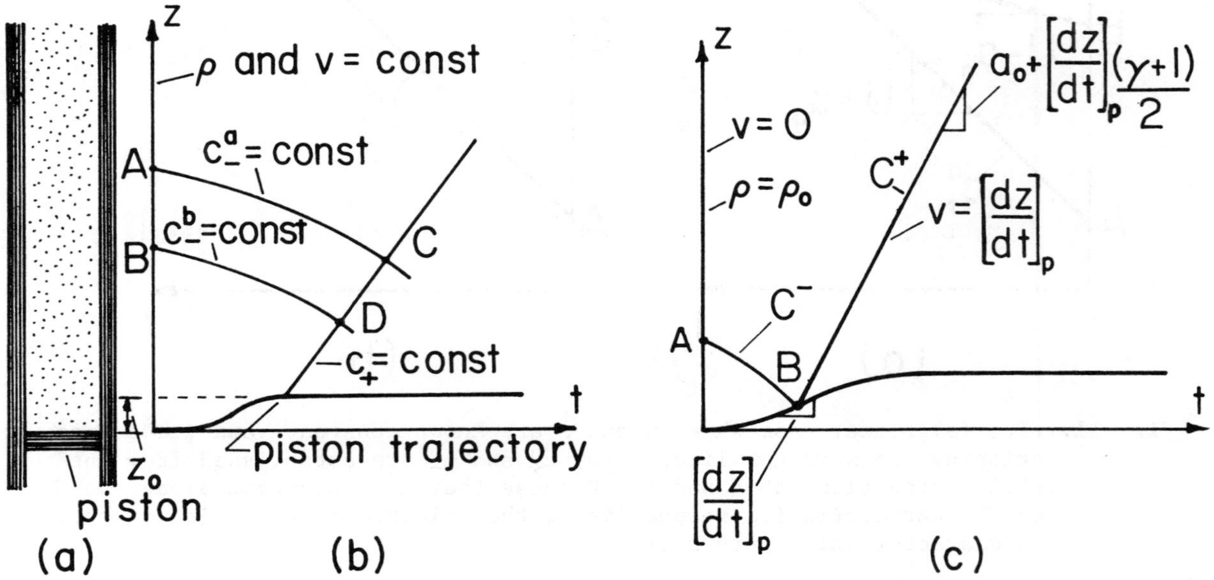

Fig. 11.7.3. (a) Gas filled tube driven by piston. (b) Boundary and initial value problem where the initial state of the fluid is uniform (when $t = 0$) and an excitation is applied by means of a piston which is initially at $z = 0$. (c) a and v are constant along C^+ characteristics, which are straight lines.

By definition, the initial boundary value problem described leads to simple-wave motions. This name designates the response to a boundary condition with the region of interest having a uniform initial state.

Consider the two C^- characteristics sketched in Fig. 11.7.3b. They intersect the z-axis at points A and B, where the initial conditions require that the fluid is stationary ($v = 0$), and that the velocity of sound is a_o. From this, it follows that the invariants c_-, established at point A and at point B using Eq. 11.6.16, are the same,

$$c_-^a = c_-^b = \frac{-2a_o}{\gamma-1} \tag{7}$$

Points C and D are intersections with the same C^+ characteristic. Hence, the invariant c_+ is the same at points C and D. Given c_+ and c_- along the characteristics which intersect at C, the velocity and density at that point are found by simultaneously solving Eqs. 11.6.16,

$$v(C) = \frac{c_+ + c_-^a}{2} \tag{8}$$

Similarly,

$$v(D) = \frac{c_+ + c_-^b}{2} \tag{9}$$

However, because of the special nature of the initial conditions, Eq. 7 requires that $c_-^a = c_-^b$, and it follows that v, and by similar arguments, $a(\rho)$ or ρ, are constant along any given C^+ characteristic. Even more, because v and a are constant, it then follows from Eqs. 11.6.11 that the C^+ characteristics have constant slope.

The C^+ characteristics appear as shown in Fig. 11.7.3c. Along the characteristics shown, the velocity, v, remains equal to that of the fluid at the piston (point B), where

$$v = \left(\frac{dz}{dt}\right)_p \tag{10}$$

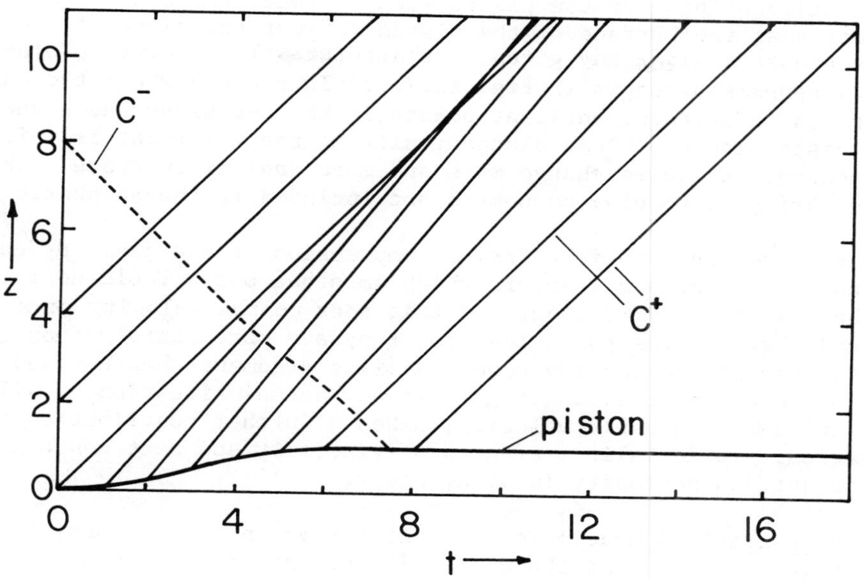

Fig. 11.7.4. Simple-wave characteristic lines initiated by piston.

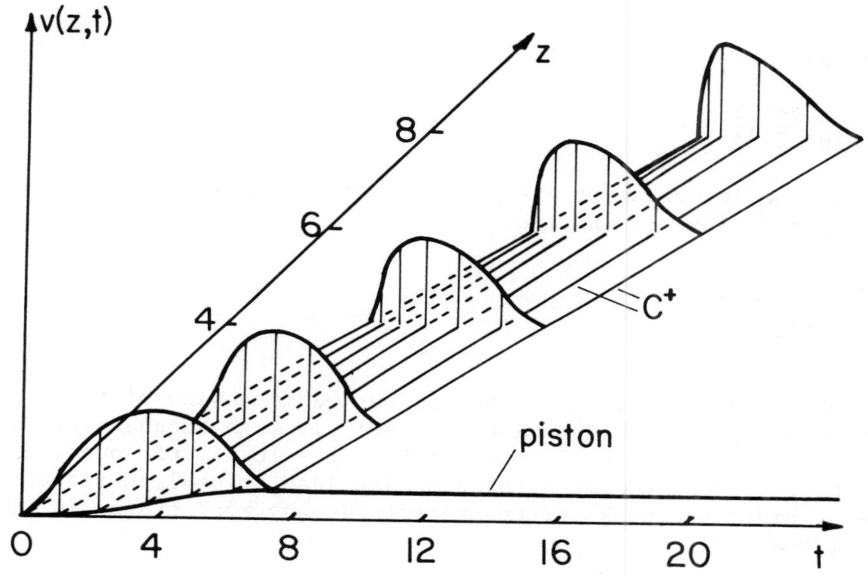

Fig. 11.7.5. Velocity of fluid as a function of z and t.

With Eq. 7, c_- is established along the C^- characteristic, and it follows from Eq. 11.6.16 that the sound velocity $a(\rho)$ at the point B on the piston surface, where v is $(dz/dt)_p$, is

$$a = a_o + \frac{(\gamma-1)}{2}\left(\frac{dz}{dt}\right)_p \tag{11}$$

This velocity of sound, a, along with the implied density ρ and velocity v from Eq. 10, remain constant along the C^+ characteristic. The picture of the dynamics is now complete, because the C^+ characteristic emerging from B has a constant slope given by the Ist characteristic equation, Eq. 11.6 .11:

$$\frac{dz}{dt} = a_o + \left(\frac{dz}{dt}\right)_p \frac{(\gamma+1)}{2} \tag{12}$$

Suppose that the piston position depends on time, as shown in Figs. 11.7.4 and 11.7.5. With no instantaneous change in velocity when t = 0, the piston reaches a maximum velocity when t = 3, and then decelerates to zero velocity by the time t = 6. The C^+ characteristics originating on the piston can be plotted directly using Eq. 12. Here, it is assumed for convenience in making the drawing that γ and

a_o are unity. Note that as the piston velocity increases, the characteristic lines increase in slope, while characteristics originating from the piston when it decelerates decrease in slope. Remember that the fluid velocity v at the surface of the piston is just the slope of the piston trajectory. This velocity remains constant along any given C^+ characteristic. Hence, a plot of the fluid velocity as a function of (z,t) appears as shown in Fig. 11.7.5. In regions where the characteristics tend to cross, the waveform tends to steepen, until at points in the z-t plane where the characteristics cross, the velocity becomes discontinuous. This discontinuity in the wavefront is referred to as a shock wave. With the steepening, variables change more and more rapidly in space. This shortening of characteristic lengths brings into play phenomena not included in the adiabatic model.

Note that a shock wave tends to form from a compression of the gas. By contrast, the decelaration of the piston tends to produce a waveform which smoothes out. Fluid near the leading edge of the pulse is moving in the positive z direction, and this adds to the velocity of a perturbation, a, in that region. Hence, variables within the pulse tend to propagate more rapidly than those nearer the leading edge, and the wave steepens at the leading edge. Similar arguments can be used to explain the smoothing out of the pulse at the trailing edge. In any actual situation, γ will exceed unity, and the increase in density, and hence acoustic velocity, makes a further contribution toward the nonlinear effect of shock formation. In actuality, effects of viscosity and heat conduction prevent the formation of a perfectly abrupt discontinuity in ρ, a, and v.

Limitation of the Linearized Model: To be quantitative in giving conditions under which nonlinearities are important, suppose that the piston is set into motion when t = 0 and reaches the velocity $(dz/dt)_p$ by the time t = T. Then the characteristics are essentially as shown in Fig. 11.7.6, where the displacement of the piston is ignored compared to other lengths of interest. From Eq. 12, the characteristic originating at t = 0, z = 0, is

$$z = a_o t \tag{13}$$

while that originating at t = T, z = 0 is

$$z = [a_o + (\frac{dz}{dt})_p(\frac{\gamma+1}{2})](t - T) \tag{14}$$

Nonlinear effects will be important at $z = \ell$, where these characteristics cross. Solving Eqs. 13 and 14 simultaneously for $z = \ell$ by eliminating t gives

$$\ell = a_o T\left[\frac{a_o}{(\frac{dz}{dt})_p(\frac{\gamma+1}{2})} + 1\right] \tag{15}$$

Hence, for a given characteristic time T, say the period in a sinusoidally excited system, there is a length (some fraction of ℓ) over which a linear model gives an adequate prediction. This distance becomes large as $(dz/dt)_p$ becomes small compared to the velocity of sound, a_o. It is clear, also, that making the period small (the frequency high) can also lead to nonlinearities. This fact is not as limiting as it seems, since the peak piston velocity in any real system is likely also to decrease as the frequency is increased.

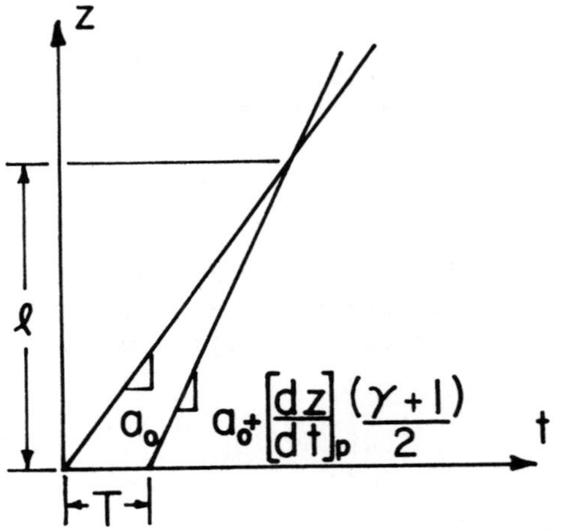

Fig. 11.7.6

An excitation at z = 0 raises the fluid velocity from 0 to $(dz/dt)_p$ in the characteristic time, T. Then nonlinear effects become important in the distance ℓ required for the resulting characteristics to cross.

11.8 Nonlinear Magneto-Acoustic Dynamics

The longitudinal motions of a perfectly conducting gas stressed by a transverse magnetic field, discussed in Sec. 8.8 for small perturbations of a slightly compressible fluid, provide an example of nonlinear electromechanical waves. The methods of Secs. 11.6. and 11.7 are put to work in many investigations of magnetohydrodynamic waves and shocks, especially in the limit of perfect conductivity considered here.[1] Motions, considered here perpendicular to the imposed magnetic field, have been considered for arbitrary orientations of the field.[2]

Equations of Motion: At the outset it is assumed that the motions are one-dimensional:

$$\vec{v} = v(z,t)\vec{i}_z; \quad \vec{H} = H(z,t)\vec{i}_y \tag{1}$$

The physical laws governing the dynamics are those of compressible fluid flow and magnetoquasistatics. Reduced to one-dimensional form, conservation of mass, Eq. 7.2.3, requires that

$$\frac{\partial\rho}{\partial t} + v\frac{\partial\rho}{\partial z} + \rho\frac{\partial v}{\partial z} = 0 \tag{2}$$

In writing the force equation, Eq. 7.16.6, viscous forces are ignored. The magnetic force density is conveniently written by using the stress tensor, Eq. 3.8.14 of Table 3.10.1:

$$\rho\left(\frac{\partial v}{\partial t} + v\frac{\partial v}{\partial z}\right) + \frac{\partial p}{\partial z} = -\mu_o H\frac{\partial H}{\partial z} \tag{3}$$

In the perfectly conducting fluid, there is by definition no electrical dissipation. If in addition effects of dissipation and heat conduction are negligible, the energy equation, Eq. 7.23.3, reduces to an expression representing an isentropic process, Eq. 7.23.7:

$$\frac{D}{Dt}(p\rho^{-\gamma}) = \left(\frac{\partial}{\partial t} + v\frac{\partial}{\partial z}\right)(p\rho^{-\gamma}) \tag{4}$$

In view of the one-dimensional approximation and Eq. 1, the field automatically has zero divergence. The combination of the laws of Ohm, Faraday and Ampère are represented by Eq. 6.2.3. The x component of that equation, in the limit where $\sigma \to \infty$, becomes

$$\frac{\partial}{\partial z}(vH) + \frac{\partial H}{\partial t} = 0 \tag{5}$$

The other components of Eqs. 3 and 5 are automatically satisfied.

Characteristic Equations: Following the technique outlined in Sec. 11.7, Eqs. 2-5, together with the relations between changes in the dependent variables along the characteristic lines and the partial derivatives, are arranged in a matrix. For convenience, $\partial\rho/\partial t \equiv \rho,_t$, $\partial\rho/\partial z = \rho,_z$, etc.:

$$\begin{bmatrix} 0 \\ 0 \\ 0 \\ 0 \\ d\rho \\ dv \\ dp \\ dH \end{bmatrix} = \begin{bmatrix} 1 & v & 0 & \rho & 0 & 0 & 0 & 0 \\ 0 & 0 & \rho & \rho v & 0 & 1 & 0 & \mu_o H \\ -\gamma p & -\gamma p v & 0 & 0 & \rho & \rho v & 0 & 0 \\ 0 & 0 & 0 & H & 0 & 0 & 1 & v \\ dt & dz & 0 & 0 & 0 & 0 & 0 & 0 \\ 0 & 0 & dt & dz & 0 & 0 & 0 & 0 \\ 0 & 0 & 0 & 0 & dt & dz & 0 & 0 \\ 0 & 0 & 0 & 0 & 0 & 0 & dt & dz \end{bmatrix} \begin{bmatrix} \rho,_t \\ \rho,_z \\ v,_t \\ v,_z \\ p,_t \\ p,_z \\ H,_t \\ H,_z \end{bmatrix} \tag{6}$$

---

1. G. W. Sutton and A. Sherman, Engineering Magnetohydrodynamics, McGraw-Hill Book Company, New York, 1965, pp. 309-339.

2. W. F. Hughes and F. J. Young, The Electromagnetodynamics of Fluids, John Wiley & Sons, New York, 1966, pp. 312-318.

To obtain the Ist characteristic equations, the determinant of the coefficients is required to vanish. The determinant is reduced by following steps similar to those that lead from Eq. 11.6.17 to 11.6.19, and then expanding by minors:

$$\frac{dz}{dt} = v \text{ on } C^p \qquad (7)$$

$$\frac{dz}{dt} = v \pm a_b \text{ on } C^{\pm}; \quad a_b \equiv \left[\frac{\mu_o H^2}{\rho} + \frac{\gamma p}{\rho}\right]^{1/2}$$

The C^p characteristics are the particle lines, and actually represent two degenerate sets of characteristics. The $C^{\pm}$ characteristics represent magneto-acoustic waves, discussed for small amplitudes in Sec. 8.8.

The second characteristic equations are obtained from Eq. 6 by solving the determinant arrived at by substituting the column on the left into the first column of the square matrix. Straightforward expansion gives

$$(\frac{dz}{dt} - v)\{dv\left[-\rho(\frac{dz}{dt} - v)\frac{dz}{dt}\right] + d\rho\left[-v(\frac{dz}{dt} - v)^2 + \frac{\gamma p v}{\rho} + \frac{\mu_o H^2 v}{\rho}\right]$$

$$-dp\,\frac{dz}{dt} - [\mu_o H\,\frac{dz}{dt}]dH\} = 0 \qquad (8)$$

The second characteristic equations along the particle lines are found from Eq. 8 using Eq. 7a. That the determinantal equations are degenerate is again reflected by the first factor in Eq. 8; remember that $(dz/dt - v)$ appeared as a quadratic factor in the denominator. The second term in brackets is zero if $dz/dt = v$, so that

$$[\frac{\gamma p}{\rho}\,d\rho - dp] + \mu_o H^2[\frac{d\rho}{\rho} - \frac{dH}{H}] = 0 \text{ on } C^p \qquad (9)$$

This equation is actually the sum of two independent expressions, as can be seen by considering Eq. 4, which on the particle characteristic can be written as

$$\frac{\gamma p}{\rho}\,d\rho - dp = 0 \text{ or } d(p\rho^{-\gamma}) = 0 \text{ on } C^p \qquad (10)$$

On the same particle characteristics, Eqs. 2 and 5 combine to give

$$\frac{d\rho}{\rho} - \frac{dH}{H} = 0 \text{ or } d(\frac{H}{\rho}) = 0 \text{ on } C^p \qquad (11)$$

These last two equations insure that 9 is satisfied, and account for the degeneracy of the two characteristic equations.

Using Eqs. 7b in 8 gives the two additional characteristic equations

$$\mp\rho a_b dv - dp - \mu_o H dH = 0 \text{ on } C^{\pm} \qquad (12)$$

Thus, the Ist characteristic equations are summarized by 7 and the IInd characteristic equations given by Eqs. 10-12.

<u>Initial Value Response</u>: To any given point in the (z,t) plane can be ascribed four intersecting characteristic lines, two of which are simply the particle line. These are illustrated in Fig. 11.8.1. In general, the solution at the given point is obtained by simultaneously solving Eqs. 10-12, which are four equations in four unknowns. The first two of these expressions can simply be integrated to give invariants along C^p:

$$p\rho^{-\gamma} = p_c\rho_c^{-\gamma} \text{ on } C^p \qquad (13)$$

$$H/\rho = H_c/\rho_c \text{ on } C^p \qquad (14)$$

The second of these states that a fluid circuit of fixed identity must conserve magnetic flux. Hence, an increase in density caused by the compression of a fluid element is accompanied by a local increase in H. The model of Fig. 8.8.1a remains a useful way of viewing the interaction.

Suppose that, at some time in the evolution of the system, the pressure, density and field are uniform and are (p_o,ρ_o,H_o). The invariants on the right in Eqs. 13 and 14 are independent of position thereafter:

$$p\rho^{-\gamma} = p_o\rho_o^{-\gamma} \tag{15}$$

$$\frac{H}{\rho} = \frac{H_o}{\rho_o} \tag{16}$$

It follows that the remaining IInd characteristic equations, Eqs. 12, become

$$\frac{a_b}{\rho}\,d\rho \pm dv = 0;\quad a_b = [p_o\rho_o^{-\gamma}(\gamma p^{\gamma-1}) + \rho\mu_o(\frac{H_o}{\rho_o})^2]^{\frac{1}{2}} \tag{17}$$

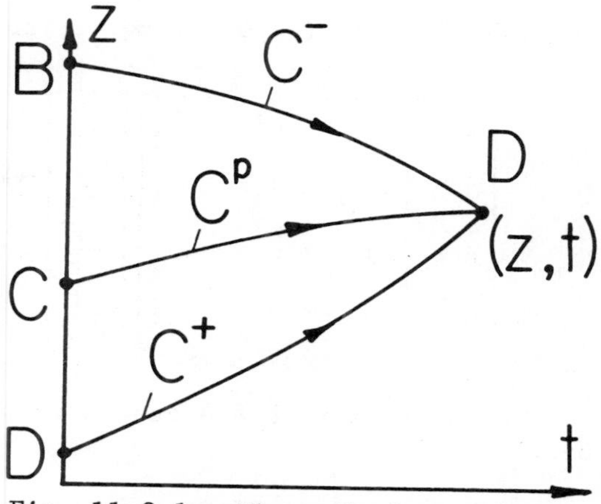

Fig. 11.8.1. The solution at (z,t) results from invariants carried along the four characteristic lines shown, with C^p representing two families of characteristics.

In effect, the dynamics now involve only the characteristics $C^{\pm}$ and two of the four original variables. Given (ρ,v) from solving Eqs. 7b and 17, p and H are found from 15 and 16. The dynamics are similar to those for the gas alone, except that $a \rightarrow a_b(\rho)$. Note that the dependence of the magneto-acoustic velocity on ρ is in part determined by H_o.

11.9 Nonlinear Electron Beam Dynamics

The nonlinear motions of streaming electrons are usually described in terms of Lagrangian coordinates. Nevertheless, an Eulerian description affords considerable insight if it is couched in terms of characteristics. By contrast with the equations of Secs. 11.7 and 11.8, those now considered are inhomogeneous.

The laws describing an electron beam neutralized by a background of ions are of the same nature as used in Sec. 11.5. Here, they are written without linearization but with the assumption that motions and fields are one-dimensional and that fields, like the motion, are z-directed. Hence, particle conservation requires that

$$(n_o + n)\frac{\partial v_z}{\partial z} + \frac{\partial n}{\partial t} + v_z\frac{\partial n}{\partial z} = 0 \tag{1}$$

where n_o is the equilibrium number density and $n(z,t)$ is the departure from that equilibrium. The longitudinal force equation is

$$\frac{\partial v_z}{\partial t} + v_z\frac{\partial v_z}{\partial z} = -\frac{e}{m}E_z \tag{2}$$

and in one dimension, Gauss' law is

$$\frac{\partial E_z}{\partial z} = -\frac{en}{\varepsilon_o} \tag{3}$$

Variables are normalized at the outset so that time is measured in terms of the plasma frequency $\omega_p \equiv \sqrt{e^2n_o/m\varepsilon_o}$,

$$\underline{t} = t\omega_p;\quad \underline{z} = z/\ell;\quad \underline{n} = n/n_o;\quad \overset{o}{\underline{e}} = \overset{o}{e}/\omega_p;\quad \underline{v} = v_z/(\ell\omega_p);$$
$$\underline{E} = E_z/(n_o\,e\ell/\varepsilon_o) \tag{4}$$

By definition,

$$\overset{o}{e} = \frac{\partial v}{\partial z} \tag{5}$$

and Eq. 3 becomes

$$n = \frac{\partial E}{\partial z} \tag{6}$$

Then, Eq. 1 and $\partial(\ )/\partial z$ of Eq. 2, combined with Eq. 3, are the first two of the four equations,

$$\begin{bmatrix} 1 & v & 0 & 0 \\ 0 & 0 & 1 & v \\ dt & dz & 0 & 0 \\ 0 & 0 & dt & dz \end{bmatrix} \begin{bmatrix} \frac{\partial n}{\partial t} \\ \frac{\partial n}{\partial z} \\ \frac{\partial \mathring{e}}{\partial t} \\ \frac{\partial \mathring{e}}{\partial z} \end{bmatrix} = \begin{bmatrix} -(1+n)\mathring{e} \\ n - \mathring{e}^2 \\ dn \\ d\mathring{e} \end{bmatrix} \qquad (7)$$

The third and fourth are expressions for dn and de, introduced following the procedure described in Sec. 11.6.

That the determinant of the coefficients vanish gives the Ist characteristic equations. There are two families of lines, but these are degenerate:

$$\frac{dz}{dt} = v \text{ on } C^{\pm} \qquad (8)$$

The second characteristic equations could be obtained by substituting the column on the right for any pair of columns on the left and setting the respective determinants equal to zero:

$$\frac{dn}{dt} = -(1+n)\mathring{e} \text{ on } C^{+} \qquad (9)$$

$$\frac{d\mathring{e}}{dt} = n - \mathring{e}^2 \text{ on } C^{-} \qquad (10)$$

These expressions are simple enough that they could have been obtained directly from Eqs. 7a and 7b by inspection.

A configuration typical of klystron beam-cavity interactions is shown in Fig. 11.9.1. The electron beam passes through screen electrodes at $z = 0$ and $z = \ell$. These are constrained to the potential difference, $V(t) = \underline{V}(t)(n_o e \ell^2/\varepsilon_o)$, which will be taken here as a given drive. In reality, $V(t)$ might be associated with a resonator that is used to either excite the beam or extract energy.

The region of interest in the z-t plane is between the electrodes, $0 < z < \ell$ and for $0 < t$. Characteristic lines that enter this region along the z-axis (when $t = 0$) are denoted by $K = N \cdots M$, while those that enter along the t-axis (where $z = 0$) are represented by $K = 1 \cdots N$. To integrate Eqs. 9 and 10, it is appropriate to have two initial conditions for the latter and two entrance boundary conditions for the former.

When $t = 0$, the velocity and electric field distributions between the screens are taken as known. As an example, if the beam is initially unmodulated, the electron velocity is constant and there is no space charge between the screens:

$$v = U, \quad E = V(0) \qquad (11)$$

For the boundary conditions, it is assumed that the beam enters with a constant velocity, U. In passing through the screen, an electron is subjected to a step in electric field, but not to an impulse. Hence, this velocity is continuous through the screen, this means that along the t-axis, where the electrons enter the region of interest, $\partial v/\partial t$ is also continuous. It follows from the force equation for an electron, Eq. 2, that $\mathring{e}$ is not continuous through the screen. Rather, if $\mathring{e} = 0$ just upstream of the screen at $z = 0$, according to Eq. 2, $\mathring{e}$ assumes the value

$$\mathring{e}(0,t) = -\frac{E(0,t)}{U} \qquad (12)$$

just downstream. The number density is continuous through the screen, so that if the beam is unmodulated upstream, then just downstream

$$n(0,t) = 0 \qquad (13)$$

Although not relevant as a boundary condition, it can be seen from Eq. 1 that the step in $\mathring{e}$ across the screen is accompanied by a step in $\partial n/\partial z$.

To make Eq. 12 a useful boundary condition, $E(0,t)$ must be related to $V(t)$. To this end, two integrations of Eq. 6, with the condition that the second intergration give $V(t)$, result in

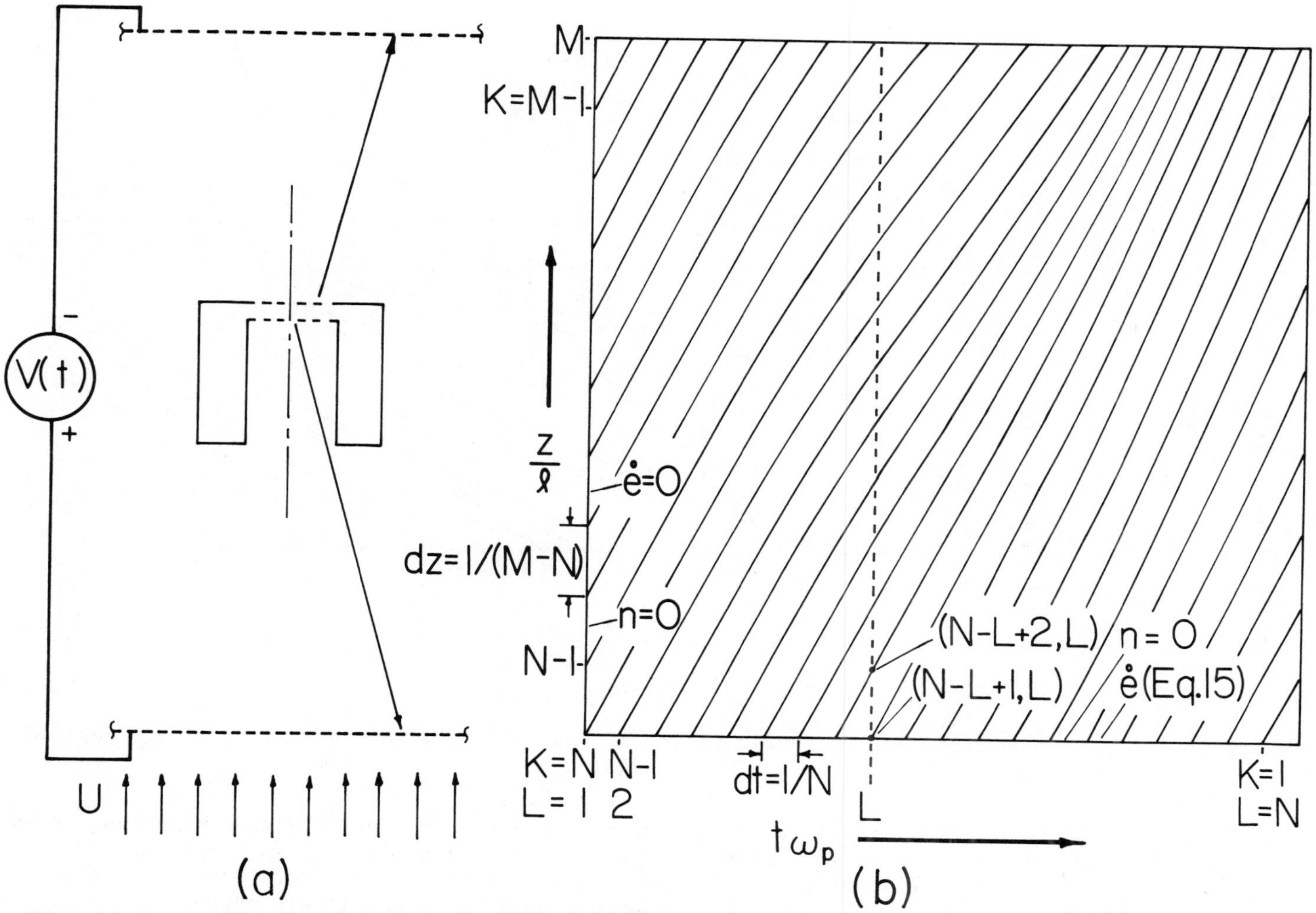

Fig. 11.9.1. (a) Beam enters region between screen electrodes with velocity U. At microwave frequencies, the screens typically provide coupling to cavity resonators, as shown by the inset. (b) Characteristic lines in the z-t plane. Coordinate (x,t) is denoted by characteristic line (K) and time (L).

$$E = V - \int_0^1\int_0^z n(z',t)dz'dz + \int_0^z n(z',t)dz' \qquad (\underline{14})$$

so that the electric field at z = 0 can be evaluated and used to express Eq. 12 as

$$\mathring{e}(0,t) = -\frac{V}{U} + \frac{1}{U}\int_0^1\int_0^z n(z',t)dz'dz \qquad (\underline{15})$$

Equations 13 and 15 comprise the boundary conditions at z = 0.

The integration of the second characteristic equations, Eqs. 9 and 10, can be carried out by treating them as simultaneous ordinary differential equations. This is possible only because the characteristic lines to which they apply are the same. Depending on whether the characteristic line of interest enters through the t = 0 axis or the z = 0 axis, the initial conditions or boundary conditions serve as "initial" conditions for this integration. However, the integration is not quite this straightforward, because superimposed on the propagational dynamics is Poisson's equation, which makes the entrance field instantaneously reflect both the net effect of the charge in the region of interest and the voltage V(t). This is why the boundary condition on $\mathring{e}$, Eq. 15, depends not only on the voltage but also on the charge throughout.

Consider the numerical steps that portray the space-time dynamics while marching forward in time. The initial conditions when t = 0 (L = 1) can be used with Eqs. 9 and 10 to establish $n(z,dt) \equiv n(K,2)$ and $\mathring{e}(z,dt) \equiv \mathring{e}(K,2)$ at the points where the characteristics $(K = N\cdots M)$ intersect the t = dt axis (L=2). Also, Eqs. 8 can be used to determine where these solutions apply, i.e., where

$$v(z,dt) = U + \int_0^z \mathring{e}(z,0)dz \qquad (\underline{16})$$

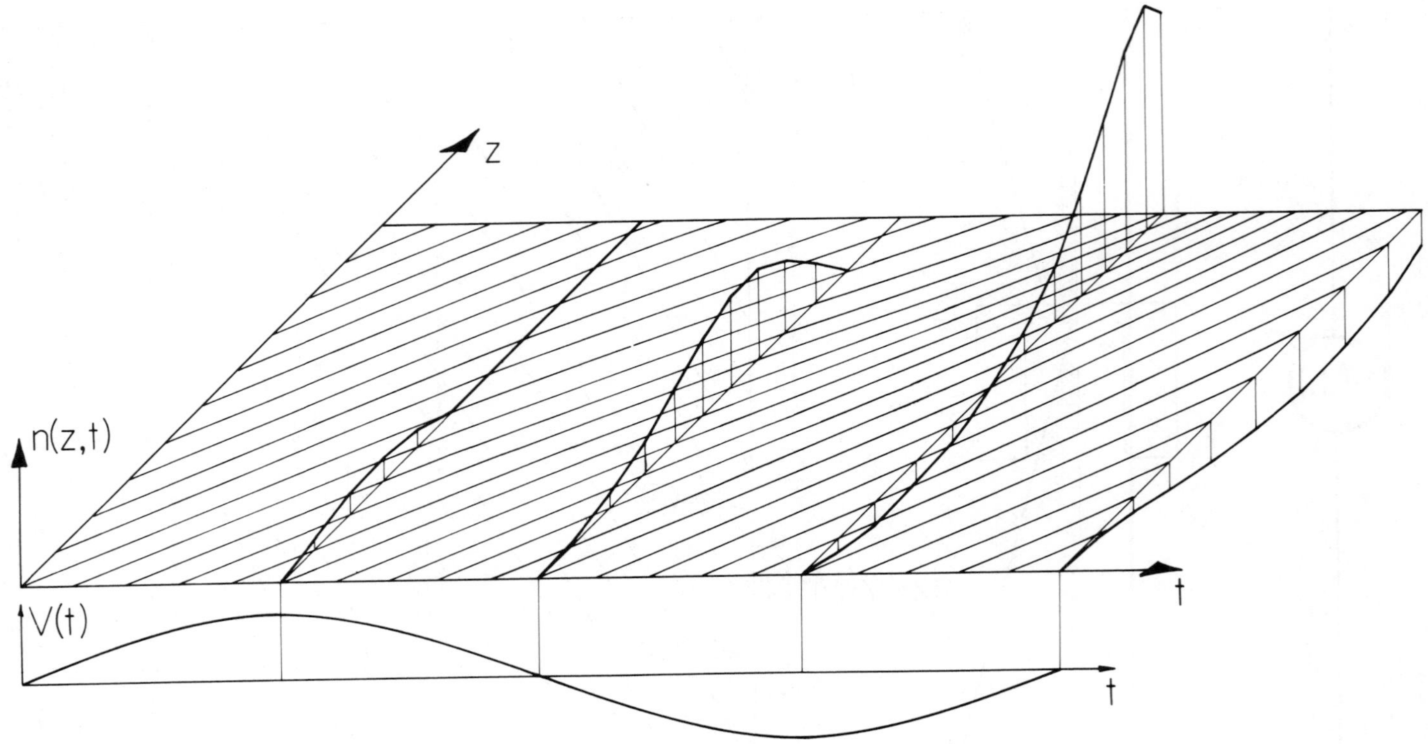

Fig. 11.9.2. Turn-on transient in configuration of Fig. 11.9.1 with sinusoidal voltage applied to screens. Normalized U = 2, V = 2 and angular frequency $\omega = (2\pi\omega_p)$.

The characteristic line K = N-1 entering at z = 0 when t = dt does so with conditions set by the boundary conditions of Eqs. 13 and 15. Note that because $n(0,dt) \equiv n(N-1,2)$ is known and n(z,dt) has already been determined at the location $K = N\cdots,L = 2$, integration called for in Eq. 15 can be carried out. Hence, $\overset{o}{e}(0,dt)$ is determined. Thus, the dynamical picture is completely established when t = dt. This process can now be repeated to determine the response when t = 2dt, and so on. The turn-on transient resulting from the application of a voltage $V(t) = V \sin \omega t$, is shown in Fig. 11.9.2.

Note that even though the transient has a well defined wave front, determined by the characteristic line passing through the origin, the characteristic lines are distorted even ahead of this wave front. This is because the applied voltage and the space charge between the screens have an instantaneous effect on the velocity of electrons throughout. Where the characteristic lines converge, abrupt changes in density occur. By increasing the driving voltage, characteristic lines can be made to cross. Electrons entering at one time are overtaken by those entering at a later time. It is to handle this situation that Lagrangian coordinates are often used.[1]

Once an electron has entered the interaction region, so that its initial conditions are established, its evolution in the state space $(\overset{o}{e},n)$ is determined. This can be seen by combining Eqs. 9 and 10 so as to eliminate time as the parameter:

$$\frac{d\overset{o}{e}}{dn} = \frac{\overset{o}{e}^2 - n}{(1 + n)\overset{o}{e}} \tag{17}$$

Given an initial position in the state space $(\overset{o}{e},n)$, numerical integration of Eq. 17 results in one of the trajectories of Fig. 11.9.3. It follows from Eqs. 9 and 10 that as time progresses, the trajectories are traced out in the direction indicated by the arrows. Thus, the number density in the neighborhood of a given electron (moving along a characteristic line) is oscillatory in nature, with a frequency typified by the plasma frequency, ω_p. For the particular initial conditions of Eqs. 13 and 15, which pertain along characteristics emanating from the t axis, the trajectories all start from the $\overset{o}{e}$ axis, but with an amplitude determined by Eq. 15. The picture is now one of particles acting as nonlinear oscillators translating in the z direction with the velocity v.

The perturbation dynamics are governed by the linearized forms of Eqs. 9 and 10, which combine to show that

1. H. M. Schneider, "Oscillations of an Inhomogeneous Plasma Slab," Ph.D. Thesis, Department of Electrical Engineering, Massachusetts Institute of Technology, Cambridge, Mass., 1969.

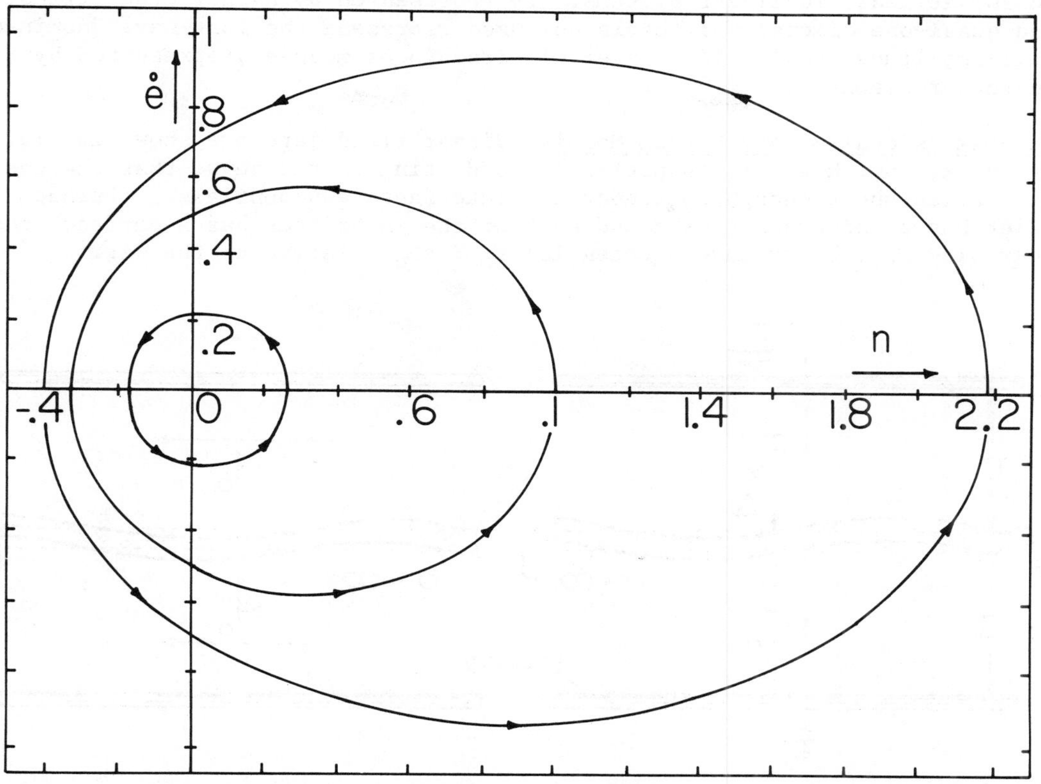

Fig. 11.9.3. Phase-plane ($\mathring{e}$-n) trajectories of oscillations of electron beam.

$$\frac{d^2n}{dt^2} + n = 0 \tag{18}$$

Thus, on a characteristic crossing the t axis when $t = t_o$ (where $n = 0$), (19)

$$n = A(t_o) \sin (t - t_o)$$

Linearized, Eq. 8 can be integrated to express the characteristic line along which Eq. 19 applies:

$$z = U(t - t_o) \tag{20}$$

Found from Eq. 20, t_o can be substituted into Eq. 19 to obtain

$$n = A(t - \frac{z}{U}) \sin (\frac{\omega_p z}{U}) \tag{21}$$

where dimensional variables have been reintroduced.

The response is the product of a stationary emvelope having a wavelength $\lambda = 2\pi U/\omega_p$ and a part traveling in the z direction with the electron velocity, U. The envelope is stationary in space because every electron oscillator passes the $z = 0$ plane with $n = 0$. The amplitude of its oscillation is determined by the initial condition on $\mathring{e}$ when it passed the screen at $z = 0$. Note that in this small-amplitude limit, the phase-plane trajectories of Fig. 11.9.3 are circles with radii much less than one. It follows that to achieve linear dynamics, $\mathring{e} \ll \omega_p$.

11.10 Causality and Boundary Conditions: Streaming Hyperbolic Systems

Objectives in this section are: (a) to develop readily visualized prototype models for streaming interactions; (b) to picture in z-t space the evolution of absolute and convective instabilities and of systems which if driven in the sinusoidal steady state would display evanescent and amplifying waves; (c) to use the method of characteristics to illustrate the crucial role of causality in the choice of boundary conditions. In terms of complex waves (and eigenmodes) a small-amplitude version of the dynamics will be considered again in Sec. 11.12. There, causal boundary conditions, as discussed here, will be essential to understanding the stability of systems of finite extent in the longitudinal direction.

Emphasized in this section is the dependence on the longitudinal (streaming) direction. Transverse dependences, at least in linear systems, are represented by higher order transverse modes. Linearized, the quasi-one-dimensional models now used represent the long-wave "dominant modes" from a complete small-amplitude model. This interrelationship of models, represented by Fig. 4.12.2, is illustrated in the problems.

Quasi-One-Dimensional Single Stream Models: Planar fluid jets are shown in Fig. 11.10.1. In the electric version, the sheet jet is perfectly conducting in the sense that charges can relax on the interface in times short enough to render the interfaces equipontials. (Perhaps a jet of water in air.) The jet has a thickness $\Delta << a$ and each of the interfaces has a surface tension γ. Electrodes to either side of the jet have a potential $V_o \equiv aE_o$ relative to the jet.

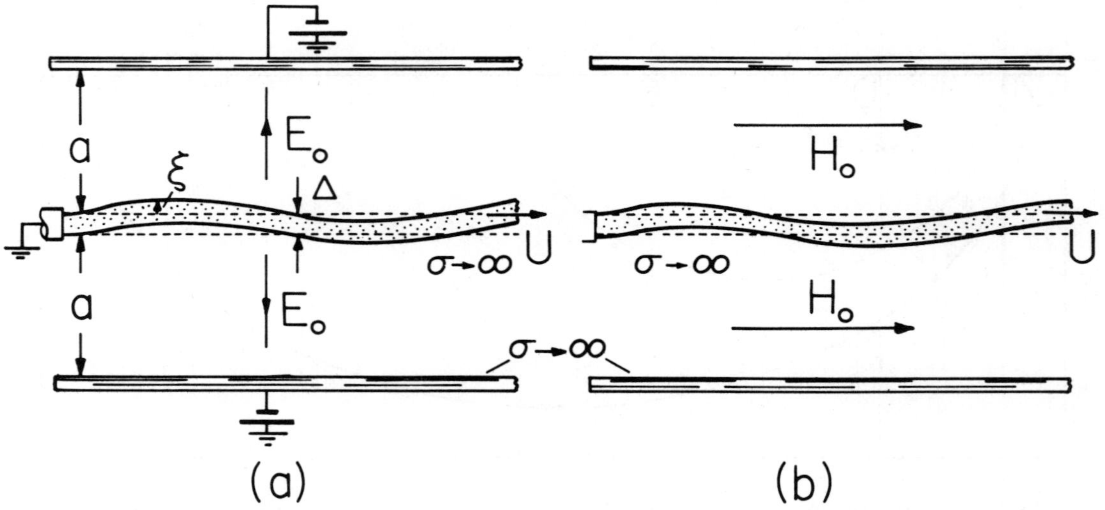

Fig. 11.10.1. Prototype single-stream systems consisting of perfectly conducting sheets convecting to the right with velocity U. (a) Potential constrained EQS configuration; (b) flux constrained MQS configuration.

For long-wave motions, the transverse electric surface force density, $T(z,t)$, can be approximated by picturing the jet as having a deflection $\xi(z,t)$ from the center line, with essentially negligible slope. Thus, perhaps by using the stress tensor on a control volume enclosing a section of the jet, it follows that

$$T = \frac{1}{2}\varepsilon_o\left[\frac{(aE_o^2)}{(a-\xi)^2} - \frac{(aE_o)^2}{(a+\xi)^2}\right] \tag{1}$$

In the magnetic version, the jet is also perfectly conducting, but now so much so that the magnetic diffusion time $\mu\sigma\Delta a >> 1$. The system is then the antidual (Sec. 8.5) of the electric one, and T obtained from Eq. 1 by replacing $\varepsilon_o E_o^2 \rightarrow -\mu_o H_o^2$. In either system, the inertial and surface tension forces acting on the sheet are now also written with the assumption that deflections are slowly varying with respect to z. With U defined as the streaming velocity, and approximated here as constant, and ρ the jet mass density, it follows that Newton's law for motions in the transverse direction is

$$\Delta\rho\left(\frac{\partial}{\partial t} + U\frac{\partial}{\partial z}\right)^2\xi = 2\gamma\frac{\partial^2\xi}{\partial^2 z} + T \tag{2}$$

The same expression would be written to describe a membrane having surface mass density $\Delta\rho$ and tension 2γ. The velocity of waves on a fixed membrane would then be $V \equiv \sqrt{2\gamma/\Delta\rho}$.

For motions having a typical time scale τ, it is convenient to write Eq. 2 in terms of the normalized variables

$$\underline{\xi} = \xi/a,\ \underline{t} = t/\tau,\ \underline{z} = z/\tau V \tag{3}$$

New variables are introduced:

$$v = \frac{\partial\xi}{\partial t}; \quad e = \frac{\partial\xi}{\partial z} \tag{4}$$

so that Eqs. 1 and 2 can be written as two first-order expressions:

$$\left(\frac{\partial v}{\partial t} + M\frac{\partial v}{\partial z}\right) + M\left(\frac{\partial e}{\partial t} + M\frac{\partial e}{\partial z}\right) = \frac{\partial e}{\partial z} + \frac{P}{4}\left[\frac{1}{(1-\xi)^2} - \frac{1}{(1+\xi)^2}\right] \tag{5}$$

$$\frac{\partial v}{\partial z} - \frac{\partial e}{\partial t} = 0 \tag{6}$$

where

$$P = \frac{2\varepsilon E_o^2}{\rho\Delta a}\tau^2 = (\tau/\tau_{EI})^2 \text{ or } -\frac{2\mu_o H_o^2}{\rho\Delta a}\tau^2 = (\tau/\tau_{MI})^2 \text{ and } M \equiv U/V$$

The last expression follows from taking cross-derivatives of Eqs. 4. Note that P is the square of the ratio of the characteristic time to an electro or magneto inertial time, while M is a Mach number. The magnetic and electric systems are respectively described with P positive and negative. With $P>0$, the transverse force acts in the same direction as the displacement, and hence promotes instability. With $P<0$, the force acts as a nonlinear spring to recenter the sheet.

Single Stream Characteristics: The characteristic representation of Eqs. 5 and 6 follows from writing Eqs. 5 and 6 in the form of Eq. 11.6.7 and using the procedures outlined in Sec. 11.6:

$$dv + de(M \mp 1) = \frac{P}{4}\left[\frac{1}{(1-\xi)^2} - \frac{1}{(1+\xi)^2}\right]dt \tag{7}$$

on

$$\frac{dz}{dt} = M \pm 1; \quad (C^{\pm}) \tag{8}$$

It follows from the definition of e, Eq. 4, that

$$\xi = \int^{z} e\, dz \tag{9}$$

where the lower limit of integration is selected as one where ξ is either known or can be related to other variables through a boundary condition.

Because the nonlinearity is confined to the second characteristic equations, Eqs. 8 can be integrated:

$$z^{\pm} = (M \pm 1)t + z^{\pm} \tag{10}$$

Thus, the characteristics are straight lines in the z-t plane, as illustrated in Fig. 11.10.2. By contrast with the situation in Sec. 11.7, where the second characteristics could be integrated, but the first not, here the z-t lines along which Eqs. 7 apply are known. It is the second characteristic equations that cause the trouble.

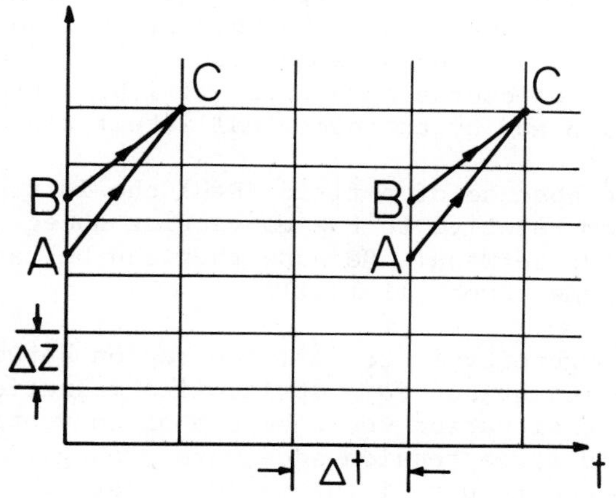

Fig. 11.10.2

Characteristic lines in z-t plane used to determine response at C given initial conditions at A and B.

There are two rewards for following the discussion now undertaken of how the characteristics can be used to give a numerical picture of the dynamics. The finite difference algorithm can be used to compute the response to initial and boundary conditions in a straightforward fashion. Perhaps more important, the implications of causality for boundary conditions becomes evident in the process.

Consider the determination of the response at C in Fig. 11.10.2, given that at B and A an instant, Δt, earlier. With the understanding that $\Delta v_A^{\pm}$ and $\Delta v_B^{\pm}$ are incremental quantities computed respectively at the points A and B:

$$v_C = \begin{cases} v_A + \Delta v_A^{+} \\ v_B + \Delta v_B^{-} \end{cases} \tag{11}$$

These two expressions must result in the same response at C. Hence, they can be simultaneously solved. The result is the first of the following four relations between the incremental variables evaluated at one or the other of the previous points on the incident characteristics:

$$\begin{bmatrix} 1 & -1 & 0 & 0 \\ 0 & 0 & 1 & -1 \\ 1 & 0 & M-1 & 0 \\ 0 & 1 & 0 & M+1 \end{bmatrix} \begin{bmatrix} \Delta v_A^{+} \\ \Delta v_B^{-} \\ \Delta e_A^{+} \\ \Delta e_B^{-} \end{bmatrix} = \begin{bmatrix} v_B - v_A \\ e_B - e_A \\ Pf_A \Delta t \\ Pf_B \Delta t \end{bmatrix} \tag{12}$$

where

$$f \equiv \frac{1}{4}\left[\frac{1}{(1-\xi)^2} - \frac{1}{(1+\xi)^2}\right]$$

The second of these equations is analogous to the first with v replaced by e. The third and fourth represent the second characteristics, Eqs. 7. Solution of Eqs. 12 results in expressions for the incremental quantities in terms of the variables evaluated at the previous time step:

$$\Delta v_A^{+} = \frac{1}{2}\{[v_A - v_B](M-1) + [e_A - e_B](M^2-1) + P[f_A(M\ 1) - f_B(M-1)]\Delta t\} \tag{13}$$

$$\Delta e_A^{+} = -\frac{1}{2}\{[v_A - v_B] + (M+1)[e_A - e_B] + P[f_A - f_B)\Delta t\} \tag{14}$$

As indicated by the superscripts, these are the incremental changes in v and e along the C^+ characteristics.

<u>Single Stream Initial Value Problem</u>: Suppose that when $t = 0$, $\xi(z,0)$ [and hence $e(z,0)$] and $v(z,0)$ are given at equally spaced points along the z axis. Further, suppose it is decided that for convenience the response is to be found when $t = \Delta t$ at points C similarly selected to fall at intervals Δz. The values of e, ξ and v at A and B can be determined from the initial conditions by interpolating between the initial values.

Then the values of e_C and v_C, e and v when $t = \Delta t$, follow from Eqs. 13 and 14 used with expressions of the form of Eq. 11. Numerical integration, as called for by Eq. 9, then gives the distribution of ξ at this time. The situation when $t = \Delta t$ is now the same as was the initial one, so the process can be repeated to find the response when $t = 2\Delta t$. Thus, the dynamics are unraveled by "marching" forward in time along the characteristic lines. Of course some error will be introduced by the interpolation required to evaluate v and e at A and B and by the numerical integration of Eq. 9.

Typical responses are shown in Fig. 11.10.3. In the absence of a field (P=0) the initial pulse divides into components propagating upstream and downstream relative to the convecting sheet. These pulses propagate without distortion, leaving a null response between. Because they can be represented analytically, this case gives a check on the numerical scheme (Prob. 11.10.1).

Regardless of the sign of P, one effect of the inhomogeneity is to fill the region between these pulses with a response. With $P < 0$, physically the sheet is subject to a spring-like magnetic restoring force. In the extreme of no tension ($V = 0$), the situation would be one of convecting nonlinear oscillators, similar to that considered in Sec. 11.9. The tension adds wave propagation effects already familiar from part (a) of Fig. 11.10.3. The combined result, illustrated in part (b) of the figure, once again shows waves propagating along the characteristic lines, but now attenuating and

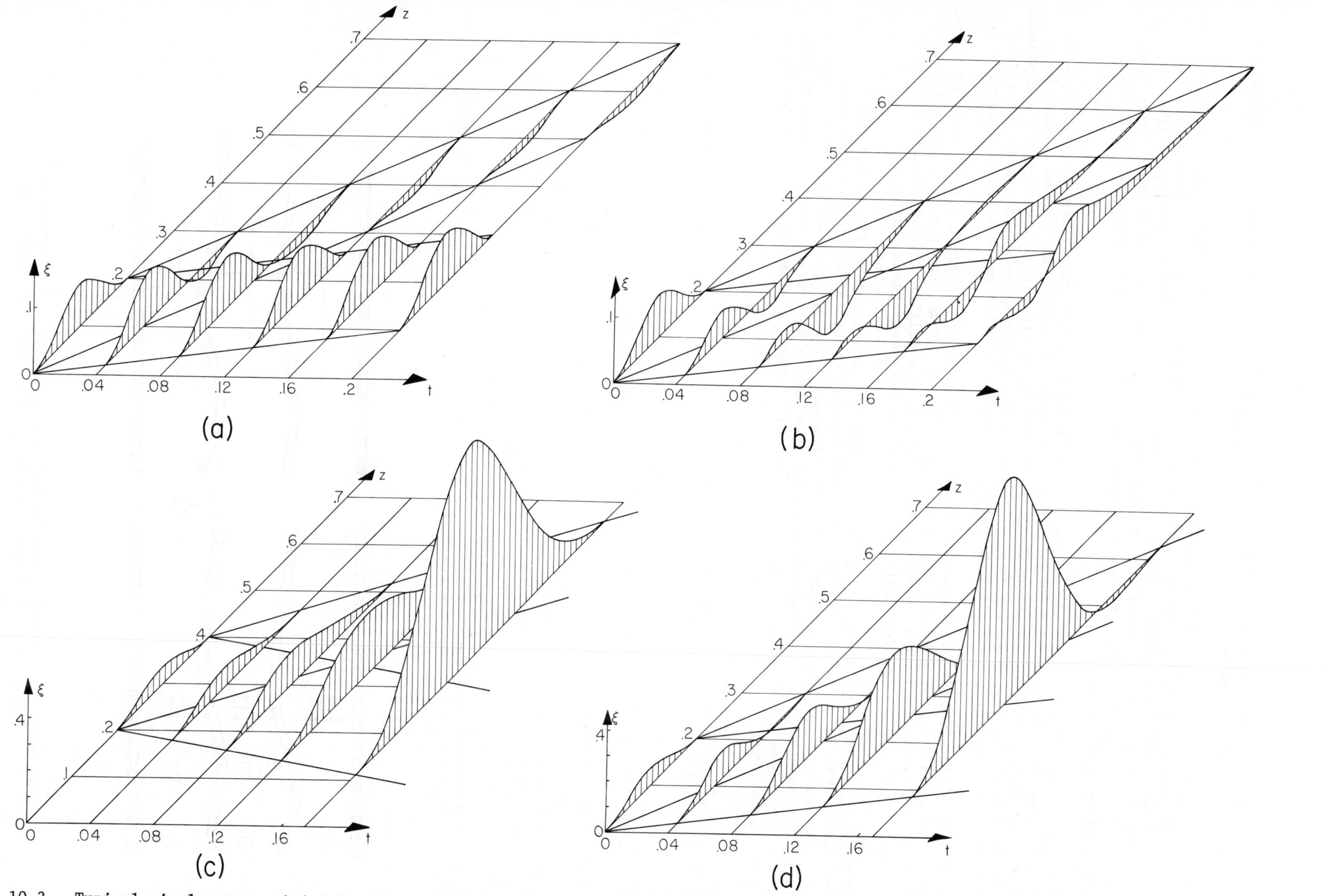

Fig. 11.10.3. Typical single-stream initial value responses determined by numerical integration with $\Delta z = .005$ and $\Delta t = .005$. The initial velocity is zero and displacement is as shown. (a) With no field, and hence no inhomogeneity, the initial pulse divides into fast and slow pulses that propagate without distortion. (b) With $P < 0$ (magnetic field), wave evanescence results. For the case shown, $M > 1$, so the response is swept downstream. (c) For $P > 0$ (electric field) and $M > 1$ ("sub"), an absolute (nonconvective) instability results. (d) For $P > 0$ but $M > 1$ ("super"), the instability is convective and the response at a given position remains bounded.

leaving behind an oscillating remnant. This oscillating part tends to be carried by the convection and have an angular frequency $\sqrt{-P}$.

As would be expected for $P > 0$, which represents a transverse electric force acting much as a nonlinear "negative" spring force, the response in parts (c) and (d) of Fig. 11.10.3 grows with time. Two types of instability are illustrated. For $M < 1$, where the flow is "sub" relative to the wave velocity V, the response becomes unbounded for an observer having a fixed location along the z axis. This is termed an absolute instability or, to distinguish it from the type of response shown for $M > 1$, a nonconvective instability.

For the convective instability of part (d), $M > 1$ and the response at a given location remains bounded. But, for an observer moving downstream it grows. Such an instability can be excited by a temporarily periodic signal at some location along the z axis and a sinusoidal steady-state established downstream in which the response takes the form of a spatially amplifying wave. At least for linear systems, such waves are best considered in the frequency domain, as illustrated in Secs. 11.11-11.13.

The nonlinear field coupling has its most pronounced effect in the electric field case. As $\xi \to a$ (its maximum possible value), the electric force becomes infinite. Thus, the peaks of the deflection tend to sharpen. In the $P > 0$ examples shown by Fig. 11.10.3, the initial deflection, consisting of a cosinusoid plus a constant in the intervals shown, tends to become a triangular pulse.

Quasi-One-Dimensional Two-Stream Models: Consider now the two-stream configurations of Fig. 11.10.4. The sheets have the respective convective velocities U_1 and U_2 and the same wave velocities V. They are now not only subject to the "self-field" effects resulting from the electric and magnetic fields, much as for the single streams, but they are also coupled to each other by this field. Thus, a given sheet is subject to "self" and "mutual" forces, represented on the right in the transverse force equations:

$$\Delta\rho\left(\frac{\partial}{\partial t} + U_1\frac{\partial}{\partial z}\right)^2\xi_1 - 2\gamma\frac{\partial^2\xi}{\partial z^2} = \frac{1}{2}\varepsilon_o E_o^2 f_1 \tag{15}$$

$$\Delta\rho\left(\frac{\partial}{\partial t} + U_2\frac{\partial}{\partial z}\right)^2\xi_2 - 2\gamma\frac{\partial^2\xi_2}{\partial z^2} = \frac{1}{2}\varepsilon_o E_o^2 f_2 \tag{16}$$

where, with the displacements normalized to a,

$$f_1(\xi_1,\xi_2) = \frac{1}{4}\left[\frac{1}{(1-\xi_1)^2} - \frac{1}{(1+\xi_1-\xi_2)^2}\right] \tag{17}$$

$$f_2(\xi_1,\xi_2) = \frac{1}{4}\left[\frac{1}{(1+\xi_1-\xi_2)^2} - \frac{1}{(1+\xi_2)^2}\right] \tag{18}$$

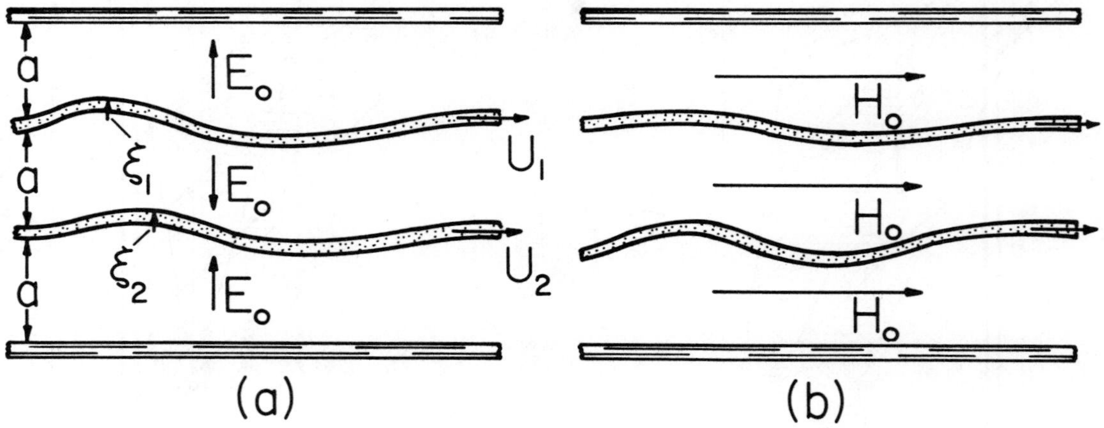

Fig. 11.10.4. Prototype two-stream configurations. (a) Potential constrained EQS configuration. (b) Flux constrained MQS configuration.

With variables defined as in Eq. 4 and normalized as suggested by the single-stream model, these equations are written as four first-order expressions:

$$(\frac{\partial}{\partial t} + M_1 \frac{\partial}{\partial z})v_1 + M_1(\frac{\partial}{\partial t} + M_1 \frac{\partial}{\partial z})e_1 - \frac{\partial e_1}{\partial z} = Pf_1(\xi_1,\xi_2) \quad (19)$$

$$(\frac{\partial}{\partial t} + M_2 \frac{\partial}{\partial z})v_2 + M_2(\frac{\partial}{\partial t} + M_2 \frac{\partial}{\partial z})e_2 - \frac{\partial e_2}{\partial z} = Pf_2(\xi_1,\xi_2) \quad (20)$$

$$\frac{\partial v_1}{\partial z} - \frac{\partial e_1}{\partial t} = 0 \quad (21)$$

$$\frac{\partial v_2}{\partial z} - \frac{\partial e_2}{\partial t} = 0 \quad (22)$$

Again, for the EQS system, $P > 0$ while for the MQS system, $P < 0$.

Two-Stream Characteristics: The same determinant approach used to find the single-stream characteristics can be applied to Eqs. 19-22. However, it is more convenient to recognize that the only coupling between streams is through the inhomogeneous terms. Thus, in view of Eqs. 7 and 8 found for a single stream, the characteristics are just what they would be for the individual streams with the inhomogeneous terms appropriately altered. Thus

$$dv_1 + de_1(M_1 \mp 1) = Pf_1(\xi_1,\xi_2)dt \quad (23)$$

on

$$\frac{dz}{dt} = M_1 \pm 1; \quad (C_1^{\pm}) \quad (24)$$

and

$$dv_2 + de_2(M_2 \mp 1) = Pf_2(\xi_1,\xi_2)dt \quad (25)$$

on

$$\frac{dz}{dt} = M_2 \pm 1; \quad (C_2^{\pm}) \quad (26)$$

The solution at some position, E, when $t = t + \Delta t$ is now determined by the response at positions A,B,C and D on the respective characteristics when $t = 1$, as illustrated in Fig. 11.10.5.:

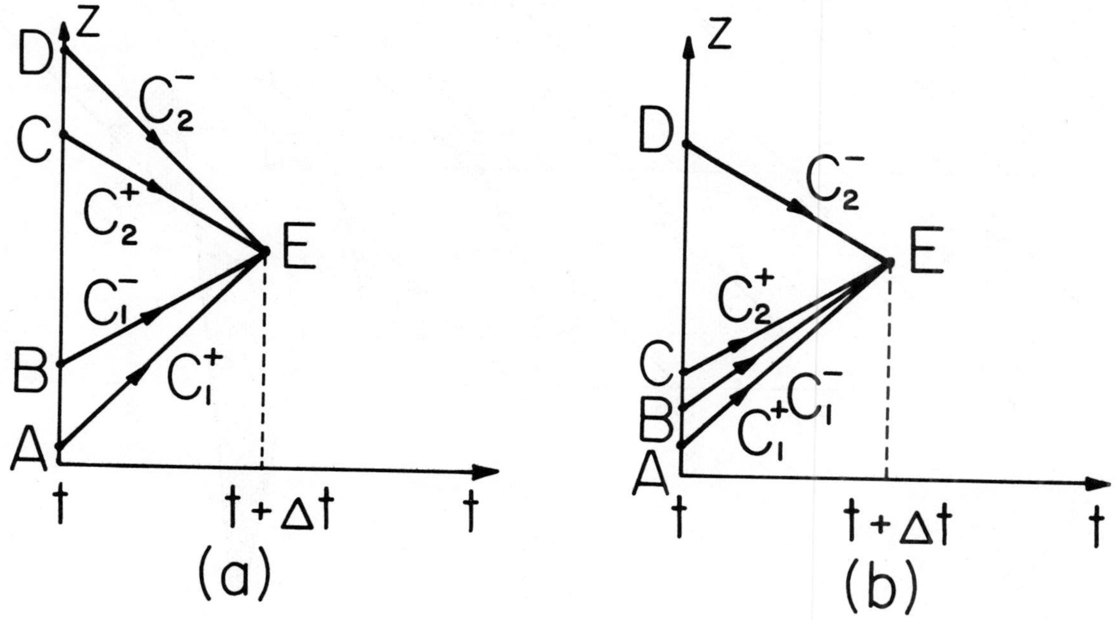

Fig. 11.10.5. Characteristics in the z-t plane illustrating (a) "super" counter-streaming and (b) "super" stream-structure interactions.

Just as Eqs. 13 and 14 follow from Eqs. 7, Eqs. 23 imply that the changes in v_1 and e_1 along the C_1^+ characteristic from A to E are given by

$$\Delta v_{1A}^{+} = \frac{1}{2}\{(v_{1A}-v_{1B})(M_1-1)+(e_{1A}-e_{1B})(M_1^2-1)+P[f_{1A}(M_1+1)-{}_{1B}(M_1-1)]\Delta t\} \tag{27}$$

$$\Delta e_{1A}^{+} = -\frac{1}{2}\{(v_{1A}-v_{1B})+(e_{1A}-e_{1B})(M_1+1)+P(f_{1A}-f_{1B})\Delta t\} \tag{28}$$

Similarly, from Eqs. 25, changes in v_2 and e_2 along C_2^+ from C to E are as given by these equations with $1 \rightarrow 2$, $A \rightarrow C$ and $B \rightarrow D$. As before, numerical integration of e_1 and e_2 gives the distributions of ξ_1 and ξ_2. The lower limits of integration in Eq. 9 should be made consistent with the entrance conditions on the respective streams.

<u>Two-Stream Initial Value Problem:</u> Given the initial distributions of ξ_1, v_1, ξ_2 and v_2, the evolution of these variables with time can be determined numerically, much as for the single streams. Given that P can be positive or negative (the two configurations of Fig. 11.10.4) and that M_1 and M_2 can be greater or less than unity (each stream can be "super" or "sub") and can be negative or positive (streaming in either direction), it is clear that there are now many physical interactions that might be considered. The super counter-streaming and super stream-structure interactions illustrated in Fig. 11.10.5 perhaps add the most physical insight.

The characteristics alone make it clear that with counter-streaming "super" streams it is possible to have an absolute instability. With $P > 0$, this instability has much the same character as for the single "sub" stream. But what might be surprising is the instability that results even with $P < 0$. In this case of magnetic field coupling, the effect of the field on the single stream is to produce decaying oscillations. With two counter-streams, oscillations are fed from one stream to the other and then back to the point of origin with a phase shift. Thus, certain oscillations build up, as the numerically computed response of Fig. 11.10.6 illustrates. Note that the instability is unbounded at a fixed position along the z axis. It takes the form of an absolute instability, in that the displacement at a given z tends to grow with time.

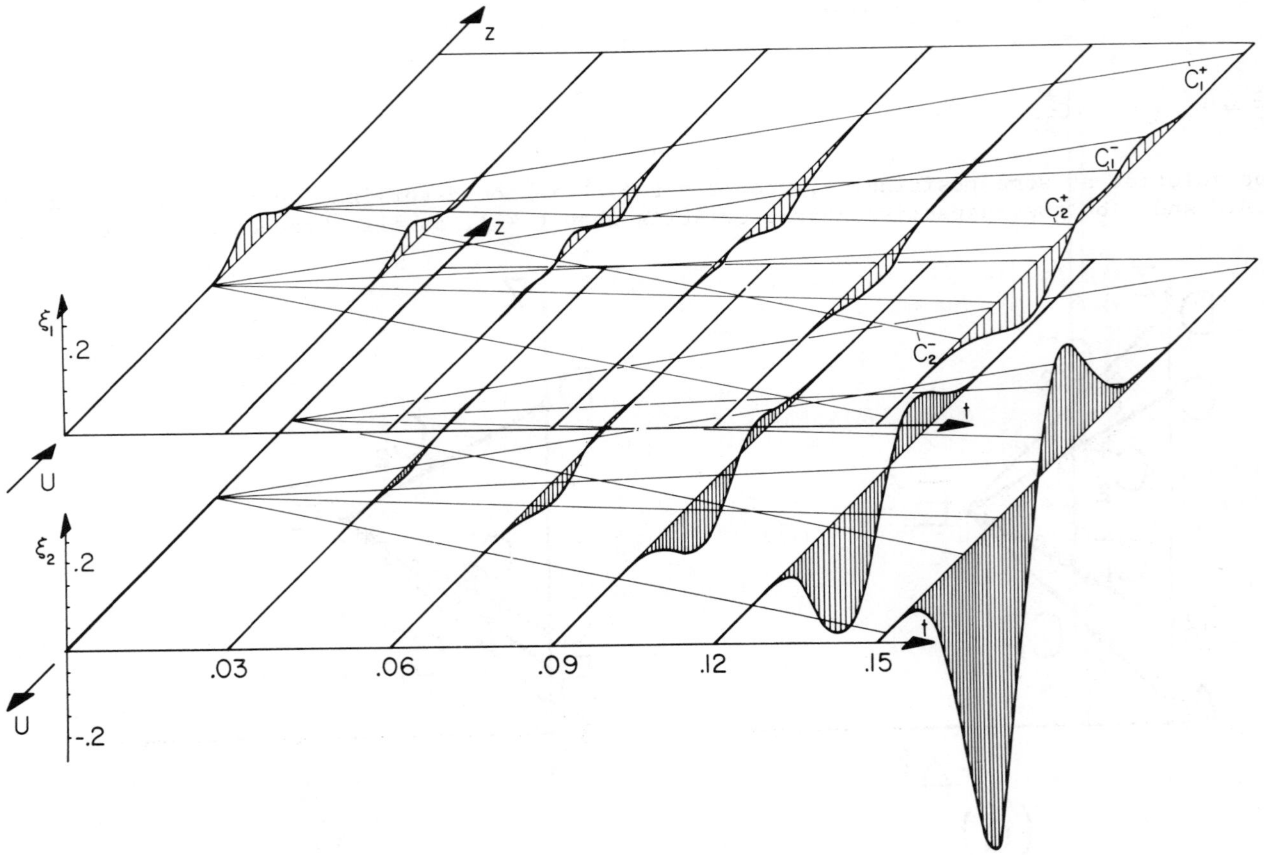

Fig. 11.10.6. Counter-streaming interactions between "super" streams coupled by magnetic field ($P = -1000$, $M_1 = 1.5$, $M_2 = -1.5$). Two-stream instability is in this case absolute.

Causality and Boundary Conditions: Any real system is of course bounded in the axial direction. One merit of the characteristic viewpoint is that causality is implicit to a specification of the conditions imposed to account for these boundaries. The dynamics unfold along the characteristic lines, always proceeding forward in time. Boundary conditions must be consistent with this requirement.

Consider first the boundary conditions for the single-stream configurations. The differential equation of motion is second-order in the longitudinal coordinate, so two conditions are required. Mathematically, these could be both imposed at $z = 0$, both at a downstream position $z = \ell$, or one at each position. But which of these is consistent with having a causal relation between the response and the initial conditions depends on whether the stream is "super" or "sub."

If the stream is supercritical, both families of characteristics are directed downstream, as illustrated in Fig. 11.10.7a. As time goes on, the response at C that depends on the initial conditions between A and B becomes one at C' that depends on both initial conditions and conditions at the upstream boundary. Finally, at points such as C", the response is fully determined by the boundary conditions at the entrance. Periodic entrance conditions clearly result in a temporally periodic reponse. The supercritical boundary conditions are equivalent to initial conditions and the response is found following the same line of reasoning as illustrated for the initial value problems. Two boundary conditions must be imposed at the upstream boundary, but none are imposed on the region $0 < z < \ell$ by the downstream boundary.

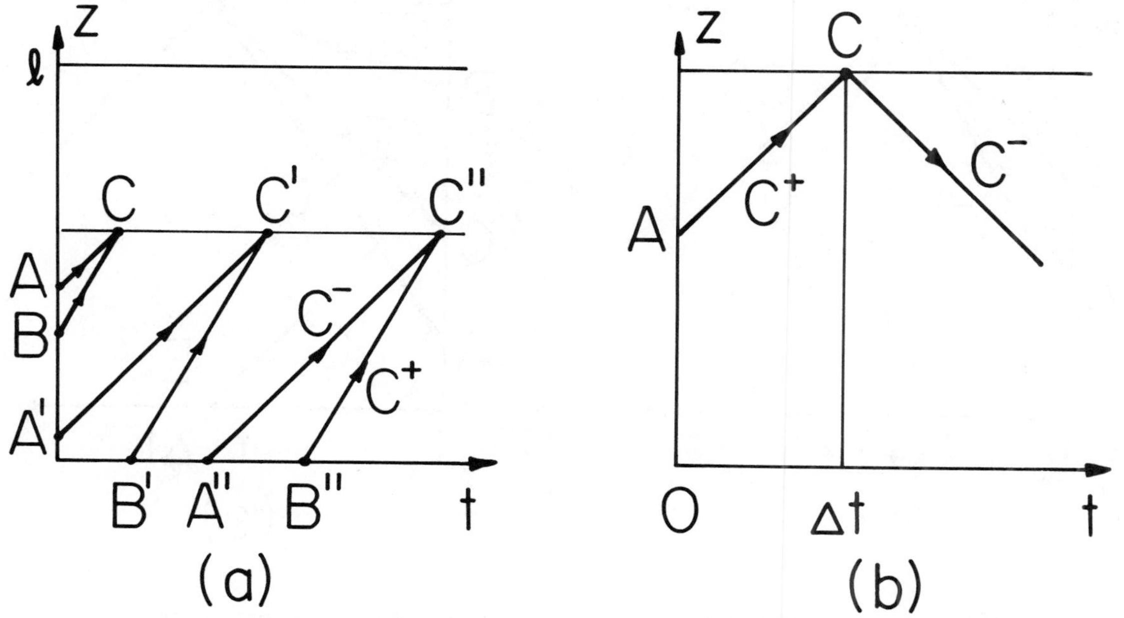

Fig. 11.10.7. Boundary conditions consistent with causality. (a) Supercritical characteristics implying two upstream conditions and none downstream, (b) Subcritical flow with one condition at each extreme.

By contrast, if the flow is subcritical, as illustrated by Fig. 11.10.7b, two conditions at either boundary leave the representation over-specified, and one condition must be imposed at each boundary. To see this, consider how the solution in the neighborhood of the downstream boundary would be found using the characteristics. To find the solution when $t = \Delta t$, the procedure is as already outlined except for the end points, like C of Fig. 11.10.7b. At this boundary point, $v = v_C$ is stipulated (say). Thus, because v_A is also known from the initial conditions, the change in v along the C^+ characteristic incident on the boundary, Δv_A^+, follows from Eq. 11a. From the second characteristic equation along C^+, Eq. 12c, the value of Δe_A^+ is then determined and hence e at the boundary, e_C, is found. Thus, e cannot be independently specified as a boundary condition. With the variables determined at the boundaries in this fashion, the stage is set for repeating the process to determine the solution when $t = 2\Delta t$.

Of course, two boundary conditions can be arbitrarily imposed, say two upstream conditions in a subcritical flow. But it is clear that the resulting solution answers the question, what initial conditions are required to make the solution satisfy the desired subsequent conditions at the boundaries? Boundary conditions are usually intended as statements made in advance to predict future events. If not causal, they place requirements on what must have taken place before to have certain conditions at the boundaries now.

Consider now causal boundary conditions for the two-stream configurations. The system is now fourth-order in z and therefore four boundary conditions are required. With the understanding that characteristics have a direction determined by increasing time, one condition must be imposed where each family of lines enters the region of interest.

As an example, consider again the supercritical counter-streaming configuration. The characteristics $C_1^{\pm}$ enter at $z = 0$ while the $C_2^{\mp}$ characteristics enter at $z = \ell$. To see that the imposition of these conditions is consistent with marching forward in time, consider how the solution is determined when $t = t + \Delta t$, given the solution when $t = t$. Provided that Δz and Δt are selected so that the characteristics passing through every interior point on the grid when $t = t + \Delta t$ pass through the line $t = t$ in the interval $0 < z < \ell$, the solution at each of the interior points is found by the same procedure as for the initial value problem. The solution at an end point, such as E in Fig. 11.10.8a, is then found by using Eqs. 27 and 28 to find v_1 and e_1 at E. The boundary conditions provide the values of v_2 and e_2 at E. In this way, the response is determined over the entire interval, including the end points, and the stage is set for the next time step. The example of Fig. 11.10.6 is in fact computed taking into account boundary conditions $\xi_1(0,t)$, $v_1(0,t)$, $\xi_2(\ell,t)$, $v_2(\ell,t)$ all zero. However, time has not progressed far enough to make these conditions significant.

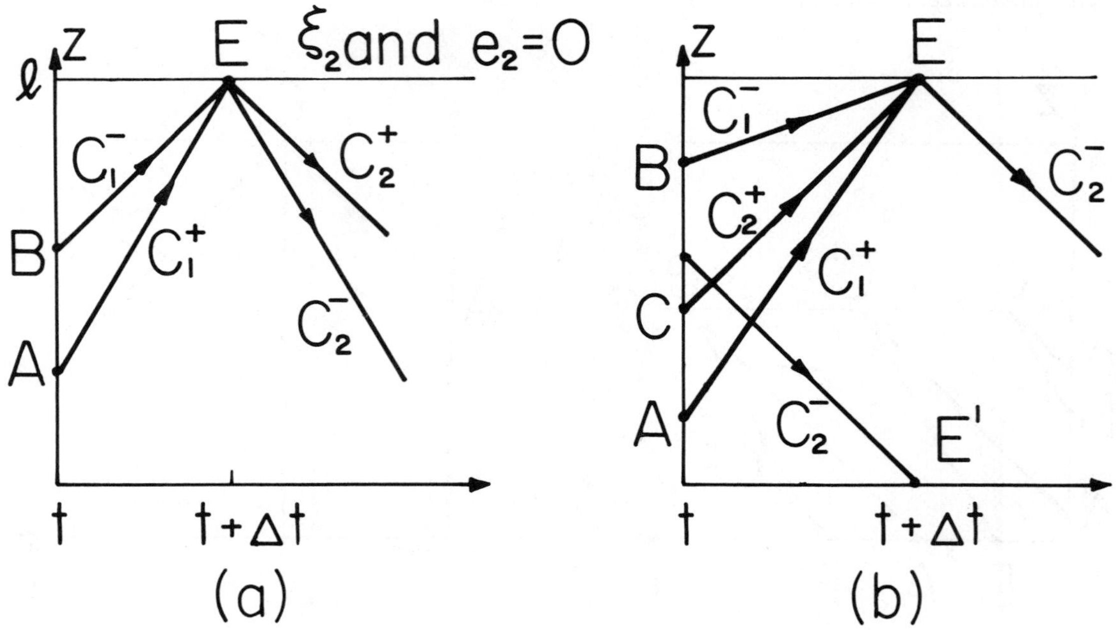

Fig. 11.10.8. (a) Downstream boundary of counter-streaming configuration at which two conditions, on ξ_2 (and hence v_2) and e_2, are imposed.

It is because coupling between characteristics for the two streams occurs only through the inhomogeneous terms that this simple procedure takes into account boundary conditions on the countersteaming supercritical streams.

In Fig. 11.10.8b, the stream-structure interaction makes more evident what is in general required. Stream (1) is supercritical while (2) is not only subcritical but is not streaming at all. To find the response at a downstream boundary, like point E of Fig. 11.10.8b, Eqs. 27 and 28 again provide (v_1, e_1) at E. One boundary condition is imposed, say v_2 is given. Because v_2 at C is also known, Δv_{2C}^+ follows. In turn, Eq. 25a (the second characteristic equation on C_2^+) can be used to solve for Δe_{2C}^+. Thus, e_{2E} is determined and all conditions at E are known.

At the upstream boundary, v_1 and e_1 are imposed as is a third condition, say that v_2 is known. From this last condition the second characteristic equation along C_2^+ can be used to determine e_2 at E'. Again, all conditions at the end points of the grid when $t = t + \Delta t$ are established and the stage is set for the next iteration.

In the absence of longitudinal boundary effects, the "super-sub" streaming interaction with $P > 0$ is convectively unstable. That is, the response to initial conditions at a fixed position is bounded in time. In this case, boundary conditions have a profound effect. Those just described turn the convective instability into an absolute one that builds up in an oscillatory fashion.

11.11 Second Order Complex Waves

The remaining sections in this chapter continue a subject begun in Sec. 5.17. There, Fourier transforms are used to represent spatial transients in terms of the sinusoidal steady-state spatial modes. The example considered there, of charge relaxation on a moving sheet, is typical of a wide class of linear systems that are uniform in the longitudinal direction, z, and excited from transverse boundaries, perhaps over an interval $0 < z < \ell$. In Sec. 5.17, the resulting temporal sinusoidal steady state consists of responses, shown in Fig. 5.17.8, that spatially decay upstream and downstream from this range. These are a superposition of the appropriate spatial modes. Within the excitation range, the response is also a superposition of spatial modes. But in addition, in this range there is the driven response having not only the same temporal frequency as the drive, but the same wavenumber as well. The Fourier transform provides a formalism for "splicing" the modes and driven response together in the planes $z = 0$ and $z = \ell$.

As pointed out in Sec. 5.17, there are two questions left unanswered in the process of finding the spatial transient. First, it is assumed there that the sinusoidal steady-state complex waves decay away from the excitation region. Thus, those spatial modes having positive imaginary k are excluded from the downstream range $\ell < z$, while those with negative imaginary parts are left out in the region $z < 0$. The examples introduced in this section include the possibility that the downstream response in fact grows with increasing z. In Sec. 11.12, the objective is to have a means of distinguishing such amplifying waves from those that are evanescent, or decay away from the drive.

Any discussion of a sinusoidal steady state is predicated on having an answer to the second question. Is the system absolutely stable, in the sense that the response is bounded with increasing time at a fixed location in space? Only then will the temporal sinusoidal steady state have a chance to establish itself. The difficulty here is that a system is not necessarily absolutely unstable even if it displays temporal modes with negative imaginary frequencies. Temporal modes, having frequencies given by the dispersion equation evaluated using real values of the wavenumber, are the response to initial conditions that are spatially periodic. These extend from $z = -\infty$ to $z = +\infty$. Thus, in an infinite system, temporal modes do not suggest whether the instability grows with time at a fixed location, z (absolute instability), or rather grows only for an observer that moves with some velocity in the z direction (convective).

The identification of an absolute instability is taken up in Sec. 11.13.

Second Order Long-Wave Models: It is the purpose of this section to set the stage for the next two sections. Although the ω-k picture of the evolution of a system in space and time is widely applicable, it is helpful to have in mind simple situations that make it possible to establish a physical rapport for what the mathematics represents. In Fig. 11.11.1, sheets of liquid stream in the z direction between plane-parallel plates or electrodes. (These same configurations are considered from another point of view in Sec. 11.10.) With the stream in steady equilibrium there is no transverse deflection, ξ, and there are uniform fields between the plane-parallel perfectly conducting walls and the sheets, as shown. Over the range $0 < z < \ell$, the transverse boundaries are driven either by electric or magnetic potentials superimposed on the uniform bias potentials which give rise to the equilibrium fields.

The models now developed highlight the dominant modes of systems actually having an infinite number of spatial modes. The higher order modes that are left out of the models come into play if "wavelengths" of interest are as short as the spacing, a, or the sheet thickness, Δ. The magnetic configuration is the antidual of the electric one, as defined in Sec. 8.5. Thus, the equation of motion follows directly from the electric case now derived, with $\varepsilon_o \rightarrow \mu_o$, $E_o \rightarrow H_o$, $\Phi \rightarrow A/\mu_o$.

With the stream modeled as a membrane having a tension twice that due to the surface tension, the equation of motion is Eq. 11.10.2. Taking into account the excitation potentials, the transverse surface force density in the long-wave limit is simply the difference between the electric stresses acting on the top and the bottom of the sheet:

$$T = \frac{1}{2}\varepsilon_o\left[\frac{(-E_oa + \Phi_d)^2}{(a - \xi)^2} - \frac{(-E_oa - \Phi_d)^2}{(a + \xi)^2}\right] \qquad (1)$$

For small amplitudes of the drive and response,

$$T \simeq -\left(\frac{2\varepsilon_oE_o}{a}\right)\Phi_d + \left(\frac{2\varepsilon_oE_o^2}{a}\right)\xi \qquad (2)$$

This expression is now used to complete the transverse force equation, Eq. 11.10.2, which takes the

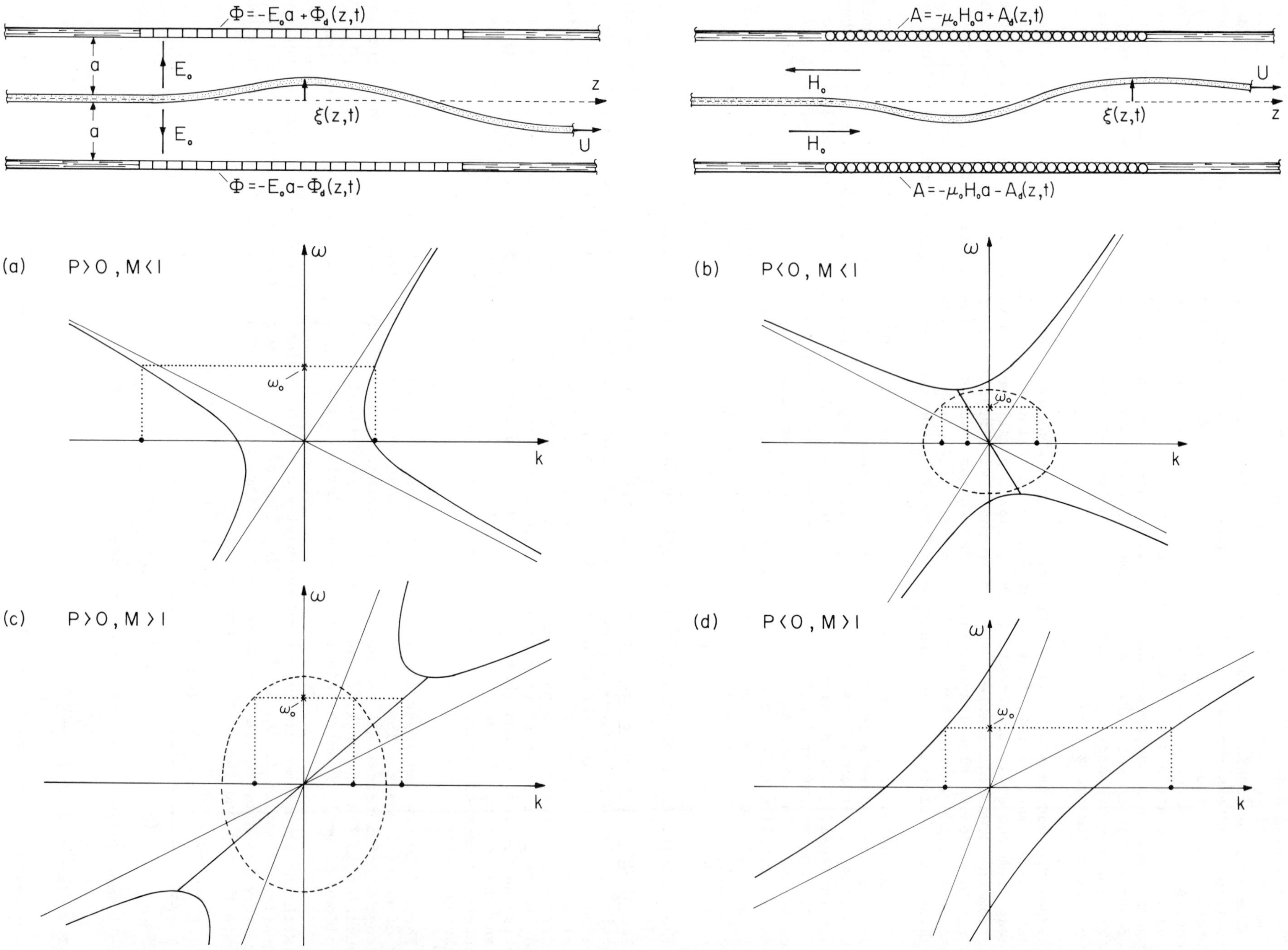

Fig. 11.11.1. Electric and magnetic long-wave models that are described by Eq. 3 together with dispersion relations. Complex values of k are shown for real values of ω: k_r ——, k_i ----. (a) Subcritical electric, (b) subcritical magnetic, (c) supercritical electric, and (d) supercritical magnetic.

normalized form ($\xi = \underline{\xi}a$, $t = \underline{t}\tau$, and $z = \underline{z}\tau V$):

$$\left(\frac{\partial}{\partial t} + M\frac{\partial}{\partial z}\right)^2 \xi = \frac{\partial^2 \xi}{\partial z^2} + P\xi - Pf(z,t) \qquad (3)$$

The membrane wave velocity and Mach number are defined as

$$V = \sqrt{\frac{2\gamma}{\Delta\rho}}; \quad M = \frac{U}{V}$$

and "pressure" parameters and forcing functions making the equation applicable to the electric and magnetic configuration, respectively, are

$$P = \frac{2\varepsilon_o E_o^2 \tau^2}{\Delta\rho a} \quad \Bigg| \quad P = -\frac{2\mu_o H_o^2 \tau^2}{\Delta\rho a} \qquad (4)$$

$$\underline{f} = \frac{\Phi_d}{aE_o} \quad \Bigg| \quad \underline{f} = \frac{A_d}{\mu_o a H_o} \qquad (5)$$

Thus, with $P > 0$ the configuration is electric, and the self-field part of the transverse force is destabilizing. That is, a deflection results in a field intensification and hence a transverse force on the stream tending to further increase the deflection. In the magnetic field configuration, $P < 0$, and it is as though there were a continuum of magnetic springs between the stream and the walls. A deflection leads to a force tending to return the stream to its equilibrium.

Whether the underlying method of characteristics from Sec. 11.10 has been followed or not, it is useful at this point to review the response to initial conditions, shown in Fig. 11.10.3, for these streaming configurations. It is the Mach number, M, that determines if the initial pulse can propagate upstream. For $M < 1$, wave fronts propagate in both directions, whereas if $M > 1$, the entire response is washed downstream.

If P is positive, the amplitude becomes unbounded. But, whether the growth is at a fixed location or for an observer moving with some velocity depends on M. Thus, in this electric case, the infinite system is absolutely unstable if $M < 1$ and convectively unstable if $M > 1$.

If P is negative, the response consists of forward and backward waves. The magnetic field results in their leaving an oscillation in the region between. If $M < 1$, this oscillation decays with time at a fixed location, while if $M > 1$, the response falls abruptly to zero as the wave front that is trying to propagate upstream is swept downstream.

With these predispositions as to what should be expected, consider now the representation of the dynamics in terms of complex waves.

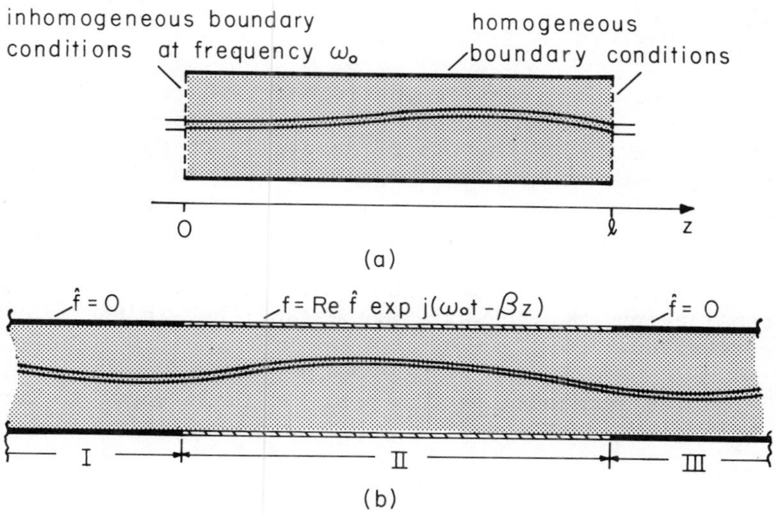

Fig. 11.11.2. Typical systems that are uniform in the z direction; (a) bounded longitudinally by planes where boundary conditions are imposed; (b) unbounded in the z direction with drive from a transverse boundary over the interval $0 < z < \ell$.

<u>Spatial Modes</u>: Consider first the response to excitations that are in the sinusoidal steady state, having a frequency $\omega = \omega_o$. Because they involve the same manipulations, but contrasting issues, two types of problems are now considered. In the first, the system is bounded by the planes $z = 0$ and $z = \ell$. The transverse boundaries are not driven. Rather, the drive is through one of the longitudinal boundary conditions which varies at the angular frequency ω_o.

The second type of problem is one extending from $z = -\infty$ to $z = +\infty$ with the excitation from the transverse boundaries over the range $0 < z < \ell$. These bounded and unbounded situations, pictured in Fig. 11.11.2, are similar enough that they are now considered at the same time.

To be consistent with normalization of Eq. 3, deflections are taken to be of the form

$$\xi = \text{Re}\,\hat{\xi}e^{j(\omega t - kz)} \qquad (6)$$

where frequencies and wavenumbers are normalized such that $\omega\tau = \underline{\omega}$, $k\tau V = \underline{k}$. The inhomogeneous solution to Eq. 3, caused by the drive on the transverse boundaries and applying over the range $0 < z < \ell$, follows by substituting this form of solution into Eq. 3 with $\omega = \omega_o$ and $k = \beta$. Solving for $\hat{\xi}$ gives

$$\hat{\xi} = \frac{P\hat{f}}{D(\omega_o,\beta)} \tag{7}$$

where the dispersion function is

$$D(\omega_o,\beta) = (\omega_o - M\beta)^2 - \beta^2 + P \tag{8}$$

For solutions of the form of Eq. 6 to satisfy the homogeneous form of Eq. 3, k must satisfy the dispersion equation $D(\omega,k) = 0$. Thus, with amplitudes $\hat{A}$ and $\hat{B}$ at this point arbitrary, the solution over any range of z is

$$\xi = \text{Re}\left[\frac{P\hat{f}e^{-j\beta z}}{D(\omega_o,\beta)} + \hat{A}e^{-jk_1 z} + \hat{B}e^{-jk_2 z}\right]e^{j\omega_o t} \tag{9}$$

where

$$D(\omega,k) = (\omega - Mk)^2 - k^2 + P \tag{10}$$

Thus, the wavenumbers k_1 and k_2 in Eq. 8 are given by solving the quadratic expression $D(\omega_o,k) = 0$:

$$k_{\substack{1\\-1}} = \eta \pm \eta \tag{11}$$

where

$$\eta \equiv \frac{\omega_o M}{M^2 - 1}, \quad \gamma \equiv \frac{\sqrt{\omega_o^2 + P(1 - M^2)}}{M^2 - 1}$$

As a graphical representation of the spatial modes, these roots are displayed in Fig. 11.11.1. For a particular driving frequency ω_o, the roots of $D(\omega_o,k)$ are represented by the intersections with the horizontal line. The solid curves indicate the real part, k_r, while the broken lines are the imaginary part, k_i. Where one k is shown, it is in common to both roots.

Four possibilities are distinguished in Fig. 11.11.1. The configuration can be electric or magnetic ($P > 0$ or $P < 0$) and it can be subcritical or supercritical ($|M| < $ or $|M| > 1$). As can be seen from Eq. 1, two of the four have ranges of frequency over which the wavenumbers are complex:

$$\omega_o^2 < P(M^2 - 1) \tag{12}$$

One is the subcritical magnetic case, where $P < 0$ and $|M| < 1$. The other is the supercritical electric case, where $P > 0$ but $|M| > 1$. In each of these, one spatial mode apparently "grows" with increasing z while the other "decays." Of course, in the magnetic subcritical case the spatial mode that appears to grow in the z direction is really an evanescent mode decaying upstream from a downstream drive. The supercritical electric case actually does involve a wave that is amplifying in the z direction as it moves away from an upstream source.

Section 11.12 shows how the distinction can be made between evanescent and amplifying waves by considering how the waves are established in the sinusoidal steady state subsequent to turning on the excitation.

The remainder of this section is intended to develop a physical understanding of evanescent and amplifying waves and of absolute and convective instabilities.

Driven Response of Bounded System: If $|M| < 1$, boundary conditions can be imposed at $z = 0$ and $z = \ell$. That these conditions are consistent with causality can be established by the method of characteristics (Sec. 11.10), or by using the arguments of the next section to determine that one spatial mode propagates in the +z direction (and hence can be used to satisfy a boundary condition at $z = 0$), while the other propagates in the -z direction (and can be used to satisfy the condition at $z = \ell$). As an example, suppose that the sheet is given a sinusoidally varying excitation at $z = 0$ and fixed at $z = \ell$. Also, make the transverse boundary excitation zero, so $f = 0$. Then, the coefficients $\hat{A}$ and $\hat{B}$ in Eq. 9 are determined and the solution becomes: (Note that in the normalized expression, $\underline{\ell} = \ell/\tau V$.)

$$\xi = -\text{Re}\,\hat{\xi}_d \frac{\sin\gamma(z - \ell)}{\sin\gamma\ell} e^{j(\omega_o t - \eta z)} \tag{13}$$

If the frequency is below the cutoff frequency, Eq. 12, γ is imaginary and the evanescent nature of the response is made more evident by writing Eq. 13 as

$$\xi = -\mathrm{Re}\,\hat{\xi}_d \frac{\sinh|\gamma|(z-\ell)}{\sinh|\gamma|\ell} e^{j(\omega_o t-\eta z)} \tag{14}$$

Some features of this steady-state response are illustrated by the experiment shown in Fig. 11.11.3. Here, the sheet is replaced by a wire under tension. In the absence of a magnetic force, it too has a deflection described by the wave equation. There is no longitudinal motion, so $M = 0$ and hence $\eta = 0$ in Eqs. 13 and 14. By passing a current through the wire and imposing a magnetic field that is all gradient along its zero-deflection axis, a magnetic force is produced that is proportional to ξ. This force tends to restore the undeflected wire to its original position. The configuration is described in Prob. 11.11.2, where it is shown that the equation of motion is again Eq. 3 with $P < 0$ and $M = 0$.

In the first picture of the sequence, the current is zero and what is seen is the standing wave resulting from the interference of two oppositely propagating ordinary waves. (In these pictures, the z direction is to the left, so the excitation is to the right.) The frequency is such that the wire is very nearly in the lowest resonance condition that prevails if $\gamma\ell = n\pi$. As the current is raised, the magnetic force tends to counteract the inertial force (that makes the wire bow outward). The current is reached where these forces just balance, and the deflections decay away from the excitation. The rate of decay is largest at zero frequency (a static deflection).

Consider next the dramatic effect of having the continuum not only stream, but be supercritical, so that $|M| > 1$. Then, two boundary conditions must be imposed at the inlet, where $z = 0$, and none that influence matters in the range of interest are imposed at the exit. For example, the deflection is again the sinusoidal one assumed before, but the spatial derivative is constrained to be zero. Then, the coefficients $\hat{A}$ and $\hat{B}$ are determined and the solution is

$$\xi = \mathrm{Re}\,\hat{\xi}_d \frac{(k_1 e^{j\gamma z} - \underline{k}_1 e^{-j\gamma z})}{2\gamma} e^{j(\omega_o t-\eta z)} \tag{15}$$

The case of most interest has the electric configuration of Fig. 11.11.1 as a prototype and hence $P > 0$. If P is raised high enough that $\omega_o^2 < P(M^2 - 1)$, γ is imaginary, and the space-time picture of the deflections given by Eq. 15 is more apparent if it is written in the form

$$\xi = -\mathrm{Re}\,\hat{\xi}_d \frac{(\underline{k}_1 e^{|\gamma|z} - k_1 e^{-|\gamma|z})}{2\gamma} e^{j(\omega_o t-\eta z)} \tag{16}$$

In the case of this supercritical stream, a demonstration is made by letting the continuum be a jet of water, with capillarity providing the (surface) tension (see Prob. 11.11.3). The drive is provided by spherical electrodes positioned just upstream of ($z = 0$) on each side of the stream and biased by a constant potential relative to the stream with a superimposed sinusoidally varying voltage having the angular frequency ω_o.

With $P = 0$, so that γ is real, the response is illustrated in Fig. 11.11.4. (Again, streaming is from right to left with the excitation at the right.) The fast and slow waves carried downstream by the convection interfere to form "beats." That is, the envelope of the deflection is a standing wave having wavelength $2\pi/\gamma$. In Fig. 11.11.4b, the frequency has been raised to the point where about one half-wavelength of the envelope appears within view. In a slow motion picture (Complex Waves II, Reference 11, Appendix C), the phases propagate through this envelope with velocity ω_o/η.

With a field applied to the jet, the kinking motions of the jet are very similar to those of the planar sheet. Thus, raising the voltage is equivalent to raising P, and has the effect on the dispersion equation and jet that is illustrated in Fig. 11.11.5. Over most of its length, the response described by Eq. 16, is dominated by the growing exponential. Again phases have the velocity ω_o/η with the exponentially growing envelope.

Instability of Bounded Systems: The importance of imposing boundary conditions that are consistent with causality is made dramatically evident by considering implications for stability of correct and incorrect conditions. For a bounded system, it is not meaningful to envision a convective instability. Once boundary conditions have been imposed, there remains only the possibility of an absolute instability in the response to initial conditions. This transient response is represented by a superposition of modes satisfying homogeneous transverse and longitudinal boundary conditions ($\hat{f}$ and $\hat{\xi}_d \to 0$).

For the subcritical system, it can be seen from Eq. 13 that the eigenvalues for these modes are

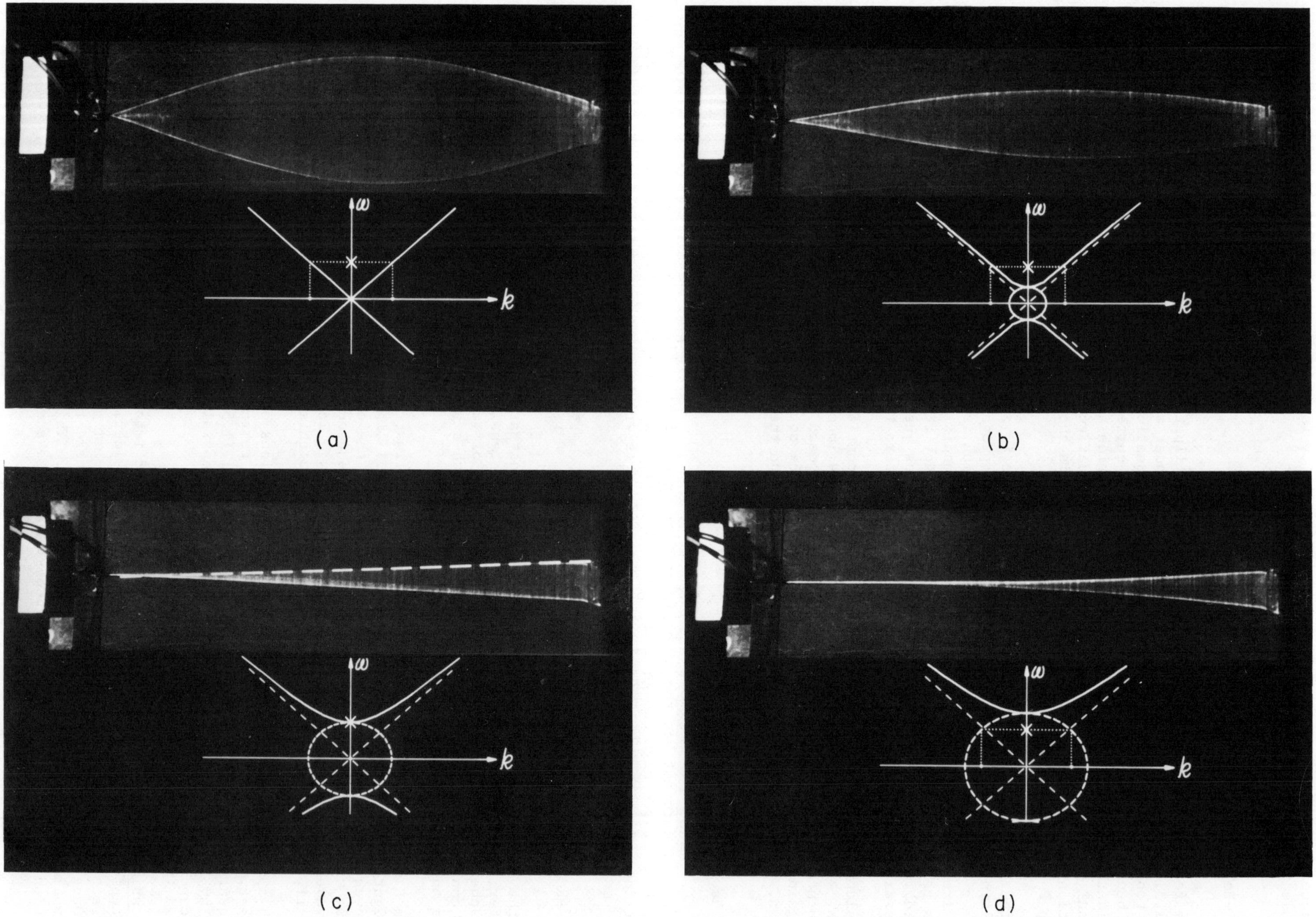

Fig. 11.11.3. Time exposure of sinusoidal steady state for a "string" represented by Eq. 3 with $M = 0$, $P < 0$ and the excitation at the right. (a) Magnetic field is off ($P = 0$) and frequency is such that response is near the first resonance. (b) Magnetic field raised somewhat, but frequency still above cutoff. (c) Field raised to cutoff condition. (d) Evanescent response resulting as field is raised to a point where frequency is less than cutoff. (From Complex Waves I, Reference 11, Appendix C.)

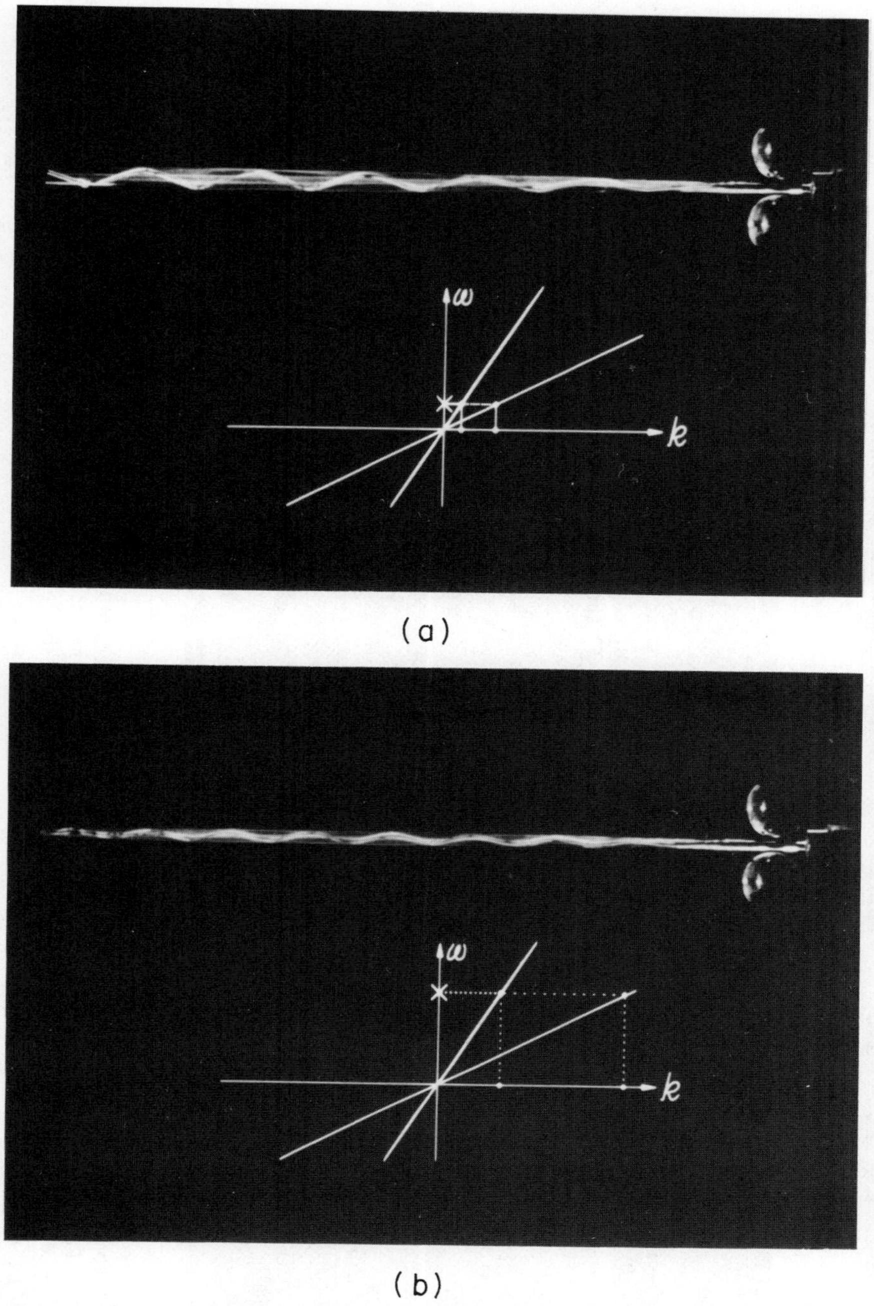

Fig. 11.11.4

Supercritical stream ($M > 1$) with no field ($P = 0$) and excitation to the right. Raising the frequency just brings one half-wave-length of "beat" into view. (From Complex Waves II, Reference 11, Appendix C.)

given simply by

$$\sin \gamma \ell = 0 \Rightarrow \gamma = n\pi/\ell, \; n = 1,2,3\cdots \tag{17}$$

In evaluating this expression, using the definition of γ given by Eq. 11, the frequency is now the eigenfrequency, conveniently represented here as $j\omega \rightarrow s_n$. Thus,

$$s_n^2 = -(\frac{n\pi}{\ell})^2(M^2 - 1)^2 + P(1 - M^2) \tag{18}$$

Because $|M| < 1$, it follows that P must be positive if there is to be instability. As P is raised, the $n = 1$ mode is the first to become unstable and that occurs if

$$P = (\frac{\pi}{\ell})^2(M^2 - 1) \tag{19}$$

At this threshold, the deflection has a shape given by Eq. 13, with an envelope having the shape of a half-wave of a sinusoid. This instability is illustrated in the limit $M \rightarrow 0$ in Complex Waves II (Reference 11, Appendix C).

Consider the consequence of an unjustified use of Eq. 18. Suppose that it is valid for the supercritical case, $|M| > 1$. It would then be concluded that the system is unstable with P made sufficiently negative (the magnetic case in Fig. 11.11.1). Of course, with $|M| > 1$, one boundary condition underlying the identification of these eigenmodes is not consistent with causality. From

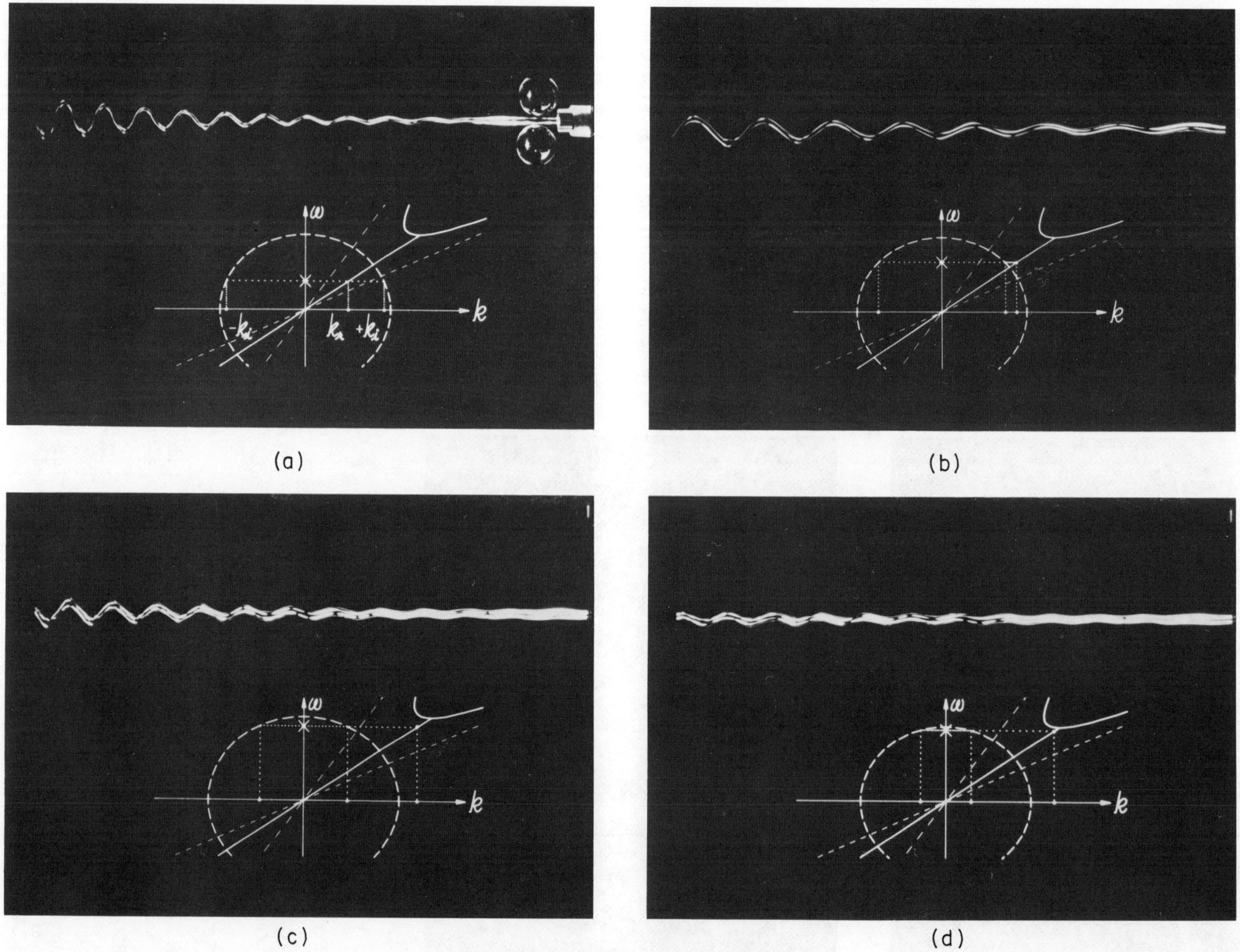

Fig. 11.11.5. Liquid jet in electric field is excited in the sinusoidal steady state by the spheres to the right. Amplifying waves, modeled by Eq. 3 with $P > 0$ and $M > 1$. Parts (b), (c), and (d) show the effect of raising the excitation frequency almost to the cutoff frequency. (Strictly, waves are short enough here to make the long-wave model of questionable validity. From Complex Waves II, Reference 11, Appendix C.)

the correct solution, Eq. 15, it is clear that in this supercritical case there are no eigenmodes, never mind modes that are unstable.

Driven Response of Unbounded System: Consider now the sinusoidal steady-state response to a drive from transverse boundaries in the unbounded configuration, Fig. 11.11.2b. For the quasi-one-dimensional model, solutions are piecewise continuous in the z direction. In regions I and III there is no drive and hence f = 0, while in region II there is a drive. With an appropriate assignment of f, the general solution, Eq. 8, can be applied to each region. There are two coefficients, $\hat{A}$ and $\hat{B}$, associated with each region. These represent the amplitudes of the spatial modes and are determined by boundary conditions at infinity and by the conditions prevailing where the regions meet at z = 0 and $z = \ell$.

A picture of the sinusoidal steady-state response in the four regimes illustrated in Fig. 11.11.1 is given in Fig. 11.11.6. First, consider the subcritical situations. Here, boundary conditions must exist at $z \to \infty$ and $z \to -\infty$ that have an effect on the asymptotic response. So long as the waves are propagating (P > 0 and P < 0 but the frequency above cutoff), it is necessary to specify conditions at infinity. One such specification might be a "radiation condition," which requires that boundaries are far enough removed that waves reflecting their presence have not returned to the region of excitation, or that these boundaries absorb the incident wave without there being any reflected wave. In either case, for $|M| < 1$, the response is

$$\xi = \text{Re}\, e^{j\omega_o t}\begin{cases} \hat{A}_I e^{-jk_{-1}z}; & z < 0 \\ \dfrac{P\hat{f}e^{-j\beta z}}{D(\omega_o,\beta)} + \hat{A}_{II}e^{-jk_{-1}z} + \hat{B}_{II}e^{-jk_1 z}; & 0 < z < \ell \\ \hat{B}_{III}e^{-jk_1 z}; & \ell < z \end{cases} \tag{20}$$

where modes representing conditions at infinity have been excluded from regions I and III.

The four coefficients are determined by making the displacement and its spatial derivative piecewise continuous. That is, requiring that ξ and $\partial\xi/\partial z$ be continuous at z = 0 and $z = \ell$ gives four conditions allowing the four amplitudes to be determined in terms of $\hat{f}$.

For P > 0 and for P < 0 with the driving frequency above cutoff, the response outside the excitation region consists of purely propagating waves. Thus, the envelope of the response in the exterior is constant in z. In the magnetic case, where P < 0, the response below the cutoff frequency consists of evanescent waves, as sketched in Fig. 11.11.6. As suggested by the general form of the solution, Eq. 20, in the excitation region the response is a sum of the spatial modes representing end effects and a driven response that has the same wavenumber as the drive. For operation below the cutoff frequency, the response in the mid-range of region II at distances removed by several decay lengths from the ends would be just the part having the same spatial periodicity as the drive. This type of behavior is familiar from Sec. 5.17, and also illustrated in detail by Fig. 5.17.8.

The case P > 0 and $|M| < 1$ is absolutely unstable. Sooner or later the response to initial conditions would dominate the sinusoidal steady state.

Now, consider the effect of having a supercritical stream, $|M| > 1$. The response in region I is entirely determined by the upstream boundary conditions. If those conditions are homogeneous, or that boundary is too far upstream to have had an effect during times of interest and initial conditions on the stream are zero, then the solution in region I is known to be zero. With this understanding, the response then takes the form

$$\xi = \text{Re}\, e^{j\omega_o t}\begin{cases} 0; & z < 0 \\ \dfrac{P\hat{f}e^{-j\beta z}}{D(\omega_o,\beta)} + \hat{A}_{II}e^{-jk_{-1}z} + \hat{B}_{II}e^{-jk_1 z}; & 0 < z < \ell \\ \hat{A}_{III}e^{-jk_1 z} + \hat{B}_{III}e^{-jk_{-1}z}; & \ell < z \end{cases} \tag{21}$$

The response continues to evolve in the direction of streaming. In region II, the amplitudes are fully determined by the requirement that ξ and $\partial\xi/\partial z$ be zero at z = 0. In turn, the downstream response in region III follows by requiring continuity of ξ and $\partial\xi/\partial z$ at $z = \ell$.

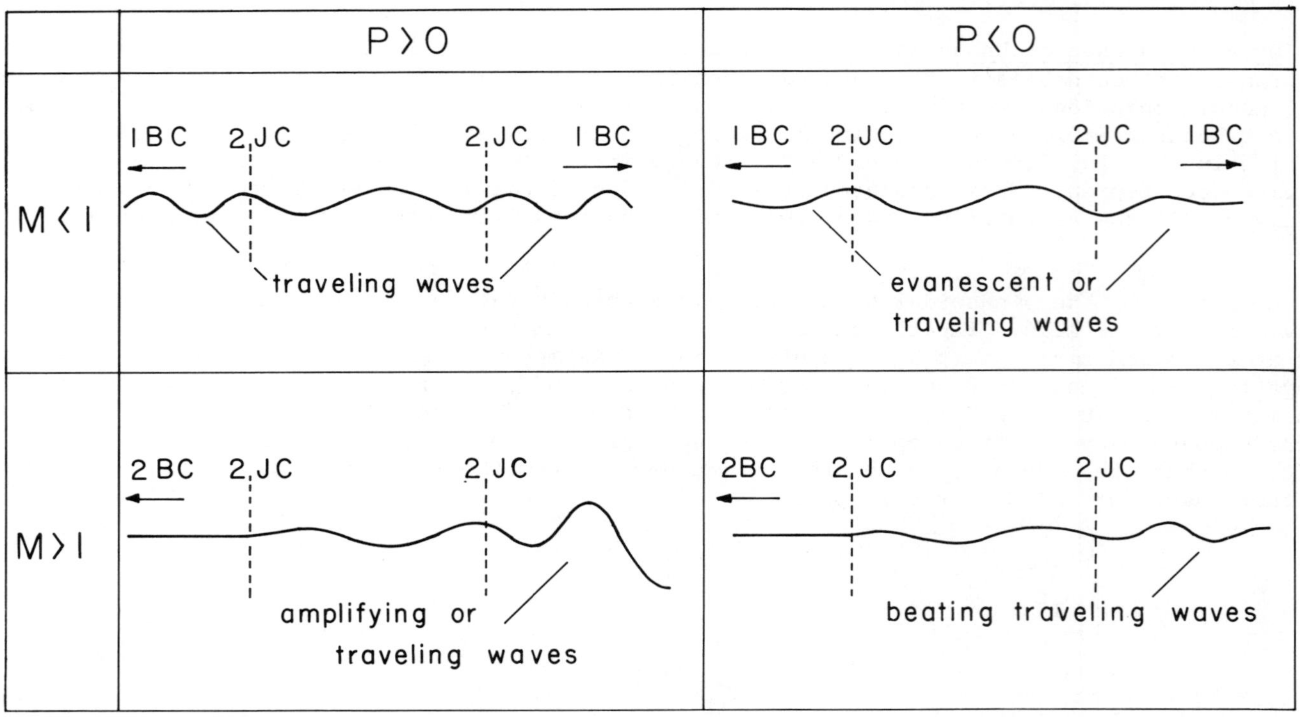

Fig. 11.11.6. Boundary conditions (B.C.) and jump conditions (J.C.) for second order unbounded systems.

In the case $P > 0$ (the unstable configuration), operation below the cutoff frequency results in an amplifying wave. Thus, it is possible that even within the excitation region the spatially periodic response will not prevail. The spatially amplifying wave certainly dominates in the downstream region, since the only other contribution to the response in that region is a decaying wave.

The downstream responses for the stable and unstable supercritical cases of Fig. 11.11.6 are illustrated experimentally by Figs. 11.11.4 and 11.11.5, respectively.

In retrospect, what is the intellectual basis for the association of the spatial modes with boundary conditions at infinity that made the difference between the supercritical and subcritical solutions? (Certainly it is not an identification of the direction of propagation of the phases.) In fact, left at this point, it is necessary to fall back on the method of characteristics, Sec. 11.10, to justify the association of modes with boundary conditions. In the next section, the objective is to have a method of relating the modes to the conditions of causality. The excitation will be turned on and the appropriate solution found as the asymptotic response. This approach can be used in systems where the method of characteristics is not applicable.

It has been presumed in this discussion of the response for the infinite system that at a given point in space, the response remains bounded as time increases. It will be the purpose of Sec. 11.13 to identify conditions for an absolute instability and to discriminate between it and a convective instability.

11.12 Distinguishing Amplifying from Evanescent Modes

Whether the excitation is from transverse or longitudinal boundaries, the sinusoidal steady-state asymptotic response is a superposition of waves having complex wavenumbers, k, for a real frequency, ω. To understand how these modes are to be combined in this long-time limit, it is necessary to picture the response in relation to the turn-on transient from which it arises.[1,2]

---

1. For a more complete exposition of criteria for identifying the types of complex waves in infinite media, see R. J. Briggs, Electron-Stream Interaction with Plasmas, The M.I.T. Press, Cambridge, Mass., 1964, pp. 8-46. Also, A. Bers, Notes for MIT subject "Electrodynamics of Waves, Media, and Interactions."

2. A. Scott, Active and Nonlinear Wave Propagation in Electronics, Wiley-Interscience, New York, 1970, pp. 27-44.

Laplace and Fourier Transform Representation in Time and Space: In Sec. 5.17, it is assumed from the outset that the temporal dependence can be represented by a complex amplitude, for example $\Phi = \text{Re}\overset{\blacktriangle}{\Phi}(x,y,z,\omega)\exp(j\omega t)$. The spatial transient is in turn represented by a Fourier transform. To permit a representation of the transient which joins the initial conditions to a possible sinusoidal steady state, this temporal complex amplitude is now replaced by the Laplace transform pair

$$\Phi(x,z,t) = \int_{-\infty-j\sigma}^{\infty-j\sigma} \overset{\blacktriangle}{\Phi}(x,z,\omega)e^{j\omega t}\frac{d\omega}{2\pi} \Leftrightarrow \overset{\blacktriangle}{\Phi}(x,z,\omega) = \int_{0}^{\infty}\Phi(x,z,t)e^{-j\omega t}\,dt \tag{1}$$

As in Sec. 5.17, the longitudinal dependence is in turn represented by the Fourier transform

$$\overset{\blacktriangle}{\Phi}(x,z,\omega) = \int_{-\infty}^{+\infty}\hat{\Phi}(x,k,\omega)e^{-jkz}\frac{dk}{2\pi} \Leftrightarrow \hat{\Phi}(x,k,\omega) = \int_{-\infty}^{+\infty}\overset{\blacktriangle}{\Phi}(x,z,\omega)e^{jkz}\,dz \tag{2}$$

The Laplace transform, Eq. 1b, starts when $t = 0$, and so the transform of temporal derivatives brings in initial conditions. For example, the Laplace transform of the first derivative is integrated by parts to give

$$\int_{0}^{\infty}\frac{\partial\Phi}{\partial t}e^{-j\omega t}\,dt = \Phi e^{-j\omega t}\Big|_{0}^{\infty} - \int_{0}^{\infty}(-j\omega)\Phi e^{-j\omega t}\,dt = -\Phi(x,z,0) + j\omega\overset{\blacktriangle}{\Phi} \tag{3}$$

Thus, if a variable is zero when $t = 0$, then the Laplace and Fourier transform of its temporal derivative is simply $j\omega\hat{\Phi}$. Of course, the Laplace and Fourier transform of the derivative with respect to z is $-jk\hat{\Phi}$. That is, relations between complex amplitudes apply also to Laplace-Fourier transforms, provided that rest conditions prevail when $t = 0$ for the transformed variables. If there are finite initial conditions, then care must be taken to include them in transforming all relations.

As an example, consider the second order systems represented by Eq. 11.11.3 in an unbounded configuration. The excitation is from transverse boundaries (Fig. 11.11.2b) where the forcing function is imposed over the interval $0 < z < \ell$ as a traveling wave

$$\begin{aligned} f &= \underline{u}_1(t)[\underline{u}_1(z) - \underline{u}_1(z-\ell)]\text{Re}\hat{f}_o e^{j(\omega_o t - \beta z)} \\ &= \underline{u}_1(t)[\underline{u}_1(z) - \underline{u}_1(z-\ell)]\tfrac{1}{2}[\hat{f}_o e^{j(\omega_o t-\beta z)} + \hat{f}_o^* e^{-j(\omega_o t - \beta z)}] \end{aligned} \tag{4}$$

Substitution, first into Eq. 1b and then into Eq. 2b, gives the transform of the forcing function as

$$\hat{f} = \frac{\hat{f}_o[1 - e^{j(k-\beta)\ell}]}{2(\omega - \omega_o)(k - \beta)} + \frac{\hat{f}_o^*[1 - e^{j(k+\beta)\ell}]}{2(\omega + \omega_o)(k + \beta)} \tag{5}$$

Note that the second term is obtained from the first by substituting $\hat{f}_o \to \hat{f}_o^*$, $\omega_o \to -\omega_o$, and $\beta \to -\beta$. With the understanding that the real response is the sum of the one now found and a response formed by making this substitution, it will be assumed in what follows that the drive is just the first term in Eq. 5. (Note that $\hat{f}_o$ is not a transform, but rather simply a complex number expressing the phase and amplitude of the drive.)

With the understanding that when $t = 0$, ξ and $\partial\xi/\partial t$ are zero, the Laplace-Fourier transform of Eq. 11.11.3 gives an expression for the transform of the sheet deflection:

$$\hat{\xi} = \frac{P\hat{f}}{D(\omega,k)};\quad D(\omega,k) \equiv (\omega - Mk)^2 - k^2 + P \tag{6}$$

This expression is obtained either by treating variables as though complex amplitudes are being introduced or by starting from scratch by multiplying Eq. 11.11.3 by $\exp[-j(\omega t - kz)]$ and then integrating both sides of the expression from 0 to ∞ on t and from $-\infty$ to ∞ on z. Integrations by parts of the terms involving derivatives and the definitions of the transforms, Eqs. 1b and 2b, then also result in Eq. 6. (Note that consistent with the normalization used in Sec. 11.11 are $\underline{k} \equiv k\tau V$ and $\underline{\omega} = \omega\tau$.) Thus, in view of Eq. 5, the desired Laplace-Fourier transform is written in terms of the specific traveling-wave excitation, turned on when $t = 0$, as

$$\hat{\xi} = \frac{\mathcal{f}(\omega)g(k)}{D(\omega,k)} \tag{7}$$

where

$$g(k) = \frac{P\hat{f}_o}{2} \frac{[e^{j(k-\beta)\ell} - 1]}{j(k - \beta)} \qquad (\underline{8})$$

$$\mathcal{f}(\omega) = \frac{1}{j(\omega - \omega_o)} \qquad (\underline{9})$$

The model represented by Eqs. 8 and 9 is long-wave. But, the form of the response transform taken by Eq. 7 is representative of a much wider range of physical situations that are uniform in the z direction. The details of a transverse dependence are determined by solving differential equations and boundary conditions over the transverse cross section. This amounts to representing the transformed variables in terms of transfer relations and boundary conditions, as exemplified many times in Chaps. 5, 6, 8, 10 and in Sec. 11.5.

<u>Laplace Transform on Time as the Sum of Spatial Modes; Causality</u>: The evolution in z-t space is determined by using Eq. 7 in evaluating the inverse transforms, Eqs. 1 and 2:

$$\hat{\xi}(z,\omega) = \mathcal{f}(\omega)\int_{-\infty}^{+\infty} \frac{g(k)}{D(\omega,k)} e^{-jkz}\frac{dk}{2\pi} \qquad (\underline{10})$$

C_F

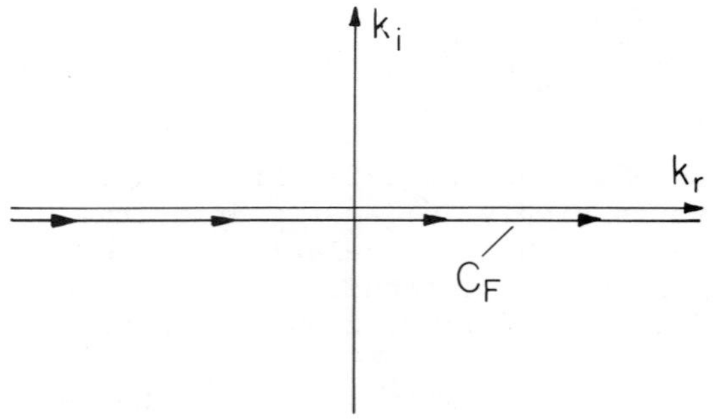

Fig. 11.12.1. Fourier contour.

$$\xi(z,t) = \int_{-\infty-j\sigma}^{\infty-j\sigma} \hat{\xi}(z,\omega)e^{j\omega t}\frac{d\omega}{2\pi} \qquad (\underline{11})$$

C_L

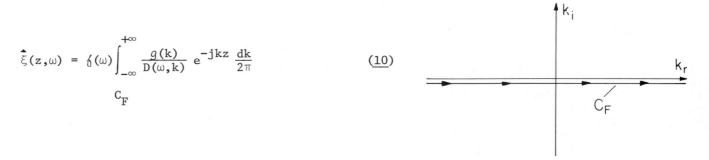

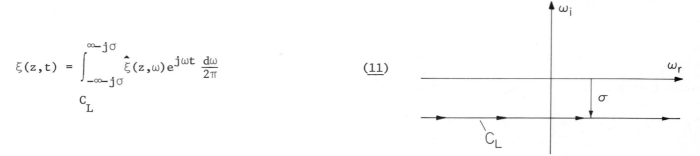

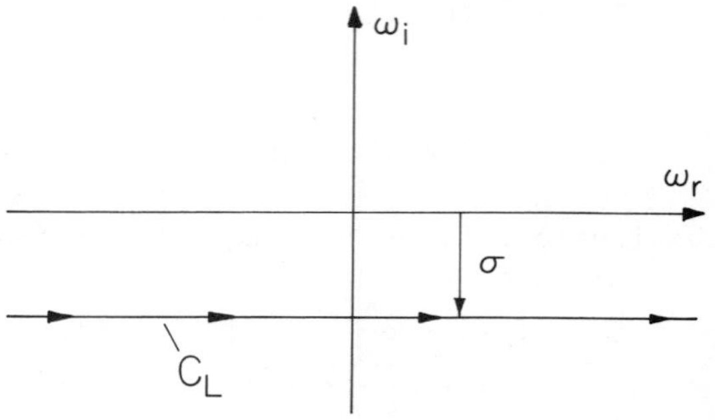

Fig. 11.12.2. Laplace contour.

The ω that appears in Eq. 10 is any one of the frequencies on the Laplace contour, C_L of Fig. 11.12.2. That is, the integral on k is carried out with ω each of the (generally complex) frequencies required to subsequently carry out the integration on ω.

Causality is built into the inversion of the transforms through the choice of the Laplace contour. On this contour, $\omega = \omega_r - j\sigma$, where σ is constant. Thus, in Eq. 11, $\exp(j\omega t) = \exp(j\omega_r t)\exp(\sigma t)$. Thus, for $t < 0$, the integrand goes to zero as $\sigma \to \infty$, and the integrand along the Laplace contour can be replaced by one closed in the lower half plane. That the response for $t < 0$ must be zero is therefore equivalent to requiring that this integral over the closed contour C_{-t} vanish. The closure is illustrated by Fig. 11.12.3. Cauchy's theorem makes it clear that this causality condition will prevail, provided that the Laplace contour is below all singularities of $\hat{\xi}(z,\omega)$ in the ω plane.

For any given frequency on this contour, the situation for inverting the Fourier transform as specified by Eq. 10 is no different than in Sec. 5.17, except that the frequency is in general some complex number. In Sec. 5.17, where it is assumed at the outset that sinusoidal steady-state conditions prevail, the frequency is the real frequency of the drive.

Note that $g(k)$ is not singular at $k = \beta$. Further, for the second order system, and for others having dispersion equations that are polynomial or transcendental in k, the roots of $D(\omega,k_n) = 0$ represent poles in the k plane. Thus, if the integration on the Fourier contour can be converted to one that is closed, then Cauchy's residue theorem

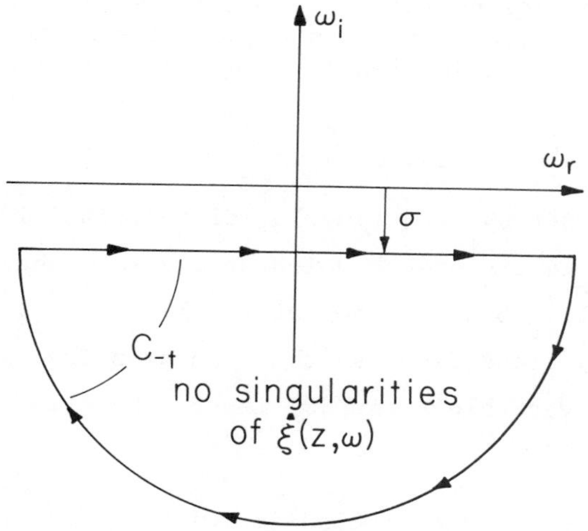

Fig. 11.12.3. Closure of Laplace contour to identify C_L consistent with causality.

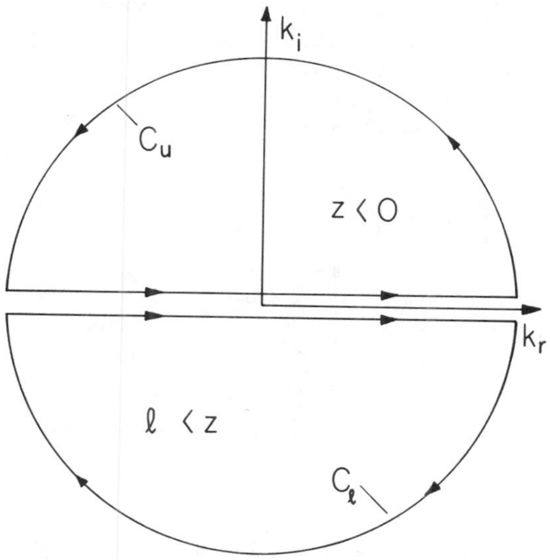

Fig. 11.12.4. Closures to evaluate Fourier integral.

$$\oint_C \frac{N(k)}{D(k)}\,dk = \pm 2\pi j(K_1 + K_2 + \cdots); \quad K_n = \frac{N(k_n)}{D'(k_n)} \tag{12}$$

provides a simple evaluation. The positive and negative signs are for counterclockwise and clockwise directions of integration. For a given frequency on the Laplace contour, the Laplace transform is the sum over the spatial modes.

To see what closed contour can be used to replace the open one that is to be evaluated, observe that in Eq. 10

$$[e^{j(k-\beta)\ell} - 1][e^{-jkz}] = e^{-j\beta\ell}e^{jk_r(z-\ell)}e^{k_i(z-\ell)} - e^{-jk_rz}e^{k_iz} \tag{13}$$

If z is in the range $z < 0$, the entire term goes to zero if $k_i \to +\infty$. Thus, the Fourier integration can be replaced by one on the closed contour C_u in the upper half of the k plane, as shown in Fig. 11.12.4. If z is in the range $\ell < z$, the entire term goes to zero if the contour is closed in the lower half plane, on C_ℓ. In the excitation range, where $0 < z < \ell$, the terms in Eq. 13 must be treated separately. The individual functions are singular at $k = \beta$, so the response in this range includes not only the spatial modes, but a "driven response" having the wavenumber β. This is considered in detail in Sec. 5.17, and will not be highlighted here. With the understanding that the summation is made appropriate to the range of z being considered (to left or right of the excitation range), Eq. 10 is integrated to give the Laplace transform

$$\hat{\xi}(z,\omega) = \pm\hat{\delta}(\omega)\sum_n \frac{jg(k_n)}{D'(\omega,k_n)}e^{-jk_nz} \tag{14}$$

where the upper and lower signs pertain to the upper and lower contours in Fig. 11.12.4. For example, in the case of the second order systems, where Eq. 6 gives the dispersion equation,

$$D(\omega,k) = (M^2 - 1)(k - k_1)(k - k_{-1})$$

$$k_{\substack{1\\-1}} = \frac{\omega M \mp \sqrt{\omega^2 + P(1 - M^2)}}{M^2 - 1} \tag{15}$$

It follows that in Eq. 14

$$D'(k_{\substack{1\\-1}}) = \pm(M^2 - 1)(k_1 - k_{-1}) = \mp 2\sqrt{\omega^2 + P(1 - M^2)} = \mp 2\sqrt{(\omega - \omega_c)(\omega + \omega_c)} \tag{16}$$

<u>Asymptotic Response in the Sinusoidal Steady State</u>: The Laplace contour, C_L, lies below all singularities of $\xi(z,\omega)$. In the second order systems, there are singularities of the individual terms in Eq. 14 at $\pm\omega_c$. Whether or not they represent singularities of $\xi(z,\omega)$ depends on the yet to be determined appropriate summation in Eq. 14. These branch poles of the individual terms are illustrated for the submagnetic case in Fig. 11.12.5. They are designated as branch poles because the function $D'(\omega)$, where $k_n = k_n(\omega)$, is double-valued if ω is allowed to pass through the branch line joining these poles (see Fig. 11.12.5). Because of $\mathcal{J}(\omega)$, there is clearly a pole of $\xi(z,\omega)$ at $\omega = \omega_o$.

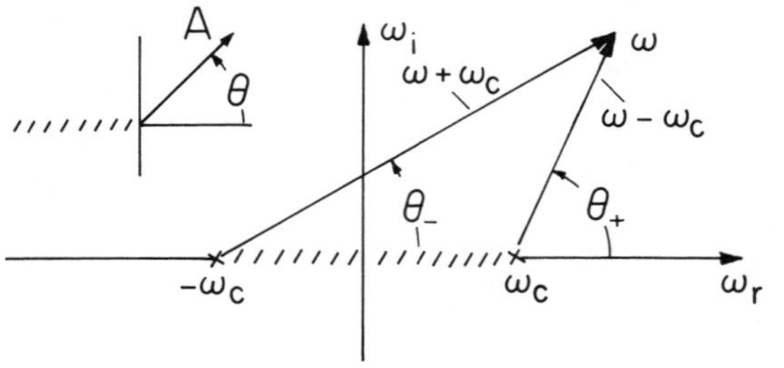

Fig. 11.12.5

For second order system, singularities of individual terms of Eq. 14 are branch poles at $\omega = \pm\omega_c$, where $\omega_c \equiv \sqrt{-P(1-M^2)}$. Branch lines of $(M^2-1)(k_{+1}-k_{-1}) \equiv \sqrt{(\omega-\omega_c)(\omega+\omega_c)}$ are defined so that the principal value of a complex variable $A\exp(j\theta)$ is $-\pi < \theta < \pi$.

The objective here is not to carry out the second integration called for with Eq. 11, but rather to discern the response when $t \to \infty$. At a given location, z, there are two long-time possibilities. The response can either reach the sinusoidal steady state, or it can become unbounded. To achieve the former, it must be possible to move the Laplace contour, C_L, so that it is as shown in Fig. 11.12.6. This is possible if there are no singularities below the (open) contour. The part, C_L', runs parallel to the real axis with ω_i slightly positive. Thus as $t \to \infty$, the integrand of Eq. 10 on C_L' goes to zero, and this part of the integration gives no asymptotic contribution. The contributions to the integral along the oppositely directed segments, C_L'', cancel. Thus, the integral reduces to a closed integral on C_L'''. The only singularity within this contour is in $\mathcal{J}(\omega)$, at $\omega = \omega_o$. Thus,

$$\xi(z,t) = \mp\sum_n \frac{g(k_n)}{D'(k_n)} e^{j(\omega_o t - k_n z)}\quad ; \quad k_n = k_n(\omega_o) \qquad (\underline{17})$$

where upper and lower signs pertain to closures of the contour in the upper and lower plane.

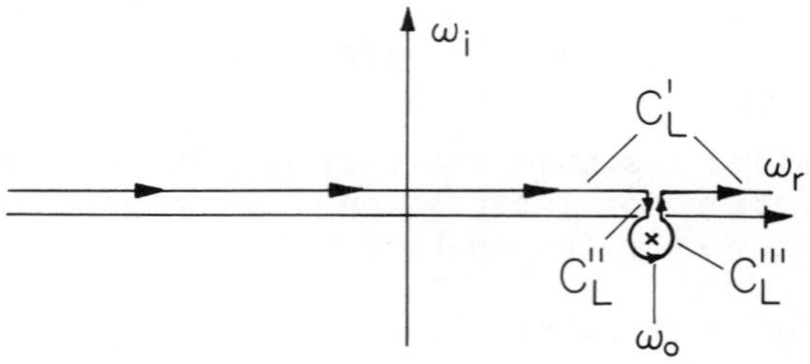

Fig. 11.12.6. Laplace contour in limit $t \to \infty$ with no singularities below contour.

As an example, consider the submagnetic second order system (with $|M| < 1$ and $P < 0$). In this case there are no singularities in the lower half of the ω plane, but there is the embarrassment of branch poles on the ω_r axis, as illustrated in Fig. 11.12.5. However, these occur on the axis because there is no damping in the system. If a force term is added to the equation of motion (Eq. 11.11.3) having the form

$$f = -B\left(\frac{\partial}{\partial t} + U\frac{\partial}{\partial z}\right)\xi \qquad (\underline{18})$$

where B is a positive damping coefficient, the branch poles are displaced into the upper half plane.

It is now possible to identify the proper contributions to the Laplace transform, Eq. 14, and hence to the sinusoidal steady-state response given by Eq. 17. For a given Laplace contour, C_L (σ a finite positive constant), the poles, k_n, form a locus of points in the k plane. Again, for the submagnetic case ($P < 0$ and $|M| < 1$), one locus is in the lower half plane and the other in the upper half plane. As the Laplace contour is displaced upward to the ω_r axis, these loci terminate in contours of complex k for real ω shown as dashed lines in Fig. 11.12.7. More prominently shown in this figure are the contours followed by the poles, k_n, as the Laplace contour is displaced upward to the ω_r axis. On these contours, ω_r is constant and σ is decreasing to zero. For example, holding $\omega_r = 0.8$, as σ is reduced to zero gives one pole that moves down and to the right in the second quadrant ($k_n = k_{-1}$) and a second that moves upward and to the left starting in the fourth quadrant ($k_n = k_1$). As σ reaches zero, these poles become complex conjugates. In general, one pole comes from above and one from below, each terminating on the locus of k for real ω. In this example, no pole starting in the lower half plane reaches the upper half plane and vice versa.

It is now clear that the $n = -1$ pole constitutes the only term in the sum in Eq. 17 with closure of the Fourier contour in the upper half plane (for $z < 0$) while $n = 1$ and closure in the lower half plane is appropriate for $z > 0$. Thus, Eq. 17 becomes

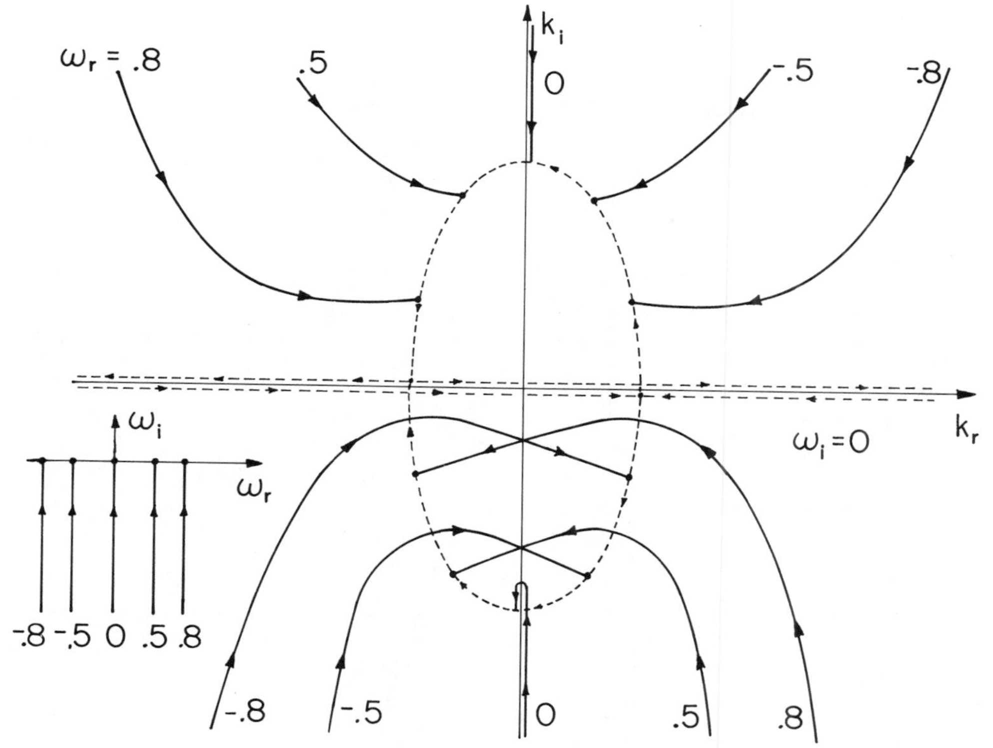

Fig. 11.12.7. Loci of k for loci of ω shown by inset. Broken curves are values of k for real ω. Submagnetic case ($|M| < 1$, $P < 0$) displays evanescent waves ($P = 1$, $M = 0.5$).

$$\xi(z,t) = -\frac{P\hat{f}_o}{2}\frac{[1 - e^{j(k_{\mp 1}-\beta)\ell}]e^{j(\omega_o t - k_{\mp 1} z)}}{(k_{\mp 1} - \beta)(k_1 - k_{-1})(M^2 - 1)}\,; \quad z \begin{matrix} < 0 \\ > \ell \end{matrix} \tag{19}$$

That is, if the frequency is in the range $\omega_o < |\omega_c|$ where waves are cut off, these waves are evanescent. They decay away from the excitation. For example, that k_{-1} has a positive imaginary part does not mean that it represents a wave that grows in the +z direction, but rather that it is a wave decaying in the -z direction. Converted to a real function of time in accordance with the discussion following Eq. 5, the result given by Eq. 19 is consistent with the sinusoidal steady-state response deduced in Sec. 11.11 for this case (Eq. 11.11.20).

Consider by contrast the establishment of a sinusoidal steady state in which there is an amplifying wave. As the Laplace contour is pushed upward toward the ω_r axis in the ω plane, one or more roots, k_n, of $D(\omega,k_n) = 0$ move across the k_r axis in the k plane. This is illustrated in Fig. 11.12.8 by the superelectric second order system ($P > 0$ and $|M| > 1$). Here, for σ large both poles, k_n, are in the lower half of the k plane. As σ is decreased keeping ω_r constant, one of the poles terminates in the lower half plane while the other passes through the k_r axis into the upper half plane.

Remember that "pushing" the Laplace contour to the ω_r axis in the ω plane is no more than a way to approximate the inverse Laplace integral in the limit $t \to \infty$. The function represented by the inverse Laplace integral must be the same, regardless of the integration contour. This requires that as the Laplace contour is moved upward, the Fourier integration contour must be distorted so that when it is evaluated by closing the contour, that contour includes the same poles of $D(\omega,k_n)$. That is, it must continue to include those that have passed through the k_r axis into the upper k plane. This "analytic continuation"[3] of the Laplace transformation is therefore obtained by using Fourier contours revised as illustrated in Fig. 11.12.9. By allowing the Fourier contour, Fig. 11.12.9, to be distorted so as to enclose the same singularities, the domain of the ω plane over which $\hat{\xi}(\omega,z)$ is analytic has been extended so that the Laplace contour can be that illustrated in Fig. 11.12.6.

For the second order superelectric system, there are now no poles enclosed by the Fourier contour closed in the upper half plane and hence appropriate to evaluating $\hat{\xi}(z,\omega)$ in the upstream region $z < 0$. Thus, the response $\xi(z,t)$ for $z < 0$ is zero. Note that this is true at any time and not just for the asymptotic response. Moreover, for closure in the lower half plane and hence $z > \ell$ the

3. P. M. Morse and H. Feshbach, Methods of Theoretical Physics, McGraw-Hill Book Company, 1953, p. 392.

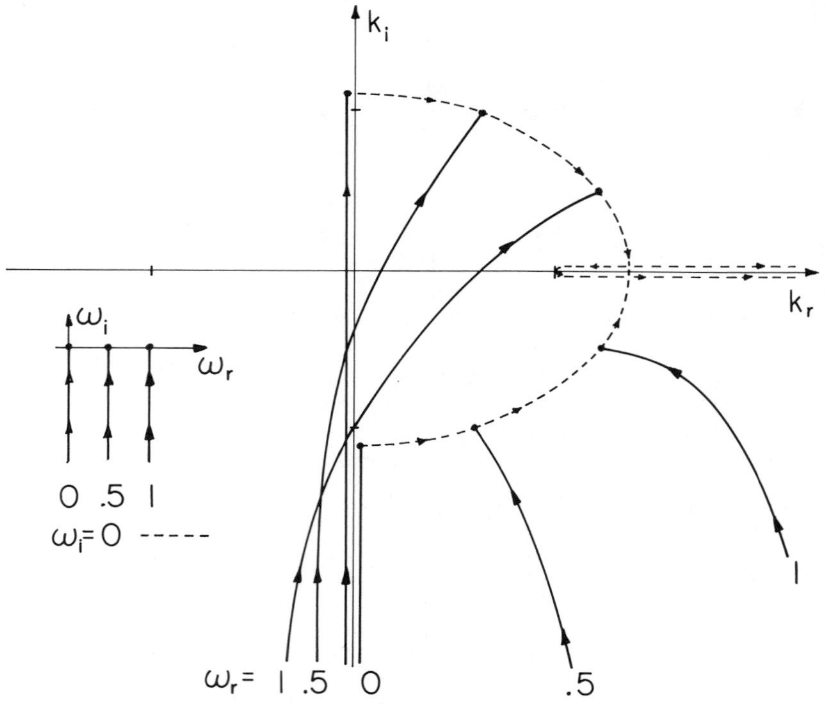

Fig. 11.12.8. Loci of k for loci of ω shown by inset. Broken curves are values of k for real ω. Superelectric case ($|M| > 1$, $P > 0$) displays amplifying waves ($P = 1$, $M = 1.5$).

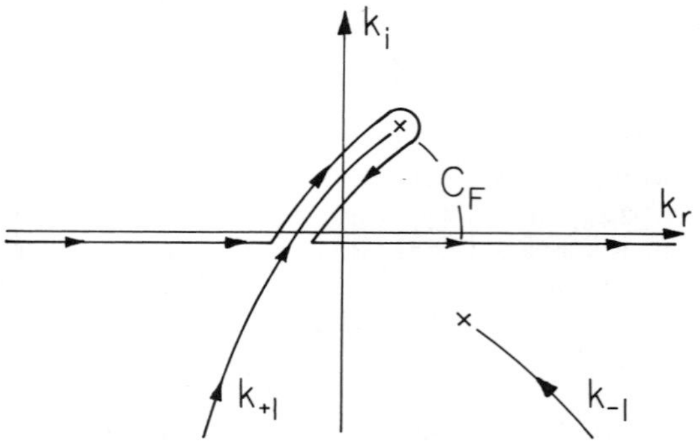

Fig. 11.12.9

As Laplace contour is pushed to ω_r axis, pole k_{+1} crosses k_r axis, indicating wave amplification.

summation for the Laplace transform is over both spatial modes. That is, Eq. 14 becomes

$$\hat{\xi}(z,\omega) = \frac{[g(k_1)e^{-jk_1 z} - g(k_{-1})e^{-jk_{-1}z}}{j(\omega - \omega_o)(M^2 - 1)(k_{-1} - k_1)} \qquad (\underline{20})$$

Thus, in the long-time limit the Laplace integration along the contour shown in Fig. 11.12.6 results in

$$\xi(z,t) = \frac{j}{(k_{-1} - k_{+1})(M^2 - 1)}[g(k_{+1})e^{j(\omega_o t - k_1 z)} - g(k_{-1})e^{j(\omega_o t - k_{-1}z)}] \qquad (\underline{21})$$

Again, remember that the real expression is obtained from this expression as described following Eq. 5.

What has been obtained for the second order example is the same sinusoidal steady-state solution summerized by Eq. 11.11.21c. If the driving frequency is less than ω_c, one of the downstream waves decays away from the source while the other amplifies.

Criterion Based on Mapping Complex k as a Function of Complex ω: Causality requires that the Laplace contour remain below all singularities of $\hat{\xi}(\omega,z)$. In the process of pushing the Laplace contour upward in the ω plane to discover the asymptotic response, singularities of $D'(\omega,k_n)$ must be identified as possible singularities of the Laplace transform, Eq. 14. In the second order system, it is possible to solve explicitly to determine the frequencies, ω_s, at which

$$D'(\omega_s,k_n) \equiv [\frac{\partial D}{\partial k}(\omega_s,k)]_{k=k_n} = 0 \qquad (22)$$

In general, these possible singularities are more difficult to identify. However, they can be determined by examining the dispersion equation.

Remember that Eq. 22 is really an expression for ω_s, because by definition

$$D(\omega_s,k_s) = 0 \qquad (23)$$

In the neighborhood of (ω_s,k_s), ω and k are related by

$$D(\omega,k) = 0 = D(\omega_s,k_s) + \frac{\partial D}{\partial \omega}\Big|_{\omega_s,k_s}(\omega-\omega_s) + \frac{\partial D}{\partial k}\Big|_{\omega_s,k_s}(k-k_s) + \frac{1}{2}\frac{\partial^2 D}{\partial k^2}\Big|_{\omega_s,k_s}(k-k_s)^2 + \cdots \qquad (24)$$

In view of Eqs. 22 and 23, Eq. 24 approximates the dispersion equation in the neighborhood of (ω_s,k_s) as

$$D(\omega,k) \simeq \frac{\partial D}{\partial \omega}\Big|_{\omega_s,k_s}(\omega-\omega_s) + \frac{1}{2}\frac{\partial^2 D}{\partial k^2}\Big|_{\omega_s,k_s}(k-k_s)^2 = 0 \qquad (25)$$

This expression makes evident what is happening in the k plane as ω approaches ω_s in the ω plane, for Eq. 27 is equivalent to the expression

$$k - k_s = \pm\sqrt{\frac{-2\frac{\partial D}{\partial \omega}\Big|_{k_s,\omega_s}(\omega-\omega_s)}{\frac{\partial^2 D}{\partial k^2}\Big|_{k_s,\omega_s}}} \qquad (26)$$

It is concluded that the coalescence of a pair of poles in the k plane is the result of having the frequency $\omega \to \omega_s$.

Candidates for poles of $\hat{\xi}(\omega,z)$ in the ω plane can be identified by mapping loci of the roots to the dispersion equation in the k plane resulting from varying $\omega = \omega_r - j\sigma$ to cover the lower half ω plane. This is conveniently done by holding ω_r at fixed values and decreasing σ from ∞ to zero.

In retrospect, for the submagnetic and superelectric second order systems, the coalescence of roots k_{+1} and k_{-1} does not occur in the lower half k plane. These mappings were illustrated by Figs. 11.12.7 and 11.12.8, respectively. But, consider the supermagnetic second order system ($|M|>1$, $P<0$). The map, illustrated in this case by Fig. 11.12.10, shows that at $\omega = -j\sqrt{-P(M^2-1)} \equiv -j\sigma_c$ (on the ω_i axis) there is a coalescence of the roots of k, and hence a singularity in the terms of Eq. 14 comprising the Laplace transform. However, because both roots are below the Fourier contour, they both contribute to the Laplace transform for $\ell < z$. In fact, as the roots coalesce, the pair of contributions to the Laplace transform are together not singular. That is, the denominator of the individual terms can be evaluated by taking the derivative of Eq. 25 and then using Eq. 26 to substitute for $k - k_s$,

$$\frac{\partial D}{\partial k} = \frac{\partial^2 D}{\partial k^2}\Big|_{k_s,\omega_s}(k-k_s) = \pm\sqrt{-2\frac{\partial^2 D}{\partial k^2}\Big|_{k_s,\omega_s}\frac{\partial D}{\partial \omega}\Big|_{k_s,\omega_s}(\omega-\omega_s)} \qquad (27)$$

where the upper and lower signs refer to the respective roots. Thus, the pair of terms resulting from the residues for the respective roots of k have opposite signs and, in the limit, equal magnitudes. Rather than being singular, the pair tend to the form 0/0, and can be shown to remain finite. It follows that the Laplace contour can be pushed to just above the ω_r axis (Fig. 11.12.6)

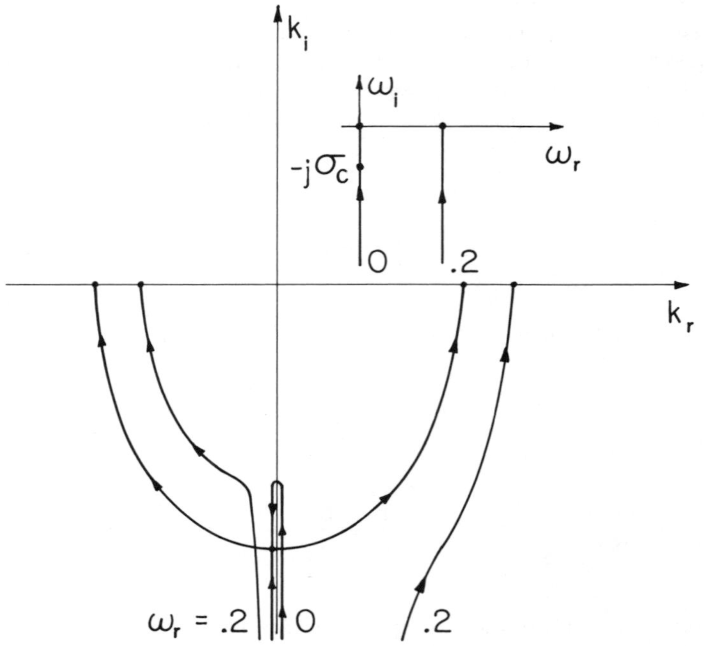

Fig. 11.12.10

Loci of k for loci of ω shown by inset. Supermagnetic case ($|M| > 1$, $P < 0$) displays ordinary waves, both propagating downstream ($M = 1.5$, $P = -1$).

to evaluate the asymptotic response as before. In the upper half k plane, there are no roots of k for ω on the Laplace contour and hence the response is zero for $z < 0$. For $\ell < z$, closure of the Fourier integral in the lower half k plane gives contributions from both k_{+1} and k_{-1}, so the Laplace transform, Eq. 14, has two terms. However, as already pointed out, the only singularity is at $\omega = \omega_o$, and so integration on the Laplace contour leads to Eq. 21. Now, $k_{\pm 1}$ can only be real and the downstream response takes the form of beating traveling waves. This is the same result as given by Eq. 11.11.15 and illustrated by Fig. 11.11.4.

In summary, the key to understanding the physical significance of a wave having complex values of k for real ω is a map of the loci of k that result from varying σ from ∞ to 0 for all values of ω_r. If the loci of a given root terminate in complex k without crossing the k_r axis, then it represents an evanescent wave. That is, in the sinusoidal steady state, the complex k represents a wave that decays away from the source. On the other hand, if loci cross the k_r axis, the mode represents an amplifying wave. At a point where a locus crosses the real k axis, k is obviously real and σ is still positive. Thus, for a crossing of the k_r axis, the dispersion equation must display "unstable" values of ω for real values of k. It is concluded that a necessary condition for existence of an amplifying wave is that wavelengths exist for which a temporal mode is unstable. That is, for k real there must be roots of $D(\omega,k)$ for which $\omega_i < 0$. As a type of instability that grows spatially rather than temporally, the amplifying wave is also termed a convective instability.

11.13 Distinguishing Absolute from Convective Instabilities

In Sec. 11.12, examples were purposely considered which were absolutely stable. Thus, the asymptotic response was in the sinusoidal steady state. A necessary condition for an amplifying wave was found to be "unstable" values of ω given by the dispersion equation for real values of k. How is this response, which is bounded at a given location, z, to be distinguished from one that grows with time at a given location?

Criterion Based on Mapping Complex k as Function of Complex ω: Suppose that a mapping of the dispersion equation, showing loci of k in the complex k plane resulting from varying σ from ∞ to 0 in the complex $\omega = \omega_r - j\sigma$ plane, results in a double pole of the type illustrated in Fig. 11.13.1. Here, as $\omega \to \omega_s$, a pair of roots coalesce, one from the upper half plane and one from the lower half plane. When it is closed, the Fourier contour can no longer be distorted to include the same poles. Before σ reaches zero, C_F is caught between the coalescing poles, one coming from above and the other coming from below.

As for the supermagnetic case from Sec. 11.12, coalescence of the roots, k_n, once again indicates the possibility of a pole of $\xi(\omega,z)$ in the lower half plane. This time the pole in fact exists, because the roots, k_n, are on opposite sides of the Fourier contour. Thus, pushed upward so that most of it is just above the ω_r axis, the Laplace contour is as shown in Fig. 11.13.2. The Fourier contour is caught between the coalescing poles, always including one or the other when closed in the upper or lower half k plane. Integration along the Laplace contour on the segments C_L' just above

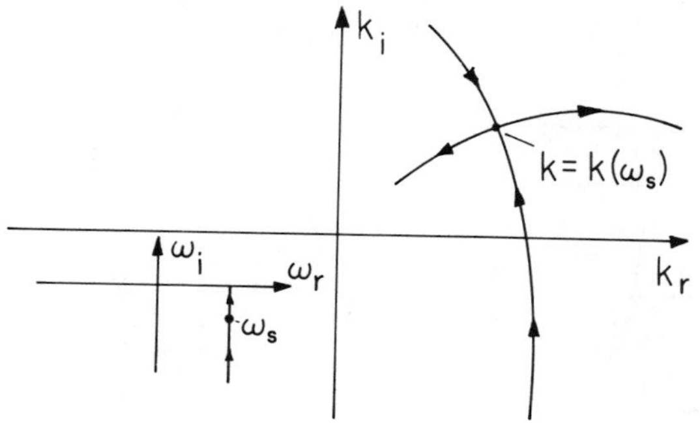

Fig. 11.13.1. Mapping of loci of k in the k plane for trajectory of ω passing through singularity in Laplace transform, showing coalescence of roots from upper and lower half plane indicating absolute instability.

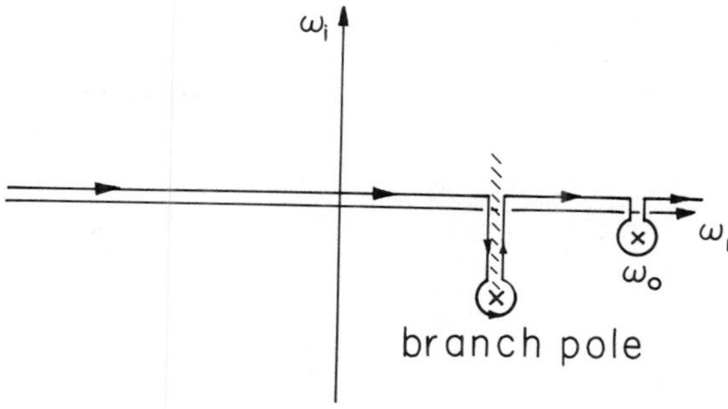

Fig. 11.13.2. Laplace contour for evaluating response as $t \to \infty$ with branch pole in lower half plane.

the ω_r axis gives no contribution as $t \to \infty$. However, integration along the segments, C_L'', adjacent to the branch-cut and around the pole do give contributions. (Note that the segments to either side of the branch-cut do not cancel, because the integrand is of opposite sign on one side from the other.) In the long-time limit, the Laplace integration reduces to

$$\lim_{t\to\infty} \xi(z,t) = \lim_{t\to\infty} \pm \int_{C_L''} \mathcal{S}(\omega) \sum_n \frac{j\mathcal{G}(k_n)}{D'(\omega,k_n)} e^{j(\omega t - k_n z)} \frac{d\omega}{2\pi} \tag{1}$$

On the contour segment C_L'', $\omega = \omega_s - j\sigma$ where σ is positive, so the asymptotic response is dominated by a part that grows with time at a fixed location, z.

It is concluded that if the mapping of the dispersion equation from the lower half ω plane into the k plane discloses a pair of coalescing roots, k_n, with one root from the upper half plane and one from the lower half plane, then the pair represent an absolute instability. This is sometimes also called a non-convective instability.

Note that if roots, k_n, of $D(\omega,k_n)$ are to come from the lower and upper half plane and coalesce, one of them must cross the k_r axis. As it does so, σ is positive. Hence, a necessary condition for absolute instability as well as convective instability is that there be "unstable" frequencies for real values of k. That is, for a wave to be a candidate for either spatial or temporal growth, it must have temporal modes that are unstable.

One consequence of this observation relates to numerous situations that can be envisioned as configurations from the sections and problems of Chap. 8 set into a state of uniform motion. For example, consider the jet of liquid described in Sec. 8.15, but now having a stationary equilibrium in which the jet streams in the z direction with velocity U. The dispersion equation is Eq. 8.15.25 with $\omega \to \omega - kU$ and takes the form

$$(\omega - kU)^2 = f(k) \tag{2}$$

Convection or not, there are only two temporal modes and only one of these can be unstable. That is, for real values of k, at most one of the two roots, ω, can have a negative imaginary part.

If the dispersion equation were to be solved for the real frequency spatial modes, there would be an infinite number, about half appearing to "grow" in the z direction. What is clear from the above result is that only one of these is a candidate for being an amplifying wave. The others are evanescent waves.

Second Order Complex Waves: The subelectric second order system ($|M| < 1$ and $P > 0$, Fig. 11.11.1) exemplifies an absolute instability. The $\omega \to k$ mapping is shown in Fig. 11.13.3. In this case, the mapping indicates a branch pole at $\omega = -\omega_c \equiv -j\sqrt{P(1 - M^2)}$, as can also be seen directly from Eq. 11.12.16. The contour to be used for inverting the Laplace transform as $t \to \infty$ is indicated in Fig. 11.13.4.

In more complicated situations, the mapping is carried out numerically by having a routine for determining the roots, k_n, of $D(\omega,k_n)$ given the complex frequency $\omega = \omega_r - j\sigma$. The pattern of the loci in the neighborhood of the coalescing roots, typified by Fig. 11.13.3, can be used to disclose the coalescing roots.

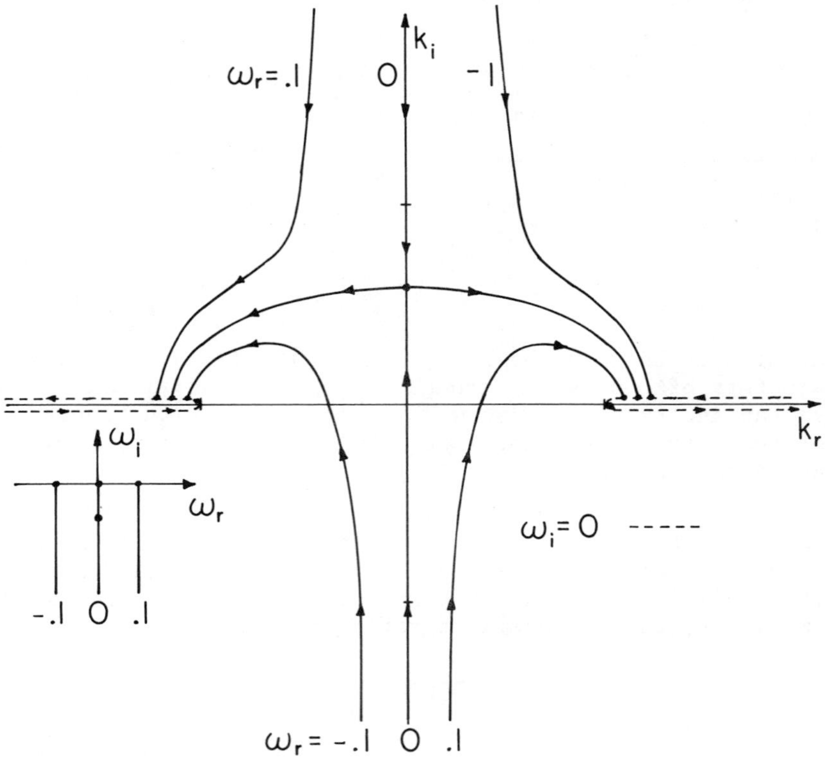

Fig. 11.13.3. Loci of k for loci of ω shown by inset. Subelectric case ($|M| < 1$, $P > 0$) displays absolute instability ($M = 0.5$, $P = 1$).

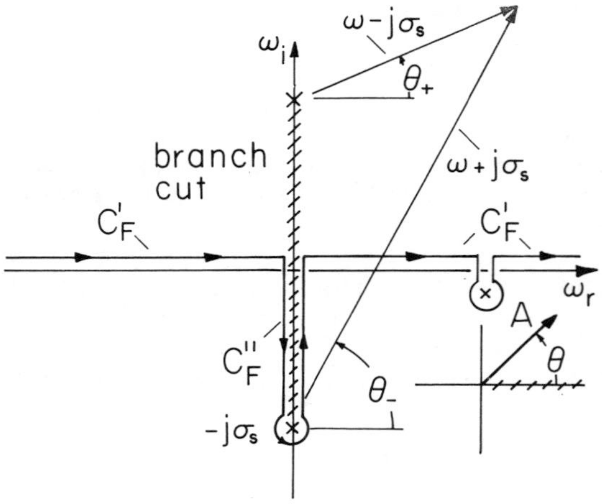

Fig. 11.13.4

Laplace contour, C_F, appropriate for evaluation of asymptotic response of subelectric ($|M| < 1$, $P > 0$) second order system. Branch cut of $\sqrt{(\omega - j\sigma_s)(\omega + j\sigma_s)}$ is defined consistent with branch line for $A\exp(j\theta)$ shown by inset.

11.14 Kelvin-Helmholtz Type Instability

Perhaps the most often discussed instability resulting from the interaction of streams is modeled by contacting layers of inviscid fluid, one having a uniform velocity relative to the other.[1] In Fig. 11.14.1, the lower layer is initially static while the upper one streams in the z direction with a uniform z-directed velocity, U. That the interface is subject to what is sometimes also called Bernoulli instability suggests why there is a critical velocity above which the stationary equilibrium is unstable. An intuitive picture makes clear the mechanism. For interfacial deformations that have a long wavelength compared to a or b, mass conservation requires that

$$v_z(a - \xi) = Ua \tag{1}$$

1. G. K. Batchelor, Introduction to Fluid Mechanics, Cambridge University Press, London, 1967, pp. 511-517.

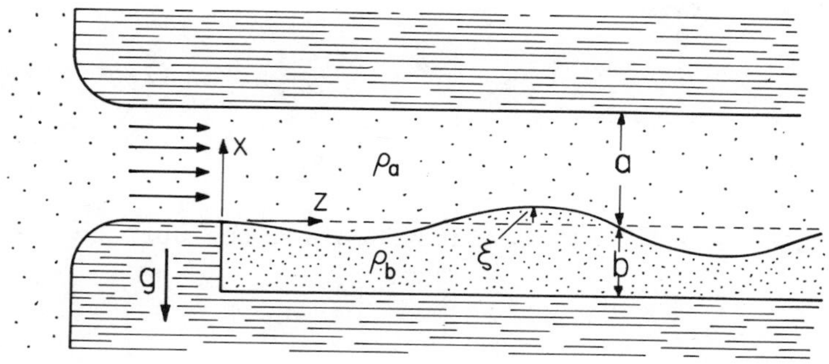

Fig. 11.14.1

In stationary state, lower fluid is static while upper one streams to the right with uniform velocity U. At the interface there is a discontinuity in fluid velocity, and hence a vorticity sheet.

Suppose that an upward directed external surface force density, T, could be applied to a section of the interface. At that point on the interface, normal stress equilibrium for the control volume of Fig. 11.14.2 requires that

$$T = p^d - p^e \tag{2}$$

where, for now, surface tension is ignored. With T applied slowly enough that the effect of the interfacial deformation on the flow can be pictured as a sequence of stationary states, the steady-state form of Bernoulli's equation is applicable to the regions above and below:

$$\begin{aligned} p + \frac{1}{2}\rho_a v^2 + \rho_a g x &= \Pi_a; \quad x > 0 \\ p + \rho_b g x &= \Pi_b; \quad x < 0 \end{aligned} \tag{3}$$

Initially, T = 0 and the interface is flat, so substitution of Eqs. 3 into Eq. 2 evaluates $\Pi_a - \Pi_b = \rho_a U^2/2$. Thus, the difference of Eqs. 3 becomes

$$p^d - p^e = \frac{1}{2}\rho_a(U^2 - v_z^2) + g(\rho_b - \rho_a)\xi \tag{4}$$

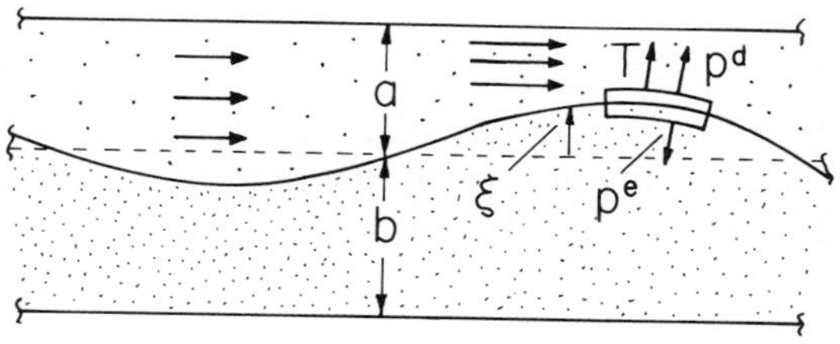

Fig. 11.14.2

Where the interface moves upward the streaming velocity increases. Thus, the pressure above drops and the interface is encouraged to further deform.

The combination of this expression of Bernoulli's equation with mass conservation, Eq. 1, and interfacial stress equilibrium, Eq. 2, gives an expression for the dependence of the surface deflection on the externally applied surface force density:

$$T = \frac{1}{2}\rho_a U^2\left[1 - \frac{1}{(1 - \frac{\xi}{a})^2}\right] + g(\rho_b - \rho_a)\xi \tag{5}$$

What is on the right is an equivalent surface force density representing the combined effects of the fluid inertia and gravity. (Rayleigh-Taylor instability is not of interest here, so it is assumed that the heavier fluid is on the bottom, $\rho_b - \rho_a > 0$.) If positive, it acts downward. The equilibrium is stable if a positive upward deformation results in a positive (restoring) surface force density. Thus, for instability,

$$\frac{dT}{d\xi}(\xi = 0) > 0 \Rightarrow \frac{g(\rho_b - \rho_a)a}{\rho_a} < U^2 \tag{6}$$

The effects of surface tension, finite wavelength and unsteady flow left out of this intuitive picture are brought in by a small-amplitude model that is little different from that developed in Sec. 8.9. What has changed is the transfer relation for the streaming layer, which in view of Eq. (c) from Table 7.9.1, is

$$\begin{bmatrix}\hat{p}^c\\ \\ \hat{p}^d\end{bmatrix} = \frac{j(\omega - k_zU)\rho_a}{k}\begin{bmatrix}-\coth ka & \dfrac{1}{\sinh ka}\\ \\ \dfrac{-1}{\sinh ka} & \coth ka\end{bmatrix}\begin{bmatrix}\hat{v}_x^c\\ \\ \hat{v}_x^d\end{bmatrix} \tag{7}$$

and the linearized relation between the vertical fluid velocity at the interface and the surface deformation (Eq. 7.5.5):

$$\hat{v}_x^d = \hat{v}_x^e = j(\omega - k_zU)\hat{\xi} \tag{8}$$

These expressions replace Eqs. 8.9.4a and 8.9.6 in an otherwise unchanged derivation resulting in the dispersion equation

$$\frac{(\omega - k_zU)^2\rho_a \coth ka}{k} + \frac{\omega^2\rho_b \coth ka}{k} = \gamma k^2 + g(\rho_b - \rho_a) \tag{9}$$

This expression is quadratic in ω but transcendental in k. Thus, there are only two temporal modes, but an infinite number of spatial modes. In the limit $U \rightarrow 0$, these are discussed in Sec. 8.9.

Consider first the temporal modes, having frequencies that follow from Eq. 9:

$$\omega = \frac{k_zU\rho_a \coth ka \pm\sqrt{(\rho_a \coth ka + \rho_b \coth kb)[\gamma k^3 + g(\rho_b - \rho_a)k] - \rho_b\rho_a k_z^2U^2 \coth ka \coth kb}}{\rho_a \coth ka + \rho_b \coth kb} \tag{10}$$

It follows that the temporal mode exhibits an oscillatory instability if the U is large enough to make the radicand negative. For instability,

$$U^2 > \left[\frac{\tanh kb}{\rho_b} + \frac{\tanh ka}{\rho_a}\right]\left[\frac{\gamma k^3}{k_z^2} + g(\rho_b - \rho_a)\frac{k}{k_z^2}\right] \tag{11}$$

Deformations with wavenumbers in the direction of streaming are most unstable, so in the following, $k = k_z$. The first wavelength to become unstable as U is raised can be found analytically in two limits. First, in the short-wave limit, where ka and kb >> 1, Eq. 11 reduces to

$$U^2 > \left[\frac{1}{\rho_b} + \frac{1}{\rho_a}\right]\left[\gamma k + \frac{g(\rho_b - \rho_a)}{k}\right] \tag{12}$$

This expression exhibits a minimum at the Taylor wavenumber

$$k^* = \sqrt{\frac{g(\rho_b - \rho_a)}{\gamma}} \tag{13}$$

which is familiar from Secs. 8.9 and 8.10. Again, this is the wavenumber of incipient instability and Eq. 12 gives the critical velocity as

$$U^* = \left\{\left[4g(\rho_b - \rho_a)\gamma\right]\left[\frac{1}{\rho_a} + \frac{1}{\rho_b}\right]^2\right\}^{1/4} \tag{14}$$

In the opposite extreme of long waves (ka and kb << 1) Eq. 11 becomes

$$U^2 > \left[\frac{b}{\rho_b} + \frac{a}{\rho_a}\right]\left[\gamma k^2 + g(\rho_b - \rho_a)\right] \tag{15}$$

In this limit, the first wavenumber to become unstable as U is raised is zero and the critical velocity is

$$U^* = \sqrt{\left(\frac{b}{\rho_b} + \frac{a}{\rho_a}\right) g(\rho_b - \rho_a)} \qquad (16)$$

Note that, in the limit $b/\rho_b << a/\rho_a$, this condition is the same as obtained by the intuitive arguments leading to Eq. 6.

With the understanding that the temporal modes so far considered here represent the response to initial conditions that are spatially periodic, or the response of a system that is reentrant in the z direction, note that the salient difference between the Kelvin-Helmholtz instability and the Rayleigh-Taylor instability of Sec. 8.9 is that the former appears oscillatory under the critical condition and as a growing oscillation for conditions beyond incipience. This is by contrast with the Rayleigh-Taylor instability which is static at incipience and grows exponentially. Note that the dynamic state at incipience is not included in the arguments leading to Eq. 6.

In connection with instabilities of the Rayleigh-Taylor type, it can be argued that because incipience occurs with a vanishing time rate of change, effects of fluid viscosity are ignorable. Here, the dynamic nature of the incipience points to inadequacies of the inviscid model.

To demonstrate the instability without there being more dominant mechanisms of instability coming into play, the configuration shown in Fig. 11.14.1 is arranged so that the upper fluid enters from a plenum to the left through a section that is perhaps of a length several times the channel height, a, to establish the essentially uniform velocity profile. This distance cannot be too large, or viscous effects from the wall and interface will have expanded into the mainstream. What is not desired is anything approaching a fully developed flow for $z > 0$. In fact, the viscous boundary layer will continue to expand over the interface, and what has been described is deformations of the interface that have effective lengths large compared to the dimensions of the boundary layer.

Complications that are not accounted for by the model include the following. First, if the interface is clean, the lower fluid will be set into motion by the viscous shear stress from the streaming fluid. Thus, the postulated static fluid equilibrium for the fluid below is not strictly valid. One way to avoid this complication would be to "turn on" the flow abruptly and establish the instability before there is time for much effect of the viscous shear stress. Another is to have an interface supporting a monomolecular film.[1] This would compress in the z direction, until the interface would be brought to a static state by the gradient in tension of the film. For certain types of films the incremental dynamics from this static equilibrium would then be as described here, with a surface tension consistent with incremental deformations about this static equilibrium.

What is actually observed at the interface could also be unrelated to the Kelvin-Helmholtz instability as modeled here because the developing boundary layer has become unstable over the interfacial region being observed. Boundary layer instability results in growing perturbations having a scale on the order of the boundary layer thickness.

Certainly, if the fluid entering from the left has a fluctuating component to begin with, interfacial motions would result that had little to do with the model. These are the types of difficulties that often make ambiguous association of the Kelvin-Helmholtz model with effects of wind and water.

In using the Kelvin-Helmholtz model to explain and predict phenomena, it is important to know whether it predicts absolute or convective instability. Does the interfacial deflection at a given position in Fig. 11.14.1 tend to grow in amplitude until it reaches a state of saturation, or is it capable of responding to an upstream sinusoidal excitation as a spatially growing wave?

Of the infinite set of spatial modes, only one exhibits a crossing of the k_r axis, and then only if Eq. 11 is satisfied. Thus, there is only one candidate for an amplifying wave. The other spatial modes must be evanescent, and are present to satisfy longitudinal boundary conditions.

Consider the (first four) lowest order spatial modes in the long-wave limit (ka and kb << 1). The dispersion equation, Eq. 9, is then quartic in k and quadratic in ω:

$$k^4 + k^2[G - U^2] + rU\omega k - \omega^2 = 0 \qquad (\underline{17})$$

where the wavenumber is normalized to an arbitrary length, ℓ, so that

$$k = \underline{k}/\ell; \quad \omega = \underline{\omega}\sqrt{\frac{\gamma}{\left(\frac{\rho_a}{a} + \frac{\rho_b}{b}\right)\ell^4}} \qquad (18)$$

---

1. J. T. Davis and E. K. Rideal, Interfacial Phenomena, Academic Press, 1963, Chap. 5.

$$G \equiv \frac{\ell^2 g(\rho_b - \rho_a)}{\gamma}; \quad r \equiv 2/\sqrt{1 + \frac{a\rho_b}{b\rho_a}} \qquad (18 \text{ cont.})$$

$$\underline{U} \equiv U\ell\sqrt{\frac{\rho_a}{a\gamma}}$$

The loci of the four roots to this expression in the complex k plane obtained by letting ω_i increase from $-\infty$ to 0 at fixed values of ω_r is shown in Fig. 11.14.3. It is clear that for the parameters summarized in the caption, there is a coalescence of roots coming from the lower and upper half planes. Thus, it is concluded that the instability is absolute and grows with time at a fixed location. This same conclusion is reached by Scott[2] in the short-wave limit ($|ka|$ and $|kb| \gg 1$).

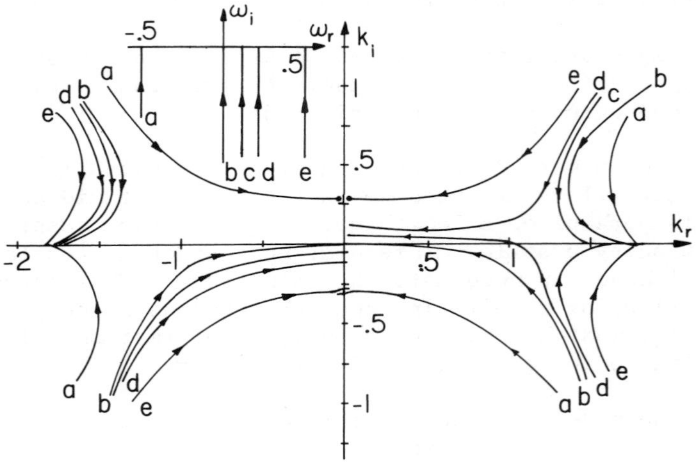

Fig. 11.14.3

For Kelvin-Helmholtz instability, loci of complex k resulting from varying ω as shown by inset. Coalescence of poles from upper and lower planes indicates that for parameters chosen, instability is absolute ($G = 1$, $U = 2$, $\gamma = 0.1$).

Spatially growing waves associated with wind blowing over water appear to require a model with more physical content than the simple one considered here.

11.15 Two-Stream Field-Coupled Interactions

Electric or magnetic coupling between streams having different velocities can occur without the necessity for physical contact between the streams. Thus, the long-wave models developed in Sec. 11.10 (Fig. 11.10.4) make a more satisfying representation of streaming interactions than does the traditional Kelvin-Helmholtz model of Sec. 11.14.[1] Transverse deflections of the respective streams are described by Eqs. 11.10.19-11.10.22, where e and v are as defined by Eq. 11.10.4. With f_1 and f_2 linearized, these expressions become

$$\left(\frac{\partial}{\partial t} + M_1 \frac{\partial}{\partial z}\right)^2 \xi_1 = \frac{\partial^2 \xi_1}{\partial z^2} + P\xi_1 - \frac{1}{2} P\xi_2 \qquad (1)$$

$$\left(\frac{\partial}{\partial t} + M_2 \frac{\partial}{\partial z}\right)^2 \xi_2 = \frac{\partial^2 \xi_2}{\partial z^2} + P\xi_2 - \frac{1}{2} P\xi_1 \qquad (2)$$

Normalization is as given by Eqs. 11.10.3 and 11.10.6. Remember, what is described is a pair of "strings" respectively convecting in the z direction with Mach numbers M_1 and M_2. With an electric field coupling the streams, P is positive, whereas with a magnetic field, P is negative. The "super" and "sub" responses to initial conditions and boundary conditions consistent with causality are discussed in detail in Sec. 11.10. There, the method of characteristics is used to describe the dynamics in real space and time.

The discussion is now limited to a description of the interaction of a stream with a fixed "string." That is, $M_2 = 0$ and $M_1 \equiv M$. The model is typical of interactions between electron beams

2. A. Scott, Active and Nonlinear Wave Propagation in Electronics, Wiley-Interscience, New York, 1970, pp. 83-87.

1. For a detailed discussion of the many possibilities for the model considered in this section, see F. D. Ketterer and J. R. Melcher, "Electromechanical Costreaming and Counterstreaming Instabilities," Phys. Fluids 11, 2179-2191 (1968) and "Electromechanical Stream-Structure Instabilities," Phys. Fluids 12, 109-117 (1969).

and transmission lines and plasmas. Counterstreaming and costreaming cases are considered in he problems.

What are the conditions for instability of the temporal modes, and is a given mode absolut ly or convectively unstable?

Substitution of complex amplitudes in Eqs. 1 and 2 results in the dispersion equation

$$[(\omega - Mk)^2 - k^2 + P][\omega^2 - k^2 + P] - \frac{1}{4}P^2 = 0 \qquad (3)$$

This expression is quartic in either ω or k. Subroutines for numerically finding the roots of polynomials are readily available.

Consider first the superelectric case, where $M > 1$ and $P > 0$. The temporal modes are summerizec by the plots of complex ω for real k illustrated in Fig. 11.15.1. For $M > 2$ and $M < 2$, respectively, Parts (a) and (c) illustrate the dynamics of the stream if the structure were held fixed, and of the structure if the stream were fixed. These are the superimposed dispersion relations for an absolute and a convective instability.

The self-consistent solutions to Eq. 3 are illustrated by Parts (b) and (d). For $M > 2$, Pt. (b), very small k results in four roots having the same real part and forming two complex conjugate pairs. At a midrange of k, two complex conjugate pairs result with the pairs having different real parts but imaginary parts of the same magnitude. At high k, waves are formed that are typical of those on the convecting string and fixed string respectively.

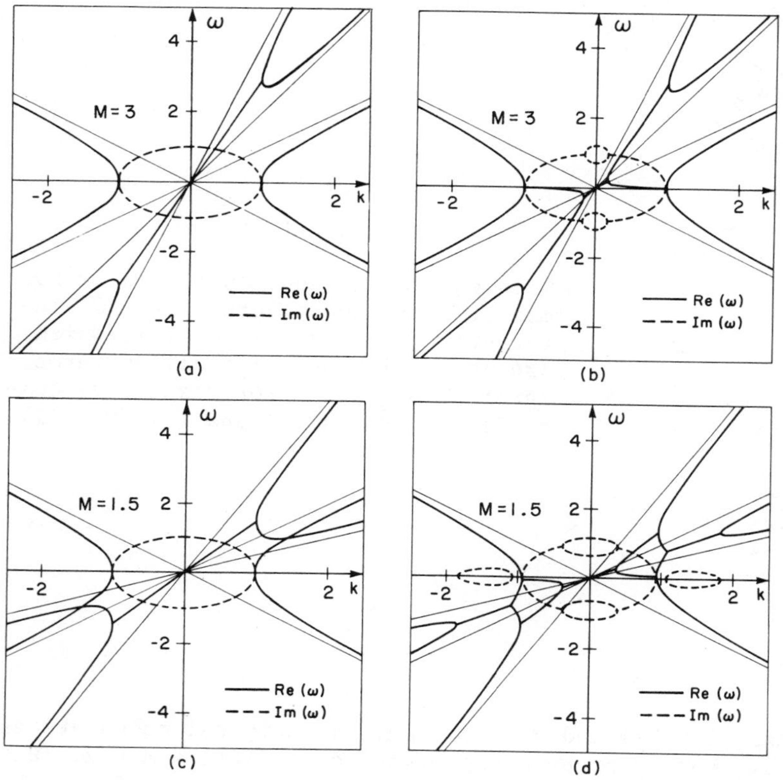

Fig. 11.15.1

Complex ω as a function of real k for electric stream structure ($P > 0$) interaction. Parts (a) and (c), which are drawn with mutual coupling terms ignored, are shown for respective comparison to Parts (b) and (d) ($P = 1$).

For $1 < M < 2$, an overstability results over a range of k where the uncoupled systems show no instability. This has a character that would not be expected from either the stream or the structure alone.

The plots of complex k for fixed ω_r as ω_i is increased from $-\infty$ to 0, shown in Fig. 11.15.2, disclose the character of these instabilities. One pair of roots move from above and below the k_r axis to coalesce in the upper half plane. Thus, this pair represent an absolute instability. The other pair migrate upward from the lower half plane, one extending into the upper half plane. This represents an amplifying wave.

It might be expected that the absolute instability eventually dominates the response. In the next section, a physical demonstration illustrates how this same system, confined to a finite length, can exhibit an oscillatory overstability that can be traced to the wave amplification.

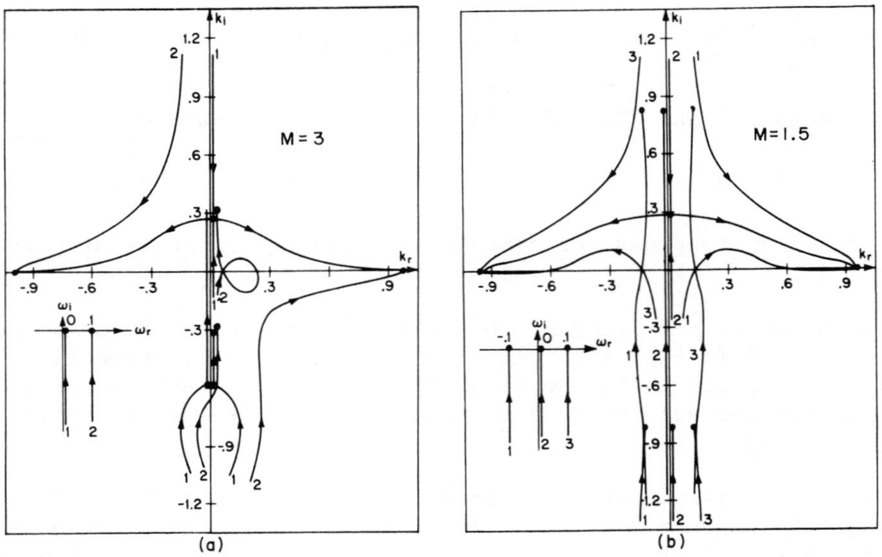

Fig. 11.15.2

For superelectric stream interacting with fixed "string" ($P > 0$) mapping of complex k plotted for fixed ω_r as ω_i is increased from $-\infty$ to 0.

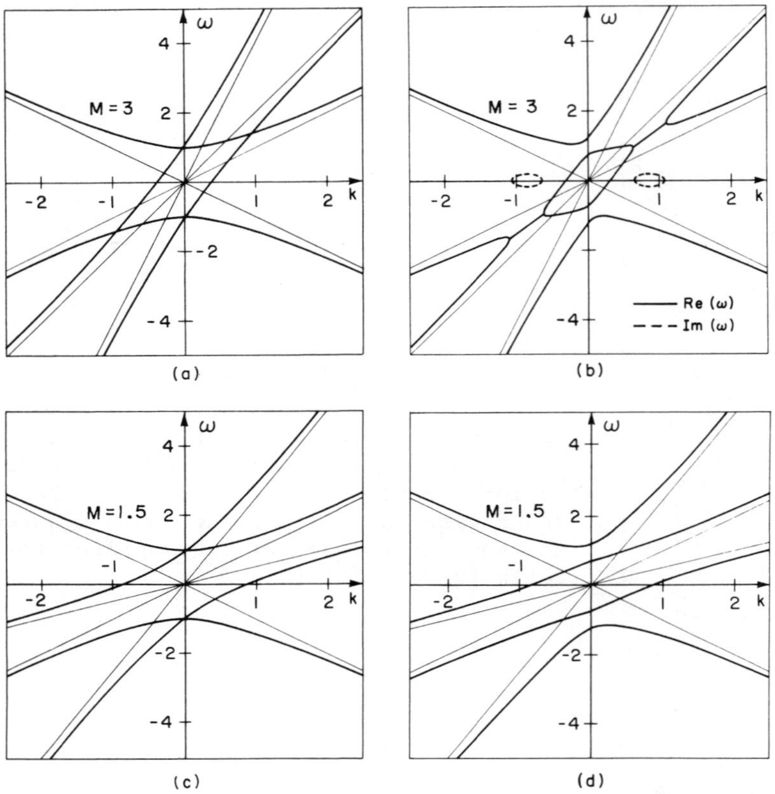

Fig. 11.15.3

Complex ω as a function of real k for magnetic stream-structure ($P < 0$) interaction. Parts (a) and (c), which are drawn with mutual coupling terms ignored, are shown for respective comparison to Parts (b) and (d) ($P = -1$).

Consider next the supermagnetic case, where $M > 1$ and $P < 0$. Now, the temporal modes are as shown by the dispersion plots of Fig. 11.15.3. Purely propagating waves result if $1 < M < 2$. However, if M is in the range $2 < M$, there is a range of k over which overstability is exhibited. Because the uncoupled stream and structure are completely stable, this temporal mode instability has some of the character of the Kelvin-Helmholtz instability. Note that it is similar to the high wavenumber overstability exhibited for the superelectric case by Fig. 11.15.1d. Whether the force on one stream due to a deflection of the other pushes or pulls, the result is a growing oscillation.

By contrast with the Kelvin-Helmholtz instability, this one is convective. This follows from the mapping of complex k for complex ω in the lower half plane illustrated (for this $2 < M$ case) by Fig. 11.15.4. The instability is indeed convective. The example is also worthy of remembrance because it illustrates how longitudinal boundary conditions, which have not yet come into play here, can have a dramatic effect. In Sec. 11.16, it is found that with boundary conditions (consistent with causality) this convective instability can become absolute.

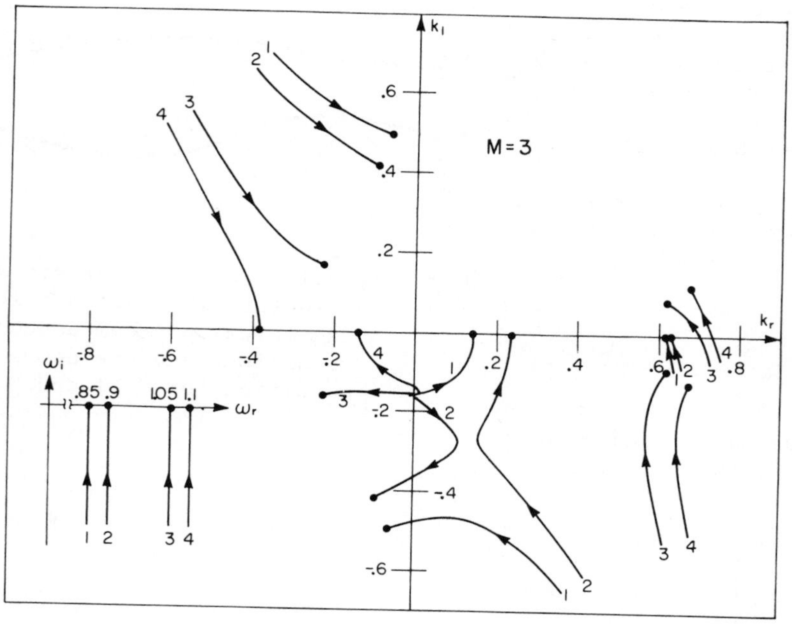

Fig. 11.15.4

For supermagnetic stream interacting with fixed "string" ($P < 0$), mapping of complex k plotted for fixed ω_r as ω_i is increased from $-\infty$ to 0.

11.16 Longitudinal Boundary Conditions and Absolute Instability

Quasi-one-dimensional linearized models, which retain the fundamental modes of a two-dimensional system having an infinite set of modes, can be obtained from the exact model by taking the long-wave limit. The coupled second order system used as a prototype model in Sec. 11.15 is one example. The following steps are generally applicable to determining the eigenfrequencies of such systems subject to homogeneous longitudinal boundary conditions. From a point of view somewhat different from that taken in Sec. 11.10, this section also emphasizes the importance of being sure that the boundary conditions are consistent with causality.

Suppose that there are N spatial modes. That is, with the frequency specified, there are N roots, k_n, to the dispersion equation $D(\omega,k_n) = 0$:

$$k_n = k_n(\omega) \tag{1}$$

Then, the response can be written as

$$\xi_2 = \mathrm{Re}\Big(\sum_{n=1}^{N} A_n e^{-jk_n z}\Big)e^{j\omega t} \tag{2}$$

where A_n are arbitrary coefficients. Although it remains to be found, the frequency of each term in Eq. 2 is the same. Thus, for initial conditions having the z dependence of the function in brackets in Eq. 2, the response would have the one (generally complex) frequency, ω. All other dependent variables can be written in terms of these solutions, and hence in terms of the N coefficients, A_n.

There are N homogeneous boundary conditions imposed either at $z = 0$ or at $z = \ell$. Evaluated using Eq. 2 and the other dependent variables written in terms of these same coefficients, these boundary conditions comprise N equations that are linear in the coefficients, A_n. The condition that the coefficient determinant vanish,

$$\mathrm{Det}(\omega,k_1,k_2,\cdots k_N) = 0 \tag{3}$$

together with the dispersion equation, Eq. 1, is the desired eigenfrequency equation.

To be specific, consider the stream coupled (either by the electric field or by a magnetic field) to a stationary continuum. The configuration is as shown physically by Fig. 11.10.4 and described by Eqs. 11.15.1 and 11.15.2 with $M_2 = 0$. Longitudinal boundary conditions, illustrated schematically by Fig. 11.16.1, are

$$\xi_1(0,t) = 0;\quad \frac{\partial \xi_1}{\partial z}(0,t) = 0;\quad \xi_2(0,t) = 0;\quad \xi_2(\ell,t) = 0 \tag{4}$$

That these are consistent with causality is demonstrated in Sec. 11.10 by appealing to the method of characteristics. Here, the same conclusions might be reached by identifying three of the four spatial modes with the upstream boundary where $z = 0$ and the other with the downstream boundary where $z = \ell$. Whether the coupling is electric (Fig. 11.15.2) or magnetic (Fig. 11.15.4), the mapping of the dispersion equation with ω_i increased from $-\infty$ to 0 at constant ω_r into the k plane gives rise to three loci originating in the lower half plane and one in the upper half plane.

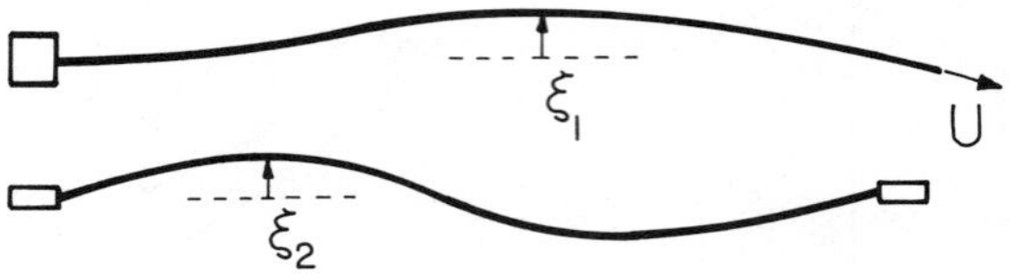

Fig. 11.16.1. Schematic of stream-structure interaction, showing three upstream boundary conditions and one downstream condition for $M_1>1$ and $M_2=0$.

The stream deflection is given by Eq. 2 with N = 4. It follows from Eq. 11.15.1 that the stationary continuum has a deflection that can be written in terms of these same coefficients as

$$\xi_1 = \text{Re}\Big(\sum_{N=1}^{4} Q_n A_n e^{-jk_n z}\Big)e^{j\omega t}; \quad Q_n \equiv \frac{2}{P}(\omega^2 - k^2 + P) \qquad (5)$$

The four linear equations obtained by using Eqs. 2 and 5 to express the homogeneous boundary conditions, Eqs. 4, give

$$\begin{bmatrix} 1 & 1 & 1 & 1 \\ e^{-jk_1\ell} & e^{-jk_2\ell} & e^{-jk_3\ell} & e^{-jk_4\ell} \\ Q_1 & Q_2 & Q_3 & Q_4 \\ k_1Q_1 & k_2Q_2 & k_3Q_3 & k_4Q_4 \end{bmatrix} \begin{bmatrix} A_1 \\ A_2 \\ A_3 \\ A_4 \end{bmatrix} = 0 \qquad (6)$$

The determinant of the coefficients in this expression set equal to zero takes the form of Eq. 3 and is a complex equation for the complex variable, ω. Remember that the k_n's are determined in terms of the frequency from the dispersion equation found from Eqs. 11.15.1 and 11.15.2.

In general, finding the roots of the complex transcendental combination of Eqs. 3 and 1 is difficult.[1] Carrying out a numerical search for the roots is as straightforward as using the root-finding techniques illustrated by Fig. 5.17.5. However, unless the computer is guided, it can easily fail to converge on certain roots. One way to obtain solutions in a relatively automated way is to start by finding the roots in a limit in which one of the parameters is small enough to make an analytical approximation possible. Then, these roots can be followed as that parameter is raised to the desired value.

Numerically determined normalized eigenfrequencies are shown as a function of the normalized length in Fig. 11.16.2. Illustrated here is the electrically coupled system ($P > 0$). The first mode reflects the destabilizing effect of the electric field. At low fields, the effect of the coupling is to produce damping, but as the field is raised, there is a threshold for static instability in this mode, much as if the stationary continuum were coupled to rigid walls. The real part of the frequency below this threshold and the condition for incipience are little different from what would be obtained if the stream were rigid.

The effect of the coupling on the second mode is something new. According to the model (which ignores viscous drag), this mode is overstable with the application of even the slightest electric field.

In the magnetic case, with eigenfrequencies shown in Fig. 11.16.3, the static instability of the lowest mode is replaced by a damping that only becomes larger as the field is raised. But, reflecting a regenerative feedback mechanism, the higher order modes exhibit overstabilities similar to those for the electric case. The eigenfunctions for the first three modes, illustrated by Fig. 11.16.4, give some idea as to the origins of this spontaneous oscillation of growing amplitude.

The theoretical second eigenfunction for the electric case, having eigenfrequencies given by Fig. 11.16.2, has an appearance little different from that of the demonstration experiment shown by Figs. 11.16.5 and 11.16.6. In the first of these, a time exposure is shown of the water jet coupled electrically to a stretched spring. A sequence of instantaneous exposures of the same system following the dynamics through one oscillation is shown in the second figure. Deflections that amplify spatially on the jet are fed back upstream by the spring with a phase that makes the feedback loop regenerative.

---

1. F. D. Ketterer, Electromechanical Streaming Interactions, Ph.D. Thesis, Department of Electrical Engineering, Massachusetts Institute of Technology, 1965.

That the system with magnetic field coupling displayed the same instability makes it clear that this mechanism of instability has little to do with the nature of the coupling between stream and structure. It is the reflection of a wave on the spring by the downstream boundary condition that turns the magnetic field configuration from an infinite system exhibiting an amplifying wave into one that is finite and absolutely unstable.

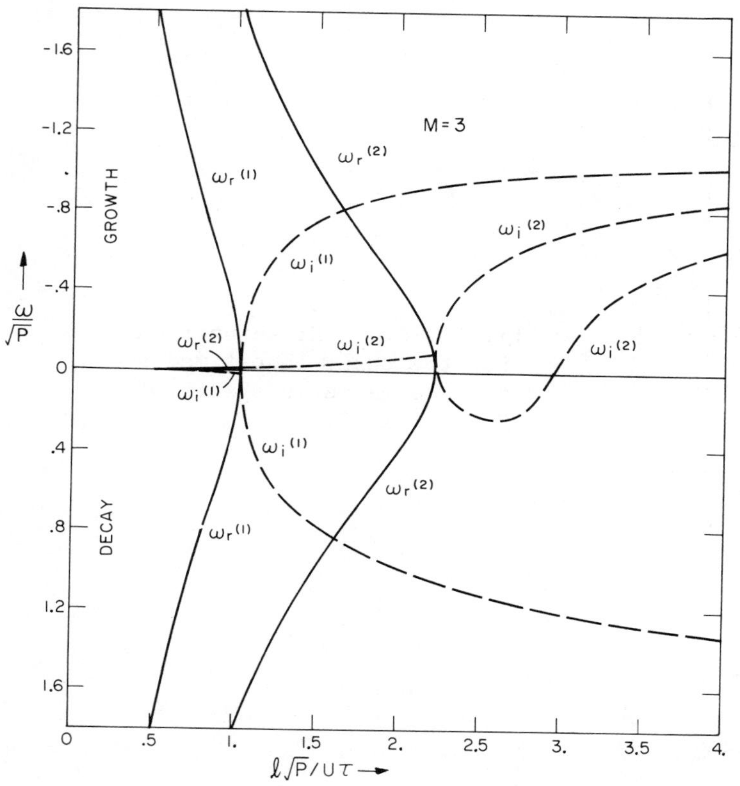

Fig. 11.16.2

Complex eigenfrequency as a function of normalized length for superelectric stream-structure interaction.

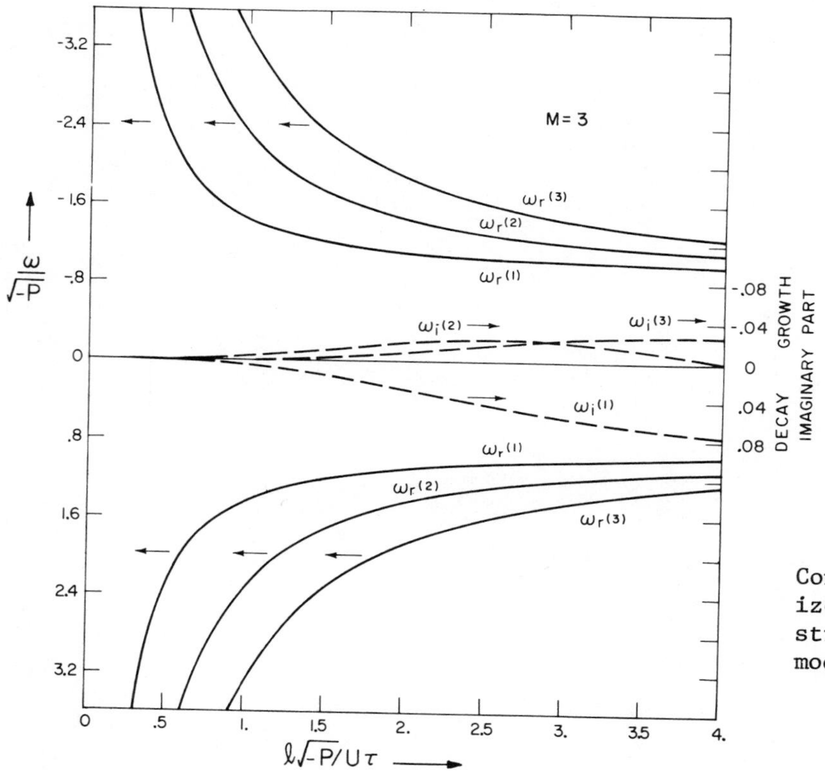

Fig. 11.16.3

Complex eigenfrequency versus normalized length for a magnetic stream-structure system for the lowest three modes.

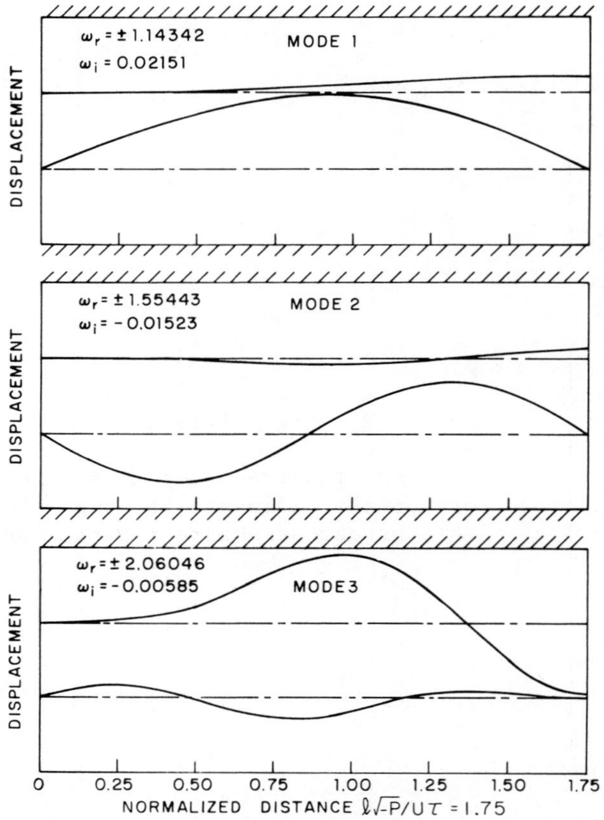

Fig. 11.16.4. Eigenfunctions for the three lowest modes for the magnetic system of Fig. 11.16.3.

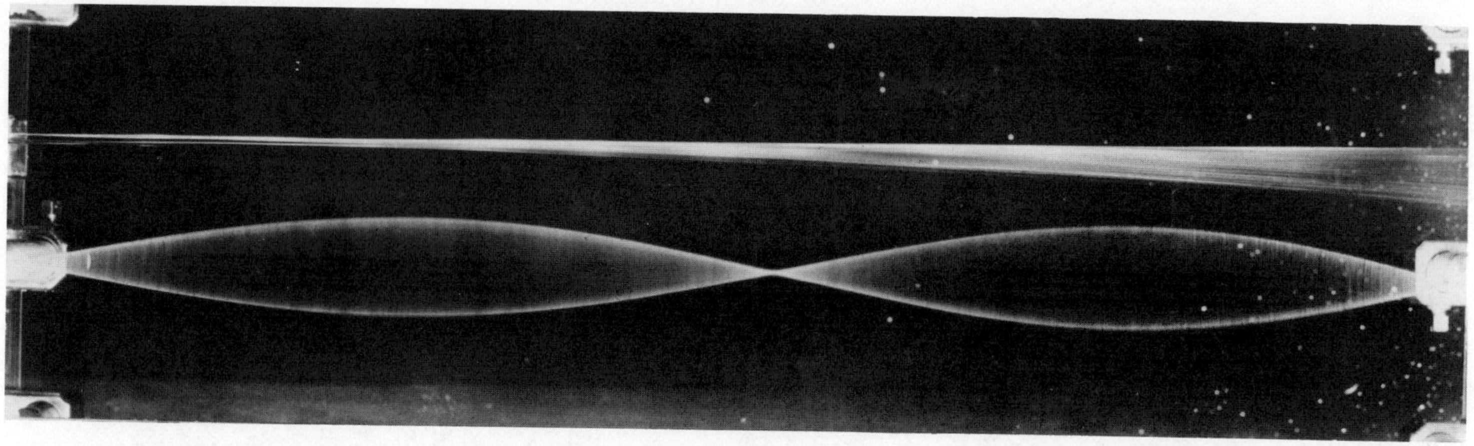

Fig. 11.16.5

Time exposure over several periods of oscillation during build-up of electric-field-coupled streaming overstability ($P>0$, $|M|>1$). Spring (bottom) and jet (top) are essentially in second mode (from film Reference 11, Appendix C).

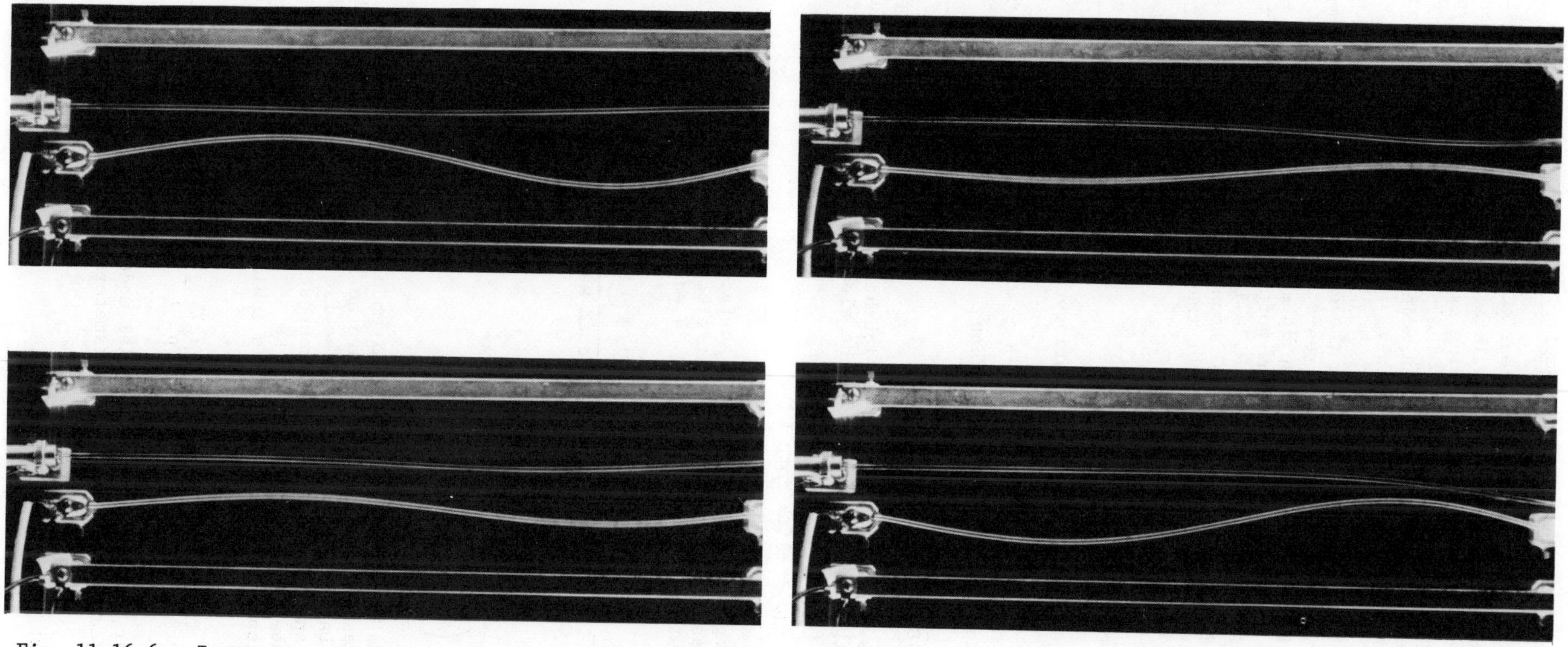

Fig. 11.16.6. Instantaneous photographs of the streaming overstability in Fig. 11.16.5. The time interval between exposures is equivalent to about 60° in phase. The frequency of oscillation is about 7 Hz (from film Reference 11, Appendix C).

11.17 Resistive-Wall Electron Beam Amplification

Either as a limitation in a linear accelerator or as the basis for making an amplifier,[1] the convective instability induced by the interaction of an electron beam with charges induced on a neighboring wall provides an interesting example of how lossy and propagating systems can interact to produce a spatially growing wave. The planar version shown in Fig. 11.17.1 incorporates the beam described in Sec. 11.5. What has been added is walls if finite conductivity, σ, themselves backed by much more highly conducting material.

With the objective of obtaining the dispersion equation for the beam coupled to the resistive layer, observe that fields in the layer (described in Sec. 5.10) are represented by the flux-potential transfer relations [Eq. (a) of Table 2.16.1]. In view of the highly conducting material bounding the resistive layer, $\hat{\Phi}^a = 0$, and it follows that on the lower surface of the layer,

$$\hat{D}_x^b = \varepsilon k \coth kd\, \hat{\Phi}^b \tag{1}$$

At this surface, the potential is continuous

$$\hat{\Phi}^b = \hat{\Phi}^c \tag{2}$$

and conservation of charge requires that

$$(j\omega + \frac{\sigma}{\varepsilon})\hat{D}_x^b - j\omega\hat{D}_x^c = 0 \tag{3}$$

In view of Eqs. 1 and 2, this expression becomes one representing the layer as "seen" by the electron beam:

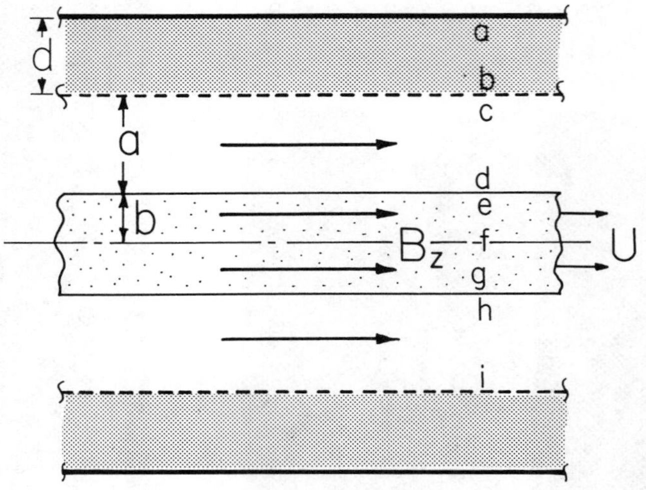

Fig. 11.17.1. Cross section of planar electron beam coupled to resistive wall.

$$\hat{D}_x^c = \frac{1}{j\omega}(j\omega + \frac{\sigma}{\varepsilon})(\varepsilon k \coth kd)\hat{\Phi}^c \tag{4}$$

The beam is represented at this same surface by Eq. 11.5.11 (with $\varepsilon = \varepsilon_o$). (Remember that it has already been assumed in Sec. 11.5 that electron motions are even.) Thus, the dispersion equation is obtained by setting D_x^c as given by Eq. 4 equal to D_x^c as given by Eq. 11.5.11:

$$\frac{1}{j\omega}(j\omega + \frac{\sigma}{\varepsilon})(\varepsilon k \coth kd) = \frac{-\varepsilon_o k(k + \gamma \coth ka \tanh \gamma b)}{k \coth ka + \gamma \tanh \gamma b} \tag{5}$$

where

$$\gamma^2 = k^2\left(1 - \frac{\omega_p^2}{(\omega - kU)^2}\right) \tag{6}$$

The discussion of temporal and spatial modes for the beam coupled to an equipotential wall given in Sec. 11.5 makes it clear that there are an infinite number of either temporal or spatial modes. Because of the Laplacian character of the field that couples the wall to the beam, the interaction between wall and beam tends to be strongest for the longest wavelengths. Thus, the long-wave limit of Eq. 5 is now taken by considering $ka \ll 1$, $kb \ll 1$, and $kd \ll 1$. Thus, Eqs. 5 and 6 become a polynomial dispersion equation that is cubic in ω and quartic in k,

$$(j\underline{\omega}R_e + 1)\left\{(\underline{\omega} - \underline{k})^2[\frac{b}{a} + \underline{k}^2] - \underline{\omega}_p^2\underline{k}^2\right\} = -j\underline{\omega}R_e r\underline{k}^2[(\underline{\omega} - \underline{k})^2(1 + \frac{b}{a}) - \frac{b}{a}\underline{\omega}_p^2] \tag{7}$$

and the higher order transverse spatial modes are neglected. Here the frequency and wavenumber have been normalized and dimensionless parameters introduced:

$$\underline{\omega} = \frac{b}{U}\omega; \quad R_e \equiv \frac{U\varepsilon}{b\sigma}; \quad r \equiv \frac{d\varepsilon_o}{b\varepsilon}$$

$$\underline{k} = kb; \quad \underline{\omega}_p \equiv \frac{b\omega_p}{U} \tag{8}$$

---

1. C. K. Birdsall, G. R. Brewer and A. V. Halff, "The Resistive-Wall Amplifier," Proc. IRE 41, 865-875 (1953).

The modes that have been retained by this long-wave model can be identified by considering two limits of Eq. 7. In the first, $\sigma \to \infty$ ($R_e \to 0$) and the expression reduces to one for space-charge oscillations superimposed on the convection:

$$\omega = kU \pm \frac{\omega_p k}{\sqrt{\frac{1}{ab} + k^2}} \tag{9}$$

These are the lowest modes from Sec. 11.5. In the second. the beam is removed by setting $\omega_p \to 0$. Then, Eq. 7 requires that

$$j\omega = -\frac{\sigma}{\varepsilon}\left[1 + \frac{d\varepsilon_o}{b\varepsilon}\frac{(1 + \frac{b}{a})k^2}{(\frac{1}{ba} + k^2)}\right] \tag{10}$$

This is the damped charge relaxation mode resulting from the ohmic loss in the layer and energy storage both within the layer and in the region of the gap and beam.

What can be expected for the coupled modes? First, to consider the temporal modes, Eq. 7 is written as a cubic in ω:

$$\omega^3\left\{jR_e[(\frac{b}{a} + k^2) + rk^2(1 + \frac{b}{a})]\right\} + \omega^2\left\{\frac{b}{a} + k^2 - 2jR_e k[(\frac{b}{a} + k^2) + rk^2(1 + \frac{b}{a})]\right\}$$

$$+\omega\left\{jR_e k^2[\frac{b}{a} + k^2(1 + r + \frac{br}{a}) - \omega_p^2(1 + \frac{rb}{a})] - 2k(\frac{b}{a} + k^2)\right\} + k^2(\frac{b}{a} + k^2 - \omega_p^2) = 0 \tag{11}$$

Numerical solution of this expression is illustrated by Fig. 11.17.2, where complex $\underline{\omega}$ is shown as a function of real $\underline{k}$. For the given parameters, the growth rate, $\underline{\omega}_{2i}$, is small compared to the other frequencies, and so it is shown on a separate plot.

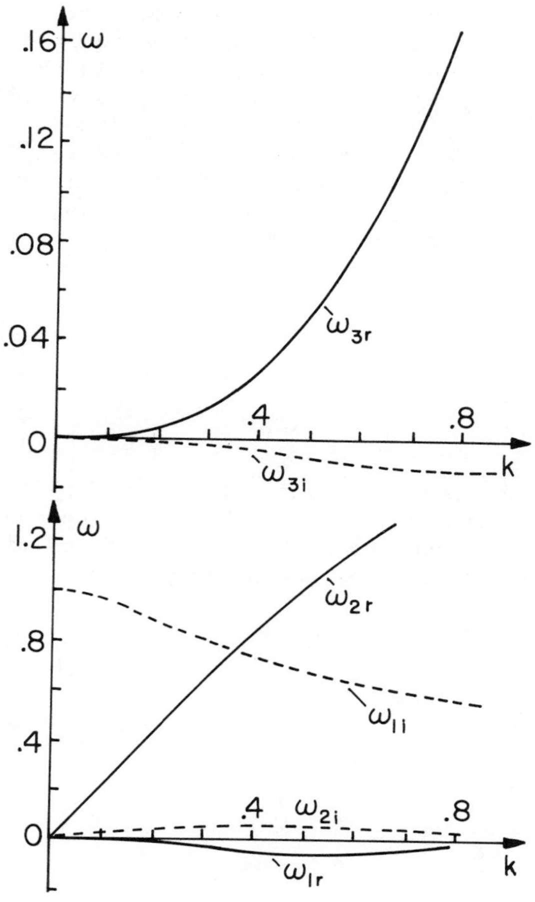

Fig. 11.17.2

Plot of dispersion equation for long-wave coupling of electron beam to resistive wall, showing normalized complex ω for normalized real k ($b/a = 1$, $r = 1$, $R_e = 1$, $\underline{\omega}_p = 1$).

That the system is unstable raises the second question. Would this instability result in a beam oscillation that grew in time at a fixed location, or would it result in a spatial amplification of any disturbance? The limitation on a particle accelerator, or the possibility of using the interaction to make an amplifier, hinge on whether the instability is convective or absolute.

In order to plot loci of complex k as ω is varied at constant ω_r from $\omega_i = -\infty$ to 0, Eq. 7 is expressed as a polynomial in $\underline{k}$:

$$k^4\left\{j\omega R_e[1 + r(1 + \frac{b}{a})] + 1\right\} + k^3\left\{-2j\omega^2 R_e[1 + r(1 + \frac{b}{a})] - 2\omega\right\}$$

$$+ k^2\left\{j\omega R_e[\omega^2 + \frac{b}{a} - \omega_p^2 + r\left(\omega^2(1 + \frac{b}{a}) - \frac{b}{a}\,\omega_p^2\right)] + (\omega^2 + \frac{b}{a} - \omega_p^2)\right\} \tag{12}$$

$$+ k\left\{-2\omega\,\frac{b}{a}(j\omega R_e + 1)\right\} + \frac{\omega^2 b}{a}(j\omega R_e + 1) = 0$$

The loci of the four roots to Eq. 12 obtained by varying ω_i from $-\infty$ to 0 with ω_r fixed are shown in Fig. 11.17.3. Of course, all values of ω_r must be considered. The ones shown are typical. Apparently the instability is convective. Thus, fluctuations on the beam as it enters the interaction region would amplify in space. To contend with an undesired instability on a beam having a given entrance fluctuation level, the length of the interaction region would have to be limited.

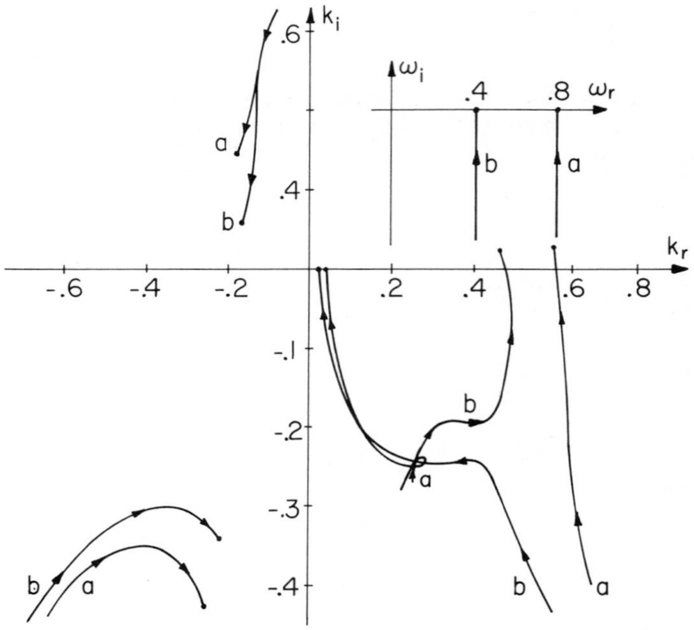

Fig. 11.17.3

For resistive wall interacting with electron beam, the loci of the four long-wave modes in the $\underline{k}$ plane are shown as $\underline{\omega_i}$ is varied from $-\infty$ to 0 holding $\underline{\omega_r}$ fixed. The inset shows the associated trajectories of $\underline{\omega}$ in the complex $\underline{\omega}$ plane. The instability is apparently convective.

The problems give some hint of the wide variety of physical situations in which amplifying waves result from coupling between charged particle streams and various types of structures and media. Electron beam interactions and plasma dynamics are rich with examples of the various forms of complex waves. So also is fluid mechanics, where boundary layer instabilities and other precursors of turbulence are often amplifying waves.

Problems for Chapter 11

For Section 11.2:

Prob. 11.2.1 Starting with the Lagrangian description of the particle motions afforded by Eq. 2, show that if $\vec{E} = -\nabla\Phi$ and Φ is independent of time, the sum of potential and kinetic energies, $\frac{1}{2}m(\vec{v}\cdot\vec{v})+q\Phi$, is invariant.

For Section 11.3:

Prob. 11.3.1 A planar version of the magnetron configuration considered in this section makes use of planar electrodes that play the role of the coaxial ones in Fig. 11.3.1. The cathode, which is at zero potential, is in the plane $x = 0$ while the anode is at $x = a$ and has the potential V. Thus, ignoring space-charge effects, the electric potential is Vx/a. A uniform magnetic flux density, B_o, is imposed in the z direction.

(a) Write the equations of motion for an electron in terms of its position $\vec{r} = x\vec{i}_x + y\vec{i}_y + z\vec{i}_z$.

(b) Combine these expressions to obtain a single second-order differential equation for $x(t)$.

(c) Under the assumption that V is constant, integrate this expression once (in a way analogous to that used in obtaining Eq. 11.3.2) to obtain a potential well that is determined by the combined effects of the applied electric and magnetic fields. Use this result to generate a potential-well picture analogous to Fig. 11.3.2 and qualitatively describe how the particle trajectories are influenced by the applied potential. (Assume that particles leave the cathode with no velocity.)

(d) Sketch typical trajectories in the x-y plane.

(e) What is the critical potential, $V = V_c$?

For Section 11.4:

Prob. 11.4.1 An electron has an initial velocity that is purely axial and moves in a region of constant potential and uniform axial magnetic field. Show that Eqs. 11.2.3-11.2.5 are satisfied by a subsequent motion in which the electron continues to move in only the z direction. How is it that Eq. 11.4.8 can predict the focusing effect of the field even though it only involves the axial component of $\vec{B}$?

Prob. 11.4.2 In a magnetic lens, the potential, Φ, is constant. Use Eq. 11.4.8 to show that all magnetic lenses are converging, in the sense that dr/dz will be less as a particle leaves the lens than as it enters.

Prob. 11.4.3 If the focal length is long compared to the magnetic length of the magnetic lens, it is said to be a weak lens. In this case, most of the deflection occurs outside the field region with the lens serving to redirect the particles. Essentially, r remains constant through the lens, but its spatial derivative is altered. Use Eq. 11.4.8 to show that in this case the focal length is

$$f = \frac{8\Phi m}{e}\left(\int_{z_-}^{z_+} B_z^2\, dz\right)^{-1}$$

Show that this general result for a weak lens is consistent with Eq. 11.4.12.

Prob. 11.4.4 An electron beam crosses the $z = 0$ plane from the region $z < 0$ with the potential V_o. As it crosses, a given electron has radial position r_o and moves parallel to the z axis. In the $z = 0$ plane, a potential $\Phi = V_oJ_o(\gamma r)$ is imposed so that for $z > 0$, $\Phi = V_oJ_o(\gamma r)\exp(-\gamma r)$ (See Eq. 2.16.18 with $jk \to \gamma$ and $m = 0$.) Use the paraxial ray equation, Eq. 11.4.8, to determine the electron trajectory. Assume that $\gamma r_o \ll 1$.

For Section 11.5:

Prob. 11.5.1 Under what conditions does an axial magnetic field suppress transverse electron motions? To answer this question take the potential distribution as having been determined from Eq. 11.7.7 and write the transverse components of the force equation with the potential terms as "drives." Argue that transverse motions are small compared to longitudinal ones provided that the frequency is low compared to the electron-cyclotron frequency.

Prob. 11.5.2 An electron beam of annular cross section with outer radius a and inner radius b streams in the z direction with velocity U.

(a) Show that the transfer relations are as summarized in Table 2.16.2 with $k^2 \rightarrow \gamma^2 \equiv k^2[1-\omega_p^2/(\omega-kU)^2]$ in the coefficients.

(b) As an application of these transfer relations, consider a beam of radius b (no free-space core) with the potential constrained to be $\hat{\Phi}^c$ at the radius a. The region $b < r < a$ is free space. Find $\hat{D}_r^c$

(c) Find the eigenfrequencies of the temporal modes with a perfect conductor at r = a.

For Section 11.6:

Prob. 11.6.1 A physical situation is represented by m dependent variables x_j

$$x_j = x_1, \; \ldots\ldots \; x_j \; \ldots\ldots \; x_m$$

which satisfy m first-order partial differential equations

$$\sum_{j=1}^{m} [F_{ij}\frac{\partial x_j}{\partial t} + G_{ij}\frac{\partial x_j}{\partial z}] = 0$$

(For example, in Sec. 11.6, m = 2, $x_1 = \rho$ and $x_2 = v$.) The coefficients F_{ij} and G_{ij} are functions of the x's as well as (z,t).

(a) Use the method of undetermined multipliers to find a determinantal equation for the first characteristic equations.

(b) Show that the same determinantal equation results by requiring that the coefficient matrix vanish.

For Section 11.7:

Prob. 11.7.1 There is a complete analogy between shallow-water gravity-wave dynamics and the one-dimensional compressible motions of gases studied in this section. Use Eqs. 9.13.11 and 9.13.12 with V = 0 and b = constant to show that if $\gamma = 2$, analogous quantities are

$$\rho \leftrightarrow \xi, \quad v \leftrightarrow v, \quad \frac{a^2}{\rho} \leftrightarrow g$$

The analogy is exploited in the film "Waves in Fluids," which deals with both types of wave systems. (See Reference 10, Appendix C.)

Prob. 11.7.2 The quasi-one-dimensional equations of motion for free surface flow contained by an electric field (Fig. 9.13.3) are Eqs. 9.13.4 and 9.13.9, with A and ρ given in Fig. 9.13.3. Find the associated first and second characteristic equations.

Prob. 11.7.3 An inviscid, incompressible fluid rests on a rigid flat bottom, as shown in Fig. 9.13.1. Consider the motions with V = 0.

(a) Show that the first characteristic equations are

$$\frac{dz}{dt} = v \pm \tfrac{1}{2}R(\xi) \text{ on } C^{\pm}$$

and that the second characteristic equations are

$$v = \mp R(\xi) + c_{\pm} \quad \text{on } C^{\pm}$$

where $R(\xi) = 2\sqrt{g\xi}$.

(b) A simple wave propagates into a region of constant depth ξ_c and zero velocity. At z = 0, $\xi = \xi_s(t)$. Find v and ξ for z > 0, t > 0. Show that only if $d\xi_s/dt \gtrless 0$ will a shock form.

(c) Consider the initial value problem: when t = 0, $v = v_o(z,0) = 1$ and $\xi = \xi_o(z,0)$, where

Prob. 11.7.3 (continued)

$$\xi_o(z,0) = \begin{cases} 1 & \text{for } z < -3 \\ -0.3|z| + 1.9 & \text{for } -3 < z < 3 \\ 1 & \text{for } 3 < z \end{cases}$$

For computational purposes, set $g = 1$ and use seven characteristics in each family originating at equal intervals along the initial pulse. Evaluate the Riemann invariants c_+ and c_- along the seven C^+ and seven C^- characteristics.

(d) Find v and $R(\xi)$ at each of the characteristic intersections.

(e) Find the characteristic slopes at each of the intersections and draw the characteristics in the z, $t > 0$ plane.

(f) Plot the velocity as a function of z when $t = 0$, $t = 2$ and $t = 4$.

Prob. 11.7.4 The region between plane parallel conducting plates having essentially infinite extent in the y-z plane is filled with material described by the constitutive laws

$$\vec{B} = \mu_o\vec{H}; \quad \vec{D} = \varepsilon\vec{E} + \delta(\vec{E}\cdot\vec{E})\vec{E}$$

where μ_o, ε and δ are constants. The plates form a transmission line for transverse electromagnetic plane waves propagating in the z direction.

(a) Show that if $\vec{E} = E(z,t)\vec{i}_x$ and $\vec{H} = H(z,t)\vec{i}_y$, Faraday's and Ampere's laws require that

$$\frac{\partial E}{\partial z} = -\mu_o\frac{\partial H}{\partial t}; \quad \frac{\partial H}{\partial z} = -(3\delta E^2+\varepsilon)\frac{\partial E}{\partial t}$$

(b) Show that the first characteristic equations are

$$\frac{dz}{dt} = \pm[\mu_o(3\delta E^2 + \varepsilon)]^{-\frac{1}{2}} \quad \text{on } C^{\pm}$$

and that the second characteristic equations are

$$H = \mp R(E) + c_{\pm} \quad \text{on } C^{\pm}$$

where $R(E)$ is

$$R(E) = \sqrt{\frac{3\delta}{4\mu_o}}\left[E\sqrt{E^2 + \frac{\varepsilon}{2\delta}} + \frac{\varepsilon}{3\delta}\ln\left(E + \sqrt{E^2 + \frac{\varepsilon}{3\delta}}\right)\right]$$

(c) With no electric or magnetic fields between them when $t = 0$, the plates are driven at $z = 0$ by a voltage source that imposes the field $E(0,t) = E_o(t)$. Prove that the problem is over-specified by imposition of $H(0,t) = H_o(t)$. Show that all C^+ characteristics are straight lines. Take μ_o, ε and δ as unity and the drive

$$E(0,t) = E_o(t)$$

where

$$E_o = \begin{cases} 0 & \text{for } t < 0 \\ \dfrac{1 - \cos(t\pi/2)}{2} & \text{for } 0 < t < 2 \\ 1 & \text{for } 2 < t \end{cases}$$

and draw the C^+ characteristics. (Use 7 lines originating at equal intervals in time from $0 < t < 2$.) Will a shock form? Use a sketch to show how the transient appears as it passes positions $z = 0$, $1/3$ and $2/3$ and sketch to show how E depends on z when $t = 0$, 1 and 2.

Prob. 11.7.5 An axially symmetric deformation of a theta-pinch plasma is shown in the figure. The plasma column has a radius $\xi(z,t)$ and is perfectly conducting. The total flux of magnetic field Λ_o trapped between the plasma and the surrounding perfectly conducting wall (radius a) is constant.

Prob. 11.7.5 (continued)

(a) Model the plasma as a perfectly conducting incompressible fluid and write quasi-one-dimensional equations of motion that include non-linear effects (see Sec. 9.13.)

(b) When t = 0, the plasma is static and has the "bulge" shown in Fig. P11.7.5b. Show that the interface never subsequently has a radius greater than ξ_{peak}.

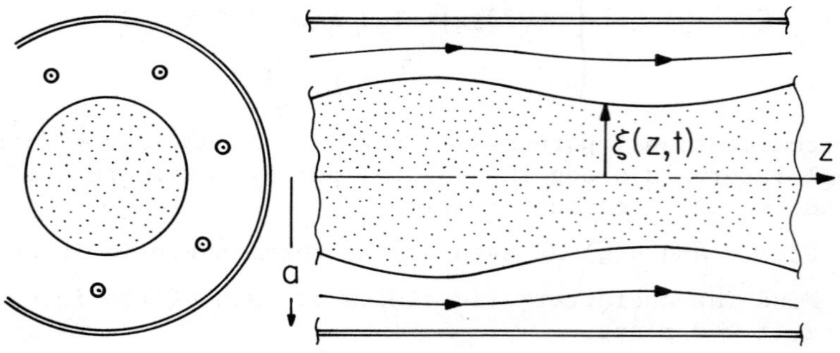

Fig. P11.7.5a

For Section 11.8:

Prob. 11.8.1 A perfectly conducting gas subject to isentropic dynamics is modeled as in Sec. 11.8. However, one-dimensional motions now considered are arbitrary in form. Consider again the one-dimensional space-time dependence of all variables; e.g. $\vec{v} = \vec{v}(x,t)$ and $\vec{H} = \vec{H}(x,t)$. Determine the first characteristic equations.

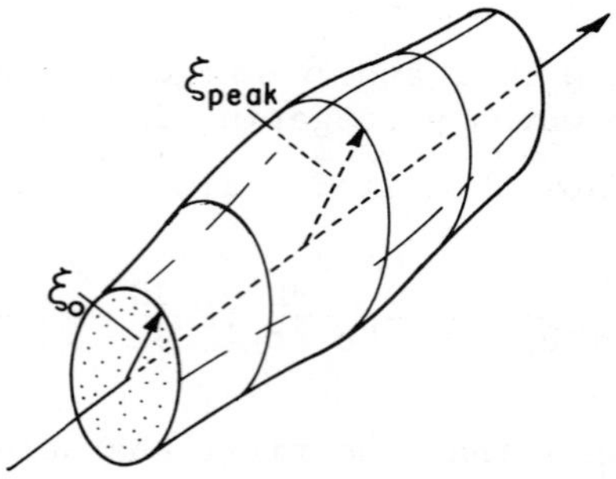

Fig. P11.7.5b

For Section 11.9:

Prob. 11.9.1 Show that for small amplitudes the phase space trajectory for the configuration of Fig. 11.9.1b is represented by an ellipse.

Prob. 11.9.2 In the configuration of Fig. 11.1.1b, the driving voltage is $V(t) = \text{Re}\,\hat{V}\exp j(\omega t)$, where $\hat{V}$ and ω are given. Under the assumption that V(t) is small enough to give only a linear response, use the characteristic equations to find n(z,t). Assume that the system is in the sinusoidal steady state.

For Section 11.10:

Prob. 11.10.1 Use Eqs. 11.10.7 and 11.10.8 to derive an analytical expression for the P=0 case illustrated by Fig. 11.10.3a.

Prob. 11.10.2 Consider the limit of the single-stream systems of Fig. 11.10.1 in which $\gamma\to\infty$ and hence $V\to 0$. In this case, the normalization velocity should be U rather than V.

(a) Show that the normalized equation of motion is

$$\left(\frac{\partial}{\partial t} + \frac{\partial}{\partial z}\right)^2 \xi = \frac{P}{4}\left[\frac{1}{(1-\xi)^2} - \frac{1}{(1+\xi)^2}\right]$$

where P is as defined after Eq. 11.10.6.

(b) In this limit the situation is similar to that of Sec. 11.9, in that the characteristics degenerate into the same lines. Show that with $\underline{v}\equiv\partial\underline{\xi}/\partial\underline{t} + \partial\underline{\xi}/\partial\underline{z}$ the normalized characteristic equations are

$$\frac{d}{dt}\left(\frac{1}{2}v^2 + E\right) = 0; \quad E \equiv -\frac{P}{4}\left(\frac{1}{1-\xi} + \frac{1}{1+\xi}\right) \quad \text{on} \quad \frac{dz}{dt} = 1$$

(c) In the phase-plane (v,ξ) a given element of the system starts with $(v,\xi) = (v_o,\xi_o)$. Sketch typical phase-plane trajectories for P > 0 and P < 0 and relate to the physical situations of Fig. 11.10.1.

Prob. 11.10.3 Starting with Eqs 11.10.19 to 11.10.22, use the determinant approach to deduce Eqs. 11.10.23 to 11.10.26. To deduce the solution at E in Fig. 11.10.5 from that at points A through D, the procedure is similar to that for the single stream situations (illustrated by Eqs. 11.10.11 to 11.10.14.) For example, because the solution at E must be the same whether obtained along the C_1^+ or C_1^- characteristics, $v_{1E} = v_{1A} + \Delta v_{1A}^+ = v_{1B} + \Delta v_{1B}^+$. Write eight equations, linear in $(\Delta v_{1A}^+, \Delta v_{1B}^+,$

Prob. 11.10.3 (continued)

Δv_{2C}^+, Δv_{2D}^+, Δe_{1A}^+, Δe_{1B}^+, Δe_{2C}^+, Δe_{2D}^+), that can be used to find these quantities in terms of the initial data (v_{1A}, v_{1B}, v_{2C}, v_{2D}, e_{1A}, e_{1B}, e_{2C}, e_{2D}, f_{1A}, f_{1B}, f_{2C}, f_{2D}). Check Eqs. 11.10.27 and 11.10.28, and the analogous expressions for (Δv_{2C}^{+}, Δe_{2C}^+).

Prob. 11.11.1 For the magnetic configuration of Fig. 11.11.1, follow steps paralleling Eqs. 11.11.1 and 11.11.2 to show that the long-wave linearized equation of motion is Eq. 11.11.3 with parameters defined by Eqs. 11.11.4b and 11.11.5b.

Prob. 11.11.2 A wire having tension T and mass per unit length m is stretched along the z axis. It suffers transverse displacements $\xi(z,t)$, in the x-z plane and carries a current I, as shown in Fig. P11.11.2. External coils are used to impose a magnetic flux density $\vec{B} = (B_o/d)(y\vec{i}_x + x\vec{i}_y)$. ($B_o$ and d are given constants) in the neighborhood of the z axis. This is the configuration for the experiment shown in Fig. 11.11.3. Show that the equation of motion takes the form of Eq. 11.11.3 with M = 0 and f = 0. What is P?

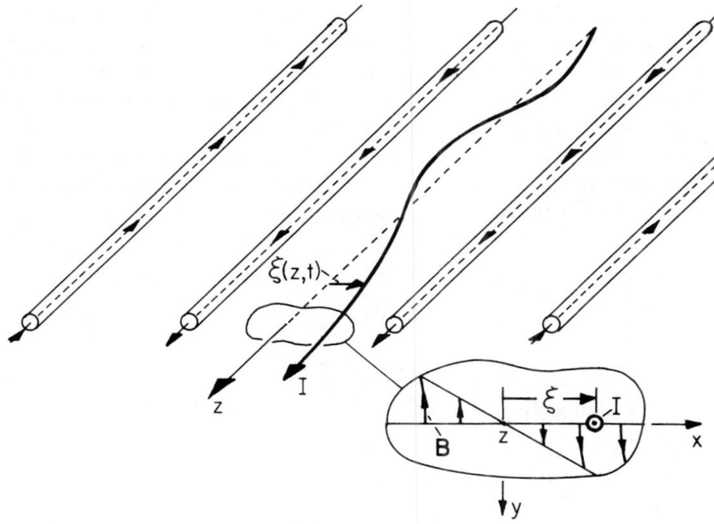

Fig. P11.11.2

Prob. 11.11.3 Derive Eqs. 11.11.13 and 11.11.14 starting with the roots of the dispersion equation, Eq. 11.11.11. Sketch the response for ω_o less than and greater than the cut-off frequency, indicating the physical significance of η and γ.

Prob. 11.11.4 Derive Eqs. 11.11.15 and 11.11.16 and sketch the response for ω_o^2 greater than and less than $P(M^2-1)$. Indicate the physical significance of η and γ.

Prob. 11.11.5 A liquid jet having equilibrium radius R streams in the z direction with velocity U. Coaxial with the jet is a circular cylindrical electrode having inner radius a and constant potential, -V, relative to the jet. The stream, perhaps tap water, can be regarded as perfectly conducting. Except for the coaxial electrode, the configuration is the same as described in Prob. 8.13.1.

(a) Determine the dispersion equation for perturbations having m = 1 (kinking motions).

(b) Take the long-wave limit of this relation ($ka \ll 1$ and hence $kR \ll 1$) to obtain a quadratic in ω and k representing the dominant modes of the system. Normalize this expression so that it takes the same form as Eq. 11.11.10 and define M and P accordingly.

Prob. 11.12.1 The equation for the deflection $\xi(z,t)$ of a "string" is the forced wave equation

$$\frac{\partial^2\xi}{\partial t^2} = V^2\frac{\partial^2\xi}{\partial z^2} + \frac{f(z,t)}{m}$$

where V^2 is the tension divided by the mass per unit length, m. The force per unit length, f(z,t), is an impulse in space and a sinusoidal function of time turned on when t=0.

$$f(z,t) = [(\Delta z)f_o]u_o(z)\cos\omega_o t\, \underline{u}_{-1}(t)$$

(a) Determine the Fourier-Laplace transform of the deflection, $\hat{\xi}(k,\omega)$. Assume that when t = 0 the deflection is zero.

(b) Invert the Fourier transform and use the residue theorem to find $\hat{\xi}(z,\omega)$.

(c) Invert the Laplace transform by using the residue theorem and causality to determine $\xi(z,t)$. Check the result to see that it satisfies initial and boundary conditions and the equation of motion.

Prob. 11.12.2 In Sec. 5.17, the spatial transient of charge induced on a moving semi-insulating sheet is considered. It is assumed that in the sinusoidal steady state considered the complex waves decay away from the region of excitation. Show that all spatial modes are indeed evanescent, thus justifying the identification of spatial modes used in Sec. 5.17. Consider in particular the two

Prob. 11.12.2 (continued)

dominant modes represented by the long-wave dispersion equation, Eq. 5.17.11, and show the loci of complex k as ω is varied ($\omega = \omega_r - j\sigma$, σ increasing from $\infty \to 0$).

For Section 11.13:

Prob. 11.13.1 For the subelectric second order system ($|M| < 1$, $P > 0$), evaluate the asymptotic response by integrating Eq. 11.13.1 on the contour shown in Fig. 11.13.4.

Prob. 11.13.2 A flexible tube is the conduit for liquid having an average velocity U. The tube is immersed in a viscous liquid which dampens transverse (kinking) motions. Its walls are elastic and (perhaps due to the viscous shear stresses from the liquid flowing within) under tension. A model for transverse motions pictures the tube as a string under tension. Because most of the inertia is due to the moving fluid within, the inertial term in the equation of motion is a convective derivative. But, because damping of transverse motions is largely due to the stationary external fluid, the damping force per unit length is proportional to the rate of change of the deflection at a given location, z. Thus, the equation of motion is

$$\left[\frac{\partial}{\partial t} + U\frac{\partial}{\partial z}\right]^2 \xi = V^2\frac{\partial^2 \xi}{\partial z^2} - \nu\frac{\partial \xi}{\partial t}$$

where V is the wave-velocity associated with the tension and mass per unit length and ν is the damping coefficient per unit length divided by the effective mass per unit length.

(a) Show that, if $|U| > V$, the tube is subject to kinking motions that are unstable.

(b) Show that if $|U| > V$, this instability is convective.

(c) Argue the result of part (b) using the method of characteristics.

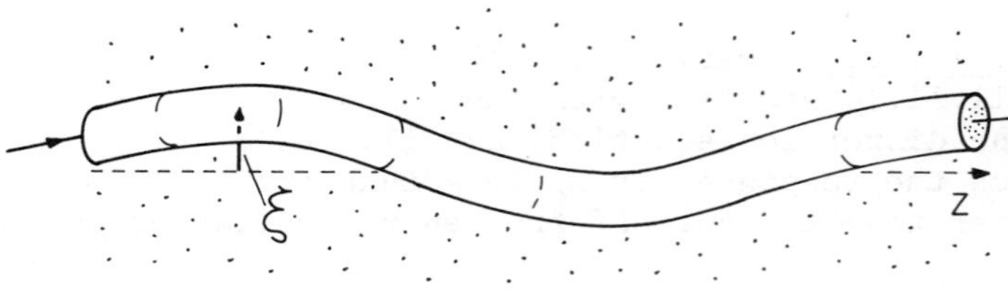

Fig. P11.13.2. Mechanical system exhibiting resistive wall instability.

For Section 11.14:

Prob. 11.14.1 In the configuration of Fig. 11.14.1, the lower fluid has a charge relaxation time that is short compared to times of interest, while the upper one is a good insulator, having uniform permittivity ε. The channel walls are metal and have a potential difference, V. Thus, with the interface flat, there is an x-directed uniform electric field in the upper fluid, $E_o = V/a$.

(a) Show that a stationary equilibrium exists, much as for the mechanical configuration. What is the stationary pressure difference across the interface under equilibrium conditions?

(b) For a perturbation having a given real wavenumber, k, what streaming velocity is required to produce instability? Discuss the conditions for instability in the short-wave and long-wave limits.

(c) Is this electromechanical Kelvin-Helmholtz type instability absolute or convective?

Prob. 11.14.2 In the configuration of Fig. 11.14.1, the lower fluid is perfectly conducting in the MQS sense while the upper fluid is a perfect insulator. The channel walls are also perfectly conducting. In a state of stationary equilibrium, much as for the mechanical system considered in Sec. 11.14, the region occupied by the upper fluid is also filled by an initially uniform z-directed magnetic field intensity H_o.

(a) What are the static pressures in each fluid under the stationary equilibrium conditions?

(b) For a given perturbation wavenumber, k, what velocity is required to produce instability? Does the magnetic field tend to stabilize the interface against Kelvin-Helmholtz instability?

(c) Are instabilities convective or absolute?

Prob. 11.14.3 A z-theta pinch has the configuration shown in Fig. 8.12.1. However, rather than having a static equilibrium, the perfectly conducting column now has a stationary equilibrium in which it streams in the z direction with a uniform velocity, U. Also, it is now surrounded by an insulating fluid having a mass density, ρ_v, that is appreciable. In the stationary equilibrium, the column has a uniform radius, R, and the surrounding fluid is static. For temporal modes, having given real wavenumber and mode number (k,m), what velocity U is required to produce instability?

Prob. 11.14.4 The configuration shown in Fig. 8.14.2 is revised by having the upper uniformly charged fluid stream in a direction parallel to the equilibrium interface with a velocity U.

(a) Determine the dispersion equation for perturbations of the interface.

(b) Determine the velocity required to induce instability of a perturbation having a given wavenumber, k.

For Section 11.15:

Prob. 11.15.1 Streams described by Eqs. 11.15.1 and 11.15.2 have equal and opposite velocities. That is, $M_1 = -M_2 = M$.

(a) Determine the dispersion equation and show that it is biquadratic in both ω and k.

(b) To describe temporal modes, make plots of complex ω for real k in the electric and magnetic cases.

(c) Are these instabilities absolute or convective?

For Section 11.16:

Prob. 11.16.1 Figure 11.16.4 shows the eigenfunctions implied by the first three longitudinal modes. Given the eigenfrequencies, describe how these functions would be obtained.

Prob. 11.16.2 Counterstreaming continua described by Eqs. 11.15.1 and 11.15.2 have equal and opposite velocities, and hence a dispersion equation as derived in Prob. 11.15.1.

(a) For $M < 1$, argue that boundary conditions consistent with causality are $\xi_1(0,t) = \xi_2(0,t) = \xi_1(\ell,t) = \xi_2(\ell,t) = 0$. What is the eigenfrequency equation? In the limit $M \to 0$, what are the solutions to this expression?

(b) For $M > 1$, argue that boundary conditions consistent with causality are $\xi_1 = 0$ and $\partial\xi_1/\partial z = 0$ at $z = 0$ and $\xi_2 = 0$ and $\partial\xi_2/\partial z = 0$ at $z = \ell$. What is the eigenfrequency equation?

For Section 11.17:

Prob. 11.17.1 Figure P11.17.1 shows distributed series of coils, each having n turns, lying in the y-z plane. Each coil is of dimension Δy in the y direction; Δy is small compared to other dimensions of interest shown. The system of coils is connected to a transmission line. Insofar as distances in the x direction are concerned, the coils comprise a thin sheet and are connected to the line outside the volume of interest. Hence, their effect on an MQS system is to be represented by the boundary conditions. Represent the circuit by an inductance and a capacitance per unit length, L and C respectively, and show that (with the understanding that B_x is continuous through the sheet) the coils are represented by the boundary conditions

$$-\frac{\partial v}{\partial y} = L\frac{\partial i}{\partial t} - nw\frac{\partial B_x}{\partial t}$$

$$-\frac{\partial i}{\partial y} = C\frac{\partial v}{\partial t}$$

$$[\![H_y]\!] = -n\frac{\partial i}{\partial y}$$

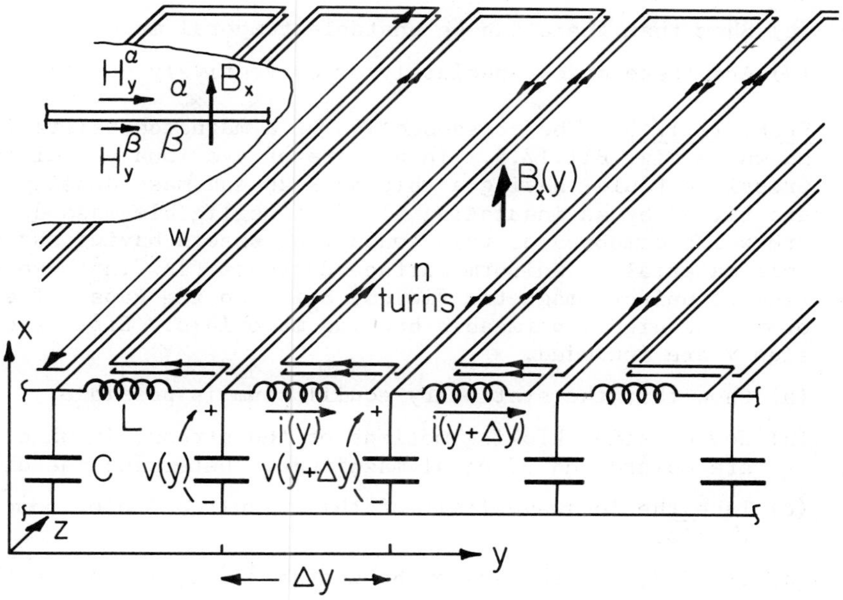

Fig. P11.17.1

Prob. 11.17.2 In what might be termed a magnetoquasistatic resistive wall interaction, the resistive wall moves and the continuum of "oscillators" remains fixed. Thus, the thin conducting sheet of Sec. 6.3 is backed by a perfectly permeable material and moves in the y direction with a velocity U. In a plane parallel to that of the sheet is the system of distributed coils of Prob. 11.17.1, connected to the transmission line as shown in Fig. P11.17.1. These coils comprise the "stator" structure, and are backed by an infinitely permeable material. There

Prob. 11.17.2 (continued)

is an air gap of thickness a between the moving sheet and coils.

(a) Find the dispersion equation describing the coupled transmission-line waves and convective magnetic diffusion modes.

(b) Show that the equation is cubic in ω, and discuss the temporal modes. Instability implies that, in a rotating configuration, the system could be used as a generator. Indeed, the foregoing is a distributed form of a self-excited induction generator. More conventionally, such generators are made by tuning the stator windings of a machine like that of Sec. 6.3.

(c) In a system that is infinite in the y direction, is the instability absolute or convective?

Prob. 11.17.3 Electrodes lying in the y-z plane, as shown in Fig. P11.17.3, have a length w in the y direction and are segmented in the z direction. Connected to the ends of the segments is a transmission line, having capacitance and inductance per unit length C and L respectively. The width of the segments in the z direction is short compared to lengths of interest in that direction. For an EQS system in which these electrodes appear as a plane having upper and lower surfaces denoted by α and β respectively, and hence where $\Phi^\alpha = \Phi^\beta \equiv v$, show that the jump conditions relating variables on the two sides are

$$-\frac{\partial v}{\partial z} = L\frac{\partial i}{\partial t}$$

$$-\frac{\partial i}{\partial z} = C\frac{\partial v}{\partial t} - w\frac{\partial \sigma_f}{\partial t}$$

$$\sigma_f = [\![D_x]\!]$$

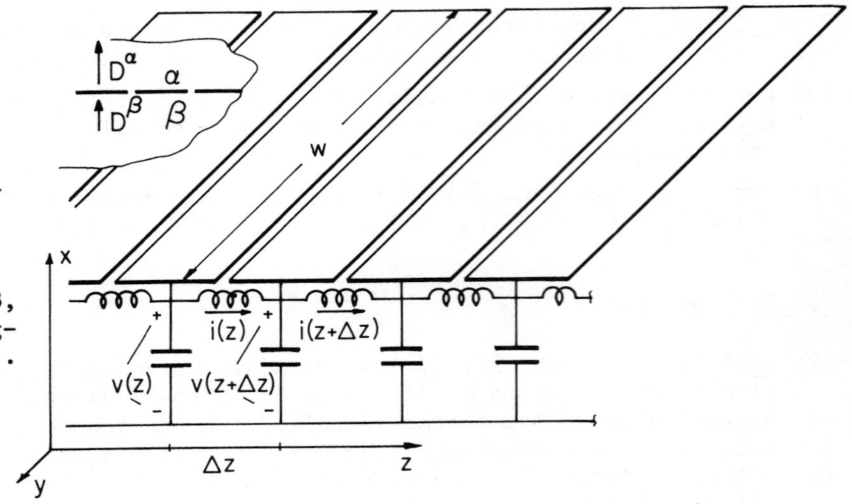

Fig. P11.17.3

Prob. 11.17.4 As a model for a traveling-wave electron-beam amplifier, consider the planar electron beam as shown in cross-section by Fig. 11.5.1. Two traveling-wave structures, described in Prob. 11.17.3, are coupled to this beam by placing the segmented electrodes in the planes (c) and (i). Respectively above and below these planes are parallel perfectly conducting sheets having spacings d from the planes of the segmented electrodes.

(a) For the even motions described in Sec. 11.5, determine the dispersion equation for the interaction.

(b) Take the long-wave limit of this expression to obtain a polynomial in ω and k.

(c) Show that there can be unstable temporal modes.

(d) Are these modes absolutely or convectively unstable?

Prob. 11.17.5 The cross-section of a magnetoquasistatic candidate for a resistive-wall amplifier is shown in Fig. P11.17.5. In a state of stationary equilibrium, a perfectly conducting inviscid incompressible fluid, having a thickness 2b and mass density ρ, streams to the right. The adjacent gaps are filled by an insulating fluid of negligible mass density. At distances a from the jet surfaces are walls composed of thin conducting sheets having surface conductivity σ_s, backed by highly permeable materials. Uniform surface currents flow in these sheets and on the free surfaces so as to give rise to uniform magnetic fields, $H_o\vec{i}_y$, in the gaps. There is no equilibrium magnetic field in the jet or in the highly permeable backing materials. Effects of gravity are absent, but those of surface tension γ are included.

(a) Show that the stationary equilibrium is possible.

(b) Now consider kinking motions of the stream, in which the transverse displacement of both interfaces are upward and of equal magnitude. Determine the dispersion equation.

(c) Take the longwave limit of this result and plot complex ω for real k. Assume that $U > V$ where $V \equiv \sqrt{2\gamma/\Delta\rho}$.

(d) If $U > V$, can the system be unstable and, if so, is the instability convective or absolute?

Prob. 11.17.6 Electron beams having equilibrium velocities U_1 and U_2 in the z direction share the same space. They have equilibrium number densities n_{o1} and n_{o2} respectively, and only interact through the macroscopic electric field $\vec{E} = -\nabla\Phi$.

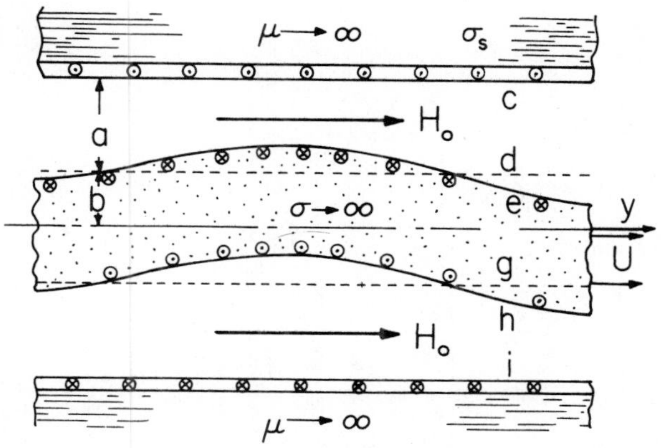

Fig. P11.17.5

(a) Use the EQS equations of motion for the two species suggested by Eqs. 11.10.1 - 11.10.4 to show that the dispersion equation for one dimensional motions is

$$1 = \frac{\omega_{p1}^2}{(\omega - kU_1)^2} + \frac{\omega_{p2}^2}{(\omega - kU_2^2)}$$

where

$$\omega_{p1} = \sqrt{n_{o1}e^2/\varepsilon_o m} \quad , \quad \omega_{p2} = \sqrt{n_{o2}e^2/\varepsilon_o m}$$

(b) In the limit $U_2 = 0$, this is the dispersion equation for an electron beam interacting with a stationary cold plasma. Show that the system is unstable and that the instability is convective.

(c) With $U_1 = -U_2$, what is described is the interaction of counterstreaming beams. Show that in this case the system is absolutely unstable.

Appendix A

Differential Operators in Cartesian, Cylindrical and Spherical Coordinates

APPENDIX A. Differential Operators in Cartesian, Cylindrical and Spherical Coordinates

Operator	Cartesian coordinates	Cylindrical coordinates	Spherical coordinates
	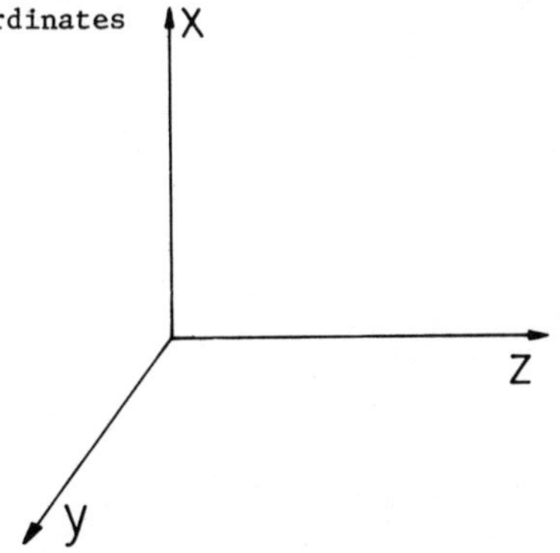	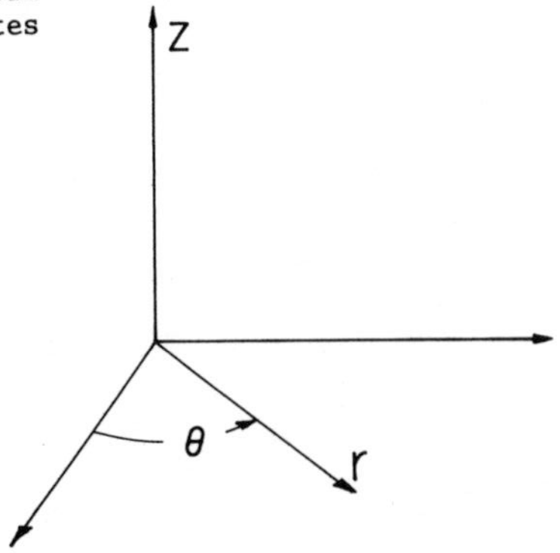	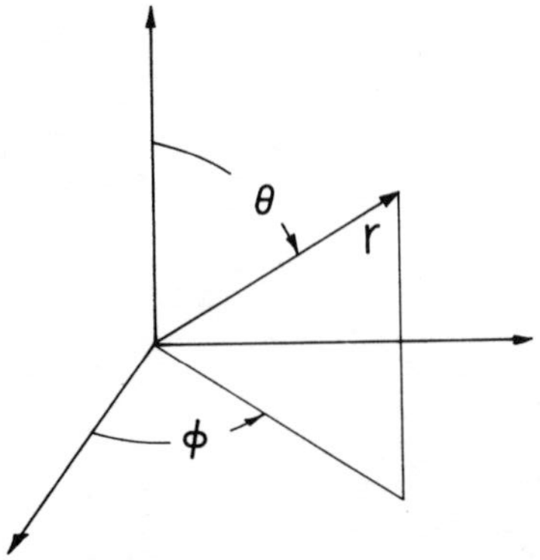
$(\nabla\cdot\vec{A})$	$\frac{\partial A_x}{\partial x} + \frac{\partial A_y}{\partial y} + \frac{\partial A_z}{\partial z}$	$\frac{1}{r}\frac{\partial}{\partial r}(rA_r) + \frac{1}{r}\frac{\partial A_\theta}{\partial \theta} + \frac{\partial A_z}{\partial z}$	$\frac{1}{r^2}\frac{\partial}{\partial r}(r^2 A_r) + \frac{1}{r\sin\theta}\frac{\partial}{\partial\theta}(A_\theta \sin\theta) + \frac{1}{r\sin\theta}\frac{\partial A_\phi}{\partial\phi}$
$\nabla\Phi$	$\frac{\partial\Phi}{\partial x}\vec{i}_x + \frac{\partial\Phi}{\partial y}\vec{i}_y + \frac{\partial\Phi}{\partial z}\vec{i}_z$	$\frac{\partial\Phi}{\partial r}\vec{i}_r + \frac{1}{r}\frac{\partial\Phi}{\partial\theta}\vec{i}_\theta + \frac{\partial\Phi}{\partial z}\vec{i}_z$	$\frac{\partial\Phi}{\partial r}\vec{i}_r + \frac{1}{r}\frac{\partial\Phi}{\partial\theta}\vec{i}_\theta + \frac{1}{r\sin\theta}\frac{\partial\Phi}{\partial\phi}\vec{i}_\phi$
$(\nabla^2\Phi)$	$\frac{\partial^2\Phi}{\partial x^2} + \frac{\partial^2\Phi}{\partial y^2} + \frac{\partial^2\Phi}{\partial z^2}$	$\frac{1}{r}\frac{\partial}{\partial r}\left(r\frac{\partial\Phi}{\partial r}\right) + \frac{1}{r^2}\frac{\partial^2\Phi}{\partial\theta^2} + \frac{\partial^2\Phi}{\partial z^2}$	$\frac{1}{r^2}\frac{\partial}{\partial r}\left(r^2\frac{\partial\Phi}{\partial r}\right) + \frac{1}{r^2\sin\theta}\frac{\partial}{\partial\theta}\left(\sin\theta\frac{\partial\Phi}{\partial\theta}\right) + \frac{1}{r^2\sin^2\theta}\frac{\partial^2\Phi}{\partial\phi^2}$
$(\nabla\times\vec{A})$	$\left(\frac{\partial A_z}{\partial y} - \frac{\partial A_y}{\partial z}\right)\vec{i}_x$ $+\left(\frac{\partial A_x}{\partial z} - \frac{\partial A_z}{\partial x}\right)\vec{i}_y$ $+\left(\frac{\partial A_y}{\partial y} - \frac{\partial A_x}{\partial y}\right)\vec{i}_z$	$\left(\frac{1}{r}\frac{\partial A_z}{\partial\theta} - \frac{\partial A_\theta}{\partial z}\right)\vec{i}_r$ $+\left(\frac{\partial A_r}{\partial z} - \frac{\partial A_z}{\partial r}\right)\vec{i}_\theta$ $+\left(\frac{1}{r}\frac{\partial}{\partial r}(rA_\theta) - \frac{1}{r}\frac{\partial A_r}{\partial\theta}\right)\vec{i}_z$	$\left(\frac{1}{r\sin\theta}\frac{\partial}{\partial\theta}(A_\phi\sin\theta) - \frac{1}{r\sin\theta}\frac{\partial A_\theta}{\partial\phi}\right)\vec{i}_r$ $+\left(\frac{1}{r\sin\theta}\frac{\partial A_r}{\partial\phi} - \frac{1}{r}\frac{\partial}{\partial r}(rA_\phi)\right)\vec{i}_\theta$ $+\left(\frac{1}{r}\frac{\partial}{\partial r}(rA_\theta) - \frac{1}{r}\frac{\partial A_r}{\partial\theta}\right)\vec{i}_\phi$

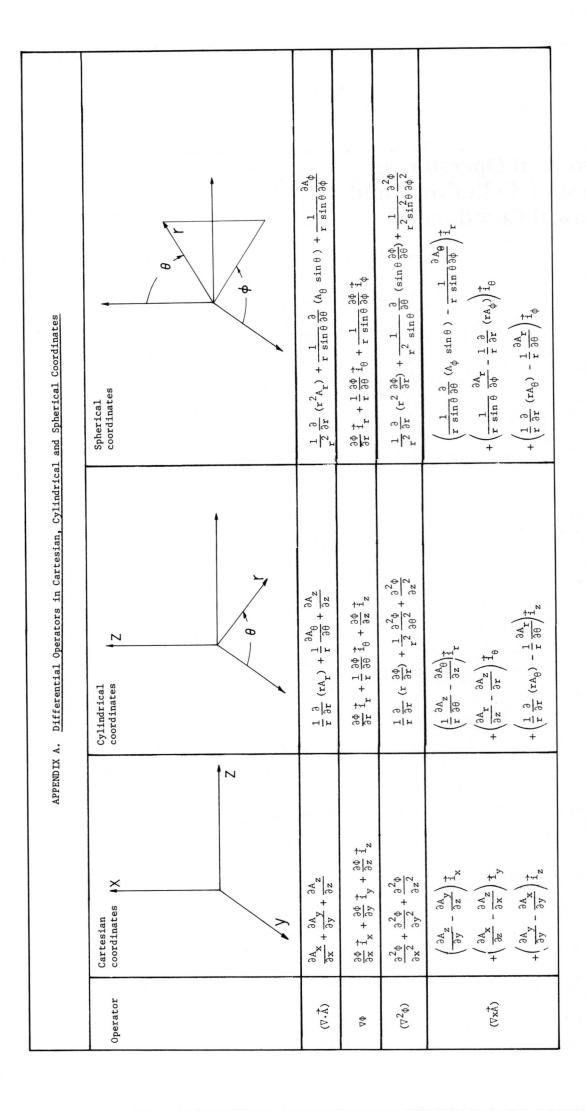

$\nabla^2\vec{A}$	$\left[\frac{\partial^2 A_x}{\partial x^2}+\frac{\partial^2 A_x}{\partial y^2}+\frac{\partial^2 A_x}{\partial z^2}\right]\vec{i}_x$ $+\left[\frac{\partial^2 A_y}{\partial x^2}+\frac{\partial^2 A_y}{\partial y^2}+\frac{\partial^2 A_y}{\partial z^2}\right]\vec{i}_y$ $+\left[\frac{\partial^2 A_z}{\partial x^2}+\frac{\partial^2 A_z}{\partial y^2}+\frac{\partial^2 A_z}{\partial z^2}\right]\vec{i}_z$	$\left[\frac{\partial}{\partial r}\left(\frac{1}{r}\frac{\partial}{\partial r}(rA_r)\right)+\frac{1}{r^2}\frac{\partial^2 A_r}{\partial\theta^2}-\frac{2}{r^2}\frac{\partial A_\theta}{\partial\theta}+\frac{\partial^2 A_r}{\partial z^2}\right]\vec{i}_r$ $+\left[\frac{\partial}{\partial r}\left(\frac{1}{r}\frac{\partial}{\partial r}(rA_\theta)\right)+\frac{1}{r^2}\frac{\partial^2 A_\theta}{\partial\theta^2}+\frac{2}{r^2}\frac{\partial A_r}{\partial\theta}+\frac{\partial^2 A_\theta}{\partial z^2}\right]\vec{i}_\theta$ $+\left[\frac{1}{r}\frac{\partial}{\partial r}\left(r\frac{\partial A_z}{\partial r}\right)+\frac{1}{r^2}\frac{\partial^2 A_z}{\partial\theta^2}+\frac{\partial^2 A_z}{\partial z^2}\right]\vec{i}_z$	$\left[\nabla^2 A_r-\frac{2A_r}{r^2}-\frac{2}{r^2}\frac{\partial A_\theta}{\partial\theta}-\frac{2A_\theta\cot\theta}{r^2}-\frac{2}{r^2\sin\theta}\frac{\partial A_\phi}{\partial\phi}\right]\vec{i}_r$ $+\left[\nabla^2 A_\theta+\frac{2}{r^2}\frac{\partial A_r}{\partial\theta}-\frac{A_\theta}{r^2\sin^2\theta}-\frac{2\cos\theta}{r^2\sin^2\theta}\frac{\partial A_\phi}{\partial\phi}\right]\vec{i}_\theta$ $+\left[\nabla^2 A_\phi-\frac{A_\phi}{r^2\sin^2\theta}+\frac{2}{r^2\sin\theta}\frac{\partial A_r}{\partial\phi}+\frac{2\cos\theta}{r^2\sin^2\theta}\frac{\partial A_\theta}{\partial\phi}\right]\vec{i}_\phi$
$\vec{C}\cdot\nabla\vec{A}$	$\left(C_x\frac{\partial A_x}{\partial x}+C_y\frac{\partial A_x}{\partial y}+C_z\frac{\partial A_x}{\partial z}\right)\vec{i}_x$ $+\left(C_x\frac{\partial A_y}{\partial x}+C_y\frac{\partial A_y}{\partial y}+C_z\frac{\partial A_y}{\partial z}\right)\vec{i}_y$ $+\left(C_x\frac{\partial A_z}{\partial x}+C_y\frac{\partial A_z}{\partial y}+C_z\frac{\partial A_z}{\partial z}\right)\vec{i}_z$	$\left(C_r\frac{\partial A_r}{\partial r}+\frac{C_\theta}{r}\frac{\partial A_r}{\partial\theta}+C_z\frac{\partial A_r}{\partial z}-\frac{C_\theta A_\theta}{r}\right)\vec{i}_r$ $+\left(C_r\frac{\partial A_\theta}{\partial r}+\frac{C_\theta}{r}\frac{\partial A_\theta}{\partial\theta}+C_z\frac{\partial A_\theta}{\partial z}+\frac{C_\theta A_r}{r}\right)\vec{i}_\theta$ $+\left(C_r\frac{\partial A_z}{\partial r}+\frac{C_\theta}{r}\frac{\partial A_z}{\partial\theta}+C_z\frac{\partial A_z}{\partial z}\right)\vec{i}_z$	$\left(C_r\frac{\partial A_r}{\partial r}+\frac{C_\theta}{r}\frac{\partial A_r}{\partial\theta}+\frac{C_\phi}{r\sin\theta}\frac{\partial A_r}{\partial\phi}-\frac{C_\theta A_\theta}{r}-\frac{C_\phi A_\phi}{r}\right)\vec{i}_r$ $+\left(C_r\frac{\partial A_\theta}{\partial r}+\frac{C_\theta}{r}\frac{\partial A_\theta}{\partial\theta}+\frac{C_\phi}{r\sin\theta}\frac{\partial A_\theta}{\partial\phi}+\frac{C_\theta A_r}{r}-\frac{C_\phi A_\phi\cot\theta}{r}\right)\vec{i}_\theta$ $+\left(C_r\frac{\partial A_\phi}{\partial r}+\frac{C_\theta}{r}\frac{\partial A_\phi}{\partial\theta}+\frac{C_\phi}{r\sin\theta}\frac{\partial A_\phi}{\partial\phi}+\frac{C_\phi A_r}{r}+\frac{C_\phi A_\theta\cot\theta}{r}\right)\vec{i}_\phi$
$(\overset{\Rightarrow}{T}:\nabla\vec{A})$	$T_{xx}\left(\frac{\partial A_x}{\partial x}\right)+T_{yy}\left(\frac{\partial A_y}{\partial y}\right)+T_{zz}\left(\frac{\partial A_z}{\partial z}\right)$ $+T_{xy}\left(\frac{\partial A_x}{\partial y}+\frac{\partial A_y}{\partial x}\right)$ $+T_{yz}\left(\frac{\partial A_y}{\partial z}+\frac{\partial A_z}{\partial y}\right)+T_{zx}\left(\frac{\partial A_z}{\partial x}+\frac{\partial A_x}{\partial z}\right)$	$T_{rr}\left(\frac{\partial A_r}{\partial r}\right)+T_{\theta\theta}\left(\frac{1}{r}\frac{\partial A_\theta}{\partial\theta}+\frac{A_r}{r}\right)+T_{zz}\left(\frac{\partial A_z}{\partial z}\right)$ $+T_{r\theta}\left(r\frac{\partial}{\partial r}\left(\frac{A_\theta}{r}\right)+\frac{1}{r}\frac{\partial A_r}{\partial\theta}\right)+T_{\theta z}\left(\frac{1}{r}\frac{\partial A_z}{\partial\theta}+\frac{\partial A_\theta}{\partial z}\right)$ $+T_{rz}\left(\frac{\partial A_z}{\partial r}+\frac{\partial A_r}{\partial z}\right)$	$T_{rr}\left(\frac{\partial A_r}{\partial r}\right)+T_{\theta\theta}\left(\frac{1}{r}\frac{\partial A_\theta}{\partial\theta}+\frac{A_r}{r}\right)+T_{\phi\phi}\left(\frac{1}{r\sin\theta}\frac{\partial A_\phi}{\partial\phi}+\frac{A_r}{r}+\frac{A_\theta\cot\theta}{r}\right)$ $+T_{r\theta}\left(\frac{\partial A_\theta}{\partial r}+\frac{1}{r}\frac{\partial A_r}{\partial\theta}-\frac{A_\theta}{r}\right)+T_{r\phi}\left(\frac{\partial A_\phi}{\partial r}+\frac{1}{r\sin\theta}\frac{\partial A_r}{\partial\phi}-\frac{A_\phi}{r}\right)$ $+T_{\theta\phi}\left(\frac{1}{r}\frac{\partial A_\phi}{\partial\theta}+\frac{1}{r\sin\theta}\frac{\partial A_\theta}{\partial\phi}-\frac{\cot\theta}{r}A_\phi\right)$
$\nabla\cdot\overset{\Rightarrow}{T}$	$\left(\frac{\partial T_{xx}}{\partial x}+\frac{\partial T_{xy}}{\partial y}+\frac{\partial T_{xz}}{\partial z}\right)\vec{i}_x$ $+\left(\frac{\partial T_{yx}}{\partial x}+\frac{\partial T_{yy}}{\partial y}+\frac{\partial T_{yz}}{\partial z}\right)\vec{i}_y$ $+\left(\frac{\partial T_{zx}}{\partial x}+\frac{\partial T_{zy}}{\partial y}+\frac{\partial T_{zz}}{\partial z}\right)\vec{i}_z$	$\left(\frac{1}{r}\frac{\partial}{\partial r}(rT_{rr})+\frac{1}{r}\frac{\partial}{\partial\theta}T_{r\theta}-\frac{1}{r}T_{\theta\theta}+\frac{\partial T_{rz}}{\partial z}\right)\vec{i}_r$ $+\left(\frac{1}{r}\frac{\partial T_{\theta\theta}}{\partial\theta}+\frac{\partial T_{r\theta}}{\partial r}+\frac{2}{r}T_{r\theta}+\frac{\partial T_{\theta z}}{\partial z}\right)\vec{i}_\theta$ $+\left(\frac{1}{r}\frac{\partial}{\partial r}(rT_{zr})+\frac{1}{r}\frac{\partial T_{z\theta}}{\partial\theta}+\frac{\partial T_{zz}}{\partial z}\right)\vec{i}_z$	$\left(\frac{1}{r^2}\frac{\partial}{\partial r}(r^2T_{rr})+\frac{1}{r\sin\theta}\frac{\partial}{\partial\theta}(T_{r\theta}\sin\theta)+\frac{1}{r\sin\theta}\frac{\partial T_{r\phi}}{\partial\phi}-\frac{T_{\theta\theta}+T_{\phi\phi}}{r}\right)\vec{i}_r$ $+\left(\frac{1}{r^2}\frac{\partial}{\partial r}(r^2T_{r\theta})+\frac{1}{r\sin\theta}\frac{\partial}{\partial\theta}(T_{\theta\theta}\sin\theta)+\frac{1}{r\sin\theta}\frac{\partial T_{r\phi}}{\partial\phi}+\frac{T_{r\theta}}{r}-\frac{\cot\theta}{r}T_{\phi\phi}\right)\vec{i}_\theta$ $+\left(\frac{1}{r^2}\frac{\partial}{\partial r}(r^2T_{r\phi})+\frac{1}{r}\frac{\partial T_{\theta\phi}}{\partial\theta}+\frac{1}{r\sin\theta}\frac{\partial T_{\phi\phi}}{\partial\phi}+\frac{T_{r\phi}}{r}+\frac{2\cot\theta}{r}T_{\theta\phi}\right)\vec{i}_\phi$

Appendix B

Vector and Operator Identities

$$\vec{A} \times \vec{B} \cdot \vec{C} = \vec{A} \cdot \vec{B} \times \vec{C} \quad (1)$$

$$\vec{A} \times (\vec{B} \times \vec{C}) = \vec{B}(\vec{A} \cdot \vec{C}) - \vec{C}(\vec{A} \cdot \vec{B}) \quad (2)$$

$$\nabla(\phi + \psi) = \nabla\phi + \nabla\psi \quad (3)$$

$$\nabla \cdot (\vec{A} + \vec{B}) = \nabla \cdot \vec{A} + \nabla \cdot \vec{B} \quad (4)$$

$$\nabla \times (\vec{A} + \vec{B}) = \nabla \times \vec{A} + \nabla \times \vec{B} \quad (5)$$

$$\nabla(\phi\psi) = \phi\nabla\psi + \psi\nabla\phi \quad (6)$$

$$\nabla \cdot (\psi\vec{A}) = \vec{A} \cdot \nabla\psi + \psi\nabla \cdot \vec{A} \quad (7)$$

$$\nabla \cdot (\vec{A} \times \vec{B}) = \vec{B} \cdot \nabla \times \vec{A} - \vec{A} \cdot \nabla \times \vec{B} \quad (8)$$

$$\nabla \cdot \nabla\phi = \nabla^2\phi \quad (9)$$

$$\nabla \cdot \nabla \times \vec{A} = 0 \quad (10)$$

$$\nabla \times \nabla\phi = 0 \quad (11)$$

$$\nabla \times (\nabla \times \vec{A}) = \nabla(\nabla \cdot \vec{A}) - \nabla^2\vec{A} \quad (12)$$

$$(\nabla \times \vec{A}) \times \vec{A} = (\vec{A} \cdot \nabla)\vec{A} - 1/2\ \nabla(\vec{A} \cdot \vec{A}) \quad (13)$$

$$\nabla(\vec{A} \cdot \vec{B}) = (\vec{A} \cdot \nabla)\vec{B} + (\vec{B} \cdot \nabla)\vec{A} + \vec{A} \times (\nabla \times \vec{B}) + \vec{B} \times (\nabla \times \vec{A}) \quad (14)$$

$$\nabla \times (\phi\vec{A}) = \nabla\phi \times \vec{A} + \phi\nabla \times \vec{A} \quad (15)$$

$$\nabla \times (\vec{A} \times \vec{B}) = \vec{A}(\nabla \cdot \vec{B}) - \vec{B}(\nabla \cdot \vec{A}) + (\vec{B} \cdot \nabla)\ \vec{A} - (\vec{A} \cdot \nabla)\vec{B} \quad (16)$$

Appendix C

Films

Developed for educational purposes with the support of the National Science Foundation at the Education Development Center, films cited fall in one of two series.

Produced by the National Committee for Fluid Mechanics Films and distributed by Encyclopedia Britannica Educational Corp., 425 N. Michigan Ave., Chicago, Illinois (60611) are:

(1) Channel Flow of a Compressible Fluid
(2) Current-induced Instability of a Mercury Jet
(3) Eulerian and Lagrangian Descriptions in Fluid Mechanics
(4) Flow Instabilities
(5) Fundamentals of Boundary Layers
(6) Low-Reynolds Number Flows
(7) Magnetohydrodynamics
(8) Pressure Fields and Fluid Acceleration
(9) Surface Tension and Fluid Mechanics
(10) Waves in Fluids

Produced by the National Committee for Electrical Engineering Films and distributed by Education Development Center, 39 Chapel Street, Newton, Mass. 02160 are:

(11) Complex Waves I and Complex Waves II
(12) Electric Fields and Moving Media

Index